PETROLEUM GEOLOGY

TITLES OF RELATED INTEREST

Physical processes of sedimentation
J. R. L. Allen

Petrology of the sedimentary rocks
J. T. Greensmith

A dynamic stratigraphy of the British Isles
R. Anderton, P. H. Bridges, M. R. Leeder & B. W. Sellwood

Microfossils
M. D. Brasier

Introduction to small-scale geological structures
G. Wilson

The poetry of geology
R. M. Hazen (ed.)

Sedimentology: process and product
M. R. Leeder

Sedimentary structures
J. D. Collinson & D. B. Thompson

Aspects of micropalaeontology
F. T. Banner & A. R. Lord (eds)

Geology and man
J. Watson

Statistical methods in geology
R. F. Cheeney

Geological howlers
W. D. I. Rolfe (ed.)

The dark side of the Earth
R. Muir Wood

Principles of physical sedimentology
J. R. L. Allen

Paleopalynology
A. Traverse

PETROLEUM GEOLOGY

F. K. North

Professor Emeritus,
Carleton University, Ottawa, Canada

Boston
ALLEN & UNWIN
London Sydney

Allen & Unwin Inc.,
8 Winchester Place, Winchester, Mass. 01890, USA

George Allen & Unwin (Publishers) Ltd,
40 Museum Street, London WC1A 1LU, UK

George Allen & Unwin (Publishers) Ltd,
Park Lane, Hemel Hempstead, Herts HP2 4TE, UK

George Allen & Unwin Australia Pty Ltd,
8 Napier Street, North Sydney, NSW 2060, Australia

First published in 1985

Library of Congress Cataloging in Publication Data

North, F. K.
Petroleum geology.
Bibliography: p.
Includes index.
1. Petroleum–Geology. I. Title.
TN870.5.N59 1985 553.2'82 85-1211
ISBN 0-04-553003-3 (alk. paper)
ISBN 0-04-553004-1 (pbk. : alk. paper)

British Library Cataloguing in Publication Data

North, F. K.
Petroleum geology.
1. Petroleum
I. Title
553.2'82 TN870
ISBN 0-04-553003-3
ISBN 0-04-553004-1 Pbk

Artwork drawn by Oxford Illustrators Ltd
Set in 10 on 12 point Times by A. J. Latham Ltd.
and printed in Great Britain
by Butler & Tanner Ltd, Frome and London

Preface and acknowledgements

This book is founded on the double proposition that *geology* is an integrative science, and that among its many constituent aspects *petroleum geology* is the most integrative of all. It is dependent upon data of its own, most of them observational or deductive, and upon the data of nearly all the other sciences, many of them experimental or theoretical.

Petroleum geologists perform a remarkable variety of functions. A book on *petroleum geology* must cover a broad array of topics if it is to make any claim to be comprehensive in the eyes of such people. Not since Arville Levorsen's famous work appeared has any geologist really tried to do this. But Levorsen's book antedated a wide spectrum of today's dominant thinking and practice, not only in petroleum geology in its strictest sense but in geology in general, and in petroleum geophysics and petroleum geochemistry as well. Today it is impossible to find any one individual who could duplicate Levorsen's achievement through his own knowledge and experience. The comprehensive text now must either be under multiple authorship, or have a single author able to call upon colleagues who will let him write through their minds.

I was invited to attempt the second route. My principal qualification for this task appeared to be that I have constructed a 33-year experience sandwich, with a long spell as a professor between two shorter ones in the oil industry. I was therefore presumed to know more than most professors about petroleum geology and more than better-qualified petroleum geologists about the students to whom my book is directed. The fortunate circumstance of spending the years 1980–83 in Chevron Oil Field Research Company (COFRC), in California, provided me with access to experts in all aspects of my subject, most importantly to those in fields in which I am myself inexpert or out-of-date (or both).

Most of my indebtedness is therefore to friends in COFRC, but I must include others with them. For instruction and correction in aspects which I could not possibly have faced with my own resources, I am grateful to Al Brown, Bruce Davis, Tom Edison, Jan Korringa, Peeter Kruus, Kenneth Peters, Giorgio Ranalli, Sol Silverman, and Ralph Simon. For much stimulation and commentary on aspects for which I had more faith in my own resources, I thank Chuck Haskin, Bob Jones, Clayton McAuliffe, George Moore, Doug Pounder, Jack St John, Lee Stephenson, and Ray Yole. For providing materials and permission to use them, and for assistance with the text, I thank Jack Allan, Roger Burtner, Charlie Everett, and several of the Chevron companies to whom text figures are attributed.

For the initial recommendation of me as potential author of this book, I thank (or blame) King Hubbert. For many helpful comments during its preparation, and for encouragement throughout, I am grateful to the late Earl Cook and to John Harbaugh (who also kept me aware of the standards set by my predecessors). As I have called upon the entire literature of the subject for my material, many practising petroleum geologists will recognize their own influences on parts of my text.

For the arduous word processing of the text, I thank Jessica Ezell and Mary Armstrong; for arranging the expenses of this lengthy process, grateful acknowledgement is due to Alice Myers of COFRC and Deans George Skippen and John ApSimon of Carleton University. The drafting of the numerous text figures involved extensive simplification and total metrication of those taken from most journals and nearly all non-periodical publications. Much of this work was done by Catherine Revell, Dan Teaford and Linda Antoniw. For permission to reproduce figures from their publications, I thank the publishers and copyright holders acknowledged on the figures themselves.

Finally, the editors will join me in much more than ritual gratitude to my wife, Kathleen, who is more interested in great art than in dirt science, but who displayed marvellous patience while my author efforts dominated my attention for more than three years. When the word processing of the text was finally complete, she spent scores of hours in the proofreading and indexing of a book which would not rank high on her own reading list. She faces no written or oral examination of what she learned from it. I hope the students who do face such tests will learn a great deal.

Ottawa, April 1984

Contents

List of tables

List of plates

Part I

INTRODUCTION

The literature of petroleum geology is as vast as befits a major science in its application to one of the world's greatest industries. Before the young geologist can hope to find a way through it, or even through a single comprehensive textbook like this one, he (or she, as we shall emphasize in a later chapter) should acquire some grasp of the vocabulary that he will need and with which he will have to become routinely familiar.

Before we embark on our consideration of what petroleum is, how it occurs, and how and where we exploit it, we may briefly review this vocabulary in its principal aspects: the literature that deploys it, the jargon that dominates it, the statistical vocabulary that cannot be avoided by the professional, and the historical development that has brought all of these to their present scope and complexity.

1 Petroleum geology as a field of study

1.1 Introduction

Titles of novels are just names for books. Their authors choose names for their books much as they choose names for their prize pets or their sailboats. The names they choose may become more familiar to the general public than the names of the authors; but they need not give any indication of the books' natures.

Titles of textbooks should not be like that. They are supposed to indicate unequivocally what the books are about. Unless a textbook's topic is very restricted in scope, however, an excessively succinct title may give a misleading idea of its contents. There have been at least a dozen textbooks published since 1928 with the title *Petroleum geology* or *Geology of petroleum*. Yet many geologists with long and successful careers in the petroleum industry maintain stoutly that there is no such subject as *petroleum geology*.

In their view, there is a science of *geology*, independent of any concern with petroleum. What geologists in the petroleum industry do is practise that science with the aid of a variety of technologies that are not themselves intrinsically geological at all. One of the best short explanations (for its time – 1942) of *geology applied to petroleum* was given that very title, by Vincent Illing. Ironically, in the light of the public's association of petroleum with profits, the paper was written for the *Proceedings* of the world's best-known association of geological amateurs, the Geologists' Association of Great Britain.

Unless individuals calling themselves petroleum geologists are first and foremost geologists in the *general* sense, they are technologists and not scientists. The philosophy behind this viewpoint has dictated the content of this book and the manner of its presentation. The reader should know this at the outset!

1.2 Relation of petroleum geology to sciences in general

Geology itself is an *integrative* science. Founded upon the *observation* of all inanimate features of the Earth – rocks, fossils, rivers, volcanoes – it seeks their explanation with the assistance of the theories and experiments of other sciences. In essence, we may think of geology as a spectrum extending between one end-member immediately related to the exact sciences of chemistry, physics, and mathematics, and another end-member related to the life sciences (zoology and botany). The first end-member is *mineralogy*, the second *paleontology*; each is a science in its own right.

Everything in between these two sciences is either physical or historical *geology*, or a combination of the two (Fig.1.1). Close to mineralogy are igneous and metamorphic *petrology*, which involve professional knowledge of chemistry and of thermodynamics. Close to paleontology is *stratigraphy*, involving an understanding of biology, meteorology, and oceanography. Close to both petrology and stratigraphy are *structural geology* and *sedimentary geology*, requiring knowledge of mechanics, hydraulics, and descriptive geometry. Any or all of these subdisciplines may adopt not only the material of the physicists and the chemists but their techniques in addition, giving us the distinctive interdisciplinary sciences of *geophysics* and *geochemistry*.

A modern university curriculum in *geology* has to embrace all these topics. As geology is more dependent upon the other sciences in the curriculum than they are upon it, the geologist's program of study must also include those sciences themselves. It is a fortunate department that can find room and time in such a program for anything specifically called *petroleum geology*. The new graduate may enter the petroleum industry armed with the core courses in geology and the other sciences alone.

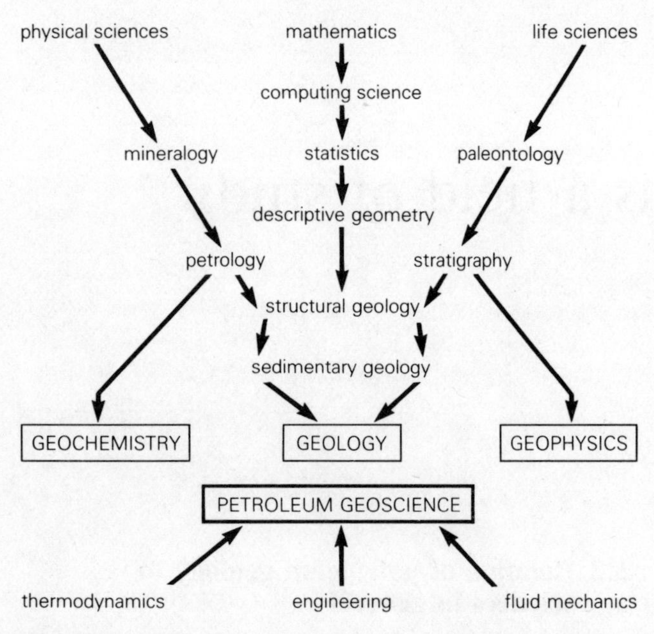

Figure 1.1 The interrelationship between petroleum geology and other sciences.

For many exploration managers and chief geologists, that is all they want in their new recruits. A well trained *geologist*, without prefix, will be able to apply his knowledge and abilities to the petroleum industry's concerns once they are augmented by a few techniques (not themselves intrinsically geological, as we have observed) that only experience in the industry can teach.

Geology, already an integrative science for the student, becomes still more integrative for the practitioner in the industry. Because oil and gas are fluids, a knowledge of *fluid dynamics* must be acquired. Because geology in industry is fundamentally practical, a knowledge of some branches of *engineering* will quickly become necessary. And because everything the geologist does will eventually come to a test in the marketplace of the real world, an understanding of *economics* is a big help.

In the modern industry, no one individual will have the opportunity to acquire experience or to use skills in all of these aspects of the subject. The modern petroleum geologist is as likely to become a specialist as any other kind of scientist is. The best petroleum geologists, nonetheless, are among the most versatile of scientists.

This book is intended to help make them so. At the risk of trying to include too much for any textbook, I have devoted sections of the book to every topic of which the petroleum *geologist* is likely to be required to demonstrate command. Excluded only are those topics falling within the pure end-members of our geological spectrum; clay mineralogy and micropaleontology, for example, are subjects far too vast to be more than touched upon here.

1.3 The literature of petroleum geology

In addition to the material available to the author from his own experience, and from the experience of colleagues whose help he gratefully acknowledges, a vast array has been gathered from library sources.

To attempt to list all the references used would add scores of pages to the book's length. To acknowledge them all within the text, as is proper in papers in scientific journals, would make the text unreadable to the students for whom it is designed. Reference lists become out-of-date very quickly, especially as review and summary articles and books continually appear. I have therefore chosen an unorthodox method of acknowledgement of my sources. All *figures* taken from others are individually attributed to their sources. All *periodical* sources from which material has been garnered are discussed here in an effort to help the beginner find his way through the jungle of the literature.

There are now more geological periodicals for libraries to accumulate than there were scientific periodicals of all kinds when the author was a student. A Russian geologist commented during the 1960s that continuing expansion of scientific literature at the present rate would lead by the end of the 20th century to its volume exceeding that of the crust of the Earth. Nobody can possibly keep up with more than a small fraction of it even by reading it 12 hours a day. The principal standard works, covering the topics presented in this book, are listed in the brief bibliography at the end of this chapter. The author acknowledges his debt to the literature he has consulted by setting out for his readers this review of his sources.

1.3.1 Bibliographies, abstracts, indices

There are abstracting, indexing, and information retrieval services for geological literature as there are for that of other sciences. The American Geological Institute (AGI) sponsors the GEOREF system; Geosystems (London) produces the equivalent GEOARCHIVE, covering most of Europe. The annual *Bibliography of Geology*, both for North America and for the world, is a valuable asset for all geologists, regardless of speciality. *Petroleum Abstracts*, produced at the University of Tulsa, is available to subscribers only. Russian titles are abstracted monthly, in English, in the *Russian Oil and Gas Bulletin*.

1.3.2 Textbooks of petroleum geoscience

Textbooks having the terms *petroleum geology*, or *geology of petroleum*, as the whole or part of their titles have been authored or edited by a number of geologists: E. R. Lilley, W. H. Emmons, C. G. Lalicker, E. N.

Tiratsoo, K. K. Landes, W. L. Russell, G. D. Hobson, A. I. Levorsen, and R. E. Chapman (all in English); M. Robert and A. Perrodon (in French); and several in other languages. The more restricted topic of *petroleum development geology* has been treated by P. A. Dickey. The geology of oil and gas *fields* has been the subject of books compiled or edited by V. C. Illing, K. K. Landes, E. N. Tiratsoo, and W. A. ver Wiebe, to mention only works in English.

Equivalent treatments of *petroleum geochemistry*, designed for geologists as much as for chemists, are by O. A. Radchenko, J. M. Hunt, and B. Tissot and D. H. Welte. Textbooks of *petroleum geophysics* include those by J. F. Claerbout, M. B. Dobrin, R. E. Sheriff, and K. H. Waters. There are several textbooks devoted to *well logging*, but the standard references for this are the publications of Schlumberger Ltd.

1.3.3 Journals and periodicals for petroleum geologists

The basic library of the petroleum geologist is easily summarized. The publications of the American Association of Petroleum Geologists (AAPG) constitute the essential part of it without assistance from any other source. No other professional organization in any science so totally dominates its field on a worldwide basis. The monthly Bulletin, the Special Volumes issued before 1960, the Memoirs since 1960, the Reprint Series since 1970, the Studies in Geology (since the mid-1970s), and the Continuing Education Series Course Notes constitute between them a complete library of petroleum geology on their own. The scope of the material published by the Association is both truly international and genuinely all-embracing within the field.

Numerous state, regional, and local affiliates of the Association publish excellent journals and semi-periodicals of their own, but of course their coverage is regional and not international. The *Transactions of the Gulf Coast Association of Geological Societies* and the volumes issued by the Rocky Mountain Oil and Gas Association (RMOGA) are excellent examples. State and local publications by governments, universities, and professional societies in Texas and Oklahoma alone constitute a substantial reference resource.

The second indispensable literature source is the Proceedings of the Quadrennial World Petroleum Congress. The eleventh of the Congresses was held in 1983. One volume of the proceedings of each Congress is devoted primarily to papers in petroleum geology.

The third vital source of references is the various serial publications of the Society of Economic Paleontologists and Mineralogists (SEPM). The fourth is the quarterly *Journal of Petroleum Geology*, published in England but of international outlook.

Many oil-producing countries publish excellent periodicals on petroleum geology which make less claim to be of international scope. The Canadian Society of Petroleum Geologists publishes a quarterly Bulletin, occasional Memoirs (of distinctly international flavor), and a variety of Special Publications. The Australian Petroleum Exploration Association (APEA) publishes a Journal. The Boletins of the Asociacion Venezolana de Geologia, Mineria y Petroleo (AVGMP) and the Asociacion Mexicana de Geologos Petroleros, mostly in Spanish, are of more local interest. An unexpectedly rich source is the *Bulletin des Centres de Recherches Exploration-Production Elf-Aquitaine*, most but by no means all of it in French. An associated Bulletin is that of l'Institut de Géologie du Bassin d'Aquitaine, published by the University of Bordeaux.

The only other widely available journal devoted entirely to petroleum geology has that title and contains very brief papers on Russian oil and gas regions, in English. Other Soviet journals in the field, notably *Geologiia Nefti i Gaza*, are in Russian.

1.3.4 Geological journals of direct concern to petroleum geologists

Journals devoted entirely to aspects of sedimentary geology are of immediate concern to petroleum geologists even though they are not designed for them. *Sedimentology, Sedimentary Geology*, and the *Journal of Sedimentary Petrology* are the principal representatives. A less obvious one is *Maritime Sediments*.

The array of journals devoted to paleontology, micropaleontology, palynology, and biostratigraphy is too wide and varied for discussion here.

1.3.5 Vital petroleum periodicals, not geological

Two groups of long-established periodicals cater to wide readerships in the petroleum industry. The first group includes journals of scientific and technological character which contain occasional (but important) materials especially for petroleum geologists. The most important in English are the *Journal of Petroleum Technology*, the *Journal of the Institute of Petroleum*, both the *Journal* and the *Transactions of the Society of Petroleum Engineers* (of the American Institute of Mining, Metallurgical, and Petroleum Engineers, or AIME), and the *Journal of Canadian Petroleum Technology*. The Institute of Petroleum, in particular, also publishes invaluable non-periodical compendia on geological topics. An easily accessible counterpart not in English is the *Révue de l'Institut Français du Pétrole*.

The second group comprises the "trade journals," which commonly contain short articles specifically for petroleum geologists. The *Oil and Gas Journal* and

World Oil, published in the USA, are read by oilmen all over the world. Less familiar but still useful are *Oil Gas – European Magazine*, published in Hamburg (and the German language *Erdoel Erdgas Zeitschrift*, also from Hamburg), the *Petroleum Gazette*, published by the Australian Institute of Petroleum, and a number of British and Canadian weeklies and monthlies.

1.3.6 Short-article periodicals

There are several long-established journals catering to scientists of all stripes, offering short articles on topics of both general and special interest. Obvious examples in English are *Nature, Science, Scientific American,* and *American Scientist*. One never knows when an article of direct relevance to the petroleum geologist might appear in such a journal.

There have sprung up more recently an assortment of similar journals specifically for geologists. Again, obvious examples are *Geotimes* (the news magazine of the AGI), *Episodes* (that of the IUGS), *Geoscience Canada* (that of the Geological Association of Canada), and *Geology* (for short papers submitted to the Geological Society of America). Papers of immediate interest to petroleum geologists occur in all of them. Of direct interest is the *Pacific Petroleum Geologist Newsletter* (of the Pacific Section of the AAPG).

1.3.7 Standard geological periodicals or series

The geologist of any specialization must choose the journals that will serve his or her interests best. We have briefly reviewed in the foregoing the publications most likely to do this for petroleum geologists. There remain the great number of purely geological, geochemical, and geophysical journals or series that may in any issue publish a paper or article of interest to any or every geologist. All that can be attempted here is a general guide to the nature of this vast library.

(a) Publications of national, state, or provincial geological surveys are likely to appeal to the petroleum geologist if the place of their origin is a producer of oil or gas. The Professional Papers of the United States Geological Survey (USGS) are a prime example.

(b) Some geological journals have a deliberately economic bent without being restricted to hard-rock economic geology. Examples of widely separated origin include the BMR's (Bureau of Mineral Resources) *Journal of Australian Geology and Geophysics*, the Dutch *Geologie en Mijnbouw*, and the *Bulletin du Bureau de Recherches Géologiques et Minières* (much of which deals with African geology).

(c) Some of the more famous national journals are determinedly international. Three examples from different oil-producing countries are the Bulletins and Memoirs of the Geological Society of America (GSA), the *Journal of the Geological Society of London*, and the *International Geology Review* (English translations of papers originally in Russian).

(d) Some much less venerable and widely known journals regularly carry papers on petroleum geology because of their geographic origin. The *Bulletin of the Geological Society of Malaysia*, the *Israel Journal of Earth Sciences*, and *Pacific Geology* (published in Tokai, Japan) are examples unfamiliar to most geologists in the English-speaking world.

(e) Established geological journals published in countries whose languages are far from international commonly carry many papers in English. The principal journals from China, Hungary, Poland, and Sweden are examples carrying some such papers on oil-geological topics. Journals published in France, Germany, Italy, and Spain are normally in their own languages, though there are exceptions (especially from Germany).

(f) The majority of long-established, nonspecialist geological journals, consciously preferring noneconomic topics and aware of the proliferation of specialized journals in the petroleum and mining fields, make it a point (acknowledged or not) to avoid papers designed for practitioners in those fields.

(g) We come finally to the world-famous national academies and learned societies whose publications are available on a regular basis only to their members and to libraries. Most such bodies admit geological sections or divisions, whose highly authoritative publications include the occasional one directed towards petroleum geologists. The *Philosophical Transactions of the Royal Society* of London (and the publications of other Royal Societies in the Commonwealth countries) and the *Comptes Rendus* of the French Académie des Sciences are outstanding examples.

Selected bibliography

Asquith, G. B. and C. R. Gibson 1982. *Basic well log analysis for geologists*. Methods in Exploration Series. Tulsa, Oklahoma: AAPG.

Bally, A. W. (ed.) 1983. *Seismic expression of structural styles*. Studies in Geology, 15, 3 vols. Tulsa, Oklahoma: AAPG.

Bebout, D., G. Davies, C. H. Moore, P. A. Scholle, and N. C. Wardlaw 1979. *Geology of carbonate porosity*. Continuing Education Course Note Series 11. Tulsa, Oklahoma: AAPG.

Chapman, R. E. 1983. *Petroleum geology*. Developments in Petroleum Science, 16. Amsterdam: Elsevier.

Chilingarian, G. V. and T. F. Yen (eds) 1978. *Bitumens, asphalts, and tar sands*. Amsterdam: Elsevier.

Collins, A. G. 1975. *Geochemistry of oilfield waters*. New York: Elsevier.

Conybeare, C. E. B. 1979. *Lithostratigraphic analysis of sedimentary basins*. New York: Academic Press.

Dahlberg, E. C. 1982. *Applied hydrodynamics in petroleum exploration*. Berlin: Springer.

Dickey, P. A. 1981. *Petroleum development geology*. Tulsa, Oklahoma: Pennwell.

Dobrin, M. B. 1976. *Introduction to geophysical prospecting*, 3rd edn. New York: McGraw-Hill.

Dott, R. H., Sr, and M. J. Reynolds 1969. *Source book for petroleum geology*. Tulsa, Oklahoma: AAPG Memoir 5.

Dunnington, H. V. 1967. Stratigraphical distribution of oilfields in the Iraq–Iran–Arabia basin. *J. Inst. Petroleum* **53**, 129–61.

Haun, J. D. (ed.) 1975. *Methods of estimating the volume of undiscovered oil and gas resources*. Studies in Geology, vol. 1. Tulsa, Oklahoma: AAPG.

Haun, J. D. and L. W. LeRoy (eds) 1958. *Subsurface geology in petroleum exploration*. Golden, Colorado: Colorado School of Mines.

Hobson, G. D. 1977. *Developments in petroleum geology*. London: Applied Science.

Hubbert, M. K. 1969. *The theory of ground-water motion and related papers*. New York: Hafner.

Hunt, J. M. 1979. *Petroleum geochemistry and geology*. San Francisco: Freeman.

Illing, V. C. 1942. Geology applied to petroleum. *Proc. Geol. Assoc.* **53**, 156–87.

King, R. E. (ed.) 1972. *Stratigraphic oil and gas fields – classification, exploration methods and case histories*. Tulsa, Oklahoma: AAPG Memoir 16.

Klemme, H. D. 1975. Giant oil fields related to their geologic setting: a possible guide to exploration. *Bull. Can. Petroleum Geol.* **23**, 30–66.

Landes, K. K. 1970. *Petroleum geology of the United States*. New York: Wiley Interscience Series.

Levorsen, A. I. 1967. *Geology of petroleum*, 2nd edn. San Francisco: Freeman.

Miall, A. D. (ed.) 1980. *Facts and principles of world petroleum occurrence*. Canadian Soc. Petroleum Geol. Memoir 6.

Payton, C. E. (ed.) 1977. *Seismic stratigraphy – applications to hydrocarbon exploration*. Tulsa, Oklahoma: AAPG Memoir 26.

Perrodon, A. 1983. *Dynamics of oil and gas accumulations*. Pau: Elf-Aquitaine Mem. 5.

Pratt, W. E. 1942–4. *Oil in the earth*. Lawrence, Kansas: University of Kansas Press.

Roberts, W. H. III and R. J. Cordell (eds) 1980. *Problems of petroleum migration*. Studies in Geology, vol. 10. Tulsa, Oklahoma: AAPG.

Tiratsoo, E. N. 1976. *Oilfields of the world*. Beaconsfield: Scientific Press.

Tiratsoo, E. N. 1979. *Natural gas*. Beaconsfield: Scientific Press.

Tissot, B. P. and D. H. Welte 1978. *Petroleum formation and occurrence*. Berlin: Springer.

Weimer, R. J. (ed.) 1973. *Sandstone reservoirs and stratigraphic concepts*. Reprint Series 7 and 8. Tulsa, Oklahoma: AAPG.

2 Basic vocabulary

The vocabulary common to all geologists is used throughout this book without labored definition. It is the vocabulary employed in the sciences and subsciences of physical and historical geology, especially in stratigraphy, structural geology, sedimentology and sedimentary petrology, paleontology, and geological mapping, whether applied to the petroleum industries or not. It is extended to include the vocabularies of geochemistry, geophysics, thermodynamics, fluid mechanics, and microscopy, again without their restriction to petroleum science.

The petroleum industries have of course spawned vocabularies of their own in addition. The parts of these vocabularies in universal use throughout the petroleum industries of the English-speaking world may be briefly identified as we begin this book; all will be enlarged upon in appropriate chapters later.

Petroleum (rock oil) is a naturally occurring complex of hydrocarbons widely distributed in the sedimentary rocks of the Earth's crust. Though its strict meaning requires that the hydrocarbons occur in liquid form, the simpler compounds among them are gaseous under normal conditions and the more complex ones are commonly found as solids. In the absence of any other widely understood term, "petroleum" should be used for the spectrum of solid, liquid, and gaseous members of the common series (see Ch. 5). The liquid members constitute *crude oil* until artificial refinement modifies their original compositions. The gaseous members constitute *natural gas*. The solid members are variously called *asphalt*, *bitumen*, or *tar*, and their nomenclature is sadly confused (see Ch. 10). Neither oil, nor gas, nor solid hydrocarbon satisfies the definition of a mineral, and they should not be called minerals by geologists.

Commercially exploitable petroleum occurs underground in the *subsurface*. The rock containing it is a reservoir rock, or simply a *reservoir*. The feature of the rock that restrains the fluid petroleum from moving out of the reservoir is called a *trap*. Petroleum becomes commercially exploitable when it is naturally gathered into a *pool*, which is a single, discrete accumulation of oil or gas in a single reservoir with a single trap. Several pools may lie in a vertical succession within a single area, or they may lie side by side or overlap laterally so as to constitute an areally continuous accumulation called a *field*.

A three-dimensional geological entity containing a number of oil- or gasfields is a sedimentary basin, or simply a *basin*. A geographical region containing petroleum fields having some geologic characteristics in common is a *province*. A province may be synonymous with a basin, or it may embrace several basins sharing clear similarities but separated by barren, nonbasinal tracts (Sec. 17.1). A geographic concentration of fields within a province or a basin is a *district*. A small area within a basin, province, or district which may contain oil or gas but has not yet been proved to do so is a *prospect*. A larger area within which the drilling of prospects has established success and pointed the way for further drilling provides a *play*.

Oil and gas occurring under these conditions are said to be *conventional*. They are discovered and exploited by drilling boreholes, or boring drillholes, into the ground. A hole which yields any fluid is a *well*. A well drilled in search of a new accumulation of oil or gas is an *exploratory* or *wildcat well*. If it is successful, it constitutes a *discovery*. Many early discovery wells spurted oil or gas high into the air above the *derrick* which had drilled the well. These strongly flowing wells were called *gushers*, but modern wellsite engineering has made the term almost obsolete. If a discovery well shows promise of being commercial it is *completed* as a *producing well*. If a well yields no recoverable oil or gas, it is a *dry hole* (despite the fact that it may yield a lot of water). A dry hole is *abandoned*. The process of recording the data derivable from the drilling of a well (whether the recording is done mechanically or as a wholly human operation) is called *logging*.

Oil which is recoverable from surface operations, and oil or gas which requires some recovery procedures other than conventional drilling, are said to be *unconventional*. Unconventional petroleum resources are of vast extent (see Sec. 10.8), but the ready availability of conventional resources has prevented their exploitation until recent years except on a local scale.

The search for new sources of petroleum constitutes *exploration*. The sources discovered by successful exploration become *reserves*, which are those portions of the total *resource* that have been shown to be accessible and recoverable under current economic and technologic conditions. Because no recovery technique can extract all the oil or gas from a field, the reserves of the field are only some fraction of the *in-place oil* which the field actually contains. The process of recovering the reserves, by drilling wells within a field and operating them successfully, is called *development*.

The techniques of exploration are not a function of geology alone but of *geophysics* also. Geophysics, like geology, has within its orbit far more than the search for and provision of petroleum. Petroleum geophysics is so dominated by its seismic branch that the term *seismic* has become almost synonymous with *geophysics* for the petroleum industry.

Other terms having nothing directly to do with petroleum geology, but a great deal to do with the duties of petroleum geologists and with the conditions under which they work, refer to the organizations on whose behalf the work is done. Organizations and individuals seeking or producing oil or gas are called *operators*. They may be individuals, many of whom provide services as *consultants* to bigger operators. They may be privately owned companies engaged in particular aspects of the petroleum industries requiring the services of geologists; these companies are *independents*. They may be privately owned companies of great size and complexity involved in all aspects of the industry, in most cases in many areas of the world. These are the *majors;* their names are household words. Involvement by one company or corporation in all phases of the industry, from exploration of virgin territory to marketing an array of petroleum products, constitutes what is called *vertical integration* of the company. A somewhat different pattern of activity, referred to as *horizontal integration,* involves significant effort in several competing fields of energy or other resources (a common example is a company having active interests in oil, gas, coal and uranium, and perhaps in renewable energy sources such as geothermal or solar power in addition). Finally there are many government-controlled or *national companies,* several of them vertically integrated and one or two horizontally integrated also.

3 Basic statistics

This book is not about numbers. Geology is not one of the exact sciences, and geologists, like other natural scientists, tend to be wary of numbers, which they do not always understand. The geologists of the oil and gas industries should constitute an exception to this familiar observation. The industries employ a large number of scientists, spend enormous amounts of money, and supply the world with enormous volumes of products. Petroleum statistics have become commonplace in newspapers and other publications quite unconnected with the industry. Oil and gas prices, OPEC production rates, pipeline costs and throughputs, national import figures, and so on have become routine topics for the information media.

All petroleum geologists should make themselves familiar with the fundamental statistics of their industry (and their science), in generalized terms, because the numbers involved are so large and so critical that misquotation or misunderstanding of them has serious consequences. As we began the 1980s, oil supplied about 50 per cent of the world's energy; oil and gas between them supplied more than two-thirds. The volumes produced and consumed are therefore very large, but the units in which they are measured are relatively small (exceedingly small, in some cases). So the words billion, trillion, and even quadrillion are matters of routine conversation and communication in the oil and gas industries; figures of this order are both easily misunderstood and easily misquoted.

3.1 Unit quantities of oil and gas

Quantities of oil have traditionally been expressed in *barrels*, because this is the unit of measure used in the USA. One barrel contains 42 US gallons, or 35 Imperial gallons. Many countries, including the UK and the USSR, have expressed oil quantities in terms not of volume but of weight: in either tons or tonnes (metric tons, of 1000 kg). So the relation between a tonne and a barrel of oil is not constant; it depends on the relative density (specific gravity) of the oil. One tonne of average world crude oil is equivalent to 7.33 US barrels, but for a heavy oil the conversion factor is 6.8 or less, and for a very light oil it is 7.6 or more. Only a few Latin American countries have traditionally expressed oil output and reserves in cubic meters. Now everybody should do so.

Quantities of natural gas have of course always been expressed in terms of volume: cubic feet in North America and many other countries, cubic meters in most European countries. Now all jurisdictions and all organizations should use the metric measure. Because oil and gas are commonly produced in the same regions, in many cases by the same producing organizations, it is often necessary to express quantities of oil and of gas in combination. The expression is then in terms of *oil equivalent*, according to heat-producing capacity of the two fuels. On this basis, one barrel of average light oil is equivalent to 6000 or 6500 cubic feet of natural gas ($1\ m^3$ of light oil $\equiv 1000-1120\ m^3$ of gas). Obviously 1 cubic foot of gas is too small a unit to be useful, and $1\ m^3$ is not much better. The unit used has therefore been 1 Mcf or even 1 MMcf (where M indicates a factor of 10^3 and MM a factor of 10^6); 1 MMcf is approximately equal to $3 \times 10^4\ m^3$; 1 tcf, a still larger unit, where t indicates one trillion, is approximately equal to $3 \times 10^{10}\ m^3$.

3.2 Very large and very small numbers

These numbers are not unmanageable for laymen. The difficulty arises because of the need to express huge quantities in terms of these very small units. The units are small because they were designed to satisfy domestic or laboratory scales of magnitude – for shopping lists, light bulbs, bicycles, experiments. The unit of work (1 erg, representing a force of 1 dyne acting through a distance of 1 centimeter) loses manageability for most people when it is used to express the work expended during a major earthquake (about 10^{25} ergs). The traditional units of heat energy are likewise too small for quantities beyond the individual or laboratory scale. In the metric world the calorie, and in the nonmetric world the BTU (British thermal unit), require multiplication by several powers of 10 to acquire any relevance to quantities actually used. The *practical* unit of heat in the nonmetric world has been the *therm* (1 therm = 10^5 BTUs, or about 25×10^6 calories). The modern metric unit of energy, the *joule*, is no more manageable for practical purposes (1 joule is the work done to move 1 kg through 1 m with an acceleration of 1 m s^{-2}). Hence 1 joule is exactly equal to 10^7 ergs and is approximately equivalent to 10^{-3} BTU; the already minute BTU is approximately equal to 1 kilojoule (kJ).

Let us now apply these minute units of energy to modern outputs of oil and gas. Approximate equivalents are as follows:

$$1 \text{ m}^3 \text{ of light oil} \equiv 35.6 \times 10^6 \text{ BTU}$$
$$1 \text{ barrel of light oil} \equiv 5.7 \times 10^6 \text{ BTU}$$
$$1 \text{ m}^3 \text{ of natural gas} \equiv 31.6 \times 10^3 \text{ BTU}$$
$$1 \text{ Mcf of natural gas} \equiv 0.9 \times 10^6 \text{ BTU}$$

Production and consumption rates are now measured in the millions of barrels (or in units of 10^5 m^3) of oil, and in trillions (10^{12}) of cubic feet (or in units of 10^9 or 10^{10} m^3) of gas. An elementary illustration is provided by the world's production of oil and gas in 1981: about 3.5×10^9 m^3 of oil and 1.65×10^{12} m^3 of gas.

Petroleum scientists will have to accustom themselves to expressing the energy output of their products in terms of *terajoules* (TJ) or of Q:

$$1 \text{ TJ} = 10^{12} \text{ joules}$$

1 m^3 of average crude oil, or 10^3 m^3 of natural gas = 0.038 TJ approximately.

So the 1981 world oil production represented the energy equivalent of more than 1.3×10^8 TJ.

$$1 \text{ Q} = 10^{18} \text{ BTU (1 quintillion)}$$
$$\cong 10^{15} \text{ cf of natural gas}$$
$$\cong 2.7 \times 10^{10} \text{ m}^3 \text{ of crude oil}$$
$$\cong 4 \times 10^{13} \text{ kg of coal}$$

The statistics of the industry of which petroleum geologists should acquire at least a general grasp are discussed below.

3.3 Numbers of people

There are probably between 40 000 and 50 000 petroleum geologists in the noncommunist world. There may be as large a number in communist countries also. The figure is more than guesswork, as there are (in 1984) more than 40 000 members of the American Association of Petroleum Geologists. By no means all of these are in fact petroleum geologists, but there must be many more active petroleum geologists who are not members of the Association (principally because their languages are not English).

At the beginning of 1984, there were about 1250 seismic crews operating in the noncommunist world. Nine out of ten of them are land crews; one-tenth operate from vessels. Nearly 60 per cent of these crews operate in the USA. Also available in 1984 in the noncommunist world were some 6500 drilling rigs and their crews, 5000 of them in North America. The proportion of these rigs active at any one time is dependent more upon economic conditions than upon drilling successes. The distribution of the rigs is of some interest in this connection.

In the world's most productive petroleum province, the Middle East, only about 200 rigs are normally active; only about 30 of these are in Saudi Arabia. It does not take a great number of wells to discover and operate fields like those in the Middle East. Of the somewhat larger numbers of rigs available in Europe, Africa, and the Asia–Pacific region, in 1984 almost one-third were employed in the North Sea, Algeria, and Indonesia. The largest concentration of rigs outside North America and the Soviet Union is in Latin America; about 300 are employed in Mexico and Brazil alone.

3.4 Numbers of wells

By late 1982, the oil and gas industries had drilled about 3.6 million wells. More than 2.8 million of these had been drilled in the USA. The industry in the USA drilled the extraordinary total of 700 000 wells between 1950 and 1964, inclusive; it surpassed even this rate of drilling in 1981, with more than 75 000 wells in the single year. These totals are about ten times the totals drilled in all the rest of the noncommunist world.

Again, the reason for this lopsided activity is primarily economic. In most countries, a small number of large companies are active, or a single government company operates without competition. In the USA there are a great number of independent operators as well as several scores of large companies, and competition between them results in the drilling of a far larger number of wells than are likely ever to be drilled in other areas of comparable size. These lopsided comparisons would undoubtedly be less lopsided if the Soviet Union were included in them. We do not know how many wells the Russians drill annually, but the numbers are probably about 20 percent of the American total. In very round figures, the Americans drill about 12×10^7 m of hole annually, the Russians about 3×10^7 m.

The USA has about 600,000 producing wells. The whole of the Middle East, including Egypt and Syria, has fewer than 7000; western Europe has fewer than 6000. Whereas typical individual oil wells in the Persian Gulf countries, or in Mexico, produce more than 10^3 m^3 daily, the average US well produces less than 3 m^3 a day. Some 75 percent of all US wells are what the industry calls *stripper wells*; each yields less than ten barrels a day (1.6 m^3). Though the actual average production of all US stripper wells is less than 0.5 m^3 of oil per well daily (plus an equal or greater amount of water), their total production is about 16 percent, and the remaining reserves they represent are about 20 percent, of the national totals.

It will be apparent that it is difficult to generalize about drilling success ratios. In the 15-year drilling spree of 1950–64, about 25 percent of the wells drilled in the USA could be described as "exploratory" wells rather than as field development wells. Of these "exploratory" wells, fewer than one in 10 000 discovered an oilfield of 10×10^6 m^3 recoverable reserves or larger. Only one or two wells out of the 700 000 discovered what would rank as giant fields outside North America. In the record drilling year of 1981, only about 10 percent of the wells were "wildcat" wells (see Sec. 23.1). Of these 8000 or so truly exploratory wells, fewer than 1 percent found fields capable of being profitably operated; none found one having 10×10^6 m^3 of oil or equivalent in gas in recoverable reserves. As a general average, US operators reported about 1000 new fields discovered each year during the late 1970s and early 1980s. About 40–45 percent of them are oilfields, 55–60 percent gasfields. Perhaps one of the 1000 is in fact a substantial field, and the total recoverable reserves represented by the 1000 "fields" may be no more than equivalent to the reserves of a single giant field in world terms. Outside the USA such success ratios would be unimaginable. A success rate better than 10 percent is normally expected; rates better than 25 percent have rewarded the drillers and the geologists during the heyday of exploration in many newly proven areas.

3.5 Numbers of fields

The several million wells that have been drilled have resulted in the discovery of about 33 000 oil- and gasfields. This is necessarily a rather rough figure; many of the 33 000 were small fields that have long ago been abandoned. About 18 000 fields have produced oil or gas, or both, in the continental USA, more than 3000 in the Soviet Union, and about 1000 in Canada. In stark contrast, there are fewer than 150 producing fields in the Middle East.

3.6 Actual reserves discovered and produced

To 1984, the total recoverable, conventional oil discovered in the world was about 2.0×10^{11} m^3 (1.2×10^{12} barrels). About 42 percent had been produced, leaving some 1×10^{11} m^3 (0.7×10^{12} barrels) as proven, recoverable reserves.

The total recoverable natural gas discovered is impossible of estimation because great quantities of it were flared or wasted before its discoverers recognized its value as a fuel in its own right. The cumulative conventional production of gas to 1984 has been about 36×10^{12} m^3 (nearly 1300 tcf). Remaining reserves were then a little less than 100×10^{12} m^3, but very much more will certainly be discovered in the future.

Of cumulative oil production to 1984, the USA and the Persian Gulf countries had each provided about 27 percent, the Soviet Union a little over 15 percent, and Venezuela a little over 8 percent. Canada, Mexico, Libya, Nigeria, and Indonesia shared 11.5 percent. Of cumulative conventional gas production, North America had provided about 60 percent, the Soviet Union and eastern Europe nearly 20 percent. Of current annual oil production (1984), approximately 3.2×10^9 m^3, 54 percent comes from the Soviet Union, the USA, Saudi Arabia, and Mexico. Of current annual gas production of some 1.6×10^{12} m^3 (55 tcf), the USA, the Soviet Union, Canada, and the Netherlands provided nearly 75 percent. The USA alone provided more than 50 percent of world oil production until 1950 and much more than 50 percent of world gas production until 1970. Even as this is written, Texas has produced more oil than Saudi Arabia, and California and Louisiana combined have produced almost as much as Venezuela.

Of remaining known reserves, however, the distribution is very different. Five Middle East countries (Saudi Arabia, Kuwait, Iran, Iraq, and Abu Dhabi) possess nearly 55 percent of the oil, the Soviet Union and China about 13 percent, and Mexico nearly 8 percent. Of the gas reserves, the Soviet Union possesses about 40 percent and Iran 16 percent; these two countries plus the USA, Algeria, Saudi Arabia, and Canada have about 75 percent of known gas reserves.

3.7 Numbers of basins

Oil- and gasfields, and their output and reserve figures, are conventionally tabulated by country because it is to national treasuries that much of their wealth is contributed. For geologists, oil and gas are better assigned to geological than to political entities. The geological entities from which petroleum is derived are *sedimentary basins* (see Ch. 17). Many productive basins lie of course wholly within the boundaries of single countries; many more extend from a single nation's territory on to that nation's continental shelf. Other basins, however, are shared by more than one country; the unique Persian Gulf Basin is shared by 12 producing countries.

There are about 700 readily distinguishable basins in the world, and finer subdivision could no doubt increase this number considerably. It has been said that China alone possesses 200 sedimentary basins. Of the world's 700, half have been intensively or partially explored for oil and gas; the remaining 300–400 are essentially unexplored at the time of this writing. About 150 basins are (or have been) commercially productive. Very few petroleum geologists (or oil people in general) could name more than half these productive basins unaided by maps or references, because many of them are of no more than trivial significance. The highly unequal distribution of reserves between the basins, and between the individual fields in them, is so important that it needs extended explanation.

3.8 The significance of lognormality

It seems to be a law of nature, though it has never been adequately explained, that size distributions of natural phenomena display a characteristic form of lopsidedness. The lopsided distributions are said to be *lognormal*, but they deviate in an important respect from true lognormality.

A lognormal distribution is one which becomes essentially *normal* (i.e. Gaussian) if the *logarithm* of the size factor is plotted against the number of its representatives. The relation therefore becomes essentially *linear* if plotted on probability paper, on which the logarithm of the size factor is plotted against the *cumulative* frequency of its occurrence.

In the case of the natural phenomena considered here, the plotting of the logarithms of the size factors against their populations results directly in a distribution which is approximately *linear* and not normal. Most of the population (or money, or power, or whatever) is contributed by a very small proportion of the representatives. The explanation of this universal phenomenon is probably to be sought in the circumstance that numerous complex controlling factors behind the magnitudes of the phenomena are multiplicative in their effect rather than additive.

Among phenomena of immediate concern to Earth scientists that illustrate lognormality of size distribution (in our sense of the term) are the elements making up the Earth's crust, the magnitudes of earthquakes, and the sizes of clasts in a detrital deposit like that of an entire stream bed. To remark only upon the first of these: of the 92 naturally occurring elements, only eight make up more than 1 percent of the crust each, by weight. Only two elements, oxygen and silicon, make up 75 percent of the total by weight; only ten make up 99 percent. The lognormality of the sizes of economic mineral and fossil fuel deposits is equally heavily lopsided.

About 95 percent of the world's known, conventional oil occurs in only 50 basins. Between 70 and 75 percent occurs in only ten of them, and almost half of the total is contributed by the Persian Gulf Basin alone; the other nine are the Maracaibo, Ural–Volga, and West Siberian basins, the Reforma–Campeche region of Mexico, the Gulf Coast and Permian basins of the USA, the Sirte Basin, the Niger delta, and the northern North Sea. If conventional gas is considered as well as oil, and converted to oil-equivalent at 1000 m^3 of gas to 1 m^3 of oil (6000 cf per barrel), calculations by Lytton Ivanhoe indicate that the top ten basins contain 68 percent, and the top 25 basins 84 percent of the total known reserves.

Of all known conventional oil, 25 percent is contained in only ten fields, seven of them in the Middle East. Fifty percent is contained in fewer than 50 fields, or no more than one-fifth of 1 percent of discovered fields. In world terms, a giant oilfield is defined as one containing initial recoverable reserves of 500×10^6 barrels (about 0.8×10^8 m^3) of oil. Some 300 fields (about 1 percent of the total found) are known to qualify for this category. These 300 fields contain about 140×10^9 m^3 of ultimately recoverable reserves of oil (about 8.8×10^{11} barrels), or 73 percent of the total.

For years, nine or ten fields have contributed 25 percent of the world's annual production of oil. Whereas more than 100 fields had each produced more than 0.1×10^9 m^3 (about 625×10^6 barrels), by the end of 1982, only nine had produced ten times this quantity. In order of output, they are Ghawar in Saudi Arabia, the Bolivar Coastal field in western Venezuela, Greater Burgan in Kuwait, Romashkino in the Ural–Volga Basin of the Soviet Union, Kirkuk, Safaniya–Khafji, and Agha Jari in three separate Middle East countries, Samotlor in Siberia, and Abqaiq in Saudi Arabia. All of these fields except Romashkino are capable of maintaining positions in the top dozen through the 1980s at least. It is only the contributions of these supergiant fields that made the growth of oil consumption in the 1960s and 1970s possible.

The historical output of natural gas has been less strikingly lopsided because few countries produced their gasfields to capacity until the 1970s. Almost 100 fields contain more than 140×10^9 m^3 (5 tcf) of recoverable reserves apiece; a number of them are also giant oilfields (Prudhoe Bay in Alaska and A. J. Bermudez in Mexico are outstanding examples). Twenty contain more than 10^{12} m^3 each (35 tcf). Eight of these are in

Siberia, five in the Persian Gulf region, three in Turkmenia, two in the Netherlands–North Sea Basin, one in Algeria and one in the southern USA.

Fifteen individual gasfields have produced more than 100×10^9 m^3 (3.5 tcf), which is approximately the gas equivalent of a giant oilfield. Twenty-five percent of cumulative gas production has come from about 75 fields. In the future, the lopsidedness will become much more pronounced. Of the 12 largest gasfields known in 1982, only Groningen in the Netherlands was discovered before 1965. Half of them are not yet in production as this is written. If markets and delivery systems were available, each of these supergiant fields could individually supply 5 percent or more of world demand. If pipelines are completed for it, the Urengoi field in Siberia may alone supply 15 percent of it by 1985. Several fields in the Persian Gulf Basin are quantitatively capable of a comparable level of production, but as they lie in a hot desert region they may never be called upon to do so.

The early wastage of natural gas has already been alluded to. Even if no wastage is permitted, the volume of gas produced at the wellheads is very different from the amount that enters the consuming world's energy stream. Of the total gross production, at least 15 percent is still either vented, flared, re-injected into the reservoir for pressure maintenance (see Sec. 23.8), or is lost through shrinkage during processing before delivery through pipelines to consumers. The net volume is called *marketed gas*. Of the original gross production, almost 40 percent is consumed by the US market and 25 percent by the Russian market, leaving only 35 percent to be consumed by the rest of the world outside the two superpowers.

3.9 Rates of additions to reserves

The lognormal size distribution of oil- and gasfields means that a single giant or supergiant discovery overshadows hundreds of routine or "average" discoveries. A graph showing annual additions to reserves, through new discoveries plus extensions and revisions of existing discoveries (see Sec. 24.2.2), therefore consists of fluctuations above and below a general average discovery rate but punctuated by huge peaks representing discoveries of the supergiant fields (Fig.3.1).

To smooth out these violent variations, it is convenient to consider graphs showing additions to reserves as five-year moving averages, with the ultimately recoverable reserves for each field dated to the year of the field's initial discovery (not to the year in which the volume of reserves was first recognized and included in reserve totals). From such a graph (Fig.3.2), it is clear that (if we exclude the war years of 1940–5) the average annual rate of additions to world oil reserves between 1935 and 1970 was a little under 3×10^9 m^3 (about 18 billion barrels). There has been a marked decline since the end of the 1960s, and world production overtook the growth of reserves at that time.

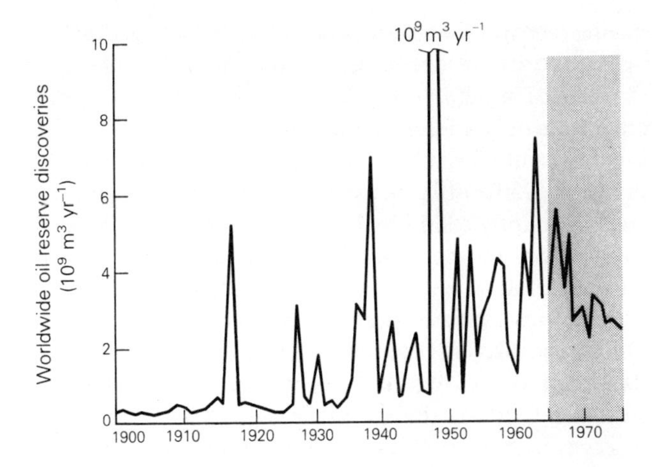

Figure 3.1 World oil discovery history, with reserves of each field assigned to the year of its discovery. Note the peaks representing the discoveries of the Bolivar Coastal field (1917), Burgan (1938), Ghawar (1948), and several supergiant fields in the 1960s. (After H. R. Warman and M. Halbouty.)

These bare statements, however, obscure a vital point. The number of "ordinary" oilfields discovered each year continues to increase year by year (even in North America). Discoveries of giant fields (those

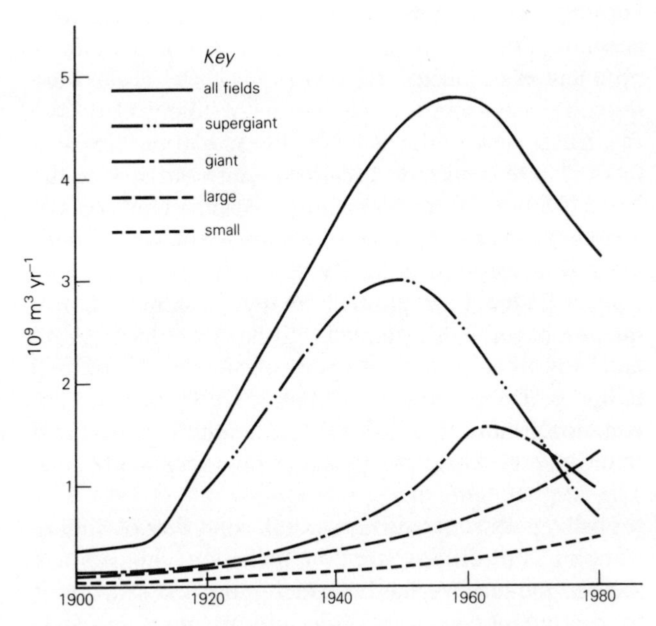

Figure 3.2 Smoothed graph of oil discovery history by field size, in 5-year intervals since 1900 (in 10^9 m^3 yr^{-1}). Note that whereas the number of small and large field discoveries is still on the increase, the number of giant and supergiant discoveries has been declining since a peak in the 1950s and 1960s. (After T. A. Fitzgerald, *AAPG Memoir* 30, 1981.)

having recoverable reserves of 500×10^6 barrels (8×10^7 m^3) or more) are becoming scarcer, however, and discoveries of supergiants (5×10^9 barrels or 8×10^8 m^3) have been declining since the 1960s (Fig.3.2). By far the greatest number of giant oilfields was discovered between 1950 and 1970, especially during the decade of 1955–64, aptly called by Harry Warman the heyday of exploration in the Middle East and North Africa. Although the six largest known oilfields were all discovered before 1955, 50 out of 150 billion-barrel (10^9 bl) fields were discoveries of those ten years. Because of the dominant role in total reserves of the relative handful of supergiant fields, the overall rate of additions to oil reserves therefore declined between 1963 and 1983.

The contrast between discovery rates for oil and for gas is very striking. The value of natural gas as a fuel independent of its association with oil was not widely recognized outside the USA until the mid-1950s (see Sec. 4.4). Of the top 50 oilfields (representing 50 percent of total known reserves), 40 were found before 1965, only ten since. Of the top 50 known gasfields, only 12 had been identified before 1965; 38 were discovered in 1965 or later. It is unlikely that giant gas discoveries have peaked even now, though no future decade is likely to surpass that between 1965 and 1974 in the number of supergiants discovered.

3.10 Costs of finding and producing oil and gas

Until the end of the 1960s, the petroleum industry's concern about the costs of finding and producing their products was concentrated on *drilling costs*. These were quoted as costs per foot (or meter) of hole drilled, and the figures were of the order of US $30–50 m^{-1} (see Sec. 23.6). These costs have escalated enormously since the early 1970s, for three elementary reasons. Unit costs of all operations have risen everywhere; exploratory wells are now drilled much deeper than formerly, and much deeper fields are exploited; finally, drilling and production have been extended into deep waters offshore and into other regions of operational difficulty in high latitudes. Deep drilling costs exceed $300 m^{-1} and are commonly more than twice that amount. The total cost of drilling the world's wells is now in the realm of $$20 \times 10^9$–$50 \times 10^9$ annually.

Drilling wells, of course, is not the only cost of finding oil or gas. Pre-drill exploration, including both geologic and seismic surveys, has been increasing in cost at about 15 percent per year for a decade or more. In 1981, seismic surveys on land in the noncommunist world cost about $$3 \times 10^9$ annually; marine surveys cost an additional $$0.75 \times 10^9$. Total seismic costs, including processing of the data, are between $$4 \times 10^9$ and $$5 \times 10^9$ annually; one-third of these costs is incurred in the USA.

The industry's attention has consequently been shifted from the costs of drilling to the *costs of finding oil* (or gas). So long as these costs were of the order of $1 per barrel ($6 m^{-3} approximately), they were of little concern. In the Middle East, the cost of finding oil was measured in cents per barrel. By 1981, the average cost of finding oil in the noncommunist world had become almost $55 m^{-3}, and this figure can only become larger.

Costs of *producing* oil and gas have risen proportionately. The operation of a single producing well, years after its field has been discovered, runs from hundreds of dollars monthly in the continental USA to tens of thousands monthly in deep waters or polar regions. Costs of production are further complicated by laws governing allowable *rates* of production. These considerations go beyond the immediate concern of the geologists (until they become members of management committees). With outlays of these magnitudes, however, it is obvious that concentration on the costs of exploration, of drilling wells, of operating wells once drilled, or of any downstream operations (transporting and refining the oil, for example) is futile, though each item is of course of critical individual concern to the industry (and to the consumer).

The industry's quantitative fiscal concern is now best expressed, therefore, in terms of the *cost per unit of daily production*. This cost represents the total capital and operating costs required for the discovery and production of the oil or gas, divided by the volume of production achievable or allowable under the physical, legal, and political conditions governing the operation. The costs considered are technical costs only, and do not include royalties or taxes (though these have to be included in the larger, overall "costs of doing business"). The actual figures, of course, vary from region to region for the physical conditions, and from jurisdiction to jurisdiction for the legal and political conditions; they also, again of course, vary with time (though invariably increasing). At the beginning of the 1980s, however, we may consider sources of oil as falling into four unit-cost categories (with costs in 1980 US dollars):

(a) *Low-cost oil*, such as that from the Persian Gulf, Indonesia, or Nigeria. This averages perhaps $3000 per daily barrel ($$2 \times 10^4$ m^{-3} of daily production). In simple terms, the achievement and maintenance of a production rate of 10^5 m^3 per day (about 600 000 barrels per day) requires investment of two billion dollars ($$2 \times 10^9$).

(b) *Medium-cost oil:* Costs average, say, $9000 per daily barrel. The larger fields in the North Sea or southeastern Mexico, and fields in the deep interiors of continents other than North America or Europe, fall into this category.

(c) *High-cost oil:* Costs average $18 000 per daily barrel or more. Arctic oil, and oil from the smaller fields in the North Sea or off eastern Canada, will be in this category. Those with difficult individual problems, such as a recalcitrant reservoir rock, are in the $30 000 per daily barrel range.

(d) *Excessive-cost oil:* Most unconventional oil sources require still higher costs per unit of output. The Athabasca tar sands, for example, will require investments variously estimated at $20 000 to more than $50 000 per daily barrel of output. Shale oil may cost between $50 000 and $100 000 per daily barrel.

Gas production may be similarly categorized. Gas from the continental USA, western Canada, or onshore western Europe is low-cost gas. Gas from the eastern Canadian offshore is high-cost gas. Gas from the Bering Sea would be of excessive cost.

As is observed in Chapters 23 and 26, the development of oil or gas resources in any remote, hostile, or deep-water region is conceivable only if output is to be large; a minimum of 50 000 m^3 per day in the case of oil, or of 5×10^7 m^3 per day in the case of gas. Minimum outlays are therefore in the billions of dollars. There is general recognition that these huge costs will approximately double by the year 2000.

4 Historical development

Commercial oil production, in the modern sense, has been carried on for more than a century. Petroleum geology, as a practice deserving to be called a science, is a wholly 20th century subject, however. Its historical development, from rudimentary and sometimes wild beginnings to its present proud status, falls naturally into five clearly demarcated periods, which have become shorter with time as the pace of all events has quickened. In some senses, we have already entered a sixth phase of the progression, but it is one brought about more by political and economic developments than by any that can be called geological.

4.1 The pre-geological years (1842–1901)

There was no such subject as "petroleum geology" before the year 1900. All areas producing oil before that time had been discovered through *seepages*. Oil in North America came from the Appalachian and Great Lakes regions (both from Paleozoic strata), and from Tertiary sandstones in southern California. Outside North America, it came principally from the Baku and Groznyy regions of the Russian Caucasus, Ploesti in Romania, Digboi in Assam, Sanga Sanga in eastern Borneo, and Talara in Peru. All of these regions yielded oil from Tertiary sandstones; it was to be 1937 before a large oilfield in Paleozoic strata was found outside North America.

Two observations should be recorded on the lack of geological contribution to this first half-century of the oil era. The first seminal, genuinely geological contribution to the development of oil finding was the *anticlinal theory* (see Sec. 16.3). It was not until the Spindletop discovery in the Texas Gulf Coast had been made, in 1901, that the anticlinal theory became a practical influence in oil exploration; yet the theory existed long before 1901. In fact it antedated the Williams and Drake wells, in Ontario and Pennsylvania, credited with being the first successful wells drilled with the recovery of oil as the objective.

The evidence for the anticlinal theory was first observed and recorded in Canada. In 1842, when William Logan became the first director of the new Geological Survey of Canada, he noted oil seepages from anticlinal structures in the Paleozoic rocks of eastern Gaspé, south of the mouth of the St Lawrence River. Thomas Sterry Hunt, an American employed by Logan's survey between 1847 and 1872, formalized the anticlinal theory in 1861. E. B. Andrews offered the same proposal in the same year, on the different reasoning that the crests of anticlines should be the sites of fractures that would permit subterranean oil and gas to reach the surface. A much-quoted observation by I. C. White, on the gas-producing region of the American Appalachians, made the same point succinctly in 1885. Yet it was not until after the turn of the century that geologists began to influence oil and gas exploration by actually looking for anticlines.

The second observation of interest is that all pre-drilling recovery of oil – from hand-dug pits and trenches in Burma, China, Peru – and the initial "oil well" itself (the Drake well, 1859), unknowingly exploited a type of trap that was to be the last formally identified and given a name. They are *geomorphic traps*, created by the truncation of the reservoir rock by a very recent erosion surface. It was not until 1966 that this important trapping mechanism was accorded individual distinction by Rudolf Martin (Sec. 16.6).

4.2 From 1901 to 1925

Surface geological surveys specifically aimed at the identification of oil structures began in 1901, following the discovery at Spindletop in Texas. Under powerful encouragement from the United States Geological Survey, the anticlinal theory was to dominate petroleum exploration until about 1932. It was indeed believed by many that *all* oilfields were actually located on some form of anticlinal structure even if no such structure could be mapped.

This simple type of structural search was spectacularly successful in the USA, but many of the successfully drilled anticlines lay within areas in which seepages were known: in the Mid-Continent, especially in Oklahoma, where the Cushing and Glennpool fields dominated production at the start of World War I and made Oklahoma the oil capital of the world; in the Rocky Mountain basins, where the Salt Creek field (see Fig. 16.94) led to the scandal surrounding nearby Teapot Dome; the Texas Panhandle; Illinois; and California. Outside the USA, many more discoveries were made through the combination of seepage and anticlinal structure: in the Maracaibo Basin of Venezuela, in Trinidad, southern Argentina, the Dutch East Indies, eastern Borneo, Burma, and, casting the longest shadow of all, at Masjid-i-Suleiman in Persia, the discovery well for the Middle East oil province. Five of the most spectacular oil provinces in the American hemisphere saw the acme of their success in this period: the Oklahoma sector of the Mid-Continent (1910–25), the San Joaquin Valley of California (before 1920), the Los Angeles Basin (1910–25), Trinidad (1912–20), and the Tampico region of Mexico (1908–20).

The distribution of world oil production exercised a profound influence on the progress of World War I. Production in the post-war period provides a vivid example of lognormality (see Sec. 3.8). The USA in 1925 produced more than 120×10^6 m^3 of oil, nearly seven times as much as the second country, Mexico. If individual American states were ranked with other countries, California, Oklahoma, and Texas occupied the first three places; Arkansas was fifth (following Mexico), Kansas seventh (following the Soviet Union), Wyoming ninth (following Persia), and Louisiana eleventh (following the Dutch East Indies). Venezuela, shortly to become the world's leading exporting country, produced little over one-twelfth of California's volume. The top four American states produced twice as much oil as all the non-American world combined. It was to be the end of the 1940s before a second country (Venezuela) reached 25 percent of America's oil production. It was not an accident of history that petroleum geology was an essentially American science until then.

Two events of 1917 were to have far-reaching effects upon the science. The first sector was discovered of what was to become the Bolivar Coastal field on the east shore of Lake Maracaibo in Venezuela. It was to become the most productive oilfield in the world during and following World War II, making an immeasurable contribution to the war supplies of the Allies; it was the first great field in a homoclinal, nonanticlinal trap, antedating East Texas by 13 years; it was the first field requiring extensive development drilling off dry land; it was to be a leader in the development of techniques for the prolific recovery of very heavy oil; and it was to establish Standard Oil Company of New Jersey (now Exxon) as the largest oil company of all.

The second important development in 1917 was the establishment of the American Association of Petroleum Geologists (the AAPG), with headquarters in Tulsa, Oklahoma, not far from the Cushing and Glennpool fields. The association has become one of the largest and most successful scientific organizations in the world, and its array of publications constitutes the library of every petroleum geologist (see Sec. 1.3.3).

4.3 From 1925 to 1945

At the onset of this period the anticlinal theory of oil and gas accumulation was still in the ascendant. The first great field in a homoclinal, stratigraphic trap had been discovered in 1917 (in western Venezuela), but it had not yet been shown to be a single field and its discovery was clearly attributable to seepage drilling. It was the East Texas discovery in 1930 that drew the attention of oil geologists to the potential of homoclinal traps. The discovery was a matter of pure chance; there was no seepage and no surface evidence of change of dip. The change of thinking set in motion by the discovery was due principally to its size; for 20 years the East Texas field was considered by American oilmen to be the largest oilfield in the world, effectively by definition (despite the fact that its earlier counterpart in Lake Maracaibo is four times as large).

The recognition of nonanticlinal traps spurred a rash of oiltrap classifications, almost all of them by American geologists (F. G. Clapp in 1929; W. B. Wilson, and E. H. McCollough in 1934; W. B. Heroy in 1941; O. Wilhelm in 1945). An outsider was V. C. Illing, who, in 1942, eschewed rigid compartmentalization and assigned oil occurrences to generalized structural styles.

A comparable change of traditional emphasis occurred in the contemplation of reservoir rocks. Reservoir rocks had been thought of as sandstones. True, there had been a number of scattered discoveries in carbonate rocks before 1925. Most of them were in the eastern or southern USA, and most were also in Paleozoic formations. An important exception was the Smackover field in Arkansas, discovered in 1922 in a Jurassic limestone and drawing attention to the rocks of that age in the upper Gulf Coast (see Sec. 26.2.4). Outside the USA there were the spectacular fields in the Mexican Golden Lane, in Cretaceous rocks; the relatively insignificant but hope enhancing Turner Valley field near Calgary, later to become the oil capital of Canada; and of course the discovery field for the Persian Gulf Basin at Masjid-i-Suleiman, in the Tertiary.

Despite these finds, carbonate reservoirs continued to be regarded as curiosities. In the first few years of the 1925–45 period, there were further discoveries in carbonate reservoirs, several much larger than any discovered before, and of much wider geographic distribution: La Paz in Venezuela (1925), Yates in western Texas (1926), Kirkuk in Iraq (1927), and Poza Rica in Mexico (1930). There followed the grandiose development of carbonate reservoir fields in the Middle East and the Permian Basin, between 1932 and 1948. But the real understanding of carbonate reservoirs was not yet achieved. An important side development, however, was the support that fossiliferous carbonate reservoirs gave to the increasingly popular hypothesis that oil and gas were of biogenic rather than inorganic origin.

The period under review was that of the greatest innovation in petroleum exploration, with the introduction of a series of ancillary aids quickly adopted as routine. The torsion balance, which detects excesses and deficiencies of gravity, had been introduced in Mexico about 1920, and into the US Gulf Coast in 1923. Its replacement by the gravity meter led to the peak of discoveries in salt dome traps there during the 1930s and 1940s. Reflection seismology, which maps unseen structure in the subsurface, was started about 1927 and enjoyed a remarkable success record in the 1930s, especially in the southern USA. The electric well log (see Ch. 12) was introduced in the early 1930s, and quickly revolutionized fluid evaluations in wells and lithologic correlation procedures between them. Rotary drilling rigs began to displace the old cable tool rigs in the late 1920s. Apart from enabling the drilling process to be much faster and more efficient, rotary rigs permitted much deeper drilling. Before 1925, most wells were shallower than 1000 m. With the new rigs, wells were soon taken to 3000 m (1931), 4500 m (1938), and 5000 m (1945). This brought the geologists new understanding of basin structures and histories; the potentials of deeply buried formations; the need to re-evaluate theories concerning porosities and fluid behavior; and an appreciation of the significance of evaporites.

Two new disciplines vital to the petroleum geologist became widely employed during the 1930s and 1940s. The decades saw the real beginnings of petroleum geochemistry, its pioneers including Benjamin Brooks and Parker Trask in the USA, W. F. Seyer in Canada, Karl Krejci-Graf in Romania, and A. P. Vinogradov in the Soviet Union. They also saw the blossoming of the new science of micropaleontology, developed as an aid to correlation and horizon identification in Tertiary basins of tip-heap bedding, especially in Trinidad and the Caucasus but spreading quickly to the Gulf Coast and other producing regions. The principal triumph of the new discipline during the 1940s was the zonation of the standard Tertiary succession by means of planktonic foraminifera.

The World Petroleum Congress held its first session in 1933 in London and its second in 1937 in Paris. World War II interrupted the new series of scientific, technical, and industrial gatherings, which were not resumed until 1951. The Proceedings of the Congresses constitute important contributions to the literature of all aspects of the industry's activities. The decades of 1925–35 and 1935–45 also saw the publication of four standard reference works for the petroleum geologist: the first two volumes of *Structure of typical American oil fields,* issued by the AAPG in 1929; the sister volume on *Stratigraphic-type oil fields,* issued in 1941; *Problems of petroleum geology,* published in 1934; and the first of the six volumes of the *Science of petroleum,* under the imprint of the Oxford University Press and appearing first in 1938.

During the period, world oil production was totally dominated by seven relatively small regions: southern California; Texas (especially the single field of East Texas, and the Permian Basin in the west, which enjoyed its greatest years between 1926 and 1936); the Mid-Continent region in general; the fields on the east side of Lake Maracaibo in Venezuela; the declining fields of eastern Mexico; Baku in the Caucasus; and the three giant fields of Masjid-i-Suleiman, Haft Kel, and Kirkuk in the Iran–Iraq foothills belt.

4.4 From 1945 to 1960

As the world began its recovery following World War II, the oil industry adjusted to a new circumstance. The USA was to continue to produce more than half of the whole world's oil until the early 1950s, but it had become clear that the center of gravity of oil availability was irreversibly shifting to other regions, most of them in the eastern hemisphere. Recognition of this circumstance was slow in coming to all but a few far-sighted individuals, but it brought with it several vital developments.

First came an extraordinary surge of drilling activity. By 1954, oilmen were drilling over 50 000 wells a year in the USA alone, more than 10 000 of them exploratory wells, and total distance drilled there exceeded 60×10^6 m yr^{-1}. In 1949 the first well to 6000 m (20 000 feet) was drilled in Wyoming, of all places. Drilling was extended to all accessible sedimentary areas, many of which were to prove petroliferous. By the end of the 1950s, most of these basins had wells drilled to the 4000–4500 m range. This drilling surge brought a revolution in the handling of the mass of new subsurface geologic data. It also brought the publication of a most authoritative and influential textbook on *Subsurface geologic methods,* published by the staff of the Colorado School of Mines in 1949.

To make advance appraisals of volumes of oil or gas that might be anticipated in newly investigated regions, geologists made use of a method devised by Lewis Weeks and first published (though not by Weeks himself) in 1948. This was based on the expected ultimate yield of oil from much-drilled and mature basins in the USA, expressed in barrels per cubic mile (or cubic kilometer) of sedimentary rock (see Sec. 24.3.2).

By the end of the 1940s, exploratory drilling had been extended to shallow waters offshore. Drilling had been carried on in Lake Maracaibo much earlier, but this was development drilling progressively extended into a very shallow lake. Early offshore exploratory drilling was concentrated upon areas adjacent to onshore production, in the Louisiana segment of the Gulf of Mexico and in the southern Caspian Sea. By the end of the 1950s, it had been attempted in the shallow shelf areas of many coastal nations.

Concurrently with the surge in drilling came enormous expansion of seismic activity, and increasing mutual understanding between geologists and geophysicists. With the drilling crews, the seismic crews moved offshore (in 1947), devising new energy sources and operating procedures suitable for their new environment. The geophysicist's equivalent of *Subsurface geologic methods,* the first major compilation of *Geophysical case histories,* was published by the Society of Exploration Geophysicists in 1948.

The principal development of this period that could be called exclusively geological was the introduction of modern sedimentology as a new petrological discipline. For the first time, sedimentary rocks were treated the way igneous rocks had been treated for 50 years. The new emphasis was brought about by two comparable but quite distinct achievements. In 1950, laboratory experiments by Kuenen and Migliorini established the mechanism for the emplacement of coarse sands in deep water, leading to the recognition of the turbidite facies. The first adaptation of this development to a prolific reservoir rock was made by Manley Natland in 1952 on the Pliocene sandstones of the Californian basins.

The second decisive development was the recognition of the great variety of carbonate sediment types that constituted productive reservoir rocks. Two almost simultaneous discoveries in Paleozoic reefs, at Leduc in western Canada and Scurry County in western Texas, brought quick attention from both a new breed of carbonate petrologists and the geophysicists. The recognition of clastic carbonate types, initiated for the oilmen by Frank Henson in the Middle East and Leslie Illing in the Bahamas and Caribbean, was equally influential.

The dependence of hydrocarbon migration and accumulation on the laws of fluid dynamics was given its clearest expression during the 1950s in the work of King Hubbert and Gilman Hill. Research in this and a number of other fields stemmed from observations from petroleum geologic experience that demanded simulation by mathematically controlled models. In several such fields of research, notably in aspects of structural geology, the oil geologists enlisted the help of distinguished academic geologists, geophysicists, and engineers.

Finally, a peculiarly American development was to lead not only to a further great increase in drilling, but to major advances in organic geochemistry and the geological study of the generation and migration of hydrocarbons. Natural gas was acknowledged as a valuable and plentiful resource in its own right, and not simply a nuisance to the seekers for oil. Recognition that some sedimentary rock facies (including the nonmarine facies), and some entire sedimentary basins, provided gas provinces and not oil provinces led to geological and geochemical appreciation of the two fuels as distinct products of distinct processes.

The period 1945–60 saw the elevation to the front ranks of producing regions of two great Paleozoic basins, that of the Russian platform (the Ural–Volga Basin) and that of western Canada.

4.5 The 1960s and after

In retrospect, the year 1960 represented a watershed for the explorers for petroleum. From 1960 to 1967 the world enjoyed a substantial surplus of oil productive capacity. The erosion of prices had dramatic consequences for the employment of geologists, geophysicists, and geochemists in the oil industries of the world, especially in North America. The explorers had been too successful, though the belief that this was something permanent and inevitable was a calamitous illusion, soon to be shattered.

The shift in the center of gravity of oil output, foreseen at the end of World War II, was now decisive. A second period of spectacular success in the Middle East, between 1957 and 1967, coincided with one almost as spectacular in Libya, and another, less spectacular but nonetheless important, in Nigeria. The 1960s saw a succession of giant discoveries in western Siberia, later to put the Soviet Union into the lead among oil-producing nations. During the decade 1967–77, annual additions to reserves of natural gas in Siberia were comparable with additions to oil reserves in the Middle East a decade earlier. The 1970s saw equally impressive successions of discoveries in the North Sea and in the southeastern corner of Mexico.

The crucial role of very large individual accumulations (an illustration of the rule of lognormality described in Ch. 3) brought the coining of the terms "giant" and "supergiant" fields; the recognition that the great majority of these huge accumulations occurred in a mere handful of sedimentary basins, most of them in either "Third World" or Communist countries; and the parallel recognition that all or most of the "obvious" structural traps had already been found. The roles of the geologists and geophysicists underwent a profound change of direction.

In the mature regions of petroleum development, and in those approaching that condition, the search for oil and gas became concentrated upon what Michel Halbouty called "subtle and obscure traps." Great emphasis was laid upon sedimentology, especially the sedimentology of carbonate rocks. A surge of carbonate rock classifications appeared in the literature. Reservoir studies were greatly improved by field research on modern sediments, emphasizing provenance, depositional mechanics, and diagenetic processes. Sandstone reservoirs, in particular, became much more intimately understood after the introduction of the scanning electron microscope during the 1960s (see Sec. 13.2.6).

Huge strides were made in the acquisition, processing, resolution, and interpretation of seismic data. The use of magnetic tape, digital recording, and processing techniques, and the introduction of new sources of seismic energy, for both land and marine operations, made for faster and more efficient data gathering, deeper penetration of the rock section and better resolution of the response. Logging techniques and other aspects of borehole geophysics (especially the borehole gravity meter) were greatly refined and largely computerized (Ch. 21).

These advances in geophysical capacity had two impressive consequences. The first was a far greater reliability in the detection of subtle or complex trapping mechanisms. The second was the greatly expanded assurance of discovering the large accumulations (if any were present) early in the first exploration phase of a basin's history. This was especially vital in frontier or offshore areas, where only the early discovery of a large field ensures that exploration is pursued to the "subtle and obscure trap" phase. A major re-direction of exploratory thinking was brought about. For many years before it, conventional wisdom among explorationists maintained that the large trap may remain undiscovered for years while scores of holes are drilled and thousands of kilometers of seismic line shot. This is in effect what had happened in western Canada, Australia, and Algeria, as is described in Chapter 26. It is nearly inconceivable that it could have happened during the 1960s or 1970s in the North Sea, on the Alaskan north slope, in the jungles of eastern Ecuador, or in the swamps of the Sudan.

Organic geochemistry, in its applications to the understanding of hydrocarbon generation, maturation, and migration, first became thoroughly comprehensible to geologists and geophysicists in the late 1960s and early 1970s. Many of the leaders in this invaluable development came from the Soviet Union and western Europe. The semi-organized guesswork of earlier generations of geologists about source sediments, in particular, was for the first time put on a genuinely scientific foundation. Quantitative laboratory techniques for source and reservoir evaluation became widely used and understood (see Ch. 7).

From about 1967 onwards came major improvements in all forms of offshore petroleum technology: exploratory and development drilling, seafloor services, and gathering systems. These led in turn to a rapid increase in geological understanding of the continental shelves and upper slopes. All these technological and instrumental advances were of course interlocked from the outset with the technology of the computer.

Unquestionably the two most publicized geoscientific developments of the late 1960s and 1970s were those of plate tectonics and remote (satellite) imagery. The former rendered all pre-1970 concepts and nomenclatures of sedimentary basins instantly obsolete. The satellite (LANDSAT) data showed geoscientists (and indeed everyone) the Earth through new eyes. The author of this book does not believe that either of these revolutionary strides (for this is what they were) influenced the actual course of petroleum exploration in any significant way. Still less does he believe that either of them bore significantly on the likelihood of success or failure of any major petroleum exploration program, or would have done so had they become available earlier.

The sedimentary basins in which oil and gas occur are ascribed now to origins in somewhat different processes than they were ascribed to before; but they are still subsident areas containing sedimentary rocks. Plate-marginal basins still unexplored as this is written – those of the Bering and other semi-enclosed seas, for example – will be explored when technology and economics permit. Once they are drilled, their fates will be determined by hydrocarbon occurrence, organic geochemistry, and sedimentary petrology (as well as technology and economics). Plate tectonics may provide convincing explanations of the exploratory outcomes after these have become known, but it is unlikely either to instigate or to deter their achievement.

On land, the sites of oil- and gasfields may or may not be revealed on satellite images. The sites of those already discovered were also revealed by other routes long followed in the industry, and there is no reason to believe that the order in which the sites were drilled would have been different had the images been available.

Both plate tectonics and satellite imagery represent true revolutions in the way Earth scientists view their subject Earth. They are considered in their appropriate settings elsewhere in this book (Chs 17 & 20) without any claim that they have especial value to petroleum geology.

Part II

THE NATURE AND ORIGIN OF PETROLEUM

The elemental constituents of petroleum are few; they are combined abundantly in nature, in some very familiar materials of very widespread distribution. The interrelations between these familiar materials (and some not so familiar), and the routes by which they come to be formed, together or sequentially, are the subjects of Part II of this book.

5 Composition of petroleum and natural gas

5.1 The basic components of petroleum

The claim might be made, one supposes, that the word *petroleum* has a geological foundation. It does not, however, have a geological definition. *Petroleum*, *oil*, and *natural gas* can only be properly defined in chemical and physical terms, and the petroleum geologist should be aware of the chemical nature and physical properties of the materials that provide his livelihood.

Petroleum and natural gas are mixtures of *hydrocarbons*. Carbon and hydrogen are the only elements essential to their composition. Great numbers of hydrocarbons occur in nature, and many more are routinely prepared in the laboratory and the chemical plant. In crude oils, however, the principal components belong to only two hydrocarbon series:

(a) The *paraffin* or *methane* series contains those straight-chain hydrocarbons having the general formula C_nH_{2n+2}. The compounds range from *methane*, CH_4, through a *homologous series* with each heavier molecule adding CH_2 to the one below it. Ethane, propane, and *n*-butane are gases at standard temperature and pressure (STP). The first liquid at STP is *n*-pentane, C_5H_{12}; the first solid is *n*-hexadecane, $C_{16}H_{34}$. Members having molecular weights higher than that are solid paraffin waxes. In modern organic chemical terminology, this series of straight-chain compounds is called the *n-alkanes*.

(b) The *naphthene* series has the general formula C_nH_{2n}. It consists of the alicyclic hydrocarbons or *cycloparaffins*, which are saturated carbon ring compounds. Ring compounds cannot possess fewer than three carbon atoms. Cyclopropane, C_3H_6, and cyclobutane, C_4H_8, are gases at STP; their structures are represented in Fig. 5.1. The first liquid is cyclopentane, C_5H_{10}; it and cyclohexane (C_6H_{12}) predominate in most crude oils.

The *olefins* or *alkenes* are isomerous with the cycloparaffins; that is, they have the same compositions as their naphthene counterparts, C_nH_{2n}, but different structures. They are unsaturated chain compounds, not saturated ring compounds. The simplest olefin, CH_2 or methylene, does not exist in nature. The first three members are ethylene, propylene, and butylene, C_2H_4 to C_4H_8. They and their homologs are important petrochemical products, but they rarely occur in crude oils and never in more than trace amounts.

Both these series of hydrocarbons are *aliphatic* compounds. The *aromatic* compounds, based on the benzene ring and having the general formula $C_nH_{2n}-6$, also occur in many crude oils but they normally comprise less than 1 percent of the crudes and in very few crudes do they exceed 10 percent. Exceptions are a handful of highly aromatic and paraffinic crudes from basins in southeast Asia and North Africa (see Chs 17 & 25). The aromaticity (defined as the ratio of aromatic carbon to total carbon in the compound) bears an inverse linear relation to the density of the crude, in general. Among products, the aromatic component is highest in lubricating oils, lowest in gasoline. Toluene, $C_6H_5CH_3$ (i.e. methylbenzene or phenylmethane) invariably exceeds benzene as a component of aromatic crudes, because it is more soluble in electrolytes.

Paraffin-base crudes, the original staple of the refiners and still the most prized of all oils, constitute only a tiny fraction of modern world crude supplies (by 1980, about 2 percent). The standard is "Pennsylvania crude," and it and most other North American representatives are Paleozoic in age (from Michigan, Ohio, and some Oklahoma fields). Oddly, nearly all the most paraffinic

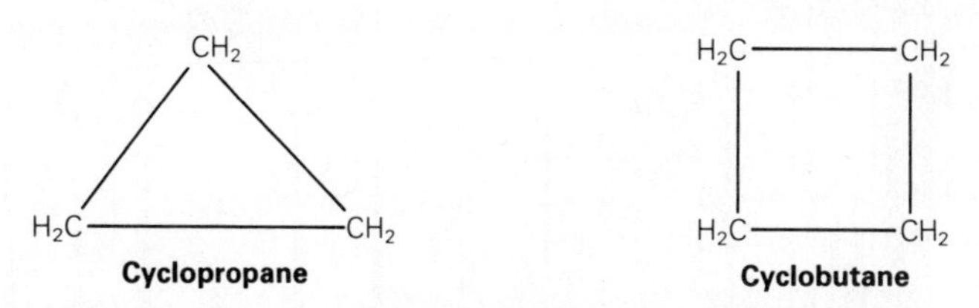

Figure 5.1 The structures of cyclopropane and cyclobutane.

crudes from other continents are much younger: Mesozoic in southern Chile, in the Reconcavo Basin of eastern Brazil, and the Mangyshlak Basin of southern Russia; and even Tertiary, in some African basins, in the northern Caucasus, Borneo, and several Chinese basins (see Chs 17 & 25).

Crudes dominated by the naphthenic components are called *asphalt-base oils*, because asphalts are closely associated with the naphthenes and not with the paraffins. Only about 15 percent of world crude supplies in 1980 were truly naphthene-based; they are the "black oils" of Venezuela, Mexico, parts of California and the Gulf Coast, and many Russian crudes. The great majority of crude oils are of *mixed base* (naphthene–paraffin); they include nearly all Middle East, Mid-Continent, and North Sea oils.

Most crude oils contain minor quantities of hydrocarbons belonging to neither the paraffin nor the naphthene series. Modern gas chromatography and mass spectrometry make the identification of these molecules rapid and accurate. Saturated hydrocarbons appear as peaks on the standard chromatograms, with the branched chain compounds as a second set of peaks adjacent to them (see Fig. 5.4). The minor hydrocarbons represent molecules synthesized by the organisms living at the time of deposition of the source sediments of the oil and preserved through subsequent geologic history; they are "geochemical fossils." In general, they are of much higher molecular weight than the molecules making up the bulk of crude oils, and have structural formulae of the type C_nH_{2n-x}; in other words, the H : C atomic ratio is less than 2.

The minor hydrocarbon molecules include long-chain paraffins derived from the waxes of higher plants; steroids, or complex ring alcohols based on cyclohexane, derived from both plants and animals; and triterpenoid alcohols, aromatic compounds based on terpene, $C_{10}H_{16}$, and having the general formula $C_{30}H_{50}OH$ (Fig.5.2). Steroid and polyterpenoid precursors give rise to naphthenes having between one and four rings in each molecule.

The most interesting and important of these minor series is that of the *acyclic isoprenoids*, based on

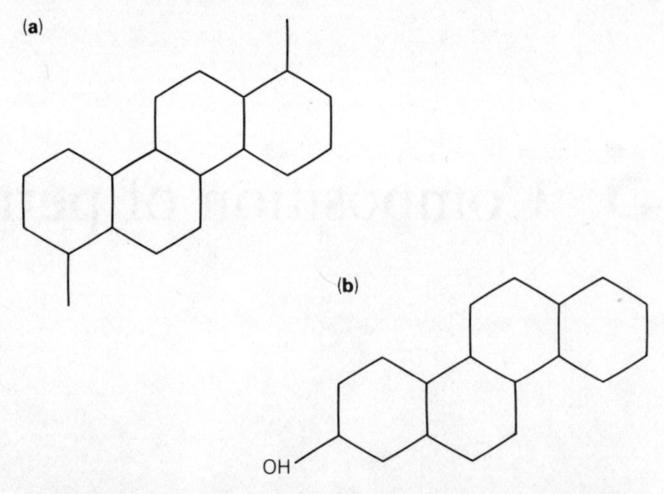

Figure 5.2 Generalized structures of (a) a steroid, and (b) a triterpenoid alcohol.

isoprene (C_5H_8), the source of synthetic rubber (Fig. 5.3). The isoprenoids form long, branched, saturated molecules having a methyl (CH_3) group linked to every fourth carbon atom. These branched molecules occur in the waxes of living plants; those members which occur in crude oils and ancient sediments are believed to be derivatives of chlorophyll, and they constitute useful indicators of an oil's origin. The commonest members are pristane and phytane. *Pristane* is 2,6,10,14-tetramethylpentadecane; it has 15 carbon atoms in its chain, with a methyl group attached to the second, sixth, tenth, and fourteenth, for a total of 19 carbon atoms. *Phytane* is a tetramethylhexadecane, and so adds its four methyl groups to a chain having 16 carbon atoms for a total of 20 carbon atoms. The ratio between pristane and phytane contents of a crude oil or rock extract, measured chromatographically, may indicate the type of organic

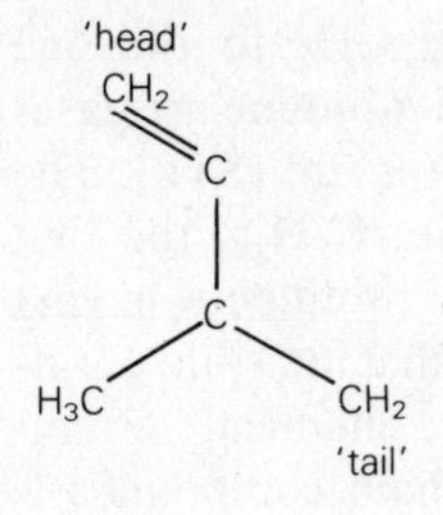

Figure 5.3 Generalized structure of isoprene.

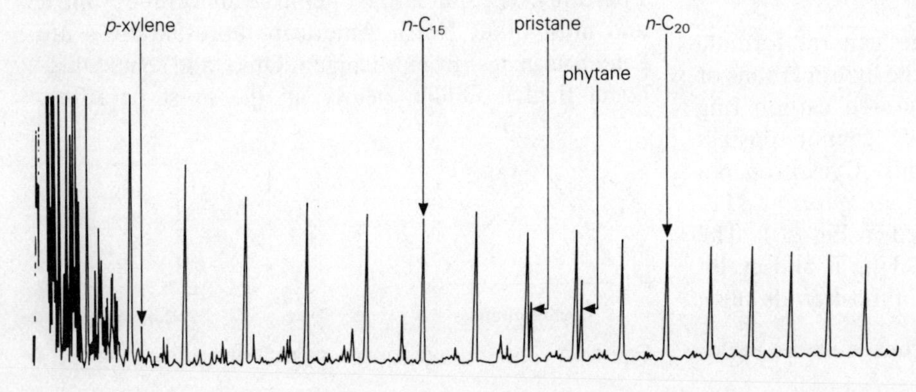

Figure 5.4 A gas chromatogram showing the range of paraffins from C_{10} to C_{25}, with much smaller intervening peaks representing the naphthenes. (Courtesy of K. E. Peters and Chevron Oil Field Research Company.)

matter from which the oil originated and/or the level of thermal maturation of the source (see Ch. 7). A high ratio, for example, may indicate a large contribution of terrestrial organic matter. The ratio is therefore a vital "fingerprint" for any crude oil containing both molecules (Fig.5.4).

5.1.1 *Natural gas*

Natural gas consists of hydrocarbons not condensable at 20 °C (68 °F) and atmospheric pressure; this effectively means the first four members of the paraffin series, methane through *n*-butane. Gas composed almost entirely of methane is *dry gas*. If the proportion of ethane (C_2H_6) and heavier molecules exceeds some arbitrary value (conventionally 0.3 US gallons of vapor per 1000 cubic feet of gas, or between 4 and 5 percent, in North America), the gas is called *wet gas*.

Natural gases consisting largely or wholly of methane may have any one of three distinct origins.

(a) *Petroleum gas* may be formed as a byproduct of the generation of petroleum. In this case the gas either accompanies oil in its subsurface reservoir, or it is contained in a separate reservoir among other reservoirs containing oil. Such gas is called *associated gas*. Petroleum gas may alternatively be formed by the thermocatalytic modification of petroleum under the temperature prevailing in deep reservoirs (see Ch. 10). This gas no longer accompanies oil and is called *nonassociated gas*.

(b) *Coal gas* is formed by the modification of coal, thermocatalytically or otherwise. Much of the world's commercial natural gas is of this origin.

(c) *Bacterial gas* is formed by the low-temperature alteration of organic matter at or near the Earth's surface. It has no direct connection with petroleum. *Marsh gas* is a familiar example occurring in stagnant waters containing decomposing vegetation.

Mineral gas is an unrelated material; the term should be restricted to gases given off during igneous activity.

The gas accompanying oil in the reservoir is in solution in the oil under the reservoir's temperature and pressure conditions. The amount of gas in solution increases with increasing reservoir pressure, and exerts great effect upon the oil's physical properties. It is also responsible for the ability of many oil wells to "flow" their oil to the surface without artificial assistance. If the gas content is sufficient to saturate the oil under the existing conditions, the amount unable to go into solution forms a *free gas cap* above the oil (see Sec. 23.8).

The volume of gas in the reservoir, in relation to the volume of oil, is called the *gas : oil ratio* (GOR). In North America, GORs have always been expressed as cubic feet (cf) of gas per barrel (bl) of oil as they exist in the reservoir. As a barrel contains less than 6 cf of oil, the beginner is surprised to encounter GORs of 1000 or more quoted for many pools. Gas is highly compressible, and the volume required to saturate oil at the high pressures existing in deep reservoirs expands enormously when the oil is brought to the surface. The oil shrinks as the gas bubbles out of solution into a separate phase. One cubic meter of oil in the reservoir therefore represents less than 1 m³ of *stock tank oil* at the surface. This phenomenon has great practical consequences for the production geologists as well as for those estimating recoverable reserves of a field; it is dealt with fully in Chapter 24.

5.1.2 *Gas liquids*

Coal gas and bacterial gas, which of course are also nonassociated gases, are invariably extremely dry, commonly containing 99 percent or more of methane and so having $C_2 : C_1$ ratios of the order of 10^{-6}. The contributions of the methane homologs to the various associations displayed by natural gas are illustrated in Fig.5.5, derived from analyses of hundreds of Russian gas deposits.

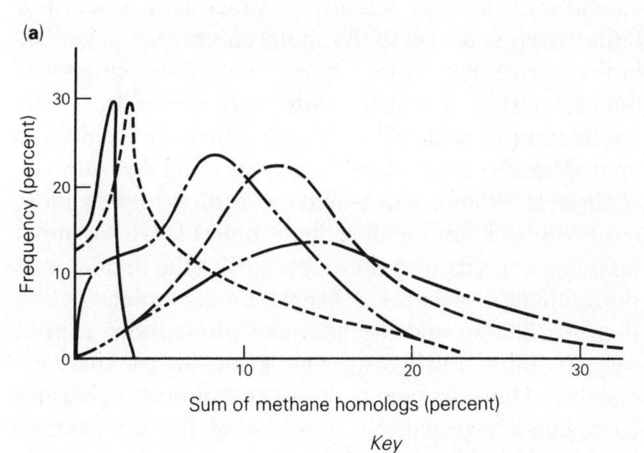

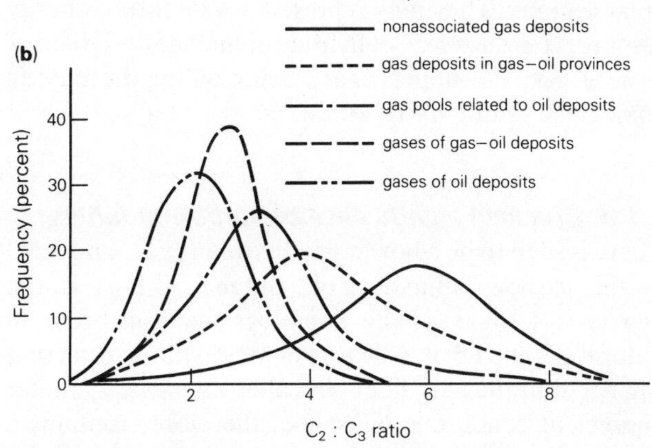

Figure 5.5 Contributions of the methane homologs to the associations displayed by natural gas. (After V. F. Nikonov, *Dokl. Akad. Nauk SSSR*, 1972.)

From 10^6 m^3 of typical wet gas, about 150 m^3 of *natural gas liquids* (NGL) are economically recoverable (about 25 bls per MMcf in the old terminology). The NGL are recovered in two forms: as *condensate*, or that portion of the wet gas that is liquid at STP (pentane and higher, and so in some quarters called *pentanes-plus*, though 10–40 percent of butanes may be included in addition); and *liquefied petroleum gases* (LPG), mostly propanes and butanes which are gaseous at STP but are readily liquefied and are then extracted from the gas for their own value. Hence "marketable gas" represents a smaller quantity than "recoverable gas" because these materials are extracted from it and sold or used separately.

The acronyms NGL and LPG are confusing to the uninitiated (especially when referred to under STP conditions), but they must not be further confused with LNG, which is *liquefied natural gas*: methane put into that condition at −160 °C (−260 °F) and atmospheric pressure. By this treatment the volume of the gas is reduced by a factor of more than 600, for transoceanic transport in specially designed cryogenic tankers. One kilogram of LNG represents about 1.5 m^3 of gas, or 1 tonne is equivalent to about 53 Mcf.

Though the liquefaction of methane requires exceedingly low temperatures under otherwise normal conditions, methane actually becomes solid when it is both frozen and *wet*. Water molecules have a cage-like lattice structure, closely resembling the pentagonal dodecahedron in which pyrite may crystallize. This configuration physically entraps other molecules of appropriately small dimensions, forming *hydrates* or *clathrates*. Ethane and isobutane form hydrates most easily (the diameter of the ethane molecule is 0.418 nm); methane and nitrogen do so less easily (the diameter of the methane molecule is 0.38 nm). Larger molecules like those of helium and the heavier hydrocarbons cannot form hydrates, and hydrogen molecules are too small for stability. Thus methane hydrates are common in permafrost zones, especially in those below the sea floor in polar regions. They may represent a vast future energy resource, because 1 m^3 of hydrate contains 50–170 m^3 of natural gas, the upper figure being about the energy equivalent of one barrel of oil.

5.1.3 *Practical significance of carbon numbers*

There is seen to be a borderline between "gas" and "oil" at the carbon number of 4. Butane, C_4H_{10}, is the heaviest of the methane homologs that can occur in natural gas at STP; it is also the lightest hydrocarbon that can occur in the very light oils called *condensates* under equivalent conditions. Gasoline, therefore, contains a range of carbon number molecules starting at the bottom with butane.

In this range and higher, it is possible to have two series of hydrocarbons with the same composition but different structures, called *isomers*. Instead of every carbon atom being attached to either one other carbon atom or to two, at least one carbon atom can be attached to *three* others, forming a *branched chain* rather than a straight chain. The straight-chain molecules are *normal* or *n*-paraffins, the branched-chain members are *iso*-paraffins (Fig.5.6).

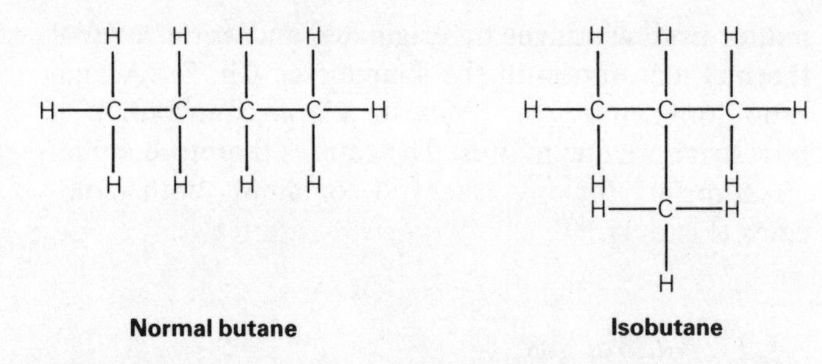

Figure 5.6 Normal butane (*n*-butane) and isobutane to illustrate straight-chain and branched-chain saturated hydrocarbons.

Both of the compounds shown in Figure 5.6 are *saturated*; they contain as many hydrogen atoms as the tetravalent nature of the carbon atoms will permit. They are therefore not particularly reactive chemically. The nearest naphthene analog, butylene (C_4H_8), is *unsaturated*. It contains double linkages between carbon atoms, and these carbon atoms may themselves be bonded to different numbers of hydrogen atoms (Fig. 5.7). Double linkages can be broken to permit reaction with other molecules. As the carbon numbers increase, the number of possible isomers increases by very large jumps; each additional carbon atom more than doubles the number that may be formed. Different isomers have different physical properties, including different boiling points.

Thus the number of hydrocarbon species in an untreated crude oil may be large. Not all of them have the properties desired by the many different users, and it is the purpose of the refiner to make the necessary modifications. The basis of refining is ordinary distillation or *fractionation*, in which the oil is heated until its various component molecules "flash" or boil to vapor in the order of their increasing boiling points. These are correlated with the molecular weights of the compounds and so also with their densities and their viscosities.

Distillation reveals that 70–80 percent of most crude oils consists of hydrocarbons heavier than C_{10}. The percentages of the products falling within the succession of

$$CH_3CH_2CH{=}CH_2$$

$$CH_3CH{=}CHCH_3$$

Figure 5.7 α-Butylene (top) and β-butylene (bottom), to illustrate different bondings of carbon atoms in unsaturated compounds having double linkages.

carbon number ranges determines the value of the oil. The order of distillation in the process is as follows:

(1) Gasoline and naphtha, with benzene and other volatile oils (4–10 carbon atoms)
(2) Kerosene and illuminating oils (11–13 carbon atoms)
(3) Diesel and light gas oils (14–18 carbon atoms)
(4) Heavy gas oil, home heating oils (19–25 carbon atoms)
(5) Lubricating oils, light fuel oils (26–40 carbon atoms)
(6) Residual and heavy fuel oils (more than 40 carbon atoms).

Fractions 2–4 are called the *middle distillates*. Remaining at the end of the process is the *residuum*, consisting chiefly of asphaltic nonhydrocarbons (see Sec. 5.2.5).

5.2 The nonhydrocarbon constituents of oils and natural gases

The nonhydrocarbon constituents commonly found in crude oils and natural gases are sulfur, nitrogen, and oxygen and their compounds (called S–N–O compounds or *heterocompounds*), and organocompounds of certain heavy metals (principally vanadium and nickel). Natural gases may also contain hydrogen, helium, or argon, the last two having nothing to do with petroleum generation.

5.2.1 Sulfur and its compounds

Sulfur provides by far the most important of the heterocompounds. Few oils are wholly without it; relatively few contain more than 3 percent of it by weight. Sulfur content is higher in heavy oils than in light oils (Fig.5.8), being concentrated in the higher molecular weight, higher boiling point, more polar fractions (resins and asphaltenes; see Sec. 5.2.5).

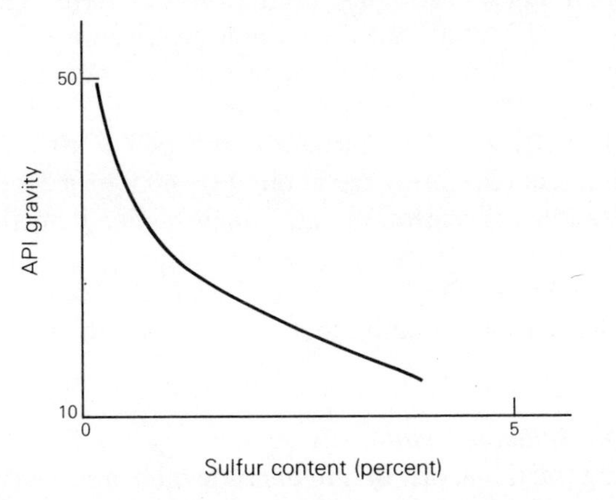

Figure 5.8 Relation between sulfur contents of oils and their API gravities.

A small proportion of the sulfur in crude oil is in the form of elemental sulfur (in solution) or of H_2S. The greater part is bonded with carbon in organic combination, mainly in polyaromatic components of the crude:

(a) Mercaptans or thioalcohols (e.g. C_2H_5SH), especially in the lower-boiling range (below 150 °C).
(b) Disulfides (such as diethyl disulfide, $C_2H_5SSC_2H_5$) and polysulfide organic compounds.
(c) Cyclic aromatic sulfides or thiophenes: $C_nH_{2n-y}S$ (e.g. C_4H_4S, which is present in commercial benzene manufactured from coal tar).

Crude oils containing detectable amounts of H_2S are called *sour crudes*. If the sulfur is in some other form than H_2S (as it usually is), the oil should be called a *high-sulfur crude* and not a "sour" crude. Very "sweet" crudes, with sulfur content as low as 0.1–0.2 percent, are remarkably concentrated in African basins (in Algeria, Angola, Nigeria). *Low-sulfur crudes* contain less than 0.6 percent sulfur by weight; intermediate, 0.6–1.7 percent (0.7–2.0 percent for some petroleum chemists); *high-sulfur crudes* contain more than 1.7 percent sulfur by weight. The highest percentages are found in reservoirs of dolomite–anhydrite facies (many fields in the Middle East – Arabian "heavy" fields, Iran, Divided Zone, Suez graben – yield oils ranging from 2.8 to 4.9 percent sulfur), and in very heavy, degraded oils (5.5 percent in the Athabasca tar sands in Alberta; 5.4 percent at Boscan in Venezuela).

Because high sulfur content is a major nuisance to refiners and to consumers of oil and gas, the industrial world has developed an excessive demand for sweet crudes. The proportion of the world's reserves that can be called sweet has declined significantly since the late 1960s. In most major exporting countries (the Middle East nations, Venezuela, Mexico) remaining reserves of high-sulfur crude are five to ten times those of sweet crude, and all synthetic crudes produced from "tar sand" or oil-shale deposits are also high in sulfur (see Ch. 10). This feature of future oil supply will dictate wholesale modification of refineries in many countries.

Sour gases contain H_2S, which must be extracted during processing of the gas for domestic or industrial use. Gases high in H_2S occur typically in carbonate–sulfate reservoirs, and are consistently associated with higher than normal concentrations of nitrogen and carbon dioxide. Concentrations of H_2S exceeding 100 p.p.m. (in gas or oil) are considered dangerous; they are also highly corrosive to drilling equipment in deep wells where temperatures are high. Yet many gases as produced contain concentrations scores of times this level: in some North American basins (the Permian, Alberta,

Tampico, and Reforma–Campeche basins for example, and the Texas Panhandle), in the Permian of the Ural–Volga region and the Jura-Cretaceous of the Aquitaine Basin in France (15–16 percent in the Lacq field), and in many others (see Chs 17 & 25). In deep Smackover (Jurassic) gases from the southeastern USA, H_2S content of 30 percent is routine, and 80–90 percent is not uncommon. In deep carbonate reservoirs of the Rocky Mountain foothills and overthrust belt it is also very high (60–65 percent below 4000 m at Moose Mountain in Alberta; nearly 20 percent in Wyoming).

5.2.2 Nitrogen

Nitrogen in crude oils is related primarily to the asphalt content. More than 0.2 percent nitrogen is considered high; such concentrations occur in some oils in the Los Angeles, Maracaibo, and Tampico basins. High-nitrogen *gases* occur especially in Paleozoic strata: up to 80 percent in some reservoirs in the San Juan Basin of New Mexico (with helium up to 3.0–7.5 percent in addition); up to 90 percent in gases of the eastern part of the Rotliegendes Basin of Europe (especially in the Danish North Sea, but more than 14 percent in the great Groningen field in the Netherlands); from 8 percent to 85 percent in many southern Alberta gases; and up to 19 percent in Pennsylvanian gases of the Uinta Basin (where the Eocene Green River oil shale is also high in nitrogen).

The association of high nitrogen content with unusual concentrations of other contaminants is illustrated from many producing regions. In the Pakistani gasfields, up to 28 percent nitrogen is accompanied by high CO_2 content; at Orenburg in the Soviet Union, sour gas in the Permian reservoir contains 5.8 percent nitrogen and 28 p.p.m. of argon in addition to 4.5 percent H_2S; in some Chinese basins, in which the gas is probably of Permo-Triassic origin, high nitrogen is again associated with high sulfur.

The nitrogen must be removed at low temperatures during reprocessing of the gas. It can be used as an injectant in the recovery of light oils (see Sec. 23.13.3).

5.2.3 Oxygen compounds

Oxygen compounds of definite structure in crude oils are acids (as in many fields in the southern Soviet Union) and phenols (hydroxyaromatic constituents of coal tars, such as carbolic acid, C_6H_5OH).

Natural gases may contain considerable quantities of CO_2. Carbon dioxide is of course itself another natural gas, but in the oxidized state and noncombustible. It constitutes up to 90 percent in the Carboniferous to Triassic reservoirs of the Paradox Basin in Utah; 80 percent in the Ordovician and Mississippian carbonates of some traps in the Overthrust Belt in Wyoming; and up to 50 percent in some fields in northern Mexico, Pakistan, and elsewhere. Both CO_2 and H_2S are high in the deep Tuscaloosa gas of the US Gulf Coast (see Sec. 13.2.7).

5.2.4 Other elements in oil and natural gas

Natural gases may also contain hydrogen, helium, or argon. The *hydrogen* content in no case exceeds 0.5 percent; usually it is less than 0.1 percent. *Helium*, which constitutes about 5 p.p.m. of the Earth's atmosphere, occurs in some "dry" gases, particularly in those with high nitrogen contents; both helium and nitrogen are inert. The He : CO_2 relation, by contrast, is inverse.

Though nitrogen-rich gases contain high helium concentrations, they do not contain the largest *reserves* of helium. Helium is associated with combustible gas, not with nitrogen accumulations as such. In fact, fields producing both oil and associated gas more commonly contain 1 percent or more of helium (a very high proportion) than do dry gas-fields. In North America, at least, all important helium-bearing reservoirs are over large basement uplifts: the Amarillo uplift, the Nemaha ridge, the Four Corners region, the Swift Current platform of southwestern Saskatchewan (see Ch. 25). The minimum commercial concentration is 0.3 percent helium. The Hugoton–Panhandle field of the USA, the greatest single producer of helium during and since World War II, averages 0.5 percent helium.

Argon, which as ^{36}Ar makes up 0.93 percent of the atmosphere, seldom exceeds 0.1 percent in natural gases. Though its atmospheric abundance is about 2000 times that of helium, its concentration in gases (as in the Panhandle field) is typically only about one-tenth that of the lighter gas. Argon in gases is enriched in the isotope ^{40}Ar (derived from the decay of ^{40}K) to the extent of 6–10 percent.

Radon, the inert, gaseous emanation from radium, occurs in many crude oils. Oilfield waters have higher contents of radon than do normal ground waters (all waters have much lower absorptions of radon than have oils). Radon activities in oils are of the order of 0.1–0.4 pCi g^{-1}. (One picocurie or 1 pCi = 10^{-12} Ci.) Radon : radium ratios are about 4.0–40.0 or more, so that radon concentrations are much higher than they should be from the radium concentrations.

Thoron is probably also present in a few crudes but its half-life is less than 1 min; that of radon is nearly 4 days.

5.2.5 Metals in crude oils

During fractional distillation of crude oil in a refinery, a heavy *residuum* is left which has to be removed from the base of the tower. It consists of some high molecular weight hydrocarbons, with 50 or more carbon atoms, but the greater part is made up of nonhydrocarbons of asphaltic character. They constitute the true impurities in the crude – sulfur compounds, asphaltenes, and metals – and must be removed by expensive catalytic hydrogenation processes (desulfurization, denitrogenation, demetallization, and so on).

Asphaltenes are high molecular weight aggregates occurring in solid bitumens (see Ch. 10). They are of variable and uncertain composition, but their basic structure is of polyaromatic sheets, plus aliphatic side-chains and some S–N–O compounds. Asphaltenes are very soluble in CCl_4, CS_2, and aromatic hydrocarbons, but not in light paraffinic hydrocarbons such as heptane. They contain very little hydrogen, and the high viscosity of heavy oils is probably a function of the size and abundance of asphaltene molecules.

Asphaltenes are also the major contributors to the total metal content of heavy oils, the metals along with much of the sulfur and nitrogen being largely bound to the asphaltene molecules. The metal-binding molecules are the *porphyrins*, the most peculiar constituents of crude oils.

5.2.6 Porphyrins

Porphyrins are hydrocarbon ring complexes containing nitrogen and a metallic nucleus (Fig.5.9). Carbon numbers are in the range 25–36, especially 30–32. In this respect crude oil porphyrins differ slightly from those occurring in bituminous coals, which lack carbon numbers higher than C_{32}. The metal is invariably either vanadium (V) or nickel (Ni). Vanadium is much the commoner; 30–300 p.p.m. are typical proportions, in the asphaltic and resinous fractions associated with high sulfur content (Fig.5.10). The vanadium content of Venezuelan crudes (up to 1100 p.p.m. at Boscan, on the western side of Lake Maracaibo) results in potential output exceeding the world's demand for vanadium (which is principally for high-strength, low-alloy steels, used especially in pipeline construction). Burning the residues of such crudes yields a vanadium-rich ash.

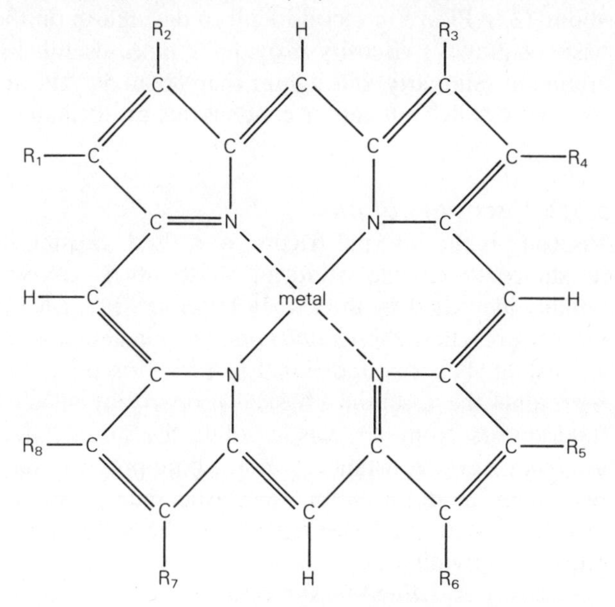

Figure 5.9 Structure of a porphyrin. R represents various carbon-containing side-chains. (Courtesy of K. E. Peters and Chevron Oil Field Research Company.)

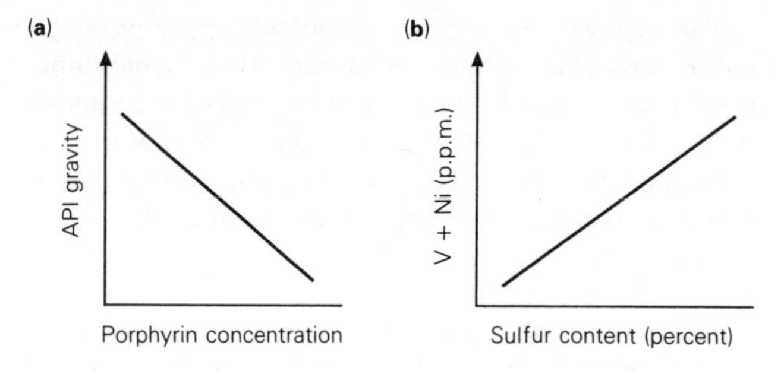

Figure 5.10 Inverse relation between porphyrin concentration and API gravity (a); direct relation between porphyrin concentration and sulfur content (b).

Nickel on the other hand is concentrated in the "oily" part of the bitumen (20–85 p.p.m. is the typical range of values), and occurs largely in crudes *low* in sulfur. Oils from the Uinta Basin of the western USA contain nickel porphyrins and are exceedingly waxy; they were derived from the famous oil shales of saline lake origin (see Ch. 10). Sapropelic matter is higher in vanadium than is humic matter, so the V : Ni ratio is highest (more than 1) in Paleozoic oils, and is less than unity in Mesozoic and Cenozoic oils because of the increasing contribution of humic matter from green plants during post-Devonian time (see Sec. 7.3).

O. A. Radchenko believed that the nickel contribution to porphyrins is primary, and the contributions of vanadium and sulfur secondary. The "oily" part of the petroleum migrates out of the source sediments, while the asphalto-resinous part remains in them. Some deep-sea sediments deposited under oxygenated conditions, and probably deriving their organic matter from terrestrial sources, contain pigments which are *copper* porphyrins. Petro-porphyrins, necessarily forming under reducing conditions in order to exist in crude oils, do not contain copper.

Total metal contents of some sulfurous crude oils are indicated here:

V + Ni (p.p.m.)	Example
300+	Gach Saran, Iran
400+	East Venezuelan tar belt
500+	Lake Maracaibo, Venezuela
1200+	Boscan, Venezuela

The nearest analogs of petroleum porphyrins in nature are chlorophyll (containing Mg), hemoglobin (Fe), and vitamin B_{12} (Co). Chlorophyll, the composition of which is represented by $C_{55}H_{72}O_5N_4Mg$, may be converted to porphyrins via intermediate products called *pheophytons*. This would require the metal nucleus of the porphyrin to be picked up subsequently, and from some source other than the source rock of the oil. Reliance on the nature of a porphyrin as an indicator of the origin of an oil has never been shown to be justified.

5.3 Physical properties of oils

The chemical compositions of crude oils are the principal determinants of their physical properties. A few of the more easily observed and measured of these properties are of significance to the geologist, and indeed to all users of oils in forms other than automotive fuel. The most readily observed properties are the specific gravity, the viscosity, and the color.

5.3.1 Specific gravities of oils

Specific gravities of oils lie in general between 0.73 and very slightly above 1.0. Paraffin-base oils are commonly light, asphalt-base oils almost invariably heavy. (The gravity is conventionally signified by the Greek letter *rho* ρ.) The gravity was formerly expressed in degrees of the European Beaumé scale, read directly on a hydrometer; this means that the degree goes up as the density goes down. A "high gravity oil" is not a heavy oil; it is one with a high scale reading and so a light oil. The Beaumé value, with the density standardized to 15.6 °C (60 °F), is given by the equation

$$\text{Bé} = \frac{140}{\rho} - 130$$

The Beaumé scale was long ago superseded by the scale of the American Petroleum Institute, called the API scale. The relation between the two scales is given by

$$\text{API value} = (1.010\,71 \times \text{Bé}) - 0.107\,14$$

In relation to the density, this is equivalent to

$$\text{API value} = \frac{141.5}{\rho} - 131.5$$

so that water under STP conditions becomes 10 °API.

In the future, the term *specific gravity* will be allowed to become obsolete. What is being measured is the *density relative to water*, and the units of measurement become kilograms per cubic decimeter (kg dm^{-3}), where 1 dm is 10^{-1} m. The density as measured approximates very closely to that expressed in these units, the absolute density of pure water at 4 °C being very close to 1 kg dm^{-3}.

The relations between API gravity, relative density, and traditional oil industry volumetric measure may be illustrated thus:

API gravity	30	33	36	LPG
density	0.876	0.860	0.845	0.570
barrels per long ton	7.31	7.45	7.58	12±

By general convention, oils with API gravities higher than 30° are considered light; those between 30° and 22° medium; and those with values below 22 °API are heavy. This convention is by no means universal, however. In a producing country with abundant heavy crudes, such as Venezuela, a 20 °API oil is considered to be of medium gravity and 26 °API might be regarded as light. The worldwide average value is about 33.3 °API. The most favored grade of crude oil is about 37 °API, equivalent to a relative density of 0.84. Crudes of about this gravity are common in the Middle East, the Mid-Continent and Appalachian provinces of the USA, Alberta, Libya, and the North Sea.

Very light crudes, above 40 °API, occur in large quantities in Algeria, southeastern Australia, and in some Indonesian and Andean fields. Very heavy crudes dominate production from California, Mexico, Venezuela, and Sicily. Some oils in fractured Miocene reservoir rocks in California's Santa Maria Basin (see Sec. 25.6.4) are heavier than 6 °API and contain nearly 8 percent of sulfur. Such oils are beyond production technology unless artificially diluted. Oils heavier than 10 °API are now called *extra-heavy*. Oils heavier than about 12 °API are in fact difficult to distinguish on the basis of gravity; viscosity provides a more useful discriminant. Similarly, oils lighter than about 50°API are not really "oils" but rather condensates or distillates.

5.3.2 Viscosities of oils

Viscosity is the internal friction of a fluid, causing its resistance to change of form. (Viscosity is conventionally identified by the Greek letter *eta*, η.) It is the ratio of stress to shear *per unit time*; shear in liquids is not a constant, but is proportional to time. Viscosity thus represents the essential physical property by which a fluid differs from an elastic solid; the latter offers *instantaneous* resistance, not time-proportionate resistance, to deformation by elastic shear, and this resistance is termed the *rigidity* of the solid. Fluids possess no rigidity.

Viscosity is defined by the ratio

$$\frac{\text{force} \times \text{distance}}{\text{area} \times \text{velocity}}$$

Its dimensions are therefore $MLT^{-2}L/L^2LT^{-1}$, or $ML^{-1}T^{-1}$. The CGS unit of viscosity is the *poise*, which is too large a unit for practical purposes in the oil industry. Viscosities of oils are therefore conventionally measured in *centipoises,* 1 cP (10^{-2} poises) being the viscosity of water at 20 °C (68 °F).

However, the CGS unit has the disadvantage of being a measure of the force required to cause the fluid to move with a particular velocity. Viscosities are more conveniently measured in some time units, such as the number of seconds needed for a steel ball to roll through a standard volume of the fluid. Such a unit is the *Saybolt universal second* (SUS):

$$\text{SUS} = \frac{\text{viscosity in centipoises} \times 4.635}{\text{(relative density)}}$$

Typical oil viscosities measured in SUS at STP are from about 1000 to 50 or less. Adherence to the SI system will require measurements of viscosities to be in *millipascal-seconds* (mPa s) (1 cP being equivalent to 1 MPa s).

Viscosities vary directly with densities, and hence viscosities of oils are a function of the number of carbon atoms and of the amount of gas dissolved in the oil. The effect of dissolved gas on both the viscosity and the API gravity of crude oils is illustrated in Fig.5.11. Viscosities of light oils are below 30 mPa s; typical values are between 5.0 and 0.6 mPa s, the latter being the viscosity of gasoline. Heavy asphaltic oils have viscosities measured in the thousands of millipascal-seconds: about 50 000 mPa s for the Miocene oil in the Bolivar Coastal field in western Venezuela, nearly 100 000 mPa s for the very heavy oil at Cold Lake, Alberta, and over 10^6 mPa s for that in the Athabasca "tar sands." Hydrocarbons having viscosities higher than 10 000 mPa s are now to be called *natural tars*.

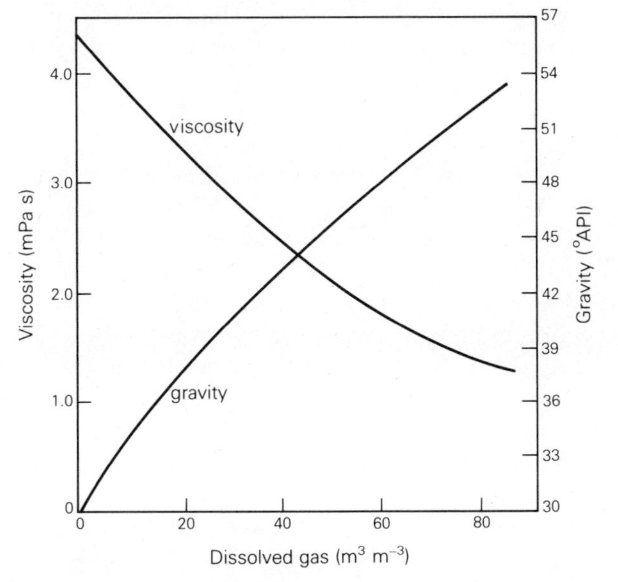

Figure 5.11 Effect of dissolved gas on viscosity and gravity of crude oil. (From *Oil Gas J.*, 13 January 1944.)

Oils with high asphaltene contents may be too viscous to be pumped through a pipeline, even in hot climates; the Boscan crude from western Venezuela, and several Californian crudes, are examples. In the cold climate of the North American interior, the first derivatives of the tar sands or the oil shales form viscoplastic masses at the surface.

A useful indicator of the viscosity of a crude oil is its *pour point*. This is the lowest temperature at which the crude will flow under prescribed, controlled conditions. Pour points above 40 °C (more than 100 °F) are relatively common among crudes having high contents of paraffin wax. Such high pour points are in startling contrast to values as low as −36 °C for some versatile Middle-Eastern and African crudes, which would flow even under Arctic conditions. Oils having high pour points because of high wax contents have a shiny appearance and are associated with formation waters of very low salinity (see Chs 9 & 17). If migration paths permit them to rise in their traps, their temperatures are lowered and the waxes crystallize out, forming a residue of high molecular weight paraffins and allowing the oils themselves to become lighter.

Very waxy crudes include those of the Uinta Basin in the western United States; the Anaco trend in eastern Venezuela; the Reconcavo Basin in Brazil and the Mendoza Basin in Argentina; the Beatrice field off eastern Scotland; the Mangyshlak fields east of the Caspian Sea (the largest of which, the Uzen field, yields a crude with 26 percent wax content); several fields in the Sirte Basin of Libya and in the Sudan; and a remarkable proportion of the fields in young sandstone reservoirs in eastern Asia and Australia (those off western India, and in the Upper Assam Basin of northeastern India, where the oils contain 10−15 percent wax; most crudes from China, Sumatra, and the Gippsland offshore basin in Australia). Prolific basins in which the original oils were paraffinic and derived oils, in younger strata, are asphaltic, include the Carpathian Basin of Romania and the Niger delta basin off Africa. These occurrences are all described in Chapters 17 and 25.

Both paraffinic and asphaltic crudes may undergo such a degree of volatilization through surface or near-surface alteration that they become totally dried up, so viscous that they are effectively solids. The drying-up process is called *inspissation*.

5.3.3 Colors and refractive indices of oils

Paraffinic oils are commonly light in color: yellow to brown by transmitted light, and the familiar green of automobile engine oil by reflected light. Asphalt-base oils are commonly brown to black; many of them are known as "black oils."

The refractive indices of oils vary with the relative density, between 1.42 and 1.48 for most examples; the lower indices are for the lighter oils. The indices are also lower at lower temperatures. Within any one molecular weight range, the refractive indices increase from paraffins, through naphthenes, to aromatics.

6 The origin of petroleum hydrocarbons

6.1 Introduction

As Sol Silverman (1971) expressed the fundamental paradox, oil and gas are incompatible with the solid and the aqueous components of the lithosphere. Petroleum hydrocarbons are a mere transition stage in the carbon cycle, constantly undergoing transformation towards products of lower free energy. The element carbon, in fact, is unstable in the Earth's crust except as graphite or as inorganic carbonate rock.

The concept of the *carbon cycle* is an expression of this constant flux. It has taken on an enlarged significance as realization has grown that the carbon fuels are far from inexhaustible. It has even been postulated (by W. O. Nutt) that the prime requirement of world resource policy should be the conservation of carbon, which, once oxidized, cannot be artificially recovered or recycled. Carbon has thus come to be thought of as an element involved in constant interchanges of form at or near the interfaces between the lithosphere, the biosphere, the hydrosphere, and the atmosphere (Fig.6.1). Because carbon compounds are *organic* substances, it is natural that hydrocarbons should be attributed to an organic origin. This attribution was not natural or widespread, however, until well into the 20th century; there are still authoritative proposals of an inorganic origin for petroleum hydrocarbons, at least for our initial endowment with them.

6.2 Theories of inorganic origin

6.2.1 Early proposals

When the incidence of fluid petroleum hydrocarbons was first recognized as being widespread, in the late 19th century, two Russian scientists put forward hypotheses of their inorganic derivation. Dmitri Mendeleev, the father of the periodic table of the elements, proposed that metallic carbides deep within the Earth reacted at high temperatures with water to form *acetylene* (C_2H_2), which subsequently condensed to form heavier hydrocarbons. The reaction is readily reproduced in the laboratory. In 1890, W. Sokoloff proposed a *cosmic* origin for petroleum; it was precipitated as rain from the original nebular matter from which the Solar System was formed, and subsequently ejected from the Earth's interior into the surface rocks. Both the *deep-seated terrestrial* and the *extraterrestrial* hypotheses have survived in a variety of forms.

6.2.2 Later variants on the deep-seated hypothesis

Solid petroleum bitumens occur relatively commonly in various igneous rock environments, though never in more than minute quantities. Micro-inclusions of petroleum hydrocarbons have been identified in the minerals of alkaline rocks associated with carbonatites, leading to the speculation that they were synthesized from hydrogen and carbon monoxide under catalysis by rare-earth minerals. This is highly implausible in view of the crystallization temperature of such magmatic rocks (about 500 °C).

More significant is the association of petroleum hydrocarbons with *hydrothermal systems*, studied by a number of Soviet scientists. Bitumens are known in the products of several active volcanoes (Katmai, Etna, Vesuvius). Gaseous homologs of methane and aromatic and naphthenic oils occur in fumaroles and ejecta in many volcanic regions, especially near "mud pots" or boiling springs associated with mercury, arsenic, and antimony minerals. Hot springs of H_2S-CO_2 type commonly deposit some bituminous materials. Ozocerite and asphalt (see Ch. 10) occur in hydrothermally altered traps and kimberlites, associated with silica, calcite, fluorite, and metallic sulfides. Hydrothermal gases commonly have carbon isotope ratios like those of organic materials (see Sec. 6.3.1).

The general postulate by proponents of deep-seated hydrocarbons is of natural chemical factories or "abyssal

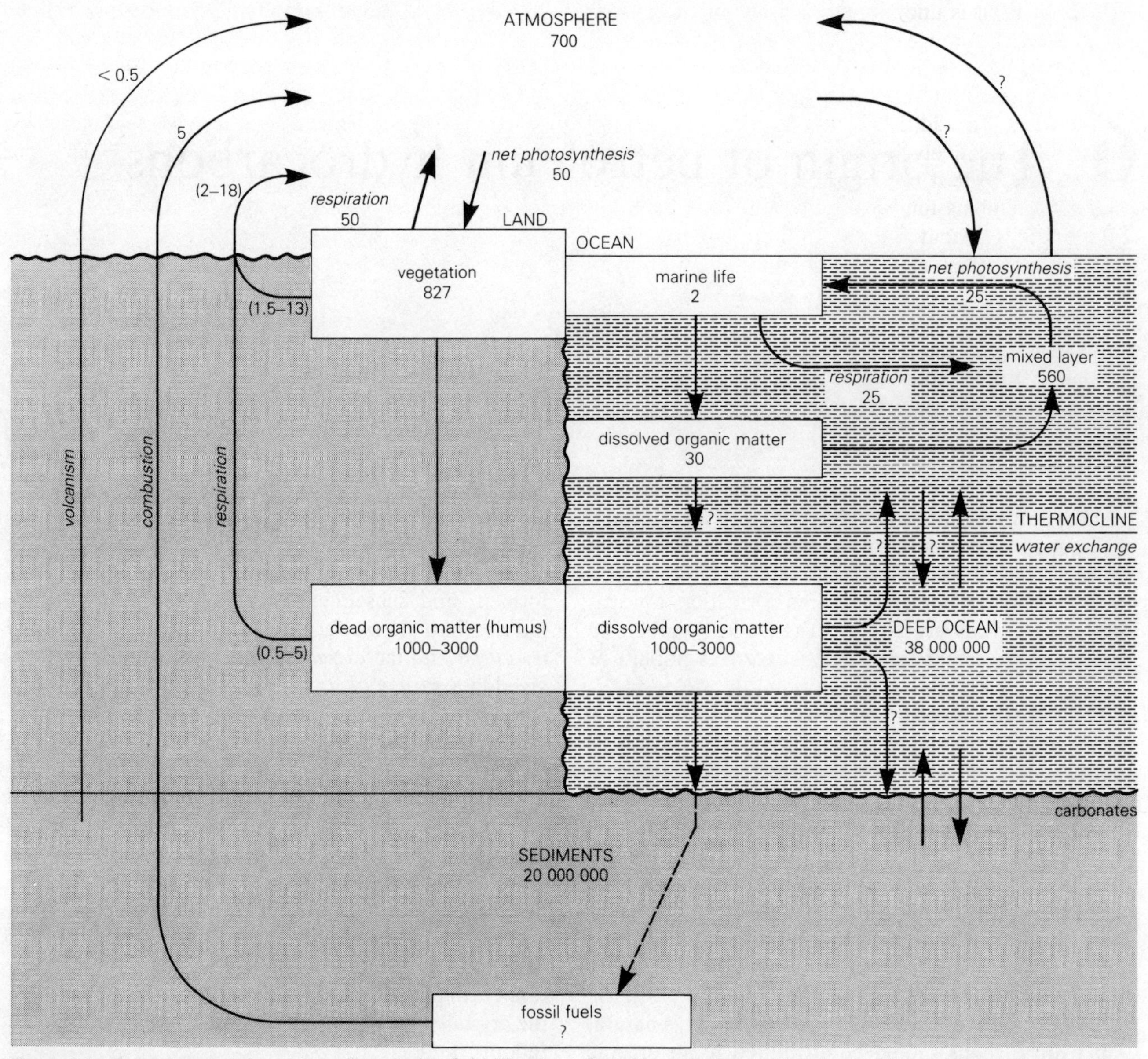

Figure 6.1 The carbon cycle in nature, as illustrated by G. M. Woodwell. Quantities are expressed in units of 10^{12} kg, or billions of tonnes. (From G. M. Woodwell, *Scient. Am.*, 1978.)

hearths" deep in the crust or in the upper mantle, in which hydrocarbons are "cooked." Telluric currents bring about the combination of water, graphite, iron, and sulfur, as in a giant chemical battery. The water would be broken down into hydrogen and oxygen; the former would hydrogenate the graphite, producing hydrocarbons. These would then be injected high into the crust via deep fractures. In 1974, V. B. Porfirev went so far as to aver that all known oilfields were formed between the early Miocene and the Pleistocene by such deep migration. Components of the oils that are thermally unstable, especially the porphyrins and waxes, are picked up during the ascent through crustal rocks.

The association of methane gas and its homologs with *volcanic rocks* is widely reported. Many oil- and gas-fields are near, or parallel to, belts of active or recent vulcanicity; Indonesia and Mexico provide obvious examples. Dying volcanoes may superficially resemble mud volcanoes powered by escaping hydrocarbons. The association is fortuitous; lines of volcanoes and compressional sedimentary basins are both aspects of young orogenic belts, and the sedimentary content of a young, petroliferous orogenic basin is very likely to include units of some volcanic derivation (see Ch. 17), but there is no closer relationship. Some CH_4-CO_2 gases appear from carbon isotope evidence to be of high-temperature origin and due to intrusive or extrusive magmatic activity (some gases from the Sacramento valley of California, for example), but only rare and volumetrically trivial cases are known.

Carbon itself is undoubtedly a constituent of many igneous rocks. A range of 50–500 p.p.m. is not unusual, the carbon being either disseminated through the rock or concentrated in diamond, graphite, carbides, or carbonatites. In addition some alkalic and ophiolitic rocks have relatively high concentrations of *organic matter*. The highest concentration known is probably that in the Khibina-tundra massif on the Kola Peninsula of the USSR, about 300 Ma old. It contains carbon compounds as high as C_{40}, predominantly of aromatic type. Where hydrocarbons are present in intrusive rocks, however, there is no proof that they themselves were of igneous source. Where the hydrocarbons are methane and its homologs, the host rock may have reacted with sea water. Where the dominant hydrocarbons are aromatics, they are probably due to passage of the intrusive through sediments containing vegetable material. Igneous intrusion into coal measures or oil shales commonly produces bituminous matter, but the phenomena are wholly local.

The volcanic origin postulate fundamentally invokes outgassing of the upper mantle via volcanic activity. An intriguing variation invokes outgassing via *earthquake activity*. Its principal proponent has been Thomas Gold. Nonassociated gas at considerable depths (below, say, 6 km) cannot be proven to be of biogenic origin, especially if it is associated with pre-Devonian rocks. Migration via deep faults may radically alter the composition of natural gas, making identification of its ultimate origin highly uncertain.

In Gold's view, the primitive, volatile constituents of the Earth were not all lost during its early evolution, chiefly as CO_2. Large volumes of methane are still being degassed from the mantle. Mass balance calculations indicate that methane is being released to the lithosphere, hydrosphere, and atmosphere at a rate of 10^6–10^7 tonnes annually. The triggering mechanism is earth movement, as testified by the sharp increases in the radon contents of ground waters and the atmosphere following many earthquakes. These hypotheses share the common basis that the Earth's store of *methane*, at least, is primitive and a feature of the origin of the Solar System.

6.2.3 Later variants on the extraterrestrial hypothesis

Sokoloff's cosmic theory has been resuscitated by many scientists, including astronomers (Fred Hoyle) and chemists (Robert Robinson) as well as geologists, because of two critical discoveries. The atmospheres of some celestial bodies (Titan, the moon of Saturn, for example) are known to contain methane. In 1950, the cosmic speculator Immanuel Velikovsky postulated that the atmospheres of some of the solar planets might consist literally of hydrocarbon "smog."

It is widely agreed that the Earth's own atmosphere has had its present oxygenated constitution for less than half of geologic time. Its original atmosphere was reducing. If it contained methane and other hydrocarbons, in addition to ammonia, hydrogen, and water vapor, various photochemical reactions brought on by ultraviolet radiation would modify it relatively rapidly. Heavier hydrocarbon compounds would be built up, and in the very low-temperature atmosphere they could accumulate a surface layer of oily, waxy materials several meters deep in, say, 10 Ma. The layer, literally a primordial oil slick, could have acted as the host for a developing variety of prebiotic compounds, possibly including the precursors of life.

The second critical discovery was that of *carbonaceous chondrites* (meteorites). These consist mainly of hydrous silicate minerals such as chlorite; carbonates and sulfates of calcium and magnesium; iron minerals; elemental sulfur; and carbon compounds, with chondrules of mafic minerals. They are distinguished within the chondrite class by having all or most of their iron in an oxidized state. This and the presence of hydrous minerals indicates an aqueous environment of the meteorite parent, at temperatures far lower than those involved in the formation of stony meteorites.

The best studied carbonaceous meteorite, that of Orgueil, was recognized from its first inspection to contain more than 6 percent of *organic matter*, which is not graphite and has $\delta\ ^{13}C$ values substantially different from those of terrestrial carbon (see Sec. 6.3.1). Hydrocarbons identified include saturated paraffins, aromatics, and familiarly associated nitrogen and sulfur compounds. There is also a large quantity of polymer-like material insoluble in ordinary solvents. The solvent-extractable lipid fraction includes optically active components with $^{13}C : ^{12}C$ ratios similar to those in crude oils and marine plants (see Sec. 6.3.1). They even contain what have been called "organized elements," meaning microstructures of simple but recognizable form such as spheres. The organic materials may be abiological, extra terrestrial biological, or terrestrial biological; the last would represent contamination, but the precautions taken against this were so thorough that it is agreed to be insufficient to account for the quantity and variety of the organic materials identified. The simplest explanation appears to be that of irradiation of light elements in a primeval dust cloud.

Robinson, Gold, and others have persuasively put the case that some undetermined proportion of our hydrocarbons could be of abiogenic or inorganic origin. What they suggest is that most oil, and all or most gas, generated before about 400 Ma had such an origin; that this early oil and gas entered the early oceans to become incorporated into early sediments. Combinations of methane, ammonia, CO_2, and water can be sources of protein-type amino acids when they are suitably irradiated. Isoprenoid and sterane hydrocarbons in Proterozoic sedimentary rocks, long antedating terrestrial plants or animals and most marine animals also, may then represent extraterrestrial evolution (not necessarily from living organisms) and capture by the early Earth. Our present stock of petroleum hydrocarbons would on this hypothesis represent biogenic additions to a fundamentally primordial endowment.

Thus insistence on an original inorganic abiogenic origin for petroleum hydrocarbons does not exclude an organic, biogenic origin for the greater part of the supply. Exclusively abiogenic origins are maintained only by a few Russian scientists, whose views have been critically summarized by one of their colleagues, M. K. Kalinko, in 1968, himself firmly in the biogenic camp.

6.3 Theories of organic origin

Before World War II, the evidence favoring a directly organic or biogenic origin for petroleum hydrocarbons consisted largely of analogies. The chemical composition of petroleum allies it with materials called "organic" because they are fundamental to life – proteins, fats, fatty acids. The general carbon cycle in nature is depicted in a variety of ways (Fig. 6.1) but all intrinsically involve living and dead animals and plants. The fact that oil and gas were inescapably associated with unaltered sedimentary rocks, which commonly contained marine fossils, led to analogies with aquatic life; whale oil and fish oil were in wide use before the industrial oil era. A further analogy was that with coal, known to originate in terrestrial plant material; an easy step was to attribute petroleum to a comparable source but of either marine or animal nature, or both.

Petroleum scientists concentrated their attention on the close association between oil and *sediments*. Just as coal is formed as a natural component of a particular sedimentary facies, so it was averred that oil and gas are born naturally in a particular cycle of sedimentation, certainly aqueous, usually marine, and requiring no abnormal circumstances such as metamorphism. Following World War II, this generalized concept was reinforced by three lines of biochemical evidence. The experimental synthesis of oils had to start with organic reagents. The detailed analysis of oils revealed the presence of so-called "biomarkers" – compounds of undoubted organic derivation constituting "fingerprints" in oils – waxes, porphyrins, and steranes. Finally, studies of carbon isotopes showed that the isotope ^{13}C is depleted, in relation to the lighter isotope ^{12}C, in all organic materials (oils, bitumens, coals, terrestrial plants) by comparison with limestones, magmatic rocks, meteorites, or atmospheric CO_2.

6.3.1 Evidence of carbon isotopes

Most of the carbon in natural compounds is the so-called "light" isotope, carbon-12 (^{12}C). If we disregard the famous carbon-14, formed by cosmic bombardment of nitrogen in the upper atmosphere, the rest of the natural carbon is the "heavy" isotope, ^{13}C.

The atomic proportions of the two isotopes, determined spectrometrically, can be expressed in two ways:

(a) $R = {}^{13}C : {}^{12}C$ ratio.

(b) $F = \dfrac{^{13}C}{^{12}C + {}^{13}C} = \dfrac{R}{(1 + R)}$

so that $F \times 10^6$ is the concentration of ^{13}C in parts per million (p.p.m.). The values of F in nature range from about 10 350 to 11 200 p.p.m.; in ratio terms, the $^{13}C : {}^{12}C$ ratio is about 1 : 92.

The conventional expression of R for any sample is by the *delta value* ($\delta\,^{13}C$), representing deviation in parts per thousand (per mil or ‰) relative to an arbitrary standard:

$$\delta^{13}C = \frac{R_{\text{sample}} - R_{\text{standard}}}{R_{\text{standard}}} \times 1000$$

The most frequently used standard R is that of a Cretaceous belemnite from Peedee, South Carolina (PDB):

$$R_{\text{PDB}} = 0.011\,237\,2$$

Positive deviations mean excesses of the heavier isotope (^{13}C); negative values mean excesses of ^{12}C.

It turned out that the calcite of the belemnite standard was at the *heavy* extreme of the scale of measured values. The only common forms of carbon yielding $\delta\,^{13}C$ values in the -4 to $+4$ range (the standard of course being zero) are unaltered, sedimentary carbonate minerals (calcite, in limestones or calcareous fossils). A typical $^{13}C : {}^{12}C$ ratio for limestones is about 1 : 90.

Carbon of mantle derivation, in magmatic rocks, volcanic gases, and diamonds, and carbon in Precambrian rocks or in meteorites, has *intermediate* δ values in the -2 to -20 range. The ratio is about 1 : 92.

Carbon in organisms or organic matter is isotopically *light*. Organic matter yields δ values in the -15 to -30 range, with the values more negative in polar than in equatorial regions. Marine plants and invertebrates yield values between -12 and -30, averaging -23. Terrestrial plants and coals, and soil humus, fall in the -23 to -28 range. The $^{13}C : {}^{12}C$ ratio is therefore about 1 : 93.

Petroleum hydrocarbons are isotopically lighter still. Crude oils fall into the δ range of −22 to −36, the light isotope being progressively depleted with decreasing age. As the oldest oils yield values similar to those of lacustrine oil shales, this depletion of ^{12}C with time is probably a consequence of the sapropelic nature of organic matter before the provision of terrestrial material from land plants. The bitumen of the Athabasca tar sands has δ = −30. Associated petroleum gases read at −35 to −55, nonassociated petroleum gases at −45 to −65, and bacterial gases at −65 to −90. The average measurement for hundreds of methane samples is −48. Gases, in fact, average about 20‰ (2 percent) lighter isotopically than their associated oils.

The evidence of carbon isotopes therefore points strongly to an origin of oils and petroleum gases in organic materials of one sort or another, and not in any magmatic or inorganic source. Attempts to use carbon isotope data to pinpoint petroleum origins more closely than this – in particular types of plant or particular ages of pelagic sediment, for example – are much less conclusive. The effects of isotopic fractionation are imperfectly understood (during photosynthesis, for example, or burial). It is now further believed that the isotopic composition of the organic carbon "reservoir" in the world ocean has itself undergone brief but significant variations through geologic time. The variations, called "excursions," may be in either the positive or negative δ directions. They are probably caused by large-scale oceanographic events, themselves no doubt consequent upon plate motions.

6.3.2 Analogy with coal

While oil was sought primarily by drilling marine sedimentary rocks, it was easy to maintain the illusion that oil and coal seldom occur together. In fact, many richly oil-bearing sandstones are members of coal-measure successions: in the Carboniferous of the Donetz Basin, the Cretaceous of western North America, and the early Tertiary of the Gippsland, Cambay, Upper Assam, and Orinoco basins, among others (see Ch. 25).

Bitumen is common in deep coal seams, and oil seepages are known from a number of them. It is therefore conceivable that oil and coal formed from the same general materials under varying conditions, one set of conditions converting carbohydrates into a wholly solid product, the other converting hydrocarbons into a wholly volatile product. The derivation of certain categories of hydrocarbon from terrestrial vegetation is familiar. Latex is a hydrocarbon derived from the *Euphorbia* tree; the jojoba bean of the Sonoran Desert in northern Mexico contains about 40 percent by weight of extractable oils, from which lubricants and heat-tolerant liquid waxes are made. The bluish haze over forests (giving the Blue Ridge and Great Smoky Mountains of the Appalachians their names) is caused by the evolution of terpenes, aromatic hydrocarbons based on the molecule $C_{10}H_{16}$.

Alternatively, coal and oil might be formed under the same general conditions (of time, climate, tectonism) but from materials of different environments. On the Russian platform, between the Ural Mountains and Moscow, about 150 oil and gas deposits are exploited from Lower Carboniferous sandstone reservoir rocks closely associated with coals. The most prolific reservoirs are in deltaic and fluvial sands. In 1966 N. I. Markovskiy proposed that the oil and coal were formed at the same time on opposite sides of a coastline.

Relatively common though the association may be between oil incidence and coal incidence in some regions, the association between natural gas and coal is very much more common (see Sec. 25.3). The large gas deposits in the Permian of western Europe and of the Ukraine, and those in the Cretaceous of northwestern Siberia, are outstanding examples. This leads to a third possibility: that coal and natural gas are natural derivatives of terrestrial vegetation, whereas oil and wet gas are natural derivatives of aquatic vegetation, like the so-called *sapropelic coals* (from the Greek σαπρος (*sapros*) = rotten, in allusion to the decomposed, amorphous, vegetable "mud" which is an important constituent of them).

Since the mid-1960s, many more geologic successions have been discovered, and exploited, in which oil, gas, and coal are intimately associated: in the Cooper Basin of central Australia, the Bohai Basin of northeastern China, the Mackenzie delta in Arctic Canada and the Mahakam delta in equatorial Borneo are important examples. They share the two characteristics of being mid-Mesozoic or younger in age, and nonmarine or incompletely marine in facies. Terrestrial or paralic organic matter contributing to coals younger than about 150 Ma seems to have contained a higher proportion of oil-prone constituents (the type II kerogen described in Section 7.3) than the more familiar Pennsylvanian coals of the northern hemisphere ever did.

Waxes and hydrocarbons similar to those in petroleum are formed in kelp and other marine algae, by conversion of the Sun's radiant energy via photosynthesis. The algal–kelp source is vast enough to have yielded all known oil (including that in oil shales and tar sands) during Pleistocene and Recent time alone if conversion and preservation were favorable. Bloom-causing algae, like the freshwater *Botryococcus* spp., yield fats, lipids, and liquid hydrocarbons. Nonmarine algae are the principal contributors to the hydrocarbon content of oil shales (see Ch. 10). The drying-up of algal colonies results in rubbery accumulations of waxy or gelatinous cell walls; the *coorongite* of Australia is an example. In a number of rich oil provinces of Tertiary age (Baku, California, Sakhalin), the oil is closely associated with (and almost certainly derived from) diatomaceous sediments. On the other hand, the most familiar marine plants, the seaweeds, cannot be an important source of oil; oils commonly contain sulfur but not iodine or bromine, and about 85 percent of seaweed is water.

Early objections to an organic origin for oil centered upon the absence of an adequate amount of life in the constitution of most petroliferous formations. But oil, being a fluid, is potentially migratory; its source may have been elsewhere than in its present host rock even if that is a fossiliferous marine limestone. Furthermore, the original fauna or flora may not have been suitable for the formation of visible fossil remains. A fish such as the lamprey, zoologically primitive, contains about 50 percent of fats and albuminoids (constituents of connective tissue) which could be a prolific source of oil, but the lamprey leaves only small teeth to be fossilized. There have been enough fish in the seas and lakes since Ordovician time to account for all the oil known, and fossilized fish remains are common in sedimentary strata believed to be oil source rocks (see Ch. 8). The fossil record is sufficient to show that soft-bodied aquatic organisms have existed throughout Phanerozoic time. Phosphatic, chitinous, and carbonaceous fossils, such as conodonts, are also commonly associated with sedimentary rocks believed to be oil sources, and may represent the only fossilizable parts of otherwise shell-less and skeleton-less organisms potentially contributing to oil accumulations.

Fish and whales are *nektonic* animals. They swim about, and if their remains are to be preserved as fossils they must fall to the sea (or lake) floor after death. Vast assemblages of fossil fish remains, such as those in the Eocene oil shales of Wyoming or the Miocene diatomite of Algeria, therefore represent environments not of life but of death (called thanatocoenoses). *Benthonic* organisms, on the other hand, live on the sea floor, and *sessile benthos* are fixed there, like corals and oysters. Their remains may be fossilized where they lived, constituting a biocoenose. Thus a biocoenose necessarily reflects an oxygenated environment, inhabited by scavengers. Higher animal organisms, including the larger benthos and all vertebrates except fish, are for this reason highly improbable candidates as sources of oil. Except for fish, therefore, we are left with *planktonic* organisms as potential sources.

The majority of marine plankton are *phytoplankton*, belonging to the plant kingdom. The most familiar phytoplankton, and volumetrically the most important, are the diatoms, siliceous unicellular plants. Minute droplets of oil accumulate in their cells late in the vegetative period. Animal plankton, collectively called *zooplankton*, include planktonic foraminifera and radiolaria, which are widely represented as fossils in young oil-bearing strata; but the most numerous and important zooplankton are the copepods, which are planktonic crustacea. Some modern foraminifera, radiolaria, and copepoda contain minute drops of oil, possibly due to the decay of the animal but also possibly taken on as a food reserve or as a floating mechanism to reduce the density of pelagic eggs or larvae. Hydrocarbons have also been isolated from bacterial protoplasm. The bulk of the evidence favors these lowly forms of aquatic life (especially marine life) as the most important sources of oil and of wet gas.

6.4 Biological and biochemical factors influencing the accumulation of organic matter in aquatic environments

Zooplankton feed upon phytoplankton, which are therefore the first item in the food chain. Phytoplankton are the original source of nearly all organic matter in sediments; very little of it comes from the land. The total quantity of organic material carried by rivers to the seas is about 7×10^{11} kg annually, or less than 1 percent of the total organic matter in the oceans.

Phytoplankton depend upon photosynthesis, which is restricted to the *euphotic zone* of penetration by sunlight, the upper 50–100 m of oceanic water bodies. In addition to sunlight, plankton growth requires mineral nutrients, and the greatest supply of these does come from the land – not merely from the inorganic sources of the rocks but also from earlier generations of plants and animals. Consequently, by far the greater part of the organic life of the seas is concentrated in the zones bordering and overlapping the continents. Shelf seas "bloom" with phenomenal outbursts of phytoplankton every spring, as the Sun climbs higher each day and the water temperature is increased.

In deeper water, below the photic zone, the existence of "marine snow" suggests that bacteria may thrive there on CO_2 dissolved in the water, converting it to organic nutrients. If this proves to be true, it would represent a further form of planktonic life not directly dependent upon photosynthesis.

The elemental attributes of organic matter favorable to its hydrocarbon source potential are of course its initial carbon content, its capacity to retain or acquire hydrogen, and as low an oxygen content as possible. A number of other elements, however, represent mineral nutrients essential to planktonic productivity. They include familiar metals (iron, copper, zinc) as well as nonmetallic elements such as sulfur. The principal limiting factor, however, is the supply of *nitrogen* and *phos-*

phorus. These are present in sea water as nitrate and phosphate ions, the natural ratio between these being fairly constant at about 15 : 1 by weight. Because of human influence, both nitrogen and phosphorus contents are increasing rapidly, and the ratio is now closer to 11 : 1. Both nitrate and phosphate reach their highest concentrations in basin depressions where oxygen content is at its lowest. A close relation between the contents of phosphorus and of organic matter characterizes almost all sediments. Fish are well known to be high in both phosphorus and nitrogen, and benthonic organisms in general are also high in phosphorus.

Among the materials synthesized by very low forms of life, both plant and animal, are the *fatty acids*, so called because they become essential constituents of animal fats and of animal and vegetable oils. All the fatty acids are monobasic, with the general formula $C_nH_{2n}O_2$, or $C_nH_{2n+1}COOH$. Familiar examples of low molecular weight are formic acid (HCOOH), which occurs in ants, stinging insects, and stinging plants, and acetic acid (CH_3COOH), which gives vinegar its taste. The fatty acids are the most stable type of organic acids, relatively resistant to hydrolysis and to bacterial action so long as oxygen is excluded. They are also the largest known source of long-chain molecules, which are almost certainly the fundamental molecules of petroleum.

We have to consider three major classes of these fundamental biomolecules which constitute potential source material for hydrocarbons:

(a) The *lipids*, mostly fats, oils, and waxes, are compounds of the fatty acids with *glycerol*, $C_3H_5(OH)_3$. (Nitroglycerine, the familiar component of explosives, is $C_3H_5(ONO_2)_3$.) Domestic beef fat, lard, and so on are essentially the glycerides of fatty acids having 18 or 19 carbon atoms. The most important is *stearic acid*, $C_{17}H_{35}COOH$, which makes the glyceride $C_{17}H_{35}COOCH_3$; common soap, used to dissolve fats, is essentially sodium stearate, $C_{17}H_{35}COONa$. Lipids, high in phosphorus and also carrying both nitrogen and sulfur, have the greatest potential as hydrocarbon generators.

(b) *Proteins* are giant molecules that make up the solid constituents of animal tissues and plant cells. They contain about 50–55 percent carbon, 7 percent hydrogen, and 19–24 percent oxygen; but they may also carry 15–19 percent nitrogen, 0.3–2.4 percent sulfur, and a small amount of phosphorus. On hydrolysis, they form amino acids of the type $RCH(NH_2)COOH$. An example is aminoacetic acid or glycine, $CH_2(NH_2)COOH$, derived from albuminoids such as glue or gelatin.

(c) The *carbohydrates*, principally *cellulose*, $(C_6H_{10}O_5)_n$, the fundamental constituent of plant tissues, are built up of molecules like that of glucose. Carbohydrates lack phosphorus, nitrogen, and sulfur.

Most zooplankton are rich in both protein and fats (lipids), but poor in lignin and in carbohydrates. Copepods, for example, contain about 65 percent protein, 13 percent fats, and 22 percent carbohydrates. For the higher invertebrates, the equivalent percentages average 70, 10, and 20 percent. Marine sediments average 40, 1, and 47 percent, but with wide variations. Marine phytoplankton, such as diatoms, contain about 25, 8, and 63 percent respectively. Terrestrial plants are rich in lignin and in carbohydrates, but very low in protein; however, G. T. Philippi showed that if their remains are sufficiently abundant they can give rise to heavy oils rich in five-ring naphthenes.

Zooplankton contain between three and eight times as much nitrogen as do phytoplankton, and therefore must consume from three to eight times their own volume of phytoplankton to account for their nitrogen content. The nitrogen comes primarily from protein, and so ultimately from animal matter. The carbon:nitrogen ratio of sediments varies between 8:1 and 12:1; their protein content is about six times their content of organic nitrogen. Fats and proteins are richer in carbon than are carbohydrates, but much lower in oxygen (18 percent in fats, 22 percent in proteins, but as much as 50 percent in carbohydrates). It will be deduced that fats and proteins are the important biomolecular precursors of petroleum.

6.5 How organic matter accumulates

Two questions immediately arise from the recognition of planktonic, aquatic organisms as the primary source of oil and wet gas. How do they accumulate in the numbers necessary to yield the vast volumes of oil and gas known; and how are such numbers preserved from destruction by either biogenic or chemical agencies? The accumulation of vast quantities of living organisms requires ample oxygen supply. Both the preservation of the organic matter resulting from the deaths of the organisms, and the subsequent enrichment of its hydrogen-rich lipid component to permit the generation of petroleum, require that oxygen be excluded.

Aquatic life depends upon the movement of oxygen-rich water. This movement depends upon the water's density, which depends in turn upon its temperature (the water density being highest at 4 °C) and upon its salinity. Oxygen-rich surface waters may sink in anticyclonic manner, moving clockwise in the Northern Hemisphere and counter-clockwise in the Southern Hemisphere. Alternatively, oxygen-rich bottom waters may rise towards the surface as they travel from colder latitudes towards the Equator; this is cyclonic motion.

Where oxygen-rich waters rise towards the surface, they bring not only oxygen but mineral nutrients into the photic zone. This leads to the phenomenon known as *waterbloom*: shallow, warm waters, penetrated by sunlight, "bloom" with phenomenal populations of phytoplankton and planktonic protista, especially dinoflagellates. The most familiar manifestation of waterbloom is "red water" or "red tide," an annual (commonly springtime) occurrence in a number of water bodies. The excessive production of plankton is called *hypertrophy*, and it creates a "biochemical fence" beyond which the natural demand for oxygen exceeds its supply. If the oxygen content falls below 0.5 ml l^{-1}, the conditions are said to be *anoxic*.

Oxygen is then not renewed in the deeper waters. Vertical mixing of the waters is inhibited, and stratification is created. Surface waters become separated by a narrow zone from deeper waters that are colder, denser, or saltier than those above them, as well as being oxygen-deficient. The dividing zone for each of those three conditions is a thermocline, pycnocline, or halocline. One consequence is the creation of *anoxic layers* within the water body, "hemmed in" by patterns of circulation. Within these layers there is mass mortality among fish and benthonic invertebrates, which are almost totally suppressed. These two groups of organisms are richest in both phosphorus and nitrogen, as we recognized earlier, and the supply of these two elements is the chief limiting factor in plankton growth. The organic carbon content of the sedimented material becomes 3 percent or much higher, and its degradation further depletes the oxygen supply.

Waterbloom and its attendant stratification of the water body are most easily observed in certain large *lakes* in humid, tropical regions without marked seasonal overturns. Very abundant dissolved plant nutrients make the lakes *eutrophic*; the surfaces become covered with green vegetable slime, exacerbated in populated regions by the introduction of phosphates from detergents. The consequence is a seasonal oxygen deficiency in the bottom waters, which become enriched in H_2S and therefore acidic (pH < 7). As bacteria are less involved in sulfate reduction here than they are in marine conditions, the organic carbon content in eutrophic lakes may become very high, exceeding 10 percent in deep lakes such as those in the East African rift valleys.

Isolated or semi-isolated bodies of saline or brackish water are likely to be *silled*; a narrow, shallow ledge or sill restricts connection with the open ocean. There are two contrasting circumstances. Where the water balance is *positive* fresher surface waters flow out as saline, nutrient-rich deeper waters flow in, creating a permanent halocline with H_2S and CH_4 generated below it. Such water bodies are nutrient traps; the Black Sea and Lake Maracaibo are obvious examples. Where the water balance is *negative*, shallow marine waters or even fresh river waters constantly flow into the basin, to be quickly evaporated in a hot, arid climate. The dense, hypersaline waters so produced are oxygenated but depleted in nutrients (and so in organic matter); they sink to the bottom or flow out of the basin as deep currents. The Mediterranean and Red Seas are obvious examples. The Caspian Sea is in principle another, though its total enclosure makes its water budget somewhat different.

A source-rock example of a semi-enclosed, silled basin, essentially synchronous with the greatest expansion of oceanic anoxia, was the mid-Cretaceous Mowry Sea of the interior of North America (Fig.6.2).

The highest concentrations of organic carbon in modern oceans occur in regions of intense *upwelling* of shelf-bottom waters. Upwelling currents bring nutrients, especially nitrogen and phosphorus, and also oxygen, towards the shallow photic zone, promoting hypertrophy and very high oxygen demand, which in turn create anoxic conditions in the deeper layers of water below the zone of upwelling (Fig.6.3). Upwelling currents are intrinsically likely to occur as cold waters

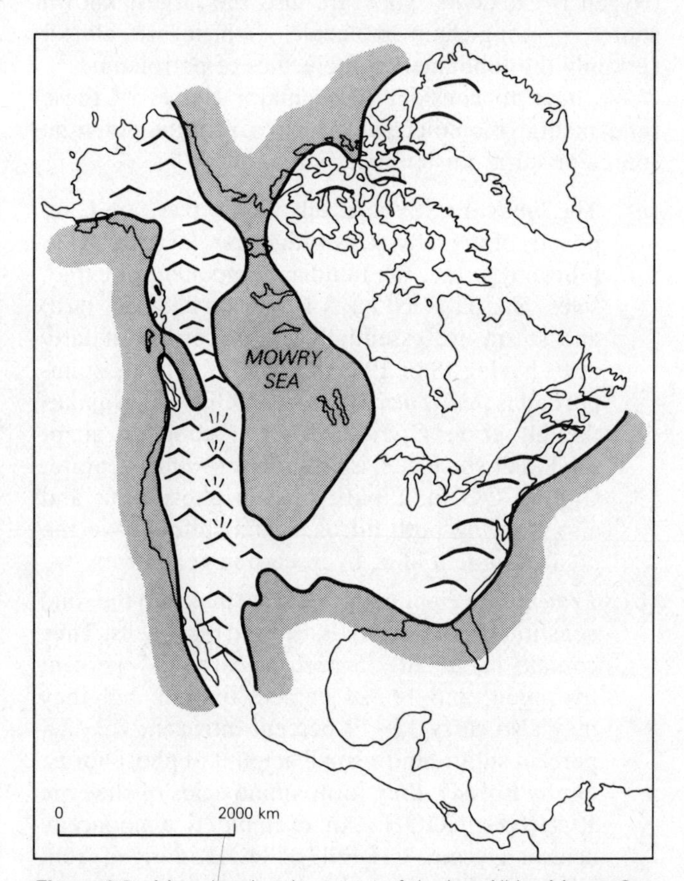

Figure 6.2 Map showing the extent of the late Albian Mowry Sea between the Canadian Shield and the rising Western Cordillera, to illustrate the typical shape of an anoxic silled basin. (From G. D. Williams and C. R. Stelck, Geol. Assoc. Canada Special Paper 13, 1975.)

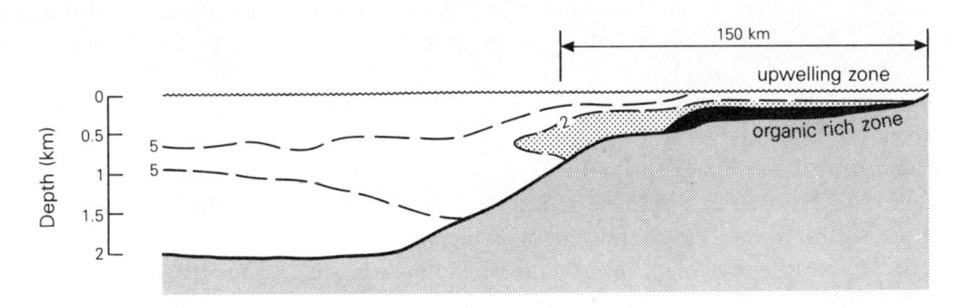

Figure 6.3 Anoxic layers at ocean margins caused by upwelling, as on the Peruvian or South-West African shelves. Numbered contours represent the oxygen content of the water in parts per thousand. Anoxic patches extend about 700 km along the shelf. The organic carbon content of the anoxic sediment is 3–26 percent and of the oxic sediment <3 percent. (After G. J. Demaison and G. T. Moore, *AAPG Bull*, 1980.)

from polar seas flow towards the Equator and become deflected westward into the equatorial drift. This elementary pattern is intensified where the orientations and shapes of the continental slope and the coastline facilitate rather than oppose it.

Upwelling by *divergence* is in large part a consequence of steadily persistent winds blowing obliquely offshore, along coasts facing to the northwest in the Northern Hemisphere and to the southwest in the Southern Hemisphere. The coasts thus face the approaching polar currents, and the coastal waters are forced to move seaward by the Coriolis force, being replaced by upwelling waters from below. The greatest effect is seen in the belts of aridity between 40°N and 40°S latitudes (the "horse latitudes"). Divergence does not normally occur along coasts facing east or southeast even within these same latitudes, but upwelling may occur locally there if the currents are deflected upwards by topographic highs on the bottom. In latitudes higher than about 40°, upwelling takes place only in this latter circumstance; more oxygen remains in solution in the colder waters.

Thus anoxic regions occur preferentially off the west sides of continents in low latitudes. A striking case, in which all favorable factors are combined, is that off the coast of Peru. The northward-moving Humboldt current from the Antarctic is uninterrupted along the meridional coast of Chile, but is forced into westward and upward deflection about 18°S by the northwest-trending continental slope off Peru.

Ancient oil-source sediments attributable to coastal upwelling are likely to be more phosphatic than normal source sediments. The Permian Phosphoria Formation of the western USA, the Oligocene Maykop Group of the Caucasus and the Miocene Monterey Formation of California are probable representatives. Now that the distributions of continents and oceans in times past can be reconstructed with some assurance, models of atmospheric circulation can be superimposed upon them and predictions made concerning probable sites of upwelling during those times.

The cores from the Deep Sea Drilling Program (DSDP) have revealed the extent of anoxic horizons in the open ocean, unconnected with any continental margin. The oxygen minima occur in water layers at intermediate depths, but they are caused by deep circulation patterns and thus indirectly by the Coriolis force. They are not, therefore, directly correlatable with high organic productivity, and certainly not with high sedimentation rates. The existence of worldwide, synchronous anoxic conditions during mid-Cretaceous time, shown by both the oceanic horizon B and the basinal oil-source sediments of that age (see Sec. 8.4), enables us to identify their fundamental geologic controls with considerable assurance. The Earth was girdled by a latitudinal, approximately equatorial ocean. The arrangement of the continents north and south of this ocean prevented the deep circulation of oxygenated, polar bottom waters. The Cretaceous climate was warm and globally equable, so that biologic productivity was high and surface waters were depleted of oxygen. The onset of rapid seafloor spreading in mid-Cretaceous time created an array of silled basins, in which anoxia was easily achieved. The spreading also led to a worldwide marine transgression, which carried the anoxic layers far landward over the upper continental slopes and shelves, interleaving organically rich sediments among ordinary shallow water deposits.

The development and decay of this particular anoxic interval can be traced in detail through the DSDP cores. In the North Atlantic Basin, for example, marine plankton became common during the Aptian–Albian and remained abundant until the Coniacian, leading to the richly organic horizon B. They then declined sharply in the latest Cretaceous, when seafloor spreading had progressed far enough for general oceanic circulation to be established, extensive new coastlines furnished great quantities of terrestrial organic matter, and kerogen content of the sediments dropped to normal levels.

The characteristic deposits of the Cretaceous anoxic realms were black, laminated, bituminous or carbonaceous shales. The high preservation of organic carbon was accompanied by higher than normal rates of accumulation of heavy metals. Organic-rich oil-source sediments of this type were widespread during the late Devonian, but they were more so between the late Jurassic and the late Cretaceous than during any other interval of geologic time (see Sec. 18.2). It seems likely that the graptolitic black shales of the Lower Paleozoic may have resulted from comparable conditions, enhanced by the absence of terrestrial organic material.

The modern equivalent of this unusual phenomenon occurs where an expanded oxygen-minimum layer intersects a suitably oriented upper continental slope – along the northern margin of the Indian Ocean, and off west-facing arid coasts such as those of Baja California, Peru, and South-West Africa (Namibia) (Fig. 6.3). The relative rarity of oceanic anoxia at present is due to the north–south (and nonequatorial) extension of the present oceans, permitting interpolar marine connections with maximum temperature gradients and excellent circulation and oxygenation of the oceans.

6.6 Organic carbon in sediments

The sedimentary basins of the Earth contain over 80×10^6 km^3 of sediments and sedimentary rocks. According to John Hunt, the *total carbon content* of the crust is about 9×10^{19} kg. Of this quantity, 1.3×10^{19} kg are in igneous and metamorphic rocks as graphite, carbonatites, and other uncommon materials. The hydrosphere and biosphere contain a much smaller quantity, of the order of 5×10^{14} kg. In the present oceans, organic carbon content decreases with depth, below about 200 m, and away from coasts.

The rest of the crustal carbon content is in sediments and sedimentary rocks. Of this portion, over 80 percent consists of the carbon in carbonates. The total *organic carbon* content of all sediments and sedimentary rocks amounts to about 1.2×10^{19} kg. This total is distributed among the principal natural media as below, in units of 10^{15} kg:

dispersed in sedimentary rocks	11 000
in coal and peat	15
petroleum in nonreservoir rocks	265
petroleum in reservoirs	1

In words, of all the organic carbon *in sedimentary rocks,* only about 0.01 percent is in the form of oil or natural gas in known reservoirs. As Hunt expressed it, the generation and accumulation of petroleum are very inefficient processes. Or, in Philip Abelson's words, in the Earth's whole history a weight of organic matter equal to the weight of the whole Earth has been synthesized, but only about one part in 10^{10} has been preserved.

In modern sediments, the content of organic carbon (OC) increases with decreasing grain size; about 75 percent of it is in clay rocks. This is because fine-grained sediments retard the rate of diffusion of oxygen and sulfate ions through the sedimentary column; these ions are the principal destroyers of organic matter. On the Atlantic coastal shelves, OC contents range from 6.4 percent in diatomaceous ooze to less than 0.3 percent in terrigenous and volcaniclastic sands. On the continental slopes, the contents are still lower, about 1 percent in oozes to 0.4 percent in volcaniclastic sands. Early work by Parker Trask suggested that the average OC content of recent sediments was about 2.5 percent, of ancient sediments about 1.5 percent. The difference represents catabolism of labile organic matter with temperature and time (see Sec. 7.4.2). More recent calculations, principally by Hunt, suggest that although about 3.5 percent is a fair figure for the *carbon* content of all sediments and sedimentary rocks, the average *organic carbon* content is more like 0.5 percent.

If OC originally constituted 1 percent of the sedimentary volume, and 0.1 percent *of this* developed into hydrocarbons, then about 800 km^3 of hydrocarbons would have been formed, or about 5×10^{12} barrels of oil equivalent. This is the right order of magnitude for the oil equivalent actually known, recoverable and unrecoverable, conventional and unconventional. It has been estimated that 2 percent of the original organic matter finally comes to rest under conditions of rapid sedimentation and favorable bottom environment; only one part in 5000 or thereabouts is effectually deposited in unfavorable environments and under very slow deposition.

We need to determine the concentration of organic matter (OM) relative to the mineral matrix of the rock. The amount of total organic carbon (TOC) is measured by total pyrolysis technique, and the OM in weight percent is derived by general conversion factors:

$$\text{OM (weight percent)} = \text{OC} \times 1.22$$
$$\text{OM (volume percent)} = \text{OM (weight percent)} \times 2.5$$

Early studies by Archangelski, Trask, and others indicated that 1.5–2.0 percent by weight of OC was adequate for the sediment to be an oil source rock. Tissot and Welte, with much better and more recent data, consider that a clastic source sediment needs at least 0.5 percent TOC by weight, a carbonate source sediment at least 0.3 percent (see Ch. 7). The actual percentage required will vary according to other circumstances discussed later (Sec. 8.1); there must certainly be a minimum value. A general consensus among petroleum geochemists categorizes source quality by TOC percentage as below:

OC (weight percent)	Quality
0.0–0.5	poor
0.5–1.0	fair
1.0–2.0	good
2.0–4.0	very good
4.0 upwards	excellent

Actual TOC may reach 20 percent or more by weight. It is highest in coals and rich oil shales, but these are not "source sediments" for oil in the practical sense.

The TOC in sedimentary rocks is separable into two fractions:

(a) The fraction that is soluble in organic solvents (such as CS_2 and chloroform) is bituminous, and loosely called *bitumen*.
(b) The insoluble, nonextractable residue from the initial transformation of OM is called *kerogen* (from the Greek κερος (*keros*) = wax). In ancient sedimentary rocks, particularly in shales, kerogen normally constitutes 80–95 percent of the total OM.

The term *kerogen* was originally coined for the OM in the oil shales of central Scotland, which on distillation yields liquid oil. It is now used for the intermediate product, formed by diagenetic transformation of the OM originally enclosed in sediments, that gives rise to petroleum (Fig.6.4). It is a complex mixture of high molecular weight organic materials, the bulk composition depending on the source and environment of the OM, as explained below. The general composition may be expressed as $(C_{12}H_{12}ON_{0.16})_x$. (Anticipating the material of Ch. 7, note that in this composition the H:C atomic ratio is 1.0 and the O : C ratio is 0.08.)

Not all the OC in sedimentary rocks, in other words, is convertible into petroleum hydrocarbons. The geochemist needs to determine the proportion of the TOC which consists of saturates and aromatics, and the ratio between these two classes of petroleum hydrocarbon. The general relation between TOC and hydrocarbon content of a sedimentary rock is shown in Fig.6.5, representing the Pennsylvanian Cherokee Group of the Mid-Continent area of the USA.

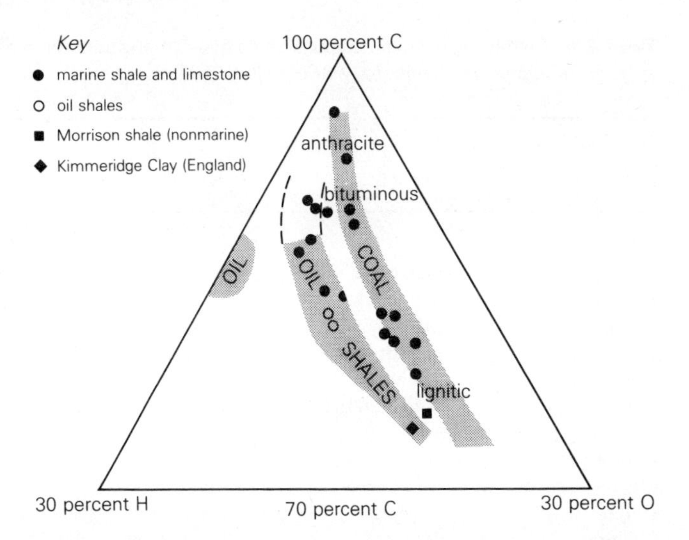

Figure 6.4 Distribution of carbon, hydrogen, and oxygen in kerogen. (From J. P. Forsman and J. M. Hunt, in *Habitat of oil*, Tulsa, Okla: AAPG, 1958.)

Table 6.1, from John Hunt, lists the hydrocarbon (HC) and organic matter (OM) contents of selected shaly and calcareous sedimentary rocks from richly petroliferous areas. An immediate observation is that carbonates in general contain much less OM than do shales, but the OM of the carbonates is much richer in hydrocarbons than is the OM of the shales. This unexpected observation was further demonstrated by H. M. Gehman, Jr in 1962; shales average about five times as much OM as limestones, but this ratio is effectively reversed for the hydrocarbon contents of the OM. The consequence is that the actual hydrocarbon contents of the two sediment types are about equal, and average 100 p.p.m. The implication is that some agent of decomposition, probably bacteria, eliminates the easily oxidized and easily degraded organic components but not the hydrocarbons (see Sec. 8.1).

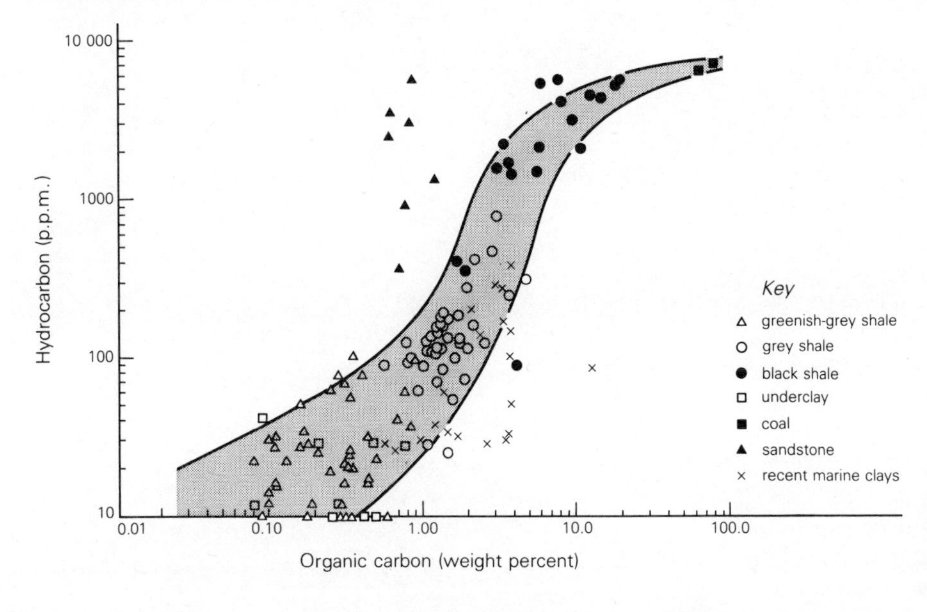

Figure 6.5 Relation between organic carbon and hydrocarbon contents for the principal rock types of the Pennsylvanian Cherokee Group in the Mid-Continent province of the USA, and (for comparison) for some recent marine clays. (From D. R. Baker, *AAPG Bull.*, 1962.)

Table 6.1 Distribution of hydrocarbons and organic matter in nonreservoir rocks. (After J. M. Hunt, *Geochim et Cosmochim. Acta,* 1961; and J. M. Hunt and G. W. Jamieson, *AAPG Bull.,* 1956.)

Formation	Locality	Age	Hydrocarbons (p.p.m.)	OM (weight %)	Hydrocarbons in OM (%)
Shales					
Wilcox	Louisiana	Eocene	180	1.0	1.80
Frontier	Wyoming	Cretaceous	300	1.5	2.00
Springer	Oklahoma	Mississippian	400	1.7	2.35
Monterey	California	Miocene	500	2.2	2.27
Woodford	Oklahoma	Devonian	3000	5.4	5.56
Bakken	North Dakota	Devonian–Mississippian	4500	13.0	3.46
Carbonates					
Mission Canyon	Montana	Mississippian	67	0.11	6.09
Ireton	Alberta	Devonian	106	0.28	3.79
Madison	Montana	Mississippian	243	0.13	16.70
Charles	Montana	Mississippian	271	0.32	8.47
Zechstein	Denmark	Permian	310	0.47	6.60
Banff	North Dakota	Mississippian	530	0.47	11.30
Calcareous Shales					
Niobrara	Wyoming	Cretaceous	1100	3.6	3.06
Antrim	Michigan	Devonian	2400	6.7	3.58
Duvernay	Alberta	Devonian	3300	7.9	4.18
Nordegg	Alberta	Jurassic	3800	12.6	3.02

7 The conversion of organic matter to petroleum

We have considered the accumulation of organic carbon in fine-grained sedimentary rocks. It is not yet kerogen, or bitumen, let alone petroleum. We now trace the transformation of the organic matter (OM), first to kerogen plus bitumen, then to petroleum hydrocarbons.

7.1 Environment of the transformation

Because the only elements essential to the constitution of petroleum are hydrogen and carbon, the transformation must be such that the oxygen and nitrogen of the original OM are largely removed, and the lipids (fats) and hydrogen-rich organic residue largely preserved. This outcome is impossible if the decomposition of the OM takes place in an oxygenated environment.

The OM must therefore be subjected to no prolonged exposure to the atmosphere, to aerated surface or subsurface waters carrying acids or bases, to elemental sulfur, or to vulcanicity or other igneous activity. Even in a continuously aquatic environment it must not be long transported, reworked, or oxidized. In many organically rich sediments, the bulk of the OM is terrestrial vegetal matter transported by rivers, depleted in hydrogen before its delivery to the site of final deposition.

If the oxygen content of the water exceeds 1 mg ℓ^{-1}, aerobic decomposition of OM is very efficient, especially at, or immediately above or below, the sediment/water interface. Though dominated by aerobic bacteria, the processes are enhanced by the presence of benthonic, metazoan predators (worms, bivalves, etc.); *bioturbation* is more or less intense under virtually all oxygenated waters, regardless of depth or type of substrate.

Under *anaerobic* conditions, with less than 0.1 mg l^{-1} of oxygen (about 1.5 percent of its normal concentration in sea water), decomposition of OM is slower and less efficient than under aerobic conditions. Anaerobic bacteria may use nitrates or sulfates in solution, until these salts are exhausted, yielding nitrogen and H_2S in addition to the CO_2 and H_2O generated by any earlier aerobic decomposition. Waters in anaerobic environments, like those in oilfield reservoirs, are characteristically lacking in the sulfate ion (see Ch. 9). As dead OM falls to the sea floor as "organic rain," the constituents necessary for hydrocarbon generation are preserved only if the water column is essentially anoxic (lacking living organisms), the fall is fairly rapid (that is, the particle size is not wholly microscopic), and bottom-dwelling predators are lacking.

Anaerobic decomposition is strongly influenced by grain size. Coarse sands permit easy diffusion of oxygen and oxy-salts, and are invariably low in OM content. In addition, the activity of anaerobic bacteria is reduced as the grain size decreases. It has been commonly thought that the process is also a function of the sedimentation *rate*: that under slow or interrupted sedimentation, complex organic substances are readily broken down by scavengers into simple salts and CO_2, whereas rapid deposition buries the OM below the reach of mud-feeding scavengers. Under normal conditions there undoubtedly is a positive correlation between the bulk sedimentation rate and the OC content of the sediments. But the correlation is less clear at very high sedimentation rates, presumably because of the dilution effect brought about by those rates; the correlation is not clear at all in anoxic waters, in which it may in fact be a negative correlation.

The transformation of OM to kerogen proceeds from shallow depths of burial to depths of perhaps 1000 m, with temperatures up to about 50 °C. On further burial and heating, the large molecules crack to form smaller, lower molecular weight hydrocarbons (*geomonomers*), in the depth and temperature ranges of 1000–6000 m and 50–175 °C. Initial products are mostly H_2O and CO_2, at higher temperatures becoming divided between volatile products (hydrogen and methane) and liquid products (from C_{13} to C_{30}). Oxygen is lost most rapidly, by dehydration and decarboxylation (loss of CO_2 from fatty acids); carbon and nitrogen are lost least rapidly. Consequently, the carbon content of the kerogen residue increases, and the H:C ratio decreases, with increasing temperature.

During this post-depositional alteration, therefore, the organic constituents of the sediments are progressively transformed by thermal processes into two fractions:

(a) a fluid product high in hydrogen, eventually petroleum and natural gas.
(b) a residue high in carbon, such as bituminous coal.

Mild transformation yields liquid products; intense transformation leads to methane plus the high-carbon residue, analogous to the formation of coke in the combustion of coal.

If this is the essence of the conversion, the quantity of residual carbon (C_r) should be a measure of the amount of conversion of the original OM to fluid hydrocarbons. Thus the progress of the transformation can be measured by means of a variety of *organic geochemical indicators*, now routinely used for the evaluation of all potential hydrocarbon source sediments.

7.2 Organic geochemical indicators

The transformation of kerogen to mature petroleum involves a progressive increase in the atomic ratio of hydrogen to carbon – the *H : C ratio.* The ratio is determined in the laboratory by combustion.

Clearly, the complete conversion process is one of carbonization and *dehydrogenation*; it produces methane and residual carbon. Consider the H : C ratios during the transformation of woody material into coal. It involves the enrichment of a fluid hydrocarbon product (methane) in hydrogen, and the depletion of that element in the source material:

Source material	H : C ratio
wood	1.464
peat	1.308
lignite	1.044
bituminous coal	0.768
anthracite	0.324–0.000

The H : C ratio of the lost methane is 4.0.

Stages in the generation of petroleum may be comparable:

Source material	H : C ratio
organisms (average)	1.920
fats, carbohydrates, proteins (average)	1.680
pyrobitumen (kerogen)	1.524
carbon	0.000

The H : C ratio of the bitumen yielded is 1.740–1.980.

Because it was for a long time difficult to determine the hydrogen content of shales in the laboratory (because of contained water, and the hydrogen combined in the clay minerals themselves), studies concentrated on a variety of expressions of the *carbon* content. As early as 1915, a paleobotanist with the United States Geological Survey, David White, recognized that he could make some estimate of a horizon's oil prospects if coal occurred nearby in a horizon of comparable age. This was the case in the Appalachian and Mid-Continent basins which at that time produced most of North America's oil and coal. Low-grade but progressive metamorphism is marked by the elimination of hydrogen and oxygen from the coals, and a consequent apparent increase in the carbon. The *carbon ratio* of the coal is then defined by

$$\text{C ratio (\%)} = \frac{\text{dry, ash-free fixed carbon}}{\text{fixed carbon} + \text{volatile matter}} \times 100$$

The values on the right-hand side are determined by pyrolysis and have always constituted the basis of the so-called *proximate analysis* of the coals.

As the sandstones which provide the reservoir rocks for oil in coal-bearing regions are affected by the same processes as are the coals, White reasoned that there would be a "deadline" of metamorphic level, expressed by his carbon ratio, beyond which no significant oil was likely to be found. The principle, with more modern elaborations, is expressed in Table 7.1.

Analogous measurements made on shales (as potential source rocks), instead of coals, include a variety of ratios. One measure of the transformation is the ratio of *retortable carbon* (the proportion fluidized at 500 °C in an inert atmosphere) to *total organic carbon* (TOC). The latter is measured by first treating the rock with acid to remove any carbonate content, and then measuring the CO_2 combustion product at about 1000 °C. Barren or metamorphosed shales yield ratios of 0.1 or less; typical source sediments measure 0.4 or 0.5; rich oil shales, like the Green River shale, 0.6 or even higher.

Some components of the organic material, notably the organic acids, will react with caustic potash (KOH) to yield potash products; the reaction is called *saponification*. During post-depositional alteration, saponifiable components of the organic material are progressively changed to nonsaponifiable products (nonhydrolyzable, in effect – hydrocarbons, steroids). The degree of transformation is then expressible by the *hydrocarbon index* (HI):

$$\text{HI} = \frac{\text{non-saponifiable carbon}}{\text{total organic carbon}} \times 100$$

Complete conversion would be represented by HI = 100. Some shales which appear to have been the sources

Reflectance, R_o (%) (see Sec. 7.5.1)	Fixed carbon (see Sec. 7.5.1)	ASTM	White (1915)	Fuller (1919)	Sandstone lithology
0.3	47	brown coal and lignite	commercial oilfields	fields of heavy oil	sands generally saturated with fresh water
0.4	50	sub-bituminous coal		principal fields of medium oils	soft sands, continuous and porous
0.5	55	bituminous coal		principal fields of light oils and gas	fairly continuous and open
0.6		high volatile		oil rare, high grade; gas common	tight with some porous beds
0.7	60				
	65		oil deadline	only shows	
1.0	70	medium volatile	no commercial oilfields; gasfields may occur	no oil or gas with rare exceptions	rocks tight and hard
1.4	80	low volatile			
1.8	90	semi-anthracite			
2.6					
6.0	100	anthracite			

Table 7.1 Rank of coal and occurrence of oil- and gasfields. (After P. A. Hacquebard and J. R. Donaldson, *Can J. Earth Sci.*, 1970.)

of prolific oil yield indices greater than 90, whereas recently deposited organic matter in which the transformation has barely begun have indices less than 40. A closely comparable ratio is that between the *organic extract* (OE) and the TOC, evaluating the "yield" of the potential source rock. The ratio is low for immature organic matter; it increases at the beginning of the stage of intense oil generation, and decreases again in the higher-temperature gasification stage. Some Russian geochemists refer to a CEB value; *chloroform-extractable bituminoids* are analyzed by luminescence. Values, however, are exceedingly low for most samples, ranging from zero to less than 0.1 percent of the total rock.

What these parameters indicate is simply some measure of the quantity of *volatile matter* remaining in the total organic matter. The quantity of oil that can be generated per unit volume of source rock is probably a linear function of the rock's organic carbon content (OC). An "oil quantity factor" can then be adjusted so that this function has a particular slope (say 45°; Fig. 7.1).

As volatiles are progressively lost, the carbon ratio increases and the H : C ratio is reduced. For peak generation, the *carbonization level* should not proceed beyond 80–84 percent of elemental carbon by weight, leaving a minimum hydrogen content of about 7 percent, preferably over 10 percent, by weight. Thus the *quality* of the kerogen as a generator of hydrocarbons may be defined by the atomic H : C ratio before thermal maturation began, and the function depicted in Figure 7.1 can be replaced by that in Figure 7.2.

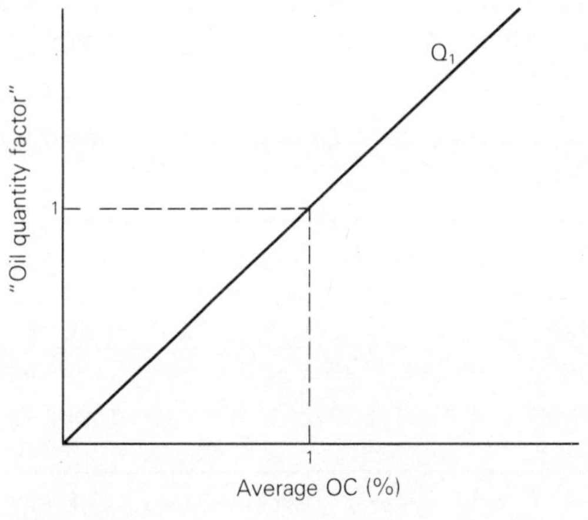

Figure 7.1 The quantity of oil generated per unit volume of source rock, shown as a linear function of the rock's content of organic carbon (OC).

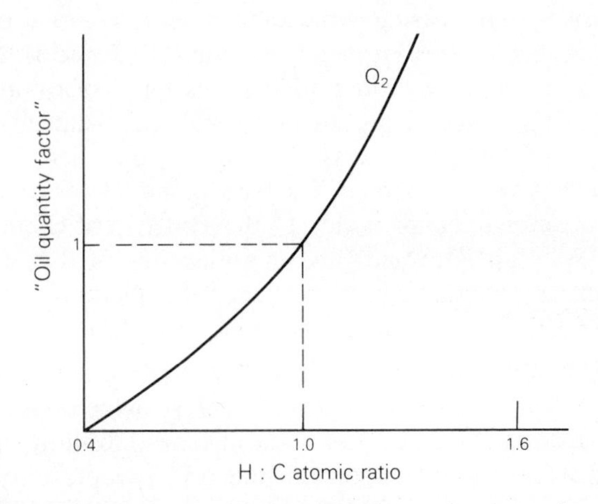

Figure 7.2 The quantity of oil generated per unit volume of source rock, shown as a function of the rock's H : C atomic ratio.

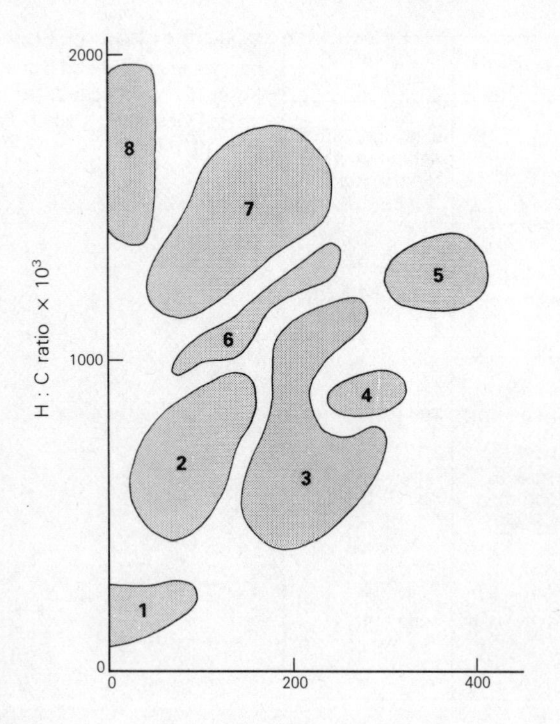

Figure 7.3 Modification of Van Krevelen diagram by K. F. Rodionova *et al.*, showing elemental composition of the combustible mass in organic fuels.

Key: 1, anthracite; 2, bituminous coals; 3, brown coals; 4, lignites; 5, peats; 6, sapropelites; 7, oil shales; 8, petroleum.

Atomic ratio	Dry gas	Wet gas	Gas condensate	Light oil	Heavy oil
H : C	0.9–1.5	1.2–1.5	1.35–1.5	1.5–2.0	1.7–1.8
O : C	0.05–0.15	0.03–0.12	0.08–0.1	0.02–0.08	0.12–0.20

(From K. F. Rodionova *et al.*, *Dokl. Akad. Nauk SSSR*, 1973.)

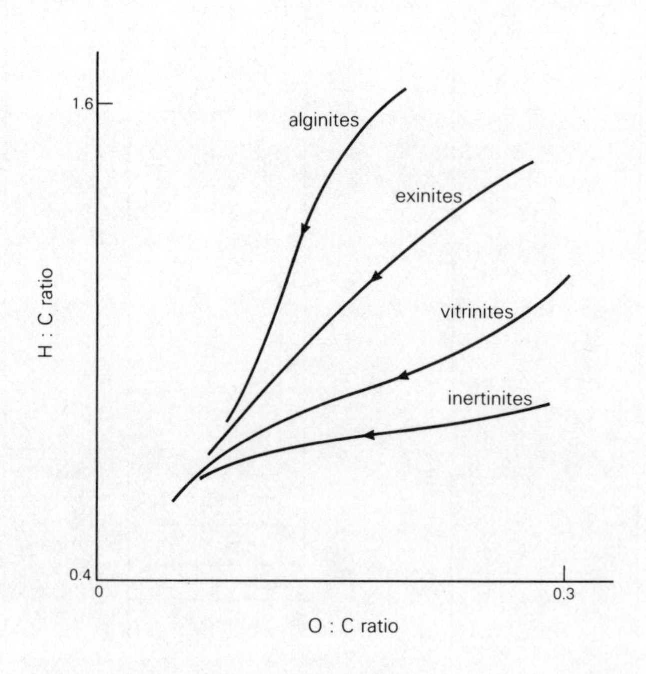

Figure 7.4 The fundamental Van Krevelen diagram, showing evolutionary paths followed by the principal maceral groups in coals.

As the H : C ratio is reduced by progressive carbonization with increasing depth and temperature, so also is the atomic ratio of oxygen to carbon (O : C ratio). By plotting these two ratios for various types of organic materials (maceral groups in coals; coals themselves; hydrocarbons), the paths of evolution and their final products can be depicted. The basic plot of H : C versus O : C atomic ratios is due to the Dutch coal chemist D. W. Van Krevelen, and modifications of the Van Krevelen diagram constitute standard depictions of the H–C–O fuels and their precursors (Fig.7.3).

The principal maceral groups in *coals* provide the basic Van Krevelen diagram (Fig.7.4), which depicts the paths of their evolution during carbonization; the paths progressively approach the origin, representing 100 percent carbon. The macerals are distinguished by their plant precursors:

(a) *Alginite* The principal organic material is of algal origin, as in sapropelic coals and oil shales. This category probably should also include resinite, the resinous components of coals. Alginite and resinite are fluorescent under ultraviolet light. Initial H : C ratios are above 1.5.

(b) *Exinite* Definite structures are recognizable. Sporinite consists of mega- and microspore exines, plus remains of plant resins and waxes. Cutinite is formed from plant cuticles. All are fluorescent under ultraviolet light. Initial H : C ratios are 1.2 or higher.

(c) *Vitrinite* Includes both telinite, in which woody structures are recognizably preserved, and collinite, the essentially structureless matrix, cement, and cavity infilling. Vitrinite is not fluorescent. Initial H : C ratio is about 0.75.

(d) *Inertinite* Includes "fossil charcoal" or fusinite; the degradation products of humic acids, lignin, and possibly proteins (micrinite); and sclerotinite, formed from fungal spores. Inertinite is not fluorescent. Initial H : C ratio is 0.5 or less.

The proportion of volatilized material yielded by the macerals under pyrolysis is a function of their H : C ratios. It ranges from more than 80 percent for algae to 5 percent for inertinite. Hydrocarbon yield ranges from 660 mg per gram of OM for amorphous exinites to 13 mg per gram of OM for inertinites. The residues after pyrolysis resemble inertinite.

7.3 The precursors of petroleum

Oil is not derived, as coal is, from terrestrial plant materials. The *humic* organic material yielded by terrestrial vegetation has to be largely or wholly supplanted by *sapropelic* material of aquatic origin, and normally also by OM of animal (zooplanktonic) derivation. The biochemical characteristics of humic and sapropelic OM are very different.

Humic organic material is largely of carbohydrate–lignin composition with some protein, relatively high in nitrogen but deficient in hydrogen (5 percent or less). It is derived from terrestrial materials and so is associated with continental sediments; it is easily identified by optical methods. Thermal alteration yields humic acids, soluble in alkalis, low in hydrogen, high in oxygen (in carboxyl groups). The yield of liquid, volatile products is very low, supplanted by CO_2 and H_2O.

Humic OM is therefore *gas-prone*. Its carbonization, exemplified by the rise in the ranks of coals, generates methane virtually without light liquid fractions. The quantities of methane so generated range from 68 kg per tonne for peat or lignite to 133 kg per tonne for hard coal. Only a small proportion of this gas accumulates in workable deposits. The rest is sorbed or occluded in the coal seams themselves, or expelled from the seams and sorbed by the host rocks, dissolved in ground waters, or lost to the atmosphere.

Sapropelic organic material is essentially structureless vegetable mud or slime (alginite), containing polymerized material of lipid and protein derivation. Lipids, derived from spores, cuticles and similar plant constituents, have high hydrogen contents (more than 10 percent). Sapropelic OM is associated with aquatic sedimentary facies, both marine and lacustrine. Its yield of volatiles is much greater than that of humic material. Thermal alteration yields water-insoluble biomolecules, especially lipids, that are derivatives of chlorophyll and are extractable from cells by organic solvents. They are efficient producers of paraffin hydrocarbons.

In palynological extracts of immature sediments, these two principal types of organic matter yield H : C ratios as below:

(a) less than 0.8, predominantly humic;
(b) between 0.8 and 1.0, mixed humic and sapropelic;
(c) above 1.0, predominantly sapropelic.

The Van Krevelen diagram was modified by Bernard Tissot (in 1974) and others to depict the maturation pathways followed by four basic types of OM in sedimentary rocks (Fig. 7.5):

(a) *Type I* A rare type of high-grade, algal sediment, commonly lacustrine, containing sapropelic OM.

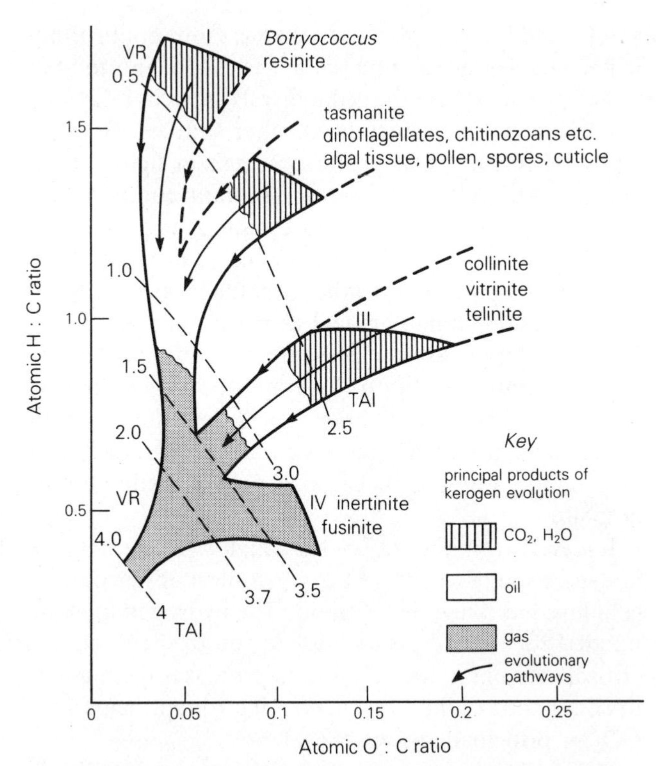

Figure 7.5 Thermal evolution pathways of four kerogen types and some of their microcomponents. The dashed lines represent values of the vitrinite reflectance (VR) and thermal alteration index (TAI). (From R. W. Jones and T. A. Edison, SEPM Publication, 1978.)

It includes oil shales, coorongite and tasmanite from Australia, boghead coals and torbanites. H : C ratio 1.6–1.8.

(b) *Type II* Type of intermediate derivation, commonly marginal marine, with admixture of continental and aquatic (especially planktonic) OM. Algal tissue, pollen, and spores are important contributors. This type includes the principal source sediments for oil, and some ancient oil shales of marine origin such as the kukkersite of Estonia (see Sec. 10.8.1). H : C ratio about 1.4.

(c) *Type III* Sediment containing primarily humic OM, of terrestrial, woody origin, equivalent to the vitrinite of coals. The material was deposited at the oxic sediment/water interface; it is gas-prone. H : C ratio 1.0 or less (that is, there are more carbon atoms than hydrogen atoms).

(d) *Type IV* OM may have come from any source, but it has been oxidized, recycled, or altered during some earlier thermal event. The inert carbonaceous material is now deficient in hydrogen (H : C ratio about 0.4 or less), has no evolutionary path left for it, and yields negligible or no hydrocarbons.

Because fluvial and deltaic sediments contain more OM of terrestrial, humic origin than do marine sediments, and because also this OM has more opportunity to become degraded or recycled, such sediments tend to be gas-prone. Where they contain oil (see Sec. 13.2.7), it is likely to be migratory from other sources. Because types III and IV had precursors high in lignin, their kerogen derivatives are high in aromatic components.

On Figure 7.5, the areas representing the production of CO_2 and H_2O constitute the zone of immature *diagenesis*. Its only hydrocarbon products are bacterial methane and sedimentary hydrocarbons (see Sec. 10.2). The unpatterned areas representing the principal liquid oil generation constitute the zone of *catagenesis* (see Sec. 7.4). The dotted areas constitute the zone of *metagenesis*, in which liquid hydrocarbons are cracked at high temperatures and the final yield is methane plus graphite.

It is clear from Figure 7.3 that a high H : C ratio is not in itself enough to ensure the generation of petroleum; cellulose has a high H : C ratio. The hydrogen must be bonded to the carbon and not to the oxygen. Thus a critical element in the maturation process is the manner of elimination of the oxygen; does it go out principally as CO_2 or principally as water?

If the organic precursors of the kerogen undergo early oxidation, much CO_2 may be generated. The yield of kerogen declines towards zero, but no hydrogen need be lost. The quantity of kerogen left as residue in the source rock may be very small even if the original OM content had been high. If, on the other hand, the oxygen is eliminated as water, most of the hydrogen is lost. The yield of kerogen is increased, but its hydrocarbon content decreases towards zero. Clearly there must not be sufficient oxygen in the original OM to allow either elimination route to proceed far enough to leave the kerogen deficient in either carbon or hydrogen.

7.4 Agents in the transformation

The kerogen we have been considering can be envisaged as a *thermally reactive* organic material. Its transformation into petroleum is to be seen as a process of initial energy loss, and its redistribution giving rise to separate molecules having a great store of energy. It is thus a thermodynamic problem. Are the energy sources required for the process intrinsic to the generative rocks, or extraneous to them?

The transformation could conceivably be *biogenic*. Bacteria and other microorganisms, in their attacks on organic matter, can produce both fats (which are easily convertible into hydrocarbons) and carbohydrates (not easily convertible). Claude ZoBell produced hydrocarbons in the C_{15} to C_{20} range in the laboratory, from naturally-occurring protoplasm. Sulfate-reducing bacteria can both make and destroy hydrocarbons; they can convert fatty acids (whether saturated, like stearic acid, $C_{17}H_{35}COOH$, or unsaturated, like oleic acid, $C_{17}H_{33}COOH$) and their hydroxy-derivatives (like lactic acid, $C_2H_4OHCOOH$) into hydrocarbons. They make organic matter more "petroleum-like," by reducing its oxygen content; they can also lower the density of hydrocarbons.

It is generally conceded that the formation of petroleum is favored not only by a negative redox potential but also by an alkaline environment, with pH between 7 and 9. The pH of sea water is generally between 7.5 and 8.4. It is highest near the surface and approaches its minimum normal alkalinity of 7.5 where oxygen is lacking and CO_2 content is high. On isolated bottoms, especially in stagnant depressions in which H_2S is produced, the pH falls below 7 and the acid environment is likely to result in types of black shale or marl in which hydrocarbons are retained and not generated. Bacteria or their enzymes help to condition the bottom environment, especially by adding CO_2; they were also thought by ZoBell to exert a greater influence than any other life forms on the pH of the water, by reducing oxy-salts such as sulfates and nitrates.

Despite all this, it is unlikely that bacterial processes can be effective in hydrocarbon generation beyond the initial breakdown of the organic matter. The numbers of bacteria decrease rapidly in the first few centimeters downward below the depositional surface; they scarcely survive at 10 m depth. Their activity is therefore exhausted before any other process has effectively begun. They may produce hydrogen at or near the sediment surface, but there is no evidence that organic matter can be directly hydrogenated to form oil. The role of bacteria in the synthesis of petroleum hydrocarbons is minor at the most.

The transformation is inevitably brought about by some kind of thermodynamic, nonbiogenic process. Such a process was called *catagenesis* by A. E. Fersman in 1922. The "zone of catagenesis" (Fig. 7.5) encompasses those levels and stages in which nonhydrocarbon material is transformed into low-order hydrocarbons, and these low-order hydrocarbons are reorganized through temperature and pressure into petroleum hydrocarbons.

The word *catagenesis* suggests a connection with *catalysis*. Catalysis of the transformation processes, by components of the mineral matrix of the source rock, has been invoked by many workers, principally by Benjamin Brooks (in 1938). The most important mineral catalysts are surfactant materials, especially those in clays, and perhaps also sulfur. The analogy with refining, in which catalysts play a vital role, has been shown by Gordon Erdman to be invalid; the water that is always present in the source sediment is more easily adsorbed onto the mineral matrix than on to any hydrocarbons, thus shielding the one from the other. The analogy furthermore implies the presupposition that the "primitive" oils are asphaltic, and that they are "broken down" into lighter paraffinic oils. As will be revealed later (Sec. 10.3), this supposition may itself be invalid in most cases. The role of catalysts is difficult to assess; it is unlikely to be critical to the transformation.

The dominant agents must include temperature and pressure as variables that increase with the burial of the source rock under younger sediments. Very high temperatures (300–400 °C, for instance) are not called for; oils contain too many components that are unstable at such temperatures, and the temperatures themselves are not achieved in sediments at depths in which oils actually occur. Lower temperatures, on the other hand, are reached at no great depths in all sedimentary basins, and they are maintained over millions of years. Similarly, the requisite pressure is derived from a few kilometers of overlying rock, later added to by gas pressure as decomposition of the kerogen progresses. This impervious cover (the lid of the retort) allows the process to go to completion, and helps to maintain the salinity of the associated water.

Early studies of the process, notably by W. F. Seyer in 1933, envisaged it as a low-temperature, high-pressure cracking process. Seyer traced the formation of oil and gas from a wax-like *protopetroleum* under these *T, P* conditions, through increasingly stable hydrocarbons of decreasing molecular weight. The successive decompositions and recombinations undergone by the petroleum hydrocarbons would then be in accordance with the laws governing unimolecular reactions, essentially independent of pressure and so dependent upon temperature alone. High pressures in some oil-bearing successions may hasten the rate of decomposition, but the rate itself, in the view of Seyer and others, must be primarily due to moderate temperatures acting over long periods. End-products are constantly differentiated, towards the methane series on the one hand and heavier hydrocarbons on the other. The lighter series builds up a great quantity of fixed gaseous hydrocarbons, creating high gas pressures if they are unable to escape. The underlying petroleum consists of the heavier hydrocarbons, grading towards fixed carbon in ancient examples.

On this theory, as on that of David White, there will be a limit to the temperature, and therefore to the depth, at which oil can occur. At greater depths, petroleum hydrocarbons will be converted to gas plus semi-solid materials high in carbon content, too viscous to flow. Seyer suggested that this limiting depth should be about 20 000 feet (6000 m; Fig.7.6). Despite disbelief in many quarters, this forecast has stood the test of deep drilling very well (see Sec. 23.5).

Later work has substantially confirmed this proposal, which is that high molecular weight components of petroleum can be converted to simpler compounds by thermal cracking. Such transformations are thermodynamically sound, because the free energies of formation of the paraffins are lower than those of naphthenes and aromatics of equivalent molecular weight, and the free energies for all classes of hydrocarbons are generally lower in the low than in the high molecular weight compounds.

Hydrocarbon generation is a *rate-controlled, thermocatalytic process*, variously referred to as catagenesis (especially by the Russians), organic metamorphism (by the Shell group and others), eometamorphism (by K. K. Landes), or simply maturation (by G. T. Philippi). The essential control is a time–temperature relation.

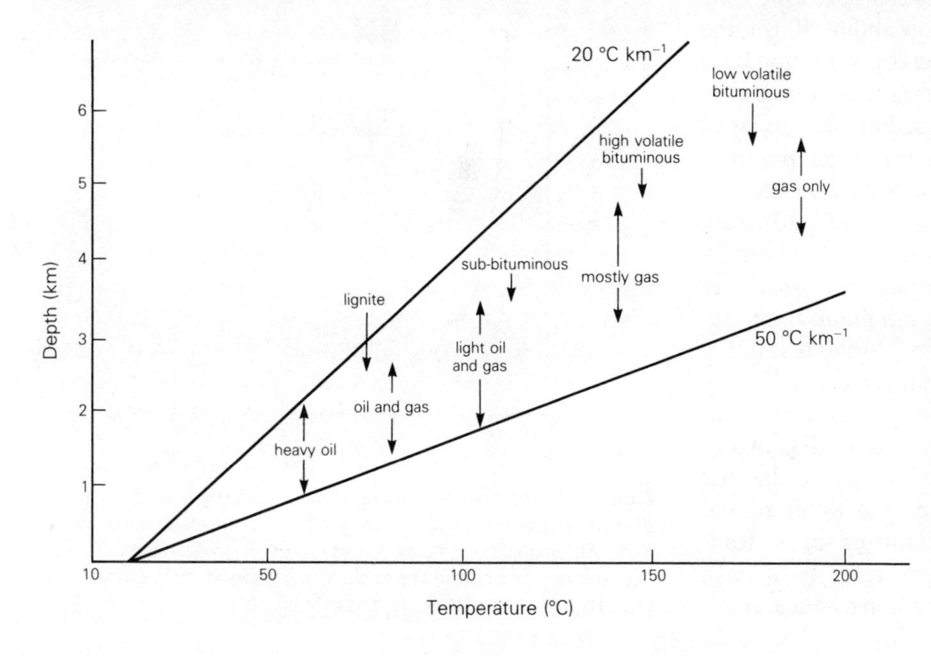

Figure 7.6 Rise in rank of coals and maturation of hydrocarbons as functions of temperature gradients. With a gradient of 50 °C km^{-1}, the light oil–wet gas stage is reached at a depth less than half that required with a gradient of 20 °C km^{-1}. (Modified from H. D. Klemme, *Oil Gas J.*, 17 July 1972.)

The dependence of chemical reaction rates upon temperature is commonly expressed by the Arrhenius equation:

$$k = Ae^{-E_a}/RT$$

where k is the rate constant, A is a pre-exponential factor, itself independent of temperature, E_a is the activation energy of the reaction, R is the universal gas constant ($= pV/T = 8.29 \text{ J K}^{-1}$), T is the temperature in kelvins.

Hence

$$\ln k = \ln A - E_a/RT$$

In words, this means that the rate of a reaction approximately doubles for every 10 °C (18 °F) rise in the temperature. Considering the generation of oil as a chemical reaction governed by this control, the amount generated increases linearly with time but exponentially with absolute temperature. The younger the source sediment, the shorter the time it has been available for generation so the higher the temperature that is required for it.

The breakthrough studies were conducted by G. T. Philippi of Shell in the Los Angeles and Ventura basins of California. Both basins are wholly Tertiary in age and wholly clastic in lithology; both have large oil columns in multiple sand reservoirs (Figs 16.15, 16.19, 16.36 & 26.15). Their hydrocarbon contents increase with depth (and age) into deeper (and warmer) shales. In the same direction, the composition of the *n*-paraffins and naphthenes in the shales, boiling above 325 °C, becomes more and more like that of the oils of the basins. In the depth range 1500–2500 m, the bitumen percentage in the source sediments rises moderately, but the concentration of paraffins increases enormously. Shale hydrocarbons having six or seven carbon atoms are especially sensitive to thermal alteration. Below about 2500 m, the bitumen percentage and the paraffin concentration both decrease markedly (perhaps because the oil which they represent has migrated out of the sediments). Most of the oils were generated at depths where the temperature now is above 115 °C, so that the shales are sterile.

Philippi concluded that the formation of petroleum from sedimentary organic matter was brought about by thermal, strongly temperature-dependent processes which must be nonbiological. The minimum temperature required appeared to be 60–65 °C (about 150 °F). The depth at which such a temperature is reached in any basin depends upon its geothermal gradient, but provided the critical temperature is reached it will promote at that depth of burial a "threshold" of intense thermal cracking. This threshold represents the onset of the *principal stage of oil formation*. Following O. A. Radchenko, many Russian geochemists refer to it as a carbonization or coalification "jump"; in essence, it is a time of rapid emission of volatiles from oil-prone macerals.

The advance of rank in coals, beginning with humic OM, is essentially smooth and continuous. It is not smooth in the carbonization of sapropelites or combustible shales. In these, dispersed OM undergoes smooth but minor carbonization to some critical depth (2000–2200 m or 6500–7250 feet), merely losing side-groupings by dehydration, decarboxylation, and similar simple processes. This prefatory episode is followed by a sharp, stepwise loss of carbon, hydrogen, and volatiles between depths of 2200 and 3500 m (7250–11 500 feet); this is the "carbonization jump." The inner bonds of the molecules are ruptured; the bulk of the kerogen is split; most of its mass is expended by the release of gas and volatile liquid products. These latter include the higher paraffins, in the C_5 to C_{14} range, which are typically absent from shallow sediments. After the "jump," the lipid content of the kerogen decreases whilst its humic content increases. Some of the kerogen's generative properties are therefore lost, and further maturation generates only methane. Radchenko's graphic depiction of this process combines the proportions, not of hydrogen and carbon, but of hydrogen and nitrogen (Fig.7.7).

There is thus a range of temperatures through which the generation of oil can take place if the source sediment is suitable (Fig. 7.8). At temperatures below the critical "jump" temperature of about 60 °C, the sediments are *immature*; at temperatures beyond some higher critical temperature (typically about 120 °C), they are *post-mature*. The temperature range of oil generation was called by Walter Pusey (in 1973) the *liquid window*; it applies to the range of true generation and

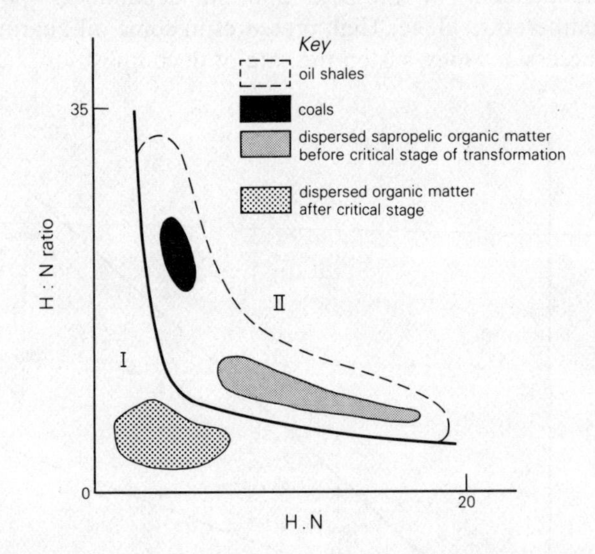

Figure 7.7 Distribution of coals, oil shales, and hydrocarbons on plot of H : N atomic ratio against H.N. In area I, humic materials dominate and hydrogen is low. In area II, lipid materials dominate, hydrogen is high and nitrogen is low. The origin (0) corresponds to graphite. (After O. A. Radchenko and V. A. Uspenskiy, *Dokl. Akad. Nauk SSSR*, 1974.)

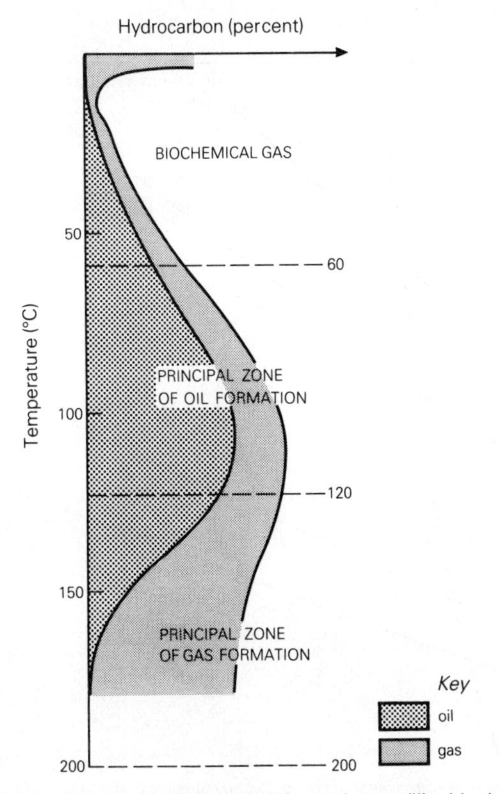

Figure 7.8 The principal zone of oil formation, or "liquid window," during the thermal generation of petroleum hydrocarbons.

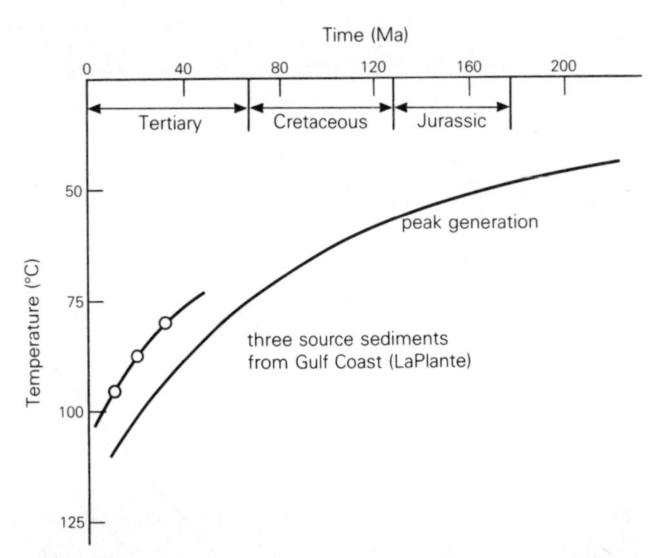

Figure 7.9 The peak generation of hydrocarbons as a function of both temperature and time.

expulsion of the oil, and should not be extended to include oils that have been subjected to extensive vertical migration into much younger and cooler rocks (see Ch. 15).

Though temperature is the critical factor in maturation of a source sediment, it is not the only one. Some basins containing organically rich sediments have attained sufficiently high temperatures yet are not known to contain oil. The *time factor* is also vital; it combines with thermal energy (rather than with actual temperature) to provide an effective exposure time, popularly referred to as the "cooking time." It constitutes the sum of the times that the rock was exposed to each incremental rise of temperature during each stage of subsidence (and/or uplift if the subsidence was interrupted) beyond the minimum threshold temperature. Conventionally it is calculated as the time that the sediment spent within 15 °C of the maximum temperature it reached.

The actual temperatures at the top and base of the liquid window therefore depend on the age of the sediments and their geothermal gradient. In the Paleozoic rocks of Algeria and Alberta, the threshold temperature or top of the window is about 60 °C. Because much of the world's oil is of Mesozoic source, a general average of 65 °C (150 °F) may be taken as the threshold temperature. Experimental work by Connan, Laplante, and others has shown that considerably higher threshold temperatures are necessary if the source sediments are of Cenozoic age. Furthermore, the threshold temperatures vary from basin to basin because of differing rock facies, types of organic material provided, and detailed geologic histories.

In the Tertiary oilfields of the US Gulf Coast, for example, temperatures for the onset of significant generation ranged from about 77 °C in Oligocene strata to 96 °C in the Upper Miocene (Fig.7.9). In Japan, the corresponding temperatures were 100 °C for the Oligocene and 105 °C for the Upper Miocene. But the threshold temperatures will always fall within a restricted range for any one time-plane and will increase as the time contribution is reduced. For Tertiary provinces in general, the range for the liquid window (that is, the tops of the oil and gas "kitchens," bracketing the zone of *catagenesis*) has been estimated by Philippi as 115 °C (240 °F) and 150 °C (300 °F). Laboratory reproduction of the equivalent transformation requires a temperature of almost 300 °C.

Because the top and bottom of the liquid window are controlled by critical temperatures, the *thickness* of the window is a function of the geothermal gradient (see Sec. 17.8). Over the drilled sedimentary areas of the world, the average gradient is 30–35 °C km^{-1} (10 °C per thousand feet); but it is much higher in Cenozoic marginal fold belts and much lower in cratonic basins. The higher the gradient, the shallower and more efficient the generation and entrapment of oil (provided geological conditions are suitable for these), and the shallower the depth at which the compositions of the shale hydrocarbons and the oil hydrocarbons are about equal. In clastic sequences, at least, a high geothermal gradient promotes higher porosity, higher permeabilities in the carrier beds, lower viscosity of the oil, and increases in the fluid pressures (Fig.7.10).

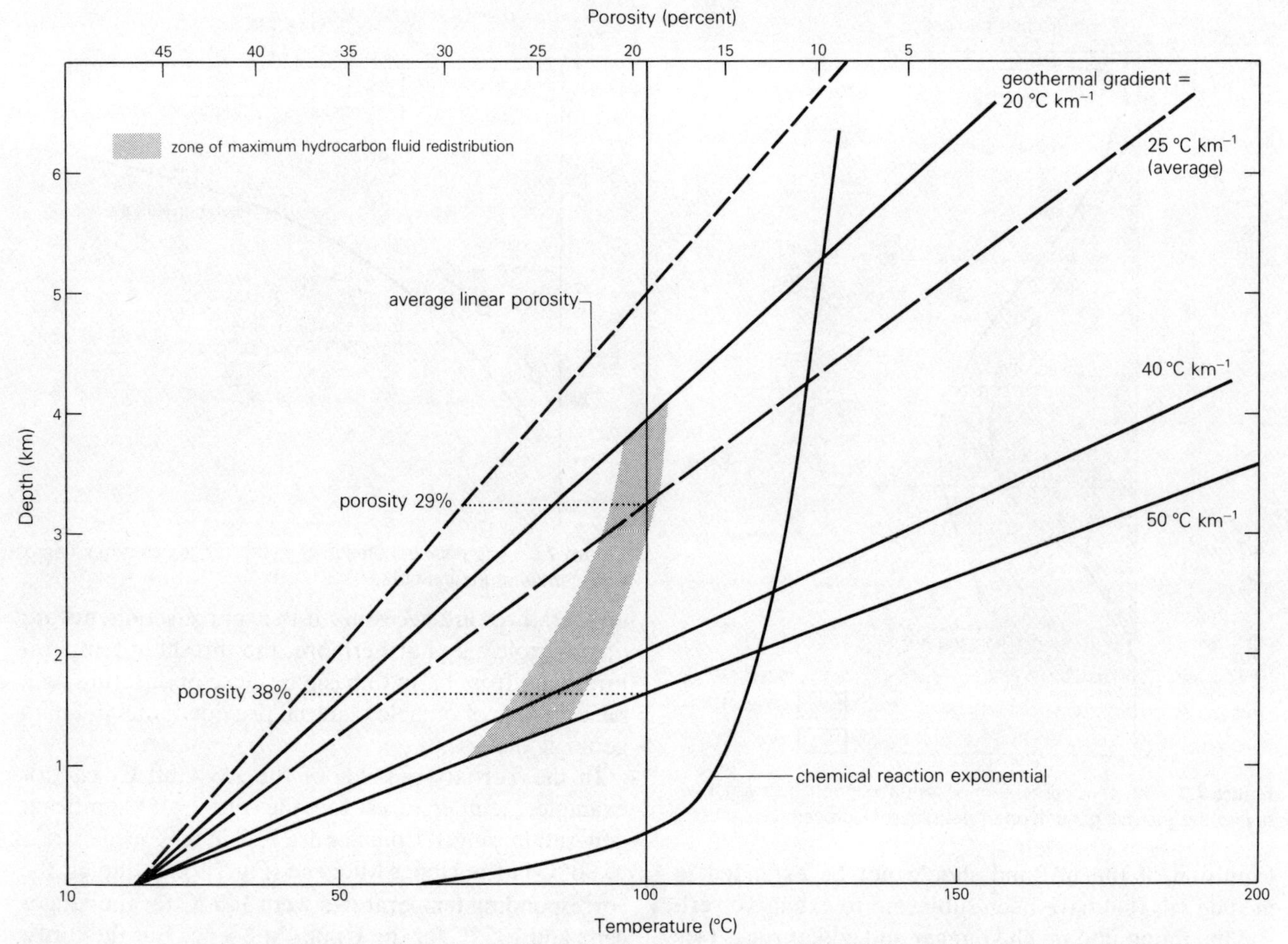

Figure 7.10 Influence of geothermal gradient on rate of reactions in the generation of hydrocarbons, and on rate of reduction of average porosity with depth. Horizontal dotted lines illustrate the difference in average porosity reduction with gradients of 50 °C km^{-1} and 25 °C km^{-1}. With the higher gradient, initial (surface) porosity of 46 percent is reduced to 38 percent at a depth equivalent to a temperature of 100 °C; with the lower gradient, porosity is reduced to 29 percent because of the much greater depth needed to achieve that reservoir temperature and thus that level of maturation. (From H. D. Klemme, *Oil Gas J.*, 17 July 1972.)

In the Tertiary basin of central Sumatra (see Sec. 17.7.5), a geothermal gradient of 90 °C km^{-1} (triple the world average) makes possible prolific production at 760 m or less; but the liquid window is thin (less than 1000 m), thermogenic methane beginning at about 1250 m depth. A normal geothermal gradient will lead to the threshold temperature of 60–65°C at depths of about 1800 m. Where this occurs, as in cratonic regions or in semi-coastal basins like the Gulf Coast, the oil column is likely to be much *deeper* but also much *thicker* (over 3000 m thick in the Gulf Coast, for example).

The bottom of the liquid window, the temperature above which oils are converted to thermogenic gas, is between 100 and 150 °C (200–300 °F). At the lower temperature, the porphyrin structure is destroyed, in part through loss of a carboxyl group (–COOH). Above the higher temperature, destruction of liquid hydrocarbons becomes dominant; they react with any sulfur present to form thioalcohols (mercaptans; see Sec. 5.2.1). Above about 200 °C processes become wholly thermal and destructive and can fairly be referred to as metamorphic, leading to the conversion of all hydrocarbons to the two end-members, methane and graphite. Optically active compounds like those occurring in natural oils are decomposed. Beyond 300 °C, many hydrocarbon-associated molecules become thermally unstable.

These widely held rules of thumb governing the maximum temperatures of petroleum generation and stability have not gone unchallenged. Leigh Price maintained that the time element is essentially insignificant in the maturation process. Hydrocarbons heavier than C_{15} are stable, however, at much higher temperatures than have commonly been accepted. This view considers destructive thermal cracking to become significant only above about 350 °C – a view having considerable impact upon hypotheses concerning the primary migration of oil, as will be discussed in Chapter 15. On the opposite hand, many geologists of the older generation still maintain that oil is generated early, at shallow depths, and at low temperatures (Sec. 15.5).

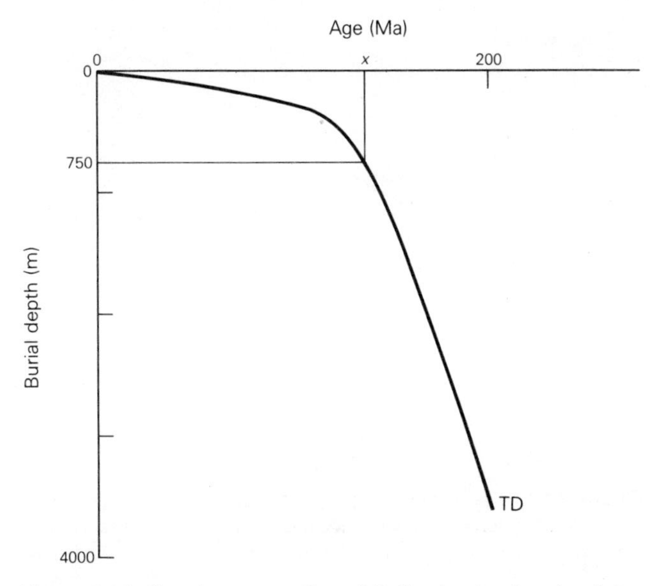

Figure 7.11 Rough computation of "effective heating time" for maturation of source sediment.

7.4.1 *Calculation of the "effective heating time"*

An approximation to the effective heating time of the source sediment can be gained through three simple steps (Fig.7.11):

(1) Plot the age of the strata penetrated against their depths in the well.
(2) From the measured geothermal gradient, calculate the depth over which the temperature increases by 15 °C. For a gradient of 20 °C km^{-1}, for example, this will be 750 m.
(3) Determine the age (x) equivalent to this depth; this is the "effective heating time," during which the sediment concerned remained within 15 °C of its threshold temperature.

The method assumes that there has been no great distance of secondary migration of the oil, and significant hiatuses in the post-reservoir section must be taken into account in converting age to time.

7.4.2 *Time and temperature in petroleum formation*

There have been a number of attempts at quantitative expression of the time–temperature relation in petroleum formation, especially by Russian geologists and geochemists. Undoubtedly the simplest is the "geochronobath" of N. B. Vassoyevich; it is simply the multiple of the depth (in kilometers) and the time (in millions of years) applicable to the source sediment. An elaboration was the "geochronothermobar" (G):

$$G = \frac{T \times t \times P}{1000}$$

(T in degrees Celsius; t in millions of years; P in atmospheres).

By far the most influential of these attempts was that by N. V. Lopatin, who, in 1971, calculated the thermal maturity of organic matter in sediments by a "time–temperature index" (TTI). As in the calculation of cooking time, the depth of burial is plotted against age; as in all the other calculation techniques, the depths have to be corrected for episodes of uplift and erosion and they may in consequence become complex (as they would in a much-faulted section with multiple unconformities). The depth–age plot is then overlain with a temperature grid, necessarily simplified because it requires assumptions about geothermal gradients in the geologic past.

Lopatin next defined two factors:

(a) An *index value* (n) for temperature intervals of 10 °C. Zero value of n was arbitrarily chosen for the interval 100–110 °C, so it becomes −1 for the interval 90–100 °C, +1 for 110–120 °C, and so on.
(b) A *temperature factor* (γ), equal to r^n where r is the factor by which the rate of maturation increases for each 10 °C rise in the reaction temperature. Thus $\gamma = r^{-3}$ for the interval 70–80 °C, and r^3 for the interval 130–140 °C, and so on. If the Arrhenius equation applies, the value of r can be taken as 2; the speed of reaction doubles for every 10 °C rise in temperature. Actual plots of r can be correlated with values derived by other methods (principally the optical methods described in the next section).

The increment of maturity, ΔM, for each time–temperature interval can then be easily calculated:

$$\Delta M = \Delta T r^{n} \text{ for that interval}$$

The total maturity index (TTI) is then the sum of all ΔM values throughout the sediment's history.

Lopatin's combination of the time and temperature factors in petroleum formation can be illustrated by Figure 7.12. Rocks of age A have been in the oil-generating zone (the liquid window) for only y Ma. During that interval, rocks of age B have been in the zone of thermogenic gas generation.

Several comparable, but somewhat more complex, logarithmic calculation indices have been proposed by North American workers. In 1977, members of the Exxon Production Research team calculated the compositions (C) of oils from the proportions of naphthenes, paraffins, and aromatics, based on the generalized equivalence

$$4N \rightarrow 3P + 1A$$

They then defined a "time–temperature integral" (tT; Fig. 7.13) by the function

$$\ln C = \text{intercept} + [\text{slope} \times (tT)]$$

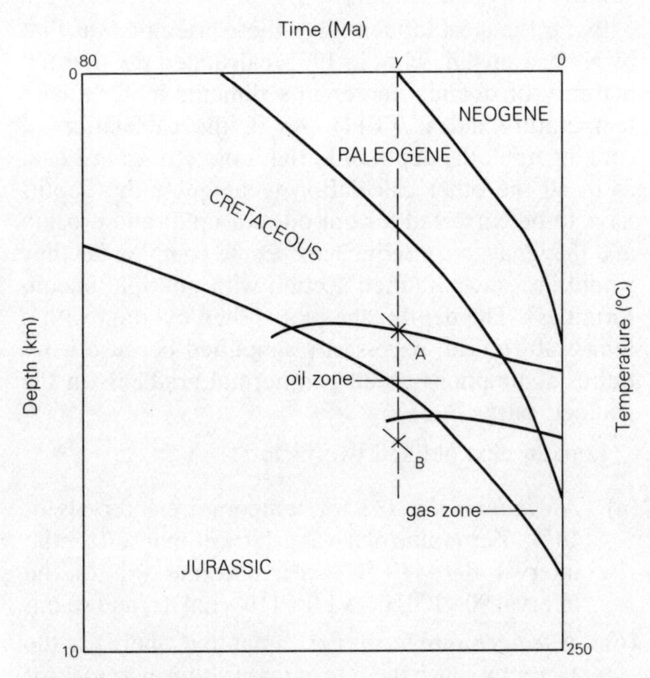

Figure 7.12 Illustration of the combined roles of temperature and time in the generation of oil and gas, as envisaged by N. V. Lopatin. Explanation of letters in text; scales arbitrary. The higher the temperature gradient, the narrower are the bands. (From N. V. Lopatin, *Int. Geol. Review*, 1980.)

A group at the Massachusetts Institute of Technology under Leigh Royden combined T and t in an equation defining the parameter C such that

$$C = \ln \int_{o}^{t} 2^{T(t)/10} dt$$

This is a variant on the Arrhenius equation. As the level of thermal alteration of the organic matter increases, the parameter C also increases. At $C \cong 10$, the process of oil generation has barely started; at $C \cong 16$ it is essentially complete, and at $C \cong 20$ the process of gas generation is similarly essentially complete.

7.4.3 *The problem of temperature measurement*

For all the calculation techniques described, it must be clear that the greatest potential source of error lies in the calculation of T; what in fact was the maximum temperature reached by the stratum in question, and when did it reach it? Actual calculations of T are probably fairly reliable for Cenozoic sedimentary rocks; present geothermal gradients may reasonably be extrapolated backwards at least to the last major orogenic episode affecting the strata. Calculations become hazardous for older strata, which may have had complex histories of subsidence, deposition, uplift, faulting, erosion, warping, renewed subsidence, and so on.

Consider a complex horst–graben structural basin, like that of the northern North Sea or those around the margins of the South Atlantic and Indian oceans. There

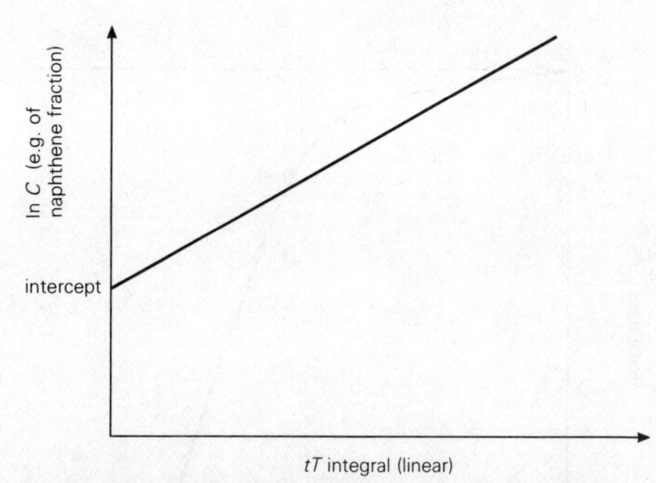

Figure 7.13 The "time–temperature integral" as a logarithmic function of the composition of crude oils. (After A. Young *et al.*, *AAPG Bull.*, 1977.)

are typically large variations in the depths to a particular value of T within short distances. These create *paleogeothermal unconformities*. We may select a single formation in such a setting, and allow it to achieve its maximum temperature, by burial, before a younger tectonic phase affects it. Uplift and erosion then cause the formation to be moved into a lower temperature regime. When further subsidence occurs, renewed sedimentation may fail to replace the section lost during the erosional episode; T_{max} may not be reached a second time, so that there will be a paleogeothermal break at the unconformity. All methods of calculation, whether geochemical or optical, will reveal higher values of thermal maturity below it than above it. Only if the post-unconformity sedimentation exceeds the amount previously lost by erosion, making it possible for younger heating to exceed the older heating, will the paleogeothermal unconformity be obliterated.

A fundamental flaw in the methods may be the implied extrapolation of pyrolysis kinetics to geologic temperature conditions. As Seyer observed in the 1930s, the high temperatures of experiments are not achieved in unmetamorphosed sedimentary sequences, but much more moderate temperatures act over exceedingly long periods. If the time–temperature relation, extended to large values of t and only small values of T, is not in fact linear, the assumptions behind the methods may be fallacious.

We may illustrate the pitfalls (which are acknowledged by the proponents of the methods) by considering the identification of source sediments by means of one of the methods, that of the time–temperature integral (tT). Using the present geothermal gradient for each petroliferous basin (with the potentials for error consequent upon that), each bed in any one stratigraphic section is assigned a unique value for its (tT) integral; the values of course become larger downwards in the section. A computer program then identifies the bed that balances the logarithmic equation adding intercept and slope. If the calculated age of the *oil* turns out to be about the same as the known age of the *reservoir rock*, the implication is that the oil and the rock have had the same thermal history and the oil can be regarded as indigenous to strata of that age. Conclusions in a significant percentage of cases are not in harmony with conclusions from geological evidence, or even of other geochemical evidence, revealing the teleological nature of the argument and procedure.

7.5 Optical parameters in source maturity studies

Use of the microscope in the study of organic materials is of much greater antiquity than the application of geochemical techniques. Microscopic evaluation of oil source materials has in a sense been an outgrowth of the microscopic study of coals, pioneered in the early 1900s by C. A. Seyler, Marie Stopes, and others. In more recent years, a great impetus came not merely from work on plant macerals in coals but from palynological research in a wider sense.

During the thermal maturation that leads to the generation of hydrocarbons, dispersed organic material (OM), viewed in transmitted light, changes color from colorless through yellow and brown to black. The color attained depends upon the temperature reached and the time during which it was maintained – the geothermal history, in fact. In artificial heating experiments, the transition from dark brown to very dark brown equates with the range of maximum generation of *n*-paraffins, with residues reaching an atomic H : C ratio of 0.80 approximately. Spore-pollen color scales have been made elaborately precise, with delightful names for some of the color varieties; a ten-point *spore coloration index* (SCI) scale is in widespread use. The scales are determined by the use of high-magnification, transmitted light techniques such as sporinite microspectrofluorescence. Organic materials of animal origin may also be utilized, provided that they are composed of chitinous or carbonaceous material and not calcium carbonate or silica. Conodonts, for example, undergo a comparable progression of color changes, from pale yellow through brown to black and, eventually, crystal clear.

The colors represent *maturity indices*, in that they are temperature- and time-dependent, progressive, cumulative, and irreversible. Color changes, of course, affect particular types of OM whether they are potentially petroleum-generative or not. To determine source quality, the changes must be studied in conjunction with visual assessment of the *types* of OM undergoing them. Finely divided OM in sediments falls into several types, recognizable and distinguishable under the microscope. In 1975 G. S. Bayliss distinguished four such types, named according to their physical appearance:

(a) *Amorphogen* Unorganized, structureless, translucent debris, finely disseminated or gathered into fluffy masses; generally absent from continental sediments.
(b) *Phyrogen* Nonopaque, herbaceous material derived from cuticles, spores, and cysts.
(c) *Hylogen* Nonopaque, fibrous material of woody (terrestrial) origin.
(d) *Melanogen* All opaque OM of coaly origin or appearance, also terrestrial.

The percentages of these types can be calibrated with the color scale to establish a *thermal alteration index* (TAI) (F. L. Staplin 1969).

The TAI is expressed according to a scale from 1 to 5, TAI = 1 representing no detectable change in color from the original pale yellow, TAI = 5 representing alteration so severe that all phytoclasts are black and the host rock offers additional evidence of metamorphism. Oil-source potential requires an abundance of amorphogen, at the expense of the terrestrial organic components, and a TAI value between 2 and 3 (Fig.7.14).

The Bayliss descriptive nomenclature was not widely adopted, because there already existed an acceptable nomenclature for the four basic types of OM, used by coal petrologists and closely paralleling the Bayliss definitions (see Sec. 7.2 & Fig. 7.5). The principal macerals in the Bayliss OM types are alginite, exinite (liptinite), vitrinite, and fusinite (inertinite) respectively.

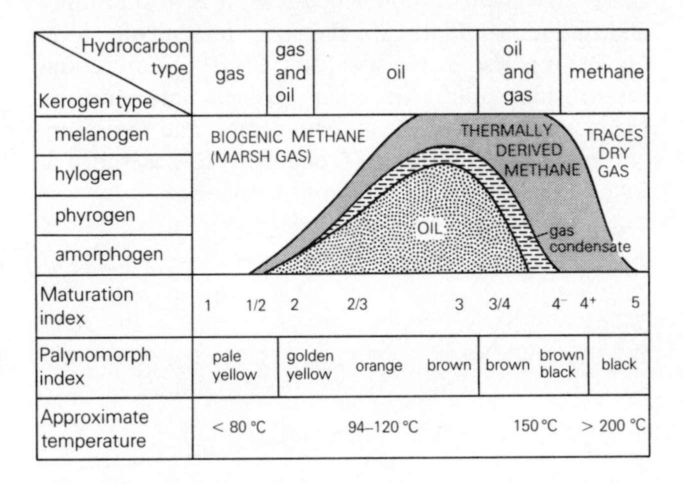

Figure 7.14 Relations between detrital kerogen type, maturation index (TAI), color index, paleotemperature, and hydrocarbon generation. (From B. Owens, *Petroleum geology of the continental shelf of North-west Europe*, L. V. Illing and G. D. Hobson (eds). Institute of Petroleum, 1981; after J. P. Bujak *et al.*, *Oil Gas J.*, 4 April 1977.)

The TAI is insufficiently versatile to be definitive of source maturity. The scale is hard to standardize with the color charts, and the range from 2.5 to 3.0 encompasses virtually the entire oil "window." Attempts have been made to extend the index to OM categories beyond spores and pollen, but a number of unfavorable circumstances can render the results misleading. The degree of opacity of the OM may be affected by changes during transport, by the redox potential in the depositional environment, or by the existence of the carbonate facies there (in the carbonate environment, OM appears to be altered less readily than in an environment of clastic deposition).

7.5.1 *Vitrinite reflectance and the value of T*

A direct descendant of David White's *carbon ratio* theory (Sec. 7.2) is the use of *vitrinite reflectance* in oil immersion (R_o, expressed as a percentage), determined by use of the reflecting microscope. The somewhat laborious technique had been developed by coal petrologists as early as 1931. It involves the concentration of the phytoclasts by acid leaching, the selection of a statistically significant sample (usually 100 specimens or more), mounting them in epoxy resin, and polishing them to yield a reflecting surface.

From histograms for each representative sample, R_o values are determined and their sequence gives the *maturation gradient*. The slope of the gradient depends on the geothermal gradient and the sedimentation rate. The values are converted to *fixed carbon* (FC) contents by reference to a chart prepared by the International Committee on Coal Petrology in 1971 (Fig.7.15).

Vitrinite reflectance provides a maturity index because, like hydrocarbon generation, it is temperature- and time-dependent. Like the spore-pollen color techniques, it is also irreversible, so it needs no corrections for erosion, uplift, or other geologic interferences. Many thousands of R_o measurements, and their conversions to FC percentages, enable us to generalize as follows the important threshold levels:

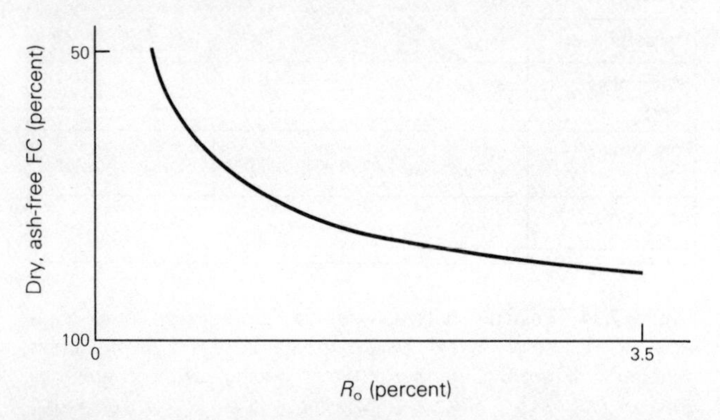

Figure 7.15 Generalized relation between vitrinite reflectance (R_o) and fixed carbon percentage. (After B. M. Thomas, *AAPG Bull.*, 1979.)

(a) For the threshold of significant oil generation from sapropelic matter, FC = 60 percent, R_o = 0.7 percent.
(b) For minor, early thermal, lean gas from humic matter, FC = 55 percent, R_o = 0.5 percent.
(c) For the threshold of significant gas generation from humic matter, FC = 65 percent, R_o = 0.9 percent.
(d) For the beginning of oil destruction, FC = 75 percent, R_o = 1.35 percent.

Though the measured values of R_o range in general between zero and about 5 percent, only a very narrow band near the lower end of this range represents the generative "window" for oil and gas.

The results of vitrinite reflectance measurements should be used with caution, because they are affected by a variety of considerations and circumstances:

(a) There is no vitrinite in pre-Carboniferous sediments; it may be lacking in any random younger sediment. Chitinozoans and scolecodonts (from worms) offer high reflectance values and may become valuable substitutes for vitrinite in Lower Paleozoic rocks.
(b) Identification of the fragment as vitrinite must be assured; some types of bitumen, and semifusinite (which is closer to inertinite than to vitrinite), may readily be mistaken for it.
(c) Vitrinite, even when correctly identified, may not be autochthonous; it may be weathered; organic matter is easily recycled or reworked.
(d) Reflectance depends in part on the surrounding lithology, possibly due to differences in thermal conductivity, or absorption of free hydrocarbons and retardation of maturation. It is lowest when in sandstones, highest (not surprisingly) in coals. It also depends on freedom from contamination, by inclusions of bitumen, pyrite, or other impurities.
(e) Above the "dead line" for oil and/or gas occurrence (Sec. 7.2), vitrinite reflectance may determine the "coal rank equivalent" for plant materials other than coal, by photoelectric recording of fluorescence spectra for sporingite or alginite. Fluorescence extinguishes at the dead line; vitrinite itself is not fluorescent.
(f) Macerals react differently to the maturation processes. In 1982 Lloyd Snowdon and Trevor Powell reported that a source sediment in which resinite (Fig. 7.5) constitutes more than about 10 percent of otherwise terrestrial OM may yield R_o levels lower than 0.6 percent and still generate significant light oil and gas-condensate.
(g) Most important of all, it must be remembered that R_o measures the *maturity of the sediment*; it does not indicate whether the sediment was an oil-prone source rock or not. Independent studies of the OM

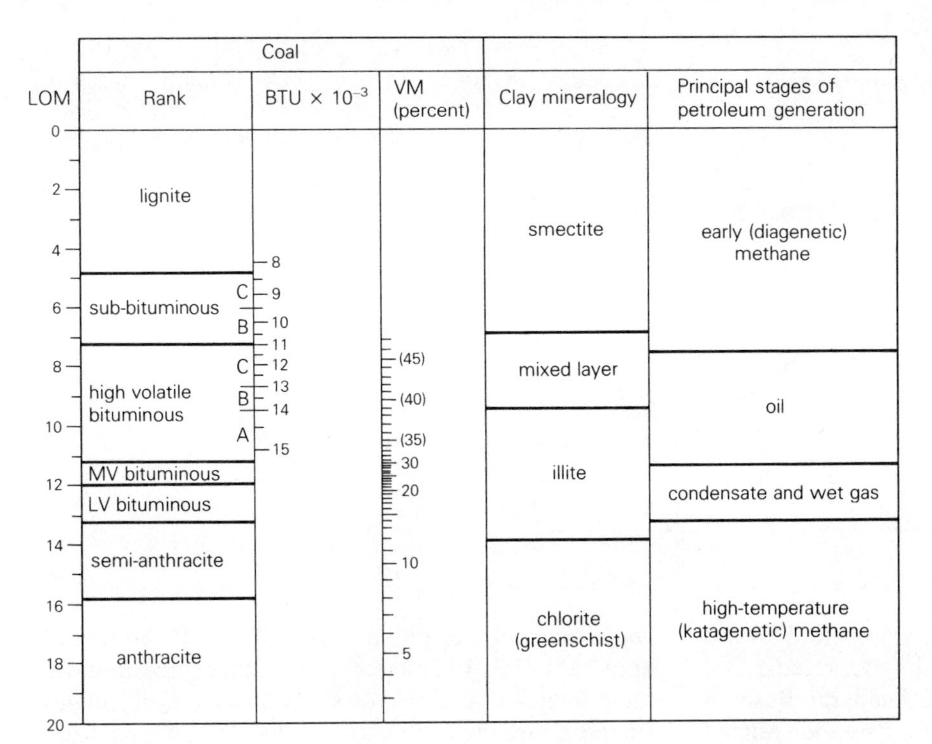

Figure 7.16 Stages of petroleum generation related to coal ranks, level of organic metamorphism (LOM), and clay mineralogy. (After A. Hood *et al.*, *AAPG Bull.*, 1975; N. B. Vassoyevich *et al.*, *Int. Geol. Review*, 1970.)

itself are necessary to settle that fundamental question.

An even more direct descendant of the carbon ratio theory was the so-called *LOM scale*; LOM stands for *level of organic metamorphism*. The scale was originally devised from studies of the Cretaceo-Eocene coal measures of New Zealand. Though rank of coals presumably advances in accordance with the Arrhenius exponential equation (see Sec. 7.4), the LOM scale was adjusted so as to become quasilinear, from zero at the surface to 20 at the anthracite/meta-anthracite boundary (Fig.7.16).

The scale reflects both the maximum temperature reached (T_{max}) and the effective heating time (Fig. 7.17). The oil "window" begins at LOM = 7–8 (R_o = 0.5 percent, temperature 100 °C or lower). Above LOM = 8, sandstone porosity and permeability are greatly reduced, chiefly by the introduction of high-temperature cements unless hydrocarbons have already been trapped. The oil "dead line" is reached where LOM = 12 or thereabouts, mean R_o exceeds 1.35 percent, and T_{max} about 140 °C (equivalent to low-volatile bituminous coals). The actual depths of these thresholds of course depend on the thermal history. Some other measurable quantity, such as the vitrinite reflectance, has to be

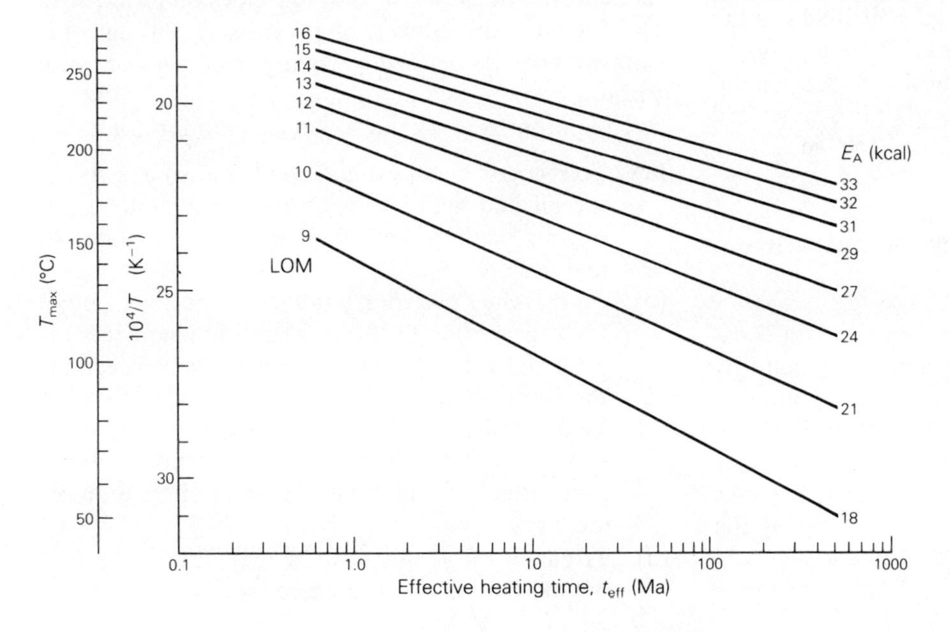

Figure 7.17 Estimation of T_{max} and effective heating time (t_{eff} = time within 15 °C of T_{max}) from measured LOM. E_A is the activation energy, in kilocalories, determined from the slope of the line. (From A. Hood *et al.*, *AAPG Bull.*, 1975.)

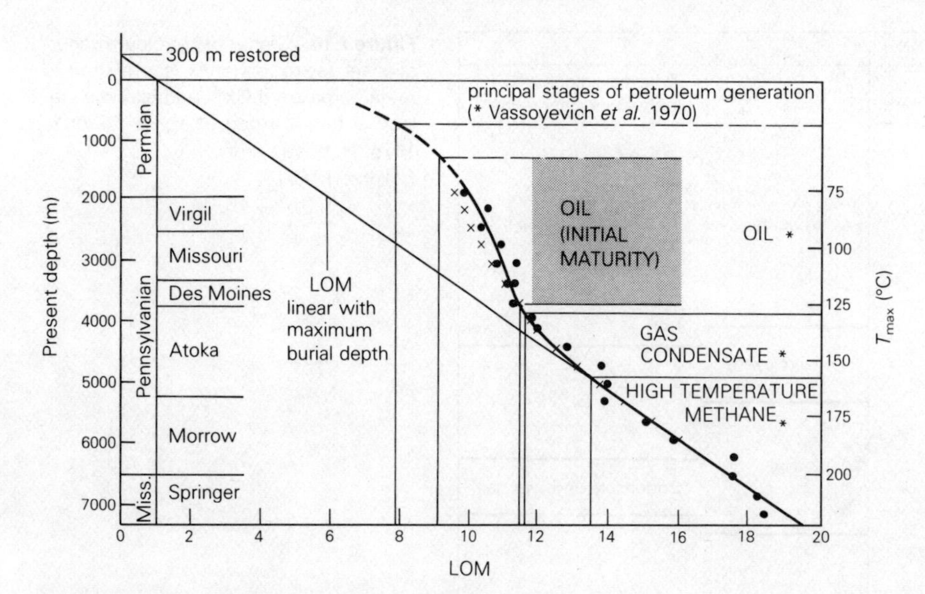

Figure 7.18 Relations between LOM, R_o, T_{max}, and petroleum generation stages for a deep well in Oklahoma. (From A. Hood *et al.*, *AAPG Bull.*, 1975.)

introduced in order to provide a "thermometer" for this history. Fig.7.18 shows a plot of the measured LOM and R_o values for a deep well in the Anadarko Basin of Oklahoma. The deduced values of T_{max} are then related to the stages of petroleum generation for the stratigraphic and depth range penetrated by the well.

The LOM scale was quickly superseded, though it enjoyed extensive usage for a time. The warning concerning reliance on material related to *coal* deserves repetition. Coal ranks must be equatable with some LOM values, some R_o equivalents, and some calculated paleotemperatures; but they may be quite unassociated with oil. Association of coal with natural gas is quite another matter. Even in the absence of free oil, the percentage of gaseous hydrocarbons in the volatile products increases with increasing metamorphism (hence depth, or rise in the rank of coal). One thousand kilograms (1000 kg) of OM yields 33 kg of dry methane in the peat stage, 77 kg in the lignite stage, and 108 kg in the bituminous coal stage. Nevertheless, the *total* organic content decreases with depth, so most methane generation is shallow; more than 50 percent of it is generated before the depth of burial has reached 1500 m.

7.6 Correlation between time–temperature and optical parameters

In 1980, Douglas Waples calibrated the Lopatin equation (Sec. 7.4.2) against a large number of maturity measurements based upon TAI and R_o techniques (Table 7.2). Lopatin's TTI was plotted against TAI and R_o on log–log scales for all plausible values of r (say from $r = 1$ to $r = 10$). Correlation is good if r falls between 1.6 and 2.5, so the value $r = 2$, obeying the Arrhenius equation, is empirically the best average. TTI may therefore now be compared with any other geochemical or optical parameter (CPI, H : C ratio, etc.), as in Table 7.3. TTI may then be further correlated with the principal stages of hydrocarbon generation and destruction, already calibrated on the R_o and TAI scales (Table 7.4).

7.6.1 Correlations between the techniques and their indicators

We are now in a position to correlate the parameters measured by the various techniques, both geochemical and optical, and to indicate what they mean in terms of the conversion of organic matter to oil, gas, or coal. Fig.7.19 shows a general correlation between depth, paleotemperature, vitrinite reflectance, spore coloration, LOM, and stages of hydrocarbon generation. Fig.7.20, prepared by Gerard Demaison in 1980, shows the values of T_{max}, R_o, TAI, and CPI representing the maturation stages for the four types of kerogen. It also includes the equivalent progression through the ranks of coal, as defined by their percentages of fixed carbon.

The purposes of all the techniques are fourfold:

(a) To recognize and evaluate potential source rocks for oil and gas, by measuring their contents of organic carbon and their degrees of thermal maturation.
(b) To correlate oil types with their respective source beds, according to the geochemical characters of both and the optical characters of the kerogen in the source beds.
(c) To determine the time of generation of the hydrocarbons, and the times of their migration and accumulation, in relation to the times of creation of the traps.
(d) To enable the geologists to estimate the volumes of oil and gas initially generated, and so to estimate

Table 7.2 Interconversion of thermal alteration index (TAI) and vitrinite reflectance values (R_o). (From D. W. Waples, *AAPG Bull.*, 1980.)

R_o	TAI	R_o	TAI
0.30	2.0	1.30	3.2
0.38	2.2	1.36	3.3
0.42	2.3	1.42	3.4
0.46	2.4	1.50	3.5
0.50	2.5	1.75	3.6
0.60	2.6	2.0	3.7
0.70	2.7	2.5	3.8
0.85	2.8	3.0	3.9
1.00	2.9	3.5	4.0
1.15	3.0	4.0	4.0
1.22	3.1	5.0	4.0

Table 7.3 Correlation of Lopatin's time–temperature index of maturity (TTI) with vitrinite reflectance (R_o). (From D. W. Waples, *AAPG Bull.*, 1980.)

R_o	TTI	R_o	TTI
0.40	<1	1.75	500
0.50	3	2.00	900
0.60	10	2.25	1 600
0.70	20	2.50	2 700
0.85	40	2.75	4 000
1.00	75	3.00	6 000
1.15	110	3.25	9 000
1.22	130	3.50	12 000
1.30	160	4.00	23 000
1.39	200	4.50	42 000
1.50	300	5.00	85 000

Table 7.4 Correlation of TTI with important stages of hydrocarbon generation and destruction, as calibrated by TAI and R_o. (From D. W. Waples, *AAPG Bull.*, 1980.)

Stage	TTI	R_o	TAI
onset of oil generation	15	0.65	2.65
peak oil generation	75	1.00	2.9
end of oil generation	160	1.30	3.2
upper TTI limit for occurrence of oil with API gravity <40°	~500	1.75	3.6
upper TTI limit for occurrence of oil with API gravity <50°	~1 000	2.0	3.7
upper TTI limit for occurrence of wet gas	~1 500	2.2	3.75
last known occurrence of dry gas	65 000	4.8	>4.0
liquid sulfur in Lone Star Baden 1 (below dry gas limit)	972 000	>5.0	>4.0

potential ultimate reserves or degree of loss of hydrocarbons from the system.

The criteria to be met for a sedimentary rock to be an *effective oil source rock* can be summarized:

(a) The TOC content should be 0.4 percent or more.
(b) Elemental carbon should be between 75 percent and 90 percent; values below 75 percent represent immaturity, values above 90 percent are due to advanced catagenesis.
(c) The ratio of bitumen to TOC should exceed 0.05.
(d) The kerogen should be of amorphous or oil-prone type rather than of structured or gas-prone type.
(e) The vitrinite reflectance (R_o) should be no lower than 0.6 and no higher than 1.3.
(f) The H : C and O : C atomic ratios of the kerogen residues should fall in the favorable sector of the Tissot plot (Fig. 7.5). The principal phase of oil formation occurs at an H : C ratio between 0.84 and 0.69.

Conflicting signals from microscopic and chemical techniques are not uncommon, and the distinction between hydrogen-rich and hydrogen-poor OM is not always easily made. In cases of conflict, chemical data are to be preferred over microscope data, but wherever possible all techniques should be employed. We may usefully end this survey of source sediment evaluation with examples of the two types of approach.

7.6.2 The microscopic technique in action

A sample yields R_o = 0.5 percent. Maceral analysis under the microscope yields these percentages of the principal macerals:

type I (say, resinite)	5 percent
type II (say, sporinite)	15 percent
type III (say, collinite)	50 percent
type IV (say, fusinite)	30 percent

On the Tissot diagram (Fig. 7.5), read off the H : C and O : C ratios at points of crossing of the 0.5 vitrinite reflectance (VR) curve and the appropriate pathways. The maceral percentages we have chosen (converted to fractions of unity) would compute about as follows:

	H : C	O : C
type I	0.05(1.5) = 0.075	0.05(0.05) = 0.0025
type II	0.15(1.3) = 0.195	0.15(0.08) = 0.012
type III	0.50(0.9) = 0.450	0.50(0.12) = 0.060
type IV	0.30(0.4) = 0.120	0.30(0.14) = 0.042
Total	0.840	0.1165

According to Demaison's chart (Fig. 7.20), these R_o and H : C values indicate that the sedimentary rock has

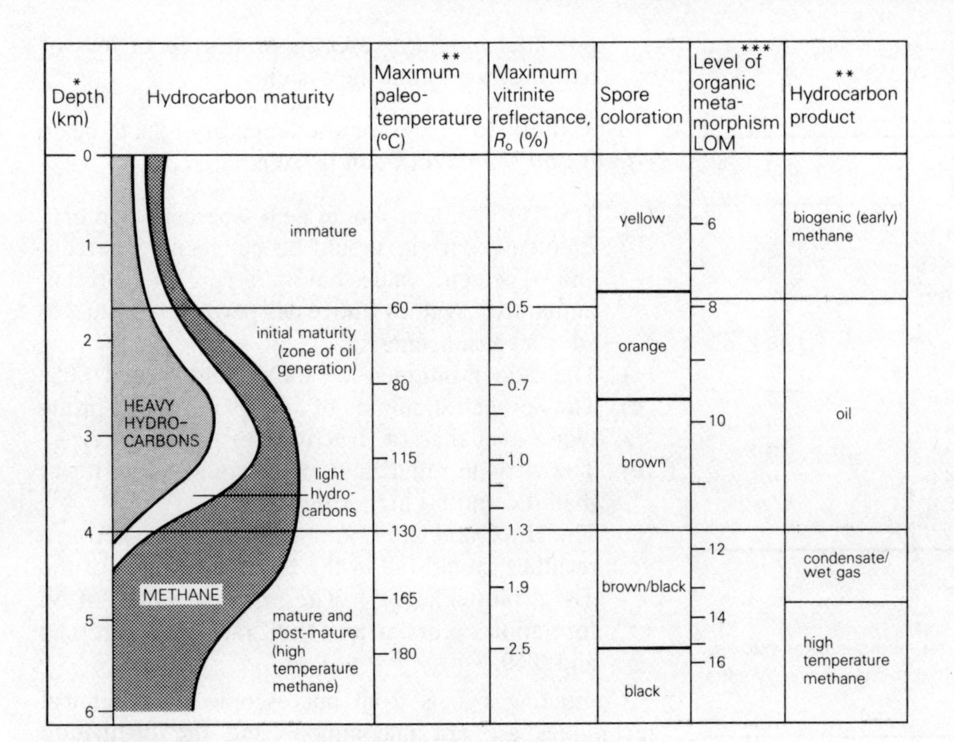

Figure 7.19 General correlation of organic maturation indices, based on the works of B. S. Cooper, A. Hood, B. Tissot, N. B. Vassoyevich, and their co-workers. (From A. J. Kantsler *et al., Oil Gas J.*, 20 November 1978.)

Figure 7.20 Maturation range chart. (From G. J. Demaison, 1980.)

Approximate temperature (°C) reached at maximum burial vs age of main overburden: 5–20 Ma	20–75 Ma	75+ Ma	Fixed carbon (percent) (coal)	Coal rank	TAI Spore color	Rock evaluation T_{max} (°C)	Kerogen type vs generative windows: IV dry-gas prone	III wet-gas prone	II oil-prone	I strongly oil-prone	C_{24+} *n*-paraffins CPI	Vitrinite reflectance, R_o (per cent)	Preservation and destruction windows in entrapment areas (staircase = migration)
implies constant heat flow through geologic time				peat	pale yellow							0.2	biogenic dry gas CO_2,N_2
65	55	50	40	lignite and sub-bituminous coals	2.0 yellow			immature			2–1.4	0.3	to minor early thermal lean gas; heavy degraded oils
93	80	65	54		2.5	~ 430	kerogen H : C = 0.6	H : C = 0.8	H : C = 1.25	H : C = 1.7	~ 1.3	0.5	undegraded medium gravity oils
					orange		oil window (generation)	mixed facies deltaic high wax crudes		oil expulsion	~ 1.2	0.6	
150	110	93	65	bituminous coals: high volatile	2.7 dark orange						~ 1.1	0.75	light oils
165	120	110	67		3.0	~ 455	0.5	0.75	0.9	1.2	1.0	1.0	mainly gas wet to dry condensates minor oil
				medium volatile	brown		gas window; DRY GAS	CONDENSATE	OIL low wax	OIL commonly high wax low sulfur; condensate expulsion			
190	150	135	75		3.3	~ 470	0.4	0.65	0.8	0.9		1.35	OIL DEADLINE lean to dry gas
				low volatile	dark brown								
205	175	150	87	anthracite	3.7 black			post-mature				2.0	CH_4, CO_2, H_2S, N_2 very low porosities
230	205	175	96		4.0	~ 500						4.0	GAS DEADLINE
				meta-anthracite				low grade metamorphics					barren

barely reached the initial stage of maturation and is wet gas-prone.

7.6.3 The chemical technique in action

An extraordinarily convenient and successful technique for the rapid evaluation of source rock potential was developed in the mid-1970s by scientists of the Institut Français du Pétrole. It is based on the selective detection of both hydrocarbon compounds and oxygenated compounds by the pyrolysis, in an inert helium atmosphere, of small, ground rock samples. The pyrolysis is performed within a rigorously controlled temperature interval in order to exclude oxygenated compounds released from mineral matter (especially CO_2 from carbonates) and not from the organic content of the rock. The analysis is completely automated.

Information derived from the analysis is the type of source kerogen, its degree of maturation, and its petroleum potential. The analysis records three peaks (Fig. 7.21), representing the volumes of three components of the organic matter (the volumes being proportional to the areas below the peaks):

(a) Free hydrocarbons, already generated and therefore volatilized below 300 °C, are represented by area S1.

(b) Hydrocarbons produced by cracking of the kerogen up to 550 °C give the residual petroleum potential or amount still capable of being generated by the sample. This is area S2. The ratio S2 : (S1 + S2) is called the *hydrogen index*, which correlates directly with the H : C atomic ratio. The ratio S1 : (S1 + S2) is the *production index*.

(c) CO_2 derived from the organic matter is represented by area S3. The ratio S3 : (S1 + S2) is the *oxygen index* and correlates with the O : C atomic ratio.

The ratio S2 : S3 gives a measure of the oil-proneness of the sample: ratios of 5 or more indicate oil, ratios of 3 or less, gas. The maximum temperature, T (in degrees Celsius), required to volatilize the OM is a function of the degree of maturation; it is given by the amplitude of the peak above S2 (Fig. 7.21).

If there has been no mobilization and accumulation of hydrocarbons for the sediments analyzed, the production index increases smoothly with depth. Accumulations of hydrocarbons interrupt the smooth curve in the direction of a higher index (A in Fig. 7.22); a zone from which hydrocarbons have been lost is represented by an interruption in the direction of lower production index (B in Fig.7.22). The three indices are plotted against depth (Fig.7.23) in curves resembling the "oil window" curve of Figure 7.8.

7.6.4 Recognition of an expanding science

White's carbon ratio theory (Sec. 7.2) has been expanded into a highly developed science – and art – in which the skills of petroleum geologists, coal petrologists, organic geochemists, palynologists, and microscopists have been very successfully combined.

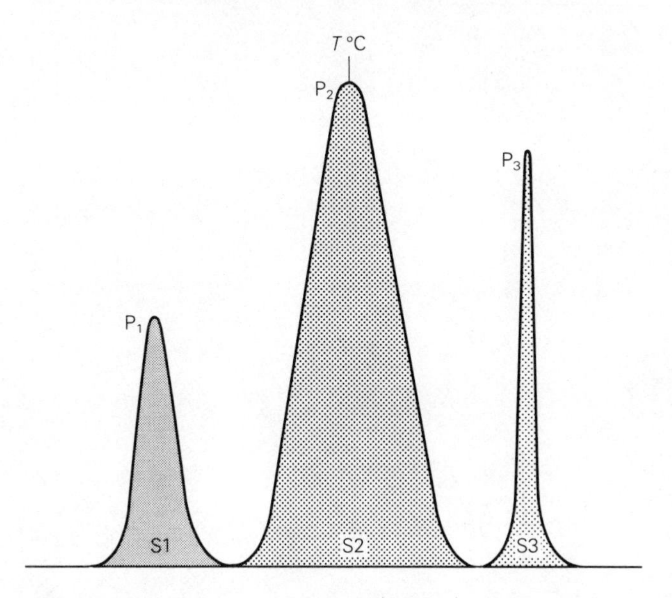

Figure 7.21 The three peaks recorded from a rock sample in the rapid rock-evaluation pyrolysis. (From J. Espitalié *et al., Offshore technology conference* 1977.)

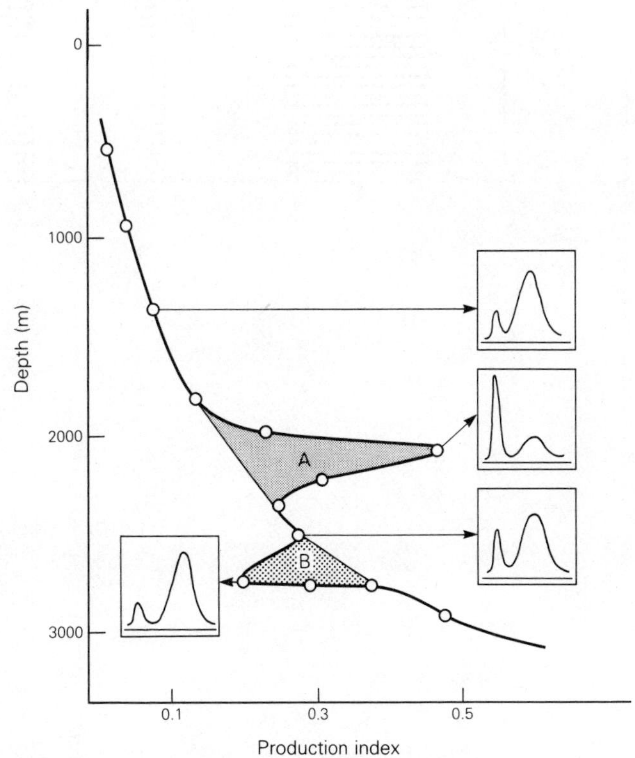

Figure 7.22 Interruptions in the production index with depth when hydrocarbons have been effectively accumulated (A) or depleted (B). (From J. Espitalié *et al., Offshore technology conference* 1977.)

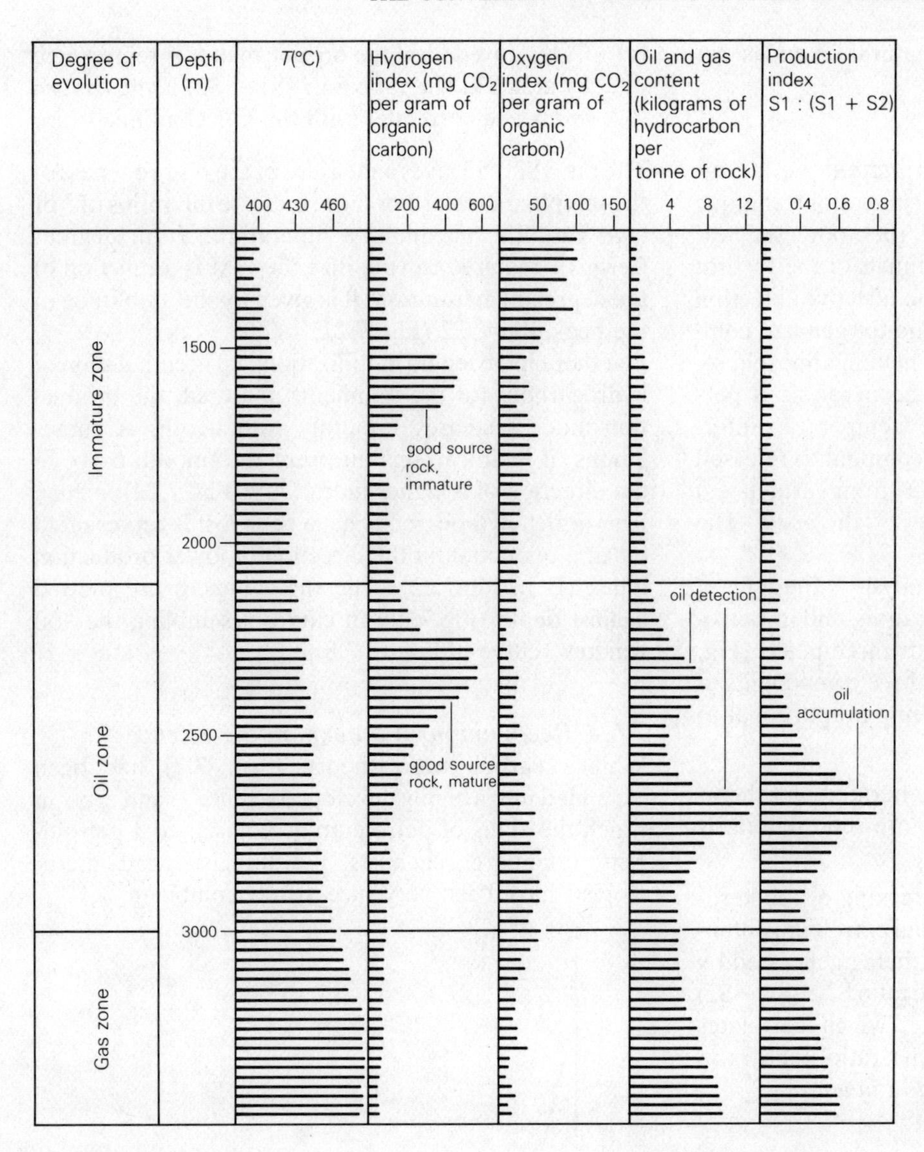

Figure 7.23 Source rock maturation log derived from rapid evaluation of well samples by pyrolsis. (From J. Espitalié *et al.*, *Offshore technology conference* 1977.)

8 Source sediments as seen by the geologist

8.1 The range of possible source sediments

We have seen that petroleum hydrocarbons are generated by thermal processes from organic material of certain restricted types entombed in sedimentary rocks. We have further seen how modern quantitative methods can determine whether a sedimentary rock contains sufficient organic matter (OM), whether it is of suitable type to generate petroleum hydrocarbons, and whether the necessary thermal processes have been operative. Any basin in which commercial hydrocarbons have been discovered must therefore contain at least one rock-stratigraphic unit that, being initially a *potential* source sediment, was converted into an *effective* source sediment, yielding and expelling at least some of its hydrocarbon content, and being then preserved as a *successful* source sediment. We have discussed its geochemical characteristics, and the conditions most conducive to its formation. What does it look like to the geologist?

The most common *reservoir rocks* are varieties of sandstone, but the sandstone clan can be dismissed as important *source rocks*. Some nearshore sandstones (in estuaries and deltas) contain minute quantities of bitumens (up to 20 p.p.m.), mostly *n*-paraffins in the $C_{20}-C_{30}$ range resembling those in crude oils. Nevertheless, the very process of coarse clastic deposition serves to dilute and degrade the organic content of the receiving basin and to disturb, not promote, its accumulation.

Nearly all nonsandstone reservoir rocks are *carbonates*, and the role of carbonates as source sediments for hydrocarbons has been a matter of enduring dispute. Organic analyses of about 1400 samples of sedimentary rocks were made by the Exxon organization's research laboratories during the 1950s and reported upon in 1962 by H. M. Gehman, Jr. The rocks came from many parts of the world and extended throughout the Phanerozoic time scale. Not surprisingly, the limestones were shown to contain much less OM than the shales (mean of 0.24 percent OM versus 1.14 percent for the shales). Surprisingly, however, the OM in the limestones was found to contain a higher proportion of hydrocarbons than the OM in the shales; in geochemical terms, the kerogens derived from carbonate OM are dominantly sapropelic whereas those derived from shale OM are mixed humic-sapropelic. As a consequence, the average hydrocarbon contents of the limestone and shale samples were almost identical at nearly 100 p.p.m.

This unexpected balance, already alluded to (Sec. 6.6), is not repeated in Recent sediments. Limes and clays contain similar quantities of OM, approximately 1 percent, and the OM in the limes is not richer in hydrocarbons than the OM in the clays. The most plausible implication, in Gehman's view, was that carbonate sediments lose much of their relatively unstable non-hydrocarbon OM during diagenesis, leaving a residue enriched in the more stable hydrocarbons. The equivalent loss of nonhydrocarbon OM from shales is obstructed by the absorptive capacity of clay minerals, lacked by carbonate minerals.

"Clean" carbonates, lacking significant argillaceous or evaporitic content, are unlikely to be effective source sediments. They are deposited in relatively agitated, oxygenated waters, to permit their biogenic component to exist; such waters lead to degradation or dissemination of the OM content. Oxidation and sulfate reduction last longer and extend over a greater vertical range in recently deposited carbonates than in other types of sediment. Carbonate sediments are also cemented and lithified more rapidly than other common sediments. Hardly any lithified, clean carbonate rock contains more than 0.5 percent OM by weight even though the percentage of this OM that consists of hydrocarbons may be 10 percent or much more. If the environment of deposition is anoxic, the organic content may be considerably higher (more than 2 percent by weight, possibly more than 5 percent), but so will be the clay content; the carbonate is not then a "clean" carbonate. If the content of clay minerals is less than about 30 per cent, the

sediment will display low catalytic activity. Moreover, the compaction process will be obstructed and the sediment is unlikely to achieve the internal overpressures necessary to expel any hydrocarbons that may be generated.

A more favorable circumstance is presented by the *carbonate–evaporite* environment of deposition, especially the mesohaline phase of it under lagoonal or sabkha conditions with 5–12 percent salinity. Surface currents then flow persistently towards the most saline parts of the basin, bringing in nutrients. Blooms of phytoplankton may occur annually, exactly as under upwelling conditions but with far fewer species of plankton represented. If the basin's setting is such as to exclude clastic detritus, the only sediment deposited consists of precipitated carbonate minerals, forming lime mud (micrite or calcilutite); there are no benthonic organisms to provide biogenic sediments.

All that is necessary to preserve the phytoplanktonic OM is density stratification of the brines in the basin. This stratification occurs in many areas where carbonate and sulfate minerals are being deposited today, notably in the Persian Gulf. Because carbonates and evaporites constitute almost the entire stratigraphic section in the petroliferous region of the Arabian platform, this process was almost certainly operative in the formation of its exceedingly rich source sediments. It may also have been the source-creating process in a number of much older petroliferous regions in which oil occurs in evaporite-enclosed reefs or banks, as in the Silurian of the Michigan Basin, the Middle Devonian of the northwestern Alberta Basin, and the Pennsylvanian of the Paradox Basin in Utah (Chs 17 & 25). The OM in these cases was supplied by algal mats and mounds rather than blooms of phytoplankton.

We come now to the third dominant category of aqueous sedimentary rocks, the *argillaceous rocks*. Argillaceous rocks are the most common of all sedimentary rocks, whether the assessment is made on stratigraphic or geochemical grounds. The proportion of sedimentary rocks that are "shales" (in the wide and inaccurate sense of "dominantly argillaceous rocks") has been variously estimated between 42 and 82 percent, but there is a general consensus that "clays" make up something like 65–70 percent of all sediments. If this huge preponderance of clay-based rocks was commonly of hydrocarbon source potential, the volume of hydrocarbons should be thousands of times what it is ever likely to have been even assuming extraordinarily inefficient conversion and preservation.

Clay-rich sediments have all the environmental and physical characteristics that a potential source rock must have; they are the dominant sediments of the deeper parts of depositional basins, where opportunities for aeration and agitation are lowest, those for stillness and stagnation highest. Of all common sediments they have the highest initial porosity and therefore the highest initial fluid content (whatever the fluid may be). Because of their compactibility (see Sec. 11.1.4), they lose volume and therefore porosity more rapidly during burial than any other sediments, and their compactibility continues for a much longer time.

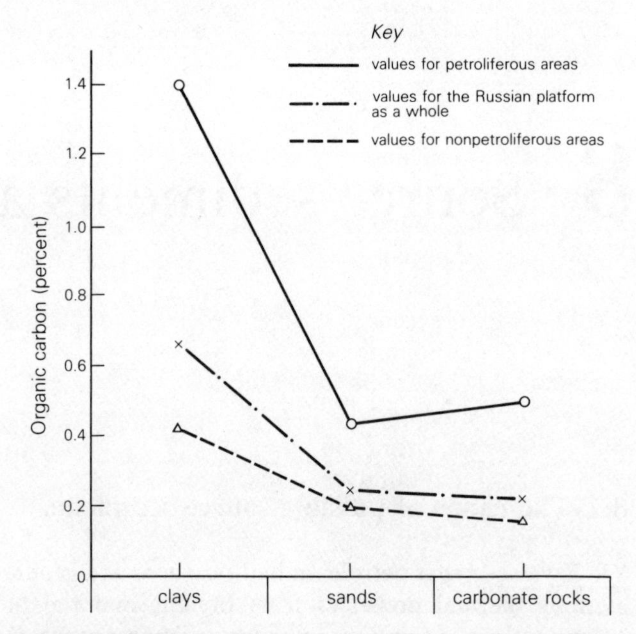

Figure 8.1 Average contents of organic carbon in different types of sedimentary rocks from petroliferous and nonpetroliferous areas of the Russian platform, based on analyses of more than 25 000 samples. (From A. B. Ronov, *Geochemistry*, 1958.)

In 1958, A. B. Ronov reported on analyses of more than 25 000 samples from the Russian platform, petroliferous and nonpetroliferous areas alike (Fig.8.1). In the petroliferous areas, organic carbon averaged 1.4 percent in "clays," 0.5 percent in carbonates, and 0.4 percent in sandstones. In nonpetroliferous areas, the equivalent figures were 0.4, 0.15, and 0.2 percent. Under comparable sedimentation rates, black shales of anoxic depositional sites have higher TOC contents than any other rock clan.

Shales and mudstones were therefore early invoked as the most probable source rocks for oil that was discovered in either sandstone or carbonate reservoirs. In reality, however, "shales" include vast volumes of fissile siltstones, blue-grey mudstones of continental shelf origin, and nonmarine, carbonaceous claystones having essentially no capacity as oil source sediments. A *rich* source sediment is no ordinary "shale." Rich source sediments of very different ages are much alike in gross appearance, the principal variation being in color. The most common color is brown, especially very dark brown, but it may be black or green. Rich source sediments are almost invariably laminated, the thinner laminae being typically 0.5–2.0 mm in thickness but with many thicker bands up to 1 cm or more. The thicker bands are commonly paler in color, and contain appreciable silt-size quartz grains. In the richest sections, alternate bands consist almost entirely of the closely packed remains of pelagic organisms, commonly of a single type and even a single species. The organisms most commonly found in young source sediments are foraminifera, especially the multitude of globigerinids. In the Kimmeridgian source sediments of the North Sea, the principal organisms in the rich bands are coccoliths; in the Miocene Monterey Formation of California they are diatoms; in other formations they are dinoflagellates or radiolaria. The presence of these bands makes a very rich source rock virtually a *primary organic sediment*.

The faunal and floral elements are characteristically microscopic, except for fish remains, which are ubiquitous, and (in many Paleozoic representatives) conodonts. In many Jurassic and Cretaceous source sediments, the laminae are interrupted by flat concretions, commonly calcareous, containing ammonites. The Kazdhumi Formation of the Iran–Iraq oil belt, a rich source rock of Albian age, was known as the "ammonite shales" before it was accorded a formal formation name. On the other hand, benthonic faunas are largely or wholly absent.

If the dominant microfauna is calcareous (as in the case of globigerinids and most other foraminifera), the carbonate content of the rock is moderate to high and the rock is a calcareous shale or marlstone. It may be sufficiently calcareous to be fairly called a limestone. Two of the most prolific source sediments known – the Jurassic Hanifa Formation in Saudi Arabia and the Cretaceous La Luna Formation in western Venezuela – are fundamentally limestones, though they were deposited in anoxic waters at least 200 m deep. If organisms secreting opaline silica are present in any quantity, discrete bands of dark chert may be interbanded with the argillaceous or calcareous bands. If the dominant organisms are siliceous (diatoms or radiolaria), the entire rock becomes siliceous instead of calcareous. The argillaceous content is always in the form of mixed-layer clays, presumably formed from original smectite ("montmorillonite;" see Sec. 15.4.1) and not from original illite.

The conspicuous laminae, presumed to be varves, are typically parallel and undisturbed over much of the basin. Minor soft-sediment deformation may yield flowage or pinch-and-swell structures, but evidence of bioturbation is normally absent; such activity would consume the OM and reduce the effectiveness of the source sediment. Fissility is highly developed. Scanning electron microscope studies show that the clay mineral content is arranged with strong preferred orientation parallel to the laminae, with closely packed, overlapping flakes. On the laminar surfaces themselves, the fabric of the clay particles is interrupted by a network of irregular pores so that porosity and permeability are relatively high (6–12 percent pore volume) parallel to the lamination but essentially zero at right angles to it.

The darker laminae are visually opaque and immensely rich in OM. They contain fluffy residual kerogen, with vague structures presumably derived from algae and anaerobic bacteria visible under the microscope. Scratching with a knife produces a shiny brown groove; rubbing with dilute acid leaves a brown residue of organic material; striking a thin slab cut from the rock results in a dull resonance more like that from wood or cannel coal than that from most rocks.

Besides the fissility there may be pronounced microscopic fracturing. This is caused by the generation and expulsion of fluids, both water from the clay minerals and oil from the kerogen, leading to shrinkage and overpressuring. A rich source sediment may be said to be inherently self-fracturing (Sec. 15.4.2).

In addition to the argillaceous, calcareous (or siliceous) and bituminous contents, most source sediments are phosphatic (though not commercially so) and uraniferous (the uranium commonly being concentrated in phosphatic nodules); some are manganiferous or seleniferous (the latter indicating contemporaneous submarine vulcanicity). Sulfides are common, especially pyrite. Carbonaceous material of terrestrial origin may be abundant, but it is not an essential component.

Most features of this extended description fit the Ordovician Simpson shales of Oklahoma (about 450 Ma old) and the early Miocene Cipero Formation of Trinidad (about 25 Ma old) equally well. Altogether, a rich oil-source sediment is a striking, "idiosyncratic" rock. From one point of view, it is the most important sedimentary rock in the world, yet one which has never been given a petrographic name of its own.

8.2 Two rich source sediments described

Following are brief summaries of the geologic characters of two long-acknowledged oil-source sediments of very different ages that have been exhaustively described by many geologists.

An acknowledged Paleozoic source sediment is the *Domanik Formation* of the Ural–Volga and Pechora basins in the Soviet Union. It is late Devonian (Frasnian) in age. The dark, almost black bituminous shales, marls, and limestones are commonly silicified over thicknesses of several tens of meters. Numerous bands are rich in OM but contain very little detrital material; some of them constitute true oil shales. A highly specific faunal complex indicates deposition in relatively deep water: conodonts (which are phosphatic), radiolaria (siliceous), tentaculites (which provide evidence of current directions), goniatites, thin-walled pelecypods, and phytoplankton; luxuriant algal growths occur in places; brachiopods occur rarely, and are phosphatic, not calcareous. Concretions, with or without fossil centers, occur sparingly.

The *La Luna Formation* of late Cretaceous (Turonian) age has been recognized as a rich source sediment since the earliest discoveries were made in the Maracaibo Basin of western Venezuela. It is a dark brownish-grey to black bituminous limestone, of fine to medium grain, thinly laminated to very well bedded with little or no disturbance by bioturbation, and with regularly spaced lenticular bands of dark chert. Discrete bands contain abundant fish bones and scales, and phosphate pellets. In the lower part of the formation are bands rich in fragments of *Inoceramus* and other pelecypods; elsewhere throughout the formation are flattened limestone concretions formed around fossils, most of them ammonites. There are also bands composed largely of the closely packed tests of pelagic foraminifera. These foraminifera (especially various globigerinids) are also sparsely present in the general matrix of the thicker bands.

Similarities between source sediments are of course even more pronounced if their ages are closely comparable, geographically distant though their basins may be from one another. The Kazdhumi Formation of Albian to early Cenomanian age, for example, regarded as the principal source rock for the great oil accumulations in the Cenomanian–Turonian and Oligo-Miocene carbonates of Iran and Iraq, is almost a duplicate of the La Luna Formation. In addition to the ubiquitous ammonites and globigerinids, there are bands containing immense floods of *Oligostegina*.

8.3 Position of source sediments in the depositional cycle

Consideration of the most favored source sediments for all the highly petroliferous basins of the world leads to a few supportable generalizations concerning the stratigraphic and depositional relations between the source sediments and the reservoir horizon(s). Source sediments are early deposits in most basins, forming at the peak of transgression following a transbasinal unconformity. In essentially continuous successions, the source sediments are commonly older than the main reservoir rocks, and may be separated from them by entire formations devoid of recoverable hydrocarbons. Where source rocks directly overlie unconformities, the reservoir rocks may as commonly be older as younger. Where both faults and unconformities contribute to the trapping mechanisms this uncertainty is especially obvious.

Regarding the depositional environment of the source sediments, the combination of high organic content, undisturbed laminar layering, pelagic fauna and flora, and presence of phosphorus, added to the need for anaerobic conditions to preserve the OM, points to deposition below a thermocline. There the temperature and pH of originally cold waters are increasing and the redox potential, E_h, is below zero. The deposition of rich source sediments is therefore most likely in the deepest part of the depositional basin. The absolute depth need not be great; source sediments are not to be attributed to abyssal or lower bathyal depths. On the other hand, the continental shelves are not the sites of important source sediment deposition. It was the fact of world-wide shelf *pooling* of oil that prompted disbelievers in oil migration to invoke shelf sediments as the sources as well as the containers. The favored sites for source sediment deposition are basinal "cells" within shelves, either faulted down within them (graben or coastal half-graben basins) or compressionally tucked down between them and newly rising mountains. Ideal depths are below 200 m (the upper limit of the bathyal zone by general convention) but not deeper than about 1000 m. This range encompasses the epipelagic or upper bathyal zone in Joel Hedgpeth's classification (published in 1957), plus the upper 20–25 percent of the middle bathyal zone in which more than 90 percent of foraminifera are planktonic.

The requirements of high organic carbon, pelagic fauna and flora, abundance of phosphate ions, undisturbed lamination, and anaerobic waters below a thermocline are, of course, met in many deep lakes as well as in deep "cells" in the marine realm (Sec. 6.5). Lacustrine OM leads more readily to the type I kerogen typical of oil shales (see Sec. 7.3) than to the source sediments of conventional petroleum, but effective source sediments of lacustrine origin are not rare (Sec. 17.4.7).

8.4 Relation of source sediment deposition to other geologic events

The establishment of a distinct thermocline and oxygen-minimum layer, and the equalization of temperatures and loss of circulation below it, are features of the *peaks of transgressive cycles*. A combination of orthodox and seismic stratigraphies has enabled Peter Vail and his co-workers to identify the peaks of transgression and troughs of regression with considerable accuracy (Fig.8.2). The first major Phanerozoic transgression, during the Cambrian, came before marine life was sufficiently abundant and diversified to provide rich accumu-

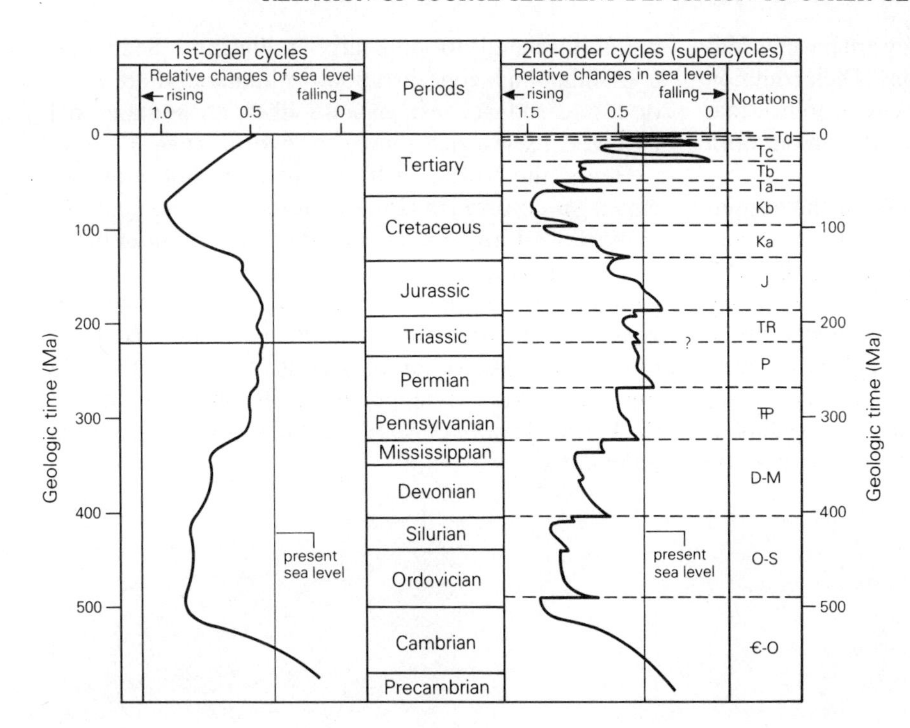

Figure 8.2 Global cycles of marine transgression and regression during the Phanerozoic. (From P. R. Vail *et al.*, *AAPG Memoir* 26, 1977.)

lations of organic matter, and when there were no land plants or animals. The second, during the Ordovician, undoubtedly contributed source sediments that became effective; a good deal of Ordovician oil has been discovered, especially in North America.

The late Devonian transgression, following the Younger Caledonian or Acadian phase of plate mobilization, contributed the principal source sediments of the Ural–Volga and western Canadian basins as well as important ones in the eastern and southern USA and northern Africa. A lesser transgression in Pennsylvanian time, following the Hercynian orogeny, saw the deposition of source sediments in the Mid-Continent and south-western regions of the USA. The greatest of all transgressions, beginning in the Middle Jurassic and continuing with minor setbacks until near the end of the Cretaceous, was responsible for the anoxic horizon B of the present ocean floors and for the principal source sediments in the Middle East, western Siberia, the North Sea, and Middle America from the upper Gulf Coast to Peru. A much smaller transgression in the Oligo-Miocene created the generative sediments of the Californian, Caucasian, Carpathian, and Indonesian provinces. The troughs of regression, in the earliest Ordovician, earliest Devonian, Permo-Triassic and earliest Jurassic, and Plio-Pleistocene, witnessed little source sediment deposition.

Parker Trask in the USA (1939) and A. B. Ronov in the Soviet Union (1958) analyzed thousands of samples from their two big and petroliferous countries to determine the organic carbon content of sedimentary rocks by age. The maxima were determined to be in rocks of Cambro-Ordovician, Carboniferous, Jurassic, and Tertiary ages, the minima in Siluro-Devonian and Permo-Triassic rocks which contain the maximum concentrations of evaporite deposits.

Helen Tappan, in 1968, and others estimated that the abundance of fossil phytoplankton reached maxima between the beginning of the Ordovician and the end of the Devonian, during the Cretaceous, the Eocene, and the Miocene. In 1968, C. B. Gregor studied the rates of erosion of the continents and formation of sedimentary rocks with time, concluding that they were at their maximum at the times of orogenic uplift in the Devonian, the Jurassic, and the Oligocene.

All these phenomena display a correlation with source sediment maxima, in some cases very close, in others more vague or uncertain. No correlation can be detected between episodes of source sediment deposition and times of great faunal or floral provinciality or cosmopolitanism. And there is a conclusive lack of correlation with episodes of numerous or of no reversals of the Earth's magnetic field.

8.5 Source sediments for natural gas

Any sediment capable of becoming an effective source sediment for oil can become one for gas also. Associated gas is very common. In addition, however, there are several depositional environments highly conducive to the accumulation of gas-generating sediments but not favorable for oil-source sediments; they are the environments that give rise to nonassociated gas. Their common occurrence accounts for the familiar circumstance that gas is very much more widely distributed than oil, both geographically and stratigraphically.

The obvious cases are those which favor the accumulation of great quantities of humic organic matter, from terrestrial vegetation, whilst excluding sapropelic OM. Nonmarine molasse deposits on the forelands of newly risen mountain ranges, brackish or freshwater fill of cratonic depressions, and the apical regions of large deltas are the common examples. The first (molasse) case is likely to be coeval with major oil-source sediment deposition but in different places. The others may be quite unrelated to oil-source sediments. Sediments containing high concentrations of humic OM are more or less carbonaceous, and are likely to contain coal (which is itself a rich source rock for methane).

Hence the widest distribution of prolific nonassociated gas occurred in Cretaceous molasse sediments, as in northwestern Siberia and the western interior of North America, and during the Permo-Triassic aggregation of the continents following the Hercynian orogeny and the wide spread of Pennsylvanian coal deposits. The Cretaceous episode is included within the time of greatest development of oil-source sediments; the Permo-Triassic episode coincides with the oil-source minimum.

9 Oilfield waters and their effects on the hydrocarbons

9.1 Why waters are found in oilfields

Nearly all sedimentary rocks are deposited in water, in layers one upon another. No matter what the composition of a sediment may be, *water* is a natural part of it. Water deposited with a sediment from the beginning is called *connate water*.

Most sedimentary rocks therefore are inherently *water-wet*. As the sediment is buried and becomes an increasingly indurated sedimentary rock, much or all of its connate water is either altered in composition, diluted or displaced by other waters, or expelled altogether. Where sedimentary rocks now contain oil or gas, the oil or gas has similarly displaced some of the water earlier occupying pore space in the rocks. The water that oil or gas is able to displace is *free water*.

Much water resists displacement by invading hydrocarbons, because it occupies pore spaces too fine for the hydrocarbons to enter (see Sec. 11.2.1) and may even adhere to the surfaces of the rock grains themselves. This is *interstitial water*. It typically occupies 10–40 percent or even more of the total pore space in the hydrocarbon-bearing reservoir rock, but it is not necessarily part of the general water column in the rest of the reservoir rock (which is an aquifer and still contains free water). Nor is interstitial water in a continuous fluid phase with the oil; wells completed in the oil column will initially yield oil free of water. The total water contained in an otherwise hydrocarbon-bearing reservoir rock is best called *formation* or *oilfield water*.

9.2 Chemistry of oilfield waters

The nature and quantity of the formation waters in an oil- or gas-bearing region are vital concerns to both exploration and exploitation of the region. Normal sea water contains about 3.5 percent of dissolved mineral matter (35 000 p.p.m.). About 90 percent of this is NaCl. Many oilfield waters contain much more mineral matter in solution than this, up to 300 000 p.p.m. or even more. Those containing more than some arbitrary percentage – say 10 percent or 100 000 p.p.m. – are called *brines*. The strongest brines occur in undeformed or little deformed basins like those overlying the stable interiors of shields, especially if the stratigraphic successions include evaporite formations. Waters in such basins not only become naturally concentrated with time; they are protected from dilution by meteoric waters entering the aquifers from the outcrop (see Sec. 17.7.1). In contrast, basins in which the reservoir rocks are close to the outcrop, or are strongly faulted, are characterized by oilfield waters deficient in salinity. Dilution by invading waters reduces the mineral content to 1 percent or less.

Oilfield waters are classified according to the dominant mineral ions present in solution. A widely used classification is that proposed by the Russian geochemist V. Sulin, distinguishing four chemical types of water according to the distribution of three cations and three anions:

type *a*: sulfate–sodium waters;
type *b*: bicarbonate–sodium waters;
type *c*: chloride–magnesium waters;
type *d*: chloride–calcium waters.

Types *c* and *d* also contain sodium, of course. Most oilfield waters, including all typical "brines," fall into type *d*; sodium dominates the cations, but both magnesium and calcium are present in addition, the Ca : Mg ratio being about 5 : 1 (Fig.9.1). Chloride is almost the sole anion. Sulfate is notably absent. Formation waters of this type characterize deep, stagnant basins in continental interiors, where there is little outcrop to permit invasion of surface waters and create a hydraulic head (see Sec. 17.7).

Type *c* water takes the place of type *d* in many evaporite-bearing sequences. In both types, the concentration of dissolved material tends to increase linearly with

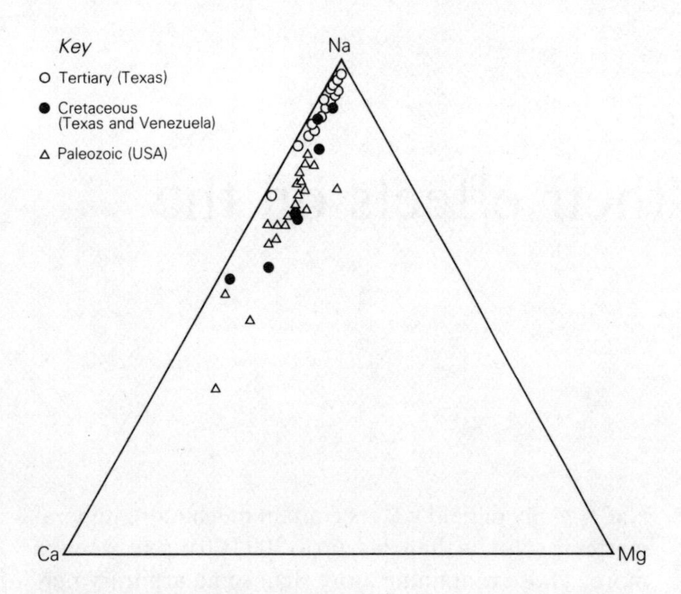

Figure 9.1 Relative amounts of cations in typical concentrated oilfield brines of connate type. (From P. A. Dickey, *AAPG Bull.*, 1966.)

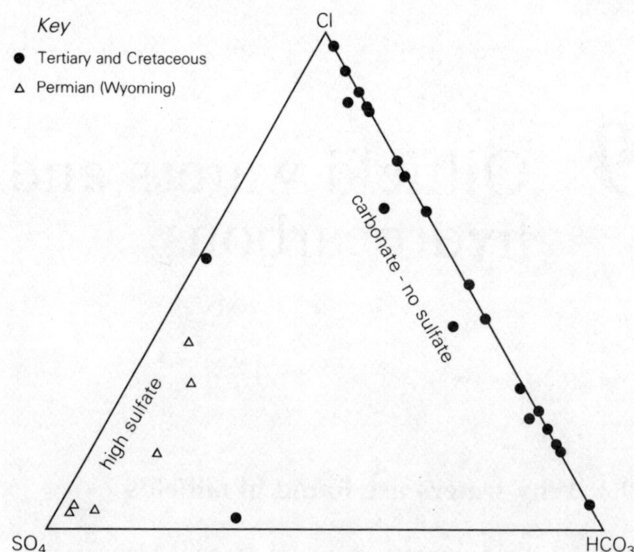

Figure 9.3 Relative amounts of anions in oilfield waters of artesian type. (From P. A. Dickey, *AAPG Bull.*, 1966.)

depth, in some basins of cratonic type as much as 10–15 percent per kilometer (Fig.9.2). This phenomenon, recognized since the early days of oil development, is probably due to reverse osmosis. As water is expelled upwards through shales and other rocks impermeable to oils, the oil and the salt remain in the aquifer below, and this therefore becomes the reservoir formation for the oil. The water in the pore spaces of the shale, in contrast, acquires higher concentrations of sulfate (SO^{2-}_4) and bicarbonate (HCO^-_3) and a lower concentration of chloride (Cl^-). The solubility of the salt increases with increasing temperature and pressure and so with increasing depth. Ultimately, of course, there must be a decrease in salinity with depth, as compaction releases the last water adsorbed on clay mineral surfaces. Very deep waters should also become more alkaline as silicate minerals are hydrolyzed by water and CO_2.

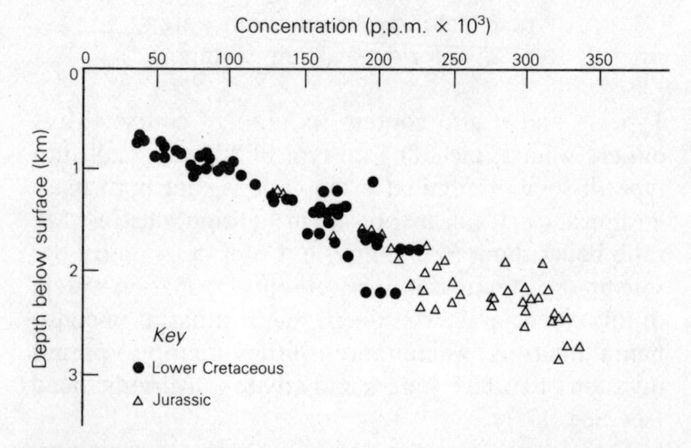

Figure 9.2 Concentration of oilfield waters versus depth in Lower Cretaceous and Jurassic strata of the US Gulf Coast Basin. (From P. A. Dickey, *AAPG Bull.*, 1966.)

Water types *a* and *b* occur in near-surface artesian conditions (Fig.9.3). Bicarbonate is the dominant anion in meteoric waters. Waters containing it and/or sulfate are oxygen-bearing and very subsaline, in some basins almost fresh. Oils associated with them become oxidized, and the presence of sulfate, as in some of the much faulted Rocky Mountain basins of the western USA (see Sec. 17.7.8), invariably means the oils are naphthenic. Combinations of water types occur in basins in which shallow reservoirs, close to outcrop, are separated by unconformities from deeper reservoirs isolated from outcrop. In the Maracaibo Basin of Venezuela, for example, the shallow Tertiary reservoirs of the Bolivar Coastal fields are associated with waters of type *b*; the deeper Cretaceous fields below the lake and to the west are accompanied by type *d* waters (Fig.9.4).

Analyses of oilfield waters are vital to both exploration and oilfield management. They are readily displayed graphically by a variety of simple constructions. A commonly used version is the Stiff diagram (Fig.9.5), in which short bars are given lengths proportional to the concentrations of the various ions, cations to the left, anions to the right.

Formation waters also contain numerous other elements in trace amounts. Russian studies of thousands of samples indicate that the halogens bromine and iodine are in higher concentration in the waters than in the associated clays; the common alkali elements sodium, calcium, magnesium, and strontium, plus boron, are in about the same concentrations in the waters and the clays; other elements are in much lower concentrations in the waters. For most of the elements analyzed, the pattern of concentrations in the waters is closer to those in the oils, and in organisms, than to those in the sedi-

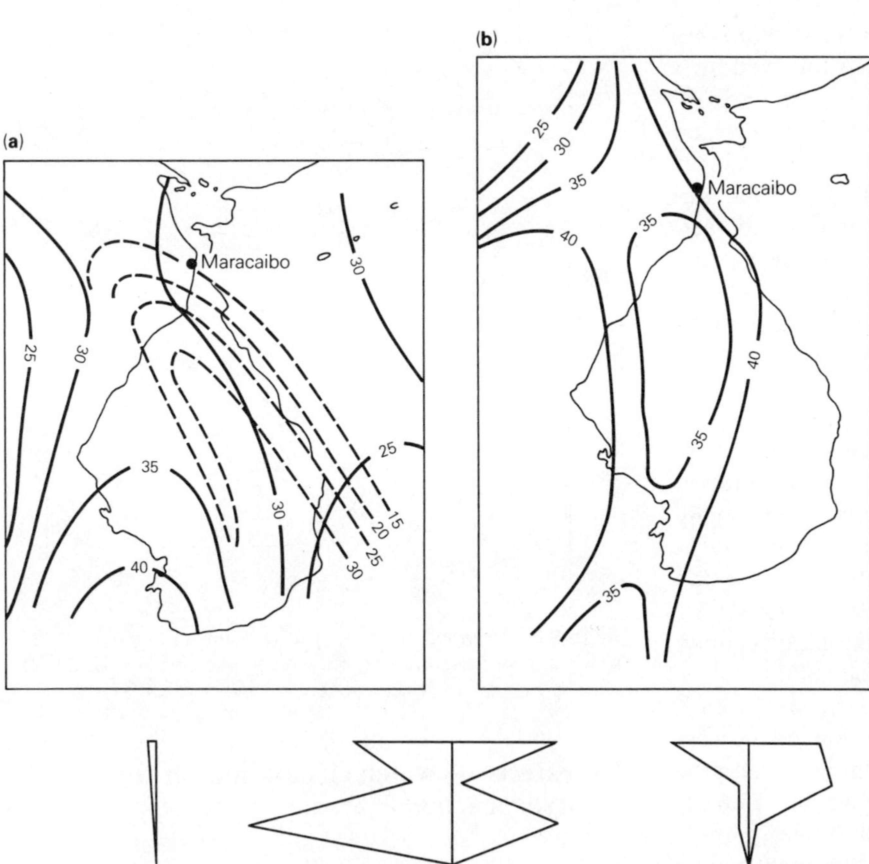

Figure 9.4 Oil gravities in the Maracaibo Basin of Venezuela: (a) oils in Eocene (solid lines) and Miocene (dashed lines) sandstones invaded from east and west by meteoric waters; (b) oils in Cretaceous strata, with no serious invasion by surface waters. (From C. Gonzales de Juana *et al.*, *Geologia de Venezuela*, Caracas, Venezuela: Ediciones Foninves 1980; after W Scherer, 1975.)

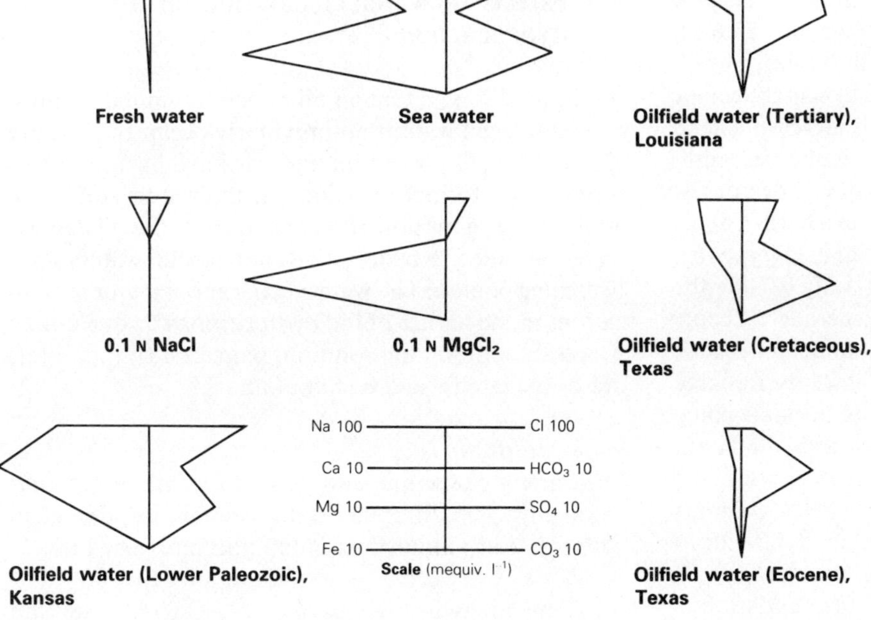

Figure 9.5 Basis of Stiff diagram for representation of compositions of oilfield waters, and common patterns of these. (From H. A. Stiff Jr, *Petrol. Trans.*, AIME, © 1951 SPE-AIME.)

ments. This seems to suggest that the trace elements were of the same organic origin as the oils. The oils contain lower concentrations of the elements than do the organisms, except in the case of vanadium, which is concentrated in petroleum porphyrins (see Sec. 5.2.6).

9.3 Reversed hydrologic and density profiles

It has been remarked (Sec. 9.2; Fig.9.2) that the salinities of oilfield waters normally increase with depth below the surface. In the same way, the oils normally become lighter with depth. In some very thick sandstone sequences, however, either of these general rules may lose application. In the most pronounced of such cases, both rules are broken together; both the hydrologic and the oil-density profiles are *reversed*. The oils become heavier with depth instead of lighter, and the formation waters become fresher instead of more saline. The lower salinity with depth is represented by reduced contents of the Na^+ and Cl^- ions; dissolved silica *increases*.

The phenomenon is characteristic of some young,

thick, deltaic sequences with multiple sand reservoirs deposited in a transgressive regime and followed by a regression. The lower sands were of brackish environment, the upper sands more marine. The reversed hydrologic profile is never smooth; instead the changes tend to be abrupt, and they may be somewhat unsystematic, each sand having its own oil/water contact. This is very clearly the case in the Greater Oficina area of eastern Venezuela, where medium-gravity oils and waters of normal salinity occur in the younger sands overlying tar oils and brackish waters in the older sands (see Sec. 17.7.5).

The simplest explanation is *density stratification*. The phenomenon occurs in a striking way only where a very large volume of oil is stored in traps having very thick oil columns – measured in hundreds of meters – in highly permeable sandstone reservoirs. The Burgan field in Kuwait is an excellent example. So is the Prudhoe Bay field in Alaska (Sec. 26.2.2), in which the Permo-Triassic reservoir contains 28° gravity oil at the gas/oil contact and 14° gravity oil at the base of the oil column.

Perhaps the most studied example is that of the Pliocene "Productive series" of the Baku region, on the west side of the Caspian Sea, from which oil has been produced for more than a hundred years (see Sec. 25.6.1). In all there are 22 productive sands in the series. The oil in the sands is in general heavier than the oil in the clayey rocks, which contains more gas. With increasing sand downwards, the oil becomes heavier, higher in naphthene content and in "tars"; water salinity decreases, and the water type changes abruptly from a hard, Ca–Cl type (*d*) to an alkaline, high-pressure $NaHCO_3$ type (*b*). The waters are sulfate-bearing (type *a*) only outside the oil-bearing region. The basin was exceedingly deeply depressed, with the Upper Pliocene and Pleistocene alone reaching a thickness of nearly 7 km. It is probable that for any single bed the situation is normal; going down the dip in a single formation, towards the South Caspian basin (Fig.9.6), the oil density decreases, the tar content decreases, the paraffin and benzene contents increase. Nonetheless the overall pattern is difficult to attribute to oxidation, water washing, or migration; it seems to be a matter of gravitational differentiation.

The concentration of dissolved material in formation waters may also undergo a sudden decrease downwards through an unconformity. The waters in the two sets of strata are simply different. Yet another circumstance leading to the same result is provided by deep basins that are abnormally pressured at depth (see Sec. 14.2). Waters in the shale members are fresher than waters in the associated sands, from which fluid escape is inhibited. All the waters become fresher downwards, in an abrupt change, as the zone of abnormal pressure is entered.

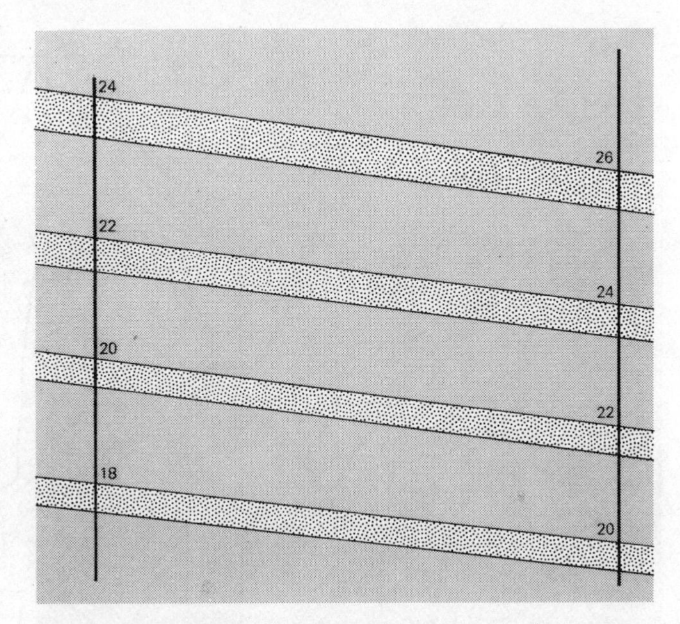

Figure 9.6 Reversed oil-gravity profile in the Baku region, USSR. Downward decrease in API gravity in any one well or field, despite *downdip* increase in any one reservoir sandstone.

9.4 Effects of water circulation on the hydrocarbons

We have observed that an oil or gas accumulation must have displaced water that previously occupied the pore space. Thus all pooled oil and gas have been in active contact with formation waters. If the waters still in the aquifers are in motion and not stagnant, the oil and gas have continued to be in active contact with waters since becoming pooled. The waters bring about major modifications in the nature of the hydrocarbons through three processes: differential solution, oxidation (if the waters are oxygenated), and bacterial attack.

9.4.1 Solution

The variety of components of a hydrocarbon accumulation have very different relative solubilities, though in absolute terms all of their solubilities are very low.

(a) The nonhydrocarbons or S–N–O compounds are the most soluble and the most easily adsorbed.
(b) The aromatic components are much more soluble than the aliphatics of equivalent carbon number; benzene, for example, is 185 times more soluble than hexane (Fig.9.7). This is especially the case in brines at atmospheric pressure, though the solubilities are still less than 0.2 percent. Because aromatics are more polar than paraffins (they have a greater attraction for water), they are more difficult to remove from the clay–mineral surfaces in the source rock. The oil in the reservoir rock is therefore less aromatic, more paraffinic or naph-

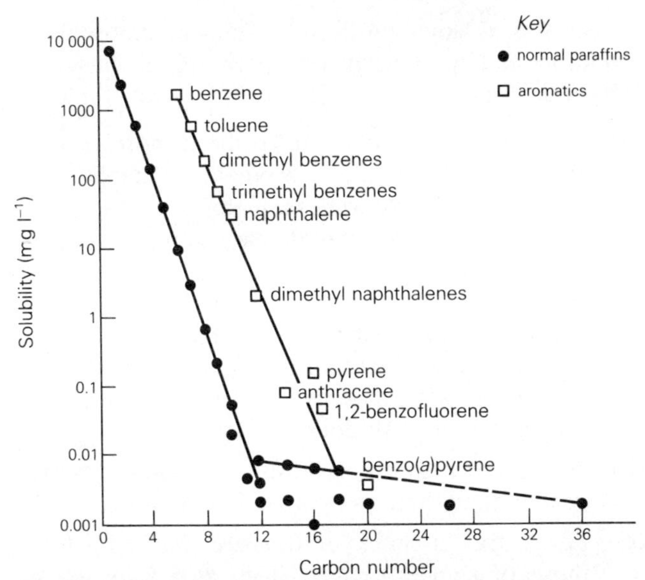

Figure 9.7 Solubilities of normal paraffins and aromatic hydrocarbons in water at 25 °C. (From C. D. McAuliffe, *AAPG Bull.*, 1979.)

thenic, than the oil that can be extracted from the corresponding source rock.

(c) Water solubilities of aliphatic hydrocarbons are chiefly dependent on their carbon numbers (molecular weights). Within any one series, solubility decreases markedly with increasing carbon number. The addition of one carbon atom decreases the solubility of paraffins by 70–75 percent (and of aromatics also). When the carbon number in *n*-paraffins exceeds 12, further increase in molecular weight brings little change in the solubility. Larger molecules are not really dissolved at all, but form aggregates with the water. Naphthenes are more soluble than paraffins of equivalent carbon number. At 25 °C, for example, cyclopentane is four times as soluble in fresh water as *n*-pentane, and cyclohexane six times as soluble as *n*-hexane.

(d) Even the lightest paraffins are only sparingly soluble. For *n*-octane and heavier, solubilities are less than 1 p.p.m. (10^{-5} percent); solubility at the C_{12} level is less than 0.01 p.p.m. (Fig.9.7). Most crude oils contain heavy molecules having these extremely low solubilities.

(e) Solubilities are increased exponentially by increasing temperature, especially above 150 °C; they may reach 10 000 p.p.m. or more (Fig.9.8). According to Leigh Price, crude oil solubilities in water at 300–350 °C range from 5 to 15 percent by weight; methane solubility becomes 140–175 m^3 per cubic meter of water at those temperatures and high pressures.

(f) On the opposite hand, solubilities are decreased by salinity of the water, and also by its increasing gas saturation. The presence of liquid hydrocarbons in the water increases the solubility of gases.

On the basis of thousands of water analyses by Russian geochemists, the average concentration of organic substances in ground waters is about 50 mg l^{-1}. Much the greater part consists of volatile nonhydrocarbons such as fatty acids, ethers, and low molecular weight alcohols. In nonpetroliferous basins, nitrogen is dominant. In petroliferous basins, the principal hydrocarbons in solution in ground waters are methane and ethylene, with ethane and propane in smaller proportions. The concentrations increase basinward, and are naturally highest in contact with oil or gas deposits, though there are many exceptions. According to the same Russian data, concentrations of methane and/or ethylene reach 800 mg l^{-1} in contact with gas-condensate deposits, and up to 370 mg l^{-1} in contact with oil deposits.

Methane is so ubiquitous in sediments that great quantities of it are held in solution in the ground waters of large artesian basins with thick aquifers. The greatest nonassociated gas deposits known are those in the northern reaches of the Siberian Basin (Sec. 25.5.7). According to V. N. Kortsenshteyn and other Russians, even these gigantic accumulations are secondary after solution gas. The gas content of the ground water is

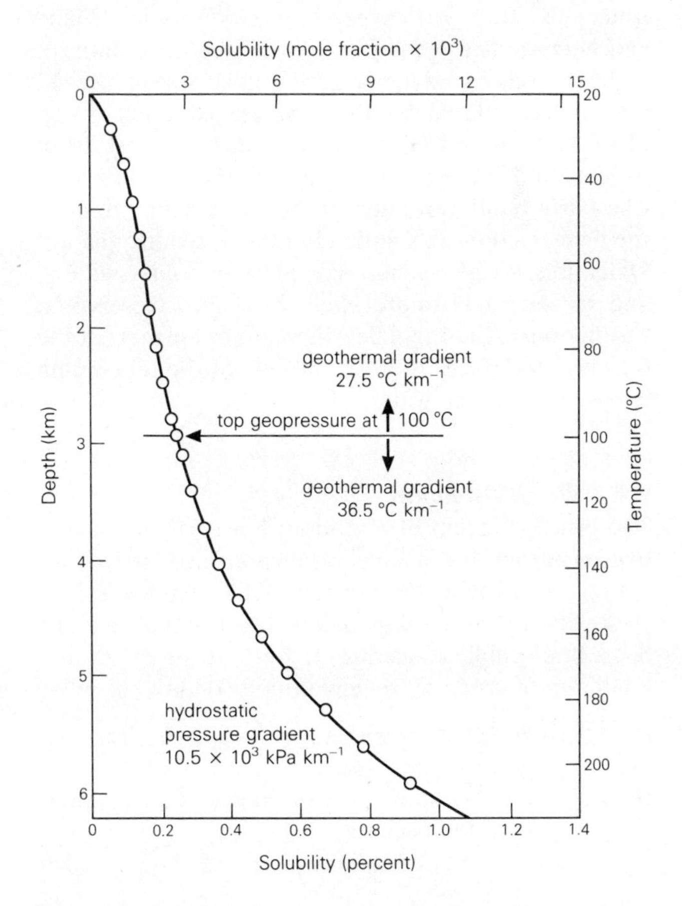

Figure 9.8 Solubility of methane in water at elevated temperatures and pressures. (After L. C. Bonham, *AAPG Bull.*, 1978.)

1.5–2.5 l l^{-1}; most of it is methane, identical to the methane in the pooled accumulations. Yet formation pressures exceed saturation pressures; the waters are undersaturated with gas, but they were earlier saturated and freed the gas to accumulate in deposits during the Pleistocene glaciation. Interstitial waters are of very low salinity and almost free of sulfate; they are of Sulin's type *d* (with the Cl ion) with a contribution of type *b* (with the HCO_{-3} ion). Temperatures along the gas/water interfaces are still very low, less than 38 °C.

Despite the huge volumes of gas freed from solution to accumulate in pools, much more still remains in solution. Dissolved gas volumes are 10 to 100 times those in the pools; the average for all Russian petroliferous basins has been estimated at 14 times the amount in pools. Ground waters are normally much undersaturated with gas, and the degree of undersaturation increases with depth; but if ratios of this order are extrapolated over the sedimentary areas of the earth the total gas potential would be of the order of 10^{10} km^3 or much more.

9.4.2 Oxidation

Apart from their capacity as solvents, surface recharge waters also transport oxygen into the reservoirs. Many geochemists and geologists doubt the oxidation of hydrocarbons by oxygen-carrying fresh waters. A. Y. Gurevich calculated that the complete oxidation of 1 kg of oil to CO_2 and H_2O would require about 3 kg of oxygen, and hence a vast volume of oxygenated water. The fairly rapid movement of hot waters may remove the light fractions in solution but fail to oxidize the oil. Oxidation would be more complete in colder waters, and so should be more likely in shallow reservoirs. Furthermore, the rapid development of tar mats at the oil/water interfaces protects the rest of the oil column from attack by the water.

9.4.3 Biodegradation

The principal agent of oxidation, however, is not free oxygen but aerobic (oxidizing) bacteria, also carried into the reservoirs by meteoric waters. The complex of processes known as *biodegradation* has become familiar because of public concern over waste disposal. Biodegradation of crude oil requires three conditions:

(a) circulating waters with about 8 mg l^{-1} of dissolved oxygen, to support the bacteria;
(b) formation temperatures of 20–50 °C, to permit growth of the bacteria;
(c) hydrocarbons essentially free of H_2S, which poisons aerobes.

Bacterial oxidation is known to occur; it is selective in its attack and is highly effective, under atmospheric conditions and up to survival temperatures of 60–80 °C (150–180 °F):

(a) Paraffinic compounds are oxidized more readily than the corresponding aromatic or naphthenic compounds. Waxy crudes are rapidly converted to heavier, nonwaxy, naphthenic–aromatic crudes, especially in faulted, young strata. The process is well demonstrated in the Niger delta fields, at Barrow Island in Australia, and elsewhere. It may account for the fact that the heavy, true paraffin ends of paraffin-base oils do not accumulate at seepages as asphalt does.
(b) Branched-chain isoparaffins are oxidized more readily than their straight-chain homologs.
(c) Long, straight-chain paraffins are attacked before those of shorter chain, which also have lower boiling points. This leads to the selective loss of *n*-paraffins above C_5, readily revealed by gas chromatograms. Though they are less *soluble* than shorter-chain compounds, the longer-chain *n*-paraffins are very susceptible to bacterial action.
(d) The attack brings about an increase in the content of S–N–O compounds, by known catalytic processes, and of asphalt. Asphaltenes are generated during biodegradation by the oxidation of hydrocarbons and the incorporation of sulfur.
(e) Biodegradation and/or water washing causes an increase of several times in the *optical rotation* of the plane of polarization, to the right or the left. The increase is due partly to the preferential removal of the *n*-paraffins and the aromatics, which are optically inactive, and the consequent enrichment of the four- and five-ring cycloparaffins (naphthenes), which are largely responsible for the optical activity of petroleums. The increased rotation is also partly due to the formation of oxygen-bearing organic solutions, especially from the steranes and terpanes (see Sec. 5.1).

The solutions may be likened to the strongly rotatory sugars: glucose or grape sugar, $C_6H_{12}O_6$, or the tartaric acid of grape wines, CH(OH)COOH. Optical rotatory power is characteristic of the so-called enantiomorphous substances, crystallizing in two habits related as the left and right hands, such as the trapezohedral-rhombohedral cinnabar or plagiohedral-isometric cuprite. The most familiar example among well-crystalline minerals is trapezohedral quartz.

Optical rotations in oils are gradually destroyed with age, and so are highest in young oils. They are commonly measurable in fractions of a degree per gram of saturated hydrocarbon in either the whole oil or the distillate fraction.

Laboratory experiments indicate that the most obvious consequence of water attack on oils – the loss of *n*-paraffins – is *reversible*. The famous floating asphalt blocks in the Dead Sea may present an example of this in nature. The asphalt is agreed to be a degraded material, yet it contains abundant *n*-paraffins. Israeli geochemists A. Bein and O. Amit suggested, in 1980, that the original oil suffered early leakage and degradation, but was re-buried during rapid subsidence of the Dead Sea graben. Re-heating at depth revived the paraffin content, and the oil was subsequently leaked to the surface a second time during renewed faulting along the rift.

9.5 Immature and senile oils

We thus see that a heavy oil, strongly naphthenic or asphaltic, may be immature and lacking the natural cracking to convert it into a light oil; but it is much more likely to be a senile, degraded oil. Crude oils from very young to mature are formed essentially by the process of natural cracking which, if not interrupted, will increase the oil's API value and its contents of gasoline and of aromatic components. Aging and senile crudes are formed by aqueous and bacterial degradation, which reduces the API value and the percentage of gasoline and benzene, and removes the porphyrins. Degraded crudes are therefore heavy, sulfurous, and high in oxygenated compounds.

We may illustrate the degradation process by two examples from contrasting geologic environments: the undeformed Mississippian carbonate reservoirs of the Williston Basin in Saskatchewan and the strongly deformed Tertiary sandstones of the Californian basins.

The Williston Basin oils were traced by Nigel Bailey and his colleagues along a trend of increasing biodegradation and water washing (from southeast to northwest). As the crudes became heavier and acquired higher S–N–O contents, the quantitative changes in properties and compositions were found to be as given in Table 9.1. The only surprise is the increase in the aromatic components. Clearly the principal process at work was biodegradation and not simple water washing; the lighter members of the aromatic series were preferentially removed.

Crudes from three fields in California's San Joaquin Valley were studied by Keith Chave. All the crudes were from Miocene or Pliocene reservoir sandstones; those from the west side are in the deep basin, those farther east are close to the outcrops on the backslope of the Sierra Nevada (see Table 9.2).

Gases can be degraded exactly like oils. They lose their gasoline and ethane contents, and their percentage of nitrogen increases, probably from the air dissolved in invading meteoric waters. Gases associated with fresh formation waters therefore have these characteristics.

Table 9.1 Quantitative changes in Williston Basin crude oils

	Least altered oil (SE)	Most altered oil (NW)
API gravity	37.6	15.2
percentage of sulfur (whole crude)	1.13	2.99
percentage of paraffins in C_{15+} saturate fraction	46.5	6.9
percentage of saturates in C_{15+} oil fraction	47.1	19.1
percentage of aromatics in C_{15+} oil fraction	37.5	43.3
percentage of asphaltenes in C_{15+} oil fraction	5.4	16.2

Bacterial and water degradation may theoretically take over at any stage of maturation of the oil. An oil now heavy and degraded may never have had the chance to become a light, sweet oil before degradation set in. However, the process must take place essentially during the secondary migration of the oil (see Sec. 15.8). Once the oil is pooled in its final reservoir, formation water cannot flow upwards through it. Water from above can flow downwards through it, but meteoric waters entering from the outcrop pass down the dip of the aquifers rather than downwards across them.

The end result of degradation is the formation, in highly petroliferous regions, of wide bodies of heavy, residual oil, either at the surface or closely below it. These *tar mats* or *tar belts* mark the edges of basins where the reservoir rocks reach or approach the surface. Notable examples are the Hit tar mat on the west side of the Persian Gulf Basin in Iraq, in a Tertiary limestone;

Table 9.2 Quantitative changes in crude oils from three fields in the San Joaquin Valley

	West (South Coles Levee)	(Fruitvale)	East (Kern River)
API gravity	43.3	21.0	14.1
percentage gasoline	52.1	5.0	0.9
gasoline gravity	59.7	42.6	44.1
percentage kerosene	6.0	6.6	1.0
benzene content	high	absent	absent
residuum gravity	15.0	13.4	11.4
saponification number†	0.1	0.8	3.6
associated water	saline	fresh	fresh
associated gas	condensate	dry	none

† The saponification number (Sec. 7.2) is the amount of KOH required to convert 1 g of the oil to a soap.

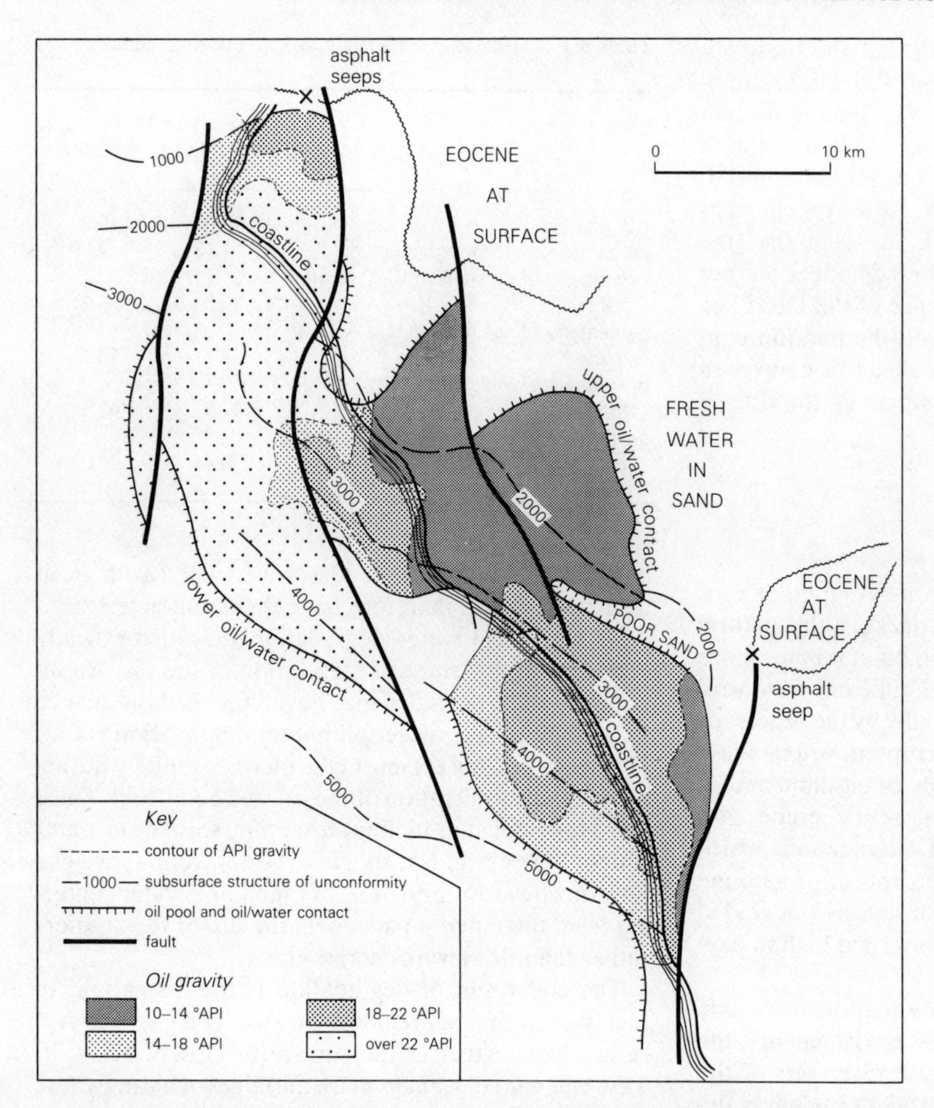

Figure 9.9 Oil gravities in Lagunillas sector of Bolivar Coastal field, Maracaibo Basin, Venezuela. Compare with structural cross section (Fig. 16.2). Water overlies heavy oil on eastern (updip) side; heavy oil overlies lighter oil, which lies on bottom water. Oil/water contact on southwest drops more than 300 m in 20 km, corresponding to post-Miocene regional tilting. (From P. A. Dickey and J. M. Hunt, *AAPG Memoir* 16, 1972; after F. E. Rubio, 1959.)

the inspissated cover of the Bolivar Coastal field on the east side of Lake Maracaibo in Venezuela (Fig. 9.9); the Bermudez or Guanoco pitch body and the Orinoco "tar belt," on the northern and southern margins of the eastern Venezuelan Basin (Fig. 10.16); and the Santa Maria tar mat in California. Where tar belts occur on both sides of a single basin, the two belts may have quite distinct oil types.

10 The variety of petroleum hydrocarbons

10.1 The variety we have to consider

We now know what petroleum consists of, and how it is formed. We also know that petroleum in the Earth's rocks is engaged in a fight for space with much more plentiful water, and that both the petroleum and the water acquire unmistakable signs of the competition.

There are two contrasting perspectives on the variety of petroleum hydrocarbons discovered in the Earth. The total number of natural hydrocarbon compounds, though unknown, is undoubtedly enormous and perhaps beyond computation. Remembering that the shift of a single atomic bond may be the only difference between two isomers (Sec. 5.1.4) having the same molecular composition but different properties, imagine the number of isomers possible in hydrocarbons with 40 carbon atoms. Several hundred hydrocarbon compounds are detectable in *individual* crude oils; more than 200 sulfur compounds are known in crudes.

From the opposite perspective, however, the hydrocarbon compounds commonly predominating in crude oils lie within the range from C_5 to C_{20}, and the dominant constituents number no more than about 50. Of these, about half are normal or isoparaffins and about one-quarter each are cycloparaffins and aromatics.

Petroleum types of course represent a continuous spectrum from pure gaseous methane to complex solid bitumens and asphaltenes. The spectrum is conventionally distributed among a relatively small number of petroleum types with names familiar to all petroleum geologists and geochemists. In this chapter we consider the range of natural hydrocarbons commonly found in rocks and the variety of petroleum types to which they give rise, from the earliest-formed to the products of their maturation or degradation.

10.2 Nature of the first-formed oil

The simplest hydrocarbon is methane, CH_4. Its common occurrence as marsh gas illustrates the ease with which it can be produced directly from organic matter. The greater part of it is formed by the reduction of CO_2 by hydrogen, which in turn is produced by the anaerobic oxidation of organic matter. The principal agents are bacteria, and all bacteria that can produce methane can do so from CO_2:

$$CO_2 + 4H_2 \rightarrow CH_4 + 2H_2O$$

Thus the first requisite step is the generation of *hydrogen*.

Considerable quantities of methane are also produced from the fatty acids, by *decarboxylation* (breaking down of the $-COOH$ group):

$$CH_3COOH + H_2O \rightarrow CH_4 + H_2CO_3$$

The generation of higher, liquid hydrocarbons is not so simple. Can they in fact be generated directly from organic matter (OM) by any process at all, or must some preliminary transformations to intermediate products take place? Minute accumulations of paraffins and aromatics are well known in modern sediments that have undergone no burial or raising of temperature. Paul Smith, in 1954, therefore proposed the spontaneous origin of oil as a fluid, directly from the OM in the sediments. Exploration of the sea floors has revealed volumes about the nature and composition of these very young, unaltered sediments. Cores from the Santa Barbara Basin off California contain indigenous hydrocarbons having molecular composition more like that of crude oil than is the composition of the hydrocarbon content of associated clays; but it is not like that of the oil in the underlying sandstone reservoirs.

The *early* generation of petroleum hydrocarbons is insisted upon by many geologists and geochemists, but their *direct* generation from a sedimentation source is widely discredited. If oil is generated directly from newly deposited OM, and is fluid and mobile from the outset, what is to prevent its quick migration and loss? The word *paraffin* means "having little affinity;" early users of paraffins discovered that they are indifferent to common reagents such as acids and oxidizing agents. In the inescapable presence of water, they also have no affinity for the clay minerals of the source sediments, and aromatic hydrocarbons have not much more. The initial, immature petroleum therefore stays in its enclosing OM until its volume exceeds both the sorption capacity of the OM and the reluctance of the clay-rock host to accept it. It must then be expelled as a separate phase, and would inevitably be lost if the surrounding rocks still possessed their original permeabilities. Loss of permeability from freshly deposited sediments involves processes such as compaction and lithification which are far from instantaneous and may be long delayed. If the amount of "primitive" petroleum generated never exceeds the sorption capacity of the clay-rock host, it is never expelled and even an ancient "source" rock still possesses a high hydrocarbon content (unless it has been metamorphosed).

How could the amount of methane associated with petroleum be deposited with the original material? It has to be *generated* by decomposition of organic precursors. Even if these decomposition reactions are as simple as those of decarboxylation described above, the decarboxylation must be of thermal stimulus and most probably affects anions of the aliphatic acids. These anions (acetates, propionates, etc.) are themselves formed by the thermocatalytic degradation of some OM in the source sediment; the anions are highly soluble, and are made available in solution during dehydration of the clays.

The original bitumen content of OM – especially of the plankton – is low in hydrocarbons heavier than methane. Early geochemical studies by Meinschein and others demonstrated that natural hydrocarbons in otherwise nonpetroliferous recent, marine sediments – the so-called *sedimentary hydrocarbons* – may represent an intermediate stage in the formation of petroleum but they do not constitute petroleum. Sedimentary hydrocarbons include similar *heavy* hydrocarbons, C_{15} and upwards; but compounds in the C_3 to C_{14} range, which constitute up to 50 percent in crude oils, are typically absent from sediments. The organic content of recent sediments is also less negative in its carbon isotope composition (i.e. it is isotopically heavier) than is typical petroleum (see Sec. 6.3.1).

An especially marked difference between sedimentary hydrocarbons and petroleum hydrocarbons is found in their *carbon preference indices* (CPI). The CPI expresses the ratio between odd- and even-numbered carbon atoms in the molecules constituting the upper range (C_{24} to C_{34}) of the normal paraffins as determined by gas chromatography. An approximate expression of the CPI is given by the ratio

$$2nC_{29} : (nC_{28} + nC_{30})$$

If *n*-paraffins having even numbers of C atoms predominate, CPI is less than unity; if odd numbers predominate, CPI exceeds unity; if odd and even numbers are essentially equal, CPI $\cong$ 1 (as it is in marine organisms). In recent muds, there is a strong predominance of odd numbers; CPI values for very young sediments are therefore high (from 2.5 to more than 5.0). In sedimentary rocks in general the values are commonly more than 2.0. Lipid derivatives of the fatty acids, on the other hand, have a strong even predominance (values for plant waxes are as low as 0.2), but decarboxylation during maturation results in odd predominance for the resulting paraffins. Further maturation then reduces the CPI towards unity by random cracking. CPI values for mature crude oils fall in the range 0.92–1.12. CPI values below 0.92 probably represent thermal degradation of the organic content of the sediment.

Light aromatic hydrocarbons, such as benzene or the double-ring compound naphthalene, $C_{10}H_8$, are much more abundant in oils than in any living matter or in sediments. Older source sediments retain high concentrations of these compounds in proportion to their total organic carbon (TOC) contents. The light aromatics must therefore have formed subsequent to any sedimentary hydrocarbons, from unknown precursors. An increase in aromatic content, and approach towards the constitution of crude oil, requires the retention of the fatty constituents of the original bitumens and their decarboxylation under reducing conditions. Under oxidizing conditions, on the other hand, the bitumens are enriched in acid components, increasing the humic content and decreasing the hydrocarbon content.

10.3 The variety of oil types

The geographic and geologic distribution of known oilfields about the year 1910 pointed to a seemingly elementary principle of oil occurrence. There were two kinds of "rock oil." One kind was light in both color and gravity, and rich in paraffinic constituents. The other kind was black or nearly black, much heavier, poor in paraffinic constituents and high in "asphalt."

The most important producing areas in the Americas at that time were the eastern interior states of the USA, California, and Mexico. All eastern American oils were of the first kind; their early discovery led to the use of "Pennsylvania crude" as the ideal standard oil for refining purposes (see Sec. 5.1.1). Such crudes were described as "sweet" crudes. Virtually all Californian and Mexican crudes were of the second kind; they were described as "sour" crudes because of their sulfur contents (see Sec. 5.2.1).

In nearly all oil-producing areas outside the Americas, the same dichotomy was apparent. The oils of Baku, for example, were black oils; those of the Dutch and British East Indies were sweet oils. The great regions of mixed-base crudes – the Middle East, the US Mid-Continent, eastern Venezuela, western Canada, the North Sea, western Siberia – had not yet become important producers.

The elementary principle referred to in the first paragraph of this section was that there were two fundamentally different kinds of natural petroleum. Another principle was becoming accepted about the same time: that oil is of organic and not inorganic origin (see Ch. 6). As these two principles came to be more closely considered, they led to two diametrically opposed conclusions. Some believed that the two oil types are fundamentally different because they are derived from fundamentally different organic ancestors (animals and plants, perhaps); others concluded that there is only one fundamental or "primitive" type of rock-oil and the other type is somehow derived from it.

If the second hypothesis is correct, which of the two oil types is the "primitive" one? The first widely accepted answer was deduced from the stratigraphic ages of the oils. Nearly all crudes heavier than 20°API are Tertiary or Cretaceous. Hundreds of analyses showed that crudes from the shallowest and youngest horizons (Pliocene, in both the Baku and most of the Californian fields) are heavy, unstable, naphthenic, asphaltic, resinous, and nonwaxy; they contain few or no light liquid fractions, their contents of natural gasoline being especially low. They are commonly associated with dry gases or no gases at all, are low in hydrogen, and high in S–N–O compounds and trace metals.

All the oils of the eastern USA, in contrast, occurred in Paleozoic reservoir rocks. They might in general have been more deeply buried than the young oils (though they are of very shallow occurrence now), and they had had a much longer time to undergo changes in their nature. Oils came to be regarded as something like wines or cheeses; they mature with age. The *maturation hypothesis* was most clearly formulated by Donald Barton in 1934. Surprisingly, Barton's evidence came not from the Paleozoic oils but from those of the US Gulf Coast, at that time one of very few regions in the world producing oil over considerable ranges of both stratigraphy and depth. Adherents of the maturation hypothesis included some of the most influential American petroleum geologists (notably Lewis Weeks); the great majority of Russian geologists and geochemists; and (later) their Chinese counterparts, whose experience was with lacustrine source sediments in post-orogenic basins (see Sec. 17.7.9).

Maturation in Barton's sense means that with increasing temperatures and pressures through increasing depth of burial, initially heavy oil rich in polar compounds is changed to a less heavy oil rich in resins, thence to a naphthenic oil of moderate gravity and finally to lighter and less cyclic, paraffinic oil at about 150 °C. This light oil is low in nonhydrocarbon constituents such as sulfur and resins. With still further maturation, the yield becomes principally wet gas, and the final end-product is probably pure methane.

The chief evidence that the heavy, asphaltic oils are "primitive," and that the lighter, paraffinic oils and wet gases are derived from them by "maturation," thus came from thermodynamic considerations. Transformations lead from large molecules of high free energy level to small molecules of lower free energy, by twin processes:

(a) Cracking, producing smaller molecules of increasing volatility, mobility, and hydrogen content; the ultimate product is *methane*.
(b) Condensation, producing carbonaceous residue of decreasing hydrogen content; the ultimate product should be *graphite*.

Barton mentioned David White's *carbon ratio theory* (see Sec. 7.2 & Fig. 7.20) without invoking it, but its echo is clear in the maturation hypothesis. In some senses, modern maturation theory in turn echoes Barton's, but with the important difference that the maturation affects the kerogen in the source rock and not the oil once it is in the reservoir rock.

Weeks enlarged greatly on Barton's hypothesis, from the foundation of a vastly greater variety of experience. In Weeks' view, lighter oils develop later, and migrate and accumulate later, than heavy, more cyclic oils. The latter were thought to be generated very early in the sedimentation process, in shallow shelf areas. The shales of such areas are therefore likely still to be "wet," waxy, and rich in hydrocarbons; those in deeper trough areas are likely to be brittle and much "drier" with respect to both water and oil.

The evidence from geologic occurrence of oils, however, was quite different. Asphaltic oils occur typically in shallow structures, and near to either unconformities or faults. In many basins, oils become progressively heavier towards the margins, especially towards the forelands where the reservoir rocks crop out (Figs 10.16,

13.75 & 17.29). The lowest API values are found in the shallowest reservoirs. Nearly all natural seepages are asphaltic. "Tars" and "asphalts" in considerable variety are far commoner than ozocerite (see Sec. 10.6). Paraffinic crudes, in contrast, may occur in traps at any depth, but they are well sealed, their reservoir rocks have no nearby outcrops, and they are commonly little faulted or unfaulted. Most oils in little-deformed cratonic basins are light oils; most oils in young, actively diastrophic basins are heavy, asphaltic oils; occurrences of both environments are described in Chapter 17.

Karl Krejci-Graf, in 1929 and 1932, and W. F. Seyer, in 1933, made early *chemical* proposals that petroleums are likely to be originally paraffinic, and that more cyclic, aromatic, and complex molecules develop later, during or after primary migration, from reproducible transformations of *n*-paraffins. The principal process causing these transformations is *microbial alteration* of originally light oils.

Chemical studies pioneered by G. T. Philippi in 1977 showed that paraffinic crudes in young reservoir rocks have compositions closely similar to or identical with those of the hydrocarbons extracted from the corresponding source rocks; they are therefore the "primary" oils. The heavier naphthenic crudes on the other hand revealed compositions quite different from those of the hydrocarbon mixtures in either mature or immature shales associated with them. The carbon preference indices (see Sec. 10.2) for both paraffins and naphthenes are closely correlatable in both crude types. As these indices are strongly depth- and temperature-dependent, it seems to follow that the paraffinic precursors of the naphthenic oil were generated in, and expelled from, the same mature source beds, at the same general depths, as the crudes which have remained paraffinic. The medium to heavy-gravity naphthenic crudes into which some of the paraffinic crudes were transformed were never found above a certain abrupt cut-off temperature in any of Philippi's basins; the temperature is commonly about 66 °C, presumably that above which the microbial transformations are ineffective.

Hydrocarbon-ingesting microbes flourish in the presence of both water and oxygen. These two agents have their own further effects of solution and oxidation (see Sec. 9.4). Whereas *microbial alteration* preferentially removes the *n*-paraffins, *solution* selectively removes the aromatics. Liquid paraffin hydrocarbons are incapable of large-scale solution–migration without elevated temperatures and are therefore likely to be restricted to deeper parts of the basins and absent from the shallow parts. *Oxidation* produces a heavy resinous oil, low in paraffins, rich in cyclic components, with high percentages of sulfur and trace metals. Oxidation is likely to be the least important of the three processes (see Sec. 9.4.2), but, in the event of its dominance, the end-products of its attack on hydrocarbons would be CO_2 and a solid bitumen residue.

There are numerous productive basins from which both light, pristine oils and heavy, naphthenic or asphaltic oils are produced almost side by side, the second oil type being considered to be the degraded derivative of the first – Alberta, eastern Venezuela, Nigeria, Indonesia among them (see Ch. 17). There are also many basins displaying intimate association between light oils (or even oil shales) and solid bitumens – in the Uinta Basin of the western USA, in Angola, Argentina and Iraq very notably.

10.4 The argument redirected

Both geologists and geochemists have recognized that this lengthy dispute was misdirected. Light paraffinic oils associated with wet gas undoubtedly represent a process of *maturation*, but it is maturation of the kerogen in the source sediment, not maturation of an older, heavier, asphaltic crude oil. It is improbable that any crude oil was originally asphaltic. Asphaltic crudes are formed by the degradation (commonly the biodegradation) of originally paraffinic, naphthenic, or aromatic crudes. Paraffinic crudes are primary crudes derived from organic materials that happen to yield paraffinic products. The variety of crude oil and natural gas types that we find in nature is not due to different stages of a descent from a common ancestral type, but to a mosaic of other causes:

(a) differences in the original organic material;
(b) different environments of its deposition and transformation;
(c) physicochemical conditions at the time of hydrocarbon generation;
(d) physicochemical conditions since generation and migration.

10.4.1 Differences in original organic material

Some who early believed that the two different crude types were different in kind and not merely in degree suggested that paraffinic oils were derived from *gelatinous plants*, asphalt-base oils from *animal remains*. The obvious source of paraffin waxes is the leaves of tropical plants, in which the cuticles and resins survive after the woody tissues have decayed.

Asphaltic oils, more complex than paraffinic oils, presumably originated in organic material high in protein, and so probably of animal rather than plant origin. Protein molecules are larger ($C_{50\pm}$) and more complex than those of fats ($C_{17\pm}$) or carbohydrates ($C_{6\pm}$). Asphaltic oils also contain an array of extraneous elements (nitrogen, sulfur, metals) most common in heavy, proteinaceous molecules. The source sediments yielding asphaltic oils are therefore also likely to be *phosphatic* (see Sec. 8.1); kerogens from phosphorus-rich source sediments are high in both nitrogen and sulfur and yield low levels of *n*-paraffins.

Despite these plausible proposals, doubt has been cast on their validity by proponents of both sides of the old argument. The nature of the parent organic material cannot be decisive. What matter are post-burial transformations of all the original OM by a variety of bacterial types. There are too many petroleum types to be accounted for by different organic precursors.

10.4.2 Differences in the environment of origin

Oils high in *aromatic components* may be of continental origin, derived from ligno-humic substances. Plausible precursors are the phenols of land plants (see Sec. 5.2.3). If this postulate were sound, aromatics should be commonest in oils of nearshore or lacustrine source, but such oils are commonly low in aromatics. Those richest in aromatics are nearly all young (Tertiary). Many of them are also sulfurous; but sulfate is notably absent from the oil-bearing successions and the sulfur is acknowledged to be secondary.

Hollis Hedberg, taking examples from all productive regions of the world, showed that *high-wax oils* occur in sandstone–shale successions of nonmarine (lacustrine) or marginally marine (brackish) environment. None of them is older than Devonian, when land plants first became abundant. This association indicates derivation of high-wax oils from kerogen types II and III, rich in dispersed exinite (spores, cuticles, etc.) and woody material (see Sec. 7.3). It also suggests a correlation between high paraffin wax content in crudes and high aromaticity. Crudes having both these properties are indeed familiar – in Indonesia and North Africa particularly. But the association between high aromatics and high waxes cannot be extended to that between high aromatics and high sulfur, referred to above. High-wax oils are low in sulfur, and are not common within entire basins having many sulfur-rich crudes (many of which occur in carbonate and not clastic reservoirs).

10.4.3 Physicochemical conditions during generation

The *redox potential* (E_h) in the source sediment is likely to exert some control on oil type. Aerated muds, with positive E_h, should yield paraffinic crudes with few or no aromatics or asphaltenes. Source muds only weakly oxygenated should yield naphthenic oils with moderate asphaltenes. Euxinic muds, with free H_2S and other sulfides, should produce asphalt-base oils rich in aromatics and asphaltenes but lacking paraffin waxes. In 1954, John Hunt suggested that this influence of the redox potential is illustrated by the upward progression through the nonmarine Cenozoic strata of the Uinta Basin in Utah.

Determination of the redox potential operative during the deposition of sedimentary rocks now lithified (and possibly altered) is necessarily uncertain, but the present oxidation state of a potential source rock may be revealed by its color, the condition of any iron minerals such as chlorite, the presence or absence of bottom-dwelling organisms, or the sedimentary evidence of water agitation.

The *temperature* reached by the source sediment during generation has been thought to be a factor in the type of oil produced. Moderate temperature and slow change were thought to lead to light, paraffinic oils, especially if they were accompanied by high pressures (permitting retention of the volatile constituents of the oil). High temperature and rapid change were conversely thought to yield asphalt-base oils, with light constituents expelled.

These factors are unlikely to be effective. Temperature is the principal control of the *phase state* of the hydrocarbons produced (Fig.10.1), but not of their chemical *type*. Highly paraffinic oils occur in basins of all ages and of such varied tectonic styles that their geothermal gradients could not all have been low. The waxy crude from the supergiant Minas field in Sumatra occurs

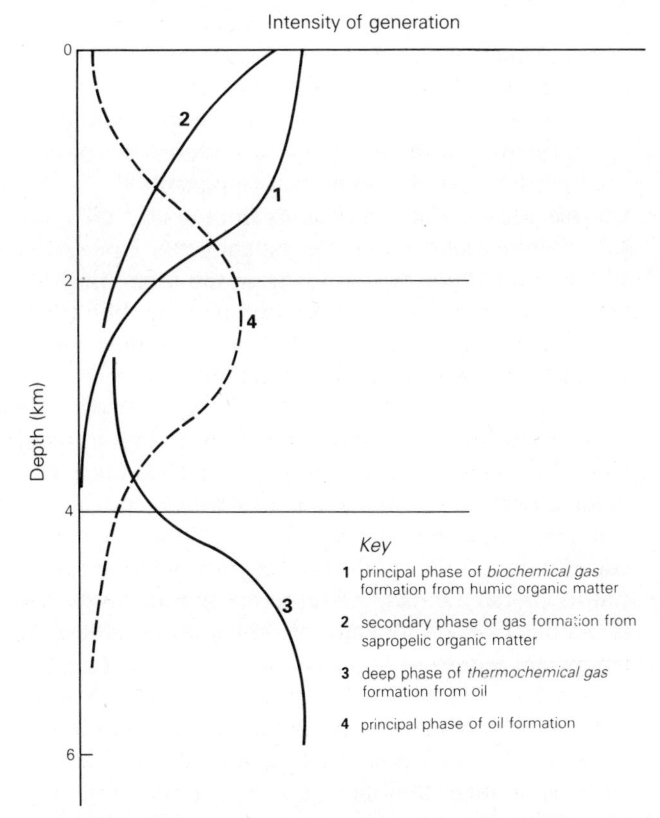

Figure 10.1 The principal phases of hydrocarbon formation, controlled by increasing temperature with increasing depth. (After N. B. Vassoyevich and others.)

in a basin having one of the highest geothermal gradients known (see Sec. 17.8), yet which did not exist before the Miocene, and the oil occurs at depths as shallow as 600–750 m. Whatever the temperature may have been at the time of generation, it is not likely to have been "low" and the generative process cannot have been "slow." Temperature and pressure are not important controls in the formation of asphalt-base oils either; the more asphaltic the oil, the more likely it is to have been degraded at very modest temperatures.

10.4.4 Physicochemical conditions during migration

In multiple-reservoir deposits, with numerous clay partings which are not seals, the oil permeates through the clays in the direction of the fluid pressure gradient (see Ch. 15). This permeation may bring about a natural fractionation of the oil into a light fraction that can get through and a heavier one that cannot. Adsorption and diffusion processes controlled by clays have a variety of effects on oils:

(a) A decrease in the components most readily adsorbed through the chromatographic effect: asphaltenes, resins, porphyrins, sulfur compounds.
(b) Enrichment in the lower molecular weight hydrocarbons, which have lower adsorbabilities and higher diffusion coefficients. The density decreases, as does the content of high-cyclic compounds. Most intensely adsorbed on to clay particles are the aromatic components.
(c) Segregation of condensates from denser oils, and their enrichment in the components having the highest vapor pressures (again the aromatics).
(d) Enrichment in the heavy isotope of carbon (^{13}C), increase in the GOR, and decrease in the proportion of methane homologs and of CO_2.

Once the oil has entered a porous reservoir rock, *secondary migration* within it (see Sec. 15.8) leads to a holding back of the high molecular weight, polar constituents but a freer passage for the lighter, nonpolar constituents. A plot of the former (asphaltenes, resins, sulfur, metals) across the reservoir should reveal the directions both of the entry of the oil into it and of its movement within it. Similarly, the gas content and its percentage of H_2S should decrease, and those of nitrogen and helium increase, going farther from the source. Plots of these variations have permitted valuable conclusions on migration directions for a number of large basins (those of western Canada and Ural–Volga notably).

If the secondary migration takes place in water-solution, on the other hand, the first components to be segregated should be the least soluble (the *n*-paraffins). Those carried farthest from the source should be the aromatics and cycloparaffins. Oils lacking *n*-paraffins may then owe this characteristic to having left them behind during the migration process because they were not soluble enough to be transported. Changes during secondary migration are considered further in Chapter 15.

10.4.5 Physicochemical conditions since hydrocarbon accumulation

The variety of physicochemical conditions to which fluid oil and gas can be subjected after their accumulation all tend ultimately in the same direction – towards the destruction of the hydrocarbons. The conditions can be thought of as involving two opposing sets of processes: processes of *maturation*, essentially through increasing temperature, and processes of *degradation*, essentially by bacteria, water, and oxygen. These processes were described in Chapters 7 and 9. The influence of temperature is further considered here only in connection with the occurrence of natural gas as the dominant hydrocarbon, rather than oil.

The successive phases of hydrocarbon formation, made familiar by the work of Bernard Tissot, N. B. Vassoyevich, and many others, and described in Chapter 7, reflect a progression through increasing temperature, depth of burial, and time. Emphasizing the temperature factor at the expense of the other two, Nigel Bailey, Calvin Evans, and other geochemists (especially many Russians) have drawn attention to the significance of *thermochemical methane* (Fig. 10.1).

Natural gas is more easily generated than is oil, requiring a lesser degree of transformation of the parent organic matter; it is therefore generated from a wider variety of source beds, and is inherently more widespread in its occurrence than is oil. Much shallow dry gas is simply *biochemical*, generated by an immature source sediment and enriched in the light isotope of carbon (^{12}C). It is the dominant hydrocarbon at depths shallower than about 1000 m, as in many deep-sea sediments studied under the JOIDES program. Medium-gravity oil is common at shallow depths, but it is likely to have migrated upwards into them; it is not indigenous to them. With increasing temperature, the mature facies yields indigenous medium-to-light oil and condensate. Excessive heating, due to high geothermal gradients, deep burial, or igneous activity, creates a metamorphic facies in which petroleum liquids are destroyed, leaving dry, *thermochemical methane* with $\delta\ ^{13}C$ values much less negative than those in biochemical methane (−35 to −40‰ rather than −60‰). Thermochemical methane, accompanied by acid gases such as H_2S and CO_2 formed

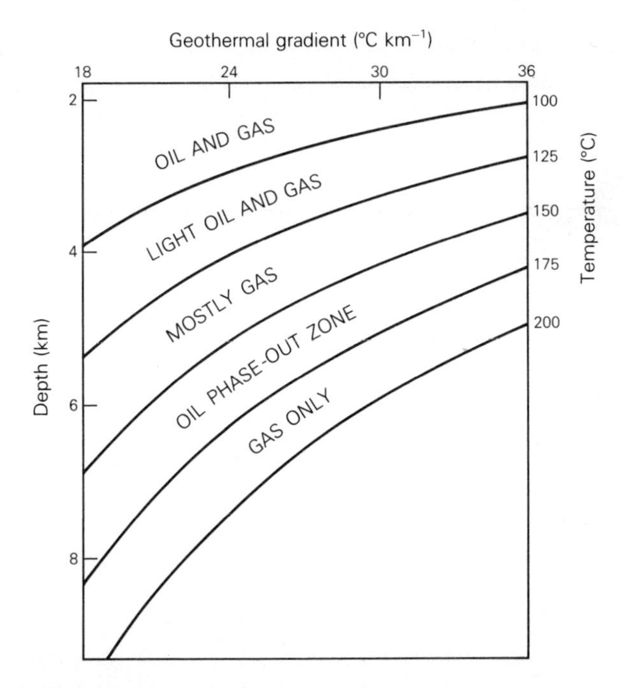

Figure 10.2 Earth temperatures and occurrence of oil and gas, based on the example of the Permian Basin.

by the reaction of elemental sulfur with the methane at high temperatures, is essentially the sole fluid below depths of about 5000 m in nearly all basins; liquid oil not "matured" to dry gas has been degraded to reservoir bitumens (see Sec. 10.6).

This downward change from oil to gas occurrence is still found in basins having had complex geologic histories. Fig.10.2 illustrates the general depth–temperature distribution of oil and gas in the Permian Basin of West Texas, a Paleozoic basin producing prolific oil and gas over a depth range in excess of 6 km and a stratigraphic section involving three major unconformities.

10.4.6 Proneness to gas rather than oil occurrence

It follows that in all basins having sedimentary fill exceeding 4–5 km in thickness, and in many much shallower than this, there is a general depth (actually an undulating but essentially simple *surface*) below which the reservoirs yield only gas or condensate, not oil. Within any one restricted area, this depth is called the *gas-condensate point*. Its actual depth depends upon the nature of the source sediments; humic OM favors proneness to gas even at moderate depths. It also depends upon the geothermal gradient (Figs 7.10 & 10.2) and upon the detailed geologic history. A long episode of erosional unloading, for example, will result in the gas-condensate point being at a depth shallower than its original depth. Furthermore, the depth varies for different reservoirs within the same basin.

Natural gas predominates among the reservoir fluids in regions of high induration and high carbon ratios (see Sec. 7.2). Methane is very stable; its stability increases with pressure and hence with depth. It is also highly compressible and mobile, so it can pass through smaller pore openings than oil can. The greater the depth (to the limit of porosity), therefore, the more gas can be stored in a given reservoir; at 2 km, for example, a reservoir can hold about 200 times as much methane as it could at sea level. The case is exactly opposite for oil; the greater the depth for any one reservoir, the smaller the pore spaces and the lower their permeability to oil.

V. F. Raaben and others have generalized about gas-condensate depths in basins of various ages and tectonic styles. In very large basins, like those of West Siberia or western Canada, there may be considerable overlap of levels, with gas occurring at much shallower horizons in some parts of the basin than in others. Generalized depth figures are nonetheless surprisingly uniform, with a world average of about 4.5 km (Fig.10.3):

(a) In Paleozoic successions on *old platforms*, gas occurs dominantly below 4 km and exclusively below 6 km. In the Appalachian Basin, for example, gas is almost the sole hydrocarbon below 1.5 km. In the Permian Basin, dry methane is essentially all that is producible below 4.5 km (Fig. 10.2). Gas or gas-condensate dominates at all depths in the Anadarko and Arkoma basins of the southern USA and the Cooper Basin of Australia (see Chs 17 & 25).

(b) On *younger platforms* with much continental sediment (in many cases including coal) gas or gas-condensate dominates below 4 km in both Mesozoic and Upper Paleozoic reservoirs. In the West Siberian Basin, there is very little oil above 1.5 km or below 3 km, and only gas or condensate is found below 4 km. Gas dominates at all levels in the San Juan Basin of New Mexico and the Sverdrup Basin of Arctic Canada.

(c) On *young platforms* having largely or wholly marine sections, oil of late generation may occur at almost any depth down to 5 km, associated with gas-condensate. In the Sirte Basin of Libya, light oil dominates down to 4.25 km. In the US Gulf Coast, gas-condensate is dominant below 2.4 km; at 3 km only 37 percent of known pools contain any oil; at 4.8 km, only about 10 percent; below 5.7 km, essentially none.

(d) In regions affected by *younger folding*, with oil largely restricted to mid-Mesozoic and younger reservoirs, oil may dominate even in the 4–6 km range, especially if pressures are high and geothermal gradients only moderate. The gas-condensate surface is below 4.5 km in the Maracaibo Basin but no deeper than 3.5 km in the South Caspian Basin. Oil occurs at all depths down to 5 km in the deep San Joaquin Basin of California,

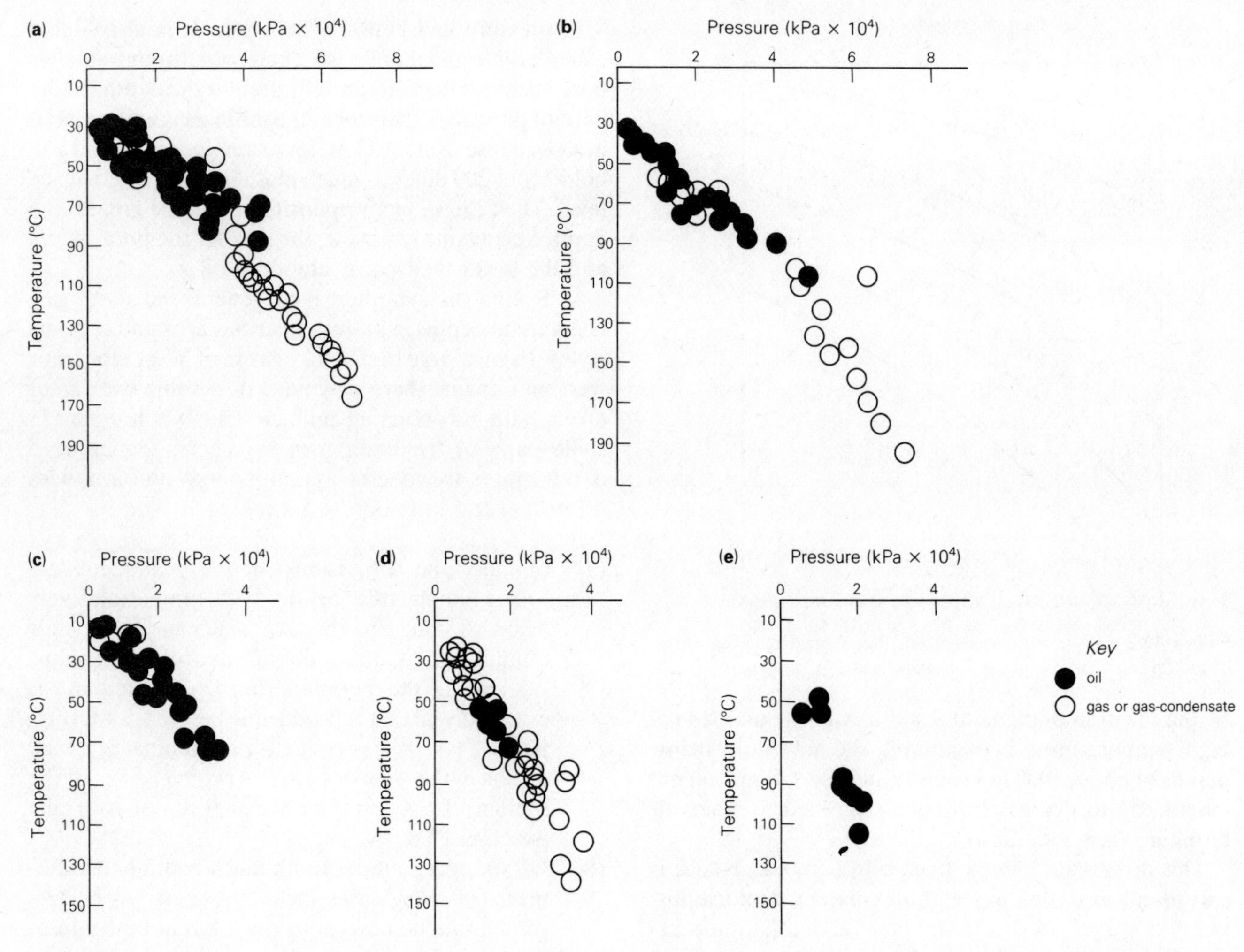

Figure 10.3 Graphs showing thermobaric conditions under which hydrocarbon phase states exist in five prolific basins: (a) Permian Basin; (b) US Rocky Mountain basins; (c) Ural–Volga Basin; (d) Western Canadian Basin; (e) Persian Gulf Basin. (From V. F. Raaben, *Int. Geol. Review*, 1976.)

but gas or gas-condensate is at all depths in the much shallower Sacramento Basin which is its present extension towards the north. The Po Valley of northern Italy is another deeply subsided basin which is gas-prone at all depths. In the Persian Gulf Basin the gas-condensate surface appears to lie at about 3.5 km, but its principal control may be stratigraphic; pre-Jurassic strata contain volumes of gas comparable with the volumes of oil in younger strata. The foothills belt of the Canadian Rocky Mountains presents the unusual case of gas and condensate-proneness in Upper Paleozoic strata in a Tertiary fold belt (Fig. 16.23).

10.4.7 Influence of age alone

As with the rise of rank in coals, there is a maturation sequence from organic material to oil or natural gas (see Ch. 7). The implication that age alone is a factor (up to a point) can be recognized in Barton's formulation of the maturation hypothesis; in the high API gravity (and high quality) of many Paleozoic oils; and in the correspondingly poorer quality of many Cenozoic, heavy, sour crudes (Sec. 5.3.1).

In 1959 the Soviet geochemist A. A. Kartsev distinguished between "paleotype" and "cenotype" oils. Hundreds of analyses showed that paleotypic oils in general contain the highest percentage of paraffins, and cenotypic oils the highest percentage of cyclic components. There are, however, very many exceptions to this generalization, many of them described in Chapter 17. In particular, Kartsev ignored the influences both of the tectonic environment of the basin and of the processes of degradation.

10.5 General classification of crude oil types

Clearly, any genetic classification of crude oils as they actually occur in reservoirs must reflect, not only the type of oil generated by the particular organic matter

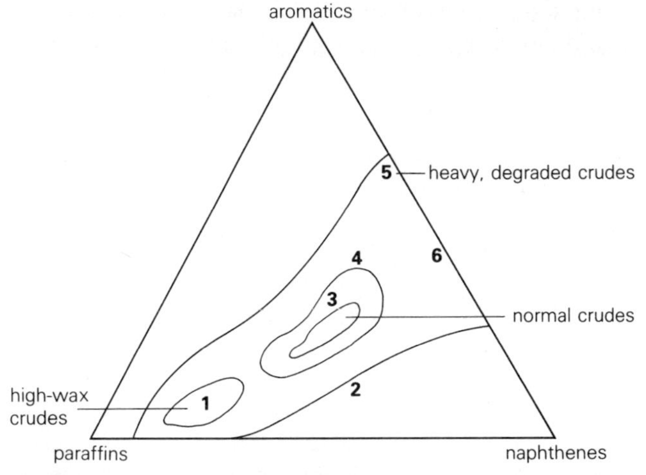

Figure 10.4 Tissot's six crude oil types displayed in a triangular diagram.

supplied to the source sediment, but also any alteration that the original oil may have undergone as a result either of further maturation or of degradation. The classification below is due to Bernard Tissot:

(1) Paraffinic crudes: less than 1 percent sulfur; density generally below 0.85.
(2) Paraffinic–naphthenic crudes, with less than 1 percent sulfur.
(3) Aromatic–intermediate crudes, with more than 1 percent sulfur.
(4) Aromatic–naphthenic crudes, with less than 1 percent sulfur and more than 25 percent naphthenes.
(5) Aromatic–asphaltic crudes, with more than 1 percent sulfur but less than 25 percent naphthenes.
(6) Asphaltic crudes, if such truly exist outside the definition of type 5.

Types 4, 5, and 6 are heavy crudes normally attributable to degradation. The Athabasca "tar sands" deposit, the largest single accumulation of this class of oils, falls into type 5 (see Sec. 10.8.4).

The six crude types are conveniently displayed on a triangular diagram with the three principal hydrocarbon series as the controls (Fig.10.4).

10.6 Natural solid hydrocarbons

Hydrocarbons of a considerable range of compositions occur in nature as solids rather than fluids. They are common in productive petroliferous regions, in "oil shales" (see Sec. 10.8.1), and even in some regions not commercially petroliferous (including both coal- and metal-mining regions).

The solids may be brittle or friable, or plastic and malleable; the latter types are really fluid hydrocarbons of very high viscosity. The brittle varieties occur as dykes, sills, and vein or fracture fillings; the plastic varieties are more common as "lakes," sheets, or tar "mats" at or near the surface or at old surfaces now represented by unconformities.

Natural solid or semi-solid hydrocarbons are of three distinct origins:

(a) *Native bitumens* are formed directly from bituminous OM in sediments (especially in clay source rocks). Because they are a form of altered kerogen, native bitumens are called *kerobitumens*; they have also been inappropriately called *protopetroleum*. They consist of high molecular weight chain hydrocarbons (C_{22} and heavier), with heteroatoms.
(b) *Reservoir bitumens* may be formed by the *in situ* alteration of conventional liquid hydrocarbons already pooled in reservoirs. The alteration is due either to thermal cracking or to de-asphalting through the agency of natural gas (see below). These bitumens may consist of remarkably pure hydrocarbons with very little contribution from S–N–O compounds. They are higher in aromatic constituents than most native bitumens, being really large aromatic ring structures (including asphaltenes). They are relatively insoluble in organic solvents. As they are presumed to represent stages in the complete dehydrogenation of hydrocarbons, they are called *graphitic bitumens*. They occupy vuggy or intergranular porosity in ordinary reservoir rocks.
(c) *Reservoir bitumens* may alternatively be formed at or near the surface under low temperature conditions, through degradation of pooled oil by inspissation, water washing, or microbial attack. These changes lead to surface plugs or mats of the plastic, tar-like material called *pitch* or *asphalt*, and this class of hydrocarbons is collectively called the *asphaltic bitumens*. They are soluble in organic solvents.

The carbon isotope ratios of reservoir bitumens are commonly identical to those of the coexisting oils. Only insignificant isotopic fractionation can have been involved in the alteration processes. These processes are therefore unlikely to have been either exclusively thermal or exclusively bacterial. Highly insoluble and infusible *pyrobitumens* in dry gas reservoirs below the gas-condensate surface (in the zone of metagenesis) are plausibly formed by primarily thermal processes. Their H : C atomic ratios indicate the maximum temperatures to which they were subjected, as carbon ratios do for coals (see Ch. 7). Conversely, reservoir bitumens close to or at the surface, overlying reservoir rocks now containing only degraded oil or stained by "dead oil," have been formed by primarily aqueous and bacterial processes, especially if the oils closest to the outcrop are more degraded than those farther from it.

The most common development of reservoir bitumens, however, consists of bands or zones of heavy oil and asphaltic "tar" at the oil/water interfaces of productive oilfields or immediately below them. The bands are called *basal tar mats*. These mats may reveal direct evidence of aqueous or bacterial attack: secondary pyrite segregations, corroded grains, or cements of reprecipitated silica. If such features are not detectable, however, a more likely degradation process is attributable to *de-asphalting* of the oil by the introduction of excess natural gas. The gas may migrate from below into an oil pool previously undersaturated with it, or increasing burial may put excess pressure on an oil pool having a gas cap (see Sec. 23.8). The excess gas causes the precipitation of the asphalt content of the oil, and the precipitated asphalt is subjected to *in situ* polymerization. The reservoir bitumen so produced is rich in asphaltenes and in S–N–O compounds (see Sec. 5.2).

The precipitation of this residuum by excess natural gas improves the quality of the remaining oil, but it may cause production problems by plugging the reservoir pore space with bitumen. It may also prevent natural water drive from maintaining the reservoir pressure necessary for production (see Sec. 23.8.3), by acting as a barrier to the entry of water from below. Nevertheless, thick tar mats mark the bases of the oil columns in some of the world's largest oilfields, in both sandstone and carbonate reservoirs. At Ghawar in Saudi Arabia and Prudhoe Bay in Alaska, the tar mats are confined to the east sides of the fields, probably because the oil/water interfaces have eastward tilts. Thick mats are also present below Burgan and many other Middle East oilfields, the Greater Oficina area of eastern Venezuela, the west side fields of the San Joaquin Valley in California, and the supergiant Sarir field in Libya, all of them described in Chapter 25. It should be emphasized that the process of natural gas de-asphalting is nowhere *proven* to be the cause of the mats.

10.6.1 Types of solid hydrocarbon and their identification

The most widely used classification of natural solid hydrocarbons is that devised by Herbert Abraham in 1960 (Fig.10.5). Abraham considered the spectrum of natural materials between liquid petroleum at one extreme and coal at the other. The solid substances in the spectrum are distinguished by the degrees of their solubility in CS_2 or in chloroform, and of their fusibility when heated. Both methods of differentiation are crude, and they do not lead to a uniform classification. Distillation chromatography may eventually lead to more clear-cut distinctions, and a combination of organic geochemistry and microscopy is also helpful (Fig.10.6).

Soluble, fusible bitumens Solid hydrocarbons that are both soluble in organic solvents and fusible below about 150 °C are *bitumens* or *petrobitumens*. They include both native and reservoir varieties.

Among native solid bitumens, the most readily fusible are the mineral waxes, effectively solid, straight-chain paraffins in the C_{22} to C_{50} range. The best known is *ozocerite*, originally from the Carpathian region of Poland (now in the Ukrainian SSR). Another notable occurrence is on the east shore of the Dead Sea. Ozocerite is a firm bitumen which breaks easily when twisted; it melts at 65–85 °C without burning. Yellow to brown in color, it consists mainly of hydrocarbons of microcrystalline structure with only small quantities of liquid components; it is nearly free of sulfur.

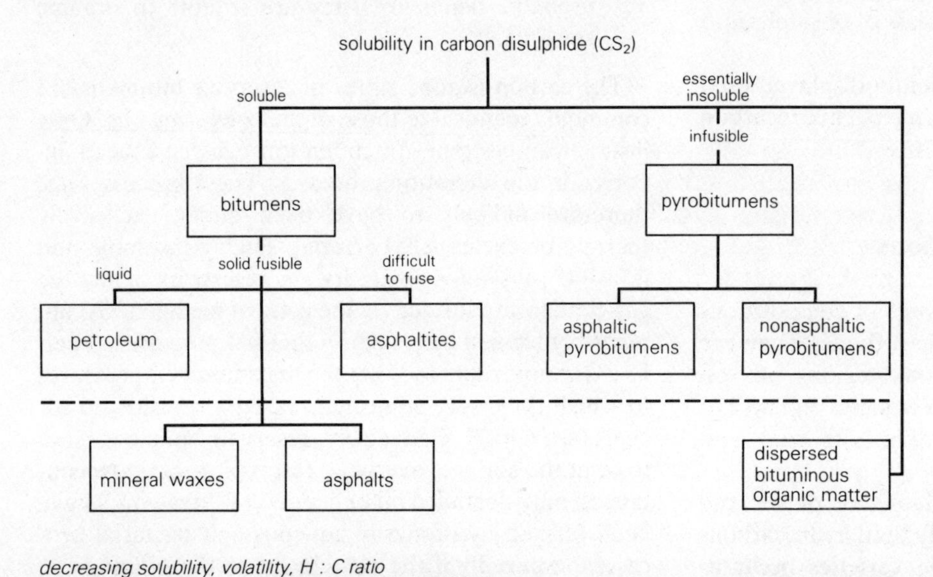

Figure 10.5 Abraham's classification of natural solid hydrocarbons. (Modified by M. A. Rogers *et al.*, *AAPG Bull.*, 1974, after K. G. Bell and J. M. Hunt, 1963.)

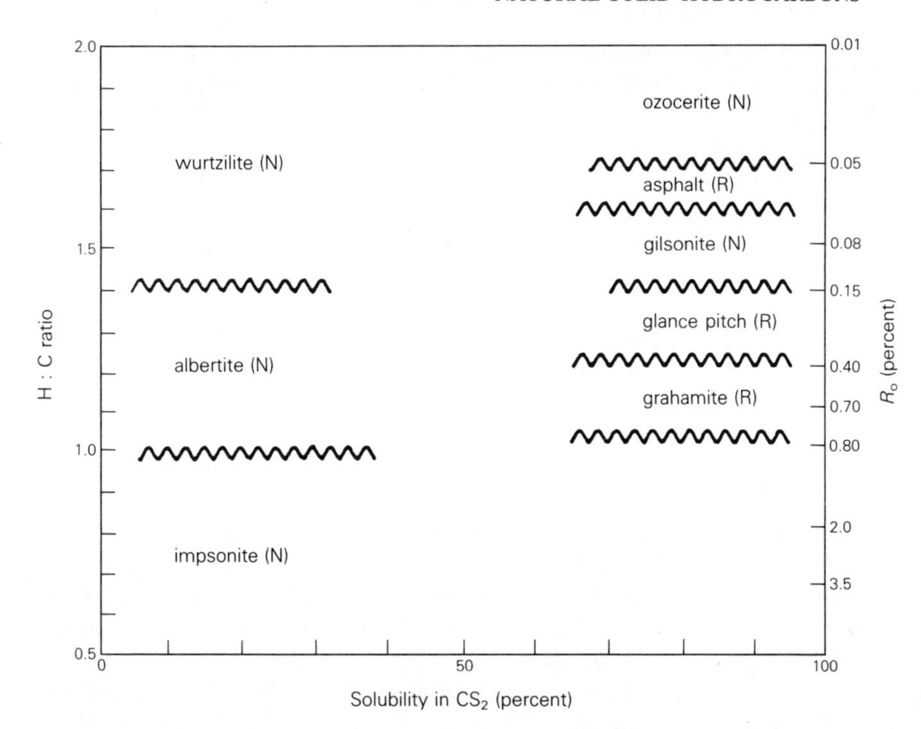

Figure 10.6 Generalized distinctions between solid hydrocarbons on geochemical and optical criteria. Native bitumens indicated by (N), asphaltic reservoir bitumens by (R). (Courtesy of D. K. Baskin and Chevron Oil Field Research Company; H : C ratios from L. H. King *et al.*, 1963; R_o values from H. Jacob, 1975.)

Somewhat less easily fusible is *natural asphalt.* Asphalts are arbitrarily and approximately distinguished from heavy crude oils by their rheological properties; asphalts soften only above 20 °C. The principal components of asphalt have a formula approximating C_nH_{2n-4}. Examples include the famous *pitch lakes* of Trinidad and eastern Venezuela, obviously degraded reservoir bitumens.

Still more difficult to fuse, melting above 110 °C, are the *asphaltites*. They are black or dark brown, soluble in CS_2 and in chloroform but only partly so in CCl_4. The fraction soluble in CS_2 but not in CCl_4 is called *carbenes*. The most familiar asphaltite is *manjak*, originally described from the filling of a mudflow vent in Barbados. Also known as *glance pitch*, it is used in the manufacture of varnish and shellac. As a generally understood term, manjak is to an asphaltic oil what ozocerite is to a paraffin-base oil; both contain 80–85 percent carbon. In North America, the most familiar native bitumen in this hard-to-fuse category is *gilsonite*, from the Uinta Basin of Utah, a predominantly aromatic substance which ignites only with great difficulty and crumbles easily.

Insoluble, infusible pyrobitumens Solid hydrocarbons insoluble in CS_2 or chloroform, and essentially infusible, are called *pyrobitumens*. They include the great majority of native bitumens or *kerobitumens*. Pyrobitumens are infusible in the sense that they decompose before they melt. The criteria of infusibility and insolubility that distinguish them from other bitumens are also met by carbonaceous materials, including coals, but bituminous and lower-rank coals contain much more oxygen.

Asphaltic pyrobitumens are all black or very dark brown in color and they occur widely under a variety of local names. *Albertite* is a very pure hydrocarbon injected as a vein into Mississipian oil shales in New Brunswick, Canada. *Elaterite* is "mineral rubber," an elastic hydrocarbon from the base metal deposits of the Carboniferous limestone in northern England. *Impsonite* is the name applied in Oklahoma, *wurtzilite* in the Uinta Basin, and other names elsewhere.

The bitumens and the pyrobitumens may be further distinguished by other geochemical means and by their vitrinite reflectances (R_o) (see Sec. 7.5 & Fig. 10.6). Native pyrobitumens have a range of H : C ratios from less than 0.5 to more than 2.0, and of R_o values from 0.01 percent or less to more than 5.0 percent. Asphaltic reservoir bitumens have H : C ratios more like those of oil-source sediments (higher than 1.0) but R_o values well below them.

10.7 Summary of petroleum and bitumen types

Gaseous, liquid, and solid hydrocarbons are all therefore merely different stages in the creation of petroleum from organic matter in sediments. Given any one type of OM provided to the sediments, the transformation of its kerogen into the hydrocarbon we find today may be *influenced* by a variety of agents and conditions, but it is *dominated* by one of three (Fig.10.7):

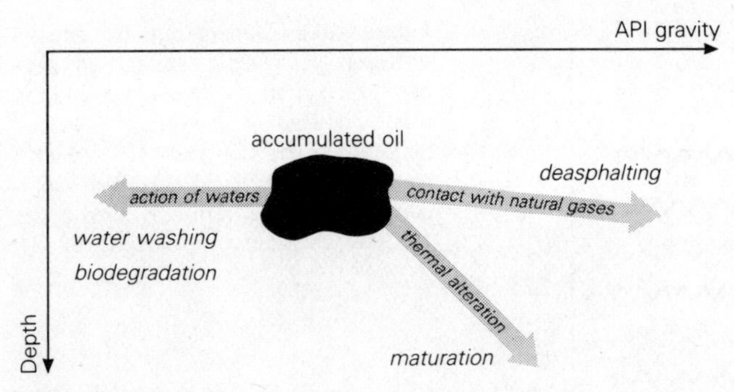

Figure 10.7 Processes causing variations in quality and type of both oils and reservoir bitumens. (From C. R. Evans *et al.*, *Chem. Geol.* 1971.)

(a) Temperature, leading to maturation and possibly to *destruction* of oil and creation of one type of reservoir bitumen.
(b) Natural gas, leading to the rejection of the aromatic–asphaltic components, consequent *improvement* of oil, and creation of a second type of reservoir bitumen.
(c) Water, with attendant oxygen and bacteria, causing *degradation* and possible *destruction* of oil and creation of a third type of reservoir bitumen.

10.8 Unconventional petroleum and natural gas deposits

We now recognize that "oil," which we conventionally think of as a *fluid* used to power and lubricate motors and heat buildings, may nonetheless occur in natural forms having such high viscosities and densities that they are no longer "fluid." In a similar way, natural gas in the presence of water may be subjected to such a low temperature that it forms a *clathrate* (see Sec. 5.1) and becomes effectively a solid.

Because our conventional uses of oil and gas depend upon their fluidity, engineers have devised mechanisms to extract, treat, and transport them as fluids when they occur in that form. Petroleum which is not sufficiently either liquid or gaseous to be extracted, transported, and used in this conventional manner is said to be *unconventional* petroleum (or fuel), and to constitute an unconventional resource occurring in unconventional deposits. The term "unconventional" now refers simply to naturally occurring oil or gas deposits of potentially commercial significance *not recoverable by conventional drilling methods*.

Because we are dealing with *petroleum geology*, we must consider the geology of oil shales and tar sands, from which oil can be recovered by mining and a variety of subsequent extractive techniques, and of certain shales and overpressured stratigraphic intervals from which natural gas can be recovered. We will not consider any aspects of oil or gas that is *manufactured* from a nonhydrocarbon source, such as coal, biomass, or living vegetation.

10.8.1 Oil shales

Long before liquid oil was recovered from rocks by digging or drilling into them, it was extracted from shaly rocks by burning them. "Shale oil" was a commercial item before "rock oil." *Oil shales* have not given up their oil content as orthodox source sediments have done. It remains part of the rock, and is recoverable only by distillation or retorting of the rock itself.

Oil shales are much commoner than is generally realized, occurring on every continent and in every geological system. They are misnamed, but it is unlikely that any more appropriate name will be acceptable either to geologists or to operators. Few important deposits actually consist of shales, and even fewer have yielded any oil in the conventional sense. Most "oil shales" are actually bituminous, nonmarine limestones or marlstones containing kerogen. Only the few marine examples can be properly described as shales. The commonest composition involves about 50 percent of mineral carbonates by weight. A variety of silicates may also be present, possibly derived from the reaction of circulating alkaline waters with volcanic debris.

The common oil-shale rock is brownish-black or yellowish-brown, very well laminated (as rich source sediments are also), in some cases of papery texture, tough, resinous, and giving off a "woody" sound when struck. Commonly associated rocks include coals, plant-bearing sandstones, cherts, and tuffs. The oil shale itself contains from 10 to 50 percent of algal organic matter (OM); in commercially significant deposits, the OM content is at least 30 percent, in the form of a high molecular weight mineraloid of indefinite composition but largely derived from algae, spores, or pollen. The OM content is therefore much higher than it is in most oil-source sediments, but it has attained such a low level of maturation that liquid oil has never been expelled from the kerogen.

The OM may be either sapropelic or humic, or any mixture of the two. In all true oil shales, however, sapropelic OM and its constituent alginite and exinite macerals are dominant, providing Tissot's Type I kerogen (Sec. 7.3). In 1981, Alan Cook and his Australian colleagues accorded oil shales a classification based on the type of contained OM exactly as Tissot classified kerogens:

(a) Ultra-rich varieties include an assortment of rocks intermediate between shale and coal. They belong among the infusible kerobitumens of Abraham's classification of those materials (Fig. 10.5), and are of freshwater origin, commonly lacustrine. *Cannel coal* contains abundant sporinite and bituminite from higher plants, as well as resinite, vitrinite, and inertinite. *Torbanite* or boghead coal contains abundant alginite. The H : C ratio is typically 1.25 or higher, the O : C ratio less than 0.10 (Fig. 7.3).

(b) The most important commercial oil shales, also lacustrine, have been called *lamosites*. They are mostly Tertiary in age, and include the famous Green River shales of the western USA. H : C ratios are typically between 1.3 and 1.7; O : C ratios 0.18 or less.

(c) Marine oil shales are rich in alginite derived from marine algae. They are important in the Paleozoic (examples include the *kukkersite* of Estonia and the *tasmanite* of Australia) and in the Mesozoic (especially in the Kimmeridgian). H : C ratios are 1.4 or higher; O : C ratios between 0.1 and 0.2.

(d) *Mixed oil shales* contain higher proportions of humic OM, and numerous dinoflagellates, acritarchs, and other organisms unimportant in richer varieties. Many such shales are called simply *black shales* or *brown shales*, like those in the Devonian of eastern North America and the Jurassic of western Europe. Their atomic ratios are too variable for specification.

In addition to the algae and other plant residues, oil shales commonly contain abundant fish remains. "Graveyards" of fish remains are to oil shales what "battlefields" of planktonic organisms are to oil-source sediments (see Sec. 8.1); they represent episodes of mass mortality during hypertrophy of the waters.

The organic constituents of a rich oil shale, such as the Green River shale, span the spectrum of natural hydrocarbons (paraffins and cycloparaffins in the C_{12} to C_{33} range, including pristane and phytane, and aromatics), carbohydrates, carboxylic and amino acids, and porphyrins. The organic matter consists of three fractions. The largest fraction is insoluble kerogen; a smaller fraction consists of soluble bitumen; the smallest is inert, insoluble material which yields no oil on pyrolysis. Associated sands may be impregnated with bitumen, or interbedded with solid or semi-solid hydrocarbons which are liable to flowage or to injection as dykes and veins into other sediments.

Oil shales are also susceptible to spontaneous combustion, though not so readily as coal. The combustion probably begins with the production of sulfuric acid through the oxidation of pyrite. The acid oxidizes the shale, raising the temperature and leading to combustion. A spectacular burned oil shale forms the "Mottled Zone" extending across much of Israel, Jordan, and Syria. A Maestrichtian-to-Paleocene succession of highly bituminous and phosphatic marls and chalks, 200 m thick, has lost its bedded character almost entirely and become a red, yellow, green, and black mass shot through by great numbers of mineral veins. The mineral assemblage indicates temperatures of formation of 600–800 °C; yet there is no geologic source of thermal metamorphism. Furthermore, the underlying cherts and overlying limestones show no sign of equivalent alteration. Where unaltered, the marls and chalks of the Mottled Zone contain 10–25 percent of OM, and spontaneous combustion of this material is the only plausible explanation of the formation's present condition.

Environments of deposition of oil shales Oil-source sediments are most readily created by marine upwelling, causing phytoplanktonic blooms and "red tides" in seas teeming with dinoflagellates or ciliates (see Sec. 6.5). Very similar blooms are caused in some lakes by freshwater green algae such as *Botryococcus* spp. Provided that the lakes contain bacteria capable of inhibiting the decay of the waxy or gelatinous algal cells, the OM is able to accumulate as a rubbery residue. One such residue, the *coorongite* of South Australia, derived from *Botryococcus* spp., has been aptly described as "the peat stage of torbanite," the sapropelic "coal."

Marine oil shales are simply immature marine source sediments; the environments of deposition of these were described in Chapter 8. The much more important lacustrine oil shales require some additional special circumstances. Their deposition is not dependent on the depth of water in the lake and it is not prevented by the existence of bottom currents. Once the conditions for the protection of algal OM are achieved, the only common interfering factor is the introduction of terrigenous clastic sediments (by rivers or turbidity currents).

Reconstruction of the lacustrine environment of oil-shale formation has been most exhaustively undertaken for the Green River shales of the western USA. Conclusions from these studies may serve as a model for all rich oil-shale deposits. The visually obvious properties demanding explanation are the prevalence of features indicative of very shallow water origin, the interbedding of a variety of saline (evaporite) minerals, and the presence of a remarkable array of authigenic carbonate and silicate minerals. The features indicative of very shallow waters include desiccation breccias (flat-pebble carbonate conglomerates), lag deposits of mud-cracked ostracode limestones, coquinas of nonmarine, air-breathing gastropods such as common land snails, algal bioherms and domal stromatolites, ooliths, pisoliths, and ubiquitous cross bedding. Perishable microfossils were preserved in the rubbery organic gel (like the Australian coorongite) when it was exposed to the atmosphere during low water levels. Shale laminae are discontinuous; barren marlstones are disrupted by flute casts.

The evaporite minerals include nahcolite (the bicarbonate of soda) and trona (the sesquicarbonate of soda) in addition to halite. The saline minerals occur as nodules in the oil shales as well as in discrete interbeds. The other inorganic constituents include many authigenic mineral species otherwise known only in alkaline igneous rocks and pegmatites, such as complex titanosilicates; also abundantly present are a variety of Ca–Mg–Fe carbonates, especially ankerite.

In a series of classic papers published between 1966 and 1973, Wilmot Bradley and Hans Eugster attempted to formulate a model lake in which oil shale, trona, and halite can be deposited in alternation. Their model was a chemically stratified or *meromict lake*, with the strong brine of the lower levels (the saline hypolimnion) overlain, above the chemicline, by a circulating, oxygenated mixolimnion carrying abundant calcium bicarbonate in solution. Such a lake could precipitate the oil shale–trona–halite triplets, but the model fails to account for the dolomite content in the shales that invariably underlie the beds of trona. If the mineral association is envisaged as consisting of *evaporites*, the degree of evaporation required to precipitate protodolomite would be such that dense algal growths would be impossible.

Eugster and Ronald Surdam, in 1973, therefore proposed as an alternative model an *alkaline desert lake* or "continental sabkha." The lake would lie within a vast alkaline-earth playa. In this surrounding playa, and not in the lake itself, calcite and protodolomite were deposited as "rains" of microcrystals. These deposits were then washed into the lake during periodic rains. The trona and the oil shales were the deposits of the very dry seasons.

Those oil-shale deposits associated with coals rather than with carbonates and evaporite minerals were the deposits of swamps, lagoons, and peat bogs. Examples are the deposits in the Carboniferous of Kazakhstan and the Tertiary of northeastern China and of New Zealand.

Oil shales and oil-source sediments The relation, if any, between "oil shales" and the source sediments for conventionally pooled oil has been the subject of much argument. Before the recognition of the biological and biochemical differences between the several types of kerogen, it seemed that the oil shales had merely failed to release their oil content for pooling because of some peculiarity in their clay content. The bituminous matter simply remained adsorbed on to clay particles of a particular type, deposited in fresh or brackish waters rather than marine waters. The advent of clay mineralogy reinforced this viewpoint. The clays deposited in freshwater lakes or swamps or in brackish lagoons were presumed to be of illitic or kaolinitic composition from the beginning, and not the smectitic type transformable into "mixed layer" clays by loss of bound water (see Sec. 15.4.1). Clays never containing bound or interlayer water were unable to provide the flushing mechanism to expel their hydrocarbons for migration and accumulation in pools.

This explanation has never been verified and its efficacy in accounting for the retention of nearly all the hydrocarbons in Paleozoic rocks having ten times the organic carbon content of an acceptable source rock must be seriously doubted. The discrimination among the four basic types of kerogen (see Sec. 7.3) leads to a much more plausible explanation. Type I kerogen is derived almost entirely from algal and spore precursors, with trivial contribution of animal OM. Its paucity in heterocompounds creates the necessity of breaking carbon–carbon bonds throughout the mass, requiring much higher temperatures than are needed for the cracking of Type II kerogen. It is unlikely that any orthodox source sediment – certainly no Paleozoic source sediment – was ever subjected in nature to temperature levels anywhere near those required in the laboratory (about 500 °C) to retort oil from oil shales. Within the range of temperature gradients measured in sedimentary basins (see Sec. 17.8), a burial depth of the order of 20 km would have to be attained (and maintained) for such temperatures to be applied to the shales to release their oil. These are the *T, P* conditions of the greenschist facies of metamorphism or higher.

Type II source sediments are *basinal* sediments; they are likely to be relatively deeply buried relatively quickly. Very many have been strongly folded (the source sediments in Tertiary intermontane basins, for example). Most oil shales, on the other hand, are among the prototypical *platform* sediments – the deposits of lakes, swamps, and shallow lagoons. They are unlikely ever to become as deeply buried as basinal sediments habitually become; the Estonian kukkersite, the outcropping Irati shale of Brazil, and the Mahogany Ledge of the Green River shale are famous oil-shale deposits that have never been significantly buried. Their high percentages of organic matter – from more than 10 percent to as much as 77 percent organic carbon – "suffocate" the sediments and inhibit their own thermal maturation below temperatures unattainable in nature except under metamorphic conditions. We have no deposits of oil slate.

Oil-shale resources Except for coals, oil shales constitute the most abundant fossil fuel resource. However, they are far more variable in quality than coals are. In sapropelic shales, like the Green River shale, up to

65–70 percent of the kerogen content is convertible to oil on distillation; in humic oil shales, less than 30 percent is convertible. Ultra-rich shales, like torbanites and tasmanites, have volatile contents in excess of 40 percent, or over 100 US gallons of oil per ton (about 0.4×10^{-3} m^3 kg^{-1}). Most oil shales are capable on distillation of yielding between 0.060×10^{-3} and 0.125×10^{-3} m^3 kg^{-1}; those yielding less than the lower amount (in traditional terms, less than half a barrel of oil per ton of shale) are not normally considered to be oil shales.

In 1976 an authoritative survey by T. F. Yen and George Chilingarian estimated total world "potential" shale-oil *resources* at 30×10^{12} barrels (4.8×10^{12} m^3). More cautious estimates by the US Geological Survey, the United Nations, and several individuals average about 500×10^9 m^3 of *recoverable reserves*. Of this amount, about 20 percent (or 2 percent of the Yen and Chilingarian figure) may be assigned to the highest commercial category available for exploitation in the near future; 80 percent falls into less rich but still exploitable categories.

Oil produced from high-grade, sapropelic shales is highly paraffinic, high in sulfur but low in trace metals; the latter are likely to be dominated by nickel. Oil from dominantly humic shales is less paraffinic and much higher in trace metals.

Table 10.1 Principal oil-shale deposits of the world tabulated according to age

Cambrian: northern Asia; southern Sweden (Kolm shale)
Ordovician: Estonia (kukkersite); Southampton Island (Hudson Bay, Canada)
Silurian: eastern Algeria
Devonian: Amazonas Province, Brazil (Curua Formation); eastern USA (Ohio and Antrim shales)
Mississippian: Scotland (Midlothian); New Brunswick, Canada (Albert Formation)
Pennsylvanian or *Carbo-Permian:* Kazakhstan (Kenderlyk shale); Spain (Puertollano Formation); South Africa (Ermelo Formation); Australia (kerosene shale)
Permian: France (Autun); Brazil (Irati Formation); Tasmania (tasmanite)
Triassic: Svalbard; Zaïre
Lower Jurassic: France (Crevenay); USSR (Kashpirian Formation); Argentina (Neuquen)
Jura-Cretaceous: northern Alaska (tasmanite); Angola; Brazil (Alagoas); Siberia (Irkutsk)
Upper Cretaceous: Wyoming (Frontier Formation); southern Saskatchewan–Manitoba, Canada; Morocco (Atlas Mountains); South Korea; western Queensland, Australia
Paleogene: Israel; Rhine graben (Messel); Liaoning, China (Fushun); western USA (Green River shales); eastern Queensland (Rundle shale); Argentina (Foyel, western Chubut); South Island of New Zealand (coal measures); France (Aix-en-Provence)
Neogene: Yugoslavia (Aleksinac shale); Brazil (Tremembe); Thailand

Principal occurrences of oil shales The most important known oil-shale deposits are listed in Table 10.1. The oldest large deposits, in the Lower Paleozoic of Eurasia and North Africa, are marine sediments but nearly all important post-Devonian examples are of lacustrine origin. Historically, geologically, and economically, the most important deposits are those of the western USA, Brazil, Scotland, Estonia, and China.

Green River shales, western USA More than 50 percent of the world's established exploitable oil-shale resources are in the Eocene *Green River Formation* of Colorado, Utah, and Wyoming in the western USA. The Laramide uplift of the basement fault blocks of the US Rocky Mountains isolated two Eocene lake basins north and south of the latitudinally trending Uinta Mountains. Subsequent uplift and erosion resulted in the preservation of four separate basins having a total area of some 42 000 km^2 (Fig. 10.8). Average recoverable oil content of the "shale" – actually a fissile bituminous limestone famous for its fossil fish remains – is 15–20 US gallons

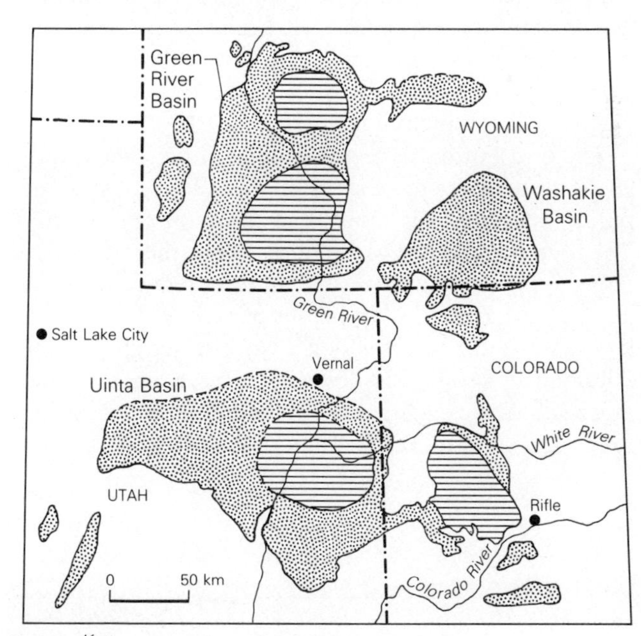

Figure 10.8 Basins containing the lacustrine, Eocene, Green River Formation in three states of the western USA. The east–west separation across the center of the map is caused by the Uinta Mountains, the north–south margin of the basins on the west by the Wasatch Mountains, both major Laramide uplifts. (From H. Wilson, *Oil Gas J.*, 29 June 1981; after Union Oil Co. of California.)

per ton (0.06×10^{-3}–0.09×10^{-3} m^3 kg^{-1}). However, a single member called the Mahogany Ledge, 25 m thick, flat-lying, and cropping out extensively in canyon walls in Colorado's Piceance Basin, contains 60–90 US gallons per ton (2.5×10^{-4}–3.6×10^{-4} m^3 kg^{-1}) and is the prime target for a surface mining operation. The OM in the Mahogany Ledge, manifestly derived largely from blue-green algae, contains about 80 percent carbon, 10 percent hydrogen, 6 percent oxygen, 2.5 percent nitrogen, and 1 percent sulfur.

The variety of hydrocarbons and other high molecular weight compounds making up the organic content of the Green River shale must have been derived by the polymerization of the hydrogen-rich components of the kerogen at temperatures of 90–125 °C. Some of the oil and gas content has been released for orthodox pooling in sandstones of the associated shoreline facies, most of it in small accumulations (by comparison with the quantity still in the shales) in the Uinta Basin of northeastern Utah.

Figure 10.10 Oil shale occurrences in Brazil. The squares in the south outline the outcrop areas of the Irati and Paraiba shales discussed in the text. (From V. T. Padula, *AAPG Bull.*, 1969.)

Irati oil shales, Brazil Much of southern Brazil is underlain by the vast Parana Basin, one of the great "continental geosynclines" of the Gondwana continent. As marine deposition gave way to continental deposition in mid-Permian time, a relatively thin but exceedingly constant and widespread oil shale was deposited. The Irati Formation consists of regularly repeated bituminous and non-bituminous shales with cherts and limestones, giving way upwards to the deposits of a freshwater, intracontinental lake-sea (Fig.10.9).

The Irati Formation is traceable as a curvilinear outcrop belt between the coastal shield and the edge of the basin, extending for more than 1500 km (Fig.10.10). The total oil resource in the formation is therefore very large, 7–14 percent distillable oil representing an average yield of about 23 US gallons per ton (0.1×10^{-3} m^3 kg^{-1}). Beyond the formation's eastern margin, in the Paraiba rift west of Rio de Janeiro, there was a second much smaller development of oil shale in lacustrine sands and clays of Tertiary age (Fig. 10.10). In neither the Permian nor the Tertiary oil shale has any episode of conventional oil release been discovered.

Midland valley of Scotland The so-called "Oil Shale Group" of the Midlothian district, in the Midland valley of Scotland, is of early Carboniferous age and part of the Carboniferous Limestone Series. The group consists of shallow-water, nonmarine limestones, shales, and sandstones, including "ganisters" or very pure silica sands penetrated by plant roots and associated with the fireclays that are the seat-earths of coal seams. The oil shales themselves are finely laminated, rhythmic deposits of lagoons formed along the western margin of a large delta system. Oil shales constitute only about 3 percent of the total succession.

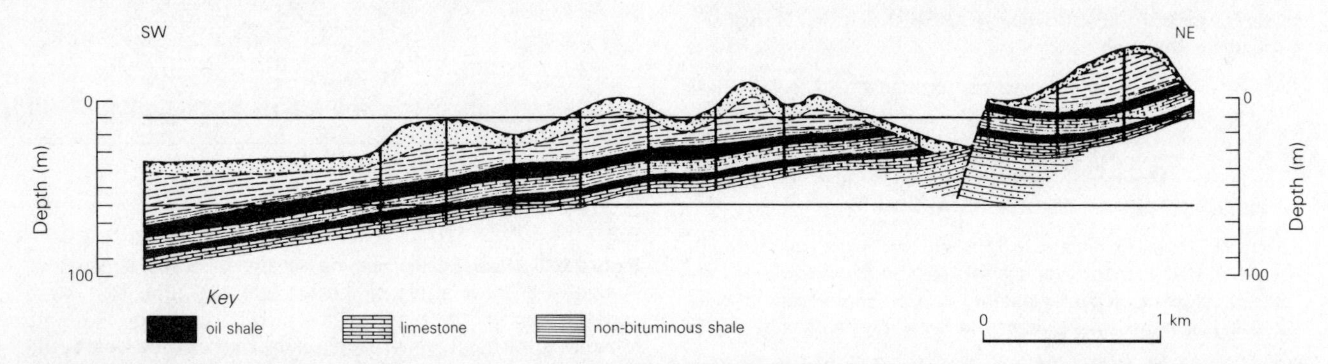

Figure 10.9 Cross section through the formations containing the Irati oil shale in São Mateus do Sul district, southernmost Brazil. (From V. T. Padula, *AAPG Bull.*, 1969.)

Commercial operations began in the mid-19th century, and the area would remain of historic interest if only because of its contribution to the chaotic nomenclature of the oil industry. It is the type locality not only of torbanite and boghead coal but of kerogen itself. The richest of the shales (not particularly rich by world standards, yielding about 8 percent of oil by weight on retorting) are so waxy and thinly laminated as to be flexible and soft enough to be cut into strips with a knife.

Deposits under active exploitation Since World War II, the principal oil-shale deposits actually under exploitation have been in communist countries. The *Chinese* deposits occur around Fushun, near the Korean border, and are associated with coal measures. The oil shales, of Tertiary age (and some Jurassic), are lean by comparison with other potentially commercial deposits (less than 10 percent organic carbon and an oil yield of less than 5 percent by weight).

The kukkersite oil shale of *Estonia* provides a remarkable contrast with the Chinese deposits. It is the lower part of the thin, marine, Ordovician carbonate-shale section which forms the escarpment or "glint" facing northwards across the Gulf of Finland towards the Baltic shield. The oil-shale rock consists mainly of the remains of microscopic marine algae; it is interbedded with a Lower Ordovician limestone so little cemented that it has been called a chalk. Kukkersite is one of the richest oil-shale rocks known, containing from 48 to 86 US gallons of oil per ton (16–30 percent oil yield). Its organic carbon content of more than 75 percent by weight makes it comparable with the type tasmanite of Tasmania, an exemplar of Tissot's Type I kerogen rock, and far richer than the average Green River or Irati shale.

Recovery of oil from oil shales The contribution of shale oil to world oil supplies during the 20th century has been negligible. Despite the great size of the resource, the technological, economic, and environmental obstacles to its utilization have frustrated years of pilot-plant experimentation. The economic and environmental problems can be touched upon only very briefly in this book, but the technological obstacles stem directly from the geologic characteristics of the shales and must be considered in more detail.

The ineluctable deterrents presented by the nature of a typical oil shale are the low percentage of the rock represented by its oil content, and the necessity for enormous input of heat to distil the oil out of the shale. Even a rich oil shale yields on average less than one barrel of oil per ton of rock (about 10^{-4} m^3 kg^{-1}). Hence 10 kg of shale must be processed to yield 1 kg of oil. To make a contribution to oil supply in any way commensurate with the size of the resource means that every oil-shale mine will be individually the biggest mining operation in the world. An oil-shale plant on the scale of a tar-sand plant – putting out about 6.0×10^9–7.0×10^9 kg per year – must therefore *process* at least 75×10^9 kg of shale annually. If the shale is *mined* before being retorted (as distinct from being retorted in the ground), the amount of rock to be *moved* will be at least double this amount because of the need to maintain slope stabilities and to dispose of overburden. To give a specific example: if Japan, anxious to diversify its sources of imported fuels, elected to contract for just 20 percent of its oil supply from the Rundle oil-shale deposit in Queensland (a very natural source), the operators of that deposit would need to create an open pit 1 km^2 in area and 10 m deep *every week*.

The problem of retorting such volumes of shale is stupefying. Whether it can be achieved with a net gain in energy† and without depriving large territories of their entire water supply has yet to be demonstrated. Even if these daunting obstacles can be overcome, the operators will still have to dispose of the spent shale, which expands during retorting to about 120 percent of its original volume – the wryly named "popcorn effect." The volumes to be disposed of would fill entire canyons and spent shale is very resistant to revegetation.

In oil-shale mining operations, the room-and-pillar pattern designed for coalmines is used. The mined shale is crushed and retorted in a closed vessel above ground. The burned shale becomes coated with a carbon residue which is itself combustible. Above 500 °C, the kerogen is broken down; the oil formed is then withdrawn as vapor from the top of the retort, and the vapor is condensed into oil, leaving only gas behind.

An ingenious variation uses a horizontal retort. In the ASPECO and TOSCO processes, preheated ceramic thermospheres and crushed shale are fed into a rotating kiln at 900 °C. The shale is converted to shale-coke, which is then burned separately in a fluidized bed to reheat the thermospheres. A plant putting out 6×10^9–7×10^9 kg of oil per year requires about 3×10^9 thermospheres.

It has been acknowledged for years that it will be necessary to develop techniques for doing the retorting underground (*in situ* techniques), so that no actual mining of the shale is necessary and the waste disposal problem is very much alleviated. Boreholes are drilled through the shale in five-spot patterns. The shale is hydraulically fractured and afterwards shot with nitroglycerin to break it up and allow the passage of heated vapors. All holes above the ignition zone are sealed with

†A net gain in energy is achieved only if the energy expended to recover the resource (in operating machinery, creating steam, etc.) is less than the energy of the resource eventually reaching the consumer.

high-temperature cement. Gas and compressed air are injected into the central input well under pressure, and ignited, creating a self-sustaining combustion zone which moves horizontally through the rubblized shale towards the output wells. A mist of shale-oil vapor is produced, which on cooling (after 24 hours or thereabouts) yields liquid oil capable of being pumped to the surface. Also produced is a gaseous mixture of CO_2, oxygen, nitrogen, and water vapor, with small residual amounts of hydrocarbons. In alternative versions of this technique the rubblized shale may be burned vertically in clusters of underground retorts, the liquid products moving by gravity to the bottoms of the retorts and being pumped from there to the surface.

The proportion of the OM which is converted to oil (commonly 10–15 percent) is related to its hydrogen content; part of the product is gas (not methane) and part is residual fixed carbon. The oil produced is black, viscous, waxy, and heavier than conventional crude oils. With a pour point of about 40 °C, it is not readily pumpable. Most shale oils are high in unsaturated hydrocarbons but highly aliphatic (that is, naphthenes dominate over paraffins plus aromatics); they contain more olefinic hydrocarbons than any conventional crudes. Finally, shale oils have much higher contents of sulfur, nitrogen, and oxygen derivatives than do conventional crudes. They are therefore not usable as they come from the retorts, but must be "upgraded" by being first *coked* (thermally cracked) and then *hydrogenated* (mixed with hydrogen at high T and P in the presence of a catalyst) to make them more "oil-like."

A major irony of the pilot engineering efforts to determine the best process for extracting oil from oil shales was a consequence of the geography and climate of the western USA, the site of the greatest and most investigated oil-shale deposits. Water is a relatively scarce commodity there, but natural gas was plentiful and blithely believed to be in assured long-term supply. The hydrogen for the experimental hydrogenation process was therefore produced by dissociating methane, until it was realized that the methane is a more valuable material in its own right than any fuel recoverable from the shales.

10.8.2 Black shales as sources of natural gas

As already noticed, many so-called oil shales are actually black or brown marine shales which have served as lean to moderate source sediments for pooled oil. This is especially true of Paleozoic representatives. They are in no case rich oil-source sediments, and few of them are ever likely to become commercial shale-oil producers. However, they are in places sufficiently thick and areally widespread to constitute a vast potential source of *methane*.

As Paleozoic shales carry negligible pore space in the conventional sense, it was originally believed that their gas content was held as free gas in fractures, like much of the gas in coal seams. In this event, the gas should be producible through these fractures in accordance with Darcy's law, though the fractures would no doubt need artificial widening. According to the so-called dual-porosity theory, there are two fractions to the gas content of the shales. Some is held as free gas in fracture porosity, but much the greater part is in an adsorbed phase in the shale matrix. This adsorbed gas diffuses slowly into any fractures wide enough to receive it, joining the free gas, and is producible as the pressure is reduced. Production of the gas on a commercial scale therefore requires physical break-up of the shale, by massive fracturing or rubblizing procedures (MHF: see Sec. 23.13). It is not producible directly by retorting, as shale oil is, but the pulverized rock still needs hydro-retorting at high temperatures.

The most comprehensively studied area for gas production from black or brown marine shales is in the Appalachian Basin of the eastern USA. Shales occur at several levels in the Middle and Upper Devonian section in a belt west of the Allegheny front (Fig. 10.11). They thin out westward into a belt offering no commercial prospects, but westward again into Ohio and Kentucky, towards the Cincinnati Arch, they thicken up in the Upper Devonian only, as the principal marine deposits offshore from the prograding Catskill delta. Here they are called Ohio or Chattanooga shales. A 1982 study by the US Geological Survey resulted in a mean estimate of 23.5×10^{12} m^3 for the in-place gas resource in the basin.

10.8.3 Natural gas from geopressured reservoirs

The subject of overpressuring is dealt with at length in Chapter 14. Waters in geopressured aquifers have abnormally high temperatures, as well as abnormally high pressures, for their depths. Temperatures up to 270 °C have been measured at less than 3000 m, and the geothermal gradient increases sharply at the top of the zone of abnormal pressure (to about 90 °C km^{-1} in the US Gulf Coast). The geopressured hot waters rise to the higher parts of structures caused by growth faulting or diapirism.

If the hot waters in geopressured sandstones, on structure, were saturated with methane in solution, the sandstones could yield 40–50 cf of gas per barrel of water (7–9 m^3 of gas per cubic meter of water), though the principal objective of development would be the geothermal energy in the water. The US Geological Survey estimated that the geopressured zone of the Gulf Coast Tertiary alone would contain almost 30×10^{12} m^3 of gas (in addition to an enormous amount of thermal energy in the hot waters) within attainable drilling

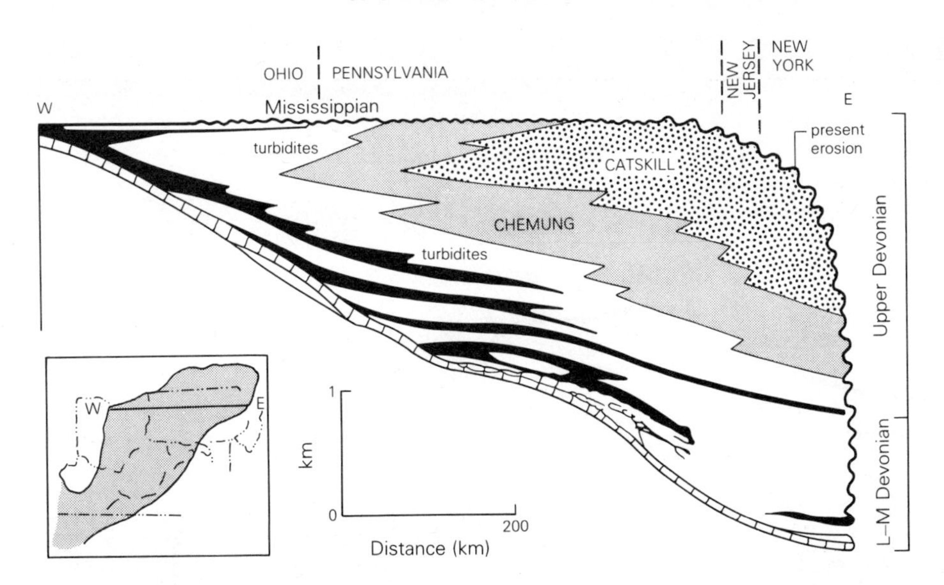

Figure 10.11 Diagrammatic stratigraphic section through the Devonian rocks from New York State to northern Ohio, showing the development of black shales between the Catskill Mountains on the east and the Cincinnati Arch on the west. Note the westward progression of black shale deposition as the nonmarine Catskill clastic sediments encroached in that direction. (From P. E. Potter *et al.*, *Oil Gas J.*, 25 January 1982.)

depths. To the time of this writing, no potentially large recoveries of gas have been supported by test drilling, and the prospects of significant supply in the short term are not high.

10.8.4 Bituminous sands

Bituminous sands are those impregnated with oil too heavy and viscous to be extracted by conventional drilling techniques. They are popularly misnamed *tar sands*, a term so entrenched in the literature and lore of the oil industry that the US Geological Survey has wisely proposed a definition for it. A *tar sand* is any consolidated or unconsolidated rock that contains a hydrocarbon material with a gas-free viscosity, measured at reservoir temperature, greater than 10 000 mPa s, or that contains a hydrocarbon material that is extractable from the mined or quarried rock. This definition excludes other fossil fuel materials of very high viscosity – the coals, oil shales, and solid bitumens. On the other hand, it seems to include tarry oil deposits in rocks other than sandstones. The large Hit "tar mat" in Iraq, for example, is in a Tertiary limestone, and the Melekess deposit north of the Caspian Sea is partly contained in a Permian dolomite. In two of California's basins, and in coastal Algeria, deposits resembling "tar sands" are actually oil-saturated Miocene diatomites (Sec. 13.4.9).

Many subsurface sandstone reservoirs contain oil too viscous to be commercially produced even under artificial stimulation. They are commonly at shallow depths (less than 600 m) and not far above major unconformities. A North American example is the group of pools of 10° gravity oil in Pennsylvanian sandstones in western Missouri and eastern Kansas, west of the Ozark uplift. They have a close parallel in the Ural–Volga Basin of the USSR, where some 80 fields of sulfurous reservoir bitumens occur (at depths of 400 m or less) in Permian strata above oil pools in Carboniferous sandstones.

Many more deposits of closely similar type, containing equally heavy oil, are nonetheless exploitable with the aid of artificial stimulation – usually cyclic steam injection under pressure (see Sec. 23.13.3). These deposits should be called *heavy oil sands*, not tar sands. Examples include the Cold Lake and Peace River deposits in Alberta, Canada, above the sub-Cretaceous unconformity (see Fig. 10.18 below). In-place oil in these two deposits amounts to some 2×10^{10} m^3. The Boscan field in western Venezuela has succeeded in putting out about 100×10^6 m^3 of oil of about 10° gravity because of the excellent porosity and permeability of the Eocene sandstone reservoir, which wedges out above an unconformity. When the light oil fields of the northern North Sea Basin approach depletion, production like that from Boscan may be achievable from the Paleocene sands which overlap on to the pre-Mesozoic basement along the western edge of the Viking graben (see Fig. 16.26); their oil is about 11.7° gravity, at shallow depths.

Surface and near-surface bituminous sands By far the largest and most prospective tar-sand deposits, however, are exposed at the Earth's surface or buried beneath no more than thin, little-consolidated surficial sediments such as glacial drift or alluvium. Provided that their bitumen content is not less than 20–40 US gallons per ton (0.8×10^{-4}–1.6×10^{-4} m^3 kg^{-1}), such deposits are potentially exploitable by either mining or steam-soak techniques.

Mining of heavy oil sands was given an early start in Europe, principally at Pechelbronn on the west side of the Rhine valley, and in Romania. In 1976 it was begun in the original sector of the Baku oilfields, on the Caspian Sea, from which 100 years of conventional production had succeeded in recovering only about 20

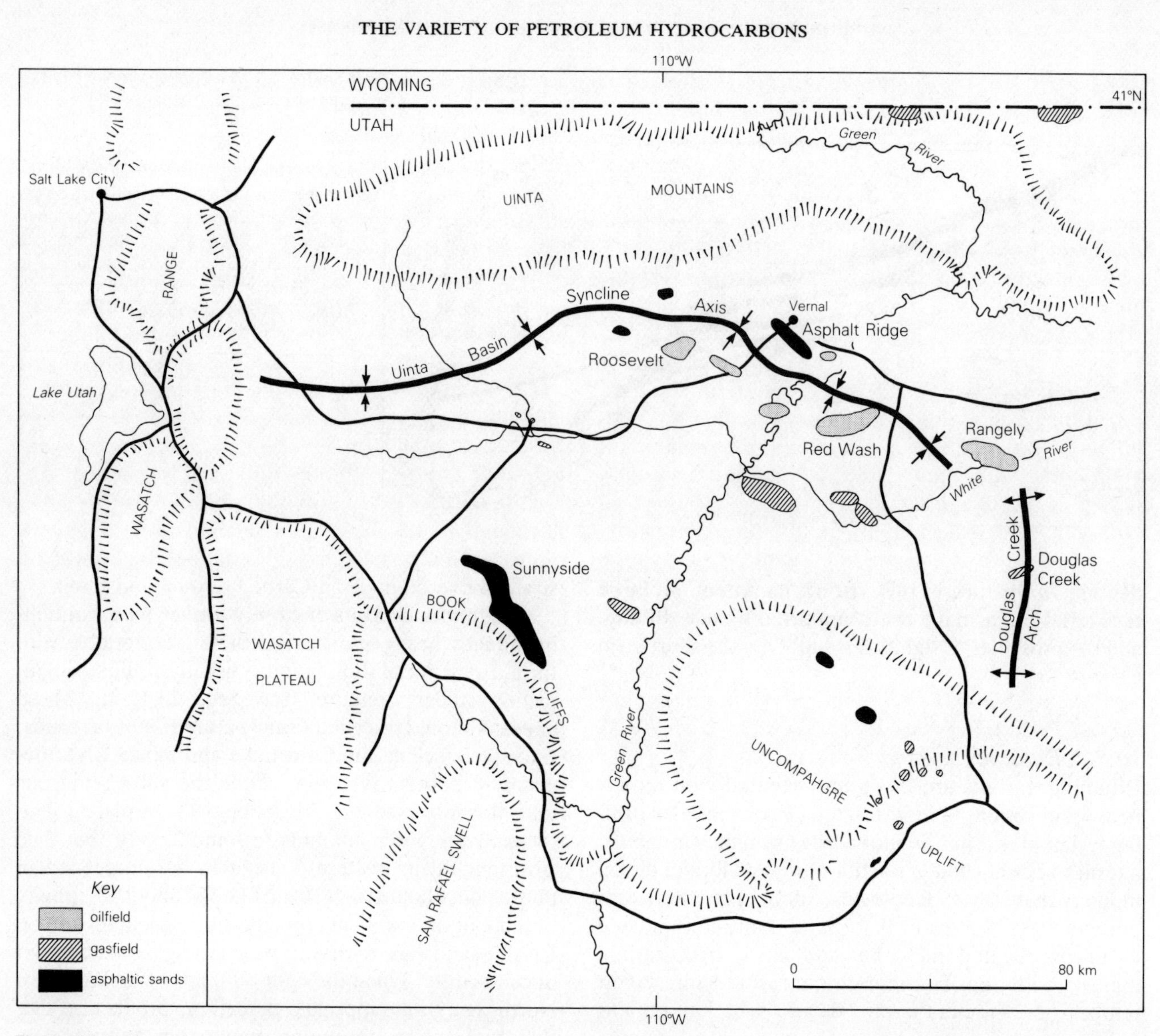

Figure 10.12 Locations of the major tar sands of the Uinta Basin, Utah, with conventional oil- and gasfields included. (From P. H. Phizackerley and L. O. Scott, 7th World Petroleum Congress, 1967; after R. E. Covington, 1963.)

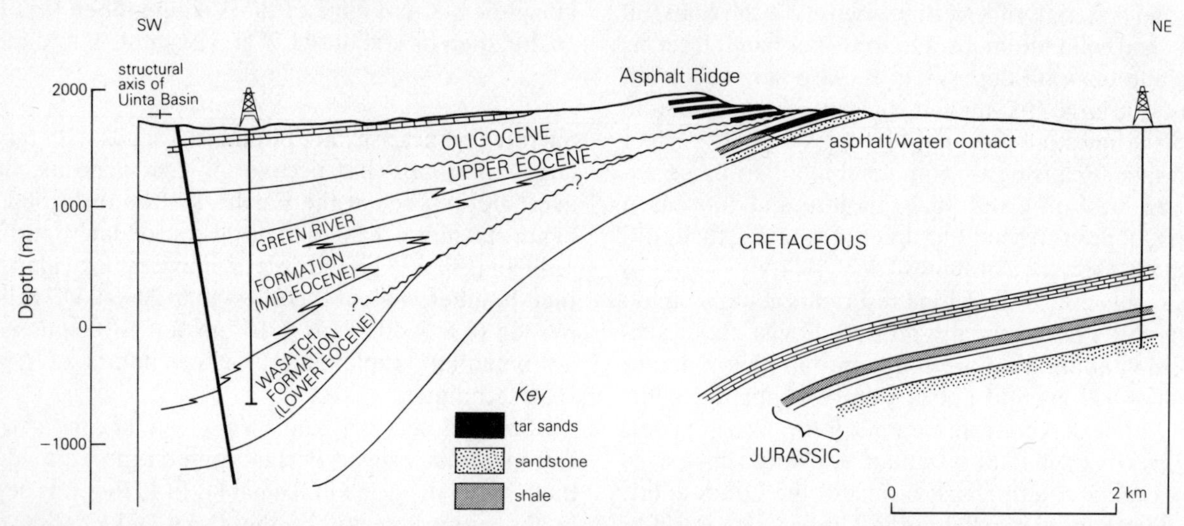

Figure 10.13 Cross section through the Asphalt Ridge deposit on the northern margin of the Uinta Basin, Utah, showing its relation to the Green River oil shales and to the Cretaceous/Tertiary unconformity. (After P. H. Phizackerley and L. O. Scott, 7th World Petroleum Congress, 1967; and R. E. Covington, 1963.)

percent of the oil in place. In these districts, the minable sands are merely the exposed parts of productive, conventional oil reservoirs. In many other regions, the "tar sands" contain nearly all the oil, with subsurface production trivial or nil. They commonly involve an outcropping sandstone immediately overlying or underlying an exposed source sediment.

Virtually all the "tar sands" of the USA are in this category. Of about 8×10^9 m^3 of 8–15° gravity oil estimated to be in place in US sands, nearly 50 percent is in two regions of Utah. In northern Utah, the sands lie close to the Cretaceous/Tertiary contact around the margins of the Uinta Basin (Figs 10.12 & 10.13), which contains the early Tertiary Green River oil shales. In southeastern Utah, the so-called "Tar Sand Triangle" contains saturated sands in a Permian arkose shed westward from a basement uplift to interfinger with the Phosphoria source sediments; the smaller Circle Cliffs deposit is in a Triassic sandstone overlying the Phosphoria. Most of the American tar sands not in Utah lie around the landward margins of the Santa Maria Basin in California, the saturated sands immediately overlying the rich Miocene source sediment of the Monterey diatomite (see Fig. 13.75).

Very large tar-sand deposits Compilations of the largest known tar-sand deposits, by Peter Phizackerley and Laura Scott (in 1967), E. J. Walters (in 1974), Gerard Demaison (in 1977), and other authors, indicate that about 20 deposits contain between them more than 3×10^{11} m^3 of oil in place (Fig.10.14). About 98 percent of this volume is contained in sands in Venezuela, the Canadian province of Alberta, and the Soviet Union. The true tar-sand deposits accessible to surface exploitation are dominated by three gigantic accumulations, those of the Orinoco oil belt, the Athabasca tar sands in Alberta, and the Olenek tar sands in Siberia. There are important differences between them.

The *Olenek deposits* are in north-central Siberia, between the Lena–Anabar trough and the Vilyuy Basin. Permian sandstones contain a volume of in-place heavy oil estimated at about 10^{11} m^3. Because of the remote location and harsh climate, there seems little likelihood of this large accumulation ever being exploited in a country possessing such large reserves of coal and natural gas.

The *Orinoco oil (or tar) belt* of eastern Venezuela is perhaps the largest continuous accumulation of oil in the world. From Guarico in the west to Cerro Negro in the east (Fig. 10.15), the belt is about 600 km long and averages 50 km wide and is easily accessible via the Rio Orinoco along its southern edge. All sands of Eocene or Oligocene age are saturated within the belt, as well as some in the underlying Cretaceous. The entire belt is covered by Quaternary and Recent alluvium, but net

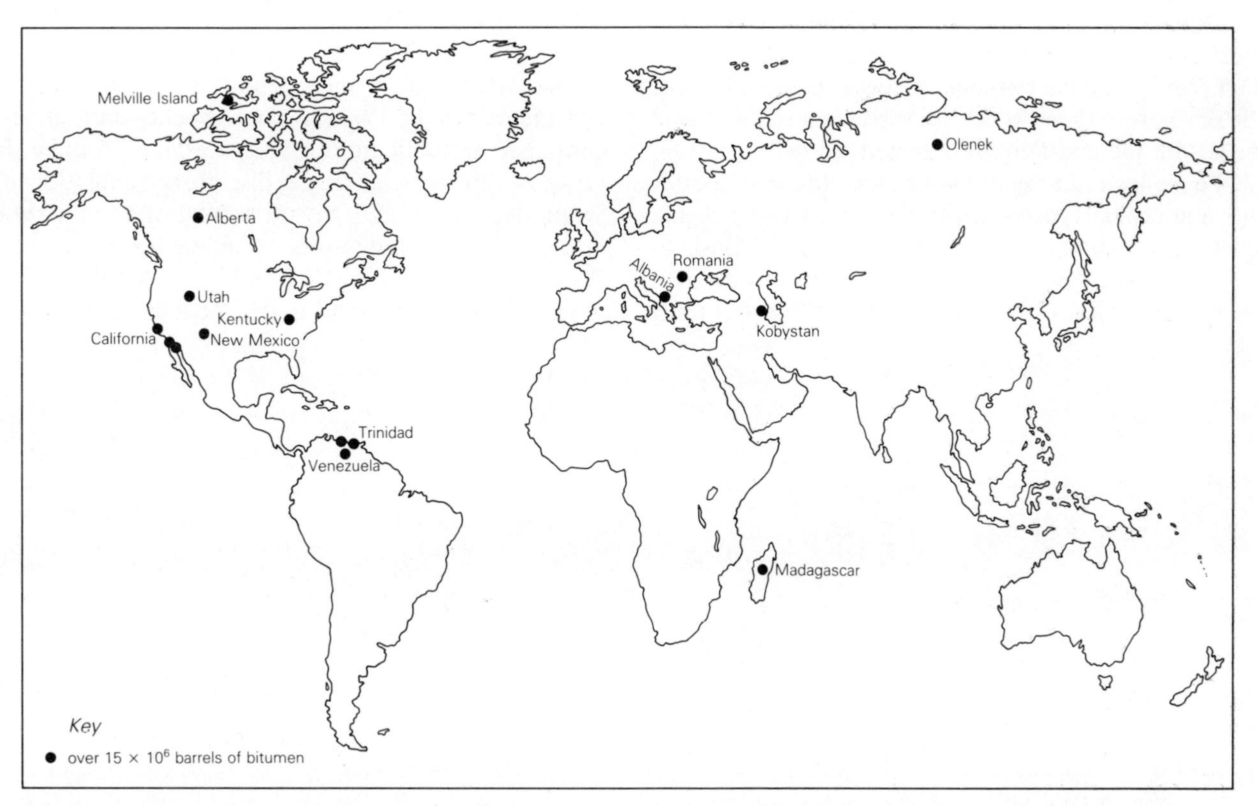

Figure 10.14 Location map of major deposits of bituminous sandstones. (From E. J. Walters, *Can. Soc. Petrolm Geol. Memoir* 3, 1974.)

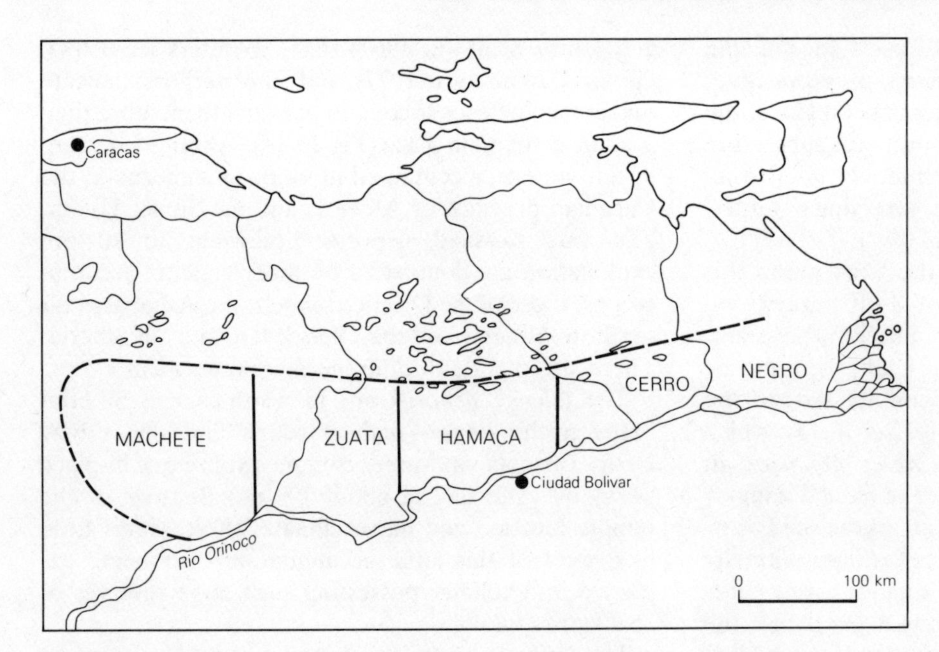

Figure 10.15 The Orinoco heavy oil belt in eastern Venezuela, showing its spatial relation to conventional oilfields of the basin north of it. (From *Oil Gas J.*, 4 February 1980.)

pay thicknesses of 40–80 m, between 180 and 1800 m depths, represent oil in place of at least 1.2×10^{11} m³, and at least 5×10^{10} m³ recoverable. The oils are mostly in the 8–12° gravity range, with 3–5 percent sulfur and high metal content, but they are of remarkably low viscosity – only a few thousand millipascal-seconds – apparently because the asphaltene content is less than 10 percent. On the US Geological Survey criterion (see above), the Orinoco sands are heavy oil sands, not true tar sands. Combined with the unconsolidated nature of most of the sands, the high geothermal gradient (about 35 °C km^{-1}), and the hot climate, these characteristics of the oils permit them to be exploited from conventional wells with the assistance of injected steam.

The geologic setting of the Orinoco heavy oil belt is the homoclinally dipping shelf between the Guyana shield to the south and the eastern Venezuelan Basin to the north. The oils become lighter northward towards the basin (see Sec. 17.7.5), and the "tar belt" is arbitrarily limited to the area south (shelfward) of the 15° isogravity (see Figs 10.16 and 17.29). Along the northern fringe of the tar belt, consequently, the oil is produced from conventional wells without artificial stimulation; many wells actually flowed for some time, and the fields in which this occurred had produced more than 100×10^6 m³ of oil by conventional means by the end of 1983. Recent forecasts of the ultimate potential of the belt are almost certainly too optimistic, but an output of 500 000 barrels per day (about 0.3×10^8 m³ yr^{-1}) by the end of the century is not beyond achievement.

The *Athabasca tar sand* deposit has been described (by Demaison in 1977) as "the world's largest self-contained accumulation of hydrocarbons." Unlike the Orinoco deposit, it is a single, discrete accumulation in a single reservoir rock. It is the largest of a remarkable group of heavy oil deposits extending across northern Alberta in western Canada (Fig.10.17). All of them are in Lower Cretaceous sandstones, the Athabasca deposit

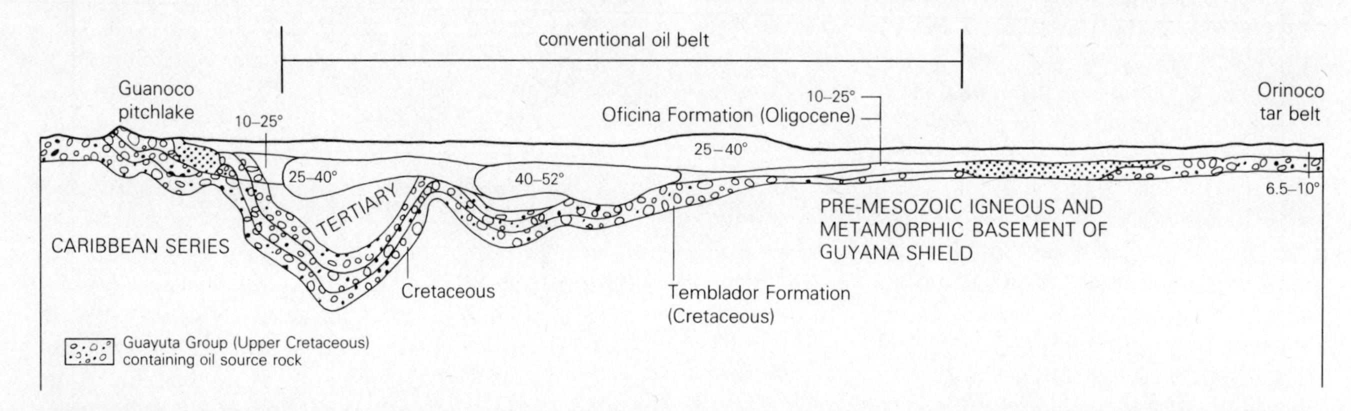

Figure 10.16 Diagrammatic cross section of the Eastern Venezuelan Basin, showing the distribution of "tar" belts, heavy oil belts, and the light oil region of the central basin. Vertical scale greatly exaggerated. (From E. J. Walters, *Can. Soc. Petrolm Geol. Memoir* 3, 1974; after J. Dufour, 1957.)

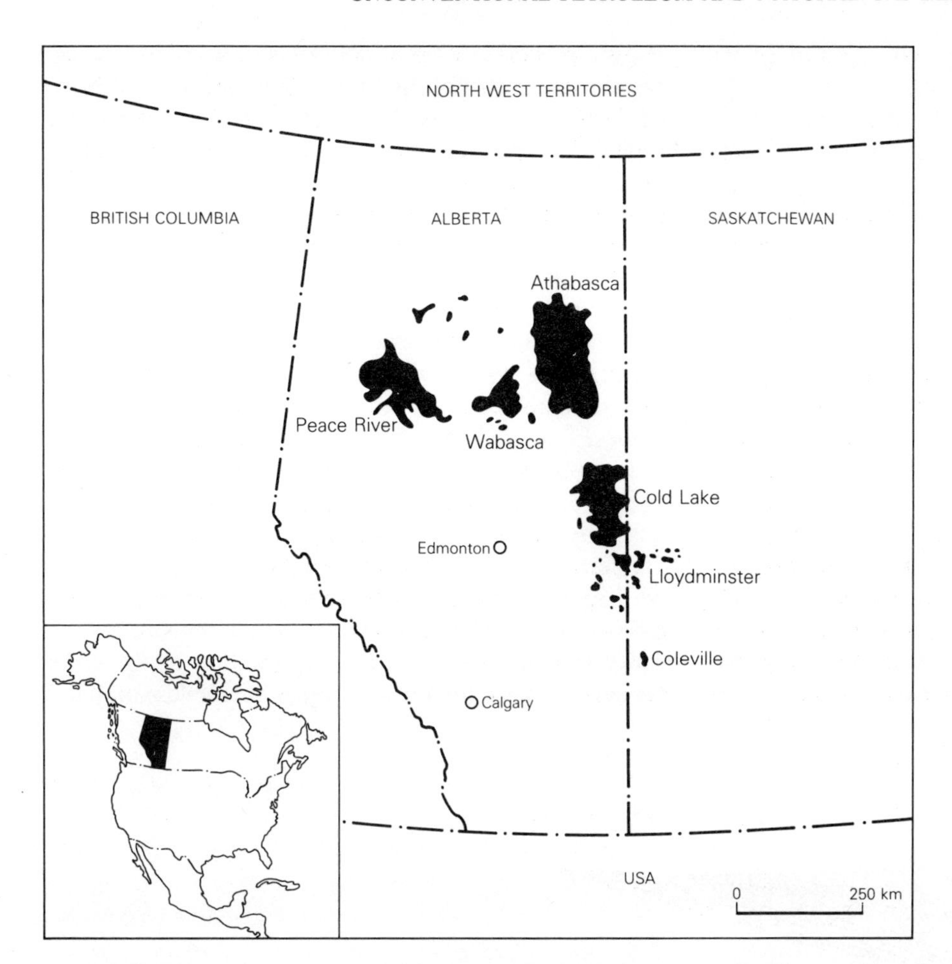

Figure 10.17 Heavy oil deposits in the Lower Cretaceous sandstones of Alberta, Canada. The Athabasca, Peace River, and Cold Lake deposits are described in this chapter. (From D. Jardine, *Can. Soc. Petrolm Geol. Memoir* 3, 1974.)

being in the lowest sandstone resting unconformably on Devonian limestones and evaporites. As the sandstone body containing the "tar" is called the McMurray Formation, and the Athabasca sandstone is a Precambrian formation on *Lake* Athabasca, almost entirely in the province of Saskatchewan, the tar sands should be called the McMurray tar sands, but efforts to promote this change of name have foundered.

The host rock is a cleanly washed, white quartz sand formation 50–80 m thick, with net thickness of about 20 m saturated with "tar." Some of the sand contains 17–19 percent by weight of bitumen, but the grade is very variable; in general, the outcropping parts are the richest. There are some interstratified *beds* of bitumen, from a few centimeters to 0.5 m thick.

The lower part of the reservoir sandstone is an alluvial complex, the upper part consists of tidal flat and lower delta plain deposits (see Sec. 13.2.7). The overall reservoir complex is therefore coastal and transgressive, principally nonmarine but with marine components at the top. The highest oil saturations are found in the fluvial and tidal channel sands, which without the "tar" to bind them would be essentially unconsolidated. Porosities up to 35 percent and permeabilities of several darcies (see Sec. 11.2.3) are recorded, but the "tar" is too viscous to take advantage of them. The "tar" is actually an asphaltic residuum of about 10.5° gravity. Its relation to its reservoir rock is unique, and has been the chief cause of the extraordinary extraction difficulties encountered by would-be operators of the deposit. Much of the oil is not in the pore spaces of the sand, but is wrapped as a coating around individual, water-wet sand grains. The extraction problem has been aptly said to be one, not of getting the oil out of the sand but of getting the sand out of the oil.

The present structure of the tar-sand region is a wide, flat dome caused by drape reversal of the regional dip due to solution of underlying Devonian salt (see Sec. 16.3.8). The salt solution created a wide depression in which a shallow lake formed, only slightly above sea level. Streams flowing into this lake built deltas towards a seaway which was transgressing the region from the northwest, between the Precambrian shield on the northeast and a ridge of resistant Paleozoic carbonate rocks to the southwest. Only the seaway received the sandstones of this earliest Cretaceous invasion.

The tar sands were first described in 1884, first worked about 1921, and became the object of vigorous discussion among geologists following the publication of an opinion by Max Ball in 1935. The discussion has circulated around three points: is the "tar" immature oil or degraded oil; what was its original source; and how did it get into its present reservoir as a coating on its grains instead of a filling of its pore spaces?

Ball thought the "tar" was highly immature "proto-petroleum," lacking natural cracking to make it mature. A low level of thermal maturation has been subsequently confirmed, but this type of immaturity does not in itself account for the characteristics of the "tar." Clifton Corbett considered the tar's present state to be essentially original, and due to *in situ* generation by humic acids from superabundant land vegetation delivered to the coal-generating Cretaceous delta.

In opposition to the hypothesis of immature *in situ*, heavy oil, geologists and geochemists have subsequently reached a measure of consensus (by no means unanimous) that the oil is senile, having suffered biodegradation and some water washing, plus possible oxidation. The effects of these processes become progressively more pronounced up the dip, from the merely heavy oils of the Cretaceous basin to the immobilized "tars" of the Athabasca deposit. Nonetheless, it has been emphasized by Douglas Montgomery and others that degradation processes cannot themselves account for the present chemistry of the "tar." The high concentration of unstable compounds – of vanadium porphyrin (360–550 p.p.m., in contrast with the 30–300 p.p.m. more characteristic of heavy crude oils), of sulfur compounds (up to 5.5 percent by weight), and of asphaltenes (responsible for the tar's infamous viscosity) – prove that the oil *in its present form* has not been extensively weathered, evaporated, or polymerized.

There is near-universal agreement that the oil has undergone long *lateral* migration. There is conspicuous lack of agreement on whether it has also undergone *vertical* migration, from pre-Cretaceous source sediments. R. G. McConnell proposed before the turn of the century that the "tar" represented oil inspissated near the surface after migrating from the underlying Devonian. T. A. Link was notable among geologists who agreed with this interpretation following the oil discoveries in Devonian reefs, down the dip to the southwest, in 1947–9. The areal relation between the tar-sand deposit and the subcrop of the Devonian shales

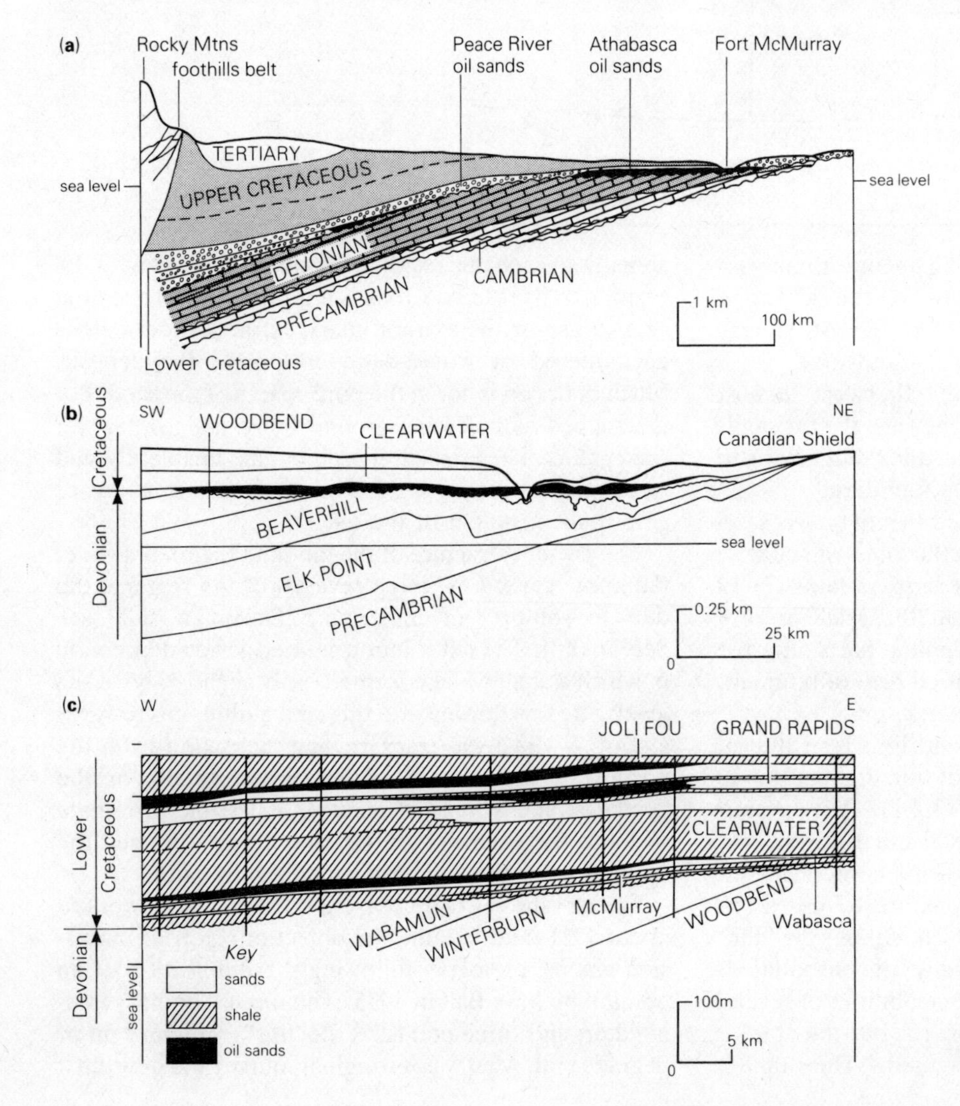

Figure 10.18 Cross sections of (a) the Alberta Basin, (b) the Athabasca area and (c) the Wabasca area, in western Canada, showing the essentially homoclinal structure of the heavy oil sand deposits and their relations to the sub-Cretaceous unconformity. (After D. Jardine, *Can. Soc. Petrolm Geol. Memoir* 3, 1974; and G. A. Stewart, First Unitar Conference, 1979.)

and carbonates is direct (Fig.10.18); in particular, a large, tabular, reef dolomite occupying much of the subcropping interval is in facies contact downdip with the principal source sediment for the productive Devonian reefs, but is itself truncated at the unconformity directly beneath the tar sands. The dolomite there contains no oil or tar, but along the so-called "carbonate trend" to the northwest the Devonian carbonates are saturated with heavy oil comparable with that in the Lower Cretaceous reservoirs. W. C. Gussow, writing before the discovery of the "carbonate trend," considered the McMurray deposit to be in a gigantic stratigraphic trap, containing oil that had migrated from post-Devonian but pre-Cretaceous rocks formerly occupying most of the Alberta Basin; the clean, tar-free sands at the updip edge of the reservoir (right-hand end of Fig. 10.18) were thought to represent an original gas cap.

The geochemists are unanimous that the "tar" oil is like other Cretaceous oils and unlike Devonian oils in the Alberta Basin. In particular, the heavy Cretaceous oils contain virtually no paraffins but much higher proportions of cycloparaffins and aromatics than any of the Devonian oils. Lewis Weeks, Laurence Vigrass, and many others therefore postulated long migration from Cretaceous shale sources far to the west, preferentially directed around barriers on which no Lower Cretaceous sandstones were deposited.

There are severe obstacles to acceptance of this conclusion. Marine conditions in the Alberta Basin lasted for only about 30 Ma in all post-Jurassic time. Most of the deposits of those 30 Ma are undoubtedly argillaceous, but they have none of the attributes of *rich* oil-source sediments. Their organic carbon content does not average anywhere near 2 percent; the kerogen in the shales equivalent to or overlying the tar sands is of Tissot's Type III; and none of the shales in the eastern half of the basin (where the tar sands are) have ever reached generative maturity. These shales are required, according to the geochemical data, to have generated enough oil to allow at least 15×10^{10} m^3 of it (equivalent to the total quantity of recoverable, orthodox oil found in the oil industry's entire history) to survive migration across the basin and accumulation in pools *after* the losses inevitably sustained in the destruction of all the light fractions.

The dilemma presented by this greatest of self-contained hydrocarbon accumulations is multifaceted. Of the various migrational mechanisms that have been proposed, and that are discussed in Chapter 15 of this book, it seems that two must be rejected out of hand as applicable to such very lean source sediments. Single-phase fluid flow, and transport via a continuous kerogen network in the carrier rocks, are out of the question. To escape this aspect of the dilemma, Laurence Vigrass suggested that the long transport of the oil had been in micellar solution (see Sec. 15.4.2). Robert Jones called upon meteoric water to transport the oil, requiring all the precipitation that has fallen upon the large drainage basin since the uplift of the Rocky Mountains 50 Ma ago to have entered the rock column. If only 1 percent of it passed through the reservoir sandstones, an oil solubility of only 3 p.p.m. would permit the trillion (10^{12}) barrels in the heavy oil deposits to be transported in aqueous solution.

There remains the possibility, maintained by Gallup, Gussow, and the present author, among others, that the source of the oil is not in the Cretaceous section at all, despite the conclusions drawn from geochemical data. The "tar," at least, cannot possibly have migrated over hundreds of kilometers in its present state. It must therefore have earlier been in a quite different state, and inferences drawn from its present chemical state are not definitive of its origin. Many of the puzzles would be removed if the oil originated in any or all of several rich, pre-Cretaceous source sediments in the basin to the west, and migrated towards the foreland in the normal way before the first Cretaceous sediments were laid down. The truncated Devonian reservoir might then lose the oil as seepage droplets into the newly formed Cretaceous lake and shallow sea; the droplets adhered to the sand grains of the developing fluvial–deltaic–coastal complex, and were deposited as part of the sediment. Thus no heavy, viscous oil has "migrated," in the conventional sense, and entry into the present reservoir would precede the cohesion of the reservoir's components. This concept, that the "tar" acquired both its present physicochemical condition and its relation to its host rock through being of "sedimentary" derivation, was probably first developed by T. A. Link, and was later elaborated upon by others, notably by J. B. F. Champlin and H. N. Dunning.

Despite its superficial attractiveness, the postulate is not without serious drawbacks. In particular, no-one has found a satisfactory explanation for the oil surviving below an erosion surface which was developed over some 150 Ma before it was finally buried by Cretaceous sediments; and it is experimentally impossible to get viscous asphaltic oil to adhere as a coating around sand grains in an aqueous environment. It is among the greatest ironies (and, for some, among the greatest embarrassments) of the oilman's science that such a large-scale and geologically simple phenomenon should continue to defy unanimity on so many vital points.

Settings for giant tar-sand deposits Nearly all large tar-sand deposits share a number of obvious common features, enumerated by many authors but most recently by Gerard Demaison in 1977. The reservoir sands themselves are in almost all cases areally widespread fluvial–deltaic–coastal complexes deposited between emergent landmasses and large, offshore sedimentary basins. The reservoirs were therefore able to draw oil by migration from large source areas.

Between source and reservoir are extensive, interfingering, permeable carrier sandstones, little interrupted by faults or other barrier-forming structures. The carrier beds and the reservoir bed occupy a simple homoclinal slope with updip stratigraphic convergence towards the edge of the basin. The common trap is therefore stratigraphic in the American sense, but it is commonly somewhat modified by either gentle arching or local reversals due to shelfward-dipping faults (see Sec. 16.7). Thus the traps are such as would be termed *lithologic-stratigraphic* or *structural-stratigraphic* by the Russians (see Sec. 16.2).

The extensive development of the reservoir rock must be followed by an equally widespread regional seal, commonly a transgressive marine shale. This combination permits long-continued, long-distance migration of the oil, focused towards a single target reservoir at the updip extremity of permeability. The migration is facilitated by a hydrodynamic gradient towards the foreland. If this updip gradient is maintained, much or all of the oil may be driven completely out of all traps and into a tar mat at the surface, unless the updip shutoff of permeability is very effective. If on the other hand the hydrodynamic gradient weakens with the reversion of the basin to the cratonic stage, meteoric waters encroach upon the oil-filled reservoir rock from foreland outcrops, causing degradation of the oil and sealing the accumulation with asphalt.

Demaison included the observation that large tar-sand deposits are likely to be Cretaceous or younger in age. This is true of the two largest known examples, but so it is of most conventional oilfields. True tar sands, at or close to the surface, are unlikely to escape destruction by erosion for many geological periods, but the actual stratigraphic distribution of bituminous sandstone deposits is not essentially different from that of conventional oil deposits.

Recovery and reserves of oil from tar sands Though tar sands and heavy oil sands represent a vast *resource* of hydrocarbons, the potential of that resource for conversion to *reserves* is a function of technology, economics, and environmental impact. *Heavy oils*, like those at Cold Lake and Peace River in Alberta, or the Orinoco oils, are recoverable from wells provided that their viscosities are reduced by some form of preheating.

The commonest method of preheating is by cyclic, high-temperature steam stimulation at about 325 °C. The steam takes a long time – several years – to sweep through the reservoir rock at a typical depth of 500–600 m. The oil is recovered from the producing wells by conventional pumping.

The hindrances to wide adoption of this process are several and obvious. The capital cost of a single plant is measured in the billions (10^9) of dollars. The demand for water is so high that it cannot be met in a region of low rainfall (or snowfall). The energy input, requiring abundant local supplies of either coal or cheap natural gas, is such that the net energy gain from the process is severely reduced. Producing wells have shorter lives than they do in conventional oilfields, because the oil in the reservoir rock is too viscous to move towards depleted wells; thus great numbers of wells have to be drilled. They may be directionally drilled, in clusters, with a certain proportion of them (say 50 percent) on the steam-injection cycle whilst the rest are in the production cycle. Operating costs therefore quickly become comparable with the initial capital costs, and the cost per daily barrel of production is very high compared with that for any other form of oil production (see Sec. 3.10).

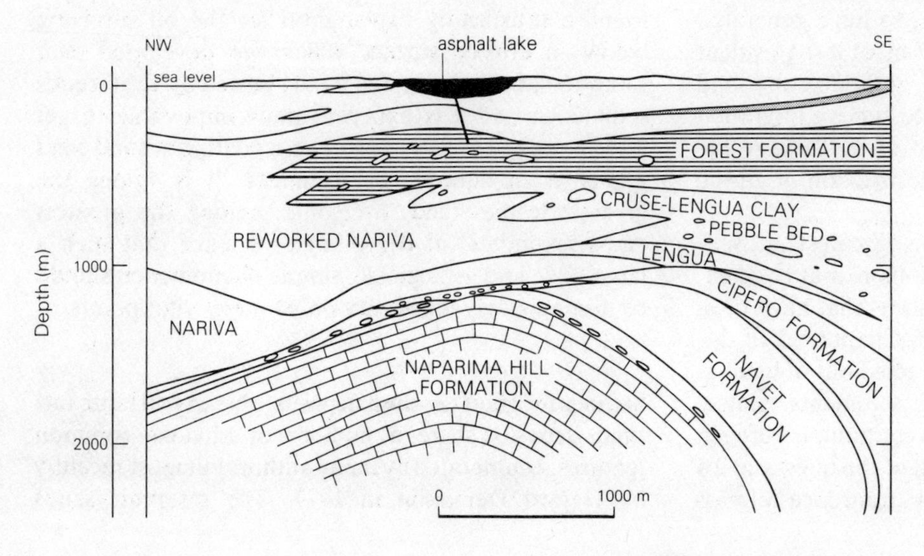

Figure 10.19 The asphalt ("pitch") lake of Trinidad in relation to rich source sediments (Lengua, Cipero, and Naparima Hill Formations). (After H. G. Kugler, Petrolm. Assoc. of Trinidad, 1961.)

No simple, *in situ*, subsurface recovery process has yet been satisfactorily developed for the *Athabasca tar sands*, because the relation between the "tar" and the sand grains of the reservoir rock precludes simple release of the oil by preheating. The tar sands consequently present a *mining* problem, not a drilling problem. Mining by gigantic toothed-wheel power shovels, or by draglines, is very inefficient in northern Alberta's harsh winters. Each month's operation by a large facility produces the world's largest open pit. Because the sand is loose and incoherent once deprived of its bitumen, it cannot be released into the interior drainage system and must therefore be put back into the pits. As it is incapable of regenerating a soil cover, the thin original overburden must be preserved and carefully replaced on top of the sand.

Clearly there is a shallow depth limit for surface mining – about 60–75 m. Below that depth, a vast volume of the tar sand is too deep for surface mining but not deep enough for *in situ* recovery (which requires some minimum thickness of overburden pressure above the steam-soaked reservoir rock). The total oil in place in the deposit is at least 100×10^9 m^3. Of this large volume, about 45×10^9 m^3 are believed to be "recoverable" – if some acceptable *in situ* extraction process can be devised. *Reserves*, accessible to the mining process, are about 4.25×10^9 m^3 – an amount representing some 4 years of supply for North America at 1980's consumption rate.

10.9 Origin of the nonhydrocarbon constituents of petroleum

The application of geochemical techniques to the range of organic materials in rocks has greatly clarified our understanding of the true nature of petroleum and natural gas (see Chs 5 & 7). The origin of their nonhydrocarbon constituents has proved to be a much more refractory problem.

Though hydrogen and CO_2, at least, are easily explicable as minor byproducts of organic transformation to petroleum, they are as readily ascribable to other origins. So are nitrogen and H_2S. Helium, argon, and radon (to say nothing of the metals in the porphyrins) *must* be ascribed to independent origins.

This conflict has led to a strange polarization of views, between an almost unanimous majority of geochemists and geologists and a persistent minority holding the opposite view, nearly all Russians. The majority inclines towards the closest plausible relationship between the nonhydrocarbons and the organically derived hydrocarbons. Until the 1960s, however, Russia's oil and natural gas came almost entirely from either Paleozoic rocks not far above Precambrian basement, or from strongly deformed Tertiary strata punctured by mud volcanoes and chopped up by faults (see Chs 17 & 25).

The outcome was a predilection among many Russian scientists for a deep-seated origin for the oil and gas themselves, and the same origin *a fortiori* for the nonhydrocarbon constituents. Because the evidence for organic origin is far less conclusive for the nonhydrocarbons than for the hydrocarbons, we will consider the alternatives for each of the former.

10.9.1 Origin of sulfur and its compounds in oils and gases

As already indicated, sulfur occurs in oils in a variety of compounds, some of them complex. They are not closely similar to the common biological compounds of sulfur, which is one of the elements essential to life. Though some of the sulfur compounds in crude oils may be formed at the same time as the oils, and possibly from the same parent material, they must undergo changes as the hydrocarbons themselves undergo changes. The sulfur contents of the oils as we now find them are therefore a function of three controls:

(a) the initial content of the organically combined sulfur in the source material;
(b) the stage of maturation of the resulting petroleum hydrocarbons;
(c) subsequent changes that may have affected both the hydrocarbons and their sulfur contents, during or after migration and accumulation (see Sec. 10.4).

The amount of sulfur initially incorporated into the sedimentary organic matter (OM) is determined by several circumstances:

(a) the amount of the sulfate (SO_4^{-2}) ion in the water (see Sec. 9.2);
(b) the redox potential (E_h) at the surface of sedimentation; this controls the intensity of microbial reduction of the sulfate in the water, forming H_2S which is both soluble and highly reactive, especially with metallic cations;
(c) the abundance of these metallic cations, especially of the thiophile elements like iron;
(d) the nature and abundance of the OM itself, as this has to provide groups which are reactive to the sulfur.

Native, *elemental* sulfur is easily formed in the supergene environment, especially where the carbonate–sulfate facies prevails. Elemental sulfur is also present in dispersed OM in far greater quantities than in oils, in which it is an exceedingly minor contributor to the total sulfur content. Both the OM and the elemental sulfur imply very low stratal temperatures, possibly below

25 °C. The *total* sulfur content of very many oils, however, is too high for mere survival from sulfur-containing biochemical compounds; the sulfur compounds characteristic of crude oils are secondary products. Their formation requires the action of sulfurizing agents – native sulfur, H2S, or the variety of polysulfides – and the simplest means of forming these agents at moderate temperatures in sediments is via microbial reduction of soluble sulfates.

The critical question is whether these reactions take place early in the generative processes leading to oil formation, or long after them when the oils are already migrating or even have formed their final accumulations. If the early case is operative, the sulfur and the oil were closely connected in their origins. If the late case is operative, the origins of the oil and of the sulfur may be quite independent of one another and the sulfur tells us nothing about the ultimate origin of the oil.

Attempts to reach an answer to this question have accorded great emphasis to studies of the *isotopes of sulfur*. Sulfur, with atomic number 16, has four stable isotopes, their mass numbers being 32, 33, 34, and 36. None of them is radiogenic. ^{33}S and ^{36}S are present in insignificant quantities (less than 1 percent); about 95 percent of all sulfur is ^{32}S. The ratio $^{34}S : ^{32}S$ is therefore important and is easily measured in relation to some arbitrary standard, which is taken to be the ratio of 1 : 22.22 in the mineral troilite (FeS) from the Canyon Diablo meteorite. Deviations from this standard value, called *delta values*, are calculated from the equation

$$\delta\ ^{34}S \text{ (per mil)} = \frac{(^{34}S : ^{32}S)_{\text{sample}}}{(^{34}S : ^{32}S)_{\text{standard}}} - 1 \times 1000$$

Thus $\delta\ ^{34}S$ (‰) is zero by definition for the meteoric sulfur. The "hard rock" geochemists reached an understanding of the true significance of sulfur isotope data well before their organic geochemistry counterparts. The sulfur-isotopic composition of any hydrothermal mineral is controlled by aspects of the ore solution chemistry: the pH level, the fugacities of oxygen and of sulfur, the temperature, and probably other quantities also. Very small changes in any of these variables may result in much larger changes in the $\delta\ ^{34}S$ values at equilibrium, and sulfur isotope data in themselves cannot be used to determine the ultimate origin of the sulfur in the deposits.

Delta values for sea water show considerable fluctuation but are always strongly positive, averaging about +20‰. Marine sulfates, such as anhydrite, are therefore also strongly positive. Bacterial reduction of sulfate in muds leads to enrichment of the heavy isotope in the sulfate but its depletion in the sulfide. The $\delta\ ^{34}S$ values in crude oils are therefore some 15‰ or so lighter than is sea water of the same age; but the values are still positive. This enrichment in the light isotope reflects isotopic fractionation during deposition of the sulfur-bearing sediment. The enrichment reaches its maximum, and $\delta\ ^{34}S$ its lowest values, in organisms, organic sulfur, sedimentary sulfides, and H_2S.

Within individual oil provinces, and particularly in reservoirs of comparable ages, $\delta\ ^{34}S$ values are remarkably homogeneous, even though the total sulfur content of the oils and gases may vary widely. The homogenization is most easily explained by isotope exchange reactions, which are temperature-dependent. This implies early incorporation of the sulfur. However, the associated H_2S is not homogenized with respect to sulfur isotopes. In shallow, low-temperature reservoirs, microbial H_2S does not react sufficiently with the associated oil to alter the $\delta\ ^{34}S$ of organic sulfur, and the $\delta\ ^{34}S$ of the oils and the H_2S are essentially unrelated. The oils, on the other hand, may be sulfurized late, after their migration, through sulfate reduction; oilfield waters are very commonly deficient in sulfate ions (see Sec. 9.2). This assumes that petroleum is about as reactive with H_2S as the predecessor OM; the latter is certainly easily sulfurized, but it is not certain that crude oils are.

Many S-organic compounds of oils and bitumens have lower thermal stabilities than do hydrocarbons; they decompose readily at 100°C or even lower, giving off H_2S. Most H_2S-rich gases are found at depths below 3–4 km, where stratal temperatures are much in excess of 100°C. Biogenic oxidation–reduction processes are excluded at these temperatures; abiotic reduction of sulfates by methane at high temperatures would produce an equivalent concentration of CO_2, but this is not found to occur. In reservoirs having these high temperatures, the thermal maturation of oils (see Ch. 7) is inevitably accompanied by *desulfurization*; the sulfur is lost partly as H_2S and partly through the precipitation of reservoir bitumens (see Sec. 10.6). According to Wilson Orr (1974), thermal desulfurization of organic sulfur compounds occurs with negligible isotope fractionation. Nonmicrobial reduction of sulfate ions, however, continues very slowly and without any fractionation, and the sulfide so formed is incorporated into both the oil and the H_2S. Organic sulfur thus becomes a dynamic system, with competing sulfurization and desulfurization leading to changes in $\delta\ ^{34}S$ towards that of the reservoir sulfate, about 15‰ heavier isotopically than the sulfur initially in the oil.

At still greater depths and higher temperatures no elemental sulfur normally survives if hydrocarbons are present. In the views of Russian geochemists, notably of L. A. Animisov and N. B. Valitov, the sulfur reacts with the hydrocarbons, forming sulfur-rich oils and gases with very low concentrations of methane homologs (which are consumed during the reaction). Any excess of sulfur sulfurizes any available OM, forming polysulfide organic compounds. These enrich the oils in sulfur. The hydrocarbon most readily reacting with sulfur is cyclohexane; it and its near relatives are dehydrogenated. Experiments by Valitov, involving reactions between gasoline and sulfur at 175–200 °C, showed that H_2S forms much earlier if limestone (not dolomite) is present as a catalyst.

It is apparent that sulfur isotope data are not particularly useful, in themselves, as indicators of the sources of the sulfur, still less of the hydrocarbons. Before the work of the hard-rock geochemists, it was maintained that sulfur isotope ratios in oils did not change materially with time, but were indicative of the nature of the source. Thus similar ratios in the oils and the sulfate of the reservoir succession (where this consists essentially of carbonate and anhydrite) meant that the oils were indigenous to the reservoir succession and had not migrated into it from any distant source. This viewpoint is no longer tenable. For some mature oils, especially oils of Paleozoic age, $\delta\ ^{34}S$ has been shown to increase during maturation, even though the total sulfur percentage decreased. Oils that have been subjected to high temperatures take on foreign sulfur. Oils that have been subjected to alteration by bacteria and oxygenated waters, at much lower temperatures, may acquire high sulfur content both by the introduction of new sulfur and the concentration of their own sulfur by the preferential removal of nonsulfur compounds.

10.9.2 Origin of nitrogen in crude oils and natural gases

Nitrogen, an inert gas, has had a number of possible sources assigned to it when it is present in crude oils:

(a) It may be of atmospheric origin, literally "fossil air." This seemingly obvious possibility is not supported by the proportions of helium or argon accompanying the nitrogen; their proportions in the atmosphere are known, and they also are inert.

(b) The nitrogen may have come from the OM in the rocks, like the oil itself. During the coalification process, starting with OM of humic type (Sec. 7.3), the amount of nitrogen released is insignificant (less than 1.5 percent). Sapropelic OM, in contrast, has a high content of protein, which readily yields ammonia and this may then be oxidized to nitrogen. A. P. Pierce in 1960 invoked the slow decomposition of the chitins, porphyrins, and amino acids contained in sedimentary rocks, which invariably have nitrogen contents higher than those of igneous rocks. Oils at the same time contain much more dissolved nitrogen than do their associated formation waters.

(c) The nitrogen may be of juvenile (abyssal) origin, having seeped upwards from the upper mantle via deep faults. Many geochemists maintain that the nitrogen in the atmosphere originated in this way, but the possibility of a direct contribution from the mantle to an oil deposit is beyond evaluation.

The much more frequent association of nitrogen with natural gases, and its much higher absolute proportion in them than in oils, suggest that the maximum output of gaseous nitrogen coincides with the late stages of the thermocatalytic generation of methane (see Sec. 10.4.5). Nitrogen content of stratal waters increases with the age of the strata. In most petroliferous basins of Tertiary age, the nitrogen content is less than 50 ml l^{-1}, but it rises to 200–300 ml l^{-1} in Paleozoic petroliferous basins and to 1200 ml l^{-1} in the formation waters of the Permian Rotliegendes Formation in northwestern Europe. The implication is that nitrogen in gas pools is enriched during migration through strata in which the waters are saturated with that gas. Studies of changes in the nitrogen's content of the heavy isotope ^{15}N support this enrichment. A parallel possibility is the enrichment in nitrogen during the second generation of a gas, the first generation being low in nitrogen or essentially without it.

Nitrogen in gases may have a different origin if the stratigraphic section contains coal at depth. Maturation of coals releases both methane (about 85 percent of the yield) and ammonia (about 15 percent). Ammonia is soluble in water, and reacts with the ferric iron in the overlying red beds to yield nitrogen and ferrous iron compounds. The nitrogen, consisting of very small molecules, migrates easily even through evaporites, and its presence may indicate deeper reservoirs filled with methane which is unable to follow it.

This separation of nitrogen from methane does not appear to have taken place in the Rotliegendes reservoir of the European gasfields. The lowest nitrogen contents among these are found in the gases of the British North Sea; contents increase eastward to their highest values in East Germany and Poland. The organic content of the largely red Permian strata is too low to account for the amount of nitrogen; isotope data confirm that the nitrogen is not biochemical in origin. The only explanation for the very high nitrogen content seems to call for its derivation from the metamorphic and radioactive Variscan basement rocks, which are enriched in the heavy ^{15}N isotope.

10.9.3 Origin of hydrogen and carbon dioxide in natural gases

Hydrogen and CO_2, like nitrogen and H_2S, can plausibly originate through the organic transformations themselves. For the hydrogen, we may need to look no further than the water present in all sedimentary and organic systems. The trivial percentage of gaseous elemental hydrogen in natural gases (0.5 percent or less) may merely reflect the circumstance that hydrogen is the constituent most easily lost from any system, and is also very easily captured, especially by oxy-salts such as sulfates.

Carbon dioxide is very soluble in petroleum (to seven volume times or more), which is why it is useful in miscible tertiary recovery from depleted oilfields (see Sec. 23.13.3). It is produced in a wide array of permanently progressing reactions: original carboxylation of OM; its subsequent oxidation; and low-temperature reactions of formation waters with carbonate rocks. In the experimental pyrolysis of OM, CO_2 is greatly dominant over methane in the early stages.

Some minor contribution of CO_2 may be made by volcanic sources. Even metamorphic basement rocks may provide a share; both CO_2 and helium occurred profusely in the deep levels of the super-deep well drilled to below 12 km into Precambrian rocks on the Kola Peninsula (see Sec. 23.5).

10.9.4 Origin of helium and argon in natural gases

Helium and argon differ from the other constituents of natural gases in having nothing directly to do with the generation of the gases. Both of them are *radiogenic*. Radiogenic ^{40}Ar is supplied mainly through the decay of radiogenic potassium in potassium-bearing minerals in the sediments (chiefly feldspars and glauconite). The ratio $^{40}Ar : ^{36}Ar$ reveals the degree of enrichment in the radiogenic isotope.

^{4}He is the radiogenic isotope of helium; ^{3}He is atmospheric, formed by the cosmic bombardment of nitrogen. The atmospheric isotope makes a trivial contribution to the helium found in natural gases; the $^{3}He : ^{4}He$ ratio is between 0.5×10^{-7} and 5.0×10^{-7}. The content of radiogenic helium in gases is regularly of the order of ten times the content of radiogenic argon, in volume percent. Both are inert gases, and they are enriched in natural gases (along with another inert gas, nitrogen, which however is not radiogenic) during the thermal decomposition of hydrocarbons by such processes as intrusive magmatism.

The origin of the helium is uncertain:

(a) It is unlikely to be formed by the oxidation of the hydrocarbons themselves; helium-bearing gas deposits invariably contain relatively complex methane homologs.

(b) It might originate jointly with the nitrogen, through α-particle bombardment, as proposed by S. C. Lind as early as 1925. If this were the case, He : N ratios should be approximately constant among gas pools, regardless of their tectonic environments. In fact, the ratios (invariably less than unity) are far from constant, and their principal variability appears to be related to the basement rocks.

(c) Does the helium then come from the basement rock, or from other country rock? No direct relation is apparent between the helium content of gases and uranium or other radioactive elements in neighboring rocks or waters. Some relation is apparent with tectonic protrusions of igneous rock. In the Dineh-bi-Keyah field in Arizona, for example, in which the reservoir rock is an intrusive sill, the helium content is about 6 percent.

(d) Finally, the helium may be of abyssal origin, generated by the degassing of mantle rocks. This origin has been favored by Russian geochemists, who call upon it for the provision of the nitrogen and other gases, for the salt in deep saline basins, and even, in a few instances, for the oil.

10.9.5 Origin of radon in crude oils

The radioactivity of crude oils, where it is measurable, is due to radium emanation and not to the presence of radium salts in solution. As we have seen, however, the radon concentration in crudes is commonly a good deal higher than it should be for the radium concentration. The excess radon must therefore come from some external source, probably from the sedimentary rocks, among which shales have the highest capacity to contribute radon and carbonates the lowest.

The radon content may affect the hydrocarbons themselves. Alpha-radiation releases hydrogen from hydrocarbons in the process of condensing them to higher homologs. The liberated hydrogen may conceivably help towards the hydrogenation of unsaturated hydrocarbons, but it can scarcely be of significance in the generation of petroleum; most natural gases lack hydrogen, and many lack both hydrogen and helium.

Part III

WHERE AND HOW OIL AND GAS ACCUMULATE

We have tracked a succession of processes, from the deposition of an organically rich source sediment to the transformation of its organic matter to petroleum hydrocarbons. In our narrative so far, the hydrocarbons are still in the sediment in which they were generated.

This is not how the geologists and drillers find petroleum hydrocarbons now. Commercial hydrocarbons are recovered from *pools* occupying spaces within rocks of the Earth's outer crust. Pools are able to accumulate in rocks having enough space to accommodate them. This space, no matter how it was created, is called *porosity* (see Sec. 11.1).

The provision of space (porosity) to a rock enables that rock to *contain* hydrocarbons (or any other fluid). It does not enable the rock to *release* the hydrocarbons for delivery to the surface so that they can be used. The property enabling a medium to release its fluid content, as distinct from merely holding on to it, is its *permeability* (see Sec. 11.2).

We have emphasized (Sec. 8.1) that a *source sediment* must possess both porosity and permeability in order for it to contain substantial amounts of OM, thence kerogen and thence, ultimately, hydrocarbons. These hydrocarbons, however, are not *pooled* in an accumulation capable of commercial exploitation. In order for pooling to take place, the source sediment must permit the *expulsion* of its hydrocarbons and their transfer to some other rock which *receives* them. In petrologic terms, the hydrocarbons must move from a rock (the *source rock*), which initially possesses porosity and permeability but loses them (and therefore also loses its hydrocarbons), to a rock which gains or retains porosity and permeability and can therefore both receive hydrocarbons from the source rock and make them available to the driller and the consumer. This receiving rock is the *reservoir rock*.

In Chapter 8, we constructed a picture of the source rock. In Chapter 7, we followed the transformation of its OM into hydrocarbons available for transfer to the reservoir rock. In the chapters of Part III, we will consider the properties of the reservoir rock; how the oil and gas get into it from the source rock; and how they are prevented from leaving it again and becoming lost at the Earth's surface.

11 Porosity and permeability

11.1 Porosity

11.1.1 Definitions

Porosity is the percentage of the total volume of the rock that is pore space, whether the pores are connected or not.

Effective porosity is a measure of the void space that is filled by *recoverable* oil or gas; the amount of pore space that is sufficiently interconnected to yield its oil or gas for recovery. It lies commonly in the range of 40–75 percent of the total porosity, except in unconsolidated sediments.

Porosity, conventionally denoted by the Greek letter *phi* (ϕ) is then given by the equation

$$\phi = \frac{\text{bulk volume} - \text{grain volume}}{\text{bulk volume}} \times 100$$

In terms not of volumes but of *densities*, and expressing the porosity in relation to unity instead of percentage, then

$$\rho_{bd} = (1 - \phi)\rho_g \quad (11.1)$$

where ρ_{bd} and ρ_g are the dry bulk and grain densities. Hence

$$\phi = 1 - \rho_{bd}/\rho_g \quad (11.2)$$

If we consider the rock not as dry but as saturated with fluid, of density ρ_f, and thereby having bulk density ρ_{bw}, then

$$\begin{aligned} \rho_{bw} &= \phi\rho_f + (1 - \phi)\rho_g \\ &= \rho_g - \phi(\rho_g - \rho_f) \quad (11.3) \end{aligned}$$

Range of porosity values For the common reservoir rock types under average operating conditions, porosity values may be viewed thus:

ϕ (percent)	Qualitative evaluation
0– 5	negligible
5–10	poor
10–15	fair
15–20	good
20+	very good

11.1.2 Nature of porosity

Porosity may be *primary* (original) or *secondary*. Primary porosity is that which the rock possesses at the end of its depositional phase, on first burial; it is the void space that would be present if the grains had not been altered, fractured, or dissolved. Primary porosity depends upon several factors:

(a) the degree of uniformity of grain size;
(b) the shapes of the grains;
(c) the method of deposition, and so the manner of packing;
(d) the effects of compaction, during or after deposition.

Secondary porosity is additional void space due to post-depositional or diagenetic processes, but the *total porosity* may be much less than the original porosity. In sandstone reservoirs, modifications of the primary porosity are due principally to the interlocking of grains through compaction, contact-solution and redeposition, and to cementation. In carbonates, the principal modifications are by solution, recrystallization (especially dolomitization), fracturing, and cementation. As will be discussed in Chapter 13, sandstones and carbonates have very different porosity characteristics.

Porosity may be visible or invisible. In size, individual pore spaces are described by a variety of somewhat casual terms, ranging from *pinpoint* to *vuggy*. The commonest porosities are interparticle (especially in sandstones) and intercrystalline (especially in dolomites), either of which may be of pinpoint dimensions.

11.1.3 Principal factors controlling porosity

Grain or particle size The actual particle size is theoretically immaterial. However, all ordinary depositional mechanisms are such that the coarser the average grain size the greater the overall variety of sizes. A rock may easily consist of very fine sand grains and little else; it will not long survive consisting of tennis-ball-sized cobbles and nothing else. Hence finer-grained sediments in general have higher porosities than coarser-grained sediments because there are invariably other factors in play. For example, freshly deposited clays have porosities of 50–85 percent. Fine sandy loam may attain 52 percent, and fine sand 48 percent, but coarse sand without cement is unlikely to surpass 40 percent.

In sandstones, the sizes of the pores and pore-throats commonly retain a close correlation with particle size. In unaltered carbonates, there is also a general relation (Fig.11.1), the pore diameter (μ_p) being smaller than the grain diameter (μ_g) (Fig.11.2); but carbonates undergo such a variety of modifications that the relation is commonly obscured or destroyed.

Grain or particle shape In sandstones, the *shapes* of the pores are obviously strongly dependent upon the shapes of the grains. In carbonates this is seldom the case unless complete dolomitization has occurred. However, the *amount* of pore space is ambiguously related to grain shapes.

The greatest porosity is theoretically possessed by a rock consisting of spherical grains of uniform size. Cubic packing of uniform spherical grains results in porosity of 47.6 percent, orthorhombic packing of 39.5 percent, and rhombohedral packing of about 26 percent (Fig. 11.3). The lowest porosity is theoretically provided with unassorted angular grains. In practice, grain shapes have unexpected effects. Grains of high sphericity tend to pack with minimum pore space; as absolute uniformity of grain size is never achieved, even in windblown sands or oolitic limestones, bimodal and polymodal size patterns lower the theoretical porosities. The best actual porosities are often found in rocks consisting of well sorted angular or subangular grains, as in many calcarenites.

Method of deposition Poorly sorted sediments are less porous than well sorted sediments; the ultimate porosity is highly dependent upon the degree of *sorting*. *Packing* helps to sort the grains according to size, but it tends always to make the rock as tight as possible. Depositional packing of course is continued by post-depositional packing and then by compaction under an increasing load of overlying sediments.

Effects of compaction Compaction may be defined as the process by which porosity is reduced below the water-content boundary between the plastic and the semi-solid state (the so-called Atterberg plastic limit). The process is induced by stress, the grains being deformed in a manner which is both inelastic and irreversible; they do not recover elastically when the load is removed.

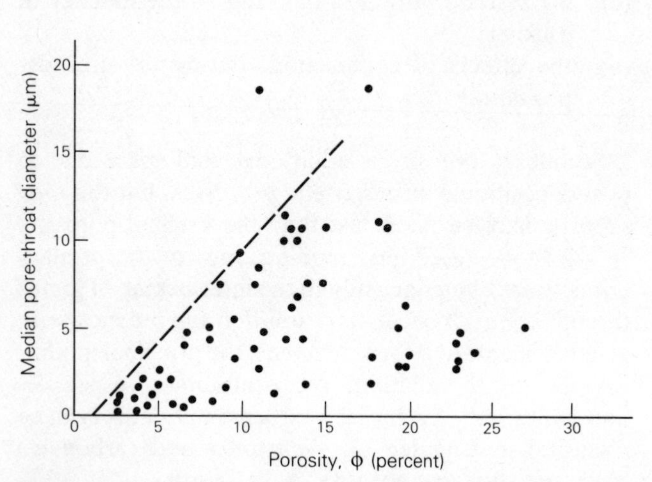

Figure 11.1 Tendency toward positive correlation between median pore-throat diameter and porosity in dolomites. For a given porosity there is a maximum throat diameter (broken line) that is not exceeded. (From N. C. Wardlaw, *AAPG Bull.*, 1976.)

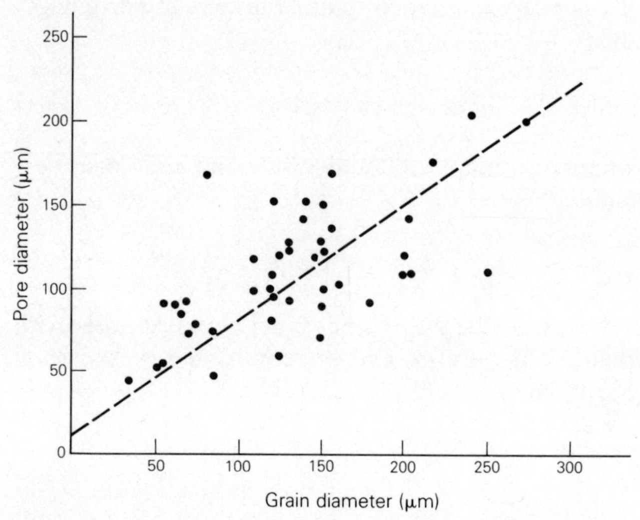

Figure 11.2 Increase of pore diameter with grain diameter, both measured in micrometers. Pore size tends to be less than grain size. (From N. C. Wardlaw, *AAPG Bull.*, 1976.)

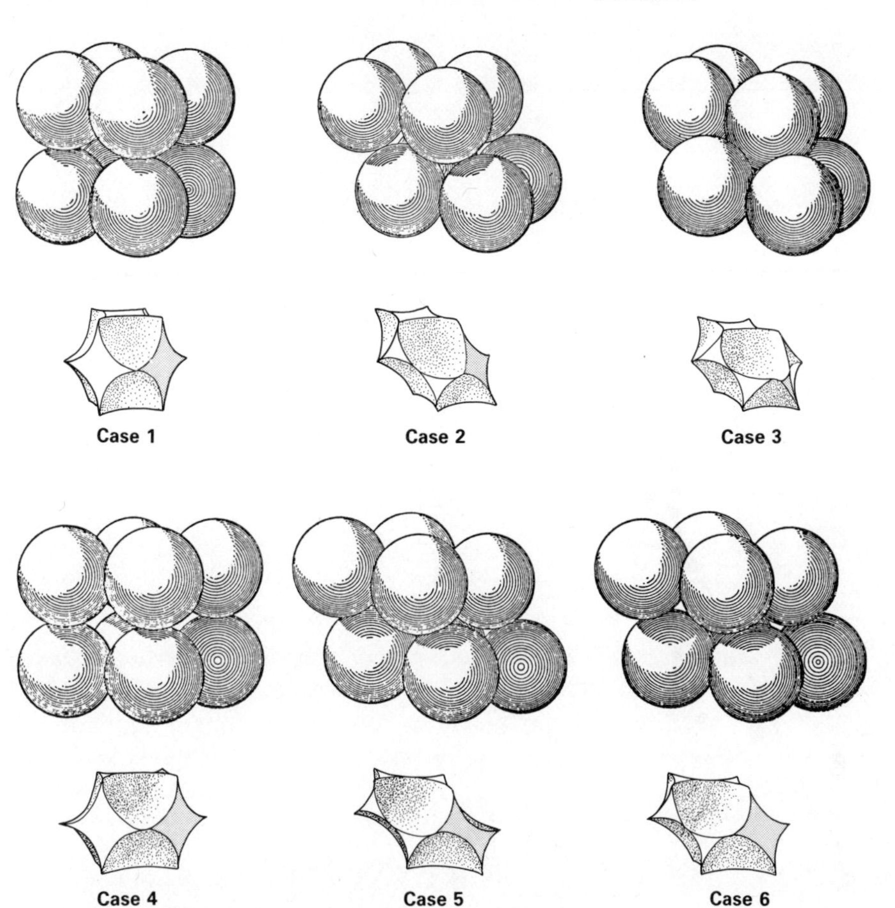

Figure 11.3 Six different packings of spheres and their resultant unit voids, on the same scale and orientation. (From L. C. Graton and H. J. Fraser, *J. Geol.*, 1935; ©The University of Chicago.)

Sands compact only about 2 percent under 25 000 kPa (equivalent to about 1000 m of burial). As the reduction of porosity with burial is essentially exponential, it becomes exceedingly slow at greater depths, but the *durations* of burial are so great that sandstones of different ages can be broadly discriminated by their porosities (Figs 11.4 & 11.5).

For most ancient sandstones it is difficult to know what the original porosity was and how much of its reduction has been due to compaction. For Tertiary sandstones, however, and especially for Neogene sandstones, this handicap is much less severe. Tertiary sandstones from the Great Valley of California, having initial porosities of 35–40 percent, lost 0.5–0.6 percent of their porosity per 100 m of burial, regardless of their actual age. Gulf Coast sandstones lost 0.4–0.5 percent per 100 m of burial. By plotting these gradients, "best

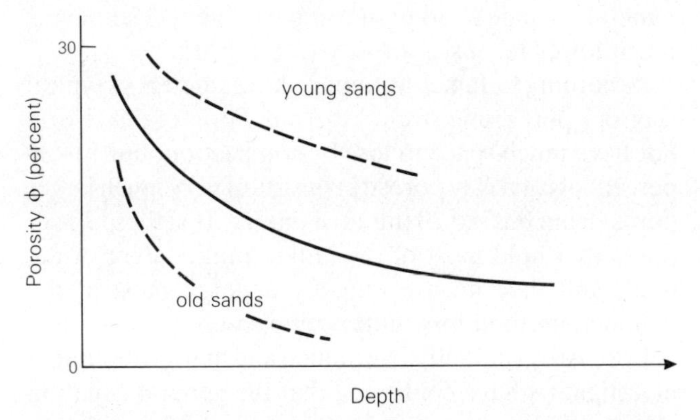

Figure 11.4 Exponential loss of porosity with depth of burial for typical sandstones. A plot of permeability against confining pressure is closely similar.

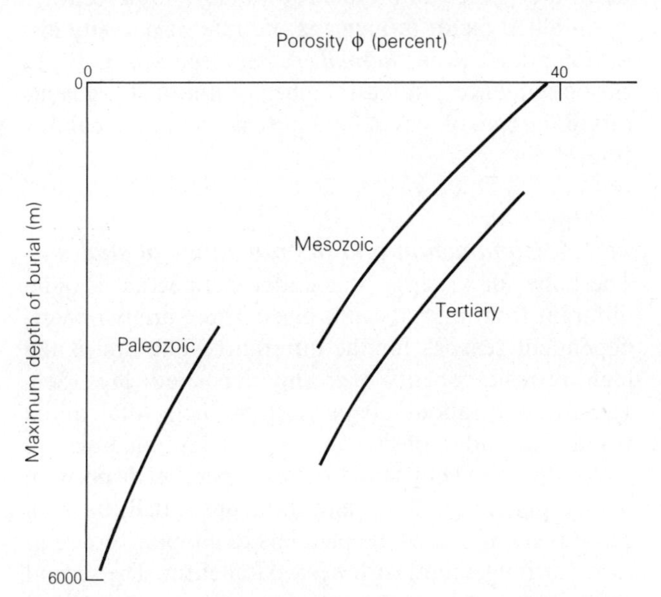

Figure 11.5 Loss of porosity with *maximum* depth of burial (not necessarily present depth) for sandstones of different ages.

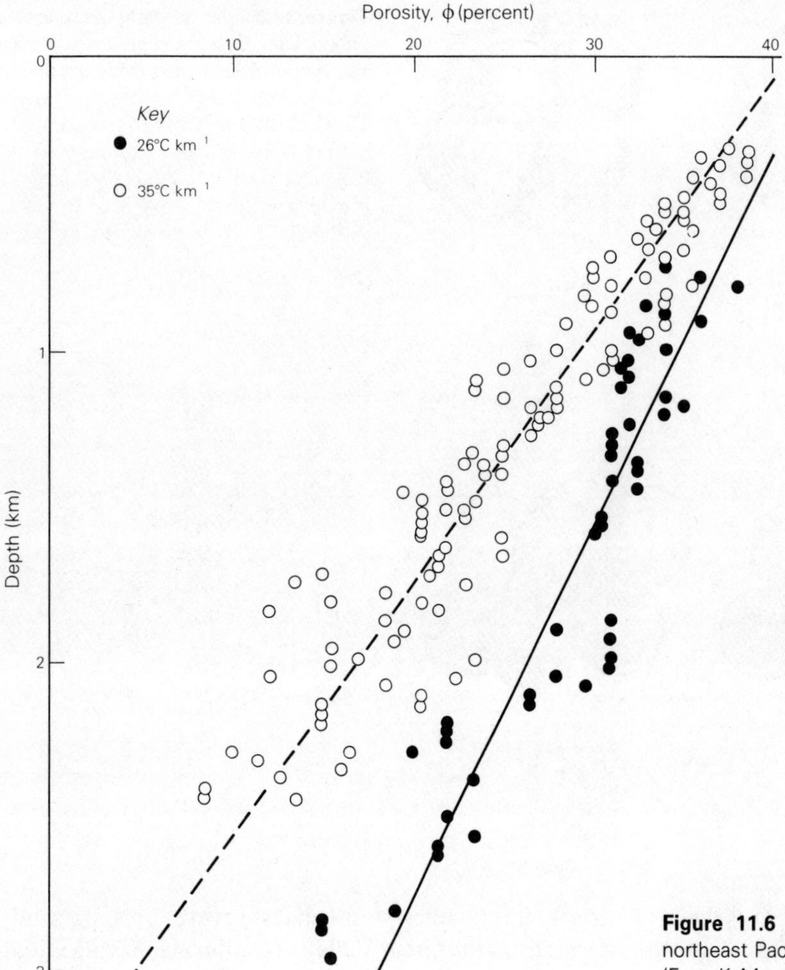

Figure 11.6 Porosity–depth relations of sandstones from the northeast Pacific arc, reflecting two different geothermal gradients. (From K. Magara, *J. Petrolm Geol.*, 1980; after W. E. Galloway, 1974.)

reservoir limits" can be determined for each Tertiary basin, represented by the depth at which the porosity may be expected to be reduced to, say, 15 percent.

An allied factor influencing the rate of porosity loss with depth is the *geothermal gradient* (see Sec. 17.8). In clastic sequences, at least, higher geothermal gradients retard the rate of loss of both porosity and permeability (Fig.11.6).

11.1.4 Compaction and the porosities of shales

The behavior of clay rocks under compaction is quite different from that of sandstones. There are two interdependent reasons for the difference. The first is the high intrinsic porosity of freshly deposited clays (Sec. 11.1.3); only about 50 percent of their total initial volume actually consists of solids. Clay minerals are *phyllosilicates*. They tend to acquire parallel disposition of their platelets under compaction, but initially the high water content causes the platelets to adopt a "house of cards" arrangement, with weak parallelism. The second reason for the compactibility of clay rocks is that not all their initial water content is held in the pore spaces.

In Chapter 15, clay mineralogy is discussed in greater detail. Here we need merely observe that some clays as originally deposited contain high proportions of *expandable clay*, especially smectite, and these clays contain "bound" water, between the layers of their crystal lattices, in addition to the water in their pores (see Fig. 15.6). As both portions of the water content of expandable clays are expelled under increasing burial, the clay minerals change to other, nonexpandable species having much lower intrinsic porosity (see Fig. 15.7).

According to James Momper, the diameters of typical clayrock pores range from 1 to 3 nm. Pores of this size do not have much space to lose by compaction, but 10–30 percent of total clay porosity consists of very much larger pores, from 0.05 to 20 μm in diameter. It is these larger pores that hold most of the initial fluid content of the mud, and they must eventually undergo most of the volume and fluid loss under compaction.

L. F. Athy (in 1930) determined and many subsequent investigators have confirmed that the general equation relating shale porosity to depth of burial is logarithmic:

$$\phi = \phi_0(e^{-cy}) \tag{11.4}$$

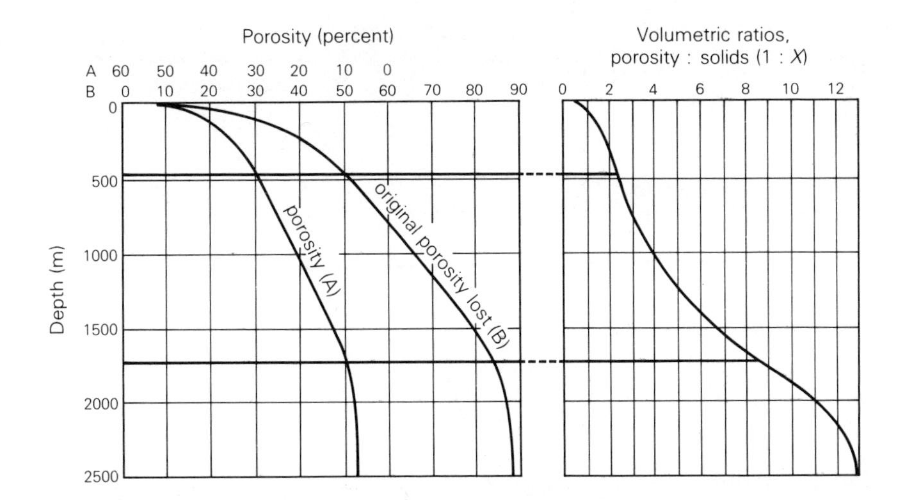

Figure 11.7 Variation of mean porosity with depth for clays and claystones (shales), after L. F. Athy (1930) and H. D. Hedberg (1936). Right-hand curve shows volumetric ratio between porosity and solids. Note changes of direction of curves at about 450 m and 1700 m; explanation in text. (After C. E. B. Conybeare, *Bull. Can. Petrolm Geol.*, 1967.)

where ϕ_0 is the average porosity of surface clays, y is the depth of burial (so $y = 0$ for ϕ_0), and c is an exponential factor called the compaction coefficient. The value of c is 1.42×10^{-3} m^{-1} (4.33×10^{-4} ft^{-1}).

When the curves of porosity decrement with increment of depth are examined in detail, however, it becomes apparent that the decrements take place in three definable stages (Fig.11.7):

(1) Down to about 450 m, *pore water* is expelled at a rapid but exponentially decreasing rate.
(2) Between 450 m and about 1700 m, the water expelled is *bound water*, from the mineral lattices. This expulsion takes place at a uniform rate, and porosity therefore decreases linearly.
(3) At about 1700 m of burial, the porosity has been reduced below 15 percent. Individual pore sizes become smaller than 10^{-3} μm (= 1 nm). The expandable clay content has been greatly reduced and the diagenetic expulsion of the remaining bound water becomes a very slow process. The remaining water saturation is *irreducible* (see Sec. 11.2.1).

The actual depths at which the porosity reduction curve changes slope depend upon the geothermal gradient (see Fig.15.8).

The practical effect of the porosity reduction process should not be underestimated by the geologist. We have already noted the relation between porosity and density (Sec. 11.1.1). Taking the averages of many measured values for shale densities at depth (ρ_g) and at the surface (ρ_0), we may write

$$\phi_0 = \frac{\rho_g - \rho_0}{\rho_g} = \frac{2.7 - 1.4}{2.7} = 0.48 \qquad (11.5)$$

Each 100 m packet of clay sediment deposited therefore has an initial porosity of about 48 percent. The packet is progressively reduced in thickness as it is buried, until at 1000 m it has lost 65 percent of its original 48 percent porosity (Fig.11.7). Its new thickness t is now given by

$$\frac{t - 52}{t} = \frac{48 - (65 \times 0.48)}{100}$$

or a little more than 60 m. A thick section of "shales," therefore, represents a considerably greater thickness of original sediment, a matter which must be borne in mind when correlating such a section with sections containing much less "shale."

Substituting densities for porosities in Equation 11.4, we reach

$$\rho_{bd} = \rho_0 + (\rho_g - \rho_0)(1 - e^{-cy}) \qquad (11.6)$$

Even under simple vertical compaction without tectonic complication, the reduction of shale porosity may have been partly caused by rock formerly overlying the shale but since eroded away. If we call the amount eroded X and the present depth Y, then

$$\phi = \frac{100Z}{X + Y + Z}$$

(Z is some constant).

When layered rocks are folded, the shales tend to become attenuated on the flanks of the folds and accumulated in the crests and troughs. Thus their porosities are still further reduced on the flanks. According to Rubey,

$$\phi_u = 100 - \cos d\,(100 - \phi_p) \qquad (11.7)$$

where ϕ_u and ϕ_p are the original (untilted) and present (tilted) porosities, and d is the angle of dip. The rather complex-appearing equation may be more readily grasped if expressed in a different form, thus:

$$\frac{\text{Original solid percentage}}{\text{Present solid percentage}} = \text{cos angle of tilt}$$

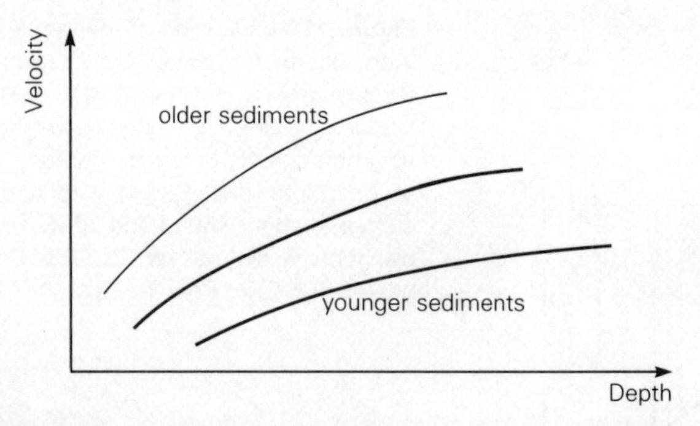

Figure 11.8 Variation in velocity of seismic waves with depth in sedimentary rock successions. (After W. H. Lang Jr, *Oil Gas J.*, 28 January 1980.)

The porosity of a shale may be reduced by about 30 percent by involvement in dips of about 50°.

As we have seen, shale porosity decreases exponentially with depth of burial. Density increases concomitantly. Age is a minor factor in density and porosity changes in shales. Unless it has been further reduced by tectonism, porosity in shales is more reliable than density as an index of depth of burial. The velocity of seismic waves through shales is inversely related to the porosity; the velocity–depth curve (Fig.11.8) is therefore the inverse of the porosity–depth curve shown in Figure 11.4. Flat pores, as in shales, reduce velocity more effectively than spherical pores do, as if the sound waves have to detour around them.

The reciprocal of the interval velocity is the *interval transit time*, which is therefore linearly related to porosity. The interval transit time (Δt) is easily measured from density or sonic logs (see Sec. 12.2.2). As it should decrease exponentially with depth of burial (like porosity), a plot of its logarithm against linear depth should be a straight line (Fig. 11.9). This is a "type compaction gradient," from A, the average transit time of modern surface clays (about 650 μs m^{-1}), to the maximum depth of practical penetration.

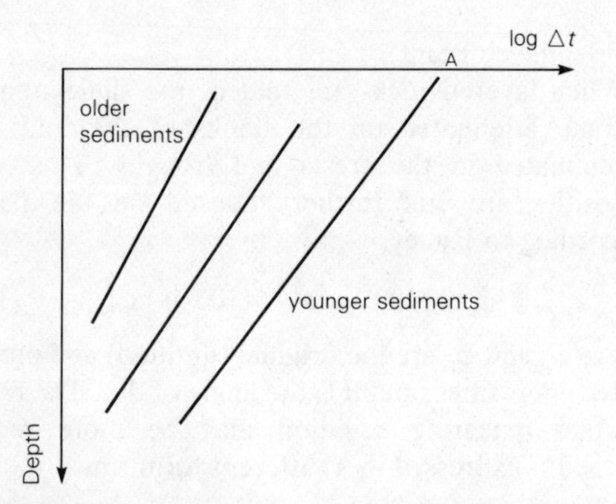

Figure 11.9 Interval transit time (Δt) of seismic waves through sedimentary rocks, decreasing exponentially with depth of burial. (After W. H. Lang Jr, *Oil Gas J.*, 28 January 1980.)

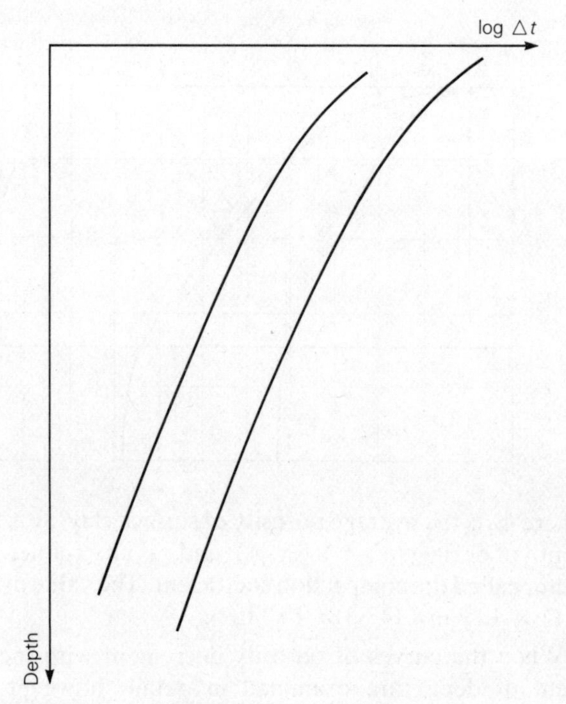

Figure 11.10 Interval transit time (Δt) of seismic waves through sedimentary rocks, decreasing as it actually appears to do in nature. (After W. H. Lang Jr, *Oil Gas J.*, 28 January 1980.)

This implies that the rate of velocity increase itself increases with depth; in fact it normally decreases. The plot is not linear under shallow burial, and the true Δt versus depth profile looks more like Figure 11.10. *Shale porosities* can therefore be calculated from density logs (Sec. 12.2.3), assuming certain values for the grain and fluid densities (say 2.68 and 1.05 g cm^{-3} respectively).

Because sedimentation seldom continued at uniform rates, or even continuously, whilst great thicknesses were accumulated, one might expect to find "stair-step" decreases in transit time, reflecting stair-step compaction. In fact this has not been shown to occur, though a major step may occur at a major unconformity. Regional uplift should, however, result in an anomalously shallow depth to a particular transit time (say 100 μs). In addition, it is known that a high temperature gradient results in a greater decrease in interval transit time with depth.

11.1.5 Compaction and porosities of carbonate reservoirs

Carbonates also undergo compaction when buried, but the process is quite different from that in clastic sediments. Compaction may reduce the thickness of a carbonate formation by as much as 20 percent if the rock contains sufficient nonskeletal material (especially lime mud and pellets). Wholly skeletal limestones undergo much less compaction, reefs scarcely any. The act of compaction, however, instigates *pressure solution*, which is much more important in limestones than in clastic sediments and quickly becomes dominant over simple mechanical compaction.

It takes place especially around fossils, flints, and other objects interrupting the bedding. As it progresses, layers of insoluble residue, including organic matter, are concentrated into seams. Familiar cases are stylolite seams and the nodular, castellated segregations called "false breccia." Advanced pressure solution can eliminate whole beds of limestone. The combination of it and compaction in carbonate reservoirs is dealt with further in Section 13.3.6.

11.1.6 Macropores and micropores

We have seen that pore sizes are commonly bimodal or polymodal, especially in sandstones. We shall further see that, in the water-wet environment of nearly all water-laid sedimentary rocks, capillarity ensures that water normally occupies the finer pore space even if oil or gas fills all the coarse porosity (Sec. 15.7). The coarse porosity may yield water-free oil, but pores smaller than 0.005–0.010 mm (5–10 μm) are almost certain to contain water both through capillarity and through imbibition. Larger pores cannot hold water by capillarity. This high *irreducible water saturation* in the finer pores means that *effective* porosity is invariably lower than *total* porosity.

This reduction of effective porosity below the measured total porosity is brought about by the partitioning of the larger pores by authigenic clay–mineral cements. The ratio of fine pore space to coarse pore space consequently commonly increases during diagenesis. The critical dimension of the finer pores is not their volume but their smallest linear dimension; the pores themselves may be equant, planar, tubular, or any variant on these shapes. If we define *microporosity* (ϕ_m) as the percentage of apertures with radius less than 0.5 μm, and *macroporosity* (ϕ_M) as the percentage of apertures larger than that, then high permeability, low water saturation, and good reservoir performance are likely only if ϕ_M is substantially higher than ϕ_m. If the two are about equal, only the macropores will yield hydrocarbons, and measurements of total porosity will exaggerate the effective porosity by as much as 50 percent. If ϕ_M is much less than ϕ_m, low permeability and high irreducible water saturation will result in poor reservoir performance.

Macroporosity may be converted to microporosity by mechanisms other than clay–mineral partitioning. Other types of cementation are equally possible. Various replacement reactions, such as those converting micas to clays, may lead to cracks (either of shrinkage or of expansion) and so to micropores. The effects of compaction, deformation, solution, and grain corrosion may also contribute to the conversion.

The existence of micropores may not be evident in thin section examination. It may have to be inferred operationally because of anomalous permeability measurements or production performances. The actual pore-size distribution in a reservoir, as distinct from its apparent distribution, is difficult to determine. Experimental measurements of pore diameters would be impossibly time-consuming, and it is commonly necessary to deduce the approximate distribution by calculating backwards from capillary pressure curves (Secs 13.3.6 & 15.7).

11.1.7 Determination of reservoir porosity

There are a number of techniques for measuring the porosity of a reservoir rock. The common basis is the withdrawal of air from the pore spaces under vacuum, and the measurement of the volume of the displaced air at atmospheric pressure. For clastic reservoirs, the measurements are commonly made on small-diameter core "plugs." For carbonate reservoirs, the nature of the porosity is insufficiently homogeneous or predictable for such a small sample to be adequate; in particular, a small plug may completely miss the large pores like vugs. Carbonate porosity is therefore better determined by full core analysis.

11.2 Permeability

11.2.1 Definitions

Permeability is the property of a medium of allowing fluids to pass through it without change in the structure of the medium or displacement of its parts. The *permeability constant*, K, is defined by the equation which expresses *Darcy's law* (Henri Darcy 1856). The equation is stated in various forms.

From the standpoint of experimental or practical hydraulic engineering, we may consider its form as stated by King Hubbert in 1940:

$$Q = \frac{-K\rho A(h_2 - h_1)}{\eta l}$$

where Q is the total discharge of fluid per unit time ($cm^3\ s^{-1}$); A is the cross-sectional area of flow path (cm^2); l is the length of flow path (cm); ρ is the density of fluid ($g\ cm^{-3}$); η is the dynamic fluid viscosity (mPa s); $h_2 - h_1$ is the hydraulic head, or pressure drop across the flow path ($g\ cm^{-2}$); K is the permeability constant in *darcies*. If the fluid is fresh water, therefore, the *specific discharge* (q) or linear rate of flow in centimeters per second is given by

$$q = \frac{Q}{A} = -K\frac{dh}{dl}$$

From the standpoint of reservoir sediments, we may instead take the form stated by Kenneth Hsü:

$$q = (Nl^2)\left(\frac{\rho}{\eta}\right)[-\text{ grad }(gh)]$$

where N is a dimensionless number which involves a group of the rock's characteristics such as grain shape and packing – it may be taken to be a constant for a particular rock; l is the length of the pore-structure of the solid (a measure of pore size and tortuosity, and hence related indirectly to grain size, sorting and compaction); $[-\text{grad }(gh)]$ is the potential function representing the amount of work required to move the fluid through length l.

Absolute permeability is independent of the nature of the fluid; it depends only on the medium (not counting induced openings). Hence the *coefficient of permeability*, k, becomes

$$k = Nl^2$$

If gas, oil, and water are all present, the absolute permeability is of little practical significance. The fluids create complex mutual interference, and there is an *effective permeability* for each fluid in the presence of the others (k_g, k_o, and k_w). The sum of these three is less than $k_{absolute}$, because the mutual interferences are retardative and not enhancing.

The *fluid saturation* of the reservoir rock is the fluid volume expressed as a fraction of the total pore space. The *relative permeability* of the rock to any one of the fluids is then the ratio of the *effective* permeability at the given fluid saturation to the *absolute* permeability at 100 percent fluid saturation. Most reservoir rocks are naturally water-wet, and contained water before the hydrocarbons entered them, preferentially occupied the coarser pore spaces, and drove the water into finer and

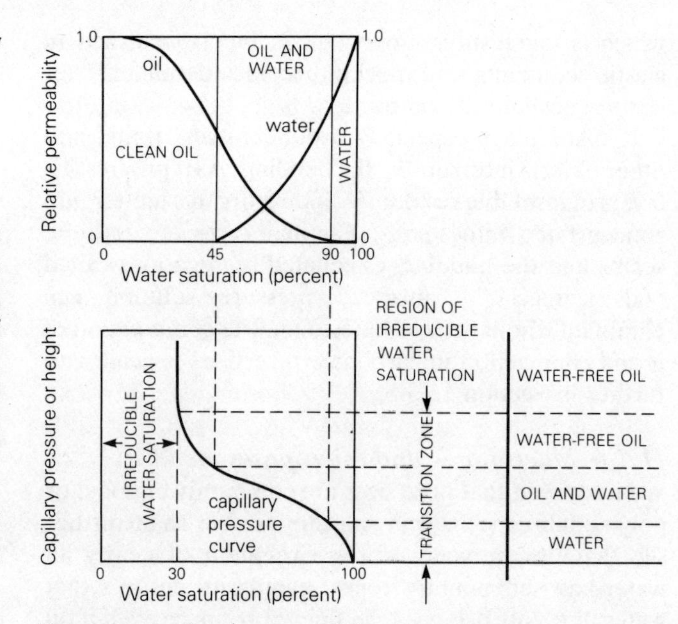

Figure 11.11 Relative permeability and capillary pressure curves. (After J. Arps, *AAPG Bull.*, 1964.)

finer spaces where it was held by capillary forces (see Sec. 15.7.1). As oil begins to enter it, the reservoir is capable of yielding only water. If we could discover and exploit a reservoir in which the displacement of water by oil had reached only (say) 20 percent, the reservoir would yield water plus some oil.

As the fluid saturation by oil increases, the permeability of the reservoir relative to water decreases towards zero, and that to oil increases towards 100 percent (Fig. 11.11). At some percentage water saturation (45 percent in Fig. 11.11, but of course dependent upon the nature of the reservoir rock as well as on the physical properties of the water and the oil), the permeability relative to water becomes effectively zero. The reservoir is now able to produce oil free of water, which no longer has any permeability available to it. The permeability relative to oil rises rapidly and the capillary pressure

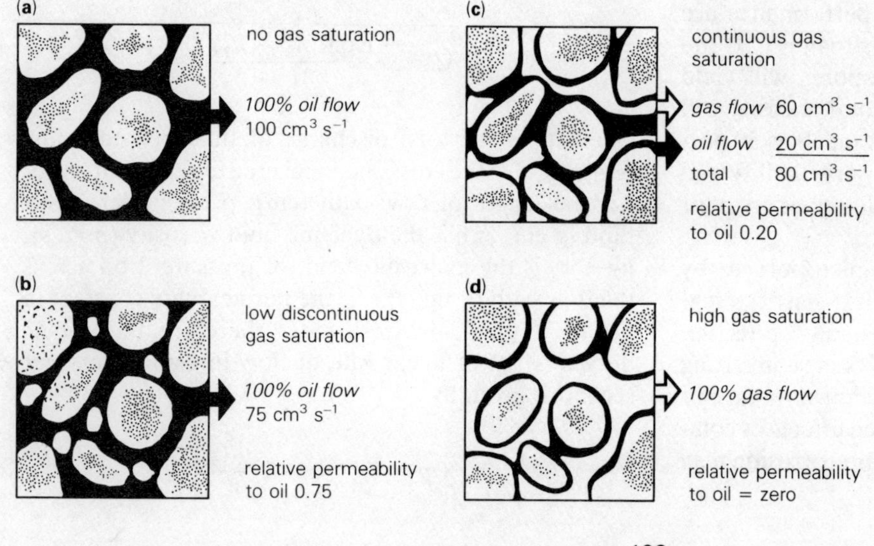

Figure 11.12 The concept of relative permeability, showing that the presence of more than one fluid in the reservoir pore space changes the capacity of each fluid to flow relative to its flow capacity if it were the only fluid present. (a) High pressure; (b) high–intermediate pressure; (c) low–intermediate pressure; (d) low pressure. (From N. J. Clark, in *Elements of petroleum reservoirs*, New York: Society of Petroleum Engineers, AIME, 1960.)

increases abruptly to a maximum. At this stage the oil has displaced all the water it is capable of displacing; the remaining water is held by capillarity in pore spaces too fine for oil to enter, and constitutes the *irreducible water saturation* (30 percent in Fig. 11.11).

If the pore space is occupied by both oil and gas, the presence of the gas reduces the relative permeability of the reservoir rock to oil (Fig. 11.12b). If the gas saturation is high enough to become continuous (Fig. 11.12c), gas and oil will move together out of the reservoir and into the well bore. If still higher gas saturation results in the *oil* saturation becoming discontinuous, the relative permeability of the reservoir to oil is reduced to zero and oil is no longer recovered from the well (Fig. 11.12d).

The validity of Darcy's law is restricted to single-phase, homogeneous or laminar fluid flow, and to circumstances in which no interaction takes place between the rock and the fluids. The law is therefore not valid for natural petroleum conditions dominated by both water and crude oil. The law's significance lies in its elucidation of the role of *permeability*.

11.2.2 Effect of rock characters on permeability

There is a general theoretical relation between porosity and permeability; it may be expressed by the familiar exponential function

$$\phi = a + b \log k$$

(see Fig. 11.13). In general, permeabilities within any one section (of sandstones, at least) change by a factor of 10 for each x percent change in porosity. In thick Tertiary successions, $x \cong 7$. Primary permeability may be no more than connected porosity, but even in cases where this is so permeability is very much more variable than porosity.

The percentage of porosity does not define drag opposing fluid movement past the small-scale roughness of the pore walls, nor the path length the fluid must follow (the so-called *tortuosity factor*). Changes in either or both of these can substantially alter permeability without any change in porosity.

An even greater divergence between permeability and porosity is observed when both are *secondary*. Especially in carbonate reservoirs, the principal permeability may result from cracks, fissures, or recrystallization (see Sec. 13.3.6). It then becomes strongly *directional*. Enlargement of bedding separations by solution facilitates horizontal permeability (k_h); fracturing may facilitate vertical permeability (k_v). Under the commonest circumstances, $k_h/k_v = 1.5–5.0$ or much more (Fig. 11.14). Under extensive fracturing across the bedding, however, $k_h/k_v \leqslant 1.0$. Widened fractures in any kind of reservoir rock, or solution channels in carbonates, lead to permeabilities hundreds of times

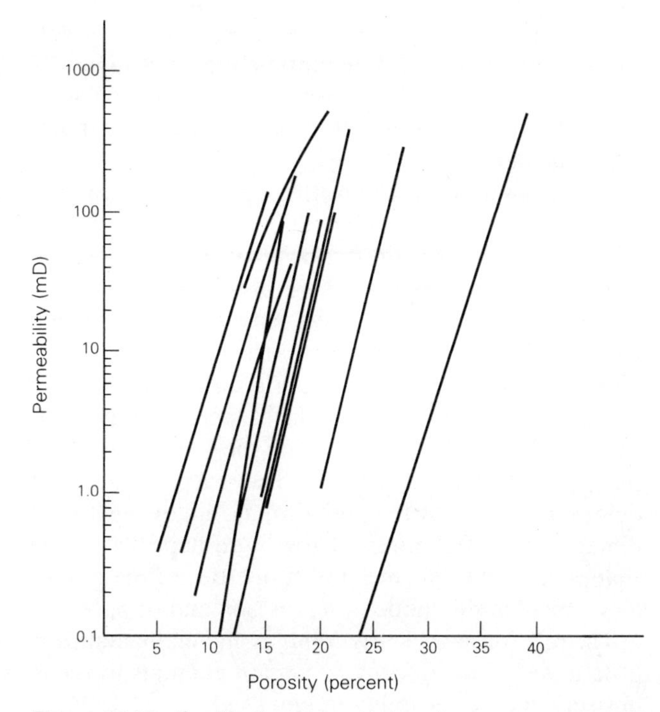

Figure 11.13 Empirical average relations between porosity and permeability for 11 American reservoir rocks. Three lines farthest to the left, indicating low porosities for given permeabilities, are for Early Cretaceous and older carbonates. Three lines farthest to the right are for Late Cretaceous and younger sandstones. (After G. E. Archie, *AAPG Bull.*, 1950.)

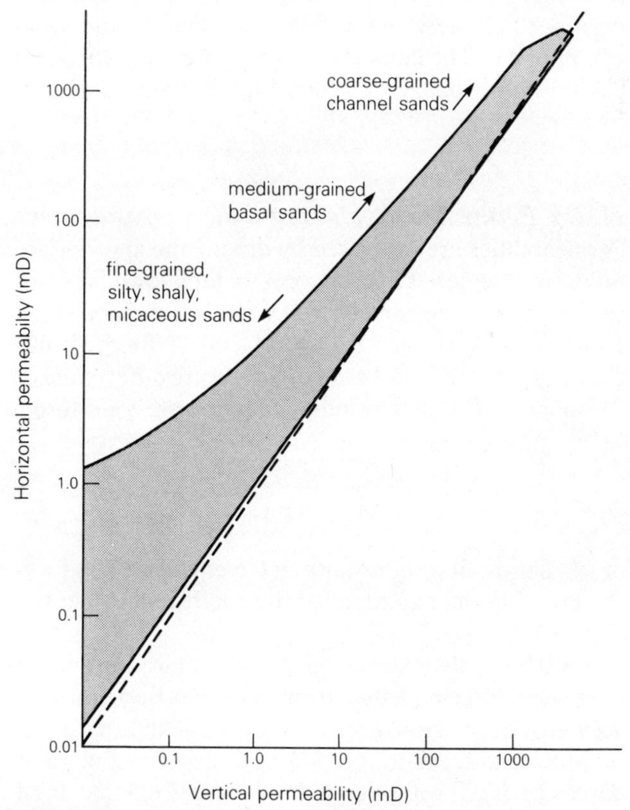

Figure 11.14 Relation between horizontal and vertical permeability in sandstone reservoirs. (From W. H. Fertl, *Oil Gas J.*, 22 May 1978.)

those of any unaltered reservoir rock. Through a slot-shaped fracture of width w centimeters, $k = 8w^2 \times 10^6$ darcies, approximately. Through a tubular solution channel of diameter d centimeters, $k = 3d^2 \times 10^6$ darcies, approximately.

The rock fabric itself also affects permeability in ways different from those affecting porosity. The permeability varies as the fourth power of the average pore radius, and hence as the square of the grain diameter, other things being equal (porosity constant, and no abnormalities of grain shape or sorting); in this it is quite unlike porosity. In general, poor sorting in sands of equal average grain size reduces permeability. The permeability is also dependent on the *geometries* of the pore spaces, not just their sizes. The real permeability is determined more by the pore throats, not by the largest dimensions of the pores. It therefore depends on the shapes and other geometrical properties of the grains, and hence on the mode of deposition and of packing.

The relations are clearly complex and mathematically little understood. *Kozeny's equation* attempts to relate porosity and permeability in any solid:

$$k = \phi^3/KS^2$$

where k is the coefficient of permeability, ϕ the porosity, and S the surface area per unit volume of the solid (rock); K is Kozeny's constant. The permeability of a porous rock therefore varies according to the ratio $\phi^3 : (1 - \phi)^2$. The equation is not truly quantitative and has no more than very general application to natural rock reservoirs.

11.2.3 Permeability values and their measurement

Permeabilities are measured by driving the appropriate fluids through series of rock cores under known pressure differentials. A permeability of *1 darcy* is defined when 1 cm^2 of rock *surface* releases 1 cm^3 of fluid of unit viscosity ($ML^{-1}T^{-1}$) in 1 s under a pressure differential of 1 atm cm^{-1}. The dimensions of a darcy are therefore

$$\frac{L^3T^{-1}\ ML^{-1}T^{-1}}{L^2MLT^{-2}L^{-2}L^{-1}}$$

or L^2. This at first sight startling circumstance was foreshadowed in the equation for the coefficient of permeability, k.

It will be easily realized that a permeability permitting the release of 1 cm^3 of fluid from 1 cm^2 of surface in 1 s is a very high permeability for a rock. Permeabilities of reservoir rocks are therefore normally expressed in *millidarcies* (mD). It will be equally easily realized that an SI unit of permeability, using meters and pascals, will apply more appropriately to a big-inch pipeline than to any rock. In order to make the SI unit comparable to the darcy, the L dimension used is the micrometer (μm or 10^{-6} m). The SI unit of permeability then becomes the *square micrometer* (μm^2), and 1 μm^2 is so close to equality with the darcy (1 darcy = 0.987 μm^2) that for geological purposes the two can be regarded as the same. The abandonment of *millidarcies* by petroleum scientists is therefore not likely to take place quickly.

By general convention, the following terms are applied to permeability values:

Qualitative description	k-value (mD)
poor to fair	< 1.0–15
moderate	15–50
good	50–250
very good	250–1000
excellent	> 1000

Based on 1000 Russian determinations, the distribution of permeabilities among common sedimentary rocks is about as follows:

Percentage of rocks	Permeability (darcies)
80	$0–10^{-6}$
13	$10^{-6}–10^{-3}$
5	$10^{-3}–1$
2	> 1

The only rocks in sedimentary basins truly impermeable *to water* are evaporites and permafrost. Normally compacted shales are rather permeable to water, though they are impermeable to oil and gas. Their permeabilities typically range from about 5×10^{-5} darcies at 1000 m depths to 1×10^{-8} darcies at 5000 m, as their porosities are reduced from about 25 percent to less than 10 percent.

It is apparent from the Russian data that about 80 percent of sedimentary rocks are potentially effective *seals* for hydrocarbon accumulations, useless as reservoir rocks. Permeabilities of the order of 0.1 mD or less (10^{-4} darcies, in the range of the 13 percent of rocks in the list above) are characteristic of the so-called *tight gas sands*, incapable of producing more than 5 barrels of oil per day (less than 300 m^3 yr^{-1}) or equivalent without expensive stimulation of the wells. Experimental studies, attempting to duplicate natural subsurface conditions, have shown that the permeabilities of argillaceous sedimentary rocks from the Lower Cretaceous of western Canada range from 10^{-4} to 10^{-7} mD, decreasing as the rocks become more clayey.

At the opposite extreme are some reservoir rocks of very high permeability. Among sandstones they include representatives of all depositional environments: fluvial (examples include the Lower Triassic top of the Sadlerochit Formation at Prudhoe Bay and the Upper Triassic sand at Statfjord in the Norwegian North Sea); deltaic (the Jurassic sands at Statfjord and some of the Miocene sands in East Kalimantan, among many examples); combined deltaic and coastal (the Burgan sands); barrier bar (the Frio sandstones in many Gulf Coast fields, almost entirely without cement or matrix); combined barrier bar and littoral (Upper Jurassic sands at the Piper field in the British North Sea); submarine fan deposits (as in the Frigg gasfield, also in the North Sea). Most of the best examples are relatively young, as would be expected, but there are some surprisingly permeable reservoir sandstones of great age (Ordovician, for example, in many fields in the American Mid-Continent; these sands are at least in part aeolian).

Very high permeabilities (and so productivities) in carbonate reservoirs are more commonly from limestones than from dolomites, and nearly always of secondary origin, from either solution or fracturing. In the first category are the legendary prominences on the El Abra limestone reef in the Golden Lane of Mexico, and many parts of the Permian limestone productive trend in West Texas and New Mexico. High fracture permeabilities are exemplified by the Asmari limestone fields of Iran, within which (as in Mexico's Golden Lane) there are individual wells that have produced more than 100 million barrels (16×10^6 m^3) of oil. Permeabilities are almost as high in the older (Permian) gas-bearing reservoirs in the Khuff Formation, in the Kangan, Pars, and other supergiant gasfields. Examples from other regions, less spectacular but still very productive, include the folded Arbuckle dolomite in some Oklahoma fields; the Triassic dolomite in Sicily; the Portlandian–Neocomian, both dolomite and limestone, in the gasfields of the Aquitaine basin in southern France; and the Mesozoic and Paleocene carbonate breccias in Mexico's Gulf of Campeche fields.

11.3 Roof rocks or seals

A reservoir rock of sufficient porosity and permeability to contain hydrocarbons and release them for production is an integral part of any trapping mechanism. Equally integral is a rock of sufficiently low porosity and permeability that, under the pressure gradient in operation, it prevents the hydrocarbons from moving any further. As the tendency of the hydrocarbons in otherwise water-bearing rocks is to move upwards, the barrier rock normally lies above the reservoir rock and is called the *roof rock* or *seal*. If the disposition of the strata is such as to facilitate lateral movement of the fluids, the lateral barrier necessary to make a trap may be called the *wall rock*. Many petroleum geologists continue to call roof rocks (and even wall rocks) *cap rocks*. Use of this term should be restricted to the secondary sheath around the tops of salt domes (see Sec. 16.3.5).

No sedimentary rock is totally impermeable, but most roof rocks have measurable permeabilities less than 10^{-4} darcies. A roof rock may be permeable to water, but if it is, the fluid potential (see Sec. 15.8.1) in the roof rock must be higher than it is in the adjacent aquifer. The rock below the accumulation need not be impervious so long as it is water-bearing.

The best seals are formed by *ductile* sedimentary rocks. Clays or shales are most common in the transgressive legs of depositional cycles and form the roof rocks for most sandstone reservoirs. More than 60 percent of known giant oilfields have shale roof rocks. Evaporites, the ideal roof rocks, are common in regressive legs of cycles and are especially favorable where the reservoir rocks are carbonates. At least 25 percent of giant gasfields are capped by evaporites of Permian and Triassic ages alone. Evaporites play havoc with conventional well logs unless great care is taken with the drilling mud (see Sec. 12.2). Anhydrite can usually be identified on density logs, its relative density being almost 3.0. Potash salts are easily picked out on radioactive logs.

A third common type of roof rock or wall rock is simply a more closely cemented or more argillaceous version of the reservoir rock, creating a *permeability pinchout*. This is more common in clastic sediments than in carbonates; only a minority of limestone reservoirs have dense carbonate roof or wall rocks.

An oddity among roof rocks is provided by *permafrost*. The enormous volume of nonassociated gas in Cretaceous sandstones in northern Siberia (see Sec. 25.5.7) would probably have been dissipated had it not passed through a clathrate stage and then been freed by regasification below the thick permafrost that still remains across the taiga.

Any normally porous sedimentary rock can be converted to a seal by diagenesis before, during, or after complete burial. Cementation by clay minerals, cementation by the introduction of soluble salts percolating downwards from overlying evaporites, recrystallization, pressure solution and re-deposition, deformation of ductile grains, or the degradation of the hydrocarbons themselves, forming asphalt or "tar" seals, are all familiar means of converting reservoir rocks into seals.

The opposite process is also important, though less common and obvious. A seal may be rendered ineffective not only by such routine geological processes as fracturing or recrystallization, but also by a change in the pressure regime. Even a dense rock is a seal only if pressure conditions permit it to be. If those conditions change sufficiently to overcome the sealing rock's entry pressure, the seal becomes a sieve, in Gilman Hill's phrase, and the hydrocarbon accumulation below or behind it is lost or greatly reduced (see Secs 15.7.2 & 16.3.9).

The efficiency of a seal, measured by the thickness of the oil or gas column it can retain below or behind it, is a function of the pore size, the "integrity" of the rock (its lack of open fractures), its continuity, and its thickness. The thickness, continuity, and integrity are quickly estimated. Evaporites and permafrost have very high integrity because they have the capacity of self-healing of fractures. To calculate the seal capacity, however, the geologist needs also to know the pore size and the parameters permitting the fluids to pass through pores of that size: the fluid densities, the interfacial tension (γ) between the fluids, and the wettability. These properties are discussed in Chapter 15. They are not easy to measure or calculate at all accurately.

The interfacial tension between two fluids is the force working to reduce the area of contact between them. Interfacial tensions are lowered by rise in temperature and by the presence of surfactants which increase fluid miscibilities. A wide variety of surfactant substances may be naturally present in the fluids, especially in the formation waters. For practical purposes, values of γ can be taken to be in the range of 5×10^{-3}–30×10^{-3} N m^{-1} for oils and 30×10^{-3}–70×10^{-3} N m^{-1} for gas, against water in each case. The higher the interfacial tension the greater the seal capacity.

Seals are a function not only of the stratigraphic regime but of the structural style in addition; their efficiency depends partly on the trapping mechanism. Simple anticlinal traps may provide *sequential* sealing surfaces because many interbedded units may be involved in them. Failure of one seal may merely cause the hydrocarbons to ascend to a higher reservoir capped by a seal which is still intact. In a nonconvex, pinchout trap, in contrast, an *adjacent* seal may depend on a single rock stratum and its failure may cause the loss of the accumulation.

In thrust belts, the lithologic seals may be destroyed but the thrust surfaces themselves may create high-pressure, tectonic seals. Oils in thrust-belt fields are typically very light and volatile (see Sec. 17.7.7). Furthermore, thrust planes are preferentially located within ductile strata like clay-shales or evaporites, which provide fault-gathering zones and lead to *décollements*. These fault-favoring strata add to the seal capacity of the thrust surfaces.

12 Well logs

12.1 Introduction

The essential mechanics of an oil or gas accumulation begin to take shape. A rock possessing both porosity and permeability must be available to receive the hydrocarbons from their source sediments; a rock lacking porosity and permeability must be suitably adjacent to the porous and permeable rock to prevent the fluids from escaping. Porosity and permeability are physical properties of rocks; like density, tensile strength, and other physical properties, they can be routinely measured in laboratories. The fluids occupying the porosity also possess physical and chemical properties measurable in the laboratory.

For many scientific and economic purposes, laboratory data of high accuracy and reliability, for both the fluids and the rocks that contain them, are of course mandatory. For the petroleum geologists and the drillers, however, such data cannot be acquired quickly enough or cheaply enough to be useful. The operators in the oilfields needed a method by which the fundamental properties of the rocks and their fluid contents could be quickly and reliably determined *in the subsurface*. The requirement was satisfied by the *electric well log*.

12.2 The basis of well logging

The continuous, instrumental logging of the physical properties of the rocks penetrated in a drillhole revolutionized the exploration and production branches of the oil industry. The names Schlumberger, Halliburton, and Lane-Wells are as much a part of the industry's history and establishment as the familiar names of large oil companies. Logging techniques have advanced as rapidly as seismic or drilling techniques, and *formation evaluation* by the interpretation of a battery of logs is both an art and a science in its own right.

The invaluable functions provided by well logs fall into two categories: the continuous recording of all measurable physical properties of the rocks penetrated by the bit and of their fluid contents; and the facilitation of correlation between wells and even between drilled areas.

The basic log is an *electric log*. An insulated sonde, or assemblage of electrodes, is lowered to the bottom of the hole on an electric cable. Truck-mounted instruments record the electrical properties of the rocks and fluids passed by the sonde as it is withdrawn from the hole. For most of the first 25 years of log development and use, the electric log consisted of one single curve and one set of three curves (Fig. 12.1) representing complementary electrical properties. This log – the *E-log* – was run as routine on all drillholes, and for many wells in old fields such logs are the principal source of information for the geologists as operators try to rejuvenate the fields.

The single curve, on the left of the log, is the *spontaneous potential* or SP curve. The source of the potential is the difference in electromotive force between the fluid in the drilling mud and the fluid, if any, in the pores of the rock. This potential difference creates an electrochemical "cell," which produces a current. The SP electrode measures the potential difference in *millivolts* (mV), with respect to a reference electrode at infinity. Readings of the SP curve are negative to the left, positive to the right. In the normal circumstance in which the drilling mud is fresher than the formation water, high negative readings, or "kicks" to the left, are given by porous rocks, especially porous sands. If the mud is saltier than the formation water, the cell is reversed and the kicks go in the positive direction (to the right). Salt water in porous rocks gives higher readings than fresh water if the mud is fresh and not salty. High negative readings are also given by coal seams if they contain "pit water." High positive readings, towards the right, are given by dense limestones, shales, and evaporites. Evaporites cause problems with both curves; the SP curve is essentially worthless if the drilling mud is salty. In limestones, also, the SP curve is of little value. Its

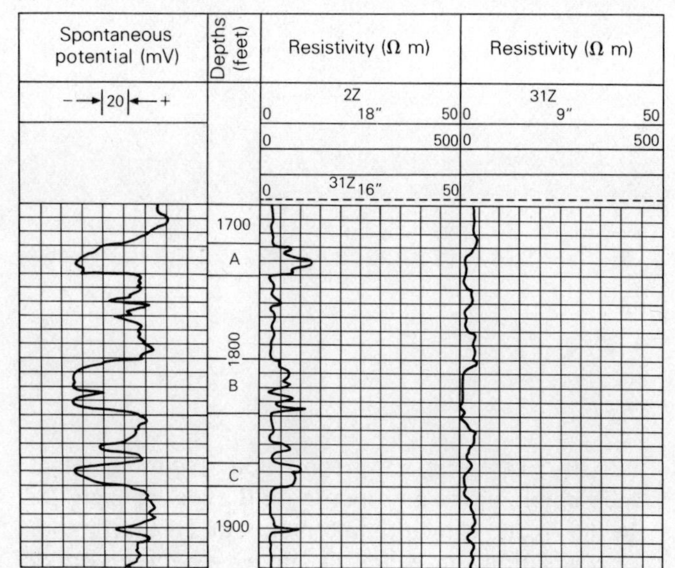

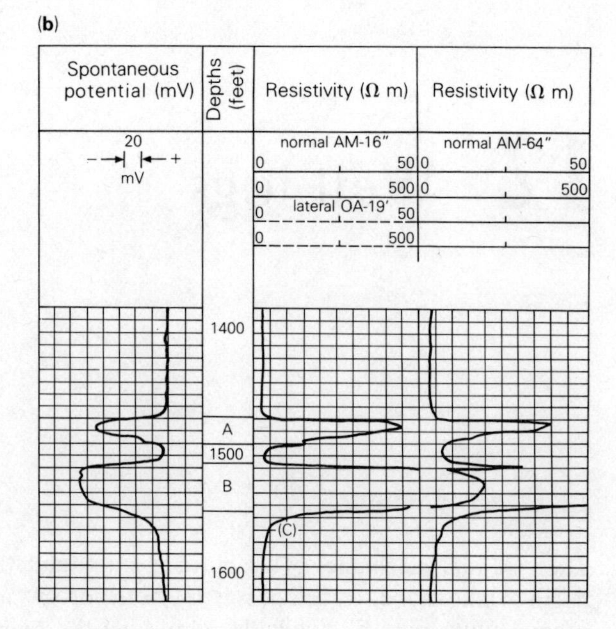

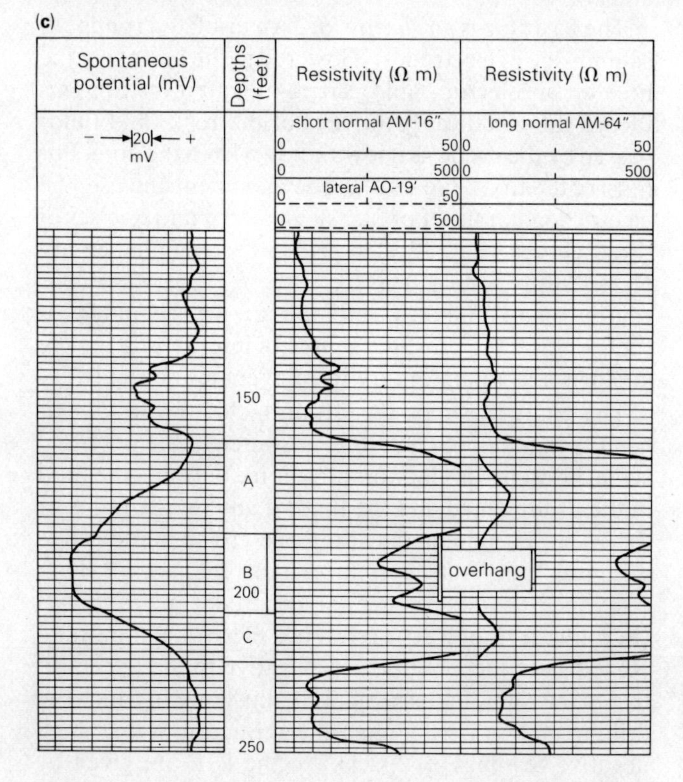

Figure 12.1 Typical SP and resistivity curves on old electric logs: (a) conductive beds; (b) resistive oil sands; (c) conductive bed (B) between resistive beds (A and C). (From E. E. King and W. H. Fertl, *Oil Gas J.*, 27 November 1978.)

principal value is its indication of the contrast between shale rocks and non-shale rocks. As most clean, porous sandstones contain thin interruptions of nonporous rock types, these interruptions appear as characteristic sharp re-entrants on the otherwise negative SP curve; the most striking case of this is provided by seams of bentonite. In sum, the SP curve reflects the *lithology* of the rock and its *fluid content* and hence, indirectly, its porosity.

The three curves on the right side of the old E-log are the long- and short-normal *resistivity* curves and the lateral resistivity curve. The short-normal curve represents the readings between electrodes spaced 16 inches (0.4 m) apart, the long-normal curve between electrodes 64 inches (1.63 m) apart, and the lateral curve between electrodes about 18 feet (5.5 m) apart. The purpose of the threefold spacing is to vary the radius from the borehole to which the device is effective; varying radii beyond the hole are invaded by drilling mud (Fig. 12.2).

Resistivity, the inverse of conductivity, is *specific resistance*, measured in ohm-meters (Ω m; but often expressed as $\Omega\ m^2\ m^{-1}$). Resistivity is high in dense

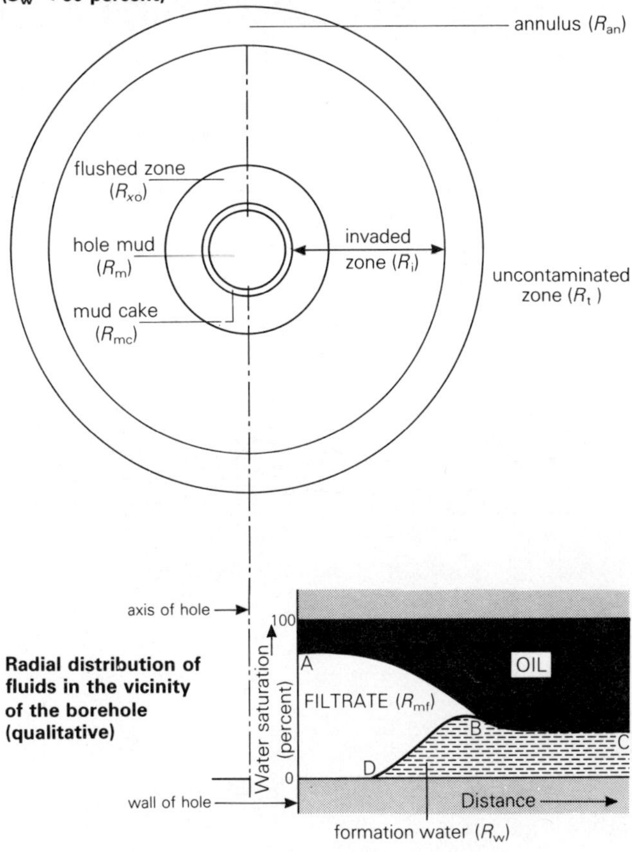

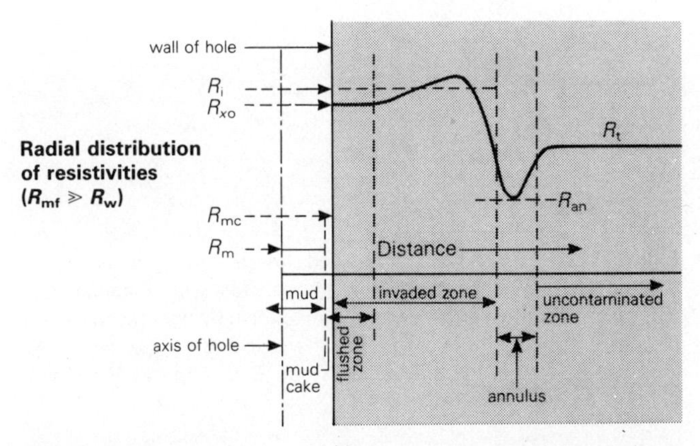

Figure 12.2 Distribution of resistivities around the borehole in an oil-bearing formation. (After Schlumberger Limited, *Log Interpretation*, vol. 1, 1972.)

limestones, evaporites (salt is almost infinitely resistive, and sulphates are also very high), metasedimentary rocks, or porous rocks containing oil or fresh water (which are effectively insulators). A thick, homogeneous oil sand (having thickness two or more times the normal electrode spacing) gives a resistivity curve kicking to the right and peaking about in its center (Fig. 12.3); an asymmetrical curve means variations in the porosity within the sand unit. High resistivity with low-to-zero SP is commonly given by tight siltstones or shales with scattered quartz grains, or coarser clastic rocks with intergranular porosity plugged by clay.

Resistivities are low in porous rocks containing salt water, which is an electrolyte. Because clay itself is a conductor, shales commonly have low resistivities. They produce no kick in either the resistivity curve or the SP curve, so that intervals clinging closely to the axis of the log on both curves constitute the *shale line*. Even in shales, however, the resistivity varies somewhat with porosity, so it is lower in clay-shales and in bentonites than in siliceous shales or silty shales. Resistivities also increase with depth for all types of shale, but they may continue to be abnormally low in shales adjacent to sands having high fluid pressures.

The range of resistivities commonly encountered is very great. Typical values (in ohm-meters) are: salt water, 0.02; fresh water, 1.0; shale and sand each about 5.0; dense carbonate, 100; coal, 150; anhydrite, 1000 or more; oil, order of 3×10^{11}; gas, nearly infinite.

Allusion has been made to the problem presented to the conventional E-log by beds of contrasting properties too thin for accurate resolution by normal electrode spacing. The *microlog* was developed to overcome this obstacle by very short spacing of the electrodes. Invasion of even thin permeable strata by drilling mud causes the build-up of *filter cake* on the bore walls at these strata. The filter cake is resistive. Two microlog curves were run, the micronormal and the micro-inverse curves. If the former yielded a higher reading than the latter, the "positive separation" of the two curves indicated a permeable bed (Fig. 12.4), and this indication should be reinforced by a negative SP kick. Lack of separation between the two microlog curves, or "negative separation," indicated an impermeable stratum. Thus the microlog was employed as a supplementary porosity log, but the degree of refinement attained was insufficient for accurate evaluation of the formation and its contents. The microlog in its original form was therefore allowed to become obsolete during the early 1970s.

12.2.1 *Focused current logs*

Conventional electric logs required the presence of a moderately conducting medium in the borehole, so that the current could flow between the electrodes. If the mud in the hole is very salty, the current is able to flow directly between the electrodes without passing through the rock with any penetration. Though satisfactory SP readings can still normally be achieved, some means had to be found of *focusing* the current directly into the formation in order to achieve an accurate reading of the resistivity. This was done by the *laterolog* (Fig. 12.5).

The efficiency of the laterolog was paid an unusual

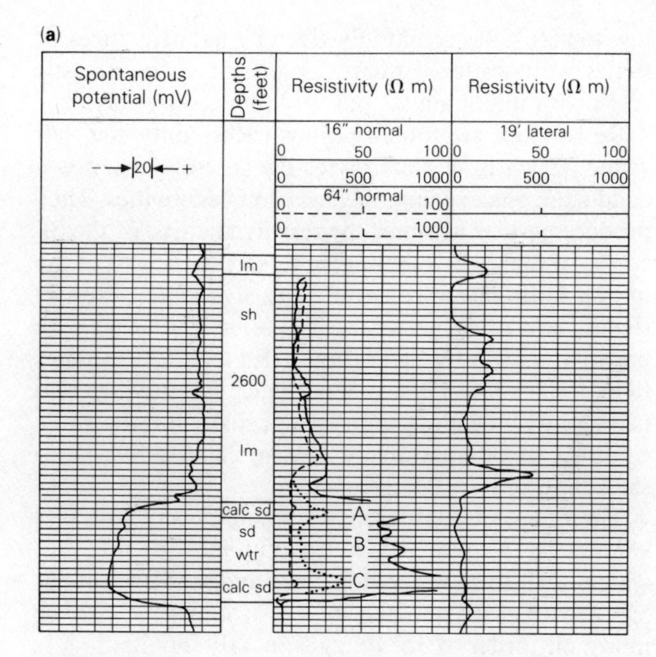

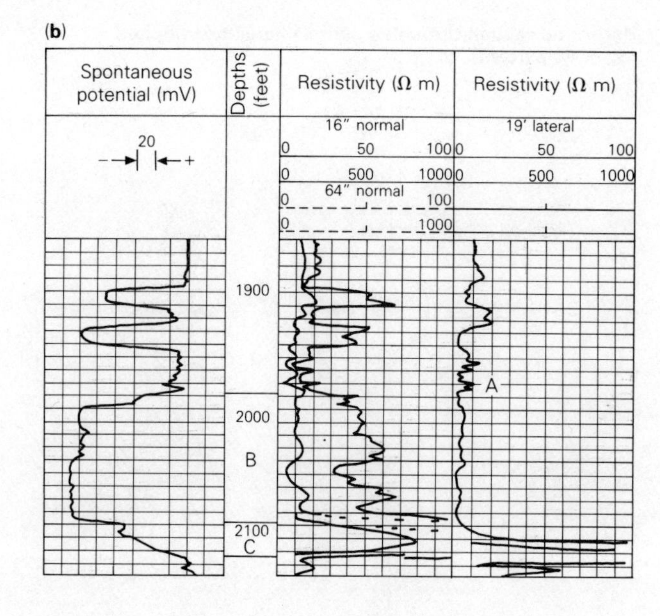

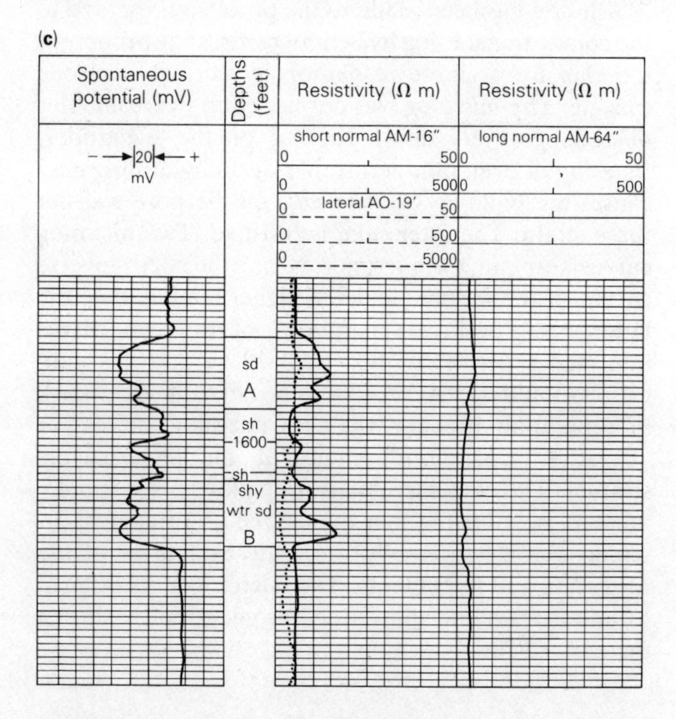

Figure 12.3 Expressions of oil- and water-bearing sands on old electric logs: (a) tight beds (A and C) bounding a water-bearing sand (B); (b) sand containing salt water (B) underlain by tight resistive bed (C); (c) oil sand (A) overlying salt-water sand (B) – low resistivity indicates that the sandstone is shaly. (From E. E. King and W. H. Fertl, *Oil Gas J.*, 27 November 1978.)

compliment by its being the only well log to give its name to a geological formation. In the Aquitaine Basin of southern France, the late Neocomian Stage is represented by black, anhydritic, unfossiliferous shales, which stand out as such a reliable marker horizon on the laterolog that they are formally referred to as the *laterolog shales*.

Serving the function of the microlog to the focused current logs is the *microlaterolog*. The current is focused in a narrow beam into the formation, by constraining it within an outer ring of a second current. The narrow beam of current opens up rapidly once in the formation. It measures the resistivity of the zone contaminated by mud filtrate. Its penetration into the formation is still not great unless the mud cake is thin, and it is otherwise replaced by a variant called the *poroximity log* (see Fig. 12.17b).

If, instead of containing salty mud, the hold is filled with nonconducting oil-base mud, conventional logs cannot be run. Even with poorly conducting water-base muds, an electromagnetic field has to be *induced* horizontally into the rock, and the resistivity is then recorded directly on an *induction log* (Fig. 12.6a), not indirectly (by passage of the current through the rock

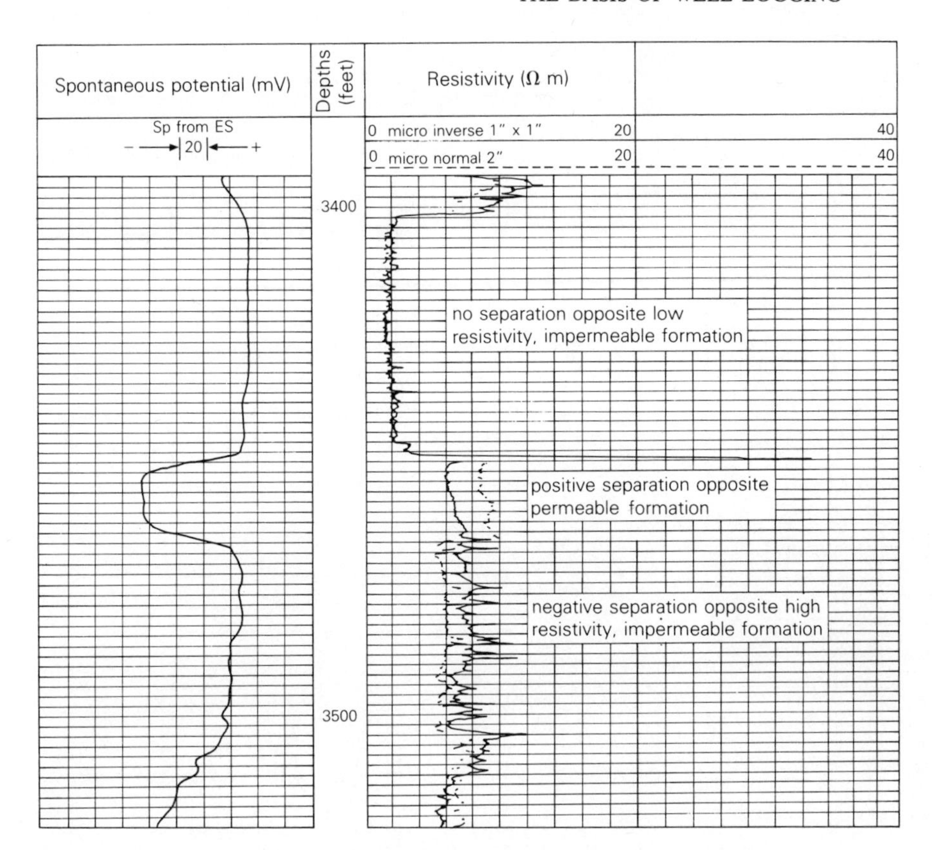

Figure 12.4 Log obtained with the microlog device. (Courtesy of Schlumberger Well Surveying Corporation.)

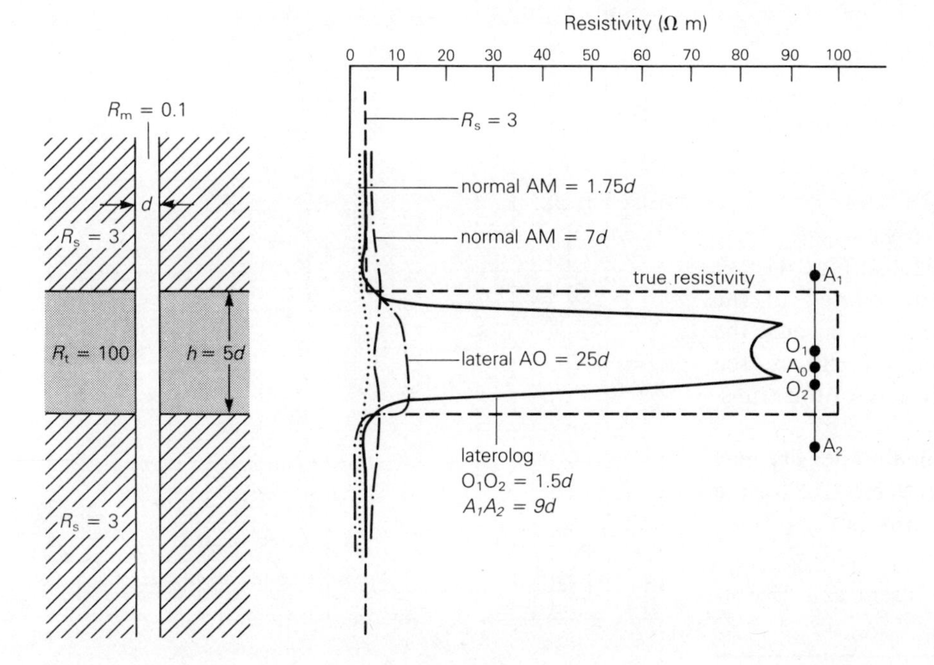

Figure 12.5 Response of the laterolog and conventional logging devices opposite a thin resistive bed of $R_t = 100$, logged with a salt mud. Note the much greater penetration of the thin bed by the laterolog current, in relation to the borehole diameter d. (From H. G. Doll, *J. Petrolm Technol.*, 1951.)

between two spaced electrodes) as in a conventional log. *Dual induction logs* provide three focused resistivity readings for deep penetration of the formations.

Standard electric logging practice by the mid-1970s replaced the old E-log or microlog with the combined *dual induction–laterolog*, still retaining the SP electrode (Fig. 12.6b). Still later practice combined the dual induction log with the *spherically focused log* (SFL), for which the currents are focused as closely as possible to a spherical equipotential front in order to overcome distortion of the current radiation by the borehole itself. The actual combination of logs is chosen according to experience in the area concerned; a battery of them can now be provided on a single run.

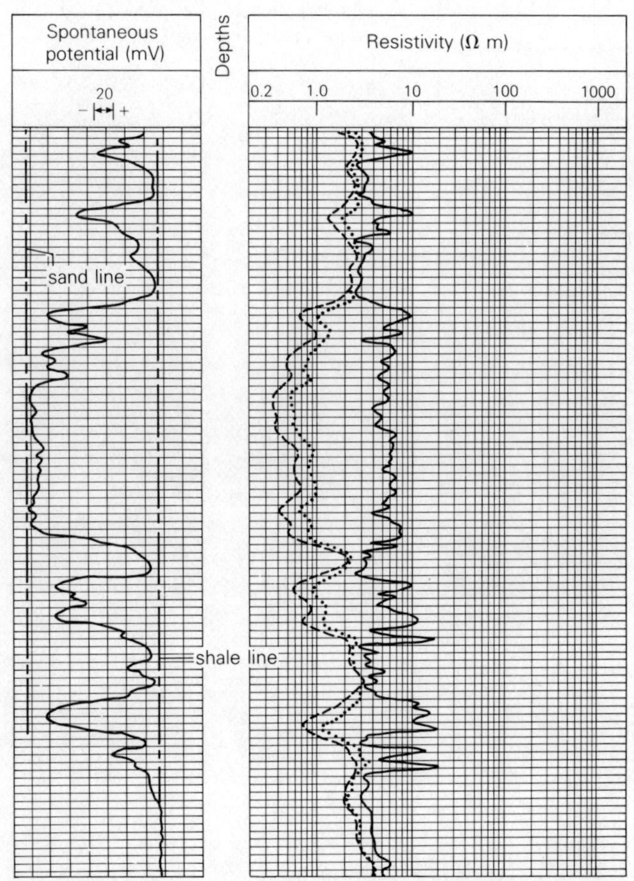

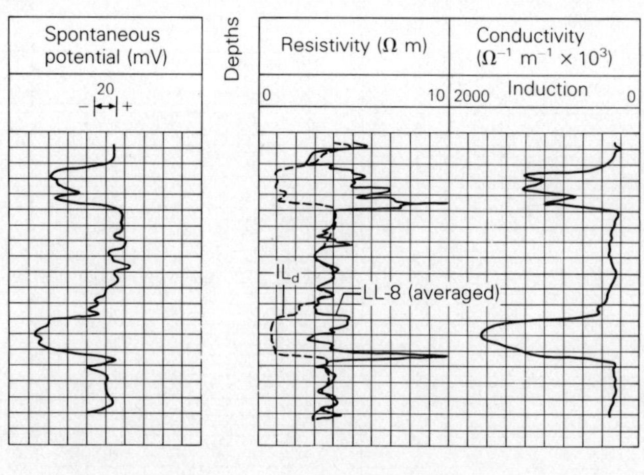

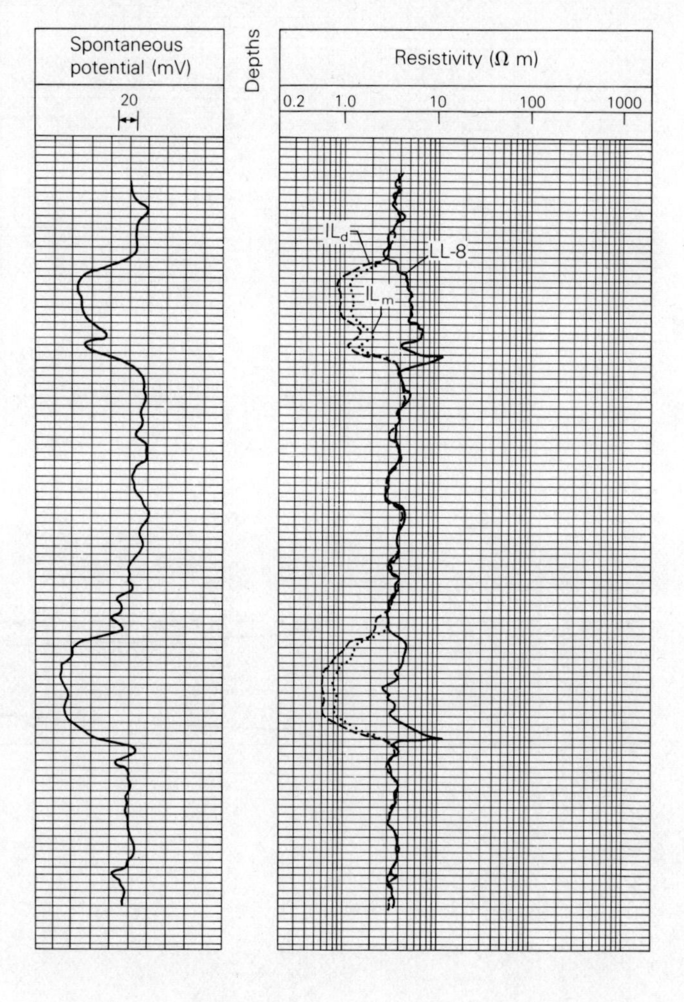

Figure 12.6 Spontaneous potential and resistivity curves as seen on (a) an induction log and (b) a dual induction–laterolog presentation. (Courtesy of Schlumberger Limited.)

12.2.2 Acoustic velocity or sonic logs

In our discussion of porosity (Sec. 11.1.4; Figs. 11.9 & 10) we observed that it is directly related to the *interval transit time* (Δt) of a seismic wave, because the transit time is inversely related to the velocity and the velocity is directly related to the density of the transmitting medium.

The dependence of sonic travel time on density may be illustrated by the values shown in Table 12.1 for the common sedimentary rocks logged in wells.

Table 12.1 Illustration of the dependence of sonic travel time on density

Rock	Bulk density (g cm^{-3})	Sonic travel time (ms m^{-1})
clean dolomite	2.88	143.5
clean limestone	2.71	160.0
clean sandstone	2.65	185.0
clean shale	2.20–2.75	230–495
clean salt	2.03	220.0

Δt is measured by the *sonic log* (Fig. 12.7), which records the transmission of a compressional sound wave through the mud, the formation, and back to two receivers below the transmitter. The sonic log is therefore an invaluable *porosity log*, and its use in conjunction with the *density log*, described below, is routine practice (see Fig. 12.12).

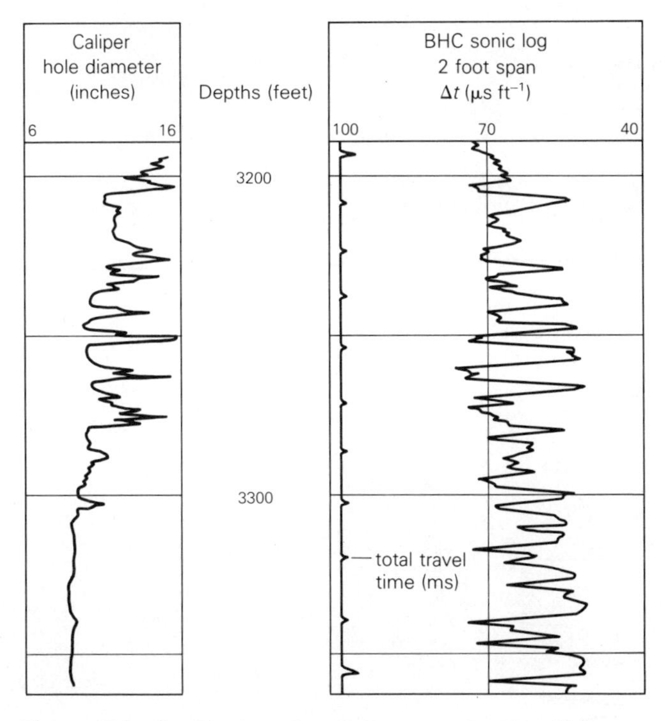

Figure 12.7 Combination of sonic log and caliper log. (Courtesy of Schlumberger Limited.)

The Wyllie equation relates the porosity (ϕ) of a *consolidated clastic* reservoir rock to the sonic travel times through it:

$$\phi = \frac{\Delta t_{\text{log}} - \Delta t_{\text{matrix}}}{\Delta t_{\text{filtrate}} - \Delta t_{\text{matrix}}}$$

where Δt represents the travel times through the various media, in μ s m^{-1} (or ft^{-1}). Because sonic velocities are measured in meters (or feet) per second, this equation expressed in velocities rather than travel times becomes

$$\phi = \frac{\Delta t_{\text{log}} - (1/V_{\text{m}})10^6}{[(1/V_{\text{f}} - \text{I}/V_{\text{m}})10^6]C_{\text{c}}}$$

where C_c is the *compaction constant*. $C_c = 1.0$, by definition, for a fully compacted formation; it is greater than 1.0 but less than about 1.6 for uncompacted formations. Do not confuse the formation's compaction constant used in Wyllie's equation with the *cementation factor* used in the calculation of the formation's water saturation (see Sec. 12.4 and Fig. 12.14).

V_m and V_f can be readily estimated. V_m ranges from about 4500 m s^{-1} for salt to almost 7000 m s^{-1} for dolomite. The general range of V_f is from 1300 m s^{-1} for oil-base mud to 1700 m s^{-1} for saline mud.

12.2.3 Radioactivity logs

Radioactivity logs have the advantage, not possessed by electric logs, that they can be run through well casing. There are three principal types of radioactivity log, normally run in association (Figs 12.8–11).

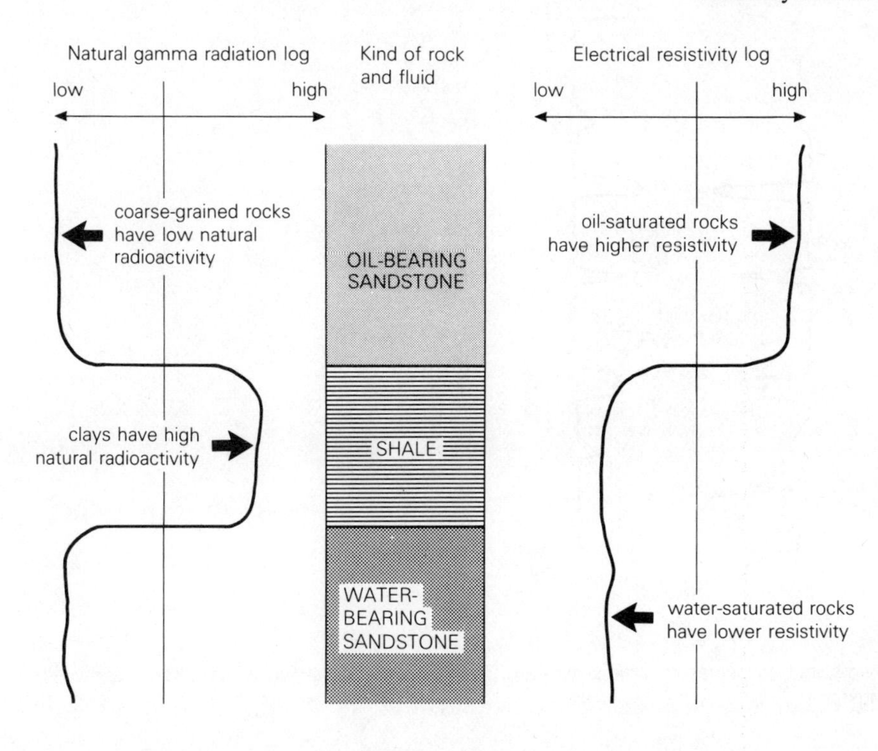

Figure 12.8 Schematic example of interpretation of combination of resistivity (from electric log) and radioactivity (from gamma-ray log). (Courtesy of Chevron USA Inc.)

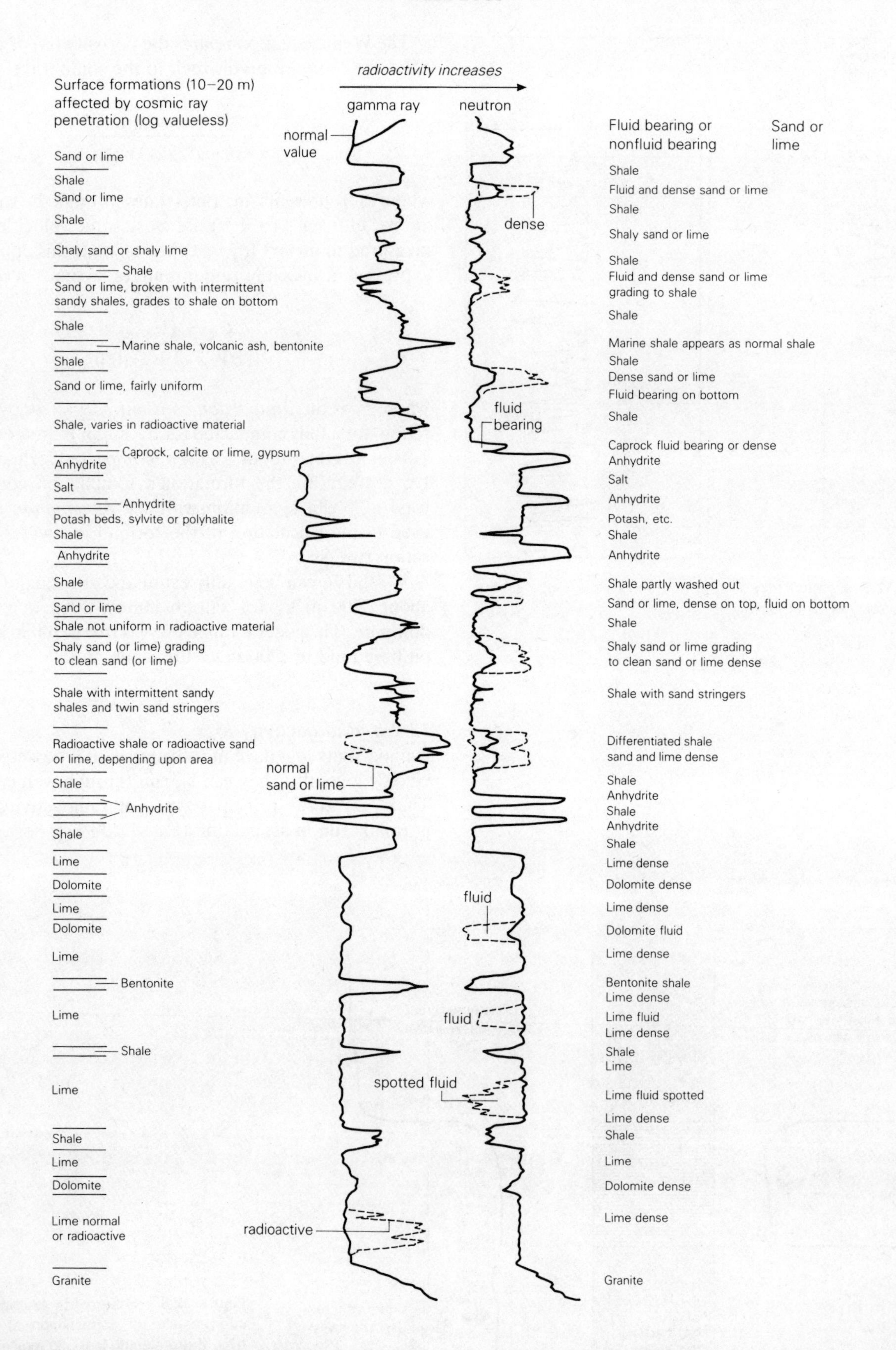

Figure 12.9 Typical responses of gamma-ray and neutron curves to various types of formation penetrated in wells. (From Lane-Wells Radioactivity Logging; Lane-Wells Company, Houston.)

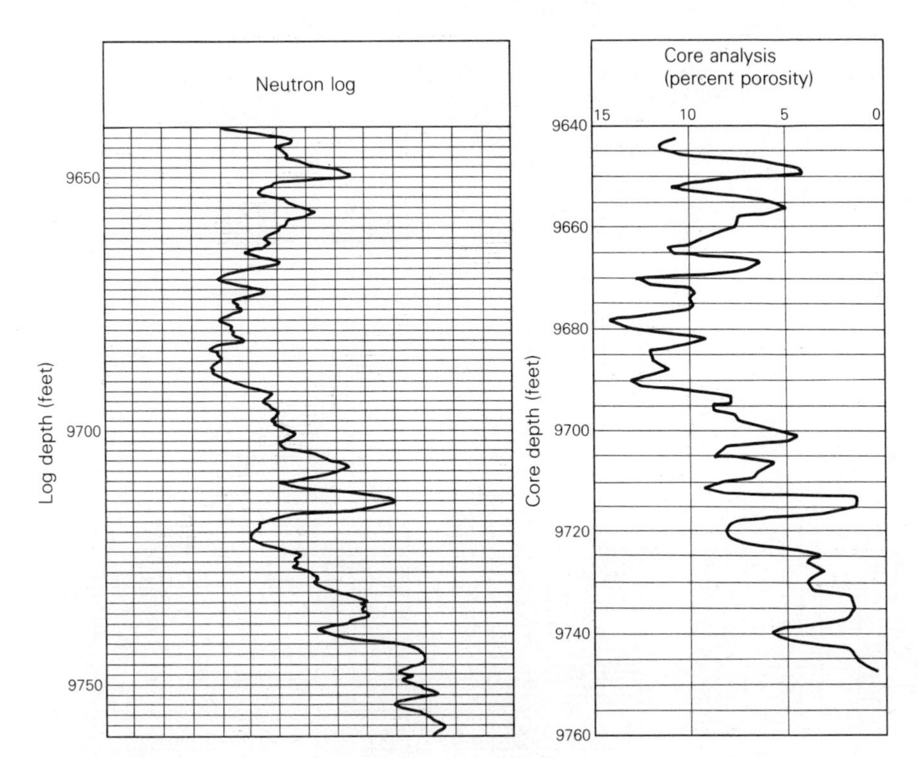

Figure 12.10 Comparison of a neutron log with porosity values determined by core analysis. (Courtesy of Chevron Standard Limited.)

The *gamma-ray log* is used to differentiate between argillaceous and nonargillaceous rocks, the former being more radioactive. However, feldspathic sandstones (including "granite wash") shed from uplifted granitic or gneissic terrains may be more radioactive than the surrounding shales, because potassium is radioactive. In clastic basins having numerous feldspathic sandstone reservoir rocks, such as the Californian basins, the usefulness of the gamma-ray curve is much reduced.

The *neutron log* reflects the abundance of hydrogen nuclei in the fluids. All fluids in and around the borehole contain hydrogen; the higher the content, the lower the reading. The hydrogen absorbs neutrons, so that porous rocks give low counts. The neutron log is especially useful for revealing variations in the porosities of carbonate rocks, which are therefore logged by compensated neutron-density logs. Unfortunately, water, oil, and mud filtrate are not distinguished by the neutron log, because all have about the same percentage of hydrogen nuclei. Dry gas, on the other hand, gives a very high neutron reading; there are fewer hydrogen nuclei in the gas phase than in the liquid phase.

The *density log* has become the most commonly used indicator of the *porosity* of a formation. A source of high-velocity gamma radiation is pressed against the side of the borehole and the rays are emitted directly into the formation. The log measures the amount of backscattering of the gamma radiation through collisions with the electrons in the rock. The reading is therefore directly related to the electron density, which is in turn related to the true bulk density and therefore inversely related to the porosity. If ρ_m, ρ_b, and ρ_f are the apparent matrix, bulk and mud filtrate densities, in g cm^{-3}, then

$$\phi = \frac{\rho_m - \rho_b}{\rho_m - \rho_f}$$

ρ_m ranges from 2.10 for halite to 2.90 or thereabouts for dolomite; the common range of ρ_f is from 0.85 for oil-base mud to 1.15 for saline mud.

12.3 Response of sonic and density logs to organic content

Organic matter has a much lower sonic velocity than any sedimentary rock. Whereas even shales transmit waves at more than 4 km s^{-1} and carbonates at about 6 km s^{-1}, velocities in oils are 1.2–1.25 km s^{-1} and in kerogens no more than 1.5 km^{-1}. Thus Δt for sedimentary rocks is of the order of 150–200 μs m^{-1}, and for organic matter about 600.

The velocity contrast between the matrix rock and its OM (about 4) is greater than the density contrast (about 2.5). The sonic log should therefore respond to the OM content of the rock more strongly than the density log. In practice, a combination of the two logs is commonly used (Fig. 12.12).

If a rock contains x percent of OM by volume, the matrix represents $(100 - x)$ percent. If the densities of

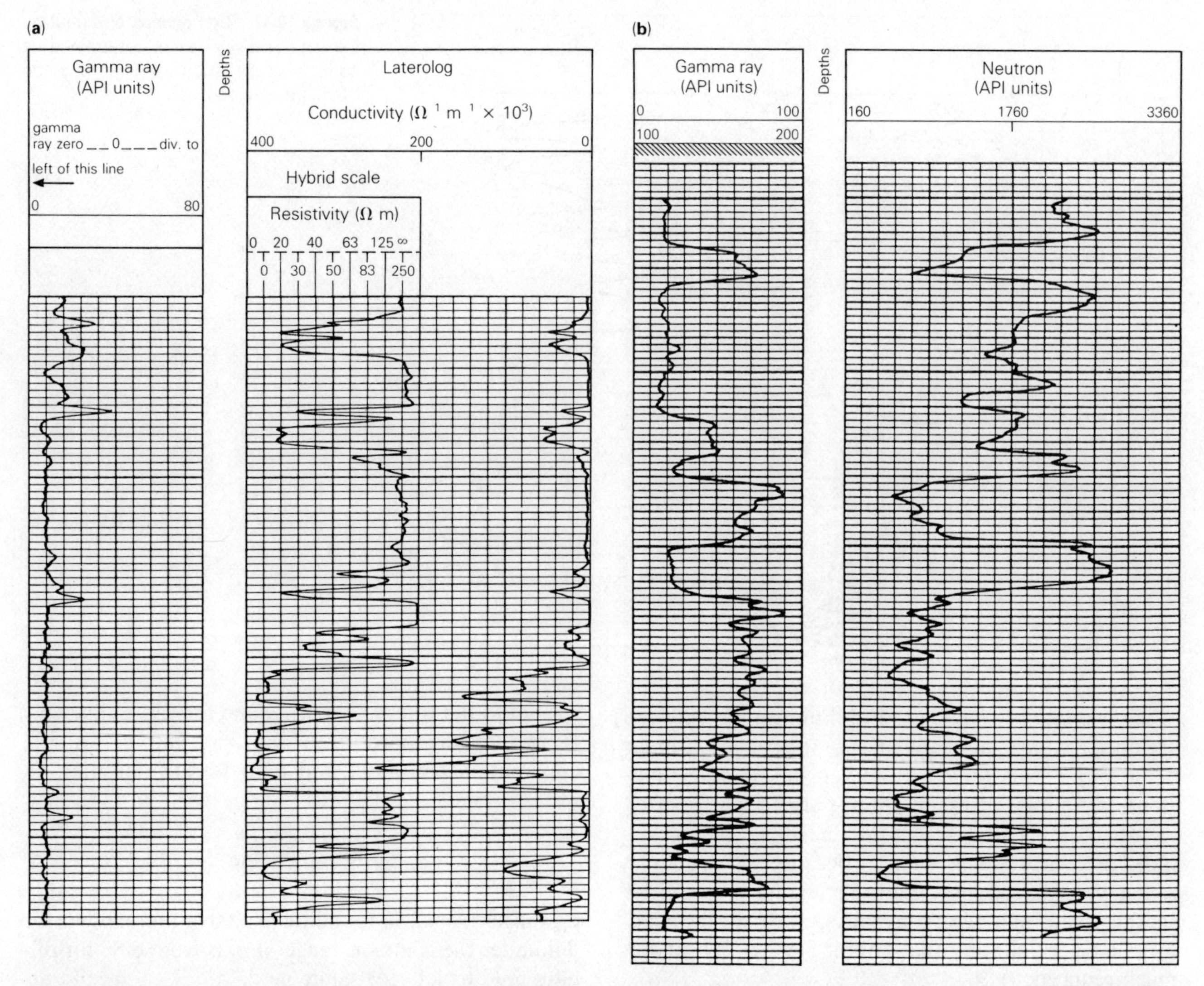

Figure 12.11 Presentations of gamma ray logs with (a) laterolog, (b) neutron log, and (c) bulk density log (forming so-called FDC log). (Courtesy of Schlumberger Limited.)

the matrix and the OM are ρ_m and ρ_0 respectively (about 2.65 and 1.10 would be good averages), then the *rock* density, ρ_r, is given by

$$\rho_r = \rho_0 x + \rho_m(100 - x)$$

and Δt can be calculated:

$$\Delta t = x\Delta t_0 + (100 - x)\ \Delta t_m$$

If the OM in a shale is accompanied by much pyrite, these two components exert opposed effects on the shale density, and calculation of the OM content can then be even roughly reliable only if the sulphur content of the shale is determined by analysis.

A firm caution is necessary for those responsible for calculating the OM contents of sediments from logs. The technique is very useful if it is used in a comparative sense (between two parts of the same formation, for example). Otherwise, the geologist and geochemist should remember that log analysis is not an exact science, and the mere fact that the data derived from this technique are quantitative does not mean that they are accurate.

12.4 Resistivity and water saturation

A vital role of resistivity curves is to provide a measure of the *water saturation* (S_w) of a reservoir rock. This is vital for two reasons. First, calculations of recoverable reserves in an oil- or gasfield will be seriously amiss if pore space occupied by water is thought to be occupied by hydrocarbons. Second, the capacity of a well to produce clean oil may lead to excessive optimism about the well's eventual decline if the water saturation is underestimated. If a reservoir sandstone contains interstitial fines (silt grains, kaolinite or other clay particles, col-

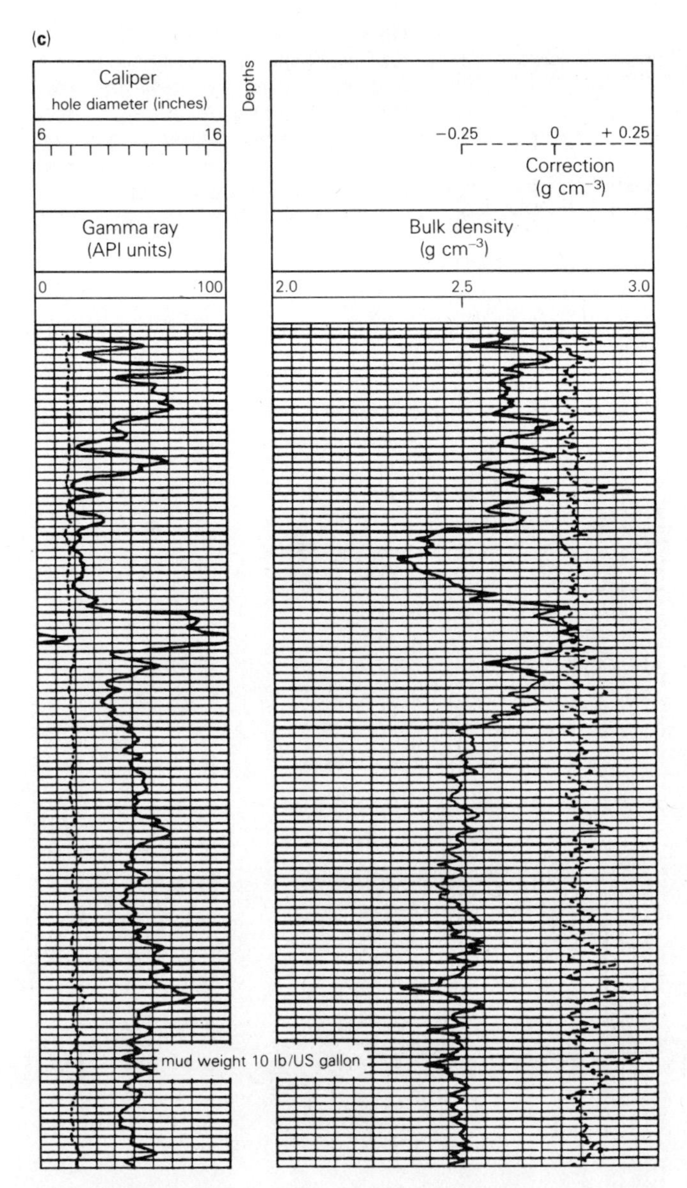

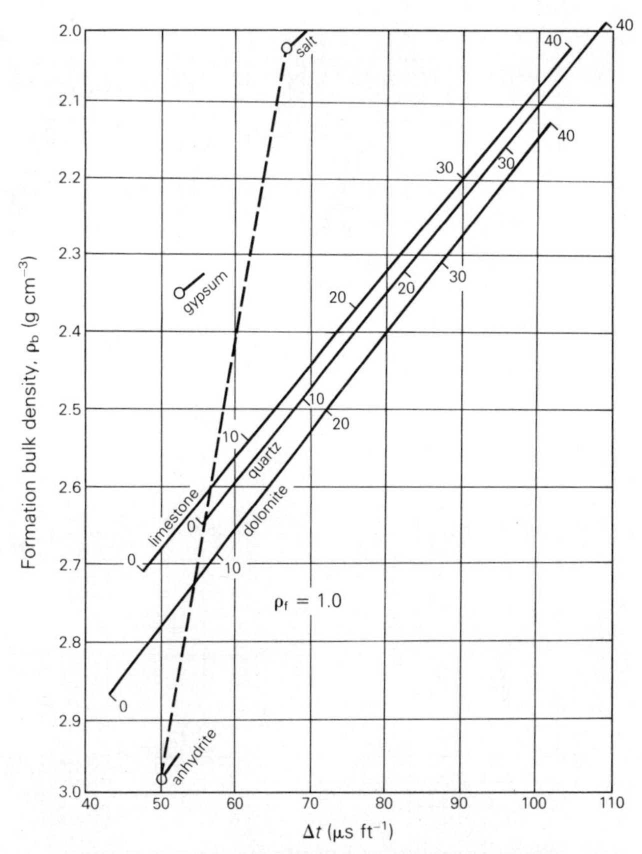

Figure 12.12 Porosity and lithology determination from density and sonic logs. Points of increasing porosity, derived from two logs, define lines which, for the common lithologies of sandstone, limestone and dolomite, are too close together to be definitive, because the densities of the rocks fall within a narrow range. The evaporite minerals, with very different densities, are more successfully discriminated. (Courtesy of Schlumberger Limited, New York, 1972.)

loidal material) between the sand grains, water may surround each fine particle, greatly increasing S_w but not releasing the water into the borehole. The sand may still produce clean oil even if S_w exceeds 50 percent, but it will not produce as much of it as a misreading of S_w would lead the operator to expect.

The measured resistivity (R) is proportional to the fluid (water) resistivity (R_w), and inversely proportional to the product of the porosity fraction and the water saturation (ϕS_w). The general relations between these quantities and the *formation resistivity factor* (F) are given by Archie's formula:

$$R = \frac{aR_w}{\phi^m S_w^{\,n}} \qquad (12.1)$$

$$F = \frac{R_0}{R_w} = a\phi^{-m} \qquad (12.2)$$

where R_0 is the resistivity of the water-bearing bed of true resistivity R_t when clean and uninvaded. F is therefore defined as the ratio between the resistivity of the saturated rock and that of the electrolyte saturating it. The value of F is very variable; it increases as porosity and permeability decrease (Fig.12.13), from 10 to 40 for most porous rocks to over 1000 for dense limestones.

The increase in F with increase in m (the *cementation factor*) is illustrated in Fig.12.14. (Note that ϕ is the porosity *fraction*, and therefore less than unity.) The value of m lies between 1.2 and 2.2; it is lowest for shaly or dirty sandstones, highest for limestones. The value of n varies from about 1.5 for dirty sands to 2.0 for clean sands or limestones. For general quantitative interpretation, n may be taken as 1.8 for sands and 2.0 for limestones.

Thus, for non-shaly rocks,

$$R = \frac{R_w F}{S_w^{\,n}} = \frac{R_0}{S_w^{\,n}}$$

and $$S_w = (R_0/R_t)^{1/n} \qquad (12.3)$$

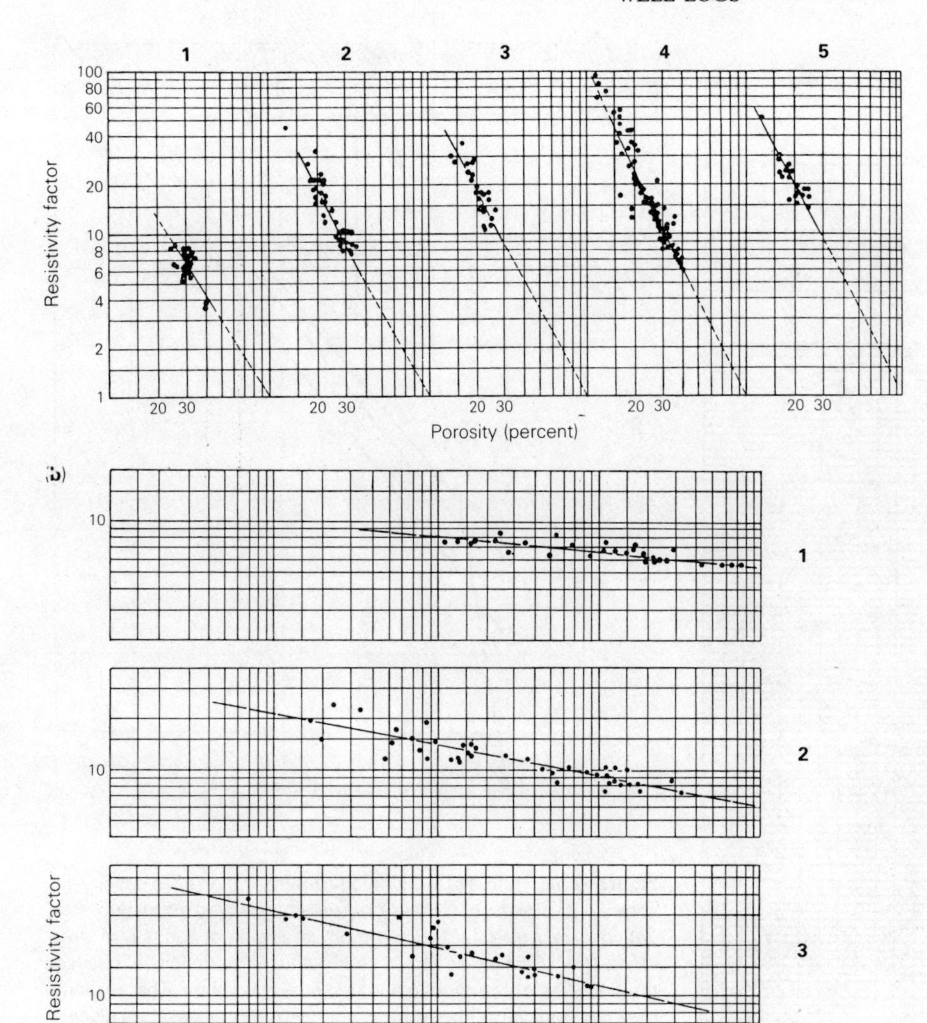

Figure 12.13 Relation of formation resistivity factor to porosity (a) and permeability (b), for five reservoir sandstones, respectively Miocene (1), Oligocene (2), Eocene (3), latest Cretaceous (4) and late Mississippian (5) in age. (From G. E. Archie, *AAPG Bull.*, 1950.)

For shaly rocks, the relation is more complex. Values of S_w are read from nomographs or the equations are programmed into calculators.

12.5 Continuous graphic logs

Selected logs are run in all wells, some of them several times (e.g. before casing is extended). Compound graphic logs should be maintained in order to keep continuous records of all observable properties of the rocks and fluids encountered by the drill: sample descriptions, hydrocarbon content of the mud as it returns to the surface from the borehole, fluorescence, drilling time, and any incidents during drilling which might have explanations valuable to the geologist. These topics are dealt with in detail in Chapters 19 and 23.

If multiple logs are run in a single well, values of the

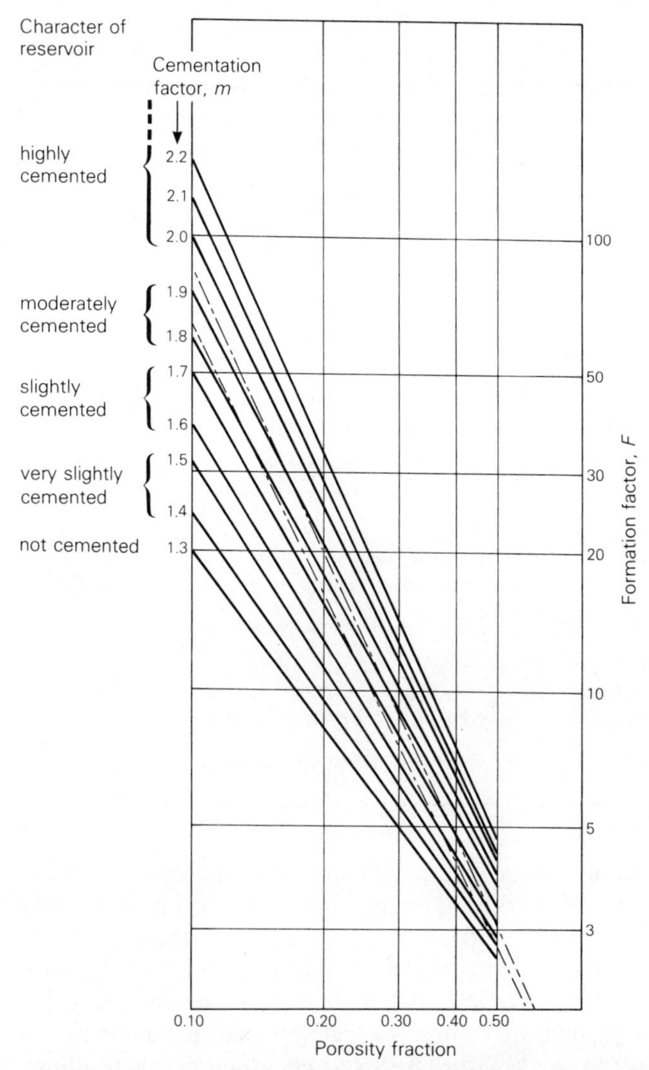

Figure 12.14 Relation between porosity (expressed as a fraction, not as a percentage) and formation resistivity factor (*F*) for sands of varying degrees of cementation, showing increase in value of *F* with increasing value of *m* (cementation factor). (From R. I. Martin, *Oil Gas J.*, 29 June 1953.)

various parameters calculated from the appropriate logs may be cross-plotted to derive fields of points (Fig. 12.12). These cross-plots can then be used to compute and display continuous analyses of critical properties such as porosity, water saturation, or clay content. The two most commonly used interpretive techniques are the SARABAND, for sand–shale sequences, and the CORIBAND, for more varied lithologies (including carbonates). Examples of the two graphic presentations are shown in Figures 12.15 and 12.16.

12.6 Use of well logs in stratigraphic correlation

Petroleum geologists all over the world spend much of their time trying to convince themselves of the correctness of the stratigraphic and structural correlations they have made between the drillholes that are their control points. In much-drilled and well-understood sedimentary basins, correlations across wide areas – even across entire basins – are accepted by everyone working there. In less thoroughly understood regions, this acceptance may be far from assured.

Because all wells are logged, whereas few are cored and even fewer provide cuttings trustworthy beyond the wellsite, stratigraphic correlation across broad areas becomes almost wholly a function of log interpretation. Where "layer-cake" geology prevails, as in many pre-Tertiary platform regions, long-distance correlation may be simple. In regions of orogenic "tip-heap" geology (Trinidad, for example, or Assam), correlation between adjacent wells may be perplexing even to old hands.

The types of well log conventionally used for correlation purposes are listed in Table 12.2. No two log types respond in identical ways to comparable conditions in drillholes. A few basic rules may help to avoid disastrous miscorrelations:

(a) Reliance on logs alone should be avoided unless no other data are available. Even apparently reasonable correlations between logs can be reduced to nonsense by fossils (especially microfossils), laboratory analyses, radiometric dates, or even cross-ties to other section-lines of correlation.

(b) Correlation sections should utilize logs of a single type (preferably resistivity logs). Figure 12.17 illustrates a typical battery of logs run on a single well. The reader may imagine the difficulty facing the geologist who has one log type for some wells and other log types for other wells in his section. Fig. 12.18 shows the correlation between wells in an area of classic "layer-cake" geology (the Michigan Basin) by radioactivity rather than resistivity logs. Imagine the differences between the left-hand and right-hand log expressions extended over a much wider area with much more complex stratigraphy.

(c) Knowledge of the geology of the area is an advantage beyond valuation. What are the most likely explanations for anomalies in the correlation sections (reefs, growth faults, gaps or repetitions in the section, wedge-outs, facies changes)? Figs. 12.19–22 illustrate some of the more routine of these anomalies. The geologist must always be mindful that most oil and gas accumulations are reflected in some kinds of "anomalous" correlation at their boundaries, and it is the geologist who first correctly identifies these who finds the fields.

(d) Finally, the geologist should at all times be ready to change any correlation that is not ironclad. The geologist who knows his correlations are complete

Table 12.2 Well logs used for stratigraphic correlation. (Courtesy of Schlumberger Ltd.)

Log	Phenomenon correlated	Requirements and recommendations
short-normal resistivity	invaded porous beds, depending upon water resistivity	uncased hole; non-salty mud combine with SP or gamma-ray logs if formations of low general resistivity (largely shales)
laterolog	as above	uncased hole; fresh or salty mud resistive formations, with high $R_t : R_w$ ratio
SP log	contrast between shale (impermeable) and non-shale (permeable) beds	uncased hole low to moderate formation resistivity; sand–shale contrasts well displayed
induction log	contrast between conductivity of pore fluids (salt water–fresh water–oil) and of non conductive beds	uncased hole; fresh mud low formation resistivity (most shales)
sonic log	transit time (Δt), dependent on lithology and porosity	fluid-filled hole low formation resistivity
density log	formation density, dependent on lithology and porosity	uncased hole without washed-out levels
neutron log	hydrogen content of porous beds; shales clearly distinguished	cased hole combined with gamma-ray log
gamma-ray log	radioactivity, related to shale content	cased or uncased hole; unaffected by drilling fluid

and inviolate has found all the fields *he* is going to find in his area. The geologist who tries a different correlation may demonstrate that not all undrilled stretches between wells are barren ground.

12.7 Dipmeter logs

An invaluable outgrowth of electric logging has been the development of devices for the accurate measurement of dips of the strata penetrated in boreholes. The device is called the *dipmeter*. Its basic design utilizes the three-point method of calculating true dips from apparent dips (or from elevations above or below a known datum) that is taught in elementary classes in structural geology.

The dipmeter tool, or sonde, bears three arms set radially 120° apart; many modern models have four arms 90° apart. The arms are pressed against the walls of the hole, and electrodes within them measure the resistivities at the points of contact. In some models, the SP is the property measured. Continuous measurements are recorded as the sonde is withdrawn from the hole. Correlations of the three traces for any individual bed pinpoint three points on that bed from which the true dip is calculated – graphically (by stereonet, for example, or electronic plotter), or by computer program (for abundant data points in complex regions).

For accurate measurements, it is not of course enough merely to recognize a single bed on the traces from all three electrodes. It is also necessary to know the azimuths of the three arms, the diameter of the hole where the bed was intersected, and any deviation of the hole from the vertical. All these values are automatically recorded, by compasses, calipers, and pendulums, encased in the arms with the electrodes. The resulting dipmeter log looks like that shown in Figure 12.23. Traces on the left are *orientation traces*, showing the azimuth of one of the electrodes (remember that this will be a magnetic and not a true azimuth and will have to be corrected for the local magnetic declination), and the deviation of the hole from the vertical. The traces on the right are *focused correlation curves* for the three electrode pads, and the diameter of the hole measured by the caliper.

Calculated true dips are plotted by direction and amount on strip logs (Fig. 12.24), sometimes called "tadpole diagrams." Positions of the circles indicate the dip angles (increasing to the right); arrows on the circles (the tails of the tadpoles) show the dip directions by azimuth. Changes in the dip angles may be gradual, as they might be in holes drilled down the flanks of folds or through the strata surrounding salt domes, or abrupt, as they may be when crossing a fault (Fig. 12.25). Interruptions of normal bedding by cross-bedding or lenticular bodies may be revealed by reversals of dip direction (Fig. 12.26). Changes in both angle and direction of dip commonly reveal unconformities (Fig. 12.27).

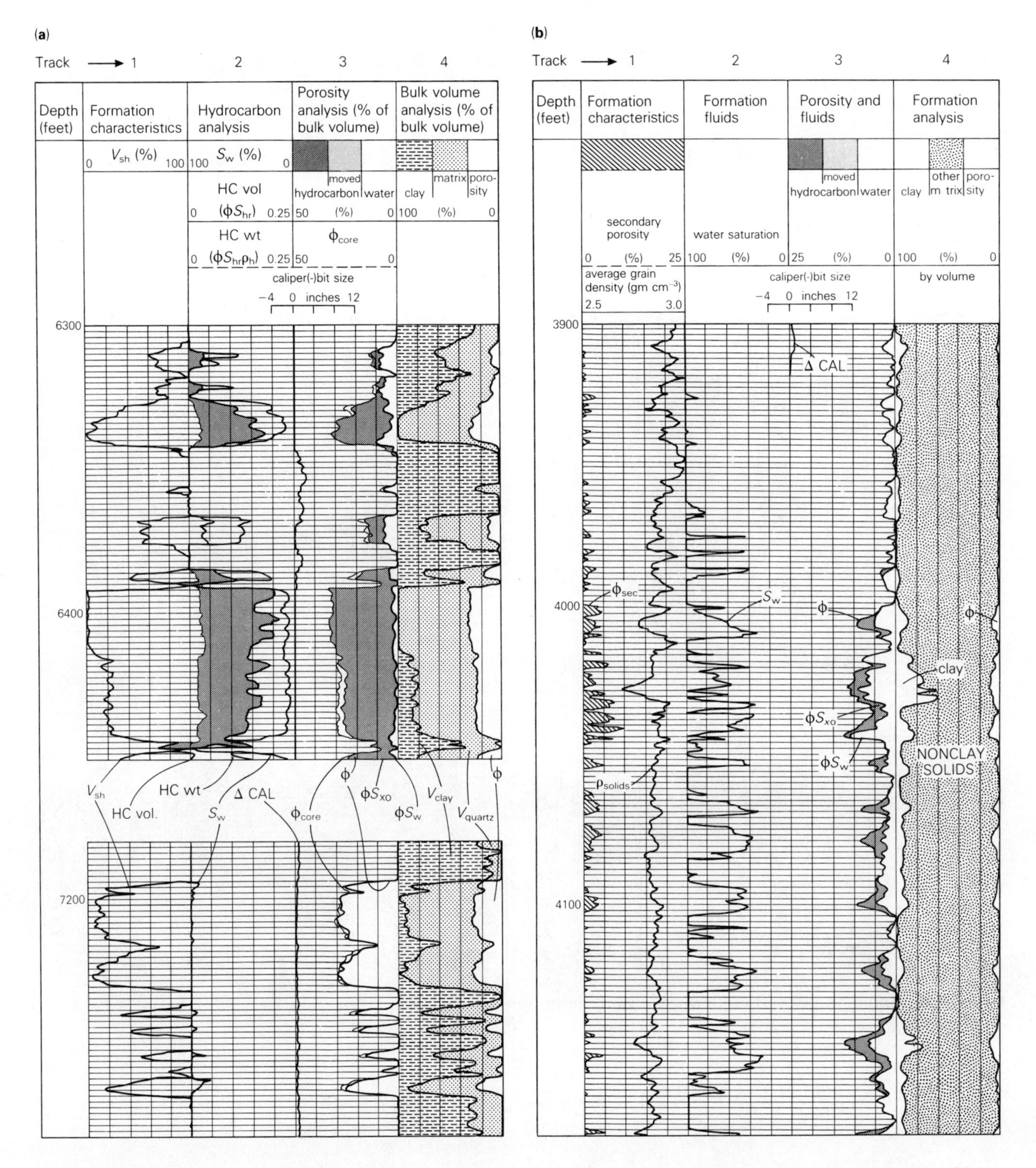

Figure 12.15 (a) Example of SARABAND interpretation format in hydrocarbon-bearing and water-bearing sandstone formations. (b) Log presentation of results computed by CORIBAND technique, showing porosity zones and their fluid contents, including water saturation. (Courtesy of Schlumberger Limited.)

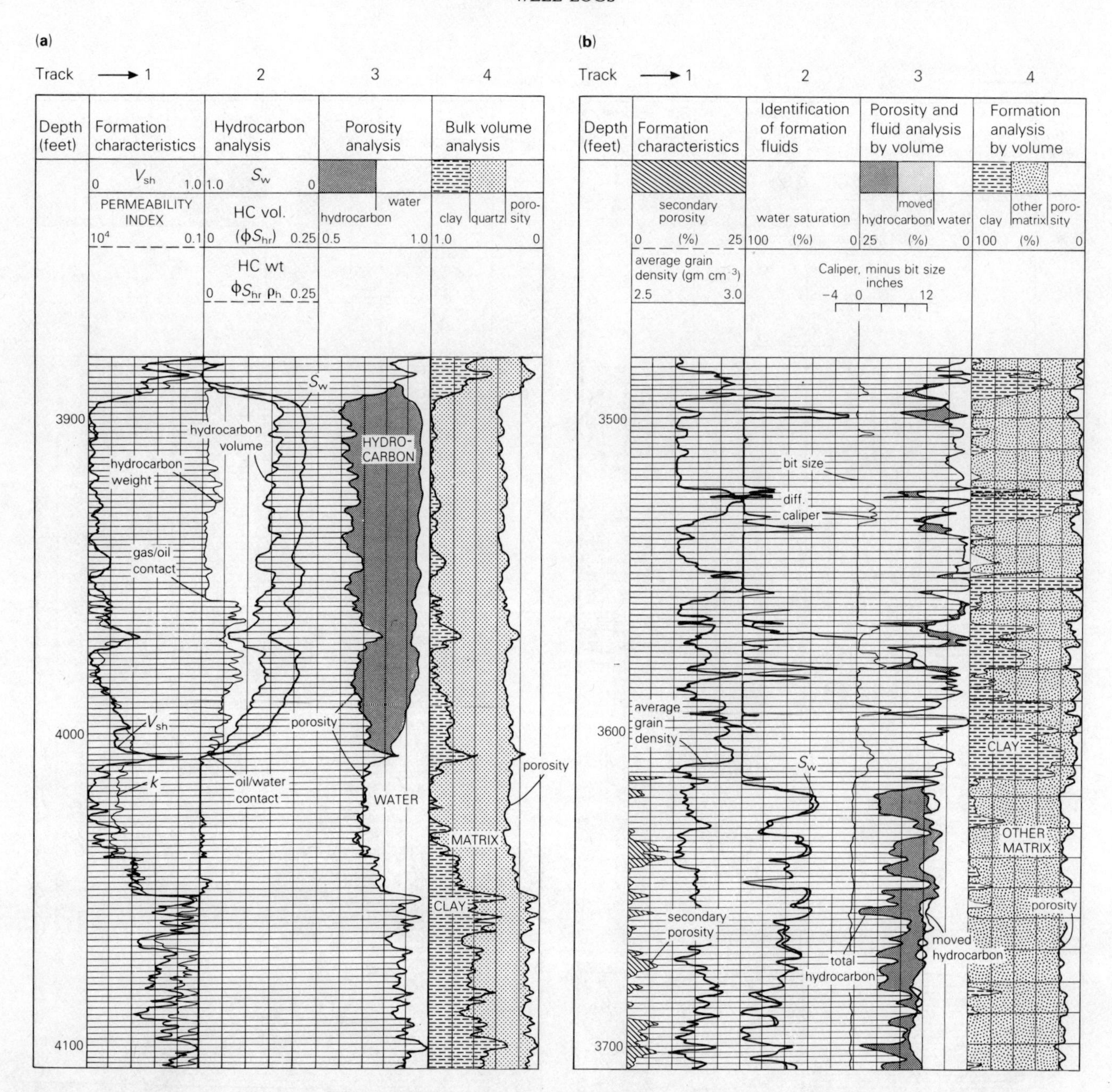

Figure 12.16 SARABAND (a) and CORIBAND (b) answer formats with identifications of all data. (Courtesy of Schlumberger Limited.)

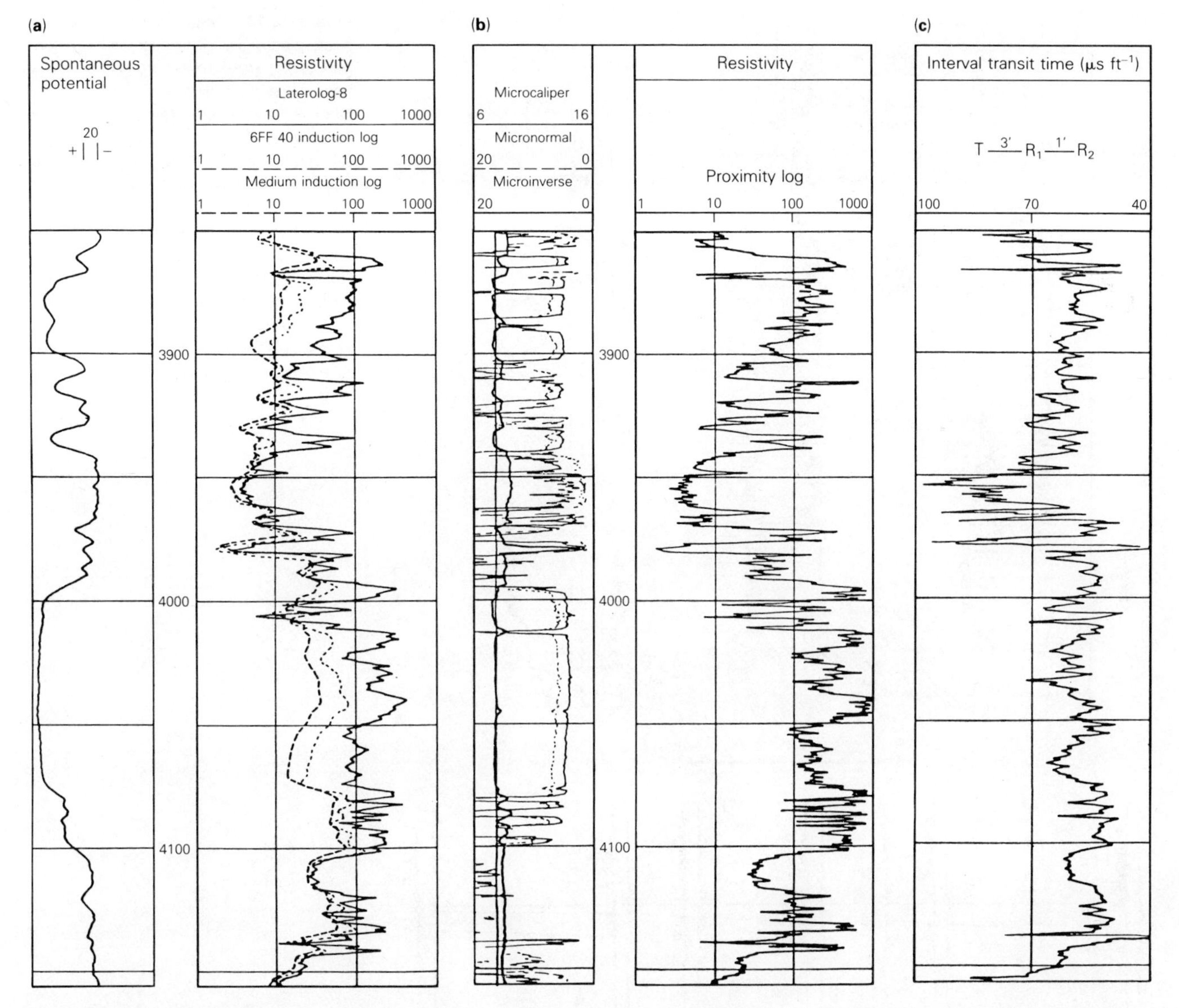

Figure 12.17 Typical grouping of logs for a single well (in the Permian Basin of West Texas): (a) dual induction–laterolog; (b) microlog–proximity log; (c) sonic log. (From H. G. Doll *et al.*, 6th World Petroleum Congress, 1963.)

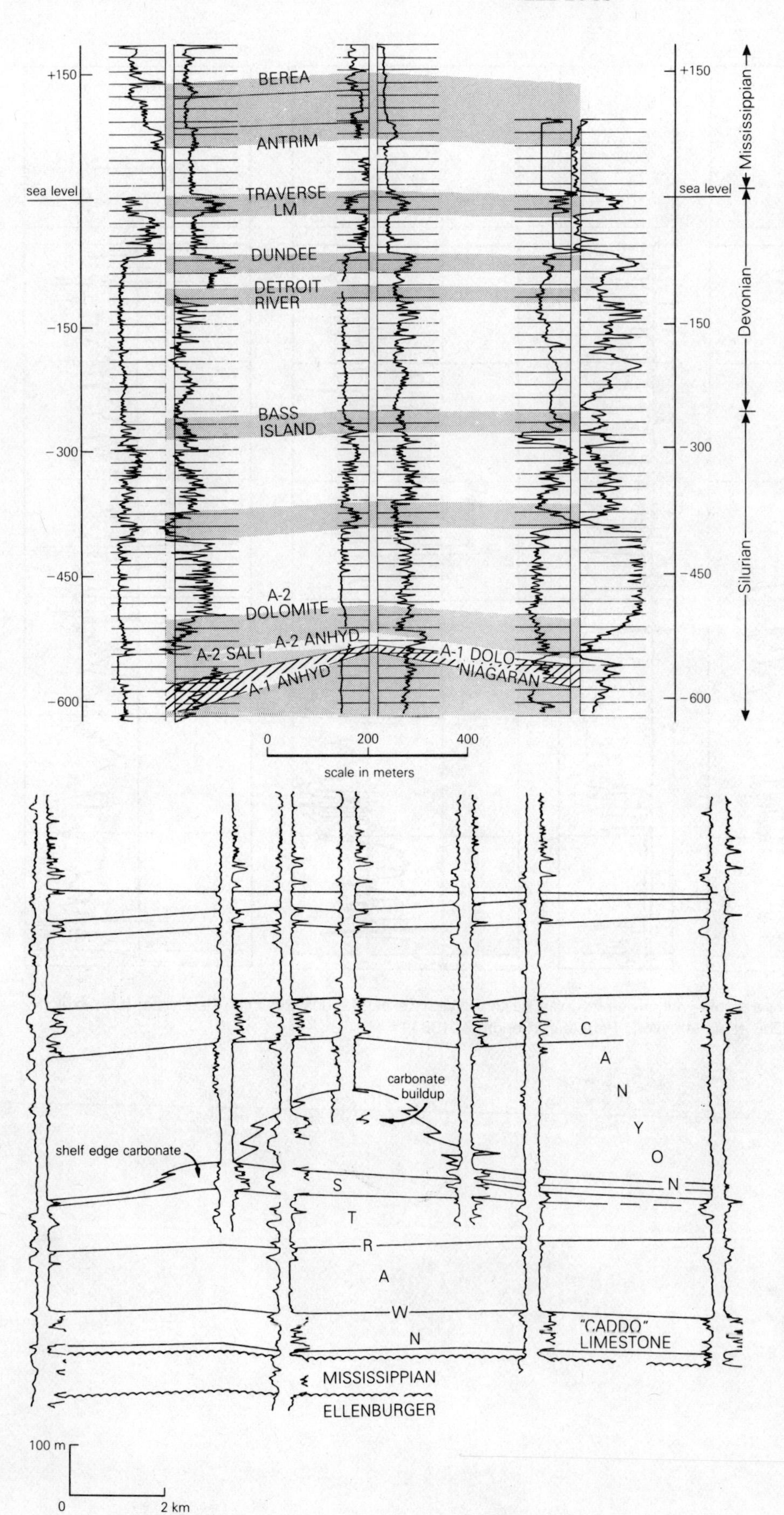

Figure 12.18 Gamma ray–neutron log section across a Silurian reef in the Michigan Basin (reef marked "Niagaran" at base of section). Illustrates simple correlation in "layer-cake" stratigraphy of post-reef beds. Note expression of Berea sandstone and Antrim shale (Mississippian); Traverse and Dundee limestones (Devonian), with lower Traverse shale between them; salt, anhydrite, dolomite, and shale of the Detroit River Group and the Salina Group (above the reef), with the Bass Island dolomite (topmost Silurian) between them. (From C. Ferris, *AAPG Memoir* 16, 1972.)

Figure 12.20 Electric log section through Middle Pennsylvanian strata in part of northern Texas, showing simple correlations through essentially parallel carbonates and shales except over the Strawn–Canyon carbonate buildup that contains the North Knox City oilfield. (From J. C. Harwell and W. R. Rector, *AAPG Memoir* 16, 1972.)

Figure 12.19 Electric log cross sections through two Tertiary sand–shale sequences, showing two common types of interruption of otherwise parallel strata. (a) Lenticular sand body, probably an offshore bar, causing draping of younger strata; Hardin field, Texas Gulf Coast. (From F. B. Rees, *AAPG Memoir* 16, 1972.) (b) Wedging-out of productive sandstone units (stippled) against post-Eocene unconformity, western Venezuela. (From P. R. Vail *et al.*, *AAPG Memoir* 26, 1977.)

(**a**)

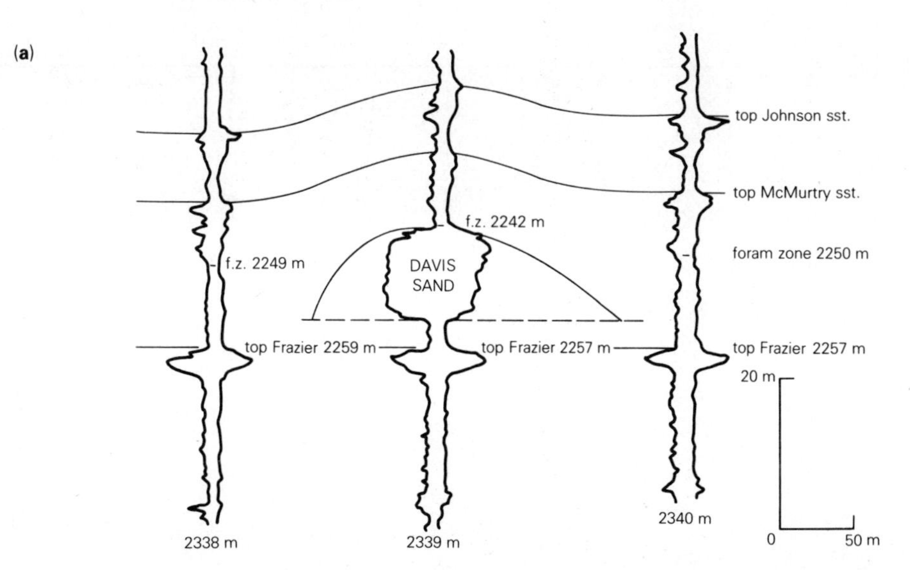

(**b**)

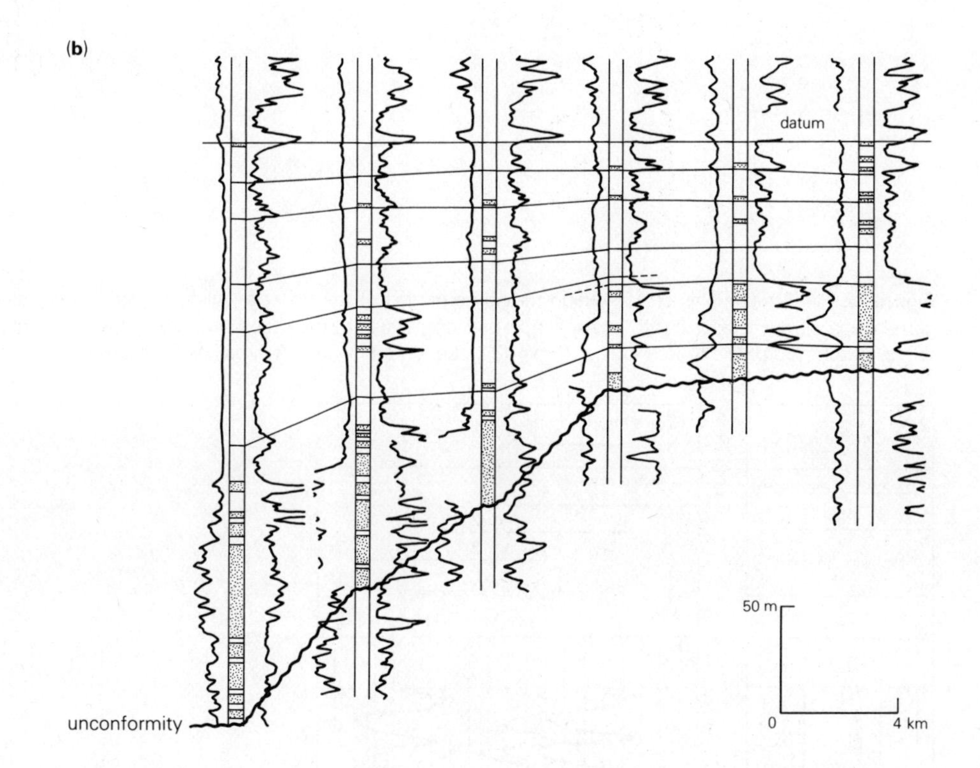

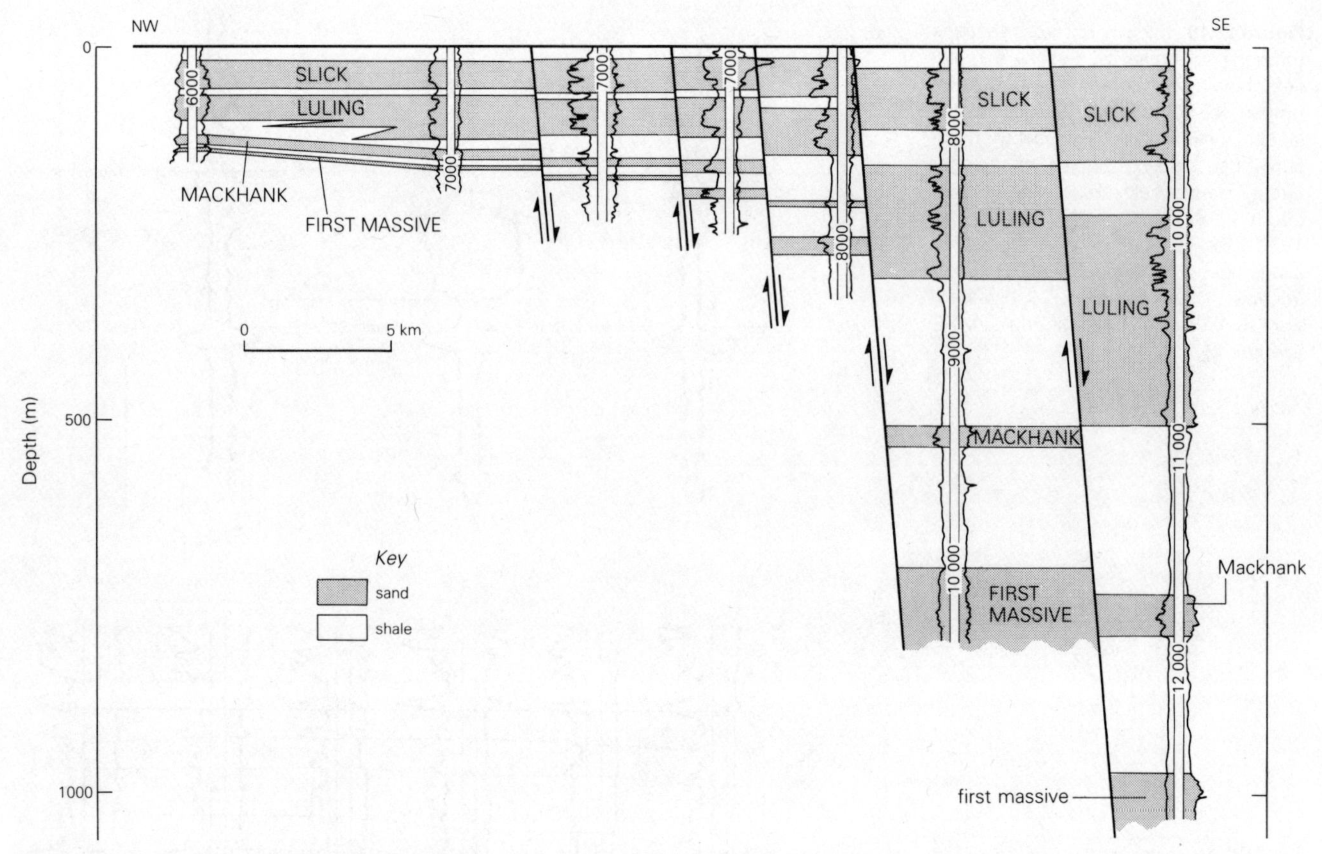

Figure 12.21 Stratigraphic cross section through a portion of the Eocene (Wilcox) trend in the Texas Gulf Coast, to illustrate electric log correlations in a succession basically very simple but complicated by thickened sections on the downdropped sides of down-to-the-coast normal faults. Datum top of the Wilcox Group. (From C. L. Lofton and W. M. Adams, *AAPG Memoir* 15, 1971.)

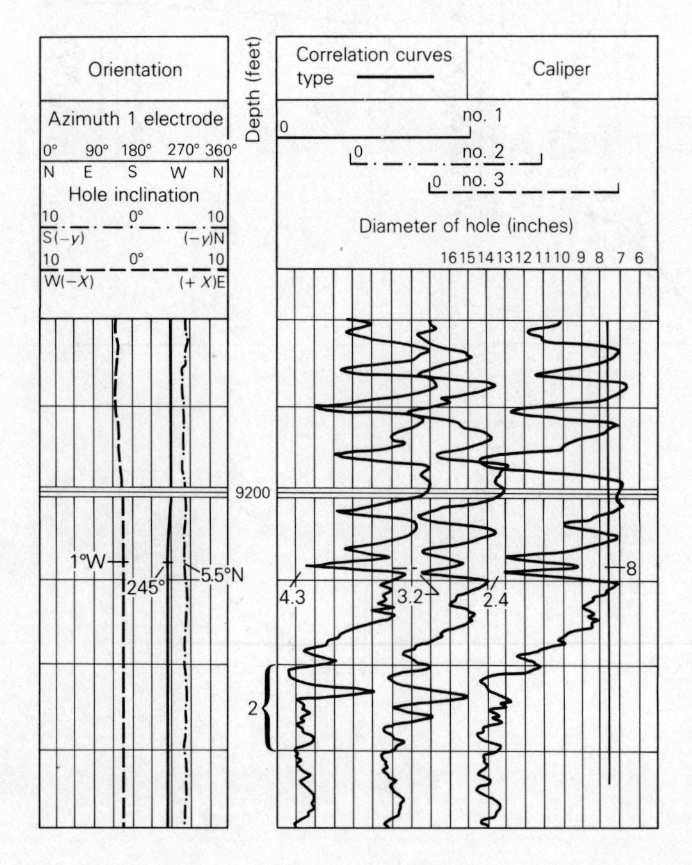

Figure 12.23 Dipmeter log, showing three components. Straight line farthest to right is caliper log, showing diameter of hole (8 inches = 20 cm). Lines on left show hole deviation from vertical. Three irregular traces are those of three pads measuring resistivity (nos 1–3), showing kicks rising in elevation from no. 3 to no. 1 (figures on curves are inches above a datum plane). These three points form the basis for the calculation of the true dip. (From A. J. Pearson, in *Petroleum exploration handbook*, G. B. Moody, Ch. 21, New York: McGraw-Hill, 1961.)

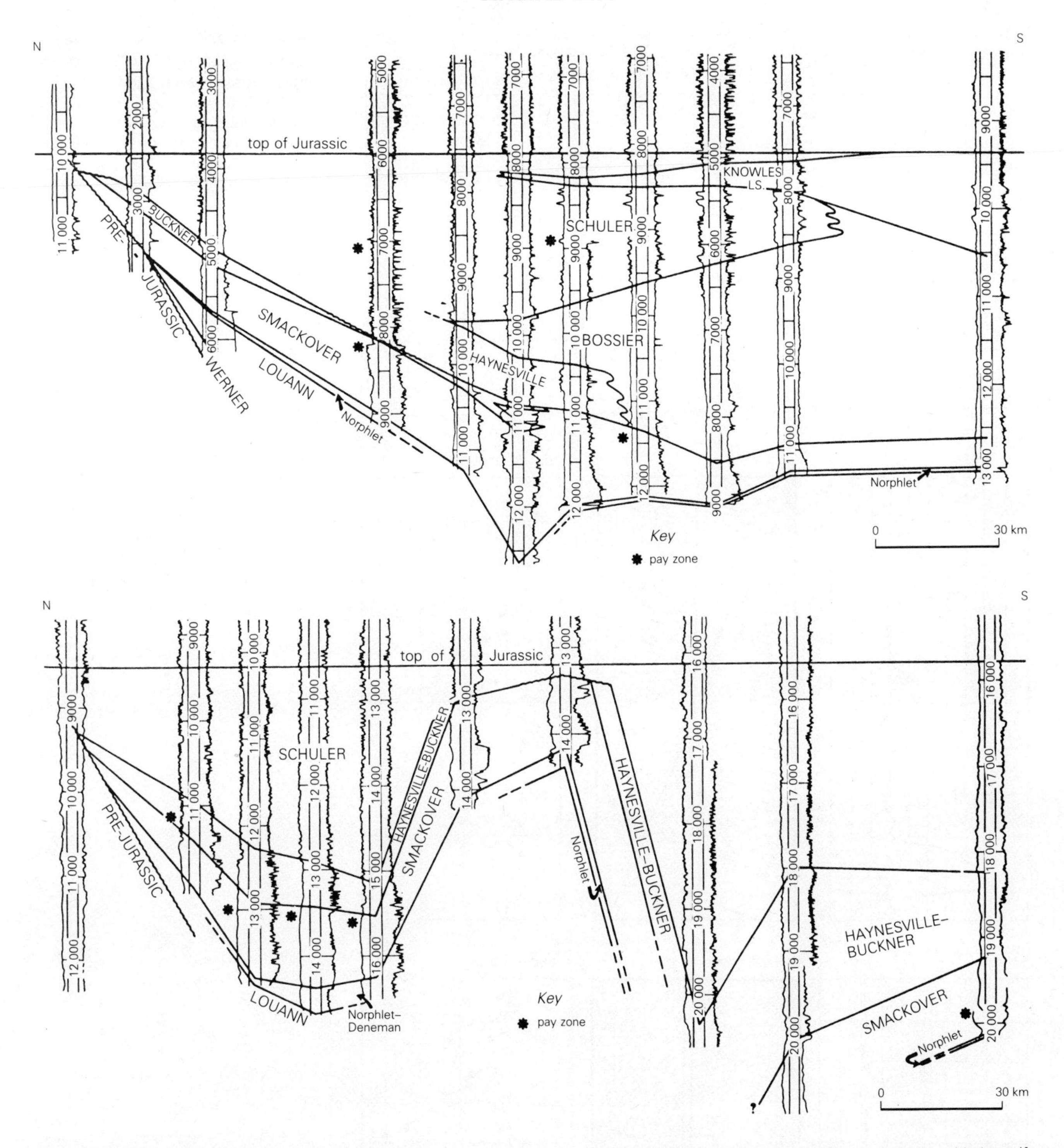

Figure 12.22 Electric log correlations along dip sections across the Jurassic succession in the faulted upper Gulf Coast salt basin, with the gulf off to the right. Complex interfingering of varied lithologies, not capable of discrimination from log data alone. The Louann Formation is salt; Buckner, evaporites and red beds; Norphlet and Haynesville also red beds; Smackover is limestone, Bossier largely shale, and Schuler largely sandstone. (From T. F. Newkirk, *AAPG Memoir* 15, 1971.)

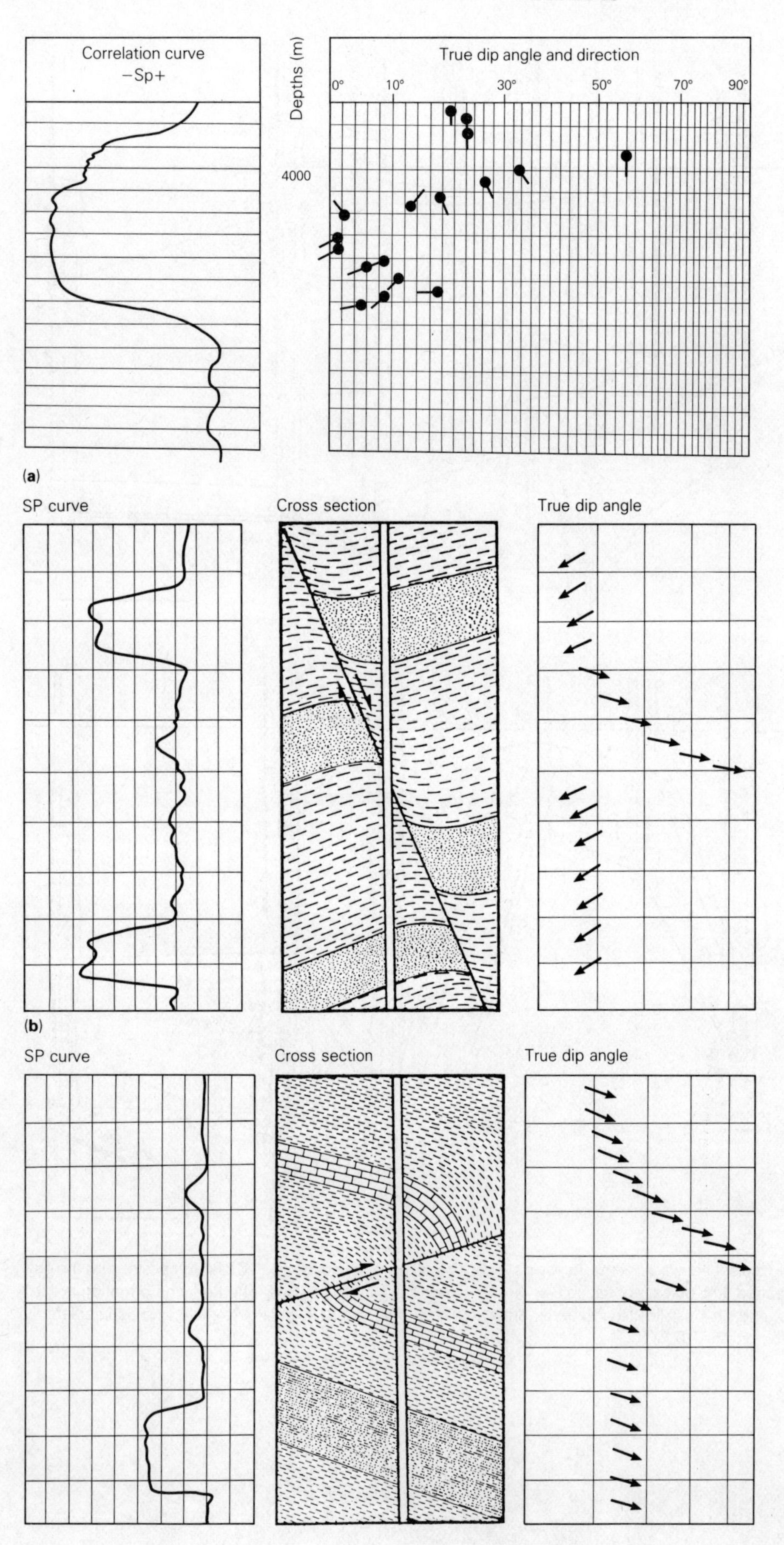

Figure 12.24 "Tadpole diagram," showing dips derived from dipmeter log plotted against depth and SP curve. Distance of body of tadpole from vertical base line indicates angle of dip; tail of tadpole shows dip direction by azimuth. (From Schlumberger Limited, *Fundamentals of dipmeter interpretation*, 1970.)

Figure 12.25 Schematic illustrations of dipmeter logs through a normal fault (a) and a thrust fault (b). (From Dresser Atlas, 1974.)

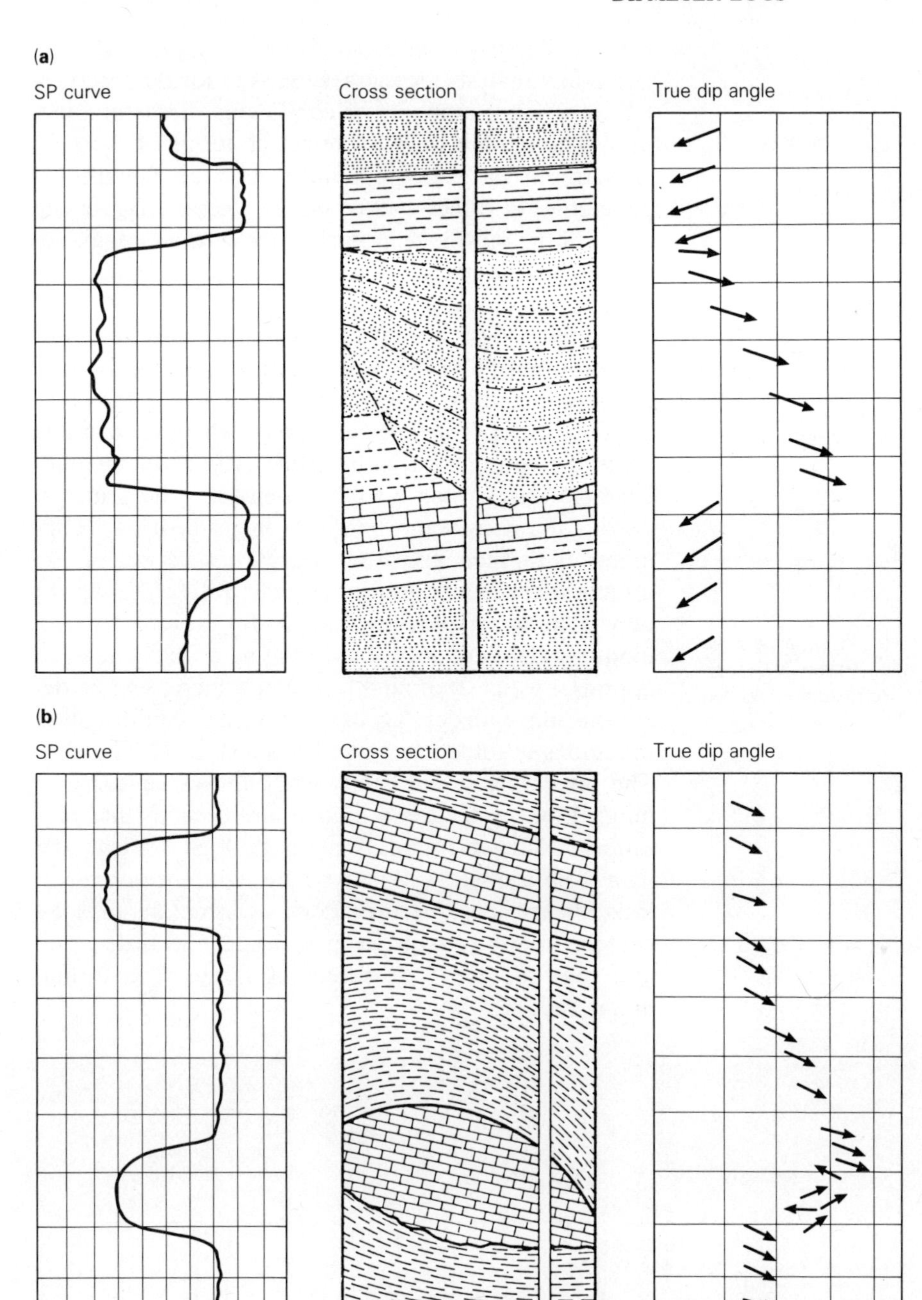

Figure 12.26 Schematic illustrations of dipmeter logs through a channel sandstone (a) and a reef (b). Note resumption of regional dips above and below the interfering rock body. (From Dresser Atlas, 1974.)

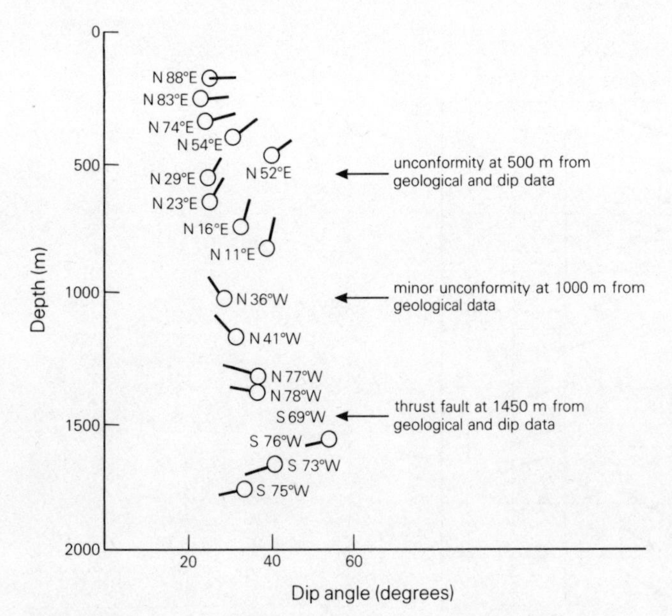

Figure 12.27 Plot of dips derived from dipmeter log, showing three discontinuities and their geological interpretation. (After P. A. Dickey, *Petroleum development geology*, Tulsa, Okla: Petroleum Publishing Co., 1979; from A. J. Pearson, 1961.)

C. A. Bengston and others have developed statistical curvature analysis techniques (SCAT) for the interpretation of large numbers of continuous dipmeter readings. The readings are expressed in terms of the bulk curvature of the geologic setting – its three-dimensional geometry, in effect – and this is then interpreted in terms of the most logical structural model among tilted, folded, and faulted structures. Examples of Bengston's models are shown in Figures 12.28 and 12.29.

Dipmeter data are invaluable in the early stages of exploration of an arèa, when the maximum amount of information must be extracted from every meter of depth of every well. The data should whenever possible be used in conjunction with seismic data. An elementary example arises when a wildcat well finds the principal porous horizons water-bearing (or gas-bearing). Dipmeter data then give the clearest indication of the direction in which to move the rig to drill farther up (or down) the dip. Following a successful exploratory test, dipmeter data are critical in the siting of delineation or appraisal wells (stepouts). Regularly increasing or decreasing dip with depth constitutes valuable information in regions of folded or faulted strata (Figs 12.28 & 29). The possibility of ambiguity must always be borne in mind; there is seldom a unique and obvious interpretation of dipmeter data. In many geologic settings, dip data from a single well may be quite unrepresentative; insignificant local perturbations of structure may be present in any setting. Even accurately recorded true dips may represent cross-bedding rather than formational bedding.

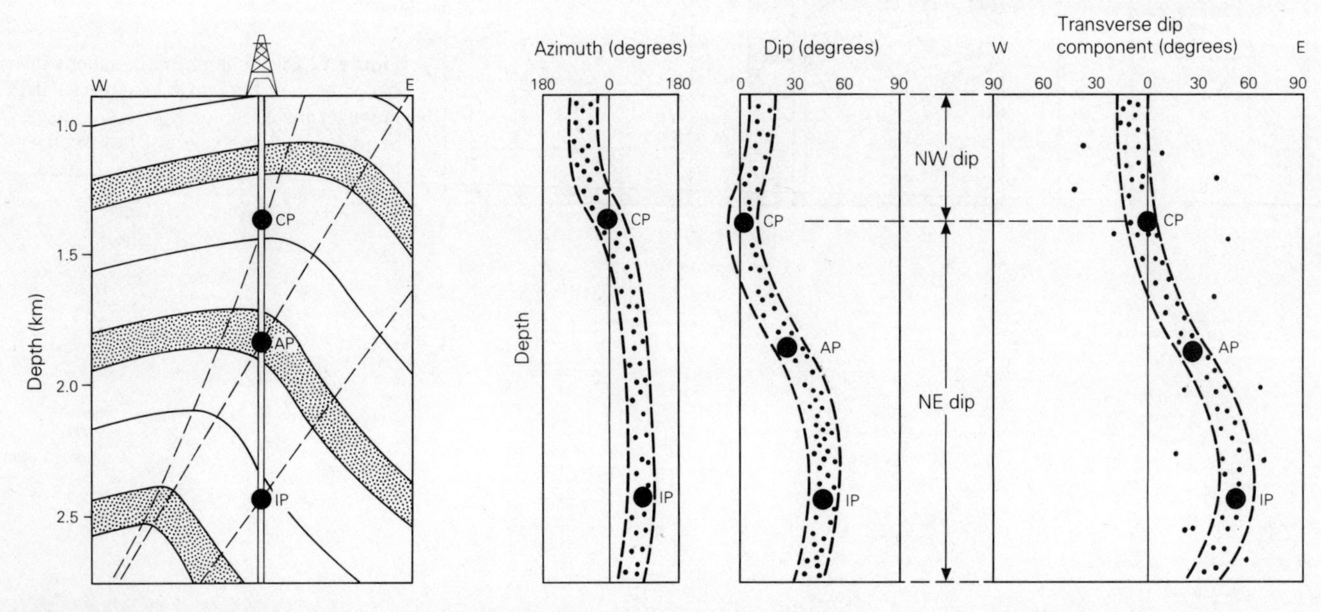

Figure 12.28 Geological cross section and SCAT plots for a plunging fold. CP, crestal plane of the fold; AP, axial plane; IP, inflection plane. (After C. A. Bengston, *AAPG Bull.*, 1981.)

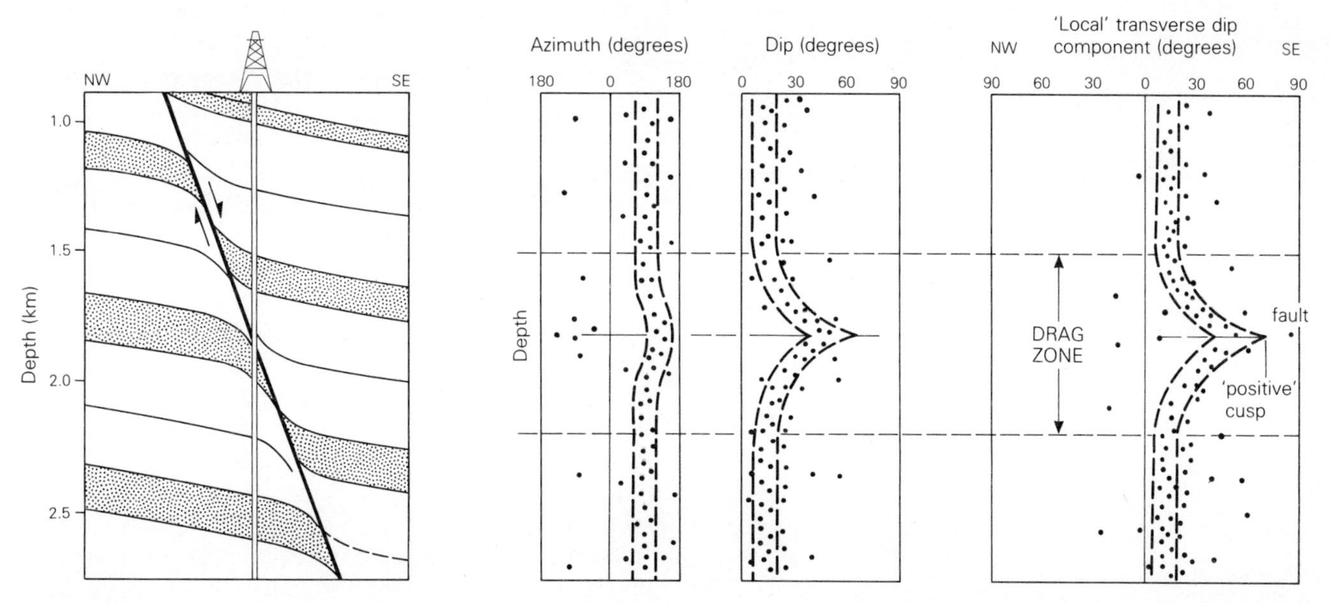

Figure 12.29 Geological cross section and SCAT plots for a normal fault with drag steepening the regional dip. (After C. A. Bengston, *AAPG Bull.*, 1981.)

Offshore exploratory and development drilling has brought new demands on the interpreters of dipmeter data. Expensive wells are drilled directionally from fixed platforms (see Sec. 23.2.4). The deviations from the vertical may be up to 45° and they differ for all the wells from any one platform. Furthermore, many petroleum traps in deep water are controlled by block faults, growth faults, or diapirs, or by unconformities themselves consequent upon one or other of these three mechanisms (see Sec. 17.9.3). Dipmeter data, crucial to the correct interpretations of structural details under these conditions, need laborious multiple checking.

13 Reservoir rocks

13.1 Introduction

Oil and gas may exist in a region in appreciable quantities and yet be inaccessible in practical and commercial terms. Though there are a number of causes of this circumstance, one of the most common and discouraging is the lack of any rock formation, in suitable stratigraphic position, possessing both the porosity and the permeability necessary both to contain and to yield oil, or gas, or both, in commercial quantities. Any rock capable of performing these functions is a *reservoir rock*.

The fundamental property of a reservoir rock is its *porosity*; for it to be an *effective* reservoir rock, the fundamental property is its *permeability*. Porosity and permeability are geometric properties of a rock, not (or at least not necessarily) genetic properties. Thus the lithologic character of the rock, especially its textural properties, is far more important than its age.

Though most sedimentary rocks possess significant porosity and permeability when freshly deposited, the rate at which these vital properties are reduced with age is, as we have seen, principally due to *compaction*. The compactibility of a sedimentary rock, like its initial porosity, depends more on texture than it does on composition. It is the textures of many familiar sedimentary rocks – geometric properties such as the sizes and shapes of the constituent grains, and the manner of their packing – that eliminate them from the effective reservoir category unless they are extensively fractured: shales and mudstones, siltstones, cherts, coals, evaporites, marlstones, dense or cherty limestones. This leaves only relatively or very coarse clastic sediments (sandstones, grits, conglomerates), and grained or crystalline carbonate rocks, as *common* reservoir rocks.

In Chapter 4, we observed that most prolific oil production before World War I came from geologic regions almost wholly clastic in sedimentary character (and nearly all Tertiary in age). In a worldwide frame, even today, most reservoirs for giant oilfields are in sandstones, certainly over 60 percent of them. North American drillers as late as the 1960s were likely to refer to their target reservoir as "the sand" (possibly pronounced "say-und") even if it were in reality a vuggy dolomite texturally and compositionally unlike any sand. In the USSR such a tradition (if it existed there) would even be appropriate. Almost 90 percent of Russian oil deposits, and considerably more than 90 percent of their known gas reserves, are in clastic reservoir rocks, even though some small proportion of these contain calcareous cements.

Some notable oilfields had been discovered and exploited in carbonate rocks before 1914, but real recognition of the importance of carbonate reservoirs came between 1925 and 1930 (see Ch. 4). Huge development followed in the 1930s and 1940s, particularly in the Middle East and the Permian Basin of the southwestern USA.

Most geologists think of sandstones and limestones (or dolomites) as being two very distinct types of rock, as of course they are *compositionally*. One is wholly inorganic and clastic, the other at least partly organic or chemical in origin even if it is now in a clastic state. The distinction between the two dominant reservoir rock types is at least partly semantic. The three familiar, archetypal sedimentary rocks were named according to different criteria: "sandstone" refers to grain size, "shale" to an induced physical property, that of fissility, "limestone" to chemical composition. Despite the spectacular performance of many reef reservoirs (see Sec. 11.2.3), it is convenient to think of other carbonates as conglomerates, sandstones, siltstones, and shales (or mudstones), but made of calcium carbonate and not of silica or clay minerals. In their capacities as reservoir rocks, many calcarenites, especially those now called grainstones, behave exactly like sandstones of comparable grain size (see Sec. 13.3.3).

This recognition of important similarities between sandstones and many carbonates, in their reservoir roles, enables us to pinpoint the essential differences between them. Porous, clastic reservoir rocks represent end products of familiar geologic processes; their essential constituents are stable, insoluble, and commonly hard. They are purest ("cleanest") when they are most extensively transported and "treated;" if they fortunately lack fine components, it may well be because these were *left behind*. Carbonates, on the other hand, even when functioning like sandstones, are composed of materials that are soft, soluble, and reactive. They are purest, and most suitable as reservoirs, when least transported; if they lack fine components, it is likely to be because these have been *removed*. A good sandstone reservoir in a sense owes its qualities to its survival during transport; a good bioclastic reservoir has escaped transport.

Reservoir rocks constitute the most valuable properties oil companies can own. Without them, oil and gas production cannot be achieved. The proper treatment of the reservoir rock is the prime task of both the development geologist and the reservoir engineer (see Ch. 19). More oil and gas may be left unrecovered through mismanagement of reservoir rocks than are left undiscovered through incompetent exploration. Understanding the properties of reservoir rocks, and utilizing rather than ignoring the differences between them, is an aspect of petroleum geology in which all its practitioners share a vital interest.

13.2 Sandstone reservoirs

The term *sand* implies a particular, restricted range of dominant grain size (conventionally accepted as being between 62.5 μm and 2.0 mm), and not a particular composition. However, the easy acquisition of that grain size during weathering, and maintenance of it during transport by water, ice, or the wind, exclude from the mineral dominance of sands and sandstones all but two or three minerals or rock-fragment species. The majority of the grains must be made of something hard, stable, insoluble, and without crystal characteristics such as cleavage that would deprive the grains of roughly equidimensional form. All reservoir sandstones therefore contain quartz as an important constituent; a great majority of important ones are essentially *quartzose* in composition.

The performance of a sandstone as a reservoir rock – its combination of porosity and permeability, in effect – depends literally upon the degree to which it is truly a sand. Concentrations of sand with effective porosity require mechanisms to exclude fine materials from both the supply (provenance) site and the depositional site. The texture of the sand-sized framework grains, whatever their composition, must not be obscured or dominated by either the *matrix* (the finer-than-sand-size material deposited with the sand grains) or the *cement* (the material subsequently deposited in the pore spaces between the grains). Furthermore, the sand grains themselves should not include an excessive proportion composed of easily decomposed or deformed materials. Thus the quality of the sandstone as initially deposited is a function of the source area, the depositional process, and the environment in which the deposition takes place. Sand exists only where there is sufficient relief to supply it.

13.2.1 The influence of provenance

The provenances of sandstones have been studied by many geologists. Familiar sandstone classifications allow for three principal types according to initial composition: those totally dominated by detrital quartz (called orthoquartzites by some); those containing significant quantities of unweathered feldspar (arkoses); and those with high contents of rock (lithic) fragments or clay matrix (loosely called greywackes). Clearly a high feldspar content is undesirable in a reservoir rock, because feldspars easily decompose to form clays. A high lithic quotient is also undesirable, because many lithic constituents are as readily decomposed as are feldspars and they may also be easily deformed under compaction, seriously reducing porosity and permeability.

Despite these cautions, it must be acknowledged that there are a great number of feldspathic, lithic, and volcaniclastic sandstone reservoirs of excellent quality, especially among young representatives, because a sandstone need not be either transported far enough or buried deeply enough to allow the nonquartz components to wreak their damage.

High-quartz sands are the dominant coarse sediments provided by three widespread continental terrains:

(a) Cratonic interiors of low relief provide the opportunity for long, slow transport of the clasts. The sands may be derived from the basement or from older sedimentary sources. Prolific reservoir representatives include sandstones of all Phanerozoic ages, many having quartz contents higher than 90 percent. The Paleozoic sandstones of the interior of the USA (in the Mississippian of the Illinois Basin, for instance), and the Cretaceous sandstones of the Arabian–Iraqi foreland of the Persian Gulf Basin, are representative.
(b) High-quartz sands are also characteristic of tectonically quiescent continental margins of the Atlantic type. According to Keith Crook, sands of this derivation contain more potash than soda. The Jurassic Piper sandstone of the North Sea and the Tertiary sands of the Niger delta are examples.

(c) Much thicker bodies of high-quartz sandstone comprise the molasse shed into foreland or back-arc basins from newly risen fold-thrust belts. They involve the recycling of sedimentary successions, and are typified by the "salt and pepper" type of sandstone, both marine and nonmarine (with the "pepper" provided by chert grains), as in the Cretaceous of the Western Canadian Basin. Also common are reddish or grey sandstones interbedded with similarly colored siltstones and clays, as in the Pliocene "Productive Series" of the eastern Caucasus in the Soviet Union.

Feldspathic or *arkosic sandstones* are derived from uplifted granitic or gneissic basement or intrusive rocks in interior basins and rift zones with rapid erosion and only moderate transport (or hardly any). Many are alluvial fan or braided stream deposits; a familiar representative is "granite wash" (Sec. 13.2.7). Examples spread across the spectrum of ages and regions include the Pennsylvanian sandstones of the basins in the Ancestral Rocky Mountain province of the western USA, the Jurassic Brent sandstone of the Viking (North Sea) graben, and the Tertiary sands of the Middle Magdalena valley in Colombia.

Sandstones with high contents of *lithic* fragments are the dominant coarse sediments derived from magmatic arc terrains, but such terrains are by no means their only source. The sands include rapidly deposited turbidites in fore-arc basins, the "dumped" infillings of successor basins in the interiors of orogens, and many sandstones properly or improperly called *flysch* or *sub-flysch* and having coal measures above. These sands, collectively called *litharenites*, are in general the least favorable as reservoir rocks.

In addition to relatively low quartz contents (still far more than 50 percent, with rare exceptions), sandstones of the arc-trench association are rich in volcaniclastic debris. This in turn is higher in soda than in potash, because the andesites and metavolcanics of arc-trench terrains are rich in plagioclase and not in K-feldspar. Examples include the Cretaceous Great Valley sequence in California, the early Tertiary reservoirs of coastal Ecuador, and the late Tertiary of Japan and Sakhalin. Volcanic ash occurs in more reservoir sandstones than is commonly realized, especially in those of young orogenic basins. Some have such high contents of volcanic material that they are difficult to distinguish from pyroclastic rocks. Such sands undergo very rapid permeability loss through lining of their pores with clays (see Sec. 13.2.6); they invariably have high irreducible water saturations (Sec. 11.2.1). Volcanic material also occurs in lesser amounts in some reservoir sandstones far removed from contemporaneous mountain building; the best-known single example is undoubtedly that of the Woodbine sand in the East Texas field, in a Cretaceous basin quite without folding or thrusting.

Sandstones deposited in collisional suture belts were commonly introduced longitudinally into remnant ocean basins undergoing closure. They contain abundant fragments of sedimentary and metasedimentary rocks (plus polycrystalline quartz), but lack volcanics.

Examples of lithic sandstone reservoir rocks include the Devonian Zilair Group of the Urals; several of the many Pennsylvanian sandstones of the Oklahoma fold belt; the Upper Cretaceous in the Caucasus; the Oligocene Barail Group of the Upper Assam valley; and the Tertiary Hemlock sandstone in the Cook Inlet of Alaska (see Chs 25 & 26).

Many reservoir sandstones are sufficiently poorly sorted to contain high percentages of both feldspar and lithic grains; they are arkoses in some places and greywackes in others. Such are the mid-Cretaceous reservoir sands of Samotlor and other fields in the West Siberian Basin. Some of these contain more than 50 percent of altered feldspar grains; others contain 20–40 percent of angular rock fragments, and up to 5 percent of mica. The matrix is clay, occasionally calcite. The slightly older Cretaceous reservoirs of the Daqing oilfield in northeastern China and in the Upper Mannville Group of Alberta, Canada, are similarly both feldspathic and lithic, and have clay matrices.

With this variety of provenance, oil sands vary considerably in natural color, but their oil contents make them commonly darker than barren sands.

13.2.2 Influence of environment and mode of deposition

Environments of sand deposition may be wholly terrestrial (aeolian or dune sands), fluvial (river deposits), tidal, deltaic, coastal (including both subaerial and submarine contributions), or deep marine (where most sand bodies are of the class called turbidites). Variations on the theme of sandstone classification have been presented by Paul Krynine, Rufus LeBlanc, Francis Pettijohn, Paul Potter, Gordon Rittenhouse, and numerous other authors; but it is the environmental and depositional criteria enumerated above that are emphasized in this book.

Individual sandstone bodies are markedly restricted in size, being lenticular or linear in overall shape and thinning out, gradually or abruptly, both along and across their strikes. Sandstone reservoirs are therefore commonly less than 25 m thick (a 1961 compilation of more than 7200 US examples set their average at 12 m), and less than 250 km^2 in area. Though productive reservoirs include representatives of all sandstone environments, most very large examples are necessarily multiple, consisting of complexes of interlocking river channels, deltas, and beaches, separated from one another by hydraulic barriers of shale.

13.2.3 Minor influence of age

All ages are represented among sandstone reservoir rocks, or among sandstones capable of acting as reservoir rocks. The oldest supergiant reservoir (in the Cambrian of Algeria) and the youngest (in the Pliocene of the southern Caspian fields) are both in sandstones. In the USA, two-thirds of sandstone reservoirs are of Cenozoic age; if the Middle East is excluded, there is no doubt that reservoirs of Tertiary age are overwhelmingly in sandstones.

13.2.4 Porosities in sandstones

The accumulation of sand requires bottom conditions suitable for localizing it. This requirement might appear to be most easily met by a low-energy, protected depositional site. However, the exclusion of fine materials from the detritus transported to and deposited at the site, or their removal following deposition, requires a high-energy agency. Water currents are by far the most common and effective such agency.

The *original* porosity of high-energy sands may be 40–55 percent, and the permeability 25–100 darcies. Loose sands have 30 percent or higher porosity, and they do not have to be young to be in this condition; the Ordovician Wilcox sandstone in Oklahoma is only weakly consolidated and has measured porosities up to 30 percent. Tight siliceous sandstones may have less than 1 percent porosity. The average porosity for all sandstone *reservoirs* is probably about 15 percent.

One square kilometer of sandstone with this average porosity (15 percent), 1 m thick and saturated with oil, would contain nearly 15×10^4 m^3 of oil in place, or nearly 1 million barrels in traditional usage. Higher oil content than this represents a very good reservoir; at 30 percent recovery, it would yield 45×10^3 m^3 km^{-2} m^{-1}, or 350 barrels per acre-foot. These figures provide a useful yardstick for the quality of a sandstone reservoir and the recoverable reserves expectable from it. Studies of reservoir sandstones in the USA indicated that the total recoverable oil from a sandstone body is proportional to the square of the average thickness, not to its cube as might be expected if the volume of pore space was the primary control. The same studies indicated an average recovery factor from sandstone reservoirs of about 27 percent. Again, the age of the sandstone has little to do with the figures; some Paleozoic sandstone fields have already yielded volumes of oil approaching 8×10^4 m^3 km^{-2} m^{-1}.

In poorly consolidated sands, even heavy oils may flow quite readily (as they do in the Orinoco "tar belt" of eastern Venezuela), carrying sand into the borehole with them; 50 percent of the flow may consist of sand. In addition to porosity values that may be high, sandstones have much better *horizontal permeability* than do carbonates, so that, as we shall see subsequently (Ch. 16), most porosity pinchout traps are in sandstones. To offset this advantage, sandstone reservoirs may have poor permeability *across* bedding, because of frequent shale breaks; such reservoirs actually consist of a number of discrete sand bodies stacked on top of one another.

The degree of *depositional sorting* exerts a double control on the eventual porosity of a sandstone. It influences not only the initial porosity but the capacity of the sandstone to receive a cement from external sources later. With good initial sorting, porosity commonly increases with decrease in median grain size (see Sec. 11.1.3). This decrease in grain size commonly also leads to a decrease in both total cement content and the proportion of the cement which is of external derivation or *allogenic* (see Sec. 13.2.5).

Geologists considering porous sedimentary rocks for reservoir qualities may overlook *conglomerates*. Conglomeratic sandstones may make excellent reservoir rocks, especially if long migration has brought water into the finer clastics. Their considerable porosity variations were treated numerically by R. H. Clarke in 1979. Clastic sediments having a bimodal distribution of clast sizes display two distinct types of packing; they may be sand-packed, with the pebbles "floating" in the sandstone matrix, or they may be pebble-packed, with many pebbles in contact with one another. In the second case, the sand content equals the porosity the conglomerate would have without the sand content. The permeabilities of conglomeratic reservoirs may be much higher than the porosities of their sand fractions would suggest, but there is a complication in this possibility. In a pebbly sandstone, the pebbles themselves increase the tortuosity factor by causing longer flow paths for fluids to follow around the pebbles.

Some extremely large fields, both oilfields and gasfields, have conglomeratic sandstone reservoirs. They include the Sadlerochit reservoir at Prudhoe Bay, the Cretaceous reservoirs at East Texas and at Pembina in Canada, the Hemlock conglomerate reservoir in the Cook Inlet fields of southern Alaska, and the Permian Rotliegendes Formation in the Groningen gasfield and other gasfields of western Europe. These fields or their reservoir rocks are illustrated in Figures 13.7, 13.12, 13.43, 16.1, and 16.10.

13.2.5 Diagenetic processes in sandstones

Principles Inorganic diagenetic processes involve the mass transfer of material into or out of the pores of sedimentary rocks by convective fluid flow. Fluid dynamics in the subsurface in turn involve the chemistries and stabilities of the fluids and the patterns and velocities of their flows. Flow transfers materials from areas of higher to areas of lower solubility; in the latter areas, precipitation takes place. Flow is therefore inevitable where strata have any significant dip, and fluids tend to become directed towards high structures. Flow velocities depend on permeabilities, fluid viscosities, temperature gradients, and *g*; pressure effects appear to be negligible. Velocities seem typically to be of the order of 10^{-8} m s^{-1}.

Diagenetic processes in sandstone reservoir rocks may therefore either introduce material into the pore spaces, or create pore space by removing material from the body of the rock. The former process is by far the more important and widespread, but we may consider the second, less important process first.

Secondary porosity in sandstones Secondary porosity in a rock is space formerly occupied by solids which are labile in the fluid environment to which the rock becomes subjected. The removal of the solids is almost always by *dissolution*, but dissolution alone is not sufficient; some products must be removed from the system. Two classes of solid are removed by solution:

(a) Those most commonly dissolved are the non-quartz constituents (carbonates, feldspars, rock fragments, shells, pellets).

(b) An introduced cement may be redissolved at depth, causing secondary regeneration of primary porosity.

Other, less significant causes of secondary porosity are the fracture, corrosion, and honeycombing (etching) of grains. The resultant is a composite of the effects of compaction, solution, and cementation.

The principal solvent is dilute carbonic acid; the waters containing it may be meteoric, or they may be squeezed from shales during decarboxylation reactions of organic matter. The solubility of *calcareous cements* is directly proportional to the pressure but inversely, exponentially related to the temperature; it therefore decreases with depth. *Aluminosilicates* are more difficult to dissolve, but their removal is facilitated by the presence of organic acids formed during the maturation of organic matter.

In sandstones, of course, it is the solution of *silica* which is critical. It requires an increase in both the temperature and the pH of the solvent, the higher alkalinity being brought about by the concentration of bicarbonate ions. The solubility of quartz is also enhanced by decreasing salinity of the waters moving through the reservoir. The net effect is that secondary porosity is significant only within the restricted depth range in which these several factors are most favorably combined. In the ideal circumstance, this depth range would coincide with the zone of principal oil formation which is similarly dependent upon temperature (Fig.7.8 & Fig.13.1).

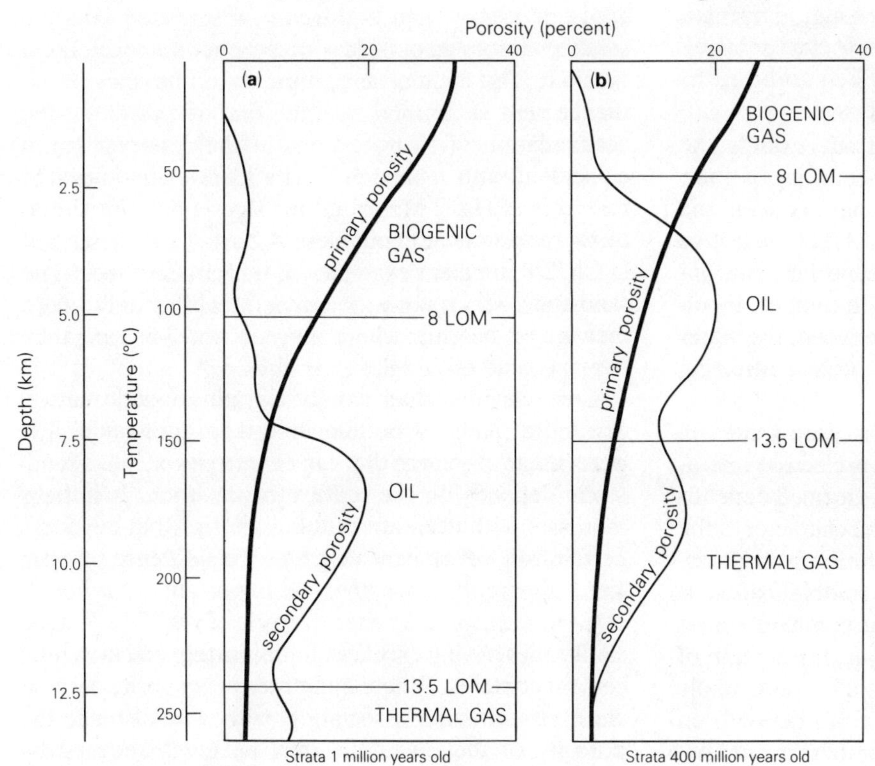

Figure 13.1 Variation in porosity with increasing depth and temperature due to reduction in primary porosity and development and reduction of secondary porosity. (a) Strata of age 1 Ma; (b) strata of age 400 Ma. Depth is based on a geothermal gradient of 18.5 °C km^{-1}. All values are approximations. LOM, level of organic metamorphism (see Sec. 7.5). (After Zuhair Al-Shaieb and J. W. Shelton, *AAPG Bull.*, 1981.)

Development of significant secondary porosity in sandstones must involve vast volumes of undersaturated pore waters. It seems that these must be of meteoric origin. Clay-mineral reactions and shale compaction cannot provide adequate volumes either for this process or for the opposite process of sandstone cementation.

The development of secondary porosity may create strange pores: oversized, moldic, elongate, or tortuous. All secondary pores in sandstones, however, are likely to be isolated, so they have little effect on the permeability or the production capacity. Creation of secondary porosity may not in fact increase the porosity at all, but merely enable the rate of its downward reduction to be arrested.

Cementation of sandstones Diagenetic processes are much more likely to reduce the porosity of a sandstone than to augment it, because the grains themselves may be relatively immune to diagenetic effects. Pure quartz sandstones present the smallest opportunity for diagenetic alteration, though overgrowths of secondary silica are common.

The proportion of the sandstone *which is not quartz* is therefore critical, and diagenetic modifications both by addition and by subtraction of material are most pronounced in lithic rather than quartzose sandstones. Plutonic grains tend to be monocrystalline, and volcaniclastic (lithic) grains polycrystalline, but both are dominated by silicate minerals which decompose to *clays*. Chemically precipitated clays are by far the most common and important cement-forming, diagenetic minerals. It is widely accepted that clay minerals plugging porosity in sandstones are likely to be authigenic, but this is not always easy to prove.

The properties of clay minerals that affect sandstone reservoirs are several. The most far-reaching is their capacity to adsorb and retain water, thus reducing the sandstone's permeability to oil (Fig.11.11). Selective adsorption facilitates base exchange, especially carbonation or the precipitation of a carbonate cement. If colloids are formed and certain ions introduced, the water films are disrupted and flocculation occurs, affecting permeability still further.

The various clay minerals affect the pore spaces in different ways, revealed by the scanning electron microscope (see Figs 13.4 & 5). Which clay is formed depends on the mineral being altered to it and the character of the altering fluid. The derivation of kaolinite from potassium feldspars, by removal of the soluble bases, is familiar. Its derivation from mica, a common minor constituent of sandstones, requires a high throughput of water of low ionic strength and low pH – essentially fresh water, in fact. The availability of this depends on both the environment and the climate; deltaic and alluvial sandstones are obviously more susceptible than turbidite sandstones to invasion by fresh water. The further alteration of kaolinite to illite, on the other hand, requires a source of potassium.

Feldspathic sandstones rich in plagioclase grains acquire cements high in expandable clays such as smectite. Chlorite cements and coatings are derived principally from basic volcanic detritus and are not common in feldspathic sandstones. Volcaniclastic debris in sandstones may become laumontite cement, which is ruinous to porosity (laumontite is a hydrated calcium–sodium aluminosilicate).

The conversion of detrital mica to clay is achieved artificially and unintentionally, but rapidly, through some oilfield practices. If drilling mud has excessively low pH, it forms clays in reservoir sands in the boreholes. The clays then migrate into the formation and reduce its permeability. In addition to kaolinite derivation from either feldspar or muscovite grains in sandstones, kaolinite platelets derived from feldspars may nucleate in cleavage or fracture planes in muscovite.

In addition to fine-grained, dispersed cements formed of authigenic clay minerals, nonporous cements are formed in sandstones by *syntaxial overgrowths*. These form by nucleation of quartz or feldspar in optical continuity over grains of their own kind. The grains must not already be completely coated by another mineral, especially by early diagenetic chlorite. Such coatings prevent overgrowth and serve thereby to preserve porosity in the sandstone instead of reducing it.

Clay minerals and silica cements may of course be *allogenic* rather than authigenic, introduced into the sandstone from an outside source rather than generated within it. This circumstance must always be suspected if the cement is formed so late that it postdates the accumulation of hydrocarbons in the reservoir (or is coincident with it in time). The Cambrian sandstone reservoir at Hassi Messaoud in Algeria and the Cretaceous sandstone at Pembina in Alberta (both described in Ch. 26) are giant examples of this circumstance. The sandstones retain porosity and permeability only where they are oil-bearing; where they are water-bearing they have become essentially quartzites.

Cements other than clay (or clay plus organic matter) are more likely to be allogenic than authigenic. The percentage of cement that can be introduced into a sandstone depends on the texture of the rock. It initially increases with decrease in the grain size, but beyond a certain ratio of cement to grains the influence of grain size apparently becomes unimportant. *Carbonate cements* are especially characteristic of very fine-grained sandstones having excellent initial sorting and high total cement contents. Where quantities of evaporite occur in close stratigraphic association with a sandstone, the porosity of the sandstone may be much reduced by infilling with *anhydrite*, as at Hassi R'Mel in Algeria (Fig.16.44) and in some of the sands in the Permian Basin. In regions lacking the carbonate–evaporite facies, such as the eastern Venezuelan and West Siberian Basins (and most Neogene basins), clay cements of sandstone reservoir rocks dominate over all others.

Cements in individual sandstone reservoir rocks may be remarkably heterogeneous, involving several different minerals introduced over geologically long periods. The paragenesis of multistage cementation, like that for the deposition of ore minerals, varies not only because of differences in the fluids transmitted through the pore spaces but also because of rising temperatures with increasing burial. The temperature dependence of the paragenesis for a Frio sandstone from the US Gulf Coast has been worked out as below:

(1) Quartz cement at 75–80 °C, at the maximum rates of fluid expulsion.
(2) Kaolinite over the quartz at about 100 °C.
(3) The most intense diagenesis occurs in the geopressured zone, at 120–140 °C.
(4) Albitization occurs at about 150 °C.
(5) Carbonate cements, authigenic or allogenic or both, seem to have been introduced over a wide range of temperatures.

Where silica, clay, and carbonate (or evaporite) cements are all present in varying degrees, that order of their introduction is very characteristic.

Combinations of clay matrix and both authigenic and allogenic clay cements (with carbonate in addition in many cases) pose tantalizing production problems in the widespread reservoir rocks called *tight gas sands*. Perhaps the most notable development of these is in the Cretaceous molasse east of the Rocky Mountains in the USA and Canada, in a series of basins extending over more than 20° of latitude. Because methane is highly compressible, great volumes of it may be stored in sandstone reservoirs with porosity and permeability so low that they could not contain oil.

Compaction of sandstone reservoirs Almost as critical to sandstone reservoirs as cementation is *compaction* (see Ch. 11). Its effects depend directly upon the textural and mineralogical *maturity* of the sandstone. We may express this in terms of the *grain fraction*, which is simply the proportion of total rock solids which is grains and neither matrix nor cement. A mature sandstone, called "clean" by petroleum geologists, may have a grain fraction of 1.0 and an original porosity of perhaps 40 percent (Fig. 13.2). In such a sandstone, burial induces packing adjustments by intergrain slippage; at

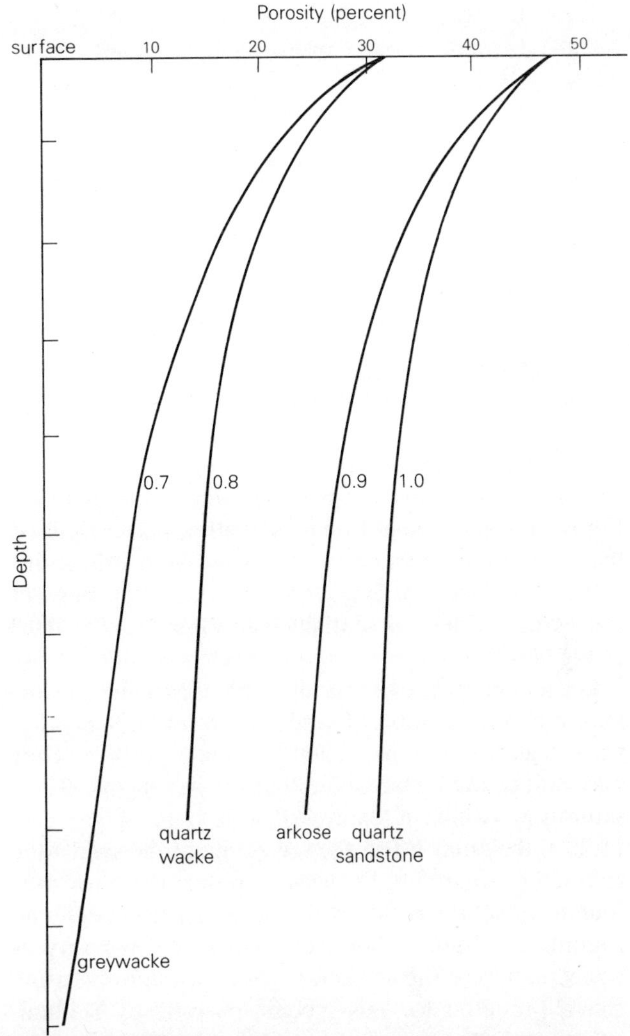

Figure 13.2 Reduction in porosities of sandstones with depth, for decreasing grain fractions. (Modified from R. C. Selley, 1978.)

depths as shallow as 2000 m these may reduce porosity by 10–12 percent. At greater depths, brittle grain deformation and pressure solution increase the lengths and suturing of grain contacts and reduce porosity still further. The loss of porosity is inversely proportional to the median grain size. In these processes, cementation must be distinguished from mere lining of pores; true cementation adds to the crushing strengths of rocks.

Mineralogically *immature* sandstones have grain fractions significantly less than 1.0 and may begin, at the surface, with porosities much less than 40 percent (say 30 percent; Fig. 13.2). In greywackes, and shaly, feldspathic, or glauconitic sandstones, the easy deformation of ductile or flexible grains brings about a more abrupt loss of porosity, by a further 5–15 percent for equivalent depths. Micaceous clasts (fragments of shale, schist, or phyllite), glauconite grains, and fecal pellets are especially able to flow into pore spaces when subjected to loads of overlying rock. In hand specimens they then resemble a detrital clay matrix for the sandstone. The

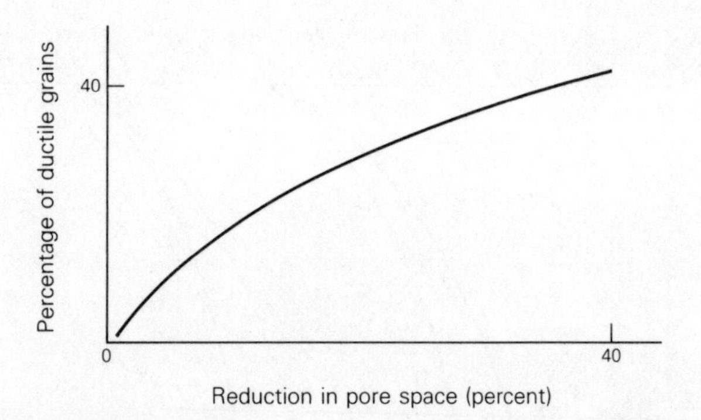

Figure 13.3 Reduction of pore space of a sandstone as a function of the proportion of ductile (nonquartz) grains. (After G. Rittenhouse, *AAPG Bull.*, 1971.)

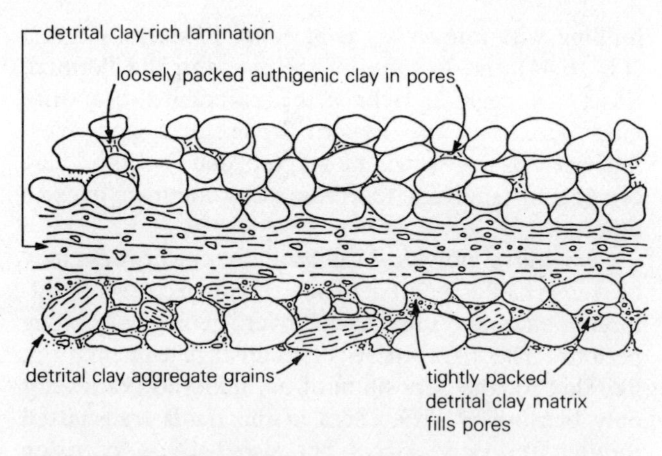

Figure 13.4 Typical distributions of clay minerals in sandstones. (From E. D. Pittman and J. B. Thomas, *AIME*, 1978.)

rate of porosity reduction with depth increases sharply with decreasing values of the grain fraction (Fig. 13.2). From studies by Jane Taylor in 1950, it was estimated that a sandstone containing 40 percent of ductile grains could suffer loss of original porosity as high as 40 percent if packing had been essentially orthorhombic (Fig. 11.3 & Fig. 13.3).

Diagenetic processes of all kinds affect the performance and evaluation of sandstone reservoirs in other ways in addition to their effects on the porosities. They may reduce the actual volume of the irreducible water saturation (which is expressed as a percentage; Sec. 11.2.1); they may affect the sensitivity of the sandstone to fracture treatment by acids or water; and they profoundly affect the rock's E-log characteristics. Under all routine conditions, however, sandstone porosity is easily read from the acoustic velocity or *sonic log*, using charts prepared for this specific purpose by Schlumberger and other logging specialists (see Ch. 12).

13.2.6 Scanning electron microscope studies of reservoir rocks

We have emphasized (Ch. 11) the important differences in the nature of porosity between sandstone and carbonate reservoir rocks, and the distinctions between porosity and permeability in all types of reservoir. Studies of these variations and their causes were enormously facilitated by the introduction of *scanning electron microscope* (SEM) techniques in the late 1960s. The SEM combines an extreme magnification range (typically from 100 to more than 10 000 times) with great depth of field, yielding a detailed three-dimensional image ideal for the scrutiny of grain and pore geometries.

The permeabilities of reservoir rocks (and hence their productivities) are substantially affected by the presence of authigenic mineral matter in the larger pores. The finer pores, or micropores, are preferentially water-wet, as we have seen (Ch. 11). In carbonate rocks, such common minerals as calcite, siderite, or anhydrite readily form pore fillings and reduce or destroy porosity and permeability; but these minerals are soluble in percolating waters of suitable composition and are as often removed as deposited. *Authigenic clay minerals* are much more common and serious reducers of pore space, and they are not readily re-soluble once deposited there from pore fluids. They especially affect sandstone reservoirs (Fig. 13.4), and it is to these that SEM studies have made their greatest contribution.

We owe to John Neasham a simple categorization of clay mineral interferences with porosity, and hence with permeability (Fig.13.5):

(a) Clay minerals may be in discrete, loosely packed particles or platelets within the pores. These normally represent low to moderate clay content of the rock, and bring about low to moderate reduction of reservoir quality. However, the particles are easily loosened and then become "rogue" grains within the rock, seriously interfering with pore-throats and so with permeability. The commonest clay particles within pores are "books" of hexagonal *kaolinite* euhedra. The euhedra, which may be rod-shaped, plate-shaped, or equant, can nucleate and grow in more than one way. They may simply grow randomly from the pore walls into open voids; this process may affect all parts of a porous rock. Alternatively they may nucleate syntaxially on detrital minerals, especially the micas, affecting only those parts of the rock that contain such minerals. These syntaxial growths wedge open the cleavages of the micaceous minerals.

(b) Clay minerals may line the pore walls, the plates growing edgewise but randomly and forming boxworks which reduce the absolute sizes of pores. The commonest mineral in this category is chlorite; illite achieves the same effect less commonly, and siderite sparingly.

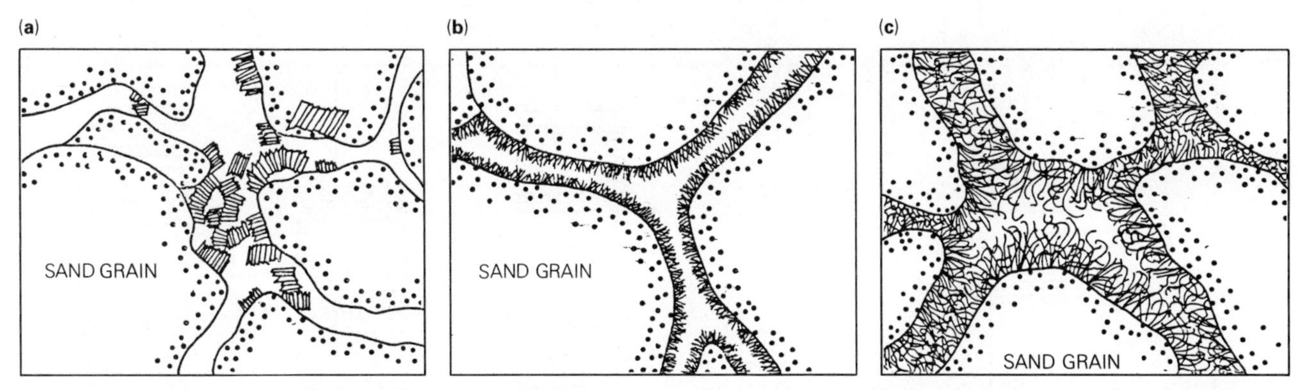

Figure 13.5 Three general types of dispersed clay in sandstone reservoir rocks: (a) discrete particles of kaolinite; (b) pore lining by chlorite; (c) pore bridging by illite. (Drawn from SEM photomicrographs by J. W. Neasham, *Trans. Soc. Petrol. Engrs*, AIME, 1977).

(c) The clay minerals may form "books" or fibers bridging across pores. This state represents the highest clay content and the greatest reduction of porosity. It also greatly increases the tortuosity factor in the permeability of the reservoir. The worst offender in this process is authigenic *illite*, but chlorite and smectite also form crystals of suitable shapes and dimensions.

The effects of the three types of clay-mineral growth may differ enormously. Figure 13.6 shows that, at constant porosity, four orders of magnitude of permeability are controlled wholly by the clay minerals in the pores.

SEM studies of porosity are by no means restricted in usefulness to sandstone reservoir rocks. They are equally valuable for diatomites, chalks, and very fine-grained limestone reservoirs. For the quite different sizes and conditions of other carbonate reservoir pores, however (see Sec. 13.3.7), SEM imagery is less useful than routine photomicrography. SEM photographs of reservoir rocks appropriate for them (in Plate A) may be compared with photomicrographs of typical carbonate reservoir rocks (in Plate C).

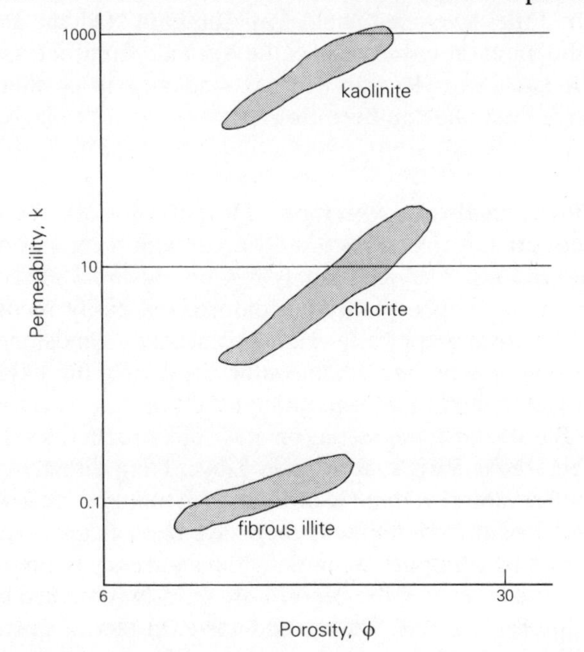

Figure 13.6 Permeabilities of sandstones ranging over four orders of magnitude without significant change in porosity, because of clay minerals in the pores.

13.2.7 The principal environments of sandstone reservoirs

Aeolian (dune) sandstone reservoirs Both coastal and true desert dune sands are represented among productive reservoirs, displaying a variety of geometries. Dune sands may be exceedingly well sorted and have adequate porosity to provide good reservoirs, but a source of hydrocarbons is less likely to be available for them than for any other important class of sandstones.

Dune sands possess characteristic cross bedding and may consist of the rounded grains created by airborne suspension. Dune formations are likely to be overlain by (or interbedded with) evaporites, and the combination easily results in complex diagenesis of the reservoir rock.

Important productive sands of dune origin are concentrated in only two intervals of geologic time. The younger interval is that of the "great drying out" during the Permian, Triassic, and early Jurassic, when the continents were assembled into the supercontinent Pangaea. Reservoir rocks representing that time include the Slochteren member of the Permian Rotliegendes Formation at Groningen and in the gas-rich area of the North Sea (Fig. 13.7); the Botucatu sandstone of the Parana Basin in eastern South America, containing very heavy "tar oil;" and the Nugget sandstone of the overthrust belt in the western USA, the surface exposures of which, called the Navajo sandstone, form the spectacular white pillars familiar to patrons of "Western" movies.

The older interval coincides with the Early Paleozoic, when there were no land plants and so no soil or forest cover to protect crystalline rocks from weathering and wind action. The reservoir sandstone at Hassi Messaoud

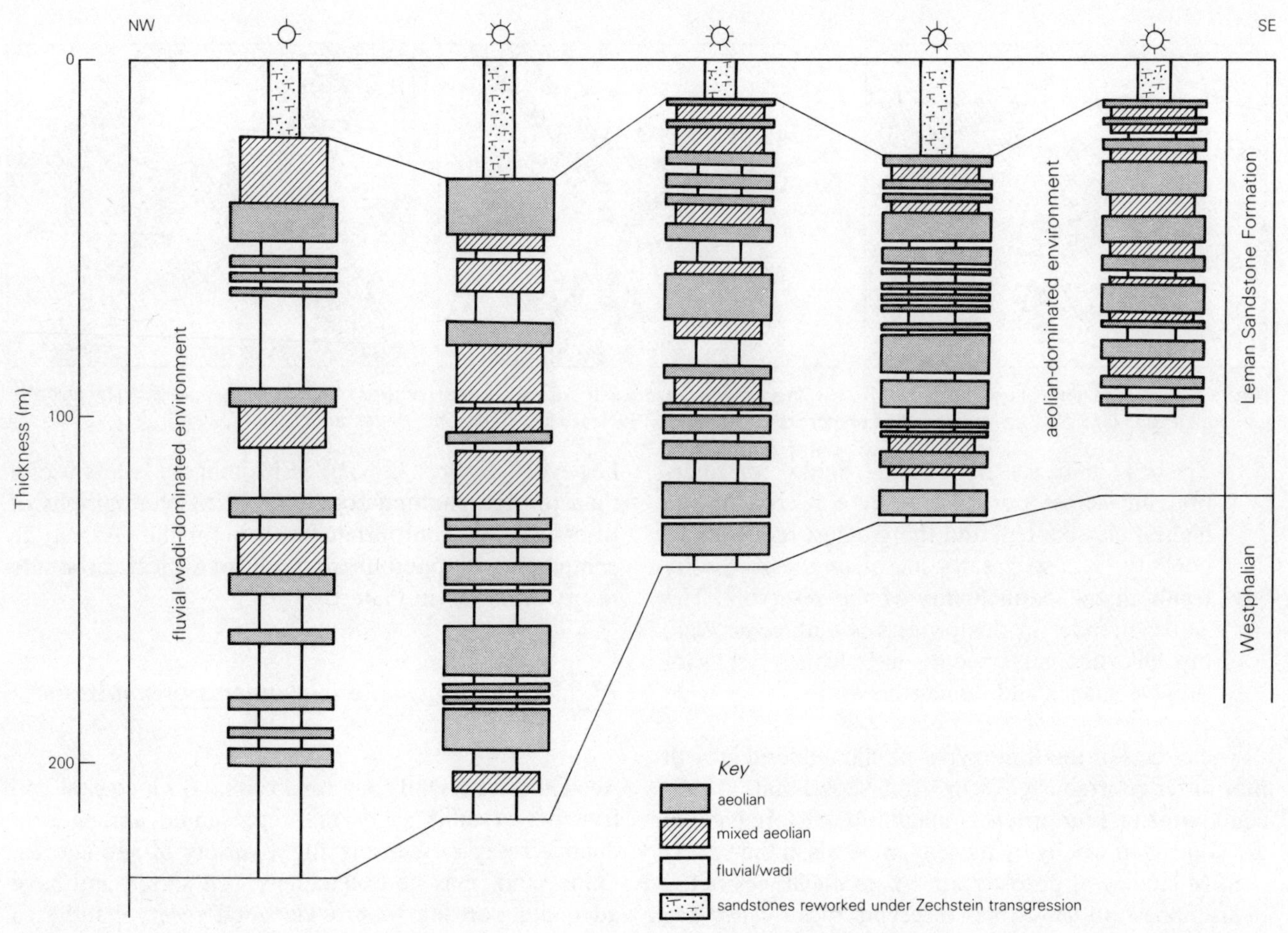

Figure 13.7 Facies variations in the aeolian Rotliegendes sandstone reservoir of the Viking gasfield, British North Sea. (From M. Gage, *AAPG Memoir* 30, 1981.)

in Algeria, which is Cambrian in age, and the Ordovician Wilcox and Simpson sands of the American Mid-Continent (typified by the reservoir of the Oklahoma City field), are representatives (Chs 25 & 26).

Glacial sandstone reservoirs A very small number of clastic reservoir rocks are safely ascribable to a *glacial* origin; they are the sand components of *tillites* (diamictites). The bulk of a tillite is a stiffly indurated rock flour, fundamentally an argillaceous greywacke, unstratified and essentially structureless. It contains randomly scattered fragments without uniformity of size or composition, but tending to have angular or tabular shapes resulting from glacial abrasion. The reservoir sands (commonly conglomeratic) constitute lenses or pods within the tillite, associated with varved argillites and merging laterally into continental, lacustrine, or marine sediments.

All examples of oil-bearing glacial sands known to the author occur in the Gondwana continents. They include the Ordovician reservoir of the Tin Fouye oil- and gasfield in eastern Algeria; a Carbo-Permian reservoir in so far little-developed fields in southern Oman and Dhofar, at the eastern end of the Arabian peninsula; and the late Pennsylvanian (Tarija) sandstone in northern Argentina and southern Bolivia.

Fluvial sandstone reservoirs Fluvial (alluvial) sands, deposited by rivers, necessarily occur within nonmarine successions. They commonly rest on erosional surface relief or on peneplaned unconformities easily recognized stratigraphically and seismically. Sandstones resting directly on such unconformities are more likely fluvial than marine, especially if the time gap is large.

Basal sandstones resting on much older rocks (like the Cretaceous Sarir sandstone in Libya, lying directly on the Precambrian) tend to be thickest in topographic lows and absent from highs. If they have been cleaned and sorted by transport, an initially fluvial agency is almost inevitable, even if the deposits are quickly reworked by transgressing seas. To be productive, in fact, a fluvial sandstone resting on basement must be at least overlapped by transgressive source sediments.

The parts of a river system in which significant sand

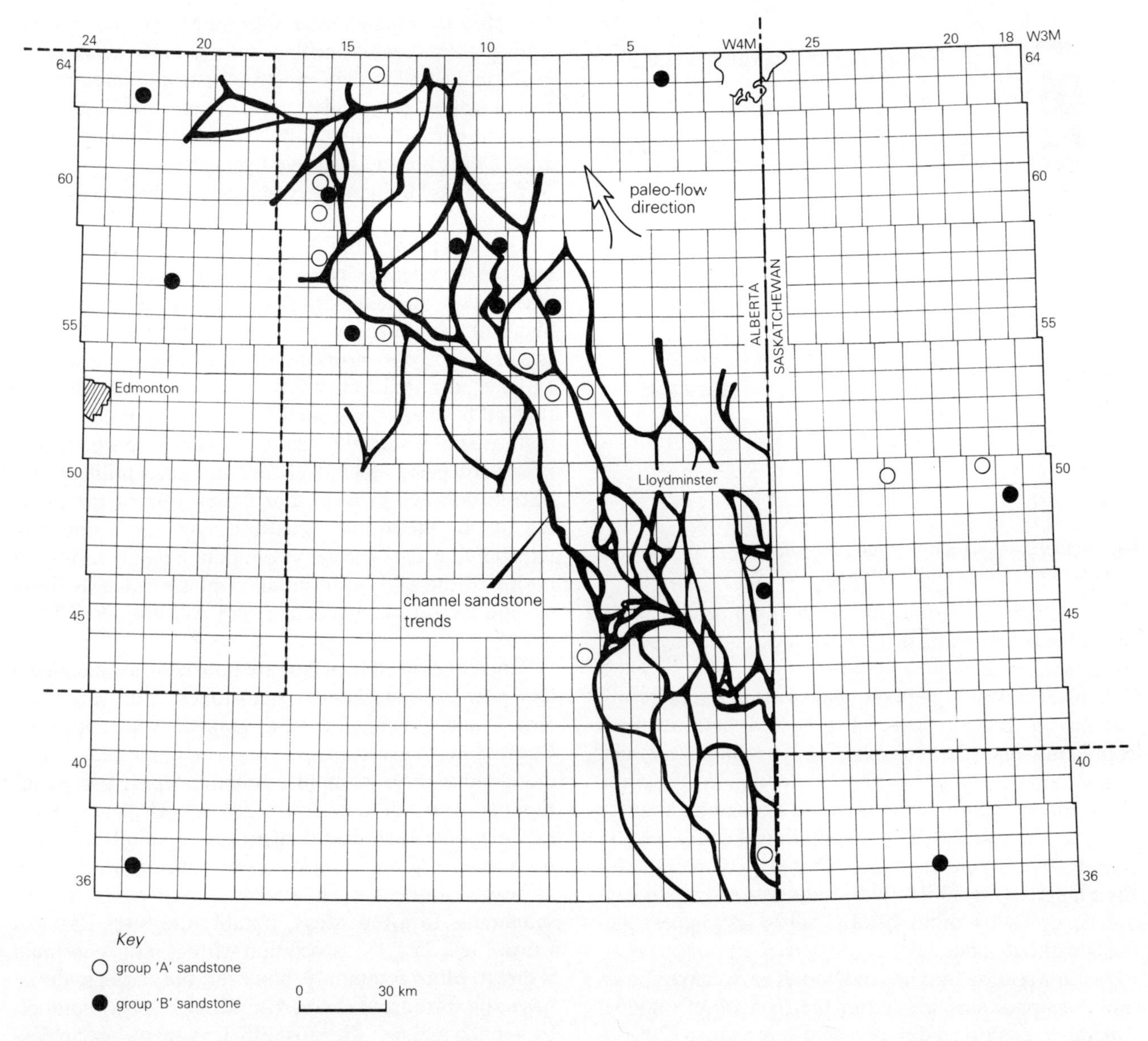

Figure 13.8 Distribution of major channel sandstones in east-central Alberta, Canada. Sandstones are of Early Cretaceous age. Note relation of channel trend to sandstone composition: group A sandstones are quartzose, group B sandstones are arkosic or lithic. (From P. E. Putnam, *AAPG Bull.*, 1982.)

bodies are deposited and preserved are the braided section, with many channels, and the meander belt in which a single master channel migrates over a wide floodplain. A braided stream deposits *longitudinal* and *transverse bars* of sand (Fig. 13.8). A variant of the braided stream is the *anastomosing stream*, with very low gradient and multiple channels. In an upper, low-amplitude meander belt, *point bars* develop on the insides of curves. In the lower meander belt, where swings of the river are much wider, multiple point bars develop, with general orientation almost perpendicular to the trend of the valley. The fluvial deposits may merge into the more widespread delta plain deposits which include *distributary channel sands* (see below); these do not normally rest on unconformities.

The vertical succession of strata in a fluvial deposit shows clear *fining upwards*, from coarse sand, gravel, and debris, through sands with tabular cross bedding; cross-laminated sands and silts on a smaller scale; ripple-laminated silts of the low-flow regime; and ending at the top with colored clays (Fig. 13.9). This succession readily distinguishes the fluvial facies from beach-sand bodies such as barrier-island or offshore bars, which typically increase in shaliness and decrease in grain size *downwards* (see below).

Fluvial sandstones are, moreover, generally less well sorted than marine sands. They contain much carbonaceous debris and are more likely than marine sands to include oxidized intervals. They lack all familiar faunas; even worm or mollusc borings indicate a marine rather

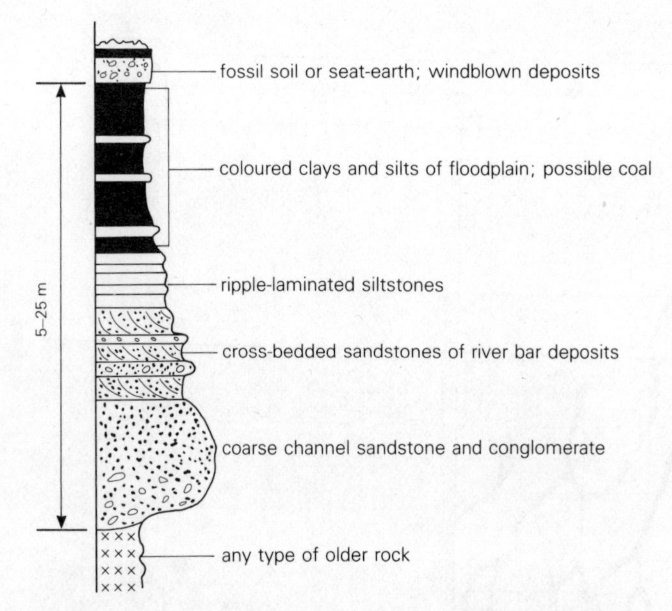

Figure 13.9 An alluvial cycle of sediments, fining upward from channel sands at the base to floodplain muds at the top.

than a fluvial environment. High permeabilities are nonetheless common in fluvial sandstones, especially near changes of flow direction.

In plan, fluvial sand bodies have abrupt lateral terminations. In adjacent wells, siltstone and shale interbeds containing thin, crevasse-splay sands but having no thick sands below may be overbank (floodplain) deposits. Repetitions of these with the fluvial sandstones are very typical. The overall shape of the fluvial sandstones in plan is one of sinuous valleys, narrow compared with their lengths (Fig.13.8). Wide areas between the valleys are occupied by thinly-bedded siltstones, shales, and coals without sands.

In cross section, fluvial sand bodies have convex bases and flat upper surfaces. After the removal of regional dip, the depositional dips revealed may be high (20–30° is not unusual). Dipmeter records tend to show uniform directions (see Fig.12.26). The sands yield smooth or bell-shaped SP curves, and abrupt lower and gradational upper contacts (see Fig.13.45). Once adequate well control is available, a feature facilitating recognition is the isopach thickness between the older erosion surface and a younger marker bed; where this is greater than the equivalent regional isopach, the extra thickness is likely to represent sandstone deposits filling incised paleovalleys.

Spatially intermediate between fluvial and coastal deposits are *estuarine* sandstones. They are hard to distinguish lithologically, as estuaries may be either tidal or nontidal. Paleocurrent directions may therefore indicate either landward or seaward flow. In tidal estuaries, the transport of sand is principally by tidal currents, which produce fairly high-angle cross bedding with common interruptions by pebbles or shale clasts. As estuaries characterize shorelines drowned during transgressions, estuarine deposits normally lie above depositional breaks; they must be restricted areally to a recognizable paleovalley; and they must have immediate marine equivalents. The most common lateral equivalents are the burrowed silty or sandy mudstones of tidal flats, unlikely themselves to provide reservoir rocks.

Also extending from the subaerial regime to the submarine, but necessarily including the fluvial, are *alluvial fans*. High-gradient, braided streams emanate from high-relief drainage areas, especially from elevated scarps during post-orogenic times (like the Pleistocene and Recent). Axial channels, active or abandoned, are flanked by sheet-flood slopes or plains on land, and by delta plains where the sea or a lake is reached. The headward parts of alluvial fans are susceptible to redistribution and erosion, but destruction of the lower reaches by marine or lacustrine agencies is minimal. Alluvial fans are especially common in lake-bearing rift basins of the early, pre-breakup phase, such as those around the Jura-Cretaceous South Atlantic (see Chs 17 & 25).

Sandstones in alluvial fans are poorly sorted and much less porous than channel sandstones, with abundant gravel and subangular sand grains, clast-supported. Because many escarpments are formed of either intrusive igneous rocks or uplifted metamorphic basement, their alluvial fan sandstones are arkosic and can be called *granite wash*. Residual arkoses grading down into unweathered granitic rocks are not normally favorable as reservoir rocks even where overlapped by source sediments. In a few areas, notably the Sirte Basin of Libya (Sec. 25.5.1), association with granitic basement is direct. More commonly, however, the clastic material has to be sorted and cleaned of most of its clay content, by winnowing, and the most efficient agents for achieving this rapidly are streams.

Most of the productive examples of granite wash consist of sands washed off uplifts and transported into *fan deltas* built into marine embayments in which source sediments were being deposited. The sands are therefore interbedded with marine rather than continental or lacustrine sediments, even with fossiliferous limestones. As reservoir rocks, they are more appropriately considered deltaic than fluvial.

Notable fluvial sandstone reservoirs Fluvial sandstone reservoirs occur throughout the stratigraphic record. In particular, numberless stratigraphic traps consist of lenticular channel sands pinching out laterally into shales of the alluvial plains. Many fields in the Pennsylvanian Cherokee Group of Kansas are of this type. Not greatly different are several fields in the Upper Pennsylvanian

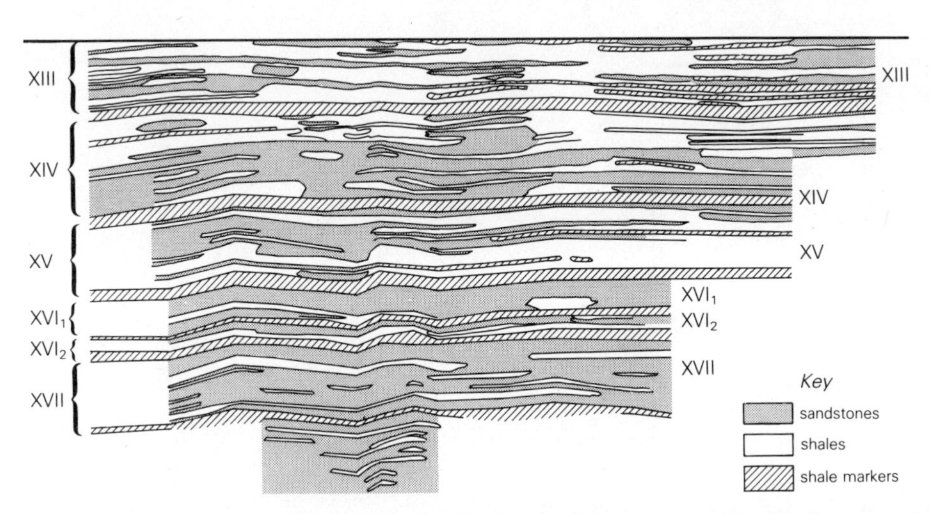

Figure 13.10 River channel sands form multiple reservoirs of Jurassic age in the Uzen field, east of the Caspian Sea in the USSR. (From G. Ulmishek and W. Harrison, *Oil Gas J.*, 24 August 1981.)

and Lower Permian on the eastern shelf of the Midland Basin in West Texas.

Much larger examples include the upper part of the Sadlerochit Formation at Prudhoe Bay, Alaska (see Fig. 13.43); the Upper Triassic sands (with some representing the deltaic facies) in the Rankin gasfields on the northwest shelf off Australia; and the approximately coeval sandstones in the Morecambe gasfield off northwestern England. All or most of the tar sands in the Uinta Basin of the USA are also fluvial deposits (see Figs 10.12 & 13).

In the North Sea, the Lower Jurassic reservoir sandstone in the Statfjord field extends from the floodplain facies at the base, through stream channel sands to coastal plain deposits at the top, reflecting the encroachment of the sea into a foundering graben occupied by sinuous, braided rivers (see below). Also Jurassic in age are the reservoir sandstones at Uzen, in the Mangyshlak Basin east of the Caspian Sea (Fig. 13.10). Those sands are poorly sorted and contain up to 40 percent of clay minerals; they appear to have been deposited by sluggish streams constantly shifting channels.

The lower part of the McMurray Formation in the Athabasca tar sands is of combined fluvial and estuarine origin, filling early Cretaceous valleys crossing eroded Paleozoic rocks. In the Lloydminster heavy oil area farther south, the complex of sandstones overlying the McMurray equivalent has been interpreted as the work of anastomosing rivers, now providing numerous stratigraphic and paleogeomorphic traps (Fig. 13.8). The sand bodies display hummocky cross-stratification and are completely enclosed in finer-grained sediments. The ascription to them of a fluvial origin is a matter of dispute.

The prolific Sarir sandstone reservoir in Libya's Sirte Basin is also of Cretaceous age but it rests on a surface of Precambrian granitic and metamorphic rocks with high relief. The thickest parts of the basal sand were undoubtedly alluvial fans. The sand is almost white where cleanly washed, but elsewhere it is oxidized and contains

Age	Graphic	Maximum thickness (m)	Lithologic description
Oligocene marine mudstone			
Eocene		1200	Late Eocene channeling common Interbedded sandstone, siltstone, shale and coal Sandstone, quartzose, very fine–coarse Siltstone, argillaceous, carbonaceous, micaceous Shale, brown to grey, micaceous and carbonaceous; coal
Paleocene		1500	Interbedded sandstone, siltstone, shale and coal Sandstone, quartzose, fine–coarse, carbonaceous, micaceous
Late Cretaceous		3000	Sandstone, quartzose, fine–coarse, with lithic fragments and feldspars Siltstone, argillaceous, carbonaceous and micaceous Shale, brown, carbonaceous and micaceous Coal
Jura-Cretaceous arkosic-lithic clastics, nonmarine			

Figure 13.11 Lithologic description of Latrobe Group strata, containing the reservoir rocks in the Gippsland Basin, Australia; delta plain and braided stream deposits. (From E. H. Franklin and B. B. Clifton, *AAPG Bull.*, 1971.)

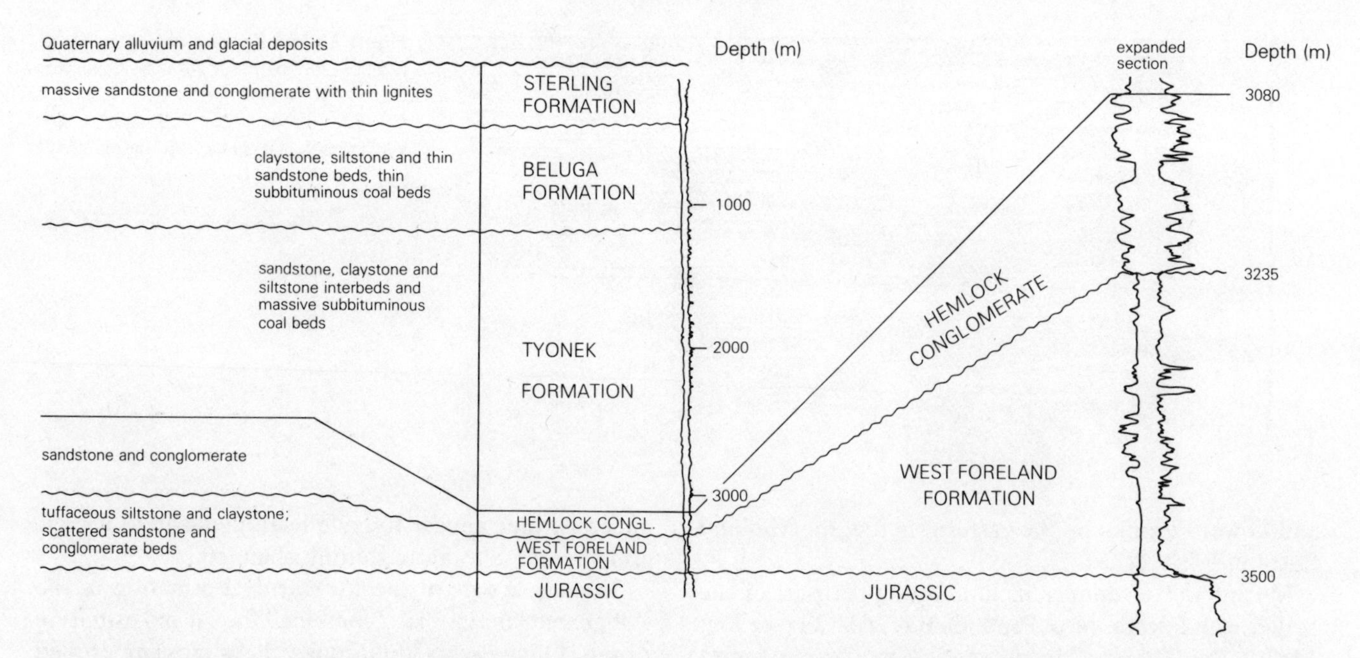

Figure 13.12 Standard section of the Kenai Group, Cook Inlet Basin, Alaska. *Left:* lithologic description of interbedded stream-channel deposits and floodplain shales with coal. *Right:* expanded log of the two lowest formations, which constitute the principal oil reservoir rocks. (After K. W. Calderwood and W. C. Fackler, *AAPG Bull.*, 1972.)

some red shale. This sand is everywhere followed by a second unconformity, above which come variegated anhydritic shales of a tidal flat or sabkha, representing thc beginning of the Cretaceous marine transgression.

The Latrobe Group, containing the reservoir rocks in Australia's Gippsland Basin, represents the deposits of braided streams crossing a delta plain (Fig. 13.11 & Ch. 26). In the lower Kenai Formation of Cook Inlet, Alaska (Fig. 13.12 & Ch. 26) a thick succession of fluvial, deltaic, swamp, and estuarine sediments includes many sands deposited by braided streams as cut-and-fill, with abundant pebbles. The sands are interbedded with floodplain shales containing coal, and the whole succession rests on a major unconformity. In California's Ventura Basin (Sec. 25.6.4), the Eo-Oligocene Sespe Formation provides reservoirs of arkosic sandstone and conglomerate with strong cross bedding, again the deposits of braided streams.

Deltaic sandstone reservoirs Many oil- and gasfields are in sedimentary deposits *associated with deltas*. Located near the boundaries between marine (or lacustrine) deposits, which include source sediments, and nonmarine deposits which represent the supply zone for reservoir rocks, deltas are ideally situated for the maximum interplay between source and reservoir facies.

A large delta needs a large drainage basin. Its identification in the subsurface therefore implies recognition of a landmass of its time, and helps to delineate the ancient shoreline. Large deltas are especially characteristic of the seaward ends of failed rifts, with thick sedimentary sections headward of them and large marine re-entrants (gulfs) seaward.

Deltas are of wholly clastic facies; large deltas inhibit carbonate deposition. A well-developed delta provides the whole gamut of clastic sediment types from carbonaceous mudstones to conglomerates. The proportions of the sediment types are controlled chiefly by the interaction of the fluvial agency supplying the material and the marine agency receiving it. This interaction leads to four general delta patterns:

(a) *High-destructive deltas,* in which marine effects are dominant and sand content is high. Beaches and barrier bars become more prominent than the deltas themselves. Productive trends are defined by local, cuspate, coastal sands of these environments, and by channel-mouth bars prograding farther downdip. The principal traps are stratigraphic. Modern examples of high-destructive deltas include those of the Po and Rhone Rivers in the Mediterranean; both rivers appear to be attempting to build birdfoot deltas.
(b) *High-constructive deltas* of birdfoot type, like the modern Mississippi delta. Numerous distributary channels build *bar fingers* seaward. Reservoir sands are largely restricted to these fingers; the deltaic deposit is otherwise dominated by muds (Fig. 13.13).
(c) *High-constructive deltas* of lobate type (horsetail deltas, like those of the modern Nile and Niger Rivers), in which a small number of active distributary channels build the delta forward along a front

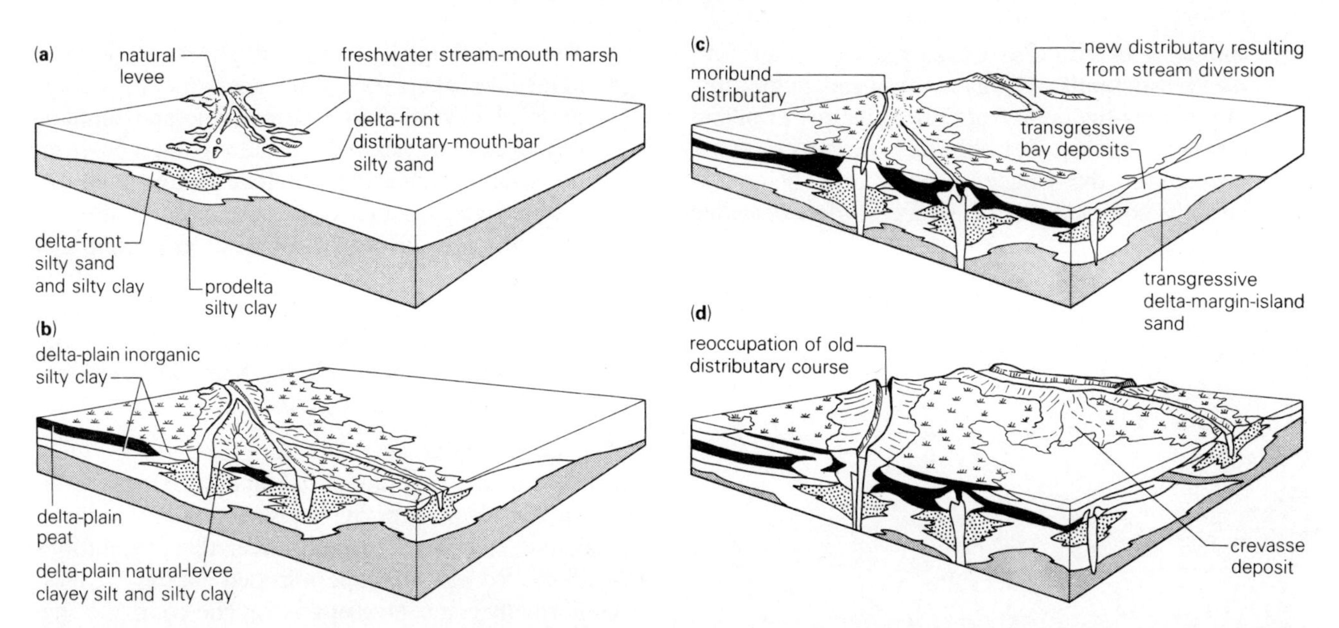

Figure 13.13 Stages in the formation of a birdfoot delta (like that of the modern Mississippi), showing interfingering and overlapping facies (a) Initial progradation; (b) enlargement by further progradation; (c) distributary abandonment and transgression; (d) repetition of cycle. (From R. R. Berg, *AAPG Continuing Education Course Notes* Series 3, 1976; after D. E. Frazier, 1967.)

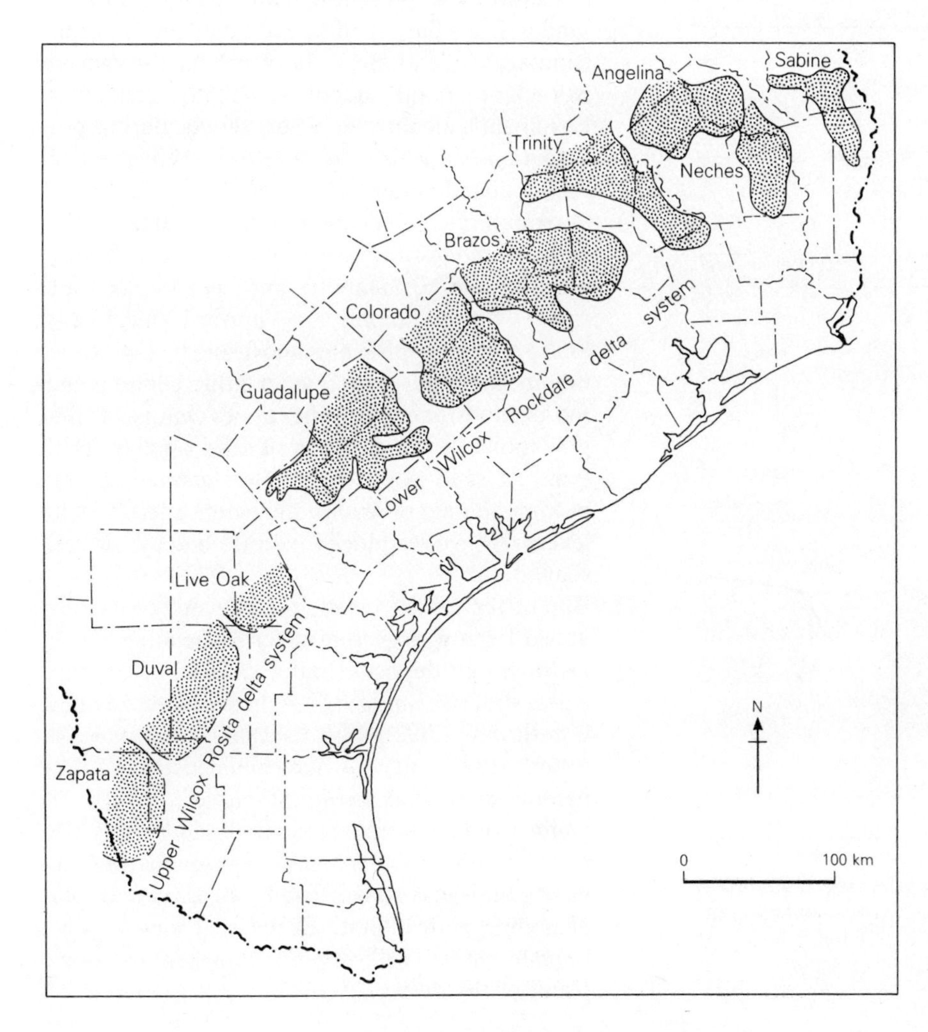

Figure 13.14 Compound delta system of Lower Eocene Wilcox Group, Texas Gulf Coast. Delta-front and pro-delta sediments prograded over a shelf-edge flexure marked by active growth faults, causing thickening of the component strata by a factor of five or even more. (From M. B. Edwards, *AAPG Bull.*, 1981.)

convex seaward. The overall regime is therefore *regressive*; sands are again thick and numerous, those along the delta front providing the principal reservoir rocks. Petroliferous examples occur throughout the Phanerozoic record, being especially common in the strata of post-orogenic episodes (Fig. 13.14).

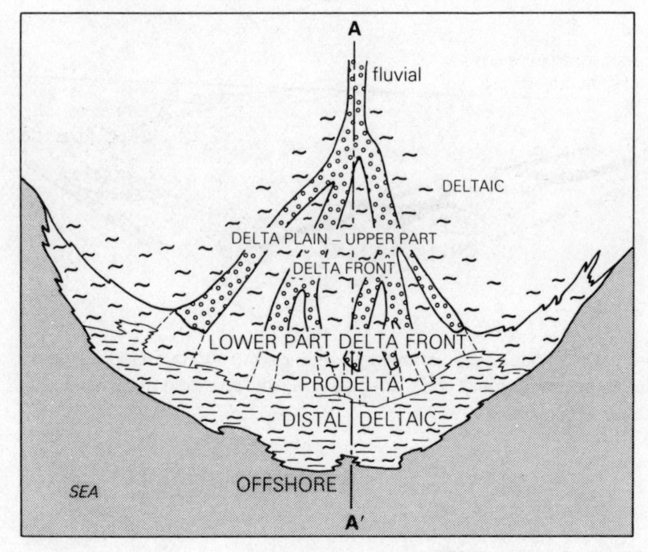

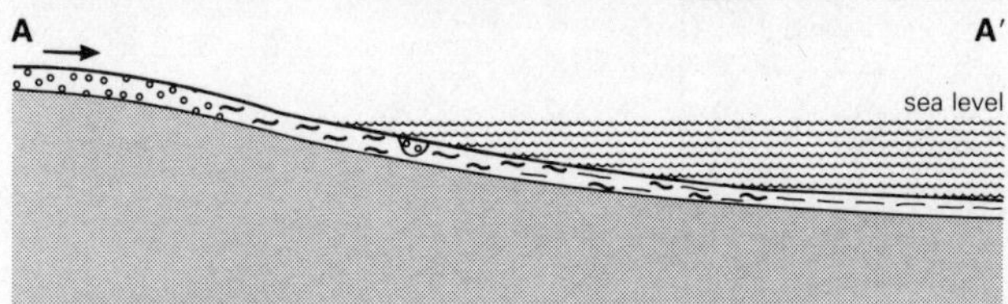

Figure 13.15 A complete fluvio-deltaic depositional model, especially appropriate for fan deltas built so rapidly that sand deposition is continuous from the fluvial to the marine environment. (From A. V. Carozzi, *J. Petrolm Geol.*, 1979.)

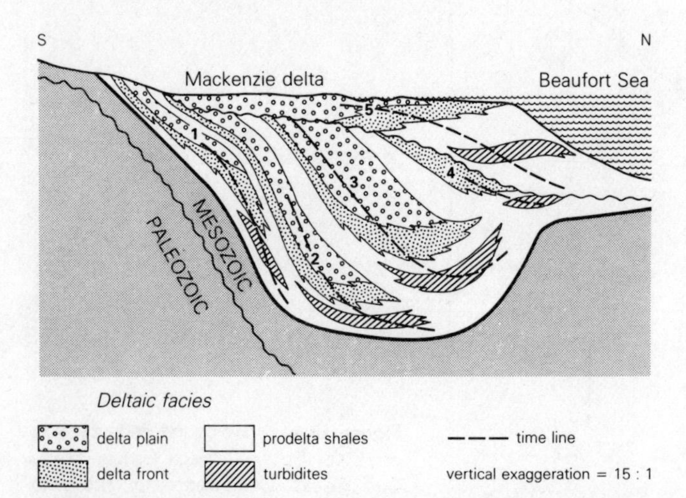

Figure 13.16 Schematic cross section of the Mackenzie River delta, Arctic Canada, showing five Tertiary deltaic cycles containing between them more than 4 × 10⁵ km³ of sediments. (From P. S. Willumsen and R. P. Cote, in *Studies in continental margin geology*, Tulsa, Okla: AAPG, 1983.)

(d) *Fan deltas* prograde into essentially standing water in lakes or inland seas. They are in fact the deltaic parts of *alluvial fans* (see above), and uninterrupted sand transport from the subaerial environment via the fluvial to the deltaic and eventually to the normal marine causes most petroliferous examples to be of composite environmental category (Fig. 13.15). Modern examples include the type Gilbert delta in Pleistocene Lake Bonneville, the large deltas of the Volga and the Danube, and numerous smaller ones built into back-arc basins around the Pacific margin.

The largest deltas are necessarily long-lived, high-constructive, and progradational, and therefore *regressive*, whether their shape in plan is horsetail or birdfoot. The deltaic complexes formed under these conditions (Fig.13.16) consist of superimposed clastic wedges, each normally coarsening upwards. The complete succession contains these components, from the top downwards:

(6) Delta plain deposits: lenticular, organic-rich, carbonaceous clay or shale, bioturbated, with brackish fauna, plant fragments, peat, and rooted sands. Tidal flat, shallow subtidal, and interdistributary bay and marsh, in which sandstones are storm or flood deposits, highly lenticular. Dominantly freshwater when active, during progradation; brackish to marine when inactive, during abandonment.

(5) Distributary-channel and mouth-bar sands, within delta plain. Former are coarse sands with good porosity and permeability and small-scale ripple marks, but commonly fine upward (Fig.13.17). Sharp contacts with shales above and below. River-mouth bar sands have higher lithic contents and more mudstones; abrupt upper contacts, but gradational lower contacts of sand on clay. Both types of sand body have linear geometries and become thinner downdip; they are associated with levees. SP curves blocky, serrate-blocky, or bell-shaped.

(4) Bar finger sands: clean, nonmarine, coarsening upward and with increasing cross-bedding.

(3) Delta fringe deposits: outer margin of exposed delta, with river-mouth bars, beaches, etc. Transition through brackish-water shaly sandstones between delta front and delta plain deposits. Low permeabilities and serrate SP curves.

(2) Delta front deposits: laminated sands, silts and clays, commonly burrowed. Sands coarsen upward, and may have good porosity and permeability, with digitate SP curves; they become thinner *updip*. These deposits represent the active phase of the delta.

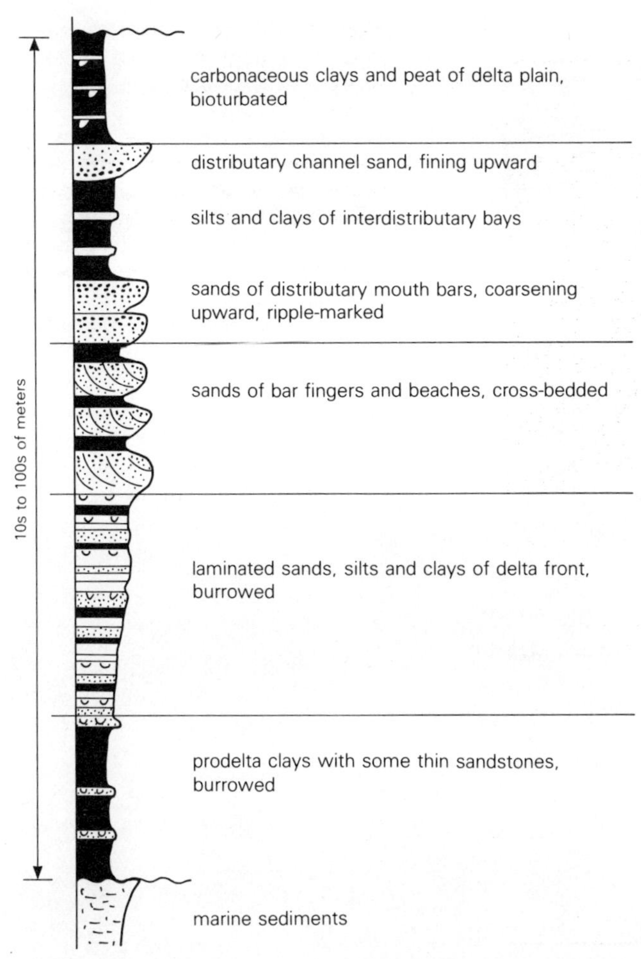

Figure 13.17 Ideal regressive (river-dominated) deltaic sequence.

(1) Prodelta clays, with a few thin sandstones or siltstones, invariably burrowed. Clays readily become undercompacted and overpressured when overlain by rapidly deposited sands of (b). Prodelta deposits rest unconformably on marine shelf deposits, which may include nonclastic components such as algal reefs.

Erosion of this sequence produces reworked quartz sands in tidal channels and inlets, lying unconformably on (e) or (f). The principal reservoirs are in delta-front sands. Productive trends are controlled both by the distribution of these sand bodies and by their geometries, which may be lobate or elongate. Trends are discontinuous along strike, the best productive areas lying around the margins of prograded deltaic lobes which are separated by embayments containing chiefly muds and tight sands. Distributary channel sands of the delta-plain facies in high constructive deltas are seldom highly productive, because initial silica cementation at the boundaries of channel and delta-front sands (due to differences in water compositions) prevents updip migration of the hydrocarbons to the topmost units of the deltaic complex.

Downdip from the delta-front sand trend are further productive trends of the same age in barrier-bar and shelf sands, but these are not strictly parts of the delta (see below). In deltaic systems in general, however, vertical stacking of sands is common, creating *multipay fields* (Fig. 13.18).

Regressive conditions are not of course uninterrupted. Brief transgressive episodes interrupt the progradational sequence and cause destructive phases. The alternation of regression and transgression causes incomplete sequences like this (and Fig. 13.19):

Offlap (regressive) complex

(7) Delta plain swamp deposits with roots and peat, and a barrier sand body marking the new coastline.
(6) Tidal channel sands, cut into
(5) Clays and silts with shallow marine fauna, burrowed at the base.

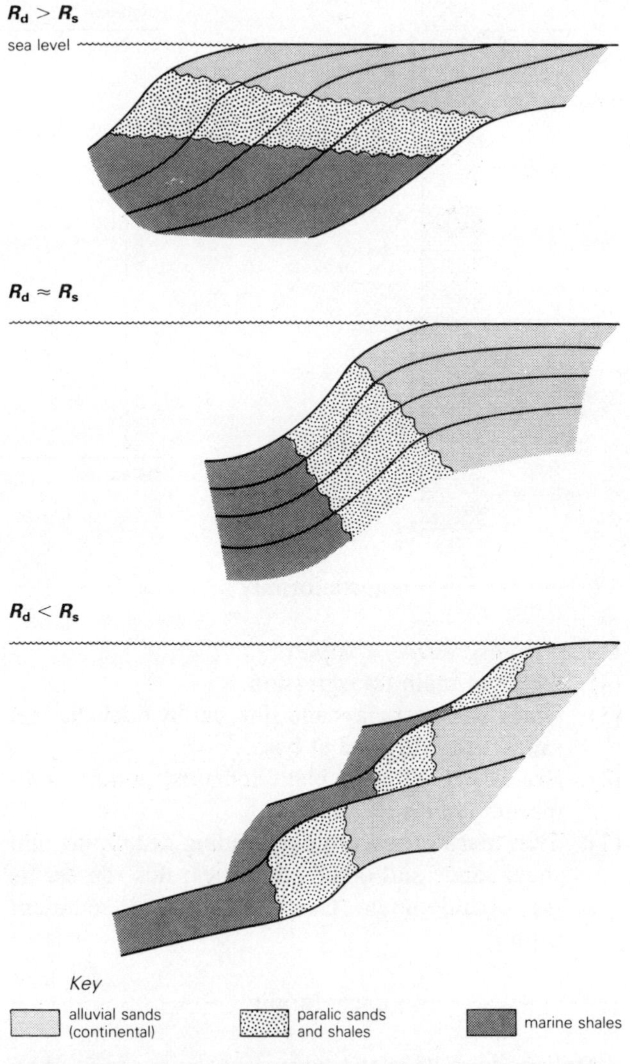

Figure 13.18 Conceptual models of deltaic sedimentation. R_d and R_s are rates of deposition and of subsidence respectively. (From D. M. Curtis, *SEPM Special Publication 15*, 1970.)

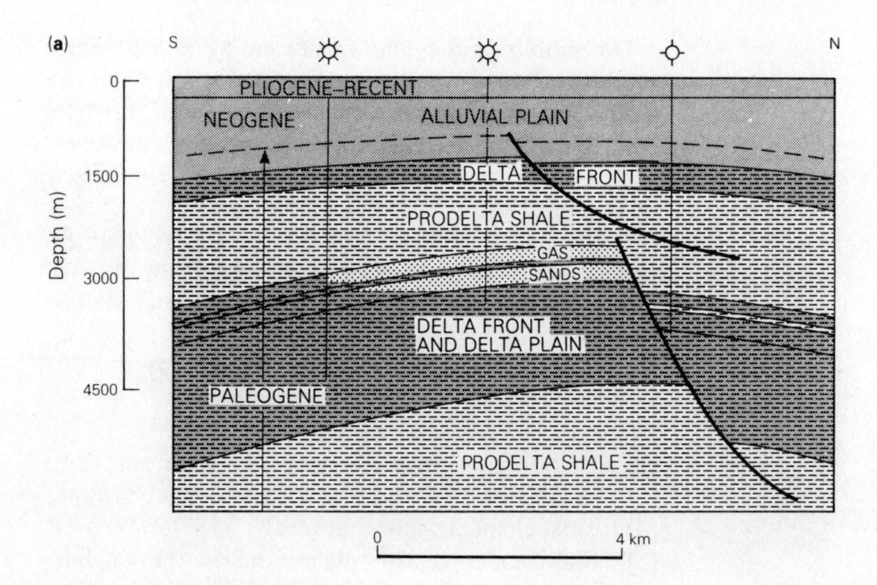

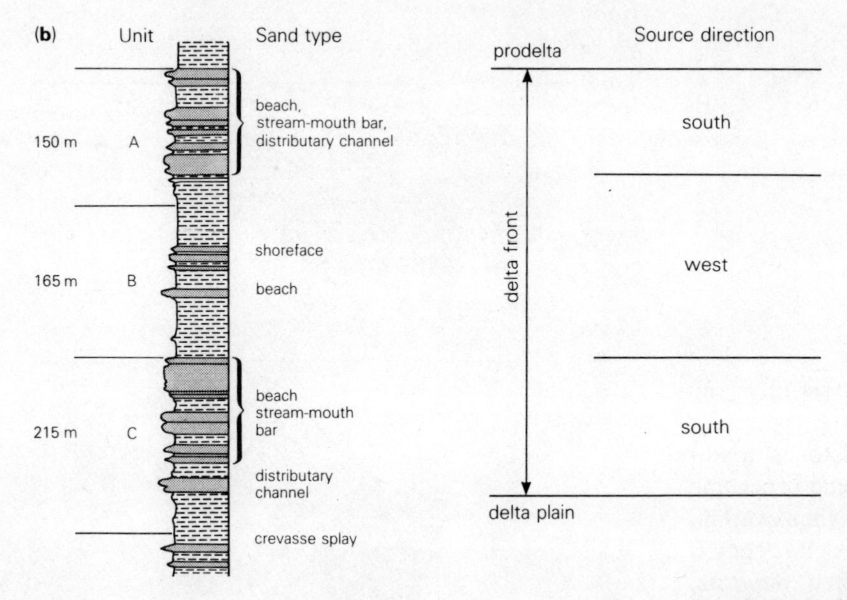

Figure 13.19 Taglu gasfield, Mackenzie River delta, northern Canada. (a) Structural cross section through field, showing succession of facies in a large Paleogene delta undergoing progradation (regression) followed by a transgression and a second regression. Main gas sands in stream-mouth and distributary channel sands at top of delta-front facies. (b) Detail of the transgressive sequence, showing reservoir sandstones. (From T. J. Hawkings *et al. AAPG Memoir* 24, 1976.)

~~~~~~~~~~~ unconformity ~~~~~~~~~~~

Onlap (transgressive) complex

(4) Sands of main transgression.
(3) Brackish-water clays and silts, cut by tidal channel sands, cross-bedded at base.
(2) Brackish beds with plant remains, pollen, non-marine faunas.
(1) Thin marker bed of peat, leading distally to thin sheet sands and other lithologies; this represents the abandonment facies of reduced sediment supply.

~~~~~~~~~~~ unconformity ~~~~~~~~~~~

Deltaic deposits in the subsurface are recognized by gross interval isopach and net sand maps (see Ch. 22), and by their SP log profiles as indicators of vertical trends in grain size (see Fig. 13.46). In the regressive sequence described in this section, the prodelta clays at the base would cause SP curves close to the shale line; delta-front sands coarsen upward and their SP kicks become greater; delta-plain deposits of interbedded sands and clays show digitate patterns; and distributary channel sands fine upward, returning their SP curves towards the shale line like those for point bars.

Trapping mechanisms in deltaic systems *Stratigraphic (lithologic) traps*, due to sand pinchouts and isolated destructional sand bodies, are very common in delta systems, especially in areas of maximum intertonguing of delta-front sands and marine muds.

Structural traps are also very common features of deltas, and they are of concern to the study of the reservoir rocks because they influence their deposition by simultaneous growth. *Growth faults* are activated by the progradation of delta-front sands over unstable, subsiding prodelta muds at the shelf edge. They are therefore especially active in high-constructive lobate delta systems. They cause numerous interruptions in the deltaic sediment sequence, with thickened repetitions of the sequences on the downdropped sides. The thicknesses may increase five- or tenfold going across an active growth-fault system. *Salt domes* coincident with delta-front facies also influence the thicknesses and geometries of sands, commonly leading to the formation of multipay reservoirs. Domes in other delta facies (delta-plain or interdeltaic, with much mud deposition) are generally unproductive. Domes may, however, enhance the generation of hydrocarbons from the prodelta sediments by their introduction of high heat flow.

Examples of productive deltaic complexes Deltaic deposits are especially typical of post-orogenic depositional successions, formed when land masses stand high and offer large drainage areas. They are therefore best developed in the Devonian and Pennsylvanian systems, in the Cretaceous–Paleocene and from the Miocene to the present. Many are parts of larger complexes, involving fluvial and coastal marine deposits in addition (Fig. 13.15). These composite sand bodies provide the reservoirs for some of the largest petroleum accumulations known, and deserve separate treatment here (see below).

Productive sandstones of truly deltaic origin are found, however, essentially throughout the oil- and gas-bearing stratigraphic column. A geologically old example discovered and exploited very early is provided by the Silurian "Clinton" sandstones of Ohio, part of the Appalachian clastic wedge that is continental in the east, deltaic in the center, and marine to the west. Multiple lenticular sandstones have produced both oil and gas since the 1890s, from stratigraphically controlled porosity variations. Most of the Devonian and Lower-to-Middle Carboniferous reservoirs in the giant oilfields of the Ural–Volga Basin are deltaic (Fig. 13.20) and closely associated with coals. The Upper Mississippian Chester Group, the principal producing interval in the Illinois Basin (see Fig. 17.19), exhibits the entire range of deltaic sub-environments: delta-front sands coarsening upward, distributary-mouth bar sands, bioturbated tidal flat sands, and massive sands of subtidal origin forming linear ridges with abrupt bases and tops.

The prolifically productive Pennsylvanian sandstones of the southern USA include deltaic representatives in a great number of fields, especially in Oklahoma. They

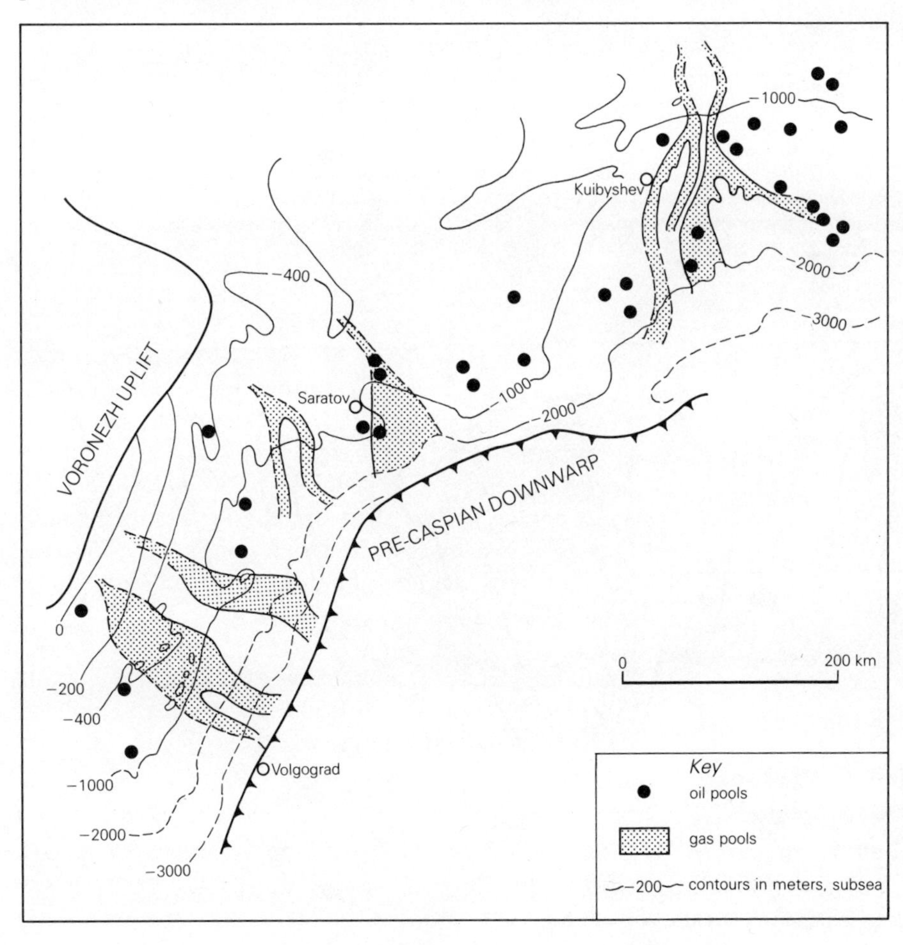

Figure 13.20 Southern portion of the Ural–Volga Basin in European Russia, showing oil- and gasfields in channel sandstones and deltas of the Vereyan Formation (Carboniferous). Deltas drained from the north and northwest into the Pre-Caspian (Emba) Basin. The three cities shown are all on the Volga River, which turns to the southeast at Volgograd where it enters the Pre-Caspian Basin. (After G. P. Ovanesov *et al.*, 9th World Petroleum Congress, 1975.)

were derived from the newly uplifted Ouachita–Amarillo fold belt (Ch. 25). The most notable examples are the channel and distributary-mouth sandstones forming lenticular, stratigraphic traps within the source-rich Cherokee Group; their overbank and interdistributary equivalents contain too much shale, and are too plugged by cements, to be good reservoirs. Similar sandstones are productive, largely of gas, in the basal Pennsylvanian of the Delaware Basin of West Texas.

Not the most productive but perhaps the most unusual sandstone reservoirs in the southern USA consist of the *granite wash* of huge *fan deltas*, which were repeatedly

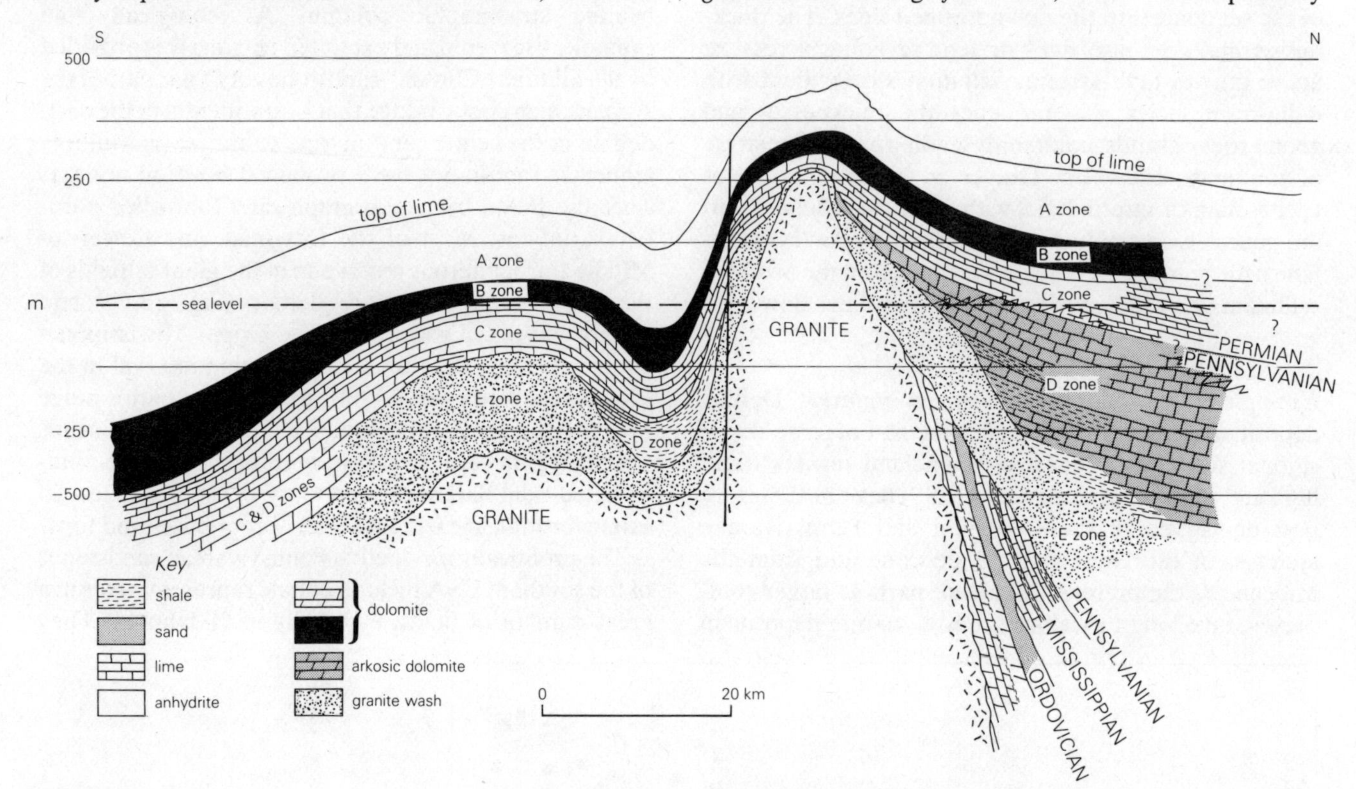

Figure 13.21 Section from south to north across the Amarillo uplift in the Texas panhandle, to illustrate the provision of thick, porous granite wash (E zone) during Pennsylvanian time. The Palo Duro Basin to the south (left) did not start to subside until Mississippian time; granite wash rests directly on Precambrian. The Anadarko Basin to the north was a major early Paleozoic basin and the granite wash lies within the Pennsylvanian succession except in immediate contact with the uplift. (After H. Rogatz, *AAPG Bull.*, 1935.)

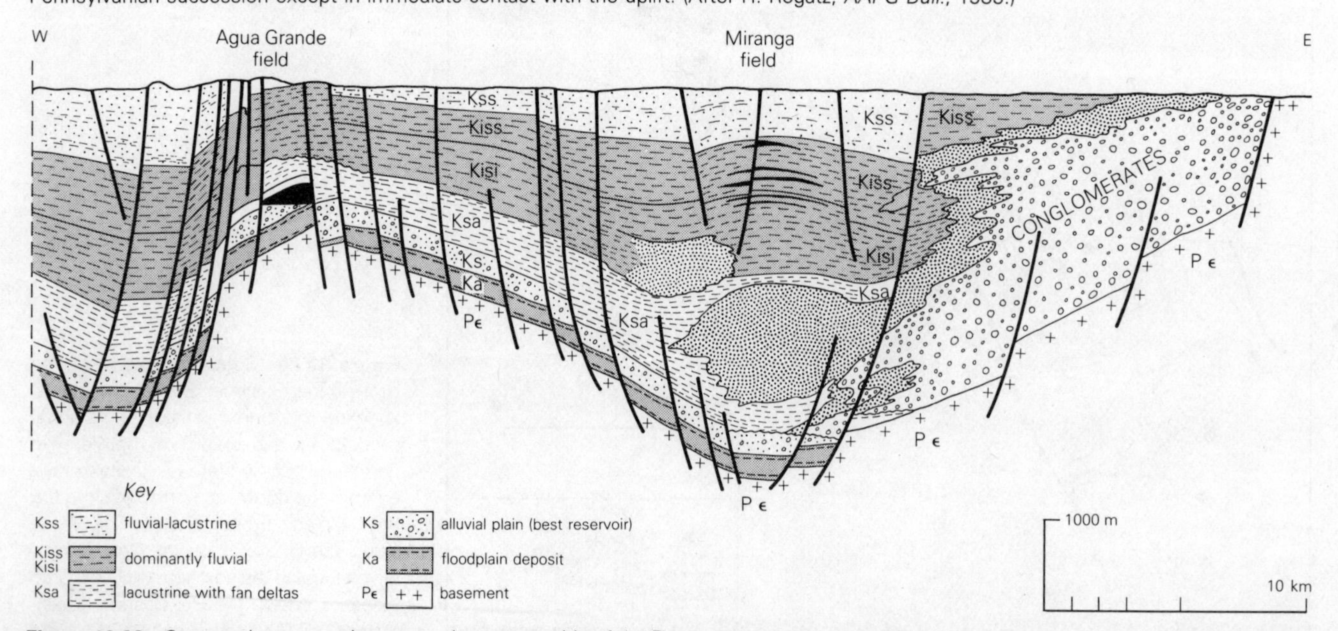

Figure 13.22 Structural cross section across the eastern side of the Reconcavo graben, coastal Brazil, showing Lower Cretaceous fan deltas and other sandstone reservoirs. (After J. I. Ghignone and G. de Andrade, *AAPG Memoir* 14, 1970.)

prograded during late Pennsylvanian and Permian times from abrupt basement uplifts. Figure 13.21 illustrates the granite wash derived from the Amarillo uplift in the Texas Panhandle and swept into the southwestern side of the Anadarko Basin. There it replaces sediments of the basin and slope systems and overlies a considerable thickness of older Paleozoic strata. The granite wash reservoir sands have surprisingly high porosities and permeabilities, up to 20 percent and 500 mD, despite the reduction of both by *in situ* alteration of feldspar grains to clay minerals. Other enclosed basins of the Ancestral Rocky Mountain province (see Ch. 25) contain similar sand wedges. Some of Utah's tar sands, for example, are in the Permian Cutler arkose in the Paradox Basin (Ch. 10).

Lower Cretaceous deltaic sandstones are richly productive in many basins. They include the Reconcavo Basin in Brazil (Fig. 13.22); the West Siberian Basin, in which many of the large oil- and gasfields occur in multilayered sandstones associated with coals (Fig. 13.23); and the Denver Basin, in which two sandstones with an organically rich shale between them constitute a marine and deltaic complex nearly 3 km thick. In other Rocky Mountain basins of North America, numerous sandstones occur throughout the Cretaceous succession, and many of them are fundamentally deltaic (the Frontier Formation of Wyoming, for example, with the widespread Pierre shale as the prodelta deposit).

The Lower Eocene (Wilcox) sand reservoirs of southern Texas are deltaic (Fig. 13.14). All thick producing sands in the Eocene of Venezuela's Maracaibo Basin include deltaic components (see Ch. 17 & Fig. 16.2). The late Miocene–early Pliocene fields of eastern Kalimantan (Borneo) and central Sumatra are in deltaic sandstone reservoirs (Fig. 13.24). Multiple pay sands are typical of them; the Attaka field, for example, is productive from 34 sands. The still younger multiple sandstone reservoirs of the Baku and Cheleken fields, in the South Caspian Basin, were deposited in a series of Pliocene fan deltas (Fig. 13.25).

Many very productive sandstone reservoirs remain enigmatic as to origin even with the aid of hundreds of well records. The large Minas field in Central Sumatra provides an excellent example (see Fig. 16.119). The early Miocene sandstone is the oldest conventional sediment in the vicinity, resting on basement on the crest of the Minas structure and on terrestrial, regolith-type clastics around its flanks. The sandstones are rather massive, very variable in grain size and poorly sorted, with some thin conglomerate beds. They are erratic in distribution, the only areally continuous sand being the one at the base. The sands in general become thinner upward, the intervening shales thicker and carbonaceous stringers more numerous. The sands contain some siderite and some burrowed structures, but no

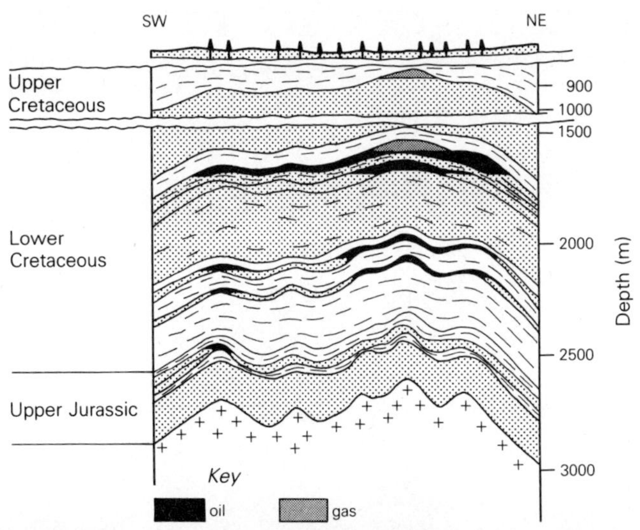

Figure 13.23 SW–NE section through the Megion (left) and Samotlor (right) oilfields, West Siberian Basin, showing multiple Lower Cretaceous sandstone reservoirs, largely of deltaic origin. (From L. I. Rovnin *et al.*, 9th World Petroleum Congress, 1975.)

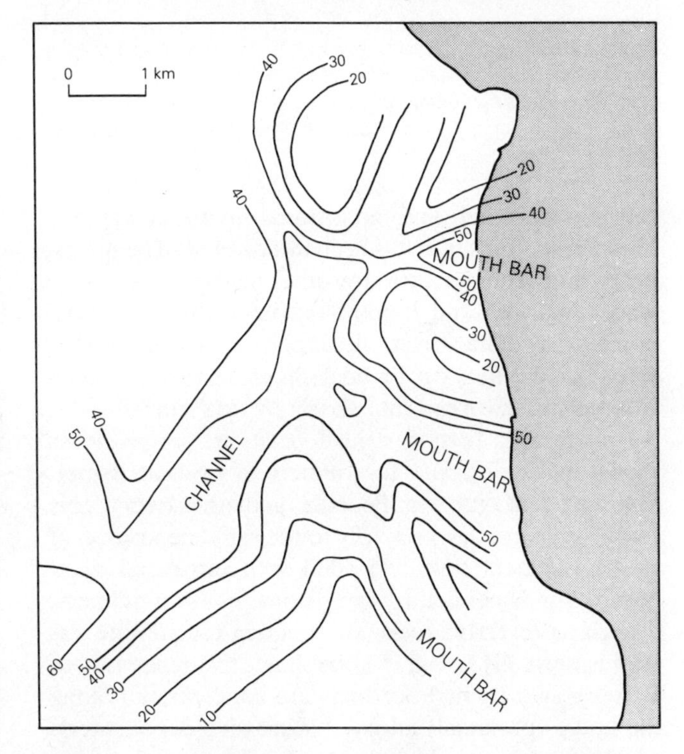

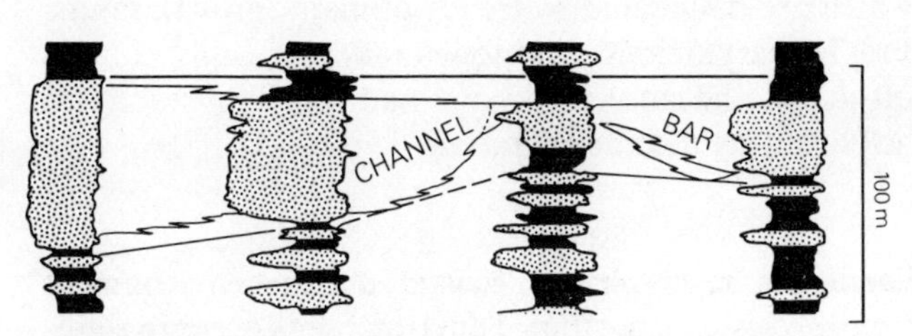

Figure 13.24 Isopach map and cross section of one of the Neogene sandstone reservoirs in the Badak field, within the Mahakam delta of eastern Kalimantan. Channel and mouth-bar sand fingers are clearly identifiable. (From R. M. Huffington and H. M. Helmig, *AAPG Memoir* 30, 1981.)

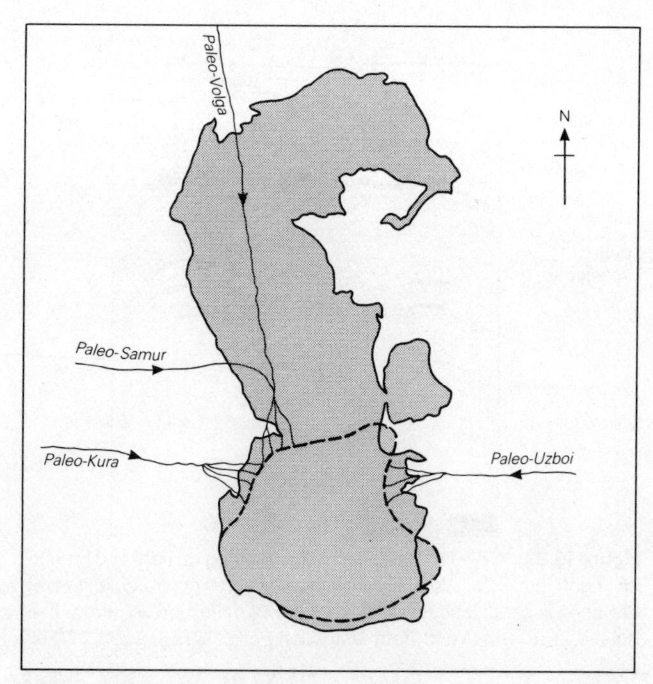

Figure 13.25 South Caspian basin (dashed outline) at end of Pliocene "Productive Formation" time, showing delta system in which productive sands of the Baku and Cheleken oilfields were deposited. (After V. P. Baturin, XVII International Geol. Congress, Moscow, 1937.)

fossils and no coal. The depositional environment of the Minas reservoir sands has been interpreted to be almost everything from terrestrial to inner neritic. It cannot be wholly marine, and is most probably that of a tide-dominated delta, from delta-plain mudflats updip, through tidal flats cut by tidal channel sands, to delta-front sands reworked into shoals by tidal currents.

Undeformed Tertiary delta deposits are *gas-prone*, as shown by the Tertiary forerunners of the Mississippi, Mackenzie, Magdalena, Po, Nile, and other rivers. This characteristic is due primarily to the rapid deposition of clastic sediment heavily loaded with terrestrial plant debris. The Niger delta is exceptional in its oil richness, though its reservoir rocks also contain much more gas than is normally found in nondeltaic sand reservoirs of Tertiary age. Its hydrocarbons are concentrated along the updip (proximal) edges of successive depocenters, which are delineated by trends of major growth faults. Pre-Tertiary deltaic sandstones may as readily contain oil as gas, because they have had the opportunity to acquire it by distant migration.

Sandstone reservoirs of coastal marine environment Coastal clastic depositional environments occur in significant variety. Coastal sandstone bodies contributing important reservoir rocks, however, are principally beach and offshore bar deposits, and the accumulations of tidal channels cutting through them. They are therefore best developed along gently shelving shorelines of emergence, subject to cyclic transgressions and regressions of the sea. They are unlikely to be well developed on strongly faulted or abrupt shorelines. They are consequently distributed in roughly linear or arcuate trends and groups demarking the margins of sedimentary basins of the past; they cannot occur in the centers of marine shale basins.

The coastal zone susceptible of sand accumulation extends from the freshwater, littoral zone to the shelf-slope break. The wider this zone is, the more extensive and more various its sand accumulations are likely to be. The inner part of the zone is at least in part subaerial, and may include dune sands; it also includes hypersaline bays and lagoons, mudflats, and salt marshes characteristic of the interdeltaic environment (Fig. 13.26). Association with deltas (commonly in their inactive phase) is therefore common. The outer part of the zone is entirely marine and neritic, extending from low tide level to the shelf break. Deposition in this outer zone is dependent upon water movements, both of waves and of offshore bottom currents. Sands deposited above wave-base display high-energy characteristics, including the vital one of good sorting; they are the familiar beach and barrier bar deposits. In sands deposited below wave-base, as shelf shoals and banks, bedding may be largely destroyed by bioturbation.

Reservoir sandstones may be deposited on any kind of substrate, but the depositional regime itself must be sand-prone. Reworking of older deposits is a constant feature, and the influence of storms in this process is much underrated. Depositional patterns migrate until they finally become buried. Sand bodies consequently tend to merge or coalesce with one another, and large reservoirs from this regime are more likely to be composite than single, isolated sand bodies. Accumulation is clearly diachronic, rates of accretion being higher laterally than vertically:

(a) If deposition exceeds subsidence, high-energy deposits prograde over lower-energy deposits; hence grain size coarsens upwards.
(b) If deposition and subsidence are in general equilibrium, stacked or multistorey sand bodies result.
(c) If subsidence exceeds deposition, a series of isolated narrow, linear bars of sand become separated by wide strips having less sand. The sand bodies may still coarsen in grain upwards.

In general terms, the transgressive and regressive coastal sequences are inverses of one another, but variations are limitless. The two sequences are not easy to identify or distinguish with certainty from logs, but petrographic study is commonly quite definitive. A hypothetical complete sequence developed under transgression would be like this:

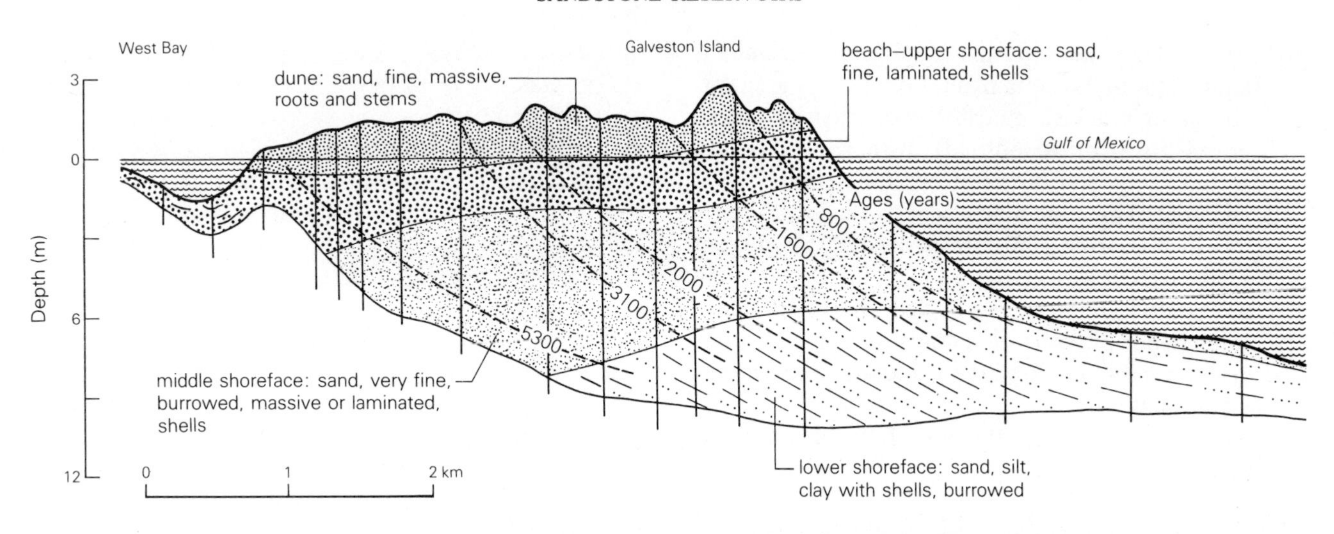

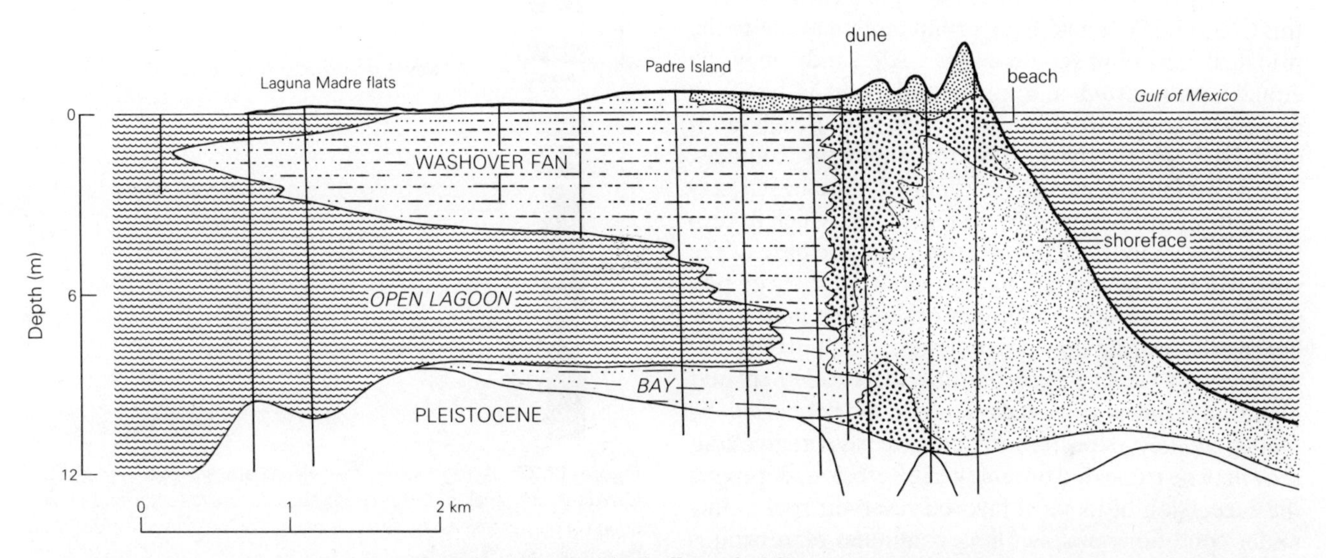

Figure 13.26 Comparative cross sections through two barrier islands, Texas Gulf Coast, to illustrate modern coastal (interdeltaic) sandstone environments. Sands extend from washover fans, washed into back-bar lagoon during storms, to shallow neritic zone or shoreface offshore from the beach. (From R. R. Berg, *AAPG Continuing Education Course Notes* Series 3, 1976.)

(8) At the top, marine shales, with alternating silts or sands below; minor bioturbation.

(7) Barrier island deposits: lower shoreface shales, silts, and sands with much clay and silt matrix (equivalent to prodelta deposits, with seaward limit perhaps 20–25 m depth). Bedding clean-cut, with shales normally thicker than sands (20–30 m, against 10–15 m); the shales may carry microfossils. Sand bars and spits are represented by lenticular sands having better sorting. This section of the sequence is cut by tidal channels, marked by rapid currents and characterized by irregularly cross-bedded sandstones. All units are burrowed, both vertically and horizontally. Sands increase downward, grading into

(6) Barrier island deposits: upper shoreface and beach sands, equivalent to the delta-front deposits of the deltaic environment. These are wave-dominated accumulations, representing the highest rates of sedimentation in the system and much influenced by longshore drift. The sands are well sorted, quartzose, and typically become coarser and cleaner upwards before being abruptly cut off at the top by muddier transgressive sediments. The sands may also be rich in shell material.

(5) Possible dune sands of the back-barrier zone: massive, fine-grained, cross-bedded sandstones containing roots and stems and occasional drifted shells. This zone and its associated beach zone may be interrupted by tongues of back-bar or lagoonal sediments encroaching from behind.

(4) Back-bar deposits: more muddy but with common intercalations of fine-grained sandstone. Many of the latter are washover fans carried in during storms. Fossils both whole and broken; bioturbation common.

(3) Bay or lagoonal deposits: mudflat or salt marsh, equivalent to the delta plain. Low-energy accumulations of brackish mudstones or siltstones (red, purplish, olive, or mottled), with minor, poorly sorted sands and layers of organic-rich clay, silt, and peat. This zone may be cut by fluvial channels in which coarser sands are deposited. Fossils are scarce except for specialized taxa like oysters; few are marine and none are of deep-water species.

(2) Inner beach deposits: thin sandstones, mostly fine-grained, silty, ripple-marked, and with plane parallel laminations.

(1) Nonmarine clastic sediments of fluvial or delta-plain environment.

The regressive succession is essentially the reverse of this (Fig. 13.27), but with important modifications to the principal reservoir components. All sands may be equally well sorted and each sand body is likely to display clear upward increase in grain size. Beach sands may have more abrupt lower than upper contacts, resting on marine shales. Alternatively, the beach may be built seaward over offshore bar sands, which in turn lie on lower shoreface or prodelta clays; the base of the sand will then be gradational through thin shaly or silty interbeds. Alternating additions of sand and mud to the beach area, during slow progradation, result in the combinations of beach ridges and tidal flats called chenier plains in Louisiana. Beach sands may be buried by muddy deltaic deposits. Under extensive regression, they may be truncated or removed by erosion, depriving the succession of its most favored reservoir rock. Only under conditions of slow, long-continued regression is the entire range of environments likely to be represented, but zones 1, 3, 4, 6, and 8 (or 8, 6, 4, 3, and 1 under transgressive conditions) are very commonly drilled through in petroliferous areas of clastic character.

The shallow coastal environment is peculiarly subject to storms, and the influence of storms on coastal deposition is an underrated process (see below). Storm-generated deposits carried landward of the beach are visually obvious; those of potential reservoir interest are *washover fans*, represented by disorganized sandstone lenses and sheets within the back-bar deposits. Both these and their counterparts in the beach area are readily reworked and are seldom well preserved in their original form.

The most commonly identifiable form is as lenticular bars of sand with hummocky cross-stratification, coarsening to gravel size upwards. The bars are separated by channels lacking sand, and possibly carpeted by shale clasts. Storms also generate density currents, which carry sands seaward rather than landward and deposit them below wave-base, still as coarse,

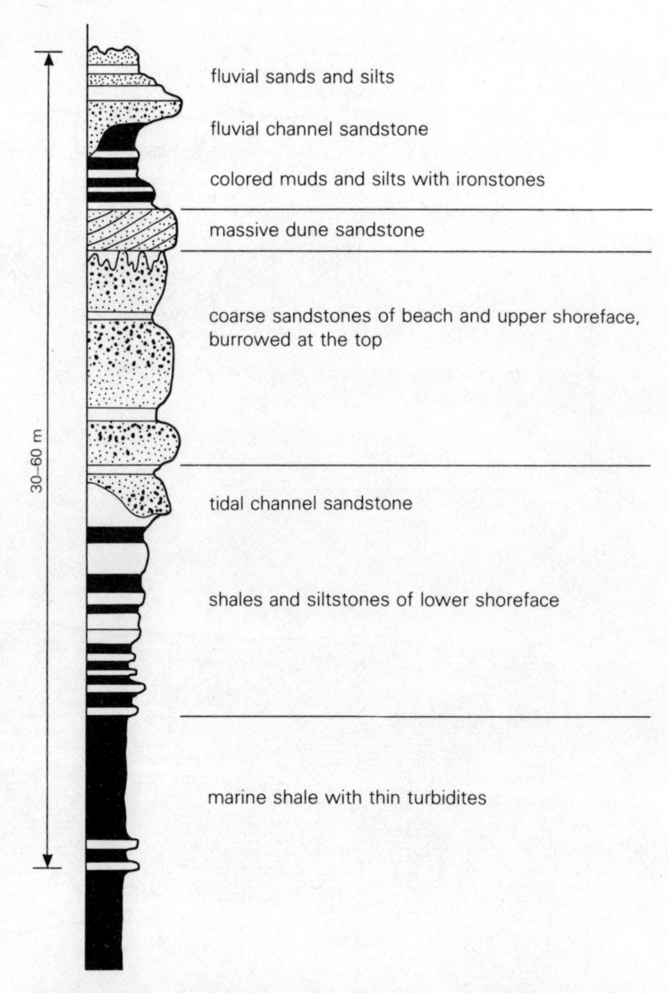

Figure 13.27 A regressive, coarsening-upward cycle with slope turbidites and shales at the base, shoreface siltstones and shales, beach or barrier bar sands, and fine clastics of the lagoonal mudflat cut by fluvial channel sands.

cross-bedded sands coarsening upward but now in an environment otherwise dominated by muds.

Petroleum characteristics of beach and barrier-bar sands
The petrographic characters of beach and barrier-bar sands make them ideal reservoir rocks. They are typically quartzose and somewhat micaceous, clean and well sorted, with good porosity and permeability. Reservoir capacity is vastly enhanced by the stacking of sand bodies of the successive belts in a heavily sand-prone environment (Figs 13.17, 19, 27). The transgressive case of this stacking may be illustrated by the Kinsale Head gasfield off southern Ireland. Its reservoir contains channel, beach, and offshore bar sands from bottom to top; the sands are of Cretaceous age, folded into an anticline during the Alpine deformation. The regressive case is well represented by the Oligo-Miocene Frio Formation of the Texas Gulf coastal plain. Between two large, fluvio-deltaic depocenters, a succession of coastal barrier and strandplain deposits was built sea-

ward in a huge, stacked wedge extending out over an older transgressive succession of prodelta muds and shelf sands (Fig. 13.28). The Frio trend has provided some of the most petroliferous sandstone bodies in North America.

Numerous examples of simple *beach sandstone* reservoirs are to be found among the Cretaceous fields in the North American Rocky Mountain province. Some of them provide the rim-rocks of eroded folds along the mountain fronts; the folds were called "sheep-herder anticlines" in Wyoming and Montana because they are so obvious that farmhands could recognize them. The beach sandstones are simply coastal-marine components of the huge, composite, clastic wedge (molasse) shed eastward from the newly risen mountains. It extends at least from northern Mexico, through the "Deep Basin" of western Alberta and around to Arctic Alaska (where the reservoir rock of the Kuparuk River oilfield is in part a beach sandstone).

All large clastic wedges provide similar coastal-marine reservoirs seaward of the deltaic and fluvial reservoirs noted in the preceding sections. In the Appalachian clastic wedge, for example, both the Devonian Oriskany sandstone and the basal Mississippian Berea sandstone span this composite depositional zone. The two Berea sandstones are particularly instructive (Fig. 13.29). Rivers from the Canadian Shield drained southward and westward into a gulf between the new Appalachian Mountains and the Cincinnati Arch. Channel sands were deposited by these rivers while a barrier bar developed in the gulf to the east. A marine transgression then laid a blanket sandstone, with both beach and deltaic components, across the older channel and barrier-bar sands. Drillers now penetrate two sand-

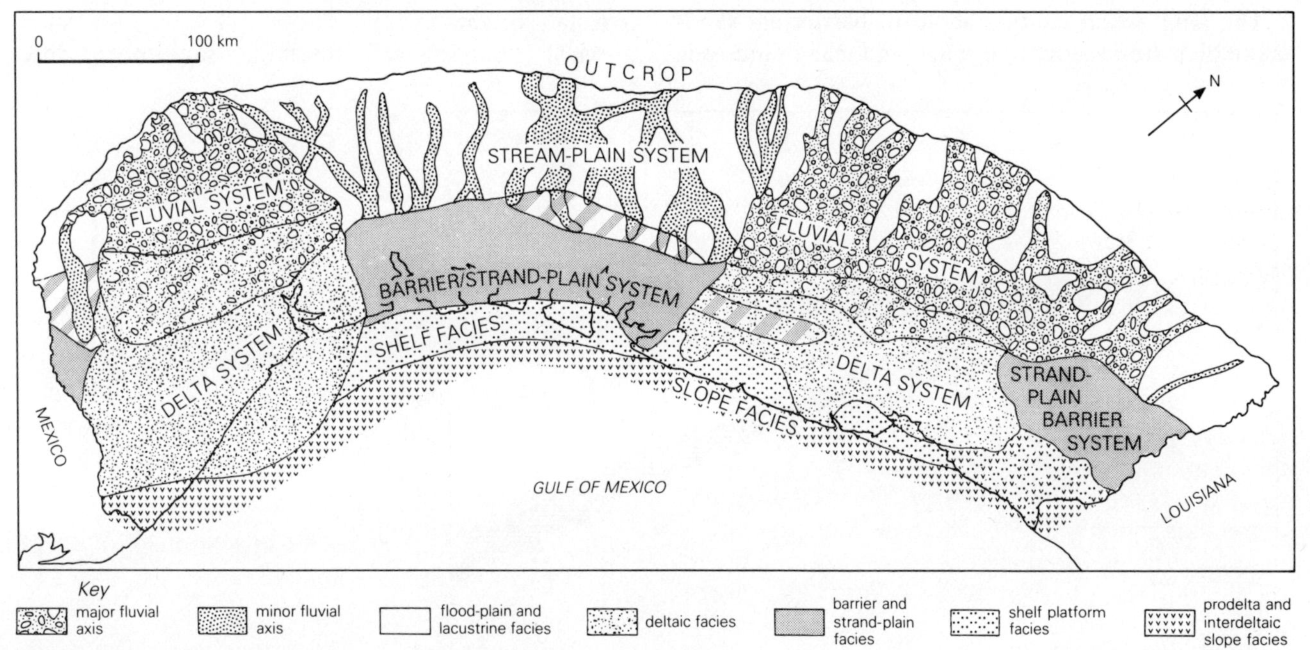

Figure 13.28 Areal relations between fluvial channels, alluvial plain, deltaic depocenters, strand-plain barrier system, shelf, and slope facies, Frio Formation (Oligocene) of the Texas Gulf Coast. (From W. E. Galloway *et al.*, *AAPG Bull.*, 1982.)

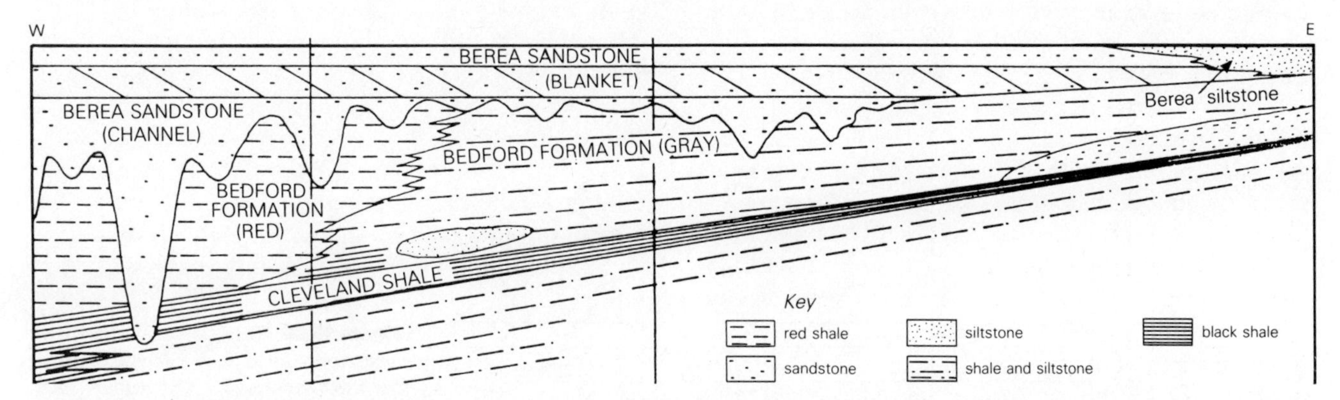

Figure 13.29 West–east stratigraphic section across northeastern Ohio, showing relation (in surface exposures) between first Berea (blanket) sandstone, of marine and deltaic facies, and second Berea (discontinuous) sandstone, of bar and channel facies. (From A. H. Coogan *et al.*, *Geological Society of America field trip guidebook*, Geol. Soc. Am., 1981.)

stones each with two facies components; all four facies have been productive.

Barrier-bar sands resemble beach sands in general shape: narrow, linear, elongate parallel to the paleoshoreline, and discontinuous laterally. They differ in being commonly more fine-grained and in grading into muds or silts in all directions. In transgressive cycles they may therefore be wholly enclosed within shales or siltstones. Oil then occurs in the tops of transgressive sands. The shoreward limit to production is caused by lack of marine shales, but the downdip limit may extend as far as permeable sands exist. They will give out in water depth such that the reduced current velocity is unable to transport and sort sands. Barrier-bar sands are less favorable as reservoir rocks in regressive cycles, because they then lack early updip seals.

The long, linear configurations of barrier-bar sands make their trends easy to predict, and many sandstone reservoirs have been interpreted as ancient barrier-bar deposits. These interpretations are based on analogy with modern barrier bars, especially those along the Texas Gulf Coast (Figs 13.26–28). Unfortunately, the analogy may not be a good one. It relies upon the doctrine of uniformitarianism, and actualistic examples of coastal phenomena are influenced by the post-Pleistocene rise of sea level. Bar sandstones in this circumstance represent brief regressive episodes during an overall transgression. They then accumulate as elongate ridges of clean, well-sorted sands, formed by tidal currents and possibly localized by subtle structural features sub-parallel to the shoreline (faults, or salt or shale ridges in the subsurface); such structural features may be effective because of their control of wave-base. The sand ridges occur in regularly spaced, parallel festoons. Individual ridges may be tens of meters high, and typically display large-scale cross bedding of very constant orientation.

Most "barrier-bar" reservoir sandstones differ

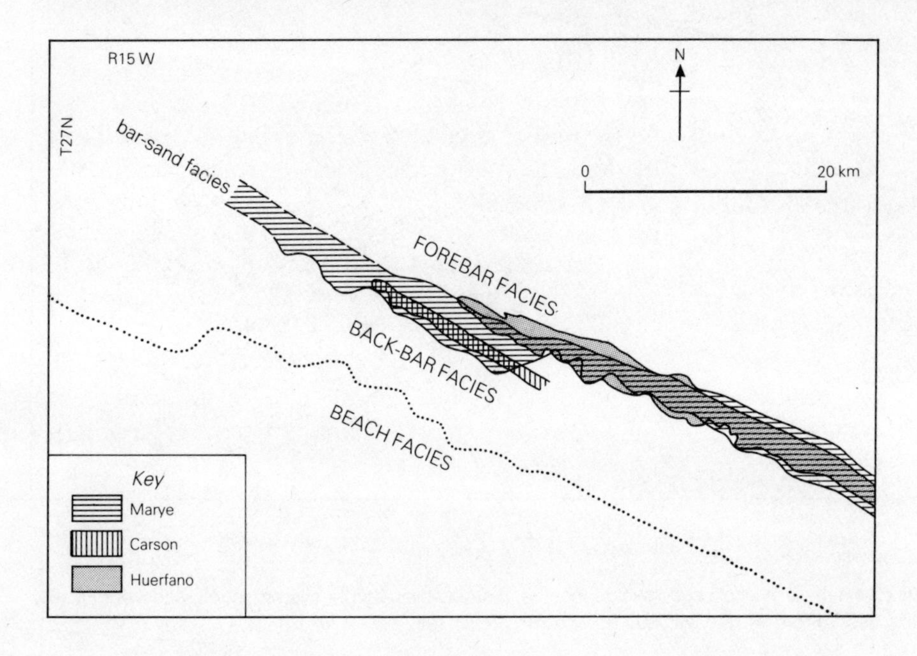

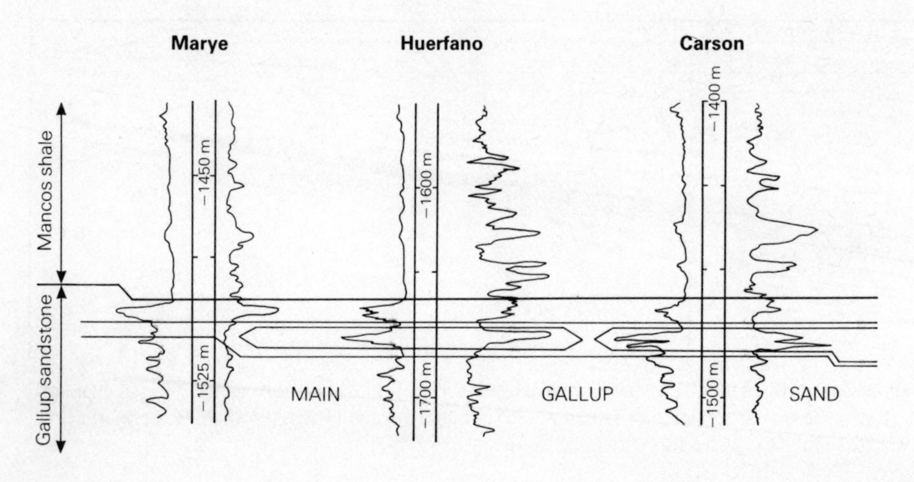

Figure 13.30 Bisti oilfield, New Mexico. Nonconvex traps in overlapping offshore sand bars enclosed in Cretaceous shale. Lower figure shows E-log relations between three named sand bars. (From F. F. Sabins, *AAPG Bull.*, 1963.)

markedly from this modern analog. They were aspects of dual transgressive–regressive cycles consequent upon orogenic episodes. In cases created by, or subsequent to, the Mesozoic drift episode, the most important reservoirs in the Jura-Cretaceous of the US Gulf Coast, the Rocky Mountain basins, and the North Sea are clearly regressive, representing upward shoaling of the sea floor. They are not long, high-crested, linear, parallel, nor grouped in swarms; they commonly coarsen upwards, and cross bedding is not of constant orientation. Their characters suggest that they were subject to more rapid changes of shelf and slope geometry than are present-day barrier bars.

Barrier-bar sandstone reservoirs provide blocky E-log patterns with both SP and resistivity kicks (see Figs 13.44 & 45). Accumulations are commonly controlled by structure, though many small stratigraphic traps also occur in bar sands. Large structures can produce from massive, nearshore sands; smaller structures produce from thinner sands interbedded with shales, usually downdip from the massive sand facies. Barrier bars are invariably divided or terminated by tidal channels. These may contain well sorted, coarse sandstones having excellent reservoir properties.

Simple barrier-bar sandstone reservoirs in North America include those in numerous fields in the Cretaceous of the Rocky Mountain province (see Sec. 25.5.10). Illustrated here are the Bisti field in the San Juan Basin of New Mexico (Fig. 13.30), in three overlapping sand bars enclosed in shale, and the Bell Creek field in Montana (Fig. 13.31), in an offshore deposit overlapped by the source rock of the Mowry shale. Offshore bars in the Lower Cretaceous of Alberta, Canada, follow carbonate cuestas on the post-Mississippian unconformity. A series of such bars in a fine-grained, glauconitic sandstone makes up the Hoadley "trend" in the south-central part of the province. The trend is about 200 km long and 25 km wide, elongated NE–SW. Open marine shales lie to the northwest, and a bay or lagoonal facies to the southeast; a deltaic complex entered this latter facies from the southeast and its deposits were redistributed to form the extensive offshore bar complex. Within the bar complex, geologists have recognized levees and ridges cut by tidal channels, and both an interbar lagoon and a back-bar belt of washover sands. The reservoirs contain gas and no oil.

Historically the most famous of productive offshore bars are the "shoestring sands" of Kansas and northeastern Oklahoma (Fig. 13.32). These are long sand bodies so narrow that a field developed in one may be only one well-location wide; but they may even so be very prolific, completely filled with oil without free

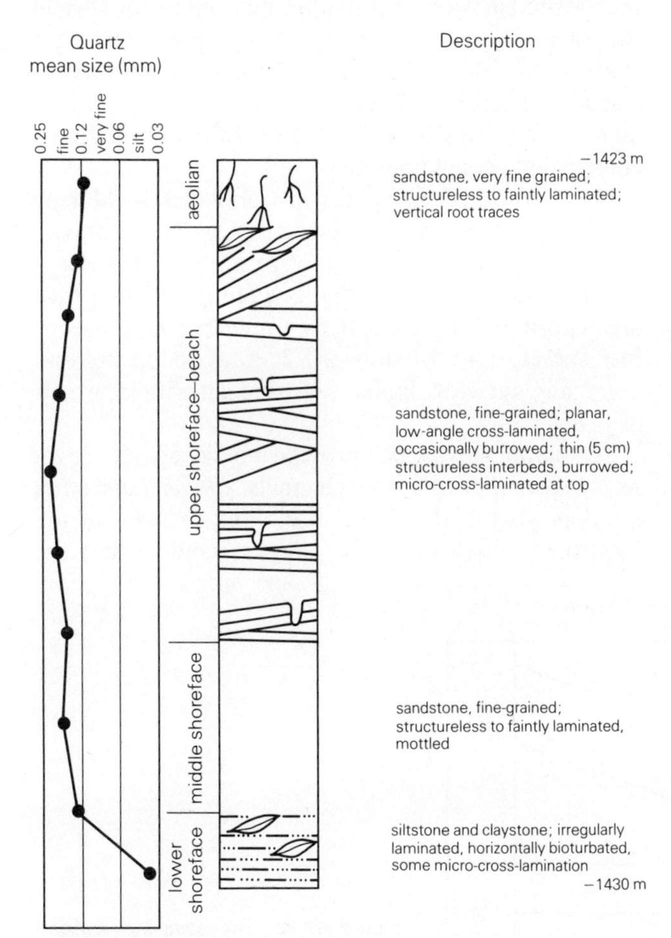

Figure 13.31 Sedimentary structures, textures, and lithologies of typical barrier-bar reservoir sediments; Cretaceous of Bell Creek oilfield, Montana. (From R. J. Weimer, *AAPG Continuing Education Course Notes* Series 2, 1977.)

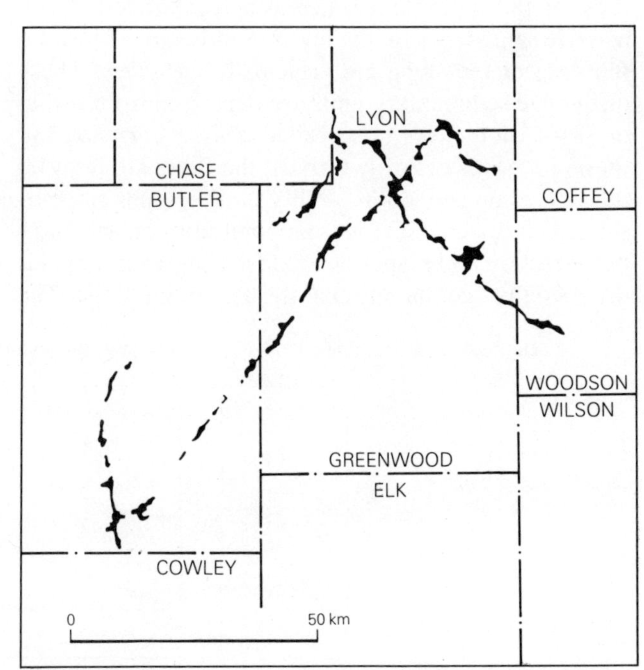

Figure 13.32 Shoestring sand pools in Kansas; sands of Pennsylvanian age, probably offshore bars. (From K. K. Landes, *Petroleum geology of the United States*, New York: Wiley–Interscience, 1970; after P. L. Hilpman, 1958.)

water and wholly enclosed in shales. Some "stray sands," similarly isolated, are river-channel deposits and not marine offshore bars; some may even be eskers; but the type "shoestring sands" are convex upward, not downward as channel sands would be. This distinction by cross-sectional shape, basic to the identification of linear sand body origin, is re-emphasized at the close of this section (Sec. 13.2.9).

Deep-water sandstone reservoirs Some reservoir sandstones were deposited in waters deeper than 200 m (possibly very much deeper), commonly associated with shales bearing pelagic fossils and other evidence of deposition far below wave-base. Such deep-water sandstones can occur only in wholly marine successions; they are not associated with shelf sediments like the common types of limestone, nor with subaerial erosional features.

The environment of their deposition extends from the continental slope, cut by submarine canyons, to the abyssal plains. Submarine fans carried beyond the foot of the slope may be productive, especially in narrow interior basins – whether extensional or compressional (see Ch. 17). The geologist must identify the deep-water facies in its relation to other, better-understood facies, with the aid of faunal-bathymetric data, continuous seismic profiles, gravity studies, and paleogeologic reconstructions once early drilling has provided the keys.

The *slope association* consists almost entirely of pelitic rocks on the upper slopes, increasingly interbedded with massive sandstones on the lower. Sands carried beyond the base of the slope are principally *turbidites*. These distinctive sedimentary units are deposited from turbid or suspension flows (also called *density currents*, but these are not necessarily turbid); the flows are induced through some sudden instability on the upper slope or shelf. The currents spread from point sources, and carry both traction and suspension loads. The traction load is dropped first, commonly near the base of the slope. The suspension load continues on a gentler gradient, beyond the channels and across the lobes of submarine fans.

Thus turbidites consist laterally and vertically of a coarse basal sand, a traction deposit, and an upper layer of suspended matrix. The matrix is critical to the recognition of a turbidite; it is the part that keeps the coarser components in motion. The so-called "complete turbidite" consists of graded beds in an ordered, five-component sequence named for the Dutch geologist Arnold Bouma (Fig. 13.33).

Normally-graded beds should not automatically be interpreted as deep-marine deposits such as submarine fans. They are common in a variety of aqueous environments, including lakes. Nor are all deep-water sands turbidites. Some lack graded textures and are low in clay. They are more probably to be attributed to non-turbid density currents (which are dense because they are cold or highly saline); these erode shallow submarine channels and permit the passage of sand and silt into the basin. Some apparent turbidites actually coarsen upwards. They appear to be storm-influenced deposits, displaced from the edges of beach or barrier sands and redeposited in water still shallow but well below normal wave-base. Such an origin has been proposed by Roger Walker for the productive Cardium sandstone in Canada's Pembina oilfield (see Ch. 26). The upward increase in grain size makes the top of that formation a chert-pebble conglomerate.

Many so-called turbidites are *gravitites* in Manley Natland's sense, carried into deep water by gravity alone, not by a moving current of water. Other deep-water sands have been called *contourites* because they are deposited by deep ocean contour currents, of energy just sufficient to winnow the bottom sediments and segregate layers of rippled sand and silt with interbeds of pelagic mud.

Submarine fans, which provide most deep marine reservoir rocks, need feeder channels. *Channel turbidites* are deposited in these incisions along the axes of the fans; they consist of coarse, massive sands, which be-

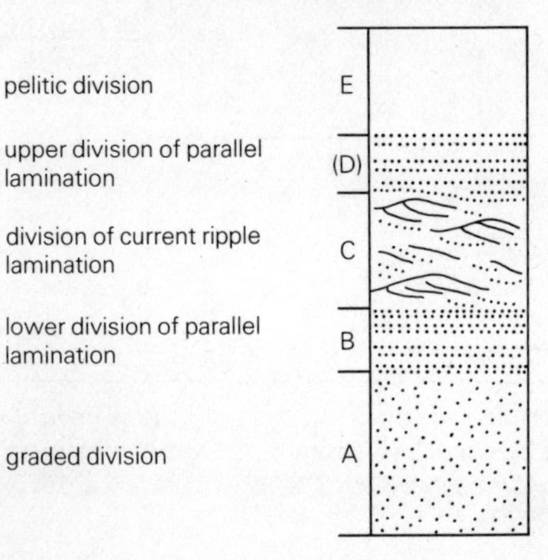

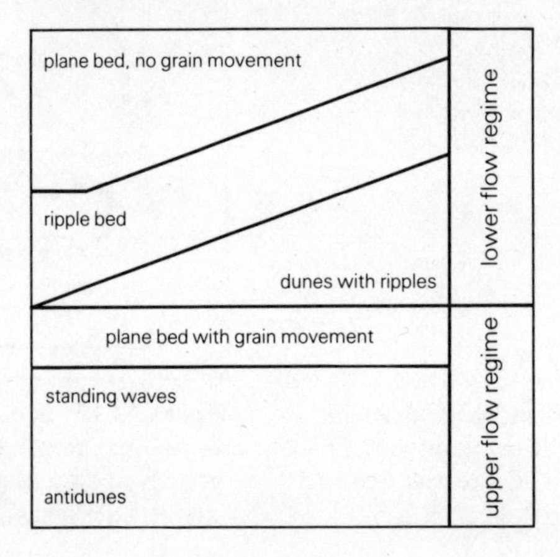

Figure 13.33 The complete turbidite unit of A. H. Bouma (1962) and its interpretation in terms of flow regimes. (From R. R. Berg, *AAPG Continuing Education Course Notes* Series 3, 1976.)

1

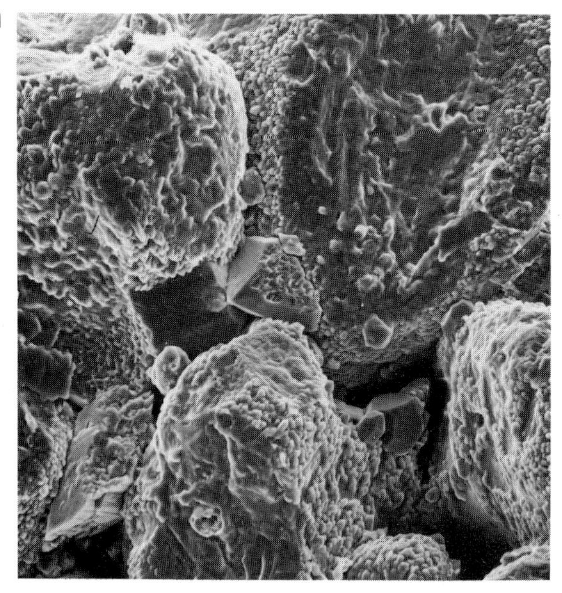

10 μm

2

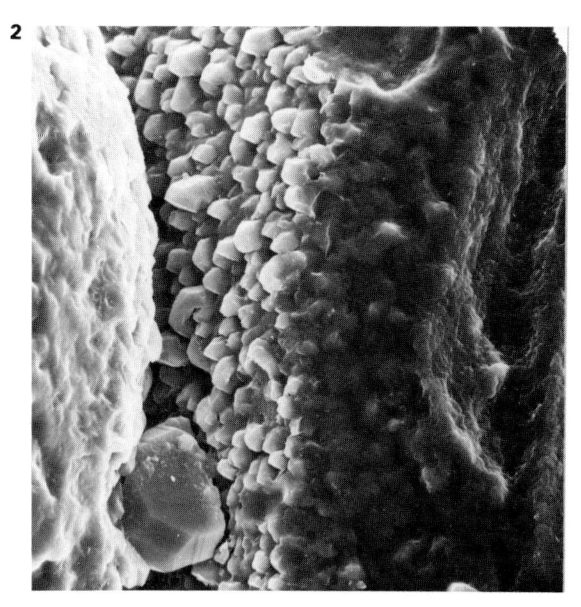

1 μm

Plate A Scanning electron micrographs of reservoir sandstones, diatomites and chalks, showing modifications of porosity by silica, feldspars, clay minerals, mica, calcite, and pyrite. Bar scales at lower left of each photograph indicate 1, 10, or 100 μm.
(From J. E. Welton, *SEM petrology atlas*, Tulsa, Okla: AAPG, 1984. Originals courtesy of R. L. Burtner and Chevron Oil Field Research Company.)

1 Authigenic quartz overgrowths; Nugget sandstone (Jurassic), Wyoming.
2 Rim of quartz crystals coating detrital quartz grains; Nugget sandstone (Jurassic), Wyoming.
3 Inner wall of foraminiferal test lined with cristobalite; Monterey Formation (Miocene), California.
4a, b Diatomite, with opaline diatom fragments partly coated with clay; Pliocene, San Joaquin Valley, California.
5 Feldspathic sandstone: detrital grain of plagioclase rimmed with overgrowth of potassium feldspar; Nugget sandstone (Jurassic), Wyoming.
6 Aeolian feldspathic sandstone: detrital K-feldspar grain in rounded quartz grains; Rotliegendes Formation (Permian), Netherlands.
7 Authigenic chlorite coating quartz grains; dark areas lack chlorite, and were formerly grain contacts; Tuscaloosa Formation (Cretaceous), Louisiana.
8 Pore-lining and pore-bridging illite coating quartz grains; dark areas are open pores; Nugget sandstone (Jurassic), Wyoming.
9 Books of kaolinite partially filling pores in Brent sandstone (Jurassic), Ninian field, North Sea.
10, 11 Detrital quartz grains (dark) partially coated by authigenic smectite; (Miocene), California.
12, 13 Pore-lining and pore-bridging smectite; (Miocene), California.
14 a, b, c Mixed-layer clays (illite–smectite) partially filling and bridging pores between detrital quartz grains; Rotliegendes Formation (Permian), Netherlands.
15 a, b Mixed-layer (illite–smectite) clays between authigenic quartz overgrowths; Brent sandstone (Jurassic), Ninian field, North Sea.
16 Detrital biotite compacted between detrital quartz grains (dark); fine-grained material is clay matrix; Stevens sandstone (Miocene), San Joaquin Valley, California.
17 Chalk: whole and broken coccolith plates; note whole coccosphere in center, and high microporosity (35 percent); (Danian), Danish North Sea.
18 Foraminiferal test containing sparry calcite infilling; Danian chalk, Danish North Sea.
19 a, b, c Pyrite octahedra and framboids, coated by clay and partly filling a burrow. Clay coatings insulate the pyrite and prevent it affecting log conductivity of the formation. Lower Cretaceous sandstone, Barrow Island, Australia.

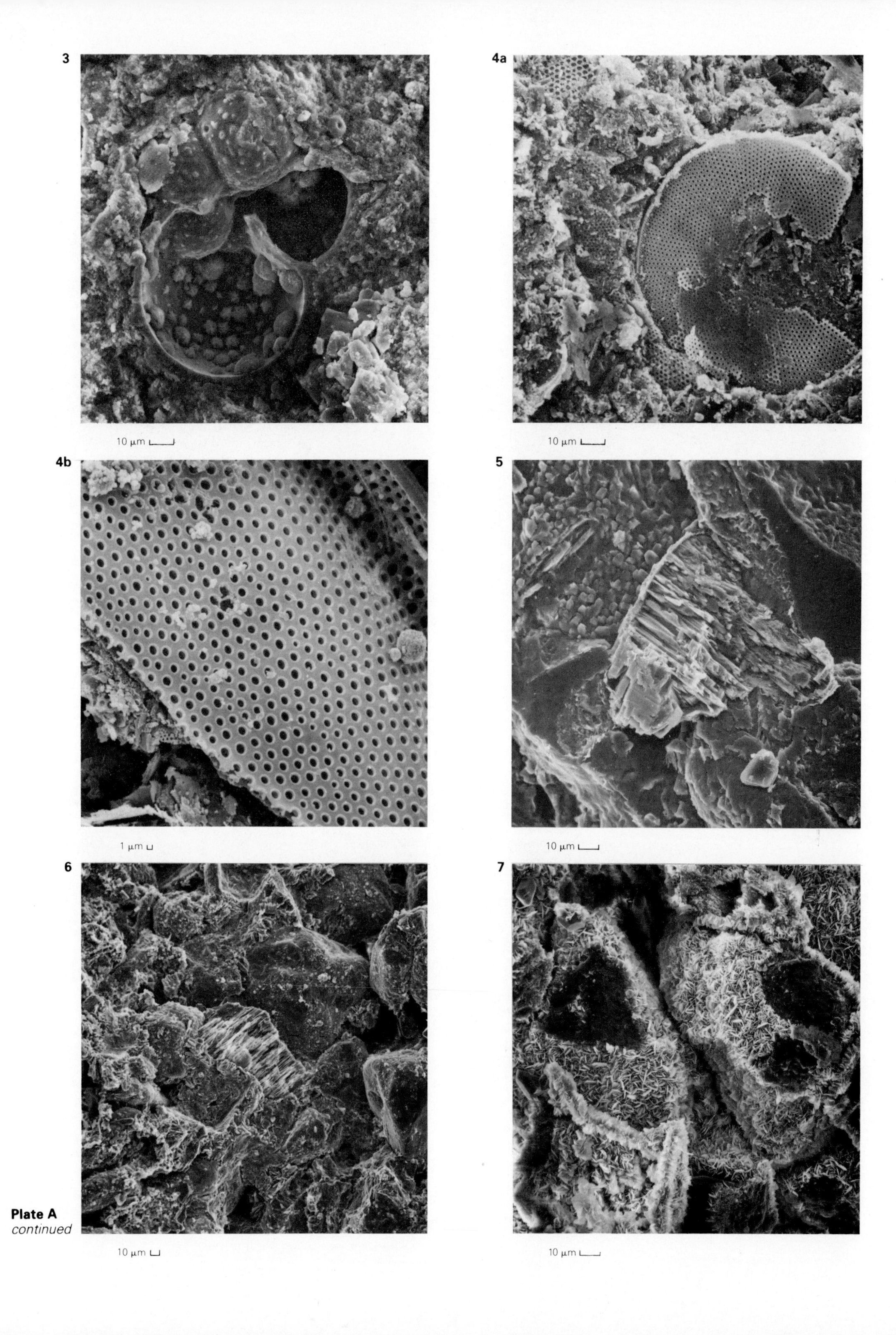

Plate A
continued

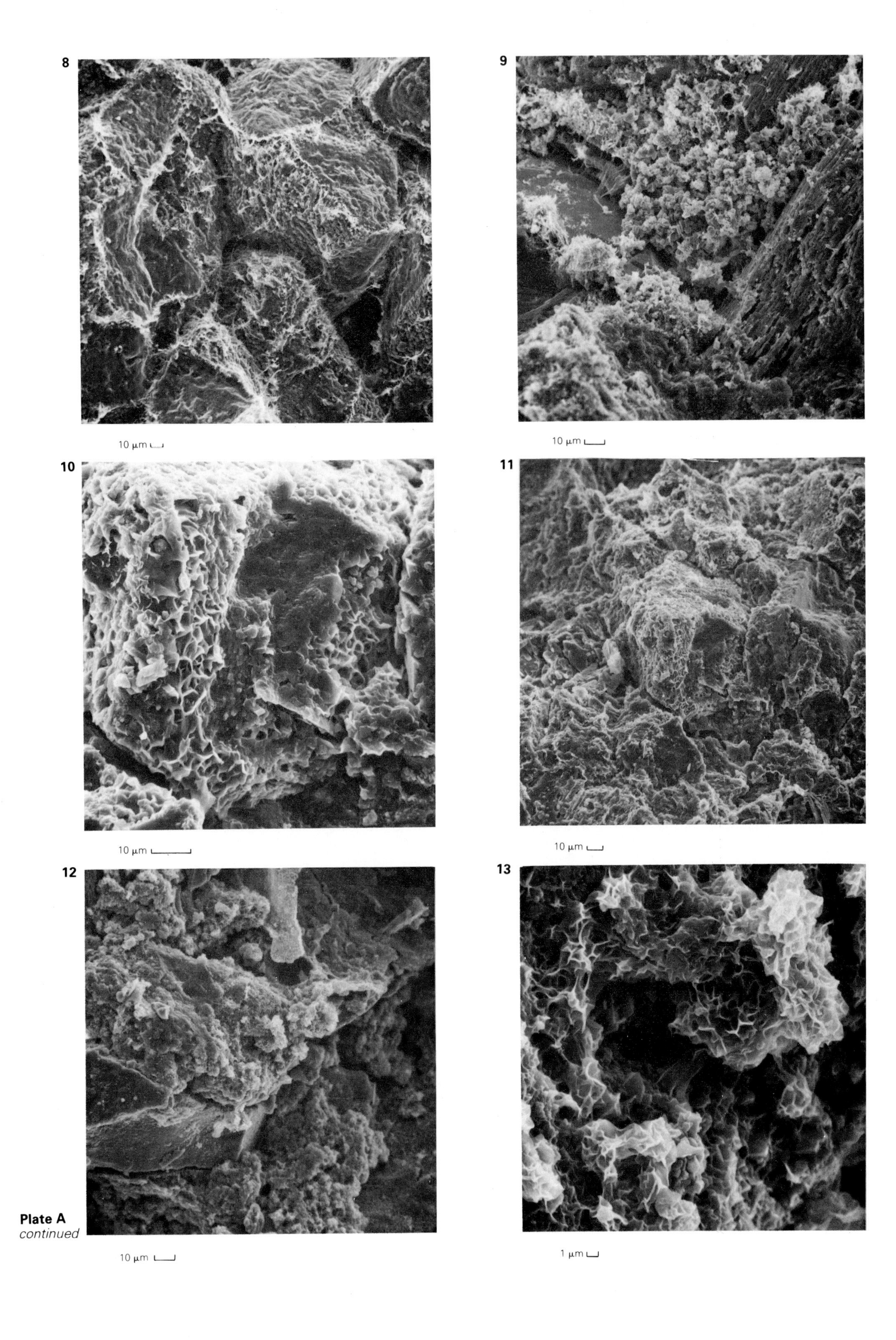

Plate A
continued

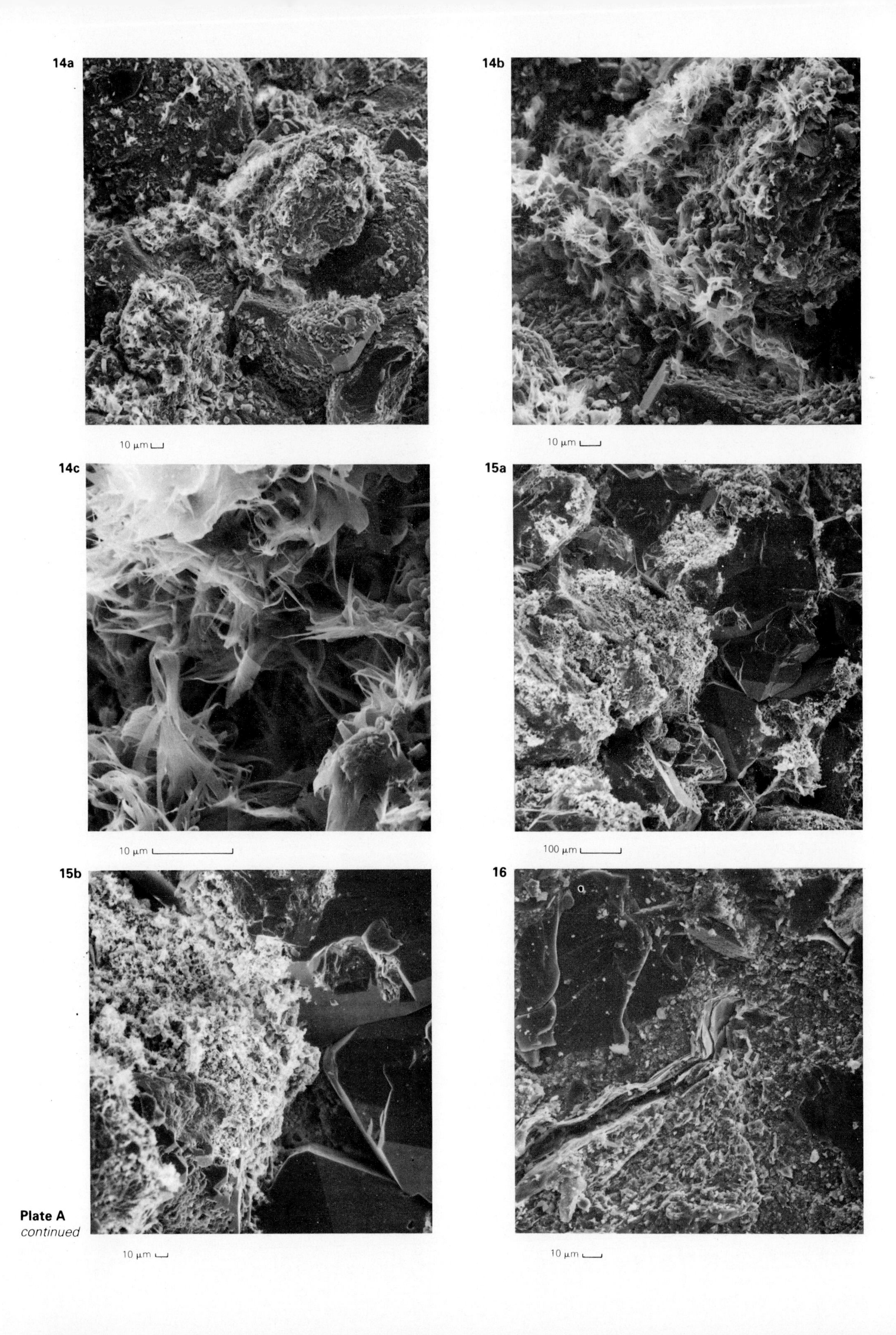

Plate A
continued

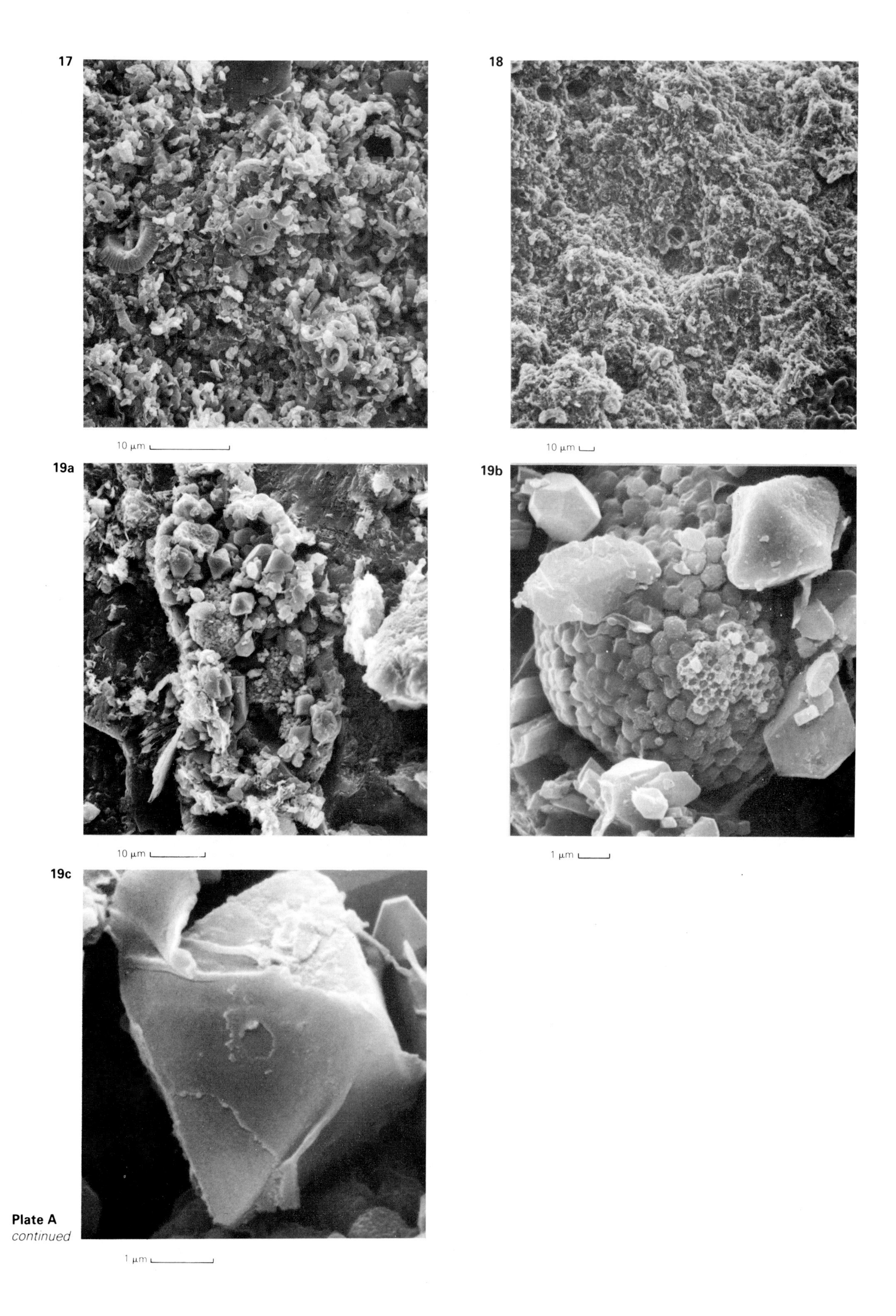

Plate A
continued

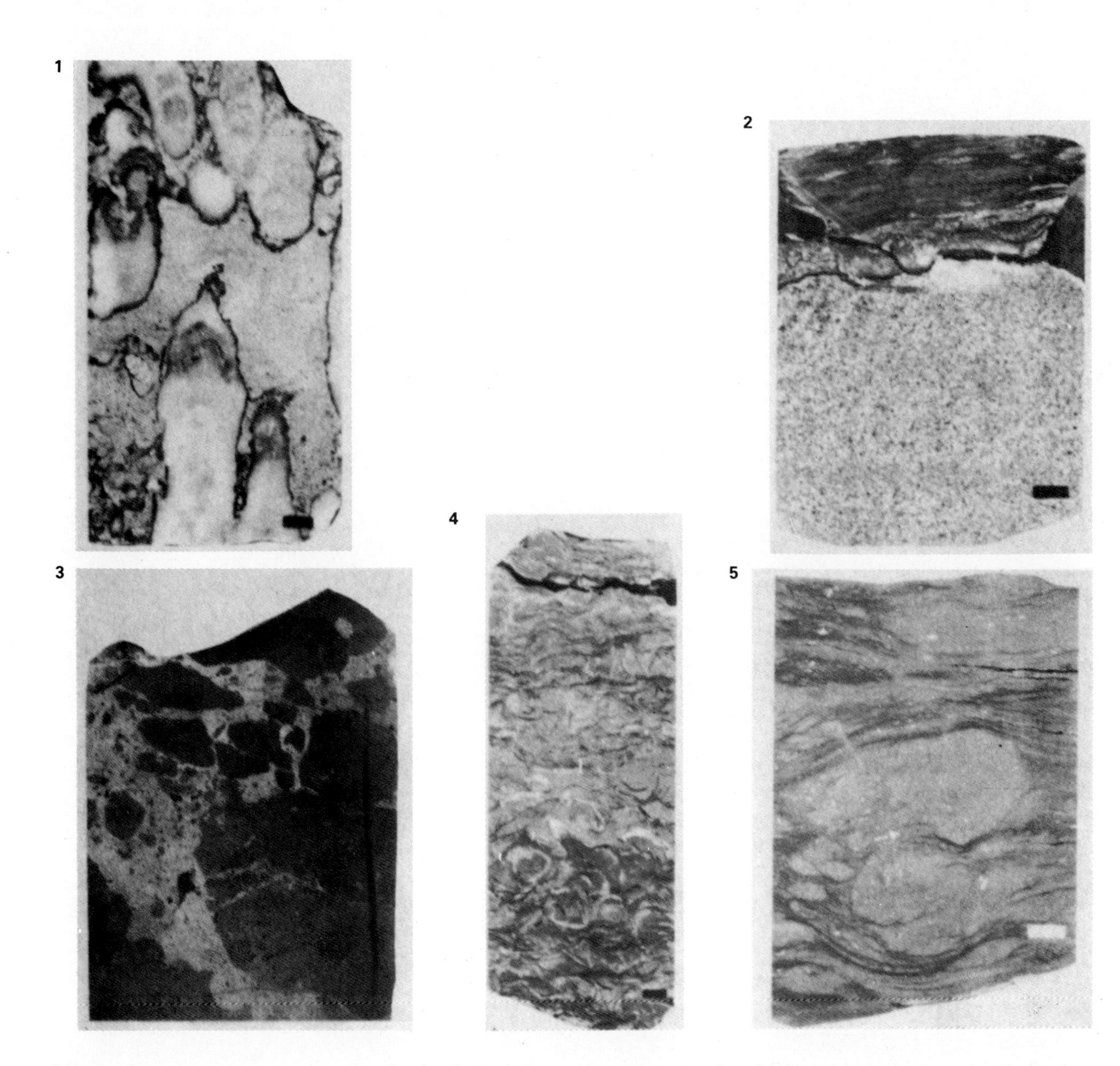

Plate B Cores from carbonate reservoir rocks showing typical examples of the categories of rock type in the Dunham classification (see p. 446).1, Boundstone; 2, grainstone; 3, mudstone; 4, packstone; 5, wackestone. (From Bureau of Economic Geology, University of Texas, *Oil Gas J*, 22 February 1982.)

Plate C Photomicrographs of common carbonate reservoir rock types taken in plane light. (Courtesy of J. R. Allan and Chevron Oil Field Research Company.)

1 Bioclastic grainstone, Jurassic, Middle East; 75×. Skeletal grains cemented by early isopachous cement. Sample contains both intragranular and intergranular porosity.
2 Sucrosic dolomite, Jurassic, Middle East; 90×. Excellent intercrystalline porosity; virtually all areas outside rhombs are pore space.
3 Pellet grainstone, Cretaceous, Texas, 40×. Grains cemented by calcite; pores filled with dead oil.
4 Oomoldic grainstone, Permian, Middle East; 40×. About half of oolite grains dissolved to form grain moldic secondary porosity. Original intergranular pores entirely filled with calcite spar cement.
5 Leached wacke–packstone, Jurassic, Middle East; 40×. Skeletal grains leached out of mud matrix to produce a good reservoir rock out of a normally tight facies.
6 Oolite grainstone, Jurassic, Middle East; 40×. Good intergranular porosity.

1

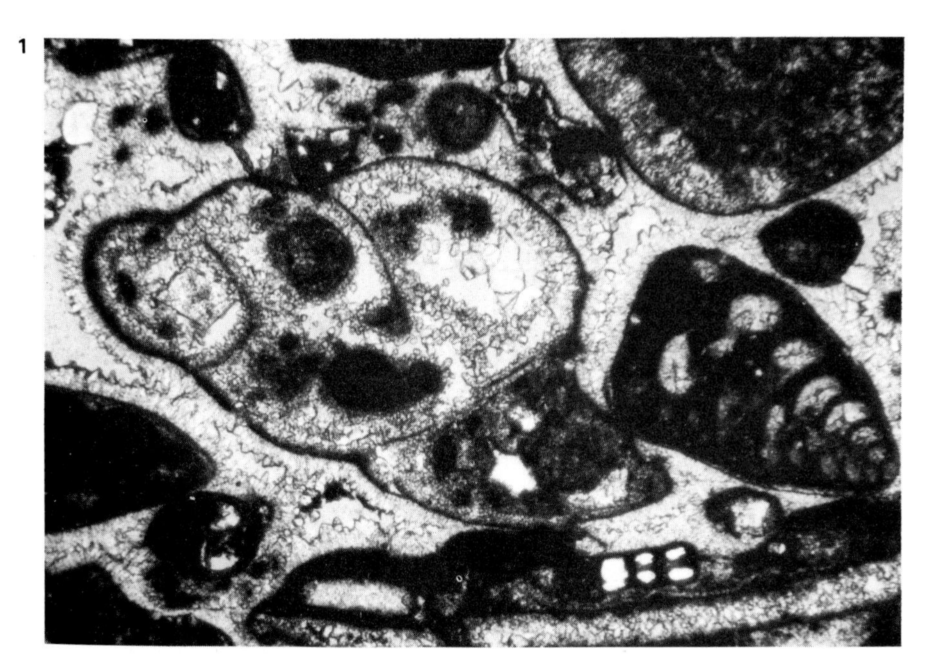

2

3

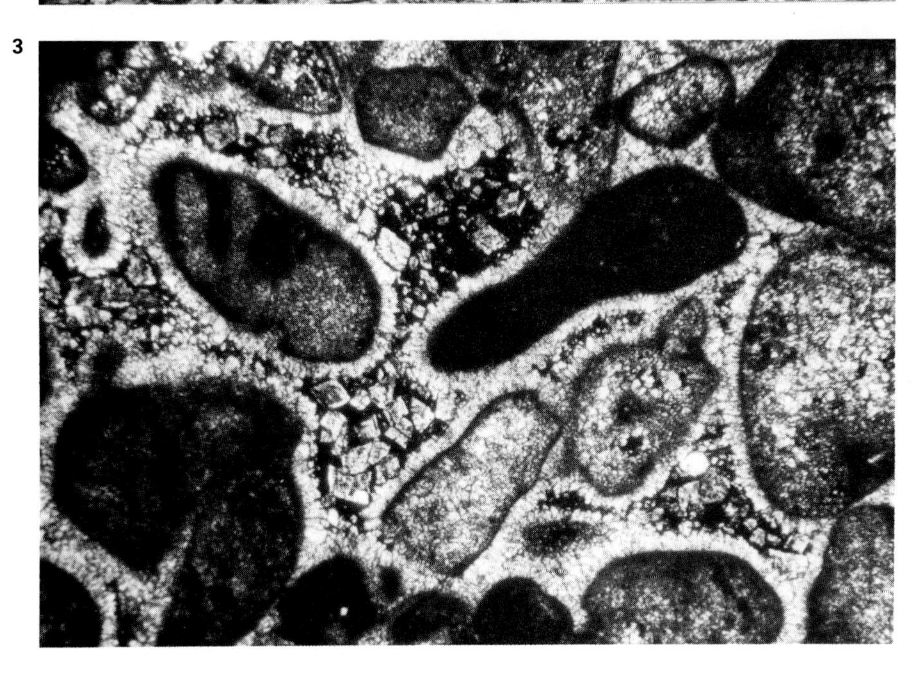

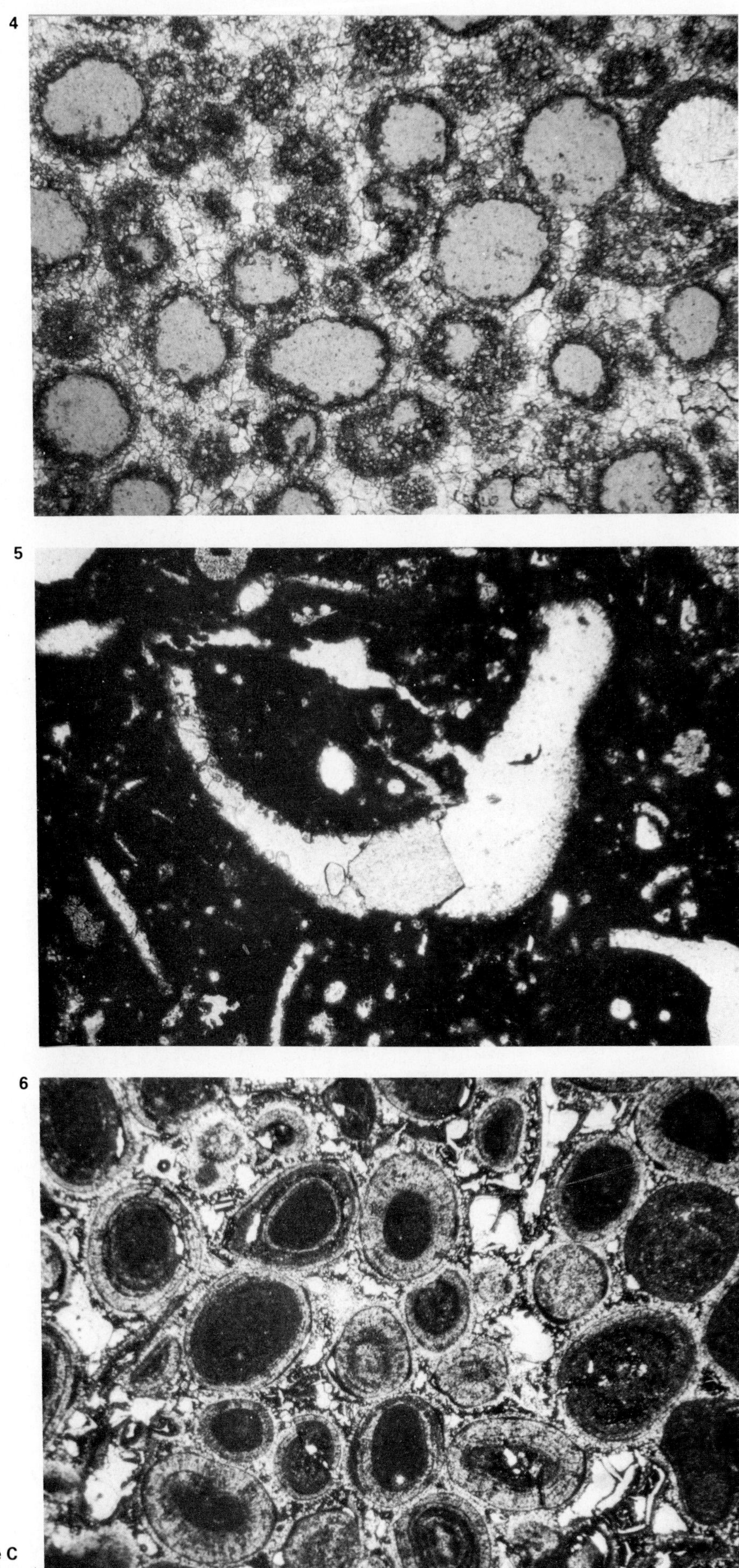

Plate C
continued

come finer and thinner upwards. The sands are poorly sorted but they are nonetheless likely to be the most porous and permeable parts of the fan. In the Bouma sequence, they commonly represent A or AE only, with the middle layers absent. These impoverished sequences, however, may lead to gross sandstone intervals of worthwhile thickness through the stacking of discrete sand lenses separated by draped, fine-grained sediments.

Fan deposits proper are habitually divided between the *proximal* or middle fan and the *distal* or outer fan (Fig. 13.34). In modern fans, the former represents principally overbank deposits of thickly bedded, ripple-marked turbidites with high sand : shale ratios but with the sands becoming thinner upwards. The upper parts of the Bouma sequence may be complete (BCDE, or even ABCD). In the outer fan, the sequence is much thinner and more thinly interbedded between sands and pelagic muds. It seldom provides good reservoir rocks. Divisions A and B of the Bouma sequence may be in the braided arrangement characteristic of distributary channels, but commonly only the upper divisions are present (CDE, for example).

Outer fan lobes have prograded farthest into the basin, and may extend laterally 15 km or more. The fine components are lost in this long transport, so the unchanneled deposits have high sand : shale ratios and the sands thicken, coarsen, and become better-sorted upwards, with relatively high porosities. The total thickness of a single lobe may be 10–50 m or more. Thickly bedded channel turbidites may prograde over the lobe deposits and overlie them. At the lobe edges the sands are interbedded with more thinly bedded, shale-rich turbidites deposited between lobes or on the abyssal plains. On the plains themselves, only muds and a few contourite sands accumulate. They have little reservoir capacity.

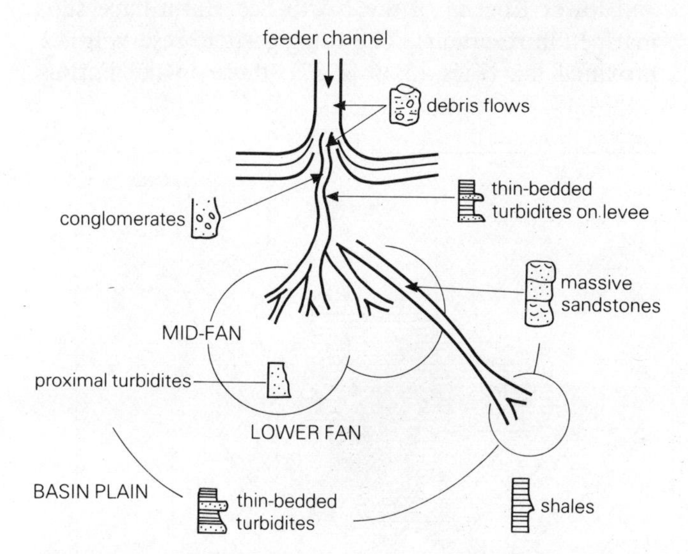

Figure 13.34 Typical facies in submarine fan deposit, interpreted for Forties field, UK North Sea, after model by R. G. Walker (1978). No relative scale implied. (From P. J. Hill and G. V. Wood, *AAPG Memoir* 30, 1981.)

Assignment of fan deposits to their proper places in deeply buried, ancient fans is not reliably made on the basis of lithofacies alone; the fan morphology must be reconstructed piecemeal from well and seismic data. As we discovered in the cases of barrier bars and reefs, ancient and modern examples are not necessarily alike. The best-studied modern fans are naturally those most complete and intact. In most ancient fans, the proximal parts are the most likely to be disturbed or destroyed by later invasions of material or by tectonism; the outermost lobes are the most likely to be preserved.

Hydrocarbons become reservoired in deep-water sands if these are introduced into generative basins. Distributary channels act as carrier beds through which hydrocarbons are channeled updip into upper-fan channel sands, or into the inner or middle fan lobes if these are adequately sealed updip. Outer fan sands are generally too thinly-bedded and contain too much mud matrix to have high permeability.

Porosity and permeability are more variable and less dependable characteristics of fan-turbidite sandstones than of any other common sandstone type. They may be exceptionally high, as they are in the North Sea examples described below (20–32 percent porosities and permeabilities ranging from only 10 mD to as high as 1.5 darcies). They may, on the other hand, be much poorer than in shallow-water sands. California's basins produce from both types. The Pliocene turbidite sands of the Los Angeles and Ventura basins and the Miocene Stevens sand of the San Joaquin Valley (Sec. 25.6.4) are cumulatively thick (because of multiple stacking) but rather poorly sorted. They provide high yields by area but relatively low yields per unit volume of reservoir (from 3.3×10^6 to almost 6.0×10^6 m^3 km^{-2} but volume ratios of 1 : 25 or even 1 : 50; these translate to 85 000–150 000 or more barrels per acre and 160–300 barrels per acre-foot). The shallow-water sands of the San Joaquin Valley, both Pliocene and Miocene without the Stevens, have much higher volumetric yields (about 1 : 20, or 385 barrels per acre-foot) but far lower yields per unit area (0.8×10^6–1.5×10^6 m^3 km^{-2}, or 20 000–40 000 barrels per acre).

Examples of petroliferous deep-water sandstones

Deep-water sandstones are obviously most likely to be deposited during low stands of sea level. Better gradients are then provided to the drainage system, so that more sands are delivered. Bottom currents become stronger, so that the sands become more thoroughly winnowed and their reservoir capacities become thereby improved.

During the time in which oil and gas were both generated and preserved, the major low stands of sea level occurred between the late Carboniferous and the early Jurassic, during the early Tertiary, and in the later Oligocene. It is in the sedimentary rocks of these epochs that we should expect to find deep-water sandstones available as reservoir rocks.

The Upper Carboniferous and Permian of the southern USA contain numerous deep-water sandstone bodies. In the basins of that region, many of these become reservoir rocks. In the Pennsylvanian of the Anadarko Basin (Fig. 13.21), scores of stratigraphic traps have been found in sandstones which grade upward from lower fan lobes, through slope-channel deposits to shallow-water shelf sands at the top; the entire sand intervals are commonly enclosed within carbonates.

Within the Permian Basin, the majority of petroliferous examples are of Early Permian age. The largest is the Spraberry Formation, confined to the Midland (eastern) Basin but oil-productive from more than 5000 wells spread over 500 km^2 (Fig. 13.35). The formation consists of several repetitions of the same cycle: 100–150 m of black, organic-rich shales and dark "carbonates", overlain by a similar thickness of very fine-grained sandstones and siltstones. Most of the "carbonates" are actually silty, dolomitic mudstones, with some limestone conglomerates; the clastics are laminated but much of the lamination has been disrupted by bioturbation. Phosphate nodules are common in the laminated siltstones. The reservoir clastics have less than 1 mD of average permeability. All the sediments were deposited about 600 m below the shelf margin; they onlap against more massive sandstones deposited on the slope.

The Midland Basin containing the Spraberry Formation in its center is rimmed by numerous fluvial and deltaic reservoir sandstones (see above). Downslope from these, the basinal fans were fed by thin, incomplete

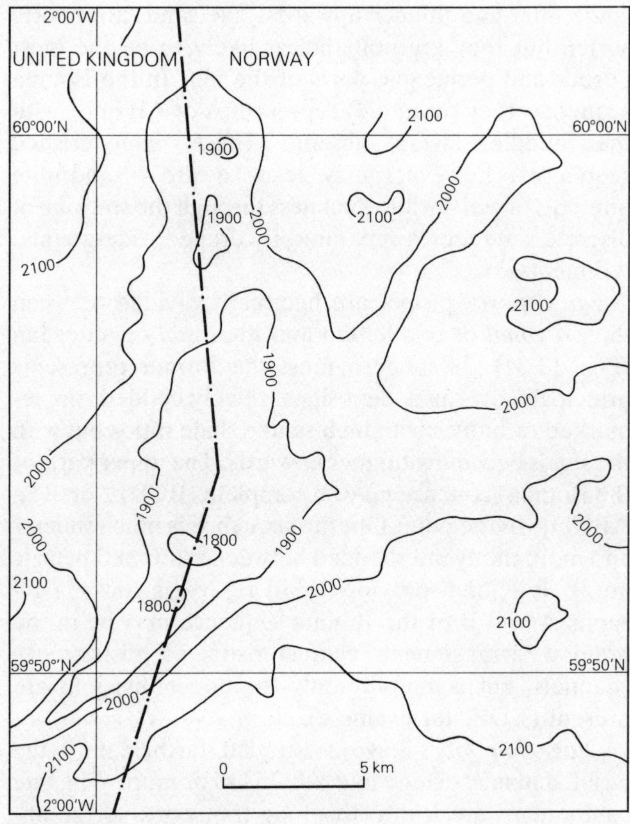

Figure 13.36 Structure contours on top of the Frigg sand in the Frigg gasfield, straddling the UK/Norway boundary in the North Sea. Contour interval 100 m. (From F. E. Heritier *et al.*, *AAPG Memoir* 30, 1981.)

(AE) Bouma sequences of numerous fan channels, and these provide a host of low-permeability reservoirs (Fig. 13.35).

Several of the large oil- and gasfields in the Paleocene and lower Eocene of the North Sea Basin have submarine fan reservoirs. The Frigg gasfield reservoir is a proximal fan (Figs 13.36 & 37); those in the Forties

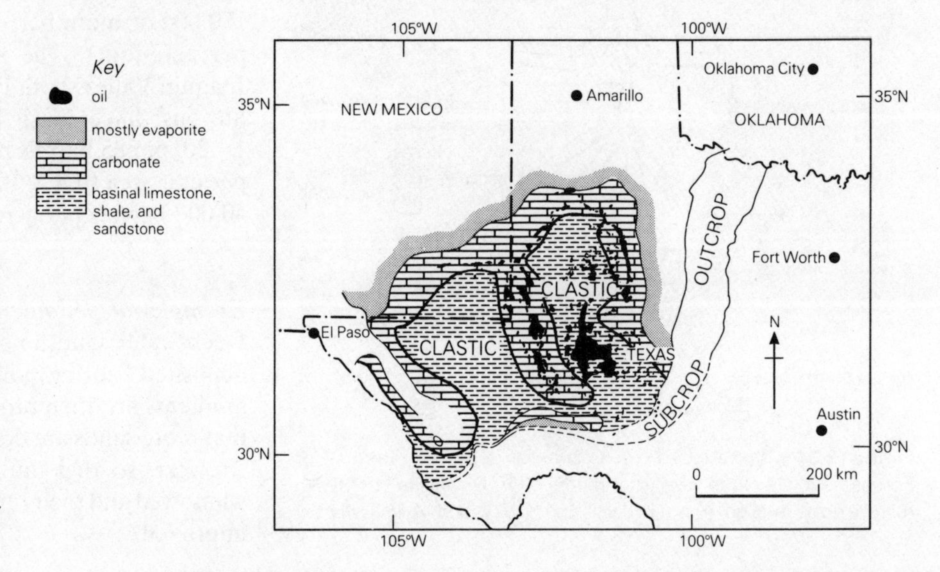

Figure 13.35 Lithofacies of Middle Permian strata in the Permian Basin of western Texas and New Mexico. The areally huge development of turbidite sands in the eastern (Midland) basin contrasts with the lack of equivalent sands in the western (Delaware) basin of this age. Oilfields shown in black. (From J. K. Hartman and L. R. Woodard, *AAPG Memoir* 15, 1971.)

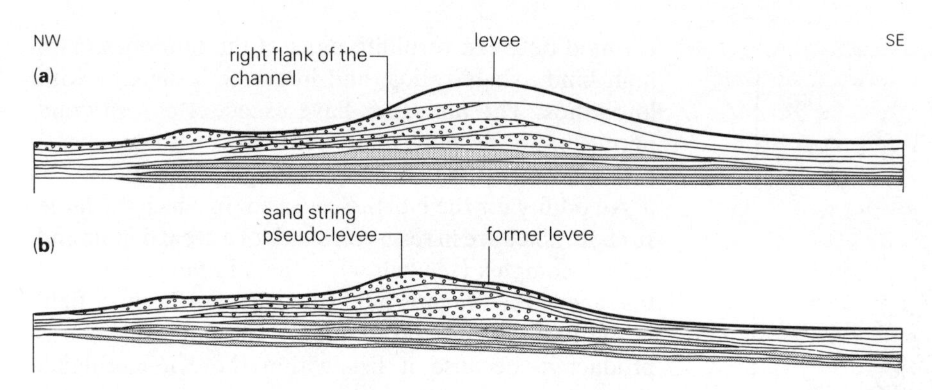

Figure 13.37 Compaction of deep-sea fan channel, illustrating development of reservoir in the Frigg gasfield, northern North Sea. (a) Levee at time of deposition, before compaction; (b) same levee after compaction; flanks of original levee are now topographic highs. (After F. E. Heritier *et al.*, *AAPG Memoir* 30, 1981.)

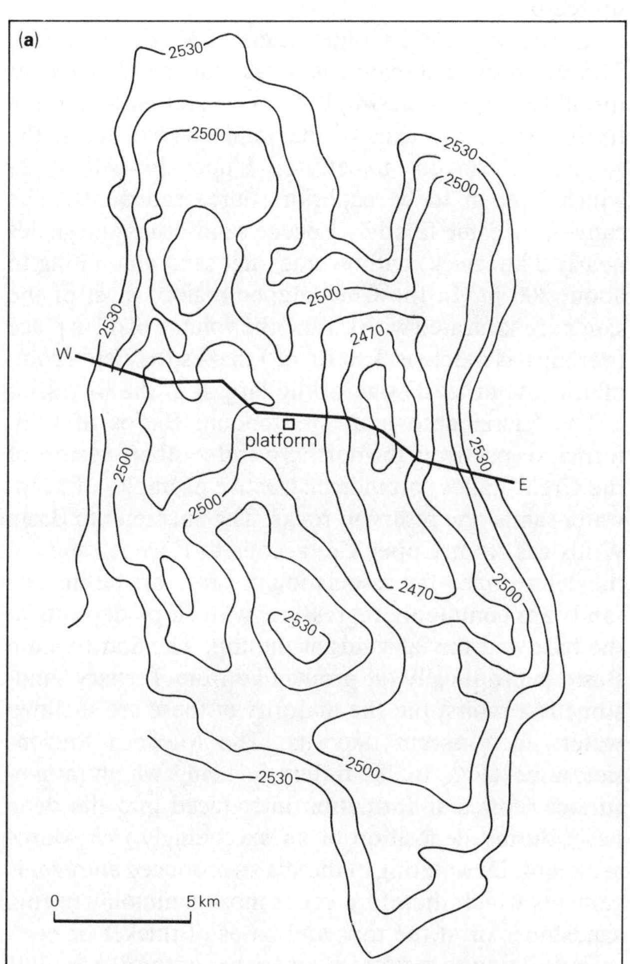

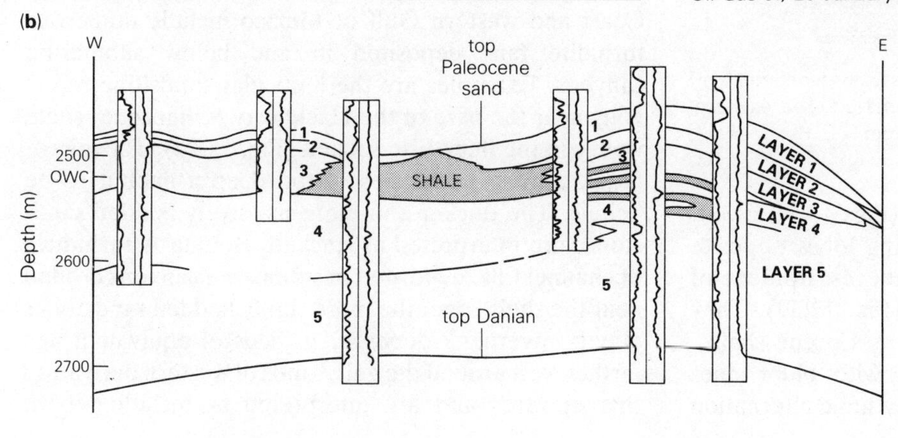

Figure 13.38 Montrose field, mid-North Sea Basin. Reservoir rock in Paleocene turbidites, with two fan lobes separated by a shalier facies, causing two separate oil columns. (a) Structure contours on top of the Paleocene sandstone reservoir, showing the bilobate pattern; (b) cross section showing the two sand lobes and the intervening shale. Contour interval 15 m (subsea). (From M. Bishlawi, *Oil Gas J.*, 26 January 1981.)

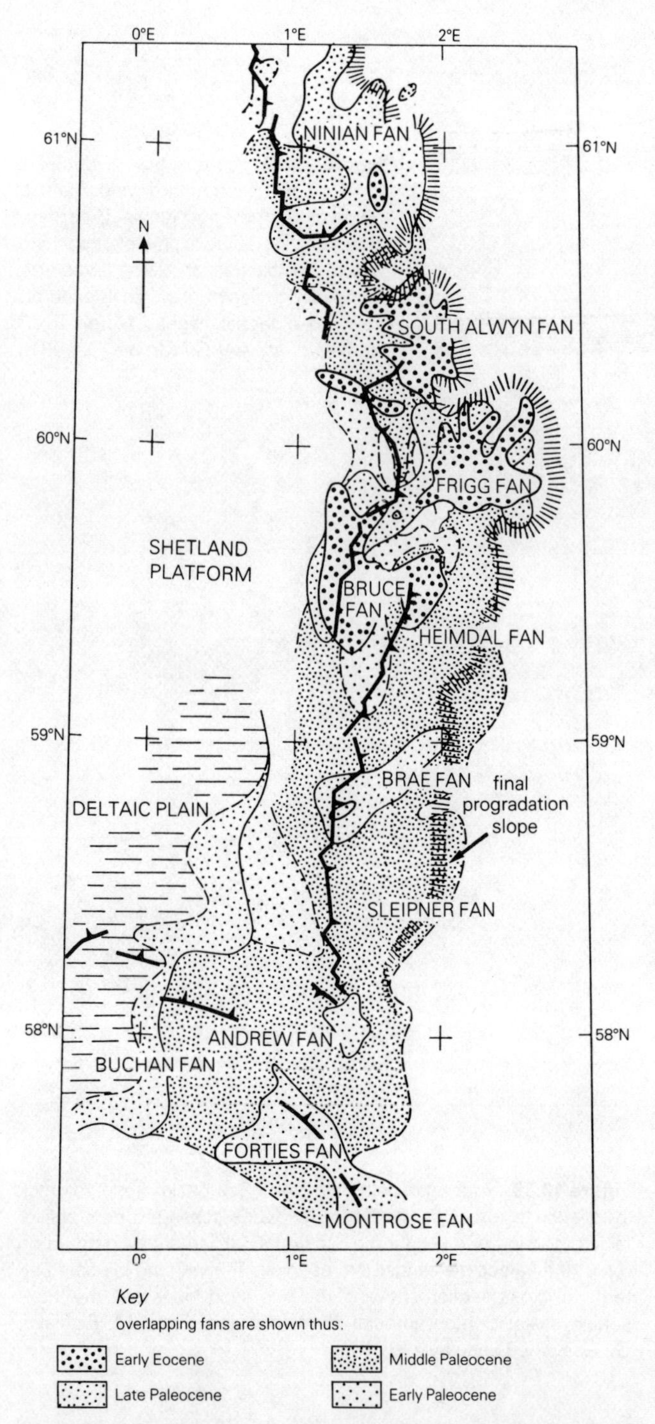

Figure 13.39 Paleocene to Early Eocene submarine fan complex off eastern margin of Shetland platform in northern North Sea, as reconstructed for the end of Early Eocene time. (From F. E. Heritier *et al.*, *AAPG Memoir* 30, 1981.)

and Montrose fields are distal fans (Figs 13.34 & 38). The fans spread as thick, overlapping lobes opposite routes of sediment descent down the escarpment of the Shetland platform to the west (Fig. 13.39). They prograde basinward and are sealed by Eocene shales. Successive sands lie side by side with older ones through lobe shifting. In places a rhythmic alternation is found between turbidite units of the fan lobes, with high sand : shale ratios, and interlobe sediments with low ratios. The fan sands have excellent porosity and permeability despite their contents of shale clasts, carbonaceous detritus, pyrite, and glauconite.

An oddity for the North Sea Basin, in which the large Jurassic fields are in reservoir sands of a great deltaic and coastal complex (see below), is the mid-fan reservoir of that age in the Magnus field (the most northerly oilfield in the basin at the end of 1982). The fan sandstone is productive because it lies within the Kimmeridgian source beds.

In part of the old Golden Lane of Mexico (see Sec. 25.5.9), erosional canyons were cut by deep-water turbidity currents during Paleocene and early Eocene times. Along the axes of maximum downcutting, the submarine canyons penetrated Upper Jurassic strata which happen to be superior source sediments. The canyons became filled by Eocene sandstones and shales nearly 2 km thick, with average net sand amounting to about 200 m. In the Chicontepec region, most of the sands are saturated with oil and the volume of oil in place (perhaps as much as 3×10^9 m^3) makes the field (volumetrically, at least) one of the largest in the world.

The Sacramento and San Joaquin Basins of California, respectively the northern and southern sectors of the Great Valley, provide instructive examples of deep-water sandstone reservoir rocks. The Sacramento Basin yields gas from Upper Cretaceous to Eocene sands of the delta-slope–fan association in a fore-arc basin. The sands are commonly regressive, with slope deposits at the base and fluvial sands at the top. The San Joaquin Basin is prolifically oil-productive from Tertiary sandstone reservoirs, but the majority of these are shallow-water, shelf-margin deposits. The Miocene Stevens sandstone (Fig. 16.38) differs in being wholly a subsurface lenticular formation introduced into the deep basin during deposition of an exceedingly rich source sediment. Dewatering of the shales produced authigenic cements which therefore occur most commonly in thin sandstones or at the tops and bases of thicker ones.

Early Tertiary sandstone reservoirs of the Texas Gulf Coast and western Gulf of Mexico include numerous turbidite fans deposited in and below submarine canyons. Examples are the lenticular sandstone reservoirs near the base of the Hackberry (Oligocene) shale wedge in the thick Frio section (Fig. 13.28). The microfauna consists of species of too deep a habitat to be deltaic. The thicker and more massively bedded sandstones are interpreted as stacked, Bouma-A turbidites of channel fills, deposited in submarine canyons eroded near the shelf edge; the more thinly bedded sandstones may be overbank deposits. In beds of equivalent age farther west around the gulf, sands of western derivation fine upwards and are interpreted as including both

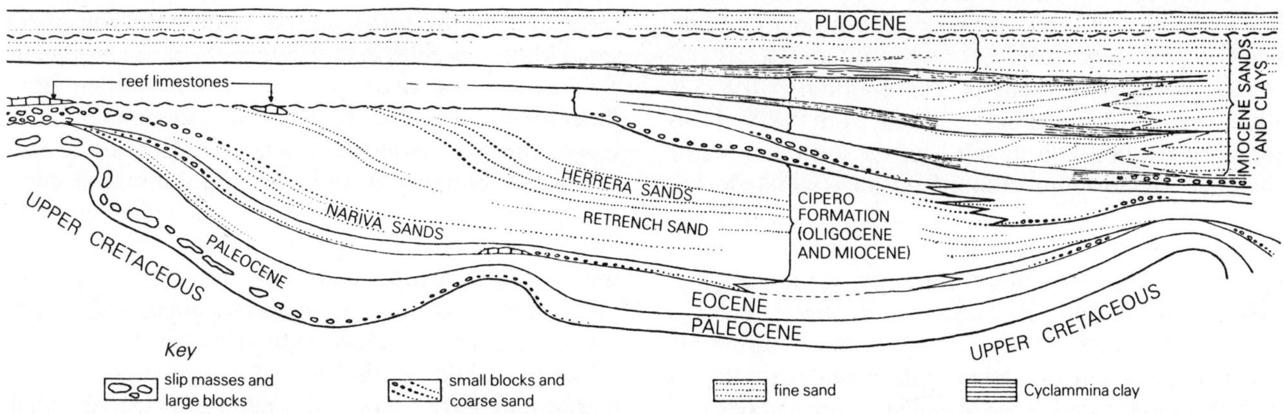

Figure 13.40 Southern basin of Trinidad, showing the introduction of turbidite sandstone reservoirs into thick, Oligo-Miocene source sediments deposited in the foredeep of the Central Range as it was progressively displaced southward. (After H. G. Kugler and J. B. Saunders, *Bol. Inf. Asoc. Venezolana Geol.*, 1967.)

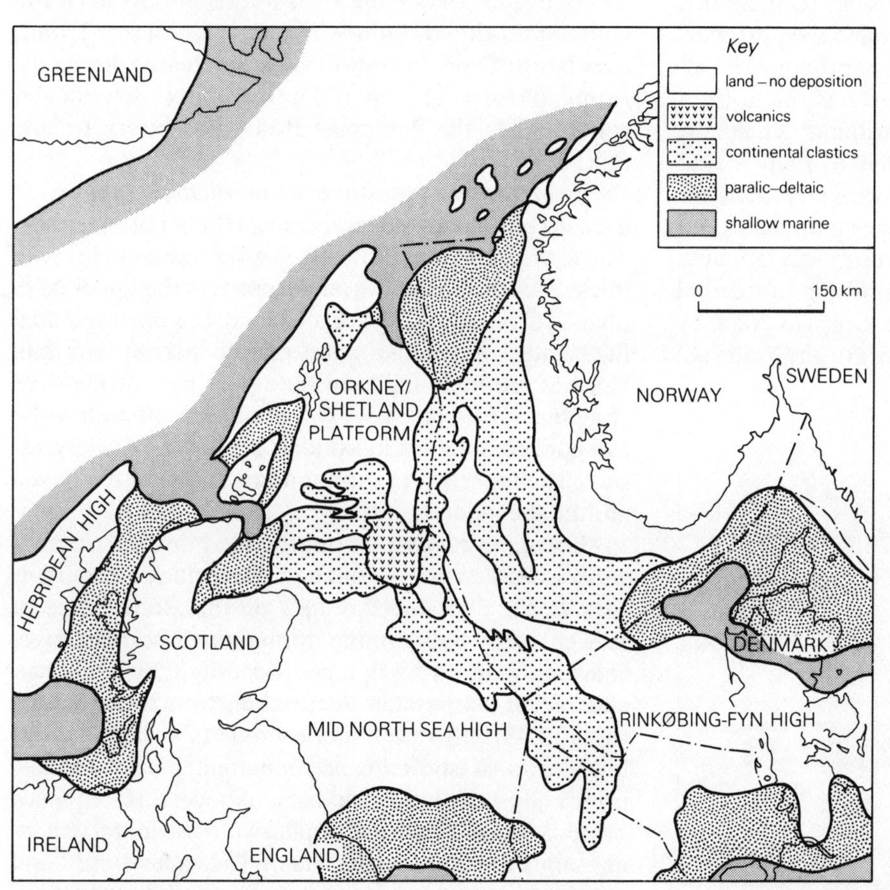

Figure 13.41 Schematic reconstruction of the paleogeography of the North Sea region in Middle Jurassic time, showing the position of the delta that was later to provide the principal oil reservoir rocks. (From G. Eynon, *Petroleum geology of the continental shelf of North-West Europe*, London: Institute of Petroleum, 1981.)

channelized mid-fan deposits and those of lobes seaward of bypassed fans.

In the Tertiary basin of eastern Venezuela and Trinidad (Chs 17 & 25), the deep-water Miocene source-shales of the Carapita and Cipero Formations enclose several thick turbidite sandstone members. The Retrench and Herrera sandstones of such Trinidad fields as Forest Reserve and Barrackpore (Fig. 13.40) and the lower reservoir sands of the Greater Jusepin area of Venezuela are examples.

Sandstone reservoirs of compound environment The four aqueous environments of sandstone deposition are intimately interrelated. The fluvial may grade down the dip into the deltaic and the deltaic laterally into the coastal. The deposits of the coastal regime may grade via uninterrupted sands through downslope channels into deep-sea fans. During transgression, the facies may succeed one another upwards, and during regression succeed one another downwards, in a single succession. Some of the very largest sandstone reservoirs are therefore compound or multiple, containing contiguous sand bodies of two or three of these depositional environments.

The Jurassic reservoirs of the *northern North Sea Basin* span the depositional spectrum from the fluvial to the deep-sea fan. During much of the period, a major river system flowed northward through a graben that underlies the present sea (Fig. 13.41). During the early Jurassic, the graben must have been occupied by an estuary; in one of the most northerly of the giant oilfields, Statfjord, the oldest productive reservoir belongs to that epoch and is a fluvial sandstone becoming marine upwards. Following a marine shale episode, a complex of deltas was created in the Middle Jurassic and reached its most northerly position, between the Shetland platform and Norway.

The "Brent group" of large oilfields occurs there in a complex of barrier-bar, deltaic, and channel sandstones. The general succession (Fig. 13.42) is progradational, coarsening upwards through micaceous delta-front sands to the coarse deposits of distributary channels. This massive sandstone body is overlain by shaly, delta-plain deposits with crevasse-splay, levee, and mouth-bar sands passing laterally into freshwater swamp shales and coals. The succession was then abruptly cut off by a renewed marine transgression depositing a second massive sand at the top. This sand in turn is overlain by Upper Jurassic marine shales which are the principal source sediments.

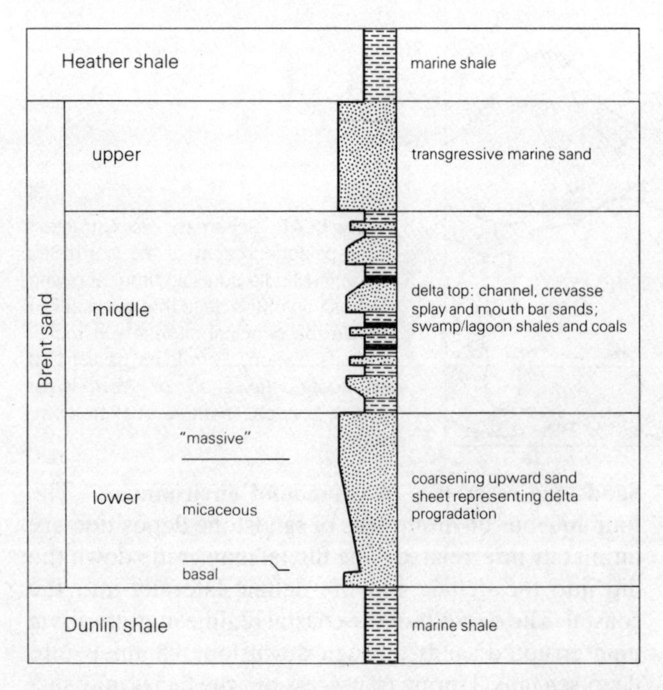

Figure 13.42 Generalized stratigraphy of the Jurassic Brent sandstone in the northern North Sea Basin. (From N. J. Hancock and M. J. Fisher, *Petroleum geology of the continental shelf of North-West Europe*, London: Institute of Petroleum, 1981.)

At the northern end of the Brent group of fields, near the mouth of the Jurassic graben, the reservoir of the small Magnus field is in an Upper Jurassic deep-sea fan. To the southwest, in a separate arm of the graben, is the Piper oilfield (Figs 16.9 & 25.3). Here the reservoir is an Upper Jurassic marine sandstone, representing a series of stacked barrier-bar and other littoral and shallow-shelf, high-energy sand bodies, laid down above wave-base and separated by thin, discontinuous siltstones and shales. The main production comes from upper offshore sheet sands and barrier-bars of the upper and middle shoreface zones, extensively bioturbated.

The probability is that Jurassic complexes of shallow-marine and barrier-bar sandstones, cut by tidal channels and backed by tidal flats, will prove to extend discontinuously from northernmost Norway, along the entire length of the continental shelf to offshore western Ireland. Gas is already known in such sandstones at Troms, near North Cape, in fault-block traps created during the rifting phase; light oil is known under very similar conditions in the Porcupine Basin off western Ireland (Ch. 25).

The McMurray sandstone of the *Athabasca "tar sand" deposit* contains all the facies except that of the deep sea. The lower part is an alluvial and estuarine complex with thick channel sandstones and deposits of the upper delta plain. This complex is overlain by tidal channel and tidal flat sands and muds; some marine barrier-bars and beaches are identifiable in the upper part of the sand complex. The complex as a whole may therefore be categorized as being mesotidal, coastal, and transgressive. The highest oil saturations are found in the fluvial and tidal channel sands.

The opposite case is represented by the Permo-Triassic Sadlerochit Formation, the principal reservoir rock in the *Prudhoe Bay field* on the Arctic slope of Alaska. The bulk of the formation is a regressive, coarsening-upward sequence deposited by a fluvial-alluvial fan-delta system encroaching from the northeast (Fig. 13.43). At the base are prodelta sediments, followed by a coastal sequence in marginal marine facies, principally stream-mouth bars. Above this sequence comes a braided stream complex with distributary channel sands crossing a delta plain. The sands have basal conglomerates and are markedly cross-bedded; those near the top of the formation provide the most favorable reservoirs. They are cut off by transgressive sediments advancing from the south.

Large productive complexes conventionally described simply as "deltaic" invariably contain important sectors properly attributable to other associated environments. In the foregoing examples, the nondeltaic components are essentially fluvial in setting. The *Niger "delta" province*, in contrast, represents the entire coastal environment. Beaches formed on top of barrier bars,

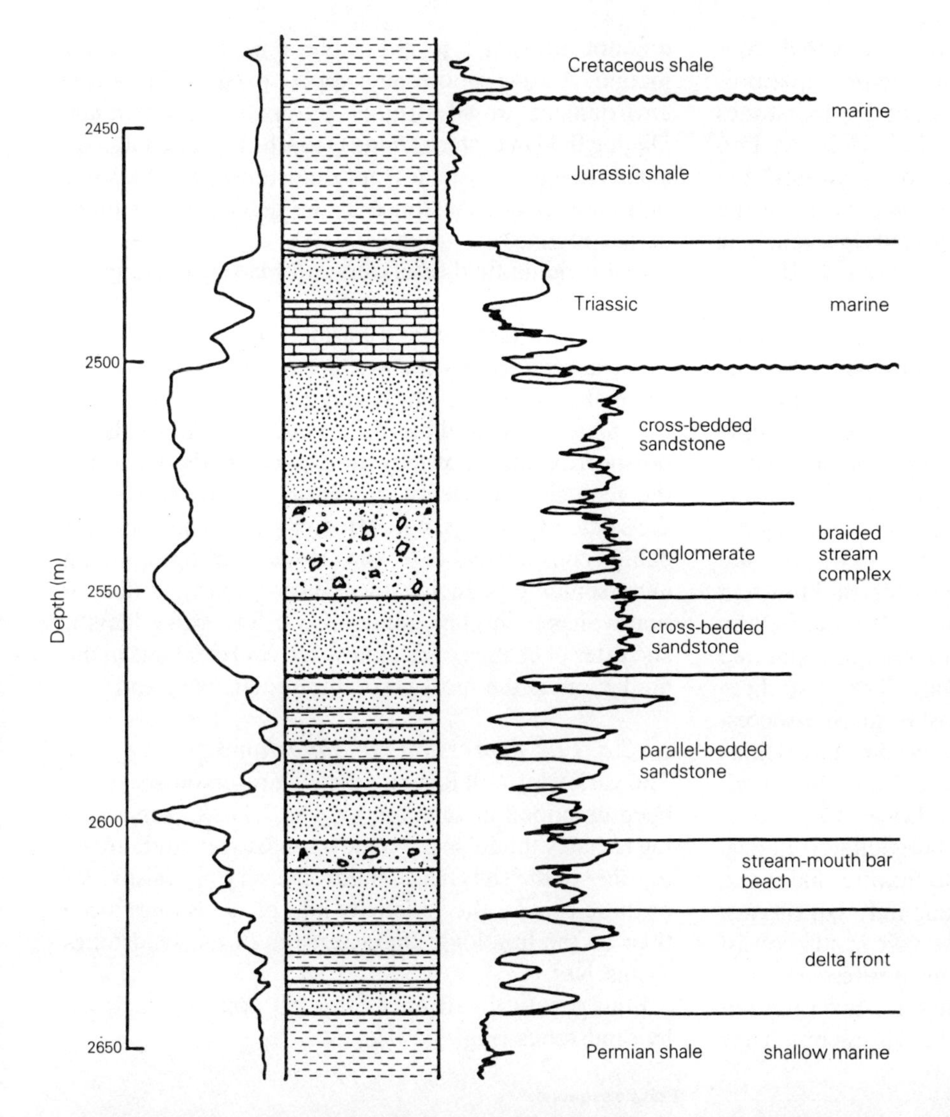

Figure 13.43 The Permo-Triassic (Sadlerochit) sandstone reservoir succession in the Prudhoe Bay oil- and gasfield, Alaska. (Based upon D. L. Morgridge and W. B. Smith Jr, *AAPG Memoir* 16, 1972, and H. C. Jamison *et al.*, *AAPG Memoir* 30, 1981.)

which had been built by waves using sand provided by longshore currents. Under regression, bars were built seaward as successions of beach ridges, with a transitional barrier foot in front and a tidal coastal plain, marsh, or lagoon behind in which clay, silt, and peat accumulated as swamp deposits. The whole complex was cut by tidal channels, which themselves contain still more sands. The principal productive sands are barrier-bar, channel, and point-bar sands, interrupted by shaly intervals deposited during episodes of transgression.

Another enormous sandstone complex occupied much of the basin of *eastern Venezuela* during Oligo-Miocene time. Sands deposited by streams flowing northward off the Guyana Shield extend continuously for more than 600 km parallel to the Orinoco River, though that river had not come into existence when the sand deposition took place. Numerous sheet- and channel-sand bodies, including point bars later filled with oil, lie within a delta-plain succession characterized by swarms of lignite beds. At the eastern end of the complex, shallow-marine shales with bar-type sands and a compound delta system form the reservoir complex in the Forest and Cruse Formations of southern Trinidad (see Sec. 25.6.5). At the northern edge of the same basin, the younger reservoir rock in the Quiriquire oilfield (see Fig. 16.83) is interpreted as an alluvial fan deposit.

The traverse from shoreline to continental slope is illustrated by the Woodbine–Tuscaloosa Groups of the *US Gulf Coast*; they are Cenomanian in age. The Sabine uplift on the Texas/Louisiana border separated two limbs of the depositional basin. On the west flank of the uplift, a strand-plain sand wedges out in the stratigraphic trap of the East Texas field. Westward, down the dip and above an unconformity, lagoonal and barrier beach facies pass into foreset and bottomset deltaic sands, then into a mud-dominated clastic wedge of the outer shelf and upper slope. Sands of this last environment are effectively turbidites, with sharp basal contacts showing sole markings and equally sharp upper contacts with organically rich shales containing foraminifera.

East of the Sabine uplift, oil occurs along the trend of

a transgressive unit of alluvial plain deposits with discontinuous, fine-grained channel sands (the principal producing horizons) alternating with grey and red siltstones and shales, passing up through deltaic deposits into marine shales. The south-flowing rivers crossed the infilled lagoon, cut across the reef complex along the edge of the older shelf, and dumped sands into a deltaic and coastal environment. There they were spread westward by longshore currents along the south front of the reef trend; these sands now constitute a quite distinct trend of deeply buried gas pools in Louisiana. Individual sands up to 300 m thick occur along growth faults.

The influence of reefs in the US Gulf Coast example reminds us that the depositional environment of compound sandstone reservoirs need not be exclusively clastic. The sands may be an alien element in a regime otherwise represented essentially by carbonates and evaporites. The largest sandstone oil reservoir known, if "tar sands" are excluded, is that in the Burgan field in *Kuwait*. The reservoir consists of four sands separated by shales and *Orbitolina* limestones. The first three sands are part of a fluctuating, coastal, marine succession; the lowest sand, a major reservoir, has deltaic components. Over a wider area, both the mid-Cretaceous (Burgan) sands and the Lower Cretaceous (Zubair) sands represent coastal–deltaic–alluvial complexes. They pass eastward into marine shales and carbonates, westward into continental sandstones. Sections above and below the reservoir sands consist almost entirely of carbonates and evaporites.

Combination sandstone reservoirs are not only not restricted to the clastic depositional environment. They are not restricted to the environment of a marine margin. A huge sandstone complex from the lacustrine environment provides the reservoirs at the supergiant Daqing field in eastern China (see Ch. 17). Fluvial sandstones merge into delta-front sandstones, and between these are sheet sandstones of the delta plain or inter-deltaic regime.

If individual field reservoirs can span such a range of clastic depositional environments, it will be obvious that petroliferous districts and basins will normally do so. The preferred reservoirs are likely to be the beach and channel sandstones, which may be interlocked in intricate patterns beyond pre-drill prediction. All sands are prospective until shown by the drill to be otherwise, and the geologist must learn to identify the components of the section from logs, cores, fossils, and whatever other data become available to him. He must also acquire an understanding of their typical dispositions and dimensions. Shapes in plan, ratios of width to strike length, and rates of change of thickness are vital elements in the prediction of the most favorable productive trends.

13.2.8 Log expressions of sandstones

The variety of well logs run through reservoir rocks has been described in Chapter 12. The "classic" sandstone log remains the SP log. As most sandstones contain one or other of the subsurface fluids, the varying resistivities of these control the resistivity logs of sandstones more than do the lithologic characteristics of the sandstones themselves.

Four genetically distinct E-log patterns are displayed by sandstones (Fig. 13.44):

Chenier beach

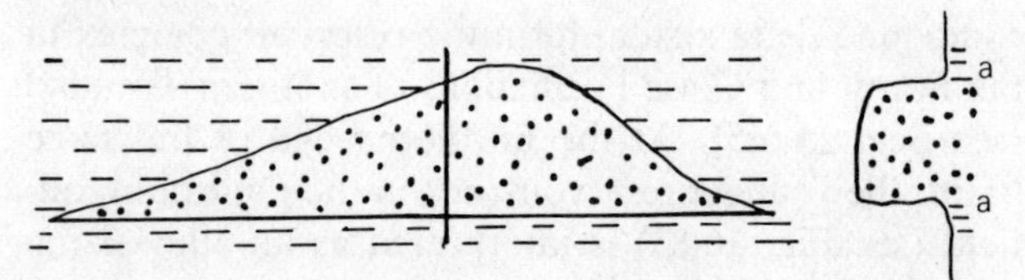

a/a contact: abrupt upper and lower contact; sands generally are well sorted by wave action

Deltaic sequence

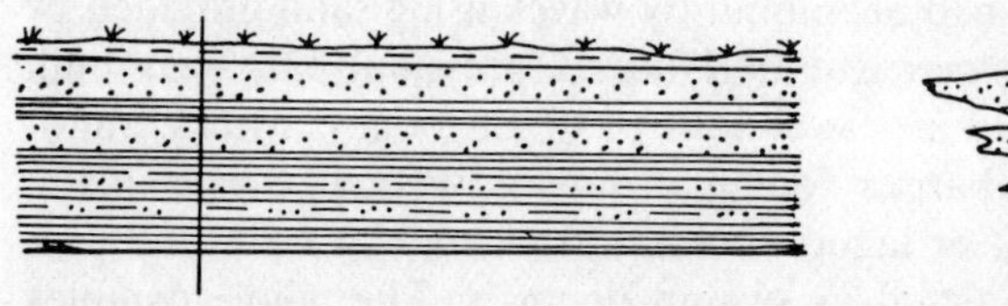

a/b contact: interbedded sand–shale sequence; the sands are fine grained

Barrier bar

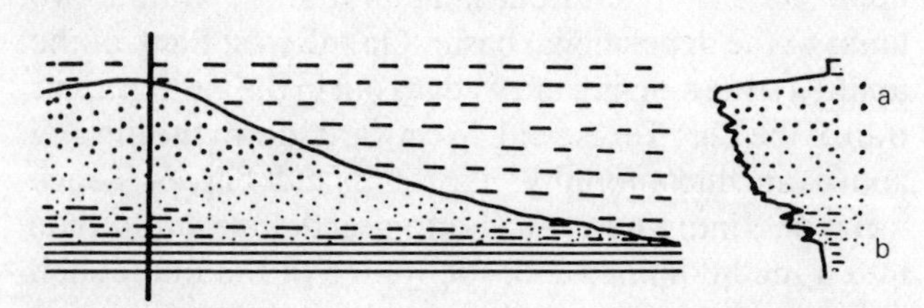

a/c or a/b: grain size increases upward in section; graduation in grain size is probably related to increasing wave energy with decreasing water depth as the bar is built up

Channel fill

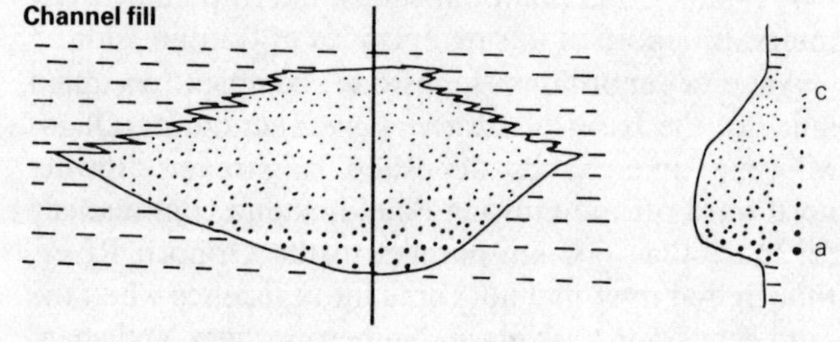

c/a contact: characterized by an abrupt lower contact and gradual upward decrease in grain size

Figure 13.44 Typical geometries of SP curves for various sandstone environments. (From P. G. Tizzard and J. F. Lerbekmo, *Bull. Can. Petrolm Geol.*, 1975.)

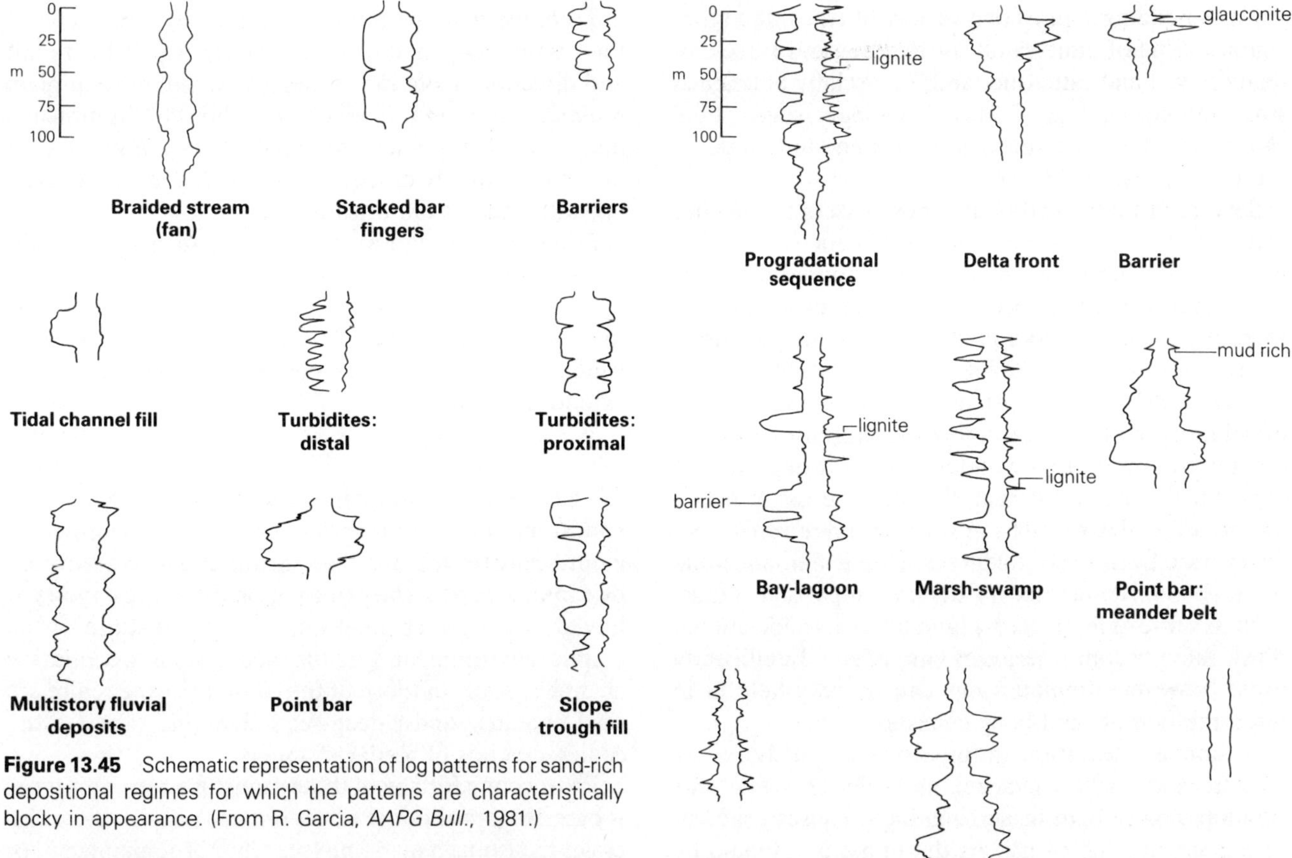

Figure 13.45 Schematic representation of log patterns for sand-rich depositional regimes for which the patterns are characteristically blocky in appearance. (From R. Garcia, *AAPG Bull.*, 1981.)

Figure 13.46 Schematic representation of log patterns of a variety of depositional environments in which sand–shale sequences are developed. (From R. Garcia, *AAPG Bull.*, 1981.)

(a) The "blocky" or bell-shaped curve (Fig. 13.45) reflects thickly bedded or massive sandstones with generally uniform lithology and few nonsandy interruptions. Some barrier-bar sands (especially those of tidal channels through the bars), fluvial channel sands, distributary-channel and bar-finger sands in the outer parts of deltas, and rapidly deposited turbidite sands close to their point sources (proximal turbidites) display blocky log patterns, especially if they are stacked one upon the other through rapid repetition of the sand-prone regime.

(b) The upward-flaring log expression is characteristic of sandstones which coarsen upwards: beach and barrier-bar sands of the upper shoreface; delta-front clastic wedges; sands of bar fingers and stream-mouth bars in deltas; and sands of some outer fan lobes deposited in deep water.

(c) The upward-narrowing expression is displayed by sandstones which fine upwards: most sands deposited by rivers (fluvial channels sands, point bars, alluvial fans, and distributary-channel sands within delta plains), and channel turbidites leading to deep-sea fans.

(d) The jagged, serrate, digitate, or "shaly" pattern reflects more thinly bedded sandstones alternating with silty or shaly interbeds (Fig. 13.46). Such alternations are characteristic of lagoonal, marsh-swamp, and some delta-front sands; of thin turbidites carried farther from their sources (distal turbidites); and of the *edges* of all nonchanneled sand bodies, such as barrier bars and submarine fans.

Combinations of the blocky and the serrate patterns are typical of progradational deltaic successions, and of floodplain deposits of low-gradient rivers.

Such combinations are exaggerated on radioactive logs, especially those from the gamma-ray device (Fig. 12.8). The sands, having low natural radiation, show large leftward deflections, sharply interrupted by rightward (positive) interdigitations for the more radioactive shales or bentonites.

13.2.9 Sandstone reservoirs on dipmeter logs

The geologist must always remain aware that sandstones are inherently likely to possess *depositional dips* high enough to be obvious. These dips may be reflected on a big scale, as in the progradational ("foreset") deposition of delta-front and shelf-edge sands (Fig. 13.18), or on the smaller scales of fine cross bedding.

Cross bedding itself is of a variety of types: the types characteristic of dune sands, of marine sand ridges, of many fluvial and estuarine sands (especially of braided stream deposits; Fig. 13.9), of bar-finger sands, tidal channel sands, barrier-bar sands, and the sands of deep-sea fan lobes (Fig. 13.37).

Several of these sand-body types possess the further characteristic of elongate, string-like shape. This shape implies a further component of depositional dip, perpendicular to the linearity. The *continuous dipmeter* is therefore a valuable tool in determining the genesis of such bodies (see Ch. 12 & Fig. 12.26). Caution in its use is essential. Dipmeter logs may provide data susceptible of solutions both unique and convincing for *structural* (acquired) dips, but they seldom do so for depositional dips. Furthermore, original dips may be confused by bioturbation, destroyed by reworking and redeposition, or reversed by tilting or folding. If a lenticular sandstone body is wholly enclosed in thick, compactible muds, even its sense of lenticularity (and hence its depositional dips) may become seriously modified. Intelligently used, however, dipmeter logs can be very helpful in interpretation of sand-body genesis.

A simple illustration distinguishes an offshore bar sand from an onshore channel sand (Fig. 13.47). A *bar*, standing higher than its surroundings, causes drape in the sediments that overlie it; the drape is revealed by downward increase in the flank dips of the post-bar sediments. Within the bar itself, the dips are outward towards its flanks. A fluvial *channel*, incised into underlying rock and convex downwards, causes no drape in overlying sediments and consequently no depositional dip there. In the channel sandstone itself, depositional dips are inward towards its center. The dips decrease upwards, from cross-laminated to ripple-laminated.

Point bars in river channels are revealed by a single direction of cross-bedding dips on azimuth–frequency plots. *Longitudinal bar sandstones* show multiple dip directions, because a stream goes around both sides of such a bar and may flow directly over it.

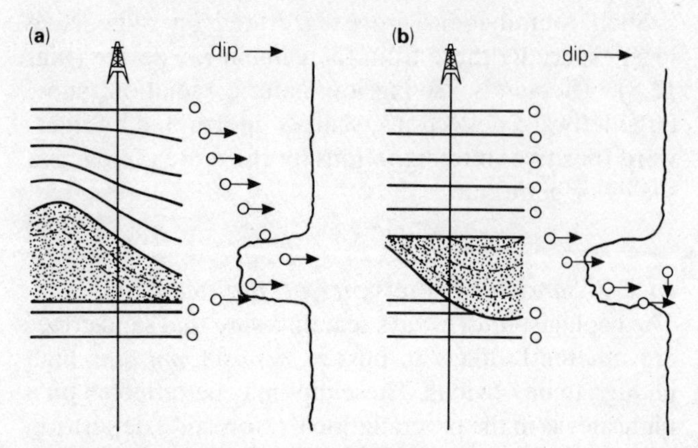

Figure 13.47 Dipmeter distinction between bar (a) and channel (b) linear sandstone bodies.

Turbidite channels normally create disconformities at their bases. Dips are likely to change, in both amount and direction, from low or negligible in the basin-plain sediments below to relatively high dips (10° or more) in the channel-fill sandstones above. *Fan lobes* create draping in overlying sediments, with dips decreasing upwards much as in offshore bars.

Proper interpretation of dipmeter data may thus indicate in which direction the axis of a sand body lies, and the direction at right angles to this direction represents the approximate strike of the linear body and provides valuable assistance to the geologist in his attempts to extend his field.

13.2.10 Prospecting for channeled sand bodies

Sandstone bodies of a variety of depositional environments share the feature of occupying channels excavated by running water. The principal sandstone reservoirs of fluvial origin are channel sands; many of those of the deltaic environment are the deposits of distributary channels; some of those of the coastal marine realm are tidal channel sands; deep-sea submarine fans are the coalesced ends of feeder channels.

Prospecting for any of these sandstone reservoir types is based on appropriate variations of a simple basic principle: the isopach mapping (see Ch. 22) of an interval of the stratigraphic section that Daniel Busch called a *genetic increment of strata* (GIS). A GIS is a vertical sequence of strata (including reservoir sandstones) within which all the lithologic components are the products of an integrated depositional cycle. It extends from the "floor" on which the sand-bearing beds were deposited (an unconformity, a facies change, or simply the base of the mapped sand interval) upwards to a time-marker horizon identifiable on well logs.

An isopach map of such an interval is in effect a paleotopographic map of the base of the sand body occupying the channels, like a plaster cast of the channel fill. Isopach maxima trace the axes of the channels, be they paleovalleys on land or submarine channels in the continental slope. The isopach data should be confirmed by structure contours on the base of the sand bodies; structurally low belts also reflect the axes of the channels.

If the only reliable time-datum horizon is *below* the sand sequence of interest, the interval mapped will extend *downwards* from the sand-base or unconformity to the datum, and the isopach *thins* will represent the channels or paleovalleys.

The isopachs of individual GISs reveal the three-dimensional shape of the river, delta, or submarine fan system during the deposition of that particular sandstone interval. Isopachs of a succession of GISs reveal the configuration of the entire depositional area during that time interval, provided that the GISs mapped are truly successive and not separated by unconformities.

It scarcely needs emphasis that many assumptions and much guesswork are unavoidable unless a good deal of subsurface control is already available. The surface over which the drainage system operated, the strand zone across which a delta was constructed, or the shelf and slope across which a fan prograded need to be reconstructed, so that the apices of distributary systems, the heads of submarine canyons, and other critical points can be located. Care must be taken to allow for any later structural deformation. Ideally, enough control should be available to permit the construction of a grid of stratigraphic cross sections, so that correlations can be assured and correlatable low points in their datum planes compared with the isopach maxima and the lows on the structure contour maps.

13.3 Carbonate reservoirs

We have observed (Ch. 4) that carbonate reservoir rocks were relative rarities until the mid-1920s. The petrographic study of carbonate rocks languished accordingly. Through the first half of the 20th century, geologists understood less of the physical and geometric properties of carbonates than of any common igneous rocks.

The most impressive and interesting aspects of limestones was their fossil content. Before the wide adoption of the modern rules of stratigraphic nomenclature, scores of limestone formations were named for their fossil contents. Some of these represented autochthonous shell accumulations, like the *Daonella limestone* of the Alpine Triassic. Many more consisted of transported shells or other skeletal fragments; crinoidal limestones are the most familiar examples. Among names coming to be accepted as "formation" names in Europe were the *Schwagerina limestone* in the Lower Permian of the Soviet Union; the *Muschelkalk*, constituting the Middle Triassic in Germany and largely the cause of the name Trias; *Aptychus limestones* in the Jura-Cretaceous of many Tethyan localities; and the *Granulatenkreide* and *Mucronatenkreide* in the uppermost chalk of northern Germany, named for two belemnite species.

Foraminifera became especially popular for this custom, and more so once thin-section studies were extended to limestones. The *Fusulina limestones* of many parts of the world during the Carboniferous and Permian; *Orbitolina limestones* interbedded with reef facies in Tethyan and Caribbean successions, especially during the worldwide Cretaceous transgression; *Globigerina limestones* in basinal or fore-reef zones; *Miliolid limestones* in the equivalent neritic zones; and *Alveolina limestones*, especially in the Eocene, were all household terms to stratigraphers and paleontologists and to many petroleum geologists. French geologists still call the Eocene Series *"le Nummulitique."*

Hence limestones were regarded essentially as organic rocks, save only for the small proportion that could be shown to be of chemical origin. The only widely recognized limestone *texture* independent of organic derivation was the oolitic. All this changed abruptly about 1950. Textural studies of young, unweathered and unaltered carbonates from the subsurface (most notably by F.R.S. Henson, working in the Iraq oilfields) and of modern carbonates from the surface (especially by L. V. Illing in the Bahamas) revealed the vast variety of limestone structures and textures, and brought recognition of the significance of wholly clastic aspects among them. Limestones, in fact, are in many cases simply shales, sands, and conglomerates made of calcium carbonate. Rapidly the studies were extended worldwide and through the whole stratified column, from ancient carbonates that had undergone significant diagenetic changes, like the Cambro-Ordovician Ellenburger Formation in West Texas, to modern carbonates in the Pacific, the Persian Gulf, and the Caribbean.

Despite the observation above on the clastic character of many carbonate bodies, carbonate rocks are not just a chemically different species of terrigenous clastic rock. Most clastic carbonate sediments are deposited at or close to the site of creation of their clasts. Mutual abrasion of the grains, a characteristic of transported terrigenous clastic sediments, is less common in carbonates. Sorting is also uncommon; the "best sorted" of modern carbonate rocks are oolites, in which the "grains" are all of essentially the same size and shape. Oolites are not in fact sorted at all; the "grains" were formed with the shapes and sizes they possess now.

13.3.1 Formation of carbonate reservoir rocks

Most carbonate rocks begin as *skeletal* assemblages. They are the accumulations of the remains of carbonate-secreting animals and plants. Though animal fossils are far more familiar to most geologists than plant fossils, plants are more important than animals in the initial creation of carbonate deposits of the types now providing reservoir rocks. Most important are the *thallophytes* or nonvascular plants, and by far the most important among these are the blue-green and red *algae*. These constitute the primary food source for all organisms and are therefore the *sine qua non* for the creation of carbonate rocks in the first place.

The dominance of carbonate-prone environments by marine plants puts some necessary constraints on the environments. As sunlight is necessary for photosynthesis, the depositional sites must be shallow and the waters warm and of essentially normal marine salinity. The waters must also be oxygenated at the surface, though they may be reducing quite shortly below it. The supply of terrigenous clastics must be low; turbid waters are inimical to carbonate accumulation and especially to reef growth. Carbonates were therefore much more widespread in the past than they are today or have been since early Tertiary time. We live now in post-orogenic, post-glacial times. Though most of the present equatorial girdle lies over ocean basins, there is no equatorially oriented ocean; the dominant marine trends and gradients run north–south.

Blue-green algae themselves do not require oxygen, but only sunlight; nor do they produce skeletons in the strict sense. Instead, they form mucilaginous mats on the floors of water bodies; these mats capture carbonate particles and grains of all types. These include a great variety of *nonskeletal* components which may come to comprise the quantitatively larger part of the final carbonate rocks. The nonskeletal components may or may not be derived directly from organisms; they are "nonskeletal" in the sense of not being visibly or detectably skeletal. Deep-sea carbonate oozes, like the widespread *Globigerina* ooze, are nonskeletal to the naked eye but are better considered *microskeletal*.

The principal nonskeletal components of carbonate reservoir rocks are described below:

(a) *Lime mud:* This is the descriptive, textural designation for all carbonate sediments of essentially clay-particle grain size (less than 0.06 mm or 62.5 μm). The "mud" may be directly organic, by accumulation of the remains of microscopic organisms; it may be wholly detrital. Commonly, however, it is precipitated directly from warm sea water as needles of aragonite. Marine waters near the surface are slightly supersaturated with $CaCO_3$ because of their content of dissolved organic matter, which inhibits precipitation of carbonate minerals. The dissolved organic matter must therefore be precipitated first, so lime muds are invariably rich in it. The "muds" themselves are then precipitated by any agency acting to remove CO_2 from the system and thereby increasing its pH – algae themselves, in sunny, warm waters, or simple rise of temperature in low latitudes, because carbonate solubilities decrease with rising temperature.

(b) *Coated grains* formed by deposition of $CaCO_3$ around any nucleus. *Oolites* consist initially of minute aragonite crystals tangentially arranged and embedded in organic matter. The crystals are later replaced by radial crystals of calcite. Oolitic limestones have an unexplained stratigraphic nonuniformity. They were common in the Mississippian and worldwide in the Jurassic, but they are not typical of pre-Mississippian systems in which carbonate reservoir rocks are exceedingly abundant, and they were nowhere near as prominent in the Cretaceous or Tertiary as they were in the Jurassic.

(c) *Fecal pellets*, most commonly formed by worms ingesting lime mud to feed on its content of organic matter.

(d) *Lumps*, a portmanteau descriptive term for any aggregation of grains. The most familiar is grapestone, an assemblage of large, spherical grains.

(e) *Detrital grains*, called intraclasts by Robert Folk. These may be abraded and redeposited, not necessarily within another carbonate. *Lime sands* are exceedingly important reservoir rocks (see Sec. 13.3.3).

13.3.2 Carbonate rock classifications

A host of carbonate rock classifications has been put forward, especially during the late 1950s and 1960s. Only a very brief commentary can be made upon a few of them here. The classification by Robert Folk in 1959 was based upon the relation between the framework elements (transported and untransported elements being separated) and the material holding them together. The principal objection to Folk's discriminating classification was undoubtedly the unwieldy nomenclature. The geologists of Aramco stressed the properties accruing from mechanical deposition including particle size. In 1966 the Russian petrologist V. P. Maslov concentrated upon "organogenic" limestones and based his classification on the environment of the organisms responsible (benthonic, planktonic, or coprolitic). Francis Pettijohn, in 1975, proposed an amalgam of genetic and descriptive limestone types in an attempt to relate them to their environments of deposition.

Among North American petroleum geologists, at least, the classification proposed by R. J. Dunham in 1962 became widely used and understood. It had three obvious features in its favor: it was designed by a petroleum geologist for use by petroleum geologists; its emphasis was upon aspects of depositional texture that control porosity and permeability, which are what really concern oilmen; and its terminology was both splendidly simple and capable of easy extension to suit special conditions. As so extended, by Ashton Embry, Ed Klovan, and others, Dunham's subdivisions are these (Plates B & C):

(a) *Boundstone:* Autochthonous carbonate rocks in which the original components were bound together during deposition. If a rigid framework is formed by massive fossils in positions of growth, the rock is a *framestone*; stromatoporoids, large corals, and rudists are the principal contributors. Tabular or lamellar fossils of encrusting or binding organisms (like many algae), supported by matrix, form *bindstone*. Stalk-shaped organisms, such as bryozoa or coralline algae, act as baffles against water flow and create *bafflestone*.

(b) *Grainstone:* The original components were not bound together during deposition, but the texture is grain-supported with essentially no mud; the nature of the grains is of secondary concern.

(c) *Packstone:* The texture is grain-supported but contains lime mud; the grain support may be either original or acquired. If more than 10 percent of the "grains" are larger than 2 mm, the rock is a *rudstone*. Rudstones are the limestone conglomerates of older terminology.

(d) *Wackestone:* The grains "float" in the mud. If the "floating" grains include many larger than 2 mm, the rock may be called a *floatstone* (another variety of lime-stone conglomerate).

(e) *Mudstone:* Essentially lime mud only.

The spatial relations of several of these rock types within a single Paleozoic reef are illustrated in Figure 13.48. Other categories will no doubt be proposed, retaining the same simple nomenclatural foundation.

All categories may be given appropriate prefixes to denote grain type, fossil contribution, or whatever. A wackestone, for example, may contain either benthonic or pelagic microfossils or both. There is complete gradation between categories, which are essentially descriptive. Carbonate deposits may change from one category to another; many packstones, for example, were originally grainstones which became infiltrated by lime mud. Carbonate units that fall into no individual category but partake of more than one may simply be described as such, without the invention of a new name. For instance, most of the carbonate rocks in Iran (including the productive Asmari limestone) are combinations of packstone and wackestone. Stromatolites may be both mudstone and boundstone. All carbonate rocks that have lost most or all of their depositional textures are simply described as *crystalline*.

13.3.3 Carbonate rock depositional systems

Favorable and unfavorable facies distributions within carbonate regimes depend in the first instance on the paleobathymetric profile of the depositional area. Recent and modern carbonates are deposited in three readily distinguishable regimes, though there exist intermediate situations involving aspects of all three.

Differentiated carbonate shelves During transgressions, gently sloping shelves and platforms become covered by shallow-water carbonate sediments with little detrital input. Highly biogenic detrital limestones of all types, including the oolitic and the pelletoid, build up the shelves. Algal and foraminiferal limestones, including patch reefs, are widely developed. Intrashelf basins contain more fine-grained carbonates which may be rich source sediments (see Ch. 8).

Relatively continuous reefs may grow along the shelf edges, facing abruptly demarcated basins. Wave-base intersects the shelf margin. The reef fronts are the loci of maximum depositional energy; grainstones are concentrated there, whereas lime mudstones accumulate close

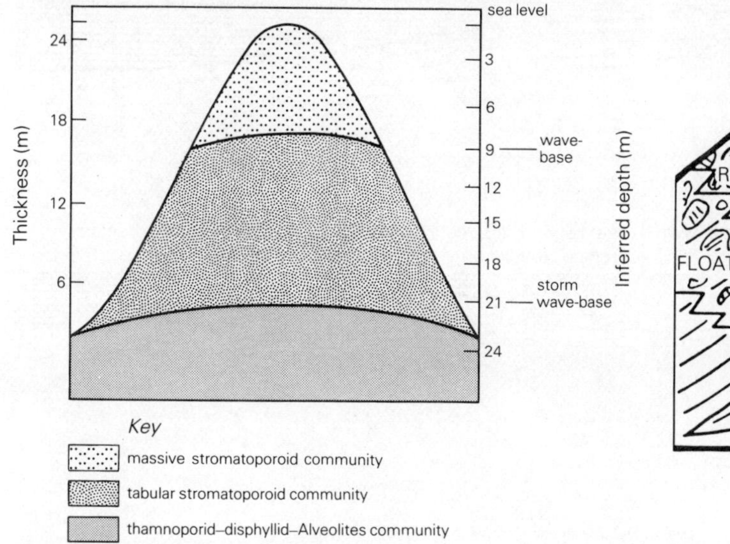

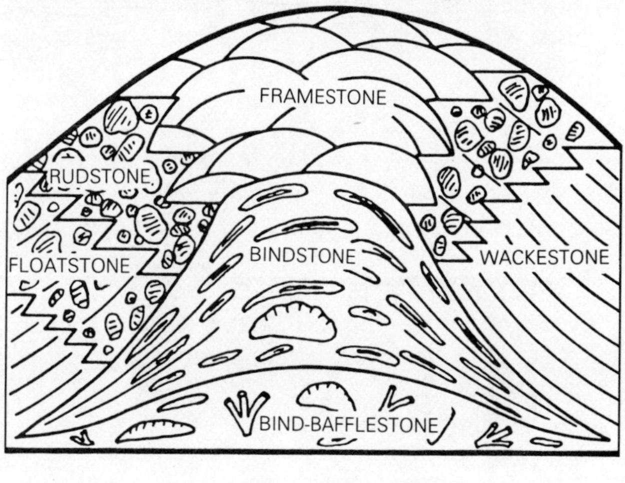

Figure 13.48 A Devonian bioherm from western Canada, interpreted in terms of Dunham's carbonate rock classification. (From J. E. Klovan, *AAPG Bull.*, 1974.)

inshore where energy is much lower. Beyond the shelf breaks, the basins are cut off from any supply of detrital sediment except for gravity-flow deposits containing carbonate clasts from the platform rocks. The basins are therefore relatively "starved" except for thin lime muds, marls, or black shales.

Depositional units are therefore interrupted and discontinuous both laterally and vertically. Packets of strata are separated by surfaces of submarine lithification (lithoclines); these surfaces may be strongly diachronous.

Carbonate banks or ramps Gently sloping carbonate platforms pass without abrupt change of slope from shoreline to basin. Wave-base intersects the depositional interface close to, and parallel to, the strandline. Sabkha or coastal, tidal flat deposits, containing "chicken-wire" anhydrite or gypsum, grade downslope through grainstones (including oolites), to wackestones and packstones (pelletoidal and bioclastic), and finally to ribbon limestones and lime mudstones. Stromatolitic or algal zones, according to age, are common. Coarse breccias or rudstones are rare.

The end result is a "layer cake" succession of mechanically deposited limestones, regularly and cyclically interbedded with argillaceous limestones, marls, shales, and perhaps a few sandstones. The latter represent phases of increased detrital input within an overall regressive regime. All units and sequences of units have wide lateral continuity, making for very easy correlation.

Reefs *Reefs,* in the strict sense, may be formed as integral parts of either the shelf or the ramp. Linear or continuous reef *trends,* however, typify the edges of true shelves. There is no equivalent preferred position in the ramp regime, in which "reefs" are isolated buildups on inherited highs (Fig. 13.49).

Reef masses consist primarily of structureless and unbedded boundstone flanked by grainstones and packstones. Reef cores tend to have vuggy porosity, but may have little permeability unless leached or fractured. In the regressive cycle, represented (for example) by the Iran oilfield belt, the reef limestone overlies its own fore-reef or basinal deposits and commonly has a depositional break at its top, with some solution; the reef is in turn overlain by back-reef deposits, commonly lagoonal evaporites and marls, then by shelf limestones and marls, and finally by continental sands and clays (Fig. 13.50). In the transgressive cycle, represented by most of the petroliferous reefs in western Canada, the succession is essentially the reverse of this.

Barrier reefs commonly margin carbonate shelves like the Bahamas Banks today. They may, however, develop across a sloping ramp during regression. In the low-energy environment behind barrier reefs, wackestones and miliolid lime-mudstones accumulate, with numerous permeability barriers. Patch reefs, banks, and debris mounds develop there, as well as grainstones washed over from the main reef. These grainstones are likely to be quickly buried by lagoonal muds, preserving their porosity and permeability and allowing them to become the most prolific reservoir rocks. Examples such as the oolitic-pelletoidal grainstones of the Jurassic Arab

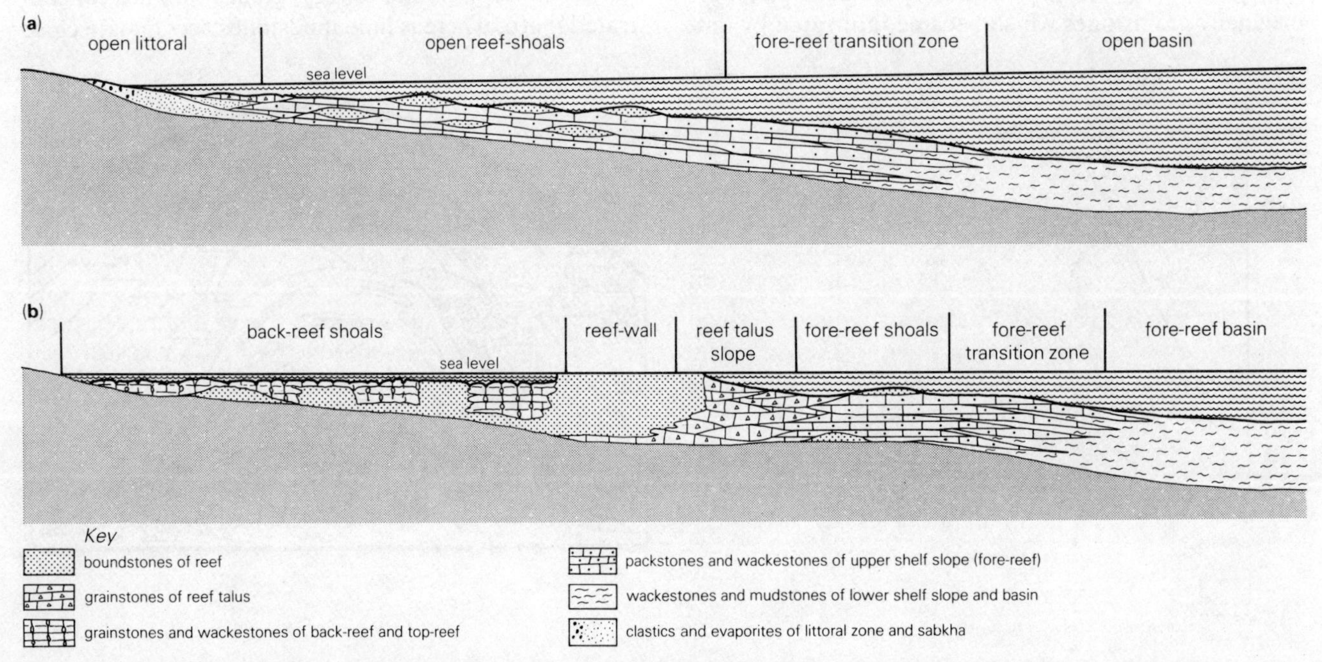

Figure 13.49 Ramp (a) and differentiated shelf (b) settings of carbonate deposition. If the basin (on the right) is generative, any carbonate facies shoreward of it may be productive. (After F. R. S. Henson, *AAPG Bull.*, 1950.)

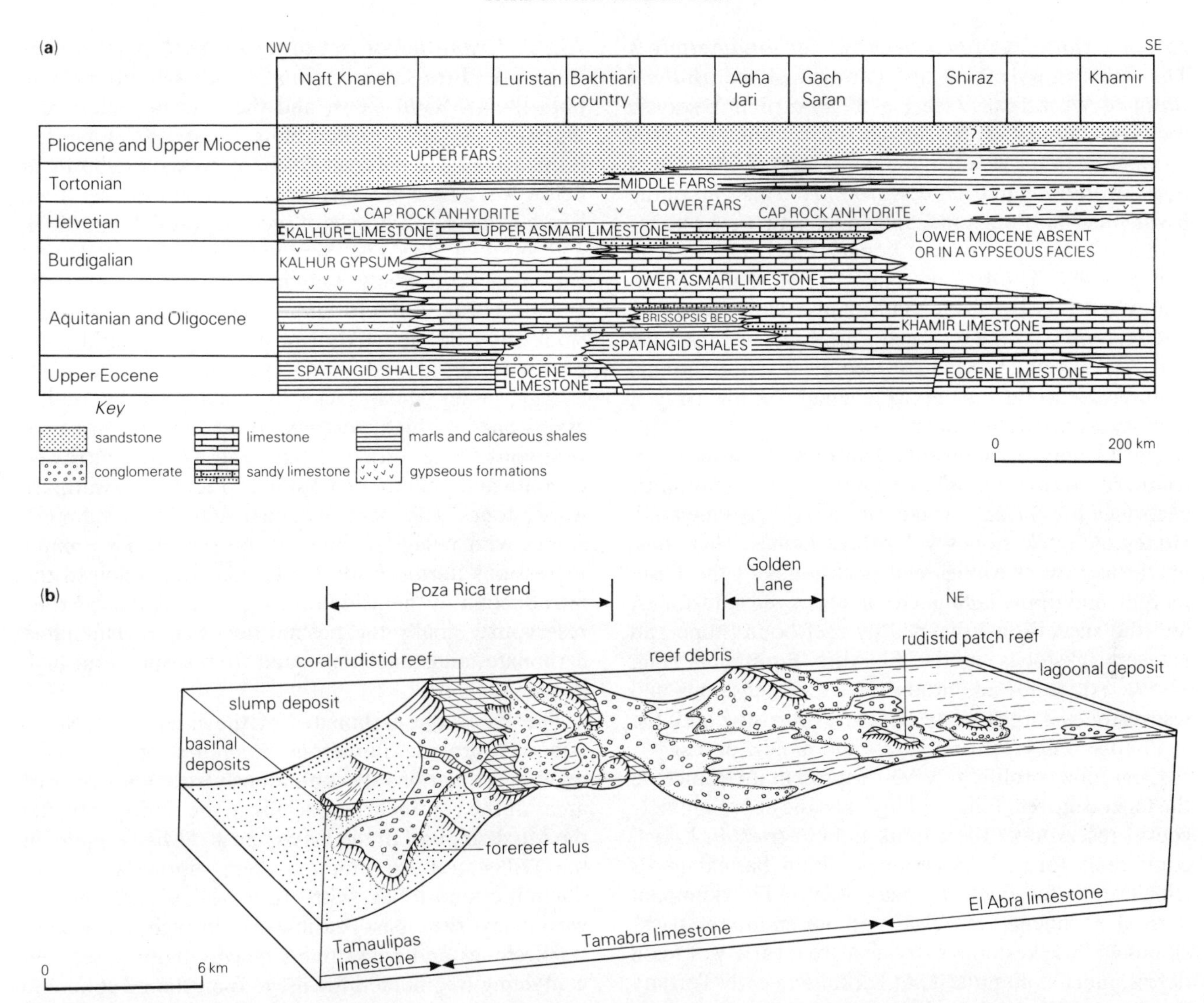

Figure 13.50 Diagrammatic reconstructions showing the great variety of carbonate and related facies types and their interrelations in two highly petroliferous successions. (a) The Tertiary succession in southwestern Iran; (b) The environments of deposition of the El Abra, Tamabra, and Tamaulipas limestones, in the Golden Lane of eastern Mexico. (From A. N. Thomas, XVIII International Geol. Congress, 1948, and after A. H. Coogan *et al.*, *AAPG Bull.*, 1972.)

Zone in Saudi Arabia and the Lower Cretaceous grainstones of the US Gulf Coast Basin are in effect the carbonate equivalents of clean quartz sandstone reservoirs.

In the high-energy fore-reef environment, pelletal grainstones accumulate in pods and shoals, with some oncolitic wackestones and packstones formed of bioclastic debris. Similar sediment types accumulate on the crests and flanks of highs on the sea floor. Numerous traps are created through facies changes were the grainstones grade updip into back-reef mudstones and wackestones. The grainstones themselves, however, are not quickly buried by mudstones in their high-energy environment, and may therefore lose their porosity and permeability during early diagenesis.

Farther downdip on the lower shelves pelagic and algal mudstones are deposited with occasional wackestones or oncolitic packstones. The lime muds may include source sediments. If basinward tilting sets in, the trap position of the reefs and their associated grainstones is enhanced.

Except for the addition of organic buildups, which have no direct counterpart in the sand-shale environment, carbonate regimes are dominated by detrital sediments just as clastic regimes are. Many of the same shallow-shelf environmental facies can be recognized in both regimes, and are subject to similar interpretations as strandline beaches, barrier bars (both with tidal channels), and shelf-edge shoals. Even carbonate dunes are fairly common today, but they have not been certainly recognized among reservoir rocks; probably their shapes and textures are destroyed by redistribution during subsequent submergence.

13.3.4 Examples of petroliferous carbonate shelves
The late Pennsylvanian and Permian of the *Midland Basin* of West Texas (see Fig. 17.3) contain reservoir rocks of almost every imaginable carbonate type. During the Pennsylvanian the type "starved basin" was here. In addition to true reefs around the margins of the basin, micritic floatstone, rudstone, and wackestones contain clasts of platform carbonates and provide excellent porosity. The porous zones are interbedded with shales and thin bioclastic sands, ripple-laminated. The basinal sediments are principally argillaceous packstone and wackestone showing soft sediment deformational features. Nearly all the Permian members of this array of carbonates are dolomitized.

The Cretaceous limestone platform of eastern and southern *Mexico* represents regressive progradation by reefs (see Fig. 17.22). At the base are cherty lime mudstones and wackestones with pelagic faunas. These pass up through wackestones and packstones of the basin margin, into ripple-laminated and burrowed mudstones, and the succession is capped by reef boundstone and coarse packstones of the El Abra Formation (Fig. 13.50). All facies are productive, the mudstones and wackestones largely from fracture porosity and the shelf-edge buildups from vuggy and moldic porosity.

Even more prolific reservoir representatives include the Paleocene reef fields of *Libya* and the approximately coeval reservoir of the Kirkuk field in *Iraq*. In Libya, coral reefs formed framestones within basinal marls which separated thick carbonate banks. The dominant detrital sediments are algal and foraminiferal packstones and wackestones carrying a great variety of fossil debris, much dolomitized. At Kirkuk, an early Tertiary shoal and shelf acquired shelf-edge bioherms between a mudflat to the north and a basin to the south (see Fig. 16.58). Repeated sabkha development and subaerial exposure resulted in dissolution, cementation, and dolomitization.

13.3.5 Examples of petroliferous carbonate ramps
The Upper Jurassic and Lower Cretaceous carbonates in both the US Gulf Coast and the Persian Gulf basins illustrate the ramp regime. In its fullest development, it is traceable from the coastal sabkha system to a basin far below wave-base in which anoxic conditions prevailed. The sabkha facies consists of cyclic supratidal, intertidal, and subtidal deposits. Supratidal deposits consist of carbonate (commonly dolomite), anhydrite, and terrigenous clastics; intertidal are cross-laminated sandstones and dolomitized algal mudstones; subtidal are burrowed fossil and pellet wackestones. The high-energy, inner shoal facies is dominated by oolite-grainstones of high porosity, providing the principal reservoirs. The outer shoal deposits are burrowed oncolite and pelletoid packstones. Farther seaward are wackestones with benthic faunas and then lime mudstones with pelagic faunas, all burrowed. Two major regressions during Early Cretaceous time allowed the introduction of the Middle East's only great sandstone reservoirs. Finally the basinal deposits are laminated carbonate mudstones deposited from suspension, high in organic matter.

The Paleozoic carbonates of the interior of North America provide numerous examples of the ramp regime. Formations are continuously traceable over vast areas of the continent, especially in the Ordovician and the Mississippian. In the Devonian and Mississippian of the Williston Basin, for example, supratidal, dense, dolomitic mudstones, with stromatolites and chicken-wire anhydrite, pass basinward through intraclastic wackestones and packstones, heavily bioturbated and containing fragmented fossils, to bioturbated coral and brachiopod packstones and wackestones with good porosity and scattered stromatoporoid framestones (Fig. 13.48).

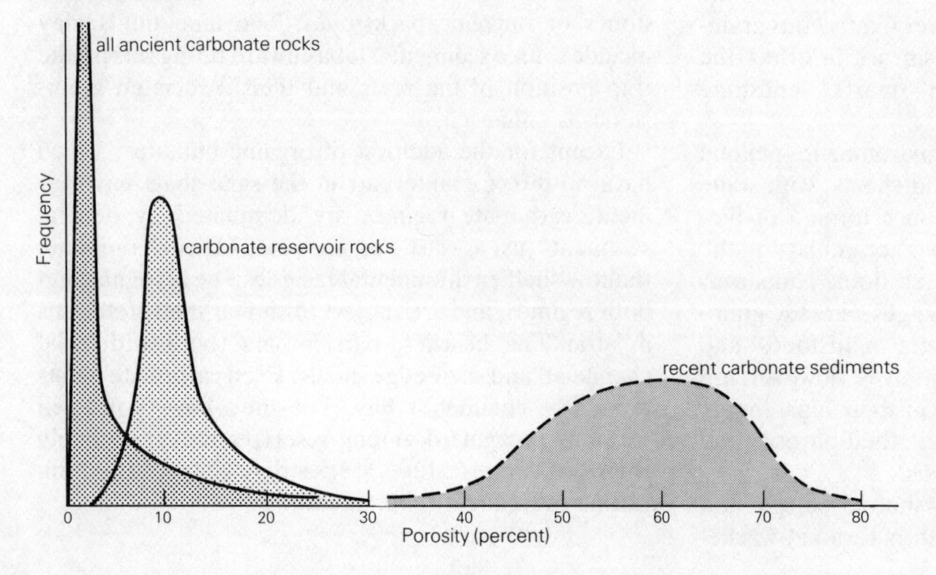

Figure 13.51 Graphical representation of porosity and permeability in ancient carbonate rocks greatly reduced through compaction and cementation but increased through leaching and/or dolomitization to reservoir values (though still far below the values for recent, uncompacted carbonate sediments). (After L. C. Pray.)

13.3.6 Carbonate diagenesis and porosity

Most shallow marine carbonates consist of aragonite or high-magnesium calcite. They are soft, soluble, unstable, and reactive. Aragonite is so unstable that, in the subsurface, it is converted to low-magnesium calcite in no more than 1 Ma. Within the phreatic zone, meteoric or shallow marine waters readily attack newly deposited limestones before they have become deeply buried. What happens to the limestones depends chiefly on the movement pattern of the waters and their degree of saturation with $Ca(HCO_3)_2$. Either cementation or the development of solution porosity will occur *early,* especially in reefs.

Recent carbonate sediments possess very high *primary porosity,* from 35 to about 75 percent (Fig. 13.51). This is the porosity the sediments have acquired by the completion of their depositional stage. Primary porosity in limestones is quite different from primary porosity in sandstones. Planar grain surfaces are rare in limestones; pores tend to be polyconcave micropores. The best primary limestone porosities are in grainstones, especially in oolites and calcarenites such as back-reef lime sands.

Packstones, wackestones, and mudstones, when consisting of pure limestone, are inherently compact in texture, and readily become further compacted during burial. The fate of carbonate porosity during burial is nearly always to be *reduced* (Fig. 13.51). *Compaction* may reduce the thickness of a limestone bed by 25–30 percent under no more than a few hundred meters of overburden, but reduction of carbonate porosity by compaction is significant only if the carbonate remains essentially uncemented. There is an inverse relation, in fact, between cementation and compaction in limestones.

Calcium carbonate is such a common substance that *cementation* of carbonate rocks by it is a very likely outcome. Other minerals – notably anhydrite – are common among cements in carbonate rocks, but by far the most common is sparry calcite *provided by the limestone itself.* A process vastly more important in carbonates than in clastic sediments is *pressure solution,* instigated by compaction and illustrated by stylolites. It is this process that causes carbonate rocks to supply their own cementing material. Cementation may proceed in stages until the cement becomes the largest single component of the rock.

Counteracting this susceptibility to pressure solution, however, is an equal susceptibility to *normal solution* in carbonated waters, which have taken CO_2 into solution from the atmosphere, from soils, or from other limestones. Solution endows the limestone with *secondary porosity* (Fig. 13.51), the ultimate development of which is represented by karst topography.

The process most favorable to leaching is marine regression, allowing early exposure to meteoric waters. Subsequent re-transgression buries the weathered, fractured, and solution-porosity zone below an unconformity, or at a depositional break. Nearly all oil in *limestone* reservoirs is pooled in this type of reservoir and trap, in the Middle East, the Williston Basin, and many Mid-Continent fields (especially in the Arbuckle limestone fields over the Central Kansas uplift; see Sec. 25.2.4). Even without early exposure, opportunities for solution porosity development may occur at any stage of diagenesis, even after relatively deep burial; so may opportunities for destruction of porosity by cementation. The critical factor is *water.* Solution porosity may come to extend over 100 m or more of thickness; it may become cavernous even at depth.

Chalks differ from most marine limestones in being initially composed largely of stable, low-magnesium calcite. They are thus less soluble than aragonitic limestones. Primary chalk porosities are intergranular, between coccoliths or their fragments. They may be as high as 70 percent at the sediment/water interface, but they are reduced by pressure solution and reprecipitation to 15 percent at depths of 2000 m and essentially to zero at 3000 m. The cementation is retarded if the pore fluids are marine and contain magnesium in solution, but it is accelerated by the entry of meteoric waters. Yet a small number of fields produce prolific oil from chalk reservoirs with primary porosity still in the 30–40 percent range below 3000 m; largest among them is probably the Ekofisk field in the Norwegian North Sea. In the opinion of Peter Scholle, preservation of this high porosity must be due either to the reduction of grain-to-grain pressures through overpressuring of the whole formation, or to the hindering of carbonate reactions by early entry of the oil into the pore spaces. Both processes are highly plausible in the setting of the Ekofisk field. It lies within a large graben, in which multiple faults, including large bounding faults, prevent lateral escape of pore fluids under compaction. Early entry of oil is known or inferred in many reservoirs, and one common carbonate reaction that it is able to hinder is dolomitization (see below).

Chalk is also among the carbonate rock types readily subject to *fracture* porosity and permeability. Its matrix permeability is very low (1–10 mD). Most limestones having secondary porosity have dense matrices like this. If they are also thickly bedded, as many limestones are, the secondary porosity most readily available to them may be fracture porosity, especially in convex traps. An important consequence is exceptionally high *vertical* permeability, as in the "crackled" Asmari limestone of Iran. This and other fractured reservoirs are considered in the next section (Sec. 13.4).

Despite the variety of routes by which limestones may acquire porosity, it remains true that good porosity in carbonate reservoirs is commonly a consequence of recrystallization, and most commonly of *dolomitization*. About 80 percent of all hydrocarbon reserves contained in carbonate reservoirs in North America are in essentially pure dolomites. This percentage is by no means representative of other regions of the world, and unquestionably reflects the number of North American carbonate reservoirs that are Paleozoic in age. In the Persian Gulf Basin, with the most prolific carbonate reservoirs in the world, the proportion of dolomite reservoirs is no more than 20 percent.

The replacement of $CaCO_3$ by dolomite involves a loss of volume of about 12.3 percent, and a consequent increase in porosity by that amount, if the replacement is molecule for molecule. It may not always be so, because volume for volume replacement is also possible. Yet it remains the case that in many fields having partially dolomitized carbonate reservoirs the oil is restricted entirely to the dolomitized portion. This portion is favorable because of *partial* dolomitization, preferentially of the finer-grained components of the limestone, and later leaching of the remaining calcitic parts which are more soluble. The most-quoted example of this phenomenon is the sprawling Lima–Indiana field south of the Great Lakes (Fig. 13.52). The oil is confined to porous dolomitic zones in the Ordovician Trenton Group where it passes over the axis of the broad, bifurcating Cincinnati Arch. Updip, porosity disappears in the unaltered limestone and only gas is recovered. Among the fields providing case histories for this book is the Jay field in Florida (Ch. 26), another example of restriction of oil to the dolomitized part of a carbonate formation. Up the dip, the undolomitized micritic limestone lacks porosity and is barren of oil. Most carbonate producing basins afford comparable examples.

An intriguing small example is provided by the Dover field, at the southwestern extremity of Ontario in Canada, like Lima–Indiana in the Ordovician Trenton Group. During the 1920s, articles on oil-bearing structures quoted Dover as an example of the rare synclinal trap. The syncline is controlled by an elongate fracture zone (Fig. 13.53) along which migrating waters have been able to dolomitize a considerable thickness of strata; the oil is restricted to the porous dolomite and therefore to the "syncline." The Albion–Scipio "trend", in Michigan's part of the same basin, is very similar and produces from the same formation (Fig. 13.52). From outside the normal concern of petroleum geologists comes the parallel case of Mississippi Valley-type lead–zinc deposits, in some of which it is established that the brines, derived from evaporite deposits, that deliver the metals also bring about dolomitization.

The selective nature of dolomitization extends to its

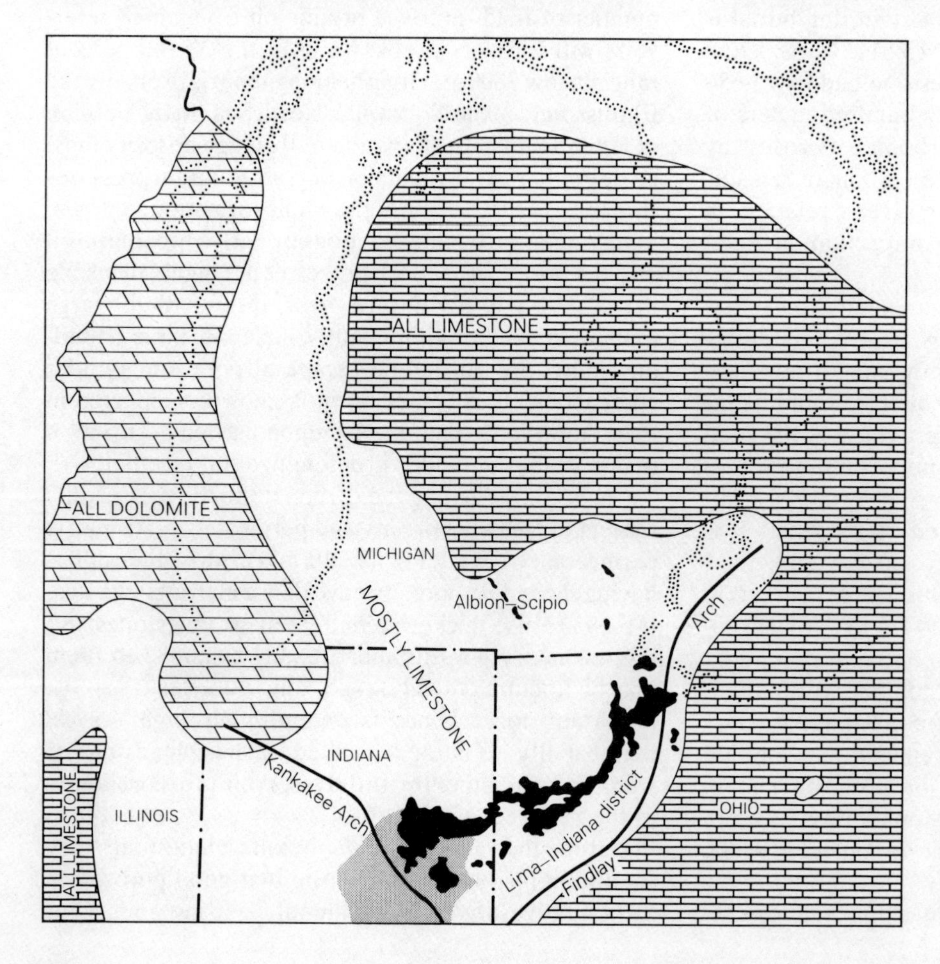

Figure 13.52 Michigan Basin in Middle Ordovician time, showing the Cincinnati Arch bifurcating into two arches as it crosses the basin from the south. The Trenton Group carbonates are dolomitized over the arches and along linear fracture zones; oil accumulations are restricted to the dolomitized portions. Dotted outlines on map delineate present Lakes Michigan and Huron. (After K. K. Landes, *Petroleum geology of the United States*, New York: Wiley–Interscience, 1970; and G. V. Cohee, US Geol. Survey Preliminary Chart no. 9, 1945.)

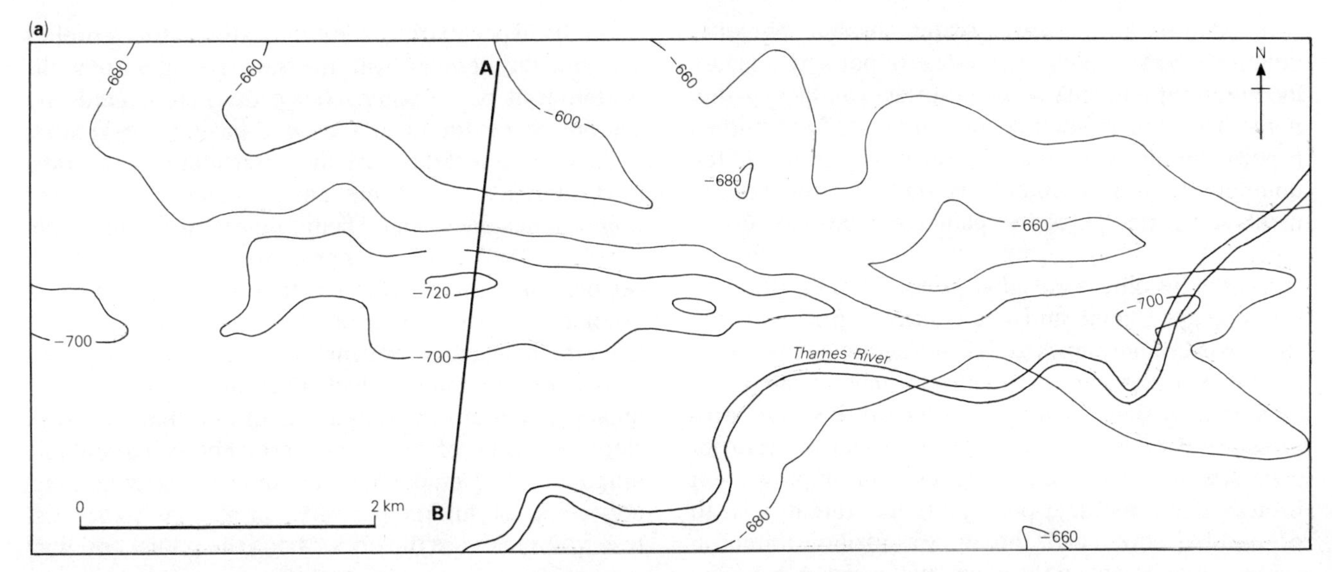

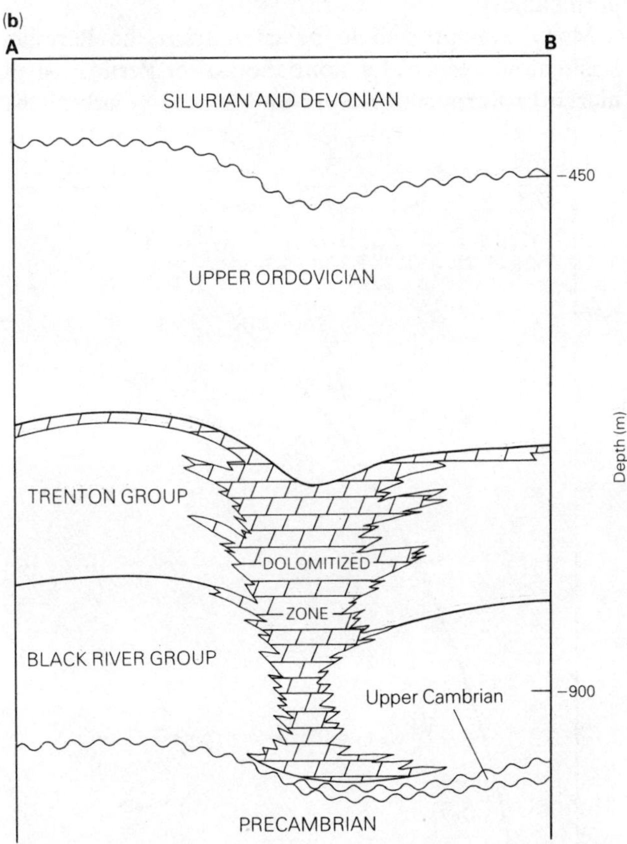

Figure 13.53 (a) Structure contours on the Ordovician Trenton Group (reservoir rock) in the Dover field, between Lakes Erie and Huron, southwestern Ontario, Canada. Contours at 20 m intervals below sea level. Note apparent synclinal structure. (b) Cross section along line A–B, to show dolomitized fracture zone creating a "pseudo-syncline." (From B. V. Sanford, Geol. Survey of Canada Paper 60-26: Fig. 10, 1961.)

effects on skeletal remains. Aragonite is much more easily dolomitized than is calcite, so shells of gastropods, cephalopods, and corals are dolomitized earlier than those of brachiopods, ostracodes, or echinoids. Well sorted crinoidal or shelly limestones are less dolomitized than surrounding rocks which contain less coarse material and more cement. Calcareous algae are easily dolomitized because high-magnesium calcite is deposited on them during their lives, and the algae themselves reduce sulfate which would otherwise inhibit dolomitization (especially of calcite). The vast mats of algae in the shallow epicontinental seas of the great Paleozoic transgressions are undoubtedly a factor in the prevalence of Paleozoic dolomites. There is very little dolomite in the stratigraphic record since the early Cretaceous, especially in the Northern Hemisphere. This may be because the present oceans, originating at that time, have a distribution and orientation different from those of earlier oceans, and epicontinental seas in low latitudes are highly restricted.

In addition to creating greater absolute porosity, dolomitization creates more *effective* porosity because its rhombs provide planar grain surfaces and *polyhedral pores*. The pore geometry of dolomites has been studied experimentally by Norman Wardlaw. The growth of the dolomite rhombs, whether by void filling or by replacement of calcite, leads to "planar compromise boundaries" (Fig. 13.54); pore sizes are thereby reduced, and the pores become *tetrahedral pores*. Because the pore surfaces are planar and not concave, growth of the rhombs eventually leads to *sheet pores,* not tubular pores as in most other porous rocks; they may be only a few micrometers thick. This pore structure becomes progressively simpler and more regular, and the ratio of pore size to throat size increases. The improvement brought about by sheet-pores and sheet-throats instead of tubular pores and throats was demonstrated in experiments involving the ejection of mercury from the pores as the pressure is reduced. Ejection decreases linearly with decrease in porosity, chiefly because of the reduction of throat sizes. The oil–water system differs from mercury, of course, in involving a wetting phase, which is pressure-dependent. For tubular pore-throats, the pressure required, P, is given by

$$P = \frac{4\,\sigma \cos \theta}{d}$$

where σ is the surface tension of the fluid, θ the contact angle for liquid upon solid, and d the diameter of the tube. For sheet-like throats, the diameter is effectively halved; it becomes the distance between two parallel, planar surfaces:

$$P = \frac{2\,\sigma \cos \theta}{d}$$

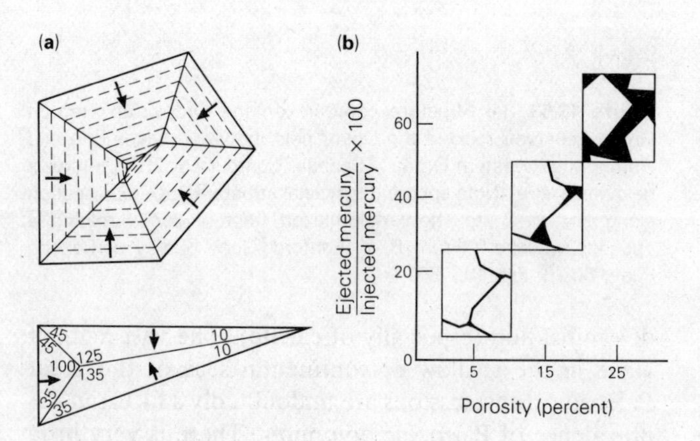

Figure 13.54 (a) Cross section of a polyhedral pore in dolomite. Growth of enclosing crystals (broken lines) reduces faces of pore from five to three. Lower figure illustrates closure of three-faced pore and development of "compromise boundaries." (b) Polyhedral pores (upper right) reduced to tetrahedral and then to interboundary-sheet pores (lower left) as growth of dolomite crystals reduces porosity. (From N. C. Wardlaw, *AAPG Bull.*, 1976.)

A critical peculiarity of dolomitization is that it prefers the mud-size grains over the sand-size grains in the limestone it is replacing. Hence the richest fields are seldom to be found at the sites having the greatest primary permeability. In the intertidal and subtidal environments, permeable units include skeletal and oolitic limestones, and calcitic limestones with secondary (leached) porosity that converted molds into irregular vugs and fractures into solution channels. If dolomitization then occurs, the calcium ions released from the limestone form anhydrite (a nearly ubiquitous associate of dolomite), which plugs the permeability and creates a dense rock from a porous one. Farther down slope, micrites of very low permeability (commonly aragonitic lime muds) are dolomitized, yielding intercrystalline or matrix porosity. In the end, the best reservoir facies are the ones having the poorest original permeability.

Many examples could be cited from the Permian Basin alone, especially along the Lower Permian shelf margin on the northwest (in New Mexico). Nearly all the

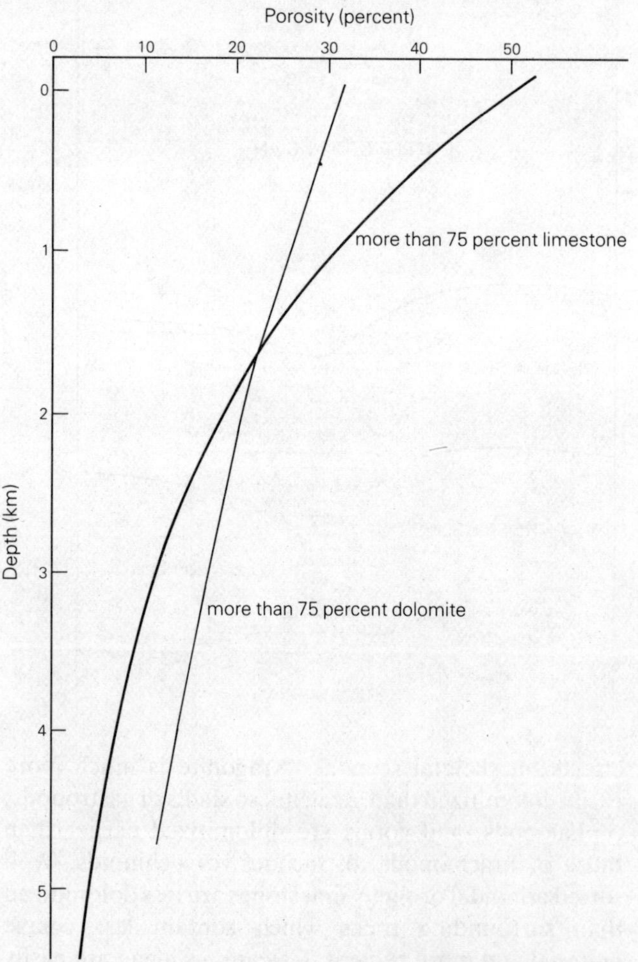

Figure 13.55 Reduction of carbonate porosity due to compaction with increasing depth, illustrating better retention of porosity by dolomites than by limestones. (From R. B. Halley and J. W. Schmoker, *AAPG Bull.*, 1982.)

Pennsylvanian production, including that from the largest single field, an atoll reef, is from limestones; nearly all the Permian production is from dolomites. In the Arab D zone in the Jurassic of Saudi Arabia, the principal production comes from detrital and bioclastic or oolitic limestones; where it comes from dolomites, the dolomitized rocks were largely micrites.

In addition to the capacity to create excellent porosity in unpromising material, dolomitization enables a carbonate reservoir rock to resist compaction. Dolomites are normally less porous than limestones at shallow depths, but they retain their porosity better during burial (Fig. 13.55). Below 4 or 5 km, porosity is much more likely in dolomites than in limestones.

Reefs of all kinds are especially susceptible to dolomitization, because the back-reef environment is a prime locale for evaporite deposition. Why some reefs are totally dolomitized whilst others close by retain their original limestone constitution is not clear. Among the multitude of reefs in the Devonian of western Canada, the limestone/dolomite distinction produced one unexpected circumstance. The giant Redwater field, one of the largest in the basin, occupies the updip (northeastern) edge of a large, lens-shaped reef, and the productive segment is undolomitized (Fig. 13.56). All the rest of the reef is dolomitized. The simplest explanation is that the oil was driven into the updip extremity of its trap by fluids that brought about the dolomitization, and could not drive the oil out and dolomitize the rest. Reefs which remain undolomitized may retain very little porosity or permeability in their boundstone cores (Fig. 13.48). Reservoir properties are then restricted to peripheral envelopes or cappings of reef detritus and lime sand (Fig. 13.56). Early estimates of reserves, based on the volume of the reef above the oil/water interface, are then vastly exaggerated.

Reefs are wave-resistant frameworks of lime-secreting organisms. The dominant frame-making organisms in any one reef province are few: calcareous algae, corals; in ancient reefs, stromatoporoids or rudistid molluscs. But the organic consortium is enormously varied, and many of the classes and orders represented have tests or skeletons that are not truly calcitic. Among contributing groups of organisms that build tests of aragonite are *Halimeda* and many other codiacean algae, scleractinian corals, and a variety of molluscs (some pelecypods and most pteropods and scaphopods). $MgCO_3$ makes up more than 10 percent of the hard parts of ostracodes, many echinoderms, alcyonarian corals, and several other coral groups; it may account for 20–25 percent of some calcareous and coralline algae; only in aragonite shells is it normally absent. The preferential

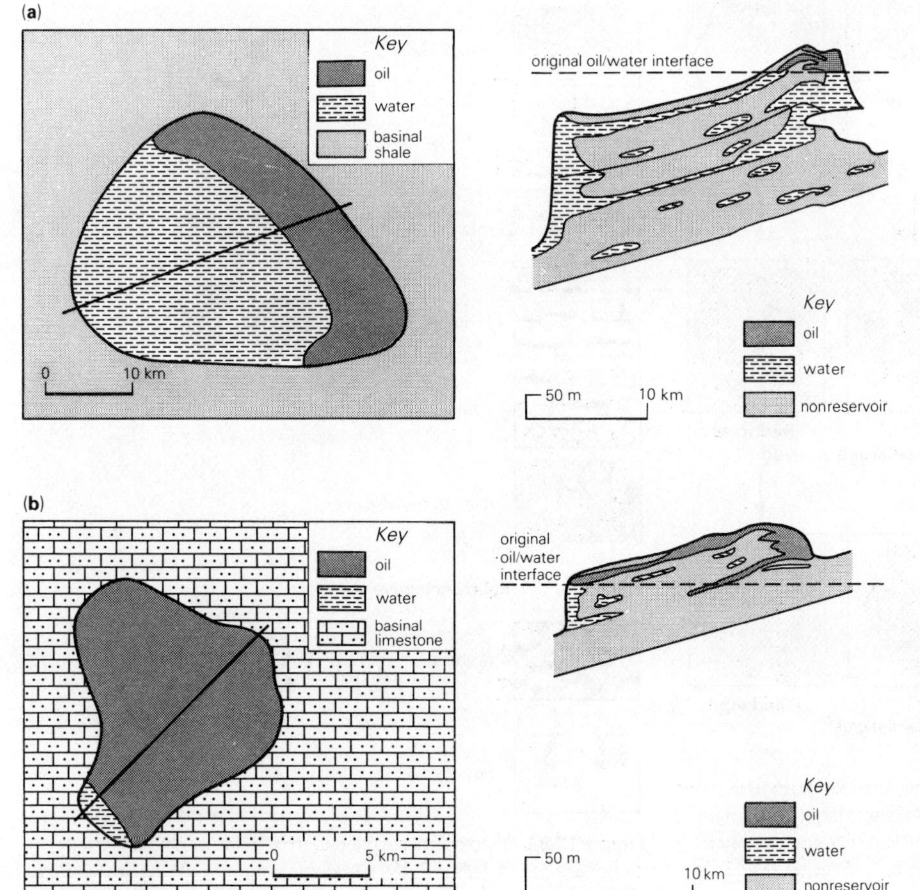

Figure 13.56 Maps and sections of two oil-bearing reefs in the Western Canadian Basin, showing essential restriction of porosity, and hence of oil, to the detrital peripheries of the reefs and not to their cores. (a) Redwater field, oil restricted to limestone reef-rim detritus; (b) Judy Creek field, oil restricted to envelope of porous reef detritus and lime sand. (From D. Jardine *et al.*, *J. Petrolm Technol.*, 1977.)

dissolution of aragonitic material, usually during the early burial stage, leads to coarse moldic and leached skeletal porosity, delightfully grouped as *crittermolds* by light-hearted paleontologists. The solution of aragonite leads to the precipitation of low-magnesium calcite, so no gain in porosity is achieved unless there is a net export of calcium carbonate. The widespread occurrence of crittermolds demonstrates that this net export is not difficult. Both the development of a mold and its refilling are illustrated in Figure 13.57. Molds, vugs, and other coarse pores in carbonate rocks can be studied visually through the medium of the medical CAT scanner (computed axial tomography). The scanner's view of the vug is plotted by computer, colored, and then studied in three dimensions by stereoscope.

For all secondary carbonate porosities, *time of development* is critical. The porosity may develop too late to receive an accumulation from a contemporaneous source. Very many carbonate reservoirs clearly acquired their hydrocarbons by downward migration from overlapping sources. The timing and mode of development of the porosity depend on the rate and manner of burial, the nature of the pore fluids and of their movements, and the availability of suitable sedimentary structures and textures. The structures and textures constitute the *fabric* of the rock, and the various forms of primary porosity are intrinsic parts of the fabric. The varieties of *secondary porosity* may then be influenced wholly, moderately, or not at all by the available fabric.

13.3.7 Summary of carbonate porosity types

The principal types of porosity found in carbonate rocks can now be summarized. The classification below is simplified from that proposed by Choquette and Pray in 1970 (Figs 13.58 & 59):

Fabric–selective porosity

(a) Interparticle porosity: most primary porosity is of this type, but so is much secondary porosity
(b) Intercrystalline porosity, typical of many dolomites
(c) Fenestral porosity, by solution along bedding or joint surfaces
(d) Skeletal, framework, moldic, or shelter porosity, by selective solution of, within, or around fossil material

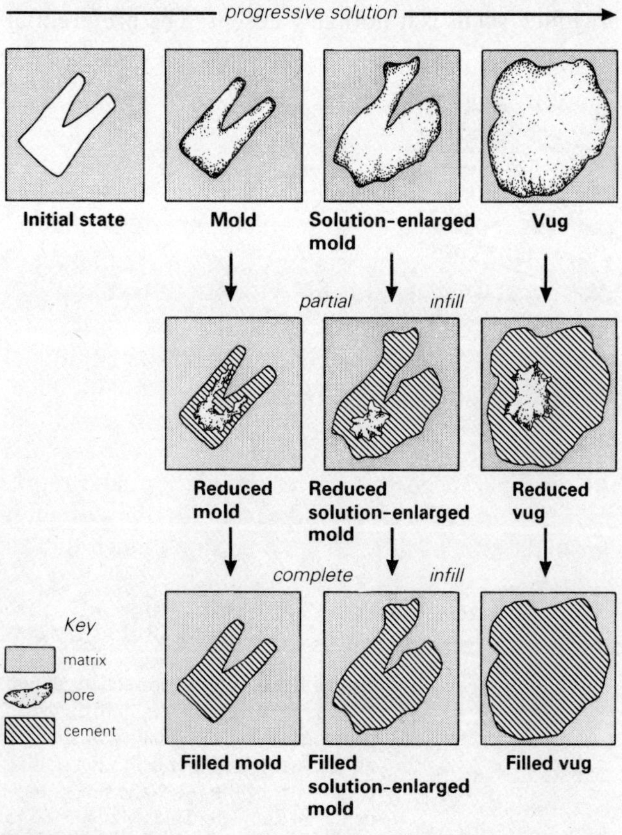

Figure 13.57 Common stages in evolution of a mold, formed in this case from a crinoid columnal (top left). At any stage of solution, leading to formation of a shapeless vug, infilling by cement may be partial or total. (From P. W. Choquette and L. C. Pray, *AAPG Bull.*, 1970.)

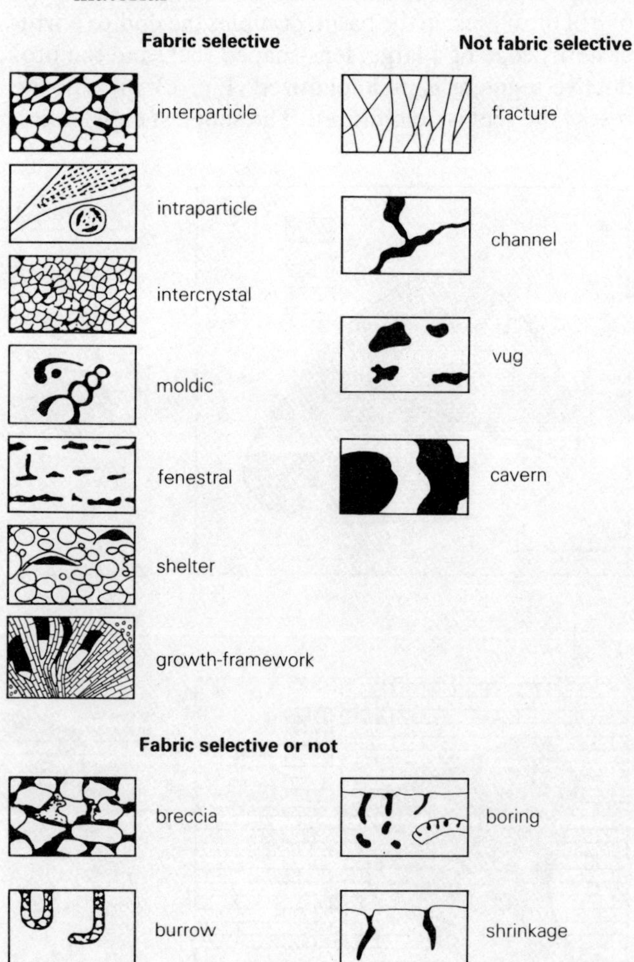

Figure 13.58 Geological classification of pores and pore systems in carbonate rocks. (After P. W. Choquette and L. C. Pray, *AAPG Bull.*, 1970.)

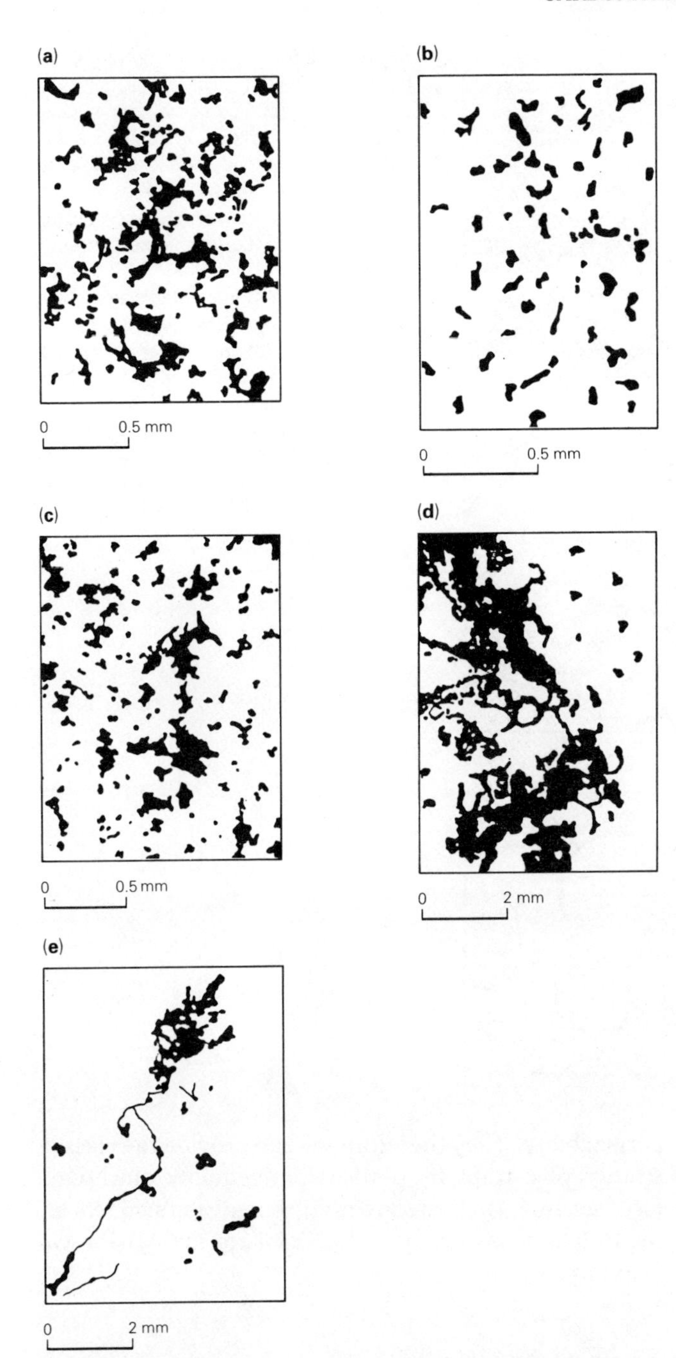

Figure 13.59 Varieties of porosity in limestone reservoirs, illustrated by drawings made from photomicrographs of plastic-impregnated thin sections; (a) and (b) show intergranular porosity, (c) and (d) show reef-type porosity, (e) shows fracture porosity. Note differences of scale. (From C. R. Stewart *et al.*, *Trans. Soc. Petrol. Engrs*, 1953.)

(e) Oomoldic porosity, a variant of (d) by selective solution of ooliths. In oolitic grainstone reservoirs (as in parts of the Arab Zone in Saudi Arabia), this process may increase the porosity to 30 percent while *reducing* the permeability to below 1 mD.

(Note that in types (c), (d), and (e), above, the induced openings may be larger than the grains.)

Porosity not fabric-selective, or not fully or necessarily so

(a) Fracture porosity, by stress or shrinkage fractures
(b) Channel porosity, through widening and coalescing of fractures or bedding or joint surfaces
(c) Vuggy or cavernous porosity, commonly through enlargement of molds
(d) Bioturbation porosity
(e) Breccia porosity, in some cases really highly developed fracture porosity.

13.3.8 Preservation of carbonate porosity

Both the original endowment of a carbonate sediment with primary porosity, and its subsequent acquisition of secondary porosity, are clearly transient phenomena. Even superior porosity is worthless if it is very short-lived. The capacity for self-cementation, facilitated especially by pressure solution (Sec. 13.3.6), must therefore be somehow denied to a limestone if its porosity is to be preserved. Compaction, the chief cause of pressure solution, must be kept to a minimum, so that grain contacts do not become "overpacked."

How this is brought about is revealed by studies of the *pre-burial diagenesis* of modern carbonate sediments (Fig. 13.60). If this takes place in the submarine environment, without emergence and exposure of the carbonates, complete saturation with water causes the deposition of fibrous aragonite or high-magnesium calcite cement as a thin, lithified crust over the rock surface. The crust becomes itself encrusted and bored by organisms, and serves as protection of the uncemented rock below. The *hard grounds* or lithoclines (Sec. 13.3.3), that interrupt carbonates in the stratified series (especially those of shelf environment), are of this origin.

In the intertidal zone on the beach, the formation of crust is less regular because saturation with water is discontinuous; the typical product is the *beach rock* behind many fringing reefs. Still higher, in the supratidal zone, the attacking waters are more meteoric than marine, so that the aragonite fibers are dissolved and reprecipitated as calcite. If complete emergence occurs, the principal diagenetic process is solution, and reprecipitation as sparry calcite, somewhere in the system, is unavoidable. The lithified crust becomes much thicker than in the submarine zone, encompassing all the sediment above the water table and much below it.

In older carbonate successions, post-burial diagenesis takes the place of surficial diagenesis and is likely to destroy all porosity except that due to dolomitization *unless the porosity is protected by early-formed surficial crusts*. These can usually be shown to have been created repeatedly where granular compaction has remained low over a considerable thickness of carbonate reservoir (as in the Arab D calcarenites in Saudi Arabia), allowing the preservation of adequate porosity and eliminating the need for the development of fracture porosity.

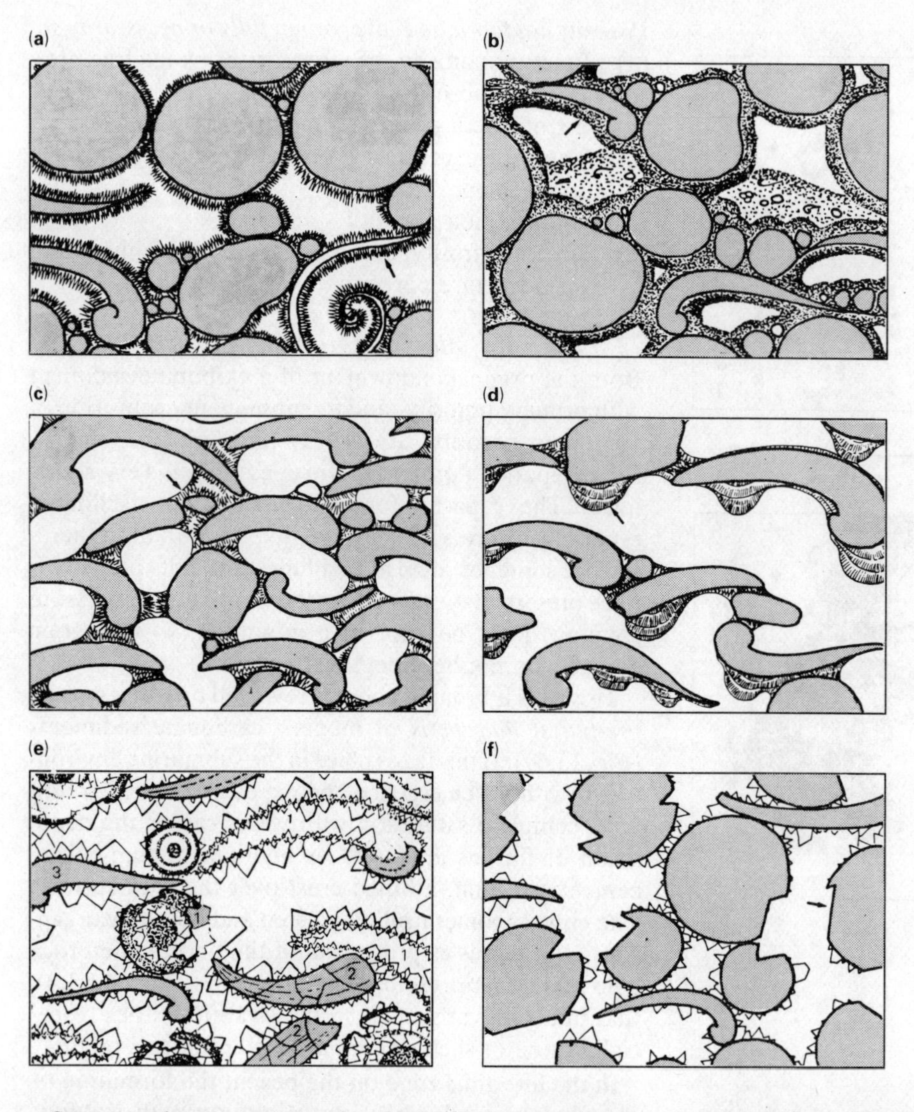

Figure 13.60 Essential petrographic aspects of Recent carbonate diagenesis. (a), (b) Submarine cements: isopachous fibrous (a) or micritic (b) aragonite or magnesian calcite cements reflect sediments saturated with marine waters. (c), (d) Intersupratidal cements: grain-contact (c) and microstalactitic (d) fibrous cements reflect partial saturation by marine waters; (e), (f) Continental diagenesis: characterized by dissolution (1) or recrystallization (2) of aragonite grains and unaltered calcite (3) particles; cement is blocky calcite. In (f) dissolution of dedolomite is indicated by angular or flat-sided secondary pore-spaces. (All drawings from thin-slides.) (From B. H. Purser, *J. Petrolm Geol.*, 1978.)

Some carbonate reservoirs have had exceedingly complex histories, creating highly heterogeneous pore systems. Rhythmic fluctuations of exposure and submergence can lead to creation or enhancement of porosity by solution, its reduction or obliteration by cementation, and repetitions of both these processes; to dolomitization, dedolomitization, anhydritization. Even mild tectonism leads readily to fracturing, and the fracture porosity may itself then be enhanced, reduced or destroyed by the forementioned processes. In prolific carbonate reservoirs that have undergone such histories there may be no primary porosity left, nor even true secondary porosity; all the porosity is *n*th cycle.

The potentiality for small lithologic traps caused entirely by porosity variations is endless. An excellent example is provided by the Permian reservoir formations of West Texas. A single formation such as the San Andres forms the reservoir rock for both a few large fields and for scores of small ones.

In contrast, carbonates (especially undolomitized limestones) suffer the critical defect of weak *horizontal* permeability. They therefore seldom provide the classic stratigraphic traps by porosity/permeability pinchouts (see Sec.16.4.4). There are no carbonate equivalents of the Bolivar Coastal field, East Texas, or the Athabasca tar sand.

13.3.9 Most favored settings for carbonate reservoirs

Finally, the geologist must remember that, despite the manifest differences between carbonates and sandstones as reservoir rocks, the two are alike in the most critical single aspect. No matter how porous, each can become an *effective* reservoir only if it has access to an effective source sediment.

The necessary combination of agitated waters (to permit the creation of porous reservoir rocks) and organic mud or lime-mud rocks (to provide the sources) is common in five settings for carbonate reservoirs. The following classification of these settings is modified from that proposed by James Lee Wilson in 1978:

(a) subtidal and intertidal complexes with updip facies changes, and commonly with downdip dolomitization (Permian and Williston basins);
(b) lime-sand bodies on shelves with incomplete cementation (Jurassic of Saudi Arabia; Smackover fields of the upper Gulf Coast);
(c) organic buildups, which may be dolomitized (Alberta, Midland, and Sirte basins);
(d) talus accumulations below carbonate banks, reefs, or buried hills (Reforma–Campeche region of Mexico);
(e) leached zones below unconformities, with or without fracturing (Paleozoic limestone fields of Mid-Continent; Cretaceous of Oman).

These settings are provided equally well by both the ramp and the differentiated shelf regimes (Sec. 13.3.3). The two regimes are illustrated in Figure 13.49.

13.3.10 Producibility of carbonate reservoirs

It is apparent that carbonate reservoirs represent a greater range of producibility than do the more common sandstone reservoirs. The most prolific sustained production rates ever recorded came from carbonate reservoirs, and most of them were from vuggy or fractured limestones, not from dolomites. The legendary wells in Mexico's Golden Lane before World War I, the first Yates well in the Permian Basin of West Texas, and the best of the Asmari limestone fields in Iran are only the most spectacular examples.

At the height of its output, Iran's Agha Jari field (see Fig. 13.72) was producing more than 10^5 m^3 of oil daily from about 36 wells. At the opposite extreme are hundreds of tight limestone reservoirs, commonly siliceous, micritic, or chalky, that will not yield their oil at all without artificial fracturing and acidization (see Ch. 23). On average, and despite the outstanding performers, carbonates are less productive than sandstones per unit volume of reservoir: 5×10^4 m^3 km^{-2} m^{-1} (500 m^3 (ha m)$^{-1}$) is exceptional and 300 m^3 (ha m)$^{-1}$ a fair average for reefs.

13.3.11 Log expressions of carbonate reservoirs

Formation evaluation in carbonate (especially in carbonate–evaporite) petroliferous successions is unexpectedly different from that in sandstone–shale successions. It is much more dependent upon laboratory study of the rocks because it is much less able to rely upon well logs.

The differences between carbonate porosities and sandstone porosities have been emphasized (Sec. 13.3.6). Both porosity and permeability are far more variable in nature in carbonates than in sandstones, and may present a much greater range of pore shapes and dimensions. The presence of vugs (even caverns), fractures, and combinations of the two permits easy invasion of carbonate reservoirs by drilling mud, confusing log interpretation (especially if the mud is salty).

The relations between sandstones and shales, with many reservoir sandstones being shaly, have provided the impetus behind a number of logging devices and the display of their expressions (as discussed in Ch. 12). Reservoir carbonates are more commonly interbedded with dense carbonates or with evaporites than with shales. Argillaceous limestones themselves are rarely satisfactory reservoir rocks (unless they are fractured, in which case wholly new formation evaluation challenges are presented).

The SP curve, indispensable in sandstone–shale successions, is of very limited value in carbonates (see

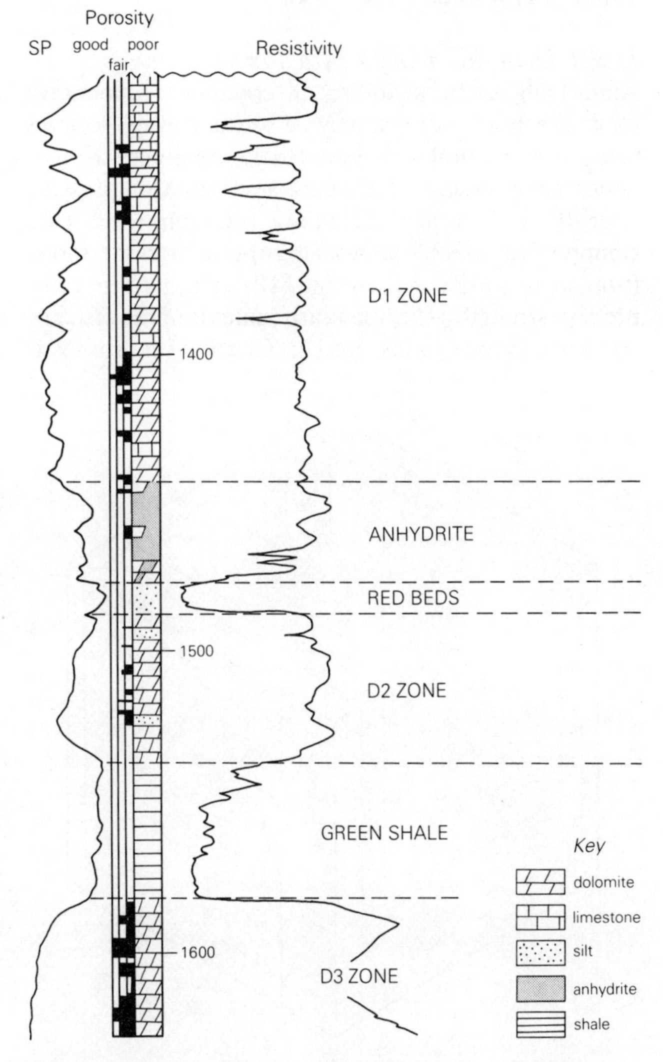

Figure 13.61 E-log (SP and resistivity) expression of massive carbonates, largely dolomites, separated by shales. The D3 zone at the base is an oil-bearing, Devonian biohermal reef in the Leduc field, Alberta, Canada. (From W. W. Waring and D. B. Layer, *AAPG Bull.*, 1950.)

Sec. 12.2). Clean, massive carbonates (especially reefs) tend to be reflected by abrupt negative SP kicks (Fig. 13.61). Resistivity is high in all carbonates except for porous limestones containing salt water.

The most useful log for carbonate reservoirs is the gamma ray–neutron log (Fig. 12.18). Clean carbonates provide sharp-shouldered kicks away from the baseline on both logs, separated by equally sharp re-entrants for the more radioactive shales. The neutron log is also helpful in the detection of porosity variations in carbonates (see Sec. 12.2.3). Dense carbonates give high readings, and evaporite interbeds even higher; the fluids in the porous members cause sharp reductions in the neutron readings (Fig. 12.8).

13.4 Fractured reservoirs

13.4.1 General theory of fracturing

Almost all rocks, including all common sedimentary rocks except evaporites, may be brittle enough to fracture during natural processes. Rocks are much stronger under compression than under tension. Compressive strengths of typical sedimentary reservoir rocks (the compressive stresses at which rupture occurs) range from about 8 MPa to about 180 MPa. The average compressive strength of lithified carbonates is of the order of 100 MPa (about 14 000 p.s.i.). That for sandstones is both lower (average about 60 MPa or 8500 p.s.i.) and much more variable. This is because sandstones may have a variety of cements (see Sec. 13.2.5) and it is a property of cemented sandstones that they break around the grains, not through them. Siliceous clastic rocks which break through grains and cement alike (because the cement is itself silica) are quartzites, and quartzites are not important reservoir rocks.

The tensile strengths of rocks are lower than their compressive strengths by factors of hundreds. This suggests that fracturing is much more likely under extension (whether regional or localized) than under compression. If fractures are to lead to effective porosity and permeability, either the body of rock must shrink (as in the cooling of an igneous rock or the dewatering of a sedimentary rock) or its overall area must increase. The latter is much the commoner cause of effective fracturing, and again it implies *extension*.

However, extension of a rock body does not imply a regional extensional regime. Fractures resulting from compression in laboratory experiments may be either *shear fractures*, formed obliquely to the compression axis (the direction of stress σ_1), or *extension fractures* formed parallel to it (Figs 13.62 & 63). A further set of extension fractures may be formed perpendicular to the axis of σ_1 on relaxation of the compression (Fig. 13.62). As the overall geometry of a rock body may be such that the axis of σ_1 is oblique to it, further fractures may result under couple or through torsion (Figs 13.64 & 65).

The results are *fracture sets*. On the large scale these are represented by regional *joint patterns* visible on the

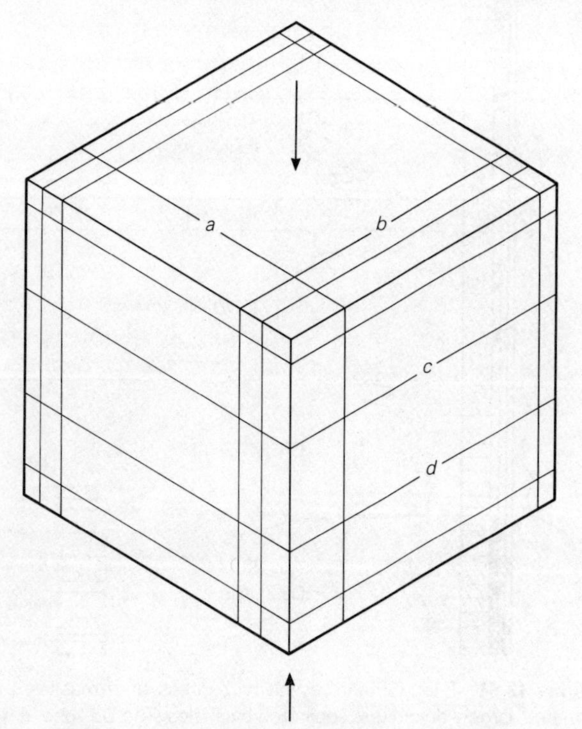

Figure 13.62 Tension fractures resulting from compression. *a* and *b* are tension fractures formed parallel to the compression axis. *c* and *d* are tension fractures formed upon release of compression.

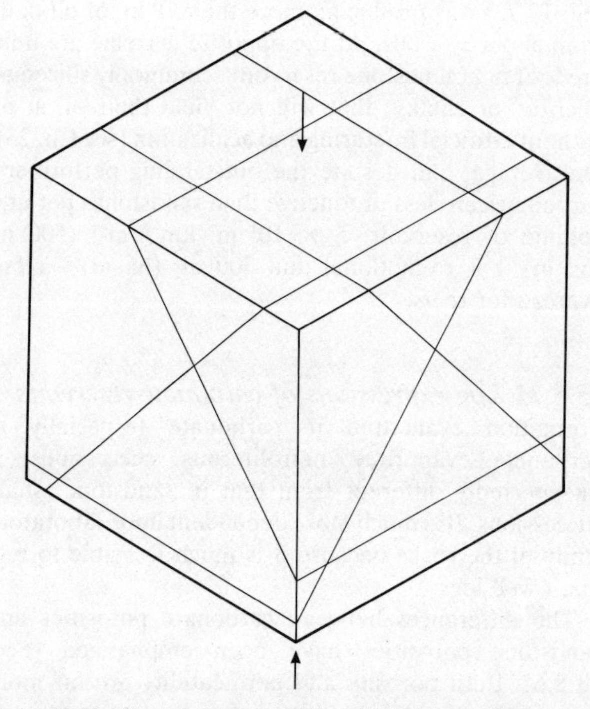

Figure 13.63 Shear fractures resulting from compression.

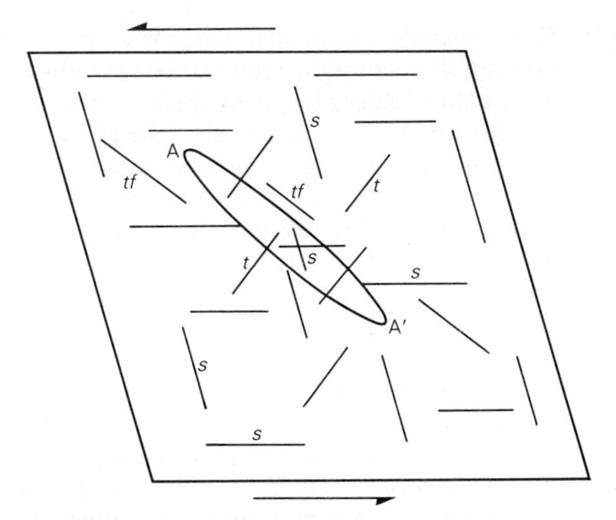

Figure 13.64 Fractures resulting from a couple. A–A′, anticlinal fold; *t*, tension fractures; *s*, shear fractures; *tf*, thrust fault.

ground and in many regions obvious on aerial photographs. Patterns are commonly orthogonal systems. They are not conjugate sets, so they are not shear fractures. Instead they seem to form essentially parallel and normal to deformed belts which may be distant from areas of well developed fractures. Major elements of the fracture systems extend downwards through considerable thicknesses of layered rock; they are apparently due to *extension*.

13.4.2 Depth of effective fracturing

There is a depth, varying according to tectonic (stress) conditions, rock types, and fluid contents, below which open fractures cannot be maintained. In the absence of the strong horizontal stress gradients which lead to folding and thrust faulting, the maximum principal stress (σ_1) is the vertical geostatic load and the minimum stress (σ_3) is the horizontal confining pressure.

At any point in a rock mass, the *deviatoric stress* (which controls change in shape rather than simple change in volume) is that component of the total stress that excludes the mean normal stress (σ_0). It is proportional to ($\sigma_1 - \sigma_3$), the spread between the maximum and minimum principal stresses that is represented by the diameter of the familiar *Mohr circle*. At shallow depths (less than 2 km), ($\sigma_1 - \sigma_3$) is large enough to permit the maintenance of open fractures under normal hydrostatic conditions; σ_1/σ_3 is of the order of 3 or more. Under conditions promoting extension, as in rift zones or at shelf edges, σ_3 is negative and open fractures perpendicular to it may be maintained at greater depths.

Sedimentary rocks contain fluids, which exert fluid or *pore pressures* which counteract the normal and deviatoric stresses (see Ch. 14). If the pore pressure is so high as to exceed the lateral confinement stress σ_3, fractures will be opened by natural hydrofracturing. At depths of 6 km or more, σ_1 and σ_3 become nearly equal; all stresses are compressional and σ_1 would exceed the crushing strengths of all sedimentary rocks if they were unconfined by the stresses exerted by the rocks surrounding them.

According to the Griffith theory of failure, fractures form through the stress-induced propagation of originally microscopic flaws. The common direction of this propagation in sedimentary rocks is *upward* from below, not downward from above. We need to consider all geological phenomena capable of causing this.

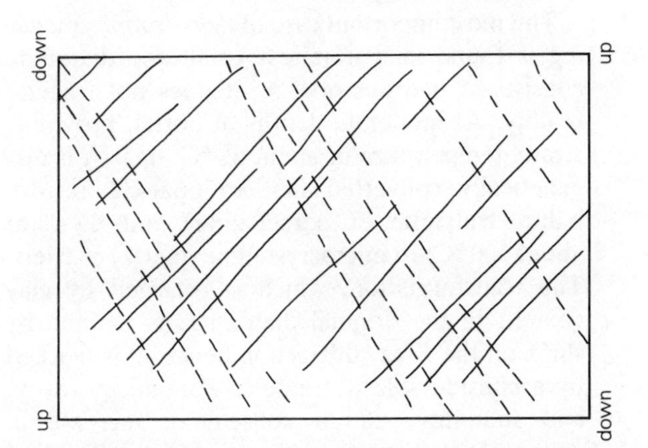

Figure 13.65 Fractures resulting from torsion. Fractures on upper side represented by solid lines. Fractures on lower side represented by dashed lines.

13.4.3 Causes of fracturing in layered rocks

The most common causes of fracturing in sedimentary rocks are set out below, in general order of their significance.

(a) *Buckle folding* is due to forces parallel to the layering of the rocks (Sec. 16.3.1). Fracturing is best developed in thick or thickly bedded formations having little capacity for flowage, like the old Arbuckle and young Asmari limestones described below (Sec. 13.4.9). Fracture formation is promoted, however, if the *underlying* rock does

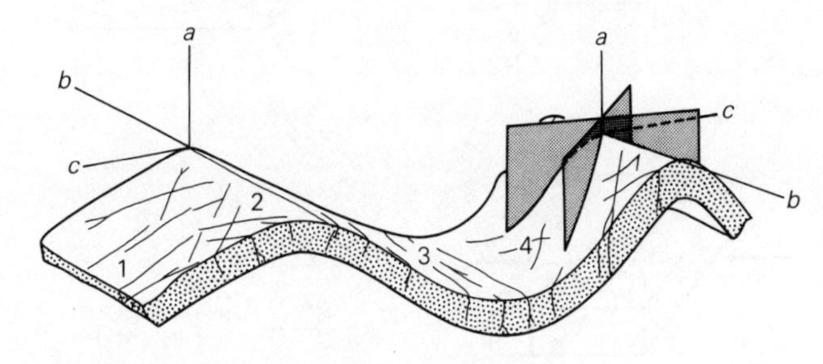

Figure 13.66 Four joint sets commonly found in folded rocks (shown in separate parts of folds for clarity only). (From B. E. Hobbs *et al., Outline of structural geology,* New York: Wiley–Interscience, 1976.)

deform by flowage; examples from the Reforma region of Mexico and the central North Sea, with salt uplifts from below in both cases, are described later in this section (Sec. 13.4.9). The most effective fractures tend to be on the anticlinal crests, which are under local extension. The fracturing is older than the final folding, which rotates the fractures and preferentially opens those parallel to the *b* direction of the tectonic strike (Fig. 13.66). Dip fractures, parallel to *c*, are likely to remain tightly closed.

(b) *Bending folding* is a consequence of forces perpendicular to the rock layering; its most important manifestation is in the vertical uplift of basement or horst blocks (Sec. 16.3.2). The principal fractures are then propagated upwards from the edges of such blocks, creating vertical fractures over the blocks and inclined (or even horizontal) fractures off their flanks.

(c) *Faulting* is itself an aspect of fracturing; most faults are merely large, discrete members of fracture systems. Longitudinal faults associated with transverse or cross faults, for example, are likely also to be associated with longitudinal and transverse fracture sets. The intensity of fracturing in general increases logarithmically with decreasing distance from a fault. Fracturing develops especially where faults and folds are intimately associated. Familiar examples affect most rollover anticlines at growth faults. In the gas-bearing black shales of the Appalachian foreland (Sec. 10.8.2), fracture porosity is best developed in flexures above detachment thrust faults or upthrusts from the basement. The general orientations of fractures related to compressional faults are illustrated in Figure 13.67.

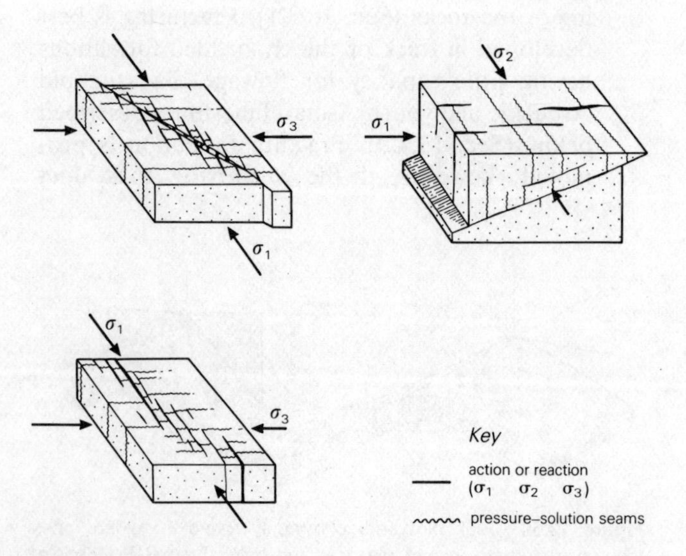

Figure 13.67 Possible development of extensional fractures in overall compressive setting. (After J. du Rouchet, *AAPG Bull.*, 1981.)

(d) *Fluid pressure* may be the commonest cause of extensional fracturing in regions lacking significant deformation. Rocks are much weaker under tension than under compression, and fluid pressure in sedimentary rocks may come to exceed the confining stress (σ_3) by a sufficient amount to overcome the rock's low tensile strength.

(e) *Relief of lithostatic pressure,* consequent upon uplift and erosion, is an important cause of fracturing both parallel and perpendicular to the layering (Fig. 13.62).

(f) *Pressure solution* is especially important in carbonate rocks that have been fairly deeply buried (Sec. 13.3.6). *Stylolites* are permeability discontinuities formed within the rock by this process. They are most common in zones of relatively low matrix porosity. The "teeth" or columns of the stylolites are aligned parallel to σ_1, and are therefore essentially vertical (and in many cases perpendicular to the bedding).

Stylolites are not open fractures, but *tension gashes* are short fractures formed by extension along the direction of minimum stress (σ_3). When associated with stylolites, therefore, they form perpendicular to them and are caused by the same stresses. The gashes are in fact extensions of the "teeth" of the stylolites (Fig. 13.67). As they are normally short, they have little effect on reservoir permeability, but they may have a large effect on laboratory measurements of it from cores.

Diatomite also readily undergoes pressure solution and stylolitization, forming *solution cleavage* which may mimic fracturing.

(g) *Loss of fluid content* is a contributing factor in the formation of shrinkage cracks, especially in the dewatering of mud rocks during compaction. Most producing reservoir rocks have undergone the replacement of one fluid by another, and this process is of little significance for them.

The most important case of *dehydration fracturing* is found in diatomites. Unaltered diatomite consists of hydrous opal-A; it does not fracture readily. At moderate depths of burial, providing stratal temperature of about 55 °C, opal-A is diagenetically converted first to opal-CT (cristobalite–tridymite or porcellanite) and then (at about 100 °C) to microcrystalline quartz, or chert. The transformation, which is retarded by clay content in the original diatomite, is defined by shifts on the X-ray diffraction peaks. It is marked by a considerable decrease in volume, porosity, and solubility, and a consequent increase in density. It also involves a decrease in ductility, and hence an increase in brittleness. Chert is therefore much more susceptible to fracture than porcellanite. The loss of volume and porosity generates excess pore pressure (see Ch. 14), which dilates existing fractures and creates new ones.

(h) *Weathering,* involving moisture, plant roots, frost, and other agents, creates fractures as well as enlarging them. Such fractures are important in reservoir rocks along old erosional surfaces now in subcrop below unconformities (see Sec. 16.5).

(i) *Cooling* of igneous rocks creates fractures, and igneous rocks form both interstratified and basement reservoir rocks in several regions (Sec. 13.4.9).

(j) *Impact craters* necessarily contain extensively fractured rocks. At least two reservoir rocks of probable impact origin are known (both in the Williston Basin), and they may be more common than imagined. Carbonate breccia there has average porosities higher than 10 percent and permeabilities of 400 mD. Breccia lenses in the fractured crystalline basement rocks below a number of well-known craters contain no known commercial reservoirs but they have measured porosities higher than 25 percent and fracture permeabilities of 10 mD or more.

13.4.4 Fracturing and deformation

Fracture porosity and permeability are therefore much more common in deformed than in undeformed rocks. This conclusion is reinforced by the phenomenon of *dilatancy*. Under compression, an initial elastic decrease of volume is followed by an inelastic increase. This results in the opening of numerous microfractures at stress levels much lower than those required for through-going fractures. Provided that the grains of the rock were originally under the closest packing arrangement appropriate to their depth of burial, the dilatant rock acquires increased porosity. Fluids then migrate in the direction of reduced pressure, which is the direction of increased porosity (in some cases, towards a vacuum) and therefore of increasing deformation. The dilatant region is structurally controlled: the crest of an anticline or the plane of a fault, for example (Fig. 13.68).

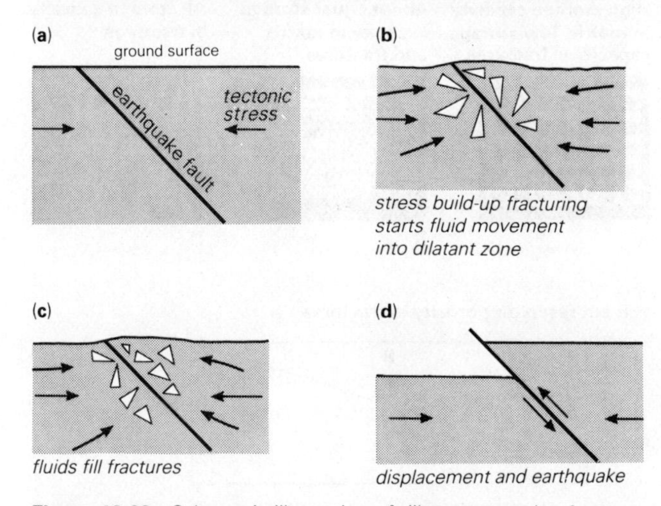

Figure 13.68 Schematic illustration of dilatancy opening fractures under compression, and flow of fluids towards them. (From R. Aguilera and H. K. van Poollen, *Oil Gas J.*, 18 December 1978.)

The relations between porosity and deformation are impossible to quantify with any assurance in nature. If the rock volumes before and after deformation and fracturing were known (V_0 and V_f), then the average fracture porosity is given by

$$\phi_f = \frac{V_f - V_o}{V_f} \times 100$$

Even if this equation were soluble it would not be much help, because we know that fracture porosity is far from uniformly distributed through a rock, so its average value has no practical meaning.

Detailed field measurements of fractures in exposed sedimentary rocks have been made in a number of areas. In 1982, Mark Grivetti reported his study of fractures in folded strata of the Miocene Monterey Formation in the Santa Maria Basin of California. Fracture porosity proved to be highest in chert beds (6.3 percent) and decreased from dolomite (3.9 percent) through siliceous mudstone (2.4 percent), phosphatic limestone (2.1 percent) to organic-phosphatic mudstone (1.8 percent). Structurally, areas close to major faults had the most fractures, folded areas had fewer, and homoclinal areas fewest. Data from equivalent formations in Sakhafin Island are shown in Figure 13.69.

13.4.5 Fracture density and bed thickness

An inverse relation between fracture density and bed thickness is indicated by both theory and observation. It suggests that spacing is controlled largely by frictional drag along bedding surfaces during deformation.

Studies by Neville Price indicate that this may be an oversimplification. In addition to dependence of fracture density on the intensity of the deformation in general and on the lithology of the fractured beds, two aspects of bed thickness appear to control it. Though the thickness of the most competent bed is important, above a certain thickness of only 1–2 m the fracture spacing becomes essentially constant. Furthermore, the spacing in the competent beds becomes wider with increasing thickness of the incompetent beds between the competent ones. These observations support the contention that most fracture sets are extensional in origin rather than deformational.

13.4.6 Oil in fractured reservoirs

The Monterey rocks in Grivetti's study are among the world's most famous *source* sediments. In the Santa Maria and some other basins of California, they are also

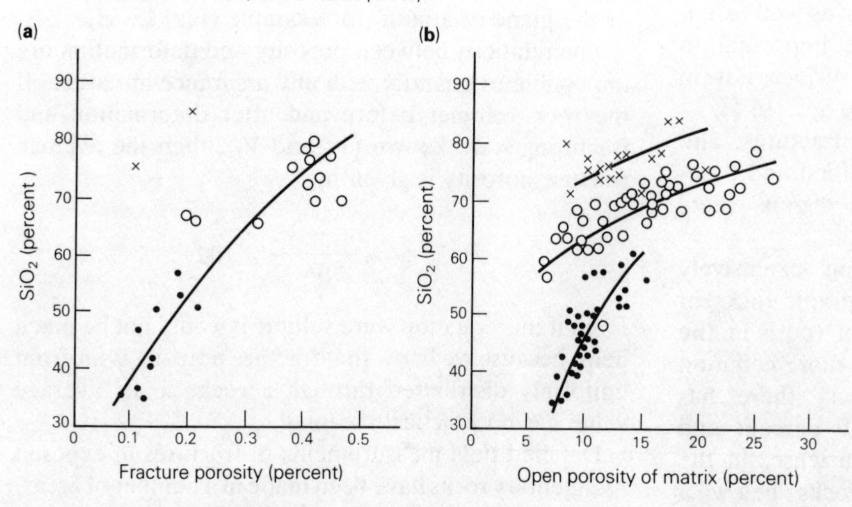

Figure 13.69 Relations between (a) fracture porosity, (b) matrix porosity, and silica percentage for three types of siliceous reservoir rock in the Miocene of Sakhalin Island. (After A. I. Yurochko, *Int. Geol. Rev.*, 1982.)

reservoir rocks (see Sec. 25.6.4), but their matrix porosities (ϕ_m) are very low and their fracture porosities (ϕ_f) obviously not high enough for commercial reservoirs under most circumstances.

If a rock is fundamentally very impermeable (with $k =$ 0.1 mD or less, as in cherts or some calcareous or clastic sedimentary rocks with very dense matrices or cements), fractures may provide all or most of its porosity and all its effective permeability. The fraction of the total porosity (ϕ) which is due to fractures is the *partitioning coefficient* (v):

$$V = \frac{\phi - \phi_m}{\phi_f + \phi_m} \times 100$$

$$= \frac{\phi_f}{\phi} \times 100$$

Fractures may be paper thin (*hairline fractures*) or up to several millimeters wide. Studies of the fractures in Iran's Asmari limestone (Sec. 13.4.9) indicated that a single fracture 1 mm wide in a reservoir rock, if intersected by a well, provides sufficient permeability to yield between 1.0×10^3 and 1.6×10^3 m^3 of oil *per day*, provided that the oil's gravity and viscosity, the formation pressure, and the attitude of the fracture plane are favorable.

Open fractures are avenues of nearly infinite permeability, and may then serve as escape channels rather than as reservoirs, causing transfer or loss of oil if they destroy the integrity of the seals. An example is afforded by oil from the Monterey source rock itself in the Lost Hills field of California's San Joaquin Valley (see Sec. 16.3.3). The most prolific producing horizon in the neighboring fields is the Temblor sandstone below the Monterey equivalent. At Lost Hills, the Temblor is unproductive because the oil has escaped through fractured Monterey into the overlying Etchegoin sands. The fractured Monterey itself yields oil. In carbonate reservoirs, fractures may be enlarged by solution so as to become essentially free channelways. The true origin of fractures found in the subsurface may be highly uncertain; it may be looked upon as irrelevant so long as the fractures are effective agents of production.

The oil may be entirely in fractures in an otherwise dense rock. A more common case has other porosity accounting for some or even most of the oil but the fractures providing the permeability essential for production. The roles of the two are infinitely variable but the general association of fracture porosity–permeability and intrinsic or matrix porosity–permeability can be summarized thus (Figs. 13.70 & 71):

High storage capacity in matrix, low storage capacity in fractures

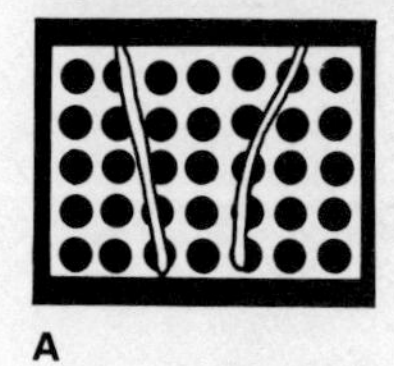

A

About equal storage capacity in matrix and fractures

B

All storage capacity in fractures

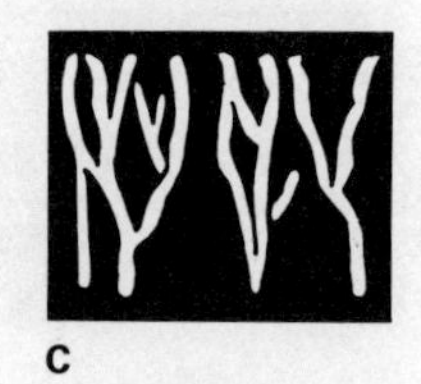

C

Percent reservoir porosity in fractures

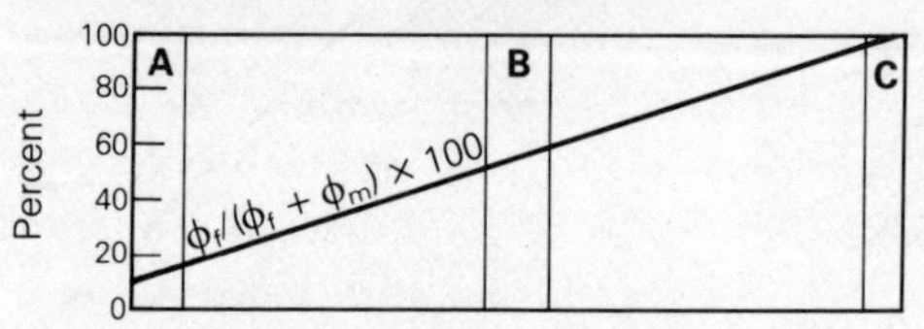

Figure 13.70 Distribution of porosity between matrix and fractures. (From R. Aguilera and H. K. van Poollen, *Oil Gas J.*, 18 December 1978.)

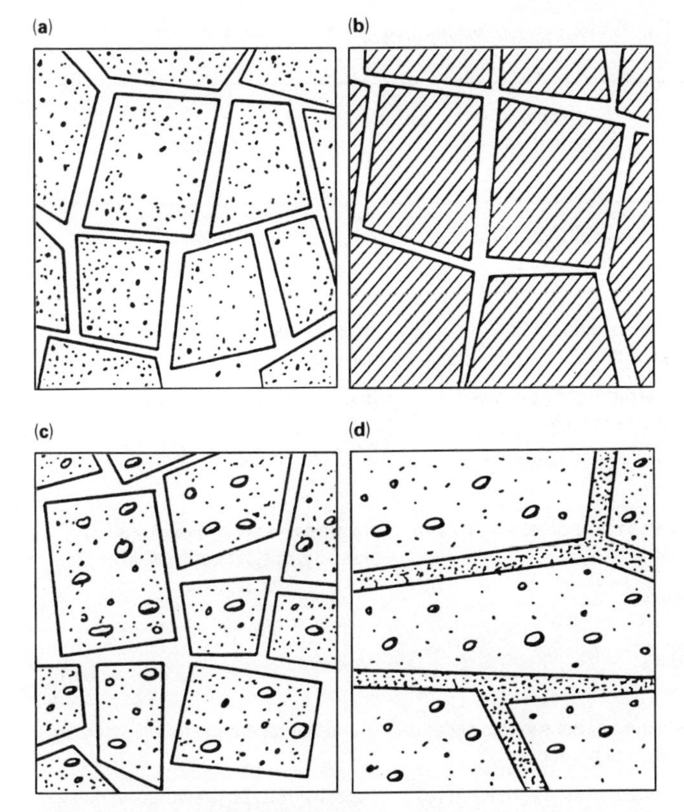

Figure 13.71 Porosity configurations in dual-porosity reservoirs: (a) fractured medium having some matrix porosity; (b) fractured medium having no matrix porosity; (c) both matrix and fracture porosity (dual porosity); (d) heterogeneous medium having porosity in both matrix and cemented fractures. (After T. D. Streltsova-Adams, *Water Resources Res.*, Am. Geophys. Union, 1976.)

(a) With high intrinsic but low fracture porosity–permeability, the latter may be more of a detriment than a help.
(b) With the two about equal an ideal combination may be achieved, with most of the porosity represented by the matrix and most of the permeability by the fractures.
(c) With low intrinsic but high fracture porosity–permeability, initial yield rates may be very high but decline is likely to be rapid.

Fracture production comes from two distinct rock regimes: from fractures in strata forming integral parts of petroliferous successions (like the Monterey Formation, or the Asmari limestone in Iran), and from fractures in basement or igneous rocks unconnected with the generative sediments and otherwise incapable of production (in the Los Angeles, Maracaibo, Sirte, and other basins; see Sec.13.4.9). Fluids produced from the fractures can be hydrocarbons only if the source rock for these is directly adjacent or not far from being so. A productive fractured stratum may be its own source rock, as in the case of the Monterey diatomaceous shales of the San Joaquin and Santa Maria basins. In fractured basement reservoirs, basement relief (as by the elevation of fault blocks) stands topographically higher than younger source strata; the traps are paleo-geomorphic (see Sec.16.6 & Figs 13.73, 13.78 & 16.89).

Not all fracture systems create or enhance effective permeability, of course. They may predate the generation of the hydrocarbons by long enough to have become closed unless rejuvenated. They may be filled by mineral matter deposited by circulating fluids. Episodic introduction of mineral matter reflects the fracture history; if the material is datable, it may reveal which fractures of a system have been rejuvenated and which have remained closed.

13.4.7 Rock types providing fractured reservoirs

Given this favorable association with source sediments, almost any rock type except evaporites may provide highly productive fractured reservoirs.

(a) Limestones and dolomites; the most spectacular fracture production comes from this class of reservoirs.
(b) Chalks and marls.
(c) Diatomites, cherts, and siliceous shales; fractured, productive Mississippian chert in Kansas and Oklahoma is called "chat" by the drillers.
(d) Bituminous shales, especially if they are also siliceous. They are both source and reservoir. The common, lowresistive clay shales are normally too pliable to provide good fractures.
(e) Siltstones (occasionally), less commonly sandstones (the principal primary reservoirs).
(f) Igneous rocks where they occur within the stratified series, as volcanics, sills, or serpentinites.
(g) Basement rocks in buried uplifts, overlapped by source sediments.

Examples of all types will be described.

13.4.8 Recognition of fractured reservoirs

The first clue to the presence of *effective* fracture porosity and permeability is well performance. Wells yield better test results and better production than the intrinsic properties of the producing horizon suggest would be possible. Identification of the fractures in their true subsurface state is not easy, however. Downhole cameras are a big help, but the geologist normally needs to rely upon his more traditional lines of evidence.

Cores might appear on first thought to be the most direct means of identification. Cores, however, are not good statistical samples of the reservoir. They may reveal that the *matrix* porosity is low and that fractures are present; but it is difficult to preserve fractures in cores in their original state and orientation. The geolo-

Table 13.1 Well logs used for fracture detection. (Simplified from J. D. Heflin and E. Frost Jr, *Pacific section*, SEPM, 1983.)

Log	Depth investigated (cm)	Recordable in cased holes	Principle of detection	Conditions
resistivity (including induction and laterolog)	75–300	no	resistivities at different depths of investigation	depth of investigation depends on vertical spacing of electrodes
microlog	3–10	no	mudcake resulting from permeability	fracture density must permit formation of mudcake
SP log	0	no	potential difference between formation and fluid	appropriate fluid must be present
microlaterolog	15	no	minor effects of fracturing on borehole	assumes effects are due to fractures
sonic log	0	no	compares total porosity and matrix porosity	fractures must significantly affect porosity, not just permeability
compensated density log	15	yes	as above	as above
compensated neutron log	20	yes	as above	as above
four-arm diplog	3.8	no	resistivity variations in four separate segments of borehole	should be repeated around borehole
caliper log	0	no	rugosity associated with fractures	rugosity must be present, and due to fractures

gist must be certain that the fractures he sees were not induced during the coring process (or by earlier drilling). They may even be created by "unloading" of vertical stress during withdrawal of the core (especially if they are more or less perpendicular to the axis of the core). Some fractures in that position may of course be natural, but if they are effective in the storing and releasing of oil they are likely to be marked by staining or by infilling with some form of hydrocarbon.

The presence of slickensides on fractures, or of tension gashes associated with stylolites (Sec. 13.4.3) is evidence of natural fracturing. So is cementation of the fractures. For cemented fractures to remain effective, of course, channelways must be left open between the cements on the opposed walls of the fractures. To determine whether such fractures are seen in the core with their true widths, *cathodoluminescence photographs* may be helpful by enabling the geologist to pick out the continuity of crystalline fracture linings (especially of well formed crystals of dolomite or silica).

Analysis of cuttings may reveal oil staining on planar surfaces which are more likely to be fractures than pore walls. If the reservoir matrix is highly permeable, however, hydrocarbons may be swept out of all porosity close to the wellbore by the drilling fluid.

Well logs present peculiar difficulties if the geologist attempts to evaluate fractures as well as both the rock itself and its fluids. Intense fracturing may cause lowered resistivity if the matrix material is normally of high resistivity (as dense limestones, common hosts of fracture porosity and permeability, normally are); the resistivity of the drilling fluid is very different from that of the rock and the fluid forms a conductive zone around the well-bore. The SP may be increased because of streaming potential, but it needs to be high for good production anyway.

Closely spaced fractures reduce the amplitudes of acoustic waves about tenfold as they pass through the rock. A modification of the sonic log (Sec. 12.2.2) to record *amplitude* has therefore been developed to aid in the detection of fractures. Unfortunately it cannot discriminate between open and closed fractures (or between useful fractures and bedding planes or slaty cleavage).

Open fractures are likely to be invaded by drilling fluid, forming filter cake on the walls of the hole and causing lost circulation (see Ch. 12). The microlog is a wall-resistivity measuring device which detects filter cake. The caliper log may locate causes of lost circulation. The rate of bit penetration may increase in fractured zones, which might then be revealed in the drill-time logs; the increase is not inevitable, however, and there are many other possible causes for it.

A potentially more trustworthy approach is to combine two logs reflecting different aspects of the rock's *porosity*. For example, the density log and neutron log purport to measure total porosity (Ch. 12). The travel time (Δt) measured by the sonic log is known to yield a lower calculated porosity than the neutron log in fractured sections. If this is because the travel time is through continuous solids, the sonic log may be reflecting only matrix porosity, without the barriers presented by fractures. In this event

$$\phi_{\text{neutron}} - \phi_{\text{sonic}} = \phi_{\text{fracture}} \quad \text{(approximately)}$$

Similarly, in the combination dual induction–laterolog (Sec. 12.2.1), the induction log registers resistivity to a field induced *horizontally* into the formation. The laterolog is affected equally by the horizontal and the *vertical* resistivity over the interval focused upon. If the horizontal resistivity is greater than the vertical, the difference may be due to mud-filled fractures closer to the vertical than to the horizontal.

If the fractured reservoir rock is a shale which is also the source rock, it will have some distinctive log characteristics. They include high natural radioactivity; non-definitive SP (because the matrix porosity is negligible and the fractures are unlikely to be closely and uniformly spaced); high resistivity because of siliceous or calcareous content and solid kerogen in the fractures; and low bulk density on the sonic log or borehole gravimeter.

Though a combination of these approaches is often helpful, none is definitive. The identification and interpretation of fractured reservoirs depend fundamentally on shrewd deduction from well behavior.

13.4.9 Case histories of fractured reservoirs

Fractured carbonate reservoirs The most famous of fractured carbonate reservoirs are in the Oligo-Miocene Asmari limestone fields of *Iran*. The reservoir limestone, more than 300 m thick, has been folded into concentric flexures of great amplitude, enhanced by the vast difference in competence between the reservoir and the seal (Fig. 13.72). The average matrix porosity is about 8 percent, the permeability 1 mD. Easy and rapid migration of fluids is made possible, however, by lateral and vertical free connection of fissures in the reservoir rock; it is said to be "crackled."

The fractures do not control the storage of the oil, but they control production rates because wells receive oil from them. The fractures themselves are replenished by a source having more limited flow. Open fractures are rare; most of them are from 0.5 to 5.0 mm wide. There is no absolute distinction in practice between fractures of this small width and pore space. At high production rates, many of the finer fractures behave as pores; at lower rates, the rock having the highest matrix permeability behaves as fractured rock.

Mexico owes much of its past and present production to fractured reservoirs in Mesozoic carbonate rocks. The old fields of Ebano-Panuco, north of Tampico (see Fig. 17.22), produced from dense and cherty but fractured Tamaulipas limestone of Cretaceous age, and from the overlying shale which is also the source sediment. The strata occupy a complex of anticlines and noses, and the fracturing may be due to a second generation of folding consequent upon vertical uplift of the Sierra Tamaulipas during the Tertiary. This led to redistribution of oil already pooled. The seals are simply more argillaceous beds of the same formation which have remained inadequately fractured or unfractured.

The much larger and newer fields of the Reforma and Campeche districts of southeastern Mexico are described elsewhere in this book (Chs 25 & 26). Laramide deformation induced salt pillowing below pre-existing horst-graben structures. The uplifts caused intense microfracturing of dolomitized, reef-derived calcaren-

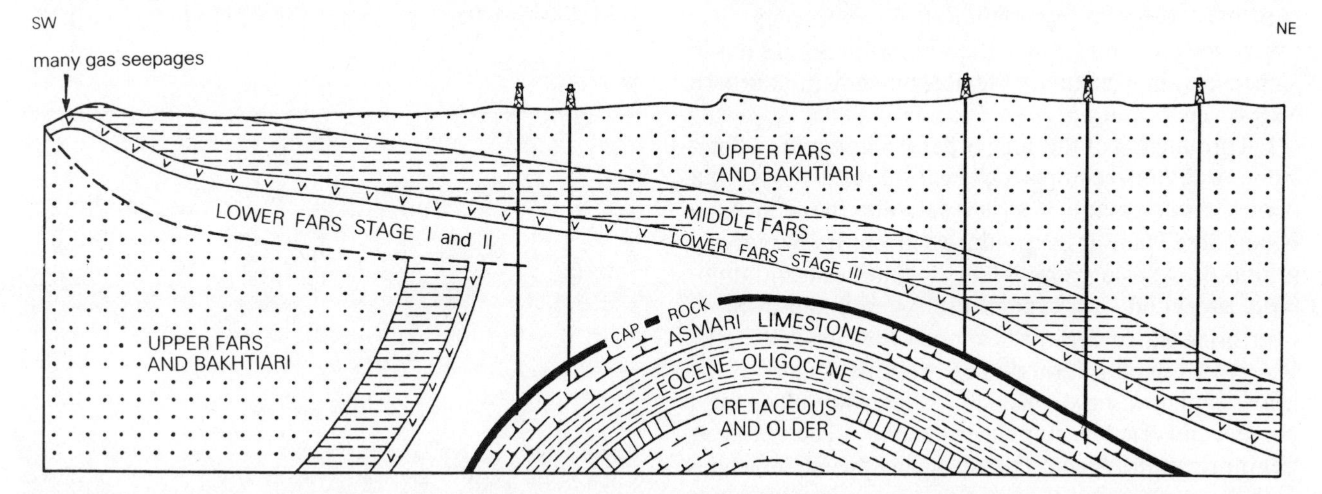

Figure 13.72 The Agha Jari oilfield in Iran. Symmetrical anticline in Asmari limestone and older rocks, disharmonically overlain by a flow-thrust sheet of weak Tertiary strata. Fissure permeability in the Asmari limestone permitted production of 4000 m³ per well per day, and 20 × 10⁶ m³ of cumulative output from individual wells. (From British Petroleum Co. Ltd, XX International Geol. Congress, Mexico City, 1956.)

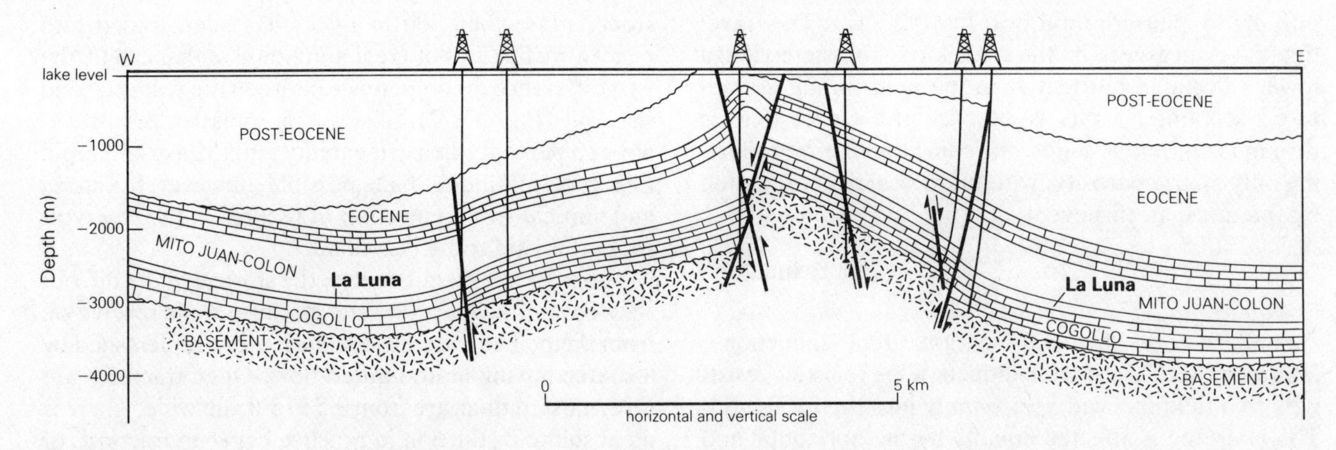

Figure 13.73 Cross section through the La Paz field, Maracaibo Basin, Venezuela. (After A. Salvador and E. E. Hotz, 6th World Petroleum Congress, 1963.)

ites of late Jurassic and Cretaceous ages. The dolomites, of very low primary porosity, acquired high secondary permeability over pay thicknesses of hundreds of meters. Of total porosity of less than 10 percent, fracture porosity represents only 1.5 percent but creates permeabilities to 7 darcies or even higher.

Along the west side of the Maracaibo Basin in *Venezuela* (Sec. 15.5.8), prolific production came from thick oil columns in fractured Cretaceous limestone and its underlying metamorphic basement. Sharp, faulted anticlines originated during the Cretaceous and continued to grow into the Miocene (Fig. 13.73). Porosity and permeability were enhanced by solution along the fractures; in places they were reduced by infilling with calcite. The best wells entered the extensional parts of the structures, especially on the upthrown sides of normal faults.

One of the many producing horizons in the Midland Basin of *West Texas* is the Cambro-Ordovician Ellenburger Formation. Dense dolomites are fractured in anticlines raised during a Pennsylvanian orogeny. The source rock is younger than the reservoir rock, as it is in Venezuela and northern Mexico but not in southern Mexico or in Iran.

Oil production of this high order from fractured limestone reservoirs requires low gas : oil ratios. Gas itself occurs less commonly in such reservoirs. In the Sichuan Basin of *China*, it is produced from fractures and solution cavities in Permo-Triassic limestone and dolomite associated with evaporites. An example from an overthrust anticline is the Savanna Creek gasfield in the Rocky Mountain foothills of western Canada; the Mississippian reservoir has matrix porosity of about 4 percent and very low matrix permeability. The Lacq gasfield north of the *French Pyrennees*, in Lower Cretaceous limestone, is perhaps the largest among many examples deeply buried below unconformities in Tertiary fold belts.

Fractured chalk or marl reservoirs Chalks and marls are uncommon reservoir rocks. The most notable are those in the Danian fieldsof the central *North Sea* in the Ekofisk complex. The reservoirs have retained remarkably high intrinsic porosity and permeability in the light of chalk's susceptibility to both compaction and pressure-solution (Sec. 13.3.6). The structures have been raised above flowage uplifts of deep salt, and crestal fractures contribute further to the effective porosity and permeability (Fig. 13.74).

Cretaceous chalks in the *United States* are normally of too low porosity to constitute reservoir rocks. Some biogenic gas is produced from the Niobrara chalk in Colorado and Kansas. The Austin chalk trend across the Texas Gulf Coast is very extensive and contains a vast volume of oil in place, especially along the upthrown side of the Luling fault zone. Yet there are no true traps along the trend, no single, distinctly defined accumulation of oil, and no oil-water contact in the conventional sense. Permeabilities are less than 1.0 mD, in many stretches less than 0.1 mD. Updip towards the

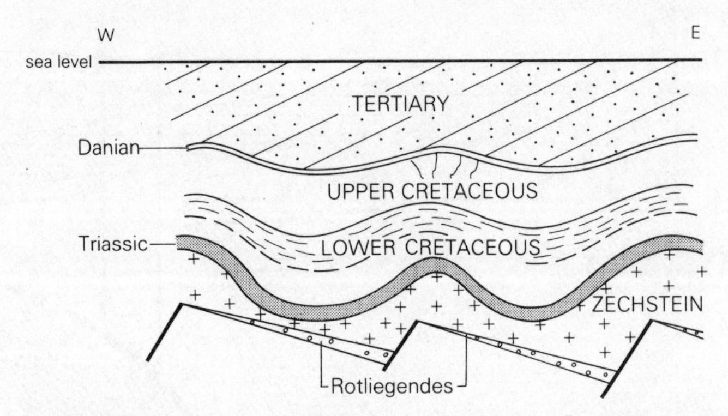

Figure 13.74 Schematic cross section of the Ekofisk sector of the central North Sea. Block faulting in Hercynian basement is proposed as the cause of Zechstein salt flowage. Upper Cretaceous and Danian chalks are fractured in domal uplifts over the salt swells. (From P. E. Kent, *J. Geol. Soc. Lond.*, 1975.)

formation's outcrop, porosity and permeability are higher, but the rock is water-bearing. Downdip towards the coast, the chalk is tighter, temperatures are higher at the greater depths, GORs increase and eventually the formation yields only gas. This characteristic of lying downdip from water is shared with a number of large-volume but impermeable reservoir rocks, including the tight gas sands in the Cretaceous molasse of the North American Cordillera (see Sec. 16.4.6). With production dependent wholly on fracture porosity and permeability, it is impossible to estimate what the chalk trend's ultimate output will be.

Fractured shale and chert reservoirs The most productive of fine-grained siliceous rocks is the Miocene Monterey Formation in *California*. It is the source sediment for most if not all of the oil produced there from orthodox sandstone reservoirs. The source formation itself is productive from fractures in several basins (Sec. 13.4.4).

The critical component of the Monterey Formation is a diatomite or diatomaceous (siliceous) shale. In the coastal basins, the opal-A of the diatomite is converted at depth to opal-CT (see Sec. 13.4.3), forming a thick porcellanous and cherty zone in the middle of the formation. Diatom tests disappear and the rock acquires lepispheres, or spherical conglomerations of small platelets. Coincident with the transformation is an abrupt loss of oil saturation. Within the oil-saturated section, oil is produced from fracture porosity.

In the *Santa Maria Basin* (Fig. 13.75), the oil-saturated section extends up into the Pliocene (still diatomaceous) and down into fractured basement rocks, giving a gross pay thickness of more than 300 m. Much of the production actually comes from dolomite horizons interbedded with the siliceous sediments; fractures are wider in the otherwise dense dolomite, providing much higher fluid conductivity. In addition to sharp anticlines raised by Pleistocene folding, traps include a large stratigraphic wedge-out of fractured diatomite, resting directly on fractured basement rock which is also productive (Fig. 13.76).

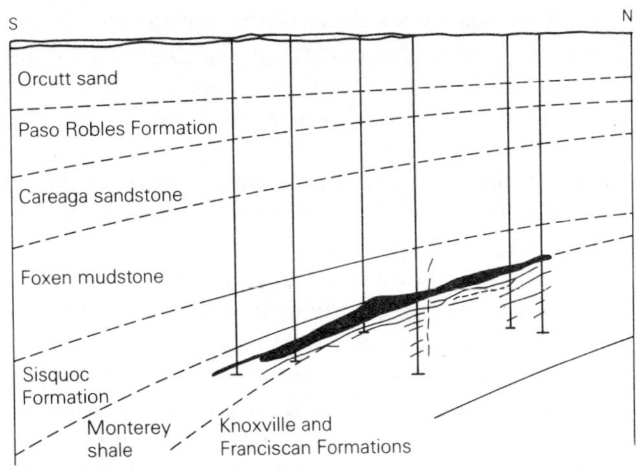

Figure 13.76 Cross section of Santa Maria Valley oilfield, California, showing productive Knoxville basement rock with well developed fracture porosity, overlain by rich source sediment of the Monterey diatomaceous shale (Miocene). (After W. P. Woodring and M. N. Bramlette, US Geological Survey Professional Paper 222, 1950.)

Along the west side of the *San Joaquin Valley* (Figs 16.37 & 38), the Monterey equivalent (here a highly diatomaceous shale) again provides both source and fractured reservoir in several fields. Fracture production is restricted to the axial regions of folds; in some fields, wells must actually intersect fractures to achieve production. Both here and in the offshore part of the *Ventura Basin*, much of the oil produced from the Monterey Formation is heavier than that produced from sandstone reservoirs. The heaviest oil is beyond production capabilities.

In the *Uinta Basin* of Utah (Chs 10 & 25), the late Cretaceous and Tertiary strata are all nonmarine. The Uinta Mountain block north of the basin was greatly uplifted in late Tertiary time. A regional fracture pattern was thereby created, extending through a great thickness of strata but failing to affect numerous ductile units against which the fractures terminate upwards and downwards. Minor production has come from thin sands and dolomites within the fractured Green River oil shale. In the axial region of one large anticline which contains a giant field (Rangely) in an Upper Paleozoic reservoir, oil is produced from fractured Cretaceous shale.

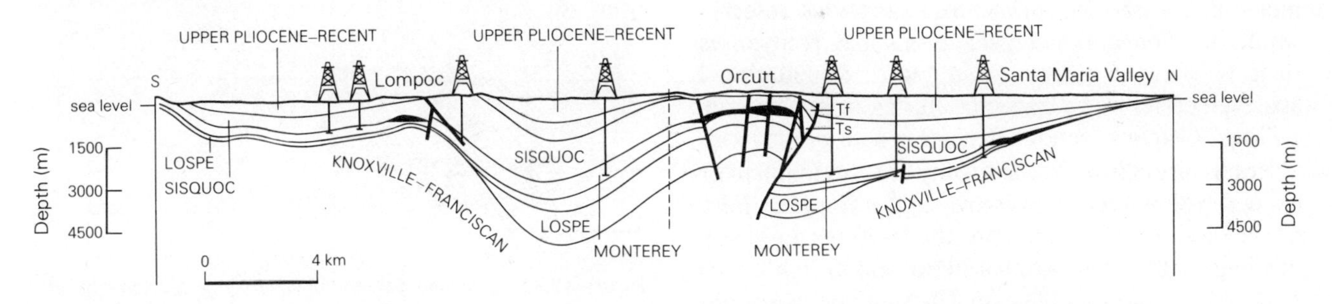

Figure 13.75 Cross section through the Santa Maria Basin, California. (From F. D. Crawford, *AAPG Memoir* 15, 1971; after K. F. Krammes and J. F. Curran, 1959.)

The Devonian black shales of the *Appalachian Basin* contain great quantities of methane gas. Some of the gas is trapped in fracture porosity in the shale; more is part of the shale matrix and diffuses only slowly into the fractures (Sec. 10.8.2). The shales are therefore source, reservoir, and seal combined. Some gas is produced from the fractures, but substantial production will require artificial fracturing on a vast scale.

In 1979 the oil world was startled by a Swedish report of the world's largest oilfield in a shale reservoir in western Siberia. It proved to be the Salym field, in which the combined source and reservoir rock is an Upper Jurassic bituminous argillite. There are no true reservoir beds. The fractured argillite is overpressured, with unfractured shales above and below it; there is no formation water. No conventional primary or enhanced recovery method has proved efficient, and the field may be a candidate for underground nuclear stimulation (Ch. 23).

Fractured sandstone–siltstone reservoirs The Devonian Oriskany sandstone of the *Appalachian* foreland provided one of the earliest producing reservoir rocks. It is actually a tight but brittle rock, fractured in faulted anticlines and thrust structures. It also produces from paleogeomorphic traps.

The largest single productive body of fractured "sandstone" is in the Lower Permian Spraberry Formation of *West Texas* (Sec. 25.2.4). Tensional fractures were formed in deep-water clastic sediments by draping and differential compaction over a large monoclinal flexure. The shale members appear to be the source sediments, because oil can be retorted from them; but fractures in the shales tend not to remain open. Most production has come from thin, tight but fractured, fine-grained siltstones, with poorly sorted, interlocking, angular grains cemented by dolomite. Average matrix permeability is 0.5 mD, but the productive intervals average 30 mD. Fractures constitute no more than 1 percent of the formation's bulk volume; their role is as channels, not as reservoirs. Capillary discontinuities are created at every fracture plane, so the sands tend to retain their oil much more than a normal, unfractured sandstone reservoir would do. Conventional mechanisms may recover less than 10 percent of the oil in place. Stimulation is achieved either by hydraulic fracturing or by "pressure pulsing" (water injection alternating with production).

Reservoirs in fractured siltstones are also found interbedded within Cretaceous shales in the *San Juan Basin* of New Mexico (Sec. 25.5.10); the fractures are due to bending of the strata over a major monocline on the edge of the Colorado Plateau. The small fields are produced by gravity drainage (Sec. 23.8.4).

The reader will have noticed that these three examples are all from the USA. Outside North America or Western Europe, such production probably could not have been economic before 1973.

Fractured reservoirs in igneous rocks Fractured igneous rocks which are not basement form commercial reservoirs in a few areas where they happen to be underlain or overlain by a source rock or intruded into it. By far the commonest cases are of fragmental rocks, especially tuffs. The largest such field is probably Tupungato, in Argentina's *Mendoza Basin*; oil is produced from fractures and pores in thick tuffs of the Upper Triassic, its source being in overlying bituminous shales containing abundant fish scales. Most other examples are Tertiary. A little oil comes from Eocene tuffs in *Cuba*, from Oligocene ignimbrites in the Great Basin province of *Nevada*, and from agglomerates, sandy tuff breccias, and intermediate lavas in the Mio-Pliocene of western Honshu, *Japan*.

Fractured lavas rarely form effective reservoirs. Oligocene lavas do so in northwestern *Java*, resting on faulted basement. Heavy, high-sulfur oil has been tested from Mio-Pliocene fractured basalt under Great Salt Lake in *Utah*. Such examples are invariably in faulted structures and they provide no significant production.

An oddity is the Dineh-bi-Keyah field on the edge of the Paradox Basin in *Arizona*. Production comes from an Oligocene syenite sill intruded into a black shale within a thick Pennsylvanian limestone. The natural fracturing provides porosity but sandfrac treatment is still necessary to achieve commercial production. Most other intrusive igneous rock reservoirs are in serpentinized rocks, in fields along the Luling fault zone in *Texas* and one or two small fields in *Cuba* (Fig. 13.77).

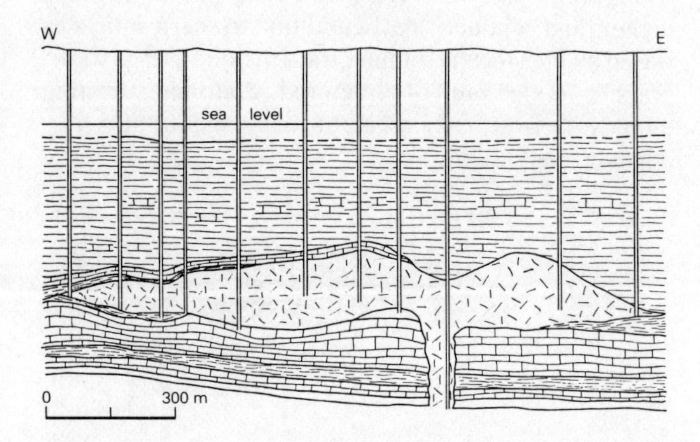

Figure 13.77 Cross section through serpentinized igneous rock reservoir of Chapman field, Texas. Actual shape of igneous body not fully known. (From K. K. Landes, *Petroleum geology of the United States*, New York: Wiley–Interscience, 1970; after E. H. Sellards.)

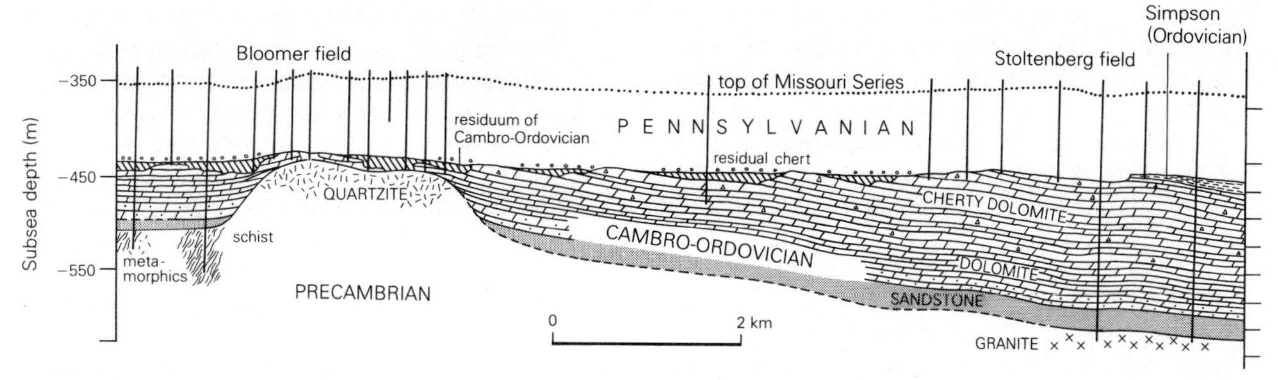

Figure 13.78 Cross section through the Central Kansas uplift, showing fracture porosity in and adjacent to buried hills of Precambrian rock, karst topography on old erosion surface of Cambro-Ordovician dolomites, and drape folds formed by compaction over basement topography. (From W. A. ver Wiebe, *North American petroleum*, 1957; after R. F. Walters, 1952.)

Reservoirs in fractured basement rocks Basement rocks, even if of sedimentary origin, have essentially no matrix porosity. If they had, they would not be basement; "basement" to petroleum geologists and drillers is not to be equated with "Precambrian." True basement rocks, however, may have considerable fracture porosity, because of deformation, weathering, or both. This porosity is prolifically productive where high basement structure or topography, forming *pseudo-anticlines*, is overlapped by effective source sediments.

Along the *Central Kansas uplift* (Fig. 13.78), rich production came from Precambrian quartzite in buried hills, overlapped by highly petroliferous Pennsylvanian rocks. In *Libya* several giant fields yield some or most of their oil from weathered and fractured Precambrian igneous rocks, mostly granites and granophyres, or from fractured Cambro-Ordovician sandstones immediately overlying them, on prominent basement highs. In the *Dnieper–Donetz depression* of the Soviet Union (Sec. 25.2.2), some production comes from jointed amphibolite and schist of the Hercynian basement, overlain by Carboniferous sedimentary rocks.

There are much younger basements also. In the *Maracaibo Basin* of Venezuela, the large La Paz and Mara fields (Fig. 13.79) produced from both fractured Cretaceous sediments and the meta-igneous basement underlying them. Fracture porosity was oil-bearing to a depth of 500 m below the top of the basement. For the *Santa Maria Basin* of California, production from fractured Mio-Pliocene diatomaceous rocks has already been described. Production extends down into fractured Jurassic sandstone which constitutes semi-metamorphic basement elsewhere in the state. Along the west side of the *Los Angeles Basin* (Fig. 16.27), several fields productive from the Neogene also produce from underlying fractured schist and schist-conglomerate, derived by the metamorphism of the ophiolitic Franciscan Group of Mesozoic age. On the east side of the *San Joaquin Valley*, some production has come from fractured basement of Cretaceous granodiorite.

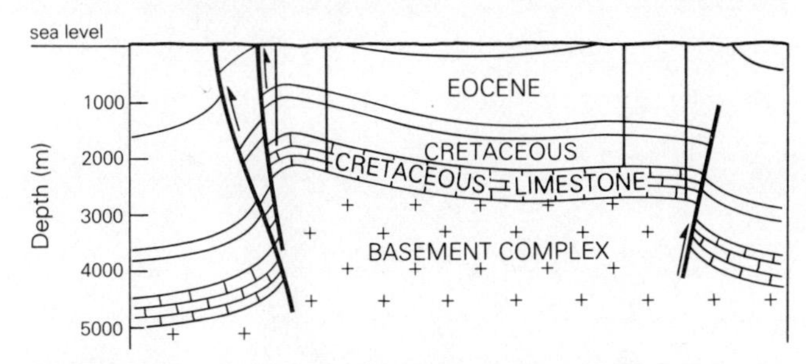

Figure 13.79 Cross section through the Mara oilfield, Maracaibo Basin of Venezuela. Production came from fracture porosity in the Cretaceous limestone and the igneous basement complex of an uplifted horst block. After E. Mencher *et al.*, *AAPG Bull.*, 1953.)

13.5 Petrophysical studies of reservoir rocks

With modern production techniques, including those of enhanced recovery (Ch. 23), and with oil and gas prices ten times what they were in the 1960s, many oil- and gas-bearing rocks have become potential reservoir rocks that could not have been so in the past. The detailed examination of such reservoir rocks is essential in judging their potential for commercial production. This is especially the case for offshore discoveries in deep water. The investment in a production platform and its ancillary services demands reliable estimates in advance of the producibility of the reservoir, in detail.

The *petrophysical* study of a reservoir rock involves computer-assisted computations from suites of well logs. In brief, the basic approach is as below.

(1) Logs are analyzed first for all *porosity* data. These data are then corrected for the effects of compaction and hydrocarbon content.

(2) The effects of *shaliness* are then taken into account. Both sandstone and carbonate reservoir rock properties are seriously affected by two kinds of shale content. Laminar shale layers within the

reservoir rock make the true, net pay thickness substantially less than the gross. Such layers, however, have been subjected to the same compaction as shale beds above or below the reservoir, so their log responses can be inferred from those of true shale units. Matrix shale dispersed in the pore spaces of the reservoir rock is subject only to hydrostatic pressure, because the reservoir rock itself has not undergone the same compaction as the bedded shales. The dispersed shale therefore retains higher water content than does the bedded shale, and this affects the reservoir porosity and consequently its log responses. The SP is reduced. Clay minerals coating grains or lining pores have high cation exchange capacity (CEC) and therefore high conductivity; the resistivity is therefore lowered for a reservoir which may nonetheless be highly productive. Density, neutron, and gamma-ray logs are also affected by the shale content of the rock, but not by the manner in which the shale is distributed within it. Shaly sands are studied with the aid of the SARABAND program, other shaly reservoirs via CORIBAND (see Ch. 12).

(3) Formation factors, water saturations and R_w calculations are made as thoroughly as possible (see Sec. 12.4).

(4) *Cross-plots* are drawn between pairs of porosity logs (e.g. the neutron and density logs; Fig. 12.12) and the lithologic analysis is completed from combinations of these.

14 Pressure conditions in the reservoir

14.1 Reservoir pressures

All rocks at depth within the crust are under the pressure of the rocks overlying them. This pressure is the *confining* or *overburden pressure*; more correctly, the confining or overburden *stress* (P_o). Sedimentary rocks normally contain fluids, and are subject to compaction (Sec. 11.1) as a consequence of the overburden stress. The overburden stress is borne by both the materials of the rock (sand grains, clay particles, etc.) and its fluid content.

The stress system in the rock, under compaction, is therefore in equilibrium when the overburden pressure (P_o) equals the sum of the compaction or grain pressure (P_c) and the fluid or pore pressure (P_f). The potential energy of the reservoir is stored in its compressed fluids, the liquid members of which (oil or water) resist compaction. Thus the present *effective stress* (σ_p) on the system is given by the equation made familiar by King Hubbert and William Rubey:

$$\sigma_p = P_o - P_f \qquad (14.1)$$

The ratio of P_f to P_o, at any depth Z, is called the *geostatic ratio* (λ). Because the term "overburden" may be ambiguous (meaning to many geologists the overlying rock only, without consideration of its fluid content), a common term for P_o is *geostatic pressure*.

Pressure increases linearly with depth (Fig. 14.1). The *lithostatic pressure gradient* depends on the densities of the rocks in the section. The *geostatic gradient* depends on the bulk wet densities of the rocks, including their fluids. A geostatic gradient of 1 p.s.i. per foot (22.6 kPa m^{-1}) would result from an average density of 2.3. In a hydrocarbon-bearing reservoir, this gradient forces more gas into solution in the oil, reducing the oil's viscosity with increasing depth (Fig. 5.11). Temperature also increases with depth, further reducing the viscosity of the oil. Porosity, however, decreases exponentially with depth (see Fig. 11.4), so that increasing volumes of fluid occupy less pore space and P_f therefore rises.

The present pressure–depth and porosity–depth curves (Figs 14.1 & 11.4–11.6) therefore diverge with depth, representing the history paths of both pressure increase and porosity decrease. The present value of σ_p may not be the highest it has ever achieved, because of the removal of formerly effective overburden by erosion; in that event, the two curves will diverge from one another more strongly.

14.2 Normal and abnormal formation pressures

The *fluid pressure gradient* is said to be "normal" if it approximates the *hydrostatic gradient*: 9.8 kPa m^{-1} (0.434 p.s.i. per foot) for fresh water, 10.5 kPa m^{-1} (0.465 p.s.i. per foot) for "standard" water containing 100 parts per thousand of total dissolved solids, and 11.3–12.4 kPa m^{-1} (0.50–0.55 p.s.i. per foot) for strong brines. Gradients 1 and 2 in Figure 14.1 represent the hydrostatic gradients for fresh water and "standard" formation water. Clearly, gradient 1 also represents a value of $\lambda = 0.434$ and gradient 4 a value of $\lambda = 1.0$. If P_f is excessive, σ_p is less than hydrostatic, and vice versa.

The concept of a *normal* formation or fluid pressure accepts as the "normal" condition the ability of the fluid content of any rock in the subsurface to equalize to the hydrostatic pressure for its depth. In nature, an array of geologic and hydrologic impediments are available to prevent this.

Reservoir formations having continuous permeability over wide areas, or very insecure seals, quickly attain the pressures of hydrostatic equilibrium, by migration of fluids out of them and the consequent loss of head. Reservoir formations consisting of poorly sorted, argillaceous sandstones, which have lost most of their fluid content through compaction, or have poor communication with the surface, also yield formation pressures differing little from hydrostatic. Their fluid pressures are *normal*.

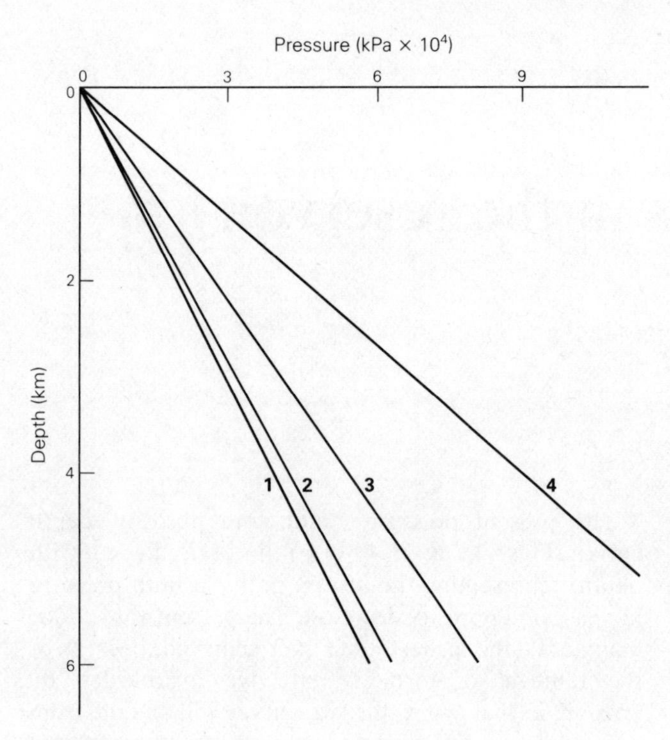

Figure 14.1 Pressure gradients under four "reference" conditions: 1, hydrostatic (fresh water); 2, hydrostatic (standard brine); 3, vertical lithostatic; 4, geostatic (2 + 3). Reservoir pressures normally lie within the envelope.

Basins in the interiors of continents, such as foredeeps and cratonic sags (Ch. 17), have been strongly uplifted and their sections involve large erosional, stratigraphic gaps. In such basins overburden pressures and formation temperatures have been reduced from earlier maxima; the present hydrodynamic regimes may still be in their infancy and the buildup towards hydrostatic equilibrium is still in progress. Fluid potential in the deeper aquifers is *subnormal*, and the pressure gradient may be much below the hydrostatic. In some basins of the North American Mid-Continent and Rocky Mountain provinces, the subnormal pressure gradients have apparently been so long-lived that effective permeability barriers to the achievement of equilibrium must exist.

Hydrocarbon-bearing reservoir formations, however, have necessarily retained open pore spaces and their fluid contents. In such formations, pressures below some critical depth must come to exceed hydrostatic during the time span required to dissipate the excess pressures and re-establish those of equilibrium. The condition of excess fluid pressure is ephemeral in terms of geologic time. Cenozoic reservoir pressures are invariably higher than hydrostatic, but pre-Cretaceous reservoirs commonly deviate little from it and Paleozoic reservoirs are as likely to be below as above normal.

Any formation pressure which exceeds the hydrostatic pressure exerted by a column of water containing 80 000 p.p.m. total solids (0.458 p.s.i. per foot or 10.3 kPa m^{-1}, equivalent to mud weight of 9.3 lb per US gallon or 1114 kg m^{-3}) is considered *abnormal*. The entrenchment of this term in the language of petroleum geology is unfortunate. The equation of "abnormal" with "abnormally high" is in fact an indirect acknowledgement that "abnormal" pressures, as formally defined, are in fact "normal" pressures in the conditions of most petroliferous basins. Pressures higher than hydrostatic are an integral aspect of the processes of generation, migration, accumulation, and preservation of fluid hydrocarbons. Many geologists and engineers speak and write not of "abnormal" pressures but of *overpressures* or *geopressures*.

Overpressures are independent of absolute depth, and also of age, though the degree of overpressuring is greatest in young basins associated with continental margins (see Ch. 17). In such basins, sedimentation has continued into recent time. If no uplift and erosion have intervened, it may still be continuing. Overpressuring requires only a *potentially dynamic fluid* (itself an aspect of both the reservoir and the migration mechanism) enclosed by a *seal*. No seal is perfect (see Sec. 11.3), but many shales have such low permeabilities compared with those of their associated sandstones that most overpressured reservoirs are *sandstones enclosed within thick shales or mudstones*. These are single-phase systems. Normally interbedded sandstones and shales may also achieve abnormal pressures as multiphase systems, with reservoir bottoms needing no seals because of buoyancy.

14.2.1 Overpressures in sandstone–shale successions

Studies of modern deltaic sediments show that thick clay formations depart from stress equilibrium, in the direction of overpressure, after a burial no deeper than 600–800 m, provided that they are loaded with *permeable* sediments (as they are likely to be during marine regressions). Fluid expulsion from the clays is then retarded; excess fluid pressure props open interparticle porosity; and compaction is prevented from achieving pressure equilibrium.

The clay formation therefore remains *undercompacted* as well as overpressured. A typical sequence of pressure conditions in a basin still lacking equilibrium is as below:

(1) at the top, normally compacted sediments;
(2) overcompacted sediments, the fluid content of which has been largely expelled;
(3) normally compacted sediments, with normal fluid pressures;
(4) undercompacted, overpressured sediments, becoming more so downwards – the fluid pressure gradient increases downwards.

The accompaniment of overpressuring by undercompaction invites the attribution of the former to the latter, but this is of course essentially circular reasoning. The fundamental feature common to the great majority of excessively pressured sections is *rapid sedimentation under a regressive regime*, in which coarse, near-shore sediments are piled on top of offshore mud-rocks. The fluid potential gradient is then upward and outward (towards the basin margins), but the loss of permeability prevents the formation fluids from escaping fast enough to maintain the normal fluid-pressure gradient. In classic studies by King Hubbert and William Rubey, measurements of deep reservoir pressures showed their approach more and more nearly towards the overburden pressure. The value of the geostatic ratio λ (defined as P_f/P_o) thus approaches unity, so that the effective stress, σ_p in Equation 14.1, becomes extremely small.

The condition is readily exacerbated by an array of geologic phenomena common in thick clastic successions. The generation of methane in the muds adds the gas phase to the overpressured liquid phase. This in turn promotes the growth of mud diapirs and mud volcanoes, and facilitates the formation of listric growth faults which permit the accumulation of still greater thicknesses of sediment on their downdropped sides (see Sec. 16.7). The faults themselves easily become lateral seals to the expulsion of fluids towards the basin margins, as in the overpressured shales of the Viking graben in the North Sea and in the US Gulf Coast.

The effects of rapid loading by sediments may be further reinforced by gravitational or tectonic loading. Masses or sheets of material may be emplaced above the section by gravity glide or by compressional thrusting. In the latter case, a component of the lateral tectonic pressure causing the thrusting is added to the overburden pressure. Conversely, the rapid removal of overburden by erosion following uplift may contribute to overpressuring if well sealed reservoirs are thereby left beneath overburdens substantially thinner than those responsible for their pressures.

Colin Barker, John Bradley, and others disputed the proposition that compressive stresses from the overburden could be the prime contributors to abnormal pressures at the relatively shallow depths at which the pressures are actually found. Of a number of contributing factors, these authors considered *temperature changes* the most important. The coefficients of thermal expansion of fluids are much higher than those of rocks; fresh pore water has a coefficient about 40 times, medium-gravity oil about 200 times, and gas 800 times those of most sedimentary rocks. All abnormally pressured zones are marked by abnormally high temperatures also. Thus the thermal expansion of the fluids in isolated packets of sediment would lead to *aquathermal pressuring*. In deeper reservoirs, the thermal degradation of petroleum itself may contribute to the pressure pump-up. Richard Chapman, William Plumley, and others have denied the adequacy of the static expansion of confined fluids as the mechanism of overpressuring; changes in P_f must result from mechanical flow of the pore fluids.

On the other hand, *underpressured gas* may indeed be caused by temperature changes. If an isolated sandstone reservoir is completely filled with gas and well sealed, the increased reservoir gas pressure due to higher *temperatures* after burial is much less than the increased *hydrostatic* pressure on the enclosing shales. If the depth of burial is decreased, because of uplift, the exactly opposite effect should result; lowering of temperature should lead to high reservoir pressure.

A third general category of explanation for overpressured shales invokes the transformation of smectite (montmorillonite) to mixed-layer clays by the expulsion of interlayer water (see Sec. 15.4.1). Interlayer water is denser than pore water, but on expulsion it merely becomes pore water, and the effective pressure pump-up will be small. It can, moreover, be effective at all only if the original clay formations contained a high proportion of expandable clay and if this high proportion was converted too rapidly (in geologic time terms) for pressure adjustment to take place. In many late Mesozoic basins particularly, abnormal pressures are met at depths considerably shallower than the 3 km or more required for complete transformation of smectite to mixed-layer clays. In deeper basins, however, overpressures may first be reached below the clay transformation zone, and the excess pressures may in such basins postdate the transformation and be in large part caused by it. Kinji Magara has shown, however, that smectite dehydration *alone* is inadequate as the cause of low bulk density and abnormally high pressure in shales.

A final category of geologic processes invoked to cause or at least to assist in overpressuring in sand–shale sequences is the variety of volume changes undergone by solid components of the system. Through diagenesis or adsorption rock materials may expand; hydrous minerals like gypsum may be dehydrated and the released water may not be free to escape; mineral coatings may be deposited from pore waters; salt crystals, gas hydrates, or permafrost may develop. None of these can be of more than local significance in the evolution of abnormal pressures.

14.2.2 Conditions in the geopressured zone

The most exhaustively studied geopressured zones are undoubtedly those of the US Gulf Coast Tertiary basin. This basin's familiar geologic characteristics (Sec. 25.6.6) have simply been summed together to yield the criteria enumerated above for the causes of overpressuring: very thick, young sediments, wholly clastic, rapidly deposited under persistent regression; abundant diapirs and growth faults; expandable clays in upper levels giving way to mixed-layer clays at depth. A belt of long, linear, geopressured aquifers 320–480 km wide extends from Mississippi around into northeastern Mexico, bounded on the north and west by the inner limit of Miocene strata and on the south and east by the edge of the continental shelf. Each sandstone–shale system has its own geopressured zone, all of them below 1.5 km depth, most of them very much below this depth, and averaging about 3 km thick. Typical pressures are about 75 000 kPa, typical temperatures about 115 °C.

The lithostatic pressure gradient, as we have seen, is linear, as is the normal fluid pressure gradient (Fig. 14.2). At some depth Z_1, varying from basin to basin and from one stratigraphic unit to another, the fluid pressure gradient departs from linearity by abruptly increasing. This is the top of the geopressured zone. Thence downwards, the difference between P_o and P_f (σ_p in Eqn 14.1) is quickly reduced, and at some greater depth Z_2 it must become equal to what it had been at some shallower depth Z_3, above the top of the geopressured zone. In the same way, the porosity ϕ at depth Z_2 again becomes theoretically equal to that at shallower depth Z_3. Porosity of course cannot in actuality increase during the burial history of its sediment; it is the fluid pressure that increases, and the porosity at any given depth remains dependent upon the value of σ_m. As Plumley expressed it, the pressure- and porosity-history paths do not coincide with the normal $P-Z$ and $\sigma-Z$ gradients (Fig. 14.2); the paths probably diverged at some burial depth, shallower than Z_1, at which the seal became effective and compaction was arrested.

The value of σ_m is the difference between P_o and P_f and so is theoretically identical with σ_p (Eqn 14.1). Because under normal pressure conditions (*above* the top of the geopressured zone) the values of P_f, σ_p, and ϕ can be read directly from the $P-Z$ and $\phi-Z$ plots, it should be equally easy to read off the reduced value of σ_p and the increased value of ϕ *below* the top of the geopressured zone (assuming that no complication has been introduced by uplift). As high P_f approaches

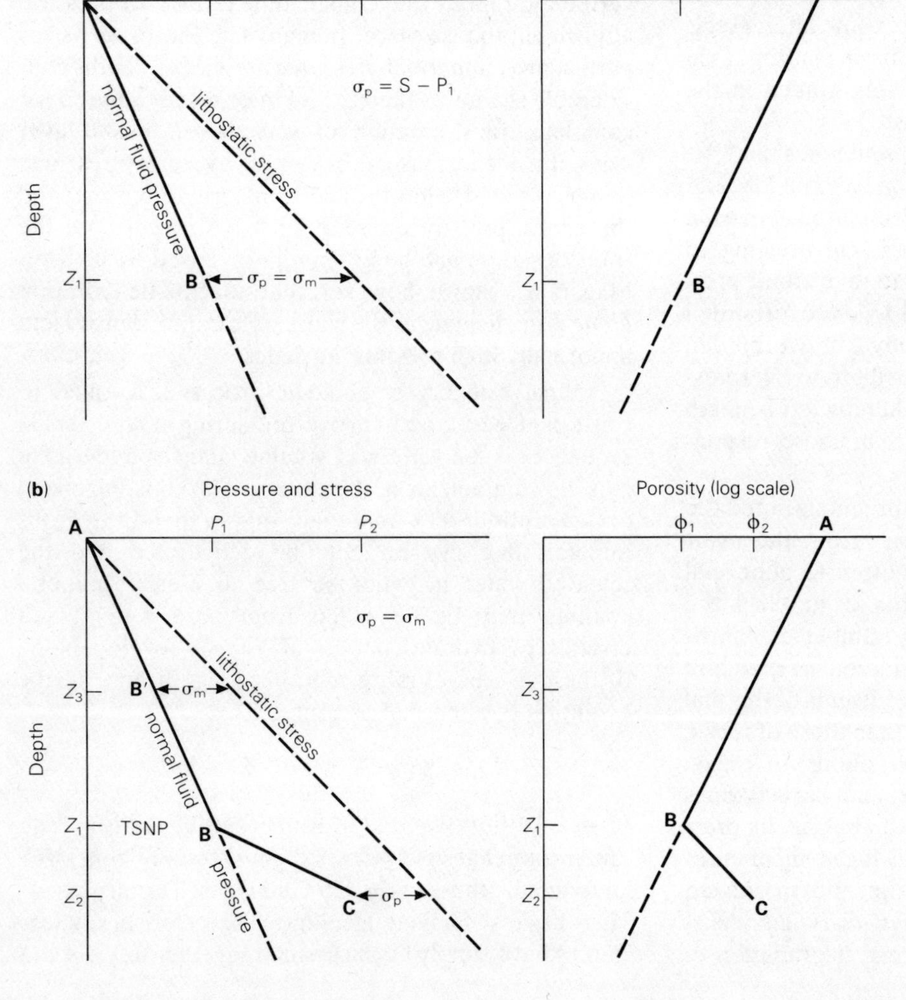

Figure 14.2 Pressure–depth and porosity–depth relations: (a) normal compaction model; (b) nonequilibrium compaction model. (From W. J. Plumley, *AAPG Bull.*, 1980.)

more and more closely the value of P_o, σ_p is reduced to a low value (Eqn 14.1) and ϕ apparently becomes very high (Fig. 14.2). But porosity, we repeat, cannot in fact increase under increasing burial, and sediments in abnormally pressured zones do not have porosities as high as the gradients would indicate; below the top of the geopressured zone, σ_p is in fact always less than σ_m. Porosities in that zone therefore do not conform with the nonequilibrium compaction model, and some "pump-up" mechanism is necessary in addition.

Plumley concluded that, though the nonequilibrium compaction model was probably the sole factor in overpressuring early in the sediment's burial history, before the clay-transformation depth has been attained, the clay transformation itself (Sec. 15.4.1) must become dominant in the later history, after that depth is passed. In comparison with these two mechanisms, the generation of hydrocarbons and the expansion of clay-water contents are of trivial significance in the creation of overpressures in clay rocks.

14.2.3 Overpressures in deep carbonate sequences

The foregoing dissertation is applicable to overpressures in sedimentary successions dominated by *shales*. The aquifers (reservoir rocks) interbedded with the shales are nearly all sandstones, and the criteria and explanations set out for overpressuring have been derived almost exclusively from studies in sandstone–shale sequences, most of them geologically young.

The highest measured *bottom-hole pressures* (formation pressures) and some of the highest *wellhead flowing pressures* (measured whilst the wells are actually in production; see Sec. 23.7) are not in sandstone–shale successions but in *carbonates*. Moreover, the carbonates are not for the most part particularly young. In the Jurassic Smackover fields of the central Mississippi salt basin (see Fig. 17.4), for example, a belt of salt "rollers" is associated with deeply buried faults that cut the Smackover limestone but fail to penetrate any younger strata, being cut off at the overlying anhydrite seal. Fluid pressure gradients approximate hydrostatic (10.5 kPa m^{-1} or 0.465 p.s.i. per foot) to depths of about 3600 m. By 6000 m, in contrast, a powerfully overpressured zone has been entered and the gradient has increased to 20.35 kPa m^{-1} (0.9 p.s.i. per foot). At 6600 m, the highest recorded bottom-hole pressure in any producing well was measured (1974) at more than 151 000 kPa, for a gradient of 22.6 kPa m^{-1} or 1.0 p.s.i. per foot (Fig. 14.1). This is equivalent to the load exerted by a column of mud weighing 20 lb per US gallon (2390 kg m^{-3}), and approximates the geostatic (overburden) stress for that depth. According to Equation 14.1, the effective stress σ_p has been reduced essentially to zero by the gross fluid pressure.

The decision to complete the highly overpressured Smackover well as a gas producer was an expensive one. A wellhead system, or Christmas tree (Sec. 23.3), built to withstand a pressure of 30 000 p.s.i. (over 200 000 kPa), stood 4.27 m high, weighed nearly 2×10^4 kg, and cost 25 times as much as the average wellhead tree. During 1980, an even higher bottom-hole pressure of 158 000 kPa was measured at 6945 m in a dry hole in the same basin; that hole is said to have cost $42 million, an almost unbelievable amount for an unproductive onshore well.

An only slightly less disconcerting and expensive case is found at the Malossa field in the Po Valley of northern Italy, in an Upper Triassic dolomite. The field produces gas and condensate of 51.8° gravity, with astonishingly low sulfur in view of its reservoir lithology and temperature. Bottom-hole pressure and temperature in the discovery well were more than 100 000 kPa (15 000 p.s.i.) and 150 °C at a depth of 5800 m. At the wellhead, the flowing pressure was 81 000 kPa and the temperature 100 °C.

Flowing pressures of nearly this order characterize the giant gasfields in the Permian Khuff limestone on the two sides of the Persian Gulf (see Sec. 25.3). Flowing tubing pressures of 60 000 kPa and shut-in equivalents of 75 000 kPa are routine in Oklahoma's Anadarko Basin wells below 5000 m depths; bottom-hole pressures below 7200 m in that basin are invariably higher than 135 000 kPa with temperatures higher than 175 °C.

All deep, overpressured carbonate formations are associated with temperatures in this range, and most of them contain sour, acid gases (25–45 percent H_2S and 3–10 percent CO_2) which are highly corrosive. The H_2S is probably formed by the *in situ* thermal cracking of sulfurous oil, and it attacks soluble minerals (as well as downhole drilling equipment). It is these late-occurring, deep, secondary geologic processes, including hydrocarbon phase changes, that create the overpressures. The compaction of interparticle porosity, fundamental to the commoner overpressuring in clastic sequences, is quite secondary in deep carbonate examples.

14.2.4 Summary of overpressures

We have considered the two commonest types of overpressure in sedimentary successions, the type due to rapid loading of clastic sediments and the type due to mineral alterations in carbonate sediments. There are other processes leading to pressures measurably higher than normal hydrostatic; they have been discussed by Hollis Hedberg and many others, and enumerated by Calvin Parker. The summary below is taken from Parker's study, with the addition of Colin Barker's category of aquathermal pressures:

(a) *Hydropressure* The driving force is the weight of mineralized water alone.

(b) *Geopressure* The driving force is the weight of the water-filled overburden:

(i) Inflated pressure, caused by fluid recharge or "pump-up";
(ii) Phase pressure, increased by mineral or hydrocarbon phase change, as in the deep carbonates described above;
(iii) Load pressure, due to rapid increase in overburden without fluid escape; this is the classic, original Gulf Coast case;
(iv) Tectonic pressure, by thrust or gravity glide sheets, tight folds.

(c) *Aquathermal pressure* The driving force is the thermal expansion of fluids.

(d) *Fossil pressure*, through removal of overburden from a sealed reservoir; it equals the hydrostatic head of the eroded overburden, formerly contributing to the pressure.

(e) *Osmotic pressure* The driving force is water of low salinity driving towards water of higher salinity through a semipermeable membrane (consisting of shale).

15 Migration of oil and natural gas

15.1 Introduction

In Part II, we considered the nature of *source rocks* for oil and natural gas. We also considered the *generation* of these fluid hydrocarbons from organic precursors which were intrinsic and original parts of these rocks.

In earlier chapters of Part III, we have considered *reservoir rocks,* and the properties of these rocks which make them natural containers for fluids, including fluid hydrocarbons, in the subsurface. Fluid hydrocarbons are not, however, intrinsic parts of these rocks that now contain them; oil and gas are discretely distinct from their host rocks.

It was early realized that the fluid nature of oil and gas, in the ground, must have enabled them to *migrate,* out of the sediments in which they originated and into the sediments in which they are now found. We may now review some of the more obvious manifestations of this capacity for migration.

15.2 *Prima facie* evidence for oil migration

15.2.1 Accumulations in structural culminations

The vast majority of oil- and gasfields occur in structural culminations of one sort or another (see Sec. 16.2). Unless one wishes to imply that the hydrocarbons in an anticlinal trap were responsible for the elevation of the anticline, or that those in fault traps controlled the faulting by weakening the rocks, the hydrocarbons must have got where they are after the structural culminations were created. The culminations gathered the hydrocarbons from the surrounding sediments by virtue of being culminations.

All properties of the culmination – folding, faulting, deposition and lithification of the cover rock – must be achieved before the accumulation can form in its present volume and position. In many fields their achievement long postdated the deposition of the reservoir rock (Sec. 15.5). Figure 15.1 shows the structure and stratigraphy of the Hassi Messaoud field in Algeria. The oil accumulation is in a major anticline involving the Precambrian basement. The reservoir rock is Cambrian, and the probable source rock Silurian, but the structure was not raised until the Carboniferous, the cover rock was not deposited until the Triassic, and the oil may not have accumulated in its pool until the Cretaceous or even later. It could not have accumulated in that particular reservoir, at that particular place, without migration, because the source rock is missing by erosion from the productive part of the structure and the structurally low areas still retaining the source sediment are barren of pooled oil.

A small number of productive structural culminations consist of the porous tops of buried hills or ridges eroded into igneous or metamorphic basement rocks (Figs 13.78 & 79). Such rocks cannot themselves have been the sources of the oil they now contain, and the three-dimensional geometries of the traps commonly leave no doubt that the oil entered them from sources that are stratigraphically higher but structurally lower than they are. The oil has taken a route into its present trap with both lateral and vertical components; this is *migration.*

Oil trapped directly below unconformities may constitute irrefutable proof of migration even if no structural culmination is involved and the oil originated in its present reservoir rock. The stratigraphic gap represented by the unconformity may be so great that erosion can be shown to have removed entire formations (or even entire geological systems) from the present roof of the pool (see Sec. 16.5).

15.2.2 Gushers

A *gusher* is a well from which oil is ejected above the derrick floor (in some cases high above the derrick itself) under natural flow. Before World War II, gushers were common sights in new oilfields, and even in established fields. Most of them came from reservoir rocks of

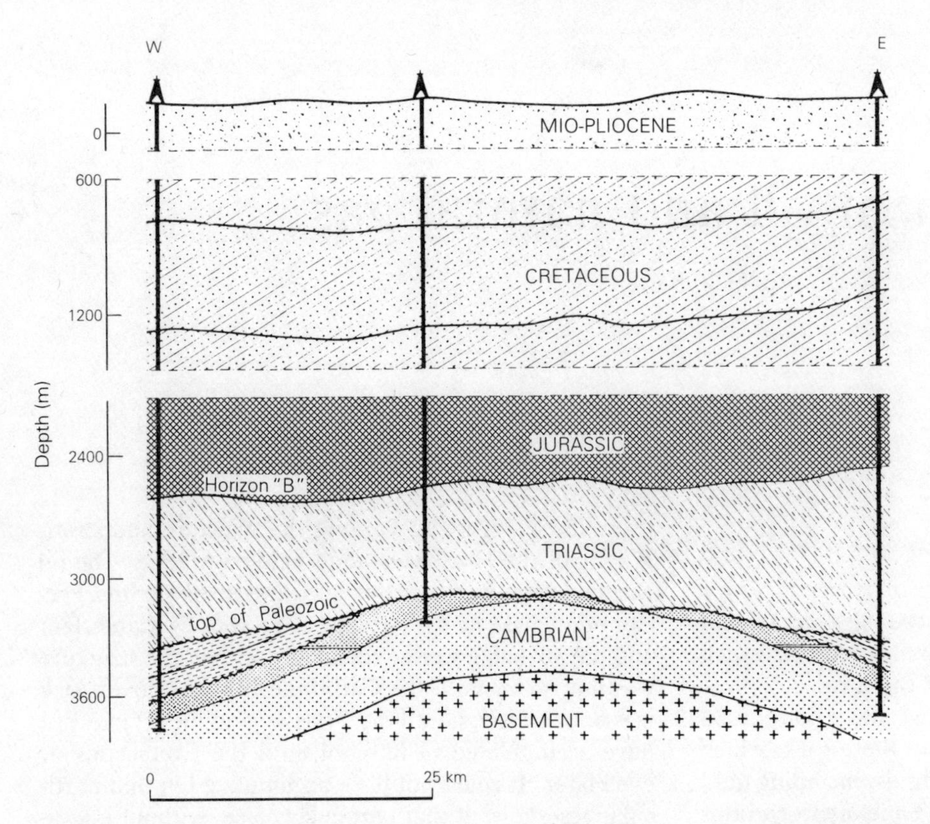

Figure 15.1 Cross section through the Hassi Messaoud oilfield in Algeria, showing relations between the Cambrian sandstone reservoir rock, Triassic evaporite seal (in unconformable contact at the top of the structure), and the Silurian source sediment (off the flanks of the structure). (From A. Balducchi and G. Pommier, *AAPG Memoir* 14, 1970.)

Tertiary age at shallow depths, but reservoir rocks of all ages have yielded gushing wells. Modern reservoir engineering has made gushers rare, but they still occur occasionally.

The oil is held under high reservoir pressure below an efficient *seal* or roof rock (see Sec. 11.3). The seal cannot be older than the reservoir rock (it may be of the same age), nor younger than the entry of the oil into the trap. Had the oil been in its discovered position before the seal was created, it would have moved elsewhere, just as it did when the drill-bit penetrated the seal and permitted the movement the drillers witnessed. The seal must therefore have interrupted movement of oil already under way.

15.2.3 *Multiple reservoir horizons in single fields*

In none of the very prolific oil basins of the world is oil restricted to reservoir rocks of a single age, even of a single epoch. In a few areally small grabens or other types of fault basin (Cook Inlet in Alaska, the Cambay and Gippsland grabens, the Songliao Basin of China, all described in Chs 17 & 25), this restriction is met in all essentials. In some areally large basins, nearly all the oil is in the reservoir rocks of a single *system*; in western Siberia, for example, it is nearly all in Cretaceous rocks, and in the eastern Venezuelan–Trinidadian Basin it is in Tertiary rocks. In the common case, however, there are many more productive reservoir horizons than there are imaginable source horizons.

In the Mid-Continent province of the USA, there is no porous formation between the initial Cambrian transgressive sand and the final Permian molasse that is not productive in one field or another. A frequently mentioned individual oilfield, the Garber field on the Nemaha uplift in Oklahoma, yielded oil of uniform character from 14 different reservoirs (some said 23, if separate sand "stringers" were distinguished) between the Ordovician and the Permian. Yet the entire succession penetrated was less than 2000 m thick. There are at least four acceptable source horizons in the Mid-Continent region, but extensive vertical migration must certainly have occurred, at least from Ordovician sources into the Siluro-Devonian and from Pennsylvanian sources into the Permian.

In the Williston Basin, in the center of North America (see Fig. 17.18), the Paleozoic section is so dominated by carbonate and evaporite strata that geologists for years were unable to agree upon *any* rich source sediment. Yet the basin yields oil from at least 18 separate horizons between the top of the Precambrian and the Triassic (Fig. 15.2). From the Cambrian to the Mississippian, essentially every formation except for the evaporites is productive; it is common to have two or three *systems* productive in a single field; yet no more than three source horizons have been identified (Sec. 17.7.2).

In the Emba Basin or Caspian depression, in the Soviet Union, commercial oil occurs in every system from the Devonian to the Cretaceous. In the Big Horn Basin of the US Rocky Mountain province (Secs 25.2.4

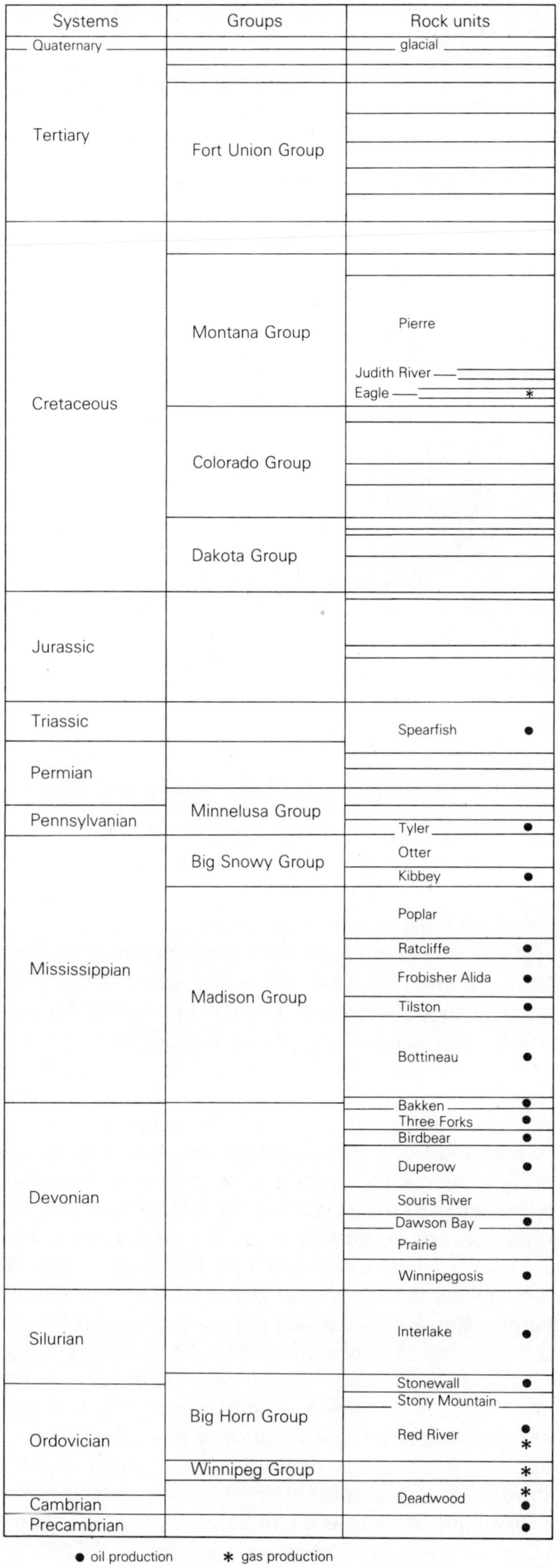

Systems	Groups	Rock units	Production
Quaternary		glacial	
Tertiary	Fort Union Group		
Cretaceous	Montana Group	Pierre	
		Judith River	
		Eagle	*
	Colorado Group		
	Dakota Group		
Jurassic			
Triassic		Spearfish	●
Permian			
Pennsylvanian	Minnelusa Group		
		Tyler	●
Mississippian	Big Snowy Group	Otter	
		Kibbey	●
	Madison Group	Poplar	
		Ratcliffe	●
		Frobisher Alida	●
		Tilston	●
		Bottineau	●
Devonian		Bakken	●
		Three Forks	●
		Birdbear	●
		Duperow	●
		Souris River	
		Dawson Bay	●
		Prairie	
		Winnipegosis	●
Silurian		Interlake	●
Ordovician	Big Horn Group	Stonewall	●
		Stony Mountain	
		Red River	● *
	Winnipeg Group		*
Cambrian		Deadwood	* ●
Precambrian			●

● oil production * gas production

Figure 15.2 Stratigraphic column for the Williston Basin in North Dakota, identifying 18 formations productive of oil and two others productive of gas. (From *Oil Gas J.*, 27 July 1981.)

& 25.5.10), oil or gas has been produced from every geological system. In the Green River Basin of the same province, oil was produced throughout the stratigraphic column, from the basal Cambrian sandstone and from every porous formation between the Mississippian and the Eocene. A single small field (Wertz) in the northeast corner of the basin produced from 11 horizons within that interval.

The basins quoted were all of Paleozoic inception, and their production extended over a vast stratigraphic range. A different circumstance is represented by multiple sandstone reservoirs, in single fields, extending over a more restricted stratigraphic range – in many cases within a single formation, almost invariably of Tertiary age. In early days, all American oil geologists quoted fields from California's Los Angeles Basin, and their Russian counterparts quoted the Baku fields; both examples have recorded up to two dozen reservoir sands within single fields, all of Pliocene age (Fig. 15.3). Later, deeper drilling in other basins discovered some still more impressive examples. About 125 separate sandstone reservoirs have produced in the Bay Marchand field in the Louisiana Gulf Coast (see Fig. 16.52), all within a depth interval of 1500 m or less. A similar number of separate reservoirs are productive at the Handil field in eastern Borneo, and at least 75 in the Oficina fields of eastern Venezuela.

The *stratigraphic* interval represented by multiple reservoir sands is small, and the clay rocks separating them are all essentially alike, so there would be no justification for invoking only one or a few of them as the contributing sources. However, all such fields involve drawn-out structural development, in some cases (like Bay Marchand) exceedingly complex – growth faults, diapirism, or anticlines growing through time – and extensive adjustment of oil reservoirs by vertical migration cannot be doubted.

An elementary conclusion is inescapable. All porous sedimentary rocks in a petroliferous basin are prospective *containers* for any oil that may be generated. Not all nonreservoir sedimentary rocks are potential *sources* of oil. Where there are many more oil-bearing reservoir rocks than oil-generative source rocks, migration of the oil must be invoked *at least in the vertical dimension.*

15.2.4 Accumulations under insufficient cover

We have seen that oil cannot be generated from its source sediments without the achievement of some minimum temperature to break down the kerogen molecules. The achievement of this temperature – barring the intervention of intrusive activity – involves some minimum depth of burial, the actual amount

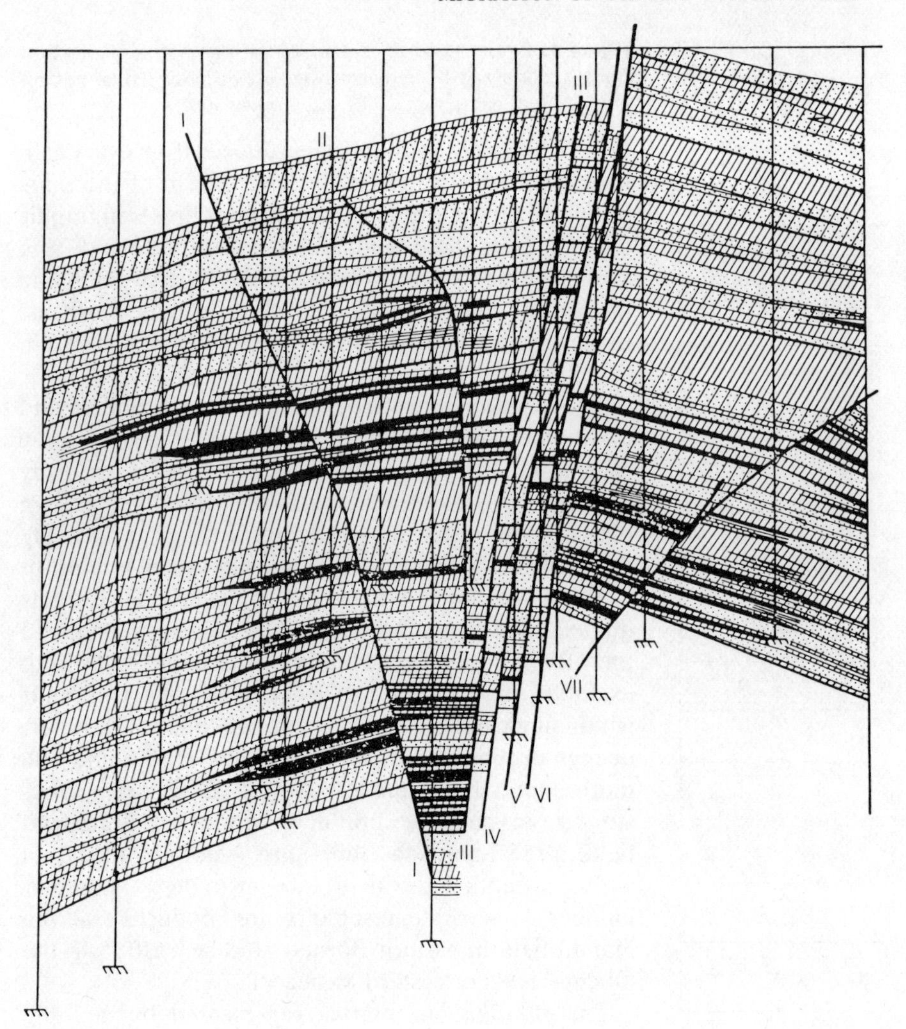

Figure 15.3 Section across the Kum Dag oilfield in the Caucasian foredeep, east side of Caspian Sea, USSR. Oil (shown black) pooled in multiple Neogene sands offset by equally numerous normal faults. Direction and scale of section not indicated in original diagram. (From D. V. Nalivkin, *Geology of the USSR* (1962); Edinburgh: Oliver & Boyd 1973, by permission of the Copyright Agency of the USSR.)

depending upon the temperature gradient (see Secs 7.4 & 17.8). This depth of burial itself requires some minimum amount of time for the accumulation of the younger sediments.

Oil pooled at shallow depths in Paleozoic strata – as in the early-discovered fields of eastern North America – has had a great deal of time for the generative process, and its present amount of cover may be much less than it was originally. These circumstances do not apply to prolific oil trapped at very shallow depths in very young rocks, especially if it is *heavy* oil at those depths. Such oil is under insufficient *present* overburden to expel it from its source sediment even if the temperature had been raised high enough for its generation.

To eliminate the risk of overlooking the erosional removal of a great thickness of earlier overburden, we may restrict our examples to fields in Tertiary strata. In California, prolific oil was (and in some cases still is) produced from less than 50 m at Kern River, on the east side of the San Joaquin Valley, and from less than 200 m at Midway–Sunset on the west side (see Fig. 16.8). In Venezuela, prolific production was achieved from about 150 m at Mene Grande, 180 m in the Bolivar Coastal field, and less than 600 m at Quiriquire (see Figs 16.2, 16.82 & 83). In Eurasia, the shallowest early fields at Baku produced for years from depths around 100 m, and at Seria in northwestern Borneo the top of the oil column was at about 250 m (see Fig. 16.121).

15.2.5 Evidence of former oil accumulations

Except in a few small, tightly-enclosed sandstones, all oil-bearing reservoir rocks contain some water as well as oil in their pore spaces (see Sec. 11.2.1). Oil entering the reservoir rock from its source rock – without regard to the spatial separation between the two rocks – did not displace *all* the water that occupied the pore spaces before. But water is clearly able to displace all the oil from its trap *if it can gain entry*. Many barren traps reveal clear evidence of flushing by moving water (see Sec. 16.4.6), which is an extremely efficient displacement agent. Porous rock may be heavily stained with oil which is not removable by ordinary organic solvents. Such rock must formerly have contained oil for a long time; the oil has migrated out of it somehow, by water flushing or some other mechanism. Tilted oil/

water interfaces show that oil moves to accommodate itself to hydrodynamic conditions (see Sec. 16.3.9). It will continue to move so long as the pressure gradient and a permeable passageway remain available to it.

15.2.6 Visual evidence of upward oil movement

World-wide seepages of oil are themselves visible evidence of oil movement. So is oil contained in mud diapirs (see Sec. 16.3.6). Light oils of low viscosity migrate very rapidly into drillholes under local pressure differentials, permitting the wells to flow for long periods. The fact that a specific area around an oil well is drained by it is a reflection of local migration of the oil. Within the effective traps of some producing fields themselves, moreover, there is impressive visual evidence of the rapid migration of both oil and gas into traps that have previously been emptied.

In the Baku fields southwest of the Caspian Sea, the oil is contained in multiple Pliocene sands of very high permeability punctured from below by numerous mud volcanoes (see Fig. 16.54). As these mud intrusions are active today, they cause constant changes in the dips of the sands they penetrate. Oil is therefore encouraged continuously to migrate farther up the steepening dips, forming "hanging traps" or one-sided accumulations, occupying single flanks of closed structures. In 1959 V. S. Melik-Pashayev reported that the migration along such flanks could actually be observed taking place. The boundaries of the trap in one field "moved with surprising speed, reaching 40 and even 100 m a month in places" during five years of the 1950s. Oil reached the highest point in the trap "during a period completely incomparable with the length of a geological period."

About ten years later, the Iranian consortium drilled a deep test on the crest of the structure at the Masjid-i-Suleiman field, the discovery field of the Persian Gulf province. The drillhole was cemented and abandoned in Jurassic sediments of basinal facies, much more plastic under stress than the brittle platformal limestones and not expected to undergo or maintain much fracturing during folding. In a very short time, the cement job in the hole was destroyed by the invasion of sour gas, which ascended from older formations and recharged the largely depleted Oligo-Miocene reservoir (in the Asmari limestone) at what Norman Falcon called the "geologically catastrophic rate" of about 70 kPa per month. This rate would have returned the reservoir to its natural pressure condition in less than 20 years. It represented an enormous transfer of gas from a deep zone of high pressure to a shallower zone of lower pressure depleted by 60 years of production.

The great anticlines of the Iran–Iraq oilfield belt – including that of Masjid-i-Suleiman – were raised during the Alpine folding episode culminating in the Pliocene. During the interval between the mid-Cretaceous unconformity and this late Tertiary culmination, thick foreland sediments overstepped a pre-Cretaceous structural pattern that was quite different from the younger one that superseded it. Harold Dunnington reasoned that the source sediments of the basin axis, east of the oilfield belt and of Oxfordian–Cenomanian age, were so rich and had been so rapidly buried that their oil had been pooled long before the late Tertiary structural traps were created. Whether the earlier pooling had been in earlier structures or in stratigraphic traps could not be determined, but the traps must have lost their oil into younger and higher reservoirs and traps as a result of the late Tertiary folding and fracturing. Basinal sediments of late Cretaceous to Oligocene age, which exist, were believed by Dunnington to have lost their oil completely, and indigenous oils are probably in negligible quantity throughout the Tertiary limestone fields. If Dunnington's persuasive reasoning is correct, more oil must have been lost than is preserved (despite the huge volume of the latter), and all the known oil must have migrated twice.

15.3 Primary and secondary migration

A moment's thought will reveal that migration leading to the phenomena we have enumerated may (indeed must) be of two distinct types, and that these types may involve at least two distinct mechanisms. The hydrocarbons first have to move, somehow, out of the source sediment into a porous host rock. The fact of the host sediment being porous merely enables it to hold the fluids, but the further fact that the fluids can be *recovered* from the reservoir (host) rock means also that they *can move through it.* It is a *permeable* medium as well as a porous one, so migration of its hydrocarbon contents may continue after they have left their source rocks through the initial migration.

Following the early discrimination between these two phases of migration, by Vincent Illing, Alex McCoy, and many others, petroleum geologists have conventionally distinguished them as the primary and secondary migrations of hydrocarbons. *Primary migration* is that from the source rock to the first porous refuge. Conditions of hydrocarbon occurrence make clear that this phase must take place largely *across the stratification* rather than along it; both observation and experiment further make clear that a prime cause of primary migration must be the *compactibility* of the source sediment (Sec. 11.1.4). *Secondary migration* takes place within the porous reservoir rock, or from one reservoir rock to another. It is the process which collects oil and gas into commercial pools. Conditions of hydrocarbon occurrence make clear that this phase takes place largely *along the stratification,* as the hydrocarbons explore the opportunities that the permeable medium provides them to find the position of least potential energy in the reservoir rock.

Because of this difference in dominant migration direction, and that between the physical characteristics of the source sediments and the reservoir rocks, it is unlikely that the two phases of migration are controlled by the same forces. Secondary migration is assuredly controlled by the differences in buoyancy between gas, oil, and water, all of them vying for space in the reservoir and all of them capable of movement so long as permeability and a pressure gradient permit.

15.4 Primary migration

Rocks below the surface are subject to *geostatic* or *overburden pressure,* P_o, equal to the weight of the column of fluid-bearing rock above them. As a general average, P_o amounts to about 230 g cm^{-2} per meter of depth. For rule-of-thumb purposes, it can be taken that about 25 kPa are added to the pressure for each meter of overburden. This stress is borne by both the materials of the rock and its fluid content (Sec. 14.1).

In sands, the constituent grains are normally settled by jarring and current action shortly after deposition; thereafter the overburden stress is borne principally by the grains of the rock and the effects of *compaction* are very small (only about 2 percent reduction in volume under 25 000 kPa; Sec. 11.1.3). Sandstone porosity is reduced much less by compaction than by the introduction of diagenetic cements, and its decrease is linear, according to John Maxwell, down to about 7500 m. At depths not much greater than this, but varying according to geologic conditions, the overburden stress exceeds the crushing strength of the rock and pore spaces are effectively closed.

During subsidence and sedimentation, in other words, reservoir sandstones are under approximate hydrostatic pressure so long as porosity remains significant. In clay rocks, in contrast, a high proportion of the overburden pressure is exerted on the rock's fluid content. Source sediments, therefore, suffer during subsidence and sedimentation under approximate geostatic pressure, in which the hydrostatic pressure is augmented by the vertical lithostatic pressure (Sec. 14.1). Their fluid contents are therefore confined under very different thermodynamic conditions from those of porous sandstones.

The surface areas of the grains in clay rocks are orders of magnitude larger than in sands, though their pore diameters are very much smaller (Sec. 11.1.4). Whereas the pore fluids in sands occupy the interiors of pore spaces that may be approximately equidimensional, those in clay rocks are adsorbed on to grain surfaces by electrostatic forces. Molecules with high dielectric constants, like water molecules, are adsorbed more readily than are those with low constants (like hydrocarbons). Shales are consequently very likely to be *water-wet,* and much of the water close to the grain surfaces is likely to be *structured water*, with more of the properties of a solid than of a liquid (see Sec. 15.4.1). The loss of shale porosity by gravitational compaction *alone* is therefore not realistically reproducible in laboratory experiments; nature reduces the porosity much more rapidly and efficiently than the experimenter.

The amount of gravitational compaction undergone by clay rocks is a logarithmic function of the burial depth (Sec. 11.1.4). The familiar density – depth and porosity – depth curves have been greatly refined during many subsequent studies, notably by Parke Dickey and Kinji Magara (Figs 15.4 & 5).

Experiments during the 1920s by Hollis Hedberg demonstrated that compaction of clay sediments, in addition to causing primary structures over pre-existing bottom highs and other interruptions, is by far the most important cause of primary hydrocarbon migration. The physics of the process was exhaustively discussed by Marvin Weller in 1959. In simple terms, *compaction* is the reduction in the bulk volume of the sediment through compressive stress. The grains are deformed irreversibly, with no elastic rebound possible if unloading takes place. The pore space of the sediment is reduced below the value corresponding to the Atterberg plastic limit, which is the water-content boundary between the plastic and the semi-solid state (see Sec. 11.1.3). Compaction is therefore a more advanced stress-induced process than *consolidation,* in which the pore space is merely reduced *to* the Atterberg limit.

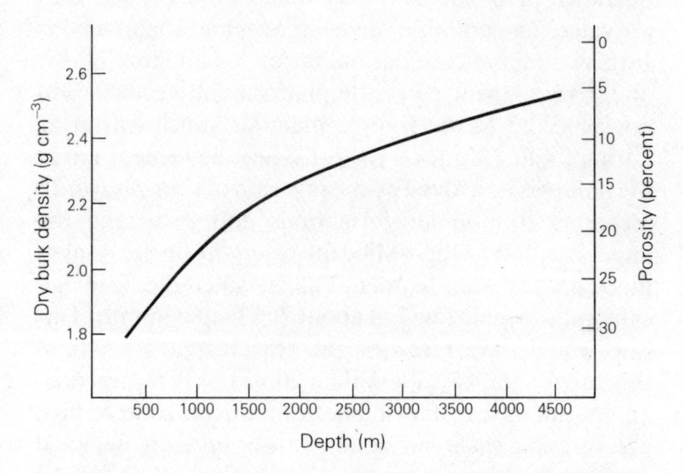

Figure 15.4 Composite curve showing relations between density and porosity of sediments and depth of burial. (Constructed from published data from numerous basins: P. A. Dickey, *AAPG Bull.*, 1975.)

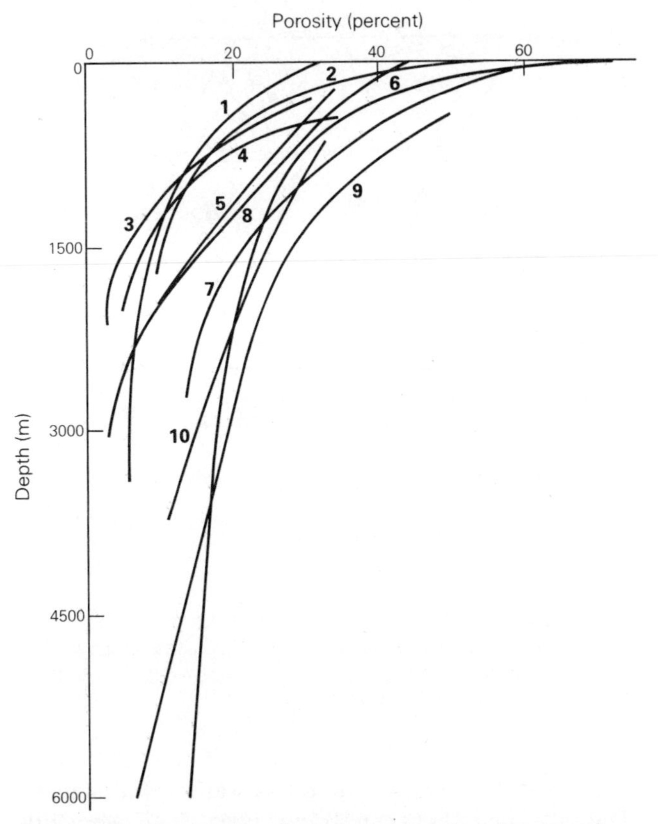

Figure 15.5 Combination of curves representing shale porosity–depth relations as measured in ten different sedimentary basins on several continents. (From K. Magara, *J. Petrolm Geol.*, 1980.)

Consolidation is approximately governed by Karl Terzaghi's equation:

$$\frac{dP_f}{dt} = C \frac{d^2P_f}{dy^2}$$

where P_f is the pore-fluid pressure; t the time of operation of the overburden stress; y the depth of burial; and C the coefficient of consolidation, a function of the void ratio.

In 1980, Kinji Magara combined the porosity–depth curve for shales with the hydrocarbon maturation "window" to adduce further evidence for primary migration. He plotted the downward increase in hydrocarbon extract (in relation to total organic carbon) with depth (that is, a maturation curve) against the normal shale-compaction trend (the porosity–depth curve). The hydrocarbon extract declines where the maximum compaction is reached, presumably because of the primary migration of hydrocarbons out of the shales below that depth.

We will accept the postulate that the principal process expelling the newly formed hydrocarbons from their source sediments is the compaction of clay rocks. The actual agency responsible for taking the hydrocarbons out of the source rocks and making them available for primary migration has been the subject of seemingly endless debate among petroleum geologists and geochemists; but it is presumably something related to the properties of the source sediments. The most obvious and measurable of these properties are those of the individual clay minerals themselves.

15.4.1 The influence of clay mineralogy

Mankind has found a variety of uses for the particular varieties of clay that swell enormously in water – in extreme cases up to 15 times their original volumes. The most familiar case for petroleum geologists is that of *bentonite,* a relatively common clay derived from volcanic ash and used in drilling muds (see Sec. 23.2) and in oil refining as well as in bleaching clays. The swelling property is a function of the clay's affinity for water.

Water in freshly deposited clay minerals is contained in three distinct fashions:

(a) Free, interparticle, or *pore water* is held in the pore spaces just as oil and gas are.
(b) Interlayer or *bound water* is held within the lattice spaces of swelling or expandable clays.
(c) *Constitutional water* is immobile water forming the hydroxyl (OH) group in the mineral lattice.

In the clay sediments of petroliferous sedimentary basins, the most important expandable clay mineral is *smectite,* commonly but mistakenly called *montmorillonite* in North America; it is the principal constituent of bentonite. The most important nonexpandable clay mineral in the same sequences is *illite.* Because smectite is slowly converted to illite during increasing depth of burial, there are important transitional members called *mixed-layer clays.*

Smectite ("montmorillonite"), $(OH)_4Al_4Si_8O_{10}.xH_2O$, constitutes the principal expanding clay. Purely smectitic clays develop most readily from volcanic ash deposits, which are not potential source sediments; they are relatively rare in recent marine sediments. Normal sedimentary clays are polymineralic, but they are dominated by nonswelling clays like illite and by swelling but mixed-layer clays. In some petroliferous Tertiary basins, however, smectite makes up 60–80 percent of the clay mineral in the upper 1500 m or so of the section – in the US Gulf Coast, for example, and in the deep basins on the two sides of the Caucasus Mountains.

Swelling or expanding clays take on far more water, as bound or interlayer water, than do other clays (Fig. 15.6). They also extract much larger quantities of organic matter from the marine environment than do non-swelling clays, and they do it much more rapidly, because of the ionic bonding of polar organic molecules, such as the fatty acids, to the interlayer sites.

Illite or *hydromica,* $(OH)_4K(Al_4Fe_4Mg_4)Si_7AlO_{20}$, is

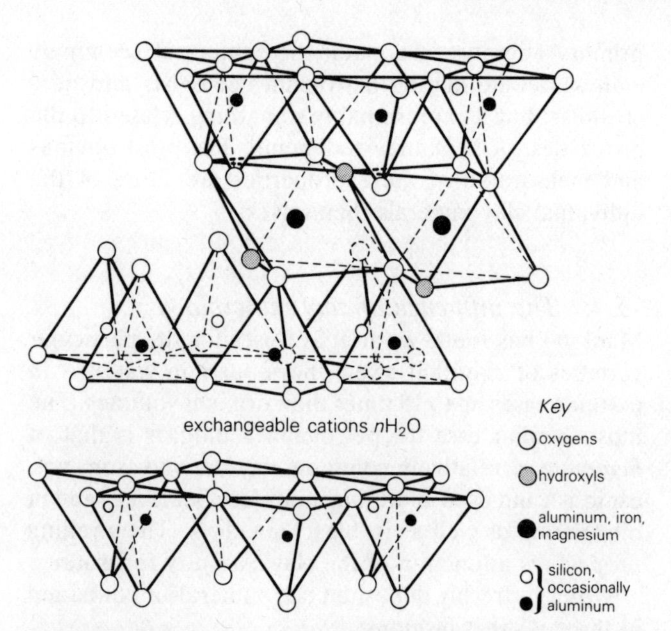

Figure 15.6 Diagrammatic sketch of the structure of an expandable clay mineral (smectite), showing both interlayer (bound) water and constitutional (hydroxyl) water. (Reprinted, by permission of the publisher, from R. E. Grim, *Applied clay mineralogy*, New York: McGraw-Hill, ©1962.)

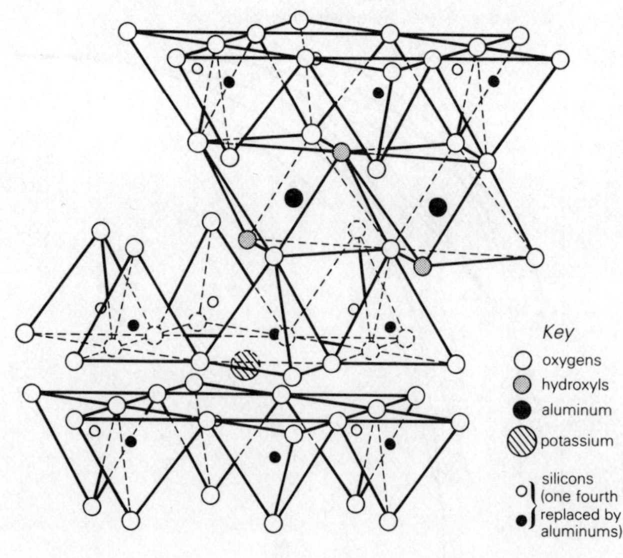

Figure 15.7 Diagrammatic sketch of the structure of a nonexpandable clay, lacking interlayer water and resembling illite. (Reprinted, by permission of the publisher, from R. E. Grim, *Applied clay mineralogy*, New York: McGraw-Hill, ©1962.)

a nonswelling clay allied to the mineral species called *sericite* by petrologists. Illite contains no interlayer water (Fig. 15.7).

The loss of smectite in Tertiary sediments below about 1500 m is due to its conversion to illite, via a progressive increase in the proportion of mixed-layer clays. The transformation, studied by Ralph Grim, Maurice Powers, John Burst, Henry Hinch, and many others, is understood in considerable detail. It is a process of staged dewatering, and not a uniform loss of water with increasing depth of burial. The pore water of both smectite and illite, and the water of excess interlayers in smectite, are lost during early, kinetic, compactional processes via vertical permeability. The fluid conductivity of most clay sediments is adequate for these processes so long as pressures are less than about 3000 kPa, porosities have not declined below about 25 percent, and pore diameters are not less than about 10 nm. Between 600 and 3000 m depths, more or less, clay-sediment porosity decreases linearly to less than 10 percent as pore diameters are reduced to about 2.5 nm. Vertical water expulsion ceases as vertical permeability of the clays is abruptly reduced, principally by a rapid increase in the rate of deposition of nonexpandable clays. Further fluid expulsion thereafter takes place laterally, in the direction of lower lithostatic pressure, rather than vertically. Fluid pressures remain hydrostatic, but water salinities are likely to increase. This is the depth range of the generation of biogenic methane.

This essentially kinetic or hydrodynamic loss of free water is overlapped and followed by the *diagenetic* loss of most of the bound or interlayer water from smectite. This takes place between 80 and about 250 °C; very little of it remains at 300 °C. The change is reproducible in laboratory experiments at pressures of 2×10^5 kPa and temperatures between 150 and 450 °C. These laboratory *T, P* conditions are equivalent to the depth range between 6 and 9 km, approximately. In natural basins, the change actually takes place between 3 and 4.5 km, more or less. The actual transfer of water is from the interlayer positions – in which the water is ineffective as an agent in migration – to interparticle positions, in which it can be an effective migration agent. The *rate* of the reaction is governed by the presence or absence of diagenetic potash and alumina (derived from the breakdown of mineral grains in the sediments), and by the temperature (and hence by the geothermal gradient; Fig. 15.8).

The progress of the transformation with increasing depth of burial can be monitored in a rather rough and ready way, by continuous measurement of the cation exchange capacity (CEC) of the drilling fluids in a borehole. The interlayer sites of the bound water in smectite are the sites of exchangeable cations (Fig. 15.6), so high values of CEC indicate sediments with high smectite contents. The transformation reaches a metastable phase (full ordering) when the illite content of the clays has reached about 80 percent; this phase still contains about 7 percent water by volume, and shale density has become about 2.5.

The slow loss of the third water fraction, the constitutional or hydroxyl water, begins at about 300 °C. It is

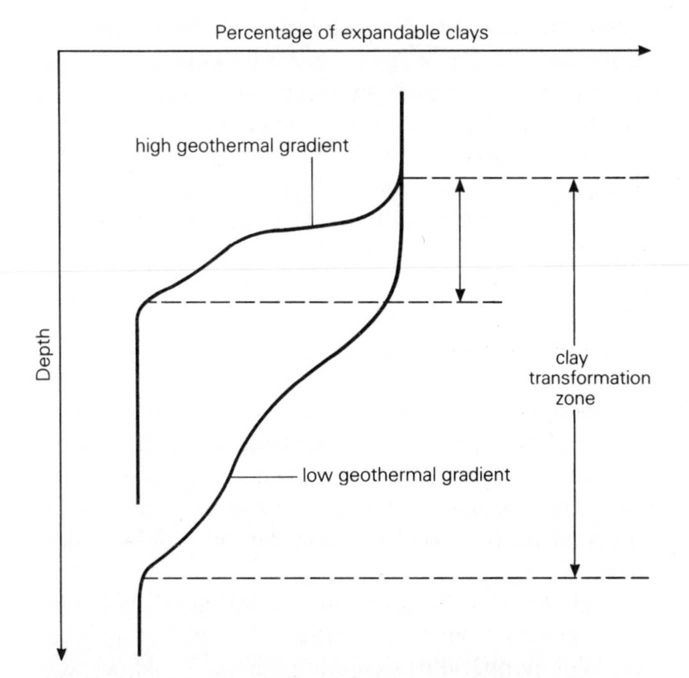

Figure 15.8 Schematic relation between the clay transformation zone (vertical arrows) and subsurface depth for high and low geothermal gradients.

lost rapidly from illite between 350 and 600 °C and from smectite between 500 and 750 °C. The loss of water from illite results in a stable modification. The complete conversion of smectite to illite requires depths of 7500 m or more; it is a change reversible only through weathering, and cannot be put into reverse by erosional unloading.

The critical stage, or clay diagenetic threshold, is reached when the depth (*P, T*) conditions are such that the bonding and hydrostatic pressure energies, which serve to *retain* clay water, are no longer sufficient to withstand the energies related to the geothermal gradient, which work to *release* it. At this threshold, no or negligible water remains in illite or in kaolinite, but the last few water interlayers still remain in smectite. The stage begins at about 1700–1800 m under average geothermal gradients. When only the four innermost layers of polar (oriented) water remain in the smectite, the clay rock undergoes an increase in density; it in essence becomes a *shale.* Between 1800 and 2800 m, virtually all the remaining "bound" water becomes "free" water, creating both porosity and permeability in the shale. In these last monolayers, the water has a density higher than normal; it is *structured water.* The hydrogen of the water molecules bonds strongly with the oxygen of the SiO_4 tetrahedra along the clay surfaces of the pores. The structuring is most pronounced in the smallest pores. The water no longer behaves like bulk liquid water, but in a sense displays some of the properties of a solid. It cannot be "squeezed out" of the rock by overburden pressure alone (unless the rock is buried to some very great depth, certainly in excess of 10 km). The last layers have an acidic character, and so are proton donors; they may help to provide the acid catalysis necessary for catagenesis of the kerogen (see Sec.7.4). Being denser than normal water, the structured interlayer water expands on its eventual release, adding to the already high fluid pressure.

The transformation ends with complete conversion to illite, at depths around 2800–3000 m. No effective porosity remains; no further compaction or fluid migration is then possible. Continuous permeability for the expulsion of water declines rapidly after the total shale-water content falls below about 10 percent. It is unlikely that the process of illitization is ever 100 percent complete unless metamorphism intervenes. The process involves a net increase in volume, as the compact water of hydration becomes free liquid water. The reaction is therefore impeded by excess fluid or overburden pressure caused by thick overlying shales or impermeable evaporites; such excess pressure reduces the reaction rate towards zero.

In some basins in which the clay transformation can be traced in detail, its completion approximately coincides with the top of the zone of overpressuring (see Sec. 14.2). Where this occurs, it has been concluded by some that the release of the bound water itself is a major factor in creating the abnormal pressures. There appears to be a sudden increase in shale porosity (to 20 percent or more) and a decrease in its density (to about 2.4). The two phenomena – clay transformation and overpressuring – are probably quite independent of one another. The overpressured zone is commonly at considerably greater depths than that of complete clay transformation; in parts of the US Gulf Coast Tertiary basin, for example, the bottom of the smectite zone is around 2500 or 2800 m and the top of the overpressured zone below 4000 m. The dehydration of mixed-layer clays probably does not result in an actual increase in shale porosity from a previously achieved lower value; instead it levels off the downward loss of porosity by inhibiting compaction.

15.4.2 Mechanisms of primary migration

An understandably persistent view is that the primary migration of oil and gas takes place in some kind of *solution* – in compaction or pore water, in clay-dehydration water, in meteoric water, in solutions of natural soaps, or in gas. We will consider the principal hypotheses in turn.

Hypothesis 1: Migration of hydrocarbons in clay-compaction water The pore water kinetically compacted out of illitic clays has been expelled after very shallow burial (a few scores of meters), long before hydrocarbons like those found in reservoirs are likely to

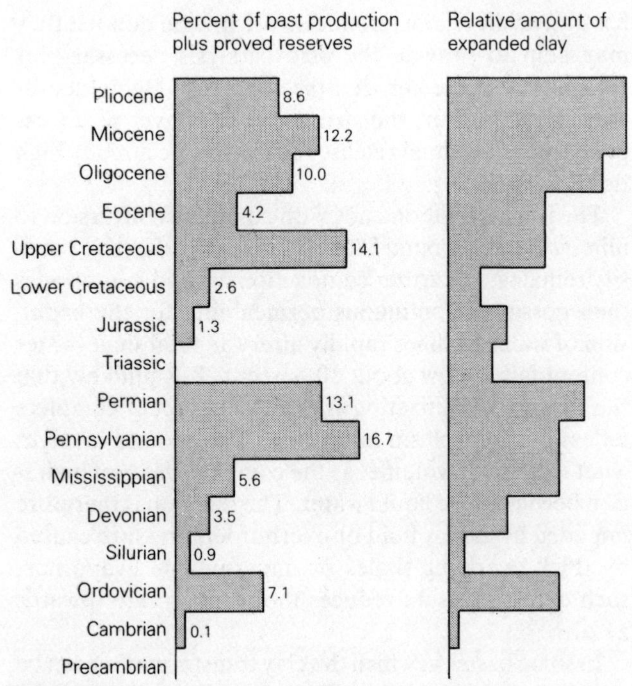

Figure 15.9 Relation of US oil production to age of reservoir rock and to relative content of expanding clays (smectite), the latter based on 20 000 X-ray diffraction patterns. (From L. C. Price, *AAPG Bull.*, 1976; after C. E. Weaver, 1960.)

have been generated. The diagenetic transformation of smectitic clay to illitic clay, however, shows some striking temperature–depth parallels with the catagenesis of organic matter. The diagenesis, as we have seen, takes place in the depth interval between 1.8 and 6.0 km, more or less, corresponding with temperatures of 80–200 °C. The transfer of clay water from bound to free state, in the upper part of that interval, creates porosity and permeability and must flush fluids out of the compacted rocks if "drainage" is good.

The question is whether the fluids available to be flushed out include hydrocarbons. Organic catagenesis also takes place in the temperature range between 80 and 200 °C, and in the depth interval variously estimated to be between 1.8 and 5.0 km according to the geothermal gradient, the rate of burial, and the activation energy required to overcome the potential barrier in changing from one state to another. Western and Russian scientists seem to have settled upon an average of 1.8–2.4 km for the upper limit of the "principal phase of oil formation" (see Ch. 7).

The organic matter taken out of the marine environment by expandable clays is not depleted after shallow burial but is retained by adsorption to depths of the order of 1.5 km or more. Decarboxylation and other carbonium ion reactions, and hydrogenation processes, then occur, forming long-chain compounds resembling petroleum hydrocarbons. In the presence of potassium and magnesium ions, clay minerals actually interact with organic compounds during the formation of bitumen.

The Russian Sergei Sarkisyan (in 1958) and the American Charles Weaver (in 1960) were early to recognize a correlation between oil production, by regions, and the contents of expanding clay in the sedimentary successions of the regions (Fig. 15.9). Within any one basin, the most productive stratigraphic intervals are those containing the highest percentages of expanding clays. Many authors have subsequently concluded that clay-compaction water is the principal agent in expelling hydrocarbons from their source sediments during primary migration. Given the necessary balance between temperature, pressure, and the chemistries of the pore water and the kerogen, hydrocarbons may be made available in quantity at the peak of the water desorption rate. Compaction alone might then displace all the interstitial oil, and much adsorbed oil film, with the expelled clay-dehydration water.

Leigh Price, who postulates solution-migration of hydrocarbons from very great depths and beginning at very high temperatures (see hypothesis 2, below), has claimed that expandable clays survive to much greater depths than other observers can accept. In basinal sediments with high original contents of such clays, 10–20 percent porosities may remain at depths of 3–6 km, and water contents of 8–10 percent by volume are feasible at 9 km or deeper. Freeing of this pore water by the smectite–illite transformation would mobilize hydrocarbons from their source beds at depths below 6 km and temperatures above 200 °C.

The several versions of the hypothesis favoring migration in clay-dehydration water are attractive and influential, but they have serious drawbacks. Oil is not ready for expulsion from the source rock until compaction (and porosity) has been greatly diminished. The clay waters, however, must move *out of* the source rocks; it is not sufficient that they move *within* them. The lack of any further porosity loss seems to imply that fluid expulsion is about at an end. The clay pore water remaining in the shales at peak hydrocarbon generation seems grossly insufficient for the expulsion of known pooled hydrocarbons, given the very low solubilities of the latter. Furthermore, much of the water remaining in the shales is highly structured (Sec. 15.4.1). Hydrocarbons will be repelled from the structured water by the electrostatic barrier, and hence will be forced directionally towards the larger pores (which may be no larger than the hydrocarbon molecules), and into bedding surfaces or fractures forced open by the fluid pressure of the hydrocarbons themselves (see hypothesis 6, below).

Hypothesis 2: Migration by molecular solution in water
Any proposal of migration by simple molecular solution in water runs up against the obstacle of the exceedingly low solubilities of oils in water (see Sec. 9.4). Aromatics

are by far the most soluble of petroleum hydrocarbons, but few crude oils have high aromatic contents. Aliphatic hydrocarbons have solubilities of the order of 1 p.p.m. or even much lower. Most crude oils contain molecules of C_{12} or heavier, and such molecules have solubilities of the order of 0.01 p.p.m. or less. If the waters are saline, hydrocarbon solubilities are even lower; hydrocarbons are "salted out" of aqueous solution at salinities higher than about 150 000 p.p.m.

The only common petroleum hydrocarbon having any significant solubility in saline water is methane. Methane undoubtedly may migrate in aqueous solution, but its presence in the solute lowers the solubilities of liquid hydrocarbons still further. Parke Dickey, Clayton McAuliffe, and many other authorities have agreed that aqueous solution cannot be the principal mechanism for the primary migration of crude oils. There simply is not enough water in the 2–4 km depth range to attain the necessary oil saturation in the pore spaces, given the exceedingly low solubilities. Any oil in solution in the pore spaces would be so far below the saturation level required for separate-phase flow that it could never exsolve to accumulate as a coherent reservoir. Mass-balance calculations by Robert W. Jones draw a stark contrast between the volume of oil expelled from unit volume of an average mature source sediment and the volume of water made available by compaction-reduction of an average original sediment porosity to an average present porosity. The hydrocarbon solubility required for transport in aqueous solution is several orders of magnitude higher than known solubilities. John Hunt, for example, estimated that the ratio of oil to water for the Western Canadian Basin would have to be about 345 p.p.m., and he did not include any of the abundant heavy oils except that in the Athabasca deposit; few constituents of crude oil have 1 percent of that solubility level in water.

Leigh Price believes that the aqueous solubilities of hydrocarbons are greatly understated in these arguments, because insufficient account is taken of *high temperatures*. Solubilities increase considerably with increasing temperature, and exponentially above 150 °C (Sec. 9.4.1). Above about 180 °C, they reach values of 20–100 p.p.m., especially for aromatic and naphthenic nuclei. The presence of S–N–O heteroatoms in the hydrocarbon molecules increases solubilities still further, and it is the *least soluble fractions* that are affected most strongly by rise in temperature. Thus the hydrocarbons in solution are less and less biased towards the light ends, and the solute becomes more nearly identical in composition to the original material. Crude oil solubilities of 8000 p.p.m. and higher are attained at temperatures of 275 °C (equivalent to depths in the 7 km range), and these solubilities are *tripled* every 600–750 m.

Two serious objections may immediately be raised to Price's conclusions. First, temperatures above 150 °C are normally sufficient to cause the conversion of liquid hydrocarbons to thermogenic gas (see Sec. 10.4.5 & Fig. 7.20). Price countered this objection by postulating a closed, water-wet system even at depths below 6 km; in such an environment, the breakage of carbon bonds is retarded until a temperature of 375 °C is reached. Price further invoked higher geothermal gradients during past structural episodes in basinal histories.

Second, at the depths required for the achievement of the high temperatures – 5 km or more – insufficient porosity remains in sedimentary rocks to provide the necessary volumes of water. Price invoked clay-dehydration waters from depths below 6 km. These waters, carrying hydrocarbons in solution, would migrate upwards, principally via deep faults and in large part because of abnormal pore pressures caused by the excess water itself. Many faults, in Price's view, are characterized by low entry pressures, which facilitate fluid ascent. The ascent in turn leads to hydrochemical, geothermal, and spore-pollen anomalies in all the fluids involved, anomalies frequently encountered (as in the oil and gas dissolved in geothermal waters). As exsolution of hydrocarbons proceeds, a continuous hydrocarbon phase is built up, eventually becoming long enough and continuous enough to overcome capillary forces. Further movement of the hydrocarbons up the faults is then in separate-phase flow. At some depth above the zone of abnormal pressure, the faults become less permeable to fluid movement than the normally pressured sandstones and carbonates, which then become potential reservoirs and are entered by the hydrocarbons via lateral (updip) migration.

It seems obvious that the source sediments in the deepest parts of a sedimentary basin must attain the expulsion threshold first, and the oil generated must move upwards through the source section until all of that section attains the threshold. The migration route must somehow open up ahead of the hydrocarbons, and faults are an obvious mechanism for this once the pressure buildup becomes adequate to keep them open. The objection to Price's hypothesis is not the necessity for fault channelways but that of excessively deep and high-temperature generation to permit migration in aqueous solution.

If we consider the solubilities of petroleum hydrocarbons in water, not as increasing with rise in temperature but as decreasing with drop in it – and to exceedingly low values (see Sec. 9.4.1) – the core of the dilemma becomes clear. Migration of hydrocarbons in aqueous solution presupposes that they exsolve as they are transferred to *cooler* stratigraphic levels. As these must normally be shallower levels, the saturated solutions must move *upwards* with respect to isothermal surfaces. Do they in fact move upwards with respect to sea level, or do the sediments pass *downwards* through an essentially fixed body of formation water?

During sedimentation, strata sink progressively with respect to sea level. Their fluid content must therefore undergo continuous heating unless the isotherms are also moving downwards at a comparable rate; in other words, the pore fluids become cooler only if the geothermal gradient decreases through time, so that the present gradient is lower in the shallow strata than in the deeper strata. The migration of oil in water solution can be a substantial factor only *after* such a reduction in geothermal gradient has occurred. Even then, the only pore waters available as solvent will be those contained in strata that have been buried deeper than the generative-temperature isotherm (T °C in Lopatin's equation; see Sec. 7.4.2).

Lawrence Bonham showed how to calculate the volume of available water in a sedimentary basin (the US Gulf Coast in his case). The calculation involves the step by step restoration of three-dimensional data from the present surface to economic basement. All thicknesses and depths must be restored, of course, to their values before compaction changed them. The essential steps in the calculation are these:

(1) Determine which stratigraphic units have passed through the generative-temperature "window" (T °C isothermal surface).
(2) By restoring the depths of key isotherms, determine the movement of this T °C surface, and hence the direction of fluid movement relative to it.
(3) By peeling off the layers of strata one by one, devise restored depth–temperature profiles for times at which each unit was at the surface as a new deposit.
(4) Using the best calculations possible for the pore volumes (by integrating the depth–porosity curves) and hydrocarbon solubilities, calculate the volumes of water and oil involved in the temperature drop from T °C to the temperature at the present depth of occurrence.

Hypothesis 3: Migration in micellar solution E. G. Baker, Robert Cordell, and others tried to avoid the difficulty presented by the very low aqueous solubilities of hydrocarbons by postulating that connate waters were solutions of colloidal electrolytes, or natural soaps. Such soaps, formed in the source sediments by the hydrolysis of glycerides (derivatives of the trihydric alcohols), form not true solutions but *stearate micelles,* or spherical clusters of soap molecules. Micelles are formed in solutions of surfactants, which are compounds that orient in a uniform manner at a fluid interface (like that between oil and water). The soap molecules are simply long carbon chains with their nonpolar ends in the centers of the spherical arrangements and their polar ends (COO^- groups) at the margins; the polar ends are soluble in water, and then constitute natural solubilizers for oils and fats.

After the micellar solutions have entered the reservoir rocks, the soaps are transformed into intermediate and high molecular weight hydrocarbons and to the asphaltenes and resins of crude oils. These are precipitated in the reservoirs when the micellar waters become diluted by meteoric or other formation waters. On this hypothesis, the relatively simple hydrocarbon composition of crude oils, when contrasted with the enormous number of possible hydrocarbon compounds, is due to preferential solubilities of a few of the compounds in the available natural electrolytes. The presence of two different crudes in a single basin (a very common phenomenon) is not necessarily due to two source sediments, but may be brought about by a change in the available solvent. Expelled connate water takes with it the hydrocarbons soluble in its particular micelles. If renewed compaction results in provision of more connate water, it may dissolve different hydrocarbons and so yield two different crudes from a single source rock.

Unable to visualize the stupendous volumes of water necessary to carry the heavy oils of Alberta in solution across the Western Canadian Basin, Laurence Vigrass invoked micellar solutions to increase the solubilities. Most geochemists, however, have objected that there is no evidence that naphthenic acid soaps are supplied in sedimentary basins in anything like sufficient quantities to be capable of transporting the volumes of oil that have actually become pooled. The micellar concentrations actually measured are far too low, and become more so at higher temperatures. Micelles undoubtedly increase the solubilities of heavier hydrocarbons in water, but they still raise them to only a few parts per million. The pore spaces in source sediments offer very low permeabilities to soap solutions, and the existence of negative electrical charges on both the clay minerals and the ionic micelles should create an electrical barrier by their mutual repulsion. A final objection to micellar migration is the lack of any obvious mechanism for removing all traces of the soaps after they have done their work. If gas, oil, and pore water all become trapped in the reservoir, why not the micelles themselves?

Hypothesis 4: Objections to a dominant role for water *Nonassociated gas* can apparently be carried out of its source rocks, and into porosity, in aqueous solution. As is emphasized in Chapter 25, many of the world's greatest gas and gas-condensate fields are pooled in relatively old rocks (Carboniferous, Permian, and Triassic); but they cannot have been so pooled before late Cretaceous time and probably not until well into Tertiary time. Why did the mobile gas not then migrate into younger strata with plenty of pore space?

Many Russian geologists maintain that their representatives of this group of gasfields (see Sec. 25.3) were created by the release of gas from solution in older formation waters, when formation pressures were reduced through rejuvenation of structures within their basins. Had the gas not been retained a long time in solution, its dissipation would have been at least likely, perhaps inevitable.

For the primary migration of *oil,* however, it is apparent that all postulates of a dominant role for water as a solvent face insuperable obstacles. No matter what the origin of the water may be, hydrocarbon solubilities are simply too low throughout the temperature range encountered in petroliferous rock successions.

Oil generation appears to be most efficient at depths where water mobility is most inhibited by loss of porosity and permeability. Waters derived from clay rocks, by whatever mechanism, are not like those actively moving through continuously permeable rocks like reservoir- or carrier-bed sandstones. The clay minerals and their water in a sense form a single system; the water behaves as part of the sediment. The flow of oil through sedimentary rocks and the flow of water are independent processes. It seems highly unlikely that petroleum can leave its source rock dissolved in water derived by the source rock's own compaction or diagenesis.

The obstacles are heightened by the still lower aqueous solubilities of oils when the waters are saline, or saturated with gas (Sec. 9.4.1). Are there *any* components of the fluids generated in source sediments that *increase* the solubilities of oils in water?

Hypothesis 5: Migration in gas-charged solution As water becomes scarcer in deeply buried sediments, the dominant fluids formed in the source sediments by thermal maturation become gaseous: CH_4, H_2S, and CO_2 (see Sec. 10.4). V. A. Sokolov was the most notable early advocate of the primary migration of oils in *gaseous solution.*

Experiments have shown that both CH_4 and CO_2 bring about vast enhancements of the solubilities of crude oils, when the waters are charged with these gases under high pressures. Russian investigators showed that solubilities of oils in compressed, liquefied natural gas, at 50 °C, increase exponentially from low levels at moderate pressures to more than 600 g m^{-3} at 5×10^5 kg m^{-2}.

Furthermore, paraffinic oils are more soluble in methane than are naphthenic oils, opposite to the aqueous solubility pattern (Sec. 9.4). Experiments reported by Leigh Price in 1983 showed that, under moderate *T*, *P* conditions, the solute is enriched in saturated hydrocarbons in the C_5–C_{15} range, but as *T* and *P* are raised the higher molecular weight compounds go into solution and the solute becomes identical to the initial crude oil. Solubilities of the order of 1000 g m^{-3} are attained at *T, P* values reached within drilling depths (Fig. 15.10).

The presence of water is critical to the solubilities of crude oils in methane. Water also lowers the *T* and *P* necessary to achieve cosolubility in which the methane and the crude oil behave as a single-phase mixture. Cosolubilities as high as 4000 g m^{-3} are then attained at about 200 °C (a representative temperature in deep reservoirs).

Methane is nevertheless unlikely to be the most important solute gas in the transfer of oils out of source sediments. Condensates are common in deep reservoirs, but even at 200 °C only a very small proportion of the gas phase consists of high molecular weight components. As the liquid hydrocarbons move upwards in gas-charged solution, moreover, these heavier components should exsolve first, and the lightest oils last. This is the opposite of the commonly observed zonation.

Experiments by E. E. Bray and W. R. Foster indicate, however, that molecules as heavy as C_{28}, at least, can be transported if the dominant solute gas is CO_2. At about

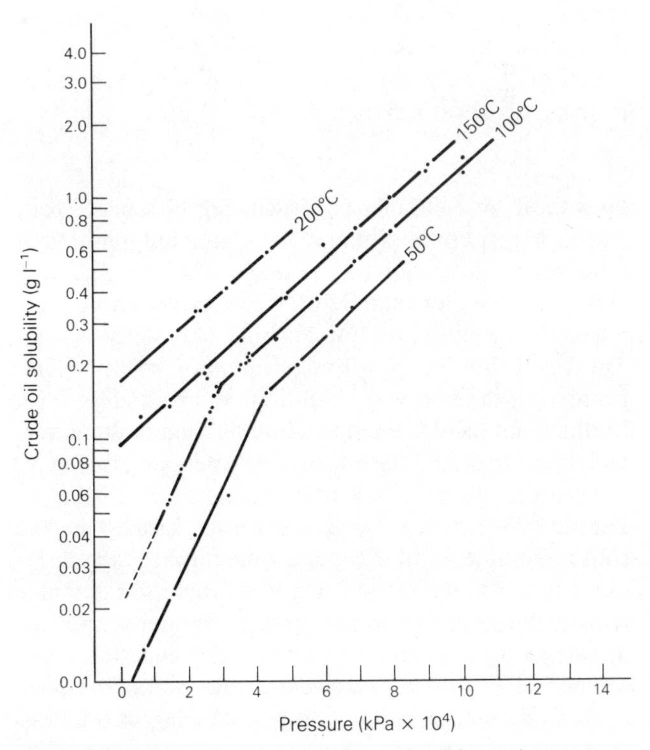

Figure 15.10 Solubility of 44° gravity crude oil in methane over a range of temperatures occurring within drilling depths. (After L. C. Price, *AAPG Bull.*, 1983.)

4×10^4 kPa, the quantity of crude oil carried in water saturated with both CO_2 and hydrocarbon gases rises to 3 percent by volume, more than 100 times the quantity taken into solution under mere rise of temperature. Nearly all the oil generated from the source rock – even tars and asphaltenes – can be accommodated in CO_2-saturated pore waters. Without CO_2 in the solute, however, the only hydrocarbons carried in the water are in the gasoline range, lighter than C_{10}.

Along the migration route, the CO_2 must normally be stripped from the moving fluid through pressure drop or by reaction with the minerals of permeable strata. CO_2 is not an important constituent of most reservoir fluids, but its reactions would produce clay minerals (especially kaolinite, which is very common in reservoir rocks) and bicarbonate ions (which are relatively common in pore waters).

The obvious question is a source of sufficient CO_2 in or near the source sediment. Some of the parent OM is converted to CO_2 during thermochemical maturation. The generation of enough of it to be an effective migration agent requires a burial depth in excess of 1000 m and some minimum weight-percent of kerogen in the source rock; both of these requirements are met among the agreed functions of a source sediment. The CO_2 is much more soluble in oil than in water. It is miscible with both oil and methane; it lowers the oil's viscosity and increases its mobility. It also precipitates unwanted products, such as asphaltenes and carbonate minerals, from the oil. It is these properties that make CO_2 useful in miscible-flood enhanced recovery (see Sec. 23.13.3). It may be the most underrated of the potential agents of primary hydrocarbon migration.

Hypothesis 6: Role of microfracturing of source rocks
James Momper determined the dominant interstitial pore spaces in shales to be smaller than 10 nm, and commonly smaller than 3 nm. These pores, moreover, normally contain water from the time of first deposition. The expulsion of *gas* from such pore spaces is no problem, whether water solution is invoked or not. Methane molecules are only about 0.4 nm in diameter, and their expulsion may reach 50–90 percent efficiency.

The expulsion of *oil* is a different matter. According to Dietrich Welte, the *kerogen particles* formed in the source sediment at the peak generation stage have diameters of about 0.1–1.0 μm. Oil droplets or globules leaving the source sediment, entrained in pore waters or dissolved in gas, cannot be much smaller than this. They are not yet *petroleum molecules,* which are only about 2–6 nm across. Even these may still be larger than the average clay pore size.

Clearly oil cannot be expelled from a clay–shale source sediment via pore spaces smaller than the hydrocarbon molecules seeking escape. It is in fact clear that the greater part of the oil generated in a source sediment never is expelled from it. The efficiency of oil *generation* from the parent OM is no doubt very low (Sec. 6.6); the efficiency of oil *expulsion* is still lower. In John Hunt's estimation, about 95 percent of the generated bitumens remain in the source rock under normal circumstances (though they may be expelled in later geologic time if *T, P* conditions change greatly). The only fraction of the fluid bitumen which is successfully expelled must find escape channels larger than normal shale pores. These larger channels must be widened partings, bedding-plane separations, or fractures.

In the absence of extensional tectonism, the creation of open fractures requires that the pore-fluid pressure (p_w, if water occupies the pores) exceeds the effective lateral confinement stress (σ_3, if σ_1 is essentially vertical and represents the geostatic load).

We have a ready mechanism for this buildup of pore-fluid pressure. It is the thermochemical generation of fluid hydrocarbons from the source rock kerogen, followed by their thermal expansion and possibly augmented by the osmotic exchange of fluids between the clay rocks and adjacent sands. The kerogen macromolecules, as we have seen, are larger than the pore spaces of the source rock, and larger also than the grains of that rock (which are mostly clay-mineral flakes). With these dimensions, the kerogen flakes are subject to the full geostatic load just as the rock flakes are. In experiments and calculations by Jean Du Rouchet, such large flakes act as microfissural flaws that can lead to the familiar Griffith-type failure.

Once the kerogen flakes become crushed under the geostatic load, they are broken down into smaller molecules by the breaking of C–C bonds. They acquire the fundamental properties of liquids, and add the extra fluid pressure of oil (p_o) to that of the water (p_w). The fluid pressure exceeds the tensile strengths of the rocks, which range from about 5×10^2 to about 150×10^2 kPa, creating microscopic, tensile hydrofractures at right angles to the flake elongations. By additionally opening up S-surfaces (bedding planes and partings), the fluids are able to migrate in the direction of σ_3.

As the fractures are propagated both laterally and vertically, the source sediments (in Momper's words) undergo natural hydrafrac. The net effect is an increase in the bulk volume of the rock (an aspect of *dilatancy*; see Sec.13.4.4), enhancing pore volume and permeability and providing avenues of pressure relief.

Microfracturing is of course of no help in primary migration from the source sediment if it occurs after the hydrocarbons are ready for migration. In nature, a dilation zone normally appears in sedimentary rocks under no great depth of burial. The actual depth is partly dependent on the geothermal gradient, being shallowest where the gradient is highest; but it may be independent of the rate of burial. As burial proceeds, dilation effects keep pace and are propagated upwards into younger and younger strata. This upward propagation is self-extending; it continues even after the original fractures have reached a critical depth at which the channelways are closed by excess lithostatic load. The flush generation of oil and gas thus leads automatically to their expulsion and migration, which may initially take place upwards, laterally, or downwards, though the first direction is likely to be upwards and the dominant later direction laterally up the dip.

Hypothesis 7: Diffusion along kerogen network The creation of the network of microfractures provides the opportunity for the already large particles of kerogen to become fused together, forming continuous, fine strands which coalesce along it. The resulting network of kerogen threads represents a *separate, continuous oil phase*; the medium is oil-wet. If the network becomes essentially continuous and three-dimensional, oil or gas, or a solution of one in the other, may diffuse along it in the direction of the pressure gradient.

The role of such a kerogen network in the primary migration of hydrocarbons out of their source rocks was first proposed by Gordon Erdman in 1965. It has been enlarged upon by Robert Jones, Clayton McAuliffe, Douglas Hobson, and others. Peak generation of bitumen from the parent kerogen occurs when the absorption capacity of the kerogen is exceeded. The bitumen is then extruded from the kerogen into the pore or fracture system of the source rock. There it first encounters water, which it displaces, causing the onset of primary migration. The oil expelled is likely to be undersaturated with gas; it is lighter, less viscous, and less polar than the bitumen from which it originated; it is further extruded as a consequence of abnormal pressure buildup during subsidence and compaction. At the interface between the source rock and a potential reservoir rock, oil droplets or gas bubbles are extruded from the ends of the network; growing to 20–30 percent pore volume, they coalesce, achieve a buoyant head (1 m or more in extent, as a general estimate), and move upwards as buoyant slugs along the upper margin of the reservoir rock. This is the beginning of secondary migration.

Douglas Hobson likened the operation of the kerogen network to that of a wick. In experiments with paraffin-soaked strands of wool immersed across the interface between paraffin and salt water, a continuous thread of paraffin formed through the wool and was expelled into the water at its end as long as a fluid pressure gradient existed in the required direction. Both experiment and observation point to a particularly vital requirement. Formation of an adequate kerogen network requires a source sediment very rich in kerogen, and hence originally very rich in organic matter. The "drainage" efficiency of the source sediment must be high. In Jones's estimation, a total organic carbon (TOC) content in excess of 2.5 weight percent is the minimum normally able to provide a major oil accumulation; many rich source sediments have much higher TOC contents than that (see Sec. 6.6). Some very petroliferous basins, however, appear to lack any source sediment so rich; the US Gulf Coast Tertiary basin is an example. It is to be doubted whether leaner source rocks are capable of providing continuous-phase oil migration.

Without the kerogen pathway, diffusion as an important mechanism for the primary migration of *oil* must be discounted. In the water-saturated pore space of the source rock, diffusion of hydrocarbons would take place along the *concentration gradient,* governed by *diffusion coefficients*. These range from 2.12×10^{-6} cm^2 s^{-1} for mobile methane to 6×10^{-9} cm^2 s^{-1} at the C_{10} level. The figure for methane suggests that diffusion is a plausible mechanism for the migration of gas; but the quantity of *oil* that could have been similarly transferred from, say, a Cretaceous source rock (see Ch. 8) is far too small to account for the quantities in known accumulations.

15.5 Evidence favoring early generation and accumulation

The reader will have recognized the correlation between concepts of primary migration – which are essentially *geological* concepts – and those of deep generation of oil and thermochemical gas – essentially a *geochemical* requirement.

If oil and associated gas form late, after some minimum burial depth and minimum maturation time (see Sec. 7.4), considerable vertical migration is inescapable. Great volumes of oil are pooled at depths much shallower than those required for generation and expulsion. Few geologists or geochemists will accept Leigh Price's implication that most oils have suffered temperatures in the 180–200°C range. Opponents of long vertical migration require that the present degree of thermal maturation of any hydrocarbons be the result of their *present* temperatures (or of the highest temperatures reached by the sediments at the hydrocarbons' present depths). Thermal maturation of petroleum in deep, overpressured sediments cannot proceed to completion because an equilibrium must intervene. For the completion of the maturation reactions, the petroleum must be removed from the overpressured source system and must enter that of normal pressures. Maturation, in other words, is a process that occurs in the reservoir rock after short migration, not in the source rock before long migration.

Persuasive advocates of early, shallow generation and accumulation abound, and include some of the most distinguished of petroleum geologists (and even some geochemists) – Lewis Weeks, Wallace Pratt, William Gussow, Hollis Hedberg, H. H. Wilson, Dietrich Welte. There are great volumes of oil in Pliocene sediments – in the Baku fields, in California, the US Gulf Coast, Borneo, Romania. There is even modest production from shallow Pleistocene sands in eastern Borneo and off the US Gulf Coast. It may be that there is much more Pliocene and Pleistocene oil and gas than we know about, because glacial and postglacial changes of sea level have caused most marine sediments of those ages – especially the basinal marine sediments – to lie far out beneath the present seas.

Many cases have been reported of free oil forming very early in the sedimentary process, before 1 Ma have elapsed, before 600 m of sediment have been deposited, before a shale dry density of 2.0 has been reached, and before the shale porosity has been reduced to 23 percent. The proto-oil must be expelled from the clay source sediments both early and rapidly, or compaction will so reduce the permeability that expulsion of the relatively large oil globules will be prevented. In Welte's opinion, some 70 percent of the liquids contained in clay sediments must be expelled before the burial depth reaches 300 m.

Such evidence is of course highly ambiguous. By far the most abundant fluid in any thick succession of clay rocks is pore water. Certainly the greater part of this is expelled at very shallow depths of burial, and this early, rapid expulsion is accompanied by decrease in porosity and permeability and increase in density of the rock. This is not evidence that light oils are expelled with this water, still less that organic matter has matured to hydrocarbons to be ready to accompany it. The commonly invoked observation that only early-formed traps are able to capture oil is readily misunderstood. If a trap is created *too late,* with respect to the timing of primary migration, it is likely to be barren. This fate is avoided if traps "grow through time" (see Sec. 16.3), and so are ready when the migration does take place; it does not mean that the migration is taking place throughout the growth of the trap. There are hundreds of fields in which the creation of the present traps occurred geologically long times after the deposition of both the reservoir rocks and the source rocks. Hardly any large field provides convincing evidence of having received its present quota of oil or gas before Jurassic time, and relatively few can have received it before early Tertiary time – even though the source and reservoir rocks may both be of Paleozoic age. The oldest supergiant reservoir-source combination is probably that at the Hassi Messaoud oilfield in Algeria (Sec. 26.2.3); many giant fields in North America and European Russia are pooled in Ordovician, Devonian, or Carboniferous strata (Ch. 25). None of these accumulations can have been formed before the Jurassic; few can plausibly have been formed before the mid-Cretaceous. Even allowing such accumulations to be of second or third generation following much earlier accumulations now dissipated, the fillings of the traps are very late, geologically, with respect to the source sediments. Enclosed, lenticular traps at shallow depths (say less than 1500 m) may never have been much if any more deeply buried than they are now. This does not mean they were filled early.

In 1976 Colin Barker attempted to reconcile the conflicting evidence surrounding this topic by proposing a two-stage model. At shallow depths – shallower than about 1500 m – the migration of hydrocarbons is potentially very efficient because source-rock permeabilities are high and active compaction makes the maximum quantities of water available. There may also be abundant biogenic gas to aid in the migration. The generation of *oil,* however, is exceedingly inefficient at such depths. Most shallow crude oils are heavy, and rich in heterocompounds; they are difficult to correlate with any potential source rocks because of their low and varying solubilities. At greater depths – below 1500 m, and extending much below – generation must be more efficient but migration much less so because of greatly reduced shale permeability. Oils must then move in separate-phase flow via pore-center networks; they undergo no chemical fractionation, so they *can* be related to their source rocks and their accumulation histories are readily deciphered. On Barker's postulate, the "generation" curves of Bernard Tissot and others (see Fig. 7.20), showing maxima in the 1500–1800 m range, are really "accumulation" curves.

15.6 Summary of primary migration

None of the mechanisms so widely discussed, for primary migration of petroleum out of its source rock, is sufficiently convincing to command majority acceptance. In some ways the most convincing of them – that of diffusion along a kerogen network, rejecting dependence upon solution – has had the shortest history of consideration. On the other hand, none of the mechanisms can be rejected for all circumstances. It is unlikely that any single mechanism is responsible for all cases. Source rocks are not all alike; the *P*, *T* conditions of generation, the availability of water, of solubilizers, of CO_2, of non-associated gas, all are variable in both time and space. The arguments will probably go on as long as new fields are discovered.

Aside from the matter of the *mechanism,* and the physical laws directing it, there are equally unresolved questions concerning the *timing* of its operation. We have discussed three processes – the transformation of the parent OM to generate the hydrocarbons, their expulsion or release, and their migration in search of porosity – as if they were in reality but aspects of a single process. Are they in fact geologically simultaneous and inseparable? If migration begins at the same time as generation, does it also end at the same time, or does it continue for some time afterwards? Does it proceed over a long or a short period, in geologic terms? And can it be resuscitated or repeated? If it can, are there any time restrictions on the resuscitation? There is probably no single answer to any of these questions, nor to many more that may have occurred to the reader. We know much more about the problem than we did before the advent of modern petroleum geochemistry; but we are not noticeably nearer to agreement and we may not be noticeably nearer to the truth.

15.7 Transfer from source rock to reservoir

In order to accumulate within a porous rock medium, hydrocarbons have to acquire the capacity for *continuous phase flow.* This in turn requires that some critical hydrocarbon saturation be exceeded. Homer Mead and others estimated that gas begins to flow when saturation is of the order of 8 pore-volume percent; oil flows at about 22 pore-volume percent. The required pore volume is relatively easily achieved if much of the water content is *structured* close to the clay surfaces (see Sec. 15.4.1). The hydrocarbons then accumulate in the position of least energy, as far as possible from the structured water, in the centers of pores.

On the kerogen network hypothesis, the continuous oil phase develops by the joining up of these pore centers. The network itself represents a continuous hydrocarbon pathway effecting continuous transfer of its own contents from source rock to reservoir. On Leigh Price's postulates, oil acquires the status of a continuous phase by exsolution from water (or gas) in fault channelways leading out of the source sediments (see Sec. 15.4.2). Under all other proposed mechanisms for primary migration, the hydrocarbons reach their first porous receptacles in a disseminated state.

At the boundary between the source sediment and a potential reservoir rock, oil can cross into the latter only by acquiring enough energy to overcome the resistance caused by the circumstance that the reservoir rock is almost certain to be *water-wet.* The oil can enter the reservoir only by displacing water. Early experiments (1926) by Alex McCoy demonstrated that this displacement occurs without regard to the direction of gravity forces. If an oil-soaked mud is placed against a water-saturated sand, water moves from the sand into the mud and oil from the mud into the sand. Oil does not push heavier water out of the way; the two fluids move in opposite directions across the interface between an initially oil-wet medium with very fine pore spaces and an initially water-wet medium with coarser ones. The fundamental force involved is that of *capillarity.*

15.7.1 Capillarity

Capillarity is the tendency of *wetting liquids* to ascend minute openings, less than 0.5 mm in diameter, through the agency of a molecular surface force acting against the force of gravity. *Nonwetting liquids,* like mercury, "climb downwards," forming a reversed meniscus. If only one fluid is present, the height of the capillary rise depends on the size of the opening, the material of the medium, and the surface tension of the fluid.

The *surface tension* is a measure of the cohesion of the molecules at a fluid's surface. In the familiar laboratory experiments, the medium into which the fluid is attempting to rise is a glass tube of radius r. The surface tension (σ) is directly proportional to the area of the tube's cross section, to the density of the liquid (ρ), and to the angle of contact (θ) between the liquid and the tube. It is inversely proportional to the length of the contact between them (the circumference of the tube, in fact):

$$\sigma = \frac{\pi r^2 h \rho \mathbf{g}}{2\pi r \cos \theta} \qquad (15.1)$$

where h is the height of rise of the liquid in the tube. In CGS units, σ is measured in dynes per centimeter. The SI units are newtons per meter. As 1 newton is 10^5 dynes, 1 dyne $cm^{-1} = 10^{-3}$ N m^{-1}.

Measured against air at 20 °C, the surface tension of pure water is 72.75×10^{-3} N m^{-1}. Brines have higher values, but most common organic compounds, such as acetone and chloroform, have much lower ones, of the order of 20×10^{-3}–30×10^{-3} N m^{-1}. So do crude oils; light crudes fall into the range of 20×10^{-3}–30×10^{-3} N m^{-1} and heavy crudes $35 \times 10^{-3} \times 40 \times 10^{-3}$ N m^{-1}. Mercury, in contrast, has a surface tension at STP of nearly 500×10^{-3} N m^{-1}. Surface tensions decline with increasing temperature, becoming close to zero at very high temperatures; this is because heat agitates the surface molecules and lowers their cohesion. Surface tensions may increase or decrease with rising pressure, with which the relation is more complex.

If more than one fluid is present, the terms *wetting* and *nonwetting* become relative. The area of contact between two contiguous fluids is kept to a minimum by the *interfacial tension* between them. The interfacial tension

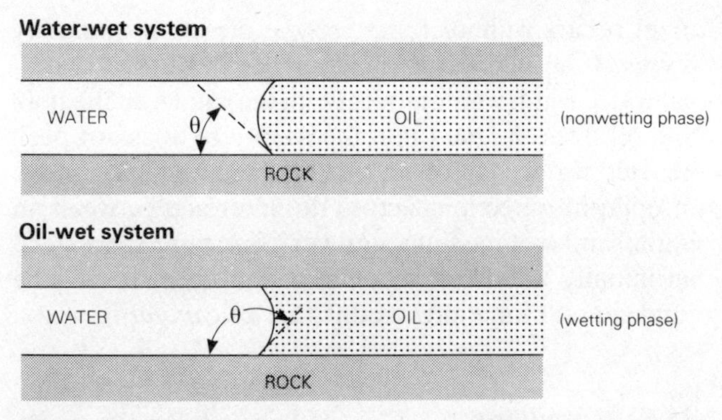

Figure 15.11 Curved interfaces between oil and water in capillary openings in water-wet and oil-wet systems. Note measurement of angle of contact (θ).

(γ) is directly comparable with the surface tension, but it depends on the densities of both fluids (ρ_1 and ρ_2):

$$\gamma = \frac{\pi r^2 h\,(\rho_1 - \rho_2)g}{2\pi r \cos\theta} \qquad (15.2)$$

Like the surface tension, the interfacial tension is measured in newtons per meter and is lowered by temperature; it is also lowered by the presence of surfactants (think of water on a clean dish and on a wet soapy dish). The wetting liquid is then the one with the lower interfacial tension; it adheres to the solid medium more easily, forcing the nonwetting liquid to find accommodation elsewhere. In the oil–water–rock system, therefore, oil is the nonwetting phase relative to water in the common case of the rock being already water-wet, but the wetting phase in the less common case of the rock being oil-wet (Fig. 15.11).

The interface between the two liquids is drawn by capillarity into a curved form, convex towards the wetting liquid. The radius of curvature of the interface depends not only on the interfacial tension between the two fluids but also on their very different surface tensions against the constantly varying materials of the pores and pore throats. In the natural water-wet, oil–water–rock system, the curved fluid interface is a reflection of the capacity of the fluid having the lower surface tension to displace the one having higher surface tension into finer pore spaces which present greater surface area.

The factor $\cos\theta$ in Equations 15.1 and 15.2 is called the *wettability*; it represents the work necessary to separate the wetting fluid from the rock medium, and is always measured through the water phase (Fig. 15.11). The factor r in the two equations is represented in nature by the radii of the largest, connected pore throats, through which the hydrocarbons must migrate in order to move from the source rock to the reservoir rock (and later within the reservoir rock itself during secondary migration).

15.7.2 Displacement pressure

To move from its source-rock pore space into its reservoir-rock pore space, the oil's capillary pressure must overcome the *displacement pressure* of the latter, which is filled with water. The displacement pressure, P_d, is defined as the smallest capillary pressure required to force hydrocarbons into the largest interconnected pores of a preferentially water-wet rock.

For cylindrical pores, the value of the displacement pressure is given by

$$P_d = \frac{2\,\gamma \cos\theta}{r} \qquad (15.3)$$

It is expressed in newtons per square meter ($1\ N\ m^{-2} = 1\ Pa = 10\ dynes\ cm^{-2}$). It may be measured directly by mercury (nonwetting) capillary pressure tests. In an oil-wet medium, the contact angle of the fluids against the pore wall is more than a right angle, because it is measured through the water phase (Fig. 15.11). The displacement pressure, proportional to $\cos\theta$, is therefore lower in an oil-wet than it would be in a water-wet medium.

The radii of the pore throats may be measured visually (in thin sections or from SEM photographs) or from pore-casts of the leached rocks. The radii are critical because the capillary pressure is greatest on the smallest fluid droplets, and is also greater in water-wet shales than in water-wet sandstones because the shales are of finer grain size. The capillary pressures in shales are of the order of 10^6 kPa; they may be 100 times higher than those in sands. A shale/sand interface is therefore a one-way street for hydrocarbons; in a water-wet system, they can move only from the finer to the coarser grain. The consequence is the selective segregation of oil from water, water occupying the pores of small diameter and driving the oil into the larger ones.

Very fine sands of course inherently possess very fine pore spaces, even though their total porosity may be high (see Sec. 13.2.4). Studies of many reservoir sandstones by John Griffiths led him to the conclusion (reported in 1952) that a sand having median grain diameter less than 62.5 μm may normally be expected to be water-bearing. If it is oil-bearing, its oil is unlikely to be recoverable from it by normal extraction methods.

The operation of the displacement pressure and its dependence on the three factors (interfacial tension, wettability, pore size) were graphically demonstrated in an experiment during the 1920s by V. C. Illing, illustrating the so-called *screening theory* of Roswell Johnson. A water–oil mixture containing 90 percent water was passed through a tube containing alternate layers of coarse and fine sand, both *water-wet.* The oil saturated the layers of coarse sand in succession, leaving the fine sand unimpregnated. The experiment was repeated with the sand layers *oil-wet* and the mixture containing only 10 percent of water. The water then forced the oil out of the coarse layers and saturated them in succession, leaving them the unimpregnated segments of the column.

15.7.3 Practical effects of capillarity in migration

We have seen (Sec. 15.7.1) that the *surface tensions* of typical formation waters are about three times those of light crude oils; the surface tension of gaseous methane is of course zero. At the boundary between a source sediment and water-wet porosity, therefore, water exerts capillary pressure about three times that of typical oil.

Interfacial tension values follow the reverse pattern, because they are measured against water and not against air. At STP, the value of γ is about 70×10^{-3} N m^{-1} for methane–water, about 30×10^{-3} N m^{-1} for light crude oil–water, and as low as 15×10^{-3} N m^{-1} for heavier crudes–water. Hence the displacement pressure is greatest for gas, rock conditions permitting; this reduces the migration potential of gas through water-saturated rocks. A given rock can seal a larger gas column than oil column because the higher interfacial tension counteracts the higher buoyant force pressure in the gas–water system (see Sec. 15.8.1). On mobilization of the hydrocarbons, therefore, the oil moves out of the source bed and into the porosity *after* the water but *before* the gas.

Thus water preferentially occupies the finest pore spaces, entry to which requires the highest capillary pressure. The hydrocarbons are driven into the coarser pore spaces, but they may not yet be in a continuous phase. A continuous, unbroken column of oil, completely saturating a fine sediment, can move into coarser sediments through capillarity alone; but such a level of oil saturation is not known to be achievable at the source/reservoir interface. How then can oil enter the coarse porosity if that is surrounded by finer porosity without oil? In many pools, reservoir sandstones are separated from source sediments by clean sandstones with no obvious staining.

Unless the high interfacial tension between oil and water is somehow reduced at elevated *T* and *P*, enabling the two fluids to move as a single phase, the oil can move through the water-filled pore space, initially, only as dispersed droplets or globules. What is more, the droplets are likely to be accompanied and separated by gas bubbles. Both the droplets and the bubbles tend to acquire the spherical shape of minimum surface area; their movement through capillary openings (pore throats, in our case) is a peculiarly difficult process, inhibited by the so-called *Jamin effect.* The spherical shapes must become severely distorted, and distortion is resisted by interfacial forces. Disseminated oil, or oil heavily charged with gas bubbles, cannot in fact migrate under capillary influences alone.

The relative sizes of the oil globules and the gas bubbles, in relation to pore diameters (values of $2r$ in Eqns 15.1 & 2), are obviously critical. Parke Dickey concluded that the oil must move as an emulsion of dispersed droplets, individually no larger than the pore spaces (1.5 μm or smaller). This kind of emulsion is easily reproducible in the laboratory, especially if a surfactant is present to lower the interfacial tension; the emulsion can be passed through a filter paper. The amount of oil it contains may be no more than 30 p.p.m. by volume. At the interface between coarse and fine porosity, water passes into the latter but the oil is screened out of the emulsion by adhering to finely particulate clay minerals. The oil therefore accumulates against the interface unless the droplets can become sufficiently distorted for it to pass through.

What is the capacity of oil globules and gas bubbles for distortion from spherical shape? As primary migration begins, we are concerned with the distortion necessary for the globules to pass through *shale* pore spaces and pore throats. In Kinji Magara's view (1981), the droplets may be *compacted* (squashed, in effect) by the structuring of the pore water in smectitic shales (Sec. 15.4.1); structured water is up to eight times as viscous as normal water. As the viscosity of the oil normally decreases with depth, there must be some depth below which the viscosity of the expelled pore water exceeds that of the oil, the oil then becoming easier to move than the water. Though there may in nature be such a depth, it seems beyond credibility that it is shallow enough to be responsible for primary oil migration in dozens of petroliferous basins having no great thickness of sedimentary column. If a continuous oil phase is achievable within the source sediment, the effect of capillary pressure on its flow through those sediments will of course be negligible.

At the source/reservoir boundary, however – and *a fortiori* once the boundary is crossed and connected porosity is finally reached – the question becomes different. Here, the size differences between the grains, the pores, and the pore throats become much magnified. The migration of oil globules (or gas bubbles) through water-wet reservoir rock demands extreme distortion of the spherical shape. In such a medium, cos θ in Equation 15.3 becomes unity. The capillary pressure in a pore (P_p) and in the throat through which the oil is being drawn (P_t) can then be expressed in the familiar Laplace equations

$$P_p = \frac{2\gamma}{r_p} \quad \text{and} \quad P_t = \frac{2\gamma}{r_t} \tag{15.4}$$

where r_p and r_t are the radii of the pore and the throat respectively (Fig. 15.12). Because the capillarity (P_c) for any single fluid (oil, in the case we are considering) is

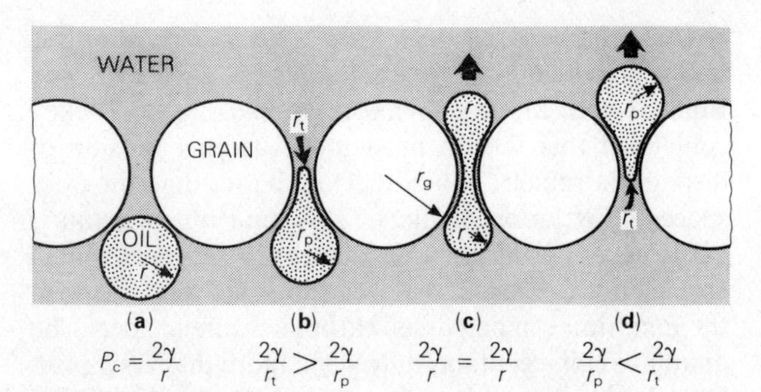

$P_c = \frac{2\gamma}{r}$ $\frac{2\gamma}{r_t} > \frac{2\gamma}{r_p}$ $\frac{2\gamma}{r} = \frac{2\gamma}{r}$ $\frac{2\gamma}{r_p} < \frac{2\gamma}{r_t}$

Figure 15.12 Oil globule of radius r, in pore space of water-wet, clastic rock, undergoing distortion as it migrates through pore throat of radius r_t. γ is the interfacial tension. (From R. R. Berg, *AAPG Bull.*, 1975.)

simply the difference between P_t and P_p, it can be expressed by

$$P_c = P_t - P_p = 2\gamma \left[\frac{1}{r_t} - \frac{1}{r_p}\right] \quad (15.5)$$

The bracketed parameter may be large, because r_p is likely to be much larger than r_t.

With perfectly uniform, spherical grains, values for r_p and r_t may be estimated as proportions of the mean effective grain radius (r_g). The values will depend upon the packing mode of the sediment (Fig. 11.3). With rhombohedral packing, for example,

$$r_p = 0.414 r_g \quad \text{and} \quad r_t = 0.154 r_g$$

The changing radii of the segments of the distorted droplet are illustrated in Figure 15.12. The maximum length, Z, achieved by the droplet as it is drawn through the pore throat (Fig. 15.12c) is proportional to $(\rho_w - \rho_o)$. Hence the capillarity can be alternatively expressed by the equation

$$P_c = \mathbf{g}\,(\rho_w - \rho_o)Z \quad (15.6)$$

When this value becomes sufficiently large, the radius of curvature of the oil/water interface equals the effective pore-throat radius (r_t), and the oil migrates through the throat until further throats are encountered with smaller effective radii.

Once out of the source rock and into a reservoir rock, the gas, oil, and water rapidly acquire separate phases. Each phase then exerts its own pressure, dependent upon its relative elevation. At the oil/water interface, for example, P_o and P_w are essentially equal. Both decrease with elevation Z above the interface, but P_w decreases faster than P_o because the oil column accumulates above the water column. Under hydrostatic conditions, therefore, $(P_o - P_w)$ increases linearly with elevation so long as permeability permits the oil to migrate through the water. Secondary migration has begun.

Quantitative studies of capillary processes and their effects are among the responsibilities of research and development geologists and reservoir engineers. Values of the fluid parameters – densities, interfacial tensions, flow rates, and so on – are read directly from nomographs supplied in oilfield manuals. Published examples for geologists have been provided by a number of authors, most recently by Tim Schowalter in 1979.

15.8 Secondary migration

A reservoir rock differs from a compacted source rock in possessing both much higher porosity and much higher permeability. Once within a reservoir rock, oil and gas migrate in search of lower energy levels; traps impede the migration and restrain the hydrocarbons because they represent *purely local* minimum values of the energy potential. The dominant parameter is the reservoir *pressure;* the migration is in the direction of decreasing hydrostatic pressure.

The availability of much larger pore spaces and much more continuous permeability in the reservoir rock than in the source rock permits the rapid achievement of a single, continuous-phase fluid flow, a flow normally taking place through water-saturated, permeable strata. The physical requirements for secondary migration are therefore three: an adequate supply of hydrocarbons to sustain it; adequately continuous permeable pathways to facilitate it; and the necessary pressure gradient to impel it. Given these three requirements, the only limit imposed on the *distance of migration* is that consequent upon the size of the basin. In a large basin of long tectonic stability (like the Ural–Volga or Mid-Continent basins), migration from the basin towards the foreland may be essentially unimpeded for scores or hundreds of kilometers. The typical *rate of the migration* was estimated by John Hunt to be about 2.5–7.5 cm yr^{-1} (25–75 km Ma^{-1}).

The migration pathways have traditionally been called *carrier beds,* and envisaged as sandstone formations of high horizontal permeability. They are better thought of as *carrier systems,* because in addition to permeable stratigraphic horizons they may be provided by uncon formities, fault or fracture systems, old weathered zones, or penetrative diapirs. Furthermore, though secondary migration is commonly a phenomenon taking place *along* the bedding direction, it may perfectly well take place *across* it, especially in young, orogenic basins containing numerous stacked, weakly consolidated sandstone members (like those of California, Indonesia, or the southern Caspian). A vital consequence is that oil or gas may now be pooled scores of kilometers laterally away from its area of generation and hundreds or thousands of meters above its horizon of generation if the latter has undergone long-continued subsidence.

In our dissertation on primary migration, the role of water had to be repeatedly both emphasized and disputed. The role of water is central to secondary migration also, but it cannot be closely similar to any possible role in primary migration. There is no serious possibility of oil, at least, undergoing secondary migration in aqueous solution, even with the assistance of solubilizing micelles (though Laurence Vigrass invoked these as the agents in the long migration of the oil now in the Athabasca "tar sand" deposit in Canada). Because of the exceedingly low solubilities of petroleum hydrocarbons, their achievement of the 25–30 percent pore-volume saturation necessary for separate-phase flow would necessitate gigantic volumes of water flowing through the reservoir rocks in all petroliferous basins. The volume of water would have to exceed the volume of the reservoir rock itself many times over. Migration as disperse globules in the water stream, another mechanism considered for primary migration, would be impossible through the water-wet sands that are by far the most likely carrier systems. Unless horizontal permeability was extraordinarily high (as along widened bedding surfaces and fractures somehow held open at depths, like underground river channels), and the disperse globules microscopically small, the hydrodynamic gradient could not possibly be high enough to overcome the forces of surface tension holding the oil globules back.

Instead, the principal role of water in secondary migration is to provide *buoyancy*. As petroleum hydrocarbons are lighter than subsurface waters, they are physically pushed ahead of water moving through water-wet rocks. Whereas capillary pressure may be envisaged as the principal force *resisting* secondary migration, the process known as *water drive,* akin to artesian flow, is undoubtedly the principal agency *promoting* it.

15.8.1 Secondary migration by water drive

Buoyancy, reflecting the difference in densities of hydrocarbons and waters, is the main mechanism of secondary migration under hydrostatic conditions, especially in coarse sands in which the pressure readily equalizes to hydrostatic (see Sec. 14.2). All crude oils float on salt water, and nearly all on fresh water.

Under hydrostatic conditions, therefore, oil (or gas) tries to migrate vertically upwards through the heavier water, displacing an equivalent volume of the water as it does so. The oil is subject to a *buoyant force* (P_b), represented by a vector directed vertically upwards and equal to the difference between the weight of a unit volume of the oil and that of a unit volume of the water that it displaced. In other words, buoyancy is inversely proportional to density and P_b increases with a rise in the API gravity of the oil. In simple quantitative terms, the buoyant force is the difference in pressure between the water phase and the hydrocarbon phase at any point in the fluid column:

$$P_b = P_w - P_o \qquad (15.7)$$

For a continuous hydrocarbon column, P_b increases vertically upwards through it. At the free water level (the level at which water would stand in an open hole), $P_b = 0$ (Fig. 15.13); at the oil/water interface, it is equal to the displacement pressure of the reservoir (see Sec. 15.7.2). All underground water is subject to "head" pressure, which determines the height to which it will rise in a borehole.

Once in a reservoir rock, oil and gas pass upwards

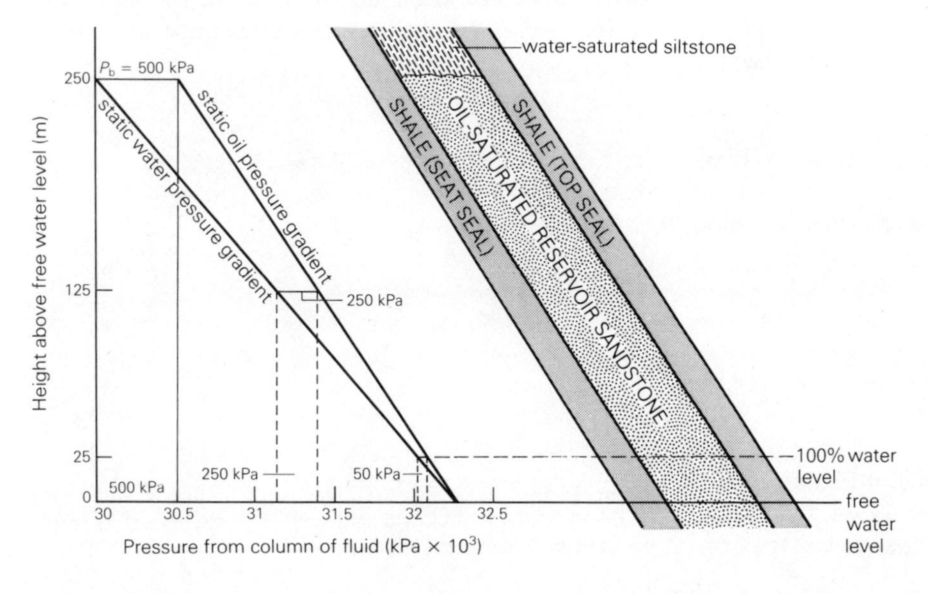

Figure 15.13 Buoyant force (P_b) in oil reservoir under hydrostatic conditions. Explanation of numbers: taking oil density = 0.795 g cm^{-3} and water density = 1.00 g cm^{-3}, then static oil pressure gradient = 0.795 × 9.8 = 7.8 kPa m^{-1}; static water pressure gradient = 1.0 x 9.8 = 9.8 kPa m^{-1}. Then P_b = 500 kPa at 250 m above free water level, where P_b = 0. At 100% water level, P_b of oil equals displacement pressure (P_d) of reservoir (= 50 kPa) and the siltstone acts as a lateral confining bed for the reservoir. (Modified from T. T. Schowalter, *AAPG Bull.*, 1979.)

through it under the control of the buoyant force until they reach its top, which is simply the level above which a stratum (the local seal or roof rock) has sufficiently lower permeability for its displacement pressure to prevent further rise. The oil and/or gas then migrates along a thin zone – possibly only the upper few centimeters – at the top of the reservoir bed in the direction of the hydraulic pressure gradient. All the rest of the reservoir rock remains water-saturated.

Under the same hydrostatic conditions, the pressure due to a column of fluid of depth h (positive upwards) and density ρ is $\rho g h$. Within the water-filled rock, surfaces of equal water pressure, P_w, are sensibly horizontal and their values *decrease* upwards. Within the hydrocarbon column, similarly, values of P_o decrease upwards, opposite to the change in the value of P_b.

Subsurface hydraulic regimes are seldom perfectly hydrostatic. Potential reservoir rocks are also both *carrier beds* and *aquifers*; in the absence of hydrocarbons, subsurface waters are constantly trying to move through them, preferably parallel to the bedding, in the direction of the hydraulic pressure gradient.

The force driving the water movement is the *fluid (hydraulic) potential,* denoted by the capital Greek letter phi (Φ). The potential of any element of fluid reflects the energy that it exerts on it surroundings as compared with the energy it would exert at some fixed datum, normally considered to be sea level; thus the dimensions of Φ are L^2T^{-2} (energy per unit of mass). The potential therefore possesses a *gradient*; the fluid tries to flow down the gradient, which may be pictured as being perpendicular to a family of surfaces of equal fluid potential, or *equipotential surfaces.*

We have seen that the *buoyant force* on the oil phase, under hydrostatic conditions, is the difference between the pressure exerted by the water and that exerted by the oil (Eqn 15.7). We have also seen that the pressure is equal to $\rho g h$. Under hydrodynamic conditions, the *fluid potential* of an oil column of height h_o, at any point of elevation Z, is related to the fluid potential of the water by the equation

$$\Phi_o = \rho_o g h_o = \rho_w g h_w - (\rho_w - \rho_o) g Z \quad (15.8)$$

In 1953 King Hubbert proposed simplifying this relationship by dividing throughout by $g(\rho_w - \rho_o)/\rho_o$. Then:

$$\left(\frac{\rho_o}{\rho_w - \rho_o}\right) h_o = \left(\frac{\rho_w}{\rho_w - \rho_o}\right) h_w - Z \quad (15.9)$$

Constant values of the left-hand side of Equation 15.9 represent the equipotential surfaces for oil; this factor may be called u. Constant values of the similar factor on the right-hand side represent the equipotential surfaces for water; this factor may be called v. Then

$$u = v - Z \quad (15.10)$$

Put simply, the difference between the equipotential surfaces for water and oil is the elevation factor, Z. Migration paths (flow lines) for each fluid are perpendicular to the equipotential surfaces.

Hydrodynamic conditions are brought about by water flow *along carrier beds,* which are aquifers. A vector in the direction of water flow, and equal to the rate of change of pressure measured along the flow path and corrected to the horizontal datum plane, is the *hydrodynamic force,* P_h, acting on unit volume of the oil as a consequence of the water flow (Fig. 15.14). The hydrodynamic force is due solely to the frictional loss caused by this flow through the permeable medium, so it may be regarded as constant for any one location in the aquifer.

The hydrodynamic force, P_h, interferes with the buoyant force, P_b. The resultant of these two forces controls the position of the oil in the aquifer in relation to the moving water. A vector equal and opposite to the resultant is the *confining force,* P_y (Fig. 15.14). The interface between the hydrocarbon and the water will become perpendicular to this resultant, and therefore no longer horizontal but *tilted*; hydrocarbon equipotential surfaces will be parallel to this tilted interface. As the buoyant force gets larger (the hydrodynamic force vector remaining essentially constant), the resultant total force becomes more and more nearly vertical, approaching the hydrostatic condition with horizontal fluid interfaces. Because buoyancy is inversely proportional to density, Pb is greatest for gas and lowest for heavy oil. Equipotential surfaces for gas therefore become tilted away from the horizontal by a smaller amount than those for oil (Fig. 15.15). The *tilt factor* for any hydrocarbon is the density ratio in the value of v: $\rho_w/(\rho_w - \rho_o)$. We shall return to this important factor

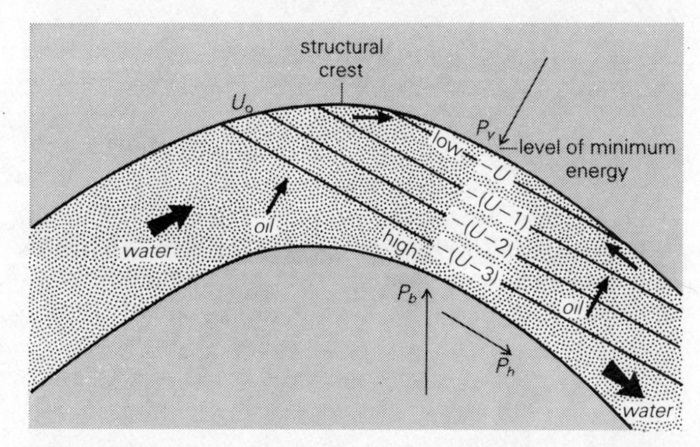

Figure 15.14 Oil isopotential surfaces (U_o) in relation to buoyant (P_b), hydrodynamic (P_h), and confining (Py) forces in a convex trap under hydrodynamic conditions.

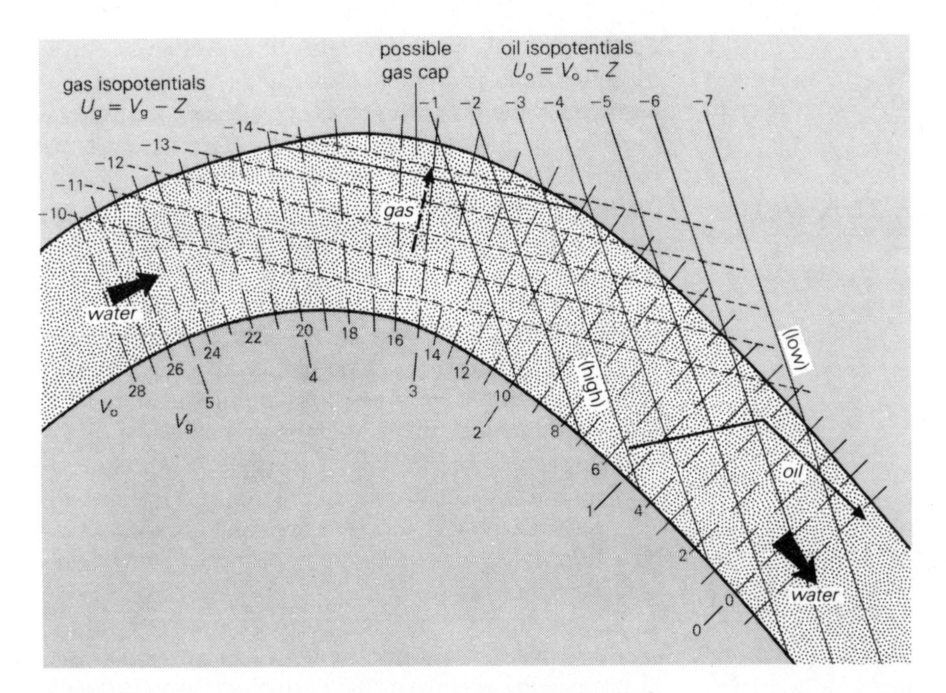

Figure 15.15 Equipotential surfaces for gas, oil, and moving water, in relation to the differential movements of the three fluids in a convex trap. Notice that the equipotential surfaces for the oil have been tilted to an angle greater than the angle of dip of the reservoir bed, so that the oil is flushed out of the trap altogether (see pp 302–3). The values on the surface traces are arbitary, to illustrate their decrease in the direction of fluid movement. (From E. C. Dahlberg, *Applied hydrodynamics in petroleum exploration*, Berlin: Springer-Verlag, 1982.)

when we consider its influence on trapping mechanisms (see Sec. 16.3.9).

The total fluid potential depends upon the *total head* (H) with respect to sea level. It was defined by Hubbert through the following relation:

$$\Phi = gH = gZ + \frac{P}{\rho} \qquad (15.11)$$

where g is the coefficient of gravity, p is the measured static pressure, and Z is the elevation of the point of measurement above or below the datum plane. Hence

$$H = Z + \frac{\rho}{pg} \qquad (15.12)$$

The total head, in simple terms, is the sum of the elevation head and the pressure head. This sum, the right-hand side of Equation 15.12, is the *potentiometric* or *piezometric surface*: it connects the highest points to which fresh water will rise, given adequate time, in wells tapping that single aquifer. A *potentiometric map* for any aquifer is simply a map of H. In *measuring* the pressure head ($p/\rho g$), remember to correct the units in which the pressure is measured to conform with the measurement of ρ in grams per cubic centimeter. For example, if the head is to be expressed in meters and the pressure is measured in kilograms per square centimeter, the latter must be multiplied by 10 to adjust the dimensions.

The potentiometric surface may be sensibly horizontal over wide areas of plain or plateau country, but faults, erosional valleys, stratigraphic discontinuities, and other geologic features create interruptions in it. The interruptions represent *lows,* or escape areas towards which water flows, and *highs,* or recharge areas away from which it flows. Under geologic conditions in which neither structure nor relief is extreme, the potentiometric surface characteristically presents a subdued reflection of the surface topography. In any one place, therefore, it has a *potentiometric gradient,* and the movement of water is down that gradient from higher to lower values of Φ. The *direction* of the movement is dependent entirely on the gradient, and not on the dip of the strata.

The gradient to depth z is dp/dz (Fig. 15.16). If this gradient is equal to ρg (Eqn 15.12), it is hydrostatic and there will be no vertical movement of the water column. If dp/dz is greater than ρg, upward migration will take place towards the outcrop or pinchout of the carrier bed; this is the circumstance we have considered so far. Most water originating in the sediments (connate water) must eventually move up the dip towards the basin margins, the direction in which the geostatic load decreases towards zero. If dp/dz is less than ρg, however, the movement of subsurface waters will be in the downdip direction and secondary migration towards the outcrop or pinchout will be obstructed or ended.

Downdip movement of formation waters creates an impenetrable barrier to the upward migration of any reservoir fluids. It therefore also creates favorable trapping conditions for hydrocarbons, by inhibiting their continued migration towards the surface (Sec. 16.3.9). The most widespread situation causing downdip water flow is that in which the ground surface slopes in the same direction as the dip of the aquifers (Fig. 15.16). Even under near-hydrostatic conditions, compacting strata may acquire a downwardly directed fluid potential gradient.

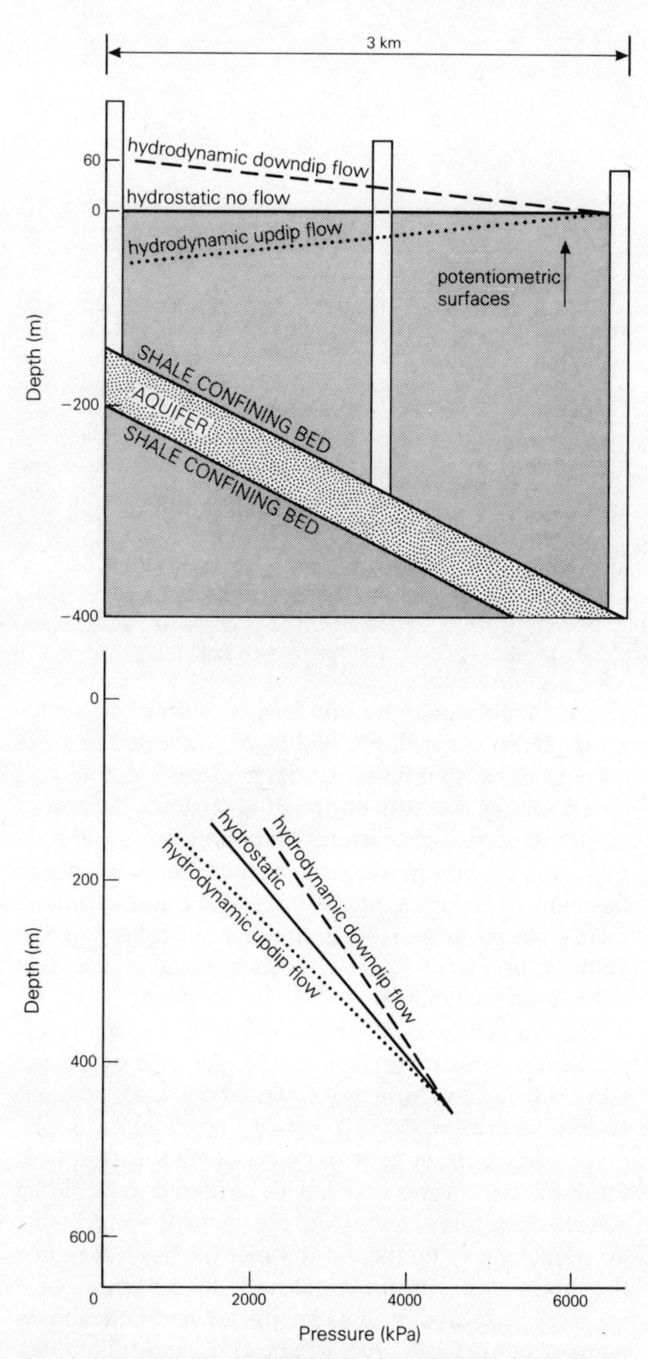

Figure 15.16 Effect of hydrodynamic conditions on pressure–depth plot of water phase, showing both updip and downdip water movement. (From T. T. Schowalter, *AAPG Bull.*, 1979.)

Downdip water movement is, nonetheless, most commonly the direct result of infiltration of meteoric waters from surface outcrops around the edges of the basin (see Sec. 17.7.1). This situation arises especially in geologically mature basins, from which most connate water has already been expelled by compaction as subsidence proceeded. Eventually a flow pattern becomes established, complex in detail but simple in overall characteristics, whereby a small number of segments of the deep basin receive converging flow systems, which ascend as cross-formational flow, and a small number of segments in the shallower parts of the basin then become the sites of diverging flow systems. Electric analogue patterns developed by Jozsef Toth showed that hydrocarbons move towards the discharge foci of the converging flow systems, and accumulate along the ascending limbs and in the stagnant zones where flow systems part. Accumulations are therefore associated with potentiometric minima, downward increases in hydraulic heads (possibly reaching artesian conditions), reduced lateral hydraulic gradients, and positions of minimum dilution of formation water by invading meteoric waters.

We discern a second form of obstruction to secondary migration of hydrocarbons up the dip of a carrier bed. This is the encounter with a permeability barrier across or within the carrier bed. The barrier may be any geologic feature presenting a *displacement pressure* (Sec. 15.7.2) too high for the water drive to overcome. If the capillary pressure exerted by the oil is lower than the displacement pressure of the barrier, the water drive is unable to force the oil through the barrier, which therefore forms a *trap*. We will consider this situation in detail in Chapter 16. At this point it is necessary to observe only that there is likely to be a limit to the amount of oil (or gas) that can accumulate behind or below such a barrier if the water drive continues to be active. If the supply of oil is great enough, the growing thickness of the column it builds up will eventually exert enough pressure to overcome the displacement pressure of the barrier, and the excess hydrocarbons will then be displaced into and through the barrier to continue their secondary migration.

15.8.2 Determinations of hydrodynamic gradients

Because the flow of subsurface waters in the direction of the hydraulic gradient can displace hydrocarbon accumulations, and create or destroy traps, it is essential to determine its pattern in any basin undergoing exploration or development. Fortunately it is not difficult to do this in broad outline, because the direction of the hydraulic gradient is essentially that from regions of high to those of low water table.

Hubbert's equation (Sec. 16.3.9) relates the tilt of the hydrocarbon/water interface to the densities of the fluids and the gradient of the potentiometric surface. The densities of the fluids must be known if reliable calculations of fluid movement are to be made (they can be read from easy nomographs). The gradient is simply the slope of the pressure–depth plot, in pascals per meter. However, a less accurate but easy alternative method can be followed if sufficient drilling has been done to provide data on *formation pressures* (see Secs 14.1 & 23.7).

The gradient of the potentiometric surface is its rate of change per unit length of flow path. It can therefore be expressed thus:

$$\Delta H = \frac{(P_1 - P_2)}{L} \qquad (15.13)$$

where P_1 and P_2 are static bottom-hole pressures (shut-in pressures, in pascals) in two wells, L meters apart. Alternatively,

$$\Delta H = \frac{144\ (P_1 - P_2)}{L} \qquad (15.14)$$

if P_1 and P_2 are measured in p.s.i. and L is in feet.

To determine the elevation and shape of any surface of equal pressure (say 15×10^6 Pa, very approximately 2000 p.s.i.) from shut-in pressures, follow these steps:

(1) The pressure–depth relation in a column of fresh water is 10^4 Pa m^{-1} or 0.434 p.s.i. per foot of column (see Sec. 14.1).
(2) The height of a column of water, of which the bottom-hole pressure is 15×10^6 Pa, is therefore

$$\frac{15 \times 10^6}{10^4} = 1500 \text{ m}$$

That is, the shut-in pressure (in pascals) measured in the well, divided by 10^4, gives the height (in meters) of the column of water above the bottom-depth of the test.
(3) Subtract the height of the column from the depth of the test, giving the depth to the 0 Pa surface.
(4) Add 1500 m to the depth of the 0 Pa surface, giving the *depth* to the 15 x 10^6 Pa surface.
(5) Subtract the KB elevation (the conventional elevation of the well above mean sea level), giving the *elevation* of the 15 x 10^6 Pa surface.

Now map the structure contours on this surface for all wells for which shut-in pressures for that particular reservoir formation are available. The map will illustrate the potentiometric surface for the tested formation over the drilled area, and will therefore indicate the directions of fluid movement.

15.8.3 Secondary migration by gas flushing

Where two or more fluids of different densities try to occupy the same trap, the *heaviest* fluid is displaced because the lighter ones move above it and force it below the *spill point* (see Sec. 16.3). William Gussow examined the case of the migration of gas, oil, and water through a succession of traps (Fig. 15.17).

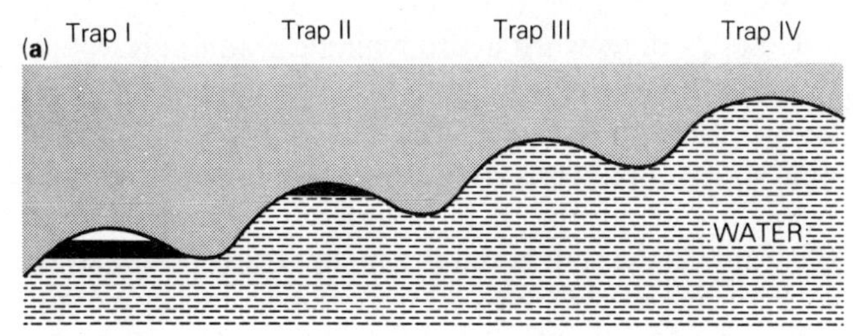

Trap I is filled to the spill point and has a gas cap; only oil is spilling updip into Trap II; Traps III and IV are full of salt water and contain no oil or gas.

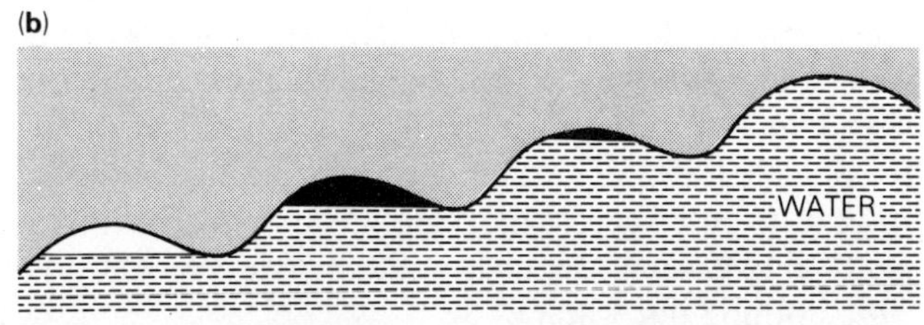

Trap I is completely filled with gas, all its oil has been flushed updip into Trap II; oil is now by-passing Trap I; Trap II is filled with oil and is spilling oil updip into III, it still has no gas cap; Trap III has a little oil, while IV is still filled with salt water.

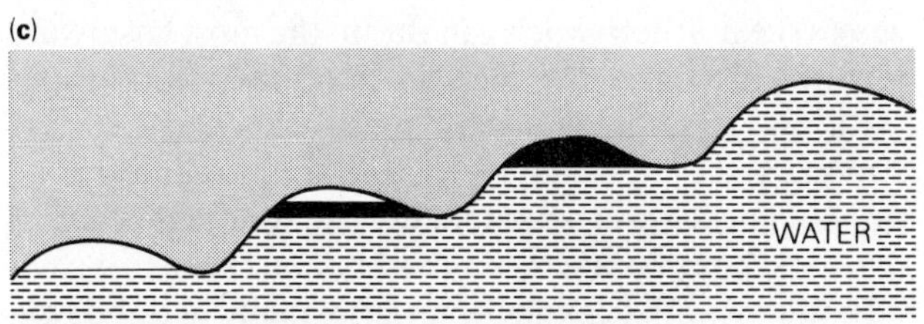

Trap I is unchanged with gas spilling updip into Trap II; oil is by-passing Trap I; Trap II now has a gas cap and is spilling oil updip into Trap III; Trap III is now filled with oil but still has no gas cap; Trap IV is dry.

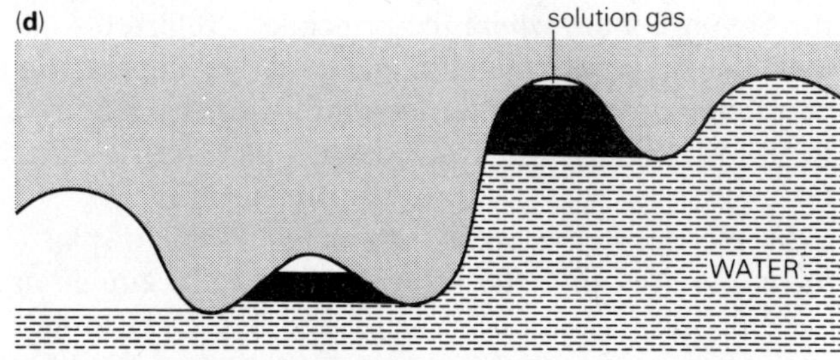

Migration same as for C, but under different structural relationships; note that height of culmination has no effect on selective trapping, elevation of spill point is the controlling feature; height of culmination above spill point determines the maximum pay section; (a solution-gas cap has formed in Trap III).

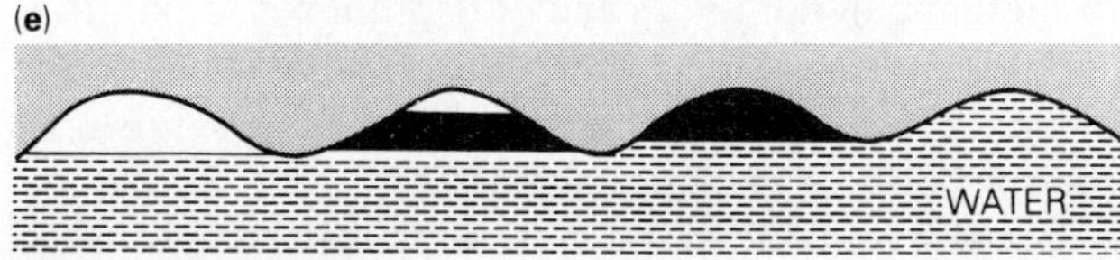

Migration same as for C; here all culminations are at same elevation; spill points control differential entrapment.

Key

Figure 15.17 Differential entrapment of hydrocarbons migrating through successive, interconnected traps. Stages (d) and (e) represent the same stage in migration but under different structural relations. (From W. C. Gussow, *J. Petrolm Technol.*, ©1968 SPE-AIME.)

Unless a downward hydrodynamic gradient is operative, the three fluids migrate up the regional dip. Gas goes to the top of the first trap encountered along the migration route, depressing the oil column and the water below it and causing them to spill out of the bottom of the trap, via the spill point, to a higher trap up-dip. Some solution gas is there released from the oil, because of the reduced pressure at the higher elevation; it forms secondary gas caps in the traps up-dip from those containing the primary gas columns. Residual oil may be forced to occupy an anticline trap if a late accession of gas occupies the main trap above it. The regionally highest anticline is very likely to end up occupied by the heaviest of the fluids, salt water. The ultimate outcome is a series of traps distributed across the regional dip, with the lowest containing gas and the highest the heaviest oil (or possibly nothing but water).

Gussow's prime example was taken from a Devonian reef trend in the Western Canadian Basin. A closely similar example is found in the Silurian pinnacle reefs across the northern Michigan Basin; the most basinward reefs contain gas, the reefs in intermediate positions yield oil, and the shelfward reefs are water-bearing. Among scores of examples in sandstone reservoirs and traps of much lower amplitudes, we may note that across the Vicksburg fault zone in the US Gulf Coast. Again, oil columns with gas caps occur principally in the down-dip rollover anticlines to the southeast; northwestward towards the foreland, oil without gas is found, and the highest, most northwesterly traps are water-bearing. In the Soviet Union, where the principle is referred to as the "multistage migration" of oil and gas, examples from petroleum provinces of all ages have been described by S. P. Maksimov, S. F. Federov, and many other geologists.

In the gas–oil system, sequential updip displacement requires that the saturation pressure equals or exceeds the formation pressure (see Sec.23.7); otherwise free gas cannot be liberated to perform the displacement. The gas : oil ratio (GOR) must decrease along the migration route. As methane is more soluble than its homologs, and naphthenes than paraffins, ratios of methane to homologs and of naphthenes to paraffins should increase along the route. These changes are evidence of differential entrapment by gas flushing; their absence is evidence against it.

In the gas–condensate system, these changes in the ratios do not take place. The opposite changes may in fact occur, because the pressure on the system should decrease isothermally as the liquid fraction condenses.

If the structures providing the traps are convex (see Sec. 16.3), the hydrocarbons migrate from lower to higher traps by *spill differential entrapment*; they "spill" up the dip via a series of spill points. If the traps are nonconvex – the "stratigraphic" traps of American terminology (Sec. 16.2) – the equivalent step by step migration may be caused by a series of local permeability barriers along the carrier (reservoir) bed. As the hydrocarbon column builds up behind any one barrier, the entry pressure of that barrier may eventually be exceeded; *gas* will then leak through it first, eventually becoming the only phase able to continue into the highest part of the updip trap. This *leak differential entrapment* therefore ends up with a situation opposite to that of spill differential entrapment.

15.8.4 Traces along the migration path

Both the water-drive and the gas-flush mechanisms of secondary migration involve the movement of oil by continuous-phase flow through the carrier system. The carrier system must in most cases be a permeable sedimentary rock (a reservoir rock, in fact). Because this process involves both the displacement of oil by water and of water by oil, the low miscibilities and solubilities of these fluids render it inconceivable that all traces of oil can be removed by the water from the rocks along the migration path. Some *irremovable oil stain* must be left as testimony of the oil's passage; unless the oil is very light and contains a high proportion of solution gas, it is likely that some *residual oil saturation* will normally be left also. Some authors, such as John Cartmill, believe that at least 20–30 percent of such saturation should remain. Such a high level is not in fact normally observed in barren strata outside the margins of petroliferous regions, but it is undoubtedly difficult to envisage the former passage of great volumes of oil through permeable strata in which no trace of residual oil can now be detected in cores.

15.8.5 Secondary migration via fractures

In Section 15.4, we considered the primary migration of hydrocarbons via kerogen networks, and recognized that such networks are consequent upon the development of *microfractures* in the source sediments. In Section 13.4, we considered the containment of oil and gas in *fractured reservoirs*. Because open fractures in the reservoir rocks are channels of nearly infinite permeability, they must constitute possible routes of secondary migration through porous strata.

Except under conditions either of tectonic extension or of reservoir undercompaction, the confining stress (σ_3) exceeds the displacement pressure of a porous stratum (which is obviously low). All oil will leave fractures and enter the coarsest available porosity. To remain in the fractures and traverse the porous bed via them, the oil must exert pressure, P_o, greater than σ_3 but less than the displacement pressure of the porous bed crossed by the fracture:

$$\sigma_3 \leq P_o < P_w + \frac{2\gamma}{r_t} \qquad (15.15)$$

where the last factor is the capillary pressure, P_c (see Sec. 15.7.3). The range of its values is about 500–3000 kPa; the higher its value, the more difficult it is for the oil to overcome P_c and enter the pore space.

Jean Du Rouchet, assuming that fractures through porous beds will preferentially follow the pore spaces, showed that fractures will be initiated when this relation obtains:

$$P_w + P_c > \sigma_3 + \sigma_t \qquad (15.16)$$

where σ_t is the tensile strength of the rock (of the order of $5-150 \times 10^2$ kPa; see Sec. 15.4.2). If $(\sigma_3 - P_w)$, the effective confining stress, is small, P_c need only be great enough to overcome the σ_t of the weakest part of the rock (say 1000 kPa).

The network of fractures so formed will spontaneously propagate in the direction of decreasing σ_3; this direction is commonly, but not necessarily, *upward*, especially towards the crests of convex structures which are under local extension.

We must recognize, of course, that fractures are a means of facilitating secondary migration; they are not a force or mechanism bringing it about. The force is still the provision of a gradient of differential pressures.

15.8.6 Influence of temperature gradients

Increased temperature reduces the viscosity of oil. As it also reduces the density, it tends to drive liquids upwards. The effect of temperature on secondary oil migration is nonetheless probably slight and local only. The thermal conductivity and specific heat of sandstones both increase, more or less linearly, with water saturation, because the water displaces air (an insulator) from the pore spaces. The thermal coefficients also increase with oil saturation, for the same reason, but at a lower rate because oil is a poorer conductor than water.

The influence of temperature may be much greater, however, on the migration of *gas*. Methane, unlike oil, has significant aqueous solubility, which increases exponentially with increase in temperature (see Sec. 9.4.1). On the other hand, the solubility decreases as the gas saturation of the water increases. Where gas pools exist in association with waters grossly undersaturated with gas, therefore, the pools very probably were formed by prolonged degassing of a formerly saturated hydrostatic system. A most striking example is provided by the great gasfields of northern Siberia, close to the Arctic Circle. The degassing is less likely to have been caused by reductions in either the pressure gradients or the groundwater salinities than by the drop in temperature during the Pleistocene glacial epoch. The region is still one of thick permafrost, in which the temperature is only about 65 °C at depths below 2500 m.

15.8.7 Influence of short-lived lithostatic loads

Because migration takes place in the direction of lower geostatic load, any abrupt increase or decrease in that load will influence either the direction or the rate of the migration, or both. Geostatic loads are not normally either increased or decreased abruptly; but they are sometimes. The most abrupt increase and decrease that can have occurred since the onset of hydrocarbon migration into the present accumulations (in Jura-Cretaceous time at the earliest) must have been those consequent upon the waxing and waning of the Pleistocene ice sheets. At their maximum, the sheets must have increased the geostatic pressure on the ice-covered strata by as much as 30×10^6 Pa (more than 4000 p.s.i.) over an interval of only 1–2 Ma.

A. A. Trofimuk considered that this rapidly acquired load was responsible for the last phase of oil and gas accumulation, and therefore for the evolution of the present petroliferous provinces in the Northern Hemisphere, at least. The increased load promoted hydrocarbon migration around the peripheries of the ice sheets, especially in areas of their confluence (such as western Siberia and western Canada). As the fluid pressures, directed outwards around the peripheries, came to exceed the overburden pressures, the sediments were naturally hydrofractured, leading to sharp but short spasms of fluid migration. Once the excess load was dissipated on recession of the ice, the migration ceased.

Trofimuk and his colleagues appeared to claim that the distribution of known oil and gas reserves over both present and paleolatitudes reflected these effects of glaciation, and that the highest pool density is to be found around the southern limits of the Pleistocene ice sheets. This claim seems to require some disregard for facts, but the migration mechanism invoked is undoubtedly a plausible one.

15.8.8 Summary of secondary migration

The deepest parts of sedimentary basins may be looked upon as *hydrodynamic divergence zones*. Oil and gas migrate away from them, up the dip of the strata and along gradients of decreasing capillary pressure, towards uplifts within the basins or the basins' uplifted margins. The latter may be looked upon as *hydrodynamic convergence zones*.

In general, the following changes take place in the reservoir fluids in the direction of migration:

(a) The API gravity of the oil decreases as the reservoir temperature and pressure decrease.
(b) Solution gas is lost upwards as reservoir pressure decreases.
(c) The sulfur content of the oil increases.
(d) The proportion of the oil content of the reservoir

that is *recoverable* therefore decreases, and the initial water content increases. Oil/water contacts may rise in a systematic fashion, and distant traps become less full as less oil is available for them.

(e) The salinity of the formation waters decreases and their oxygen content increases, because of the increased influence of meteoric waters and the bacterial activity that these permit.

It is vital that attempts be made to reconstruct the migrational histories of the oil and gas to reach any reservoirs and traps they may now occupy. The items to be taken into account in making these attempts may be summarized:

Geological conditions

(a) Disposition of strata; areas of homoclinal dip versus areas of folded strata
(b) Availability of carrier systems
(c) Presence or absence of restraining structures or strata near source-rock areas
(d) History of earth movement
(e) Form of potentiometric surface
(f) Variations in geothermal gradient

Physical and chemical characters of fluids

(a) Gravity and viscosity of oil
(b) State of association of oil, solution gas, free gas, and water
(c) State of dissemination of fluids within sediments
(d) Availability of meteoric water recharge

Lithological characters of reservoir rocks

(a) Porosities and permeabilities
(b) Proportion of induced openings (fissures, joints, solution pipes, etc.)
(c) Entry pressures and permeabilities of roof rocks
(d) Degree of saturation with water
(e) Effectiveness of cementation.

16 Trapping mechanisms for oil and gas

Because oil and gas are mobile fluids, they are able to migrate from one place to another in the natural environment. If they are to become collected into an accumulation capable of economic exploitation, their capacity for migration has somehow to be arrested. The arresting agent is called a *trapping mechanism*.

It is in the understanding of trapping mechanisms that the geologist's grasp of *structural geology* is tested. Unfortunately, experienced petroleum geologists are nearly unanimous in the opinion that structural geology constitutes the weakest part of the young petroleum geologist's training. There is an obvious explanation for this.

Petroleum geology is a "core" or required course in very few university curricula in geology. The faculty member teaching it is much more likely to be a specialist in some branch of sedimentary geology than in structural geology. The faculty member responsible for the structural courses, on the other hand, is unlikely to have any experience in the petroleum industries. Of all the "core" courses in the curriculum, structural geology is the one least amenable to understanding in the laboratory and most dependent upon field study. The need for field exposures leads the instructor and his students to places of high relief in mountain belts and of spectacular structure in Precambrian shields; there the student is introduced to the analysis of complex polyphase folding.

This is not the kind of structural geology the petroleum geologist is likely to encounter. He is concerned with unmetamorphosed, Phanerozoic, layered sedimentary rocks in the subsurface. Foreland deformation is quite different from the axial or geosynclinal deformation that gives rise to polyphase fold belts. Foreland deformation normally involves a much thinner section, in which the natural rock contrasts and structural anisotropies are *not* obliterated during deformation as they are in the metamorphosed axial zones. Heterogeneity is not erased but increased; relative elevations are enhanced, not reversed. The heterogeneous rock environment leads to heterogeneous response to deforming forces (including the force of gravity and its simplest effects in compaction). Responses are discontinuously disharmonic, with regional unconformities and *décollements* leading to critical changes in structural style downwards.

16.1 Behavior of sedimentary rocks in outer crust

In university courses and textbooks in structural geology, folds and faults are treated as different types of response to stress. At the surface in foreland areas, the geologist sees a two-dimensional view both of folds and of faults. Faults preferentially follow weak rocks because they obey the principle of least work; the weak rocks are the hardest to map.

The exploration geologist's prime concern is to find a *trap*; the development geologist's concern is to exploit it; both need to understand it. A trap is simply any geometric arrangement of strata that permits the accumulation of oil or gas, or both, in commercial quantities. The commonest traps, by far, are *anticlines*; but are they necessarily *folds*?

The term *anticline* has a geometric and not a genetic connotation. It means a structure in which the flanks slope away from one another. Student geologists see many examples of anticlines that involve faults. Folding of layered rocks undoubtedly gives rise to faulting very easily; faulting also gives rise to "folding," and in very many anticlinal traps the fundamental structure is a fault, not a fold.

In the subsurface, nearly all pre-Pleistocene sedimentary rocks are faulted. Faults and "folds" are integral parts of most "structures" as the petroleum geologist has to envisage them. Faults, however, may be much more difficult to interpret in three dimensions than are folds. Nowhere does the field geologist see a fault exposed over 5 km of relief in unmetamorphosed sedimentary rocks, but the petroleum geologist has to deal with faults in three dimensions all measured in that magnitude. The experimental geologist also finds it difficult to model faults as effectively as folds over the necessary depth range.

These concerns first came to light because petroleum geologists drilled through folds and faults at depth and recognized that they did not always behave the way surface information suggested they should. They therefore asked structural and experimental geologists to explain and model their subsurface structures. Some of these studies revolutionized the understanding of both folding and faulting; they will be referred to further in this chapter.

The petroleum geologists now have an added complication. Offshore in deep water, where well records provide the only direct information, all exploration is concentrated upon convex structures detectable by marine seismic surveys (see Sec. 21.6). Such structures on continental shelves and slopes are highly likely to be faulted; if they had been formed by compressional buckle folding they would be more likely to be parts of mountain belts than of continental shelves. Wells in deep-water fields are drilled directionally from fixed platforms (see Fig. 23.2). In a vertical drillhole on land, the bit necessarily passes downwards from the hanging wall of a fault into the footwall. Drilling vertically through a normal fault results in loss of section; if the section is repeated in the drillhole, the fault is interpreted as either a high-angle reverse fault or a thrust fault. In a deviated drillhole, the bit may as easily pass from the footwall of a fault into the hanging wall. A normal fault may repeat section, and a reverse fault may cut out section.

Complexly faulted structures offshore may never be fully worked out. Both compressional and extensional examples will be discussed and illustrated in this chapter.

16.2 Fundamental types of trapping mechanism

As we have seen, any arrangement of strata that permits the accumulation of hydrocarbons in commercial quantities is a *trap*. A trap therefore has two functions; it receives the hydrocarbons, and it obstructs their escape. If we think in terms of oil and gas being determinedly migratory, a trap is something that interrupts the migration and prevents it from continuing. Arrangements of strata capable of doing this can have a great variety of forms, but all have a single feature in common: a *porous* rock at least partially enclosed in rocks that are *relatively* impervious. "Relatively impervious" simply means that the permeability of the enclosing rock (see Sec. 11.2) must be too low for pressure and temperature conditions of the oil or gas in the trap to take advantage of it.

Water is an important agent in directing the oil or gas into its trap in the first place. Most traps are originally *water-wet*. The production histories of thousands of oil- and gasfields show that water also normally displaces the oil and gas from their traps during depletion of the accumulation by production. In addition to receiving the oil and gas, therefore, a trap must be able to expel water at depth and later re-admit it. Traps are not passive receivers of fluid into otherwise empty space; they are focal points of active fluid exchange.

The rocks above the trap (the *roof rocks*) and those alongside it (the *wallrocks*) may be impermeable not only to oil and gas but also to water under the reservoir's pressure conditions. If this is the case, the water originally in the trap must be displaced downwards by the accumulating oil and gas; the pool will contain *bottom water*. If the wallrock is not impermeable to water, it must be water-saturated, and the pool will be bounded laterally by *edge water*.

There have been scores of attempts to classify trapping mechanisms. Only those based on fundamental principles can be considered here. W. B. Heroy's classification of 1941 considered the genesis of the trap: depositional, diagenetic, or deformational. W. B. Wilson, in 1934, based his classification on the manner of achievement of closure (see Sec. 16.3):

(a) Reservoirs may be closed by local deformation of strata, for example by folding, so that contours drawn on the upper surface of the reservoir close locally.
(b) Reservoirs may be closed because of varying porosities of the rocks. Deformation is then unnecessary though regional tilting will commonly be involved. Contours may not close if drawn on the upper surface of the reservoir only.
(c) Reservoirs may be closed by any combination of (a) and (b).
(d) Reservoirs may be open, in the sense of having no detectable means of closure. No important field needs this category to accommodate it.

Otto Wilhelm, in 1945, defined traps according to the genetic causes of their boundaries:

(a) Convex trap reservoirs, formed by folding or by differential thicknesses of strata and due to convexity alone. The reservoir is completely surrounded by edge water as porosity extends in all directions beyond the reservoir area.
(b) Permeability trap reservoirs, due to changes in reservoir power. The periphery of the reservoir is

partly defined by edge water and partly by a barrier resulting from loss of permeability.

(c) Pinchout trap reservoirs, formed by lenticular structures, including reefs. The periphery of the reservoir is partly defined by edge water and partly by the pinchout, to zero thickness, of the reservoir bed.

(d) Fault trap reservoirs, in which the periphery is partly defined by edge water and partly by a fault boundary.

(e) Piercement trap reservoirs, formed by diapirs or volcanic necks. The periphery of the reservoir is partly defined by edge water and partly by a piercement contact.

By far the most familiar distinction between trap types is that between *structural* and *stratigraphic* traps. Structural traps are created by involvement of the strata in any kind of secondary, post-depositional structure – folding, draping, faulting, piercing – unless the involvement is in erosion or tilting only. Stratigraphic traps are created by any variation in the stratigraphy that is essentially independent of structural deformation (other than simple erosion or uncomplicated tilting). The distinction between the two types is obviously not very clear cut, and many fields combine aspects of both in such a way that both are vital to the trapping mechanism. Such fields are commonly attributed to *combination* traps. They include some of the world's largest fields, but it is worth noting that, in all famous cases, the term "combination trap" has conventionally been restricted to traps that are fundamentally *stratigraphic*, with only modification or enhancement by "structure." The structure may be obvious, and its identification may have led to the discovery of the field; but the oil or gas would be in the same place if the structure had never been created (and may have been there before it was created) because it is restricted to that place by the wedging-out or truncation of the reservoir bed. The Prudhoe Bay field in Alaska is a striking example (Fig. 16.1).

If stratigraphic traps do not need structural deformation, they should be available earlier than are structural traps, and one might expect that there would be more of them. We do not in fact know whether this is true, because the distribution of stratigraphic traps (if we adhere to the original definition) has two startling aspects. First, their size distribution is grotesquely bimodal. A relative handful are among the areally largest oil- and gasfields in the world, and among the most productive. The largest oilfield in the Western Hemisphere, the Bolivar Coastal field in Venezuela, contained original recoverable reserves of about 5×10^9 m^3

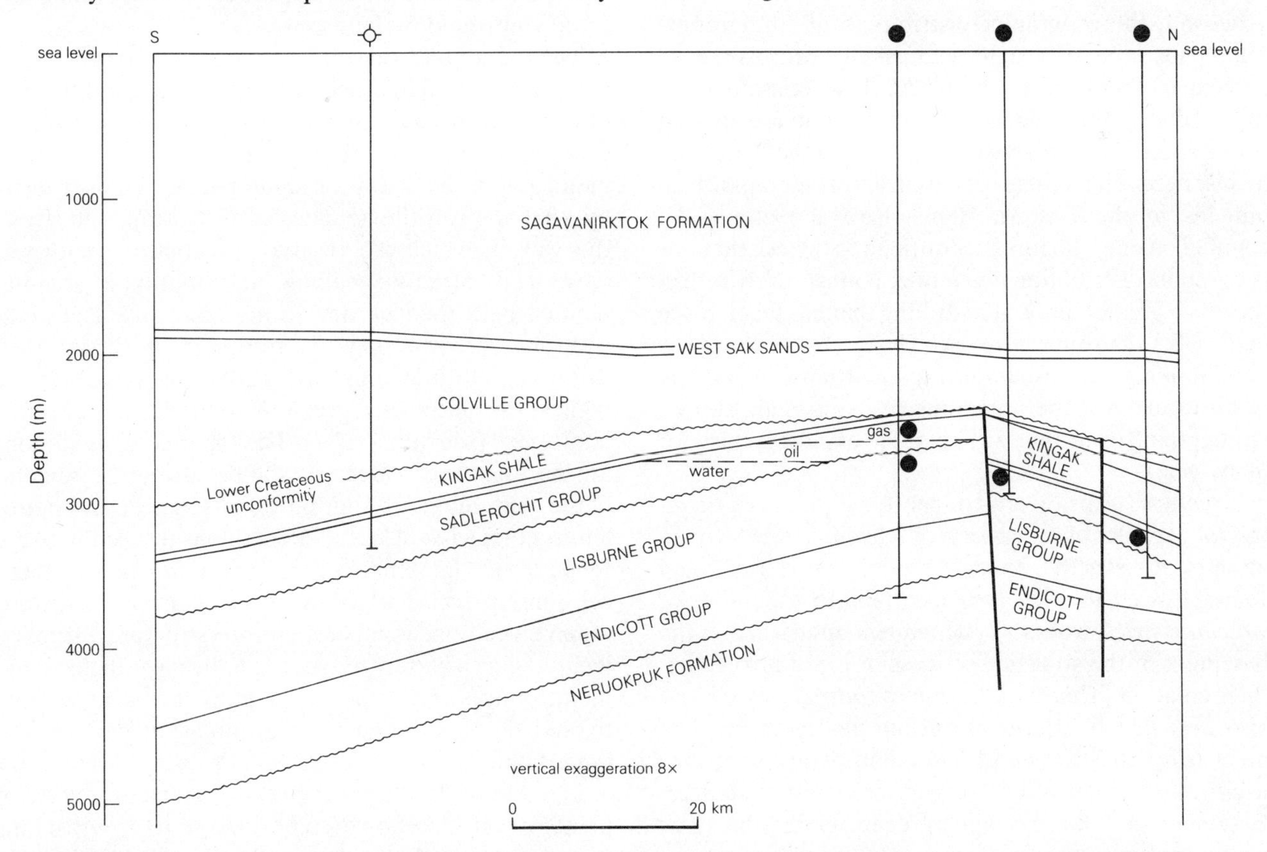

Figure 16.1 North–south cross section through the Prudhoe Bay oil- and gasfield, Arctic Alaska. Though the overall structure is a huge anticline, the actual trap for the principal pool is the product of a homoclinally dipping reservoir rock cut off up the dip by a normal fault and truncated by a major unconformity. (After H. C. Jamison *et al.*, *AAPG Memoir* 30, 1981.)

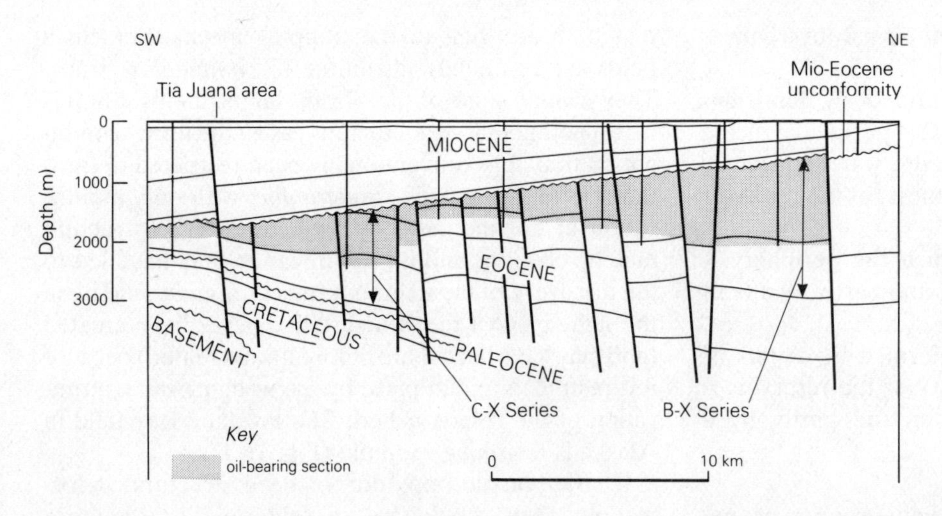

Figure 16.2 Cross section through the Bolivar Coastal field, Maracaibo Basin, Venezuela. Prolific oil production from both Miocene sands, in homoclinally dipping stratigraphic trap above the major unconformity, and widely separated sands through the much-faulted Eocene succession below the unconformity. (From H. D. Borger and E. F. Lenert, 5th World Petroleum Congress, 1959.)

(30 billion barrels) in an essentially stratigraphic trap (Fig. 16.2). The largest oilfield in the continental USA (East Texas, Fig. 23.8) occupied nearly 650 km^2; the largest gasfield (Hugoton) sprawls over more than 1.8 × 10^4 km^2. The Lima–Indiana field south of the Great Lakes was productive over 1.3 × 10^4 km^2 (Fig. 13.52). The largest onshore oilfield in Canada, the Pembina field, occupies about 1.7 × 10^3 km^2. At the other extreme are hundreds of fields in stratigraphic traps small both in area and in productive capacity.

Second, the geographic distribution of stratigraphic trap fields is at first sight astonishing. More than 90 percent of them are in the USA; if we consider only those of less than giant size, 95 percent are in that country, and more than 98 percent are in the American hemisphere. The number of important stratigraphic trap oilfields in the Eastern Hemisphere can almost be counted on one's fingers. It is difficult to believe that this geographic distribution is genuine; it must surely reflect the much greater amount of drilling that has taken place in the USA than anywhere else. The rest of the world has been looking for structural traps and must eventually begin to uncover the numerous fields in small, subtle, stratigraphic traps that will keep the explorers busy for many years.

Structural traps, even stringently defined, are themselves capable of a variety of classifications. In an attempt at a strictly genetic division, Tod Harding and James Lowell in 1979 assigned traps to a number of *structural styles*, essentially dependent upon whether the basement of the stratified series is or is not involved in their creation. There are *basement-controlled traps* and also *detached structures* having no basement involvement (Fig. 16.3). Lack of basement involvement may itself reflect either of two very different circumstances. Basement may be exceedingly deep, giving the thick cover multiple levels of potential *décollement* and enhancing the likelihood of detached structural traps. This appears to be the case in the foredeep of the Appalachians and in the Persian Gulf Basin. Alternatively, basement may be shallow but rigid, as in large parts of the covered Canadian, Baltic, and Guyana shields (Figs 10.16, 16.41 & 42). This enhances the likelihood of simple stratigraphic traps, as recognized by William Mallory in 1963. The principal controls of oil occurrence on cratons are regional facies changes, reefs, isolated sandstone bodies such as ancient offshore bars, and regional, low-angle unconformities; the only common structural traps in such environments are likely to be controlled by faults.

The influential concept of stratigraphic-type oil- and gasfields has defied precise definition except at the cost of severe restriction. Gordon Rittenhouse in 1972, for example, took a leaf from Wilhelm's book and recommended that the definition should be determined solely by what controls the boundaries of the trap, and not by the way in which the reservoir's actual capacity was developed. Strictly speaking, a stratigraphic accumulation would then be one formed by a displacement-pressure barrier along the reservoir or carrier bed. Within this terminating barrier zone, the rock with the highest displacement pressure acts as the effective lateral seal (see Sec. 15.7.2). Though precise definitions are necessary to science, it should always be remembered that geology is not an exact science, and petroleum geology is still less so. Punctilious restriction of the scope of stratigraphic traps has two unfortunate practical consequences; it enlarges the scope of nonstratigraphic (and presumably therefore structural) traps to include such a variety of types as to be meaningless, and it introduces artificial distinctions between closely related fields, or even between adjacent parts of the same field.

The case of Prudhoe Bay may be taken as the starting point for consideration of some other trap types that blur the distinctions between structural and stratigraphic traps. The oil at Prudhoe Bay is in its particular trap because its reservoir rock was truncated by erosion to

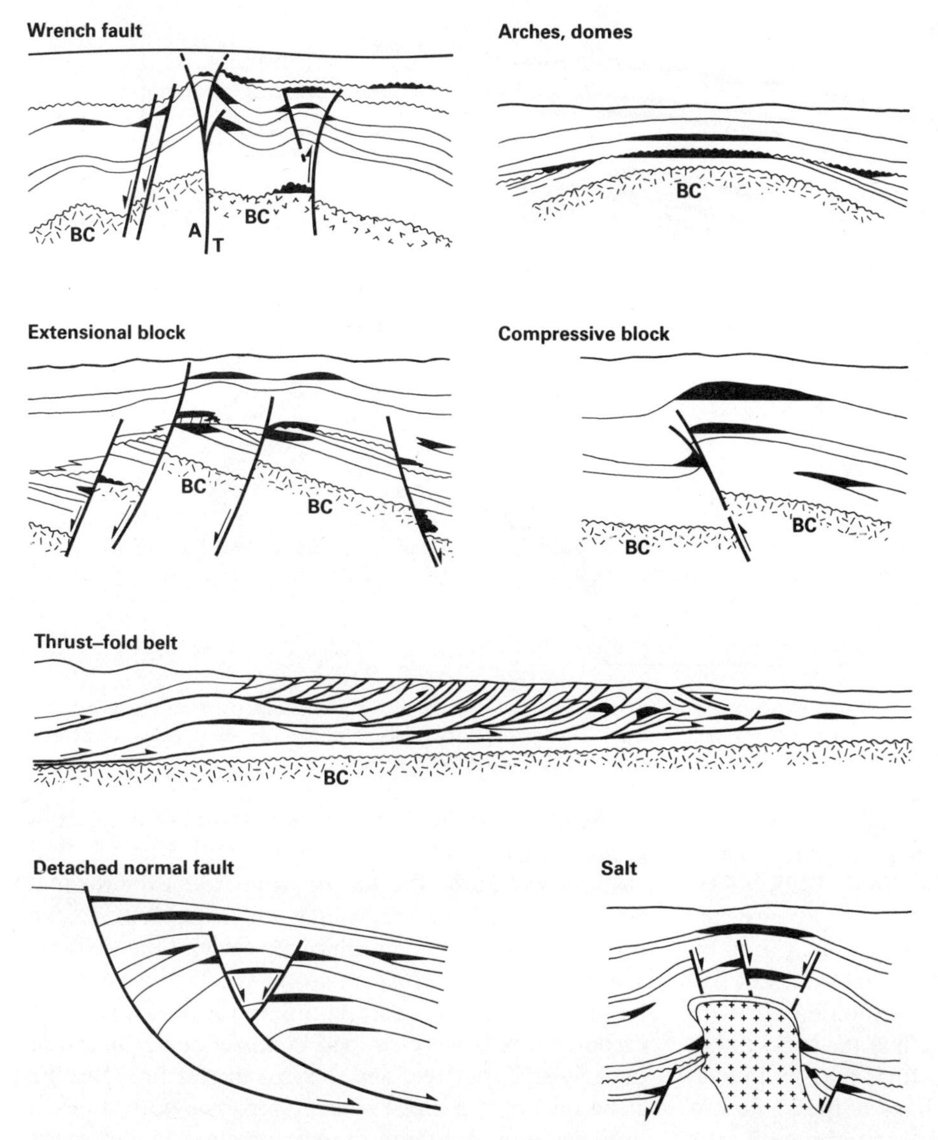

Figure 16.3 Schematic representation of hydrocarbon traps most commonly associated with the variety of structural traps. In upper left diagram, T and A indicate displacements towards and away from viewer; BC, basement complex. (From T. P. Harding and J. D. Lowell, *AAPG Bull.*, 1979.)

form an updip wedge-out (Fig. 16.1). An easily detectable structure of great size was then imposed upon the trap. On the original definition of a stratigraphic trap, the Prudhoe Bay trap was stratigraphic; on a restricted definition it may not be; in any event it is now a structural trap. In the Bolivar Coastal field in the Maracaibo Basin (Fig. 16.2), the oil in the Eocene is in a trap like the original trap at Prudhoe Bay, truncated above by erosion. The oil in the Miocene lies above the erosional unconformity, and its reservoir beds dip in the direction opposite to the dip in the Eocene beds. The Miocene beds are in a classic stratigraphic trap, but its boundaries are not controlled by the same mechanism as the boundaries of the Eocene trap below. The most famous of stratigraphic traps, in the East Texas field, was formed by erosional truncation of the reservoir sandstone, exactly like the original trap at Prudhoe Bay. It is maintained by some, however (notably W. C. Gussow), that the reservoir sandstone itself rests unconformably on older beds and pinches out depositionally against structurally controlled topography. If either or both of these erosional truncations permits the East Texas trap to remain in the stratigraphic pigeonhole, what is the proper designation for a trap in which the reservoir bed is truncated at a considerable angle by the unconformity, especially if the unconformity itself was consequent upon the raising of a major anticline? The Ordovician Wilcox sand trap in the Oklahoma City field may serve as an example (Fig. 16.4).

High-standing horsts or tilted fault blocks are undoubtedly structural in origin. If the high position permits the truncation of the beds atop the structure, oil may occupy the truncated strata for the same reason as it occupies the reservoir at East Texas. Are fields on the Brent platform of the North Sea (Fig. 16.5) structural or stratigraphic? If a tilted bed truncated up-dip by an angular unconformity is in stratigraphic trap position, what is the proper designation of a trap formed by a

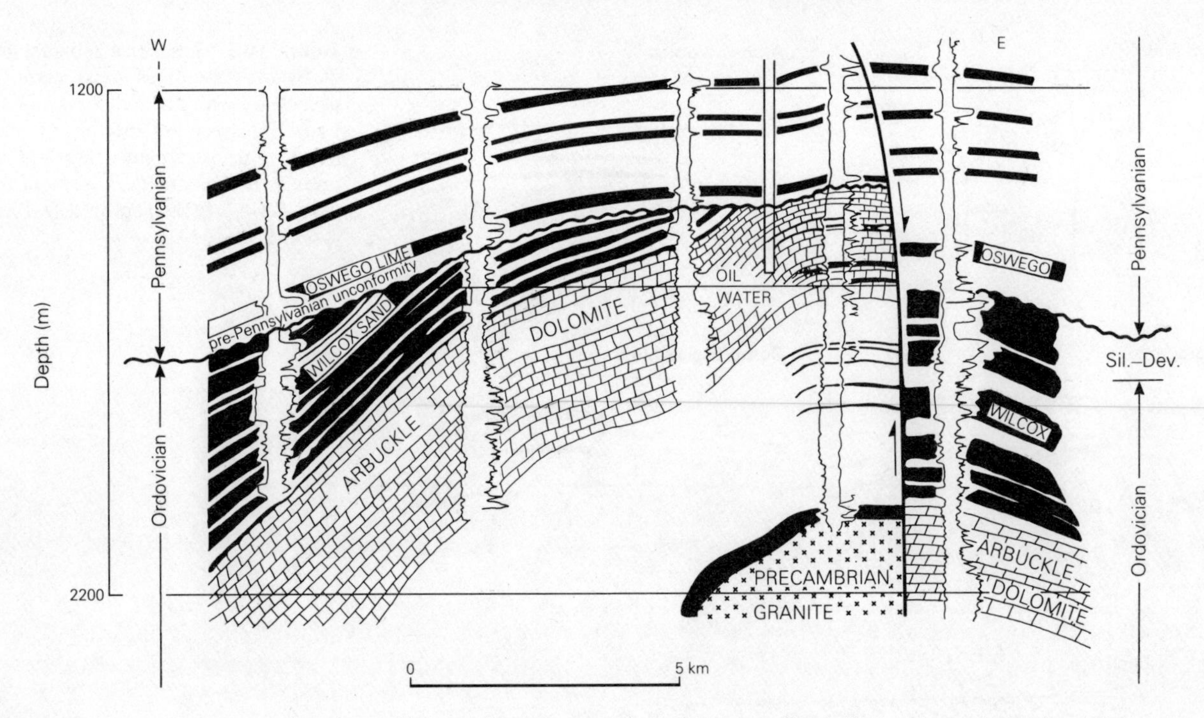

Figure 16.4 Structural cross section of pre-Mid-Pennsylvanian succession in Oklahoma City field, showing undoubted anticlinal structure but control of principal production by unconformable truncation of westerly-dipping Ordovician strata. Note much thicker Lower Paleozoic section below unconformity on downdropped side of major fault. (After L. E. Gatewood, *AAPG Memoir* 14, 1970.)

tilted bed truncated up-dip by a low-angle fault, or even a vertical fault? Hundreds of traps in complex horst-graben structures are like this (Figs 16.5, 31, 95 & 99). Traps caused by piercement salt domes also are undoubtedly structural in origin; but they may not be structural in geometry. Because of the intermittent rise of a dome, a variety of controls is exercised by it on sedimentation around it. A sand such as the M-5 sand at Bay Marchand in Louisiana (Fig. 16.6) may be in true stratigraphic trap position on one flank of the dome and in structural trap position on the other because it is truncated by the salt.

The discovery field for the modern era of petroleum production in western Canada is the Leduc oil- and gasfield (Fig. 16.7). Production came from two principal horizons in the Upper Devonian. Oil in the older, Leduc Formation is in a stratigraphic trap because the Leduc is a bioherm on undeformed base. Oil in the younger, Nisku Formation is in a structural trap because the Nisku carbonate beds were draped compactionally over the underlying Leduc reef and so form an anticline. But the same oil, from a single source, occupies both traps; it occupies both for the same reason, that they provide geometric culminations into which the oil was able to

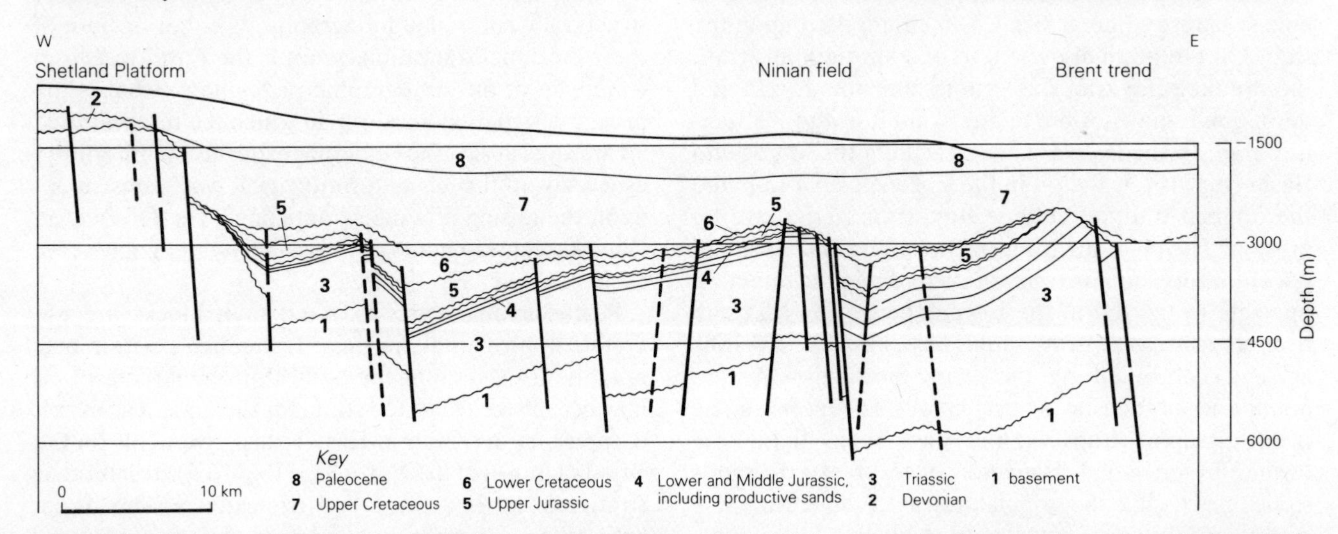

Figure 16.5 West–east cross section through Shetland basin and Brent platform, UK, North Sea, showing truncations of west-dipping strata in fault blocks. (From W. A. Albright *et al., AAPG Memoir* 30, 1981.)

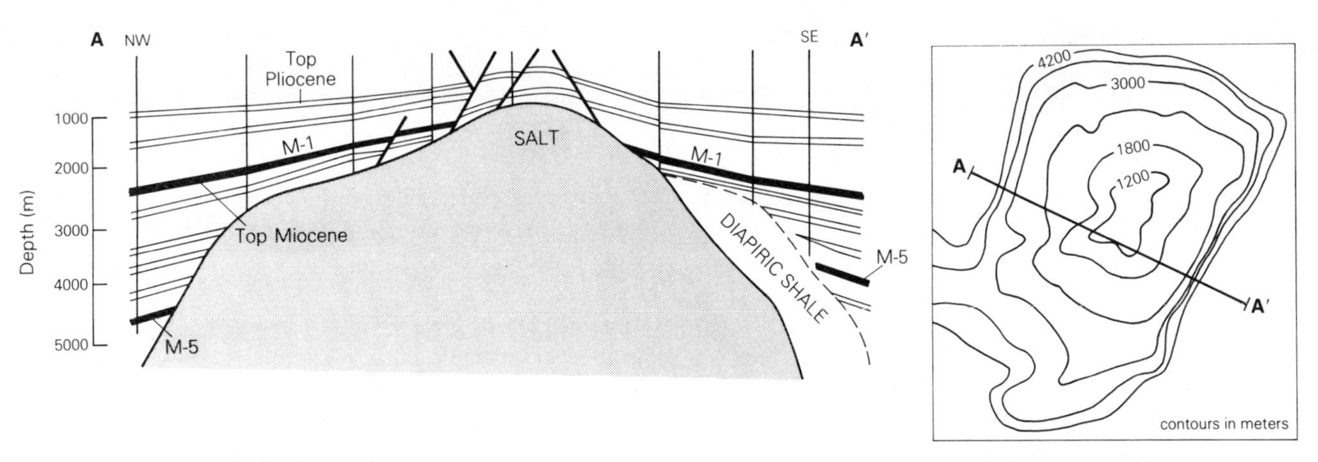

Figure 16.6 Bay Marchand field, southern Louisiana; NW–SE cross section without vertical exaggeration. Both salt and shale are diapiric, forming large convex trap. But M-1 sand is in nonconvex trapping position, and M-5 sand is in nonconvex, stratigraphic trap position southeast of the dome and in nonconvex, structural (fault) trap position northwest of the dome. (From M. G. Frey and W. H. Grimes, *AAPG Memoir* 14, 1970.)

rise; it occupied both traps at essentially the same time, and was produced from both traps by the same wells.

Any attempt to fit phenomena as various as hydrocarbon traps into a neat classification is bound to create overlaps and ambiguities of these kinds. The problem, familiar to geologists, stems from trying to avoid a hybrid or noncoherent classification which improperly combines genetic and descriptive criteria. The classifications outlined at the beginning of this chapter were all attempts at genetic discrimination. This is what all scientists aim for, but the difficulties are immense. It is true that traps of essentially identical geometry (shape) may be formed by quite different processes and under quite different conditions; numerous examples are illustrated in this chapter. But it is also true that traps formed under comparable conditions, and so justifying association in a genetic classification, may be of quite different geometries.

Russian geologists have tried to avoid this dilemma by lumping together a great variety of trap geometries in the *structural* category, because all owe their origin to some kind – any kind – of deformation of the rocks; but severely restricting the *stratigraphic* category by introducing a third fundamental category of *lithologic* traps. Most formerly "stratigraphic" traps are then in fact combinations of two or all three of these basic types.

(a) *Lithologic traps* are those formed by changes in the depositional products as results of changes in the topography of the depositional sites. The change in topography may be relatively gross – a shelf edge, or a residual "high" – or so subtle that its only visible consequence is a change in the pattern of water or wind currents. In general, however, the change is one that separates a high-energy from a lower-energy regime. Lithologic traps include reefs, offshore bars, channel sands, fan lobes, and so on. They also include a genetically quite distinct group of traps formed by diagenetic processes, such as dolomitization and cementation.

(b) *Stratigraphic traps*, which for Levorsen and others included all the traps now described as *lithologic*,

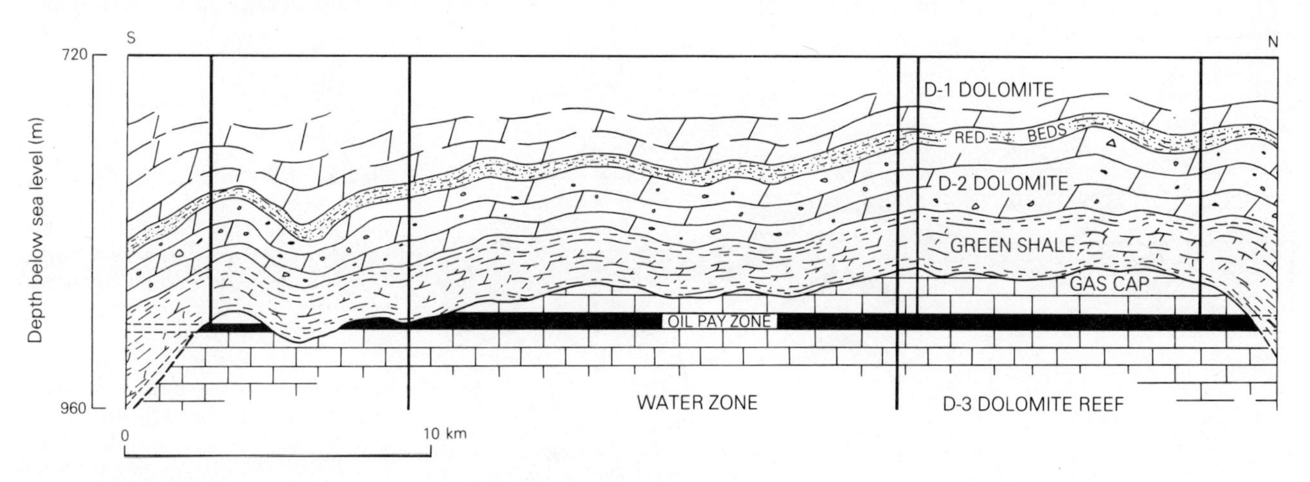

Figure 16.7 Cross section of Leduc oil- and gasfield, Western Canadian Basin, to illustrate drape folding in the Devonian D-2 reservoir over the incompactible D-3 reef reservoir. (From J. Baugh, *World Oil*, March 1951.)

are restricted by the Russians to traps created by post-depositional features of tilting and erosion; they lie below unconformities, but the development of the unconformity must not involve any active "structural" agent, such as folding, nor any kind of prominence, even if it is wholly residual in character.

(c) Traps which involve either of these two features plus a "structure" become combination traps. They are among the most common productive traps. Those in which the change in lithology occurs on the flank of a growing structural feature are *structural-lithologic traps*; obvious examples are the sand cutoffs against the flanks of salt domes (Fig. 16.49). Traps occupying the flanks or crests of truncated structural or residual "highs" and overlain by impermeable roof rocks are *structural-stratigraphic traps*. Most of the large North Sea fields occur in this kind of trap, as do such combination trap fields as Oklahoma City (Figs 16.4 & 5).

(d) A final category involves no morphologic "prominence" of any kind, but the change in lithology involves a change of stratum across an unconformity or disconformity; hence such traps can be called *lithologic-stratigraphic traps*. Those in simple supercrop position include channel sands filling erosional valleys, current-controlled sand lenses filling depressions on the basin floor, some cases of reef detritus, and so on. Those formed by progressive onlap over a tilted erosional surface are represented in some very large oilfields – the Pliocene part of the Midway–Sunset field in California (Fig. 16.8), and the Miocene part of the Bolivar Coastal field in Venezuela (Fig. 16.2) – as well as scores of small ones. A much less impressive group is that formed by erosional topography buried below unconformities; it includes fields trapped in fractured, weathered, or solution-affected carbonate and basement rocks (Figs 13.78 & 79). Many traps included in the *lithologic-stratigraphic* category by the Russians are now referred to yet another category, that of *paleogeomorphic traps*, by many Western geologists (see Sec. 16.6).

This comprehensive classification of traps has many merits, but great numbers of very efficient traps defy simple categorization. There is nothing particularly inconvenient in the fact that the Miocene pools in the Bolivar Coastal field (Fig. 16.2) are in lithologic-stratigraphic traps while the Eocene pools below the unconformity are in straightforward stratigraphic traps – except that it is only the presence of the younger reservoir that prevents the Eocene pools from being in a structural-stratigraphic trap because it is on the flank of a major basement uplift. If Gussow's interpretation of the East Texas trap is correct, it is both structural-stratigraphic and lithologic-stratigraphic. At Mene Grande in western Venezuela (Fig. 16.82), the main producing horizon (the Miocene Isnotu Formation) is in a lithologic-stratigraphic trap on the east side of the field, resting on an unconformity, but in both structural-lithologic and structural-stratigraphic traps on the west side – despite the facts that production from the Isnotu Formation was continuous across the crest of a large convex structure and the relative roles of folding, faulting, and unconformity are very difficult to isolate. The Piper field in the North Sea (Fig. 16.9) is in a structural trap on all traditional definitions of that term, but its flanks are individual lithologic-stratigraphic traps because they have homoclinal dips and rest on an unconformity. The Groningen gasfield in Holland (Fig. 16.10) is very similar; the list could be extended indefinitely.

This author has long accepted the hazards of trying to shoehorn oil and gas traps into rigidly defined slots. He believes that the simplest framework within which to consider traps is the one that separates them into fundamentally convex and fundamentally nonconvex categories. The distinction is both geometric and functional.

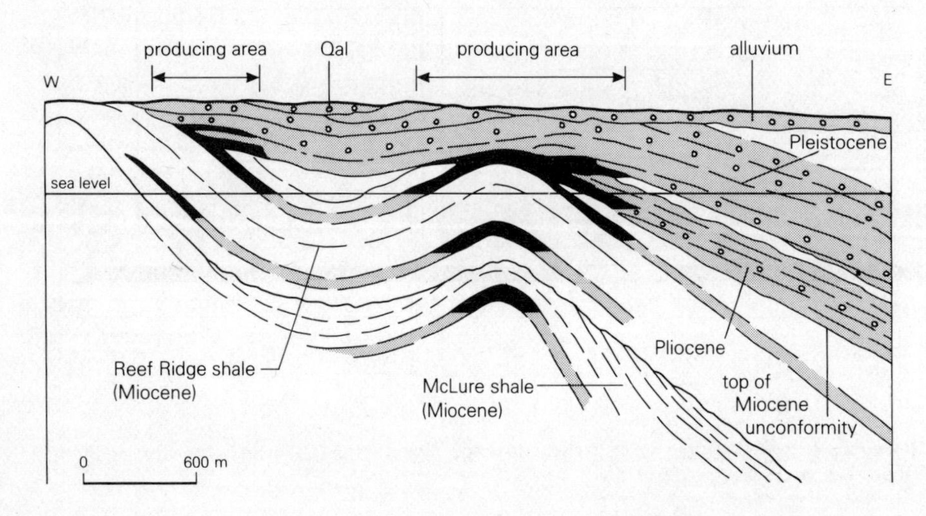

Figure 16.8 Cross section of the Midway–Sunset field, a giant oilfield in the San Joaquin Valley of California, discovered in 1901 and still among the top ten producing fields in North America 80 years later. The Coast Range is to the west, the basin to the east. Oil accumulations (solid black) show a combination of convex traps, nonconvex pinch-out traps, and nonconvex traps in both the subcrop and the supercrop of the Miocene–Pliocene unconformity – all in the same field. (From H. W. Hoots *et al.*, California Division of Mines Bull. 170, 1954.)

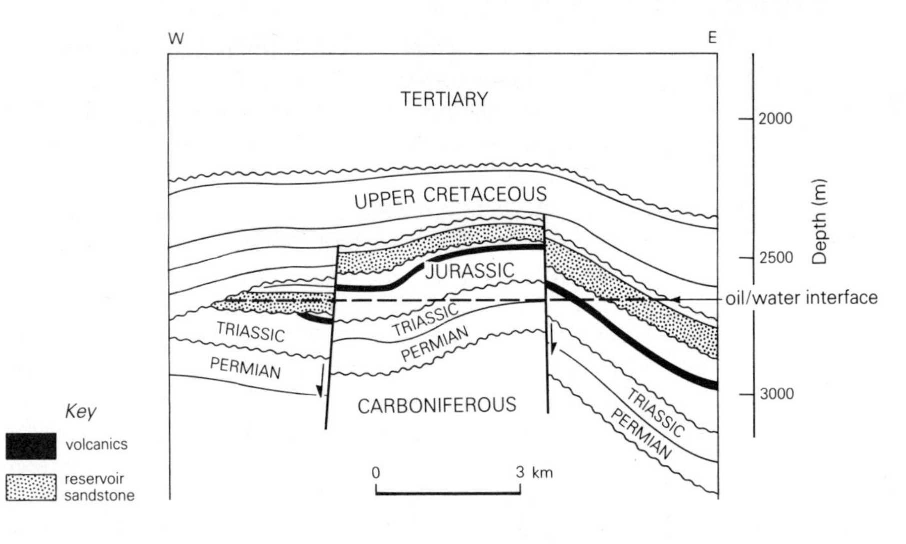

Figure 16.9 Cross section through the Piper oilfield, UK North Sea. (After P. E. Kent, *J. Geol. Soc.*, 1975.)

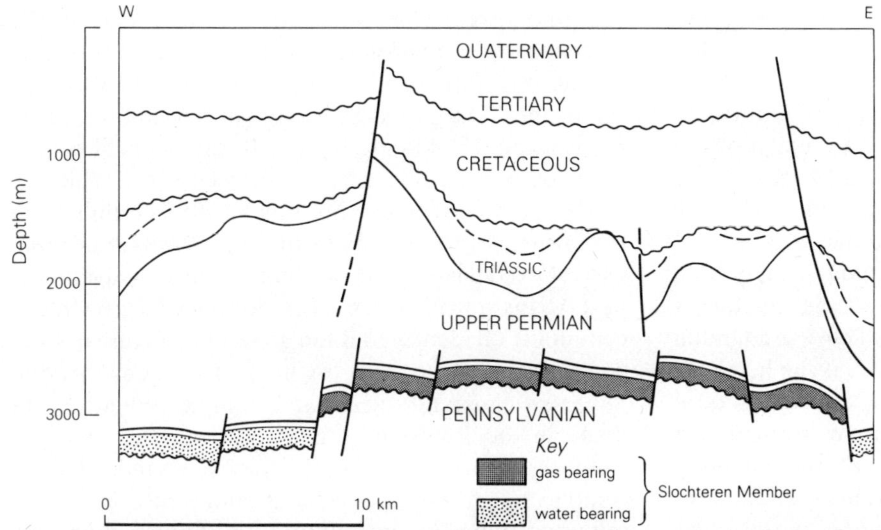

Figure 16.10 Cross section through Groningen gasfield, Netherlands. (From A. J. Stäuble and G. Milius, *AAPG Memoir* 14, 1970.)

Convex traps have static closure. As Wilhelm observed in defining his first trap category, convex traps owe their efficacy to their convexity alone. *Nonconvex traps* have no static closure, and therefore have to acquire the equivalent of closure by some dynamic means. As we shall see, the simplest and commonest dynamic means is hydrodynamic.

A great many traps, both convex and nonconvex, depend for their efficacy on the presence of *unconformities*, which may themselves be either convex or nonconvex. A probable majority of convex traps, and many that are nonconvex, are intimately associated with *faulting*. In many clearly convex traps, the convexity is due to faults and not the other way round; the faults are the essential ingredients. It will be convenient to consider most fault-controlled traps, and all unconformity traps, from those essential standpoints alone (Secs 16.5 & 7). We should note, however, that every individual trap is either convex or nonconvex even though complex fields may contain traps of both types (Figs 16.3, 6, 8, 31 & 82).

16.3 Convex traps

Given the opportunity, oil and gas get as high in their reservoir as they can. Because they are lighter than water, this is their position of least potential energy. This elementary postulate was put forward at the very beginning of the petroleum era, in 1861, and it quickly became the basis for exploration as the *anticlinal theory*. It was in fact put forward independently by two men, an American, E. B. Andrews, and a Canadian, Sterry Hunt. It is unlikely that either Hunt or Andrews ever saw an oilfield truly owing its existence to anticlinal structure, but their postulate was the most influential and successful one ever put forward in the industry.

Many geologists, if we exclude those whose speciality is structural analysis, think of anticlines as *folds*. An anticline is simply a geometric arrangement of strata such that they dip away from a central area which is relatively "high"; the term has no necessary genetic connotation, and many true anticlines have nothing to do with folding. If we accept the term in this wider and

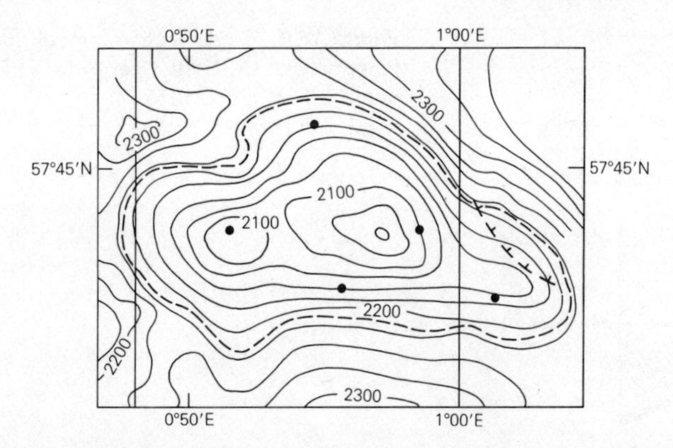

Figure 16.11 A simple domal trap to illustrate closure: the Forties field in the UK North Sea. Structure contours on the Paleocene reservoir are at 25 m intervals below sea level. The dashed contour is the oil/water interface. The lowest closing contour is −2225 m; maximum closure is about 140 m. The long dimension of the field is 18 km. Control wells are denoted by ●. (After A. N. Thomas *et al.*, *AAPG Bull.*, 1974.)

correct sense, it is certain that a majority of known oil- and gasfields are in "anticlinal" traps. Compilations of the world's *large* fields, made during the 1970s, indicate that some 85 percent of the oilfields and at least 90 percent of the gasfields are in such traps.

Convex traps, which are of tremendous variety both genetic and geometric, are all versions of the anticline as far as the oil and gas are concerned. All have in common the possession of static closure. *Closure* is the height to the crest of the structure above the lowest structural contour which closes; it is the effective depth of the structure that can contain oil or gas above the *spill point* (Fig. 16.11). Below that point, identified on a structure-contour map by a contour that fails to close, the oil or gas will spill out laterally into an adjacent structure. Closure must be three-dimensional. Under totally static conditions, oil will occupy the crest of the structure unless gas is present in excess of the amount required to saturate the oil under the reservoir's pressure and temperature conditions; the excess of gas then occupies the crest of the structure, with the oil below. The structure may be filled with oil and/or gas to the spill point. If there is insufficient oil or gas to achieve this, water (commonly saline water) occupies the flanks of the structure, the saddles, and the synclines.

The ideal convex structure is the isolated dome or *structural high*, with quaquaversal dip (dip in all directions radially away from the central high). This structural shape can gather oil or gas from a large subsurface drainage area. In steep, angular folds, oil tends to occupy a narrow belt close to the crest of each fold. The actual position of the oil or gas in the uplift depends on the movement of the underlying water, because conditions are seldom truly static but more commonly hydrodynamic (see Sec. 16.3.9). If water were completely absent, the oil would accumulate in the synclines through gravity, but this situation is rare and no field of any size could accumulate in it.

The ideal convex trap also reveals a history of *growth through time*. The structure develops in a series of pulses over a long interval, during the migration of the hydrocarbons. Thus it presents the oil or gas with a larger and larger trap. The growth of the structure is recorded in the thinning of stratigraphic units over it as it rises and their relative thickening into the off-structure areas that were not rising (Fig. 16.12). Histograms of the differences between on- and off-structure thicknesses then pinpoint the episodes of growth. The dome occupied by the Burgan field in Kuwait, the second-largest oilfield in the world, was uplifted about 220 m during Cenomanian time (causing the development of a widespread erosional surface during the Turonian), a further 150–160 m between the Maestrichtian and the Middle Eocene, and more during the early Tertiary; the oil is of Cretaceous source and the reservoir consists of Albian sandstones. The growing structure may be uplifted along faults rather than being folded. In the fields of central Oklahoma that produce from Ordovician strata (like the Garber and Oklahoma City fields), isopach thins over the structures show growth in early Ordovician time, early Caradocian, post-Devonian, and finally in the post-Mississippian. Even more episodes of growth can commonly be identified if the growing structure is a salt dome. As an illustration for his famous book, Arville Levorsen chose the cylindrical, nonpiercement Erath dome in the Louisiana Gulf Coast, overlain by gas-condensate pools in multiple Miocene sands. Highly complex, composite, piercement domes rose spasmodically and at times semicontinuously through long time spans. Some 125 separate sand bodies have yielded oil, from strata ranging in age from mid-Miocene to Pleistocene, in the domal complex of Bay Marchand–Caillou Island–Timbalier Bay straddling the coast of Louisiana (Fig. 16.6).

Just as early structural growth is ideal, providing a trap during the oil's migration, so the opposite circumstance is normally fatal to accumulation. Structures that arise long after the migration period are little more likely to contain oil than those that arise outside the migration pathways.

Convex traps come in a great variety of styles, but a general genetic distinction should be drawn between three fundamentally different types:

(a) Buckle- and thrust-fold traps, created by tangential compression without any necessary involvement by the basement. As such traps are bound to die out or be cut off downwards, they have been called "suspended folds." They may die out upwards also.
(b) Bending fold traps, caused by essentially vertical movements without necessary shortening of the

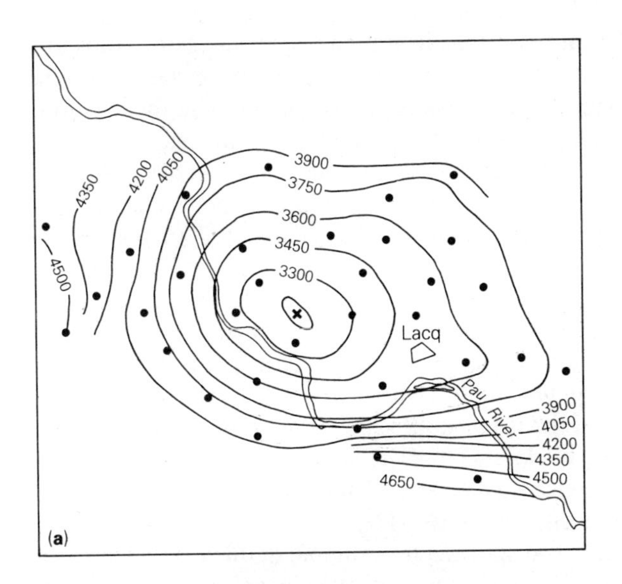

Figure 16.12 Stages in the growth of the anticlinal trap of the Lacq gasfield, southwestern France, illustrated by isopachs of the mappable Cretaceous and Tertiary stratigraphic units (contours in meters). See Figure 16.21 for a cross section through this structure. (a) Structure contour map on top of the Jurassic, as it was before deposition of the final (Tertiary) nonmarine fill; this map identifies the position of the structure; (b) isopach map of the Neocomian (earliest Cretaceous) stage; (c) isopach map of the Aptian–Albian stage; (d) isopach map of the Upper Cretaceous; (e) isopach map of the marine Tertiary. Note the great thinning of every mapped interval over the growing structure; note also that the growth caused the thinning whether the thinning was depositional or erosional. (From E. Winnock and Y. Pontalier, *AAPG Memoir* 14, 1970.)

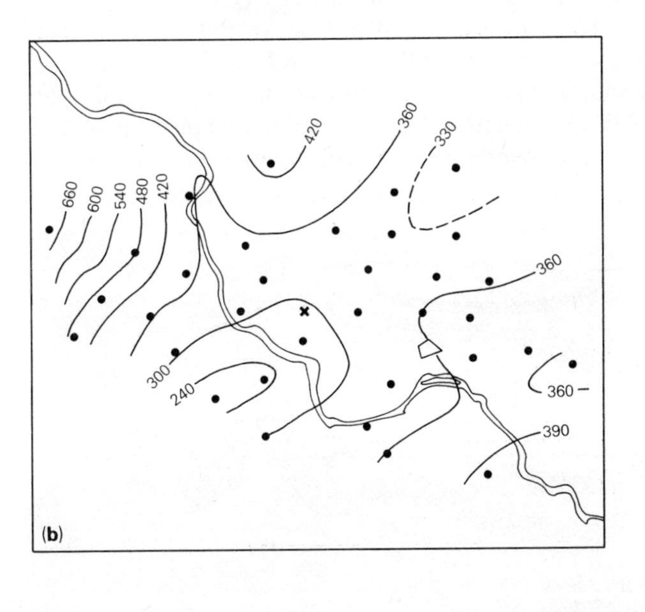

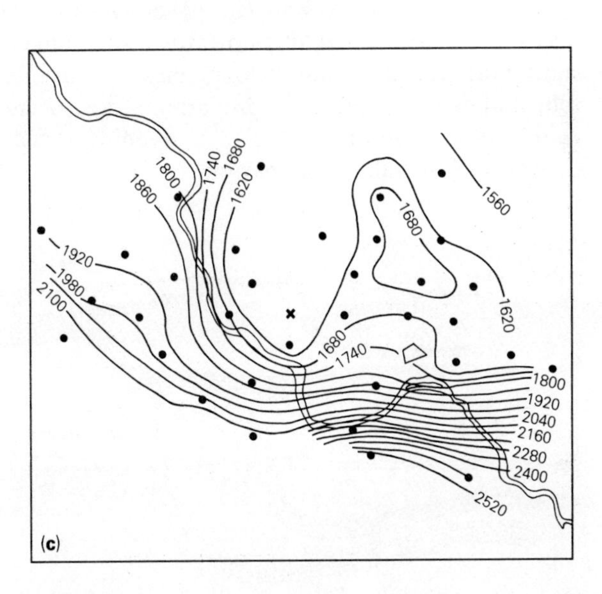

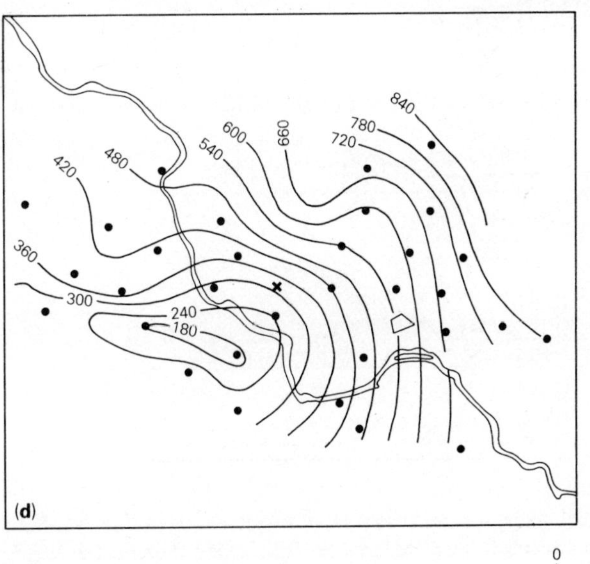

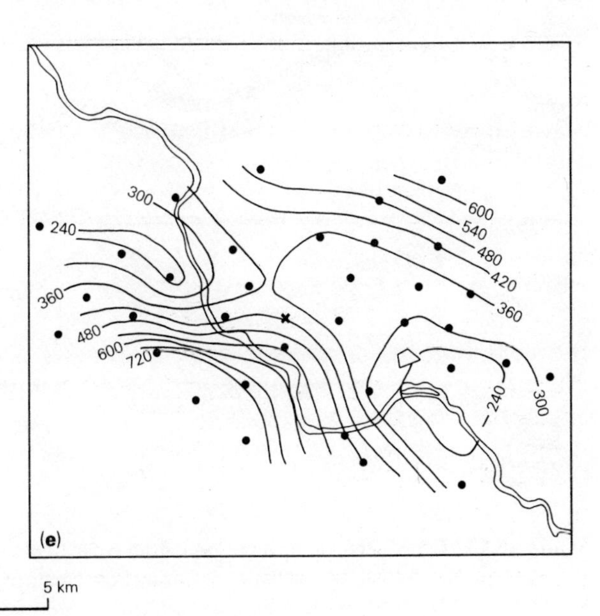

0 5 km

beds. These traps are very likely to involve basement. They may therefore die out upwards, and may be undetectable in shallow levels. If the upward movement continues, however, plastic members of the succession may become *diapiric*.

(c) Traps of immobile convexity, created by specific geologic events antedating the whole of the younger section. The convex, immobile agent may be residual (buried hills) or depositional (principally as reefs). Convexity in overlying strata is due to *drape folding*; it is inherently likely to die out upwards.

16.3.1 Buckle- and thrust-fold traps

Buckle folds are formed by tangential compression and represent horizontal shortening of the stratified cover rocks. Basement is not involved; the structures are "rootless" and die out (or are cut off) downwards. They may die out (or be cut off) upwards also. Tangential shortening typically results in long, linear, or curvilinear fold and fault festoons. Hydrocarbon traps occur in *culminations* along these lines; intervening structural depressions or saddles are barren.

Great variations in the natural sizes of the folds are consequences of the *viscosity contrasts* between the strongest and weakest beds making up the section. Though viscosity is a property of fluids (it is a fluid's internal friction, causing its capacity to resist change of shape), the very fact of being folded shows that layered rocks possess viscosity when subjected to tectonism; they are not perfectly rigid elastic solids. As the viscosity increases, so does the effect of the layering. With very great viscosity contrast, the thick, strong stratigraphic units control the shapes of the major folds, the weaker, incompetent units being "carried." Under this circumstance, the wavelengths (L) of the buckle folds depend on the thickness (t) of the master bed or beds, as well as on the viscosity contrast. The $L : t$ ratio for buckle folds is typically about 10 : 1.

The most spectacular example of oil-rich anticlines in a belt of buckle folds controlled by a master formation is the Iran–Iraq fold belt in front of the Zagros Mountains. Here, the Oligo-Miocene Asmari limestone, about 300 m thick, was thrown into huge buckle folds with wavelengths of 10–20 km and amplitudes of 2–5 km. The overlying sediments contain thick evaporites

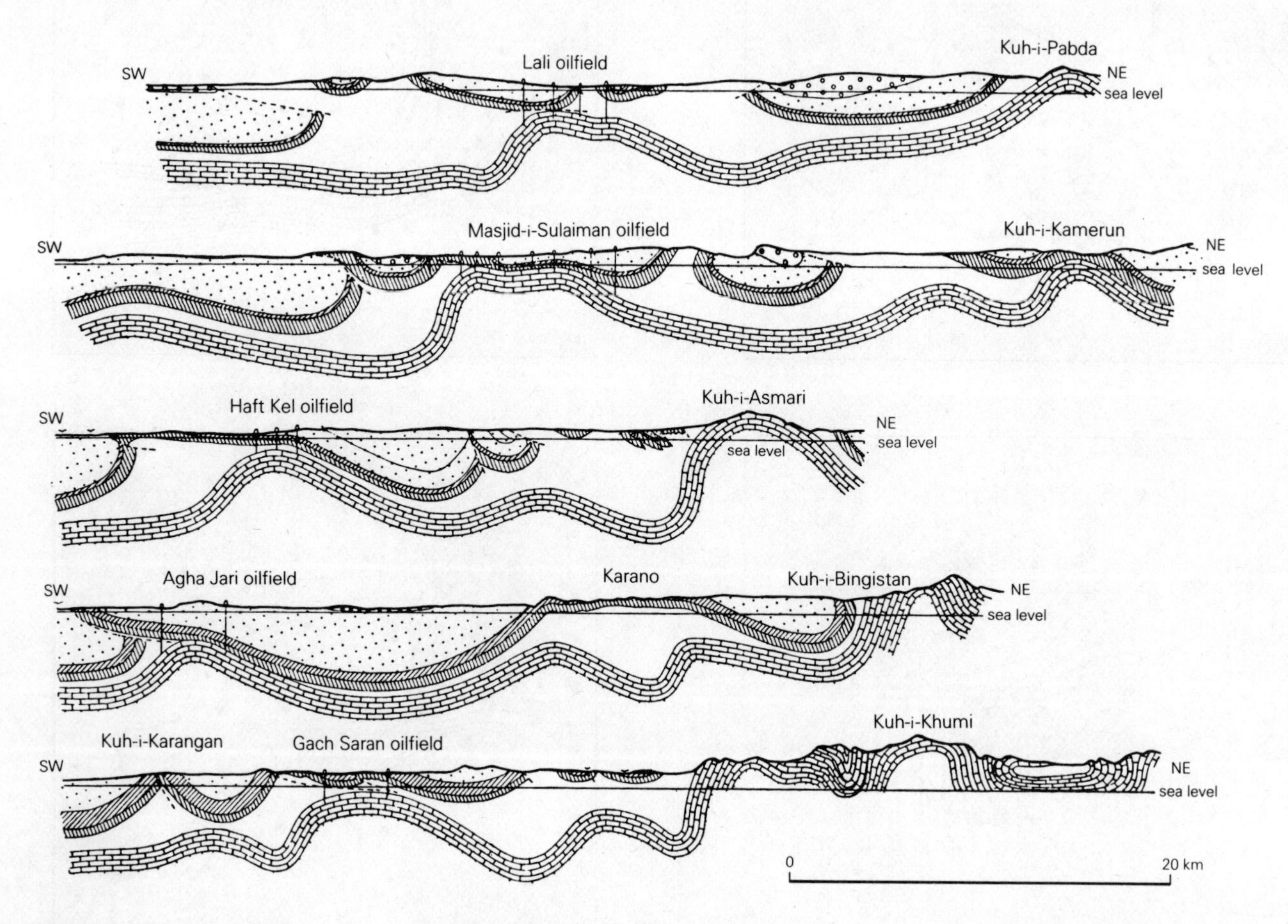

Figure 16.13 Cross sections through the Iranian oilfield belt, showing buckle folds in productive Asmari limestone (block pattern). Note flowage of evaporite roof rock (unpatterned) away from crests of underlying anticlines. (After G. M. Lees, in *The science of petroleum*, Vol. 6, Oxford University Press, 1953.)

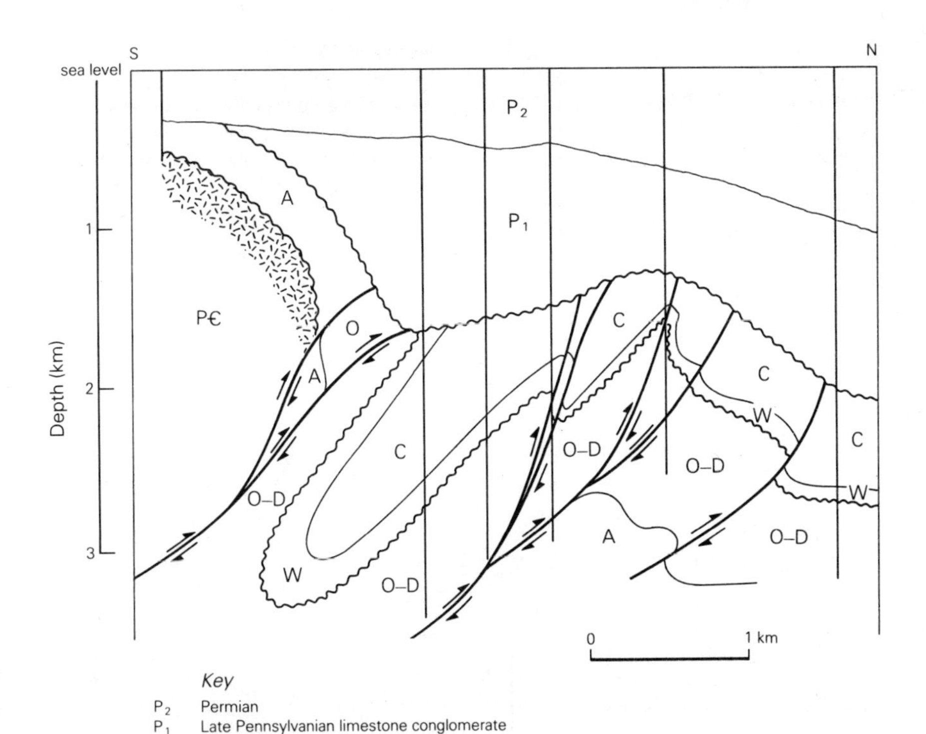

Figure 16.14 Cross section through Arbuckle uplift, southern Oklahoma, showing trap of Eola oilfield in thrust-faulted and overfolded Lower Paleozoic reservoir rocks. Note interpretation that Precambrian igneous rocks are folded in the main anticline. (After R. M. Swesnik and T. H. Green, *AAPG Bull.*, 1950.)

and are exceedingly weak under compression; they were squeezed from the high-stress areas, flowed away from the rising anticlinal crests, and accumulated in low-stress areas in front of them (Figs 13.72 & 16.13).

The forward limb of a buckle fold may of course become oversteepened, even overturned, attenuated, and sliced through by a thrust fault. Many traps in the Appalachian and Oklahoma fold belts, and along the east flank of the Timan–Pechora Basin in northern Russia, are in structures of Paleozoic age of this type (Fig. 16.14). Along the northern flank of the Orinoco Basin in Venezuela, in the Californian basins, and even along the outer Carpathians, comparable traps are found in Tertiary strata (Fig. 16.15). The great gasfields of southern Iran combine the two ages: the reservoir rock is Paleozoic (Permian), the structure Tertiary (Pliocene). The thick, competent strata are heavily fractured in the giant flexural folds (Fig. 16.16).

Buckle anticlinal traps are commonly also segmented by cross faults, transverse to their axes. Such faults may be totally irrelevant to the trap's capacity (as at the north end of the Elk Basin field in Fig. 16.17), or they may control it by dividing the structure into segments with oil or gas restricted to the upthrown sides of the faults (as at the south end of the field in Fig. 16.17).

In very young, weak sediments, with no thick competent units, both average viscosity and viscosity contrast are low. This situation is characteristic of rapidly deposited, fluid-charged clastic sections in Tertiary orogenic basins. Complexes of tight folds are created, commonly without any governing direction of asymmetry. Doubly-recumbent isoclinal folds are not unknown; they may have excellent trapping capacities but

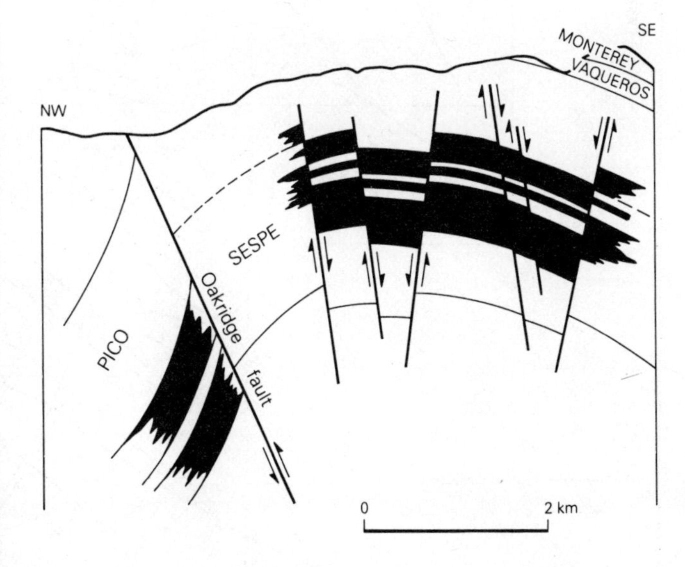

Figure 16.15 Cross section through the South Mountain oilfield, Ventura Basin, California; oil in Oligocene (Sespe) sandstones in hanging wall of thrust fault, but in Pliocene sandstones in footwall. Source in Miocene; basin is to the northwest (left). (After K. K. Landes, *Petroleum geology of the United States*, New York: Wiley–Interscience, 1970; and California Division of Oil and Gas, California Oil and Gas Fields, Part 2, 1961.)

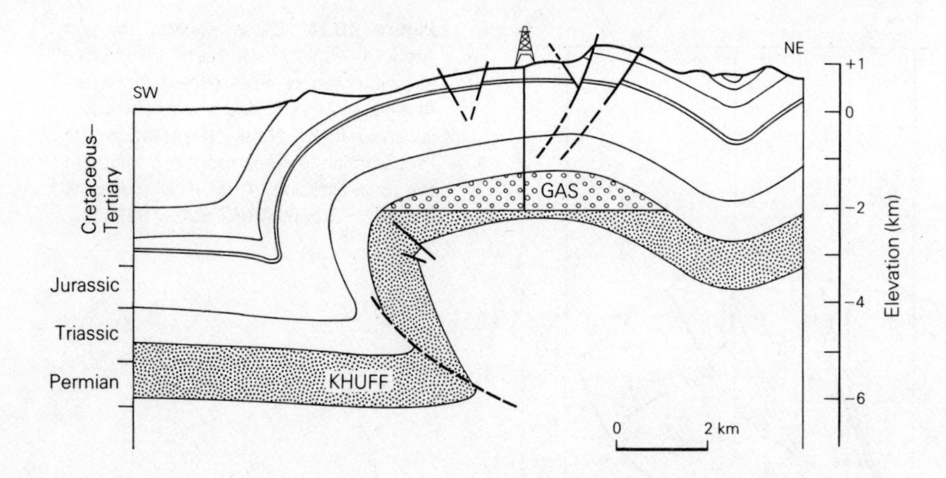

Figure 16.16 Cross section through the Kangan gasfield, southern Iran, in a thrust-faulted buckle fold having an amplitude of about 4 km. (From D. Reyre, *Rév. Assoc. française des Techniciens du Pétrole*; Paris, 1975.)

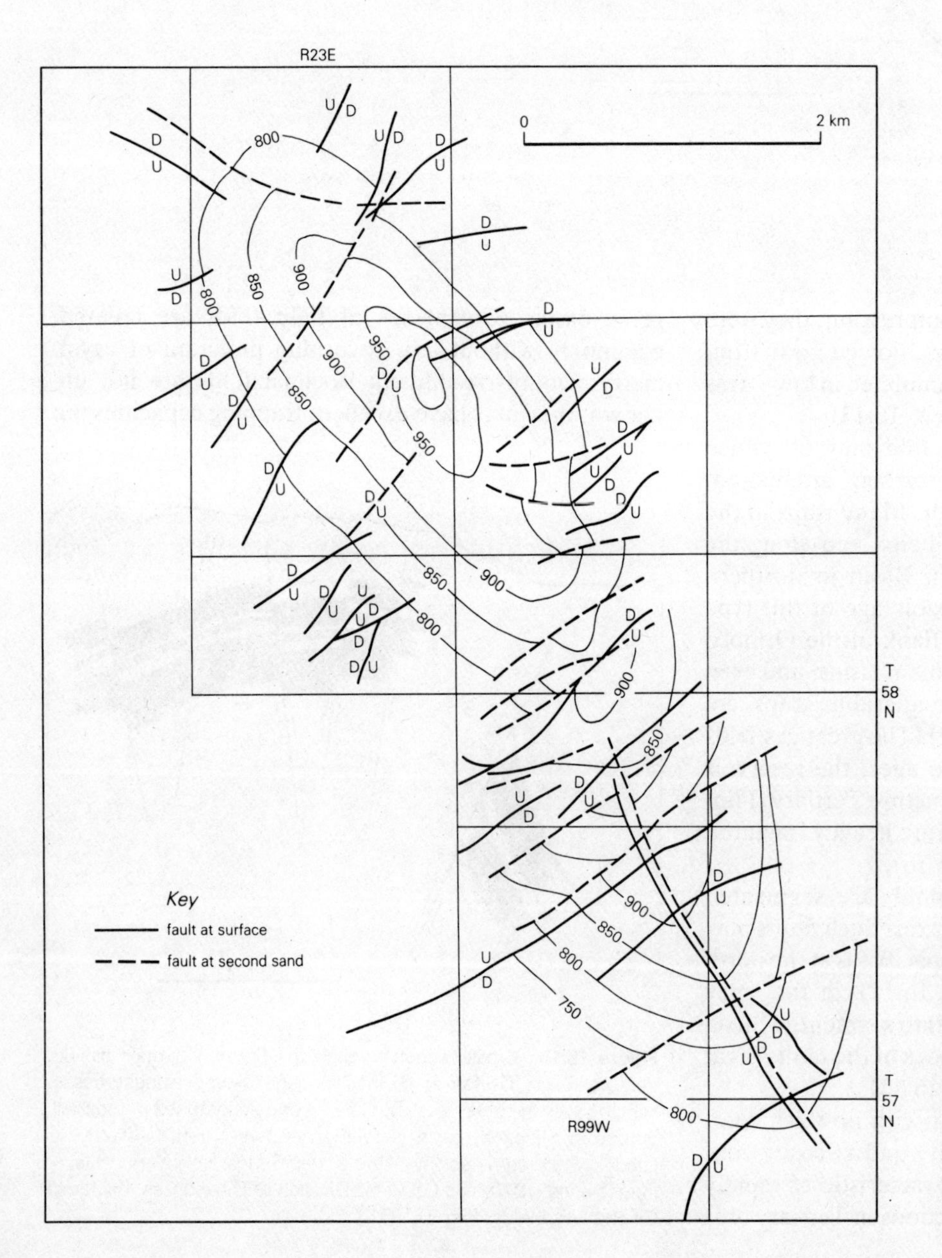

Figure 16.17 Elk Basin oilfield, Big Horn Basin, Wyoming–Montana. Structure contours (above sea level) on top of the principal Cretaceous sandstone reservoir; the field also produced oil from Upper Paleozoic sandstones. Contour interval 50 meters. To illustrate a combined buckle and bending fold, creating an elongate, asymmetrical anticline cut by numerous normal cross faults. (Modified after K. K. Landes, *Petroleum geology of the United States*, New York: Wiley–Interscience, 1970; and J. G. Bartram, *Structure of typical American oilfields*, Tulsa, Okla: AAPG, 1929.)

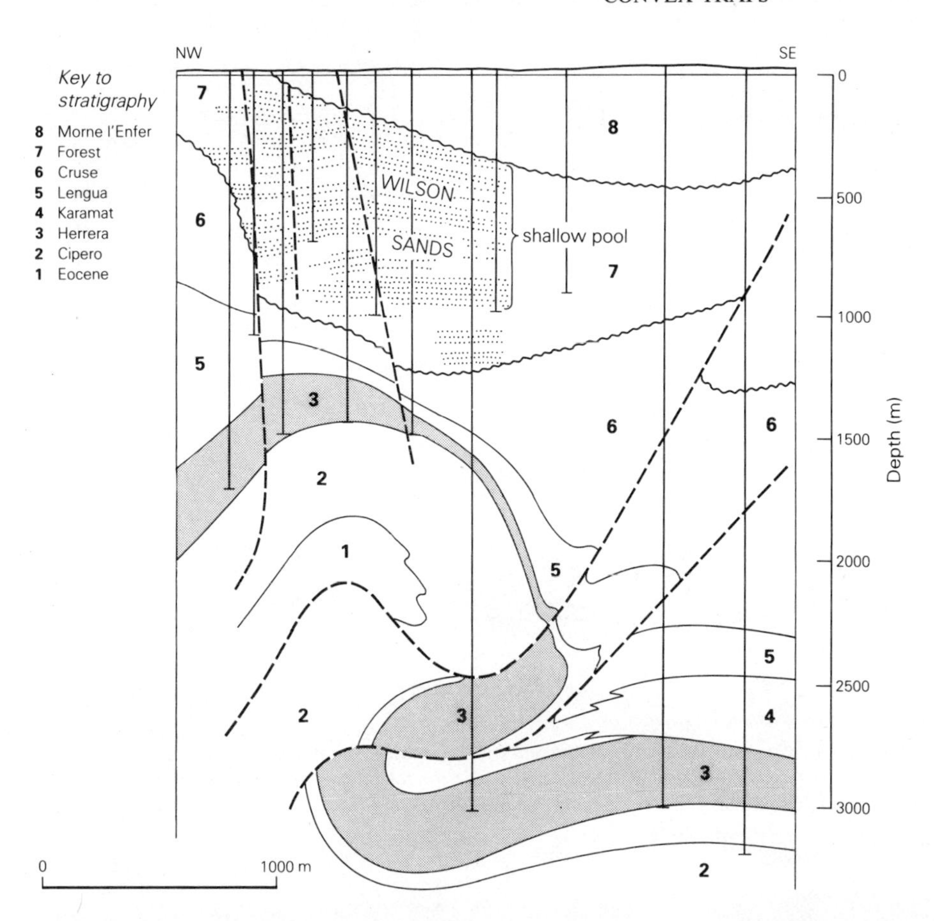

Figure 16.18 Cross section through East Penal field, Trinidad. Three intersections of productive Herrera sands (Lower Miocene) in doubly recumbent anticlinal structure with opposed thrust faults conforming to both directions of overfolding. True scale. (After K. Ablewhite and G. E. Higgins, 4th Caribbean Geology Conference, 1965 (1968); and P. Bitterli, *AAPG Bull.*, 1958.)

with lower closure upside-down (Fig. 16.18). The space problem in the tight folds is compensated by multiple thrust faults converging upwards (and commonly intersecting). Exceedingly complex traps result, and they must be rootless (Fig. 16.19). Such traps are seldom individually large, but examples occur in nearly all compressional Tertiary basins: the Ventura Basin of California (both on- and offshore), southern Trinidad, the Caucasian and Carpathian basins in eastern Europe, the Po Valley in Italy, and others. Even more extreme conditions are reached in the axial zones of orogenic belts, in which isoclinal folding is accompanied by the development of rock cleavage; neither basin nor reservoir survives.

Between these two extreme circumstances, in which viscosity contrasts within the productive strata are either enormous or negligible, is a range of moderate to strong contrasts giving rise to a quite different response by

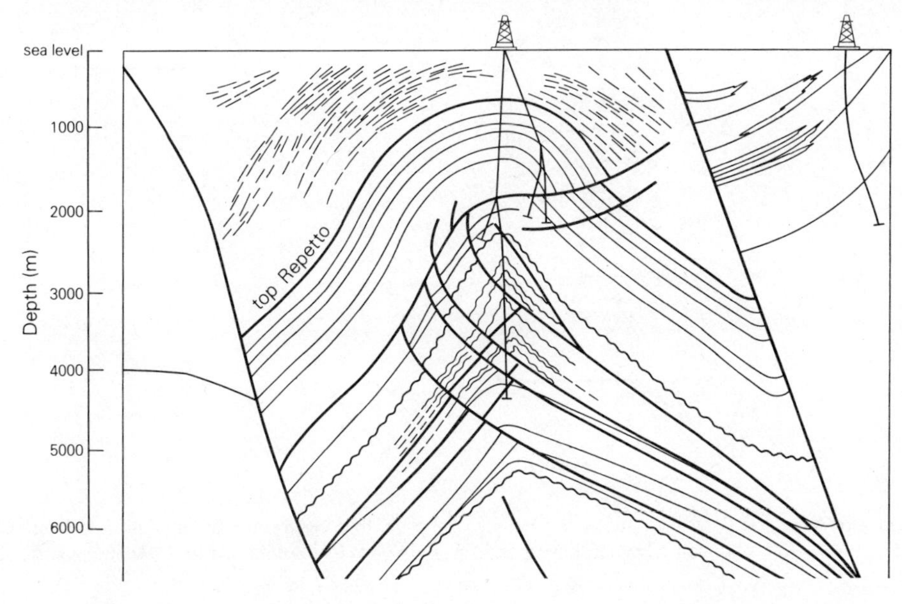

Figure 16.19 Carpinteria field in the Santa Barbara Channel, offshore California. Strongly folded anticlinal trap in weak Neogene strata, cut by opposed thrust faults. (Courtesy of Chevron USA Inc.)

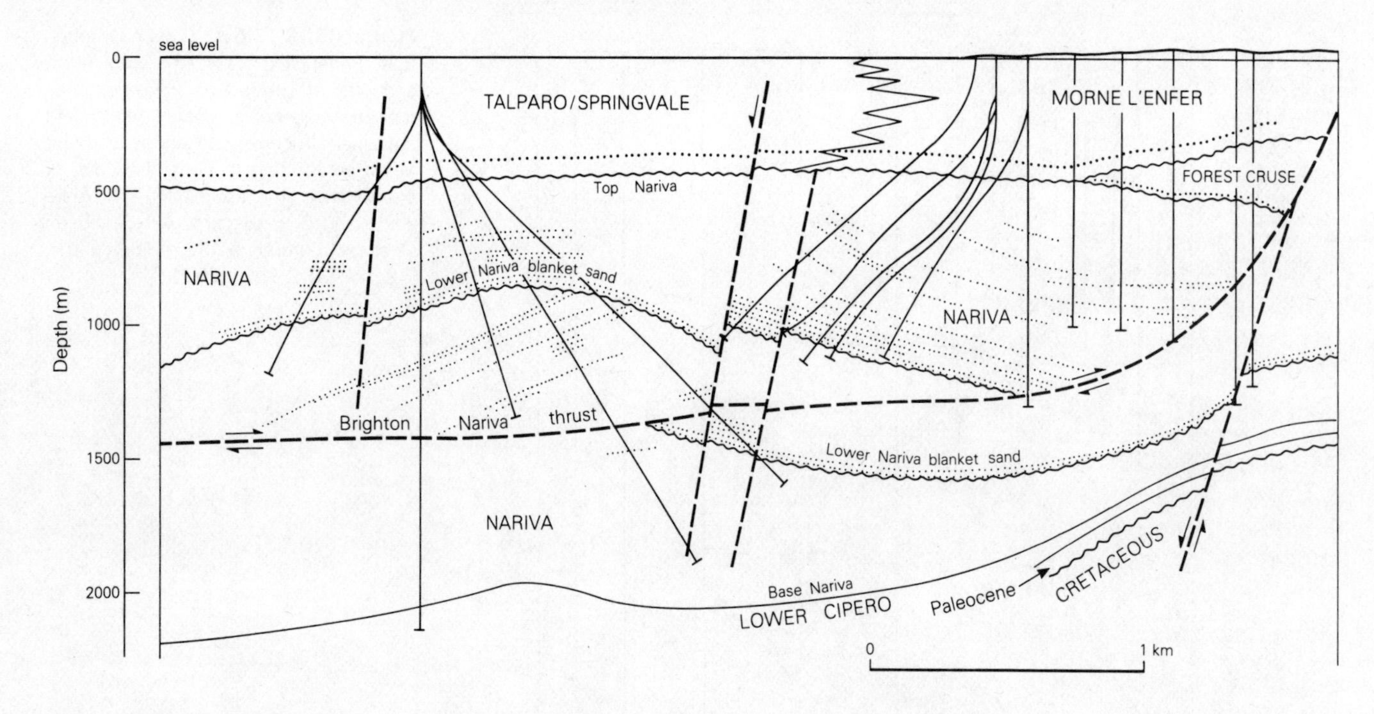

Figure 16.20 Cross section through the Brighton Marine field, southwestern Trinidad: anticlinal trap antecedent to a flattish thrust fault which transects it. The fundamental structure is the fold, not the thrust. Cretaceous strata are involved in the fold at greater depth. (From K. Ablewhite and G. E. Higgins, 4th Caribbean Geology Conference, 1965 (1968).)

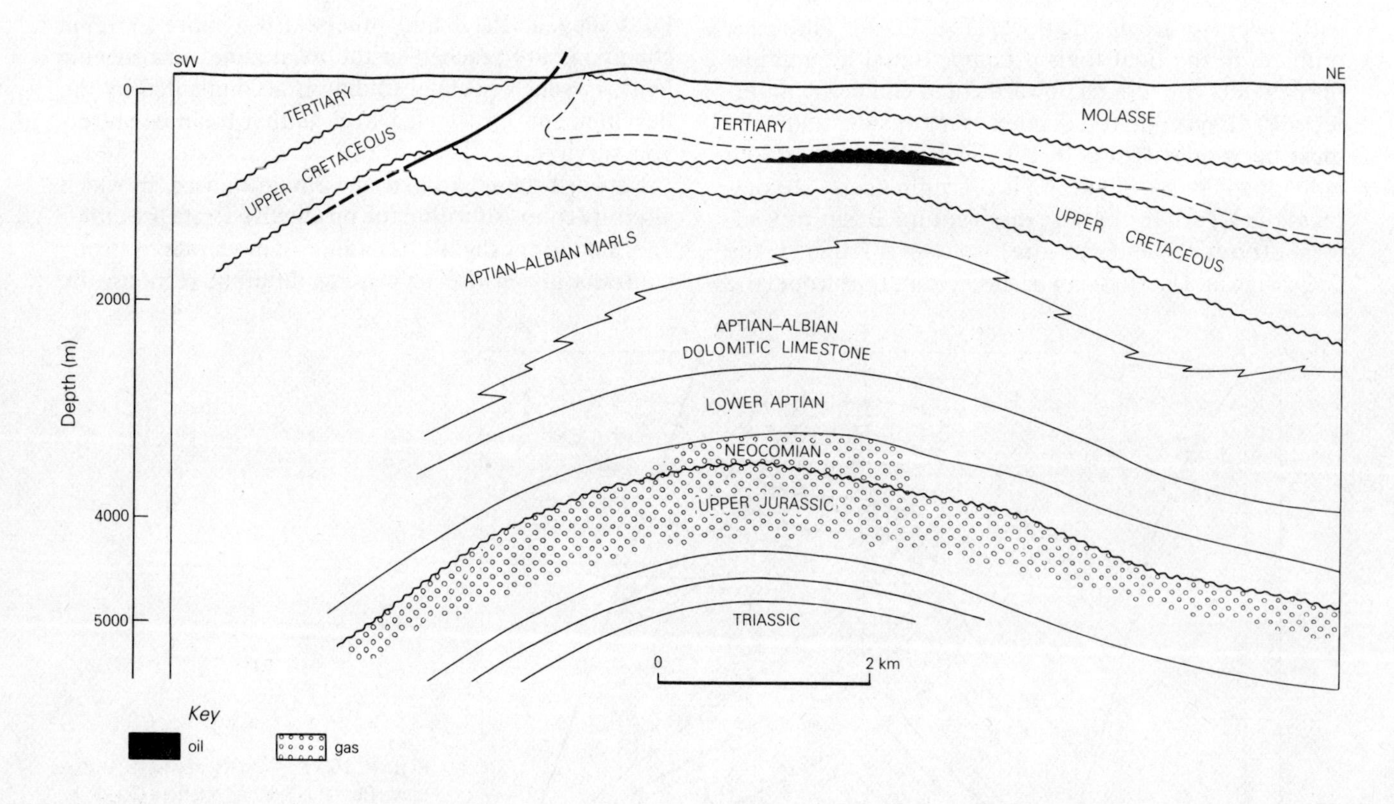

Figure 16.21 Cross section through Lacq gasfield, Aquitaine Basin, southwestern France. Anticlinal trap interpreted as bending fold over uplifted Triassic salt, but could be interpreted as buckle fold with salt pillowing below anticlines above *décollement*. (After E. Winnock and Y. Pontalier, *AAPG Memoir* 14, 1970.)

rocks undergoing tangential compression. Sinusoidal alternations of anticlines and synclines, exemplified by the Asmari limestone response in Fig. 16.13, become unlikely. Instead, the style of folding called *ejective* by Hans Stille is likely to arise: long, isolated, ridge-like anticlines rise above wide intervening "synclines" which in fact represent regional dip. Because no beds are actually downfolded, the pattern of folds comes to resemble those created by pushing a carpet over the floor; the anticlines are often called *rug folds*, and are analogous to pressure ridges.

Clearly a *décollement* or level of detachment must occur at no great depth. The cores of the narrow anticlines are likely to become sliced by minor thrust faults rising to the surface from this basal detachment fault. Some of the anticlines may be beheaded by later, flatter thrusts at depth (Fig. 16.20). There may be several detachment levels, as there probably are below the Iranian folds and certainly are below those in front of the American Appalachians. If the *décollement* takes advantage of a salt horizon, the salt may become pillowed below the rising anticlines, which may then be difficult to distinguish from structures caused wholly by rising salt (Fig. 16.21; see Sec. 16.3.5). Other petroliferous examples are to be found in southern Trinidad, the Romanian Carpathians, the Caucasus, the eastern cordillera of Bolivia, and parts of the Ouachita Mountain fold belt in Oklahoma (Figs 16.20, 24 & 53).

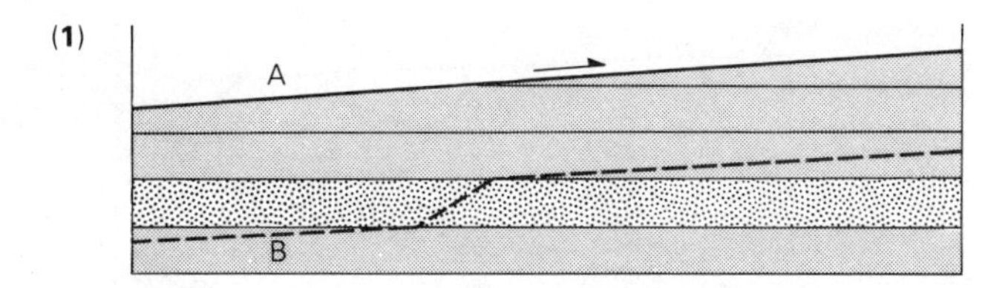

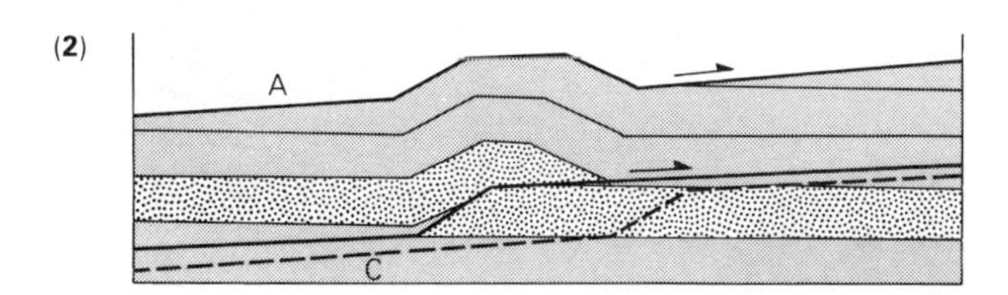

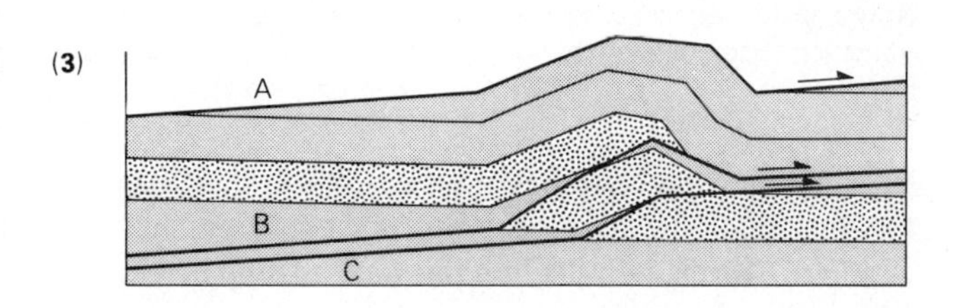

Figure 16.22 Development of folded faults from bedding-plane thrust faults, resulting in stacking of thrust plates and raising of anticlines in hanging walls where thrusts cut across competent horizons (stippled). (From P. B. Jones, *AAPG Bull.*, 1971.)

The combination of high average viscosity and strong viscosity contrast, if occurring along the external (foreland) flank of a deformed back-arc basin, results in *thrust-fold assemblages* of convex structures. The fundamental structures are *underthrusts*, by rigid, foreland basement, of thick, competent, stratified rocks, which become stripped and stacked in succession from the bottom upwards (Fig. 16.22). The stack of imbricate thrust sheets may preserve a clear, decipherable order and constitute a set of parallel strips like those in the front ranges of the Canadian Rocky Mountains and their equivalent in the Overthrust Belt of the western USA. Alternatively, their order may be modified by gravitational spreading, especially if the foreland contained marked salients and re-entrants. In either event, the characteristic sequence is from the oldest rocks, at the highest elevations in the axis of the belt, through successively younger formations in younger thrust structures passing towards the undisturbed external zone or foreland.

The thrusts begin as bedding-plane detachments at depth, commonly in weak horizons closely underlying thick, strong horizons. The underthrusting utilizes the weak horizon, or *fault-gathering zone*, until friction makes further shortening impossible; the fault then cuts obliquely upwards through the overlying competent unit until it reaches the next weak unit, which it again follows as long as possible (the principle of least work). In whatever number may be necessary of flat "treads" in the weak beds and steep "risers" in the strong ones, the thrust approaches the surface, steepening in dip as it does so because of lighter load to lift. It eventually breaks through to the surface as a relatively steep fault. Its overall shape is therefore that of a complex listric thrust merging with a bedding plane at depth. In any one segment of it, its attitude is governed by the footwall lithology.

The thrust must cut up-section as shortening progresses. As the geologist sees it now, the thrust surface must intersect progressively older beds *in both blocks* with increasing depth. Where the thrust cuts obliquely across the competent units, the resulting "riser" in the thrust surface would impede further movement *unless the competent unit itself is raised into a fold above the riser* (Fig. 16.22). Thus a sharp anticline is raised above each transection of a competent unit by the thrust. The anticlines increase in amplitude upwards as the throw of the thrust decreases in compensation. As no anticlines are raised over the "treads" in the weak strata, the anticlines over the "risers" are separated by wide, flat synclines which are in fact slabs of regional dip.

The stratigraphic throw must vary along the fault surface, being much greater at "risers" than at "treads." The amount of displacement in the dip direction cannot

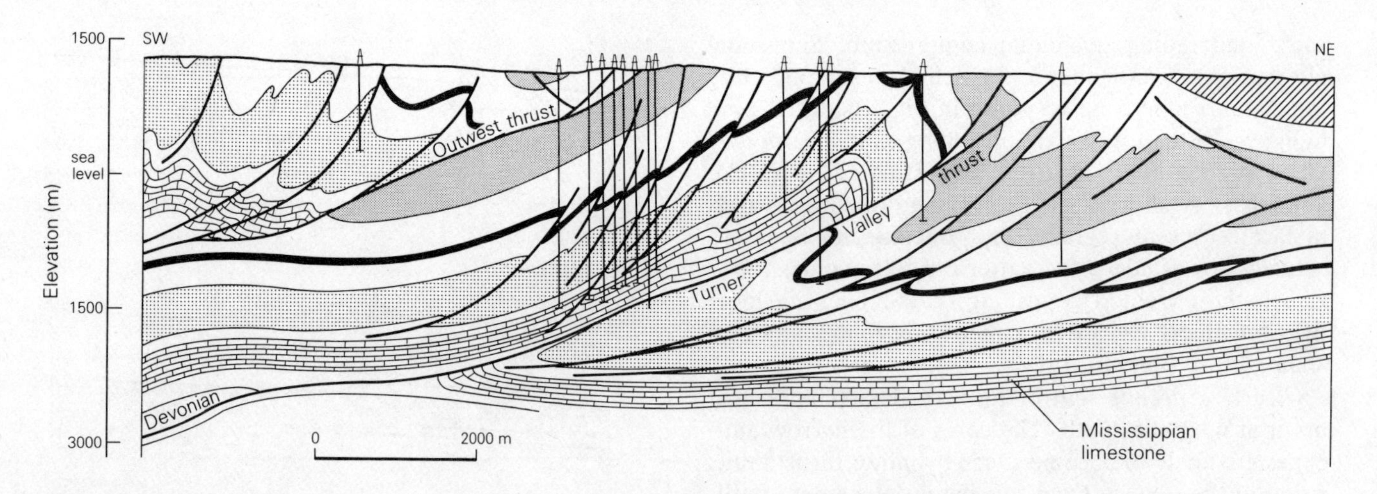

Figure 16.23 Structural section through Turner Valley oil- and gasfield, Rocky Mountain foothills of Alberta. Anticline in productive Mississippian limestone raised where bedding-plane thrust fault cuts across the competent limestone. (From F. G. Fox, *AAPG Bull.*, 1959.)

be constant; one wall must shorten or lengthen relative to the other. The basement also must shorten, but not within the thrust belt itself; it lies undisturbed below the lowest thrust.

Imbricate splay thrusts arise from the master thrusts as these cross the competent units. The splays therefore cut the folded structure, commonly through its gentler back limb. Continuing compression may fold the thrust faults themselves along with the bedding.

Virtually all the hydrocarbons are trapped in the hanging-wall anticlines, which may be very large structures. The fault itself may form one edge for a pool, either in the hanging wall or over the upturned flank of the fold in its footwall, but this is rare in *décollement* thrust anticlines. The accumulation normally postdates the thrusting, though a growing thrust-fold may carry a pre-existing trap with it (as may have occurred in the Turner Valley field in the foothills of the Canadian Rocky Mountains; Fig. 16.23).

The critical feature of *décollement* thrust-folds is that the fundamental aspects of the structures are the *thrusts*; the folds are subsidiary to the thrusts and were derived from them. Even if the folds are now overturned towards their foreland, the overturning was subsequent to translation along the thrusts and did not precede them.

The opposite case, with *folds* the fundamental structures and thrusts subsidiary to them, is illustrated in Figure 16.24. An intermediate case is represented by Figure 16.25. In the Sub-Andean Zone of Bolivia, thin-skinned detachment thrusting first raised rug folds and, after these had reached a critical amplitude, beheaded every anticline by splay faults rising steeply from the basal detachment plane.

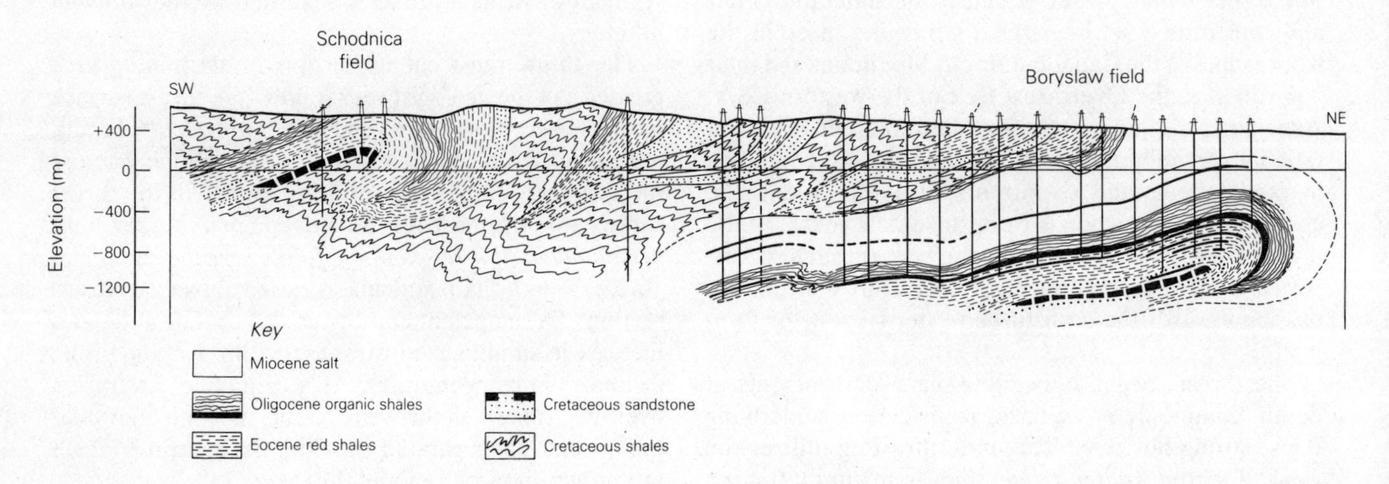

Figure 16.24 SW–NE section through the Ukrainian Carpathians, formerly in southern Poland, showing the structural conditions in the thrust-belt oilfields. Length of section, about 10 km. Oil-bearing horizons in black. (From W. van der Gracht, *The science of petroleum*, Oxford: Oxford University Press, Vol. 1, 1938; after K. Tolwinski, 1934.)

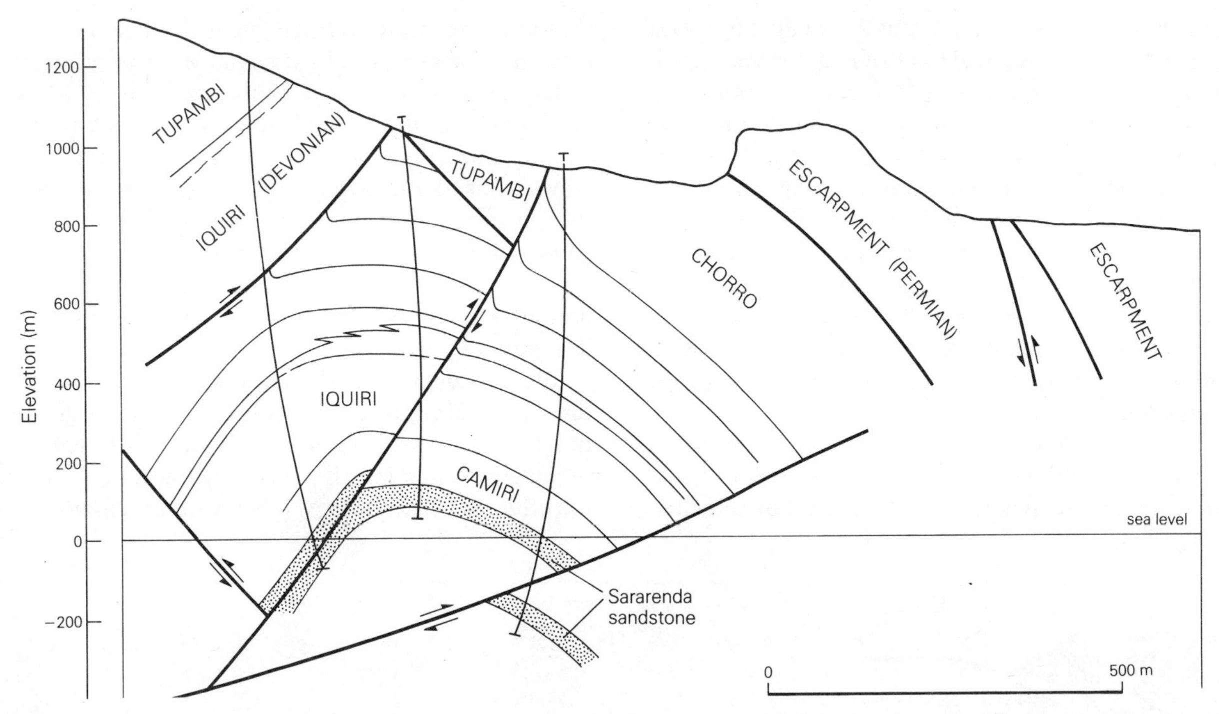

Figure 16.25 Structural section through the Camiri oilfield, in the sub-Andean zone of Bolivia, showing an anticlinal trap cut by a thrust fault rising from the master basal thrust. (From F. Ahlfeld and L. Branisa, *Geologia de Bolivia,* La Paz: Instituto Boliviano del Petroleo, 1960.)

16.3.2 Bending fold traps

The buckle folds described above fulfil the layman's essential conception of any *fold*: the width across the folded medium is smaller after the folding than before it. Geologists call this *stratal shortening*. Many "folds" in layered rocks involve no significant shortening; they are caused not by tangential compression parallel to the layering but by differential movement perpendicular to it. They are *bending folds*.

Bending folds may be much larger than buckle folds. In carbonate successions, they may have wavelengths (L) of the order of 50 km and $L : t$ ratios as high as 100 : 1. The amplitudes of the structures normally increase downwards, as they are propagated upwards from below and no *décollement* is involved.

Most oil- and gas-bearing structures are quite local. On this local scale, depression of unfoldable basement rocks is hard to envisage, whereas their *uplift* is a fam-

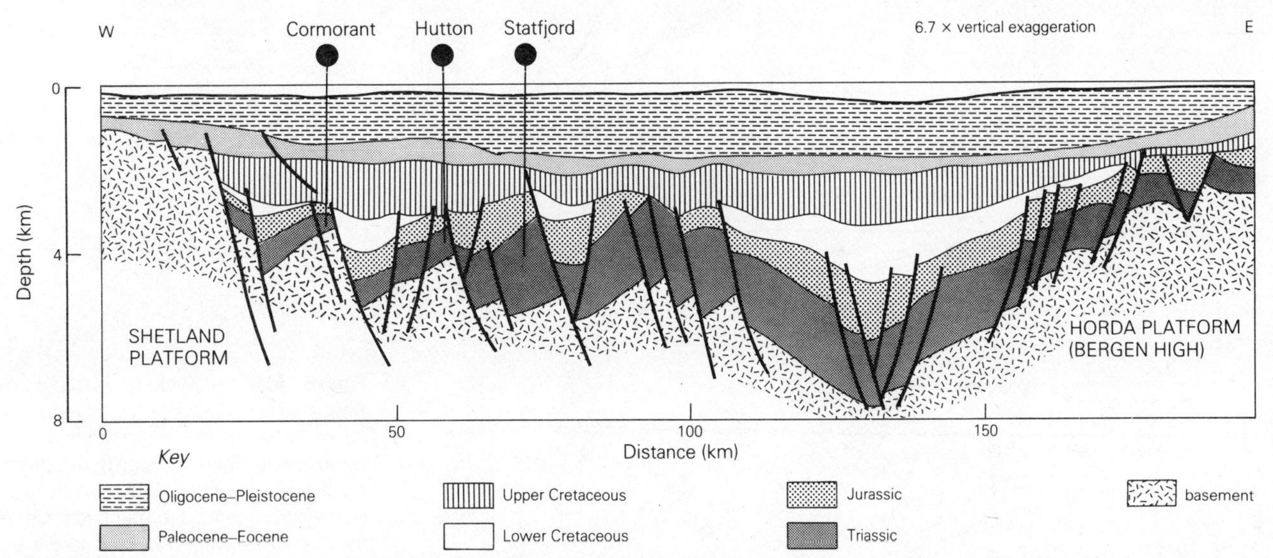

Figure 16.26 Diagrammatic section across the Viking graben, in the North Sea between Scotland and Norway. The foundered graben consists of tilted fault blocks overlain by drape-folded but unfaulted beds. The three fields shown are in bending fold traps. (After R. H. Kirk, *AAPG Memoir* 30, 1981.)

iliar phenomenon. Similarly, the ascent of light or plastic layered rocks towards the unconstrained surface is easily brought about by a variety of natural processes; the opposite circumstance, of a heavy rock sinking through other rocks, is not unknown but it is not a significant process in layered rocks on a trap-forming scale.

On the larger scale, rifts and grabens have developed within all continental plates and throughout geologic history. Differential *subsidence* is a consequence of crustal extension. The fault blocks into which the foundered crust is fragmented become rotated, most of them away from the most deeply subsided part of the graben (Fig. 16.26).

The uplifted blocks, under compression, and the least depressed of the tilted blocks, under extension, cause arching of the strata lying above them, and possibly their rupture. The strata uplifted highest, or depressed the least, may be removed by erosion. If the uplift or subsidence is long continued, it causes thinning of the strata deposited over the "highs"; some younger units may be missing altogether over them. Because the section in the "lows" thus becomes thicker than that over the "highs," the closures of the latter are increased by the greater compaction of the off-structure section. The strata are said to *drape* over the uplifts. Compaction seldom initiates structures of any size; it accentuates structures developed by other means.

Bending fold traps may therefore be considered under three general cases, all exceedingly common. The first is the compressional, fault-controlled uplift within the basement, which may be of any age so long as it forms the effective, rigid floor of the petroliferous succession.

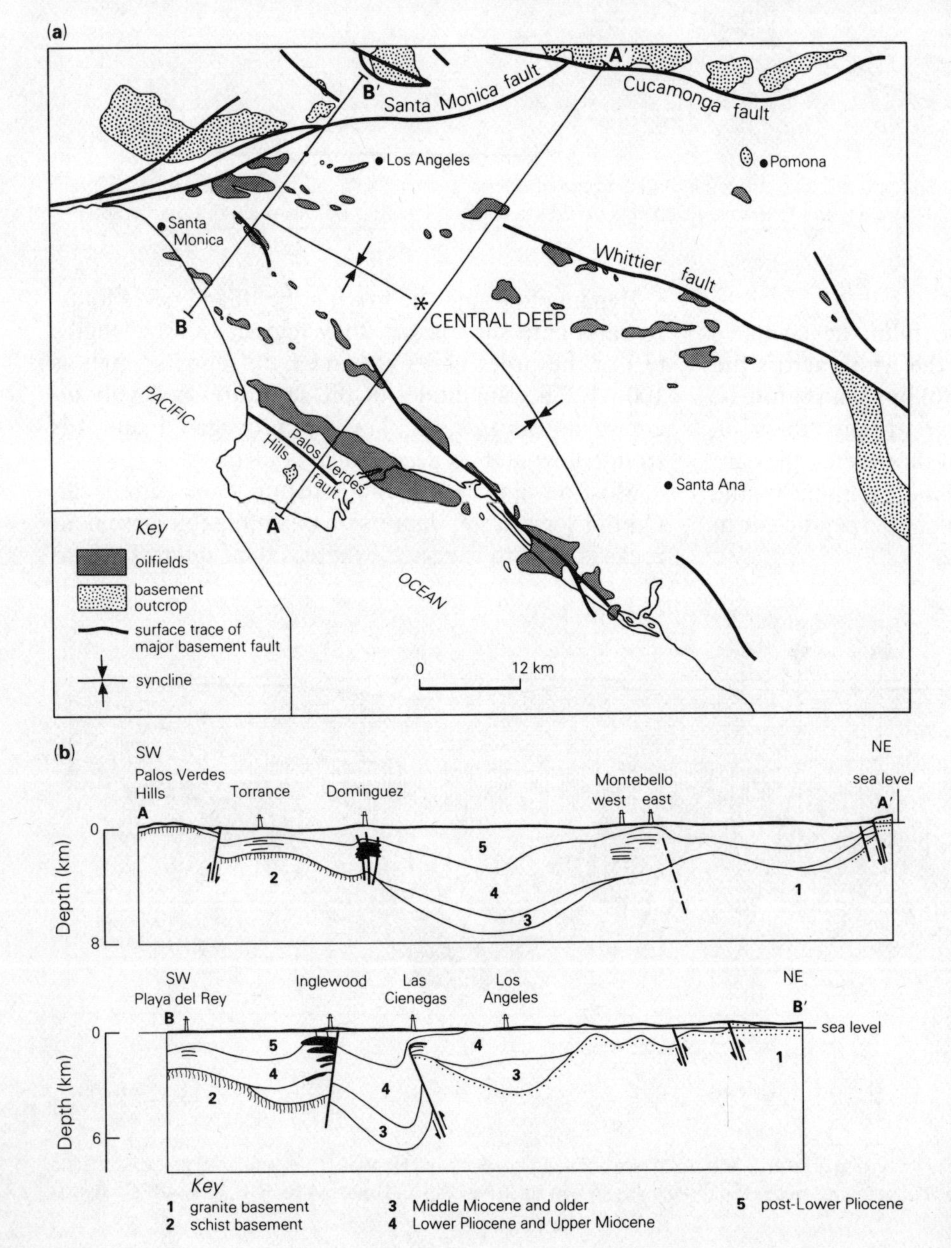

Figure 16.27 Oilfield structures in Los Angeles Basin, California. (a) Map showing alignments of fields in bending-fold traps along reverse fault zones on both sides of the basin, with unfaulted buckle-fold traps in the basin center. (b) Sections across the deep central basin and the northwest end of the basin. Pacific Ocean to the west in both sections. (After P. H. Gardett, *AAPG Memoir* 15, 1971.)

The second is the differential subsidence, under extension, of crustal blocks separated by steep faults. The third case is the restricted vertical uplift from below caused by differential movement of underlying rock that is not basement but is both stratified and plastic. The rock most commonly fulfilling this role is salt. This third type of uplift may be (and in many cases undoubtedly is) caused by either the first type or the second.

Compressional bending folds are best developed as linear trains of dome-like culminations extending along the forelands of orogenic belts (Figs 16.27 & 37). In belts of thick-skinned tectonics, huge blocks of basement are uplifted on steep reverse faults, in contrast to the low-angle thrusting and buckle folding of layered rocks in thin-skinned deformation. Individual trap structures then tend to mimic, in strike, shape, and vergence, the larger uplifts that rise above the basin beside them. The traps arise from basement fractures which are developed *en echelon* along hinge belts or breakover zones between the basins and their shelves. The fractures, essentially vertical in the basement, are propagated upwards through the cover rocks, as reverse faults having dips decreasing upwards. The faults may never reach the surface, in which case strata not cut by them form unfaulted drape folds which constitute excellent traps (Figs 16.27 & 28). The uplifts may be markedly asymmetrical in cross section, and bounded by "synclines" which are not downfolds but merely regionally dipping strata which have not been uplifted.

Examples are readily found along the mobile sides of most compressionally deformed, back-arc basins: those of the US Mid-Continent and Rocky Mountain provinces, the Maracaibo and Orinoco basins in Venezuela, the Aquitaine Basin in France, and the Timan–Pechora Basin in the Soviet Union (Figs 13.73, 13.79 & 16.4).

Not all compressional, bending fold traps conform spatially and temporally to deformed belts. Some straight, elongate anticlines assuredly ascribable to basement fault control are strikingly discordant with their surroundings. The anticlines may be markedly asymmetrical, with low flank dips at the surface but both dips and amplitudes increasing continuously with depth. The thicker the strata, the greater the difference in width of the structure between upper and lower formations; the older the beds, the smaller the area of the structure but the steeper the dips of its limbs. Because they appear to be unrelated in orientation or vergence to their nearest deformed belts, and may be far distant from them,

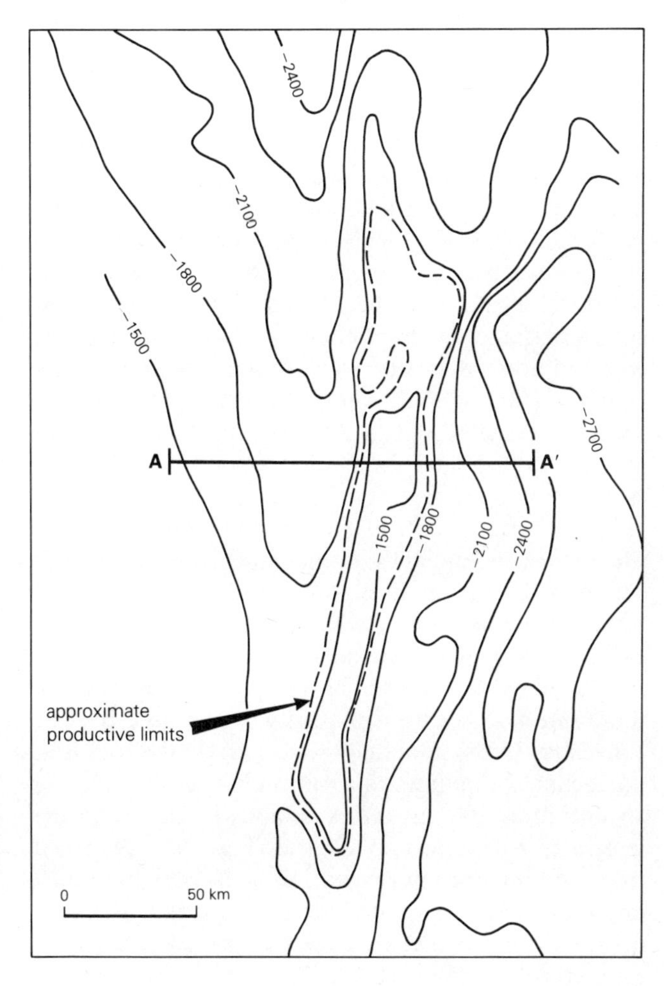

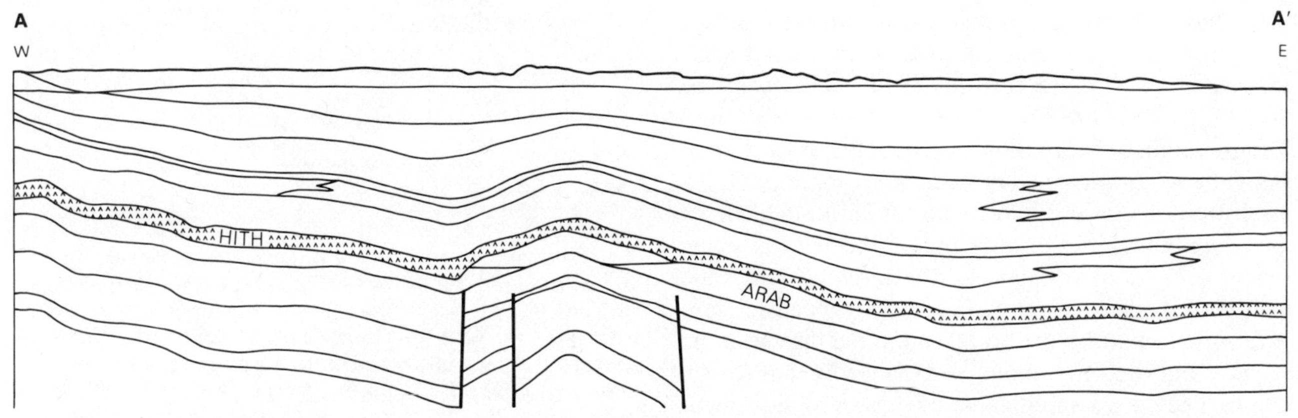

Figure 16.28 Ghawar oilfield, Saudi Arabia, the largest conventional oilfield known. Structure contours on the productive Jurassic limestone, and structure section through the field, showing elongate bending fold trap. (Courtesy Chevron Oil Field Research Company.)

these structures have been variously called *epeirogenic, plains-type*, or (by some Russian geologists) *pseudo-concentric* fold structures.

The entire Central Basin Platform bisecting the Permian Basin in West Texas may be regarded as a gigantic bending fold discordant with its deformed belt; but it is far too large and complex a structure to be compared with individual traps (see Fig. 17.3). The largest anticlinally trapped oilfield known, however, is in such a trap. The great linear anticlines of Ghawar and other fields on the Arabian foreland are quite dissimilar in shape and orientation to those in the orogenic belt far to the northeast (Fig. 25.2). Their growth is proved to have been essentially continuous over a long period and to antedate the compressional orogeny by 100 Ma or more. There is no conclusive evidence of basement block faulting below. Deep salt walls or elongate pillows may be responsible, as deep salt is known to be present; but the elongate, asymmetrical anticlines are quite dissimilar in shape and orientation to the wide domes to the east and north that are known to have deep salt cores (see Sec. 16.3.5). Other great bending fold traps like those of Hassi R'Mel in Algeria and Amal in Libya present no such uncertainty of ascription (Figs 16.44 & 84).

Extensional bending fold traps, created by horsts and tilted blocks bounded by normal faults, are exceedingly common. They occur in all graben and coastal pull-apart basins; examples from the Viking, Suez, Sirte, Reconcavo, and Dampier basins are illustrated here (Figs 16.5, 29–32 & 84). The actual trapping mechanism in these settings, however, is hardly ever caused by true convexity. In the great majority of cases, the trap is in homoclinally dipping strata truncated over the uplift by an unconformity, cut off by a fault, or closed by the intersection of these two features (Figs 16.1, 29 & 74). Because both unconformities and faults constitute such important trapping mechanisms in their own rights, with or without convexity, they are considered as separate categories here (Secs 16.5 & 7).

Graben and pull-apart basins are associated either with divergent plate boundaries or with "failed arm" extensions from these (the so-called aulacogens; see Sec. 17.2). They are also found, but are rarely petroliferous, within or behind convergent boundaries. Extensional bending folds also provide many large but enigmatic traps in cratonic interior basins far removed from any plate boundaries of their age. In the West Siberian Basin, for example, some of the large oilfields and nearly all the supergiant gasfields are trapped in elongate anticlines which are broad and rather diffuse at the surface but become more pronouncedly convex in depth. They are assuredly due to persistent movement along pre-basinal faults in the Baykalian basement.

Because there have been no significant folding

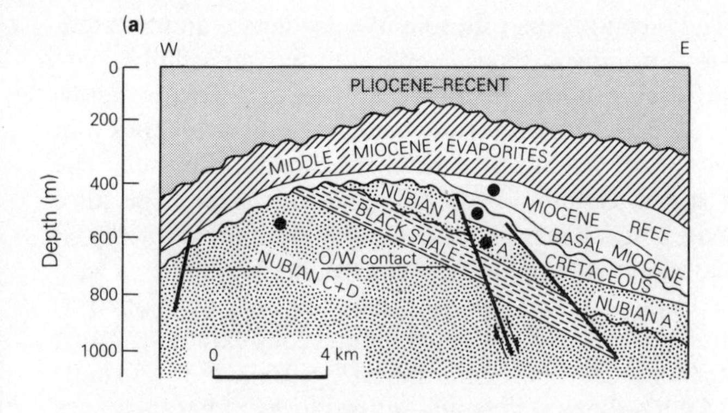

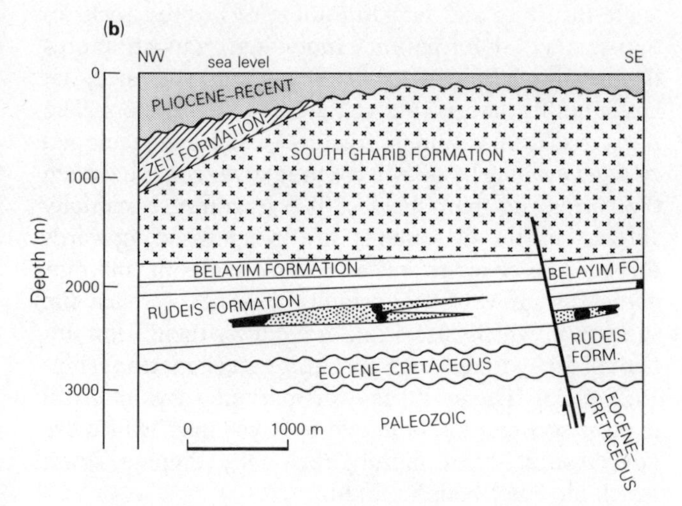

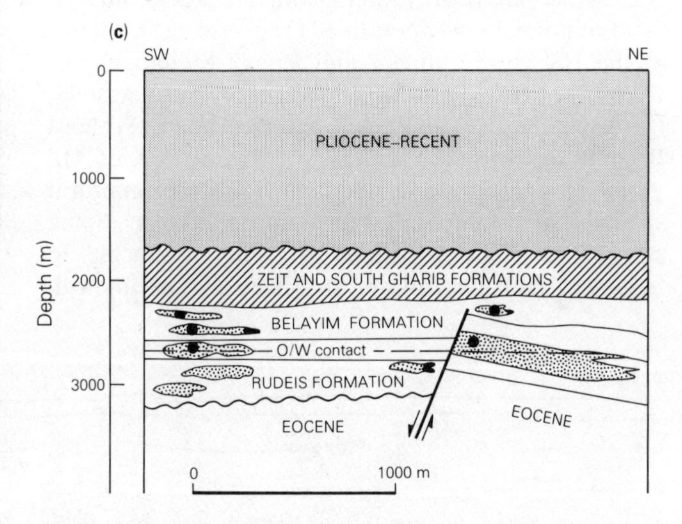

Figure 16.29 Three fields in the Suez graben, illustrating the variety of trap types in an area lacking compressional anticlines. Cross-hatched pattern identifies Middle Miocene evaporites. (a) Ras Gharib field, on the west side of the graben. Oil in Miocene reef, and in sandstone reservoirs of Carboniferous Nubian Group in both synthetic fault and unconformity traps associated with erosional convexity. (b) Amal field, in center-trough. Oil in lenses of Tertiary sandstone. (c) Belayim Land field, on the east side of the graben. Oil on shelfward side of antithetic fault, and in lenses of Tertiary sandstone. (From Y. Gilboa and A. Cohen, *Israel J. Earth Sci.*, 1979.)

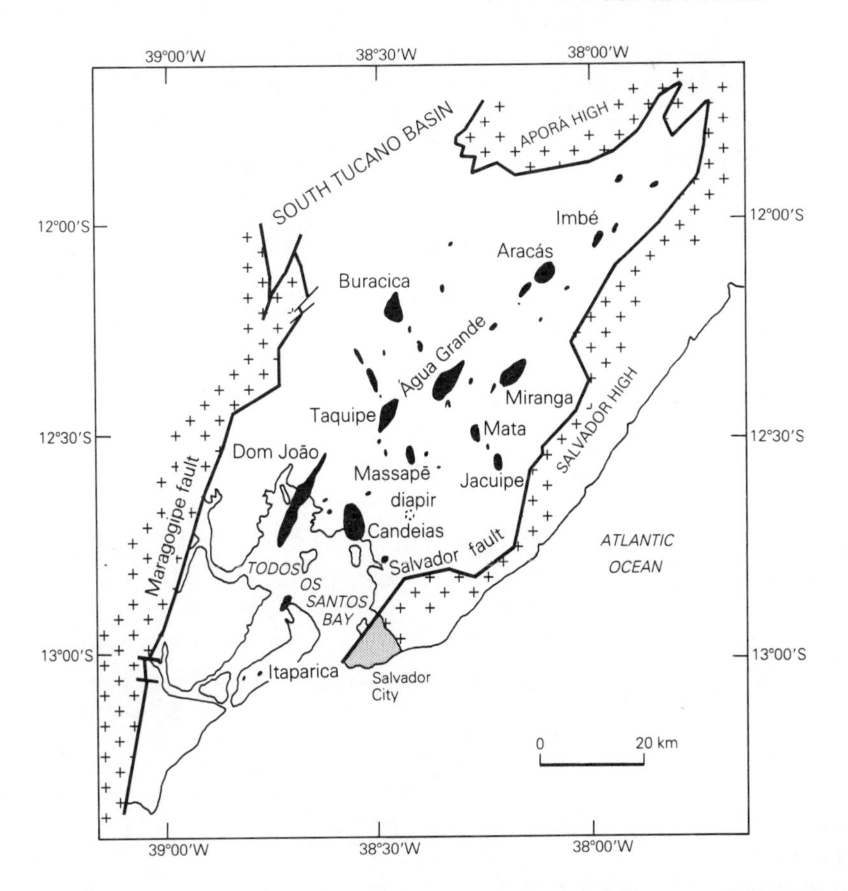

Figure 16.30 Oilfields of Reconcavo Basin, coastal Brazil, to illustrate distribution of fault-block traps in a graben deepest on its eastern side. (From J. I. Ghignone and G. De Andrade, *AAPG Memoir* 14, 1970.)

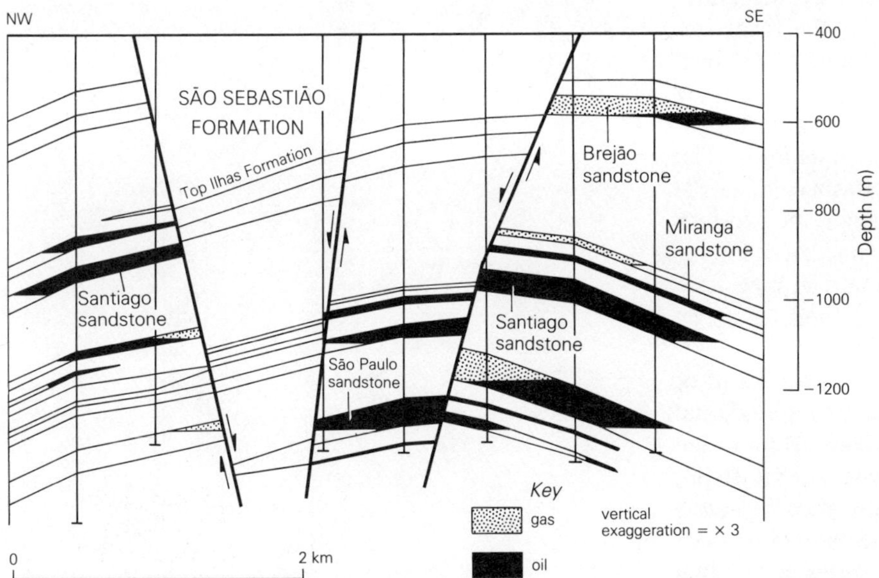

Figure 16.31 Miranga field, Reconcavo Basin, coastal Brazil: bending-fold traps and true fault traps, both convex and non-convex, all consequent upon normal faulting. (After J. I. Ghignone and G. De Andrade, *AAPG Memoir* 14, 1970.)

episodes in cratonic basins such as that of western Siberia, or those in the interiors of Australia or Africa, since the present basins were initiated, we infer that the bending folds are consequences of normal and not reverse faulting. In a number of cases, we have no direct proof of this. In the intracratonic basins of North America, in contrast, so many wells have been drilled that geologists can be much more specific about the faults creating their bending fold traps. In virtually all cases in which the behaviors of the faults have been established in detail, the bending folds turn out to be *anomalous* in more ways than one.

In the first place, they may be scores or hundreds of kilometers removed from any deformed belt of their age. Elongate fold structures like the Cedar Creek anticline in the Williston Basin (Fig. 16.33) or the Lasalle anticline in the Illinois Basin lie far out on undeformed cratons. They are roughly parallel to the Cordilleran

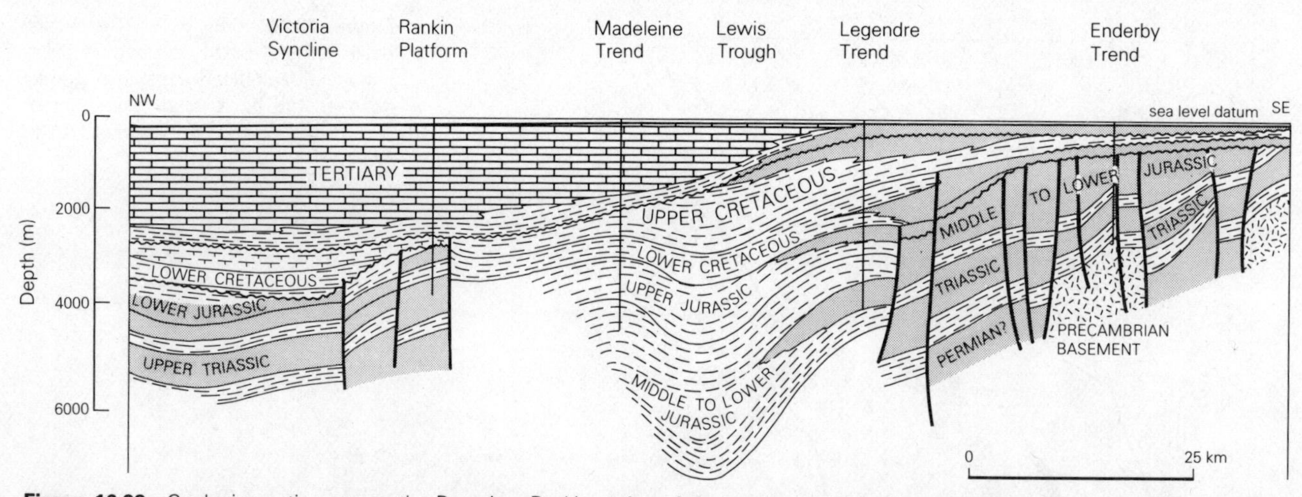

Figure 16.32 Geologic section across the Dampier–Rankin region of the northwest Australian continental shelf, illustrating extensional bending folds. (From A. Crostella and M. A. Chaney, *APEA J.*, 1978.)

trend but they long antedate it. They are also parallel to much broader, basement-controlled structures such as the Central Basin Platform in Texas and the Central Kansas uplift, and the whole of this pervasive structural alignment is perpendicular to the trend of the Transcontinental Arch, the Paleozoic backbone of North America (see Sec. 25.2.4).

A further anomaly is found in the fact that not all of the structures can properly be described as folds. The Albion–Scipio trend in the Michigan Basin (Fig. 13.52) and others similar to it in the Michigan and western Canadian basins are straight, narrow bands of dolomitization porosity in otherwise noncommercial limestone reservoirs. The fractures controlling them may be consequent upon distant wrenching or torsion.

The most puzzling feature, however, is the habit of some of the faults of changing from compressional (reverse) to extensional (normal) and back again. In the Williston Basin, the two largest anticlines (Figs 16.33, 16.34 & 17.18) are bending folds controlled by faults having this history. The Nemaha Ridge, extending from Oklahoma into Kansas and having a series of bending fold traps along it (including that at Oklahoma City; Fig. 16.4), is bounded by a major fault along its eastern side. In Pennsylvanian strata, the fault is a steep normal fault; but downward increases in both the offset of the fault and the convexity of the fold (both greatest in the crystalline basement) strongly suggest that the basement may be reversely faulted.

Anomalous bending fold traps of the kinds described present difficult problems both in the interpretation of their origin and in the prediction of their behavior at

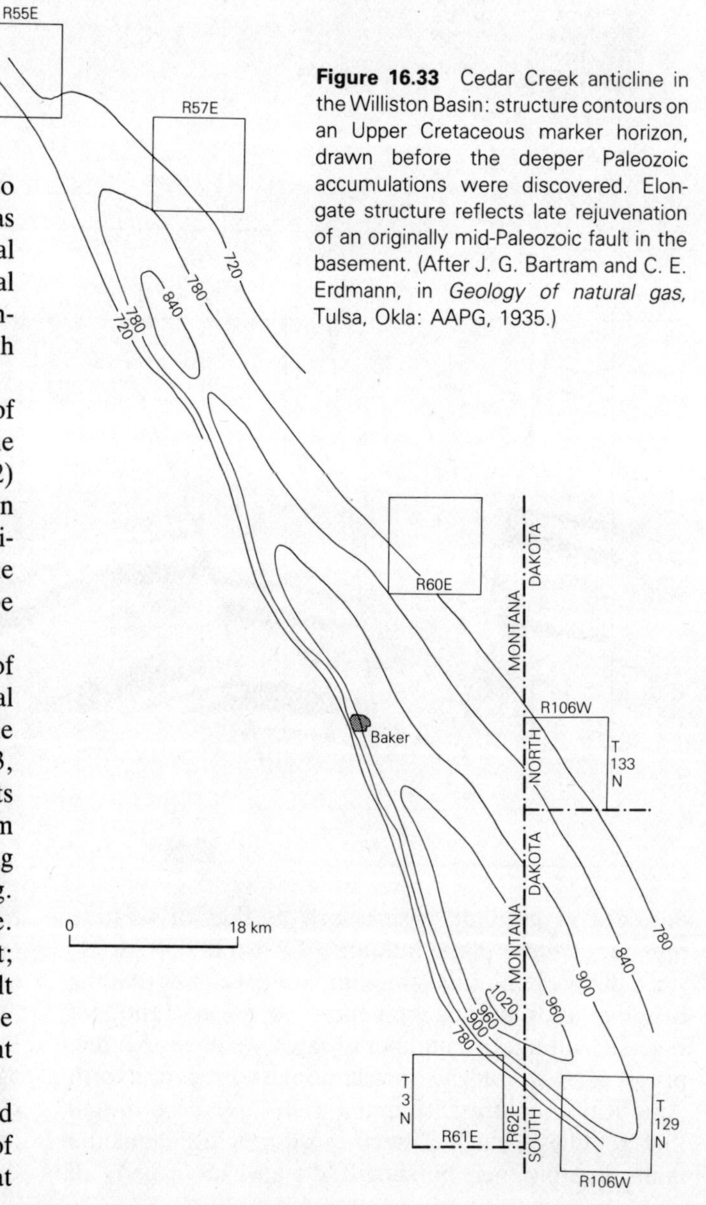

Figure 16.33 Cedar Creek anticline in the Williston Basin: structure contours on an Upper Cretaceous marker horizon, drawn before the deeper Paleozoic accumulations were discovered. Elongate structure reflects late rejuvenation of an originally mid-Paleozoic fault in the basement. (After J. G. Bartram and C. E. Erdmann, in *Geology of natural gas*, Tulsa, Okla: AAPG, 1935.)

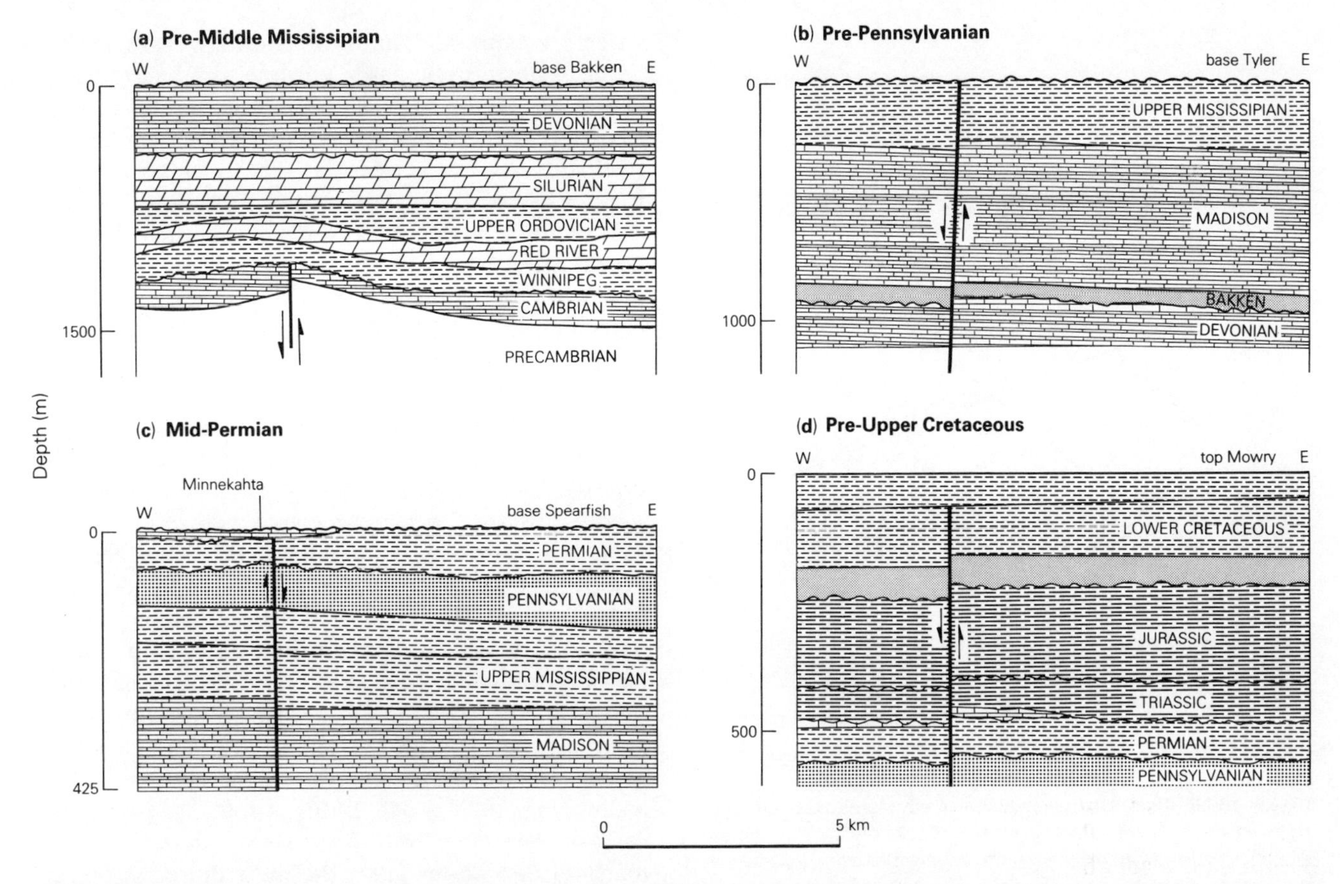

Figure 16.34 Nesson anticline in the Williston Basin, raised above a Paleozoic fault in the basement. An originally reverse fault became a normal fault in the Permian and reverted to its original reverse style in the Cretaceous. (From L. C. Gerhard *et al.*, *AAPG Bull.*, 1982.)

depth. As with any structure that "grows through time" (Sec. 16.3), the folds show changes in thickness of some strata over their crests. If these strata can be accurately dated, and if thickness changes that are regional in scope are removed from consideration, the geologist can plot the percentage of the total growth of the structure against the appropriate time interval for every available marker horizon. This may reveal sporadic or "step-stair" growth, but there are important cases in which the growth was essentially continuous and the plot is a straight line. No evidence is revealed of interruptions in the growth of the folds even though important breaks in sedimentation may be established.

16.3.3 Combination bending and buckle fold traps

True bending folds, like those in the fields along the Nemaha Ridge, and true buckle folds, like those in the Iranian fields, are really end-members of a spectrum of fold types.

Bending fold traps, as we have remarked, are genetically associated with high-angle normal and reverse faults. In the large sense, these structures are intrinsic features of the mechanism that creates sedimentary basins. Buckle fold traps, in contrast, are genetically associated with thrust faults; they are aspects of the mechanism that terminates basins. Many very large traps are created by combinations of bending and buckle folding, created as early regimes of extensional faulting give way to final episodes of compressional stratal shortening and deformation. The effective result is the preferential upwarping of the thicker section against the thinner.

The process is most clearly displayed along belts of deformation adjacent to strike-slip faults, which arise through oblique convergence of two plates or blocks of crust (the so-called *transform boundaries*). It is important to recognize that strike-slip faults themselves (the "master wrench faults" of many authors) are not intrinsic features of sedimentary basins, as both normal and thrust faults are. They have to be looked upon as "superior elements" (in Peter Misch's phrase) which may cut across all rock types and all structures with little or no conformity with their ages or attitudes. Many of the greatest of strike-slip faults have no direct association with any petroliferous basin.

Oblique extension, or *transtension*, opens rhomb-shaped basins or rhombochasms, bounded by faults having large components of normal movement. The thickest sections are deposited close to these faults on

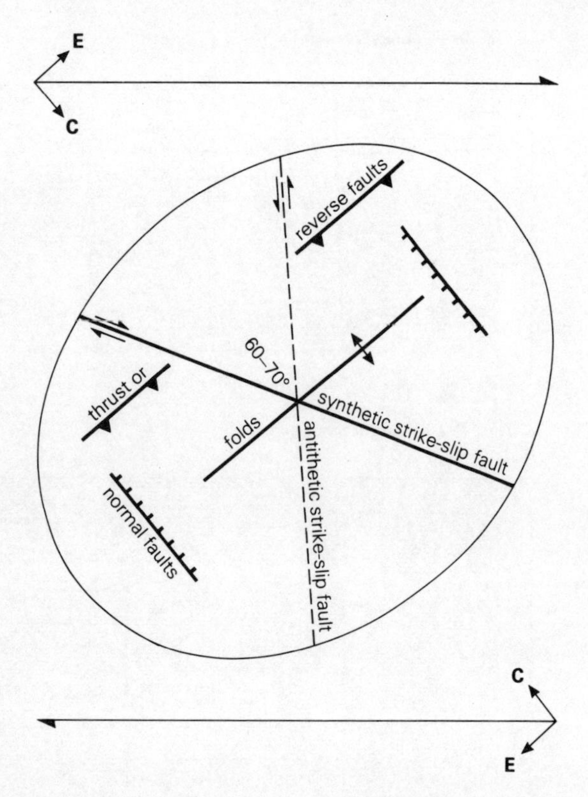

Figure 16.35 Forces and composite of structures that can result from wrench deformation, as determined by laboratory experiments and all combined schematically with the strain ellipse. Right-handed movement depicted; left-handed movement illustrated by viewing in reverse. **C**, vector of compression derived from wrenching; **E**, vector of extension derived from wrenching. (From T. P. Harding, *AAPG Bull.*, 1974.)

their downdropped sides. If oblique convergence becomes dominant, *transpression* deforms and closes the basins; originally normal faults become reversed, and the sections on their originally downthrown sides become hanging-wall anticlines.

Structures ancillary to master wrench faults are readily modelled in the laboratory (Fig. 16.35). They range from synthetic and antithetic fractures (see Sec. 16.7.3), through tensional (normal) and compressional (thrust) faults commonly associated with drape folding, to asymmetrical buckle anticlines with little fault involvement. The trap-forming structures we are considering, however, were not *created* by the strike-slip faults; they were created by extension followed by compression, and the lateral movement along the strike-slip faults merely *modified* them.

As the compression comes to be taken up by movement along the strike-slip fault, rather than on shelfward-directed thrust faults, folds not initially caused by the fault become dependent upon it. They become progressively accentuated, and rotated obliquely to the fault; they thereby acquire an *en echelon* arrangement and fold loci migrate away from the fault towards the basin. Thus the *oldest* folds, closest to the fault, are most sharply compressed and most strongly faulted. Their vertical closure decreases upwards, and may be undetectable at the surface (Sec. 26.2.3). The *youngest* folds are farthest away from the master fault and less likely themselves to be faulted. They may arise too late to reverse the original basinal dip of the strata, in which case they will be essentially surface structures, dying out downwards.

In only three regions of the world are first-order petroliferous basins bounded by deformed belts in which the principal recent tectonic displacement has been of strike-slip (transform) character: southern California, Sumatra, and (more doubtfully) eastern Venezuela and Trinidad. In California, the Los Angeles and San Joaquin Valley basins lie on opposite sides of the San Andreas fault. The Los Angeles Basin (Fig. 16.27) is itself cut by smaller, subparallel faults, with their own associated trains of *en echelon* anticlinal oilfields. These fields lie along the smaller fault zones, and not merely parallel to them. Their traps include both compressional bending folds and true fault traps against high-angle reverse faults (Figs 16.27 & 36); both types of trap resemble countless others elsewhere that have no known association with strike-slip faults. In the San Joaquin Valley (Figs 16.37 & 38), episodes of fold growth can be correlated in detail with episodes of strike-slip movement on the master fault, the most thoroughly documented large fault on Earth. There is little doubt of the effect of the strike-slip movement on the fold traps; but it did not initiate them.

A special case is encountered in which a shelfward-dipping normal fault has been active over a sufficiently long time to acquire a section of plastic sediment much thicker along its downdropped than along its stable side. During subsequent transpression, the thicker section is bowed into an anticline. The basinward dip of the older strata, however, is difficult to reverse, and the final structure loses convexity downwards and remains synclinal in the deepest levels (Fig. 16.39). Such structures are in a sense the compressional equivalents of rollover anticlines (see Sec. 16.7.3). They form important traps in several very young basins in the Tethyan belt, notably in the southeast Asian oilfield areas where they have been referred to as *Sunda folds*. In the compound Sumatra Basin (Fig. 16.40), most of the fields are in faulted anticlines of this sort, obliquely disposed to the Barisan Mountain fault.

The special case becomes still more special if the plastic sediment in the thicker section is not shale but salt. The switch to compression then gives rise to linear salt anticlines like those in the Paradox Basin of Utah (Fig. 16.73).

An enigmatic category of trap combines remarkable linearity and elongation and the downward sharpening of convexity (plus, in some cases, the anomalous ver-

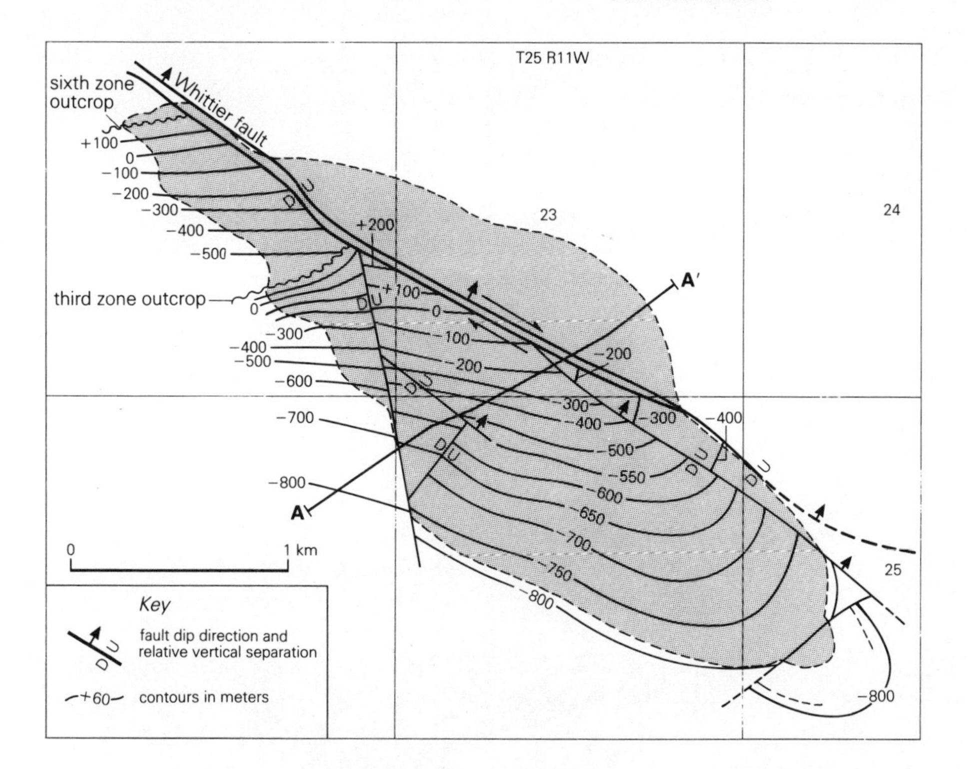

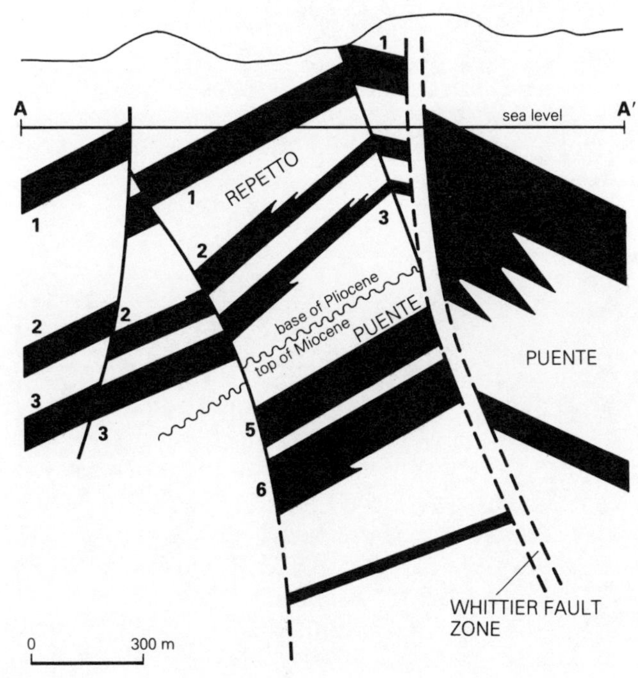

Figure 16.36 Map and cross section of Whittier oilfield, Los Angeles Basin, California. True fault trap against high-angle reverse fault, augmented by bending fold and tar seal. Reservoir sands on northeast (right) side of master fault are Miocene, and of minor significance; reservoir sands in contoured area of field are Pliocene and Upper Miocene, numbered as shown in section. On map, contours south of third zone outcrop are on top of that zone (Lower Pliocene); contours north of third zone outcrop are on top of sixth zone (Upper Miocene). Contours in meters above and below sea level. (From T. P. Harding, *AAPG Bull.*, 1974; after V. F. Gaede, 1964.)

gence) of the traps described above, with an apparently appropriate location in front of a compressionally deformed belt. In the Americas, striking examples are the Lost Hills and Kettleman fields in California's San Joaquin Valley (Fig. 16.37). In Eurasia, the Layavozh and neighboring fields in the Timan–Pechora Basin of northern Russia appear to be similar. The sharp structures rise from deep, linear troughs (possibly grabens) within basement; they are probably aligned along the junctions between different basement types, one more readily foldable than the other (see Sec. 16.3.4). Like the Arabian structures, they antedate the *present* deformed belts, which here are not far away from them. Thus they share some of the characteristics of the Arabian and Sunda types of fold.

The Californian and Sumatran basins are attributable to transtensional origins and transpressional culminations. They were never grabens broken into tilted fault blocks, like the North Sea or Reconcavo basins. Grabens may nonetheless pass through very similar transpressional histories if some element of crustal shear augments the earlier extension. The Gippsland Basin between Australia and Tasmania illustrates this condition (Sec. 26.2.5). In essence it is an aborted rift, formed by the differential collapse of basement fault blocks along a new continental margin. Right-handed shear came later. The traps are now highly asymmetrical drape folds truncated by a basinwide unconformity.

Many deep, old faults may have undergone important strike-slip movement during earlier periods even if the simplest explanation for them now is that they are high-angle normal or reverse faults. The so-called southern

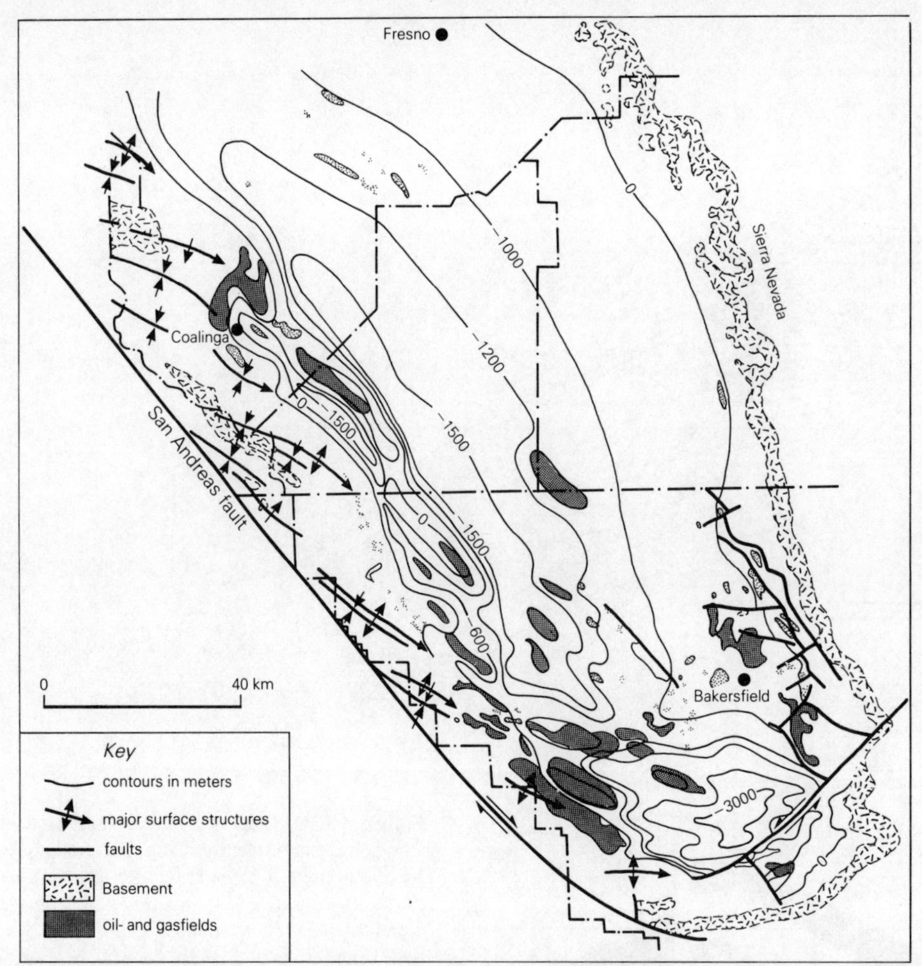

Figure 16.37 San Joaquin Valley Basin, California, showing surface fold axes in mountain ranges along the west side, productive subsurface anticlines in the basin, and the relation of these to the San Andreas fault. Note fault trap fields on the east side, east of Bakersfield. Contours on top of Lower Pliocene; variable contour interval. (From T. P. Harding, *AAPG Bull.*, 1976.)

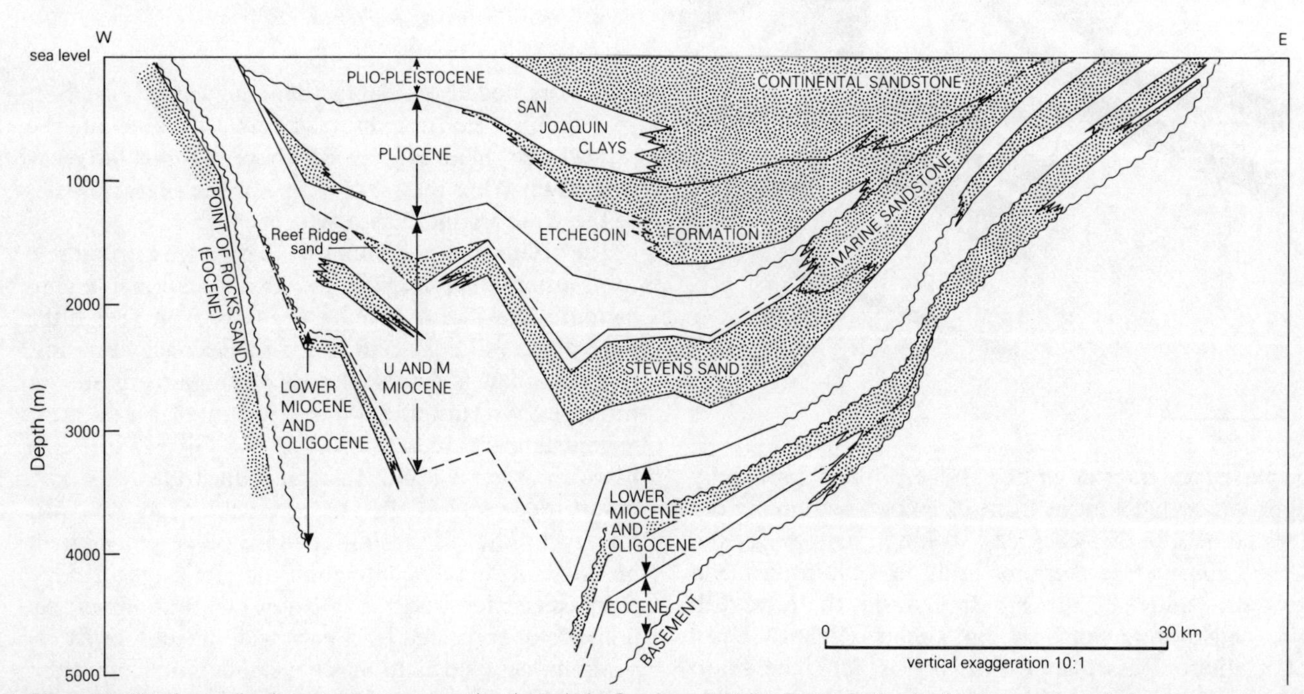

Figure 16.38 Generalized west–east cross section through the San Joaquin Valley Basin, California. West side sharply faulted; traps mostly in buckle folds over foldable oceanic basement. Traps along axis of basin probably bending folds, over basement boundary. East side traps nonconvex, over granitic basement on backslope of Sierra Nevada. Note the Stevens and Reef Ridge sands, turbidite units introduced into the deep basin during deposition of the Upper Miocene source sediments. (After W. J. M. Bazeley, *AAPG Memoir* 16, 1972.)

(a) No basement dip reversal

(b) With basement dip reversal

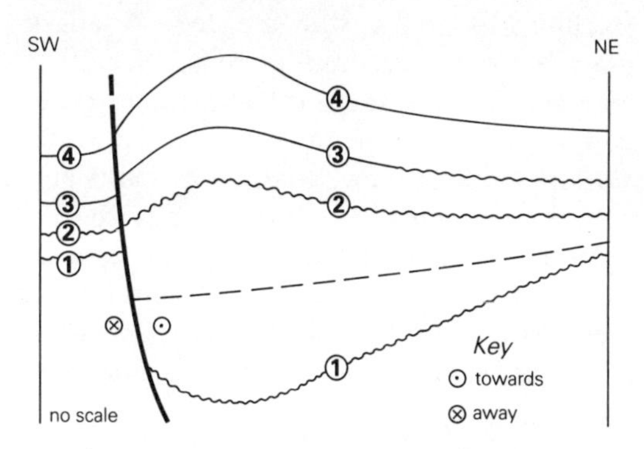

Figure 16.39 Diagrammatic development of a *Sunda fold.* Section between markers 1 and 2 was deposited on the downgoing side of an extensional growth fault. Regime then became compressional, bowing the older units into a syncline and the upper units into an anticlinal trap. Diagram is based on an example from central Sumatra. (From R. Eubank and A. Chaidar Makki, *Oil Gas J.*, 14 December 1981.)

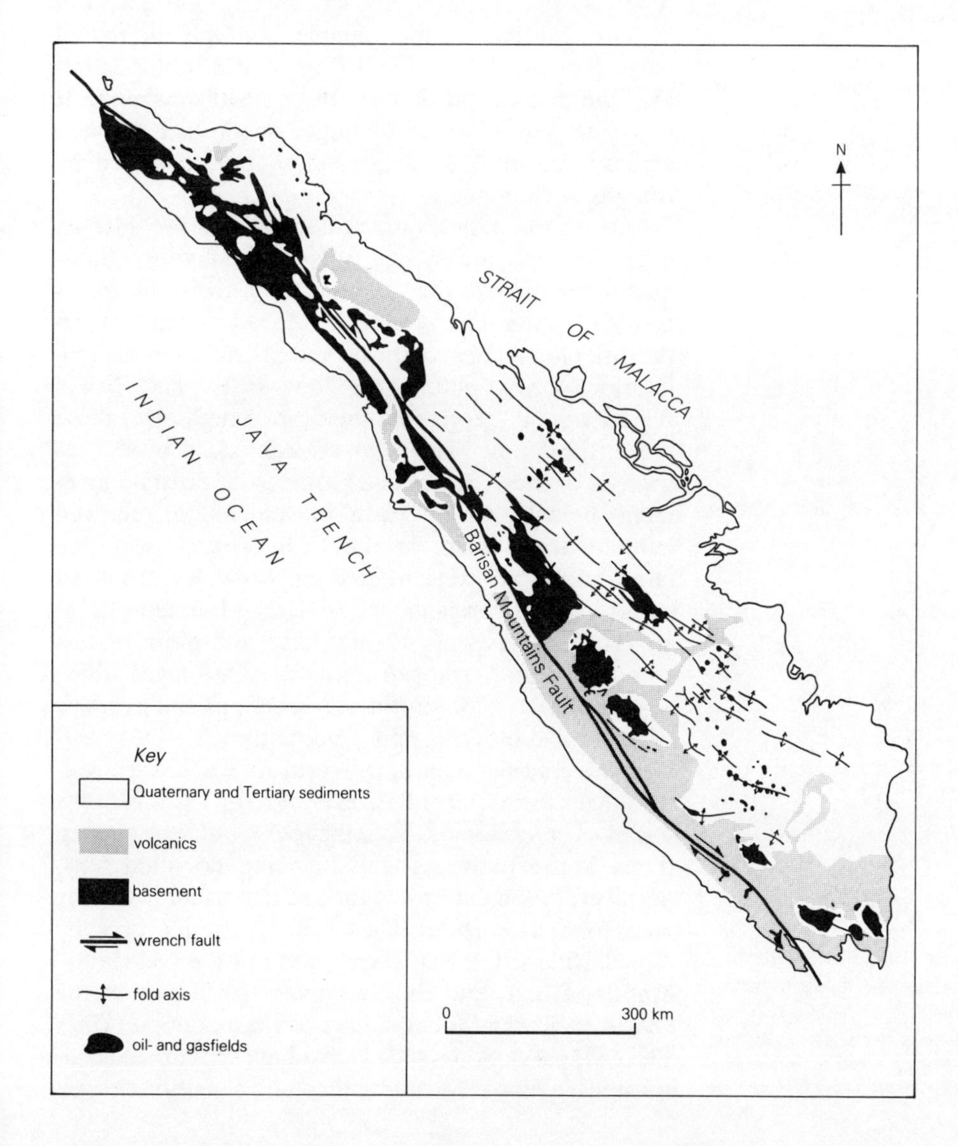

Figure 16.40 Island of Sumatra, showing anticlinal axes and oilfields of the central and southern Sumatran basins and their relation to the Barisan Mountain strike-slip fault. (From T. P. Harding, *AAPG Bull.*, 1974.)

Oklahoma aulacogen gave rise in late Paleozoic time to the Wichita Mountains, with the deep Anadarko and Ardmore basins to the north of them (Sec. 25.2.4). A group of WNW–ESE faults transects the area, which is basically one of high-angle thrust faulting. However, several strikingly straight and continuous faults appear to be "superior elements" (see above): they cut obliquely through the compressional structures, and are regarded as left-slip faults rejuvenated from Precambrian features. Scores of Paleozoic oil- and gasfields occupy traps disposed obliquely to these great lineaments.

16.3.4 Bending fold traps and basement topography

In all the bending folds so far considered, the traps were formed by localized, vertical uplift of something underneath them, either real or effective basement. There is no requirement that the basement itself should have any particular form or surface before the localized uplift takes place. It might have the form of a flat or tilted plane. The Precambrian basement of the Western Canadian Basin, at least between 48° and 56°N, is the almost unmodified continental shelf of the Churchill province of the Canadian shield. Contours on its surface are essentially parallel (Fig. 16.41) and the dip towards the southwest is about 6 m km^{-1} all the way from the Precambrian outcrop to the Rocky Mountains.

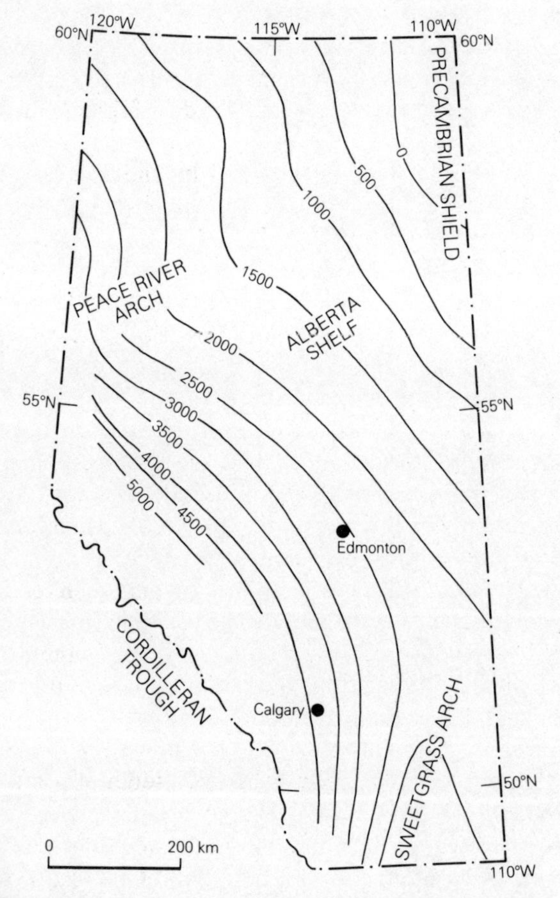

Figure 16.41 Generalized basin-fill isopach map of the Western Canadian Basin in Alberta, reflecting simple homoclinal dip of basement towards the southwest. Contour interval 500 m. Heavy lines delimiting contours on the southwest are eastern limits of the disturbed belt and the foothills of the Rocky Mountains. (After G. Deroo *et al.*, Geological Survey of Canada Bull. 262, Fig. 2-1, 1977.)

Some basements, in contrast, possess considerable relief which long antedated the localized uplifts we have been considering. Because all highly petroliferous regions occur in Phanerozoic sedimentary rocks outside the major orogenic belts, their basements may have undergone long erosion before the present cover of sediments was begun. Thus residual relief on such a basement is unlikely to be abrupt. Instead, the basement surface is widely undulating, with sags and swells which we may collectively call *warps*. Their form, size, and distribution suggest first-order vertical movements due to subcrustal processes, possibly enhanced by heterogeneities within the crust (sags where the basement rocks are relatively heavy, and swells where they are relatively light). The basement of the Ural–Volga Basin in European Russia, for example, is simply the buried portion of the Baltic shield (Fig. 16.42), but it is not like the buried portion of the Canadian shield. It is characterized instead by large, simple flexures with arches, structural terraces, and so on, separated by troughs or depressions.

Most of the oilfields in the Ural–Volga province lie over the high areas, and Russian geologists consequently emphasize these areas, which they call *domes* and North Americans call *arches*. In North America, on the other hand, most of the oil and gas in the continental interior occurs within the low areas, and North American oil geologists therefore emphasize these areas, the *basins*. The *traps* are controlled by the responses of these domes and basins to subcrustal movements below them or crustal movements far removed laterally from them. In the Ural–Volga Basin, the largest fields are concentrated on two wide, long-lived swells on the foreland of the Ural Mountains (Fig. 16.43). The fields are areally large and more or less equidimensional, trapped over subsidiary highs which also localized unconformities, truncations and overlaps of strata and other trapping mechanisms.

In the cratonic interior of North America, such residual highs form Lower Paleozoic outcrop areas *between* basins. The cratonic basins themselves do not contain them. In the foredeeps of the flanking mountain belts, however, basement arches formed important gathering areas for hydrocarbons. The Cincinnati Arch west of the Appalachians, the Bend Arch north of the Ouachita–Marathon belt, and the Sweetgrass Arch east of the Canadian Rocky Mountains are obvious examples (Figs 13.52 & 16.41). Some such arches have remained inconspicuous in amplitude and difficult of definition despite

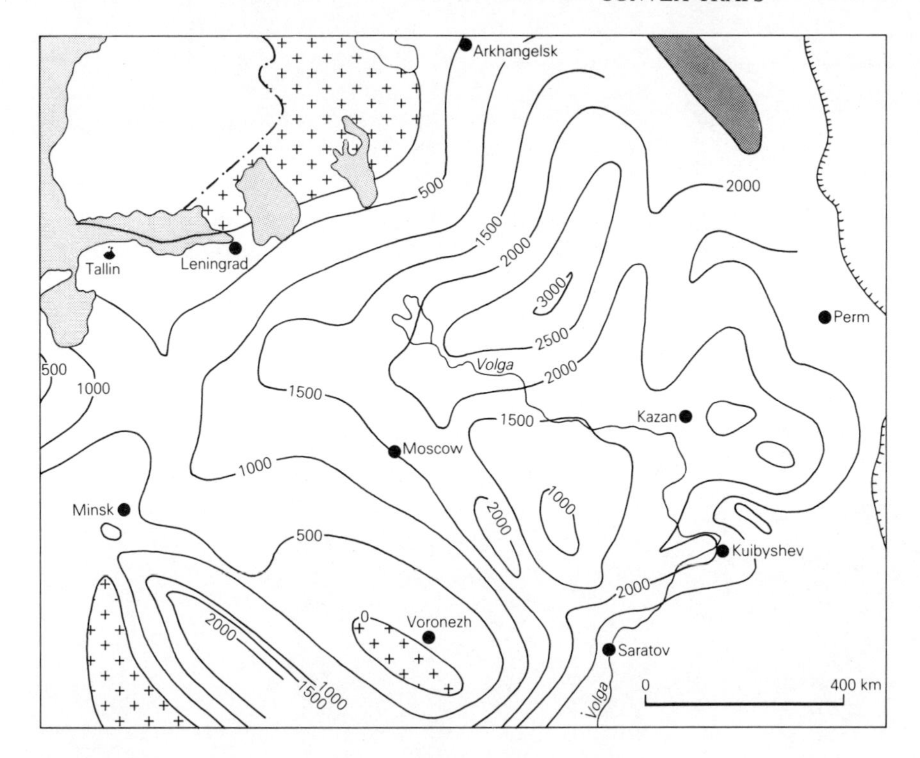

Figure 16.42 Structure contours at 500 m intervals on the Precambrian basement of the Russian platform. Patterned areas are Precambrian outcrops; hachured line on the right is the west front of the Ural Mountains uplift. Eastern third of the map area is the Ural–Volga Basin. Compare basement surface with that in the otherwise comparable Western Canadian Basin (Fig. 16.41). (After D. V. Nalivkin, *Geology of the USSR* (1962); Edinburgh: Oliver & Boyd, 1973, by permission of the Copyright Agency of the USSR.)

rejuvenation over very long time spans. Arising in the earliest stages of crustal unrest – during the Caledonian or Hercynian upheavals, in most cases – they remain unobtrusively active so long as unrest continues (into the Cretaceous or Tertiary, in the cases of features in front of the North American Cordillera). Such features may localize productive trends in reservoirs and traps of considerable variety over 100 Ma or longer.

Why some Precambrian basements respond to secular deformation by warping and some by block faulting is not clearly understood, because the distribution of rock types in deeply buried basements is not known in sufficient detail. Presumably the cause is to be found in some kind of viscosity contrast. Basement rocks in some cratons behave both ways. In Algeria, for example, the two responses are illustrated by the traps in the two largest fields. The Hassi Messaoud oilfield lies over a basement swell, the Hassi R'Mel gasfield over a basement horst (Figs 16.44 & 26.17).

Some basins have basement rocks that are relatively young. An arcuate band of petroliferous basins stretches across the Northern Hemisphere from the inner Gulf Coast of the USA, through western and central Europe, southern Russia, and across China, resting upon what Russian geologists call the epi-Hercynian platform; their basements consist of metamorphosed Paleozoic rocks as well as Precambrian. Still younger basins lie on basement rocks of Mesozoic age, especially of Jura-Cretaceous (Nevadan or Subhercynian) intrusive rocks. Where these represent the synorogenic, calc-alkaline magmatism characteristic of Cordilleran-type plate boundaries, their peripheral or fore-arc basins, between the former island arcs and their trenches, are floored at least in part by oceanic basement. If this is ophiolitic or serpentine-bearing, it is itself readily foldable under strong compression.

A superb example is provided by the San Joaquin Valley Basin in California. This highly productive basin lies between the Sierra Nevada to the east, composed of Mesozoic granitic rocks, and the Coast Ranges to the west, composed largely of Franciscan ophiolitic rocks and transected by the San Andreas fault. Beneath the axis of the basin, buried below more than 8 km of young sediments, is the boundary between the two basement types. That on the west was readily folded during the Pleistocene deformation of the Coast Ranges; that on the east was not. Hence the multitude of traps on the west side of the basin are principally convex, those on the east side principally nonconvex (Fig. 16.38). The same situation occurs to a less striking degree in the Los Angeles Basin (Fig. 16.27). Other young basins in comparable plate-boundary locations elsewhere in the world have similar basement contrasts and produce oil from similar traps (Burma, Borneo, Sakhalin).

Basins lying wholly outside or oceanward of the main arc system are floored by arc-trench sediments and/or oceanic crust, more or less metamorphosed and reduced to *mélange* during the uplift they underwent to create the basins as we see them now. The small Cretaceous and Tertiary basins of coastal Ecuador and Peru are of this type; their basements are folded in essential conformity with their cover rocks. The oceanward extensions of prograding basins on divergent (trailing) margins are also likely to lie on basement of oceanic crust (in the US

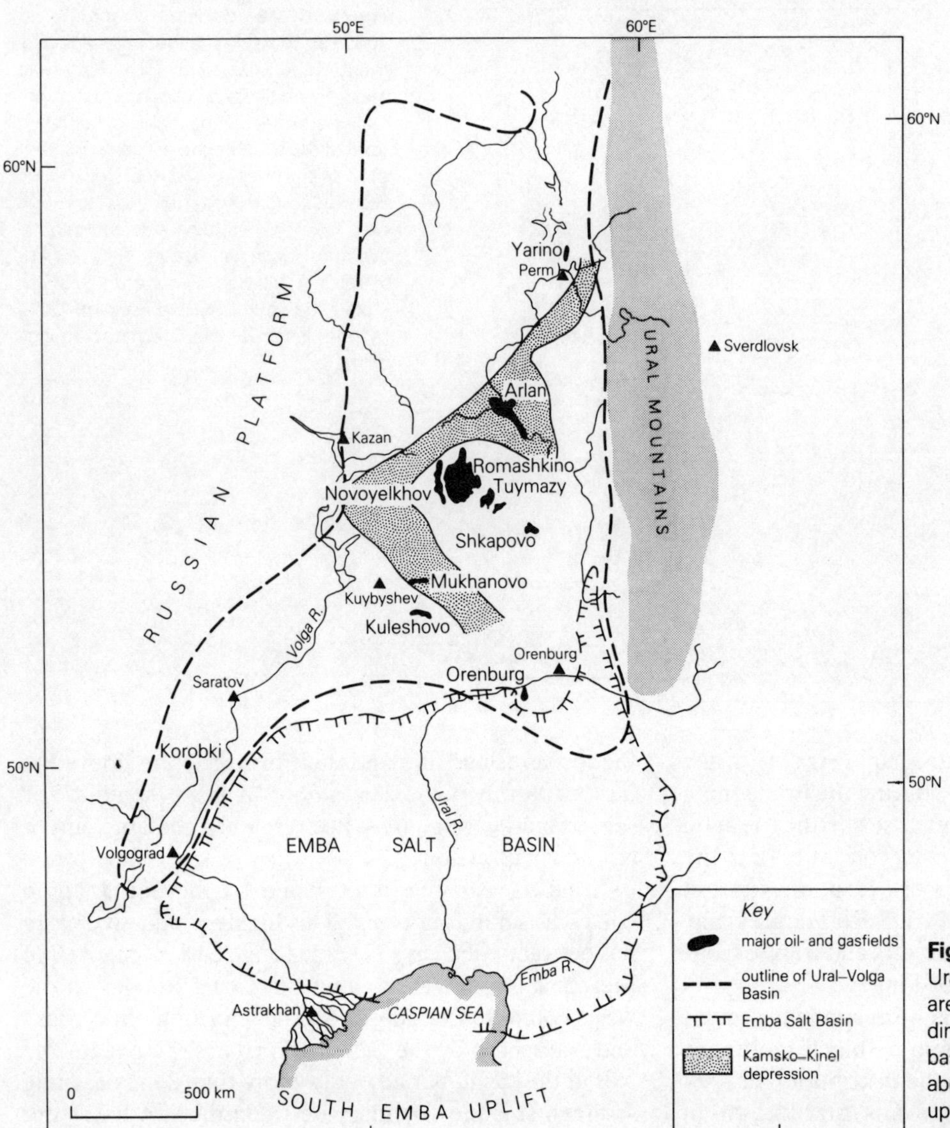

Figure 16.43 Principal oilfields of the Ural–Volga Basin, USSR. Fields are areally very large and more or less equidimensional. Traps are combinations of basement-controlled anticlines lying above broad, gently-convex basement uplifts or "domes." (After S. G. Sarkisyan *et al., AAPG Bull.*, 1973.)

Gulf Coast and Niger delta basins, for examples), but as the basement along such margins is not deformed it is apparently not directly responsible for any traps.

So, basement can be folded, and basement involvement in trap formation is not restricted to large-wavelength warping or near-vertical block faulting. Can crystalline "granitic" basement fold under foreland conditions? We know it can be folded on a huge scale in the axial zones of collision orogenies, but these are not petroliferous regions. Under nonmetamorphic, nonmobilized foreland folding, granitic basement may be involved in the folds under exceptional circumstances. The Pleistocene folding and faulting of the California Coast Ranges created such features locally, but no known petroleum traps were created as a consequence.

Along the north flank of the Arbuckle Mountains in Oklahoma, however, unmetamorphosed Paleozoic limestones are deeply infolded into Precambrian basement. The basement is involved with the cover in northwardly-directed thrusting and overfolding (Fig. 16.14). It is now known that the principal deformational style in this strip of basement is one of thrusting, and that the associated folding is of high amplitude only where the Precambrian granitic basement is overlain by thick volcanic rocks dated as Cambrian. The folding of the Precambrian granitic basement is of much lower amplitude, but it is folded nonetheless.

Thus our categories of compressionally shortened fold traps, independent of basement, and vertically uplifted fault traps, controlled from the basement, are not without intermediate members. There is also a trap category, albeit of minor significance, controlled by compressional folds and warps in the basement itself.

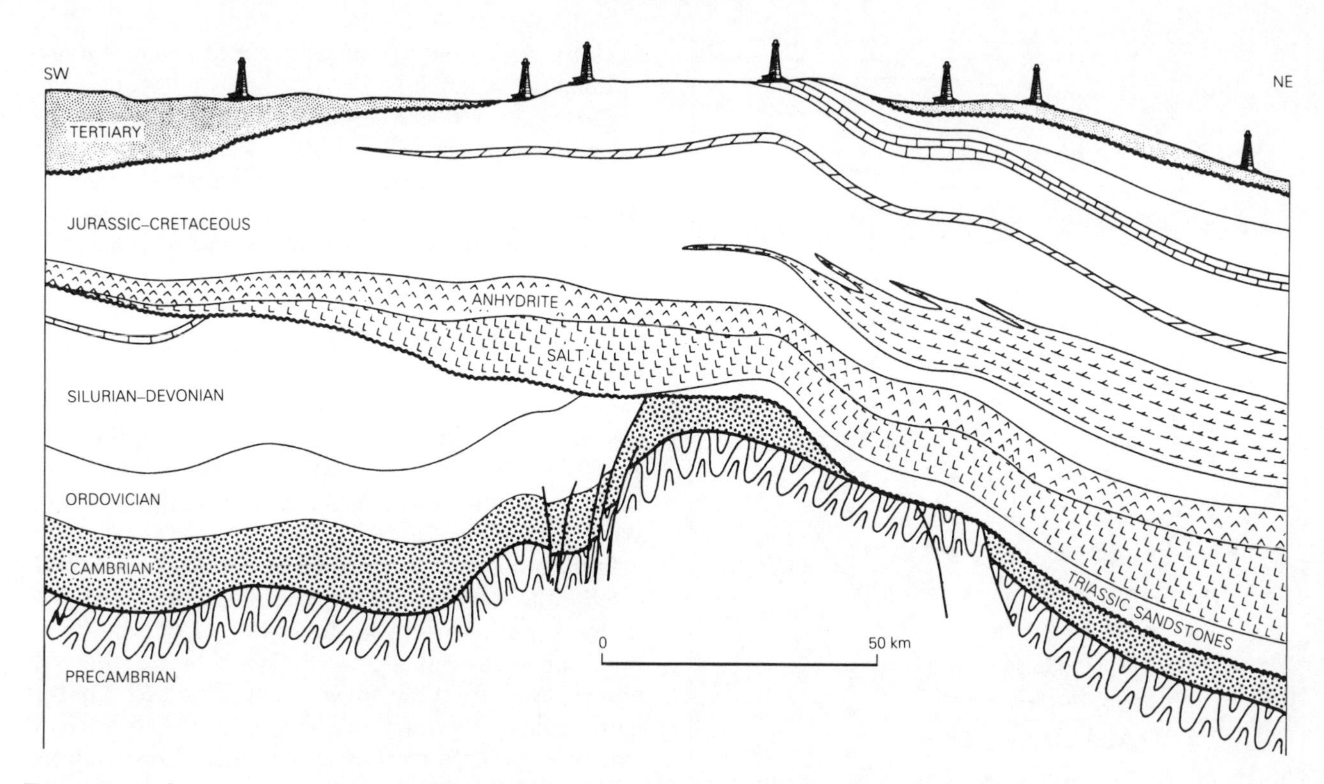

Figure 16.44 Cross section through Hassi R'Mel gasfield, Algeria. Trap formed by horst-like uplift of Precambrian basement; reservoir rocks are Lower Triassic sandstones; seal is Triassic salt and anhydrite. (After P. R. Magloire, *AAPG Memoir* 14, 1970.)

16.3.5 Bending fold traps with mobile basement

The bending fold traps considered so far involve faulted uplifts of brittle basement and stratified rocks of the lower cover, followed upwards by ductile drape folding of younger cover rocks over the uplifts. In many regions, the *effective* basement is itself ductile and not brittle.

In extensional regimes, created during crustal separation, a common early sediment is *salt*, rapidly deposited in deepening water. A common stratigraphic position for thick salt deposits is therefore close to the base of a potentially petroliferous succession. Movement of basement blocks below, whether brought on by extension or compression, easily mobilizes the salt and initiates its concentration into pillows, swells, walls, and eventually domes.

Deep-seated salt structures are especially important trapping mechanisms where much of the post-salt succession consists of competent carbonate rocks. In this circumstance the salt is unlikely to become diapiric; it is unable to pierce towards the surface as a cylindrical plug or dome. A great thickness of strata is simply arched upwards from underneath, forming broad domes with quaquaversal dips, the most favored of all traps (Fig. 16.45). Because the uplifted strata are under extension, the carbonate members at least are likely to become intensely fractured, greatly enhancing their capacities as reservoir rocks (Sec. 13.4).

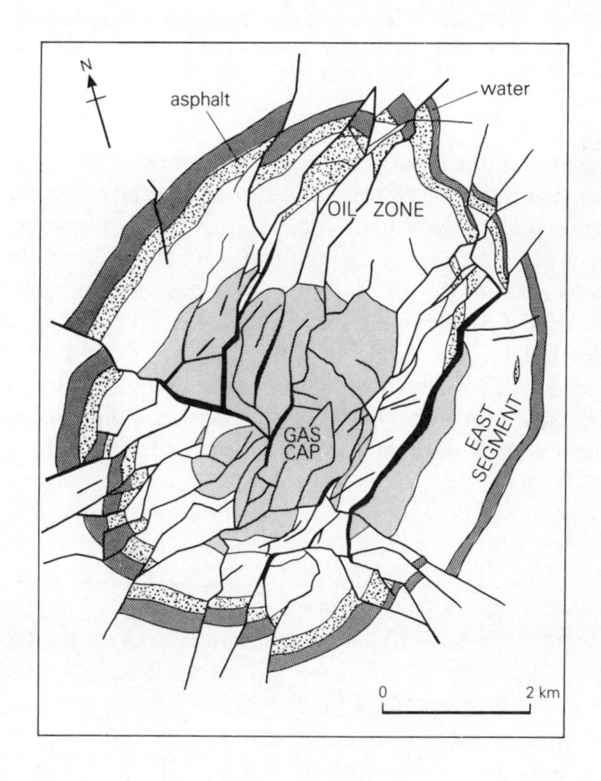

Figure 16.45 Hawkins oil- and gasfield, Texas Gulf Coast Basin. Symmetrical dome over deep salt uplift, with abundant radial and longitudinal normal faults which have only minor influence on the trap. (From R. L. King and W. J. Lee, *J. Petrolm Technol.*, 1976.)

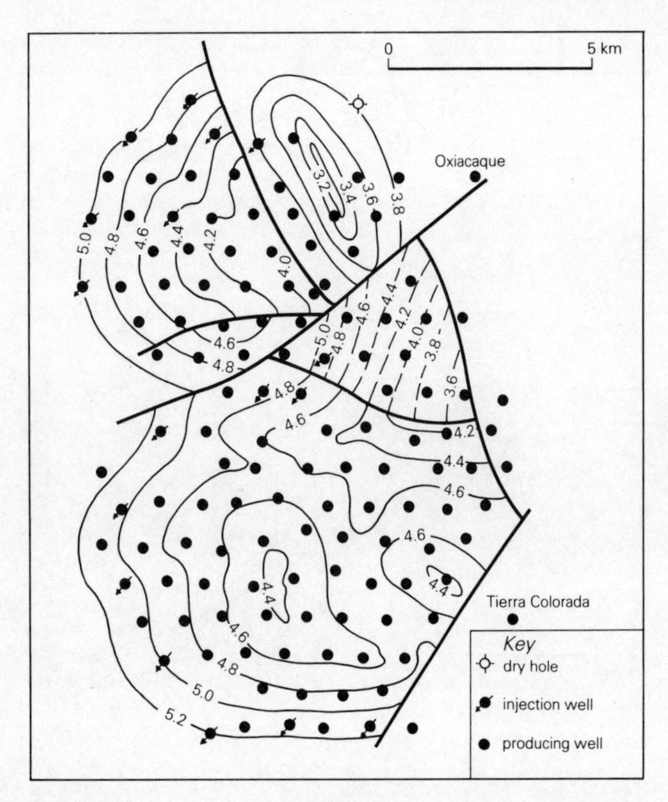

Figure 16.46 A. J. Bermudez field, Reforma region, southeastern Mexico. Structure contours on productive Cretaceous carbonate formation, in roughly equidimensional dome over an intrusion of Jurassic salt. Reservoir is much fractured and faults are undoubtedly more numerous than shown. Contours in kilometers below sea level. (From F. Viniegra-O, *J. Petrolm Geol.*, 1981; after *Oil Gas J.*, 5 June 1978.)

Some of the most productive oil regions of the world are characterized by traps due either to basement horst-block uplift, halokinesis (deep-seated salt movement), or a combination of the two. In the salt basin of the Persian Gulf, the great domes of Burgan, Bahrain, and Dukhan form obvious surface features. In the central North Sea, the fields in Paleocene strata (Ekofisk, Forties) lie over uplifts of Permian salt (Fig. 13.74). The principal fields in the Reforma area of southeastern Mexico, including the supergiant Bermudez field, lie in fractured Mesozoic carbonates over salt-caused horsts or domes (Fig. 16.46).

These salt-motivated uplifts conform strictly to their categorization; they lift up the overlying strata from below, without piercing them. Salt is unable to pierce a post-salt section of competent carbonate beds, such as occupy most cratonic basins (Michigan, Williston, and Aquitaine basins, for examples; see Ch. 17).

Where the post-salt section is of sand–shale facies, mobilized salt easily pierces upwards through great thicknesses of strata; it becomes *diapiric*. Diapirism, or piercement of strata from below, is especially characteristic of subsiding half-basins at divergent margins, such as the Gulf Coast basins of Middle America, the Emba Basin north of the Caspian Sea, and the basins of coastal West Africa (Fig. 16.47).

Initial movement of the salt is essentially lateral, instigated either by movements in rocks of the underlying floor or by gravity-induced basinward movement of the overlying sediments. A wave-like motion of the salt ensues, away from the active agent. This lateral movement quickly becomes restricted as the initial salt structures become more deeply buried. Thus the early structures begin to rise vertically because of deficiency of mass; they rupture the overlying strata and are forcefully injected through them. Structures farther basinward find piercement more and more obstructed; their areal sizes increase but their amplitudes decrease and large swells, walls, or irregular salt massifs are formed instead of piercement domes.

Four important structural features are observed in association with virtually every diapiric salt plug. The plug is surrounded by a *rim syncline*, representing the zone from which salt was withdrawn to form the plug. If the zone of early withdrawal was filled with sediments, those sediments are likely to dip inwards towards the salt pillar, forming a *turtle structure*. The pierced sediments higher in the section are turned upwards and brought into trap position where they abut against the plug. And *keystone normal faulting* occurs directly over the top of the plug. It may become highly complex, but a valuable

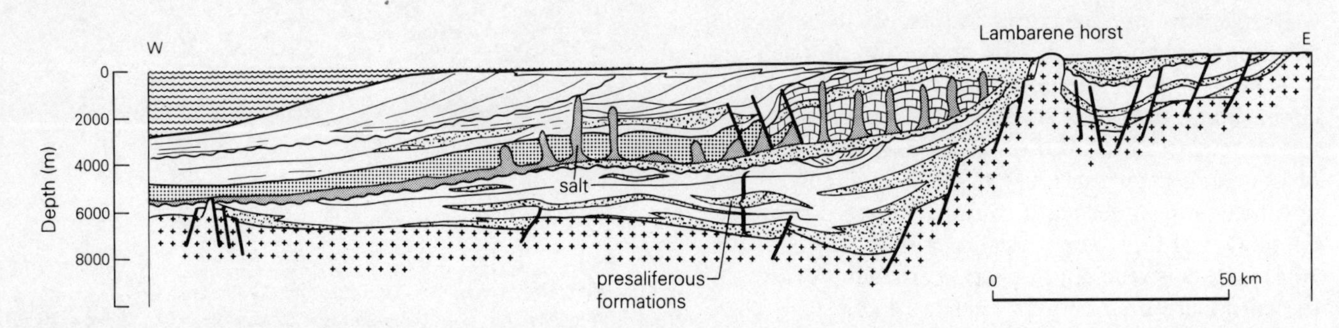

Figure 16.47 Structure of the Gabon Basin, West Africa, a coastal pull-apart basin invaded by salt domes. (From J. P. Cassan *et al.*, *Bull. Centres Rech. Explor. Prod. Elf-Aquitaine,* 1981.)

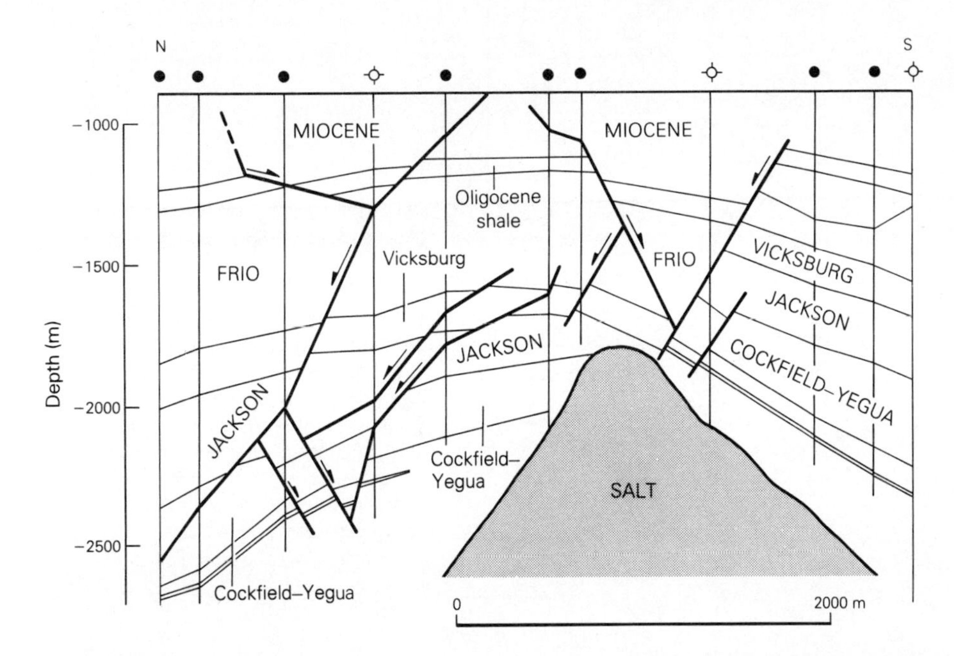

Figure 16.48 Cross section through a multi-reservoir salt dome oilfield (Esperson dome, Texas), showing fragmentation by normal faults of sediments overlying salt core. Reservoirs in Miocene, Frio, and Vicksburg are in traps formed by normal faulting and upward flexure; those in Yegua and Crockett sands (latter is lowest formation shown) are in stratigraphic pinchouts and truncations. (After J. T. Riddell, Houston Geological Society, 1962.)

generalization is that the youngest faults form a true graben directly over the crest (Fig. 16.48).

The irregular and by no means unidirectional growth of a salt dome creates numerous successive and varied traps for oil and gas. Chief among them are the following (Figs 16.49 & 50):

(a) in the closed structure over the dome, associated with sulfur formed by the bacterial reduction of sulfate accompanying the salt. The largest and most prolific traps are in many cases not pierced at reservoir level.

(b) in traps formed by extensional faults over the highest part of the dome.

(c) in the caprock of the dome, commonly gypsiferous or anhydrite-bearing limestone.

(d) in upraised flank sediments which buttress against the salt mass, or in peripheral fault blocks around it; these may be in nearly vertical or even overturned attitudes beneath a salt overhang at the top of the dome (Fig. 16.51).

(e) in stratigraphic pinchouts on the basinward side of the dome, commonly at stratigraphic disconformities.

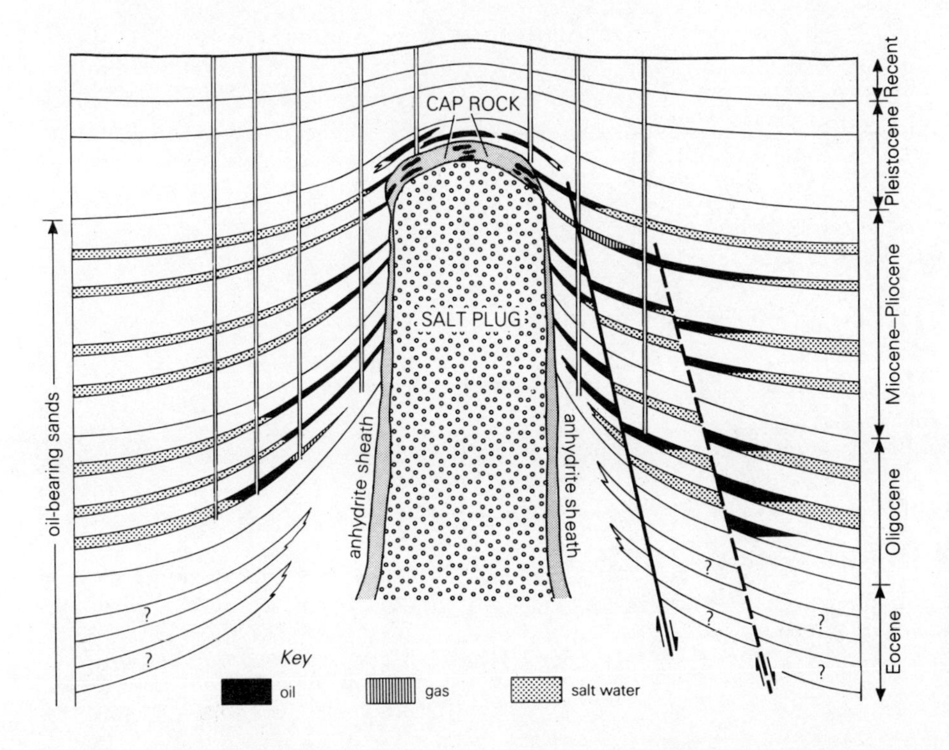

Figure 16.49 Schematic illustration of salt dome traps: in supercap, caprock, and flanking sands (abutting, pinchout, and fault sealed). (From K. K. Landes, *Petroleum geology of the United States*, New York: Wiley–Interscience, 1970; after J. A. Clark and M. T. Halbouty, 1952.)

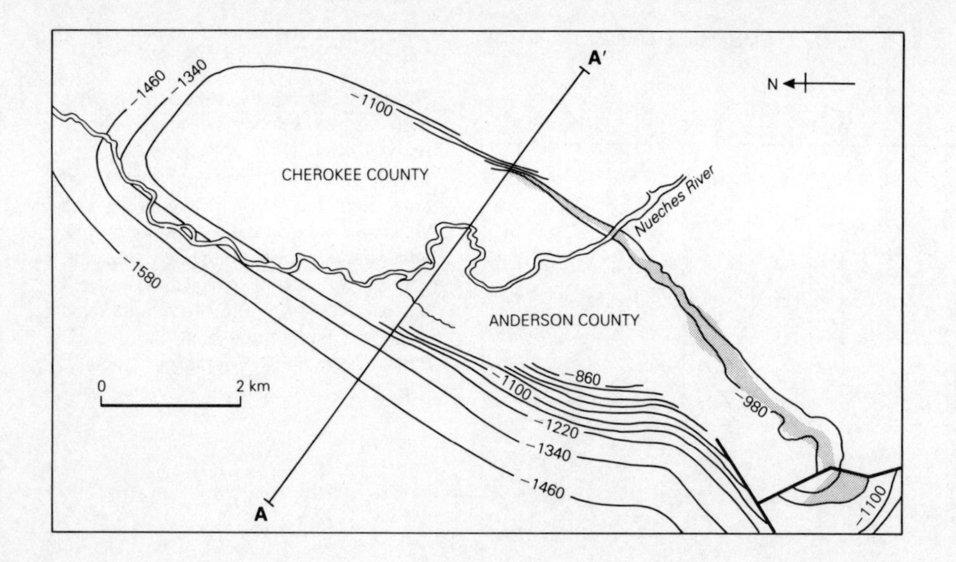

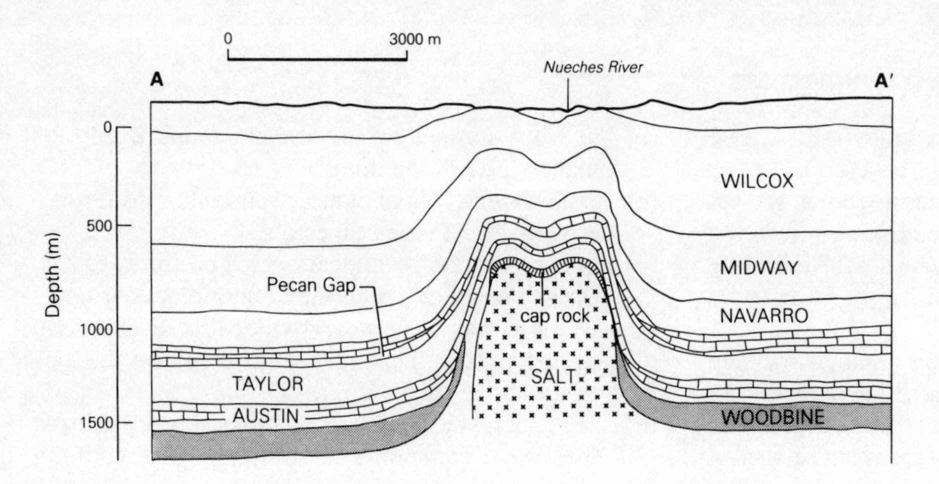

Figure 16.50 Map and cross section of the Boggy Creek oilfield, northeastern Texas, showing a nonconvex trap formed by truncation of the reservoir rock by a piercement salt dome. Structure contours at 120 m intervals on top of the Woodbine (Cretaceous) sandstone reservoir rock. Blank area in center has no Woodbine because of the piercement; the structure is "bald-headed." The productive area (stippled pattern) is restricted to the southeast flank of the intrusion, on the right-hand side of the cross section. (After R. W. Eaton, University of Texas Publication 5116, 1951.)

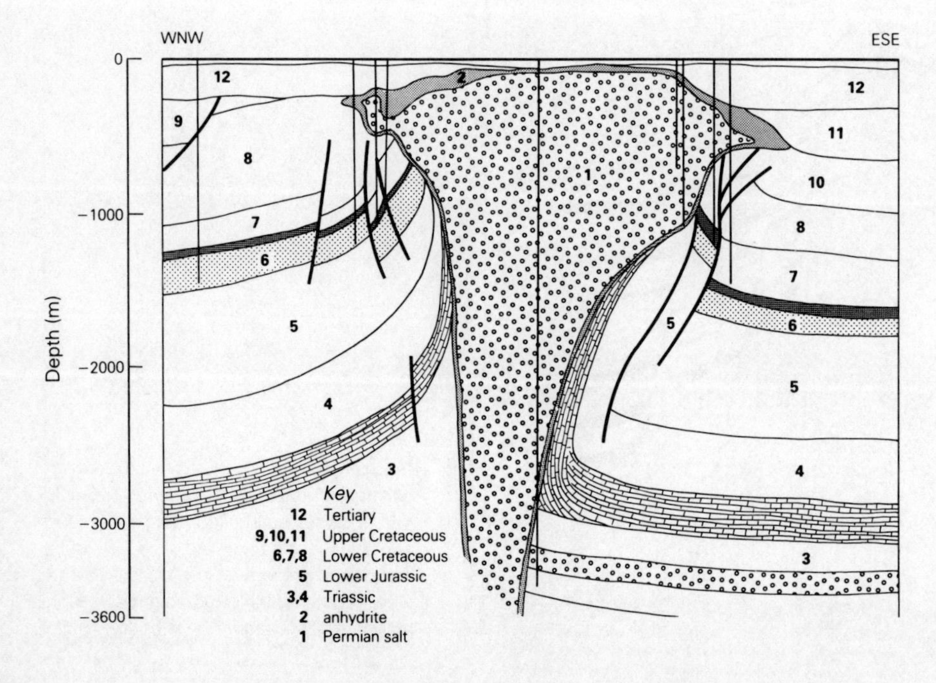

Figure 16.51 Cross section through traps below flank overhangs of the Wienhausen–Eicklingen salt dome, Hanover basin, northern Germany. (From F. Lotze, *Steinsalz und Kalisalze*, Berlin: Borntraeger, 1957; after A. Bentz, 1949.)

(f) some drape and turtle structures over residual highs (such as salt pillows left behind during the mobilization of the salt) may also be productive.

Traps in locations (d) and (e) are very likely to be non-convex traps.

Some piercement salt domes are actually spines or cupolas rising from huge, complex salt massifs that represent coalesced pillows, walls, or ridges. Such a structure has been called a *flowage anticline*. Clearly it provides multiple trapping possibilities well removed from the piercing spines themselves. The largest such complex known in the Louisiana Gulf Coast is about 50 km long and 18 km wide; its base is at a depth of at least 7 km, possibly 10 km, and its crest at only 600 m. Below about 1500 m, the mass represents intermediate piercement; above 1200 m it represents shallow or true piercement, culminating in the spines of the Bay Marchand, Timbalier Bay, and Caillou Island domes (Figs 16.6 & 52). The complex contained initial recoverable reserves of more than 500 x 10^6 m^3 of oil in more than 100 now-separate sand reservoirs.

16.3.6 Bucklefold traps with mobile basement

Salt *domes*, like those in the Gulf Coast just described, and deep salt *uplifts*, like those in the Persian Gulf and Mexico described under "Bending Folds," are fundamentally gravitational features and can be called *halokinetic*. They are independent of regional compression. Where regional compression has prevailed, however, buckle folding is imposed on layered rocks which may well include salt.

Thick salt beds, with their admixture of gypsum and clay, react plastically to moderate but continuous stress. When tangentially folded, the salt accumulates in the anticlinal axes and is withdrawn from the synclines; the flanks of the anticlines become oversteepcned and the salt is eventually intruded into the overlying sediments and out on to the surface, through the crests of the folds. *Salt diapir anticlines* may be called *halotectonic* structures; they represent the extreme case of disharmonic folding.

The cores of salt anticlines are very variable in shape: nearly round, or narrow, sinuous, arcuate, even bifurcate. They may be in very confused relationship with other strata, and chaotic structures are common. Whereas gravitational salt domes are features of extensional basins or of passive, divergent margins, salt anticlines occur along the foreland edges of compressional zones as in the Andes and Antilles. They are less commonly the cause of oil or gas traps than are ordinary salt domes, even where they adjoin generally petroliferous areas. The most notable region of productive salt anticlines is the Carpathian foredeep of Romania, where diapirs of Miocene salt pierce the anticlines from beneath the oil-bearing Pliocene sands and clays (Fig. 16.53).

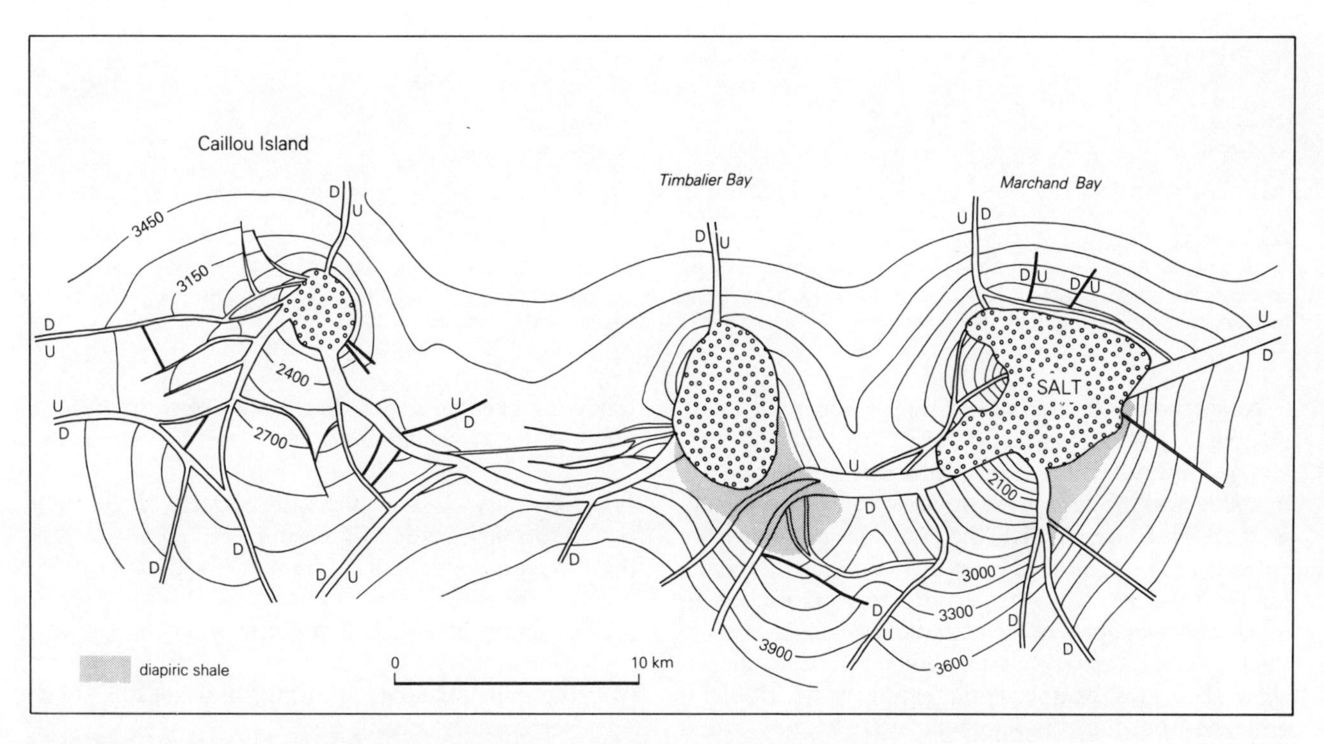

Figure 16.52 Bay Marchand and associated salt dome traps, coastal Louisiana. Structure contours, in meters, on Upper Miocene productive sandstone. Note complex fault pattern radiating from each salt culmination. (After M. G. Frey and W. H. Grimes, *AAPG Memoir* 14, 1970.)

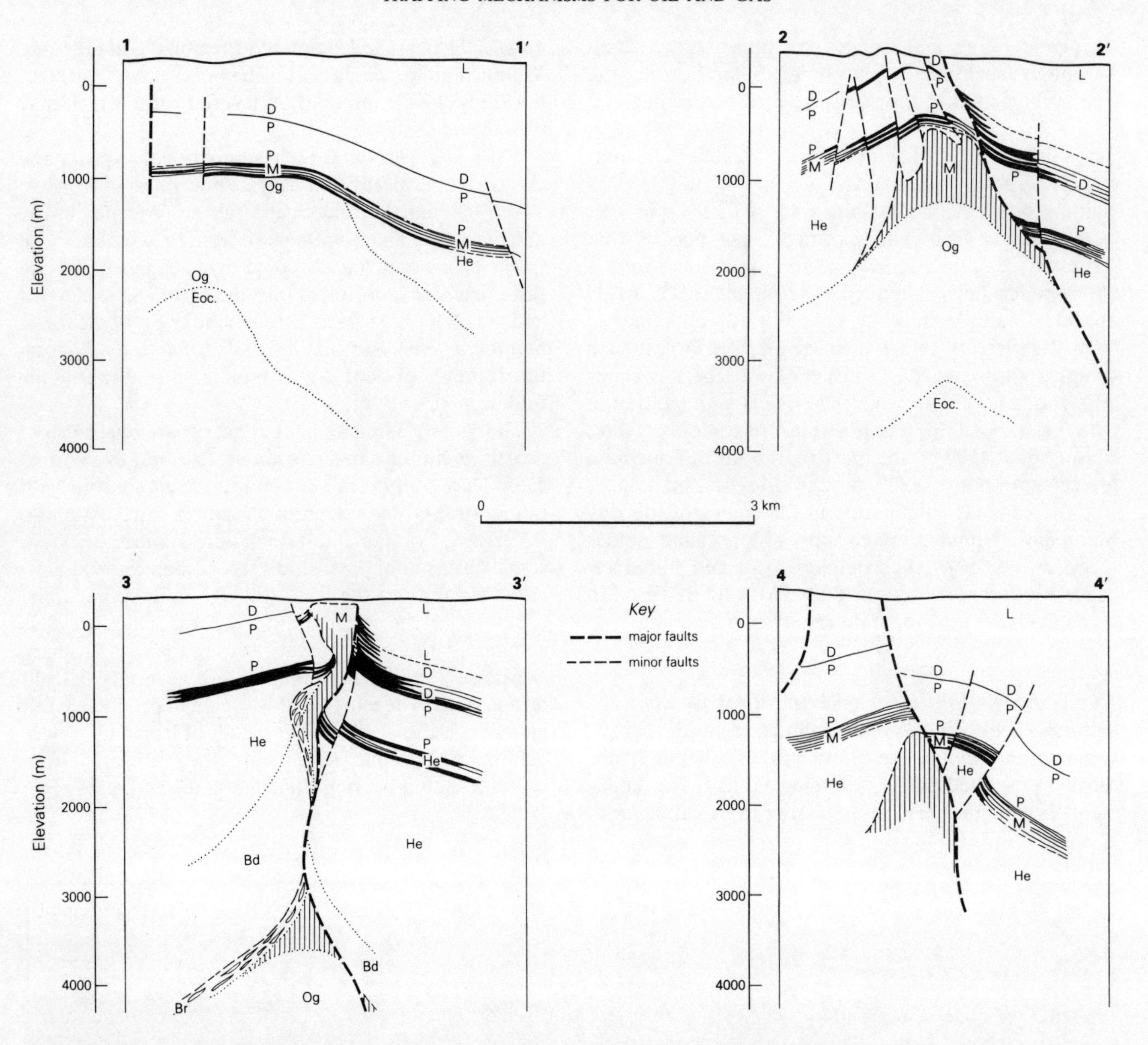

Figure 16.53 Cross sections through the Moreni–Gura Ocnitei oilfield, the largest field in the Romanian Carpathian foredeep. Productive sandstones are Pliocene in age. Note constriction and complication of fundamentally simple anticline where the Lower Miocene salt (shown in vertical ruling) pierces the crest. Note also concentration of accumulation on south (right) side of structure, which is also the downthrown side of the principal fault. Vertical scale in meters. (From D. Paraschiv and G. Olteanu, *AAPG Memoir* 14, 1970.)

Analogous to salt anticlines, and in some places accompanying them, are *clay diapirs*. Soft, water- or gas-charged muds become thixotropic under stress and are intruded into the cores of sharp anticlines exactly as salt is. If the gas content is high, the intrusions may reach the surface and form *mud volcanoes*. Active mud diapirism implies that the area is still undergoing compaction, so all notable examples of the phenomenon are in areas of thick, weak, Tertiary clastic sediments. Clay diapirs share their most characteristic features with salt diapir anticlines, but they display them better because clay is not soluble, as salt is, so it retains its layering better, and clays may carry fossils. Many clay diapirs display the following characteristics that mark them off from all other structural phenomena:

(a) very steep dips in the central parts of the folds, becoming gentler all around;
(b) weak but invariable asymmetry, with limbs unlike one another;
(c) common absence of any governing direction of overturning;
(d) internal crushing, shearing, and slicken siding, very readily seen in clays;
(e) piercement of the diapirs by clay dykes;
(f) an extraordinary assortment of microfossils,

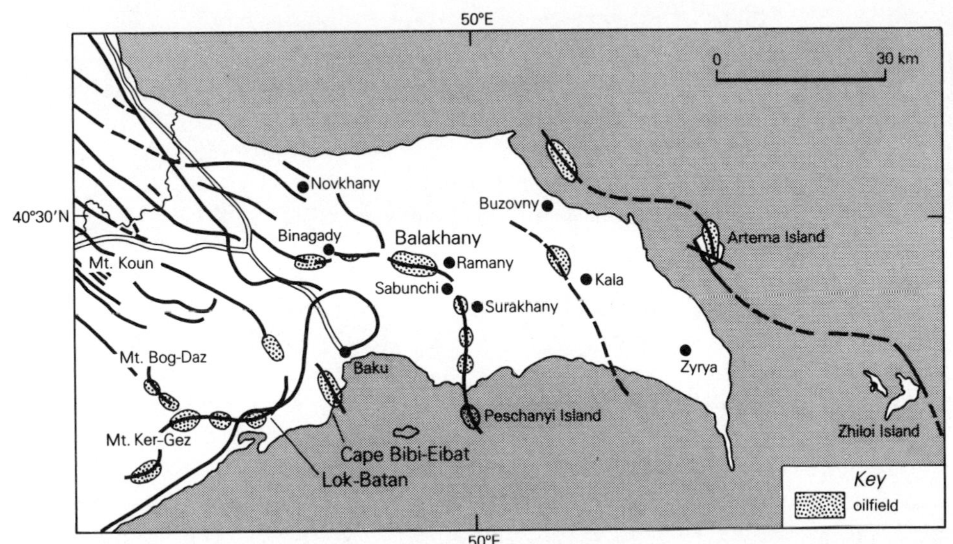

Figure 16.54 The oil-productive Apsheron Peninsula west of the Caspian Sea, centered on Baku. Note the curvilinear tectonic trends, culminating in clay-cored anticlines with mud volcanoes. The largest anticlines, Balakhany and Bibi-Eibat, have produced oil since the 19th century. (From D. V. Nalivkin, *Geology of the USSR* (1962); Edinburgh: Oliver & Boyd, 1973, by permission of the Copyright Agency of the USSR.)

representing all ages from that of the base of the clay (and even earlier, if pre-clay strata are involved in the same structures) to the present.

Clay diapirs are especially developed at structural points of weakness or of deflection, such as bends in anticlinal axes or crossings of such axes by faults.

The most famous clay diapirs and mud volcanoes are those of the Baku region, where the eastern end of the Caucasus Mountains plunges beneath the Caspian Sea (Fig. 16.54). The burning gas provided the impetus for the Zoroastrian religion; the crushed clays from which the gas comes are probably also the principal source of the oil, which has been produced in the Baku region for more than 100 years. Not much less famous are the mud diapirs and mud volcanoes of southern Trinidad, called *mornes* or *bouffes* (Fig. 16.55); the clay is of Paleogene age. Though mud intrusions penetrate many structures in the productive area, they themselves do not form important trapping mechanisms in Trinidad. Few Tertiary orogenic basins, productive or unproductive, are completely without clay diapirs.

Shale structures are to clay diapirs what salt pillows are to salt anticlines. They are nondiapiric, residual shale masses, best developed in deltas. Rapid deposition inhibits the expulsion of pore water; the muds become undercompacted and overpressured, and consequently buoyant and mobile. They then tend to creep down the delta toe or the continental slope, but they neither creep as far laterally nor rise as high vertically as do salt masses. Rim synclines are unlikely to be developed, but shale lumps accumulate below the delta toe and help to

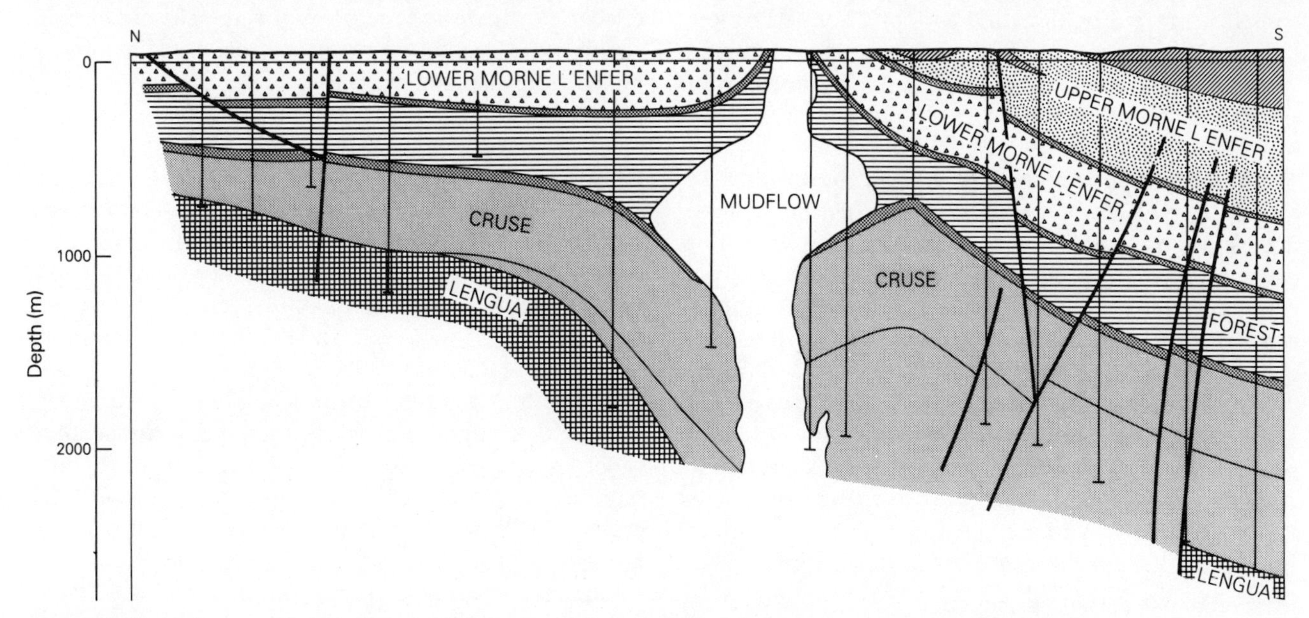

Figure 16.55 North–south cross section through Forest Reserve field, Trinidad. (From T. H. Bower, 4th Caribbean Geology Conference (1965), 1968).

initiate detached normal faults (see Sec. 16.7.3). Shale structures of this sort are seldom important in trapping oil or gas, but they are common in most oil- or gas-bearing deltas of Cretaceous or Tertiary age (in southern Texas, the Mackenzie delta area of Arctic Canada, and the Niger and Nile deltas).

16.3.7 Convex traps formed by immobile features on the surface of deposition

Horsts and rising salt domes, the subjects of preceding sections, are active trap-forming agents. They are subject to rejuvenation, and so to repeated growth. Many convex traps, in contrast, are formed by features that are fixed at or below the surface of deposition. They may be residual features, left over from some previous cycle of crustal development and antedating the cycle that creates the reservoir and the trap; or they may be deposited during that second cycle and then remain as immobile convex structures after it is completed. In either event, they form excellent traps and are very widespread.

Residual, pre-depositional features are parts of the landscape which is transgressed by a new depositional seaway. The features may be of structural origin, like rock ridges controlled by faults, or they may be simply erosional remnants, called *buried hills*. They contribute several trap- and reservoir-creating phenomena. They shed aprons of clastic sediment around their margins. If the residual is made of feldspathic igneous rock, the clastic detritus will be arkosic and be called *granite wash*; it may be very porous and permeable (see Sec. 13.2.7). The residuals also localize porous sediments derived from elsewhere, by standing high above the sea floor and so being affected by more agitated and aerated waters than the low areas around them. These more agitated waters "screen" the sands over the highs, winnowing out the finer material and rendering the sands more effective as reservoirs.

Any porous sediments deposited in the lower area around the high will wedge out towards the high or against it, and so will be in trap position from the outset (in nonconvex trap position, but made possible by a convex structure). The screened sands over the top of the structure will eventually be in trap position also, because the residual high will be incompactible (otherwise why would it be a residual high?) and the overlying and surrounding sediments will become *draped* over it by differential compaction. The draping will form a true anticline, though no folding will have taken place. As the relief of the structure is reduced by deposition around it, there will be an interval during which sediments are deposited high up the flanks of the now subdued high but not across its crest; or, if they are deposited across its crest, a very small change of sea level will provoke their removal by sea-floor scouring. Thus the high will be "bald-headed" to those particular strata, and several episodes of bald-headedness may alternate with episodes of unconformable overlap, forming stratigraphic traps.

Finally, the convex residual structure itself provides a trap if it contains an adequate reservoir rock. As most such residual structures are necessarily composed of hard rock which has survived erosion, no normal reservoir rocks are likely to be present in them. Fracture porosity, however, is actually facilitated, both by the convex shape of the structure and by the prolonged weathering to which it may have been subjected. Important production has been derived from depths of 500 m into such porous basement rock in the Maracaibo Basin of Venezuela, and in California, Kansas, Libya, and elsewhere (see Sec. 13.4.9).

Convex depositional features create two types of trap. Those consisting of sandstone may arise in almost any environment in which discrete sand bodies accumulate – dunes, offshore barrier bars, deep-sea fan lobes (see Sec. 13.2.7) – but the sands are compactible. They therefore seldom provide distinctly convex traps unless they are affected by later folding. Only in this event are they likely to remain convex enough to be detectable seismically. They are more conveniently regarded as nonconvex traps (Sec. 16.4).

Organic buildups, on the other hand, may be much less compactible than the sediments around and above them. Carbonate reefs form important traps that are conventionally regarded as stratigraphic, which in the genetic sense of course they are. But they form their own primary structures, which are upwardly convex and may be very pronouncedly so. As far as the oil and gas are concerned, therefore, reefs offer static closure and are simply stratigraphic versions of the anticline. Reefs make their own convex traps.

Organic reefs Organic reefs possess all the ancillary trapping mechanisms of buried hills: flanking clastic sediments of high porosity, facies changes over their crests, and drape anticlines in post-reef strata. They add two other vital characteristics of their own: they are themselves excellent reservoirs and they are commonly very well sealed.

The term "reef" has come to embrace a variety of structural-stratigraphic rock assemblages. Organic reefs range in size from gigantic barrier-forming structures like the one east of Australia, through large atolls, discrete bioherms, to tiny isolated bodies only a few hundred square meters in area. Most reefs, large or small, are carbonate constructions of colonies of algae, corals, molluscs, and some groups of now extinct organisms such as stromatoporoids, with symbiotic crinoids,

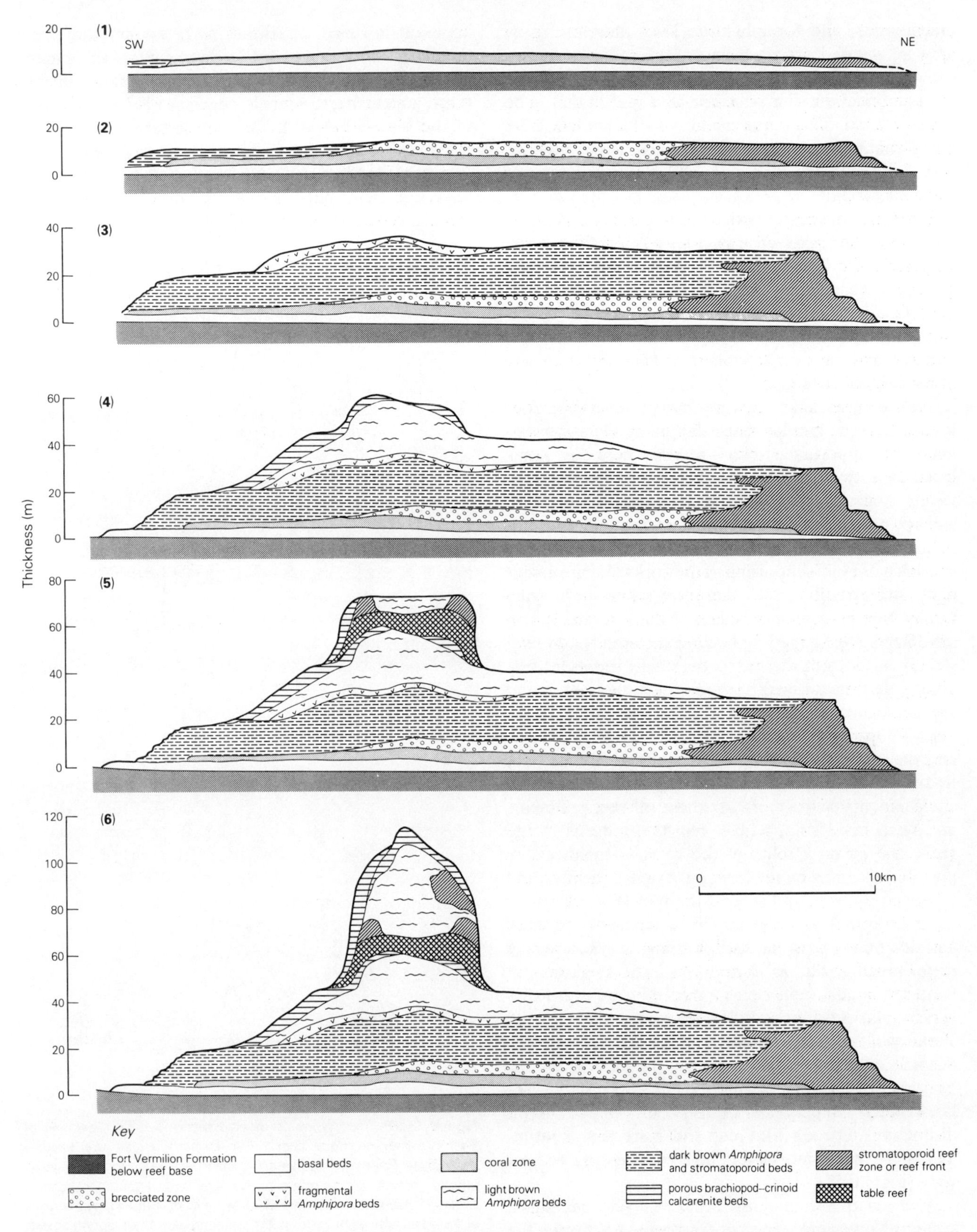

Figure 16.56 Six stages in the development of a Devonian reef in the Swan Hills field, northern Alberta, Canada: 1, gradual transgression; 2, rising sea level (reef only in the northeast); 3, eastward tilting, drowning northeast reef and exposing southwest side; 4, rapid *Amphipora* growth in central lagoon-bank; 5, circular table reef of stromatoporoids during slow subsidence; 6, return to supratidal conditions, followed by sudden submergence and end of reef growth. (From C. R. Hemphill *et al., AAPG Memoir* 14, 1970.)

brachiopods, and foraminifera. They therefore grow only in agitated waters facing nutrient- and oxygen-carrying waves; for many geologists a requirement for the identification of a structure as a reef is that it be *wave-resistant*. This requirement would eliminate from the definition many organically constructed buildups and all the so-called *reef knolls*, which may be largely of inorganic origin.

Reefs are inherently both reservoir rocks and traps. They owe their *reservoir* capacities to their lithologies, as described in Chapter 13. With high primary, skeletal porosity, enhanced by the solution of aragonitic fossils, reefs readily acquire a unique internal texture: reticulating, breccia-like networks of permeable channelways, fissures, and vugs, independent of bedding or of any other original structure.

Such textures make reefs into natural reservoirs even if they have the tabular shape that many reefs approximate. The importance of reefs as *traps*, however, stems from their twofold tendency to acquire *natural convexity*. Colonial marine organisms grow both outward and upward, in order to maintain an essentially constant relation to water depth (Fig. 16.56). The edifices they construct are self-supporting frameworks that are essentially incompactible; they therefore retain both structurally high positions and much of their original porosity. Reefs are flanked by bedded calcarenites dipping steeply away from the reef cores; these represent carbonate detritus derived from the reefs themselves and are analogous to the granite wash and other clastic deposits around buried hills (Sec. 13.3.3). Muds, marls, and the most decomposable parts of organic material find rest only in quiet waters removed from the aerated environments of the reefs. As these off-reef sediments are much more compactible than the material of the reefs, the primary relief of the reefs is enhanced by post-burial compaction, forming drape anticlines and monoclines above and around the reef (Fig. 16.7).

As important to a reef trap as its inherent convexity and its porosity is its *seal*. During a transgressive depositional cycle, as during the Late Devonian of western Canada, reefs become buried by muds or marls which make excellent seals. In regressive cycles, as in the Tertiary of the Persian Gulf Basin, barrier reefs or atolls become overlapped by evaporites or marls from the back-reef or lagoonal regime (Sec. 13.3.3). Small, isolated reefs, called *pinnacle reefs*, involve no useful distinction between fore-reef and back-reef environments and may have either shale or evaporite as roof rock (Fig. 16.57).

Reefs in the wide sense occur in every geological system; they were especially widespread during the Devonian, Permian, and Cretaceous periods and in the later Tertiary and Recent epochs. Despite this extensive distribution in both time and space, and despite their apparent intrinsic superiority as reservoirs and traps, reefs are not as important in these capacities as many geologists imagine them to be. A mental fixation on reef traps, amounting to a fetish, began in 1947–8 as a result of the discoveries at Leduc, in western Canada, and Kelly–Snyder in the Permian Basin of West Texas. Western Canadian reefs fulfilled the expectations entertained of them; but the only first-order reef-trap province discovered before 1947 was that in the Tampico

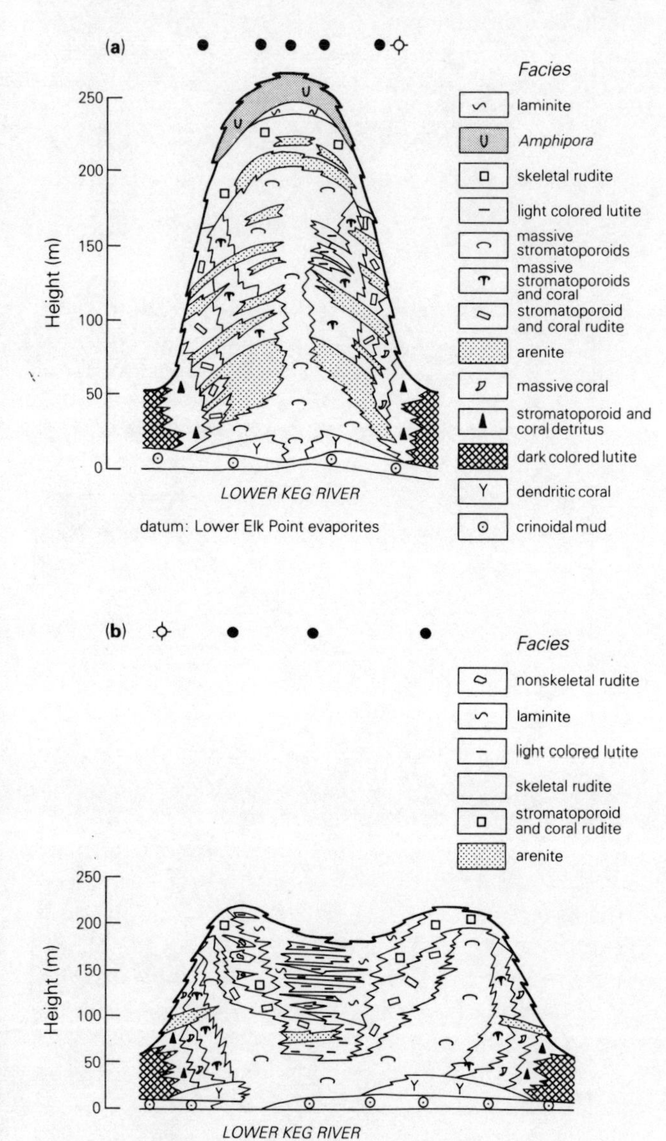

Figure 16.57 Two types of organic reef reservoir, Devonian of Rainbow area, northern Alberta, Canada. (a) Pinnacle reef, shallow bank phase. Reef grows upward from narrow base (2 km wide or less), around core of stromatoporoids. (b) Atoll reef, lagoon-bank phase. Broader base from beginning (3 km wide or more); organisms grew outward as well as upward, with formation of interior lagoon of relatively quiet water. Final reef is somewhat less elevated than pinnacle reef, but occupies a larger area; central and upper parts are composed of lagoonal detritus, not of organic framework. (From D. L. Barss *et al.*, *AAPG Memoir* 14, 1970.)

region of Mexico, and the only one discovered since has been in the Sirte Basin of Libya. These include some large oil-fields, but the largest fields are not in reef traps in any known basin except the Permian Basin, and in that basin Kelly–Snyder is the only giant field to be in such a trap.

There are a number of reasons for this apparently anomalous situation. Though reef growth is confined to low latitudes because of requirements of water temperature and sunlight, it is also essentially confined to the eastern sides of continents, which receive warm currents from the westward equatorial drift. The western side of any continent (in either hemisphere) receives a cold current from polar waters, which is inimical to reef growth; there are no reefs off the western side of the Americas, or of Africa. On the other hand, organic-rich, phosphatic, source-type sediments are best developed either in restricted interior seas, not normally reef-prone, or preferentially on the western sides of continents, because that is where the upwelling of equator-seeking cold waters takes place. Thus no significant oil has been found in reef limestones in Florida or the Bahamas, the largest area of Cenozoic–Recent carbonates and reefs on Earth.

During the most prolific and widespread development of rich source sediments, from Oxfordian time to the Paleocene, reef growth spread throughout the Tethyan region between Mexico and Indonesia. It was especially well developed during the Cretaceous reign of the rudists, in the so-called Urgonian facies. Rudistid reefs were coeval with the widest development of oxygen-deficient ocean waters in the Phanerozoic record, in which the anoxic "Horizon B" was deposited (see Secs 6.5 & 8.3). Yet there are surprisingly few large oil- or gasfields in reef traps of these ages; only in Mexico and Libya are there very large ones. The nearest approach to this source sediment ubiquity during Paleozoic time occurred in the late Devonian, especially in the Frasnian. This was contemporaneous with vast reef developments in western Canada, the Spanish Sahara, northern Europe from the Ardennes to the north Caspian Basin, European Russia, and the Canning Basin of Australia. Only in Alberta are these reefs known to be highly productive reservoir rocks.

The Ural–Volga Basin (Fig. 16.43) is one of the half-dozen richest oil basins known. The structures forming the traps in its largest fields appear to have developed over Upper Devonian atoll reefs, but the oil is in deltaic sandstone reservoirs and not in the reefs. The basin and the adjacent Ural Mountains also provide the type locality of the Permian System, with world-famous reef and evaporite development; the Permian contains less than 3 percent of the oil of the basin. The Middle East, the Gulf of Mexico, and the Caspian region are also among the most oil-rich regions of the world. The first was essentially surrounded by reefs during the Tertiary, the second during the Cretaceous, and the third during the Carboniferous and Permian. Relatively little of the oil of those three regions has been found in reef traps.

Post-Eocene reefs are very widespread; so are post-Eocene source sediments and oilfields. Yet nearly all these fields, except in Iran, are in sandstone reservoirs. The only sizable reef reservoirs elsewhere are in the Miocene of the southwest Pacific province; they are dwarfed by the reservoirs in Miocene sandstones (see Sec. 25.6.3).

Though a number of authors have invoked the reef-forming and reef-inhabiting organisms themselves as the providers of oil-source material, it is now widely accepted that this is not justified. Where reefs contain oil, they are merely containers; the oil has migrated into them from basinal sediments, which may overlie the reefs after prolonged transgression but are more likely to be their lateral equivalents. Reefs around isolated volcanic islands, as in the Pacific and Indian Oceans today, are out of contact with basinal sediments and therefore also with source material. Long-lived reef masses tend to "climb" through the section, either fore-stepping or back-stepping according to whether regression or transgression is taking place. They readily become escape conduits for any oil available to them.

A reef, or an entire trend of reefs, may of course originate because of some earlier convex structure, such as an anticline, a deep horst or a buried hill. Alternatively, a reef may become caught up in the formation of a later convex structure. In such a case, the reef may provide all or most of the reservoir, but it is the younger convex structure that forms the trap. The most spectacular example is that of the Kirkuk field in Iraq, where a thick, porous, barrier-reef limestone of Tertiary age was caught up in the Zagros fold belt during the Pliocene, and a linear anticline was raised that is productive over a length of almost 100 km (Fig. 16.58). The strike of the anticline is oblique to the trend of the barrier reef, so that the field provides an oblique section through the reef.

It was observed above that a reef needs access to a source sediment if it is to fulfil its promise as a trap. The richest source sediments are found in the basins seaward from the reefs, especially if these are intraplatformal or interior basins rather than basins of the open ocean. The basins are separated from the reefs by the clastic carbonate sediments of the fore-reef, inter-reef, and back-reef zones. *These* sediments are highly favored reservoir rocks (Sec. 13.3), but they are not in reef traps and if they are in convex traps at all it is because they have become involved in one of the types of convex trap already described. The sediments are nearly always grainstones, in the broad sense – mechanically deposited lime sands of high porosity and permeability,

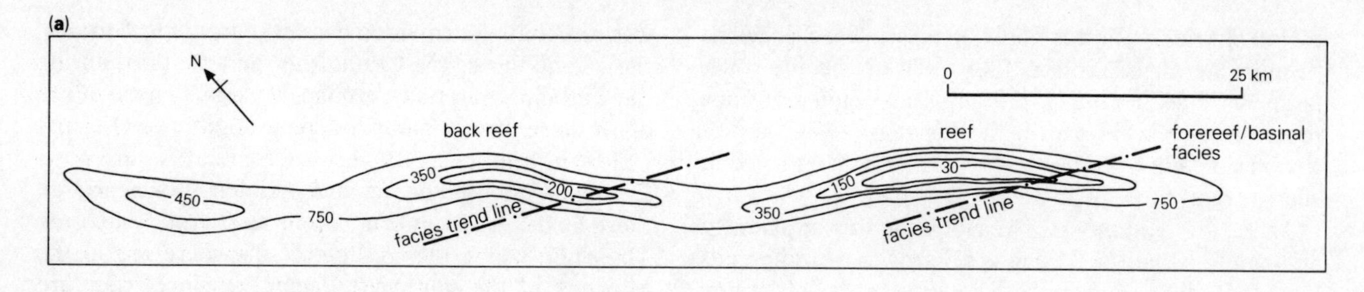

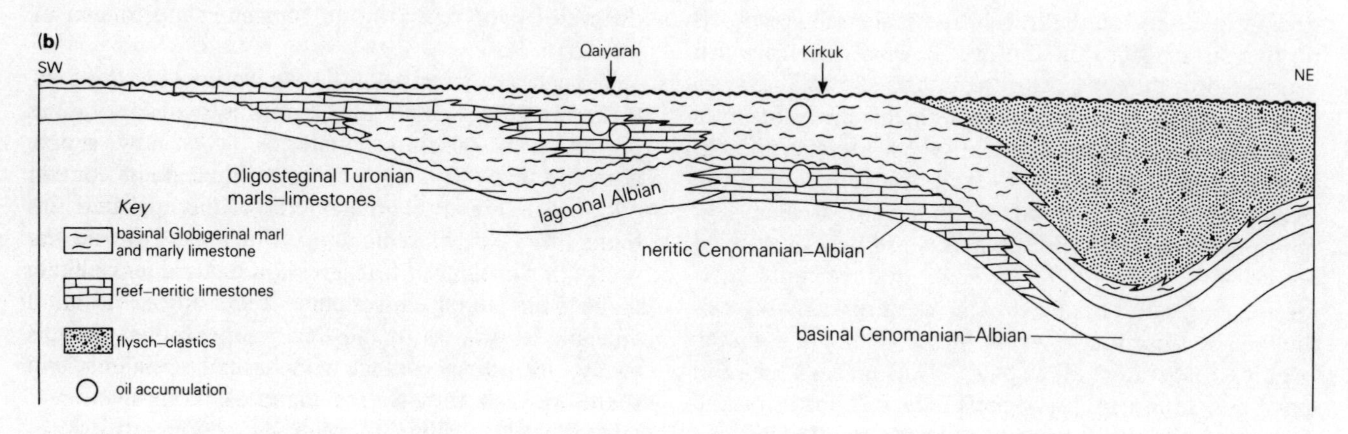

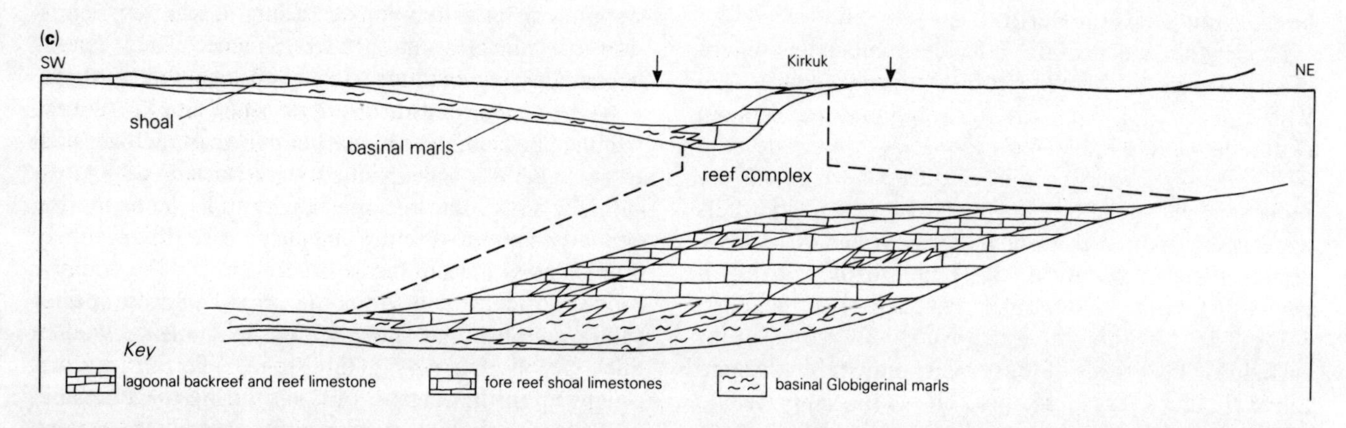

Figure 16.58 Kirkuk oilfield, Iraq. (a) Generalized structural map on top of the main reservoir limestone; contours in meters below sea level. (b) Diagrammatic cross section, not to scale, showing position of Kirkuk structure during Late Cretaceous time, when the probable source sediment was deposited. (c) Same section for Oligocene time, showing reversal of structure from Cretaceous time and development of reef trap facing intraplatformal seaway. (From W. C. Krueger Jr and F. K. North, *SEPM Special Publication* 33, 1983; after E. J. Daniel, 1954, and H. V. Dunnington, 1958.)

abundantly augmented by pellets and shells of sand-grain size.

A superb example is provided by the Permian Basin of West Texas and New Mexico, characterized by reef development through much of Pennsylvanian and Permian time (see Sec. 13.3.4). The largest and most productive single reef is a horseshoe atoll of Pennsylvanian limestone, the Scurry reef in the Kelly–Snyder field. It lies within a starved basin, but all the other major reefs rim the large, horst-like Central Basin Platform which transects the basin from north to south (Fig. 17.3). The fans, wedges, and sheets of their detritus provide the reservoirs for numerous giant oilfields, but nearly all the large ones except for Kelly–Snyder itself are in nonconvex traps formed by porosity wedge-outs, erosional truncations, or facies changes at the platform margin. Conversely, the large fields in the Reforma–Campeche region of Mexico (Fig. 16.46 & Sec. 26.2.4) are in structurally controlled convex traps, as have already been described; but the reservoir sediments were derived from a barrier reef to the east which is not itself known to be oil-bearing.

Despite the eye-catching characteristics of exposed reefs, there is surprising disagreement on which parts of productive carbonate complexes do and do not constitute true reefs. The example of the Golden Lane fields of Mexico is notorious. After the early descriptions by E. de Golyer in 1915 and J. M. Muir in 1936, the productive El Abra limestone, of Cretaceous age, was widely accepted as a prime example of a miliolid–rudistid barrier reef or atoll. The Tamaulipas limestone is its basinal equivalent. The intervening Tamabra dolomite (even the name is designed to show its intermediacy between its two counterparts) is the fore-reef deposit derived from the El Abra reef and deposited by turbidity flows. The prominent relief, in this view, is due to reefing. L. V. Illing in 1956, however, considered the El Abra to be a shelf deposit, the Tamabra a mixture of reef and fore-reef, and the Tamaulipas, as before, of basinal or open shelf origin. On this view, the relief is due to faulting along the edge of a carbonate platform. Mexican geologists also discarded the true reef or bioherm interpretation of the El Abra, considering it to be a "tilted arc" of biostromal type being extended northward. Still later interpretations, notably by A. H. Coogan and Pemex and Exxon colleagues in 1972 (Fig. 13.50), almost turned the original one backwards. No true coral-rudist reef could be established; the El Abra is a shallow shelf or lagoonal deposit with oolite banks, miliolids, rudist reef patches, and some anhydrite, resembling large areas of the Bahamas Banks today. It is in the Tamabra facies that the only real coral-rudist reefs occurred, embedded in shallow-water, reef-derived debris which is not a slide or turbidite deposit. The Tamaulipas is simply a micritic limestone lacking reef debris. In the light of the easy dolomitization of reefs and of micritic limestones, already alluded to (Sec. 13.3.6), it is interesting that neither the El Abra "reef" nor the Tamaulipas micrite is dolomitized, but the Tamabra intermediate facies is (at least at Poza Rica).

In contrast, isolated, equidimensional reefs ("pinnacle reefs") are easily identified. As they shed little carbonate detritus, and offer no useful distinction between forereef and back-reef facies, they are sharply delimited from their surroundings. Unfortunately, they are commonly small, and exploration for them is dependent upon careful seismic resolution (see Sec. 26.2.1). This is especially difficult if the reefs underlie other carbonates, or evaporites, as is normally the case. If on the other hand the reefs are *underlain* by a carbonate platform which can act as both reflecting horizon and aquifer, even small reefs may be easily detected but they are likely to be flushed by waters moving within the underlying aquifer and being funneled upward through the reefs. Despite this, a number of petroliferous provinces have been productive from these tiny targets: the Silurian reefs of the Michigan Basin, below thick salt beds; the Devonian pinnacle reefs of northwestern Alberta (Fig. 16.57) and the younger reefs, still Devonian, of the West Pembina field of that province (see Sec. 26.2.1); and the Miocene reefs of Irian Jaya in New Guinea.

16.3.8 Oddities among convex traps

Salt, as we have seen, is the immediate cause of many convex structures because of its propensity to rise. It may also cause convex structures by the ease with which it is dissolved in meteoric waters. Waters entering an aquifer closely associated with a salt horizon easily take the salt into solution, removing it from the section and bringing about the collapse of the post-salt beds over the new edge of the salt. If the removal of the salt is achieved slowly and progressively, the collapse will be represented by a drape or sag of the younger beds over the edge. If this drape or sag reverses the regional dip, a convex structure is formed.

Salt solution may be discontinuous, being initiated and re-initiated in response to diastrophic events. This may lead to very odd structures. First-phase solution creates sags or grabens in younger strata, permitting thicker sections to accumulate there. Much later, second-phase solution removes salt elsewhere, where thicknesses of younger strata are normal, and the thicker sections over the older lows now become highs, as domes or structural ridges. The highs become targets for the drill.

Unfortunately, movement of meteoric waters through the reservoir may long postdate oil migration and accumulation; it may be going on today. The solution of salt may itself lead to the flushing of the traps by the same waters. This has undoubtedly happened in the Middle Devonian reefs in the Williston Basin in Saskatchewan, Canada, where most zero-edges of the otherwise thick salt are now solution edges. Salt solution has markedly affected both convex and nonconvex traps in many basins: along the western side of the Central Basin Platform in the Permian Basin of the USA; in the Paradox Basin in Utah, where the salt is Pennsylvanian (Fig. 16.73); the Suez graben (Fig. 16.29), where the salt is Miocene; and the Emba Basin in the Soviet Union, where it is Permian.

I am not aware of any important convex trap that can be shown to have been *created* this way. There is, however, one gigantic nonconvex trap which has been provided with local convexity through solution of underlying salt. The Lower Cretaceous sandstone in the Athabasca "tar sand" deposit of northeastern Alberta has been given a reversal of its regional southwesterly dip through solution of the same Middle Devonian salt as underlies the Williston Basin in Saskatchewan (Fig. 16.59). Many geologists familiar with the "tar sands"

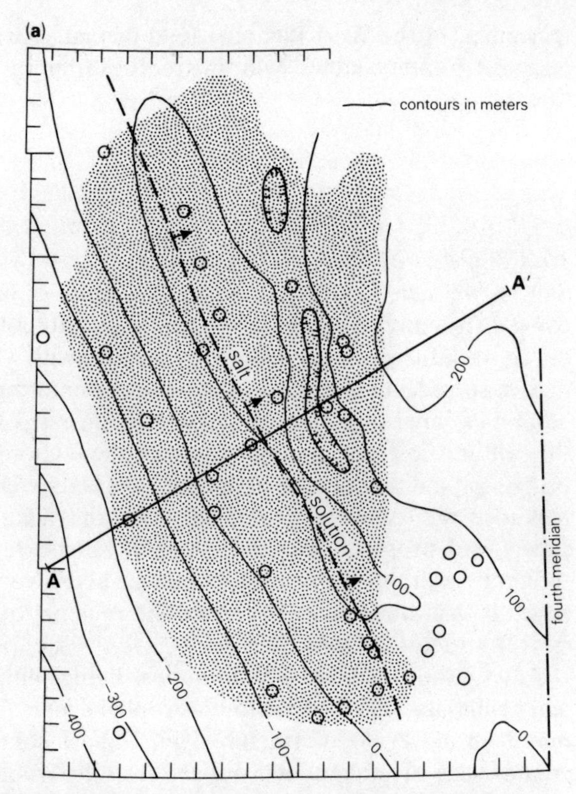

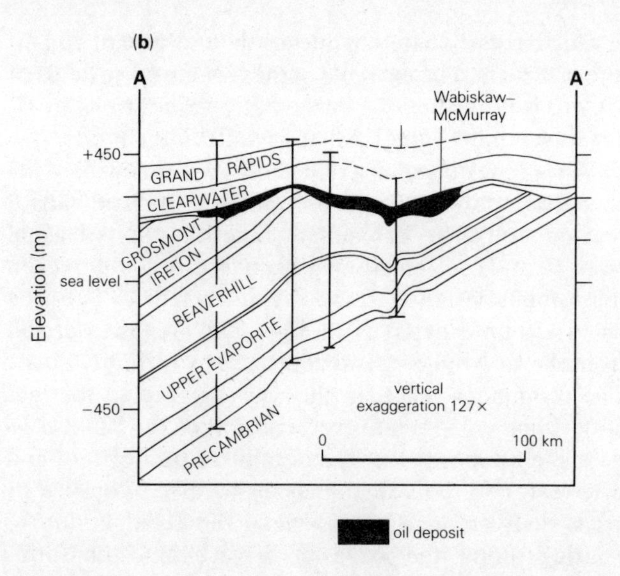

Figure 16.59 (a) Structure contour map on top of the Devonian Elk Point Group, Athabasca area, western Canada, to show the relation of the "tar sand" deposit to the area of dissolution of Elk Point salt ("Upper Evaporite") beneath the Cretaceous reservoir rock. (b) Cross section along line A–A′, showing solution-induced topography beneath the Cretaceous overlap. Vertical scale greatly exaggerated to emphasize this. (After L. W. Vigrass, *AAPG Bull.*, 1968.)

maintain that this reversal created a culmination towards which the migration of the heavy oil was preferentially funneled.

An exceedingly rare category of convex trap may be formed by *meteoritic impact craters*. At least two probable examples are known, both in the Williston Basin and both in Mississippian carbonate breccia reservoirs. The breccias are believed to have been formed by impacts during Mesozoic time.

One trap is in the central uplift of a crater, which is surrounded by a ring depression. The other trap is in the rim facies, which is absent from the center of the structure (which in this case is a cavity containing anomalously thick sediments). Underlying the rim facies are other carbonate reservoirs unrelated to the impact structure (Fig. 16.60).

As impacts typically affect crystalline basement as well as the cover rocks, many potential traps may exist in fractured porosity in such rocks. We may anticipate more identifications of hydrocarbon traps in impact structures, but it is difficult to imagine any of them being very large.

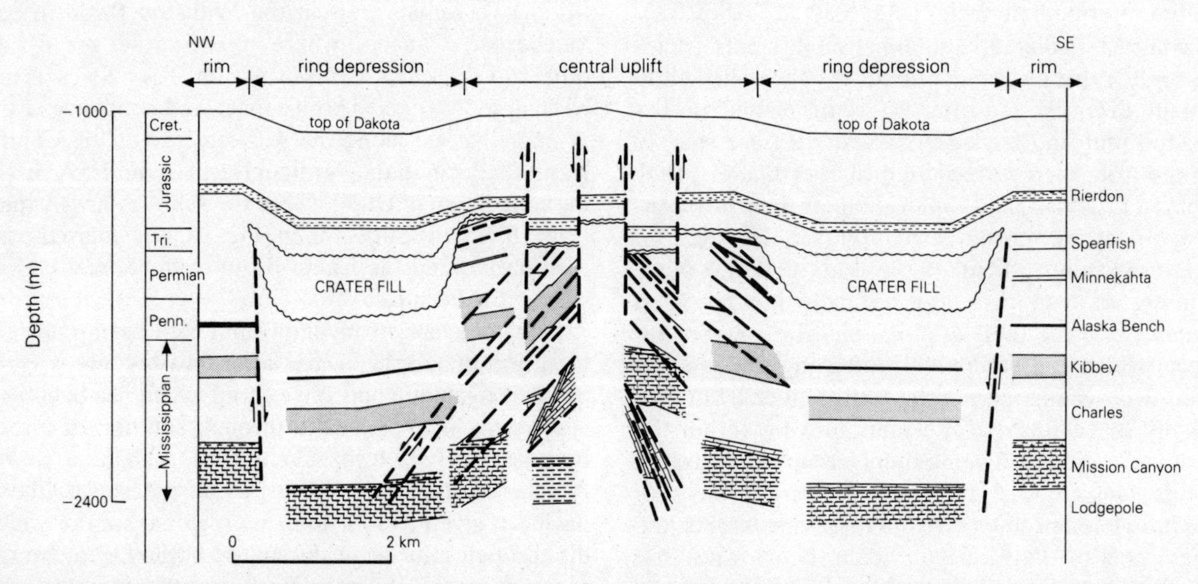

Figure 16.60 Cross section through Red Wing Creek structure, Williston Basin of North Dakota, interpreted as a meteorite-impact feature. (From L. C. Gerhard *et al.*, *AAPG Bull.*, 1982.)

16.3.9 Convex traps without static closure

It has been emphasized that all the different varieties of convex trap – compressional, extensional, depositional, residual – possess the common property of *static closure*. Yet very many convex traps in fact lack static closure and still contain oil or gas. Moreover, very many convex traps in petroliferous basins possess static closure and yet do not contain oil or gas. These two circumstances are common because the structures involved are not under static fluid conditions.

Consider a simple anticline or dome with flanking dips of no more than 3°. There are hundreds of oil- and gasfields in convex structures as gentle as that. If that anticline or dome affects beds otherwise horizontal, it has static closure. But if the beds are tilted to a *regional* dip of 4°, the anticline no longer has static closure. A reef or constructional sand bar may have convex shape, but if the dip of its shoreward flank is lower than the slope of the shelf on which it is developed it has no static closure. Yet any of these structures may provide an adequate trap for an oil- or gasfield if the conditions are *hydrodynamic*.

In Chapter 14, we considered the natures of the *pressures* under which fluids exist in the subsurface environment. We return to the influence of subsurface pressures in Chapter 23. The potential energy of the reservoir is stored in its compressed fluids. If the fluid pressure in a reservoir rock is the same as that in the overlying rock (the potential seal), there is no impetus to fluid interchange and the conditions are hydrostatic. Normally, however, a reservoir rock constitutes a reservoir and a roof rock constitutes a seal because the two have distinctly different porosities and permeabilities. Fluid pressures in the two are therefore not likely to be equal. They may be unequal but not sufficiently unequal to promote fluid interchange; the conditions then are still *effectively* hydrostatic.

If any significant pressure difference exists between reservoir and seal, however, the fluids must move in an attempt to eliminate the difference. In 1951, Gilman Hill distinguished between positive and negative *hydrodynamic* conditions. Under positive hydrodynamic conditions, the pressure in the roof rock exceeds the pressure in the aquifer; this permits oil or gas to accumulate in the aquifer (reservoir) until its volume is such as to exert a pressure greater than that in the roof rock. Under negative hydrodynamic conditions, pressure in the aquifer exceeds that in the roof rock; oil therefore tends to migrate up through the roof rock until the pressures are equalized. If two aquifers are separated by a shale barrier, and oil occurs in a convex trap in the upper aquifer, the pressure in that aquifer must be lower than the pressure in the lower aquifer. If it is higher, the oil will migrate downwards through the shale once the capillary pressure it exerts exceeds the entry displacement) pressure of the shale (see Sec. 15.7.2).

In the common circumstance, the most ubiquitous of the reservoir fluids – the water – is trying to move not vertically but laterally through the aquifers, in the direction of the hydraulic pressure gradient. Subsurface water flows along the potentiometric gradient from regions of high water table to those of low water table. The more continuous and permeable an aquifer, the more quickly will it attain normal hydrostatic pressure for the system, because the water is able to migrate out of the system, reducing the head and eventually eliminating it. As long as the water is moving, however, it can displace oil and gas accumulations and create or destroy traps.

In Chapter 15, we set out the principles of hydrodynamic theory in order to understand its control of the secondary migration of oil and gas. We must now consider how these same principles may create or destroy convex traps.

If the potentiometric surface overlying an oil pool is other than horizontal, water will move through the reservoir rock (which is also an aquifer) in accordance with the datum pressure difference (Δp) across the width (x) of the pool. Δp is positive if the water flow has a *downdip* component, negative if the flow is updip.

In the circumstance illustrated in Fig. 16.61, therefore, Δp is positive and $dZ/dx > dh/dx$. The pressure difference due to the column of water, Δp_w, is then given by

$$\Delta p_w = \rho_w g \, dZ - \rho_w g \, dh \qquad (16.1)$$

The pressure difference in the oil column, Δp_o, is of course independent of dh and is given by

$$\Delta p_o = \rho_o g \, dZ \qquad (16.2)$$

When equilibrium is achieved, Δp_w and Δp_o are equalized; thus

$$dZ = \frac{\rho_w}{(\rho_w - \rho_o)} \, dh = \frac{\rho_w}{(\rho_w - \rho_o)} \frac{dh}{dx} \qquad (16.3)$$

If the reservoir bed, shown with homoclinal dip in Fig. 16.61, has been provided (by whatever means) with a convex trap, the *direction* of water movement through it will be constrained to conform with the shape of the reservoir. Equipotential surfaces for the water, represented by the middle factor in Equation 16.3 (and identified as the parameter v in Sec. 15.8.1), change attitude as the water moves around the convex trap, pushing the oil ahead.

Equipotential surfaces for the oil (or gas) become normal to the resultant force created by the hydrodynamic gradient (Fig. 15.14). The oil/water interface

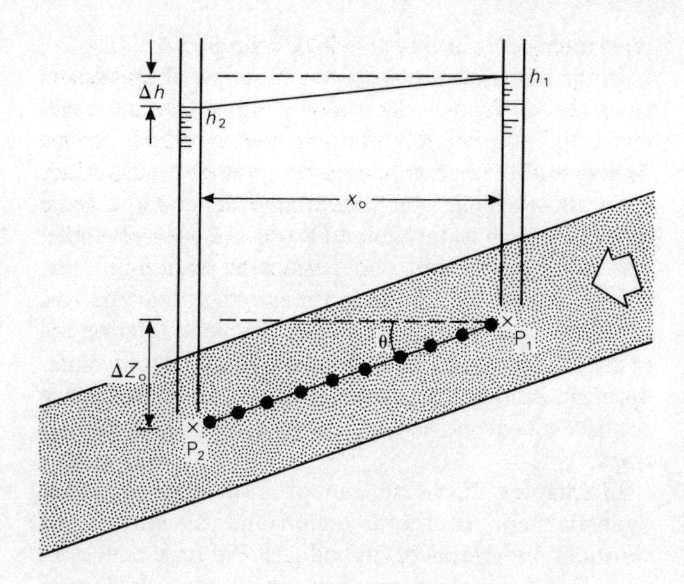

Figure 16.61 Relation between vertical height of hydrocarbon column (ΔZ_o) and drop in level of potentiometric surface (Δh), under hydrodynamic conditions. (From R. R. Berg, *AAPG Bull.*, 1975.)

conforms with the equipotential surfaces for the oil under this constraint; the tangent of the angle of tilt (θ) of the interface is dZ/x (Fig. 16.61). From Equation 16.3, therefore, we derive

$$\tan\theta = \frac{dZ}{x} = \frac{\rho_w}{(\rho_w - \rho_o)}\frac{dh}{dx} \tag{16.4}$$

This is Hubbert's equation expressing the physical conditions governing entrapment of hydrocarbons under hydrodynamic conditions. The hydrocarbon/water interface is tilted, to angle θ, in the direction of movement of formation water; the amount of tilt is a function of the hydraulic gradient across the accumulation (the rate of change of the potentiometric surface per unit length of flow path) and of the densities of the water and the hydrocarbons. Given no otherwise unfavorable geologic factors, and assuming reasonably uniform transmissibility properties in the sediments, the *vertical shape* of a hydrocarbon accumulation is the potentiometric surface (dh/dx) *magnified by the tilt factor* ($\rho_w/(\rho_w - \rho_o)$). We anticipated this conclusion in discussing the role of water flow in secondary migration (Sec. 15.8.1).

We find an apparently startling consequence. The value of $(\rho_w - \rho_o)$ is greatest when the hydrocarbons are lightest, and $(\rho_w - \rho_o)$ is the denominator in Equation 16.4. The heavier the hydrocarbons, the larger the tilt factor (Fig. 16.62). For natural gas, the factor is between 1 and 2; for 35° oil, it is about 6; for 20° oil, about 12. A structure to hold 20° oil, and to prevent its loss by forced migration, must be six to twelve times as steep as a structure to hold gas, under equivalent hydrodynamic conditions. Gas is so light, mobile, and migratory that it is difficult to realize that it is in fact much more easily retained in a trap than is weakly mobile, heavy oil, under active hydrodynamic movement.

Let us now combine the reservoir dispositions illustrated in Figs 16.61 and 16.62, by superimposing the convex trap of the latter on the regional dip of the former. Clearly the direction of the regional hydraulic pressure gradient is critical. Almost equally critical is the

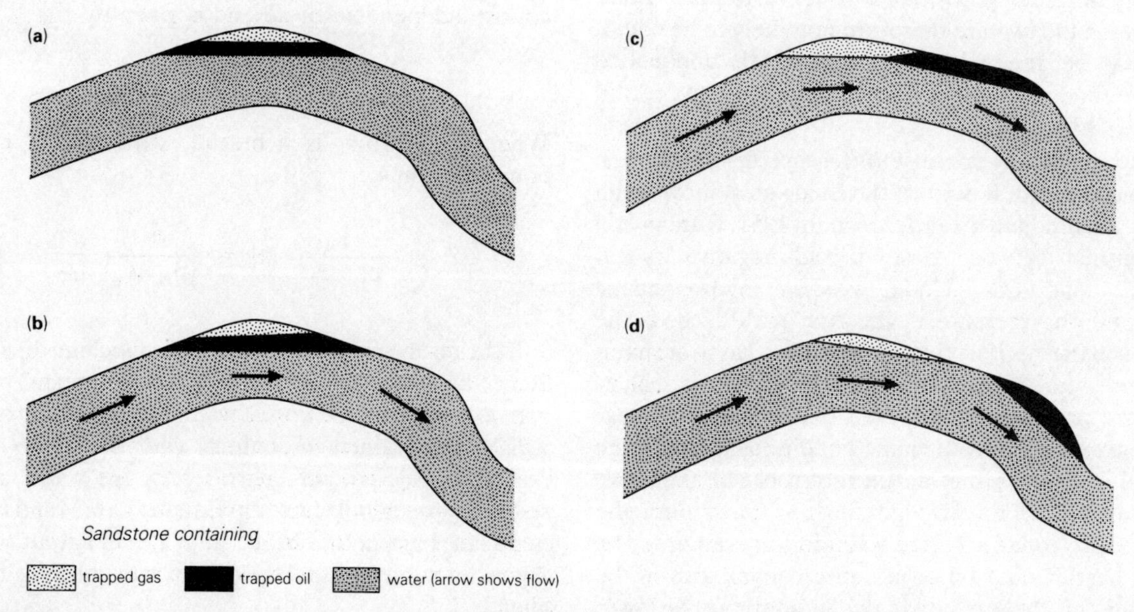

Figure 16.62 Gas and oil trapped in a thick, anticlinally folded sandstone under hydrostatic conditions (a), and modifications of the trapping capacity by the introduction of hydrodynamic conditions. Successive effects illustrate either increase in the flow rate, with hydrocarbons of constant density, or increase in the density of the oil at constant flow rate. Note increasing angle of tilt of the oil/water interface. (After M. K. Hubbert, *AAPG Bull.*, 1953.)

precise relation between the dips on the flanks of the convex structure and the regional dip (Fig. 16.63).

With simple monoclinal flexuring of the reservoir bed, increasing its dip in the downdip direction (case 1), downdip water movement creates a trap in the flexure despite its lack of static closure. Updip water movement flushes hydrocarbons out of the structure altogether. A true flexure interrupting otherwise uniform regional dip (case 2) provides static closure, but downdip water movement augments this closure and updip movement diminishes it without destroying it. The tilting of a reservoir bed on which a true anticlinal fold has been raised (case 3) leads to a similar result. A "reverse drag" fold, whether on the inverted limb of an overturned larger fold or brought about by gravitational creasing of weakly lithified strata down their dip (flap fold), has the effect of reversing the tilt of the interface (case 4), but the outcome is still more favorable under downdip than under updip flow. Such structures are in any event rare in petroliferous successions. It is a nearly universal principle, therefore, that *downdip water flow promotes hydrocarbon accumulation and updip flow reduces or prevents it.*

The tilt of an oil/water interface seldom exceeds 1 or 2°, but tilts as steep as 10° are not unknown. With flank dips on very many convex traps no more than 2° or so, such interfacial tilts can displace oil accumulations well away from the positions they would occupy under hydrostatic conditions. Theoretically, as we have observed, oil and gas try to get as high in their reservoir as they can;

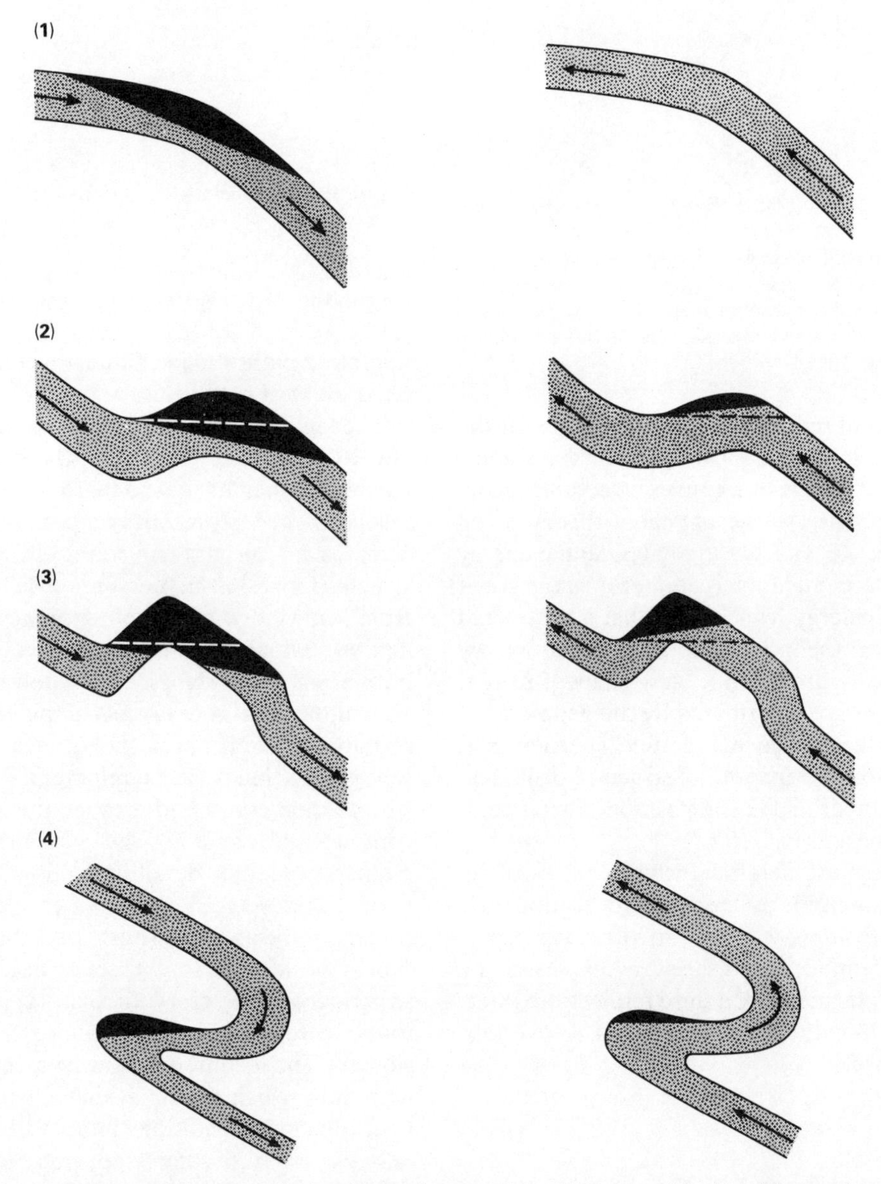

Figure 16.63 Four cases of hydrodynamic traps made possible by changes of dip of the reservoir bed involving varying degrees of convexity. Cases illustrate effects on trapping capacity brought about by downdip and updip water flow (left and right columns, respectively). No scale implied.

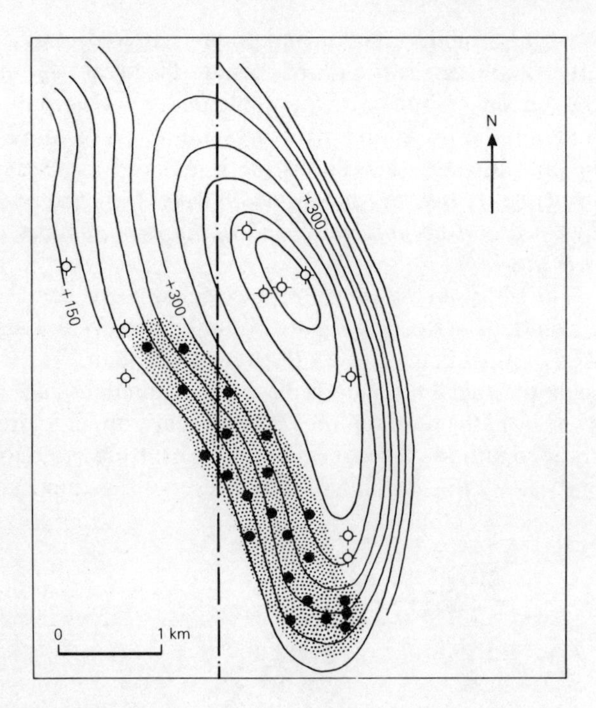

Figure 16.64 Sage Creek oilfield, Big Horn Basin, Wyoming: structure contours on the productive Tensleep sandstone (Pennsylvanian). Oil pool (black dots are wells) is displaced hydrodynamically more than 1 km down the west flank of the anticline; oil/water interface is tilted to the southwest at 150 m km^{-1}; crest of structure is barren (indicated by dry hole symbols). (After Wyoming Geological Association, 1957.)

this is the position of minimum potential energy for the trap geometry available. Under hydrostatic conditions, this position is in the crest of a convex structure; recognition of this gave rise to the anticlinal theory of oil accumulation (Sec. 16.3). But the total potential energy of a hydrocarbon accumulation is greater than the gravitational potential energy, which is all that is accounted for in the anticlinal theory. The total potential energy also includes the hydrodynamic and shape-potential energies, and these are determined by the geometry of the structure and the change in hydraulic head across it. If these contributors are ignored, a structure drilled on its crest may not be drilled in the position of true minimum potential energy (Fig. 16.64).

Area evaluation must therefore include analysis of the direction and magnitude of regional potentiometric gradients. *Hydrodynamic mapping* involves the superposition of two families of surfaces, as expressed by contours, and the deduction of a third family from intersections between the others. In Section 15.8.1 we established this relation:

$$u = v - Z \tag{16.5}$$

in which constant values of u and v represent equipotential surfaces for oil and water, respectively, and Z is the elevation of the top of the hydrocarbon column. The

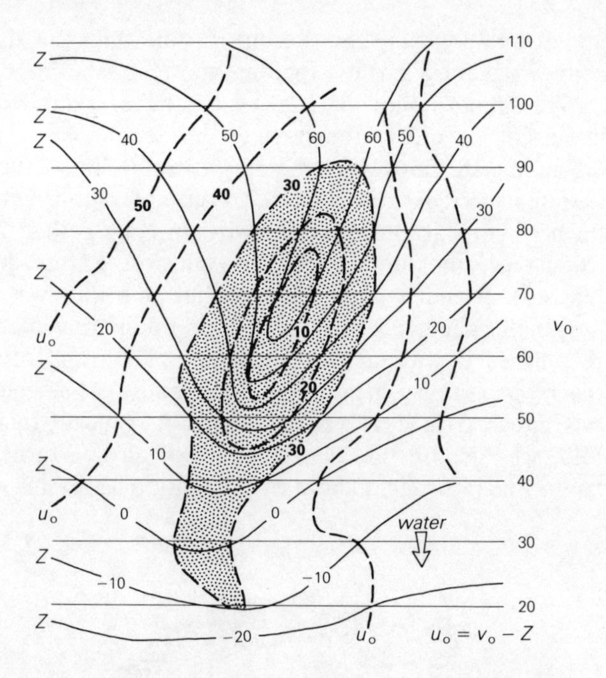

Figure 16.65 Structure contours on potential oil trap (u_o, in dotted pattern), deduced from contours on the reservoir rock (Z) and the potentiometric surface (v_o). Contours are arbitrarily diagrammatic and without implied scale. (From E. C. Dahlberg, *Applied hydrodynamics in petroleum exploration*, Berlin: Springer, 1982.)

migration paths for each fluid are perpendicular to these equipotential surfaces.

The families of surfaces are illustrated in Figures 16.65 and 16.66. For any reservoir, the hydrocarbons cannot rise higher than its top, so that structure contours on the reservoir rock represent contours on Z. Equipotential surfaces for the moving water (the factor v) are constructed from shut-in pressures of wells (see Sec. 23.7) or from knowledge of the potentiometric surface for the aquifer (which is also the reservoir rock). Until the potentiometric surface is adequately established, it must be contoured as a uniformly sloping plane; in the most primitive circumstances, it may have to be deduced from topographic maps. The equipotential surfaces for the oil (u) are then contoured through the intersections of the contours of v and Z. The mapping procedure is explained in greater detail in Section 22.8.

In Figure 16.65, the water potential decreases gradually towards the south, and the subsurface water flow is therefore in that direction at some velocity x. The structure on the reservoir rock, represented by Z contours, reveals a simple plunging nose without static closure. The u contours, however, reveal a closed minimum into which any oil available would flow under the hydrodynamic conditions imposed by the water flow.

In Figure 16.66, the water potential again decreases towards the south but the gradient is twice as steep as in Figure 16.65. The velocity of the water flow (other factors being equal) is therefore about $2x$, sufficient to flush

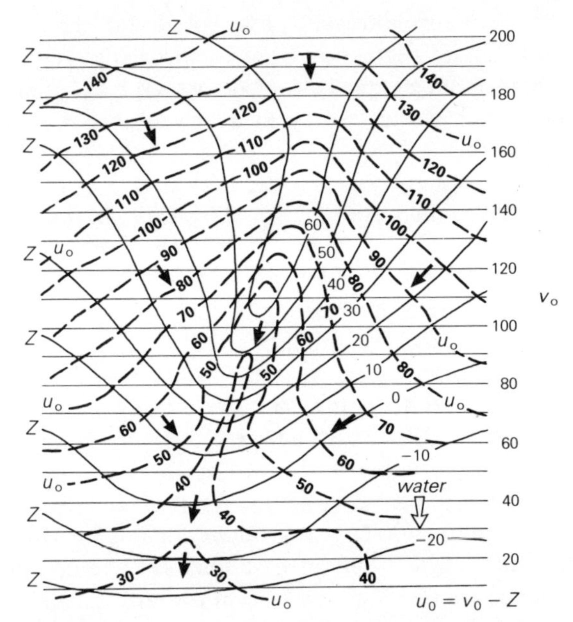

Figure 16.66 The same structure as in Figure 16.65, but with rate of southward water flow doubled. Contours on oil equipotential surface (u_o) show no closure, indicating flushing of the structure towards the south by the water flow. (From E. C. Dahlberg, *Applied hydrodynamics in petroleum exploration*, Berlin: Springer, 1982.)

all oil out of the structure. Contours of the *u* surface show no closed hydrodynamic minimum; the migration paths for the oil, at right angles to the equipotential contours, follow the heavy arrows.

As is so often the case in our inexact science, a caution is necessary. There is an inclination on the part of many development geologists to ascribe all tilted fluid interfaces which they encounter to hydrodynamic influence. Though this influence cannot be doubted, experience indicates that it cannot be the only cause of all tilts. At least two other causes must be considered.

First is the internal geometry of the reservoir. The nature and distribution of pore space or of microfractures, and the microscale nature of diagenetic effects such as cementation and differential compaction, may permit hydrocarbons to accumulate, or water to penetrate, more successfully along one side of a reservoir or trap than along the other. Diagenetic effects are especially influential if they occur after the accumulation of the hydrocarbons but before the imposition of the tilt. Second is recent tectonic history. Once subsurface fluids have achieved their proper gravitational relations within a reservoir, their readjustment to neotectonic increments in the regional dip cannot be immediate. The interfaces between them may therefore for some time retain downward tilts in the direction of the latest subsidences of their basins. These tilts will normally be *away from* the youngest adjacent uplifts, which may have continued to rise isostatically into geologically recent time. Such tilts will of course conform in direction with hydrodynamically induced tilts, if the potentiometric surface is strongly influenced by hydraulic head to surface outcrops. The relative significances of the two causes may be beyond resolution.

It must furthermore be borne in mind that reasoning about hydrodynamic conditions from interface tilts (and vice versa) may easily become teleological. The geologist must be clear in his own mind whether he is deducing a paleoflow pattern from present conditions in his traps, or interpreting his traps from the present hydrodynamic regime, which may be quite different from that in operation when the trapping took place.

Imperfectly convex traps, revisited We now see why a structure which is geometrically convex may have no static closure and yet contain an oil or gas accumulation. What is more, such an accumulation may be of much greater amplitude than any static closure the trap does possess would permit. Water creates the trap by moving down-dip from the direction in which closure is lacking or deficient.

Hundreds of oilfields, and some gasfields, are found in structural "noses" like that illustrated in Figure 16.65, extending into basinal areas or sinks from larger highs, or into embayments from those basinal areas. All that is necessary is that the slope of the oil/water interface, in the direction of water flow, should be less than the plunge of the axis of the "nose" in the same direction. Similarly in a monocline, or a structural terrace, no static closure exists (Fig. 16.63). In the direction of water flow, the slope of the oil/water interface must be less than the dip of the structure (Fig. 16.65).

Pools contained in such structures are seldom very large, for obvious reasons, but in many cases they are much larger than they would be if they were governed wholly by the static closure available to them. The Jay field in Florida and the Mene Grande field in Venezuela are examples illustrated here (Figs 16.82 & 26.20).

A vital consequence should be apparent. Given favorable hydrodynamic conditions, not only need there be no closure for a trap; there may be no change of dip whatever. We have traversed from the truly convex trap, having static closure, to the truly nonconvex trap, having no geometric closure at all.

16.4 Nonconvex traps

Traps lacking convexity have no static closure. They must therefore be closed dynamically, commonly hydrodynamically, either with downwardly directed water drive above, or with permeability pinchout (including asphalt seal) above and water drive below.

Nonconvex traps embrace all the types popularly called "stratigraphic" traps, except only reefs, which are convex structures (Sec. 16.3.7). They also include a variety of "structural" traps in which the porosity/permeability cutoff is caused or at least influenced by faulting. There are five basic types of nonconvex trap, genetically distinct though geometrically very similar.

16.4.1 Depositional wedge-out traps

Simple depositional wedge-out traps within otherwise continuous successions are characteristically in sandstones, which have much higher horizontal permeabilities than do limestones. There are two cases.

In the first case, the sand body wedges out up the depositional dip, as beach and bar sands do (Fig. 16.67). Such sands are in trap position from the outset. They may also possess optimum source–reservoir–seal relations if they merge down-dip into basinal shales and up-dip into delta-plain or lagoonal sediments of low permeability. They may instead merge down-dip into deep-water turbidite sands associated with the basinal shales. Under regression, their updip wedge-outs may remain too long exposed, or be covered by continental sands which permit escape of hydrocarbons directly to the surface.

Dozens of sand bodies in the US Gulf Coast are in the favorable configuration, particularly those in the Yegua and Frio Formations. Many traps within them have been created by faulting or by salt uplifts (Fig. 16.48), but there are a great number of nonconvex traps that have remained essentially unmodified.

In the second case, the depositional wedge-out is down-dip, resulting in a sandstone body wedged between two shales. If such a wedge-out is to acquire trap position, its original dip must somehow be reversed. This is easily achieved by further uplift of the source area of the sand, and the consequent loading of the depositional basin by molasse sediments. Large, simple examples are the Pembina field in western Canada, in a

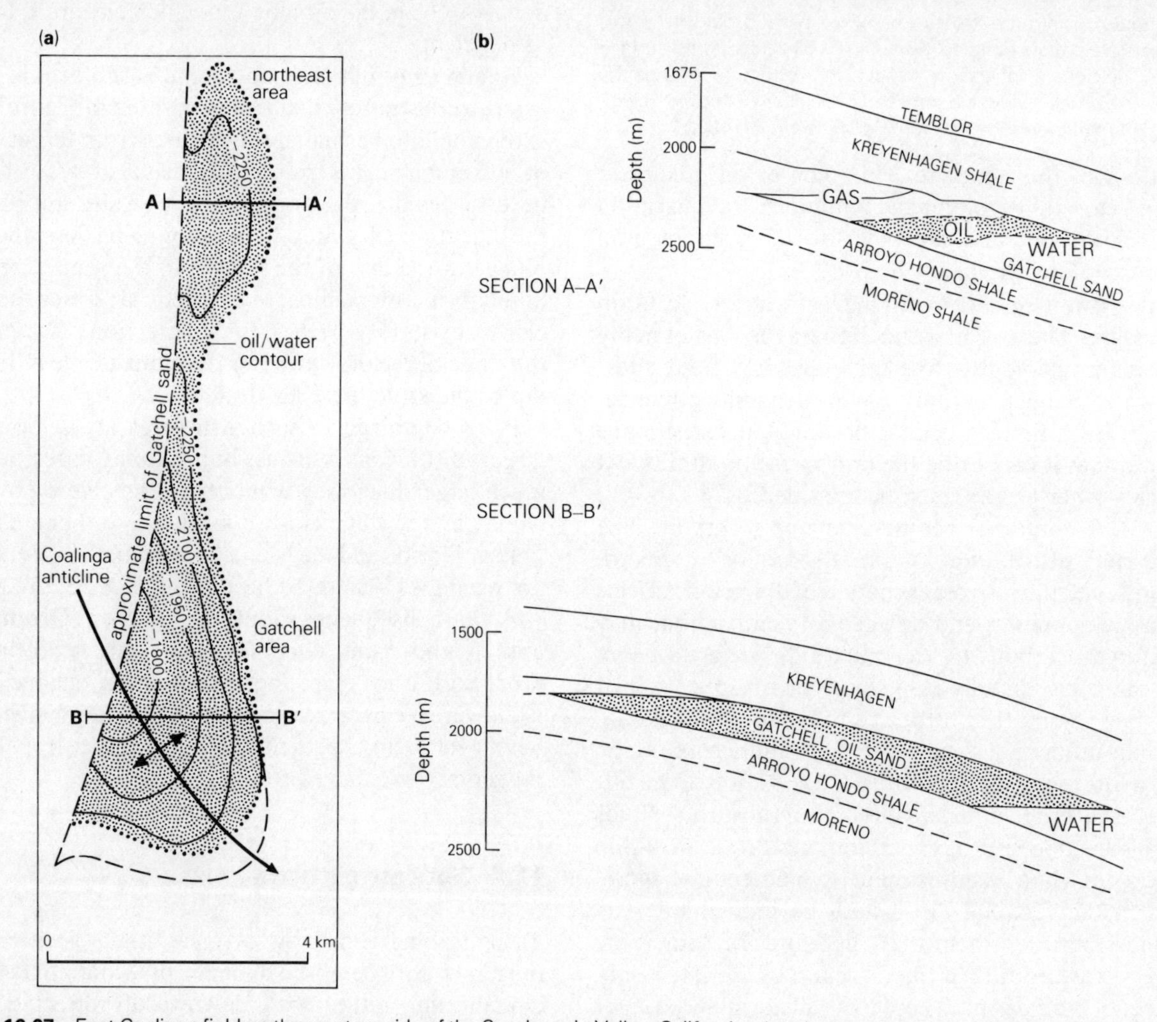

Figure 16.67 East Coalinga field on the western side of the San Joaquin Valley, California, showing nonconvex, stratigraphic trap due to updip wedge-out of a bar-type reservoir sandstone, of early Eocene age, into an impermeable siltstone. The much larger Coalinga field lies updip and westward from this field, and its principal reservoir is in a different sandstone. (a) Structure contours at 150 m intervals on top of the productive sandstone, with productive area stippled; (b) cross sections along lines shown on the map. (From L. S. Chambers, California Division of Mines Bull. 118, 1943.)

Cretaceous sandstone interrupting a succession of dark shales (Figs 26.1 & 2), and the Burbank field in Oklahoma, in a sandstone member of the thick Pennsylvanian molasse sequence.

This simple trap type is bound to be rare, because the very process of reversing the original dip involves uplift, which may involve the reservoir rock itself and impose convexity upon it. This is what happened to many productive sand bodies that in their original depositional settings must have been closely similar to the Pembina reservoir: the Forest and Cruse sandstones in the Miocene of Trinidad (Fig. 16.55), and the Mio-Pliocene sands on the west side of the San Joaquin Basin in California (Fig. 16.8).

16.4.2 Erosional wedge-out traps

Erosional wedge-outs are exceedingly common trapping mechanisms. An especially common case is provided by sandstones eroded from the tops of uplifts that become "bald-headed"; the sandstone remnants down-flank are in perfect trap position, with the unconformities above them. The Russians would call these *structural-lithologic* traps. Scores of fields in the basins of southern Oklahoma are trapped in Upper Mississippian and Lower Pennsylvanian sandstones in this way.

The erosion surface may alternatively be below the reservoir sandstone (in some cases, both above and below it). This category contains the largest of all the classic "stratigraphic" traps, such as the Bolivar Coastal field in Venezuela and the East Texas field (and in all essentials the Athabasca "tar sands" of Alberta and the Orinoco "tar belt" of eastern Venezuela; Figs 10.16, 10.18 & 16.2). In the commonest case, a basal transgressive sandstone lies unconformably on a depositional slope and retains its original dip; the sandstone may be overlain by transgressive marine shale or carbonate micrite (Figs 16.68 & 69).

Most nonconvex traps in carbonate rocks are in erosional wedge-outs underneath unconformities. Unconformities are so vital to trapping mechanisms in so many petroliferous regions that they deserve extended separate treatment (see Sec. 16.5).

The distinction between erosional and depositional wedge-outs may not be easily made in drillholes. In the depositional wedge-out of a sandstone unit, the entire unit thins progressively towards its zero isopach; in the erosional wedge-out, only the top beds of the unit need do so. The uppermost beds of the unit reach the zero isopach line of a depositional wedge-out, the lowermost beds do so in an erosional wedge-out. The rate of loss of thickness of the unit is likely to be gradual and regular in the depositional case, abrupt in the erosional case. Shallow-water features, especially increasing sand content and grain size, should become more pronounced towards the zero edge in a depositional wedge-out; there

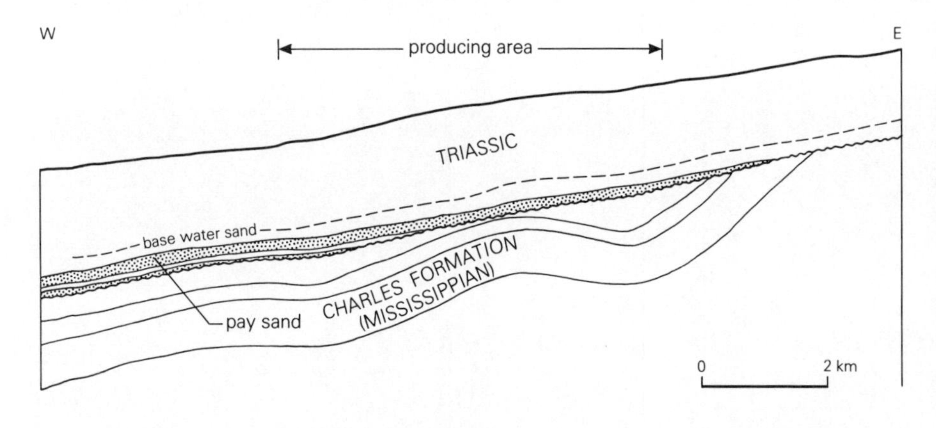

Figure 16.68 Newburg field, North Dakota, a classically simple trap of the type called lithologic-stratigraphic by Russian geologists. A basal sandstone of the Triassic transgression wedges out updip against an erosional surface of Mississippian carbonates on the east flank of the Williston Basin. The carbonates are themselves petroliferous and are the source of the oil in the Newburg sandstone. (From K. K. Landes, *Petroleum geology of the United States*, New York: Wiley–Interscience, 1970; and North Dakota Geological Society, 1962.)

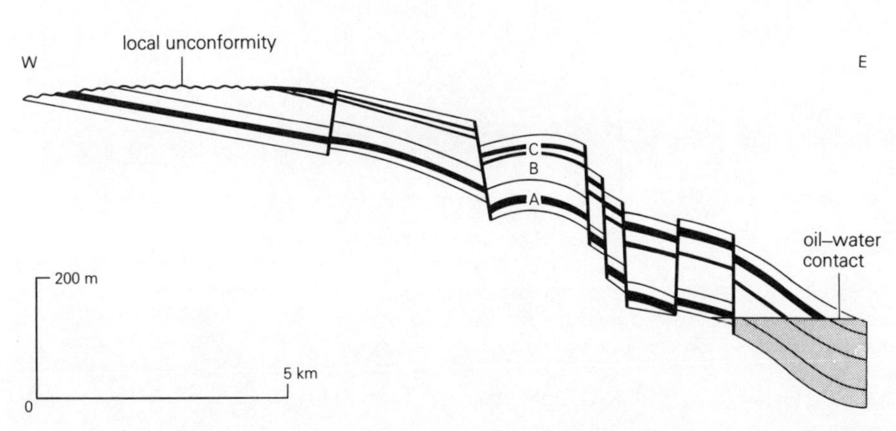

Figure 16.69 Diagrammatic cross section through the Kuparuk River field, Arctic Alaska, showing onlap of Lower Cretaceous clastic sediments on to a high to the west, followed by their partial erosional truncation. Faults give appearance of breaking up the reservoir after the trapping of the oil, but in fact the faults do not extend upwards into the younger strata. Oil pay intervals in black. (From G. J. Carman and P. Hardwick, *Oil Gas J.*, 22 November 1982.)

is no reason why they should do so in the erosional case. Finally, lithofacies, biofacies, and isopach trends are likely to be concordant with the zero edge in the depositional case and discordant with it in the erosional case.

Simple wedge-out traps, whether depositional or erosional in origin, tend to be cut up by normal faults. Mosaics of faults in the basement become extended upwards into the petroliferous cover rocks. The faults themselves may constitute important nonconvex trapping mechanisms (see Sec. 16.7).

16.4.3 Isolated or lenticular bodies

Isolated, lenticular bodies, commonly of sandstone, form closed system traps, the "isolani" of Caswell Silver (1973). Because most sandstone depositional environments (Sec. 13.2.7) are replete with such bodies, they are very common as traps. Some of them are prolifically productive because they are filled with oil without free water. They are easily subject to overpressuring (see Sec. 14.2.1). Their settings are preferentially in the structural lows of their basins.

Most traps in this category are necessarily small, and an extraordinarily high percentage of the productive examples are in the USA. This is undoubtedly in some degree due to the ease with which even very small oil- or gasfields can be economically exploited in that country. A much more important reason, however, is that operators in the USA have consciously looked for these "subtle" traps. Difficult to detect seismically, erratic in areal distribution, easily missed by widely spaced drillholes, they are discovered in numbers only under conditions of dense, possibly random, wildcat drilling.

Examples abound among the fields, and among the pools within fields, in the Cretaceous sandstones of the western states (Fig. 16.70), the Permian and Pennsylvanian sandstones of the Midland, Delaware, and Mid-Continent basins (Fig. 16.71), and the Mississippian Springer sandstones in the Anadarko and Ardmore basins of Oklahoma. As many contain gas as contain oil. Outside the USA, there are numerous examples in the Cretaceous sandstones of Alberta and the Tertiary sandstones of eastern Venezuela, especially in the heavy oil areas of both regions.

Some biostromal bodies of limestone or dolomite also provide tabular or lenticular traps, commonly passing laterally into denser limestones. They are even more difficult to detect than their sandstone counterparts.

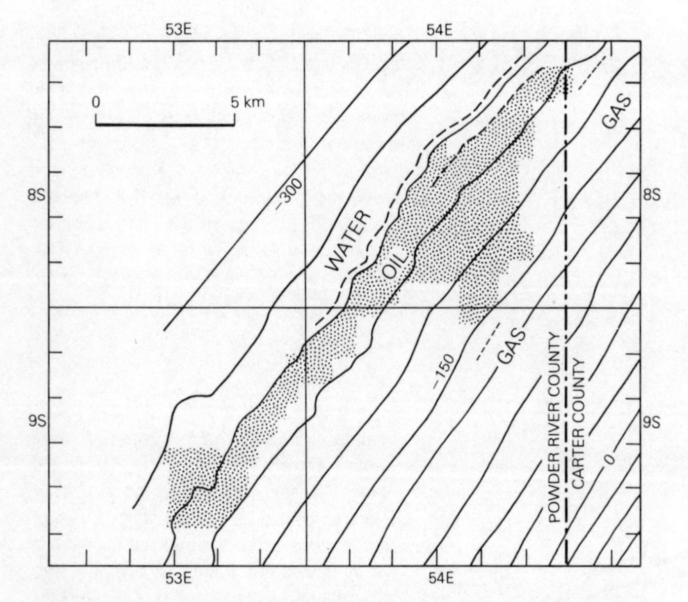

Figure 16.70 Bell Creek oilfield in the Powder River Basin, Montana. Structure contours at 30 m intervals on top of the Cretaceous producing sand, principally a barrier bar, showing remarkably uniform homoclinal dip without indication of structural closure. (After A. A. McGregor and C. A. Biggs, *AAPG Memoir* 14, 1970.)

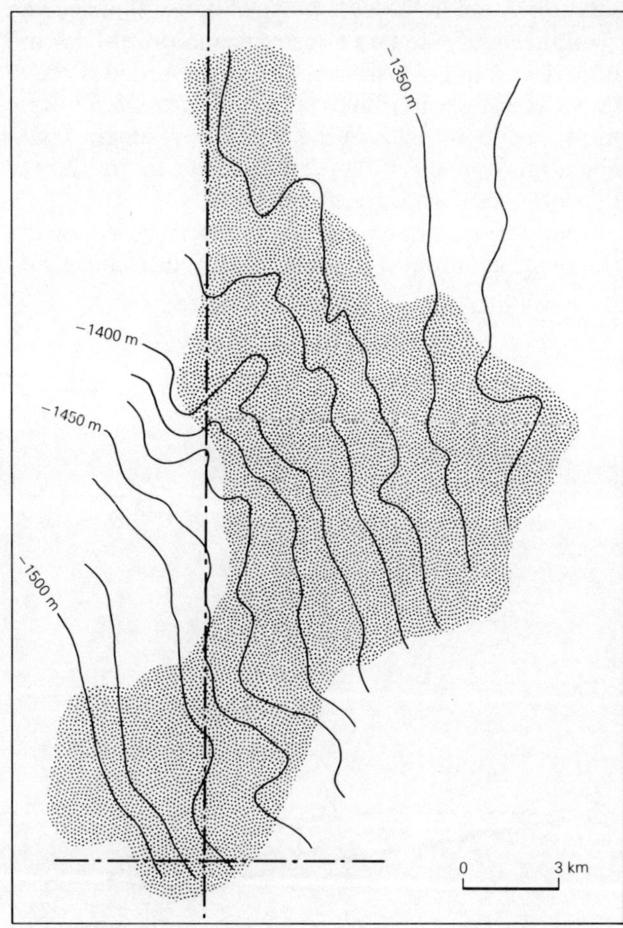

Figure 16.71 Jo Mill oilfield, in Spraberry sandstone trend of Permian Basin, West Texas. Structure contours on top of producing horizon, showing homoclinal westerly dip of about 10 m km^{-1}; trapping mechanisms are updip wedge-outs of a series of lenticular sandstones. Producing area stippled. (From K. K. Landes, *Petroleum geology of the United States*; New York: Wiley–Interscience, 1970; after R. D. Jons, 1966.)

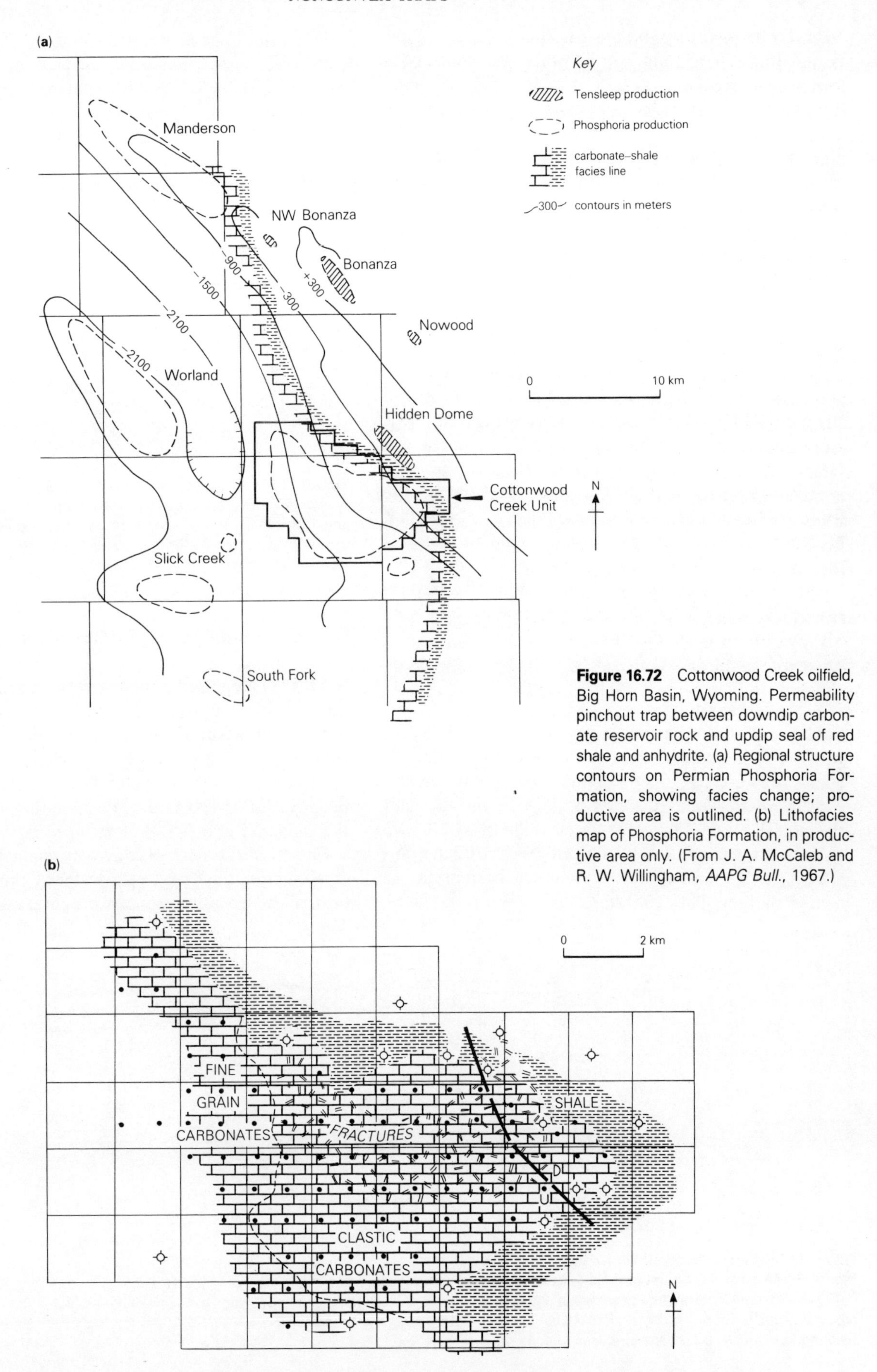

Figure 16.72 Cottonwood Creek oilfield, Big Horn Basin, Wyoming. Permeability pinchout trap between downdip carbonate reservoir rock and updip seal of red shale and anhydrite. (a) Regional structure contours on Permian Phosphoria Formation, showing facies change; productive area is outlined. (b) Lithofacies map of Phosphoria Formation, in productive area only. (From J. A. McCaleb and R. W. Willingham, *AAPG Bull.*, 1967.)

16.4.4 Permeability pinchouts

Permeability pinchouts within otherwise continuous formations provide traps in both clastic and carbonate successions. In sandstone reservoirs, the loss of permeability in the updip direction is due to increasing shaliness of the sand or to passage into a band of secondary cementation. More striking are the carbonate representatives of this trap category, about equally divided between updip transitions from dolomitized to undolomitized limestone and from limestone (occasionally dolomite) into evaporite time-equivalents.

Familiar examples of the former case occur in the Ordovician dolomites of the Cincinnati Arch and Williston Basin in the USA, overlain by shale, micrite, or anhydrite (Fig. 13.52). More prolifically productive are examples of the second case, particularly in the Permian fields of West Texas; the large Slaughter field is an example. In the Cottonwood Creek field in the Big Horn Basin of Wyoming (Fig. 16.72), the reservoir rock is another Permian carbonate; down-dip it becomes the source sediment of the Phosphoria Formation, up-dip it becomes the evaporite seal. In the Canadian sector of the Williston Basin, the Mississippian limestone reservoirs are sealed up-dip by either anhydrite or mudstone, derived from the unconformably overlying Mesozoic beds (Fig. 16.86).

16.4.5 Fault cutoffs

Fault cutoffs of homoclinally dipping reservoir beds are very common among traps. They are especially effective where they are combined with unconformities, the unconformity providing the roof of the trap and the fault the lateral termination of the permeability (Figs 16.1, 73 & 74). Both fault- and unconformity-controlled traps are considered in categories of their own in later sections of this chapter (Secs 16.5 & 7).

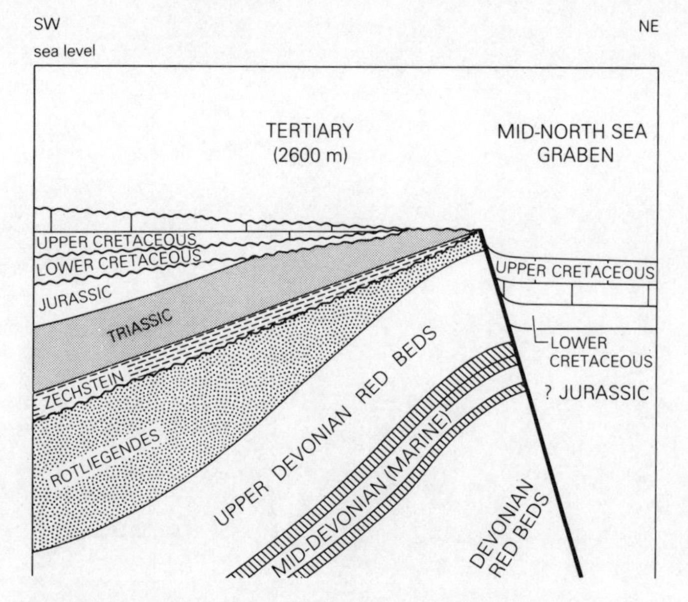

Figure 16.74 Cross section of the Argyll oilfield, UK, North Sea, in a nonconvex trap created by the intersection of a tilted fault block and an unconformity. The reservoirs are the Rotliegendes sandstone and the Zechstein dolomite, both Permian but separated by the oldest of four unconformities. (From P. E. Kent, *J. Geol. Soc.*, 1975.)

16.4.6 Hydrodynamic influence on nonconvex traps

All these categories of nonconvex trap are strongly influenced by hydrodynamic forces. We have observed in our consideration of convex traps (Sec. 16.3.9) that the change in datum pressure (Δp) across a pool is considered positive if the water flow is down the dip and negative if it is up the dip. Positive flow increases trapping capacity, negative flow reduces it.

Clearly the absence of convexity facilitates the control of accumulation by moving water. Updip flow would flush all hydrocarbons out of the reservoir unless some

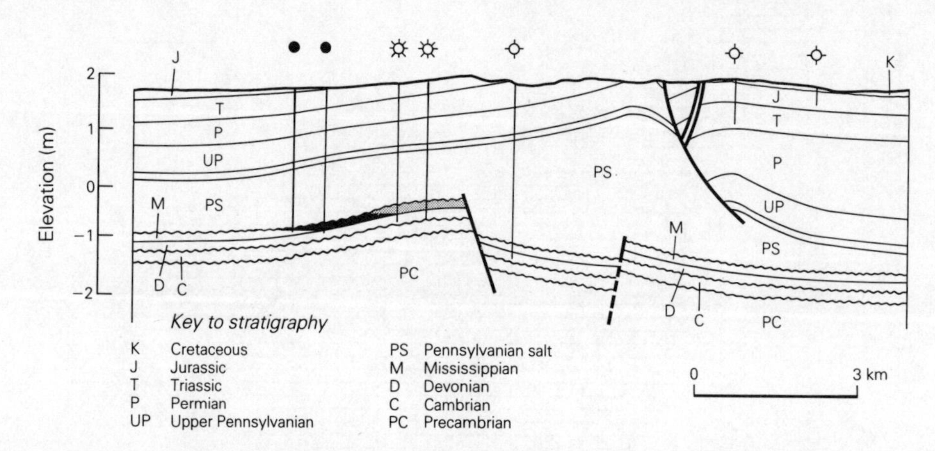

Figure 16.73 Northwest Lisbon field, Paradox Basin, Utah. Oil and gas-condensate pool in trap below unconformity overlain by thick Pennsylvanian salt. The salt thickened over the graben (center) formed during the Pennsylvanian elevation of the Ancestral Rocky Mountains by basement uplift. Thus the trap is nonconvex, in a homoclinally-dipping, truncated reservoir rock cut off updip by a normal fault; but the fundamental structure is basement block faulting, which in turn led to salt flowage. (After J. M. Parker, *AAPG Memoir* 9, Vol. 2, 1968.)

form of permeability barrier existed across it. Whether the barrier is due to wedging-out of the reservoir bed, to a facies change within it, or to a fault across it, its function is to present a *displacement pressure* too great for the hydraulic pressure to overcome (see Sec. 15.7.2).

In Chapter 15, we offered two expressions for the capillary pressure, P_c, exerted at the boundary between the permeable reservoir bed and the impermeable barrier. The pressure may be expressed in terms either of the radii of the pores or of the densities of the fluids:

$$P_c = P_t - P_p = 2\gamma \left[\frac{1}{r_t} - \frac{1}{r_p}\right] \quad (16.6)$$

$$P_c = \mathbf{g}\,(\rho_w - \rho_o)\,Z \quad (16.7)$$

The critical condition for entrapment is that P_c must not exceed the displacement pressure of the barrier. If P_c becomes sufficiently large to overcome the displacement pressure, oil migrates through the finer porosity of the barrier (represented by the factor r_t, the radius of the smallest pore throats).

The height of the oil column (Z_o) which can accumulate behind the barrier before this penetration will occur is given by Hobson's equation, modified by Robert Berg; it is simply a combination of Equations 16.6 and 16.7:

$$Z_o = \frac{2\gamma \left[\frac{1}{r_t} - \frac{1}{r_p}\right]}{\mathbf{g}\,(\rho_w - \rho_o)} \quad (16.8)$$

We may put some figures into our equations. If the displacement pressure of the barrier is 2 kg cm^{-2} (about 30 p.s.i., a fair value for a good seal, with permeability of about 0.01 mD or less); $\rho_w = 1.15$ (salt water) and $\rho_o = 0.85$ (for 35° API oil); and the hydrostatic pressure gradient for the water is a little over 0.1 kg cm^{-2} m^{-1}, then *under hydrostatic conditions*

$$Z_o \simeq \frac{2}{0.3 \times 0.1} \quad (16.9)$$

or about 65 m of oil column. Put simply, there is a linear relation between the thickness of the roof rock (the barrier) and the thickness of the hydrocarbon column it can hold behind or below it (assuming that hydrocarbons are available to fill the trap).

The oil column actually found behind permeability barriers in nonconvex traps is commonly much in excess of this calculated hydrostatic oil-holding capacity. For example, a 65 m oil column might be held under a barrier having 0.5 mD permeability, which under hydrostatic conditions would permit an oil column only about 5 m thick. The great improvement in trapping capacity is caused by *downdip water flow* (positive Δp).

Depending on the value of ($\rho_w - \rho_o$), Z_o may become two to ten times the hydraulic difference (Δh in Fig. 16.61). The most common cause of this circumstance is the entry of meteoric water into the reservoir aquifer from the surface. Becoming formation water, it flows down the dip through the oil column, increasing the pressure gradient, reducing the buoyancy (P_b), and enhancing the seal capacity (Fig. 16.75). Further updip migration of the oil becomes impossible.

Some of the largest oil and gas accumulations in the world have been strongly influenced by downdip flow of water through the reservoir or trap. They range in age and rock type from Paleozoic carbonates (as in the huge Hugoton gasfield in the southwestern USA) to Neogene sandstones (as in the Bolivar Coastal field of Venezuela; Figs 16.2 & 17.30). The trapping capacities of downdip hydrodynamic gradients are so clearly established

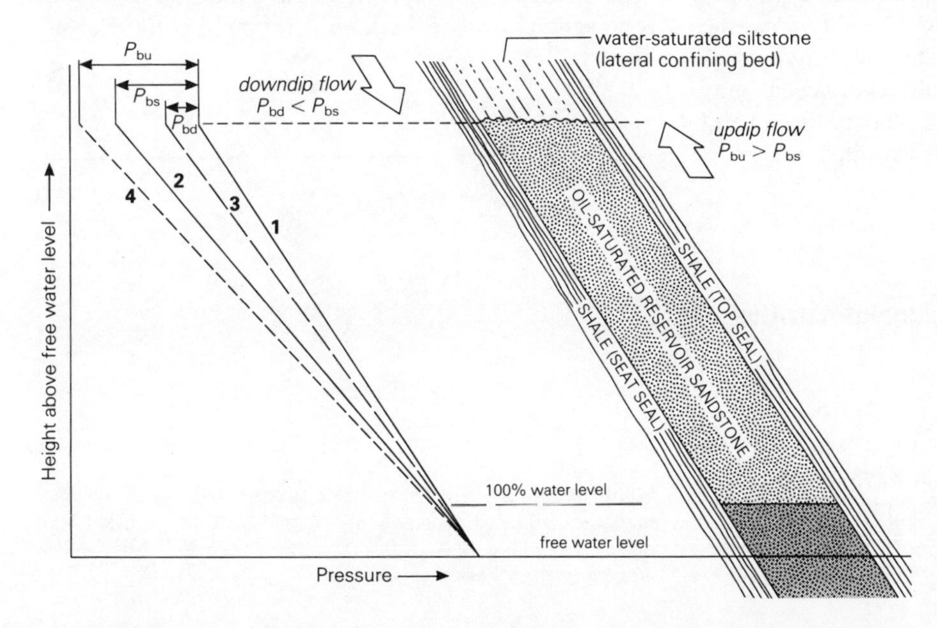

Figure 16.75 Variation in buoyant force (P_b) under hydrodynamic conditions, for constant height of oil column. Four pressure gradients are shown: 1 static oil pressure gradient (buoyant pressure = P_b); 2 static water pressure gradient (P_{bs}); 3 hydrodynamic pressure gradient with downdip flow ($P_{bd} < P_{bs}$); 4 hydrodynamic pressure gradient with updip flow ($P_{bu} > P_{bs}$). Compare with Figure 15.13. (After T. T. Schowalter, *AAPG Bull.*, 1979.)

through experience, as well as through theory, that it is a matter for astonishment that so many geologists were surprised to learn that the greatest gas deposits in the American hemisphere are trapped down-dip from water.

The "tight gas sands" of the Cretaceous of western Canada and the western USA – from the so-called "Deep Basin" in Alberta and British Columbia to the San Juan Basin in New Mexico – occupy giant stratigraphic traps resting on the post-Paleozoic unconformity along the western flank of regional arches. The sands contain gas in sandstone formations of low porosity and permeability, down-dip from more porous sands containing water. The gas is confined by a downdip flow of nearly fresh water; the waters are saline only around the outsides of the gas accumulations. Tongues of sandstone which retain high porosity down-dip into the deep basins remain water-saturated. If their porosities and permeabilities are reduced below a critical lower limit, they are saturated with gas. With porosities of 10 percent or less and permeabilities exceedingly low (say 1 mD), the permeabilities of the sands to gas are reduced sharply as the water saturation increases. At 65 percent water saturation, the sands are almost completely impermeable to gas flow.

If on the opposite hand the water flow has an *updip component*, $\triangle p$ is negative and causes a decrease in the trapping capacity. The pressure gradient is decreased, the buoyancy increased, the seal capacity reduced, and the migration potential increased (Fig. 16.75). The conditions are unfavorable for hydrocarbon entrapment. Most water originating in the sediments (connate water) tends to move up the dip (Fig. 17.17), but its flow rate is very low (25–75 mm yr^{-1} according to John Hunt). If the disposition of the strata is such that meteoric waters can readily enter the reservoir aquifer at the surface, migrate down-dip into the central parts of the basin and there possess sufficient head to continue migration up the dip on the opposite side, active water flushing of the reservoir may result. Oil and gas may then be driven out of all traps, or gas may be lost and oil reduced to mats of heavy, degraded residuum.

16.5 Traps dependent upon unconformities

16.5.1 Definitions

An *unconformity* is an interruption of the rock succession by a surface of nondeposition or erosion, with or without deformation. Some geologists restrict the term *unconformity* to surfaces separating packets of rocks having visibly different dips, making all unconformities *angular unconformities*. Surfaces separating strata in which the dips are parallel above and below are *disconformities*. The petroleum geologist should be especially careful in his use of these terms. The term *angular unconformity* is established in geological usage for a particular kind of unconformity having a single clear characteristic; it will be a pity if the term is rendered redundant by making all unconformities angular.

A break may separate parallel strata within a restricted area of outcrop or of subsurface control, and yet over a wider area be found to lie above rocks of a variety of ages. Though such a break is legitimately called a disconformity within the restricted area, on a regional scale it is an unconformity, as indeed nearly all stratigraphic breaks must be unless they represent only small intervals of geologic time. In some unfolded successions, especially in thickly bedded carbonates with somewhat irregular bedding surfaces, no clear discontinuity may be immediately apparent even if a large stratigraphic break is proven to be present by faunal evidence. This situation has been called a *paraconformity*. It is superbly exemplified in scores of outcrops in western North America, where either Devonian or Mississippian carbonates rest in near-horizontality on Cambrian or Ordovician carbonates of very similar lithology (see Sec. 16.5.2).

Even an angular unconformity may be relatively local. Common examples are those around the margins of basins (Fig. 16.76) in the centers of which the successions are continuous, and those over arches or local uplifts which interrupt sedimentation otherwise continuous off both flanks. Many classic stratigraphic or truncation traps are created by these conditions. Alternatively an unconformity may be regional, even semicontinental, in extent. Such areally widespread unconformities divide the Phanerozoic stratified series into *sequences*, which are packets of strata bounded above and below by the unconformities. Sequences are the rock records of major tectonic-depositional cycles, representing one of the largest rhythms in the Earth's history.

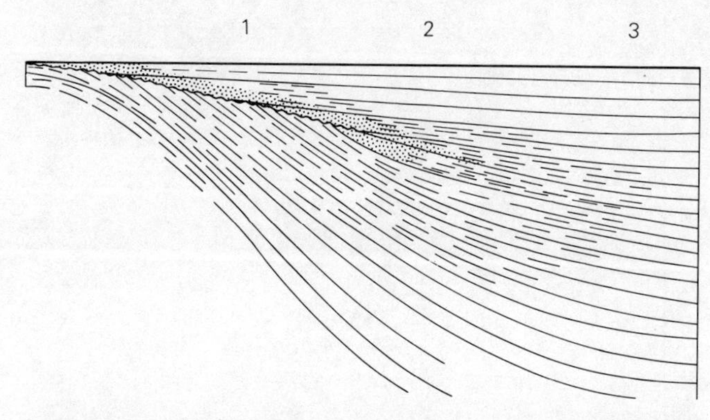

Figure 16.76 Diagrammatic section across one flank of a sedimentary basin, showing continuity of section in basin center (at 3) but angular unconformity between regressive older and transgressive younger units at basin margin (1).

16.5.2 Types of interregional unconformity

For the stratigrapher or paleontologist, the most important feature of a regional unconformity (or disconformity) may be whether it is regionally synchronous or diachronous. The former would presumably be due to eustatic sea-level changes brought about by global activity along oceanic ridges. For the seismic stratigrapher, the most important feature of a regional unconformity is its cross-sectional trace, which depends on whether the break was created wholly subaerially or at least in part in the submarine realm. The latter would favor the formation of deep-water sandstone reservoirs, such as canyon turbidites and submarine fans.

For the petroleum geologist, the prime significance of an unconformity is its trap-creating capacity. Trapping mechanisms directly associated with unconformities are exceedingly common. They may be convex or nonconvex, as indeed unconformities themselves may be, and their natures may be clarified if we consider unconformities as a structural style of their own, like asymmetrical folds or fault blocks. Interregional unconformities on forelands provide two quite distinct structural styles, which reflect two stages of the axial (geosynclinal) deformational cycle.

Axial or geosynclinal deformation is the aspect of plate interactions that raises compressional orogenic belts. As plate convergence begins, the earliest deformation involves the transfer of the rise-slope sedimentary prism on to the continental margin as nappe- and glide-slices. This propels the waters far landward over the essentially undeformed cover of the continental platform. The result is a regional paraconformity, between younger, transgressive marine strata and an older platform succession resting on a rigid foreland basement. The time-gap may be large, but it may remain relatively constant over enormous areas; the structural discordance is minor.

The transgression over the unconformity provides numerous reservoir rocks but its trap-creating capacity is largely confined to stratigraphic and lithologic traps. The early Ordovician (early Caledonian) unconformity on the two sides of the present North Atlantic provides a classic example; it is called the Knox unconformity in North America and the Tremadocian unconformity in western Europe. The sub-Devonian (in places sub-Mississippian) unconformity in the western interior of North America is an even more familiar example, perhaps because it is visible in the walls of the Grand Canyon.

Once the structural convergence has consumed the rise-slope prism, further compression involves the shelf succession also. This initiates further cratonic sags and swells on the foreland, and culminates in vast deformation, intrusion, and uplift of the basinal succession, and regression of the sea from the foreland. The entire foreland succession is then flooded with post-orogenic or molasse sediments, which may be partly or wholly nonmarine. These sediments cover all older rocks. Abrupt variations in the underlying structure and topography result in erosional gaps which may vary from negligible in downtucked belts to those across uplifted belts which may cut down all the way to the basement (which itself extends from the transitional basement of the rise-slope prism to the sialic basement of the foreland). This pattern produces an array of structural and stratigraphic traps, many of them directly below or above the unconformity.

Four prolifically productive examples of this abrupt type of unconformity may be cited: the sub-Pennsylvanian unconformity along the southern margin of the Transcontinental Arch in the southern USA (Sec. 25.2.4 & Fig. 16.4); the sub-Mesozoic unconformity on the North African craton, with Triassic, Jurassic, or Cretaceous rocks resting on all older rocks down to the Precambrian (Fig. 16.44); the sub-Cretaceous unconformity across much of the Northern Hemisphere; and the mid-Cretaceous unconformity in the Neuquen Basin of Argentina (Sec. 25.5.8). The much more varied structural style of this type of unconformity, as compared with the older type, endows it with a vastly greater array of trapping mechanisms, convex, nonconvex, and paleogeomorphic.

16.5.3 Recognition of unconformities in drillholes

The recognition of unconformities is not always easy even in outcrop. Some standard criteria for it – the truncation of faults by unfaulted younger strata, for example – are normally not revealed by drillholes, so the recognition of unconformities is still less easy there. If the stratigraphic gap or hiatus representing an unconformity is essentially constant over a wide area (as in the first of the two structural styles just described), the unconformity may be very hard to recognize in drillholes unless abundant, definitive fossils can be recovered. Ease of correlation between well logs does not imply that the sequence is complete. The rather astonishing uncertainty about the number of unconformities governing the world's most famous stratigraphic trap (in the East Texas oilfield) has been alluded to (Sec. 16.2).

Indicators of probable unconformities detectable in individual drillholes include basal conglomerates fining upwards; *remanié* beds with rolled or phosphatized pebbles and chert residua; paleosols and old regoliths, with evidence of bioturbation or plant roots; karst effects on carbonates, including secondary porosity and infilling by anhydrite or clastic detritus, especially if this is red. If the strata are fossiliferous, vital evidence may be provided by abrupt lithologic or faunal changes or by paleontologic gaps, especially among securely dated microfossils. Once a number of drillholes permit correlation of strata between them, the critical evidence is given by the identification of a peneplaned or irregular surface underlying a classic transgressive succession as described in Chapter 13.

Until an adequate number of holes have been drilled, the geologist may have trouble determining whether stratigraphic hiatuses in a few holes represent an unconformity or a zone of closely spaced faults. Some useful guides may be offered. A stratigraphic hiatus has some necessary stratigraphic sequence across it; no marker horizon may appear both above and below it in any drillhole. This is not true of a fault. A fault may pass straight through an unconformity, but it must offset it; an unconformity cannot pass straight through a fault. A fault in pre-unconformity beds may terminate upwards at the unconformity. Drillstem tests may provide conclusive clues; displacement of fluid contacts should not occur across unconformities, but they are routinely created by faults.

16.5.4 Trapping of oil at unconformities

Oil trapped at an unconformity may be in the subcrop section, the supercrop section, or both (Figs 16.2, 4, 8 & 9). Oil in the *subcrop* below an unconformity is commonly somewhat heavier than oil in the same basin removed from the unconformity; it is likely to contain little or no gas, but there are major exceptions (Figs 16.1 & 77). The reservoir may be capped by almost any

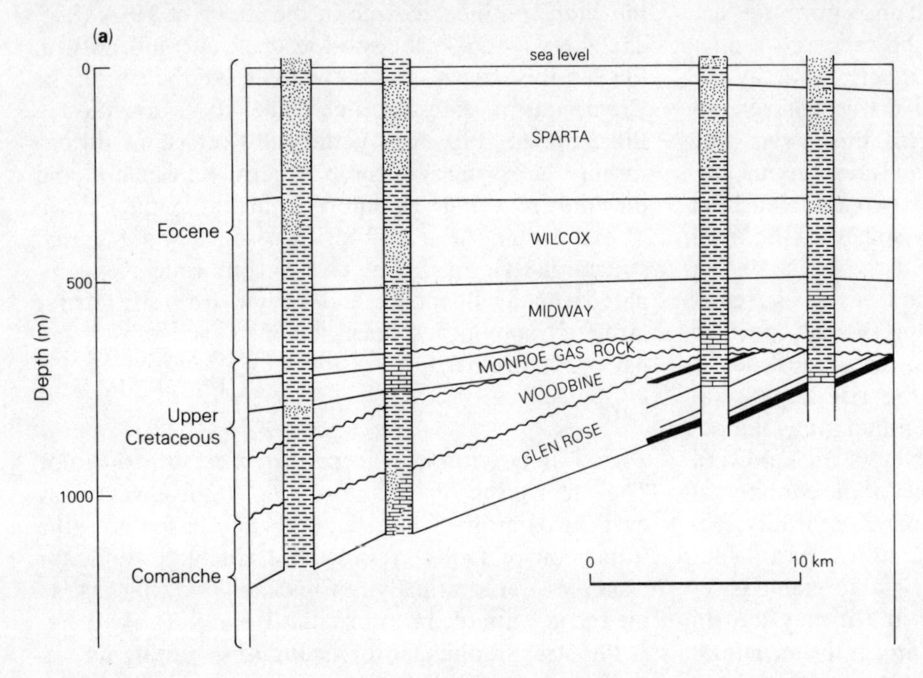

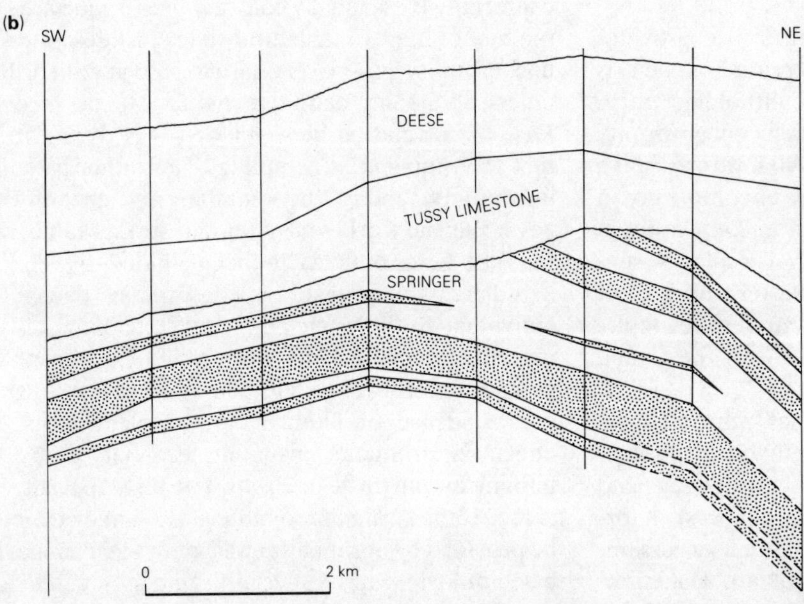

Figure 16.77 Two large fields in unconformity traps in quite different geological settings. (a) Monroe gasfield, upper Gulf Coast Basin in northern Louisiana. Principal reservoir in Cretaceous carbonate, wedging out on a truncated uplift. There is no folding, despite the triple unconformity. (b) Sholom Aleichem oilfield, now part of the Sho-Vel-Tum field in the deep Ardmore Basin between basement-cored Pennsylvanian uplifts in southern Oklahoma (see Figs 16.14, 17.1). Strong folding was directed towards the northeast. Production derived both from truncated, nonconvex traps and from true convex, anticlinal traps in Mississippian Springer sandstones, unconformably overlain by Pennsylvanian strata. ((a) After P. Fergus, *Geology of natural gas*. Tulsa, Okla: AAPG, 1935; (b) after H. R. Billingsley, *Petroleum geology of Southern Oklahoma*, Tulsa, Okla: AAPG 1956.)

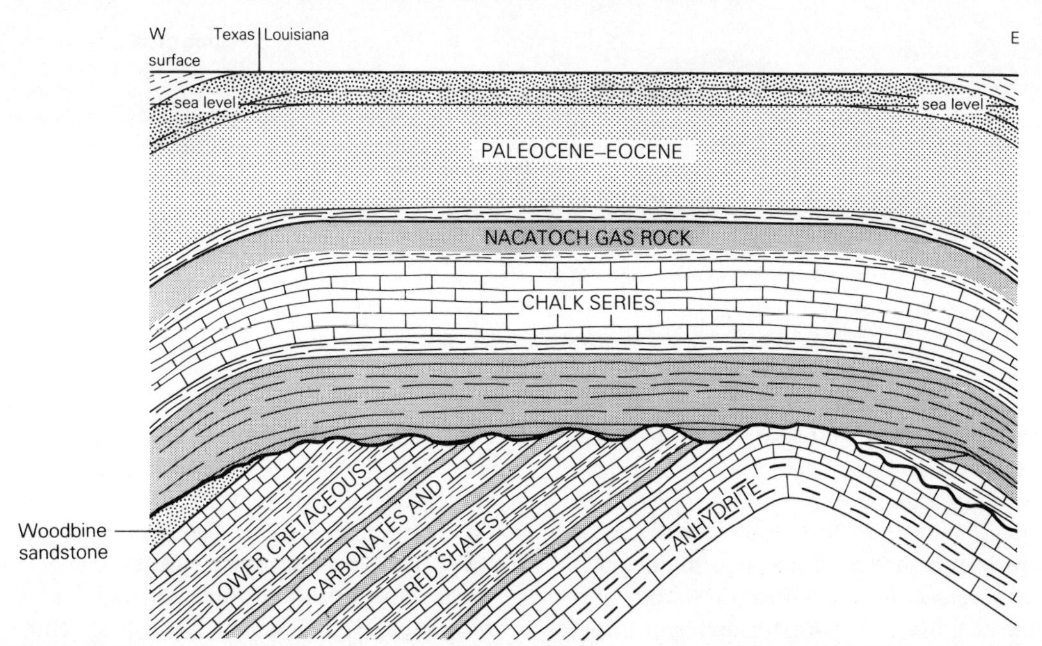

Figure 16.78 Simplified west–east cross section through the Caddo oilfield, near the top of the Sabine uplift in northwestern Louisiana. Gas occurred in the Nacatoch Formation, and oil in the "chalk rock" (in fracture porosity), both Late Cretaceous in age and arched over a buried, truncated anticline in Lower Cretaceous strata. (The amplitude of the buried anticline is exaggerated in the diagram.) The main oil zone was in the misnamed Woodbine sand, truncated on the flanks of the main anticline, which is therefore "bald-headed" to the principal reservoir rock. The field therefore combines a convex trap, nonconvex traps consequent upon convexity, and the effects of the Mid-Cretaceous unconformity. (From C. D. Fletcher, *Structure of typical American oil fields*, Tulsa, Okla: AAPG, 1929.)

sedimentary rock type, but commonly by shale, colored clay (which may represent fossil soil), red beds, or evaporites.

Among hundreds of examples of *sandstone* reservoirs in subcrop traps below unconformities, we select fields from the Suez graben (Fig. 16.29), the US Gulf Coast (Fig. 16.78), the North Sea (Figs 16.79 & 80), and the Mid-Continent region of the USA (Figs 16.4 & 77). There may be more than one regional unconformity in a petroliferous basin, and each may provide traps which may be quite unlike one another except for the dependence upon the unconformities (Fig. 16.81). If the unconformity truncates a pronouncedly convex structure, it may create a nonconvex subcrop trap far removed from the crest of that structure (Fig. 16.78).

Asphalt sealing is a special case; the oil accumulates under the surface products of its own inspissation, especially if these cover dipping beds. Very large

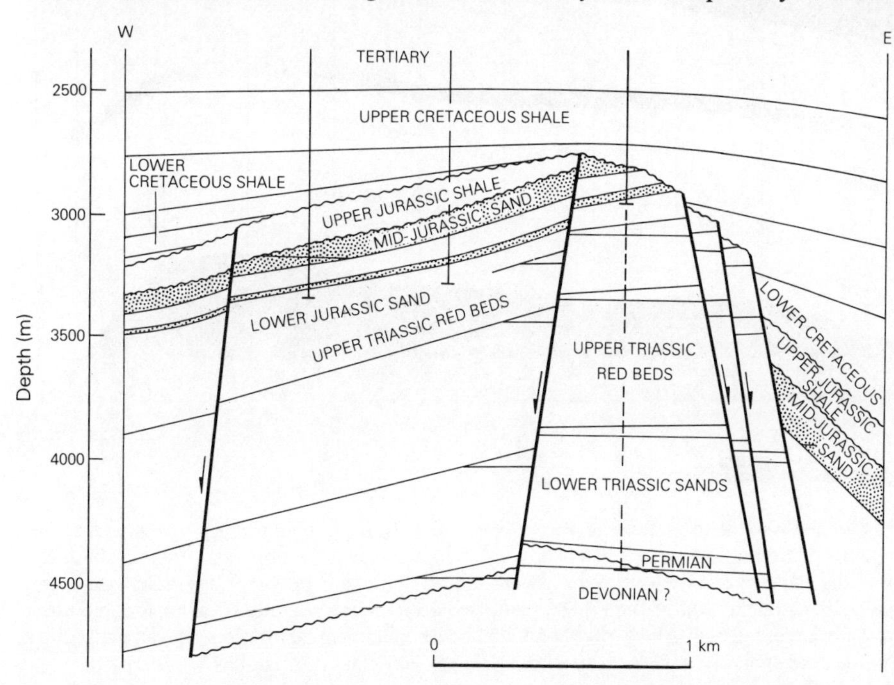

Figure 16.79 Structural cross section through the Ninian oilfield, UK, North Sea. Though the Jurassic *reservoir* is in tilted horst blocks, the trapping mechanism is an unconformity. No true fault traps are present. (Courtesy of Chevron Overseas Petroleum Inc.)

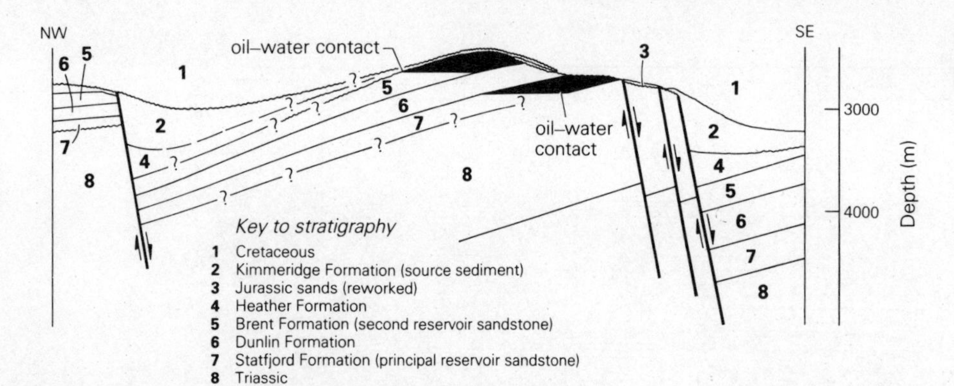

Figure 16.80 Cross section 17 km long through the Statfjord oil- and gasfield, on the UK/Norway boundary in the North Sea. Datum is 2250 m subsea. (After R. H. Kirk, *AAPG Memoir* 30, 1981.)

accumulations at least partly attributable to this type of seal occur along the updip margins of the Maracaibo, Orinoco, and San Joaquin Valley basins (Figs 10.16, 16.38 & 16.82). Over considerable distances the accumulations are under very little present cover.

Sandstones intersecting unconformities from above are called *buttress sands*. They are formed by transgression over a newly submerged erosion surface, and provide many excellent traps. Sandstone reservoirs in the *supercrop* may be illustrated by several other of the North Sea's giant fields (e.g. the Piper field, Fig. 16.9), by the Quiriquire field in eastern Venezuela (Fig. 16.83), by the Athabasca tar sand deposit (Figs 10.18 & 16.59), and by scores of small fields (Fig. 16.68). Supercrop traps also contain some enormous gasfields, including Groningen in the Netherlands and Hassi R'Mel in Algeria (Figs 16.10 & 44).

Traps in both subcrop and supercrop are found in such giant fields as the Bolivar Coastal field in Venezuela (Fig. 16.2), Midway–Sunset in California (Fig. 16.8), and Amal in Libya (Fig. 16.84). The contrast between the subcrop and supercrop traps may be extraordinary if the subcrop consists of strongly deformed strata and the supercrop is essentially undeformed. An extreme case is presented by the Vienna Basin (Fig. 16.85).

Unconformity traps in carbonate reservoirs are also very numerous, nearly all of them in the subcrop. They are especially characteristic in North America's Paleozoic: the Ellenburger (Cambro-Ordovician) fields along the Central Basin Platform of West Texas (Fig. 16.81), and the Mississippian fields of western Alberta, Kansas, and the northeastern edge of the Williston Basin in Saskatchewan (Canada). The last are of extra interest because of the role of evaporites as seals. The reservoir carbonates themselves are interbedded with salt and anhydrite units, which form lateral seals, and the entire

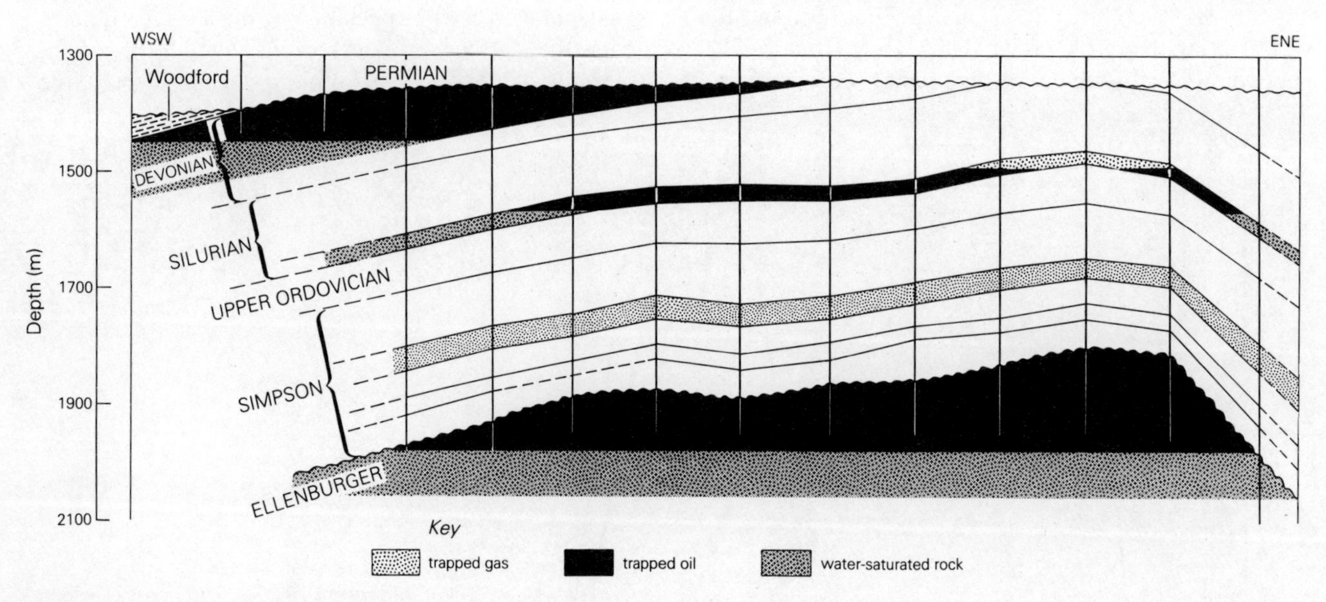

Figure 16.81 TXL field, on the east side of the Central Basin Platform, Permian Basin, West Texas. The cross section is perpendicular to the long dimension of the field, showing multiple oil pools in Cambro-Ordovician (Ellenburger), Silurian, and Devonian formations. Principal traps are below unconformities, the youngest of which is the Pennsylvanian unconformity overlapped by Permian. However, the main trap in the Ellenburger is also a true convex trap, in the center of the structure; that in the Silurian, also in the center of the structure, is a simple convex trap unassociated with any unconformity. The principal source sediments (Woodford and Simpson Formations) bear different relations to their unconformity-bounded reservoir rocks. (From C. G. Cooper and B. J. Ferris, University of Texas Publication 5716, 1957.)

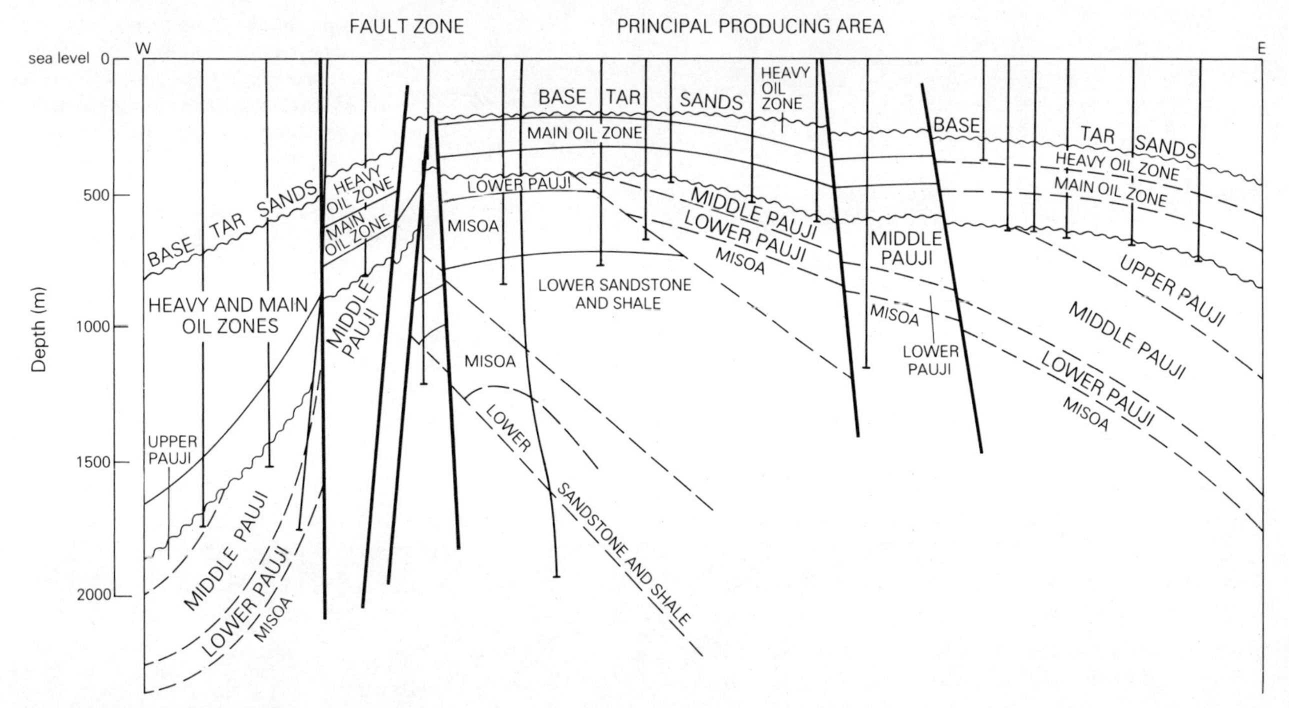

Figure 16.82 Cross section through the Mene Grande oilfield, Maracaibo Basin, Venezuela (horizontal and vertical scales equivalent). Note the ubiquitous asphalt seal and the confinement of the principal producing horizons to the interval between two regional unconformities. (From R. W. Fary Jr, US Geological Survey Open File Report 80–782, 1980; after E. Mencher *et al.*, *AAPG Bull.*, 1953.)

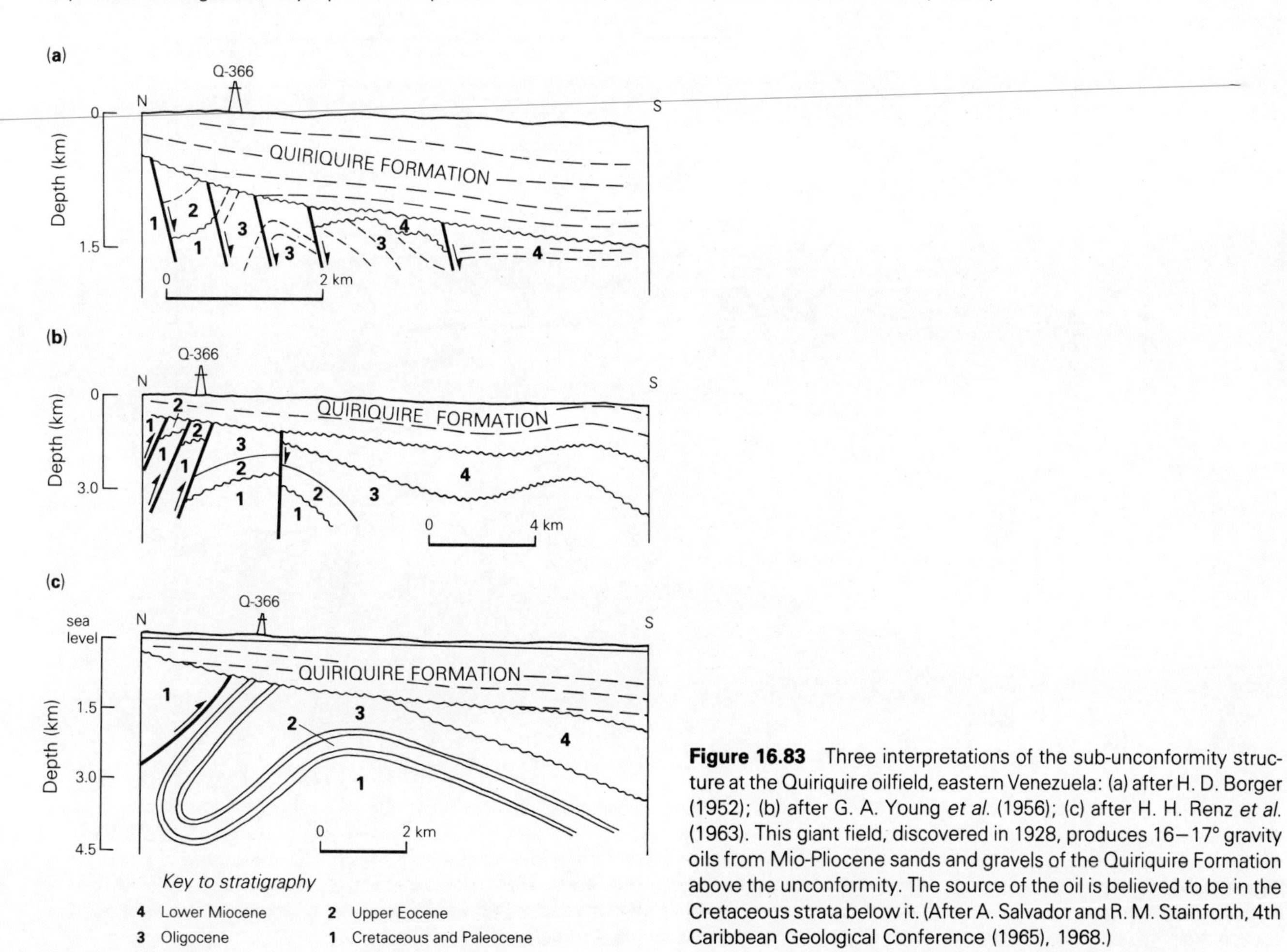

Figure 16.83 Three interpretations of the sub-unconformity structure at the Quiriquire oilfield, eastern Venezuela: (a) after H. D. Borger (1952); (b) after G. A. Young *et al.* (1956); (c) after H. H. Renz *et al.* (1963). This giant field, discovered in 1928, produces 16–17° gravity oils from Mio-Pliocene sands and gravels of the Quiriquire Formation above the unconformity. The source of the oil is believed to be in the Cretaceous strata below it. (After A. Salvador and R. M. Stainforth, 4th Caribbean Geological Conference (1965), 1968.)

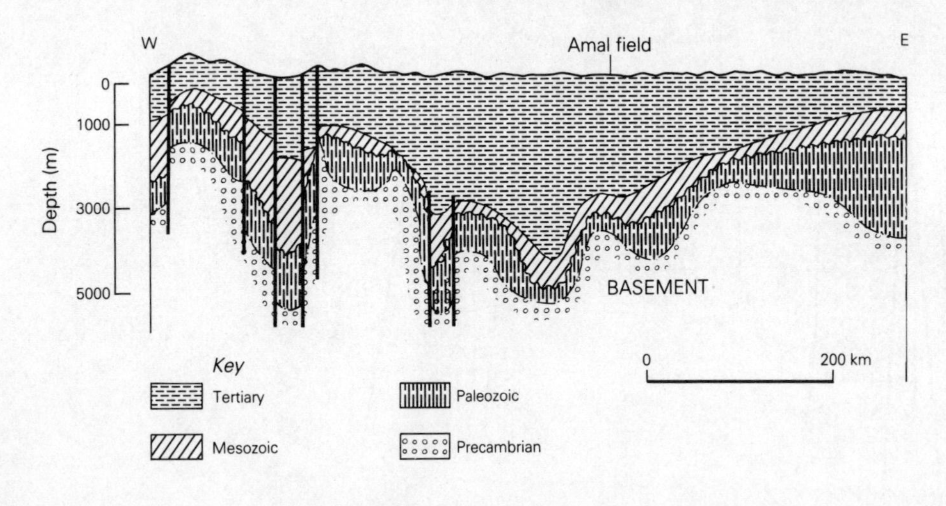

Figure 16.84 Generalized structure section through northern Sirte Basin, Libya, showing the relation of the Amal oilfield to regional unconformity between Paleozoic and Mesozoic strata. Reservoir and trap are in Cambro-Ordovician sandstone below the unconformity and Lower Cretaceous sandstone above it, where uplifted by a basement-controlled bending fold. (From J. M. Roberts, *AAPG Memoir* 14, 1970.)

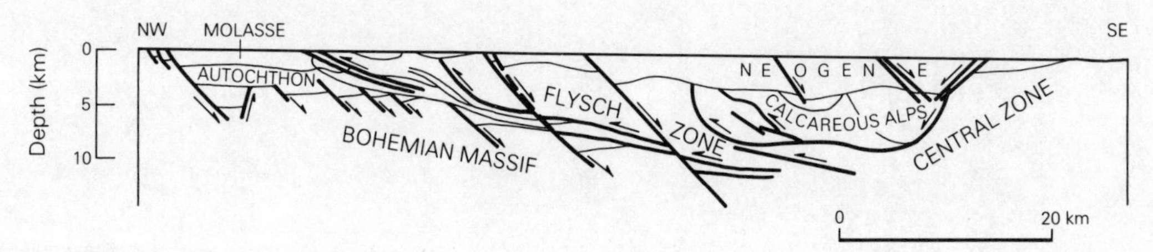

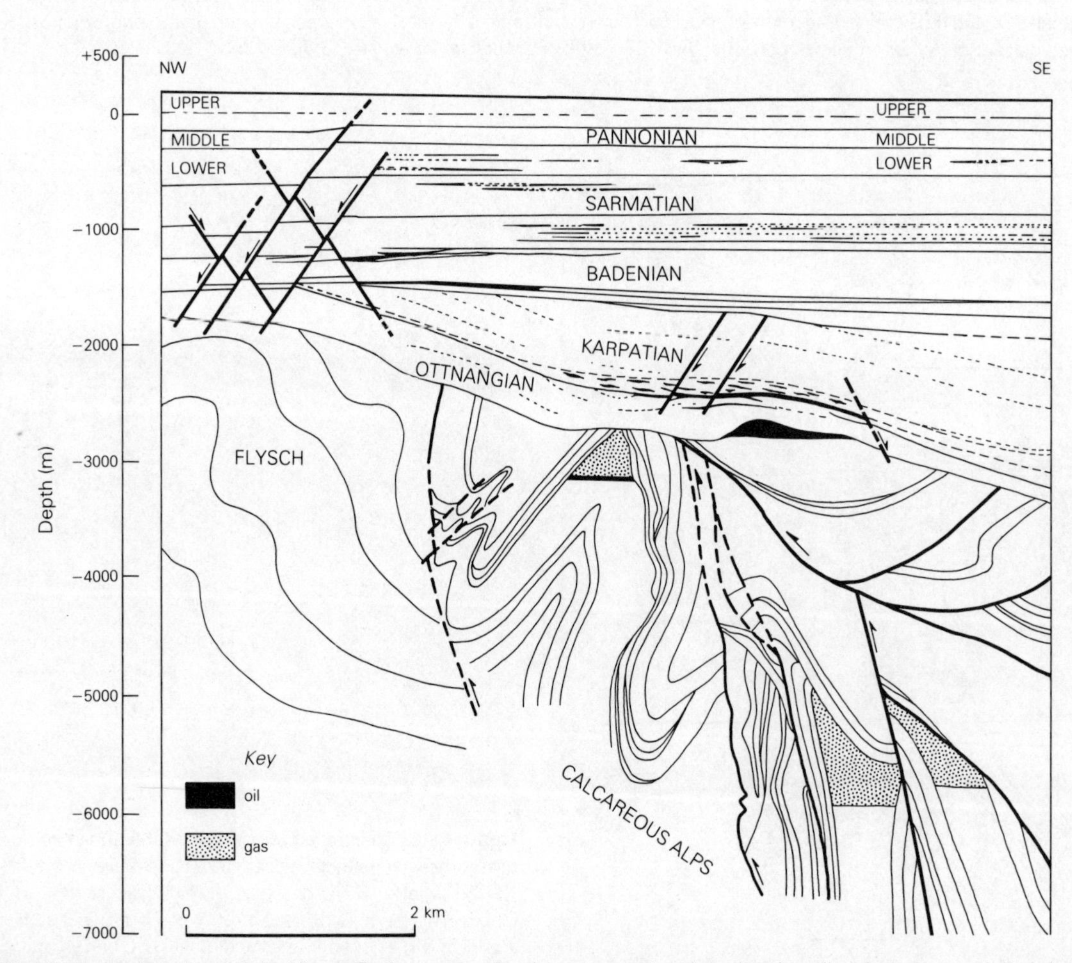

Figure 16.85 Cross sections through the Vienna Basin, at natural and exaggerated scales, showing oil and gas accumulations both in faulted (but unfolded) Neogene strata above an angular unconformity, and in violently folded and faulted Mesozoic strata overthrust against Cretaceo-Eocene Alpine flysch below the unconformity. (From D. H. Welte *et al.*, *Chem. Geol.*, 1982.)

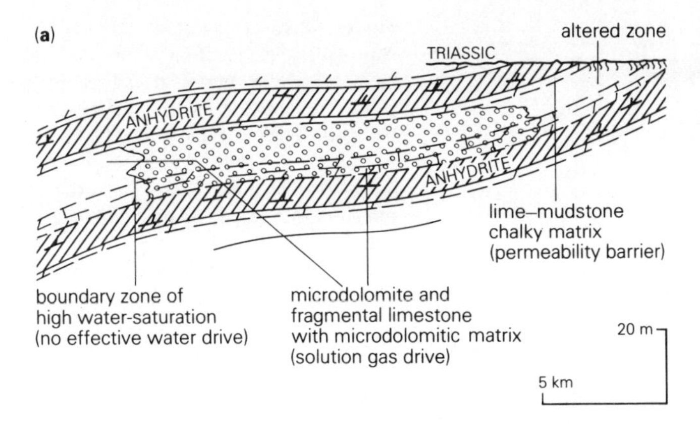

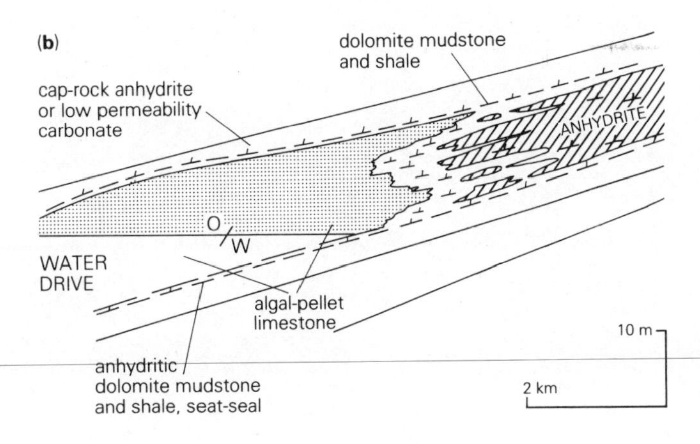

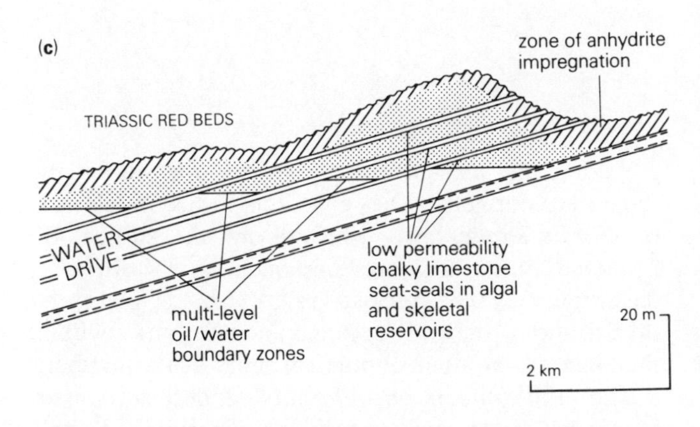

Figure 16.86 Three principal kinds of stratigraphic trap in the Mississippian strata of the Williston Basin. Effective oil-bearing reservoirs are stippled. (a) Trap in dolomitized carbonates associated with fossil sabkha deposits; (b) trap in off-sabkha pellet-limestone shoal; (c) trap in unconformity-sealed reservoir. Either of the two anhydrite deposits shown in (a) when traced downdip is found to break up into anhydrite stringers and pockets closely associated with microdolomite, which in turn passes into algal-pellet limestone. This is illustrated in (b). The complete facies change spreads over about 2 km. (From L. V. Illing *et al.*, 7th World Petroleum Congress, 1967.)

reservoir succession, after bevelling, lies in subcrop position below Mesozoic evaporites which form the top seal (Fig. 16.86).

Few carbonate reservoirs are as complexly deformed as are many sandstone reservoirs; traps like those illustrated in Figures 16.18, 19 and 53 are almost unimaginable in carbonates. The little Heide field beside the Kiel Canal in Germany is an illuminating exception (Fig. 16.87). Containing three of the classic trapping mechanisms (limestone anticline, salt dome, and permeability pinchouts in sandstones), the field's numerous individual pools are nonetheless all trapped below two Cretaceous unconformities.

The *source* of the oil reservoired above or below unconformities may itself lie either above or below, more commonly below, or laterally down-dip. In the fields on the eastern side of Lake Maracaibo in Venezuela, and those along the northeastern border of the Williston Basin in Saskatchewan, for examples, there is no doubt that the principal source rock lies below the trapping unconformity (Figs 16.2 & 86). At Prudhoe Bay in Alaska and along the Nemaha Ridge in Oklahoma, the principal source sediment is probably the one immediately overlying the unconformity (Figs 16.1 & 4). In some cases, the source sediment lies between two unconformities, as in several North Sea fields and in the Songliao Basin of China. In such cases (Figs 16.9, 16.77 & 17.24), the same source sediment may provide oil for both the subcrop and supercrop reservoirs.

In still other cases, there have apparently been two episodes of generation, the first being lost at the surface during the creation of the unconformity, the second generated after retransgression of the eroded rocks and their burial beneath younger sediments. The second generation may come from source rocks still below the unconformity, immature at the time of erosion and preserved from it (the "preservation window" of Gregory Webb, 1976), or from younger source sediments overlapping the bevelled strata.

16.5.5 The significance of unconformity traps

The significance of unconformities as trapping mechanisms is understated in most compilations of oil- and gasfields by type. In a 1970 compilation of giant fields by Michel Halbouty *et al.*, only 25 percent of the oilfields were said to have an unconformity as an *essential* factor in the trap; surprisingly, this was claimed for 42 percent of the giant gasfields. But unconformities are *important* factors in a great variety of traps. We have already surveyed their common association with non-convex or stratigraphic traps (see Sec. 16.4). They are also vital features of many traps that are normally considered structural in nature. In dozens of fields that were

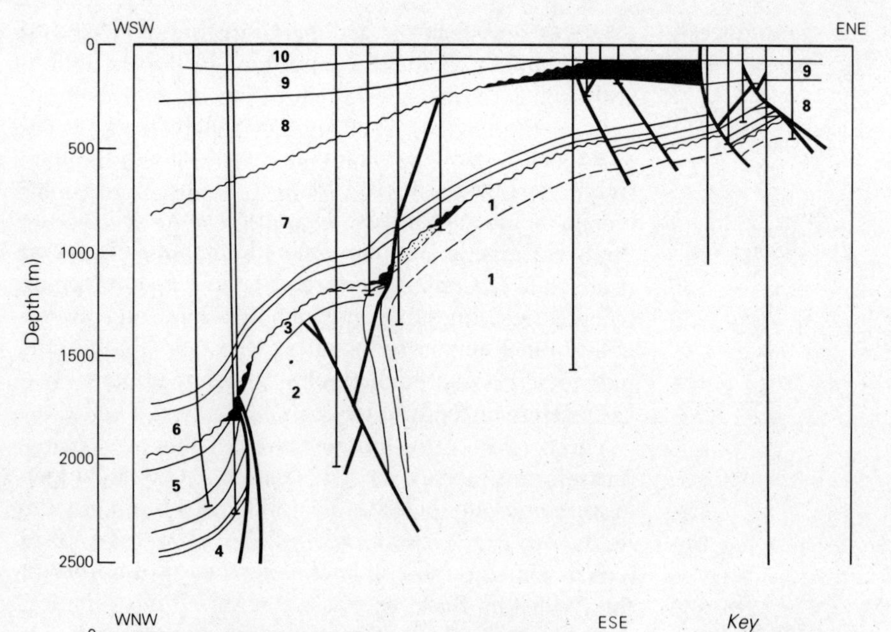

Figure 16.87 Heide oilfield, near the German/Danish border: two cross sections across the field; oil pools in black. The structure is a highly complex, elongate anticline, bounded by faults and containing a diapiric salt core. Yet the actual trapping mechanism is provided by the sub-Cretaceous and post-Cretaceous unconformities. (After H. Weber, XX International Geol. Congress, Mexico City, 1956.)

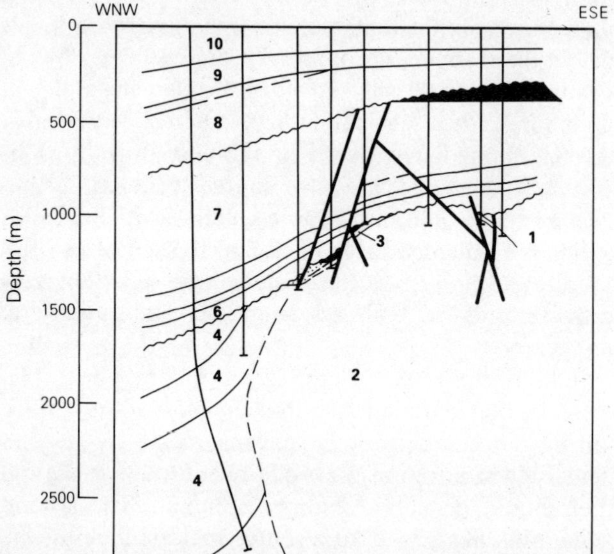

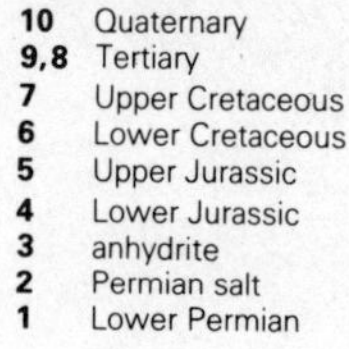

discovered because they were in anticlines, the actual trapping mechanisms are unconformities; Prudhoe Bay, Oklahoma City, and Hassi Messaoud are large, famous examples (Figs 16.1, 16.4 & 26.17).

In all petroliferous grabens, the basic structures of many individual fields are horsts bounded by faults. In only a minority of them are the actual pools sealed by faults; they are more likely to be sealed at unconformities, as in the examples from the Viking (North Sea), Suez, and Reconcavo grabens shown in Figures 16.5, 9, 29, 31, 79, and 80. The greatest "tar sand" deposits are all intimately associated with unconformities (Figs 10.16, 10.18 & 16.59).

16.5.6 The great unconformities

The significance of *interregional* unconformities is difficult to overstate. They reflect the great cratonic transgressions, and are surprisingly synchronous in widely separated continents. They constitute the essential basis of *seismic stratigraphy*, which forms the subject of Section 21.9. As they are consequences of worldwide diastrophic events, themselves aspects of plate reorganizations, the truly regional unconformities within the Phanerozoic are no more than about five in number.

The *Caledonian unconformity* underlies Upper Devonian strata (occasionally Mississippian) in the Ural–Volga, western Canadian, and eastern Australian basins. *Hercynian unconformities* underlie Permo-Pennsylvanian or Permo-Triassic rocks in the Ural–Volga and Mid-Continent basins, across most of eastern Australia and North Africa, and in Iran; in any single area this unconformity is very likely to lie within the Upper Carboniferous succession. The *sub-Jurassic unconformity* may reflect the intra-Triassic "older Cimmerian" disturbance or it may reflect delayed transgression of a Hercynian unconformity; it is present in western Siberia and in the earliest of the "drift" basins, but it is not a critical factor in trap formation.

The most important single structural feature responsible for trapping oil or gas on a worldwide basis is no doubt the anticline. But "anticlines" are not a genetically homogeneous category of structures, as we have already observed. There is a case to be made for the claim that the most important structural phenomena involved in the trapping of oil and gas on a worldwide basis are the *sub-Cretaceous* and *intra-Cretaceous unconformities*. Unconformities below Cretaceous strata are vital features of the North Sea and nearly all other graben basins, the Western Canadian Basin, the Arctic slope of Alaska, and nearly every Andean basin in South America. Still more widespread and important is the unconformity related to the worldwide mid-Cretaceous transgression; its most common stratigraphic level is post-Cenomanian, pre-Turonian, but the level naturally varies from basin to basin. In many rich basins this unconformity is informally used to separate the Lower Cretaceous from the Upper Cretaceous, or the Middle Cretaceous from the Upper Cretaceous, according to the preferred local nomenclature. It is important in the Middle East (where it is essentially the only unconformity having any significance in the oil-producing areas), the upper Gulf Coast, the North Sea, the Arctic slope of Alaska, and the circum-Atlantic basins.

The last worldwide unconformity important to our concern is the *post-Eocene unconformity*, overlain either by Oligocene or by Miocene rocks. It is important in California, the Magdalena and Maracaibo basins in the northern Andes, the Caucasus, Sicily, northern Sumatra, and in drift basins around the present ocean margins, such as the Cambay Basin in India, the Gippsland Basin in Australia, and the Bohai Gulf in China.

16.6 Paleogeomorphic traps

We have recognized a category of convex traps owing their convexities to humps that protruded above the surfaces on which the sediments were deposited. The most familiar of these humps are the buried hills or ridges (Sec. 16.3.7). In a number of nonconvex traps, the reservoir beds abut up-dip against such buried humps (Fig. 13.76); others abut against the banks of rivers or submarine channels incised into the older surfaces. A very much larger number occur in reservoir beds themselves truncated by the erosional surfaces; they have been included among our "unconformity traps."

All these trap types share the common characteristic of control by an old landscape; they were called *paleogeomorphic traps* by Rudolf Martin in 1966. Unless they happen to be convex, they are difficult to detect in advance of drilling. There must therefore be a large number still undiscovered, at least outside North America, and it is worthwhile to consider the characteristics they possess simply by virtue of their geomorphic origin.

Paleogeomorphic traps fall fundamentally into two categories, each itself twofold: those associated with prominences that rose above the levels of initial sedimentation, and those associated with erosional depressions below those levels. In the "prominence" group, the hydrocarbons may be in younger strata which abut against the flanks of the prominences. Important examples include the fields on the flanks of the Precambrian Central Kansas uplift in the US Mid-Continent (Fig. 13.78); the Panhandle oil- and gasfield in northern Texas, along the edge of the crystalline Amarillo uplift after it rose during the Pennsylvanian (Fig. 13.21); and many fields in basal sandstones above unconformities. An excellent example of these last is the Peace River heavy oil deposit in Alberta (Fig. 10.18), in Lower Cretaceous sandstones wedging out against an erosional cuesta of Mississippian limestone.

Alternatively, the accumulation may be within the erosional landforms themselves (Figs 16.88 & 89). The Mississippian limestone cuestas causing the trap of Figure 10.18 continue around the northeastern edge of the Williston Basin in Saskatchewan. There they contain oil beneath a horizontal cover of mineralized Mesozoic red beds and evaporites (Fig. 16.86).

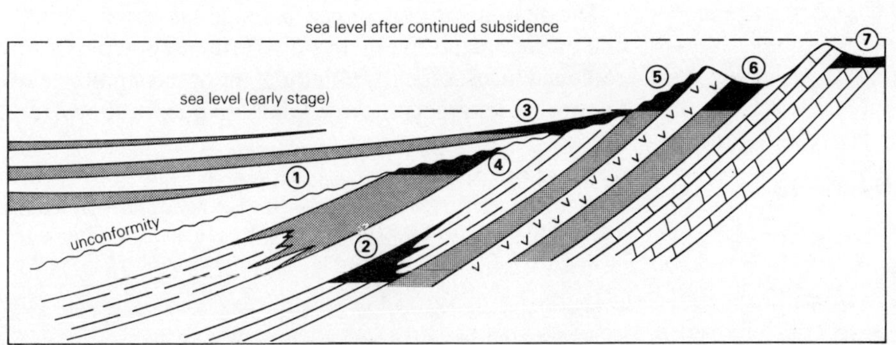

Key
1,2 true stratigraphic traps
3 unconformity traps above the unconformity
4 unconformity traps below the unconformity
5–7 paleogeomorphic traps (in buried hills)

Figure 16.88 Traps dependent on the presence of buried hills during basinal subsidence and marine transgression. (From R. Martin, *AAPG Bull.*, 1966.)

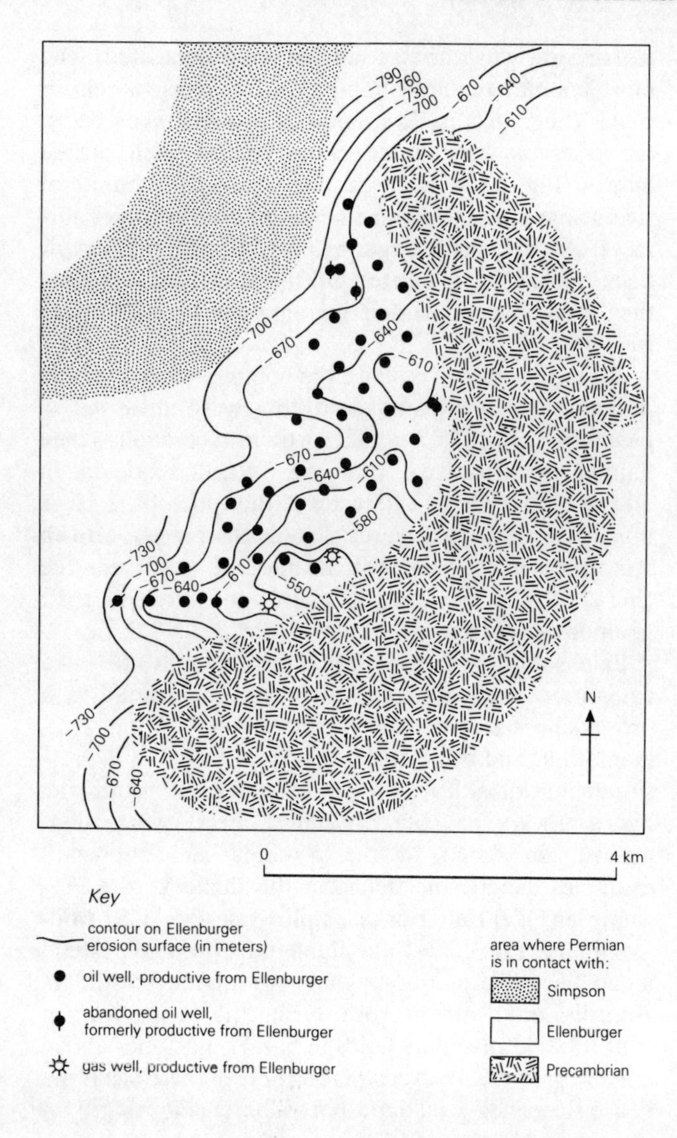

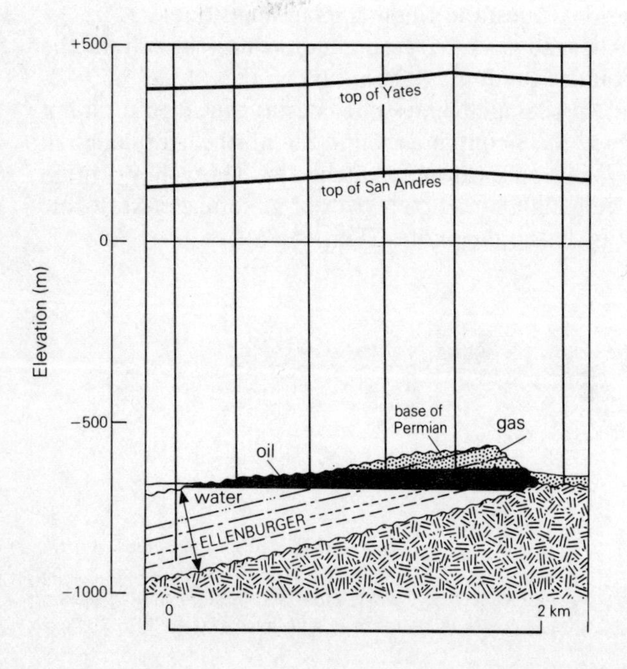

Figure 16.89 Apco oil- and gasfield, West Texas: a paleogeomorphic trap in an erosional cuesta of the Cambro-Ordovician Ellenburger dolomite, overlapped by Permian sediments. Structure contours, at 30 m intervals, are on the erosional surface. The heavily stippled area lacks the Ellenburger Formation due to truncation; the Permian there rests directly on the Precambrian. (After S. P. Ellison Jr, University of Texas Publication 5716, 1957.)

The "erosional depression" category is similarly twofold. The oil may be in undeformed reservoirs (usually of sandstone) in the older succession, cut into by river channels, submarine channels or canyons, or glacial valleys. Of historical interest is the original Drake well in the Oil Creek field in Pennsylvania, the birthplace of the American oil industry. The oil was trapped in a "stray" sandstone at the top of the Devonian, the bedrock in the region, where it was truncated by a glacial valley containing lake sediments (Fig. 16.90). In the Lloydminster region of western Canada, much of the heavy oil is pooled in Lower Cretaceous sandstones cut off by erosional gulleys now containing impervious shales (Fig. 16.91). Oil trapped in buried river terrace deposits would be under exactly the same trapping conditions. A more surprising circumstance is the trapping of *gas* in erosional remnants between incised channels (Fig. 16.92).

Alternatively, the reservoirs may be in the contents of the erosional depressions, not in the rocks into which the depressions were incised. Channel sandstone reservoirs are in this category; many examples were offered in Chapter 13. Illustrated here (Fig. 16.93) is the Cretaceous sandstone fill forming part of the reservoir in the Bell Creek field in Montana. Whereas the Lloydminster channels of Figure 16.91 were cut into reservoir sandstones and themselves contain shales, the Bell Creek channels were cut into shales and themselves contain reservoir sandstones. The Bell Creek valley channels, like those in other parts of the US Rocky Mountain province, were controlled by paleostructures brought about by recurrent movement of basement fault blocks during the Nevadan deformation. Some of the channels began as true grabens.

The Chicontepec area of Mexico, lying southwest of the old Golden Lane and Poza Rica fields (Figs 13.50 & 26.20), contains the same stratigraphic succession as is productive in those fields, but the whole succession was incised during the Eocene by deep-water, turbidite channels. The total oil in place within the channel fills may be very large, but its distribution in detail appears to be controlled by sand percentage, which is very erratic and may prevent the achievement of any high rate of production.

The search for nonconvex paleogeomorphic traps is one of the most difficult and frustrating the exploration geologist faces, especially in the light of the small size of the probable targets. Like all detective work, however,

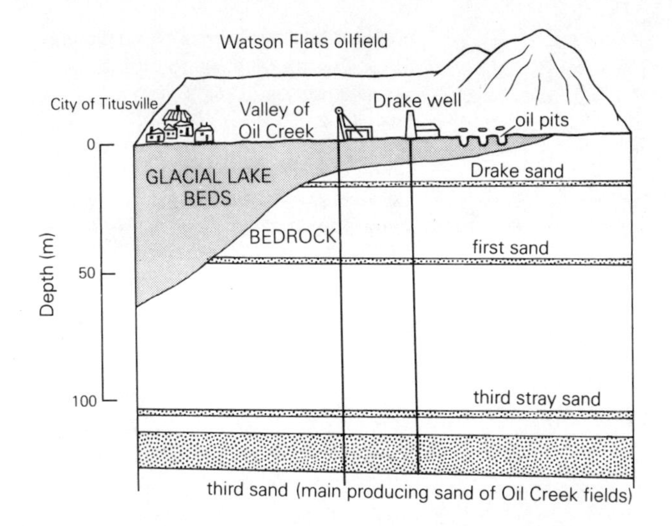

Figure 16.90 Geology of the Drake well. Pennsylvania, 1859. The wellsite was chosen at a seepage of light oil from Devonian sandstone breached by glacial erosion during the Pleistocene. The accumulation is in a paleogeomorphic trap. The well was deepened to Third sand in 1888. (From P. A. Dickey and J. M. Hunt, *AAPG Memoir* 16, 1972.)

its pursuit to a successful conclusion provides rare satisfaction. The process is a variant of that prescribed earlier (Sec. 13.2.10) for prospecting for channeled sand bodies of any type. The objective now is to reconstruct an erosional surface as it was before it became tilted, buried, and perhaps deformed.

First, sufficient drilling is necessary to identify an unconformity having some identifiable pattern of relief below it. Second, this unconformity must be carefully contoured, and then recontoured with any regional dip taken out (see Sec. 22.3). This provides a topographic map of the landscape before it became buried by younger sediments. Third, the best available time-parallel marker bed in the post-unconformity section must be chosen. A bentonite or thin volcanic ash bed is ideal, its deposition being to all intents and purposes geologically instantaneous; but a thin, persistent coal seam or grit layer will serve. Fourth, the subsea elevations of this marker are referred to some arbitrary but suitable and constant datum plane.

If the elevations so derived are not horizontal, or at

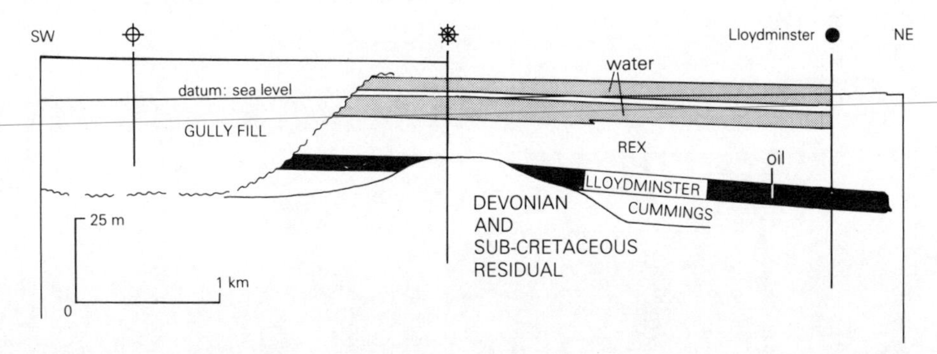

Figure 16.91 Correlation section through part of the South Chauvin oilfield, Lloydminster area, eastern Alberta, Canada, showing one of many paleogeomorphic traps in Lower Cretaceous sandstones. This location illustrates two trapping mechanisms, one against a pre-depositional high, the other against a post-depositional gully. (After L. W. Vigrass, *AAPG Bull.*, 1977.)

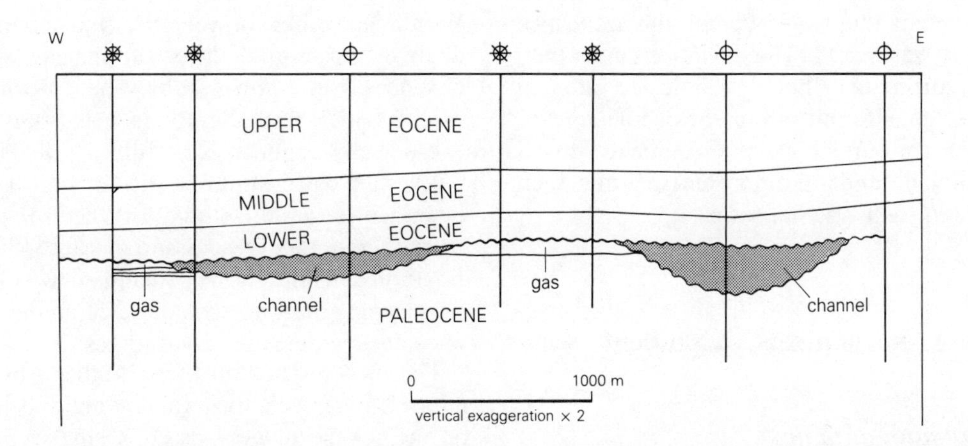

Figure 16.92 Cross section through West Thornton gasfield, Sacramento Valley, California. Paleogeomorphic traps in Paleocene sandstone erosional remnants, between submarine channels filled with clay. (After A. B. Dickas and J. L. Payne, *AAPG Bull.*, 1967.)

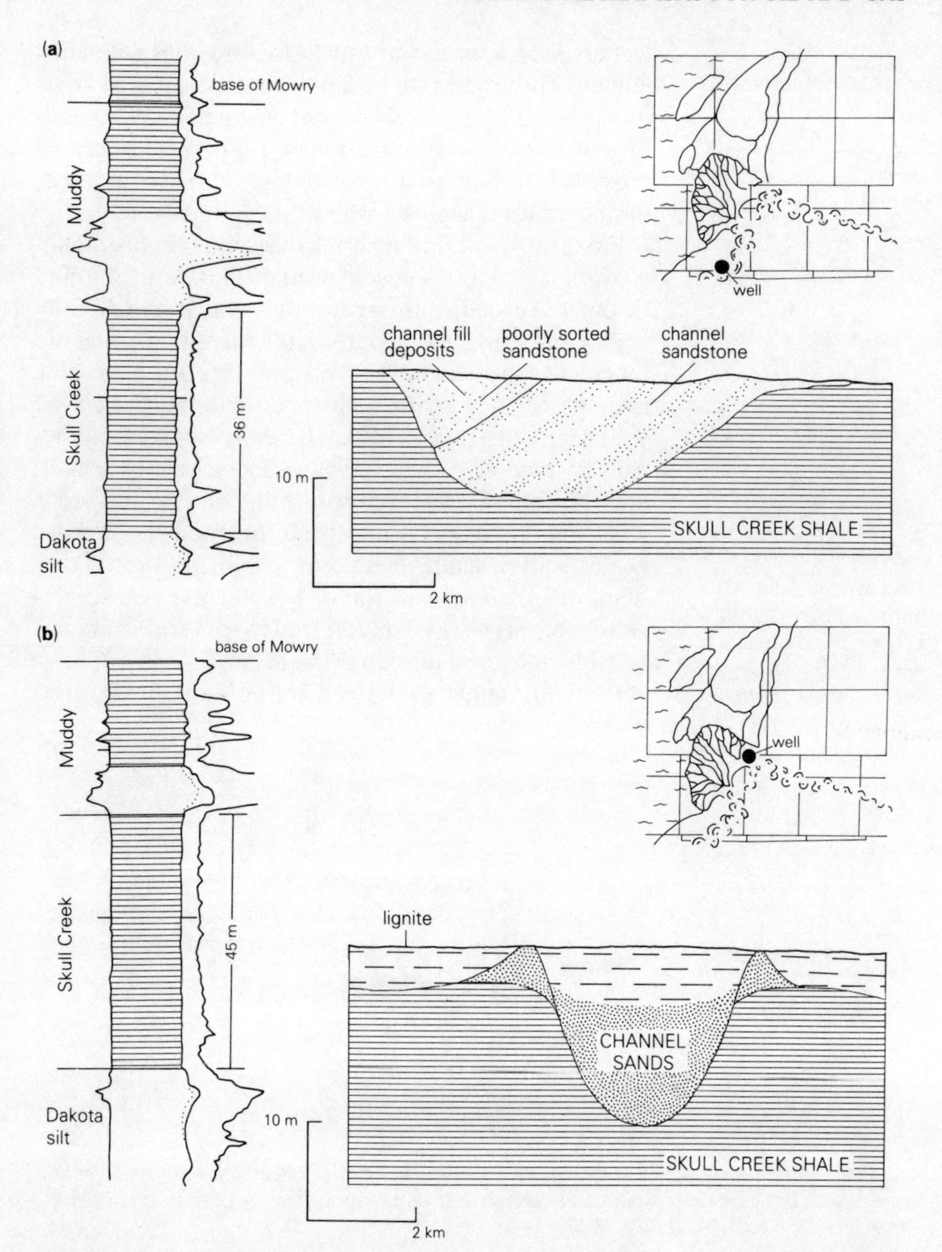

Figure 16.93 Bell Creek oilfield, Powder River Basin, Montana, in Cretaceous sandstones, to illustrate paleogeomorphic trapping mechanisms. (a) Oil in meandering channel sandstones deposited on eroded shales; (b) oil in distributary channel sandstone in delta-marsh facies, also deposited on eroded shales. (After A. A. McGregor and C. A. Biggs, *AAPG Memoir* 14, 1970.)

least planar and nearly horizontal, the deviations represent movements that have affected the area since the unconformity was buried. The deviations must then be applied as corrections, hole by hole, to the log readings of the unconformity, and the whole unconformable surface contoured for the third time to show it as it was originally. Potential geomorphic traps may then be recognizable in their original form.

16.7 Trapping mechanisms dependent upon faults

16.7.1 The ubiquity of faults

Statistical studies by Russian geologists purport to have shown a clear association between oil- and gasfields and *faults*, especially deep-seated faults on platforms. Within petroliferous regions, the percentage of productive "structures" drops off sharply with increasing distances from known faults. The control by fault densities is said to be clearer for gasfields than for oilfields; it is especially applicable to multipay fields.

Russian basin studies on fault densities were later extended to satellite studies of fracture densities, particularly over the West Siberian Basin. The maximum fracture densities were concluded to characterize unpromising areas; intermediate fault densities favored oil, and low densities favored gas.

The impetus behind these studies was the belief of many Russian scientists (and some non-Russians) that oil and gas are of deep-seated origin (see Sec. 6.2.2). Oil and gas can on this hypothesis attain their shallow crustal positions only by ascent along deep-reaching fractures.

Geologists and geochemists who believe that oil and gas migrate into their shallow traps in aqueous solution (see Sec. 15.4) require solvent temperatures of 180 °C or considerably higher. Again, vertical ascent through 6–10 km of crustal rocks implies association of the hydrocarbons with faults reaching to those depths. Within the shallow, productive section itself there is ample evidence that oil and gas can become preferentially re-pooled about faults that cut their original reservoirs; the Los Bajos wrench fault in southwestern Trinidad is an often-quoted example (Fig. 22.9).

We have both theoretical and observational justification, however, for regarding the Russian conclusions as teleological. It is impossible for any segment of the Earth's crust to subside over a long time, becoming a basin in which hydrocarbons may be generated, without acquiring some secondary structural features. It is equally impossible for a segment to be uplifted over a long time, becoming an arch or dome towards which hydrocarbons tend to migrate, without similarly acquiring such structures. One may or may not subscribe to Giuseppe Merla's dictum that rocks bend only when they cannot be faulted; but there is no doubt that supracrustal rocks have spent much time under conditions in which failure by faulting is more likely than failure by folding. The most common of all "structures" are faults of one kind or another. Any structure that develops "against the grain," becoming for example a "high" within a subsiding basin, is intrinsically likely to become faulted.

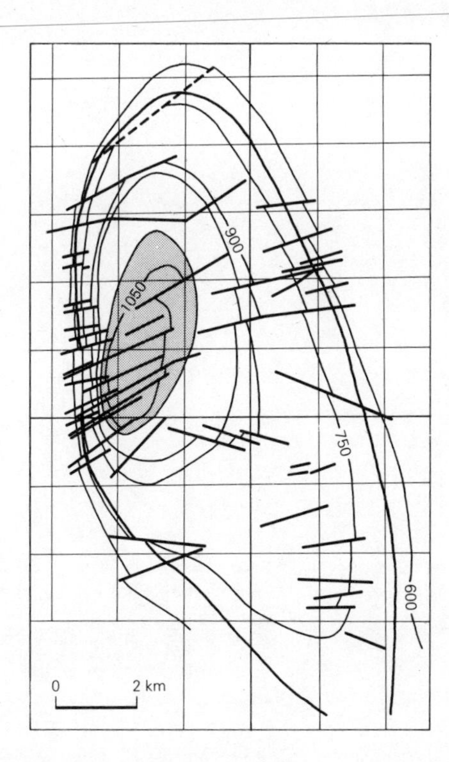

Figure 16.94 Salt Creek oilfield, Wyoming: structure contours (in meters above sea level) on the principal reservoir sandstone (one of four such reservoirs). Area of production shaded. To illustrate irrelevance to trapping mechanism of numerous faults that cut it. (From J. A. Barlow, Jr, and J. D. Haun, *AAPG Memoir* 14, 1970.)

The most favored traps are versions of the anticline, as we have seen. If we exclude reefs, all convex structures are inherently likely to be associated with faults – radial, transverse, longitudinal, or marginal. Great numbers of faults in oilfield areas are quite irrelevant to the traps. Figure 16.94 shows the Salt Creek field in Wyoming, in a large anticline mappable at the surface. The structure is cut by scores of faults, but they are irrelevant to the trapping mechanism and they would be present on that anticline even if it were utterly barren of oil or gas. A nonpetroliferous basin may contain as great a density of faults as the West Siberian or Gulf Coast basins, but their actual density and distribution will never be as well documented as they would be in a productive basin.

The most fundamental role of faults in trapping mechanisms is this intrinsically probable association with virtually every other type of trap-forming geometry: drape structures; differential deposition or preservation of sediments; growth of convex structures with time; truncations, and the formation of unconformity traps; uplifts of many kinds from horsts and diapirs to simple rug folds. Whether the fault-influenced geometry is *effective* as a trapping mechanism, or ineffective, is less a function of the geometry than of the *timing*. From the standpoint of timing, faults may be broadly assigned to four categories:

(a) "Dead" faults, antedating the petroliferous part of the column and so restricted to the effective basement.
(b) Continuously developing faults, possibly inherited from older faults but with movement continuing during the depositional phase. This is by far the most important category, especially so if movement ceases while sedimentation continues.
(c) Young faults, not formed until late in the depositional phase. These are likely to be confined to relatively shallow strata and not reflected in the gravity or magnetic fields of the region; they are unimportant as trapping mechanisms.
(d) Late-regenerated faults, moving late in the depositional phase but following long dormancy. These faults are more likely to destroy traps than to create them.

The close association between oil- and gasfields and faults implies that fault traps will be very common. Though this is true in the general sense, traps due wholly to faults are relatively uncommon. Most of them are in very young strata and none of them rank among the world's great oil- or gasfields. Nearly all fault-trap

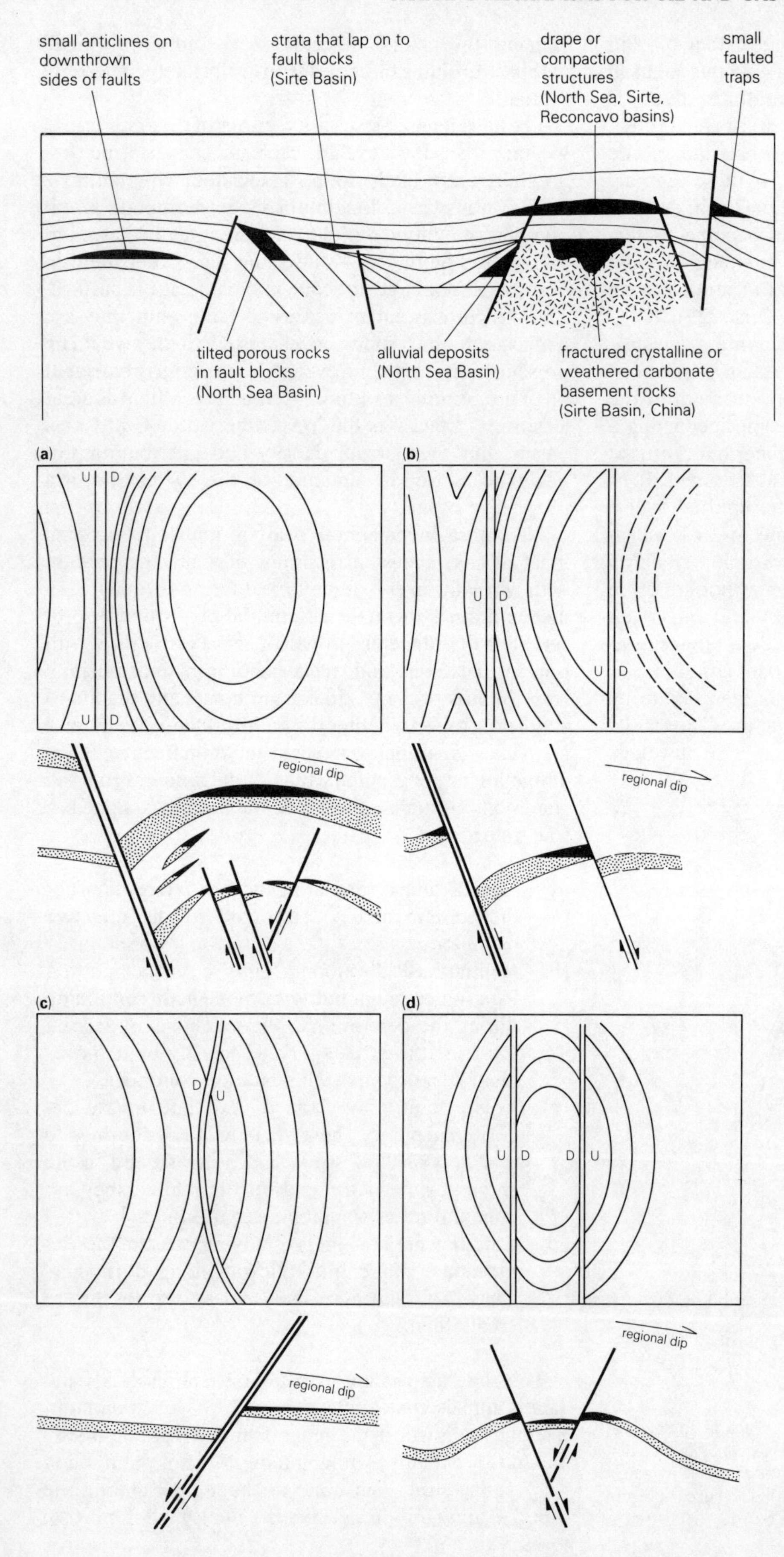

Figure 16.95 An assortment of trapping mechanisms created by faults; only two of the seven examples are true fault traps. (From M. A. Fisher, *AAPG Bull.*, 1982.)

Figure 16.96 Typical fault-associated traps in Tertiary strata of the western Gulf Coast Basin, USA: (a) both stratigraphic and structural accumulations in rollover anticline, on downthrown side of down-to-the-coast fault; (b) accumulation primarily controlled by closure on upthrown side of down-to-the-coast fault; (c) accumulation against up-to-the-coast normal fault; (d) accumulation against keystone faults over the crest of deep-seated salt intrusion. (From R. B. Johnson and H. E. Mathy, *World Oil*, June 1958.)

accumulations are restricted to the upthrown sides of their faults; most of the exceptions to this rule are caused by variations in reservoir pressure (see Ch. 14).

We must distinguish three types of association between accumulations and faults (Figs 16.3, 95 & 96):

(a) The fault itself may make the trap, by sealing the edge of the pool, without any ancillary trapping mechanism such as convexity. Faults in such cases are nearly always normal faults, dipping towards their downthrown sides. Reverse and thrust faults, which dip towards their upthrown sides, commonly form traps involving convexity in addition.
(b) The fault may be responsible for creating another structure, such as a fold or a horst block, and this fault-generated structure rather than the fault itself forms the actual trap.
(c) The fault(s) may be the *consequence* of another kind of structure (especially of a fold) which itself makes the real trap, even if individual pools within it are sealed against faults.

16.7.2 True fault traps

The simplest geometry for a fault trap that seals the pool without any other trapping mechanism involves a reservoir rock being faulted directly against an impermeable rock. If the fault responsible for this juxtaposition is a normal or extensional fault, as it is in the majority of known cases, the geologist might expect the oil or gas to migrate towards the surface via the fault zone itself. In fact, fault traces in oilfield areas are seldom marked by important seepages. In the most frequently quoted classification of seepage causes, by Walter Link in 1952, none of the five categories is attributed specifically to fault control. Faults, even normal faults, are more commonly seals for oil and gas than conduits for them towards the surface (unless, of course, the faults are rejuvenated after pooling has taken place, and so belong in the fourth age-category of the list in Sec. 16.7.1).

It is important to consider how and why this somewhat unexpected circumstance comes to be the rule. Fault zones commonly contain *gouge* or "natural mudcake," formed of attenuated, deformed shale trapped in them during early movement and smeared along them as offset proceeds. Normal faults are *fluid* seals because they are *pressure* seals. The trapping capacity of a faulted structure depends principally on the pressure conditions in the reservoir rocks *on both sides of the fault.* In the normally hydropressured zone at shallow or moderate depths, the reservoir pressures on the two sides are not likely to differ significantly and the potentiometric surface of the reservoir rock will lie well below the topographic surface. Oil cannot then rise to the surface along the fault, which functions as an effective lateral seal to the pool. This can lead to anticlinal traps being broken by sealing faults into separate pools (Figs 16.9 & 99).

In geopressured zones, on the other hand, the pressures in the aquifers on the two sides of a fault may differ substantially. Hydrocarbons may then migrate *across* the fault in the direction of lower fluid potential, provided that permeability is available for this to happen. The fluid potential $\phi = gh$, where h is the head measured on the potentiometric surface (see Sec. 15.8.1). If the pressure conditions are such that the potentiometric surface of the reservoir rock is above the topographic surface, the fault itself becomes a conduit to the surface. Oil and gas cannot then be held against the fault, and the trap cannot extend downwards below the contact between the top of the reservoir and the fault zone (Fig. 16.97).

A normal fault may, of course, have permeable rocks juxtaposed across it as easily as it may juxtapose permeable rocks against impermeable. There are endless variations, but the most common are these (Fig. 16.98):

(a) Parts of the same reservoir rock (we will take a sandstone) are juxtaposed across the fault within the hydrocarbon column. The fault is likely then to be of nonsealing capacity, at least in the interval of sand-to-sand contact. Any hydrocarbon accumulation will then extend across the fault without change in fluid interface levels.
(b) Parts of the same sandstone body are juxtaposed across the fault but not within the hydrocarbon column. Any accumulation must then be controlled by some type of closure – either vertical

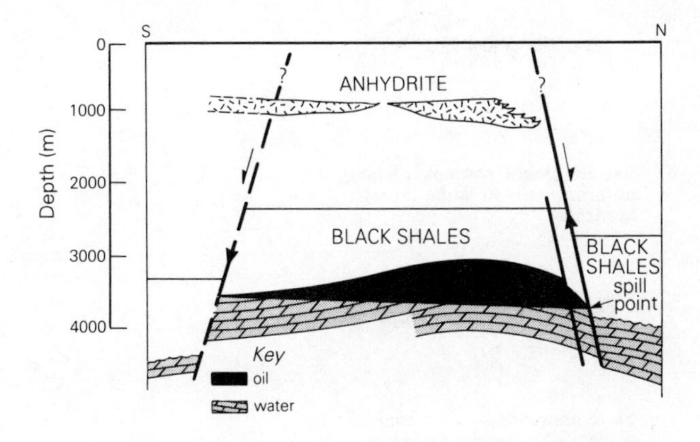

Figure 16.97 Cross section of Gela oilfield, Sicily, in a faulted anticlinal trap in a Triassic dolomite. Surface water (here marine) percolates downward along the fault on the left (south), moves northward through the reservoir, and leaves the system along the fault on the right (north). The oil/water interface is tilted in the direction of water movement, and intersects the top of the reservoir where it is cut by the faults. (After S. Neglia, *AAPG Bull.*, 1979.)

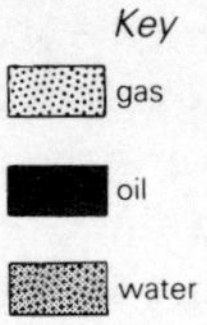

Hypothetical situation	Analysis of fault seal	
	Vertical migration	Lateral migration
(a) Sand opposite shale at the fault; hydrocarbons juxtaposed with shale 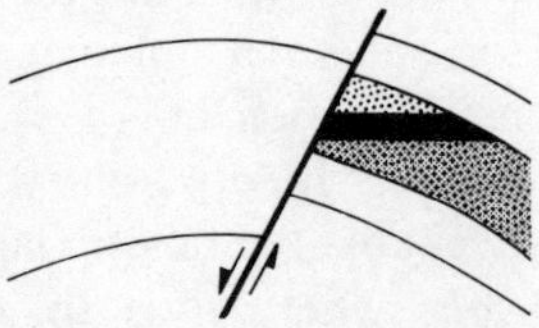	sealing	sealing reservoir boundary material may be the shale formation or fault zone material
(b) Sand opposite sand at the fault; hydrocarbons juxtaposed with water 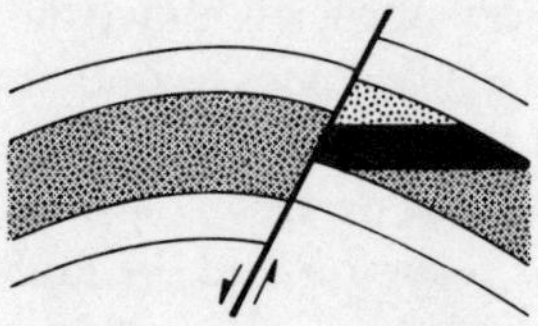	sealing	sealing seal may be due to a difference in displacement pressures of the sands or to fault zone material with a displacement pressure greater than that of the sands
(c) Sand opposite sand at the fault; common hydrocarbon content and contacts 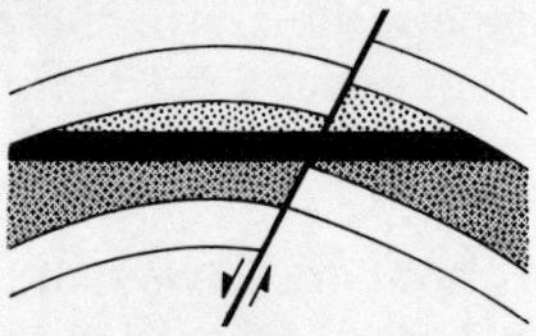	sealing	nonsealing possibility is remote that fault is sealing and the reservoirs of different capacity have been filled to exactly the same level by migrating hydrocarbons
(d) Sand opposite sand at the fault; different water levels 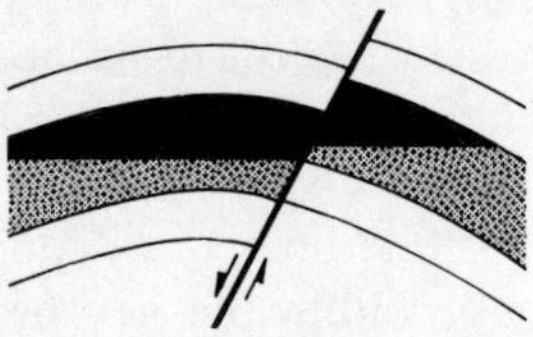	sealing	unknown nonsealing if water level difference is due to differences in capillary properties of the juxtaposed sands; sealing if water level difference is not due to differences in capillary properties of the juxtaposed sands
(e) Sand opposite sand at the fault; common gas/oil contact, different oil/water contact 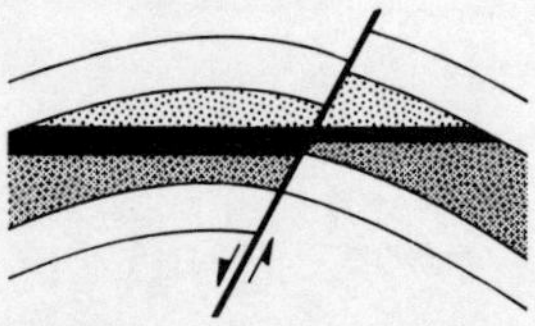	sealing	nonsealing possibility is remote that fault is sealing and migrating gas has filled the reservoirs of different capacity to exactly the same level
(f) Sand opposite sand at the fault; different gas/oil and oil/water contacts 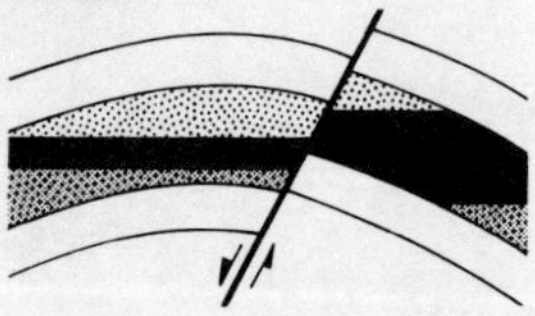	sealing	sealing a difference in both gas/oil contact and oil/water contact implies the presence of boundary fault zone material along the fault
(g) Sand opposite sand at the fault; water juxtaposed with water 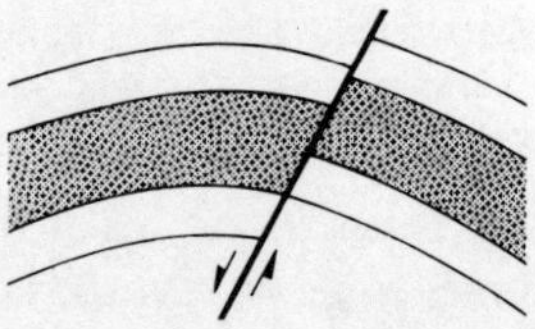	unknown	unknown

Key: gas; oil; water

Figure 16.98 Analysis of sealing and nonsealing capacities of normal faults in the American Gulf Coast, according to juxtapositions against sand bodies. (From D. A. Smith, *AAPG Bull.*, 1980.)

(structural) closure, or lateral closure of permeable against impermeable rock across the fault. This tells us nothing about the sealing capacity of the fault.

(c) Different sandstone bodies are juxtaposed across the fault, within the hydrocarbon column. In the majority of such cases, the fault forms a seal, but it may be nonsealing at different levels.

The reason for the difference between case (a) and case (c) lies in the boundary displacement pressures on the two sides of the fault (see Sec. 15.7.2). Displacement pressure gradients in sandstone reservoir rocks may differ by an order of magnitude or more between a clean sand (say 20 kPa m^{-1} or 1 p.s.i. per foot) and a sand with high clay content (several hundred kilopascals per meter). Such differences will not exist between juxtaposed segments of the same sandstone (case (a), above). The buoyant force of the oil or gas then enables it to migrate across the fault without difficulty, in the direction of lower fluid potential.

If the fault zone contains fault breccia and not gouge, the fluid potential ϕ in the zone may be higher than that in the reservoir. With good reservoir permeability, datum pressure in the bottom water then quickly reaches hydrostatic, through migration out of the reservoir and into the fault zone; hydrocarbon entrapment cannot then last very long. If permeability is poor, however, pressures take much longer to equalize with hydrostatic pressure. During the then slow process of destruction of the accumulation, oil may remain efficiently trapped against normal faults with equally permeable strata on the opposite sides of the faults being water-bearing.

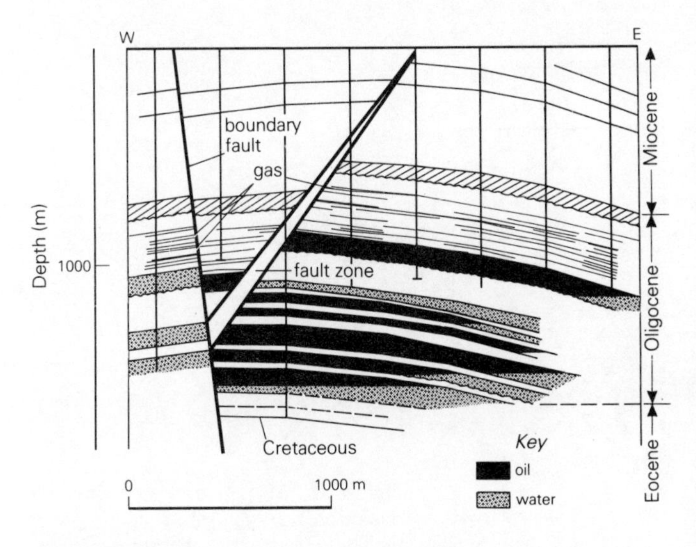

Figure 16.99 Structural cross section through the Casabe field, Middle Magdalena Valley of Colombia, showing the same sandstones oil-bearing and water-bearing across a normal fault. (After L. G. Morales *et al.*, in *Habitat of oil*, Tulsa, Okla: AAPG, 1958.)

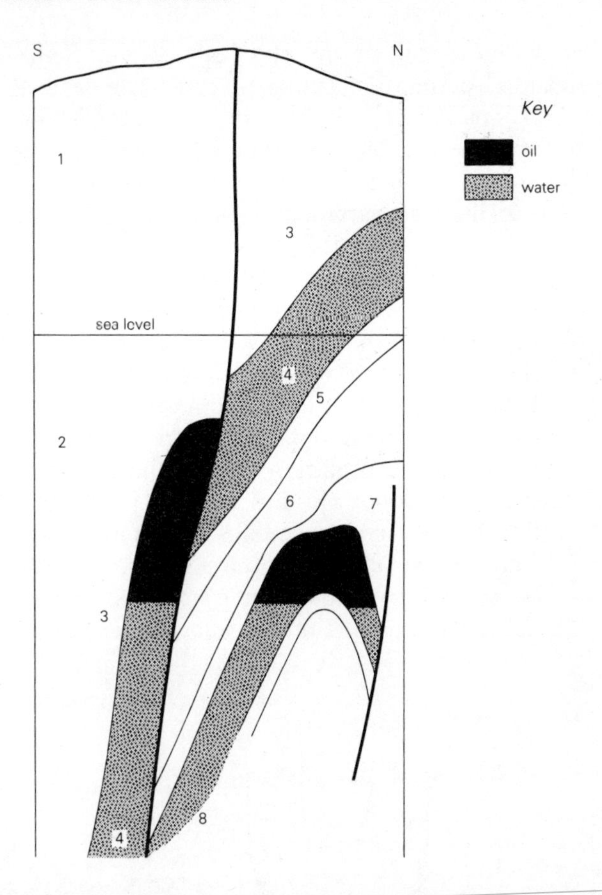

Figure 16.100 Structure section (not to scale) through the Osobniza oilfield, Polish Carpathians. True fault trap in Eocene sandstone (unit 4) on downthrown side of normal fault, with water updip across the fault in the same sandstone. (From N. A. Eremenko and I. M. Michailov, *Bull. Can. Petrolm Geol.*, 1974.)

Examples from regions of very different structure are illustrated in Figures 16.99 and 16.100.

A somewhat similar circumstance is presented by a water-bearing sand overlying an oil- or gas-bearing sand. In this case there is likely to be a thin shale "break" between the two sands, possibly the smeared-out clay filling of the fault zone separating water in the downthrown block from hydrocarbons in the upthrown block.

16.7.3 Settings for normal fault traps

Large *normal faults*, fundamentally extensional in genesis, occur in two principal geologic settings of concern to petroleum geologists.

The first setting is within fault-bounded grabens or half-grabens. The assemblages of fault blocks in this setting may be highly complex in plan, with zigzag or rhomboidal grid patterns of faults. In cross section, however, the pattern is much simpler. Faults may dip towards either boundary of the graben, but all of them are normal faults; compressional folds or reverse faults are wholly or largely absent. In the lower levels are tilted fault blocks; higher up, the beds display fault-drag flexures, perhaps including reverse drag, and these flexures pass up into intact, unfaulted drape flexures in the upper levels. Most fields in such graben settings as the North Sea, the Suez graben, or the Reconcavo Basin of Brazil illustrate variations on this profile (Figs 16.5, 9, 29 & 31).

The faulting precedes the accumulation of hydrocarbons, and in only a minority of cases do faults alone form the traps. Oil so pooled is likely to be over the top of a horst, and so in a type of convex trap; or in the erosional wedge-out of a reservoir bed below a fault-caused unconformity, and so in a type of stratigraphic trap. The most favored trap attributable wholly to the faults is in trapdoor closures at fault junctions (Fig. 16.101). In horst-graben settings, therefore, cases (a), (b), and (c) may all be represented (Sec. 16.7.2).

The second common setting for assemblages of normal faults has them cutting homoclinal sequences, on the forelands of compressional basins or in unconfined, coastal half-basins of the Gulf Coast type. The faults are especially well developed where the homoclinal sequence thickens rapidly basinward into deltaic, shelf-edge, or submarine fan clastic successions. The flexure allowing such thickening, whether marked by faults or not, is a *hinge belt*. Famous examples include those in the Los Angeles Basin (Fig. 16.27) and the Eastern Venezuelan Basin (Figs 10.16 & 17.29).

Synthetic and antithetic faults Normal faults, as we have seen, are intrinsic features of sedimentary basins; they form as part of the basin-creating process. The majority of the faults are syndepositional, listric faults dipping towards the subsiding basin. As their dips are in the same direction as the regional dip, they are said to be *synthetic faults*. They are also called down-to-the-basin faults, or down-to-the-coast (if the basin still lies off-

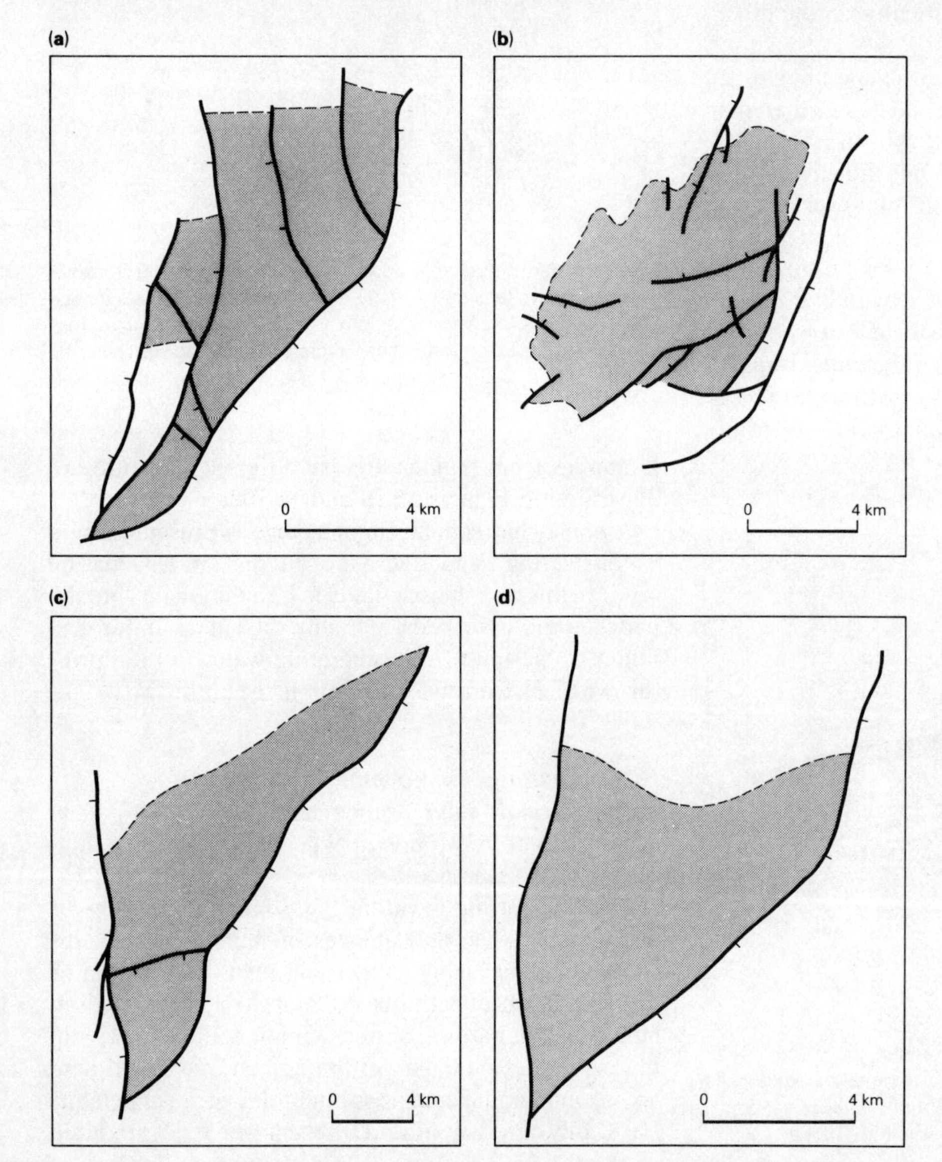

Figure 16.101 Traps in structural culminations between intersecting normal faults; Rankin trend of Dampier Basin, offshore northwestern Australia. (a) North Rankin field; (b) Angel field; (c) Goodwyn field; (d) West Tryal Rocks field. (From P. E. Playford, *APEA J.*, 1975.)

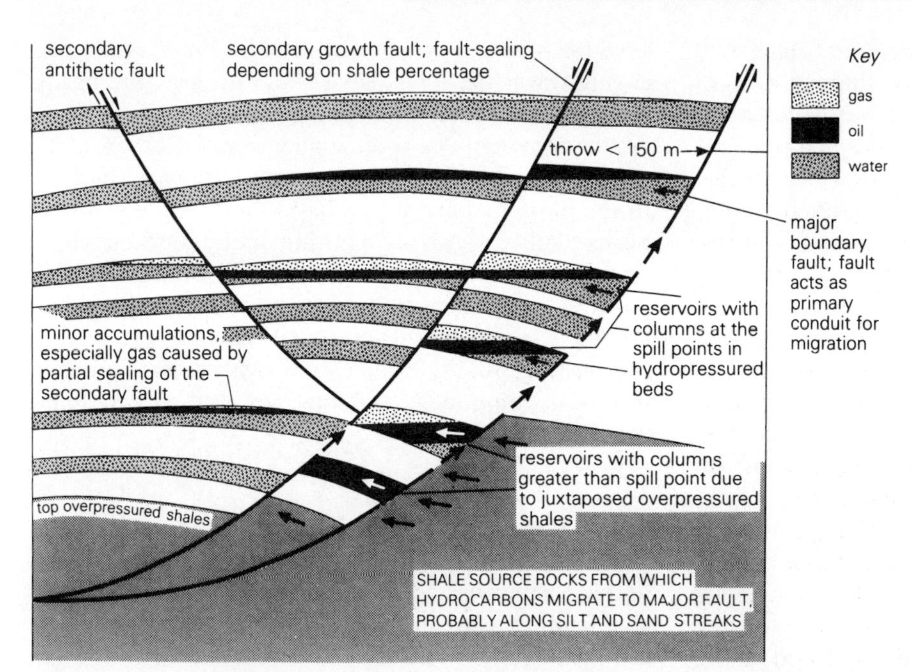

Figure 16.102 Typical traps in Niger delta oilfield, on basinward side of listric, synthetic fault. (From E. Onu Egbogah and D. O. Lambert-Aikhionbare, *Oil Gas J.*, 14 April 1980; after K. J. Weber and E. Daukoru, 1975.)

shore, as in the US Gulf Coast and the Niger delta), or down-to-the-fold-belt faults (if the basin has been partly consumed by later folding, as in the Persian Gulf and eastern Venezuelan basins).

Synthetic faults are commonly also *growth faults*. They were initiated early by flexing of the lithosphere towards the basin as the basin was loaded by sediments; they continued to develop during sedimentation. Thus their throws increase downwards (as will be seen shortly, it would be more correct to say that their throws decrease upwards). Stratigraphic thicknesses are normal on the upthrown (shelfward) sides of the faults and anomalously large on the downthrown (basinward) sides (Fig. 16.102).

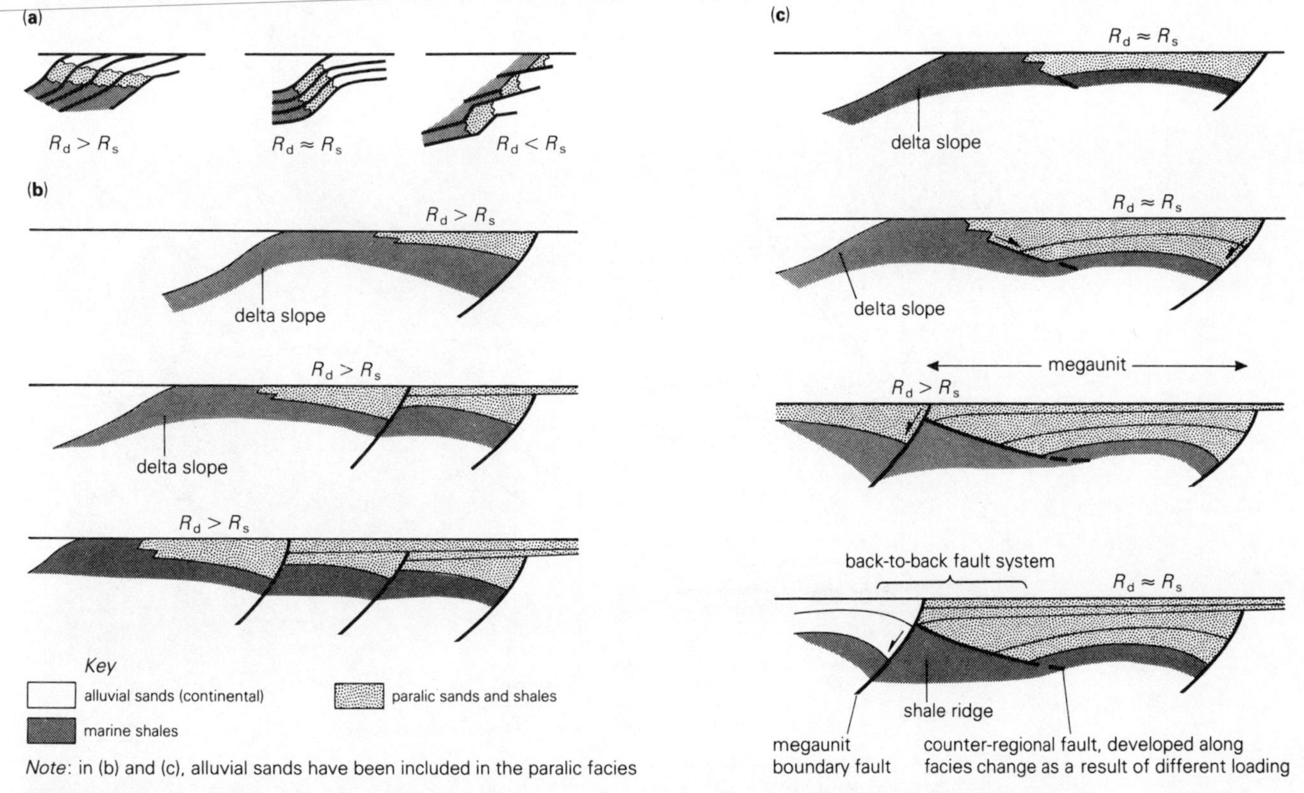

Figure 16.103 Development of synthetic and antithetic synsedimentary faults in Tertiary Niger delta. In (b), development of growth faults shown when rate of deposition (R_d) exceeds rate of subsidence (R_s). In (c), equivalent development shown when R_d is in balance with R_s. (From B. D. Evamy *et al.*, *AAPG Bull.*, 1978.)

Normal fault assemblages cutting homoclinal sequences also include *antithetic faults*, dipping against the regional dip and towards the shelf. Antithetic faults may show little or no growth with time; they may then be viewed simply as compensations for extensions of the overburden. In many cases, however, antithetic faults achieve growth status of their own, and become *counter-regional faults*. In the Niger delta, such faults form the shelfward (inner) boundaries of large shale ridges; the basinward (outer) boundaries of these ridges are major down-to-the-basin faults. Thus the shale ridges taper upwards between normal faults diverging downwards (Fig. 16.103). In the US Gulf Coast, salt ridges and diapirs take the place of shale ridges (though some of those occur also). Down-to-the-gulf faults occur preferentially along the seaward sides of rising ridges or stocks, separating them from depositional sinks into the underlying salt, which are also sand depocenters in which strata thicken on the downthrown sides of the growth faults. Faults on the landward (shelfward) sides of the ridges are mostly post- or late-depositional, antithetic, and of small displacement.

The explanation of this array of features has been demonstrated experimentally, notably by Ernst Cloos in 1968. As young, weak strata are subject to gravitational sliding down the homoclinal slope towards the basin, faults are initiated as bedding-plane detachments at depth, commonly between a glide zone of salt or mudstone and its overburden. Rock particles move parallel to the master, down-to-the-basin fault, and essentially horizontally towards an opposing, antithetic fault which moves down slope. The two opposing faults are propagated upwards, becoming steeper as they rise. They are *listric* faults; in cross-sectional view they are *shovel-shaped* (Fig. 16.102). The eventual structural regime is that of a compound, asymmetrical graben extending itself down slope.

Accumulation of oil and gas against normal faults
Where oil has accumulated against the faults themselves in this framework, so that faults seal the pools without any auxiliary trapping mechanisms, the accumulations are against shelfward-dipping or antithetic faults far more commonly than against basinward-dipping or synthetic faults. The accumulations are on the upthrown but downdip sides of the faults, which of course antedate the accumulations.

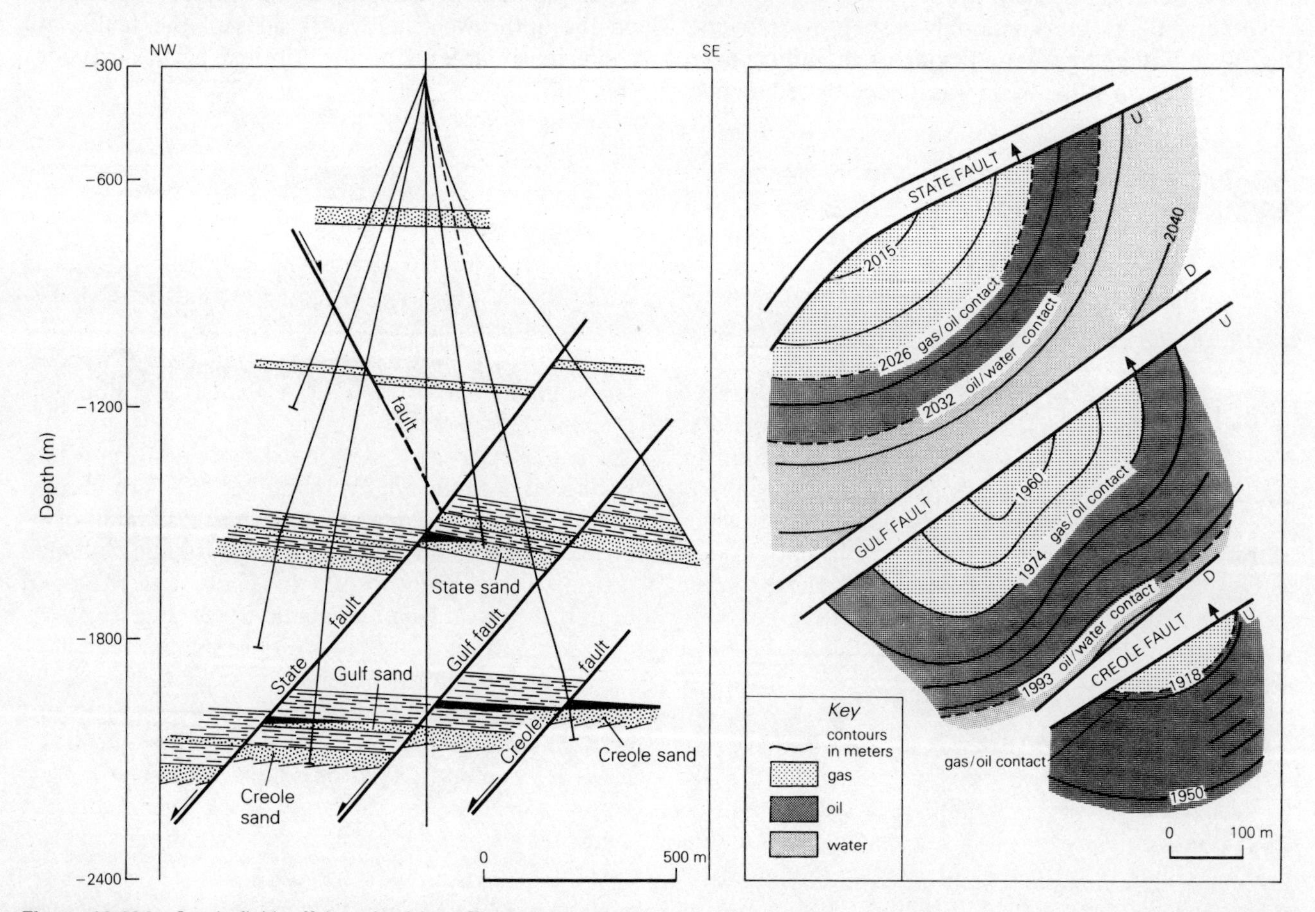

Figure 16.104 Creole field, offshore Louisiana. Three separate pools in a single Miocene sandstone, in nonconvex, true fault traps against antithetic faults (i.e. dipping against the regional dip). All faults in turn overlie a convex structure (deep salt intrusion). (After T. Wasson, in *Structure of typical American oilfields*, Vol. 3, Tulsa, Okla: AAPG, 1948.)

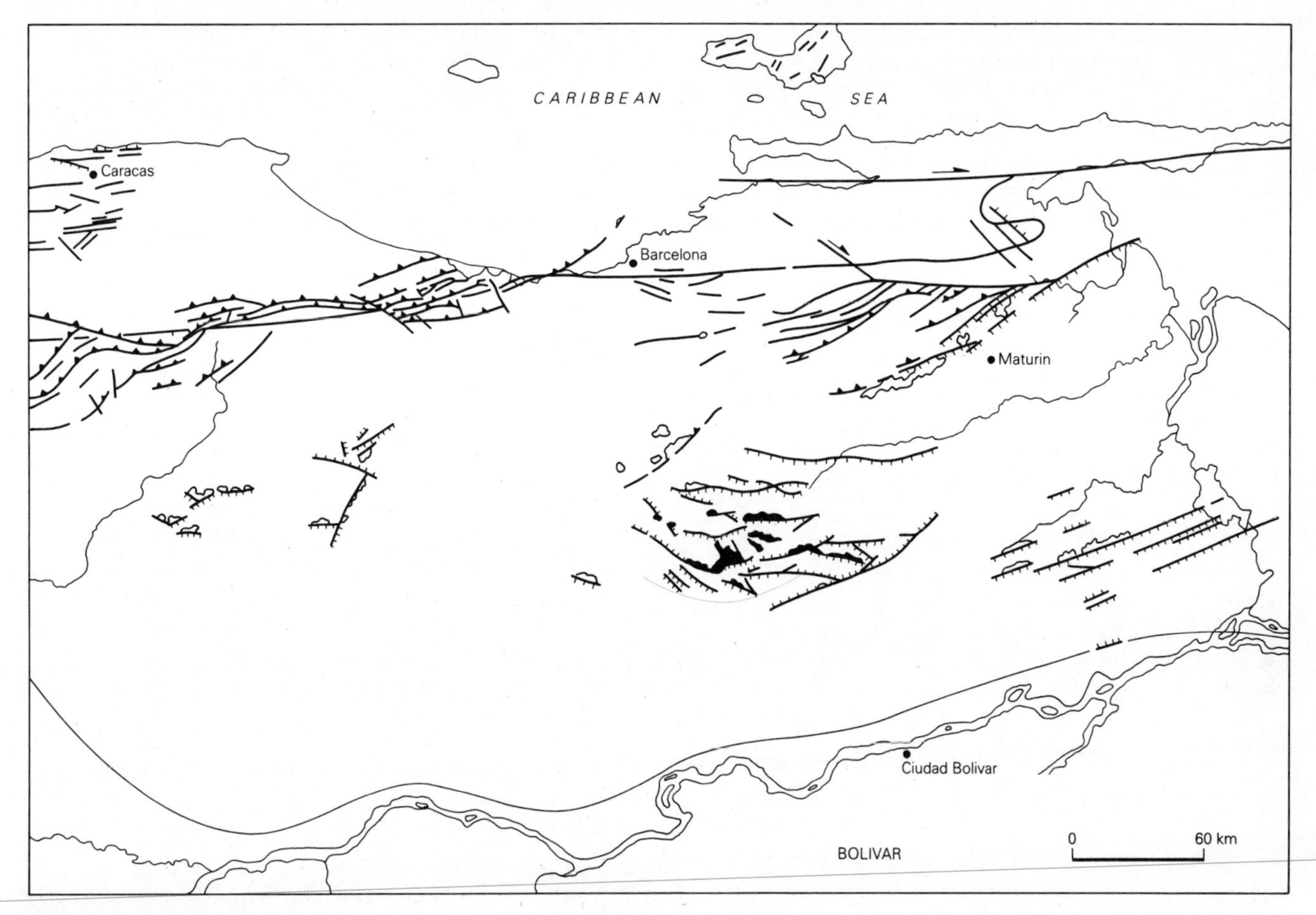

Figure 16.105 Structural map of the Eastern Venezuelan Basin during the late Tertiary. Oilfields of Greater Oficina district shown in black near the center of the figure. Nearly all fields are on the north sides of south-dipping normal faults. (From M. Alayeto and L. W. Louder, *Can. Soc. Petrolm Geol. Memoir* 3, 1974.)

In the American Gulf Coast, where the basin lies south of the shelf, most of the oil which is in fault traps (much of it of course is in quite other kinds of traps) is pooled on the upthrown, south sides of up-to-the-basin normal faults (Fig. 16.104). In the Greater Oficina area of eastern Venezuela (Fig. 16.105), a hinge-belt between the basin and the shelf is marked by a complex pattern of low-angle, normal growth faults. Virtually all fields are on the north (basinward) sides of faults which are concave towards the shelf (the south) in both plan and section. The faults are cuspate-concave in the direction of tectonic transport. At depth, they dip southward at about 25°; towards the surface this dip becomes about 45°, but many of the faults do not reach the surface.

This phenomenon had become apparent within a year or two of the discovery of the Oficina fields, while the Gulf Coast Tertiary basin was still regarded primarily as a salt-dome province. It was given a simple explanation by V. C. Illing (Fig. 16.106). Basinward-dipping faults have trough areas on their downdip sides, which oil bypasses as it migrates out of the basin towards the faults. Unless the fault is very long (and nearly all faults in this category are characteristically short), the oil continues its migration around the ends of the fault. Shelfward-dipping faults have domes or noses on their downdip sides, which are also their upthrown sides. Oil migrating from the basin towards these domes is "funneled" into them, forming traps.

The phenomenon is seen in exaggerated form where all the large faults are synthetic but they are accompanied by multiple, compensating, antithetic faults of individually trivial offset. Every small antithetic fault may provide a trap, with no oil (or almost none) trapped against the much larger synthetic faults (Fig. 16.107).

It is important to note that differential movement in the shelfward fault block alone will not in itself normally form a trap, no matter whether the fault dips shelfward or basinward. Migration of the oil is from the basin towards the shelf, so that the oil has to bypass the fault before it reaches the shelfward block, where the dip continues upwards and provides no trapping mechanism (Fig. 16.108).

Illing's explanation presupposes that migrating oil will choose a path at right angles to the structure contours; it will climb as steeply as possible. This in turn presupposes

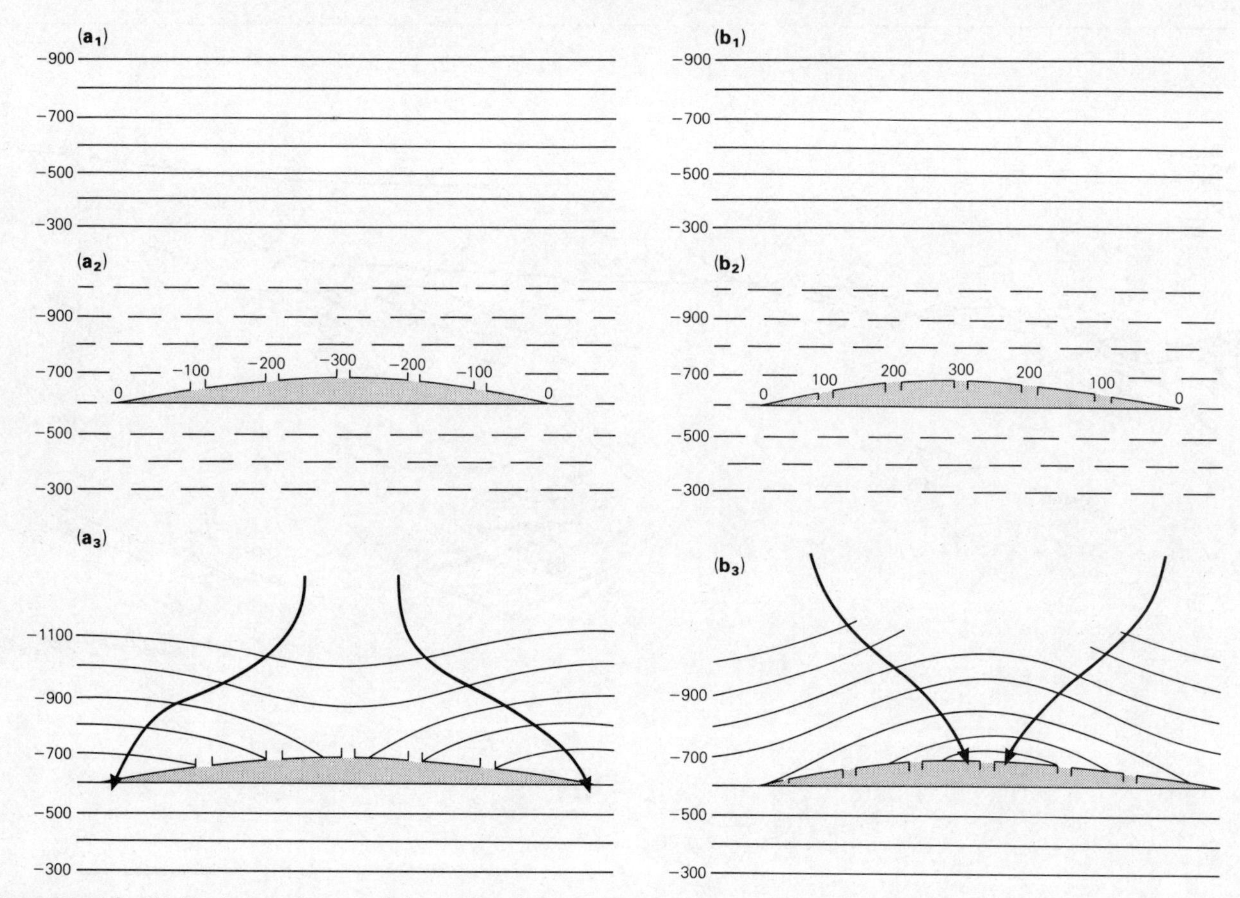

Figure 16.106 Illing's principle of accumulation against shelfward-dipping normal faults, as illustrated by L. C. Sass: (a_1) contours showing regional dip into basin before faulting occurs; (a_2) a long strike fault occurs – downthrown basinward – with displacements as indicated, resulting in the new contouring shown in (a_3); (a_3) in this case oil and gas being generated out in the basin and moving updip through permeable reservoir rock would be deflected around the edges of the fault – i.e. away from the fault – and no trap exists unless additional supplementary barriers are also present; (b_1) contours showing regional dip into basin before faulting occurs; (b_2) a long strike fault occurs – upthrown basinward – with displacements as indicated, resulting in the new contouring shown in (b_3); (b_3) oil and gas focused into fault. (From L. C. Sass, *Asoc. Venez. Geol. Min. Petrol. Bol. Inform.*, 1961.)

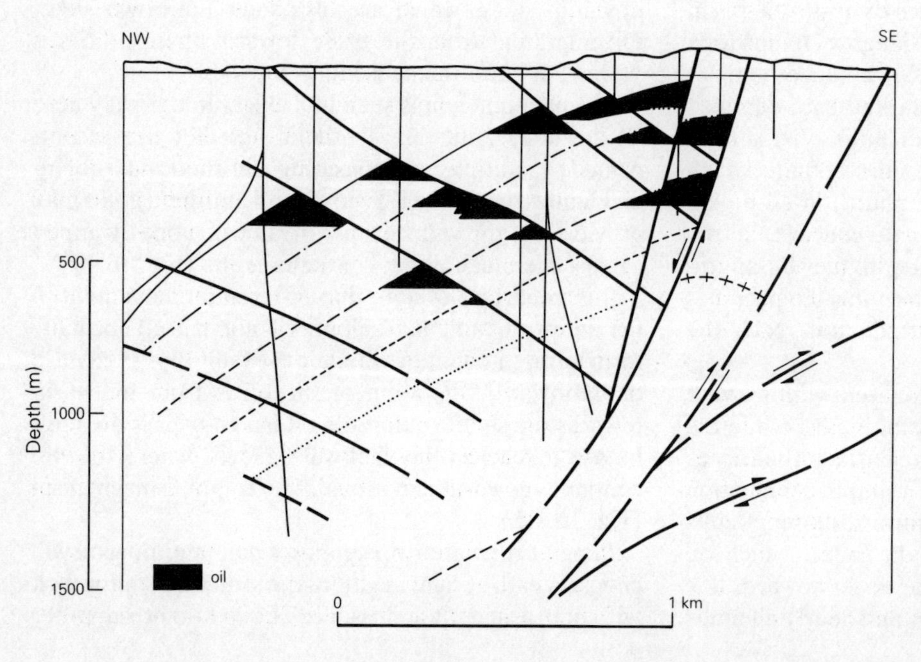

Figure 16.107 Miri oilfield, northwestern Borneo. Numerous oil accumulations in Mio-Pliocene sands on the basinward side of a diapiric ridge of Lower Miocene shale (top of ridge marked by +++ pattern). Most accumulations on basinward (upthrown) sides of antithetic faults; some on the basinward (downthrown) side of a major synthetic fault, at very shallow depth. (From E. J. H. Rijks, *Geol. Soc. Malaysia Bull.* 14, 1981.)

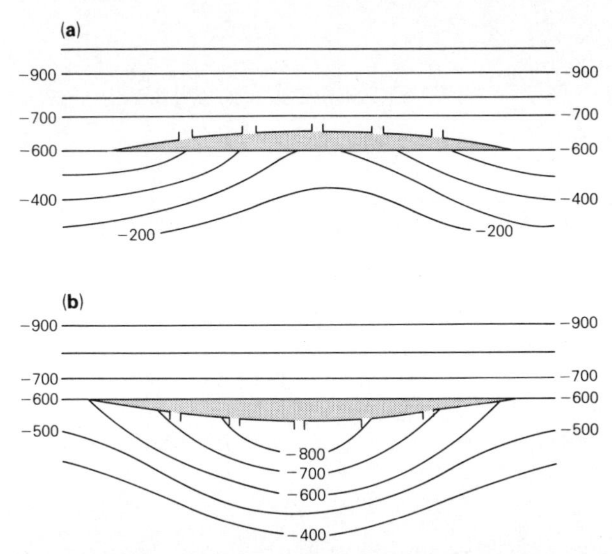

Figure 16.108 Illing's principle, continued. Differential movement in the southern (shelfward) block alone will not result in a trap, regardless of whether the fault hades basinward (a) or shelfward (b). (From L. C. Sass, *Asoc. Venez. Geol. Min. Petrol. Bol. Inform.*, 1961.)

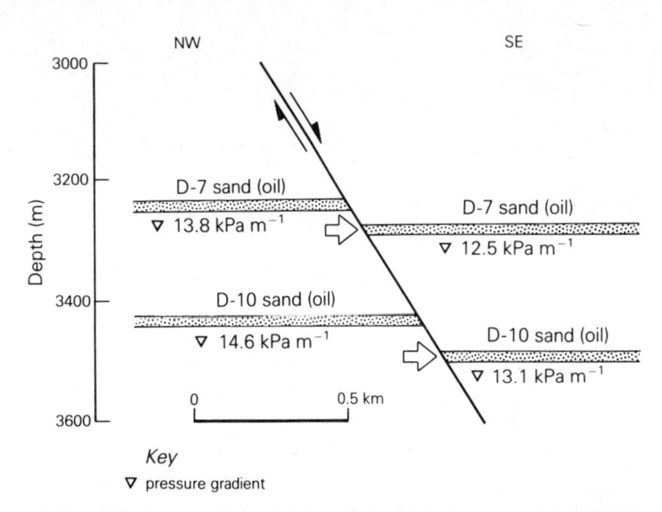

Figure 16.109 Section through oil sands in Block 21 field, South Timbalier area, offshore Louisiana; oil on both sides of a down-to-the-basin normal fault. (After J. D. Myers, *Trans. Gulf Coast Assoc. Geol. Socs*, 1968.)

essentially uniform permeability of the carrier beds. Permeability barriers, of course, may be enough to direct the oil towards the troughs or away from the domes, so there are some exceptions to Illing's rule. But as the fault systems we are considering are preferentially developed in thick, young, poorly consolidated clastic sections, the exceptions are few and they include no very large fields. The fundamental cause of nearly all exceptions is a strong *differential pressure gradient* across the fault zone. There are two general cases.

The first is a basinward hydrodynamic gradient strong enough to raise the entry pressure on the shelfward sides of synthetic faults and so permit oil accumulations in true fault traps on their basinward sides (Fig. 16.99). The hydrodynamic flow may be away from either high outcrops beyond the shelf, or diapiric ridges shelfward from the faults. Oil-bearing reservoirs on the basinward sides of synthetic faults may then abut directly against water-bearing reservoirs on the shelfward sides. In a few cases, oil or gas occurs in true fault traps on *both* sides of a fault (Fig. 16.109). This is likely only if the thicknesses of individual reservoir beds are less than the throw of the fault; otherwise the oil would inevitably leak across the fault.

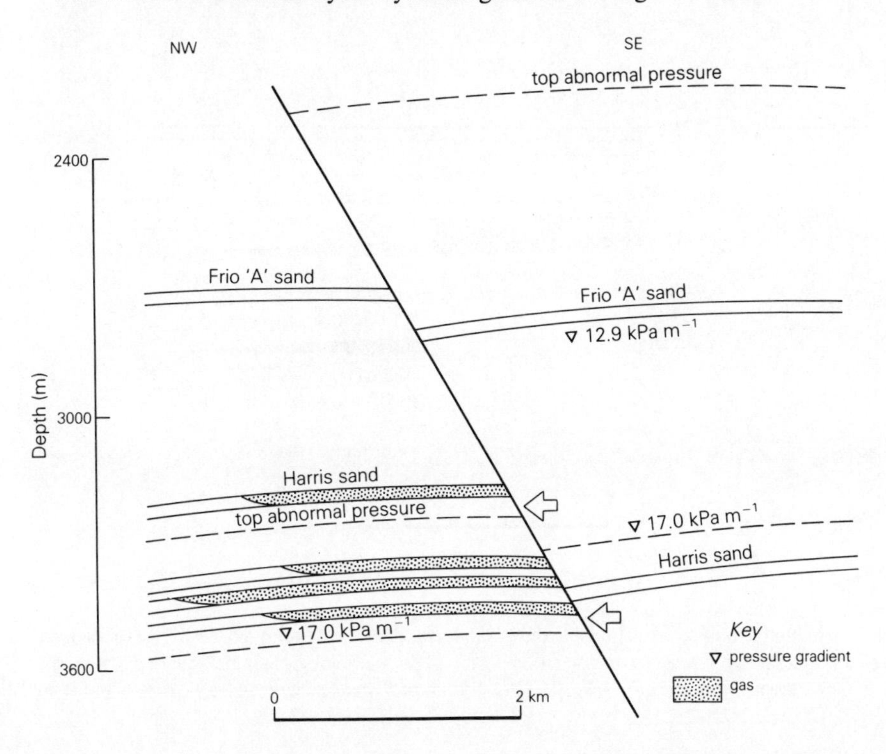

Figure 16.110 Cross section through West Alta Loma field, Texas Gulf Coast: gas in thin sandstones on *shelfward* side of down-to-the-basin normal fault; note top of abnormally pressured zone on two sides of fault. (After J. D. Myers, *Trans. Gulf Coast Assoc. Geol. Socs*, 1968.)

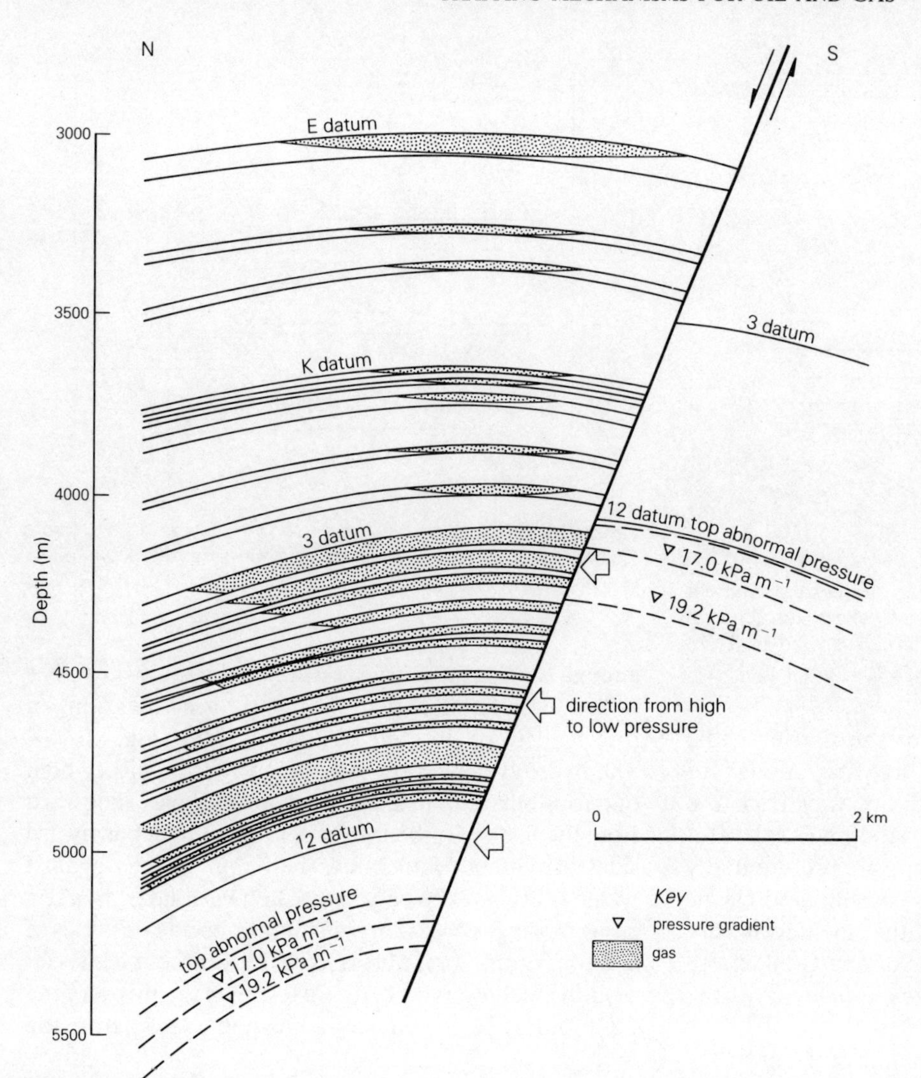

Figure 16.111 Producing Mio-Pliocene sand interval in Block 28 field, offshore Louisiana, illustrating updip pressure gradient sufficiently strong to trap oil on the *shelfward* side of an antithetic fault. (After J. D. Myers, *Trans. Gulf Coast Assoc. Geol. Socs,* 1968.)

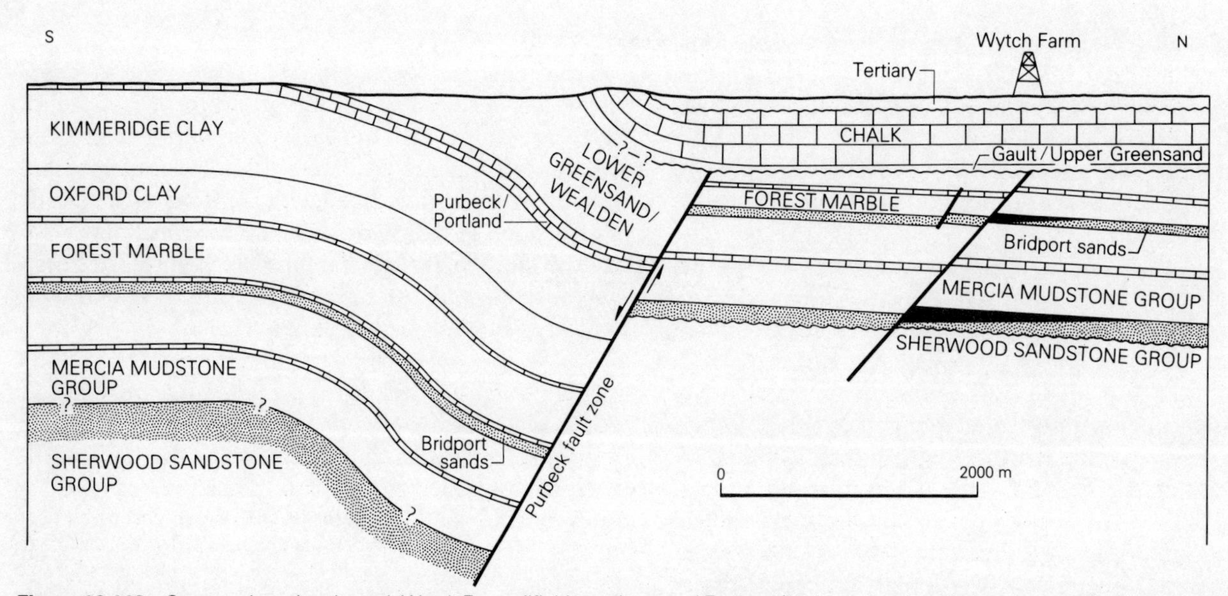

Figure 16.112 Structural section through Wytch Farm oilfield, south-coastal England (horizontal and vertical scales equal). Oil in Lower Jurassic sandstones on *shelfward* side of down-to-the-basin normal fault, following reversal of original dip of strata. Folding, of mid-Tertiary (Alpine) age, postdates normal faulting. (From V. S. Colter and D. J. Havard, *Petroleum geology of the continental shelf of North-West Europe*, London: Institute of Petroleum, 1981.)

The second case occurs below the top of the geopressured zone (see Sec. 14.2.1). The pressure differential on the two sides of a fault may then be easily sufficient to confine the oil or gas to the side having the lower pressure (invariably the shelfward side); the reservoirs are juxtaposed against higher-pressure strata. Gas (less commonly oil) may then be trapped on the *shelfward* sides of faults having either shelfward or basinward dips (Figs 16.110 & 111). The greater the pressure differential, the thicker the hydrocarbon column can be.

An unusual case occurs where the original basinward dip of the strata is reversed during later folding. Oil migrating up the new dip may now find true fault traps on the shelfward sides of synthetic faults (Fig 16.112).

Other settings for normal fault traps There is of course a spectrum of basin shapes and sizes between a one-sided, undeformed basin such as the US Gulf Coast Basin, and a closed, two-sided basin with a strongly deformed margin like that of eastern Venezuela. There are narrow basins with one compressionally deformed margin and a shelf-like foreland, like the San Joaquin Valley (Fig. 16.38). There are narrow basins essentially of graben (extensional) origin but having undergone some compressional deformation later, as in the Middle Magdalena Basin in Colombia (Figs 16.99 & 17.25). And there are true grabens with numerous extensional fault blocks, such as the Viking, Suez, and Reconcavo grabens (Figs 16.5, 26 & 29–31).

If any of these basins lacks a distinct asymmetry (as in the Los Angeles, Suez, and Gippsland basins), the terms "shelfward" and "basinward," or for that matter the terms "synthetic" and "antithetic" fault, lose much of their meaning. In much the most common case, however, the basin is strongly asymmetrical, with its greatest depression not in the center but close to one margin (see Sec. 17.4.2). The fundamental structure is then a *half-graben*, and the faults bounding the deeper side are antithetic faults because the shallower side represents the regional dip.

In the San Joaquin Valley, for example, fields on the shelfward (eastern) side that have fault boundaries are all at antithetic (east-dipping) faults (Fig. 16.38). In the Molasse Basin of Austria, oil- and gasfields trapped in fault blocks are all against antithetic faults dividing strata dipping southward towards the Alps. In the Viking graben of the North Sea, the principal boundary faults are on the west side (Fig. 16.5); the major intra-graben faults are downdropped to the east but rotated in the opposite sense. All the large fields are on the upthrown (west) sides of shelfward- (east-) dipping normal faults (Fig. 16.113). The rule is much less clearly adhered to in more symmetrical basins, and in one-sided basins having a dense mosaic of very small faults (the coastal basin of northern Peru is the classic example; Fig. 16.114); but it is a fundamental rule nonetheless.

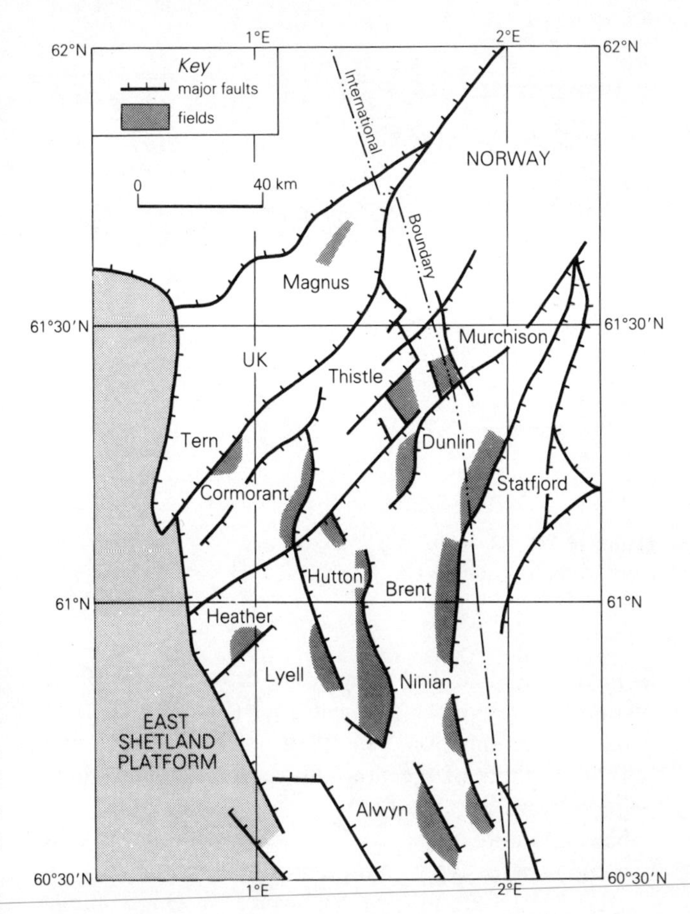

Figure 16.113 Pattern of faulting within Viking graben, northern North Sea. (From J. T. C. Hay, *J. Petroleum Geol.*, 1978.)

Traps against basinward-dipping normal faults There are a lot of oilfields on the downthrown sides of basinward-dipping growth faults, but very few of them are trapped against the faults. Some supplementary trapping mechanism is needed in addition to the faults: intersections of faults, or stratigraphic pinchouts. Unlike the fault-controlled trapping mechanisms in horst-graben settings, drape flexures are rare or absent over basinward-dipping growth faults.

By far the most common and important supplementary trapping mechanism is the *rollover anticline*. A growth fault begins at depth, in or close to bedding. It is propagated upwards, and steepens, as the basinward, down-moving side "pulls away." Thus the beds on this down-moving side are rotated into the fault (Fig. 16.115). If they are brittle, or if movement is very rapid, they must lose continuity, and become broken by step-like antithetic faults. If they are little older than the faulting (as is implied in the concept of growth faults), and have high fluid content, they retain continuity and

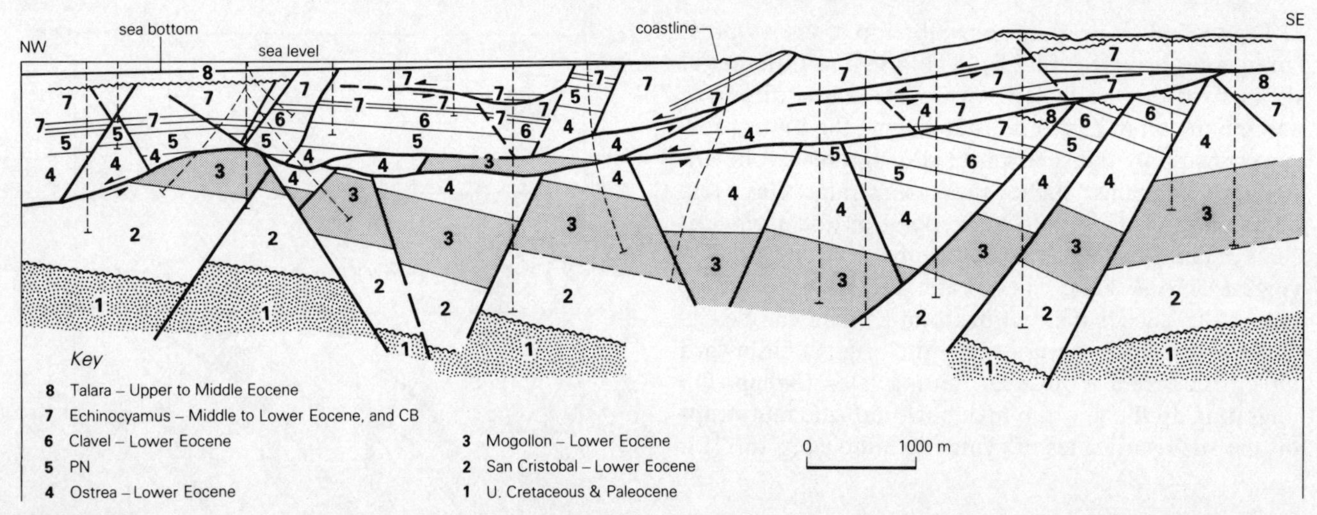

Figure 16.114 Onshore–offshore cross section through the Cabo Blanco oilfield, northwestern Peru. The Paleocene and Eocene section is broken by normal faults along the inner edge of a fore-arc basin; there is no sign of folding. Low-angle gravity slides later covered the fault blocks. The principal reservoir sandstone (3, stippled) is Lower Eocene. (From R. Moberly *et al.*, Geological Society of London Special Publication 10, 1982; courtesy of Belco Petroleum Co.)

form an anticline; the downward drag into the fault reverses the regional dip away from it (Fig. 16.103). The general phenomenon is called *reverse drag*; one normally thinks of the drag on a downdropped block being upward into the fault.

Normal faults of the type we are considering should become progressively younger in the direction of sedimentary progradation, or basinward. Younger faults may intersect rollover anticlinal traps already occupied by oil. A complex example is afforded by the Teak field offshore from Trinidad (Fig. 16.116), in which oil pools appear now to be confined to traps on the shelfward ("wrong") sides of synthetic faults whilst gas occurs on both the shelfward and basinward sides. The general rule of younger faults basinward may be obscured, however, because the forward progression itself precipitates rejuvenation of the older, more shelfward faults, and even the initiation of new faults there. Within any one *zone* of down-to-the-basin faulting, therefore, the faults may get younger landwards. This is the case in the Vicksburg fault zone, one of the more basinward of the arcuate fault zones around the northern margin of the Gulf of Mexico basin. As the faults flatten with depth, the rollover anticlines shift basinward, so that the greatest rollover is in the oldest beds (in which, unfortunately, secondary faulting also becomes more

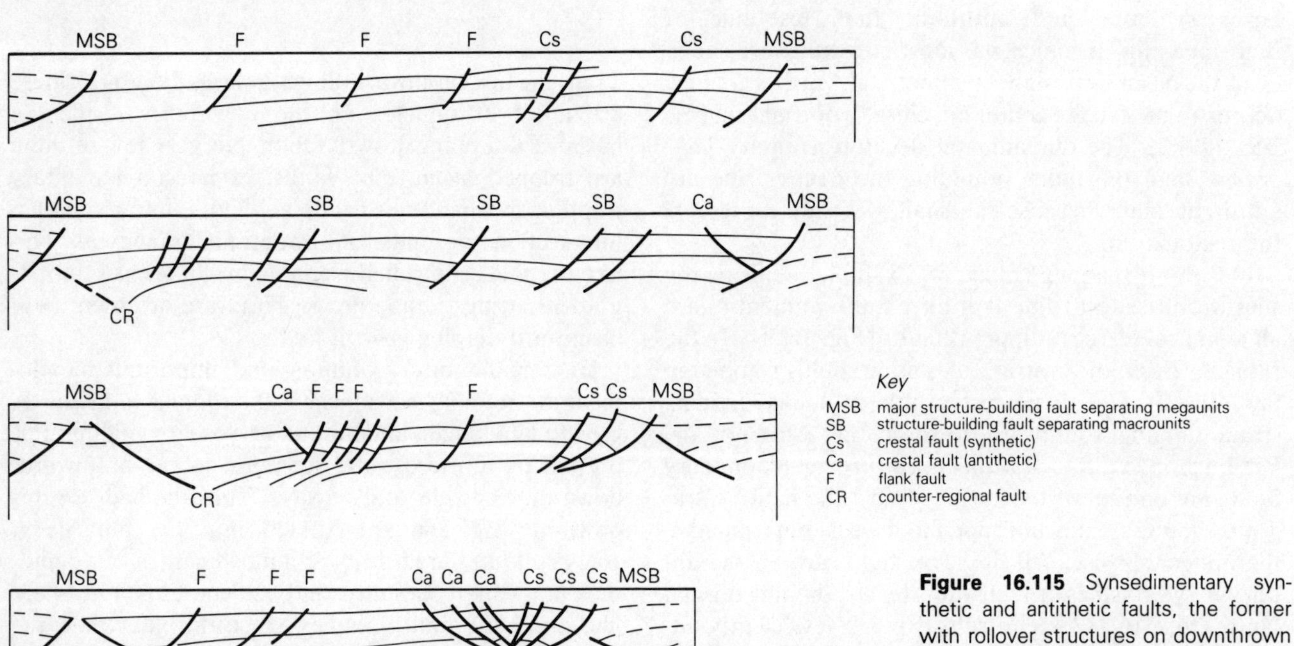

Figure 16.115 Synsedimentary synthetic and antithetic faults, the former with rollover structures on downthrown sides; Niger Delta Basin. (From B. D. Evamy *et al.*, *AAPG Bull.*, 1978.)

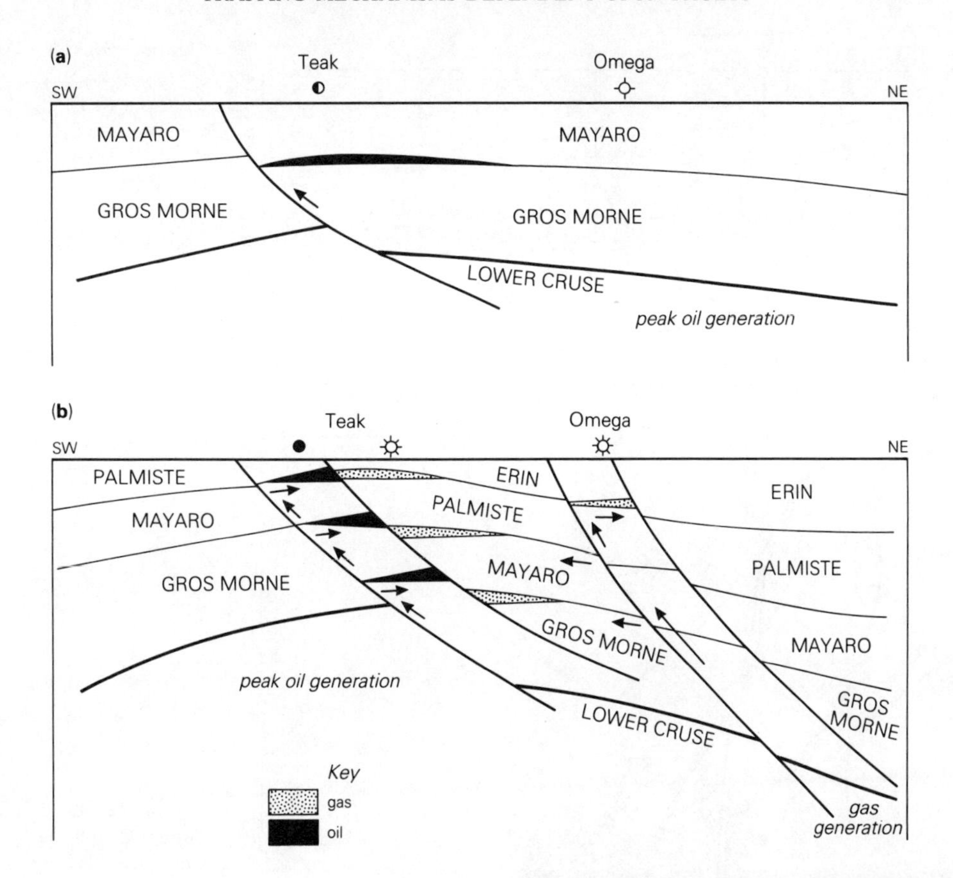

Figure 16.116 Teak oil- and gasfield, offshore southeastern Trinidad. Present oil accumulations are on shelfward side of synthetic, normal, eastward-dipping fault; gas accumulations are on basinward side (b). First accumulation of oil attributed to rollover anticlines on basinward side of fault, before faults farther basinward were formed (a); these traps now contain gas. (From R. Leonard, *AAPG Bull.*, 1983.)

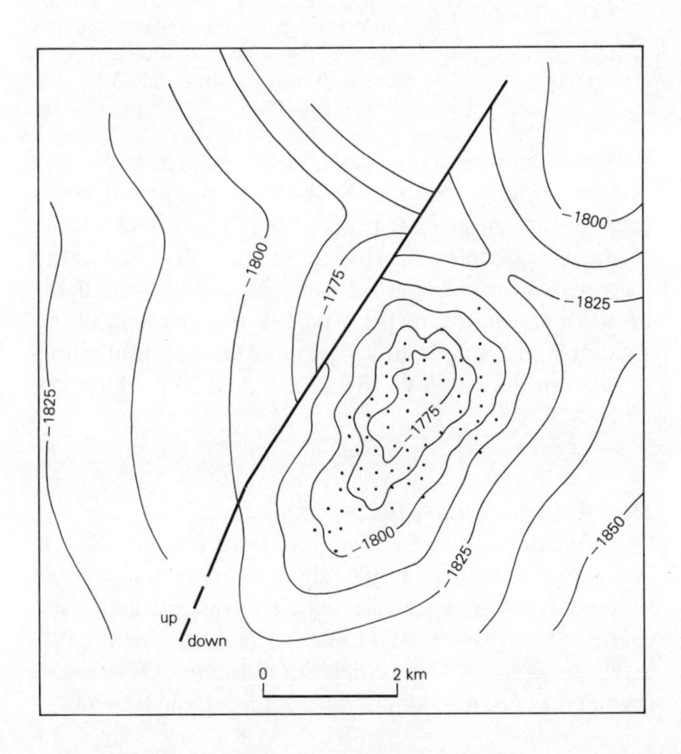

Figure 16.117 La Rosa field, Texas Gulf Coast. Structure contours (in meters) on the main producing sandstone. Note subcircular rollover anticline on the downthrown (basinward) side of the normal fault, and absence of trapping against the fault itself. (After B. Fisher, *AAPG Bull.*, 1941.)

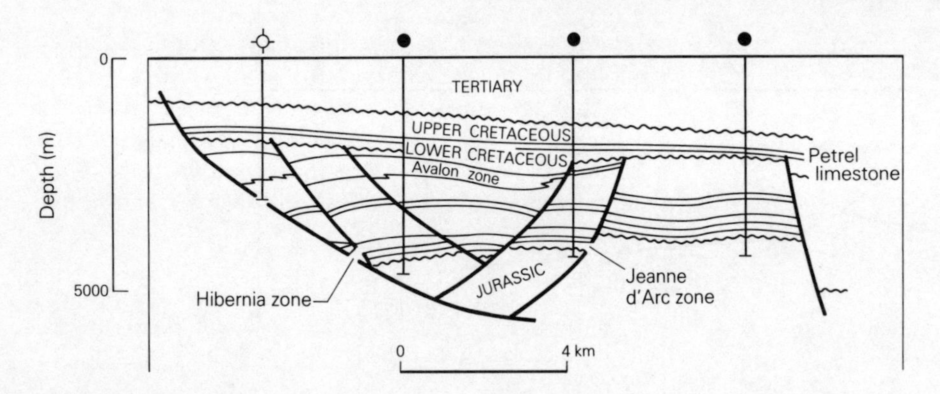

Figure 16.118 West–east cross section through the Hibernia field, offshore Newfoundland, showing rollover anticlinal trap between synthetic and antithetic normal faults. (From R. M. McKenzie, *Oil Gas J.*, 21 September 1981.)

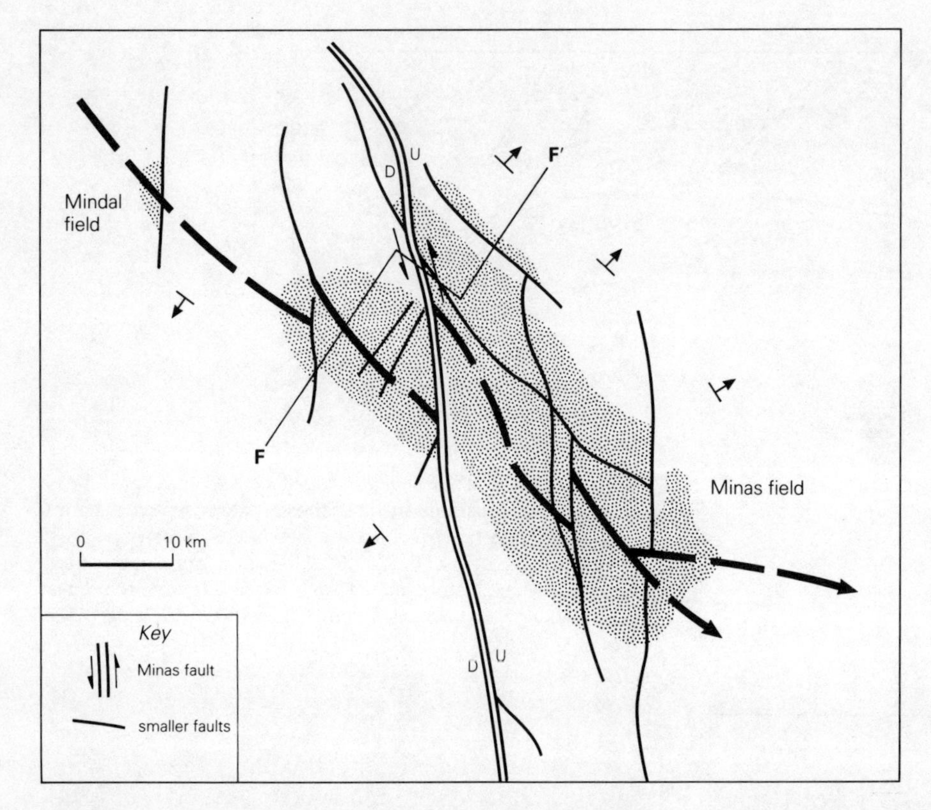

Figure 16.119 Minas oilfield, central Sumatra. Structure on top of the productive Lower Miocene sandstone. Note two sets of faults. North–south faults are extensional and antedate the reservoir rock; the large Minas fault shown by the double line underwent left-handed strike-slip movement beginning in the Paleogene. NW–SE faults are post-reservoir and have moved since the accumulation took place. Faults extend to the surface. All formations thin over the anticline; productive area is shaded. (After R. Eubank and A. Chaidar Makki, *Oil Gas J.*, 14 December 1981.)

complex). A combination of these two features (faults becoming younger landward and rollover being greatest in the oldest beds) means that the downbending decreases both landward and upward. Figure 16.117 illustrates the separation of the accumulation from its fault in a typical rollover anticlinal field.

Rollover anticlines may prove to provide the most common convex trap types among recent and future field discoveries, because they characterize divergent continental margins where offshore drilling must be concentrated. The basins surrounding Australia, India, and the two sides of the Atlantic Ocean are more likely to yield this trapping mechanism than any other (Fig. 16.118).

In this extended discussion of true fault traps (case (a) of Sec. 16.7.1), it has been necessary to note that the faulting antedated the accumulations though it may have continued during them. In a number of accumulations, however, faulting has occurred later, so that convex or nonconvex traps become broken up by normal faults into mosaics of separate traps with individual oil–water contacts. Examples illustrated here include some of the Reconcavo Basin fields in Brazil (Fig. 16.31); the fields of north-coastal Peru (Fig. 16.114), where many of the smaller faults simply offset pre-existing accumulations of oil; and the late NW–SE faults at Minas in Sumatra (Fig. 16.119).

16.7.4 Anomalous fault traps

Our consideration of *bending fold traps* took note of an anomalous category of strongly convex structures consequent upon steep faults which have changed their senses of displacement (see Sec. 16.3.2). In all well established cases, the earliest identifiable deformation involved a compressional downflexure in the basement.

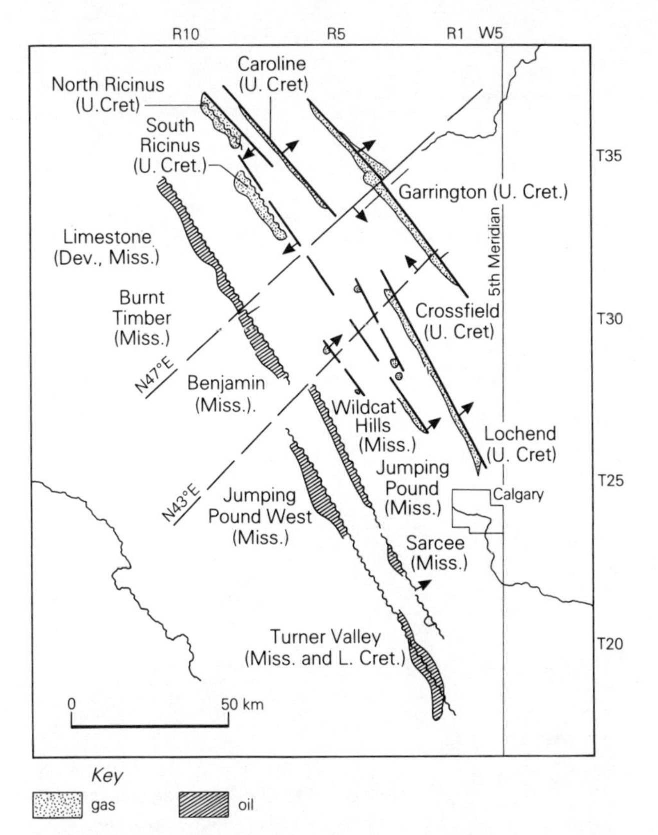

Figure 16.120 Hydrocarbon productive trends in southwestern Alberta, Canada, in linear pattern controlled by subtle faults of inconsistent movement sense. (From R. M. P. Jones, *Bull. Can. Petrolm Geol.*, 1980.)

There followed buoyant isostatic uplift and the creation of a trelliswork of normal faults, aspects of basin creation or accentuation. As they extend into the basement, these faults are not listric; they are therefore unlikely to create any of the trap types we have just described.

They may, however, readily create "stratigraphic" traps, by constituting updip cutoffs of thin sandstone or calcarenite units deposited only on their downthrown (basinward) sides. In 1980 Roy Jones postulated that numerous subtle nonconvex and diagenetic traps were formed in this way in the Western Canadian Basin (Fig. 16.120). Jones coined the term *isostatic adjustment faults* for these unobtrusive faults, which finally reveal themselves when late reversals form small compensating grabens along their shelfward sides.

16.7.5 Fault traps and faulted traps

We recognize a somewhat unexpected phenomenon. A great majority of hydrocarbon traps involve faults of one kind or another. In very many traps, faults are vital components. Yet only a relatively small minority of fields are in true fault traps, in which at least one boundary of the field is at a fault. Of our three categories of *faulted traps* (Sec. 16.7.1), the first is numerically and economically the least important.

Far more numerous and important are faulted traps of our second category, in which the faults cause other structures and these other structures form the traps. We have examined many varieties, both convex and nonconvex:

(a) Horsts and tilted blocks may constitute the essential structural style of an entire basin; but the actual trapping mechanisms are much more likely to be either unconformities or drape folds than faults.
(b) Overthrust anticlines may indeed create fault traps, especially along their footwalls, but pools in such traps are invariably overshadowed by others due wholly to convexity (Fig. 16.15).
(c) Trap structures associated with strike-slip faults are also likely to be due to convexity rather than to faults.
(d) Faulted depositional sites give rise to great numbers of depositional trapping mechanisms from sand bodies to reefs, and to a variety of paleogeomorphic traps.
(e) Rollover anticlines are the only common convex structures wholly consequent upon normal faulting; but it is their convexity that makes them traps.

There remain the faulted traps of our third category, in which the faults, though they may be vital features of the traps, are themselves consequences of quite different structures which are the real causes of the traps. The obvious case is of a fold or arch that acquires normal or high-angle reverse faults in the course of its growth. This is the converse of the convex drape anticline overlying a basement horst block; there the faults are the fundamental structures and the fold is a consequence of them. Here the fold is the fundamental structure and the faults are consequences of it; we have a fault-modified convex trap. Many hundreds of oil- and gasfields occur in such traps; those of Cook Inlet in Alaska may serve as examples (Sec. 26.2.2).

A comparable trap mechanism may be created by the faults generated along the flanks of a salt or shale uplift. Here the faults may be either cause or consequence of the mobile structure. Both the salt and the shale cases are illustrated in Fig. 16.121. The traps are formed on the upthrown sides of the faults (the footwall sides in the common case of normal faults, the hanging wall sides in the less common case of high-angle reverse faults) if permeable beds are thrown against impermeable. Accumulations are almost certain to postdate faulting if the faults actually bound the pools.

Accumulation may precede faulting, but late faulting is more likely to destroy any true fault traps in a field. An exceptional example much quoted in earlier literature is

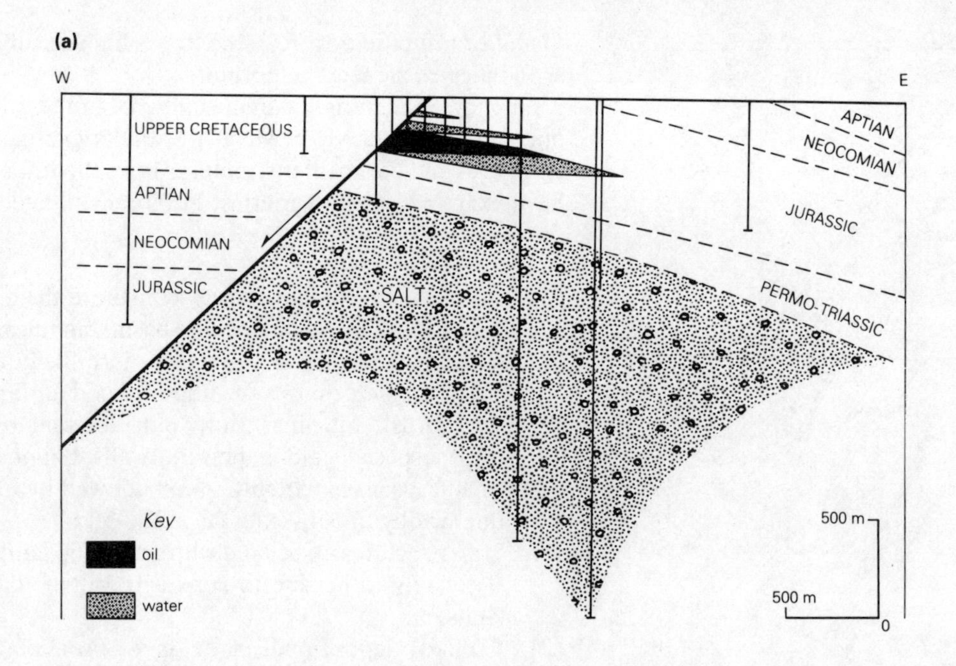

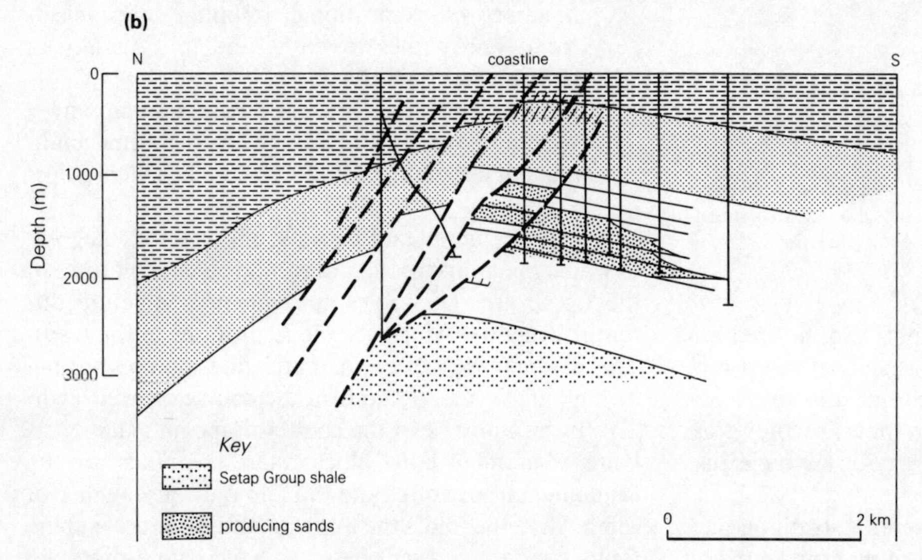

Figure 16.121 Cross sections through two oilfields, of different ages in different basins, showing great similarity even between complex composite or combination traps. Both have semi-mobilized cores uplifted along normal faults; oil is in non-convex traps with fault seals, on one flank of a convex uplift. (a) Dossor field in the Emba Basin, north of the Caspian Sea. Formations are all Mesozoic except for the semi-mobilized salt, which is Permian. (After C. W. Sanders, *AAPG Bull.*, 1939.) (b) Seria field in Brunei, northwestern Borneo. Formations are wholly Tertiary; semi-mobilized rock is shale. (After H. P. Schaub and A. Jackson, in *Habitat of oil*, Tulsa, Okla: AAPG, 1958.)

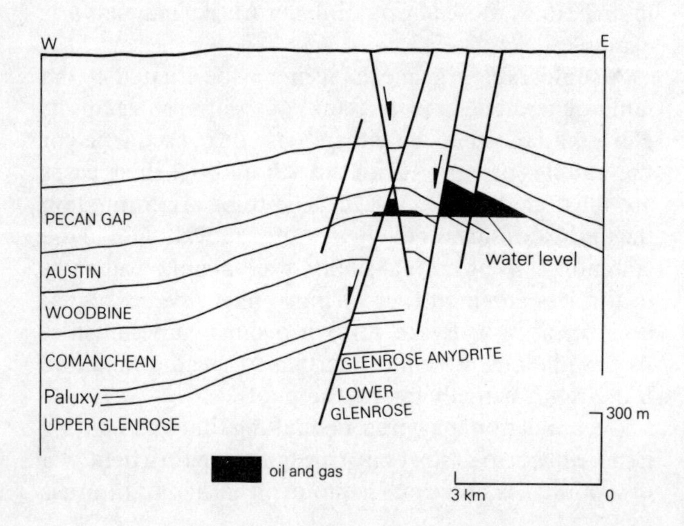

Figure 16.122 Cross section through the Van oilfield, in the upper Gulf Coast Basin of Texas. The structure overlies a deep-seated salt intrusion, with characteristic graben faulting over its crest. The oil pool on the west side originally had a free gas cap (in the small wedge between the downwardly diverging faults). The latest faulting must have postdated the accumulation, yet the oil/water contact has readjusted itself to a horizontal surface. (After R. Liddle, University of Texas Bull. 3601, 1936.)

that of the Van field in northern Texas (Fig. 16.122). Oil occurs in two pools in the Cretaceous Woodbine sandstone, separated by a large normal fault. Several other normal faults cut the structure, which overlies a deep-seated salt intrusion. The faults form trap boundaries within the field but not for the field as a whole. Yet the oil/water contact is constant throughout the field, crossing even faults which bound accumulations. There was an initial free gas cap above the oil column between the downwardly diverging faults on the western side of the field. This part of the field must therefore originally have been on the highest part of the structure, but it is now lower than the structure of the larger pool on the eastern side of the field. The faults must have moved since the accumulation was formed (though they may have moved earlier as well), and the more westerly faults underwent greater downthrow than the more easterly faults. The faults apparently introduced no permeability barriers (that is, they were nonsealing) because the oil/water contact was able to re-establish itself as a continuous surface across the whole field.

Part IV

EXPLORATION, EXPLOITATION, FORECASTING

We at last have oil and gas, in the states in which we know them, trapped in commercial quantities in accumulations ready for mankind to exploit. Mankind learned how to do this by following a succession of simple, practical steps which had begun when someone recognized that certain materials exuding from the Earth's surface were so readily inflammable that they could be gathered and used to provide heat and light. Now that we use these materials for much more than heat and light, the practical steps have become more complicated and far from simple. In this fourth part of the book, we shall consider where we have learned to look for the petroleum-filled reservoirs and traps we discussed in Part III; what criteria we seek and how we seek them; how the geologists and their industry colleagues go about the tasks of recovering petroleum in the most efficient way they can; and how they try to express in usable numbers their conviction that they can go on doing this far into the future.

17 Petroleum basins

17.1 Petroleum provinces and petroliferous basins

Oilmen have long referred with casual vagueness to petroleum (or oil, or hydrocarbon) *provinces,* oil *districts,* oil (or petroliferous) *basins,* and to other terms denoting geographic or geologic units containing oil and/or gas. These terms overlap to a confusing degree.

An oil (or gas, or petroleum) *province* is a region containing oil or gas under a possibly wide variety of conditions but with one or more critical, unifying geological factors. These can be wholly geological (all the reservoir rocks are Tertiary sandstones, for example); or they may combine geological and other factors (such as the presence of a particular chemical type of oil, or the dominance of gas). The factors should not, however, be exclusively geographical, political, or geopolitical, despite the fact that the word "province" has such connotations outside the industry.

A "North African oil province" would have no geological significance, even to OPEC. The large oil- and gasfields in Algeria are in Paleozoic and early Mesozoic sandstones. Those in Libya are in a basin that did not develop until the Cretaceous, even though a lot of the oil migrated downward into fractured Precambrian basement. Nearly all the Egyptian oilfields are in a Miocene salt basin.

An even more pointed example of a purely geographic province is the "Western Canadian oil province." More than 90 percent of the oil in western Canada is in either Paleozoic carbonates or Cretaceous sandstones. There is no semicontinuous spectrum of stratigraphic age for the reservoirs, as there is in the US Gulf Coast, for example. Canada's Cretaceous oil was generated in a basin created by the rise of the late Mesozoic cordillera to the west. That basin happens to be closely coincident, geographically, with the much older Devonian basin; so the Western Canadian Basin is really two basins, one on top of the other. The term "province" has no additional significance in such a case and should not be used.

The unifying factor(s) must be fundamentally *geological*; commonly all hydrocarbons within even a large province are contained within strata of one age span. In the American Mid-Continent and Russian platform provinces, for example, oil and gas are abundant in a great variety of traps; but all significant occurrences are in Paleozoic strata of foreland or platform facies lying in front of a Paleozoic deformed belt. Each of these provinces embraces at least four separate *basins* and the *uplifts* or *arches* separating them; but the basins and arches were established in essentially their present forms in Paleozoic times, and the Paleozoic strata in the separate basins have many features in common (Figs 16.43, & 17.1).

Additional unifying factors may be similarities of trap type, reservoir rock, and even oil characteristics. In the Californian, Carpathian, and southeast Asian provinces, for example, virtually all the reservoir rocks are of Tertiary age, and the great majority are sandstones; nearly all the traps are anticlines of one sort or another; and the oils in any one province exhibit such chemical uniformity that they must share a common source rock.

Structural unity is a third criterion by which a hydrocarbon province may be defined. Large-scale examples are the Persian Gulf province and the Gulf Coast province of the southern USA. Oil and gas occur in orderly successions of strata across each province, essentially from Jurassic to late Tertiary. There is also some uniformity of trap and reservoir type in each province. The fundamental justification for referring to each as a single *province,* however, is that both are relatively simple structural units. The Persian Gulf province is pressed down below a bounding mountain belt on one side, and its stratigraphic content tapers out to nothing against a Precambrian shield on the opposite side. The Gulf Coast province is faulted down along the back slope of an older mountain belt, and its stratigraphic content thickens abruptly toward a deep oceanic basin on the opposite side (Figs 17.2 & 22.15).

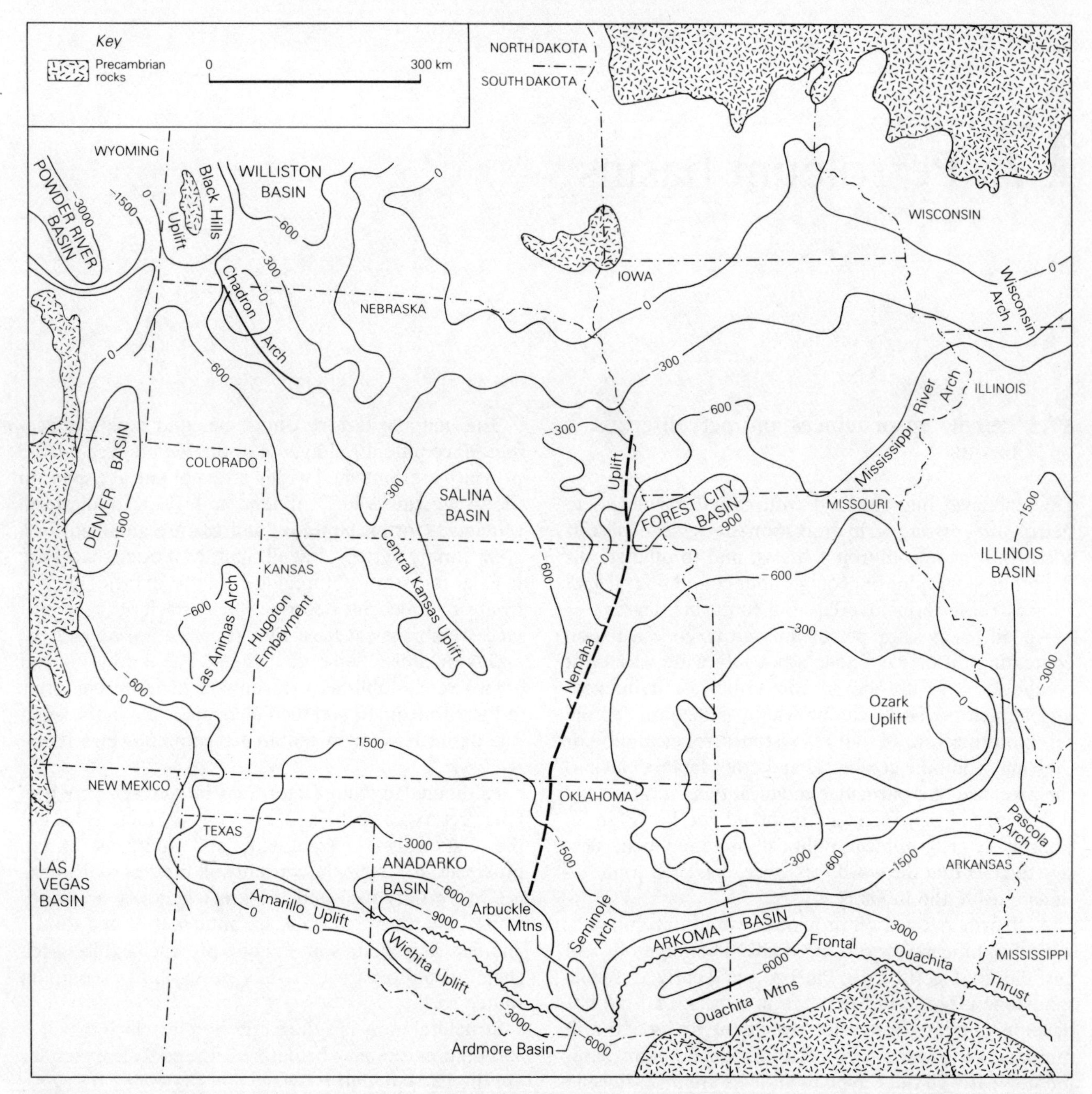

Figure 17.1 Basement structure map of the Mid-Continent petroleum province of the US, showing numerous individual basins and arches (outlined by schematic contours in meters below sea level; contour interval not constant). The basins are filled with Paleozoic sedimentary rocks under a cover of Cretaceous. All exposures of Precambrian basement rock across the northern and western parts of the map area, plus the Central Kansas uplift, were within the Transcontinental Arch of Paleozoic time. (From W. H. Parsons, *Can. Soc. Petrolm Geol. Memoir* 1, 1973; after F. J. Adler, *AAPG Memoir* 15, 1971.)

Obviously the term "petroleum (or 'oil' or 'gas') province" is too vague to be particularly useful unless we agree on the province's *boundaries*. The Mid-Continent "province," for instance, is not in the middle of North America at all, unless we allow it to include the Williston Basin which straddles the USA/Canada boundary in the dead middle of the continent (see Fig. 17.18). The Williston Basin's oil is in Paleozoic reservoir rocks, several of them having almost the same age and lithology as those in the Mid-Continent fields of Kansas and Oklahoma.

Similarly the fields in the Permian Basin of Texas and New Mexico are almost linked to those of the Mid-Continent's Ardmore Basin by the fields along the Bend and Red River arches of northern Texas (Fig. 17.3). Again, all the oil in the Permian Basin is in strata coeval with those containing the oil in Oklahoma and Kansas. But the Permian Basin was separated from the "Mid-

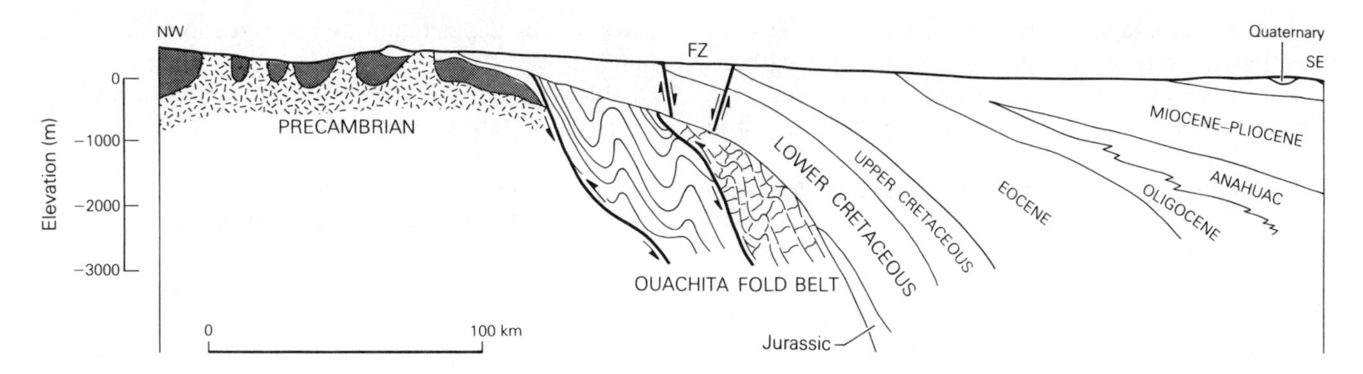

Figure 17.2 Cross section from the Llano uplift in western Texas, across the Balcones fault zone (FZ) and into the Gulf of Mexico, showing the development of the Gulf Coast Basin by Mesozoic downflexing along the back slope of the Paleozoic Ouachita fold belt. Vertical exaggeration, 21x. (From K. K. Landes, *Petroleum geology of the United States*, New York: Wiley–Interscience, 1970; after Dallas Geological Society, 1963.)

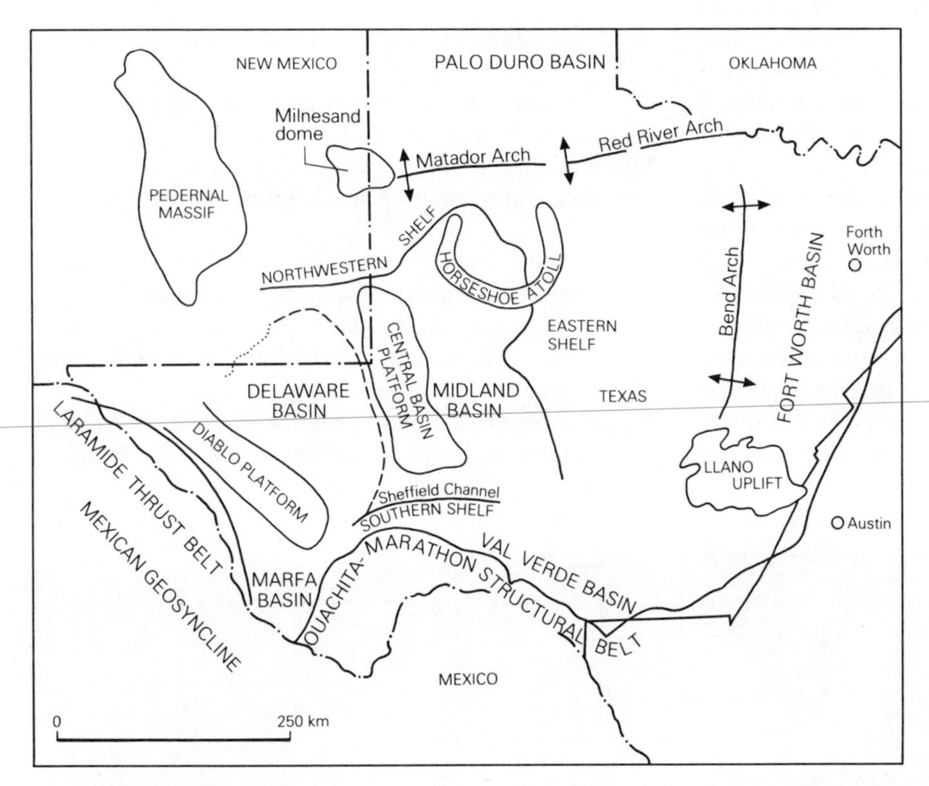

Figure 17.3 Principal structural features of the Permian Basin and adjacent elements. (After J. K. Hartman and L. R. Woodard, *AAPG Memoir* 15, 1971.)

Continent" basins during most of Paleozoic time by a broad flexure which now forms the basin's Eastern Shelf; during the Pennsylvanian orogenic episodes, a major mountain range arose between the Permian and Mid-Continent basins.

The Williston Basin, on the other hand, was an extension of the Western Canadian Basin during much of the Paleozoic, and was totally cut off from the Mid-Continent province by the Precambrian backbone of North America, the Transcontinental Arch (see Sec. 25.2.4). If the "Mid-Continent province" were to be interpreted as extending north of this Precambrian orogenic belt, or south and west of the Pennsylvanian orogenic belt, it would become co-extensive with the sediment-covered North American craton and might as well embrace northern and much of eastern Canada.

The term "province" to define a productive region has therefore never been adopted in countries relatively recently important in the oil world, such as Australia, China, and West Africa. Elsewhere, a "province" has come to equate either with a single productive *basin* or with a group of closely related basins and their separating uplifts.

A petroliferous (or hydrocarbon) *basin* must be distinguished from a hydrocarbon *province,* even though the two are sometimes the same. The Persian Gulf province is in an acceptable sense identical with the Persian Gulf Basin. Though there were distinguishable depocenters within it at various times, it was never fragmented into sub-basins until late Tertiary mountain building and uplift ended its depositional history. The Gulf Coast province of the USA and Mexico, on the other hand, contained a number of separate Mesozoic salt basins from its inception. Prolific hydrocarbons occur, under quite different geological conditions, in at least six separate basins inherited from those ancestral salt basins (Fig. 17.4).

An oil or gas *basin* is a geological structure that subsided in such a manner as to receive a unique sedimentary succession. This succession is necessarily dissimilar to that outside the basin, though it may be very similar to that in a nearby basin. For example, six separate basins rim the Guyana shield along the inner (continental) side of the Andes between northern Peru and Trinidad. Their Cretaceous and Tertiary stratigraphic successions are remarkably alike, and all contain oil source and reservoir rocks of comparable ages and lithologies. Such basins inevitably shared similar origins. Nevertheless, the manner and cause of subsidence are immaterial to the existence of a basin, but in nearly all cases a direct association with a regional diastrophic event is quite evident.

It follows that a basin has a definable life span, from the beginning to the end of its subsidence. During that life span, at least some of the time-stratigraphic units necessarily change facies as they are traced beyond the basin. At least some of the rock-stratigraphic units must also change thickness in that direction; most will become thinner (some to zero). During parts of a basin's life it may be starved of sediment while its margins accumulate thick carbonates. A basin's contents need not be marine; lakes occupy depositional basins, and a long-lived lake that becomes thermally or chemically stratified may accumulate several kilometers of sediment and become richly petroliferous (see Secs 17.4.7. & 17.7.9).

An oil or gas *district* is a restricted area within a basin in which two or more productive structures are closely similar in style and fluid content, and essentially alike in age. Examples of richly petroliferous districts are the Oficina district in the eastern Venezuelan Basin (see Fig. 17.29), the Orenburg district within the Ural–Volga Basin (Fig. 16.43), and the Ekofisk district in the North Sea Basin (Fig. 25.3).

17.2 General development of basin studies

The development of basin studies can be generalized into five principal phases, sequential in a general way but overlapping widely, at least in the days before plate tectonics. Interpretation has been in the light of

(a) eustatic theory and epeirogeny, as reflected in global transgressions and regressions of the seas;
(b) geosynclinal theory, as reflected in the basins' locations with respect to mobile belts and stable regions;
(c) fundamental geophysical calculations of thinning and/or flexing potentialities of the lithosphere;

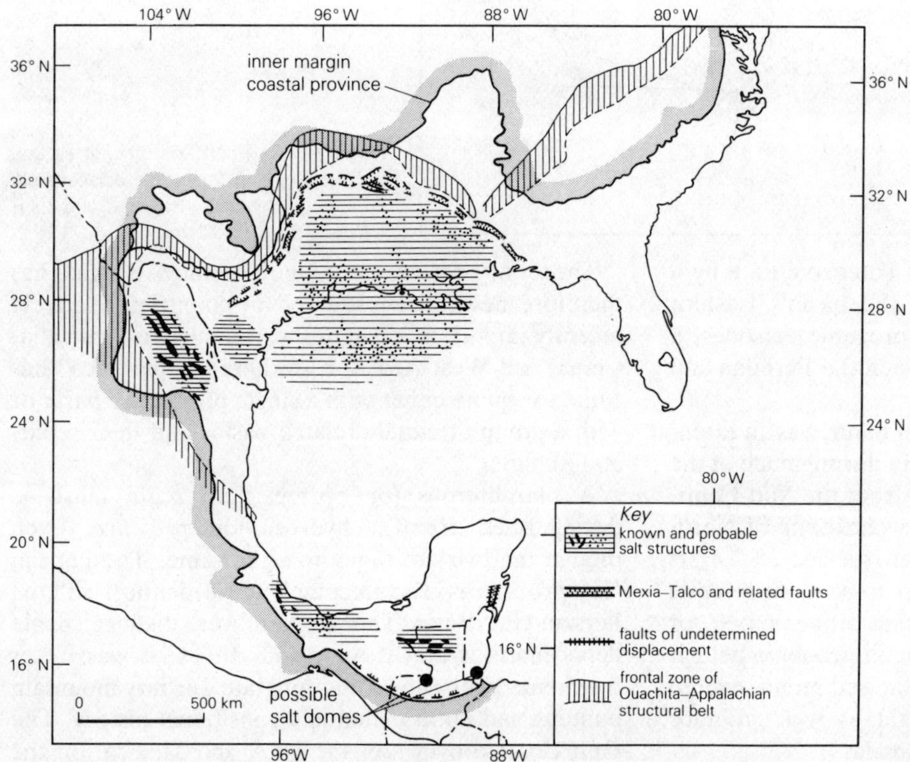

Figure 17.4 The Gulf of Mexico oil and gas *province*, containing at least six discrete *basins* as delineated by accumulations of Jurassic salt. (From G. E. Murray, *AAPG Memoir* 8, 1968.)

(d) oil-geologic experience as reflected in the structural styles of basins, determined by surface, subsurface, and seismic data;
(e) plate tectonics, as reflected in the basins' relations to crustal types and to plate boundaries or interiors.

We will consider (a), (b) and (c) briefly, and (d) and (e) in detail.

Phase 1 It was observed early that thick sedimentary sequences comprising petroliferous basins tend to begin during marine transgressions and end with nonmarine "fill" deposited during regression or withdrawal of the sea. Hans Stille connected the withdrawals with relatively brief orogenic episodes, which he regarded as synchronous on a global scale. Emil Haug contended that the reverse was the case; that orogenic episodes brought about transgressions by expelling waters from the geosynclines. This line of thinking was dominant before World War I; it ultimately led, through Laurence Sloss's *sequences* (packets of strata bounded above and below by interregional unconformities), to a modern culmination in *seismic stratigraphy,* discussed in Chapter 21.

Phase 2 Charles Schuchert classified *geosynclines* in two broad types: those already deformed into mountains (specifically the Appalachian and Cordilleran geosynclines); and those still receiving thick sediments but as yet undeformed (the Mediterranean, for example, and the Sea of Japan). All lie alongside or between continents or their "borderlands." Marine extensions of these deeply subsiding elements across continental interiors (most petroliferous basins, in fact) were called *embayments* by Schuchert.

Marshall Kay's extension of this classification became really a classification of *basins,* whether petroliferous or not. A geosyncline (hence any sedimentary basin) was conceived of as a *surface* developed at the base of extensive surficial rocks during their accumulation. Schuchert's embayments were divided into a number of basinal types lying within rather than beside continental platforms. A further category was proposed for basins created within older geosynclines as a result of their deformation.

A serious drawback to attempts to make all basins aspects of geosynclines was that some common petroliferous types were not conveniently or convincingly accommodated. It is not obvious to which of Kay's categories he could have assigned the greatest of all oil basins, that in the Middle East, without associating it with some incongruous bedfellows. Wallace Cady, in adopting Kay's categories (Fig. 17.5), was compelled to supplement them with no fewer than seven traditionally titled basin categories, including all the familiar petroliferous types.

Russian geologist Nikolai Shatsky proposed a widely adopted variant of the Schuchert and Kay classifications. Long-lived, deeply-subsiding depressions within cratons, bounded by large faults and not inherited from older depressions, deserve to rank alongside major geosynclines as sediment-filled troughs. Shatsky's term *aulacogen* for such basins became preferred by many over the equivalent American terms.

Phase 3 Geophysicists have made many studies on how segments of the Earth's spheroidal crust subside to form simple basins. This occurs within cratons (by thinning of the crust, perhaps by thermal contraction); at cratonic margins (by elastic downflexing, assisted by sediment load); and in ocean basins. These studies imply that there are limits to both the area and depth of a subsiding or downflexing basin, and upper and lower limits to the rate at which it can subside. These implications influenced the thinking of a number of petroleum geologists, notably Karl Dallmus.

Dallmus calculated that the maximum diameter for "simple" basins (whether due to subsidence or downfaulting) is 400–500 km. This is, in fact, very roughly the diameter of the Permian Basin and the productive part of the Ural–Volga Basin. It is not easy to identify any basin in which rocks of a single system are productive over significantly more than 2×10^5 km^2 (consider the eastern Venezuelan *Tertiary* basin, one of Dallmus' type specimens, or even the at first sight gigantic Ural–Volga Basin; Figs 16.43 & 17.29). Larger basins can commonly be shown either to be composite (subdivided by arches or other structural features) or to have shifted their axes significantly with age.

Oil generated in the central part of such a basin and moving at only 1 cm yr^{-1} could migrate into a marginal trap 200–250 km away (given adequate carrier beds) in only 20–25 Ma (about the duration of the Eocene Epoch). However, neither Dallmus nor his fellow investigators discriminated in any detail the variety of basin types with which the practising petroleum geologist has to contend.

Phase 4 Classifications devised specifically to accommodate *petroliferous basins* came naturally from petroleum geologists. In the days before plate tectonics, most published classifications fell into one of two general types, not inappropriately describable as the "American" and "Russian" types. The former was devised by practical oilmen to accommodate all known American examples (plus many not American); such classifications became closely similar to classifications of

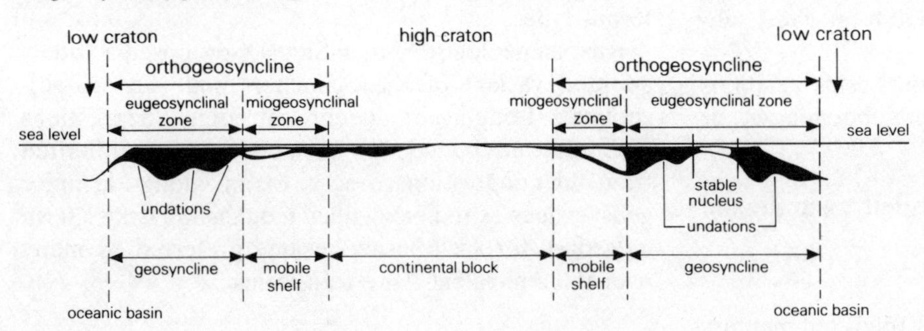

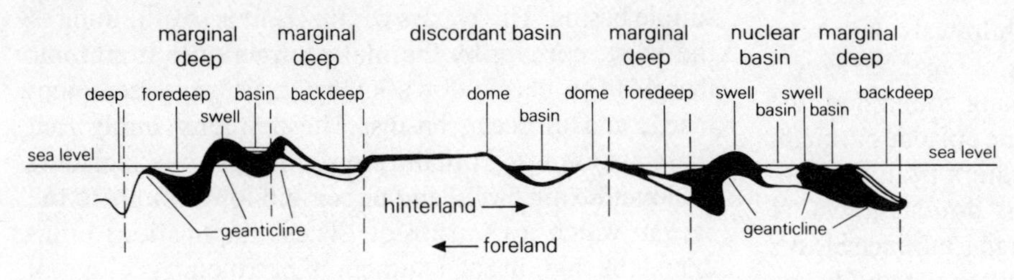

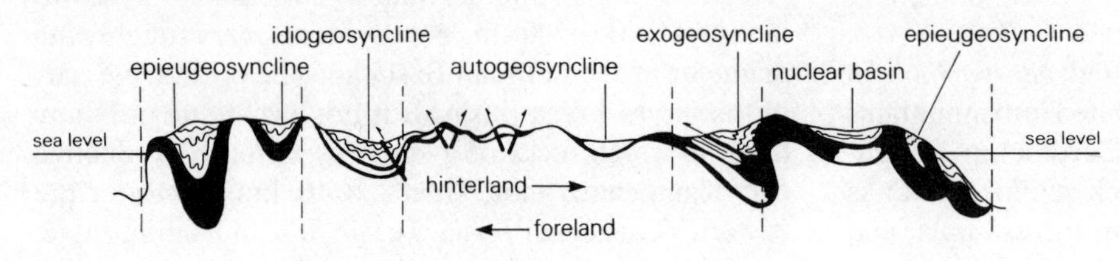

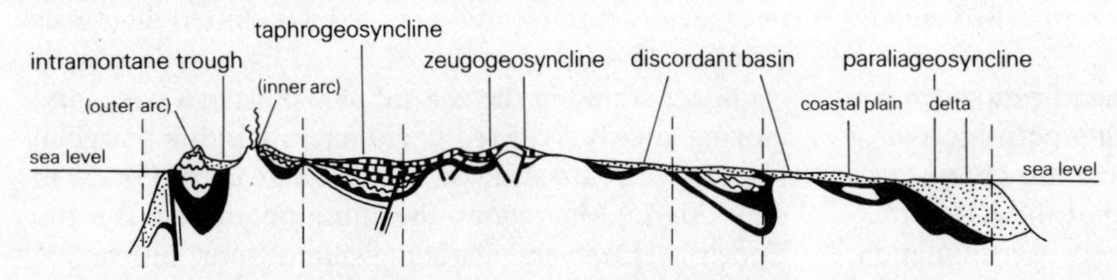

Figure 17.5 Succession of basinal developments during the diastrophic cycle. (After W. M. Cady, *Trans. Am. geophys. Un.*, 1950, ©American Geophysical Union.)

trapping mechanisms. Early Russian models were devised by structural geologists, commonly believers in the dominance of vertical over horizontal tectonics. These models comprised hypothetical, descriptive types based on structural theory, usually without clear indications concerning non-Russian basins. Productive regions were classified according to mode of basement control, and emphasis was accordingly on *uplifts* ("domes" or "arches" in Russian terminology) rather than on *basins*.

Among Western geologists the emphasis depended upon the individual's attitude towards continental drift. "Fixists" based their classifications upon the basin's position with respect to continents, ocean basins, and deformed belts, which were considered essentially immobile. The most influential classification of this type was that of Lewis Weeks. "Mobilists" tended instead to emphasize mechanisms of basin formation, as the locales were not imagined to be fixed. An example of this type of classification was proposed by North. By considering these two classifications in slightly more detail, the different emphases will be made apparent.

The *Weeks classification* was position-centered. The primary distinction between *mobile belt basins* and *stable region basins* was refined by dividing each primary group into three subgroups. Mobile belt basins can be *marginal* to the mobile belt (between it and an ocean basin, as in the American Gulf Coast Basin; or between the mobile belt and the continent, like the Cretaceous basins between the Andes and the Brazilian craton, or the Sea of Japan today). Mobile belt basins can alternatively be *intracontinental*, like the Appalachian and Donetz basins and the Persian Gulf Basin to some extent; or they can be *intermontane* (actually within the mobile belts, like the San Joaquin Valley Basin of California). Stable region basins may be *interior* (and either faulted like the Dead Sea or essentially unfaulted like the Michigan Basin) or *coastal* (normally between shield edges and ocean basins, as around western Australia and Africa in the Cretaceous and Tertiary).

North's classification, purportedly genetic, was based on the mechanism of basin creation. *Primary* (or *structural*) *basins*, due wholly to hypogene processes, can be: compressional in origin (*foredeeps* in front of compressional deformed belts, and *interdeeps* between them); extensional (*backdeeps* behind deformed belts, and *graben*, which can themselves be forms of backdeep); or apparently wholly gravitational (*cratonic sags*), presumably created by contraction of subcrustal rocks. *Secondary basins* are any of these types that have been so modified by supergene (depositional) agencies that the effective generative and productive basins must be regarded as *constructional*. Principal examples are those basins in which the isolating mechanism is caused or influenced by reef growth.

Russian classifications of basins have been changed relatively little since the advent of plate tectonics. Modern versions maintain the foundations set down by N. Yu. Uspenskaya, V. B. Olenin, and others during the 1960s. Uspenskaya emphasized the nature and setting of the geologic province within which a particular basin was formed. All basins lie within either platform provinces or mobile belt provinces. Olenin recognized "homogeneous" basins associated with each of these two major province types, but categorized most of the great petroliferous basins of the world as "heterogeneous," combining aspects of both. All important foredeep and marginal basins, for example, are "heterogeneous" in this sense. Like Uspenskaya, Olenin emphasized basinal localities with respect to platforms and fold belts, and essentially ignored processes involved in basinal formation. Olenin's scheme differs from Uspenskaya's in incorporating post-basinal history, taking account of the manner in which the hydrocarbon content was generated, accumulated, and preserved.

Even during this plate tectonics era, scientists dealing with interregional seismic correlation find it necessary to classify basins according to structural and positional criteria. These can be determined from long-distance reflection profiles. The categories employed by COCORP and COPSTOC, for example, do not differ greatly from those proposed by Weeks or North: rifted continental margins, grabens, aulacogens, foreland basins, and intracratonic basins.

Phase 5 The advent of plate tectonics did surprisingly little to alter the geological understanding of basins. It provided a unified framework within which to identify their appropriate places, and it made a chaotic basinal nomenclature even more so. Petroleum geologists continued into the early 1980s to use traditional, descriptive terms like "cratonic basin," "foredeep," and "intermont basin" for productive regions of their concern.

The old Stille–Haug controversy over the control of marine transgressions and regressions was quickly reactivated. Amid considerable disagreement in detail, a general consensus maintained that transgressions occur when plate tectonism is active. The growth of new oceanic ridges leads to eustatic rises in sea level; plate interactions culminate in vulcanism, uplift, crustal shortening, and basin subsidence. In short, transgressions synchronize with orogenic episodes. Regressions occur during plate quiescence, with epeirogenic upwarps, eustatic falls in sea level, diminished tectonism, diminished subsidence, and the filling and shallowing of basins.

Expressed in slightly different terms by James Valentine and Eldridge Moores, the assembly of continents is accompanied by regression; John Bird and John Dewey had quite the opposite view. Erle Kauffman regarded the development and decline of large basins and their sedimentary contents as reflecting the interaction of eustatic, tectonic, and sedimentologic processes.

There are only four principal modes of interaction between plates: separation, convergence, collision, and transform sliding. The simplest way to relate basin creation and development to plate tectonics is to assign basins to those four processes. Max Pitcher did this, and concluded that 75 percent of the world's basins containing giant oilfields were related to collisions and suturing; only 20 percent were related to the other three processes; and 6 percent could not be assigned to identifiable plate activity at all.

More extended discrimination between basin types in the light of plate tectonics has produced two general styles of classification. One attempts to identify all basin types created during plate activity, without regard to their relative significances to petroleum geologists. The other concentrates on petroliferous basins and assigns them – commonly by their traditional names – to appropriate niches in the plate tectonics scheme. Richard Chapman disdained both approaches and maintained that all basins are either symmetrical or asymmetrical, and faulted or unfaulted. Several authoritative versions of the two larger styles involved such fine discriminations that they are considered briefly below.

17.3 Basinal classifications in plate tectonics

The most exhaustive representation of the first style is that by William Dickinson (published in 1977). His emphasis is on the setting of the basin in relation to the nature and stage of plate activity, especially at plate margins. The latter can be *activation margins* (of Cordilleran type); *collision margins* (of Alpine or Taconic type, according to whether the collision is of continent–continent or arc–continent type); or *rifted margins* (of Atlantic type). Dickinson's fundamental division is between extensional or *rifted basins* and compressional or *orogenic basins*; this division is common to classifications of otherwise quite different styles, including North's.

Dickinson considered all types of basinal subsidence which plate interactions are capable of generating, whether any petroliferous examples are known or not. Most classifications of basins proposed or widely adopted by practising petroleum geologists have been much simpler than Dickinson's. The only one of comparably exhaustive complexity to enjoy wide influence was Albert Bally's (first published in 1975, expanded in 1980). Whereas Dickinson emphasized tectonic *processes*, Bally emphasized tectonic *boundaries*, especially compressional plate boundaries involving decoupled crust. These "megasutures" were held to occupy the whole of the Earth's surface except for the ocean basins.

There are fold belts (derived from megasutures) of various ages: Precambrian, forming the shields; Paleozoic, also now wholly in continental crust (like the megasuture of the Appalachians–Caledonides, extending from the Caribbean to the Canadian Arctic); and Mesozoic–Cenozoic, retaining much remnant oceanic crust as ophiolites (as along the Tethyan belt). The ocean basins are also of Mesozoic–Cenozoic age.

The term *basin* was interpreted by Bally as embracing all "realms of subsidence" within the lithosphere, provided that they have been preserved with some minimum content of sedimentary rock (say 1 km of thickness or more, so that deep ocean basins and thinly covered areas of continents are excluded). Petroleum content is not a criterion. Bally recognized three families of such basins:

(a) basins (of all ages) on rigid lithosphere, without association with megasutures – these are *cratonic* and extensional basins;

(b) basins (of all ages) on rigid lithosphere, flanking compressional megasutures (*perisutural basins*);

(c) basins within or upon compressional megasutures (*episutural basins*), necessarily young because they are easily destroyed by renewed orogeny.

These three "families" of basin were subdivided by Bally into 19 basinal categories.

The second general category of plate tectonics classification concentrates on assigning *petroliferous* basins to their apparently appropriate niches. One early classification, by Douglas Klemme in 1970, became widely influential. An associate of Lewis Weeks, Klemme classified basins in a manner not very different from Weeks (see Sec. 17.2) except that the basin categories are firmly identified in relation to crustal plates. One innovation was the separate category for delta basins. Like Weeks, Klemme emphasized the nature of the underlying crust and the mechanism of the basin's creation.

Like the present author, Klemme also emphasized that many basins are composites; one structural type has been superimposed on another. The Cook Inlet Basin of Alaska, for example (see Sec. 26.2.2), has been classified by most plate tectonics authors as a fore-arc basin; but the present Cook Inlet Basin is a compressional fault trough transverse to the structural grain, and it produces from a section entirely nonmarine. Similarly, little is learned by associating the Permian Basin of the southwestern USA either wholly with cratonic basins or wholly with orogenic foredeeps (see Sec. 17.4).

A closely similar division of basin types, used by Exxon, was proposed by Keith Huff. Huff adopted the more usual plate tectonics division into extensional (divergent) basins and compressional (convergent) basins, eschewing Klemme's emphasis on the subbasinal crust. Huff's four classes of extensional basin equate closely with Klemme's types 1, 3, 5, and 8, plus part of 4C. Huff took a somewhat different view of compressional basins than Klemme, distinguishing between fore-arc, back-arc, and strike-slip (transform) basins within Klemme's types 6 and 7 (plus part of 4B). He lumped Klemme's already large types 2 and 4A into the vast category of foreland basins.

17.3.1 Classification of basins by Klemme

A. *Intracontinental basins, on cratonic crust*

Type 1 Single-cycle cratonic basins, mostly Paleozoic: Williston

Type 2 Composite, multicycle cratonic basins, containing most Paleozoic oil (Ural–Volga, Western Canadian, Mid-Continent, Permian, Appalachian, Algerian), some Mesozoic oil (West Siberian), and most gas reserves

Type 3 Graben or rift basins, that may be Paleozoic (Dnieper–Donetz); Mesozoic (North Sea, Sirte, Northeast Chinese); or Cenozoic (Suez)

B. *Extracontinental, borderland basins* Formed on intermediate crust associated with plate margins; nearly all are Mesozoic and/or Cenozoic

Type 4 Downwarps into small ocean basins; these basins contain 50 percent of world oil reserves:
4A Basins closed by deformed belts: Persian Gulf, Orinoco, Caucasian
4B Troughs between deformed belts: Upper Assam, Po Valley
4C Open, one-sided basins: Gulf Coast, Alaska North Slope, northwest Borneo

Type 5 Pull-apart (stable coastal) basins on trailing edges of continents; ultimate stage of Type 3 basins: Gabon, Dampier–Rankin

Type 6 Subduction basins, second-cycle, intermontane, small, mostly Tertiary; may be fore-arc (Talara); back-arc (Sumatra); or nonarc, usually transform (California, Baku)

Type 7 Median basins, also second-cycle, intermontane, within compressed and uplifted zones: Maracaibo, Gippsland

Type 8 Delta basins, essentially Tertiary: Niger, Mississippi, Mahakam.

17.3.2 General summary of petroliferous basins

We see that there are three foundations of basin creation and style from which a classifier may choose his principal emphasis. Emphasis may be on the basin-creating process (extension versus compression, in the simplest case), on the type of underlying crust (stable craton versus mobile or transitional belt), or on the basin's situation in relation to global architecture (to a particular type of plate boundary, perhaps). In reality, all three foundations should be looked upon as different aspects of a single one.

The great majority of sedimentary basins were initiated by some form of thinning of originally thick (continental or sialic) lithosphere. How the thinning was brought about is secondary for petroleum geologists. The simplest process is stretching or "necking," permitting passive rise of hot asthenosphere from below. This leads to fracture of the crust into fault blocks which tilt in order to occupy a greater width, and consequently subside (Fig. 16.26). The end of stretching puts an end to the faulting, and slow thermal subsidence continues thereafter and covers a wider and wider area.

If the stretching and the uprise of asthenosphere proceed beyond complete rupture, a new ocean is created and therefore new plate boundaries. Developments along such boundaries depend on whether extension continues or plate convergence takes over and extension gives way to compression. It is on this point that the real distinctions between basins must be made.

Basins that remain extensional pass through a far simpler history than those that become compressional. In an extensional basin, sialic crust is the *first* to be deformed; structures develop from the margins inward; and the original basin is preserved, with the youngest rocks of its fill more or less in its center. Conversion to compression may result in an Andean style of margin, with trench and mountain belt between oceanic and continental plates; to an Aleutian style, with an offshore arc in addition; to an Alpine suture between two continents; or to a transform margin. In all cases, the wedge of sedimentary and volcanic rocks forming the continental rise and slope must be deformed before anything much can happen to the sediments lying on the craton. The sialic crust is therefore the *last* to be deformed; structures develop outward towards the far margin of the depositional area; the original basin, which had lain off the edge of the continent, is totally destroyed and a new one is created on the craton. It is this new basin which becomes petroliferous, not the original one as in the case of a wholly extensional basin.

A compressional basin is thus vastly accentuated by downflexing of the lithosphere under tectonic stress, and by either tectonic (thrust-sheet) or sedimentational loading. It will contain both extensional and compressional structures, whereas a wholly extensional basin contains no truly compressional structures. In the extensional basin, the oldest widespread sediments are nonmarine and the basin ends with a cover of transgressive marine strata deposited during post-rift subsidence. Traps are controlled by normal faults and unconformities; structural traps are chiefly diapiric. In compressional basins, the oldest widespread sediments in the foreland basin are marine and transgressive; numerous stratigraphic traps are created and structural traps are principally bending folds (see Sec. 16.3.2). These old transgressive sedimentary rocks are later covered by regressive molasse, which may be largely nonmarine and its oil or gas content is likely to be migratory into structural traps.

Petroliferous basins might therefore be viewed as transient features in a progression towards equilibrium for a segment of lithosphere containing a plate boundary:

(1) *Constructive disequilibrium* Equilibrium has not yet been reached and continuation of the movements bringing about the present state would continue the basin's life. This state is represented by extensional basins distant from consuming margins or in the interiors of plates (Gulf Coast and North Sea basins).

(2) *Destructive disequilibrium* implies that continuation of the movements that led to the present state would destroy the basin. It is represented by basins on underthrusting continental slabs, such as the Persian Gulf, south Caspian, and Upper Assam basins. This state exists only while convergence is

gentle enough to avoid serious subduction, obduction, or collapse.

(3) *Unstable equilibrium* is the state in which continuation of the most recent movement might or might not continue the basin's life. This state is represented by transform boundaries, seen along the youngest collision margin in the Biscay–Pyrenean and Anatolian stretches of the margin.

(4) *Stable equilibrium* is the state in which further movement in the same sense is irrelevant; the basin as a petroliferous entity has been obliterated. Examples include the Alpine and Himalayan collision zones, the Greater Antilles, and the eastern Cordillera of Colombia.

(5) *Unattainable equilibrium* is represented by steady-state subduction of an oceanic plate. So long as oceanic crust is the only type arriving at the convergent margin, no petroliferous basin forms there. This is why Klemme ignored several basin types listed by Bally and Dickinson. Petroliferous basins form almost exclusively on the uplifted sides of the margins, between the arcs and the continents, and they are compressional from the beginning; they are back-arc or retro-arc basins. They are more commonly petroliferous than any other class of basin; examples include the basins of northern South America, from Colombia to Trinidad; the Sumatra basins; and those of the North Caucasus (see Ch. 25).

17.4 Fundamental types of petroliferous basin

Whatever the basis chosen by the classifier, there are six or seven clearly differentiated basinal styles recognizable among the world's petroliferous basins (Figs 17.6 & 7).

There are simple cratonic sags, like the Michigan and Williston basins in North America. There are simple rift basins, such as the North Sea Basin and the Suez graben. There are basins formed by downflexing or downfaulting of the lithosphere along the backslopes of deformed belts, like the US Gulf Coast and West Siberian basins. A few basins created principally by transform faults, initial transtension being superseded by transpression, have never progressed beyond that condition; the only outstanding examples are in California.

Then there are basins in which compression has taken over completely from earlier extension. Some are wide basins bounded on one side by deformed belts and on the other by stable shields. These may be lumped together as foredeeps, or separated into two or three subcategories (partly by age); they include some of the greatest oil and gas basins known. Other highly petroliferous, compressional basins are narrow and intermontane, between belts deformed towards them and the backslopes of belts deformed away from them, as in the San Joaquin Valley of California and the Maracaibo Basin of Venezuela.

Finally there are basins in which initial tectonic control is later augmented by depositional control, as when extensive reefs come to form the edges of the basins, no matter what may have formed them originally. This category might be extended to include a variety of single-factor basins the critical features of which are exclusively supergene and depositional – deltaic depocenters and continental basins containing stratified lakes.

We have considered the origins and tectonic settings of these basinal types. What do they look like, geologically and geometrically?

17.4.1 Simple cratonic basins (Fig. 17.8)

Simple cratonic basins lack significant asymmetry; the terms "foreland" and "hinterland" have no meaning in relation to them. They are circular to elliptical sags, wholly within stable cratons and unrelated to any complementary highlands. Their subsidence may be due to: thermal contraction of the lithosphere; the presence of dense basement rock below; or activation of buried Precambrian rift faults. North American examples seem to overlie the boundaries between shield provinces.

The basins, fundamentally atectonic, form slowly but exist for hundreds of millions of years. They may never acquire either deep-water deposition or any rapidly deposited clastic sediment. A sedimentary succession about 4 km thick is all that gravitational subsidence enables the basin to accumulate. Black shales, carbonates, and evaporites are common; basement faults are almost the only important primary structures.

17.4.2 Rift or graben basins (Figs 16.26 & 30)

Petroliferous rift basins are known in strata of all eras. The best-known Paleozoic example is probably the Dnieper–Donetz Basin north of the Black Sea, Shatsky's type aulacogen. Most North American geologists now consider the Anadarko and Ardmore basins of southern Oklahoma (Sec. 25.2.4 & Fig. 17.1) to be remnants of a Paleozoic aulacogen.

Rift basins are more closely related to continental margins than to plate boundaries, but several lie well within major cratons. Some are concordant with an older structural grain; others are distinctly transverse to it; some were transtensional, formed by wrench faulting. They may lie atop the sediments of older, pre-rift basins, and these older basins may be the ones that provide the source sediments.

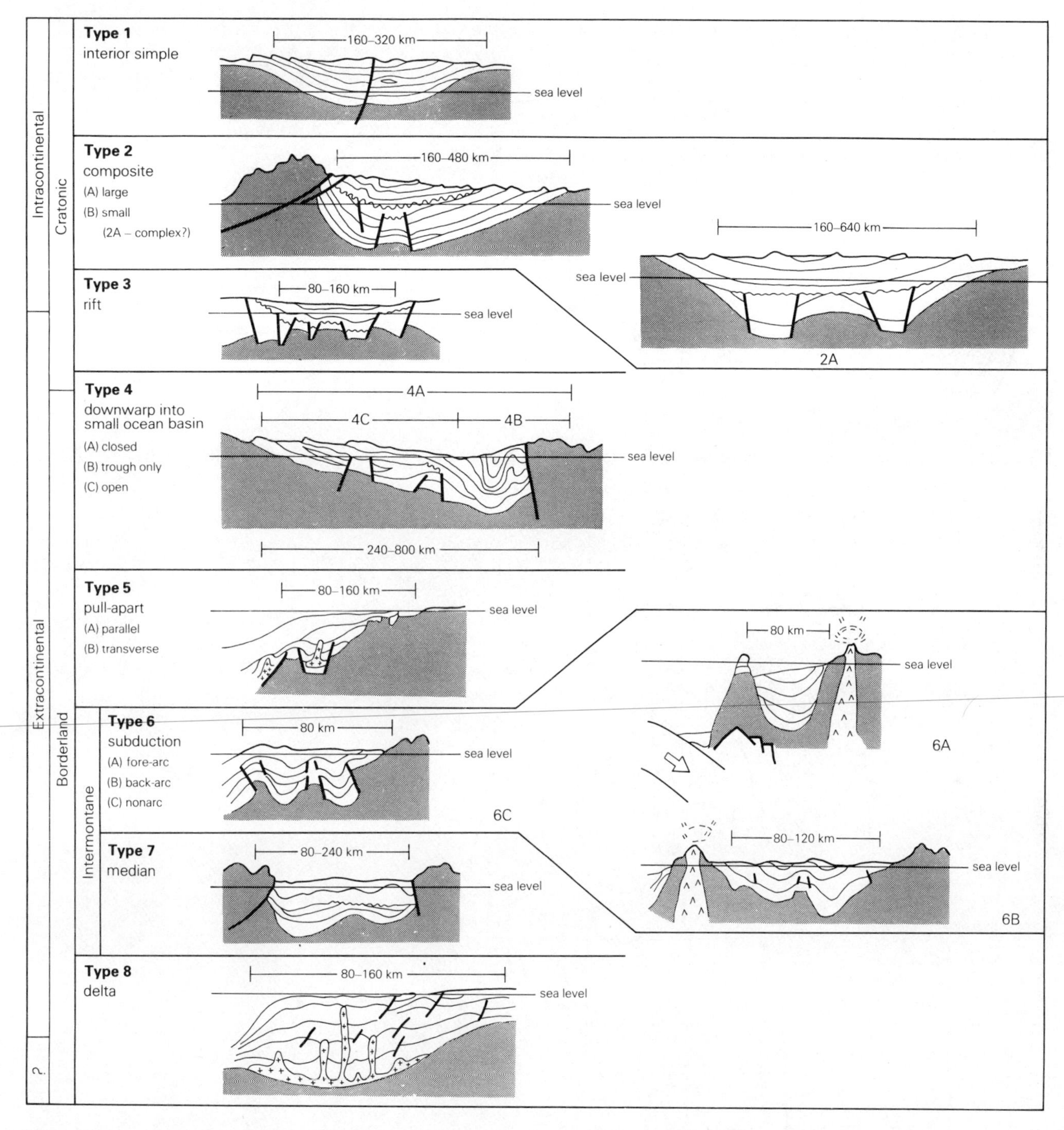

Figure 17.6 Typical cross sections of the principal structural types of petroliferous basin, named according to the classification of H. D. Klemme. (*J. Petrolm Geol.*, 1980.)

The Middle Magdalena Basin of Colombia was a Tertiary creation within a Cretaceous back-arc basin; the Cook Inlet Basin of Alaska was a Tertiary creation within a Jurassic fore-arc basin. In each case the sediments confined to the graben are entirely nonmarine and supply only the reservoir rocks. Furthermore, both basins are in the process of becoming compressional intermontane basins, with as many thrust or reverse faults as normal faults contributing to the trapping mechanisms (Figs 16.99 & 26.5).

The most petroliferous of post-Paleozoic grabens have not been compressionally deformed in this way; they have remained wholly extensional. The North Sea and Sirte basins are the most productive Mesozoic examples. The Gippsland and Cambay grabens are early Tertiary, the Suez and Vienna basins late Tertiary.

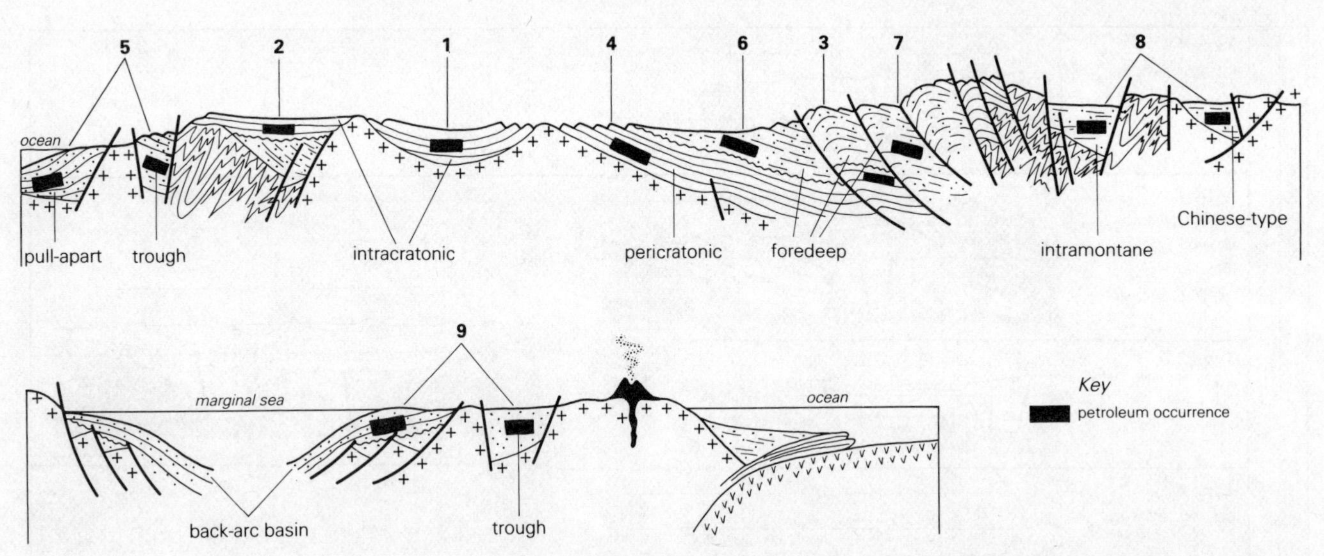

Figure 17.7 Geotectonic classification of petroleum occurrences as proposed by C. Bois, P. Bouche, and R. Pelet. (AAPG Bull., 1982.)

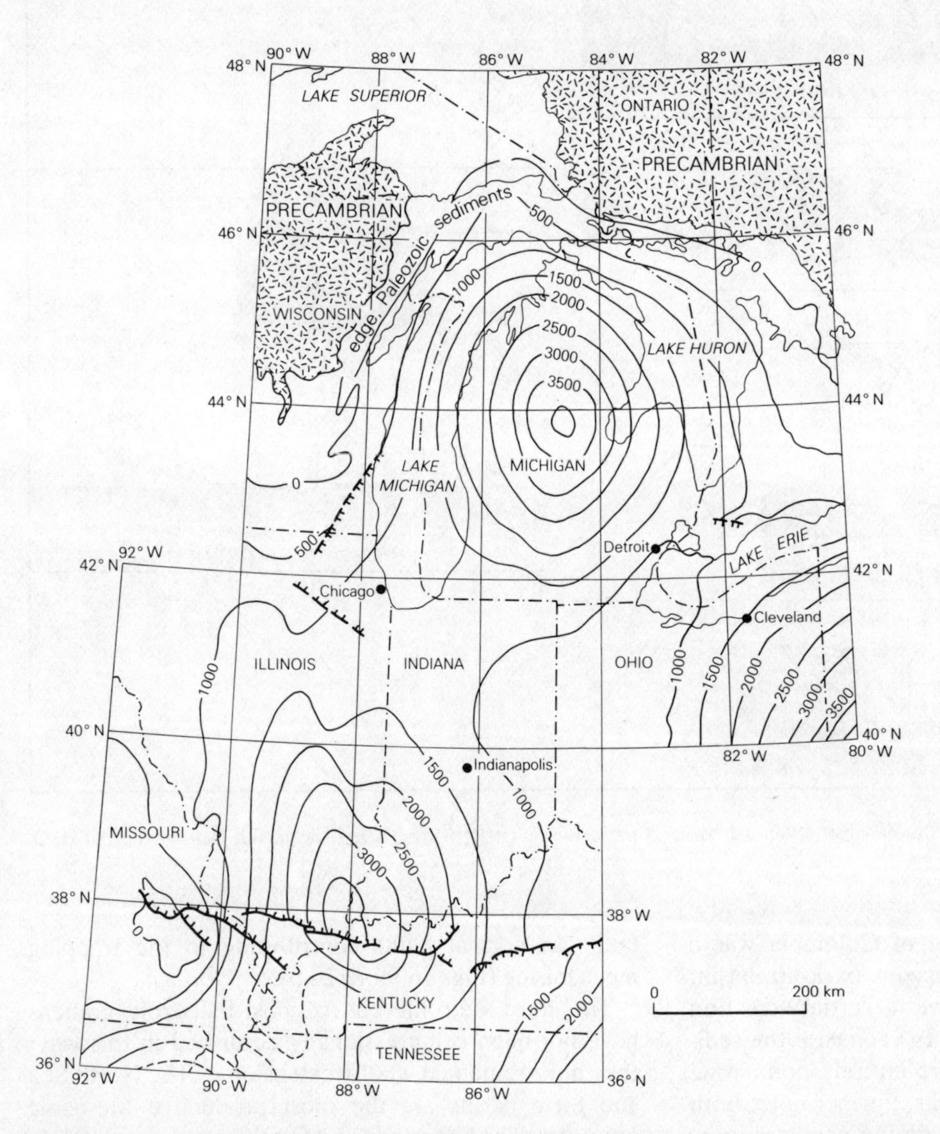

Figure 17.8 Simple cratonic basins, illustrated by the Michigan and Illinois basins in the USA and Canada. Structure contours on Precambrian basement; contour interval 500 m below sea level. (From J. E. Christopher *et al., Can. Soc. Petrolm Geol. Memoir* 1, 1973.)

Rift basins formed on continental crust (as all petroliferous examples were) are naturally likely to begin by receiving nonmarine sedimentation. Nearly all of them pass through an early lacustrine phase; the eventually marine basins around the Cretaceous South Atlantic Ocean are excellent examples (see Sec. 25.5.4). Some rift basins which never progress beyond the continental, lacustrine phase may nonetheless become richly petroliferous (see Sec. 17.4.7).

Most rift basins are actually highly asymmetrical half-grabens. They are more likely than other basin types to contain true fault traps (see Sec. 16.7), but the majority of graben oilfield traps are related more to unconformities than to faults. The youngest sediments in graben basins are likely to be quite unfaulted, because all grabens eventually give way to simple gravity-controlled sags over areas much wider than the original rifts (see Sec. 17.7.4 & Fig. 16.26). A single sedimentary cycle normally dominates the section.

17.4.3 Extensional downwarps (Figs 16.32, 47 & 17.2)

Few classifications unite in one category the American Gulf Coast Basin, the West Siberian Basin, the Magallanes and San Jorge basins of Patagonia, and such coastal downwarps as those of the Congo Basin off West Africa, the Campos Basin off Brazil, and the Dampier Basin off northwestern Australia. Yet all are fundamentally downflexed or downfaulted depressions formed under crustal extension.

The Mesozoic West Siberian Basin (Fig. 17.9) differs from the others in that its extension did not lead to a new ocean basin. However, its situation behind the Ural Mountains (which have a compressional Paleozoic basin in front of them) is directly akin to that of the Mesozoic Gulf Coast Basin, behind the Oklahoma fold belt which also has compressional Paleozoic oil basins in front of it (Fig. 17.1). The Magallanes Basin also differs from the others in its spatial relation to a young orogenic belt, making it appear to be a compressional foredeep. But the Jura-Cretaceous age of its petroliferous section long antedates uplift of the southern Andes.

Extensional downwarps that open into new ocean basins display an asymmetry as pronounced as that of compressional foredeeps, and are likely to display the same stratigraphic zonation of production. The Gulf Coast Basin is a textbook example of the zonation, with Jurassic production closest to the foreland and concentric belts of progressively younger trends seaward (Fig. 17.10). Even within the wholly Tertiary and Recent Niger delta basin this zonation is evident. Production from the northern zone, on the continental side, comes from sands of Eocene age; in the central zone, the reservoirs are Oligo-Miocene; and in the southern zone, Mio-Pliocene in age.

Asymmetrical extensional downwarps are likely to be floored by either salt or basic lava. Orthodox anticlinal traps are naturally exceptional; the commonest traps are rollover anticlines associated with normal growth faults (see Sec. 16.7.3) and facets of diapiric structures.

17.4.4 One-sided compressional basins

The wide depressions between the fronts of deformed belts and distant basement outcrops are among the areally largest of sedimentary basins. In general, they represent the undeformed or little-deformed foreland portions of miogeosynclines and their associated shelves. Initial depression is abrupt; the principal source sediment is likely to be an early deposit.

An inherent characteristic of such basins controls their potential for hydrocarbon content and its distribution. The basins possess an original and irreversible asymmetry, with an uplifted, deformed, and commonly overthrust belt on one side and a stable craton on the other. Uplift and deformation cause deposition in the basin to be strongly regressive. Hence the oldest reservoirs are nearest to the forelands, the youngest nearest to the deformed belts.

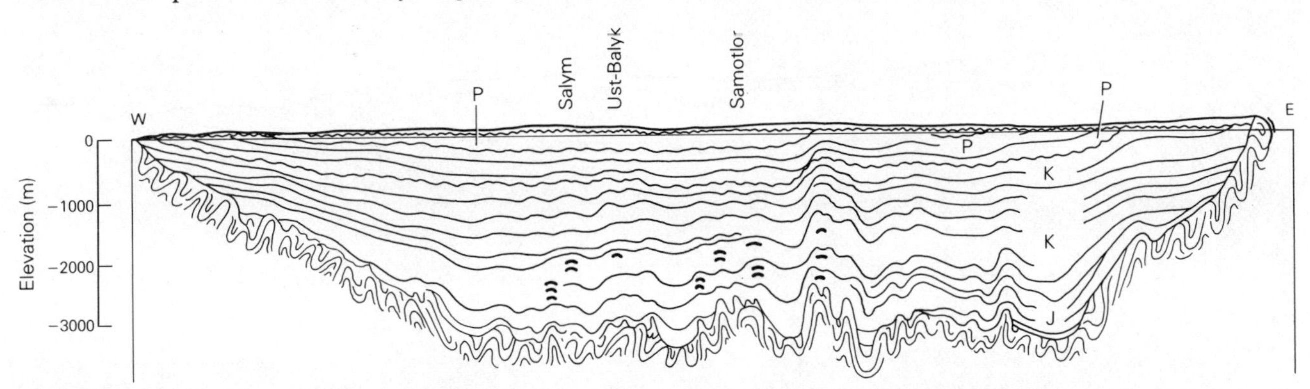

Figure 17.9 West–east cross section (length about 1600 km) through the West Siberian Basin and its largest oilfields. The basin was formed by the downflexing of the folded, faulted and metamorphosed backslope of the Hercynian Ural Mountains (off to the left of the cross section). The sedimentary section in the basin extends from Jurassic (J) to Paleogene (P). (From I. P. Zhabrev *et al.*, 9th World Petroleum Congress, 1975.)

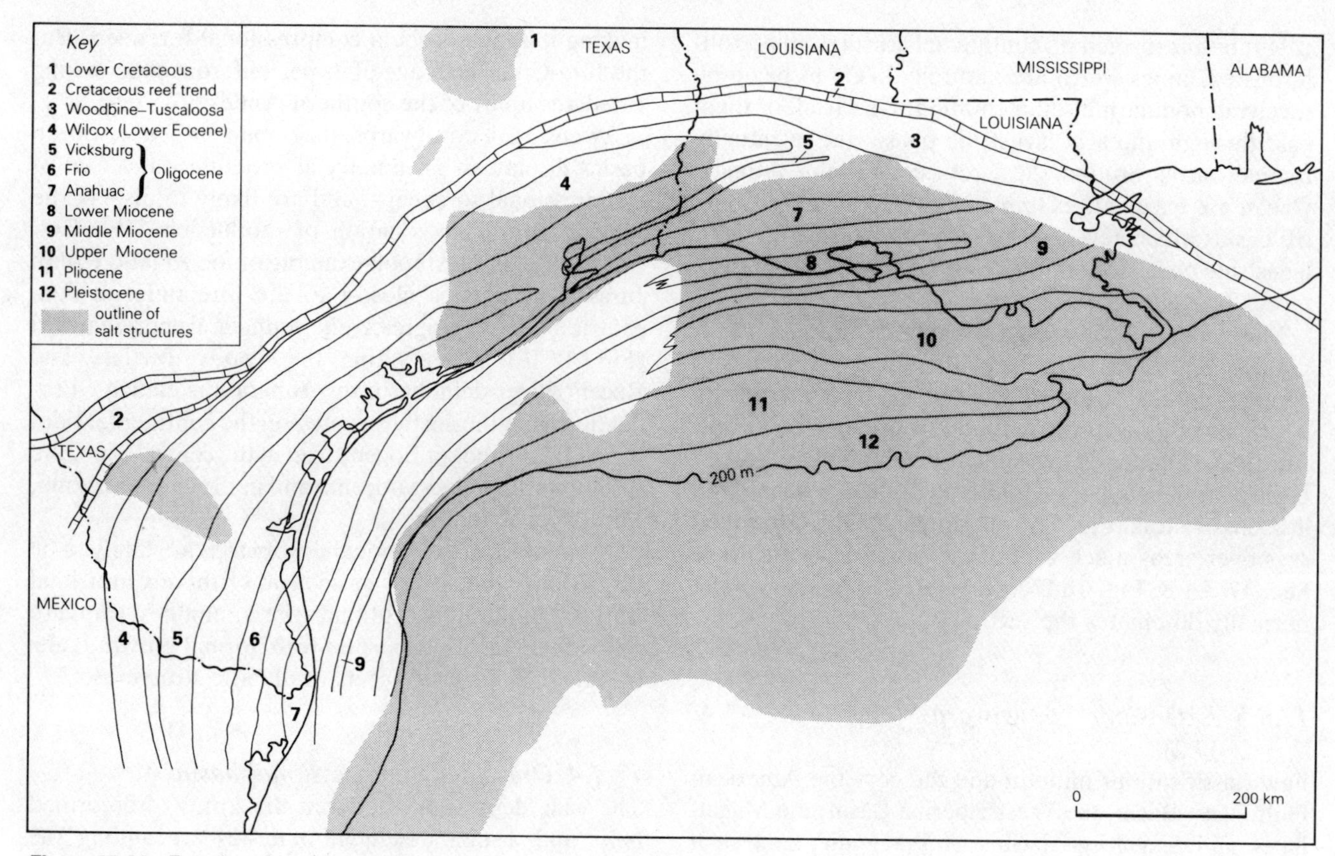

Figure 17.10 Zonation of productive trends in the Gulf Coast Basin of Texas, Louisiana, and Mississippi. (Compiled from maps by B. B. Mason *et al., AAPG Memoir* 15, 1971; by C. D. Winker, T. E. Ewing, *Oil Gas J.*, 30 January 1984.)

This phenomenon is seen in asymmetrical basins of all ages. In the Middle East province, virtually all Jurassic production comes from fields west of the Gulf; Cretaceous fields underlie the Gulf and extend through its northwestern end; Tertiary reservoirs lie along the front of the disturbed belt in Iran and Iraq. In the Mesozoic basin of western Canada (Figs 17.11 & 25.1), most Jurassic fields are in southwestern Saskatchewan; Lower Cretaceous fields are concentrated in the heavy oil belt of eastern Alberta; and Upper Cretaceous fields are in west-central Alberta near the Rocky Mountain front. The Paleozoic Appalachian foredeep and its Tertiary counterpart in eastern Venezuela illustrate the same feature to a less striking degree (Ch. 25).

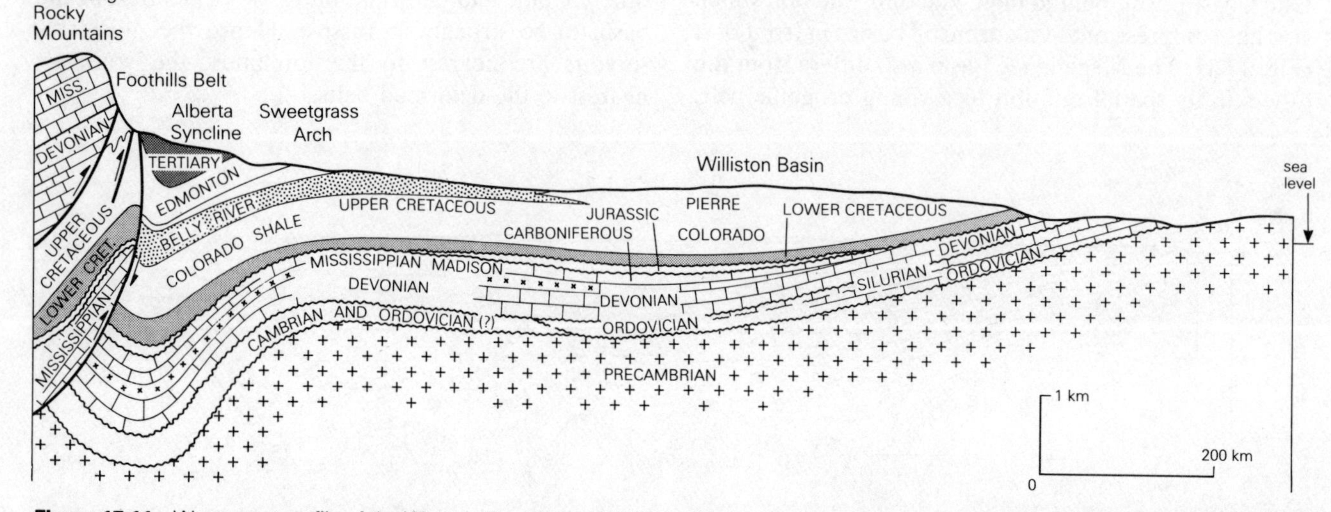

Figure 17.11 West–east profile of the Western Canadian and Williston Basins, a miogeoclinal wedge in the Paleozoic and a major compressional foredeep in the Mesozoic. (From Alberta Society of Petroleum Geologists, in *Possible future oil provinces of the United States and Canada*, Tulsa, Okla: AAPG, 1941.)

Asymmetrical foredeep basins yield hydrocarbons from all types of trap. Those of Paleozoic age are especially rich around large basement arches or uplifts like the Tataria Arch in the Ural–Volga Basin and the Bend Arch in central Texas (Figs 16.43 & 17.3). A common characteristic of such basins is the existence of one exceedingly large field in a trap at least partly stratigraphic, possibly containing heavier oil than the average for the basin.

17.4.5 Two-sided compressional basins (Figs 16.27, 37 & 38)

Compressional intermontane basins, tucked between pairs of deformed belts, are commonly narrow and likely to be short-lived. Hence most productive examples are of Tertiary age. The total volume of sedimentary fill is less than 10^5 km^3. Orogenic associations are everywhere evident: rapid downwarping (average probably 180 m Ma^{-1}); high heat flow; negative gravity anomalies; diapirism; abundant faulting; anticlinal traps; volcanic contributions to the stratigraphic section, which is almost wholly clastic. It is the combination of these features that made possible the early generation and pooling of the oil content; distant migration could not avoid leading to its loss (see Ch. 18).

A few productive interdeeps are little more than deeply downtucked synclines between opposed thrust faults. The axes of California's Ventura Basin and India's Upper Assam Valley have that form (Fig. 17.12). However, compressional intermont basins usually have one mobile side, deformed towards the basin, opposed by the backslope of a slightly older deformed belt that acts as the basin's foreland.

This structural asymmetry is unlike that of one-sided foredeeps in that it can be reversed. Successive deformations may mobilize either flank, or both flanks simultaneously. In the Los Angeles, Maracaibo, and South Caspian basins, both flanks were mobile at different times during the Tertiary, and this was exceedingly favorable for repeated oil generation and accumulation. Both flanks developed structural traps and both provided coarse clastic sediments for multiple reservoir sands.

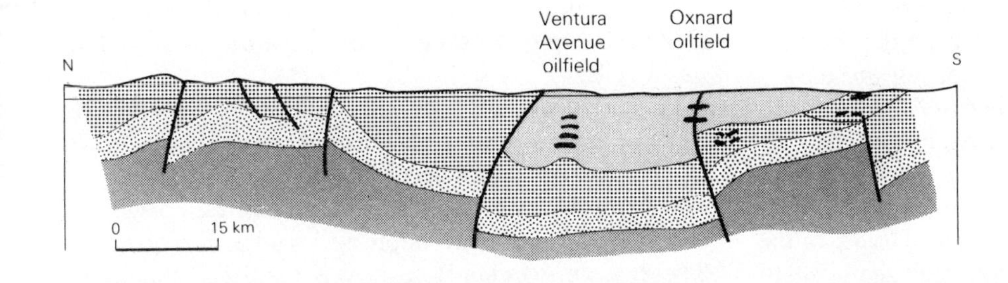

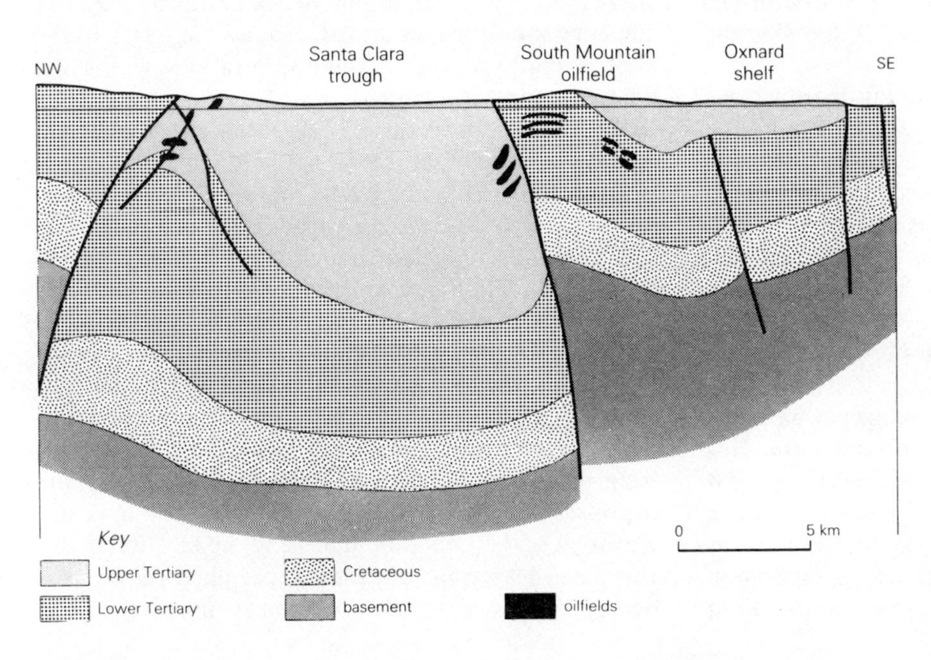

Figure 17.12 Two cross sections through the Ventura Basin, California, transverse to the structural grain, showing intermontane basin deeply downfolded between downwardly diverging thrust faults. Oilfields occur in both hanging walls and footwalls of the thrusts. (After M. C. Blake, Jr, *et al.*, *AAPG Bull.*, 1978.)

17.4.6 Combination basins (Figs 17.3 & 22.8)

We have considered some half-dozen types of structurally created sedimentary basins that are commonly petroliferous. Each has certain easily discernible and understood characteristics, but no two basins are exactly alike even in gross geologic features or generalized geologic histories. Many defy easy categorization, not because their natures are obscure but because they are complex.

The late Jurassic uplift of the western Canadian Cordillera initiated a late Mesozoic foredeep basin unconformably above a wide Devonian basin. Both basins were richly petroliferous, in virtually the same areas. They were not alike geologically at all, yet no published plate tectonics classification of basins unequivocally assigns them to different categories. The Devonian basin had more in common with the Jura-Cretaceous basin of the Persian Gulf than with the Jura-Cretaceous basin directly above it; but only Bally's classification assigns the Western Canadian and Persian Gulf basins to the same category.

The Western Canadian Basin is an excellent example of two quite different ways in which a sedimentary basin can become a *combination* (or *composite*) basin. First, a basin created by some hypogene diastrophic event can become so modified by purely supergene depositional processes that these can be said to control the basin's style; such a basin has been described as *constructional*. Alternatively, the original basin can be converted by further diastrophism into quite another kind of basin. This either overlies the earlier basin (as in western Canada), or supplants it, as the Cretaceous basins of the US Rocky Mountain province supplanted their Paleozoic precursor, and the Tertiary Middle Magdalena graben of Colombia supplanted a much larger, oil-generative Cretaceous foredeep that had been deformed out of existence by an early phase of the Andean orogeny (see Ch. 25).

The more intricate and discriminating the divisions between basin types, the more basins have to be assigned to some combination category. Among Klemme's relatively modest subdivisions, most or all pull-apart basins must have started as rift basins; and all closed downwarps (like the Persian Gulf Basin) must have begun as open downwarps. In other words, initially extensional basins may become still more extensional or they may become compressional (Sec. 17.3.2 & Fig. 17.13).

Reef growth is the only depositional agency we know of that can grossly modify a large basin's geometry. This is not likely in a mobile orogenic basin, receiving great quantities of clastic sediment. Large-scale reefs developed on cratonic forelands or hinterlands, as results of subtle seafloor topography consequent upon an immediately preceding diastrophic event. Reef basins are not created by orogenic structures; their contents may never become deformed; but they are due to deformation elsewhere and at an earlier time. The necessary framework is provided by a form of structural remote control. Once large-scale reef growth takes advantage of this control, it rapidly takes over, building up the basin margins and intrabasin structures and starving the centers. This phase justifies the word *constructional* in the description of reef-girt basins.

The combination basin can be illustrated by thumbnail histories of two richly productive basins, one combinational in the constructional sense, the other in the superpositional sense. Between them, they represent virtually every basic basinal type and all petroliferous ages, overlapping in the productive time frame only during part of Permian time.

The *Permian Basin* of West Texas and New Mexico (Figs 17.3 & 14) began as part of an early Pennsylvanian fold belt extending NNW from northern Mexico into Colorado. Numerous meridional fault blocks provided a grain for the new basin. The region then became part of the foreland of the Ouachita–Marathon thrust episode in mid-Pennsylvanian time. Acting almost at right angles to the slightly older grain, this episode accentuated that grain, producing the northward-tilting Central Basin Platform between two abruptly subsiding basin cells.

In the meantime the Ancestral Rocky Mountains were uplifted to the west and northwest. The western (Delaware) basin thus became an interdeep between the Ancestral Rocky Mountains and the Central Basin Platform, and was deeply depressed under compression. The eastern (Midland) basin was therefore starved of sediments, because it was separated from sources by shelves on which limestones were deposited, but was itself too deep for limestone deposition. The way was thus prepared for reef growth around the edges of the central platform and in suitable positions within the eastern basin. The basin as a whole became constructional, with depositional relief accentuated across it until finally filled in by salt.

By a fluke of development not completely understood, the now dead basin escaped being refolded during the Laramide orogeny, though the extensions of the basin's foundations both northward and southward were extensively re-elevated in the Laramide.

The US Gulf Coast Basin has been described so thoroughly that we choose another large basin very similar in origin but older, and having progressed to the deformation of the youngest geosyncline on what had been its oceanward side. The *Pre-Caucasus (Caspian) depression* began as a late Permian backdeep after the elevation of the Ural Mountains and the creation of the European Hercynian platform. That platform became deeply depressed by the downward hingeing of the

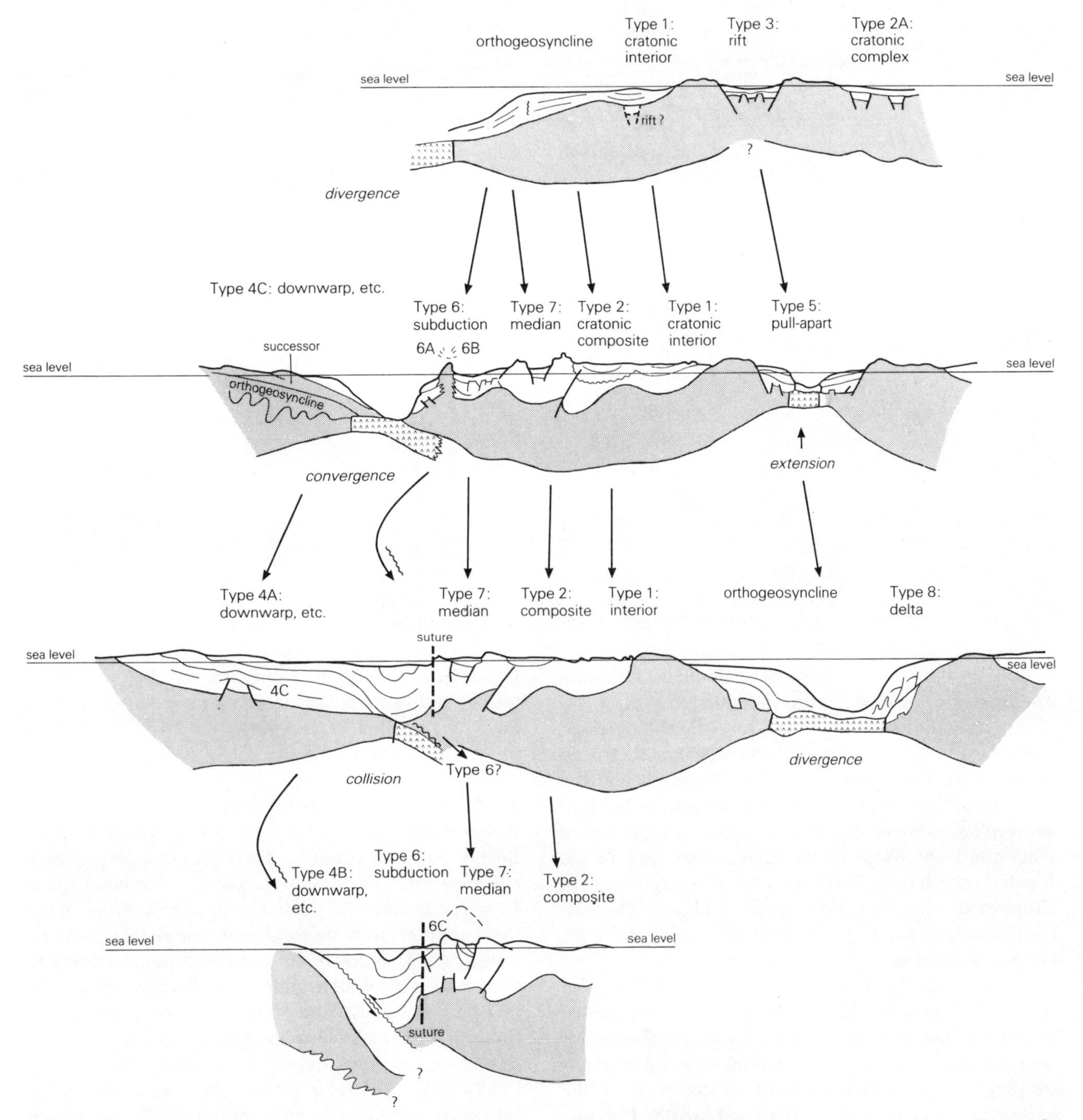

Figure 17.13 Possible patterns of basin evolution and the development of combination-type basins, on Klemme's nomenclature. (From H. D. Klemme, *J. Petrolm Geol.*, 1980.)

Russian shield's southern margin. The boundary between the two generations of crystalline basement, Precambrian and Hercynian, is now the southern limits of the Zechstein salt domes and the Emba sector of the productive basin. In conformity with the rule that the productive stratigraphy becomes younger away from the foreland, the oldest strata producing from the new basin are in the Triassic of the Emba sector, northeast of the Caspian Sea.

The first post-Hercynian development was the folding of the Mangyshlak anticlinorium, across the middle of the new pre-Caucasian basin. This yielded molasse sediments of Permo-Triassic age resting directly on metamorphosed Precambrian basement. (The region thus structurally resembles the backslope of the Ouachita Mountains, which forms the foreland of the adjoining Gulf Coast Basin.) The main Caucasus folding followed in the late Jurassic, creating a southward-sloping foredeep between the new mountains and the Emba region. This deformation accentuated the broad north–south

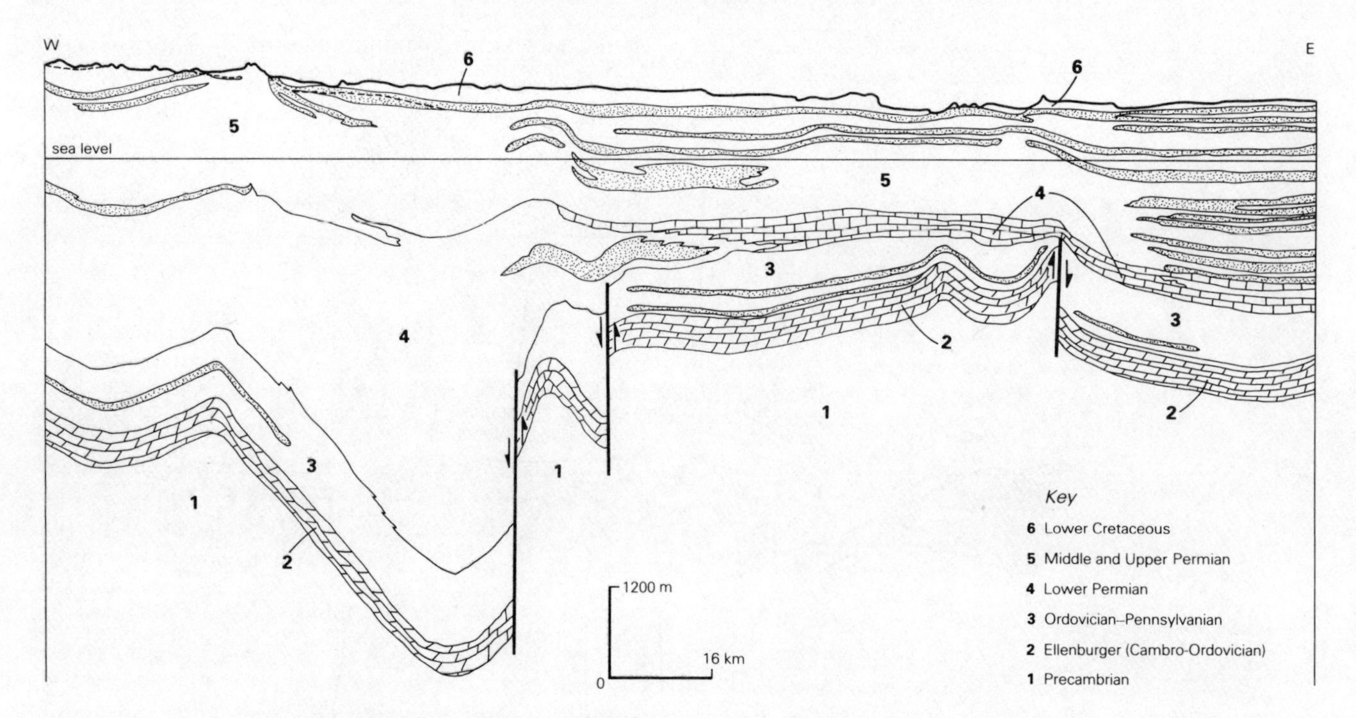

Figure 17.14 Cross section through the Permian Basin of West Texas, near its southern end; Delaware Basin on the west, with the deep Val Verde extension adjacent to the Central Basin Platform (center); Midland Basin on the east. (After K. K. Landes, *Petroleum geology of the United States*, New York: Wiley–Interscience, 1970, and West Texas Geological Society, 1962).

folds of the earlier generation, much as did the Ouachita folding in the Permian Basin, described above.

The raising of east–west structures then progressed southward, and the Jurassic basin became an interdeep between the Caucasian folding on the south and the Kara-Tau folding on the north. The Jurassic oilfields of western Kazakhstan, east of the Caspian Sea, are within that interdeep; those of the Emba Basin are in the foredeep north of it. In this new pre-Caucasian trough, Cretaceous oil occurs in the center, and large gasfields (in Cretaceous sandstones) occur at its ends, near the Black and Aral seas.

As the Alpine folding arose in the south, the belt from the Sea of Azov to the Karaboghaz Gulf became its foredeep. Oligocene oil occurs at the edge of the foredeep (at Maykop); Miocene oil in the new frontal interdeeps (in places overlying the Cretaceous oil of the earlier interdeep, as at Groznyy); and, finally, Pliocene oil within a true intrasutural interdeep at Baku.

According to many Russian geologists and geophysicists, the rise of the Caucasus and the depression of its foredeep are still continuing, and a wave-like progression of structural activity is moving southward. The resultant folding and compaction of the very young sediments activate the mud volcanoes for which the area is famous. The deep basin of the South Caspian Sea is a remnant of the Alpine geosyncline still undeformed; all pre-Pleistocene foldings went around it. The basin as a whole, of which this deep sea is the last vestige, has thus passed through backdeep, foredeep, and interdeep phases. Presumably its latitudinal position, its orientation, and its persistent tectonism were unsuitable for any constructional (reef) phase.

17.4.7 Lacustrine basins (Fig. 10.8)

Basins whose entire generative section is lacustrine and fluvial are never accorded their own category in basin classifications. They deserve at least separate mention, however, because they differ from marine basins in that their settings must be local and completely enclosed. Lakes in general can be structurally "open," needing no structural control, depending instead upon topography, bedrock petrology, and rainfall (or snow-melt). The thousands of shallow lakes on the Canadian and Baltic shields are obvious examples. They are geologically short-lived, highly changeable, and easily diluted and refreshed. Their sediments are unlikely to become petroliferous.

A petroliferous lake basin must be structurally "closed," so that it can contain for a geologically significant time a body of water (possibly but not necessarily deep) that becomes either thermally or chemically stratified. The enclosure, involving at least one steep side, is almost bound to be brought about fundamentally by faulting, in many cases associated with igneous activity. There are three common methods of fault enclosure for a lake basin. It can occupy: a downdropped rift or half-graben, formed under cratonic extension (like the lakes of the African rift valleys and the Dead Sea); a downfaulted graben or half-graben formed in the axis of an uplifted orogenic belt as an indirect result of compression (Lakes Baikal and Titicaca); or a basin isolated by the compressional uplift of surrounding basement fault blocks (Pleistocene Lake Bonneville, the predecessor of the Great Salt Lake).

It follows that virtually all petroliferous lacustrine basins fall into the categories that require high-angle normal or reverse fault control (types 3, 5, and 7 in Klemme's classification). Within type 3 are the Songliao and Vienna basins (for which Bally introduced two special subcategories). Within type 5 are the Jura-Cretaceous basins of Bahia Reconcavo in Brazil and Gabon–Cabinda in West Africa, before the South Atlantic became a seaway in which the widespread Aptian salt was deposited. Within type 7 are the Eocene basins north and south of the Uinta Mountains of the American west, hemmed in by high-standing fault blocks raised during the Laramide orogeny.

A lacustrine *source* for a basin's petroleum need not imply a lacustrine *reservoir*. The reservoir rock may be a fluvial, channel, or turbidite sandstone, or even an aeolian sandstone. Traps in lacustrine basins are commonly anticlines around the basin margins, fault and combination traps on the slopes, and stratigraphic (lithologic) traps in the deep lake centers.

17.5 Dependence of basinal style upon geologic age

The world's petroliferous basins reveal elementary geologic differences directly attributable to age differences. These need not be large, in relation to geologic time, for the dissimilarities to manifest themselves.

The productive basins of California or Indonesia or the Carpathians were developed wholly within the Tertiary Period; their pre-Tertiary rocks are basement. The basins of northern and northwestern South America, such as the Maracaibo and Orinoco basins, produce from formations essentially the same age as in the Californian basins. However, the South American basins developed during the Cretaceous and Paleocene; their common history was terminated by diastrophism before that of the Californian basins began. Though the oil produced from Tertiary sandstones in California or Indonesia is necessarily of Tertiary origin, the oil produced from Tertiary sandstones in South America is not; much of it is undoubtedly of Cretaceous origin.

Basins representing these two conditions can be distinguished by a number of geologic criteria, derived quite empirically. The indigenous Tertiary basins were created by Tertiary diastrophic movements (in most cases orogenic movements) within deformed geosynclines only slightly older than the basins. Their basements are young (commonly Mesozoic). The productive successions bear all the hallmarks of orogenic derivation: volcanic content, displaced sandstone reservoirs, compressional structures in intermont settings, common diapirism, and absence or near-absence of carbonates. Basins owing their origins to Mesozoic diastrophism may have such features in their pre-Eocene successions, but they are unlikely to have them in younger strata. Several of the older basins are true foredeeps rather than interdeeps (the Persian Gulf and Orinoco basins, for examples). Those that are interdeeps (e.g. the Maracaibo and Middle Magdalena basins) were foredeeps when their source sediments were deposited, and became fragmented later. Many basins of Mesozoic origin are completely noncompressional (Gippsland, Nigeria, Sirte). Carbonate reservoirs are common in both compressional and noncompressional types (Iran–Iraq, Sirte, and the central North Sea).

The differences between Paleozoic and Cenozoic petroliferous provinces are due to two principal factors: the late (Cretaceous or Tertiary) generation of hydrocarbons from thin Paleozoic source sediments as "cooking time" becomes dominant over "cooking temperature" (see Ch. 7); and the migration of many Paleozoic oils far towards the forelands of their original basins, large volumes of their source sediments having long ago become incorporated into the flanking late Paleozoic orogenic belts. The Mid-Continent province of the USA, with the flanking Ouachita Mountains, is a classic example. Essentially the only Paleozoic oil preserved is that sufficiently far out on the forelands to have escaped the deformation of the basin sediments. Most Paleozoic oilfields therefore lie, not in true compressional basins, but on rises within covered shields along the distal edges of Paleozoic foredeeps.

Basement control in Paleozoic provinces is exerted by Precambrian structures enormously older than the petroliferous sediments, and commonly showing little relation to the grains of the new basins. In Tertiary basins the equivalent control reflects basements little older than the basins and in many cases still foldable. Thus cratonic sags and reef-girt foreland basins are typically Paleozoic and have no close Tertiary counterparts; the intermont basins typical of Tertiary provinces have no surviving representatives in the Paleozoic.

As Paleozoic oil and gas are typically reservoired far out on the forelands, they are pooled in shelf sediments, as likely as not in carbonates. None of the characteristics of Tertiary anticlinal sandstone fields are to be expected where Paleozoic hydrocarbons are now pooled. As these hydrocarbons were generated in the foredeeps of Paleozoic orogenic belts, they are more likely to be in late Ordovician, late Devonian, or Permo-Pennsylvanian rocks than in those of any other ages.

Mesozoic oil provinces do not simply combine Paleozoic and Cenozoic characteristics. Instead, they reflect two aspects of global tectonics. Mesozoic time saw no worldwide orogenic episode until almost the end of the Jurassic, and that episode continued through the Cretaceous as plate motions exerted maximum effect. Compressional basins (especially foredeeps) created by such late Mesozoic orogenies continued into the Cenozoic; these were the basins distinguished, in an earlier paragraph, from those created wholly within the Cenozoic. In cases in which no Cenozoic fragmentation took place, the basins remained foredeeps and did not become interdeeps. The manner of pooling of their oils was similar to that in the Paleozoic provinces, even though the basins produce from both Mesozoic and Cenozoic rocks. Such Mesozoic provinces as the Arabian platform and the Alberta–Saskatchewan region of Canada illustrate these characteristics; traps are not sinusoidal buckle folds, carbonate reservoirs are common, and the oil is far out on the forelands.

The second tectonic control over Mesozoic oil occurrence was provided by the advent of continental drift, introducing crustal extension as a widespread basin-creating mechanism. Basins so formed begin as grabens or half-grabens, commonly coastal in situation. The most productive examples are the Gulf Coast, Sirte, and North Sea basins. They have few Paleozoic counterparts, because most basins created during Paleozoic extensional episodes were converted into compressional basins or deformed belts during the late Paleozoic orogenies that paved the way for Pangaea.

Two generations of continental crust were available to be extended by rifting: the Precambrian shields and the platforms of the basement-creating Hercynian orogeny. The marginal South Atlantic basins, as well as the Sirte Basin, lie over rifted Precambrian crust and exemplify the essential age restriction to the Cretaceous and Paleocene. Among productive basins developed over Hercynian crust are the North Sea, Hanover–Netherlands, Aquitaine, and Emba basins; the Songliao and Sichuan basins of China; and the Gippsland Basin of Australia.

Two of the largest and most productive extensional basins display a striking relation between age and structure. They lie on the backslopes of late Paleozoic mountain belts, which consequently have Paleozoic oil provinces in compressional foredeeps in front of them and Mesozoic provinces in extensional backdeeps behind them. The Mesozoic Gulf Coast province of the USA is separated from the Paleozoic Mid-Continent province by the Hercynian Ouachita Mountains; the West Siberian Basin is separated from the Paleozoic Ural–Volga and Pechora basins by the Hercynian Ural Mountains.

Where the continental crust was completely ruptured, the first sediment in the new basin was normally salt, rapidly deposited in deep water exactly as in newly formed rifts today. A reef phase commonly followed, unless the basin was unfavorably oriented with respect to its continent. Where abundant terrigenous detritus maintained rapid sedimentation, later deposition became wholly clastic and built outward over the new oceanic crust (as in the Gulf Coast and Nigerian basins). Whether the rupture stage was reached or not, however, there is no molasse phase to act as a single, dominant reservoir, and individual traps are unlikely to be compressional anticlines.

Crustal extension did not, of course, end in Paleocene time, though the new ocean basins had opened by then. There are a few petroliferous grabens wholly of Tertiary age – the Cambay and Suez grabens, for example – but the mid-Tertiary closure of Tethys and the plate convergence around the Pacific and the Caribbean ensured that most wholly Tertiary oil basins would be compressional interdeeps.

Petroliferous lacustrine basins have less than an even chance of surviving intact, let alone of achieving maturity of their sediments and pooling of their oil. Productive representatives are therefore unlikely to be either very old or very young. All important examples are fault-controlled and either extensional in origin or governed by essentially vertical tectonics. Nearly all therefore originated during the same time interval as the extensional basins just described: Mesozoic to early Tertiary.

The nature of a basin is thus closely dependent upon its age, because it reflects the global diastrophic regime at the time it was formed. The volume of hydrocarbons, and their position in the basin, reflect the basin's type and therefore its age. These three phenomena – the genetic character of the basin, its richness in hydrocarbons, and the hydrocarbons' position within the basin – are further reflected in the chemistry of the reservoir fluids – the gas and the water as well as the oil. Let us now consider these phenomena as related to basinal age and architecture.

17.6 Relations between basin type and hydrocarbon richness

More than 700 sedimentary basins and sub-basins have been identified, some wholly offshore (Sec. 3.7). Of the 400 or so subjected to some exploration for oil and gas, about 150 are (or have been) commercially productive. As much as 90 percent of the world's known oil and gas (excluding "tar sands" and oil shales) occurs in fewer than three dozen of these basins (Fig. 17.15). This is a dramatic reminder that hydrocarbon content among the basins is exceedingly variable.

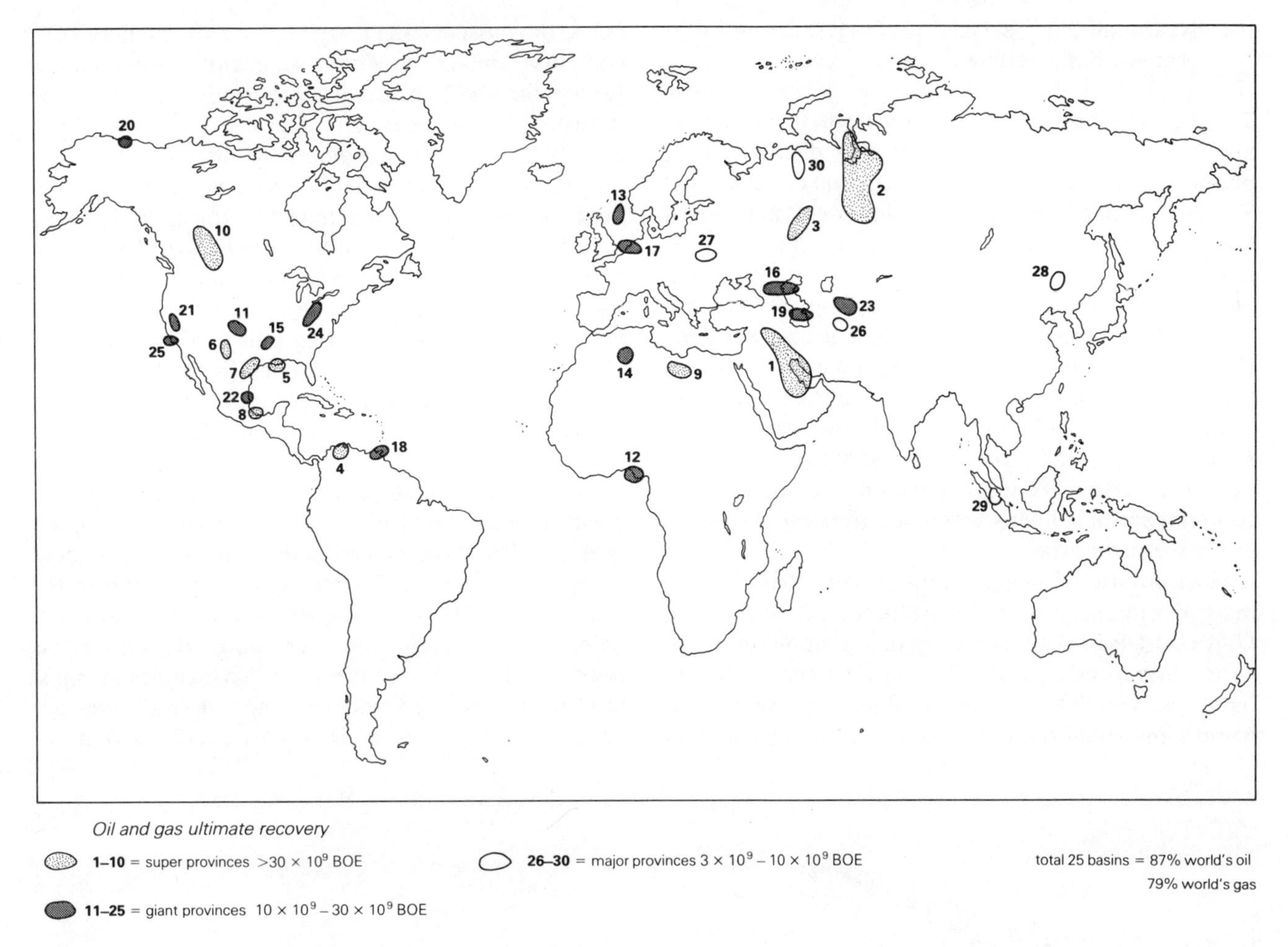

Figure 17.15 The world's giant petroleum provinces, ranked in order of ultimate estimated recovery of oil and oil equivalent in gas. (From L. F. Ivanhoe, *Oil Gas J.*, 30 June 1980.)

Indeed, three-quarters of the known oil and almost two-thirds of the known gas are pooled in only four general regions: in the Persian Gulf Basin; along the northern and western sides of the Gulf of Mexico and the south side of the Caribbean; and along the two flanks of the Ouachita–Marathon Mountains in the southern USA and of the Ural Mountains in the Soviet Union. No more than a dozen truly separate basins are involved, of at least four structural types – a spectacular illustration of lognormality of distribution (see Sec. 3.8).

The recoverable hydrocarbon richness of a basin can be expressed in various ways, the most common being hydrocarbon volume per unit area or per unit of sedimentary rock volume (Sec. 24.3.2). Following Weeks, Klemme used the volume of oil equivalent (conventional oil plus gas converted to the energy equivalent in oil), in barrels recoverable from a cubic mile of the sediments of the basin. Bally preferred the volume of oil equivalent, in cubic meters per thousand square kilometers of basin area.

Frequently quoted figures for general or average oil richness have been from 30 000 to 60 000 barrels per cubic mile of sedimentary rock (about 1200–2400 m^3 km^{-3}). These figures are heavily weighted downward by the inclusion of regions of North America with very little oil. Generalized comparison between basins of different productive capacities indicates the following ranges of hydrocarbon richness (in oil equivalents, OE):

(1) ultrarich basins with more than 10 000 m^3 km^{-3};
(2) rich basins, 3600–10 000 m^3 km^{-3};
(3) average basins, 2400–3600 m^3 km^{-3};
(4) lean basins, 1000–2400 m^3 km^{-3};
(5) very lean (but still productive) basins, less than 1000 m^3 km^{-3}.

Assignment of hydrocarbon richness to particular situations within plates and their margins permits only two conclusions:

(a) Nearly all Paleozoic oil reserves are in intracratonic (platformal) basins and nearly all Cenozoic oil reserves close to plate boundaries, with Mesozoic oil reserves more equally divided between the platforms and their boundaries.

(b) Nearly all gas reserves, of all ages, are in intracratonic basin settings.

These conclusions can be maintained whether the giant basins are plotted against the present distribution of plates (Fig. 17.16) or against palinspastic distributions. The conclusions themselves are almost certainly teleological. We do not know what has happened to the oil or gas that may earlier have been associated with Paleozoic plate boundaries; Paleozoic oil has been preserved only where it had managed to migrate high on to the cratons before the plate boundaries of the time were converted into orogenic belts. And Cenozoic marine transgressions did not extend far enough over the cratonic platforms to permit rich hydrocarbon accumulations to be developed there. We will confine our consideration of tectonic control of hydrocarbon occurrence to the basins as we see them now.

Most ultrarich basins are downwarps at continental margins, either still open towards oceanic crust (like the Gulf Coast Basin) or closed by deformation and converted into foredeeps (the Persian Gulf Basin). These basins (Klemme's type 4) contain about 50 percent of the world's known oil reserves. Their present-day counterparts, the passive (rift) margins, have the highest TOC contents among modern sediment accumulations (except for wholly inland seas); deep-sea basins have the lowest. Compressional interdeeps (Klemme's types 6 and 7) also include a number of ultrarich representatives, with a remarkable incidence of giant individual fields, even though more than two-thirds of explored interdeeps are almost or quite unproductive. The reason for the exceptional richness of some intermontane basins is the stacking of multiple turbidite sand reservoirs in narrow basins during pulses of deformation. These basins contribute about one-eighth of the world's known oil reserves. A few rift or graben basins are also highly productive, containing about 10 percent of the reserves.

The most commonly productive basin type is the composite interior basin or cratonic foredeep (Klemme's type 2). This type is productive on every continent except Antarctica, and contains about a quarter of the world's oil reserves. It includes most of the reserves in Paleozoic rocks, and well over half of the known gas reserves. However, this category also includes the most thoroughly explored and exploited basins; they are areally the largest and lie almost totally within the

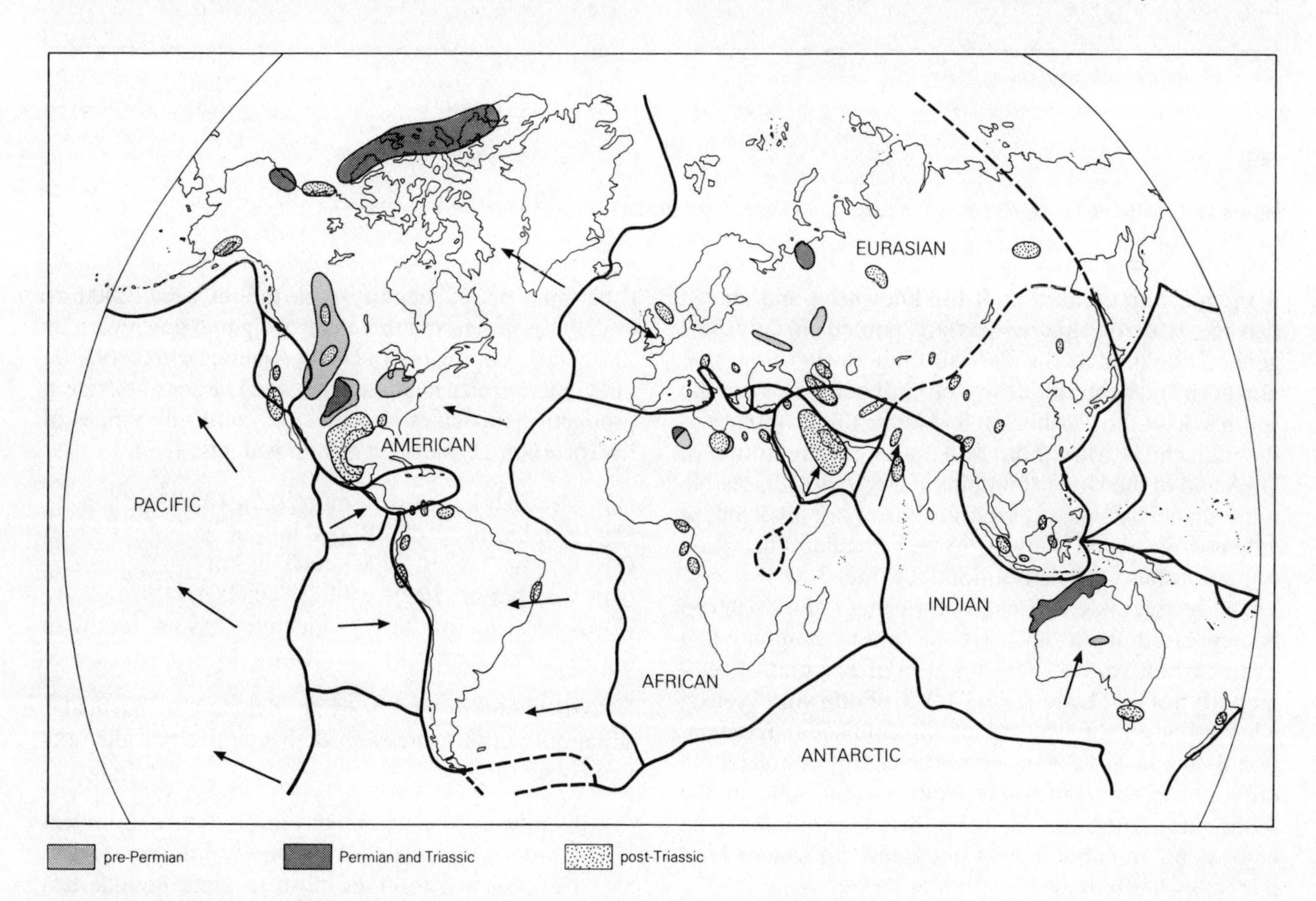

Figure 17.16 Basins containing giant oil- or gasfields of different ages, in relation to the present plate boundaries. (From D. A. Holmgren *et al.*, 9th World Petroleum Congress, 1975.)

continents. They therefore offer little expectation of additional contributions from "frontier" areas, which lie largely offshore.

Not many young deltas are oil productive; they contribute less than 5 percent to the world's oil reserves. They tend to be gas-prone, presumably because of their dominantly terrestrial organic contents. Those that are oil-prone are richly so, but not because they are well endowed with large fields. Unlike Klemme's types 2 and 4 basins, in which overall hydrocarbon richness depends heavily on one or a few supergiant fields, oil-prone delta basins contain many small to modest-sized fields. This is because the prevalent trapping mechanisms (diapirs and rollover anticlines) are individually of small capacity. The Tertiary Niger delta, an impressive example, contains some 100 fields with at least 5×10^7 barrels (8×10^6 m^3) of recoverable reserves; but fewer than a dozen have 5×10^8 barrels and none are known to have a billion barrels (160×10^6 m^3).

Our "average" oil richness, 2400–3600 m^3 km^{-3} of oil or oil equivalent, is indeed about the average for foredeep basins (Klemme's type 2), and is below the average for interdeeps and rift basins. But it is about the maximum richness reached by coastal half-grabens like those around the Atlantic Ocean (Klemme's pull-apart basins). However, large parts of these half-grabens lie under deep water; their hydrocarbon potentials have not yet been properly evaluated. They may have the highest promise in the "frontier" regions as deeper waters become accessible, but they are not expected to attain the richness of other types of basin.

The least productive basinal type is the simple cratonic basin within a Precambrian shield. These average well under 1000 m^3 km^{-3} of recoverable reserves; many contain none at all, and none contains an individual billion-barrel field. If the term "cratonic basin" is extended (as it has been by Bally and others) to include all unfolded downwarps with little-faulted boundaries (like the West Siberian Basin, or even the North Sea Basin from the mid-Cretaceous onwards), such cratonic basins range from lean to rich in hydrocarbon content.

Few petroliferous lacustrine basins have yielded their oil for orthodox pooling; most remains locked up in the kerogen shales. Large lacustrine basins normally contain no very large individual fields (the Daqing field in the Songliao Basin of Manchuria is an obvious exception), and they are of no more than average overall richness. Areally small lacustrine basins, on the other hand, are really narrow rift basins cut off from the sea, and they can possess the richness of marine rift basins. The Vienna Basin may ultimately yield about 7×10^3 m^3 of oil per cubic kilometer (and its duration as a lacustrine basin was only about 10 Ma). The eastern Chinese basins may turn out to be comparably rich.

17.7 Influence of basin structure on formation fluid chemistry

We have seen (Ch. 9) that waters moving through hydrocarbon-bearing strata affect the accumulations both physically and chemically. Physical effects can involve displacement or removal of the accumulations; chemical effects can involve degradation (causing lower producibility), removal, or destruction. The oil and gas chemistries are therefore of triple dependence: they depend upon the natures of the organic precursors; on the processes through which these were transformed into hydrocarbons; and on any subsequent chemical or physical changes caused by water movement (Sec. 10.4).

The nature of the subsurface waters is therefore important in exploration (and exploitation). Normal sea water contains about 35 000 p.p.m. of dissolved salts; its mineral salinity is expressed as 3.5 percent. Waters associated with light, unaltered oils or wet gases are commonly supersaline (or hypersaline), having 50 000 to over 150 000 p.p.m. salt content, because of post-burial concentration. These waters are *oilfield brines*. Formation waters fresher than normal sea water commonly indicate flushing by meteoric waters; oils associated with brackish or nearly fresh formation waters are likely to be heavy, asphaltic, and degraded. *Isosalinity* or *isoconcentration maps* delineating trends of water movement aid materially in exploration (see Sec. 22.6).

17.7.1 Fundamental types of hydrodynamic basin

Let us imagine the history of an initially simple, symmetrical basin that through geological time acquires a variety of complexities. The *juvenile* basin is buried and sealed by sediments without any uplift or deformation. The only effective force on the basin fluids is compaction, which establishes centrifugal gradients; the waters are expelled upwards and outwards. Unless permeabilities in the aquifer are uniformly high, fluids cannot be expelled fast enough, and pressures come to exceed the hydrostatic head. Little or no meteoric water can reach the deep parts of the basin, so waters there become hypersaline and stagnant, and any hydrocarbons accumulated with them are easily retained, unaltered.

If the basin is then uplifted vertically, sufficient erosion will let surface waters enter around the margins. These waters migrate towards the basin center, establishing an active gravitational and hydrodynamic regime, now centripetal instead of centrifugal (Fig. 17.17). At first the meteoric waters invade only the shallow aquifers, driving the original fluids upwards. Lateral fluid movement gradually supersedes vertical; fresh waters occupy the larger pore spaces, destroying

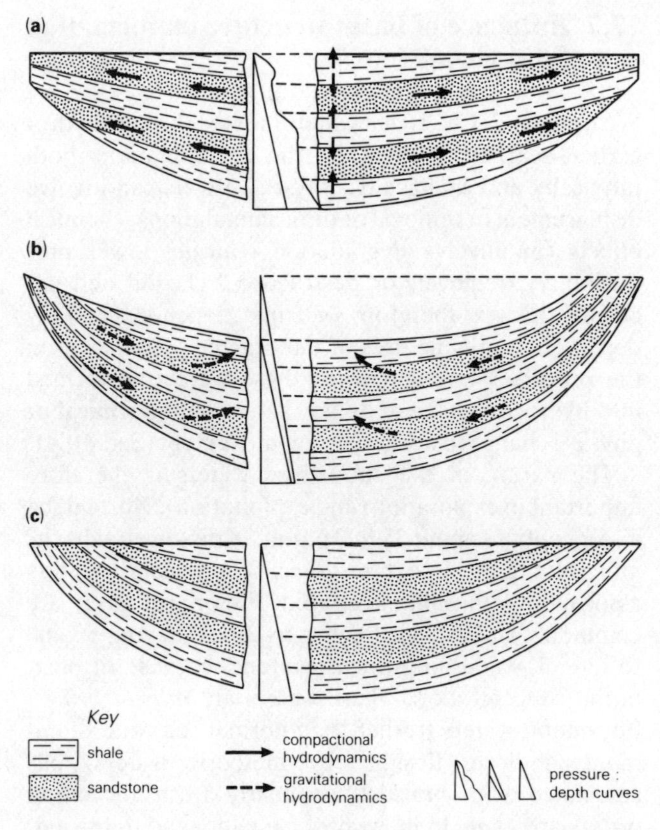

Figure 17.17 Three types of hydrodynamics in sedimentary basins: (a) juvenile basin, centrifugal water movement; (b) intermediate basin, centripetal water movement; (c) senile basin, hydrostatic conditions. (From H. Coustau *et al.*, 9th World Petroleum Congress, 1975.)

any hydrocarbons present or degrading them to asphaltic residues. Saline waters and any remaining hydrocarbons become confined to the finer pore spaces, and oil/water interfaces become tilted (see Sec. 16.3.9).

The last aquifers to be invaded are well closed enclaves distant from recharge areas and lenticular aquifers within impermeable parts of the section. The remaining saline water and oil are driven upwards and outwards into any adequately sealed traps, leaving fresh waters below and basinward of them (reversed hydrologic profiles: see Sec. 9.3). The basin becomes *senile*, with hydrostatic conditions established throughout and no hydrodynamic pressure gradients remaining.

While this concept of the progressive destruction of a petroliferous basin contains important truths, it distorts nature. Some basins remain essentially "juvenile" through hundreds of millions of years. Such are the Michigan Basin in the USA and the Illizi Basin in Algeria, described below. Other basins pass almost immediately to the stage of water invasion and oil degradation; they are *born senile*, like those of California.

Obviously basins of "eternal youth" lie in very stable areas, on ancient cratons. Basins of "instant senility" lie in highly mobile areas, often within boundary zones between continental and oceanic plates. The critical element is not *time*, but the *structural setting and style* of the basin. Insofar as hydrocarbon chemistries are controlled by water movement, they are controlled essentially by *structure*. Let us now review the principal types of petroliferous basin from this point of view. Some generalizations will emerge about the relations between structural style and contained fluids. The geologist should be prepared to find an explanation if his basin is strikingly atypical.

17.7.2 Simple cratonic basins

Basins underlain and surrounded by stable cratonic regions in the interiors of continental plates can be independent of fault control. Such basins undergo little tectonic activity during or after sedimentation. They contain many slowly deposited beds of low permeability because there are no adjacent highlands to supply coarse detritus. During transgressions, sedimentary units overlap those preceding them; as a result, many formations do not crop out. Low topographic relief offers little encouragement to invasion by surface waters. Formation waters are nearly all hypersaline, with some increase in salinity basinward and with depth.

The *Michigan Basin* (Fig. 17.8) contains some of the most saline formation waters known (several hundred thousand parts per million of dissolved minerals). Oils of 40° gravity occur there at less than 400 m depth, and shallow wet gas is common. The oils are paraffinic and aromatic, and occur throughout a Paleozoic section ranging from Ordovician to Carboniferous in age. A still older (Cambro–Ordovician) sandstone yields light oil and wet gas from 1500 m in the *Amadeus Basin* in the middle of Australia. Oils of similar gravities are found in the Devonian of the *Illizi Basin* of Algeria, at less than 1500 m. Aquifers surrounding the oil pools contain fresh or weakly mineralized waters of bicarbonate and sulfate types (Sulin's classes *a* and *b*; see Sec. 9.2), but these oxidizing meteoric waters were unable to penetrate into the traps.

The *Williston Basin,* in the center of North America (Fig. 17.18), also yields oil from reservoirs through the Ordovician–Mississippian span. However, these reservoir rocks are carbonates, and the basin's history was markedly different from those of the Michigan and Illizi basins. Ordovician and Silurian rocks are restricted to the basin, isolated from other rocks of the same ages in North America, but they crop out extensively along the basin's northeastern margin. Devonian and Mississippian rocks, in contrast, extend across the entire prairie interior and into the outcrop belt of the western Cordillera. Devonian rocks also crop out along the eastern margin of the basin, but Mississippian strata,

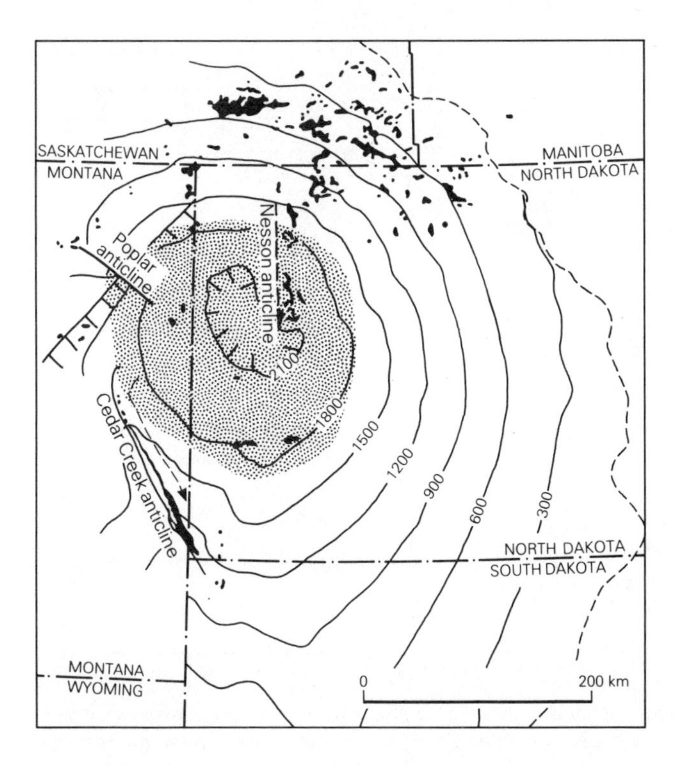

Figure 17.18 Williston Basin, straddling the USA/Canada border in the center of North America. Approximate structure contours, in meters, on top of the productive Mississippian limestone. Oilfields in black. (After L. C. Price, *J. Petrolm Geol.*, 1980.)

which include the principal producing horizons, do not. Meteoric waters drive eastward from the Rocky Mountain outcrops through the Paleozoics below the prairies. They are halted by the large fault zone on the west side of the Cedar Creek anticline (Fig. 16.33). West of this structure, Paleozoic rocks (especially Middle Devonian and older) contain nearly fresh water; there is no oil in strata older than a thick Middle Devonian salt formation. East of the fault salinities are high and Paleozoic oilfields are numerous. The pattern of their oil gravities is instructive.

Differences between the oils in the Lower Paleozoic carbonate reservoirs and those in the Upper Paleozoic (chiefly Mississippian) have been attributed to generation from two different source rocks, one Ordovician in age and the other early Mississippian. Distinctions in sulfur isotope ratios are consistent between the two oil types. The older types are highly mature, straight-chain oils with low sulfur; many are lighter than 40°API. These occur especially along the western side of the basin. Mississippian oils have more cyclic components and higher sulfur; few are lighter than 40°API. More detailed differences cannot be accounted for by source alone, but correlate well with the structural characteristics of the basin.

All productive horizons along the Nesson anticline, near the deep center of the basin (Fig. 17.18), contain light-colored, high-gravity oils. Mississippian oils at 1600 m depth are 40° gravity and higher; Silurian and Ordovician oils below 2800 m are lighter than 52° gravity. On the Poplar anticline, farther west, Mississippian oils are 40° gravity at little over 1000 m depths. Along the Cedar Creek anticline, oils are mostly black and average about 35°API, even in the Ordovician reservoirs at 2000 m depth. At the southeast end of the anticline, beyond the direct influence of the bounding fault, the oils are even heavier (28–32° in the Ordovician reservoirs).

Along the northeast margin of the basin, virtually all the oil is in Mississippian carbonates. There is no locally preserved source sediment, and the oils are agreed to be of early Mississippian (Bakken) source; yet they are pooled almost 150 km beyond the limits of such a source. The oils must have migrated laterally in the porous carbonates and thence along the unconformity below the Jurassic red shale seal. This migration itself caused no measurable change in the sulfur isotope values, but water washing and biodegradation increased the $\delta^{34}S$ markedly as the sulfur content increased by a factor of three and the oil gravity was reduced from 38 to 28° at depths of about 1400 m, closely below the post-Mississippian unconformity. In the basal Mississippian sandstone, 12–14° gravity oils are pooled along the zero erosional edge, with much fresher formation water. But down-dip into northern North Dakota, Mississippian oils are almost 40° gravity.

A striking feature of the great majority of Williston Basin oils, in view of their carbonate–evaporite environment and the manifest effects of water invasion, is their very low sulfur content. Less surprising is the small amount of associated gas.

Around the margins of the *Illinois Basin*, in the east-central USA (Fig. 17.19), Silurian reefs (including the type bioherms) and Devonian carbonates crop out extensively, especially in the north and east. Surface waters invaded those strata in two broad tongues that cross the basin contours, degrading crudes there. Producible oil in Siluro-Devonian reservoirs is restricted to patches, mostly on the west side of the basin (Fig. 17.20). The Mississippian Ste. Genevieve oolite, in contrast, has negligible outcrop area; it was widely overlapped by Pennsylvanian strata. Thus a large area of production in the center of the basin was unaffected by water invasion; the crudes in the oolite reservoir are comparable throughout the basin.

All our examples so far are cratonic basins of Paleozoic age (pre-Pennsylvanian, in fact). As testimony that it is not the age of the oil that is the critical factor, Jurassic sandstones in the *Cooper Basin* of central Australia (the southwestern extremity of the Great Artesian Basin) yield 54° gravity oil from little deeper than 1500 m.

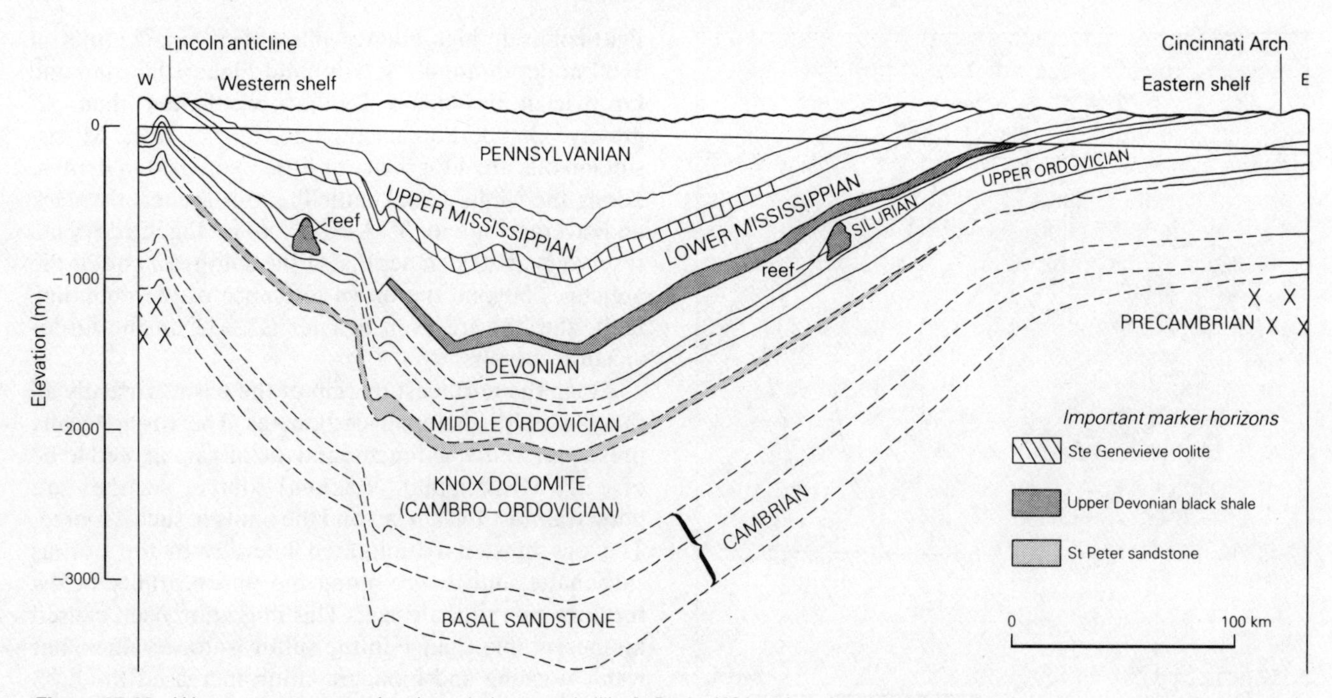

Figure 17.19 West–east cross section through the cratonic Illinois Basin, US interior, showing the stratigraphic positions of the Silurian reefs and the Mississippian Ste. Genevieve oolite. (From K. K. Landes, *Petroleum geology of the United States*, New York: Wiley–Interscience, 1970; after Illinois Geological Society, 1951.)

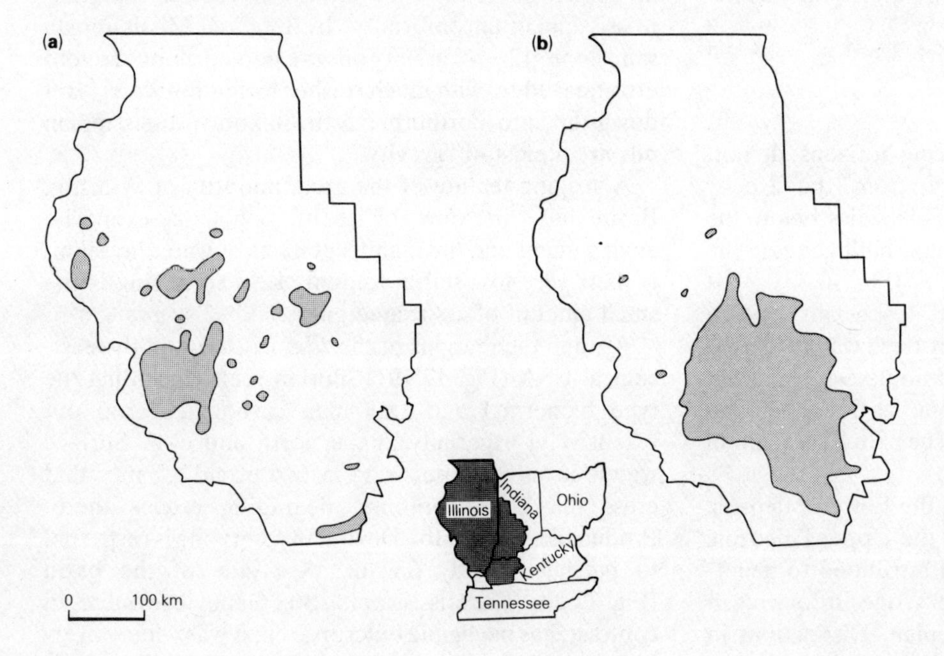

Figure 17.20 Illinois Basin, showing very different areal patterns of production from Siluro-Devonian strata (a) and Lower to Middle Mississippian strata (b). (From D. C. Bond *et al.*, *AAPG Memoir* 15, 1971.)

17.7.3 Semicratonic (passive) coastal basins

Only slightly more complex structurally than simple cratonic sags are basins formed by extensional downflexing of continental margins into deeply subsiding depressions commonly floored by salt. Most authors have subdivided these highly asymmetric basins into more than one structural type. Klemme, for instance, calls some of them *open downwarps* and others *pull-apart basins*. They should also include petroliferous *delta basins* of post-rift age, because it is into such basins that the productive deltas have been constructed. Young deltas are markedly gas-prone, because of the dominance of terrestrial over marine organic material contributed to them (Sec. 13.2.7).

The critical characteristics these basins possess in common are three. Persistently regressive deposition caused most formations to crop out in bands becoming successively younger basinward. A multitude of sandstone formations, all of late Mesozoic or Cenozoic age, include many of low induration and high permeability.

As the basins are fundamentally extensional, and compressional deformation is either absent or both moderate and very late, the structures consist principally of shallow, listric normal faults (growth faults), rollover anticlines, and salt- or shale-cored diapirs.

The extensional faulting and shallow sandstone formations permit shallow, fresh waters to mix with deep, saline waters. A characteristic consequence is the presence of at least two contrasting types of crude oil: shallow asphaltic oils in younger reservoir rocks overlying lighter and commonly paraffinic oils in the deeper levels.

The basin of the *Niger delta* (Fig. 17.21), like other semicratonic downwarps, is marked by concentric arcs of productive horizons that become younger basinward (Sec. 17.4.3). Four distinguishable crude types occur in the basin. In the eastern and inner parts of the delta, the crudes survived in their primary state: green, paraffinic, and waxy, with moderate to high pour points and gravities of 35–45° or even higher. Along the longest axis of the delta, intermediate crudes of 30–37° gravity are produced from Oligo-Miocene rocks. The central and western areas produce naphthenic, nonwaxy, biodegraded crudes of low pour point and gravity about 26°. Finally, heavy degraded oils occur in continental sandstones of Plio-Pleistocene age, and seeps of heavy oil are common. Bituminous sands are present in the Upper Cretaceous around the east and west flanks of the basin, where its stratigraphic content wedges out against the Precambrian.

The older *circum-Atlantic basins* (the pull-aparts of Klemme, on and off Brazil and West Africa) are typified by growth faults and salt flowage structures (Fig. 16.47). In the oldest reservoir rocks, latest Jurassic or earliest Cretaceous, sweet paraffinic oils of 35–40° gravity are the rule. Younger Cretaceous oils are 25–32° gravity, and those in topmost Cretaceous and early Tertiary reservoirs, truncated by a pre-Miocene unconformity, are 13–20° gravity.

In the US *Gulf Coast Basin*, 40° gravity oil is normally found only below 1250 m, often much deeper. Lighter and lighter oil occurs with depth, and all deeper zones offshore contain wet gas and condensate. This may possibly be an original feature, as in any one well each successively older, regressive formation is penetrated farther from its depositional shoreline (Fig. 17.2).

The *Mexican sector* of the Gulf Coast Basin has had a totally different history from that of the US sector (or from those of the other extensional basins considered here). It was impinged upon from the west by rising cordillera from early in Eocene time. Nearly all the oil is in Mesozoic carbonate rocks, all of which crop out in mountainous tracts at no great distance from the fields. Degradation of the Mesozoic oils is the rule. They become progressively deeper from northwest to southeast. Post-orogenic sandstone reservoirs, deformed only by salt flowage from below, contain undegraded light oils and condensates.

The northern Tampico fields yielded 12° gravity asphaltic oil, with more than 5 percent sulfur, from limestone so faulted and fractured at 400–500 m depths that it is effectively shattered (see Sec. 13.4.9). Water invasion of the wells was literally visible; reservoir pressures built up during the rainy season. The atoll that includes the famous Golden Lane (Fig. 13.50) is now elevated on the west and deeply depressed on the east. Early wells onshore yielded oil of 20° average gravity with nearly 4 percent sulfur from 550 m. Offshore oils along the east rim are deeper than 2000 m and have gravities between 30 and 40°. The Poza Rica reservoir

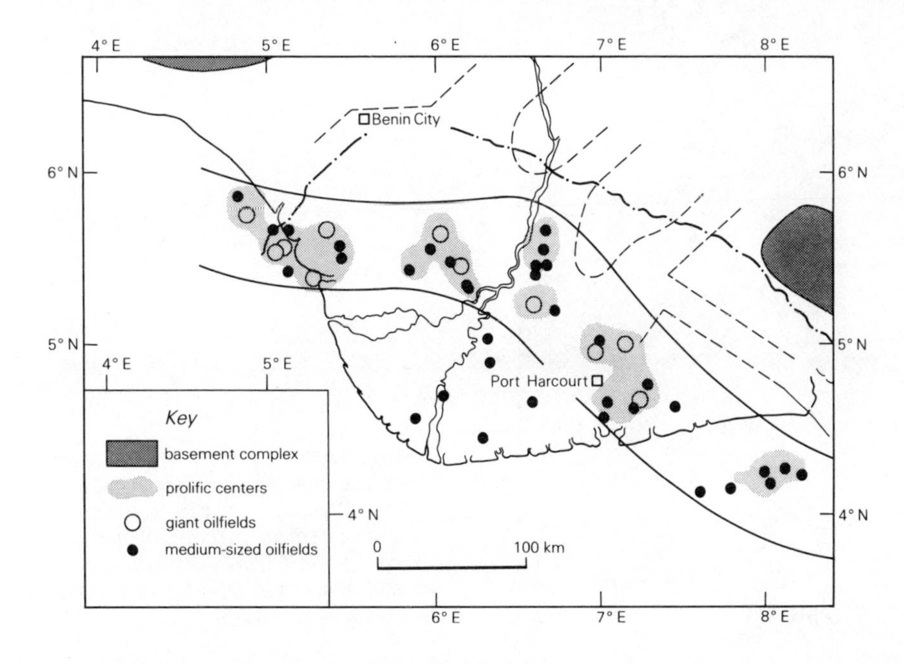

Figure 17.21 Niger Delta Basin, showing position of continental edge (dash-dot line) and three belts of oil production becoming younger seaward. Circles represent large oilfields. Cretaceous bituminous sands occur in the northwest and northeast corners of the area shown. (After J. E. Ejedawe, *AAPG Bull.*, 1981.)

also is buried below 2000 m and its oil has escaped degradation. The huge Chicontepec accumulation and the Isthmus fields are in Tertiary sandstones and most of the oil is lighter than 30° gravity. Indeed, the small Macuspana Basin yields 55° gravity condensate from depths between 2000 and 3000 m.

The Reforma and Campeche districts (Fig. 17.22) which dominate Mexico's present production yield oils of such contrasting types that they are marketed abroad under different names. The "Isthmus" crude from the Reforma fields is 33° gravity or lighter, undersaturated, and contains less than 2 percent of sulfur by weight. The dolomite reservoirs are below a regional unconformity; some are deeper than 4000 m. The "Mayan" crude from the Campeche Bank fields is of 20–24° gravity with 3 percent or more of sulfur, apparently because of salt water invasion of the heavily faulted, fractured, and brecciated limestone reservoirs above and below a major unconformity. The fractious quality of the Mayan crude presents a daunting problem to its refiners.

17.7.4 Graben basins

The US Gulf Coast Basin, considered above among the tectonically simple but one-sided basins, began as a half-graben downfaulted along its northern margin. True two-sided grabens, created during extensional regimes and controlled by normal faults, might on first thought appear to be easily invadable by surface waters.

This would be to misunderstand grabens. Most petroliferous grabens originated during the Mesozoic phase of continental extension. Furthermore (as already discussed in Ch. 16), they are normally half-grabens, highly asymmetrical but not progressing to the extension achieved by the Gulf Coast Basin. Their oldest sediments were confined to the downfaulted area; however, significant downfaulting soon ceased, and younger sediments were deposited over an area far wider than the original graben. Thus the older sediments became sealed by the younger, themselves deposited in what were essentially young cratonic basins much like the Paleozoic Michigan Basin (Figs 16.26 & 17.8).

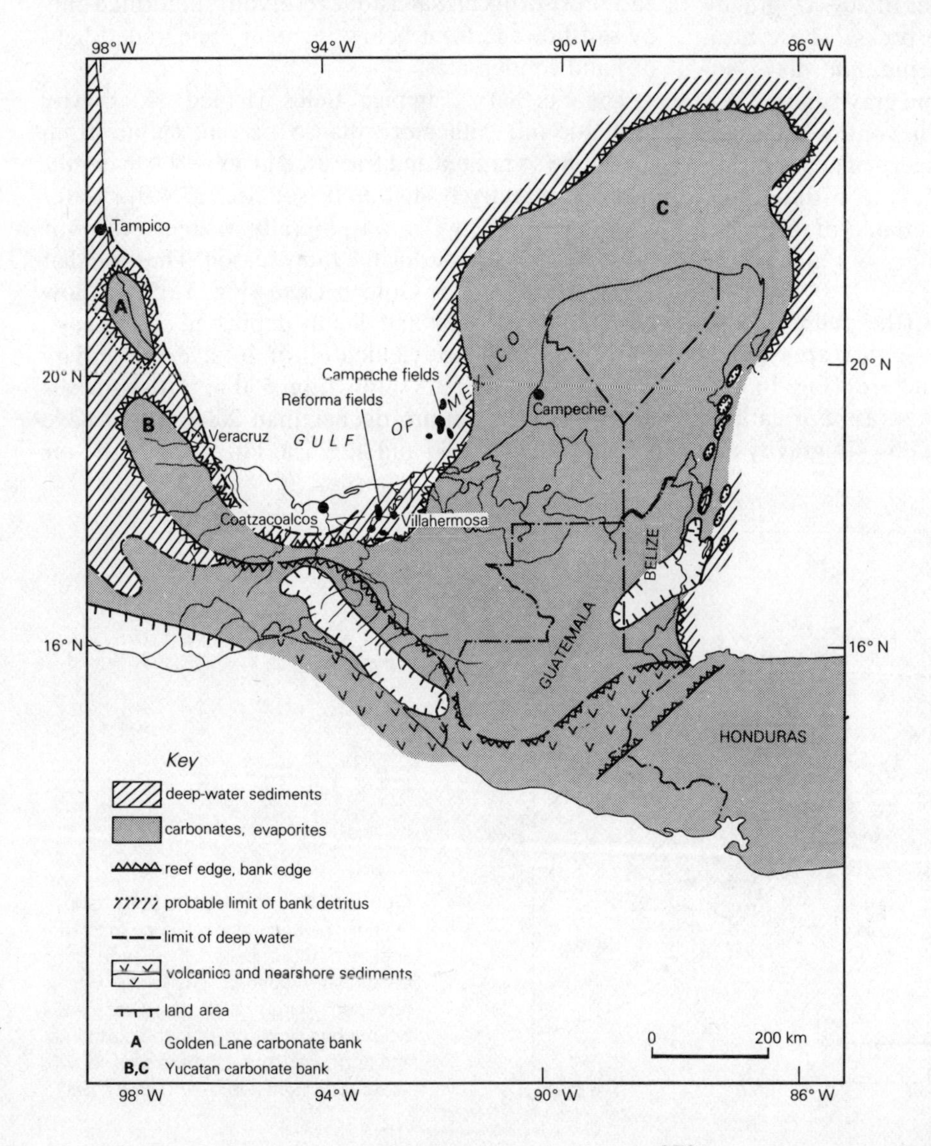

Figure 17.22 The Reforma–Campeche province of southern Mexico, in relation to the Cretaceous carbonate bank. The Reforma district (onshore) produces light crude, the offshore Campeche district heavy, sour crude. (From F. Viniegra O., *J. Petrolm Geol.*, 1981.)

In the North Sea graben and several others, this simple basinal sagging began in the Aptian–Albian. In the Sirte Basin of Libya, overstep of the original fault blocks had occurred by Senonian time and was broadened during the Cenozoic. In the Gippsland Basin of Australia, the earliest Tertiary sediments following the usual mid-Cretaceous unconformity similarly spread far outside the original graben. Almost all grabens so modified contain very light oils, especially in the faulted reservoir rocks of the rift and pre-rift stages. These reservoirs are also likely to be overpressured, though rifted basins in general have relatively low gas contents because the early source sediments contain abundant Type I kerogen and are highly oil-prone.

Some North Sea oils, many Libyan oils and all Gippsland oils are lighter than 40°API. In the somewhat younger Cambay graben of India, 45° gravity oil occurs at about 1500 m. Reservoir rocks tend to be invaded by surface waters only where the grabens are very young and have been uplifted without the wide, unfaulted sag stage being reached. Petroliferous representatives of such invaded grabens (the Vienna and Suez grabens and the Middle Magdalena valley of Colombia) are described here, along with their uninvaded counterparts.

The *Sirte Basin* of Libya was created by mid-Cretaceous block faulting, mainly along the southwest side. The blocks were completely submerged before the end of Cretaceous time. The wide present basin, containing important reservoir horizons but probably no effective source sediments, dates from the mid-Paleocene. The overall shape of the basin is somewhat like the letter W (Fig. 17.23). Along the strongly faulted western margin, oils of 30–39° gravity occur in Cretaceous and Paleocene reservoirs. In the faulted axial trend, all important oils are in the 35–39° gravity range and most are in Paleocene carbonate reservoirs. Similar-weight oils along the little-faulted eastern foreland are associated with hypersaline formation waters (having salt contents approaching 200 000 p.p.m.); they are pooled immediately above and below the Precambrian–Cretaceous unconformity. In the nearly unfaulted strips flanking the axial fault zone, the oils are all lighter than 35°, and many are lighter than 45°. Some water invasion of shallow aquifers is apparent, but oils in the deeper zones have been adequately protected from it. They are nearly all paraffinic and very low in sulfur.

The *North Sea Basin* is mainly a westerly-dipping homocline in pre-mid-Cretaceous strata. It is much cut by east-dipping normal faults and is cut off by major faults at both sides (Figs 16.26 & 113). Upper Cretaceous strata, essentially unaffected by the faults, do not extensively overlap the graben. Early Tertiary strata, however, do overlap the graben margins and occupy what is effectively a shallow, unfaulted cratonic sag. The effects of water invasion are minimal, because of a lack of renewed structural influences. Formation waters lack the sulfate ion. In the north part of the graben, most of the oil is in faulted Jurassic sandstones; gravities are in the mid- to upper 30s. The post-faulting Paleocene strata contain abundant associated gas. There is clear evidence of some local biodegradation there. Gas in the giant Frigg field is very low in homologs of methane; oil underneath it is naphthenic, of 24° gravity, and associated with tar. Apparently the mid-Tertiary uplift permitted access of meteoric waters via the mid-Tertiary and pre-Tertiary unconformities. In the southern parts of the graben (the Greater Ekofisk area), oils in the Paleocene reservoirs attain 40° and even 50° gravities below about 3000 m.

The *Songliao Basin* in northeastern China, containing

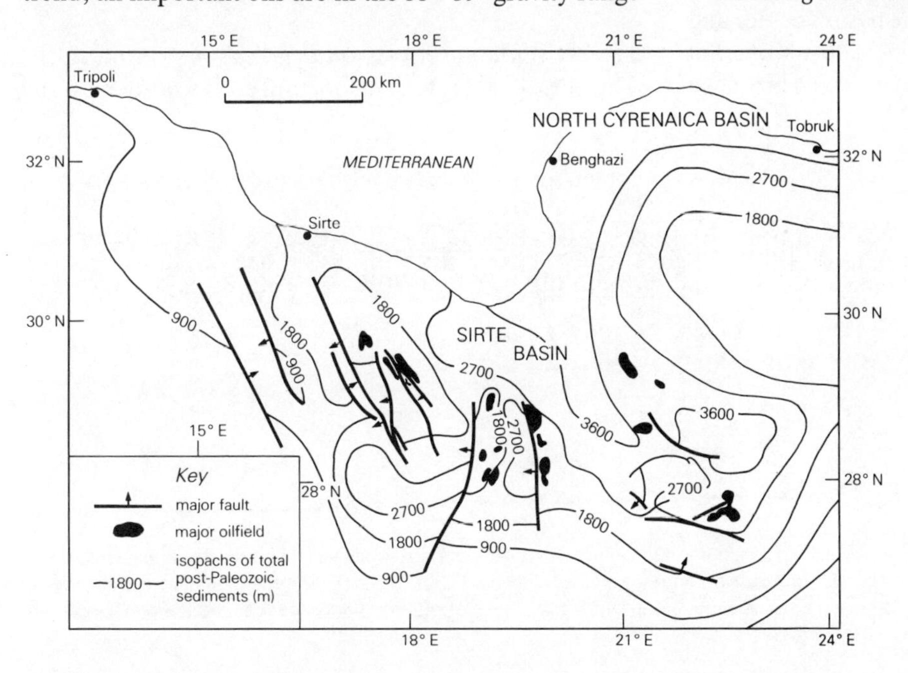

Figure 17.23 Sirte Basin, Libya, showing general structure and isopachs of total post-Paleozoic section. (From J. Gillespie and R. M. Sanford, 7th World Petroleum Congress, 1967.)

the supergiant Daqing oilfield, began as a mid-Jurassic graben (Fig. 17.24). Following the ubiquitous sub- and intra-Cretaceous unconformities, the margins of the graben were overlapped by transgressive Upper Cretaceous and younger strata, all lacustrine or fluvial. Transgressive Middle Cretaceous bituminous shales are the probable source beds, but the reservoir sandstones are in the Lower Cretaceous below the overlap and have been shielded from water attack. The oils are highly paraffinic.

The basin of the *Gulf of Suez* began as a mirror image of the Mesozoic North Sea graben. At the latitudes of its principal oilfields (but excluding its northern and southern ends), it is an eastward-dipping homocline with normal faults downdropped to the west (Fig. 16.29). Below a pre-Miocene unconformity, the block faulting created great stratigraphic relief; above the unconformity, Miocene and younger sediments are almost unfaulted. Precambrian basement blocks are at the surface on both basin flanks. Oils from close to the flanking faults are all heavier than 28° gravity, and most occur at less than 1250 m. Only in the center of the trough at its southern end do oils reach 35° gravity, at almost 3500 m depths. Isotope and geochemical studies indicate that a single (Miocene) source is likely for all the oil, even though reservoir rocks range in age from Carboniferous to Miocene. Geological considerations suggest that a Mesozoic source is also a possibility for the stratigraphically older reservoirs.

The *Vienna Basin* in Austria and Czechoslovakia is so young (wholly of Mio-Pliocene age) that it never achieved serious overlap (Fig. 16.85). The major normal faults dip towards an eastern hinterland, where the chief stratigraphic units crop out. Only gas has been found in the eastern and southern sectors of the basin. Asphaltic oils of 19–23° gravity are found in the basin's center and in the shallowest productive horizons near its western border. These oils, thought by many to be of Neogene source, obviously owe their condition to oxidation by sulfate-bearing waters. Water drive is the dominant recovery mechanism throughout the west flank fields. Older Miocene horizons there contain mixed paraffinic–naphthenic crudes of 24–28° gravity at moderate depths. The oldest Miocene horizons and the underlying Eocene flysch, having no significant outcrop in the basin, yield highly paraffinic crudes of almost 40° gravity. These crudes are of pre-Miocene source, and probably pre-Tertiary. The simplest explanation, supported by the geochemists, assigns all the crudes of the basin to a single source, below the unconformity.

The graben of the *Middle Magdalena Valley* of Colombia (Fig. 17.25) developed during the Tertiary within thick, basinal Cretaceous strata. The essential structure can be represented by an eastward-dipping wedge cut by numerous faults, not all of which are normal. The thickness of the wedge reaches about 12 km along the eastern edge, about one-third marine Cretaceous and two-thirds nonmarine Tertiary. Probably all the oil originated in Cretaceous source rocks, but seeps from Tertiary strata are very numerous. The only significant production from Cretaceous reservoirs comes from the extreme north end of the graben, where the section is thinnest. Paraffinic oil of 37° gravity is produced from the base of the carbonate section at nearly 2000 m in an unfaulted structure. None of the oils from the non-marine Paleogene sandstones is lighter than 28° gravity; in the youngest of the sands (Oligocene) gravities range from 23° to about 13°, even at 2200 m. All these oils are naphthenic or asphaltic. The formation waters are saline to brackish, but the unproductive sands above the producing zones contain essentially fresh waters.

17.7.5 Wide compressional foredeeps on cratons

The thrust faults that characterize one border of any

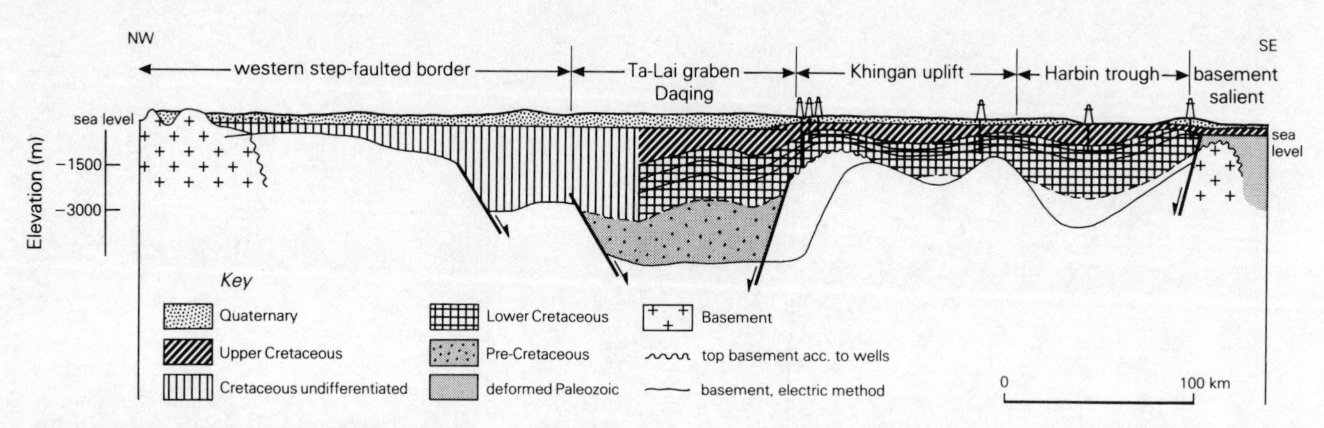

Figure 17.24 Cross section through the Songliao Basin in Manchuria, containing the supergiant Daqing oilfield. The original graben was formed after the Yenshan orogeny, during the Middle Jurassic. The principal source sediments are mid-Cretaceous in age; the principal reservoir rocks, early Cretaceous; all are lacustrine or fluvial. Following two major unconformities, the Upper Cretaceous widely overlapped the original graben in a gravity-controlled sag. (From Chin Chen, *J. Petrolm Geol.*, 1980.)

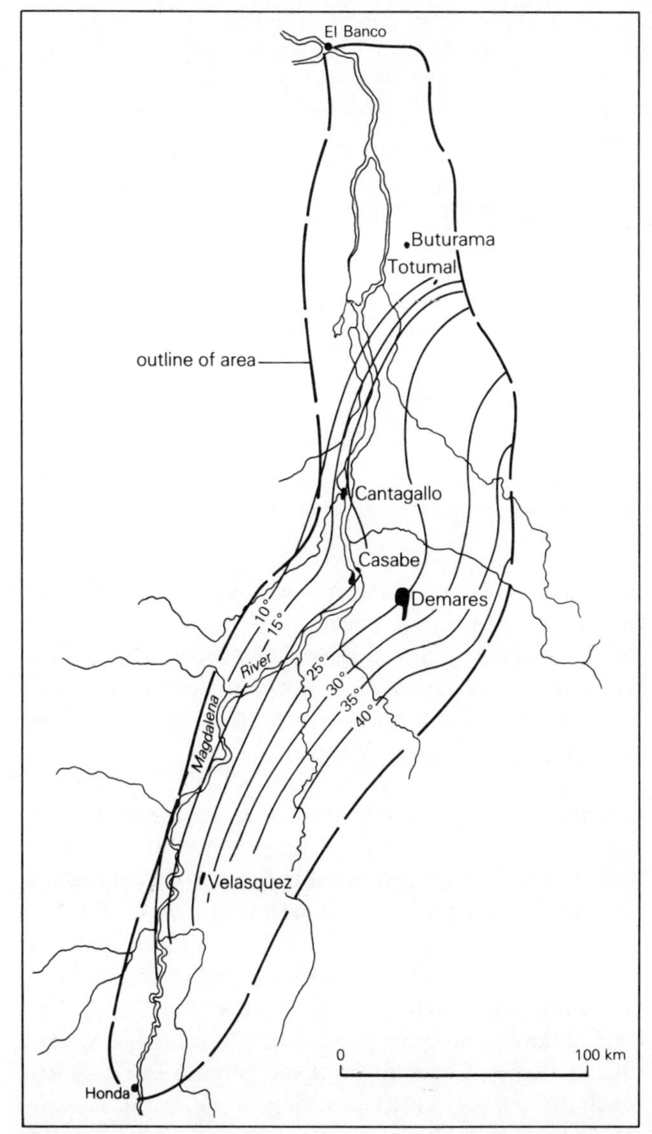

Figure 17.25 Middle Magdalena Valley Basin in the Colombian Andes. Generalized contours of the gravities of the oils in Tertiary reservoirs at intervals of 5° API. The Tertiary sedimentary fill thins from over 8 km along the southeastern edge of the basin to zero along the western edge. (From L. G. Morales *et al.*, in *Habitat of oil*, Tulsa, Okla: AAPG, 1958.)

compressional foredeep can create barriers to the entry of fresh waters from surface outcrops. The lower strata lack important normal faults and are separated from their outcrops by the thrust barriers. They also commonly have a major unconformity above them, and a covering of molasse sediments. Degradation of hydrocarbons is common only at or above this unconformity, and is therefore much more likely along the stable foreland side of the basin than along its mobile side or its axis. We can illustrate this phenomenon through a succession of basins showing advancing degradation.

In *Burma* and *northern Sumatra*, elongate intermont basins are bounded by thrust belts on the west and by backslopes of older orogenic belts on the east. Paraffinic oils lighter than 50° gravity occur at 1000 m depths along the west sides; asphaltic oils as heavy as 20° gravity along the east sides.

In *southern Sumatra*, early recognition of three distinct oil types precipitated a debate about the explanation that lasted for years. The basin contains a single depositional cycle of transgression followed by regression. There is now little dispute that there is only a single oil-source rock, a bathyal shale formation low in the transgressive leg of the cycle. Interfingering with the shale are prolific reservoir sandstones draped over highs in the pre-Tertiary basement (Fig. 17.26). These sandstones received very waxy, paraffinic oils, having a range of gravities between 27° and about 57°, by migration from the deep basin to the west. Regression began in later Miocene time, preparatory to very late folding. The shallow regressive sandstones contain exceedingly light oils (45–55° gravity) that are paraffinic but almost free of wax. These oils must have undergone vertical migration of several thousand meters, presumably via faults, from the same source as the waxy oils in the deeper sandstones. In still younger sands closest to the outcrop, asphaltic oils of 22–26° gravity are produced from about 250 m. Biodegradation by meteoric waters is not to be doubted.

The basin of *central Sumatra* was the last of the three onshore Sumatran basins to be discovered, because it provides fewer seepages than the other two. The oil there is reservoired almost wholly in the shelf facies. Waxy oil of 35° gravity is produced from large fields at less than 1000 m. Equally waxy oil of 21° gravity occurs at less than 200 m in the Duri field, on a linear uplift bringing pre-Tertiary basement rocks close to the surface. The degraded Duri crude is the target of the world's largest steam flood (see Sec. 23.13.3).

Prolific oil basins on opposite sides of *Borneo* are also wholly of Tertiary age (Fig. 17.27). Regressive sandstone reservoirs are principally of deltaic environment in the fields on the southeast side, of coastal marine environment on the northwest side. Early fields were inevitably shallow (less than 1000 m), and yielded asphaltic oils heavier than 30° gravity associated with almost fresh waters. These oils later proved to be merely caps above deeper (1800–2800 m), highly paraffinic oils of 40° gravity or lighter (Fig. 16.121). Geological and geochemical evidence points to the oils coming from a single source and being emplaced following considerable vertical migration. There are two oddities about the fluid columns in several fields. In the shallower zones, the formation waters tend to become fresher and more oxidized with depth. In the deeper zones of light oil content, the lightest and most fluid oils are shallower than somewhat heavier oils with higher pour points. Presumably the lightest oils migrated farthest upwards via extensional faults.

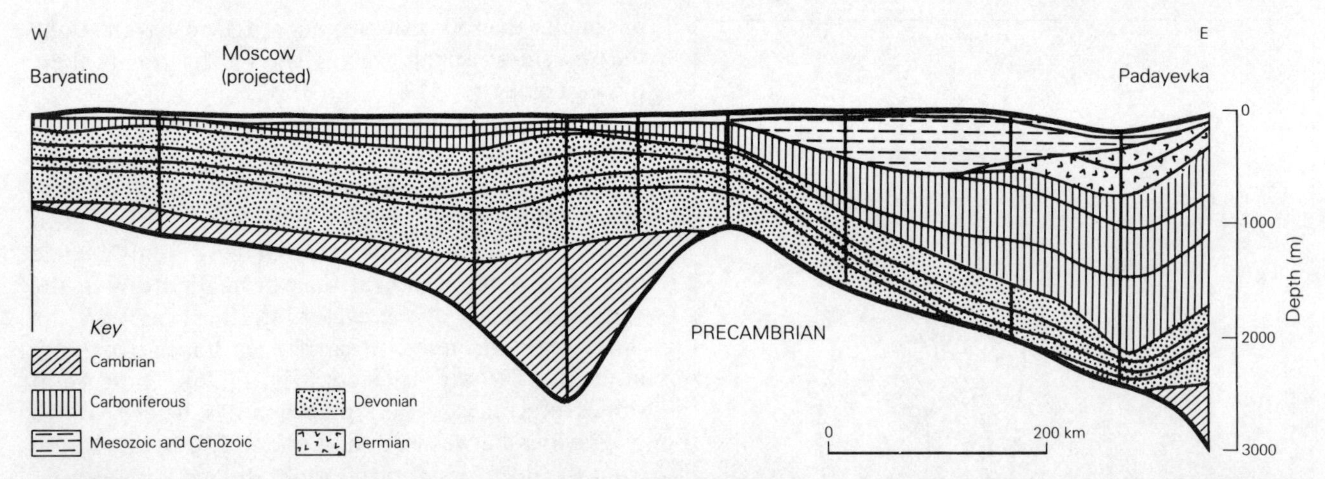

Figure 17.26 Reconstructed stratigraphic section across the South Sumatra Basin, showing pre-folding relations between transgressive and regressive depositional cycles and occurrences of three different types of oil at three distinct stratigraphic levels. (After J. H. L. Wennekers, in *Habitat of oil*, Tulsa, Okla: AAPG, 1958.)

All petroliferous southeast Asian basins were of Tertiary creation; their characteristics are not duplicated in older basins. In the *Persian Gulf Basin*, fresh waters are invading slowly around the margins; but the region has lain in its present arid belt since its emergence in Tertiary time, and the outcropping section consists almost entirely of low-permeability carbonates without sands. Almost all the oils in fields of the Iran–Iraq fold belt are between 32° and 38° gravity. Oils from the Arabian platform to the west are broadly similar, but oils heavier than 30° occur at the north end of the gulf and oils as light as 40° are common at the south end.

The *Western Canadian Basin* contains abundant light oil and gas in Devonian carbonate rocks, stratigraphically well below a regional unconformity. Oil as light as 40° gravity occurs at less than 2000 m. This enormous basin differs from the other wide foredeeps considered here in that it was *not* a foredeep during the time its principal source and reservoir rocks were deposited. The deformed belt to the west did not achieve its present state until the Laramide deformation in post-Cretaceous time; it provided the present water recharge area and its hydraulic head. There is therefore downward water movement through thick Mesozoic and Cenozoic clastic strata close to the disturbed belt, but at pre-unconformity levels the flow is up-dip towards the northeast through thick Paleozoic carbonates. There is neither oil nor gas in pre-Devonian rocks west of the Williston Basin (see Sec. 17.7.2), and no reason to believe there ever has been. Though there are no heavy, degraded oils in Devonian reservoirs, Lower Cretaceous sandstones above the unconformity contain some of the greatest accumulations of such oil known (Fig. 10.17).

The huge *Ural–Volga Basin* (Fig. 16.43) is wholly Paleozoic. Devonian oilfields are confined to an area east of a basement arch that lacks *terrigenous* Devonian rocks. Oil migrated from southeast to northwest, away from a deep basin north of the Caspian Sea. In that direction: oils become heavier within a narrow 30–33° gravity range; sulfur and asphaltenes increase; gas, light fractions, and aromatics decrease; and oil/water contacts rise so that traps become less full of oil. The central part of the province in fact lacks gas almost entirely. Formation waters are of chloride–calcium (stagnant basin) type (Sulin's type *d*; see Sec. 9.2), highly mineralized but decreasingly so to the northwest. Sulfate ions are almost absent; they probably reacted with aromatic gasoline fractions and converted them to asphaltic tars.

Carboniferous strata thicken greatly eastward towards the Urals as the Devonian thins (Fig. 17.28). The Carboniferous outcrop is far larger than the Devonian. Oils in Lower Carboniferous sandstone reservoirs are much more widely dispersed than those in the Devonian, extending far beyond the crest of the arch. They are also heavier (about 27° gravity in the largest fields), have less of the light fractions, aromatics, and naphthenes but more sulfur and asphaltic tars. The Carboniferous gasoline fraction contains more branched hydrocarbons than normal hydrocarbons. The formation waters are less mineralized than in the Devonian but contain more sulfates.

The structures topping the arch apparently gathered oil over much of Mesozoic and some of Cenozoic time, from surrounding smaller structures that lost their seals. The arch is now ringed by a wide structural depression (Fig. 16.43), which has shielded oilfields in younger Carboniferous sandstones on its outside (platform side) from surface waters entering from outcrops in the Urals. Thus these fields yield oils lighter than 40° gravity from less than 1600 m.

The much-studied basin of *eastern Venezuela* (Figs 10.16 & 17.29) is an exemplar of oil gravity variation across a basin. At the northern edge is the Guanoco

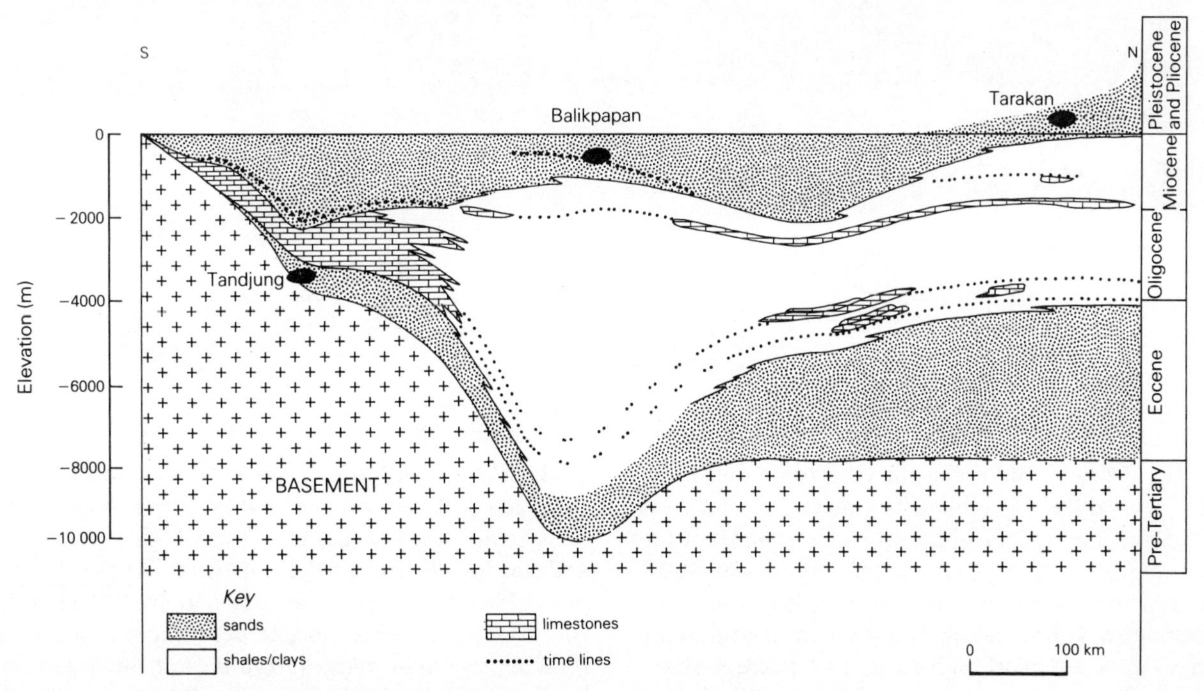

Figure 17.27 Reconstructed stratigraphic section along the east coast of Borneo (Kalimantan), showing pre-folding relations between shelf, deep basin, and final molasse cover. (After J. Weeda, in *Habitat of oil*, Tulsa, Okla: AAPG, 1958.)

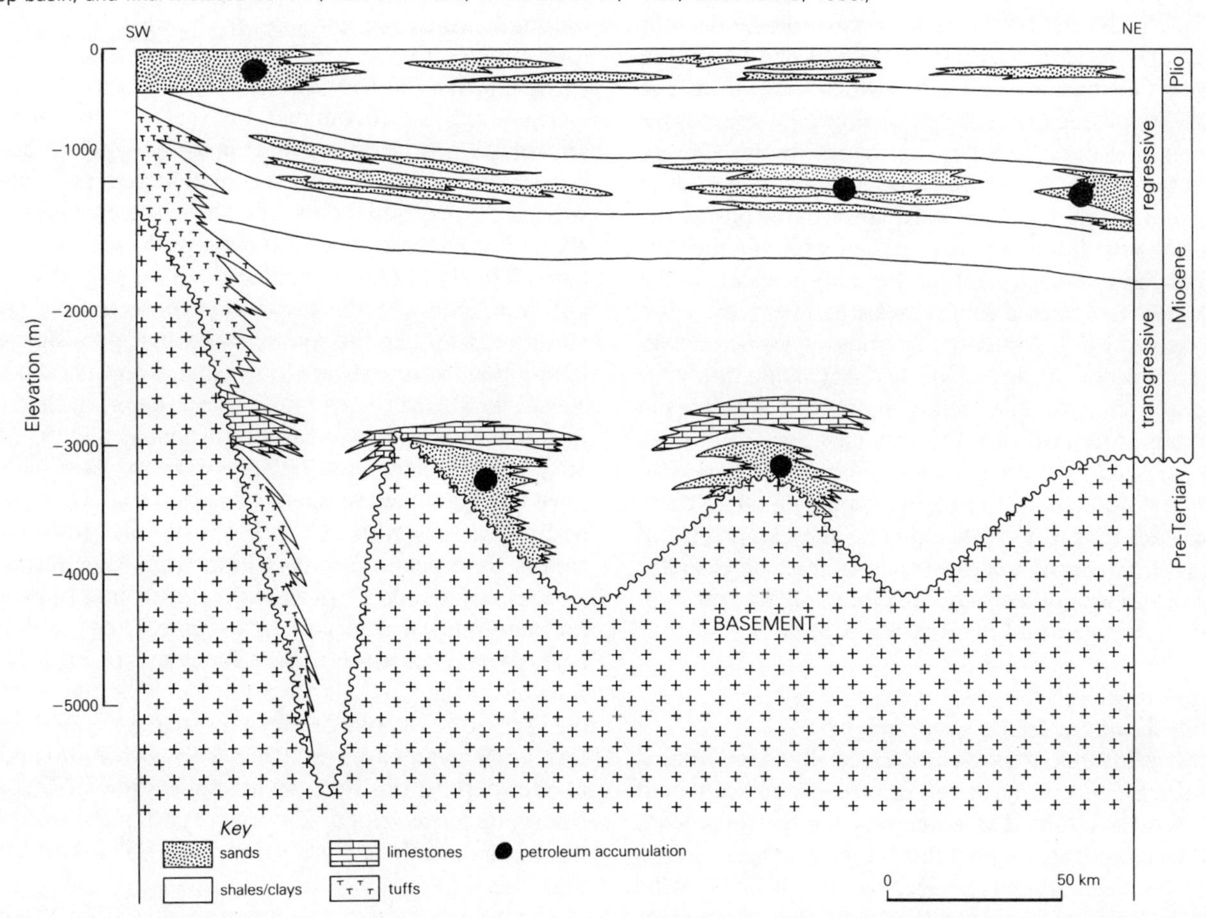

Figure 17.28 Generalized geologic profile across the Ural–Volga Basin, USSR, from southwest of Moscow to northeast of Kuibyshev, showing the considerable relief on the Precambrian basement, and the thinning of Devonian and thickening of Carboniferous strata eastward towards the Ural Mountains. (From D. V. Nalivkin, *Geology of the USSR* (1962); Edinburgh: Oliver & Boyd, 1973, by permission of the Copyright Agency of the USSR.)

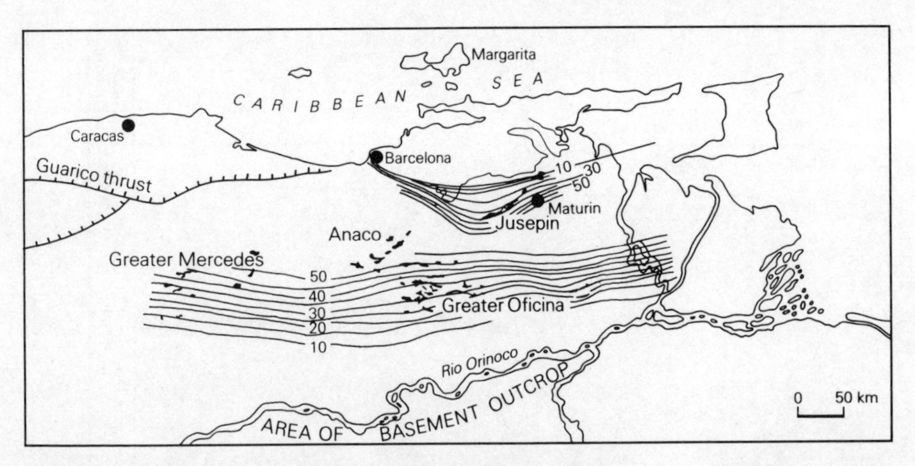

Figure 17.29 The East Venezuelan (Orinoco) Basin, showing distribution of oil gravities. Isogravity contours are at intervals of 5° API, based on the highest values found in each field. (From K. F. Dallmus, Tulsa Geological Society, 1963.)

pitch lake. Closely south of it, the Quiriquire field (Fig. 16.83) yields oil of 16° (average) gravity from buttress sands above a major unconformity. The salinity of the formation waters is very variable but goes as low as 0.2 percent. The far southern flank of the basin is lined by the Orinoco heavy oil belt (6.5–15° gravity), containing perhaps $0.2 \times 10^{12} m^3$ of oil with up to 5 percent sulfur and 500 p.p.m. of heavy metals. Oils become sharply lighter, and formation waters more saline, towards the axis of the basin. In the Anaco wax–oil fields, oils lighter than 50° gravity occur at 2000 m depths. In the Greater Oficina area, both the hydrologic profile and oil-gravity profile are reversed; lighter oils occur above heavier, and high-bicarbonate waters in the Oficina Formation, the principal producing horizon drop below normal salinity northward, towards the basin axis.

The *Denver Basin* receives strong hydrodynamic flow from the mountains to its west. Formation waters in the productive Cretaceous sandstones are of very low salinity. The sprawling Wattenberg gasfield is trapped downdip from water by the strong hydrodynamic gradient, much as is the gas in the tight Cretaceous sandstones in Wyoming, Montana, and Alberta (see Sec. 16.4.6).

In the *Great Artesian (Eromanga) Basin* of Australia, water invasion is nearly total. No significant oil accumulations remain in its center. Only in the Cooper Basin (Sec. 17.7.2), really a separate, older cratonic basin far removed from any recharge area, have significant nonassociated gas and a little light oil been found.

17.7.6 Young intermontane basins

In narrow intermontane basins, especially those created or accentuated during Cenozoic time, extensive recent tectonism has provided numerous faults and unconformities for water circulation. Topographic relief is large; seepages abundant. Repeated erosion allowed entry of fresh waters into all shallow and many deep strata. Hence nearly all waters in such basins are subsaline, and 40° gravity oils are rare; they are likely to be nonexistent above 2500 m.

The *Maracaibo Basin* of western Venezuela has a very complicated structural and stratigraphic history. During the Cretaceous, sedimentation was highest in the west and southwest, least in the northeast. At that time the present basin was a shallow carbonate platform between two troughs. The late Eocene deformation raised high, flanking mountain ranges from these troughs, converting the platform into a depression almost surrounded by deformed belts. Sedimentation has since been highest in the southeast corner and least in the west. The present foreland lies to the west, where the greatest trough existed during the Cretaceous.

The basin is surrounded by vast oil and asphalt seepages, most close to a transbasinal pre-Miocene unconformity (Fig. 17.30). Formation waters in reservoir sandstones both below and above the unconformity are of bicarbonate type (Sulin's type *b*; see Sec. 9.2). Commercial oilfields, including the third largest known and four others in the top hundred, occur in a great variety of traps. In the southwest corner, paraffinic oils lighter than 40° gravity are found shallower than 1800 m. Yet several structures there have been flushed by meteoric waters, and at least one producing field yielded 30° gravity oil associated with almost fresh water. The oil must have occupied its sharp structural trap when it still had saline formation water below it; the fresh water cannot have been there very long in geologic terms.

West of the lake, oils in faulted anticlinal traps and pre-deformation reservoir rocks range from asphaltic types as heavy as water to paraffinic types of 36° gravity. East of the lake, all oils are asphaltic. Near the outcrop on the northeast, oils are heavier than 14° gravity and formation waters are fresh (Fig. 9.9). Down-dip and at greater depths, oils become lighter but few in Tertiary reservoirs are as light as 25° gravity east of the mid-lake fields, and hardly any are lighter than 35° gravity (Fig. 9.4).

The basins of *California*, all producing from Tertiary sandstones, have been equally affected by meteoric waters. Many of the oils are heavier than 20° gravity, especially in the Los Angeles and Santa Maria basins. In

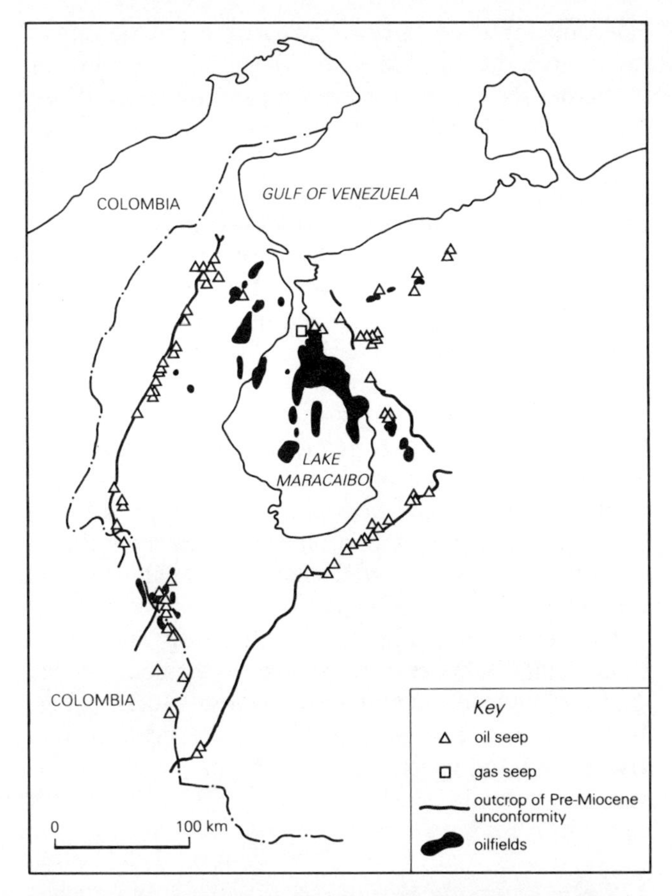

Figure 17.30 Oil seepages in and around the Maracaibo Basin, western Venezuela. The southeastern, southwestern, and northwestern margins were all strongly uplifted and faulted in early Tertiary time, and meteoric waters gained access through them as well as via the homoclinally dipping strata on the northeastern margin. Along that margin, nearly all asphalt seepages are within a few hundred meters of the outcrop of the sub-Miocene unconformity. (From P. A. Dickey and J. M. Hunt, *AAPG Memoir* 16, 1972.)

the San Joaquin Valley (Fig. 16.38), all oils near either of the marginal outcrops are heavier than 20° gravity; pools in some west-side fields have oil/water contacts tilted to more than 100 m km^{-1}. Basin-center sands like the Stevens turbidite sandstone (Sec. 13.2.7) yield 36° gravity oil. In the Ventura Basin, oil occurs in the same sandstones in both the hanging walls and footwalls of thrust faults (Fig. 17.12). Sandstones in the thrust plates crop out; their oils are oxidized, and the associated waters are again of bicarbonate type. On the shallow northern shelf are "tars" of 7° gravity. Sandstones in the footwalls of the thrusts are autochthonous and do not crop out within those structures; their oils have escaped oxidation.

17.7.7 Thrust-belt fields

California's Ventura Basin, considered among the intermontane basins above, is perhaps the closest approach to a major petroliferous basin entirely within a thrust belt. A number of second- and third-rank basins have thrust-fault fields essentially throughout them – the southern Trinidad and Bolivian Subandean basins are obvious examples. In a number of prolific foredeeps and interdeeps, production extends into the overthrust margins.

As already noted, thrust faults may destroy traps but they also create new ones. Being compressional, they may also create barriers to invasion by meteoric waters. Characteristic of thrust-belt oilfields are high-gravity, highly paraffinic oils and condensates (including retrograde condensates; see Sec. 23.7). In the Andean foothills fields of *Bolivia* (Fig. 16.25), oils shallower than 1200 m range from 40 to 65° gravity. In the fields of the complex, overthrust, and recumbent Carpathians of *Poland* and the *Ukraine* (Fig. 16.24), the oils in deep traps are light, paraffinic, and waxy; associated waters are saline and of chloride–calcium type. This is the type locality for the solid paraffin wax ozocerite (see Sec. 10.6). The exceptions to this rule occur above shallow overthrusts, in allochthonous plates containing oxidized oils and nearly fresh waters of bicarbonate–sodium type. The nearby *Romanian Carpathians* (Fig. 16.53) provide the classic juxtaposition of highly paraffinic crudes in Lower Pliocene sands and nonparaffinic crudes in the Upper Pliocene.

Oils in the aeolian sandstone of the *Wyoming–Utah* overthrust belt of the western USA reach 48° gravity and are paraffinic. Some of them are in fact retrograde condensates. The *Upper Assam Valley* fields contain waxy oils of nearly 40° gravity. The oil in the Turner Valley field, in the overthrust foothills of the *Canadian Rocky Mountains* (Fig. 16.23), is of 39° gravity and unoxidized over a vast depth range below about 1000 m. All these oils are in traps in the hanging walls of their thrusts.

The most conspicuous exception to these generalizations is the basin of *southern Trinidad*. It is a continuation of the northern flank of the basin of eastern Venezuela, which contains some highly waxy, light oils (see Sec. 17.7.5). In Trinidad, however, the longer continuation of Neogene subsidence and tectonism has resulted in sharp, linear anticlines separated by wide, nearly flat synclines and cut by thrust faults, strike-slip faults, and (in several cases) mud diapirs (Fig. 16.55). Highly discontinuous, lenticular, young sandstone reservoir rocks crop out extensively. Surface manifestations of oil and gas, found all over the basin, include the world's most famous asphalt lake. The basin now yields no oil as light as 40° gravity; only in and off its southeastern corner are the oils lighter than 30° gravity. In all the older large fields, the oils are 25° gravity or heavier. They are not waxy.

17.7.8 Multiply-deformed basins

The rarest and most startling effects of basin structure on fluid chemistry occur where an intense orogenic episode fragments an older basin already petroliferous. Classic examples are provided by the Rocky Mountain basins of the western USA, especially the Big Horn, Powder River, Wind River, and Uinta basins (see Ch. 25). A simple foredeep created during the Pennsylvanian contained oil from a Permian source. Renewed deformation in late Mesozoic time culminated in violent Laramide thrusting during the earliest Tertiary. The Precambrian crystalline basement of the old foredeep was broken into alternating strongly uplifted and deeply depressed rhomboidal blocks (Fig. 22.6).

The individual basins, wholly structural and topographic with respect to their pre-Tertiary sedimentary contents, were in no way separate before Laramide time. Structural relief from uplift to basin now exceeds 12 km. The Laramide deformation created a set of abrupt mountain outcrops exposing both the source and reservoir rocks of the Paleozoic foredeep. The uplifts were bounded by faults extending to great depths, at which intervening basins now contain almost fresh formation waters. Only at those depths and below do oils in the Pennsylvanian reservoirs attain 40° gravity. Otherwise Pennsylvanian oils are heavy, black, viscous, and sulfurous.

If such oils owe those characteristics to immaturity and shallow origin (see Sec. 10.2), the Pennsylvanian oils can be explained by their positions in the basins. The heavy oils would be shelf oils, of eastern (platform) derivation. In the deeper central basin to the west and southwest, Pennsylvanian oils should be lighter. However, the intermont basins superimposed above the Paleozoic foredeep carry 40° gravity oils in Cretaceous reservoirs at about 1500 m, overlying the heavy oils of the earlier shelf. The heavy oils apparently suffered late degradation and oxidation by waters of the new tectonic regime.

The much-studied Salt Creek field (Fig. 17.31), on the arch between the Powder River and Wind River basins, illustrates the condition perfectly because of its multiple reservoir horizons. The deepest of these, in the Pennsylvanian, lies below a regional unconformity; it contains sulfate-bearing water very low in sodium, and salinities less than 0.2 percent extend below 3000 m. The shallowest producing horizon, the Upper Cretaceous Shannon sandstone, crops out around the field and contains water of bicarbonate–sulfate type. The second shallowest sandstone reservoir contains water of bicarbonate type. All the other productive horizons, Jurassic and Cretaceous in age, yield the normal chloride-type waters with some bicarbonate.

The Pennsylvanian fields of the Big Horn Basin provided King Hubbert with the type specimens for his study of tilted oil/water interfaces (see Sec. 16.3.9 & Fig. 16.64). The tilting is exaggerated between heavy, viscous oils and nearly fresh waters, under subnormal pressures and temperatures.

17.7.9 Lacustrine basins

Petroliferous lacustrine basins are nearly all bounded by high-angle faults, extensional or compressional (see Sec. 17.4.7). This might at first suggest that their formation fluids would be generally similar to those in rift basins or in the Rocky Mountain type of basin described above. Lacustrine basins, of course, did not contain sea water in

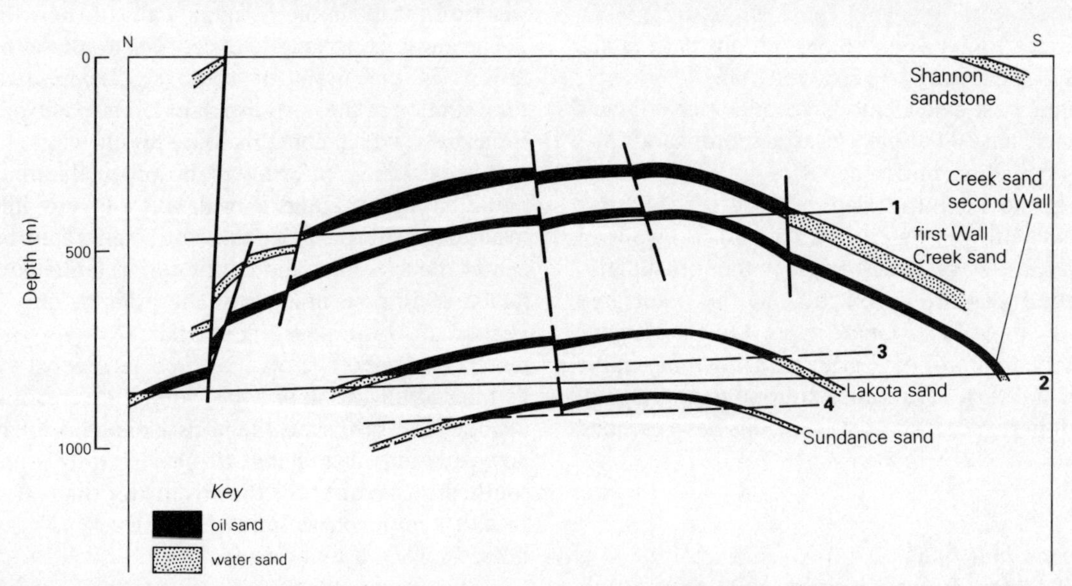

Figure 17.31 Cross section through the Salt Creek field, Wyoming, showing lack of influence of numerous faults cutting the structure. The different oil/water interfaces, numbered 1–4 for the four producing sands and all tilted towards the north (left), also reflect the lack of intercommunication between the four pools. (From E. Beck, in *Structure of typical American oilfields*, Tulsa, Okla: AAPG, 1929.)

the first place, and any oils generated were not like those of marine basins.

Lacustrine oils are mainly derived from rich algal and diatomaceous organic material, highly sapropelic and yielding Tissot's Type I kerogen (Sec. 7.3). They are therefore invariably rich in insoluble waxes and have high pour points. They are typically as light as 36–40° gravity, very low in sulfur, and contain nickel porphyrins rather than the much commoner vanadium porphyrins (Sec. 5.2.6). Distant lateral migration is improbable, but the deposits of lake margins tend to become much faulted. This leads to poor interconnection between multitudes of individually thin reservoirs, and variations in oil chemistry between these and between fault blocks are chiefly due to differential loss of paraffins.

The chemistry of the associated waters depends on the history of the lake while it still received sediment. If the waters were diluted by large rivers, as the Caspian Sea and Great Slave Lake are today, conversion of organic matter to petroleum would be very unlikely. If no persistent dilution took place, and stratification of the water body was long continued, it would become either *saline* (with waters of Sulin's chloride–calcium type, *d*) or *alkaline* (with waters of bicarbonate–sodium type, *b*). The Dead Sea and Great Salt Lake are saline; Lake Tanganyika is alkaline, as were the Eocene lakes of the western USA, in which extensive oil shales were deposited. Most alkaline lakes are in areas having (or having had) some volcanic activity; most modern playa lakes represent a phase of alkaline lake development.

The most favorable circumstances involve carbonates in the bedrock or drainage basin of the lake. This allows the buildup of the pH in the epilimnion and the precipitation of either calcite or (with increasing alkalinity) dolomite (the commonest alkaline associate), or even aragonite. The characteristic deposits are then seasonal varves of dolomitic micrite (the light-colored summer layers), evaporites (especially carbonate salts), and darker (cooler season) diatom or algal blooms.

17.8 Sedimentary basins and geothermal gradients

The role of temperature in the maturation of source sediments and the generation and migration of hydrocarbons has been emphasized (Sec. 7.4). The geothermal gradient, or rate of increase of temperature with depth, has been measured in most producing areas; the flow of heat through the Earth's upper crust has also been measured in all well-explored sedimentary basins as well as in otherwise unexplored areas. The "world mean" heat flow rate is taken to be 1.47 HFU (heat flow units), 1 HFU being 10^{-6} cal cm^{-2} s^{-1} (or in SI units, approximately 42 mW m^{-2}). If individual volcanic and hot spring centers are ignored, the flow is highest over oceanic ridges and lowest over oceanic trenches.

Hundreds of measurements collected by the Russian geophysicist F. A. Makarenko indicate that heat flow in petroliferous regions differs by 10–20 percent from the "world mean" or background value. Because petroleum geologists measure temperatures directly in drillholes, the *geothermal gradient* is a much more convenient expression for them than is heat flow. Geothermal gradients in sedimentary basins generally range from 15–50 °C km^{-1}; the average value may be taken to be 30 °C km^{-1}. Like the access of surface waters, the geothermal gradient of a basin depends largely on the basin's structural style.

The lowest values are found on Paleozoic platforms overlying stable Precambrian shields. They therefore characterize both simple cratonic basins (like those in the interior of North America) and wide Paleozoic foredeeps and constructional basins derived from them. In the Ural–Volga Basin and the Irkutsk amphitheatre of central Asia the geothermal gradient is about 15 °C km^{-1}. The Mid-Continent and Permian basins in North America and the Paris and North German–Netherlands basins in Europe are not much hotter. An exception to this generalization is the Western Canadian Basin, and the explanation of the exception is very simple. The basin began as a Paleozoic platform resting on a Precambrian shield, and much of its oil and gas are in Upper Paleozoic carbonate rocks. Doubtless the region then had a low or moderate geothermal gradient. The Laramide deformation of early Tertiary time elevated the Cordilleran belt across the western edge of the platform, creating a new hydraulic head which has caused low geothermal gradients in Mesozoic strata along the water recharge zone in front of the deformed belt, but a gradient higher than normal in Paleozoic strata over the main platform because heat is transferred towards it by the northeasterly water drive at depth.

Also yielding low geothermal gradients are Tertiary delta basins and coastal pull-apart basins like those around the Atlantic Ocean. Productive regions such as the northwest Australian and eastern Brazilian shelves probably achieved generational maturity only because of high sedimentation rates; on both these shelves the combined Cretaceous and Tertiary section exceeds 3 km in thickness. The very low gradient in the Beaufort–Mackenzie Basin may be a factor contributing to its relatively lean hydrocarbon content.

Most Mesozoic basins give values close to the mean, whether they are compressional (like the Persian Gulf Basin) or extensional (the US Gulf Coast and the West Siberian basins). Surprisingly, similar values also characterize some regions of Cenozoic mud vulcanism, like southern Trinidad. Salt diapirs, in contrast, are geothermal highs, because evaporites have high thermal conductivities.

Graben and half-graben basins have high gradients regardless of age. The Paleozoic Dnieper–Donetz Basin gradient is as high as those from the Tertiary Gippsland and Suez grabens. In the Paleocene fields of the central North Sea, the productive structures are arched over deep salt uplifts but remain essentially unfaulted. Gradients are close to 40 °C km^{-1}. In the northern North Sea, the productive structures do not lie over salt uplifts, but the principal reservoirs are cut by numerous extensional faults (Fig. 16.26) which appear to have acted as conduits for the upward transfer of heat by fluids. The gradients there are 30–35 °C km^{-1}. The overlying Cretaceous is unfaulted and gradients in it are much lower (15–30 °C km^{-1}). In the petroliferous graben basins of eastern China, the average gradient is about 33 °C km^{-1}, possibly a reflection of associated vulcanicity. The most productive of these basins, that of Songliao, is said to have a gradient of 60 °C km^{-1} in the productive Cretaceous section. This section is of lacustrine origin, and lacustrine basins must be of high geothermal gradient if their sediments are to achieve maturity before accumulation ceases.

The younger the graben the higher its gradient, in general. The Vienna and Gippsland basins and the Rhine graben all have gradients between 36 and 50 °C km^{-1}. In the Cambay–Bombay High grabens of western India, the gradient is more than 60 °C km^{-1}.

Not surprisingly, the highest of all geothermal gradients and heat flow values in sedimentary basins, up to double or even triple the mean, are found in Tertiary back-arc basins associated with the Tethyan and circum-Pacific volcanic belts. The central Sumatra basin's very high gradient of 65–90 °C km^{-1} (heat flow more than 2.0 HFU) has already been commented upon in connection with its shallow oil-generative "window" (Sec. 7.4). The basins of northwestern and eastern Borneo, geographically next door to the Sumatra basins and of the same age, are not back-arc basins but their geothermal gradients range from 32 to about 50 °C km^{-1}.

17.9 The continental shelves and slopes

17.9.1 The promise of the offshore

That the continental shelves which fringe all landmasses are simply submerged extensions of the landmasses themselves, and not shallow parts of the ocean basins, was early realized by geographers and geologists. Coalminers extended their underground workings out beneath the sea in both Britain and Nova Scotia without losing the sequence of strata they recognized on land.

If coal measures could be followed beyond the shoreline, so obviously could the reservoir rocks of such oilfields as had already been shown to reach the shore. Wells were drilled from wooden piers off Summerland, in California's Santa Barbara Channel, as early as 1896. The most obvious targets were in oil districts which extended into shallow, enclosed seaways having more oilfields on their opposite sides. Four very large fields lining the east shore of Lake Maracaibo in Venezuela produced heavy oil from shallow sands dipping directly under the "lake" (Figs 16.2 & 17.30). The structural trend of the anticlinal Baku oilfields on the Apsheron Peninsula (Fig. 16.54) headed directly into the western side of the Caspian Sea and reappeared in the Cheleken Peninsula on the eastern side. Offshore production from trellis-like platforms became important in Lake Maracaibo before World War II and in the Caspian Sea shortly after it. Both were quickly eclipsed in activity by the northern, US margin of the Gulf of Mexico.

The heyday of successful offshore drilling came between 1960 and 1975. During that short time all productive coastal areas acquired productive offshore extensions. By the beginning of the 1970s, petroleum recovered from offshore wells was worth 30 times as much as all other offshore mineral production put together; about one well in six was being drilled offshore. By the beginning of the 1980s, about 42 percent of the noncommunist world's entire continental shelf was under lease to oil companies, and was being probed by some 800 offshore drilling rigs and platforms.

A seriously misguided belief had been engendered in much of the oil exploration industry. Geological and geophysical work had demonstrated that the most familiar continental shelves had been constructed by the process called *progradation*; they had been built across the downflexed margins of the continents by sediment derived from the continents themselves. The shelves came to be looked upon as the final repository of virtually all clastic detritus, including all clastic carbonate detritus. They were regarded by many geologists as constituting a gigantic, compound sedimentary basin in their own right, discontinuously but completely circumscribing all landmasses. It was asserted that the continental shelves would contain more oil and gas than the emergent continents. Some went further and maintained that the potential of the continental slopes and rises, seaward from the shelves, would equal or exceed that of the shelves. Volumetric calculations of "potential" oil and gas resources were extended to include these zones, even into waters 3000 m or more in depth.

There were superficially good reasons for this optimistic outlook. The best studied (but then undrilled) shelves, those on the two sides of the North Atlantic, were fundamentally wedges of young sediments, comparable in age and thickness with the sediments productive in dozens of onshore regions. Even accepting the (then) hypothetical postulate that the shelves concerned were no older than Jurassic (because no pre-Jurassic sediments were known in any of the emergent coastal plains), the age interval represented was that containing two-thirds of all known oil. The lack of metamorphism or even of much induration in shelf sediments meant that porosity and permeability could be expected to be high in multiple potential reservoir horizons. The known productive examples gave ample support to these expectations. Furthermore, a number of untested offshore regions were known on both geological and geophysical grounds to be true sedimentary basins; they were, in fact, under water precisely because they were basins. Hudson Bay is a saucer-shaped cratonic sag occupied by Paleozoic sediments much like those in the productive Michigan Basin to its south. The Gulf of St Lawrence occupies the site of a Carboniferous basin, the Irish Sea of a Triassic basin. The Black Sea occupies a Tertiary basin as surely as the productive Gulf of Mexico and Persian Gulf occupy Tertiary basins. Thus the continental shelves constituted the explorer's last resort for sedimentary basins, but a resort thought to be assured of provision.

All of these considerations carry some justification, but all were clearly partial. Prograding shelves are all mid-Mesozoic or younger. Productive Paleozoic basins have no immediate counterparts offshore; Paleozoic shelves are now mountain ranges. Though many basins are submerged literally because they are basins, *compressional* basins (including the highly productive foredeeps and interdeeps) are exceedingly unlikely to be *wholly* offshore. The buoyant mountain roots alongside them render the basins at least partly emergent.

All offshore regions commercially productive before the advent of plate tectonic thinking (1968–9) were simply extensions of regions richly productive onshore. Of the seven productive offshore regions important before 1970, six (the Persian Gulf, the southern Caspian, the Maracaibo Basin, the Gulf of Paria, the offshore Ventura Basin, and the interior seas of Indonesia) were the cores of richly productive, compressional, orogenic basins which happened still to have water in them; the submerged parts are virtually surrounded by onshore production. They are merely one step short of the situation ultimately reached by the equally prolific San Joaquin Valley Basin of California, now an isolated semi-desert hemmed in by mountains but not cleared of the sea until the Pleistocene. The seventh of the early-productive shelf regions, the US outer Gulf Coast Basin, was not surrounded by onshore production but it was continuously flanked by it in reservoir rocks of all Tertiary ages dipping uninterruptedly offshore. These evidently favorable examples in reality provided no analogy for shelf areas unconnected with productive onshore flanks: the Gulf of California, for example, or the Baltic or Arafura Seas.

Prograding continental shelves composed of undeformed, seaward-dipping sedimentary rocks are virtually bound to be artesian. Their inner margins were inevitably exposed during the Pleistocene glaciation. If the strata are petroliferous, their outcrops should be marked by lines of seepages. Such shelves are not "basins" in any worthwhile sense. They are likely to be productive only where they are *interrupted* by discrete basins developed within or across them: grabens, like those of Cambay in India and Reconcavo in Brazil, or deltaic depocenters like those of the Mississippi and Niger rivers. These discrete basinal "cells" within the shelves will normally reveal some onshore evidence of their existence – the continuation of the graben, even if it is not significantly productive onshore (as in the case of the Gippsland Basin between Australia and Tasmania), or the large river providing the delta. There is one outstanding exception – the northern North Sea Basin – which lies wholly offshore and provides no onshore evidence of its existence (Fig. 25.3). To the time of this writing, the Campos Basin off Brazil is the only other case of more than trivial significance. There will undoubtedly be others to provide future discoveries (they are confidently anticipated off southeastern China, for example); but the distinction between "shelf" and "basin" is as applicable offshore as on.

Large deltas are among the thickest wedges of sedimentary rock known. Once extended seaward to the edges of their continental shelves, they also provide the optimum interrelation between multiple reservoir sandstones and potential source sediments deposited beyond the shelf edges. A great number of oilfields, including many giants, occur in deltaic reservoir rocks (see Sec. 13.2.7). These reservoir rocks, however, are not in deltas today and in few cases have they been in deltas since early Tertiary time. Most post-Eocene deltas, and some dating back to the Cretaceous, are markedly gas-prone, a point early emphasized by Klemme. The Tertiary deltas of the Niger and Mahakam Rivers (the latter in eastern Borneo) contain large oilfields, but they contain no supergiant fields and their oils are characterized by very high gas : oil ratios. The sands of the Tertiary Mississippi delta contain much more gas and condensate than oil; the widespread impression to the contrary is due to confusion between the Tertiary delta and the Tertiary Gulf Coast province. Other large, young deltas overlying older deltaic sediments that have proven to be gas-prone are those of the Ganges, Mackenzie, Magdalena, Mekong, Nile, Po, and Sacramento rivers. The tendency towards gas generation is presumably due to the dominance of the humic type in the organic material delivered to a delta from a large drainage basin (see Sec. 7.3). This observation may be carried further. *Oil-prone* source sediments are in fact rather rare in Cretaceous and Tertiary deposits of the continental shelves.

The final caution against excessive expectation from the continental shelves is economic and statistical. Fewer than 50 percent of explored *onshore* basins are commercially productive; fewer than 25 percent are sufficiently productive to be commercial if they were to lie offshore. Not only are the continental shelves not all basinal; the parts that are basinal are not all productive; and the parts that are productive are not necessarily as productive as they would be on dry land.

17.9.2 The change in appraisal of the offshore

Large deltas need large drainage basins (Sec. 13.2.7); they also need master rivers with high ratios between maximum and minimum discharges. They are thus much more likely to form on trailing-margin coasts than on consuming-margin coasts, on which young mountain belts may reverse much of the drainage. The concepts of plate tectonics, upon which the terms in the previous sentence depend, erupted into the Earth scientists' thought-frame in the middle of the heyday of offshore exploration.

To a degree unimaginable in our visualization of onshore basins, which are at least partly visible and tangible, impressions of offshore basins are powerfully influenced by plate tectonic concepts. Only those who reject these concepts can still think of the submerged continental margins as consisting of a vast, disjointed string of potentially petroliferous basins.

The typical trailing-margin shelf has an extensional early history. Discrete graben basins will be totally characteristic of it; deltas will be allowed long lives to become large. An early stage of restricted sedimentation, the so-called taphrogenic or rift stage, will provide ready-made source sediments. A long-lived, prograding accumulation of clastic sediments, with a slow rhythm of transgressions and regressions, provides numerous reservoir possibilities. The trailing or Atlantic type of continental shelf is inherently favorable for oil accumulation.

The typical consuming-margin shelf, surrounding most of the Cenozoic and present Pacific Ocean, has passed through a very different history. Extreme instability, with rapid alternations between transgression and regression and between shoal and deep-water clastic deposition, extensive magmatism, frequent deformation, persistent mineralogical immaturity of the sediments and the lack of carbonates among them, and the low geothermal gradients associated with oceanic trenches, all are features potentially inhibitive of petroleum generation and preservation. Marginal mountain ranges are much more likely to have foredeep and back-arc basins on their continental sides than fore-arc and backdeep basins on their oceanward sides (Fig. 17.7). The foredeep and back-arc basins may be partly submerged, like those of the Persian Gulf and Indonesia, but if the submerged parts are petroliferous the emergent parts are sure to be so also. The *oceanward* sides of consuming margins are inherently unfavorable for the development of petroliferous basins. Fore-arc basins are significantly productive only where fault troughs are superimposed upon or across them (as in northwestern Peru and southern Alaska).

Transform margins are inherently neither favorable nor unfavorable for the development of basins (see Sec. 16.3.3). The invariably quoted case of the Californian basins (which were offshore until the Pleistocene and are still partly so) is unique. The richly petroliferous nature of the basins is not due to the presence of a transform margin alone but to its inception *before* the deposition of the diatomaceous source sediment. The source was consequently preserved intact and not reduced to slivers in the flysch of the arc-trench gap, as most other Miocene diatomites were. Tertiary depressions along the Bartlett, or New Zealand Alpine or Anatolian faults, offshore or onshore, may have structural histories closely similar to those in California, but they are not known to be petroliferous.

17.9.3 Distinctions between offshore and onshore petroleum occurrences

Depth of production A simplistic summary of the foregoing comparisons is that the sedimentary prism off a consuming margin is inherently likely to become deformed and elevated (unless it is unfortunate enough to become consumed). The sedimentary prism off a trailing or passive margin is inherently likely to become deeply depressed. We have also observed (Sec. 8.3) that the principal source sediments in most petroliferous basins were early deposits of them. As downflexure of a margin continues, the source sediment and any reservoir rock stratigraphically close to it is carried to considerable depth. If the offshore basin is unenclosed and eventually faces open ocean, it then receives a thick pile of unpromising sedimentary fill lacking any further source rock. The *average depth of productive intervals* in offshore basins is therefore greater than that for onshore basins, and the difference can only become greater as basins on the outer shelves and slopes are explored – despite the fact that no offshore wells are likely to be drilled as deeply as many onshore wells have already been drilled.

Ages of productive intervals With a few trivial exceptions, such as the late Paleozoic production in some

North Sea fields, offshore oil production is and must remain overwhelmingly from rocks no older than Jurassic, the beginning of the creation of most of the present shelves. Like onshore production, therefore, that from the offshore is (and is likely to remain) in the order of Mesozoic first and Cenozoic second, but with the Paleozoic nearly nowhere. There is no expectation of an offshore equivalent of the Ural–Volga or Mid-Continent basins.

Reservoir rocks The narrow dominance of sandstone over carbonate reservoirs among onshore fields would be very much greater if it were not for the long-lasting carbonate–evaporite regime (Permian to Miocene) in the Persian Gulf region (see Sec. 25.4). This imbalance is repeated in the offshore. Except in the Persian Gulf, offshore reservoirs are predominantly of sandstone, and future discoveries are unlikely to cause much change in the proportion. Only a minor part of the continental shelves received carbonate sediment during much of Cenozoic time.

Types of basin All early offshore oil or gas discoveries were made in extensions of petroliferous onshore basins. Except for the US Gulf Coast Basin, all were within foredeep or interdeep basins, types 4 and 7 in Klemme's classification (see Sec. 17.3.1); they are onshore basins having some offshore production in them. They are not parts of the *present* continental margins. Truly offshore basins, within the present continental margins and facing open ocean, will be neither foredeeps nor interdeeps except where a Mesozoic or Cenozoic orogenic belt faces the ocean (as in the Beaufort Sea and the Gulf of Campeche). Instead they are semi-cratonic (coastal) downwarps, and rift basins (including simple half-grabens), both parallel and transverse to the margins; they are Klemme's types 3, 4, 5, and 7.

Delta basins (Klemme's type 8) of course also extend offshore and are responsible for a good deal of offshore production; but a delta is really a sedimentary depocenter on top of some other kind of basin, commonly of Klemme's type 4 or 5.

Types of trapping mechanism The foregoing comparisons between productive offshore and onshore basins are important geologically and geophysically. Economically and operationally, however, they are of minor significance when compared with the contrasts in trap types on- and offshore.

In enclosed, compressional basins, traps offshore are just like traps onshore. The great majority are formed by the various versions of the anticline described in Chapter 16. The true, marginal, continental shelves, however, virtually by definition have never been compressionally deformed. The pre-Jurassic shelves were deformed; they are now mountain belts. The post-Triassic shelves would no longer be shelves if they had been compressionally deformed. Thus basins within the true shelves are unlikely to contain genuinely compressional traps. To be commercially exploitable in water deeper than, say, 250 m, however, a field must have a *convex* trap. It needs the thick pay zone made possible by such a trap if it is to be exploited from a small number of expensive fixed platforms. This critical requirement is discussed further in Chapter 26.

17.9.4 Offshore basins and hydrocarbon incidence

The scaling down of expectations from the continental shelves has been an instructive process. Pre-plate-tectonic conventional maps showed shelves to the 600 foot (180 m) subsea contour to occupy an area of about 30×10^6 km². In 1965, Lewis Weeks reduced this figure to 15.8×10^6 km² by concentrating on those shelf sectors that could reasonably be expected to be "basinal." This figure compared with an area of about 46×10^6 km² of "basinal" area on land, for a ratio of about 1 : 3.

V. E. Khain amended this ratio to more than 1 : 2. He assigned an area of 148×10^6 km² to prospective areas on land and not less than 80×10^6 km² to the "submerged extension of known continental and marine petroliferous basins." Sixty percent of this "submerged extension" was assigned to continental *slopes* between 200 and 2500 m depths. Khain used the International Tectonic Map of the World to compute the volume of the sedimentary layer in the oceans by triple integration, assuming minimal thicknesses; the volume on the slopes and shelves was computed to be 163×10^6 km³.

Lytton Ivanhoe attempted much more discriminating calculations of prospective offshore territory. The total prospective offshore area having water depths no more than 200 m was calculated to be a little over 20×10^6 km², compared with nearly 35×10^6 km² onshore, for a ratio of 1 : 1.66.

Weeks supposed that 4 percent of the most prospective offshore acreage might yield 35 000 barrels per acre (about 13.75×10^5 m³ km^{-2}), and 2.5 percent of the second-class acreage 20 000 barrels per acre (about 7.85×10^5 m³ km^{-2}). The total potentially recoverable, conventional, continental shelf reserves of oil would then be about 110×10^9 m³. In 1971, Weeks inflated this estimate to nearly 180×10^9 m³ of oil, plus 54×10^{12} m³ of gas. For the shelves plus the slopes down to water depths of 1000 m, Weeks calculated the "potentially recoverable resources" at 365×10^9 m³ of oil plus the equivalent in gas. These resources were optimistically forecast to be exploitable "by 1980 or during the 1980s," despite requiring the development of techniques for drilling in deep waters that at the time of Weeks' writing were not beyond the drawing boards.

Using far more liberal assumptions, Khain predicted the minimum reserves (of oil plus equivalent in gas) to be not less than 2.6×10^{12} tons (about 3×10^{12} m^3), adding that they might be as high as 4×10^{12} tons (more than 4.5×10^{12} m^3). The smaller figure is about 12 times the actual reserves known in 1980. Khain acknowledged that the optimal zone of hydrocarbon accumulation might in actuality extend only to the foot of the continental slope; under deep waters, the resources are wholly conjectural.

Ivanhoe conducted altogether more realistic calculations. Of his 20.8×10^6 km^2 of prospective shelf area, less than 4 percent was considered to have "excellent" or "good" prospects, 95 percent to have poor or unknown prospects. If the area having favorable prospects were as productive as Weeks had proposed, the expectable continental shelf reserves would be about 42.5×10^9 m^3, not 110×10^9.

Estimates made in the mid-1970s by the National Petroleum Council of the USA, and by William Krueger and several others, were that between 55 and 70 percent of all recoverable, offshore petroleum resources would be pooled under the continental shelves in water depths of less than 200 m; 90–98 percent would be pooled beneath the combined shelves and slopes, leaving no more than 5 percent below 2500 m on the continental rises and in the deep ocean basins.

If therefore we add 50 percent to Ivanhoe's resource estimate to represent the prospective contribution from the continental slopes and rises, we reach a figure in the 60×10^9–65×10^9 m^3 range for confidently expectable offshore oil resources. By the middle of 1983, discovered offshore oil reserves totalled more than 30×10^9 m^3, of which almost one-third had been produced. Discovered gas reserves were about 16×10^{12} m^3. On the data of Ivanhoe and Krueger, about 30×10^9–35×10^9 m^3 of recoverable oil then remained to be discovered offshore. On the land : sea area ratios, confidently expectable, total offshore reserves would be between 35×10^9 and 70×10^9 m^3; about 30×10^9 have already been discovered and 10×10^9 already produced.

Offshore fields maintain between 20 and 25 percent of both the world's oil production and the world's known oil reserves; they also maintain about 15 percent of the world's gas reserves, though much less than 15 percent of the gas production. To the old standbys of the Persian Gulf, Lake Maracaibo, and the US outer continental shelf have been added the UK and Norwegian sectors of the North Sea, southeastern Mexico, and the shelves off southeastern Asia and Australia. It should be looked upon as remarkable, and disturbing, that the offshore contributions to the totals did not grow more significantly during the 1970s. Many shelf areas, and possibly many parts of the continental slope, remain to be added to the production tables, but the sanguine forecasts of shelf dominance over dry land are unlikely to be revived.

18 Factors favoring hydrocarbon abundance

18.1 The conventional wisdom

Geologists engaged in petroleum exploration need to be of optimistic outlook. They also need to acquire what amounts to a fundamental faith in certain tangible, observable criteria that are able to incite and maintain their optimism. They believe that certain simple geological factors within a region lead naturally to that region being richly petroliferous.

Many authors have tried to assess these factors. A relatively small number of much-studied basins have provided the "type material" for most of them: the Gulf Coast and Californian basins for the Americans, the Persian Gulf and Trinidad for the Europeans, Baku for the Russians. These specimen basins offer both clastic and carbonate successions; but they are all relatively young stratigraphically and all of them provide an abundance of structurally controlled traps. The favorable factors consistently invoked from their study include the following:

(a) A very large volume of marine sediments, including some deposited under stagnant, reducing conditions. These permit both an ample accumulation of organic material and maximum opportunity for its protection from chemical or biological destruction.

(b) Almost uninterrupted subsidence during deposition, and absence of abrupt or large-scale uplift or erosion until a complete depositional cycle of transgression and regression is effectively completed. These factors sustain reservoir fluids and their pressures. They also result in sufficient load compression, at least in the center of the basin, to squeeze the entrapped fluids from the fine-grained source sediments.

(c) Widespread interfingering of fine-grained mud rocks with coarser-grained carrier and reservoir sands, or with shallow water carbonates. The potential reservoir rock should be laterally persistent, and of high permeability.

(d) Large traps should be created near the margins of the basin before or during oil migration. Growth of these traps throughout the migration phase is an added advantage.

(e) The traps must involve efficient seals or roof rocks, and these must not be damaged by excessive faulting or destroyed by erosion.

(f) The prospects for prolific production are best if the area is geologically relatively young, without a long-standing load of superjacent, much younger rock.

(g) Deformation should not be more intense than results from a single, moderate orogenic episode, resulting in little post-depositional alteration of the sediments.

(h) The first-formed light oil should not be converted to either heavy tar oil or gas by any subsequent geologic events. Above all, these subsequent events should not be such as to permit the oil's loss by erosion, evaporation, or chemical destruction.

The reader will recognize that these factors in effect include all those that normally contribute to the development of any large, long-lived sedimentary basin, with the pointed exclusion of the conversion of the basin into a mountain belt during subsequent compression. The factors may be accorded weighted values, and combined into "indices of productivity" for basins of different types. Where all the factors are well developed, favorability is at its maximum. On the other hand, the absence of a single factor may mean the absence of oil; the absence of a good source rock *will* mean the absence of prolific oil.

The reader will also recognize that several essential factors may be intimately interrelated, and so occur in conjunction or not at all. High porosity, for example, is frequently a consequence of current winnowing of the sediments during the early growth of structure. The process of load compression leading to the expulsion of fluids from the source sediments is an aspect of the process that forms drape traps.

The critical factor, without any doubt, is the maturation of a rich source sediment. Without that, the provision of compressive load or of carrier beds, and the growth of traps or the deposition of seals, is in vain. Yet the maturation must not be excessive, and the absence of deformation must not be total. For some years, therefore, exploration geologists tended to "translate" the factors governing oil incidence into a few visible or measurable geologic phenomena:

(a) Generally young strata, showing no great induration and no sign of metamorphism.
(b) Thick strata deposited within a restricted span of geologic time, representing rapid and uninterrupted sedimentation.
(c) The greatest favorability lay in the transgressive legs of marine depositional cycles, in which organic productivity and preservation are both at their highest.
(d) The most favorable segment of a basin is the hinge belt between the basin and its foreland shelf. Here the interfingering of source and reservoir facies is at its maximum.
(e) Because the commonest traps are "anticlines," regions of moderate folding are to be preferred over those of severe folding or of no folding at all.
(f) High geothermal gradients enhance the maturation and migration of hydrocarbons.

We shall consider these factors in more detail.

18.1.1 Favorability of young strata

Most of the prolific early oil provinces were in Tertiary strata because Tertiary basins are easily detected, their contents are not buried and they are abundantly supplied with seepages. Of the top ten petroleum provinces (Fig. 17.15), however, two are wholly Paleozoic (the Ural–Volga and Permian basins) and a third (western Canada) is largely so. Only the rich deltas of the Mississippi and Niger Rivers are wholly Tertiary. As gauged by weight percent of TOC and hydrocarbon content, some of the richest source sediments known are Paleozoic (see Chs 6 & 8).

An oil's "age" is an ambiguous concept. The age of the reservoir rock is accurately known and is usually the one intended in a reference to (for example) "Cretaceous oil"; but the reservoir rock is merely the oil's container. The age of the source rock *may* be accurately known, but there is not always agreement on the source rock. The age of formation of the trap may also be established with considerable accuracy, but the hydrocarbons may not have got into it at that time. It is unlikely that many *present* accumulations of oil of large size formed before the late Mesozoic, regardless of the age of the source rock or of the reservoir rock (see Sec. 15.5). The greater part of the world's known conventional gas must be of relatively recent accumulation, even if both source and reservoir are Paleozoic.

It is a safe generalization that TOC content in sedimentary rocks is much more dependent upon lithology than upon age (Secs 6.6 & 8.1).

18.1.2 Rapid sedimentation

The early-found Tertiary basins of Baku, Sumatra, Trinidad, and Ventura certainly gave support to the idea that thick and rapid sedimentation favors the generation and occurrence of oil. Many geologists judged the potential of a basin by either the average or the maximum thickness of the sedimentary pile in it; the Weeks method of computing the potential productivity of a basin is based upon it (Sec. 24.3.2).

It is not the sediment that generates the oil, however. Generation is a function of organic carbon (OC) content and OC type. Many thick and rapidly deposited shale sequences are nearly or quite barren of oil (the coastal Colombian Basin, the Amazon Basin, the Tindouf Basin of Algeria, the Sverdrup Basin of Arctic Canada, and many others contain notorious examples). Sedimentation rates are much higher along active subduction margins than along passive, trailing margins, but the passive margins enjoy a much higher incidence of oil. Many cases of rapid sedimentation involve excessively coarse sediment and large quantities of terrestrial OM, which itself may have already been degraded during a prefatory sedimentation episode before the final rapid accumulation (in recycled or redeposited deltaic, turbidite, or slump deposits).

The role of thick and rapid sedimentation in *young* basins is to achieve early maturation of the hydrocarbons (the "maximum phase of oil formation" as the Russians put it), and hence early pooling of the oil. As the rate of sedimentation increases, the TOC content of the sediment also increases because the sediment passes more rapidly through the near-surface zone in which intense microbial degradation of the OM would take place (Fig. 18.1). Moderate or average rates of sedimentation may be taken to be 10–50 m Ma^{-1}; rates below this are low for subsiding basins; rates higher than 100 m Ma^{-1} are high. In many petroliferous basins of late Mesozoic or Cenozoic age (in California and Sumatra, for example), rates of 200–400 m Ma^{-1} were common for geologically brief intervals, but these are much higher than any Gulf Coast rate today.

High sedimentation rates (above about 50 m Ma^{-1}) may result in dilution of the organic input by clastic material, unless primary organic productivity is very high. Where both sedimentation rate and productivity are high (where, for example, upwelling occurs in low latitudes off an active, high-relief coast like that of Peru or Morocco), the preservation of OC is at its maximum

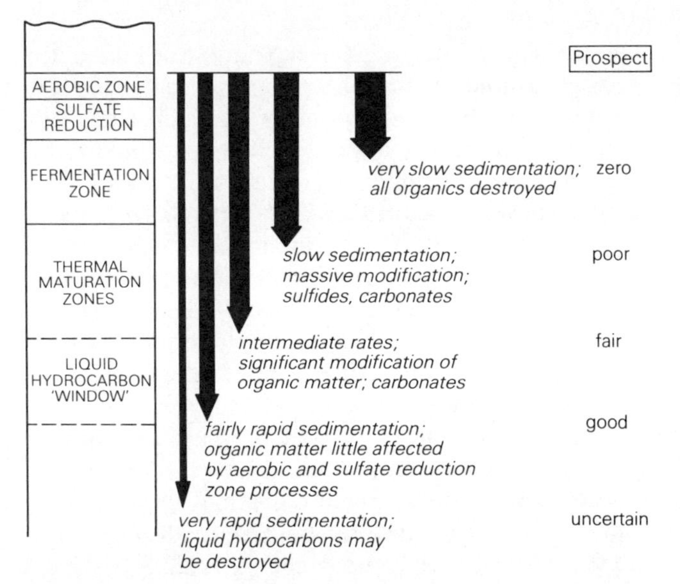

Figure 18.1 Comparison of effects of sedimentation rates on the preservation and maturation of organic matter. (From M. L. Coleman *et al.*, *World Oil*, March 1979.)

(over 4 percent or even over 10 percent by weight), and its near-surface degradation is minimized. This has been established for Mesozoic, Cenozoic, and Recent sediments.

It is a totally different matter with Paleozoic strata. Hydrocarbon maturation indices, like the TTI of Lopatin (Sec. 7.4.2), involve both time and temperature factors. Because the necessary cooking temperature is reached at a certain depth (say 2000 m) in any basin which continues to receive sediment, there is no need for that depth to be significantly exceeded if the time factor is large. The TTI is essentially independent of depth in basins older than about 150–200 Ma.

Paleozoic production may therefore be prolific in basins in which the source and reservoir sediments have never been buried more deeply than about 3 km. The Ural–Volga and Western Canadian basins are obvious examples. Paleozoic basins still containing enormous thicknesses of unmetamorphosed sedimentary rocks are normally foredeeps filled with molasse, like the Anadarko and Arkoma basins in front of the Oklahoma fold belt (Fig. 17.1). Such basins are inevitably gas-prone. Permo-Triassic strata, between 270 and 180 Ma old, may contain thermogenic gas at virtually any depth and without ever having exceeded routine sedimentation rates or thicknesses (Sec. 25.3).

18.1.3 Transgressive versus regressive cycle

The widest spreads of rich *source sediments* coincide with marine transgressions following major orogenies: in the Frasnian (late Devonian), the late Pennsylvanian, the late Jurassic, the mid- to late Cretaceous, and the Oligo-Miocene. During transgressions, OM is better preserved; during regressions, deposition overtakes subsidence and OM is liable to exposure unless it is swept out beyond the regressive facies. The transgressive facies commonly provides suitable cover rocks (shales) over suitable reservoir rocks (sands); the regressive facies tends to coarsen upwards and to lack seals for fluids.

Experience in many young, clastic successions (especially in US and Caribbean basins) gave support to the preference for the transgressive over the regressive facies. The reasoning from this experience may be faulty. *Source* sediments are best developed during the initial transgressions of basins newly created in intracratonic settings, consequent upon the deformation of plate-marginal basins elsewhere. To become *pooled* in the transgressive sequence, oil may have to find contemporaneous or older reservoirs, for which the source sediment is also the seal. Traps in such reservoirs are likely to be either residual (such as buried topography) or depositional (basal sands of the transgression, as in many North African fields, or reefs, as in the Devonian of western Canada and the Pennsylvanian of West Texas). There may be no reservoir rocks younger than the source rock until the regressive phase begins.

Once the regressive phase does begin, the basinal sediments become loaded by permeable, shallow-water sediments (including molasse). The mudrocks of the source sequence compact from the top downward, not from the bottom upward as they do in the transgressive phase. This compaction readily leads to drape folding, diapirism, and the formation of growth faults, all aspects of contemporaneous structural growth internal to the basin. Most hydrocarbon migration and accumulation takes place during this phase, and most hydrocarbon occurrences in the regressive facies are in structural traps. In clastic sequences, at least, hydrocarbon occurrences in the regressive facies are more likely than those in the transgressive facies to be of degraded oil.

The favorability of the regressive facies *for the reservoir rock* is strikingly seen in carbonate sequences. There, the source rock is more likely to lie stratigraphically below the principal reservoir rock than above it; the oil migrates upward to the reservoir, possibly via fractures. The best reservoirs may be provided by the introduction of regressive sands into the otherwise sand-free carbonate environment; the Cretaceous sands of Kuwait and Iraq are superb examples (Sec. 25.4). The best seals are provided by the evaporites of the last stages of the regression.

Prolonged retreat, of course, may be destructive. Exposure reduces hydrostatic pressure to zero at the liquid table. Volatiles are lost by inspissation, rendering the remaining oil viscous and difficult to produce. Asphalt seals the pools, as in some Maracaibo and San Joaquin Valley fields (Figs 9.9 & 16.38). The reservoirs may be destroyed by erosion. Brief retreat followed by renewed transgression is the ideal order of events. The worldwide sub-Cretaceous and mid-Cretaceous unconformities, the most petroliferous of all settings, exemplify the short-duration regressive–transgressive turnaround (see Sec.16.5.6).

18.1.4 *Most favored sectors of a basin*

The several distinct structural types of basin have different shapes both in plan and in profile. Now that we have some understanding of the deposition of the source sediment, the derivation of hydrocarbons from it, their migration in search of traps, and the genesis of trap geometries, we should be able to deduce some principles to govern the selection of the most favored sectors of the basins.

First, the source sediments accumulate in the deeper "cells" or "sinks" of the depositional basin. These sinks may not be deep in any absolute sense: hundreds of meters of water depth are more likely than thousands. Being features of basin centers, source sediments are therefore the least susceptible to tectonism of the basin's entire sedimentary accumulation.

The migration of hydrocarbons from the mature source sediment is therefore likely to be radially *outward* and *upward*. Secondary migration is a focused process, taking place as closely as possible perpendicular to the basin contours. Fluid migration therefore tends to become concentrated along routes that are upwardly concave, and to lead to upwardly convex gathering areas between concave depocenters. Thus *changes of basinal slope*, from concave below to convex above, are highly favorable. Well developed changes of slope of this sort are *hinge belts* or *hinge zones* (see Chs 16 & 17). The most ubiquitous case is the change of slope which permits the demarcation of the "basin" from the "shelf." Along such inflection zones, porosity may become concentrated (in reefs, or shelf-edge sand bodies); facies changes are characteristic, including the change from source-type to reservoir-type facies; and fault structures and flexure structures are easily created.

If folding takes place, it is likely to do so entirely on one side of the hinge belt; if it engulfs the hinge belt, the fold style is likely to change across it. The most mobile zones are those undergoing the greatest compression and the highest heat flow; maturity is reached earliest there, but the hydrocarbons migrate away from such zones and accumulate preferentially in relatively inactive zones adjacent to them. Hydrocarbon incidence is much lower over distant forelands and in shallow overlaps of stable shields. In Paleozoic basins, however, much of the original basin is very likely to have become deformed out of existence, and the oil and gas that remain have indeed migrated far on to their forelands.

18.1.5 *Most favored traps*

During the early stages in the exploration of a basin, the geologists should assume that any *oil*, at least, is likely to be pooled in its sites of primary (original) accumulation; gas is another matter. The first opportunities for accumulation presented to the migrating oil were in or around convex structures *available at the time of primary migration*. Any structure forming a "high" during deposition is to be preferred over even a large structure created much later. Sand bars, reefs, buried rock ridges, horsts, rising salt domes, all localize effective porosity. They also offer opportunities for unconformable overlaps, marginal faults, and depositional pinchouts.

Structure available at the time of migration may be quite different from the present structure. It may not be revealed by ordinary structure contour maps, and requires elucidation and reconstruction through controlled isopach and facies maps (Secs 22.3 & 22.4). Nevertheless, early structures are unlikely to have been totally obliterated, and all evidences of discontinuity or change of dip in lower stratigraphic levels should be scrutinized on the assumption that they reflect these favourable features.

18.1.6 *Most-favored temperature conditions*

The dependence of generation and accumulation upon temperature regimes was described in Sections 7.4 and 17.8. The regimes are not safely forecastable in advance of drilling. All that need be added here is the caution that the time element in the Arrhenius equation must not be over-looked in concern over the temperature element. Paleozoic intracratonic basins do not need high geothermal gradients.

18.2 Combination of all favorable factors

Anyone can conclude that assurance of success lies in finding all the favorable factors together. We noted (Sec. 18.1) that some of the factors do indeed tend to occur either in association or not at all.

The discussion is circular, of course. We have discovered what the favorable factors are through experience and deduction, not through any kind of abstract reasoning. We can offer plausible reasons for back-arc basins being more favorable than fore-arc basins because we know by experience that they *are* more favorable. Reasoning from the petroleum geologists' experience can be extended to a global, panoramic synthesis.

Episodes of maximum *oil* incidence coincide with geologically brief cycles of transgression followed by regression, brought on by orogenic events in turn reflecting the partial closures of equatorial oceans and the opening of longitudinal oceans. The episodes are then terminated by further orogenic events which lead to wide-spread uplift of continental plates and evaporite maxima.

These episodes can be identified by tracing the record of global sea-level stands through Phanerozoic time. The standard depiction of this record is by Peter Vail and his colleagues (Fig. 8.2) and its significance for our concern is discussed in Chapter 8. A simplistic recipe for successful exploration can be seriously deduced: the most favorable places to look for oil are those having well developed marine successions of late Devonian, late Jurassic, Cretaceous, or Oligo-Miocene ages.

Natural gas is of much more widespread occurrence than oil, and much easier to find (see Sec. 10.4.5); but the parallel test of experience points to most of it being in intracratonic successions of Pennsylvanian through Triassic age or in Cretaceous coal-bearing molasse.

19 The petroleum geologist in action

19.1 The functions of the petroleum geologist

We have considered everything that petroleum geologists try to find, to determine, or to achieve. How do they actually do it?

Petroleum geologists include many extraordinarily versatile individuals who can do almost everything the oil finder and oil producer need to do. With the increasing specialization of all scientific endeavors, such individuals are increasingly rare. The processes of finding and producing oil and gas, extending from the evaluation of totally undrilled acreage to the correlation of oils among many mature producing fields, are too numerous and complex for any one person, no matter how versatile. A traditional separation of the roles and duties of petroleum geologists among three basic functions provides us with a convenient breakdown for our present review.

The geologists whose primary responsibility is to discover oil and gas accumulations are *exploration geologists*. Their basic fields of expertise are stratigraphy, structural geology, and sedimentary petrology; their immediate associates are *exploration geophysicists*. Geologists whose primary responsibilities are drilling for oil and gas and evaluating, producing, and protecting the reservoirs found by the drilling are *development* or *production geologists*. Those development geologists who spend most of their time on or around wellsites are *wellsite geologists*; those who monitor entire fields or pools are *reservoir geologists*. The immediate associates of all of them are *petroleum engineers*; the job of all of them is to apply geological principles and techniques to exploit the reservoirs and their contents to maximum advantage. Finally there are *research geologists* whose task is to enhance understanding and appreciation of the geologic factors that underlie the generation, migration, and accumulation of oil and gas. Their associates are scientists of nearly all other disciplines.

Some petroleum geologists of more specialized function operate within all three realms, of exploration, development, and research. The *organic geochemists* and the *paleontologist-palynologists* are those whose work is most familiar and far-reaching. Finally, an indispensable specialist especially associated with the development geologists but overlapping all functions is the *formation evaluation geologist*, who studies the characteristics of well bores and producing reservoirs primarily from suites of logs.

When Max Ball, Arville Levorsen, and other pioneers wrote their familiar textbooks, they were able to describe a petroleum geologist's training, his characteristics, and his duties; and they were able to do the same for the geophysicist, the geochemist, the engineer, and the driller. Most drillers are still males, except possibly in China, but by no means all the others are. This author happens to be a male, but in this chapter on the petroleum geologist his unavoidable use of the masculine pronouns should be understood to encompass the feminine equivalents as well.

19.2 The exploration geologist in new territory

The exploration geologist's job is to find oil and gas deposits that can be exploited commercially. "Geologists" have been doing this for more than 100 years, but the professional exploration geologist is a creature of the 20th century, and his role now is very different from his role even at the end of World War II. Until that time, exploration geology was essentially a matter of surface geological mapping and surveying, looking primarily for structures. The geologist undertaking it was a member of a small field party just like the geologist engaged in a government survey or a search for mineral prospects.

The days of unassisted surface field work are largely gone for the petroleum geologist in most of the accessible part of the Earth. The search is now concentrated in areas already well mapped but not previously considered prospective for oil or gas; in remote and poorly accessible regions with little or no outcrop; on poorly understood strata deeply buried below unconformities and cropping out only a long way off if at all; and in offshore tracts with no rock to be seen. Exploration in such areas is very expensive and carries high risk of failure. To use money and manpower to the best advantage and to reduce the risks, a team approach is vastly to be preferred over the solitary search. This is in no way to decry the critical role of the imaginative individual who causes the team to be put to work *in the right place*; and that individual is more likely to be a geologist than any other member of the team. He is also the one who should best be able to coordinate the work of the team. Who are the members of the team, and why are their skills essential?

First, the *exploration geologist* himself; his understanding of regional, historical, and structural geology, of sedimentary facies and stratigraphic concepts, of reservoir rocks, roof rocks, and source rocks, must fit him for the job of identifying the places and the conditions in which oil and gas are most likely to be found. Second, the *organic geochemist*; he must discover from the geologist's rock samples whether the area being investigated contains source sediments of sufficient quality to have generated hydrocarbons, how any hydrocarbons found are to be correlated, and what fluid movements must have taken place if they are to be commercially pooled. Third, the *paleontologist-palynologist*, who contributes the "bio" portion of the biostratigraphic understanding of the area; the interpretation of a depositional environment suitable for hydrocarbon occurrence, the ages of the rocks deposited in it, and time correlations across the facies recognized by the geologist, are in his purview. The bio-stratigrapher is of course himself a geologist, and the exploration geologist whom we have put at the head of the team may be his own biostratigrapher and paleontologist. Fourth, a vital individual whom oilmen call the *landman* (even if she is female), who is really a combination of salesman, diplomat, and psychologist, and who may have to be a lawyer in addition; the explorers can achieve nothing if they cannot get the subsurface rights to the land they want and then maintain them against competitors, governments, farmers, and real estate developers.

Last in our roll call of the team, but equal first in its exploratory operations, is the *exploration geophysicist*. His knowledge of rock properties and stratal geometries enables him to determine in detail the three-dimensional shape of the stratified rock content of the basin, and his command of data processing is needed to coordinate the great mass of data the team builds up.

What can be done depends on how complete is the area of total expertise available in the exploration team. The geologists have always known that the absence of a single geologic factor may doom a basin otherwise seemingly highly attractive. If all the potential reservoir rocks are tightly cemented or flushed by fresh waters, or all the structures arose too early or too late, the other features he finds attractive are to no avail. Non-geologic factors may be equally decisive. The play will come to little or nothing if the organic content of the shaly rocks is of the wrong type, or if deep structures are beyond the seismograph's detection capability, or if the team cannot get title to the acreage.

Geology itself is an integrative science, combining the laws and data of other sciences with direct observation (Sec. 1.2). Petroleum exploration must go further, and integrate geology with its sister geosciences. The petroleum exploration geologist must gather all the data that can be gathered; he must then extrapolate these data into a realm in which data are inadequate or nil and try to predict *correctly* what the missing data will be. What is more, in many countries and many circumstances, he must do this ahead of his competitors; time and money constraints are unavoidable aspects of his work.

Even in a totally virgin area – as the northern North Sea, western Siberia, the Sudan, and scores of other regions were until the oil geologists got there – the exploration geologist has one indispensable weapon if he can use it correctly: the process of analogy with known productive regions elsewhere. He is engaged in a form of detective work. The deduction of the entire dynamic history of a prospective region from what can be learned of its *present* geology has been picturesquely, but inappropriately, likened to the reconstruction of a moving film from a "still" of its final scene. The simile would be more appropriate if the reconstructor also had the benefit of the fragments from the cutting-room floor. The dynamic history of any region of immediate concern to the petroleum geologist began with the creation of conditions enabling a source sediment to be deposited; its progression to the present time has left many pieces of direct, unarguable evidence, and still more of indirect evidence, which it is the geologist's task to discover, interpret and recombine. The task begins with activities common to all geological work: the investigative or data-gathering phase. This must then be followed by the establishment of multiple working hypotheses for the reconstruction itself. We will follow the principal steps from beginning to successful conclusion.

19.2.1 Sequence of exploratory steps in the exploration of a new region

Investigative phase

Step 1 The primary objective in this first phase is the *identification of a sedimentary basin*. Travellers' records, geologic literature, maps (including simple topographic maps), and air photographs are the first materials searched for leads. LANDSAT images are helpful for the identification of basins on land; the presence of layered rocks disappearing beneath a low-lying area is a sufficient early lead. Aerial reconnaissance by the geologist should follow. The basin may lie beyond present accessibility; it may no longer contain useful targets; but the geologist seeks to identify sites most likely to have contained "basinal" sediments (see Ch. 8).

Step 2 The second step is an initial interpretation of the *type of basin* identified. Does it lie in front of a folded mountain belt? Does it consist of strata dipping off the margin of a landmass and disappearing under the sea? Does it have the aspect of a graben or fault trough? Do its contents appear to be themselves deformed? Different basin types have different orders of expectation of hydrocarbon incidence (see Sec. 17.6).

Step 3 The most important single factor required for the basin to have become petroliferous is its *stratigraphy*. Descriptions by early geologists or observant travellers may be much more helpful than young geologists realize. Stratigraphy and paleontology were among the earliest geologic topics to receive widespread and intensive study in the field, and oil geologists exploring during the second half of the 20th century in the Canadian and US Rocky Mountain belts, in northern India and Pakistan, in Sicily, South America, Australia, and elsewhere, found that stratigraphic data recorded 50 or more years earlier provided excellent foundations for their own work.

The geologist tries to gauge the possible thickness and areal extent of the sedimentary section as well as its lithologic and faunal characteristics. Great thickness of sedimentary rocks implies long-continued sedimentation and improves the chances of multiple source–reservoir relations. Large areal extent increases the chances of multiple favorable trends. Differences in the observable section on opposite sides of the basin are important; asymmetrical foredeep basins are in general much more productive than symmetrical cratonic basins. Variety in the stratigraphic section is normally more favorable than monotony, and the map or air-photograph pattern of the basin may suggest variety by revealing alternations of feature-making units (which may provide reservoir rocks) and non-feature-making units, which may include source sediments or roof rocks. Strata mappable at the surface may give direct evidence of a depositional environment conducive to hydrocarbon generation, and strata from this environment may extend beneath younger rocks into the basin. The presence of marine strata, especially of marine shales lacking large, shelly fossils, or of interfingering fine-grained and coarse-grained sedimentary rocks, is a valuable indicator.

During this phase of the investigation, the geologists must always remember that no region should be written off, as unprospective, on the basis of its *visible surface stratigraphy* alone. A monotonous section of nonmarine strata may overlie thick marine beds which crop out only far away from the prospective area or not at all.

Step 4 Surface evidences of hydrocarbons, if they exist, should be recognized early in the investigative phase. Seepages, mud volcanoes, burnt clays, sulfur or gypsite in soils are among these evidences. Those at all extensive may be betrayed by tonal anomalies visible on colored aerial photographs. Other indications of petroliferous strata that are detectable from the surface, but at a later stage of the exploration, are described in Chapter 20.

Step 5 Any *paleontologic data* already available should be taken into account in this initial stage of exploration. Paleontology was a widely practised science long before igneous or sedimentary petrology, and records of fossils may be available in literature even if little or no other geologic information was recorded. Both ages and types of fossils are valuable, as are their abundance and variety. In thick shaly strata, the absence of visible, collectable fossils may give as much information as their abundance.

Step 6 The *structural setting*, though important, is secondary to the stratigraphic environment in this phase. Some aspects of it should be given early emphasis, however. What is the areal relation between the basin and the principal outcrop belt, especially if the latter is a belt of structural deformation? In coastal areas, what is the basin's relation to the present continental shelf? Are any lineaments visible which might indicate favorable trends or exploration targets? What is the apparent age-relation between detectable structures and the strata envisaged as the exploration targets? Concordance between structure and topography, with little-eroded anticlinal ridges and synclinal valleys, indicates very youthful structures, which would be an unfavorable sign if the stratigraphic targets were Paleozoic in age. The same evidence may be favorable, however, if it is found in an area in which modern seismicity or gravity data testify that structural growth is still continuing. It may then have begun early enough to trap oil or gas generated in earlier Tertiary or even late Mesozoic strata, as in parts of the circum-Pacific belt.

Step 7 With this preliminary geological assessment in his possession, the geologist must turn to logistic and quantitative considerations, some of which will more properly be the responsibility of others (tax, royalty, and labor conditions, supply channels, assurance of land tenure, and many more). Even if these considerations are judged to be decisively negative, however, this should not deter a geologic appraisal; it will merely reduce its urgency. The negative logistic assessment may change. Through his knowledge of geologically comparable basins elsewhere, the geologist must assess the probability of large fields being present; in many remote regions, and all those underlying deep water, development is feasible only if individually large fields are discovered early. From his study of the terrain, the geologist must also come to some conclusions about its accessibility. With the geophysicist, he estimates the minimum costs of pre-drill evaluation of the acreage. The two of them must recommend its acquisition in what they judge to be the most favorable areas.

Reconnaissance phase

Step 1 The primary objective in this second phase is the localization of favorable stratigraphic trends, called *plays*. The process begins with a more detailed *photogeologic reconnaissance* for regional control; this can in many circumstances be conducted rapidly and at low cost.

Step 2 A more thorough understanding of the *regional stratigraphy* and of the configuration of the strata within it comes next. This requires less control than structural mapping, and the geologic foundation of it can be quickly augmented by aerial gravity and magnetometer surveys (for the configuration of the basin and estimates of the depth to basement in its various parts) and by *reconnaissance seismic profiles* (to identify the positions of reflecting horizons, possible unconformities, and general high and low areas). This phase should lead to a decision on what will initially be regarded as basement for the drillers, its approximate depth, and sites at which it appears to stand higher or lower than the average basin floor.

Step 3 The geologist should try next to get some impression of the *structural style*. Are the detectable structures compressional or extensional, high or low in amplitude; is there a governing direction of vergence? If outcrop is adequate, an attempt should be made to identify specific *trends*, such as that of a depositional hinge belt between shelf and slope, or lines of onlap or offlap, or of any anomalous dips or topographic features. If, for example, there are alignments of streams or of bends in their courses, or of subtle elevations or depressions, the reasons for these should be considered.

Step 4 Check the initial stratigraphic and structural interpretations, including those derived from reconnaissance seismic data by drilling *stratigraphic tests* at low-cost locations. The purpose of these is to verify the stratigraphy and identify potential source and reservoir horizons. Conformity with structure is quite secondary; if the structural pattern is vague or indeterminate, the drilling of stratigraphic tests should not be hindered by the expense of searching for favorable structures. Data from stratigraphic tests must be integrated with any old well records or surface data available. Temperature measurements in all stratigraphic tests give an initial estimate of the geothermal gradient in the basin.

Step 5 From the test-well samples, cores, and logs, carefully study all horizons offering source or reservoir potential or materials capable of accurate dating. The *organic geochemist* should become intimately involved in this phase. He determines the organic carbon content and maturation characteristics of potential source beds. Combination of these data with those on the basin's shape, as determined by geophysics, and on its geothermal gradient, enables the geologists to evaluate the most probable directions of fluid migration and hence the most favorable parts of the basin for hydrocarbon accumulation.

Economic phase The single objective in this phase is the identification of drillable structures, called *prospects*. Structure now becomes of prior concern over stratigraphy. No exploration of a new region can *begin* with a search for subtle, nonconvex traps; early wildcat wells *must* test structures even if the nature of these is ambiguous. This is especially critical for offshore prospects.

Seismic study is now localized and intensified over areas revealed during the reconnaissance phase to be in any way "anomalous:" apparent highs, linear trends, unconformable truncations, abrupt discontinuities in gravity or magnetic character. Dips, unconformities, and linear trends are carefully projected to detect culminations or intersections.

Initial exploratory drillsites are thus selected; target horizons are chosen, and drilling depths planned accordingly. The duration and costs of drilling are estimated (see Ch. 23). The gathering of all data the drillhole can yield is arranged: logging, testing, temperature and pressure measurements, velocity survey for calibration between the hole and the seismic data.

Still better calibration of geology and geophysics is achieved if further detailed seismic profiles are run after completion of the hole – especially if it is a discovery. The planning of step-out (appraisal) wells is thereby controlled as carefully as possible; early estimates of reserves are more reliable; and more land may be acquired for protection in case a new trend has been identified. If the structure around the discovery well is at all complex, and most particularly if the discovery is offshore, the detailed seismic follow-up should be three-dimensional (see Ch. 21).

Consolidation phase At this stage of exploration, the exploration geologists should try to assemble the crucial information concerning the basin in which they are operating. This information should then be summarized, in terse, tabulated terms, preferably in a format suitable for poster or viewgraph display. The following items are suggested for inclusion once production has been established:

(1) Name and location of basin, and its type
(2) Nature of the basin's boundaries
(3) Ages of inception and termination of the basin (correlatable deformations)
(4) Intervening deformations (significant unconformities)
(5) Age and nature of economic basement
(6) Thickness(es) of sedimentary rocks: (a) in basin center; (b) along its margins
(7) Estimated volume of sedimentary rocks
(8) Maximum depth of deposition, and formation representing it
(9) Geothermal gradient(s)
(10) Most favored source rocks, and their maturation values
(11) Most favored reservoir rocks, and their depths
(12) Most favored trap types, and time of their creation
(13) Probable directions and times of migration
(14) Period(s) of maximum pooling
(15) Hydrodynamic characteristics
(16) Probable distribution of fields
(17) Additional observations; references.

The summary should be accompanied by a single *prospect map* and a single *correlation chart*, for display purposes. Large companies have standardized their evaluation formats for each type of prospective basin; a number of versions have been published. Lytton Ivanhoe has set forth the evaluation procedures he developed for Occidental Petroleum Corporation, and the Exxon format was described in 1983 by explorer-extraordinary Dave Kingston and his colleagues from Esso Exploration.

19.3 The exploration geologist in an established producing area

Once a region is shown to be commercially petroliferous, exploration and development proceed concurrently. As they proceed, the volume of data available to the exploration geologist increases very rapidly. His immediate tasks are three. He must try to establish the probable *limits* of the basin in which the oil and gas are likely to occur. He must determine the *controls* of their occurrence: the most probable source sediments, the most favored migration paths, the critical aspects of the structural style, the influence of particular facies changes or particular levels of truncation of the strata. And he must establish working procedures to enable him to receive and evaluate new data on his area as they become available.

In the pre-economic phases of the exploration, the data may have come entirely from the geologist's own organization, operating without active competition. Once a commercial discovery is made, this circumstance is unlikely to continue for long; the era of large, exclusive concessions is gone. As more wells are drilled and more seismic lines shot, possibly by more and more competing organizations, the geologist will acquire data of differing levels of reliability and value from different sources; all of them must be evaluated. Most of these data will be available to his competitors also. As was remarked in the preamble to this chapter, the rewards are greatest for the geologists who succeed in extrapolating their data into areas in which data are inadequate and predicting correctly where the unknown data will be favorable. All interpretations of the data must be constantly subject to critical scrutiny; no potentially favorable reinterpretation must be allowed to go untested.

The basic tools of the exploration geologist's trade in explored, maturing areas are well logs, maps, cross sections, and seismic records. *Maps* must be prepared showing structure contours on every correlatable stratigraphic horizon, and isopach contours for every stratigraphic interval (see Ch. 22). As *well logs* are received, all potentially valuable data from them (formation tops, thicknesses, unconformities, gaps) must be added to the maps, which are thereby constantly updated and revised. *Cross sections*, derived from the logs and the maps and combined into *fence diagrams*, should be refined so as to reveal accurately the configurations of formation contacts, unconformities, subcrops, and both depositional and erosional wedge-outs (Fig. 22.10). The successive attitudes of these features, from their time of origin to the present day, should be worked out, thus revealing changes of dip, growth of structures, loss of section by erosion, and other geologic events through time. This reconstruction will reveal the history of the migration paths of the hydrocarbons and of the trapping mechanisms in which they became stored. In this way, other potential loci of the trapping mechanisms, and perhaps new mechanisms, can be indicated.

Cross sections must be drawn showing both litho-stratigraphic and time-stratigraphic correlations (Fig. 22.24). The former are controlled essentially by well logs, augmented by petrographic studies. Time-stratigraphic correlations involve biostratigraphic, faunal, and floral studies; if the geologist lacks the skills necessary for these, he must work with the biostratigrapher, the paleontologist, and the palynologist. Without proper time correlation, neither facies changes nor diachronous (time-transgressive) rock units can be recognized with certainty; both may be crucial to the trapping mechanisms.

Where well data are plentiful, a most useful adjunct to the stratigraphic cross section is the *stacked log section*. Full well logs are cumbersome things, and they commonly include long intervals of little interest. The geologist should select intervals which combine potential commercial interest and distinctive log characteristics; extract one type of log from all the wells in the area which penetrated those intervals; digitize the logs on a uniform scale (by converting their curves to numerical values at fixed intervals of depth, or opposite kicks); and display the traces side by side (Fig. 12.22). The process is somewhat analogous to the stacking of seismic traces for enhancement of the reflection signals (Fig. 21.5). The actual choice of log types for digitization and display depends on what the geologist seeks to emphasize. Variations or pinchouts of porosity, for example, are best illustrated by logs designed to distinguish porous intervals (sonic, density, or neutron logs); hydrocarbon-bearing zones may be best illustrated by one of the resistivity logs; and so on (see Ch. 12).

The geologist must decide which logs he needs for his work, though of course he cannot ensure that competitor companies will run them on their wells. Individual areas have individual logging needs and problems – marginal or highly variable porosities, overpressured zones, variable dips, unusual horizons like uraniferous shales, thick cherts or evaporites, zones of chronic mud loss. These special needs are quickly learned on the spot, and particular suites of logs become standard for all exploratory wells in the area. Geologists should particularly insist on *dipmeter logs* in areas of substantial or changeable dip.

Geological structure contour maps should harmonize with *seismic structure maps*; geological isopach maps should harmonize with *seismic time-interval maps*. If they do not, the explanation for the conflict must be sought. If geological correlations are correct, the conflict means that seismic data are being misinterpreted. In the initial stages of exploration, even continuous and through-going reflectors may understandably be wrongly identified; this is unpardonable once the development stage is reached.

In cratonic regions, lacking pronounced structures, a *regional dip* is likely to be detectable once both drilling and seismic data are well distributed. Once the regional dip is established, structure contour maps should be paired with versions having the "regional" removed. Structural maps of the region thus returned to untilted position commonly reveal important new structural patterns which may have been responsible for trapping mechanisms still surviving. Many structures now closed lose their closures when the regional is removed; many unclosed structures acquire closure. The technique is particularly useful in reconstructing unconformities on which paleogeomorphic traps are developed (see Sec. 16.6).

If the exploration geologist's responsibilities extend over a large region in which drilling activity is high, he will soon find himself maintaining great numbers of maps and sections. For purposes of acreage evaluation and management decision making, this mass of working data must be integrated into a smaller number of *prospect maps*. These should illustrate the salient characteristics of every stratigraphic interval of potential commercial interest: its facies distribution (in color), thickness changes (by isopach contours), number and thicknesses of potential pay zones, known hydrocarbon occurrences, fluid recoveries from wells, and pertinent hydrodynamic data. The better integrated the data, the less likely the geologist is to be caught by surprise by developments within his area of responsibility.

Well designed prospect maps – which should be on a variety of scales according to the detail of data available – constantly point up areas or features deserving of fuller understanding. If, for example, favorable trends of more than one potential reservoir horizon converge towards an area in which seismic information is less than adequate, more detailed, common-depth-point coverage should be designed for that area. If, as another example, favorable characteristics of a deep horizon are projected beneath an area in which drilling and seismic activity has been concentrated upon a shallower horizon, the need for deeper wells, or the deepening of carefully chosen existing wells, is called for.

19.3.1 After all the big structures have been drilled

Fields continue to be found for years in mature producing basins in which one would think that all the detectable structural anomalies had been drilled long ago. Many are found by essentially random drilling, especially in the USA. Many others, however, are found by careful detective work by subsurface geologists familiar with the region, especially if their leads are supported by equally careful study of newly reshot seismic profiles.

In essence, the process involves continuous consideration of the three-dimensional configuration of every known stratigraphic unit in the area, taking account of all known and potential faults, diastems, or other discontinuities, and all available seismic data. The more data the geologists have, the more possibilities are revealed; during the same data accumulation process, of course, other possibilities are eliminated. Of the plethora of possibilities, we may consider a single very common one. Suppose that a step-out well, drilled in the hope of extending a field in a carbonate reservoir and a convex trap, misses its pool but finds oil in a small sandstone body identified as a point bar deposit and beyond detection seismically. Is it likely that a point bar is the only one of its kind? A river is unlikely to acquire a single, unique meander; a basin is unlikely to be entered by a single, unique river. The geologist therefore looks at the logs of all wells drilled deep enough to penetrate the horizon of the sand to see if any others may have penetrated an untested sand at about the same horizon. From his findings he reconstructs the apparent distribution of the sand (or sands), and this may reveal a pattern which points to features of the paleodrainage system.

The searches that a geologist in this situation may make are so many that mere tabulation of some of the more obvious ones must suffice:

(1) Recheck all logs and cores for porous stratigraphic horizons that have not been tested, or which are of uncertain distribution.
(2) Make sure that the true limits of all known pools are established beyond doubt, and that the reasons for the limits are also known.
(3) Are there areas of established production in shallow horizons which have received inadequate attention to deeper horizons productive elsewhere? The case of the Pembina and West Pembina fields in western Canada provides one of our case histories (Sec. 26.2.1).
(4) Have all known or reasonably postulated trends been followed to their productive limits? In particular, are accumulations considered to be separate pools actually known to be unconnected?
(5) If accumulations are divided up, or separated, by faults, have all potential reservoirs been tested in every fault block?
(6) If the basin contains productive, nonconvex traps which are undetectable seismically, are they of such a size that others could occur within undrilled areas in which no anomalies have attracted exploratory wells of suitable depth?
(7) Have all zero edges, of whatever cause, identified on seismic or geologic sections been fully followed out and accurately located? Are any segments of the wedge-outs in potential trap position but undrilled?
(8) Is the subcrop pattern below every unconformity properly understood? If any formation is productive from sub-unconformity traps, has its entire subcrop been explored?
(9) Is the exact nature of every trap related to an unconformity known with certainty? Might some of those thought to be structural or wholly lithologic actually be paleogeomorphic? If so, does this interpretation change the geologist's impression of the likely distribution of such traps?
(10) If there are horizons like thick evaporites which make structural conditions below them difficult to decipher, have all reasonable geologic and seismic leads been tried and have enough wells been drilled into the deeper levels?
(11) Is the geologist sure of the explanation for all known changes of dip?
(12) Are any of the known pools dependent upon fracture porosity or permeability? If so, is the fracturing necessarily consequent upon detectable structure? Might it not be present where no convex structure can be detected?

Many experienced petroleum geologists will be surprised at the omission from this tabulation of the detective procedures that were most successful in their own areas.

19.4 The geologist at the wellsite

The *wellsite geologist's* tasks are basically two: to make sure that all possible information is derived from the drilling of the well before the rig is released, and to use his geological skills to ensure that any prospective producing zone penetrated by the well is given every chance to become a real producing zone. On an exploratory well, no porous horizon should be drilled through without testing for hydrocarbons, and only if the horizon is already well known should it be drilled through without coring.

Before drilling begins, the wellsite geologist must be provided with a *well prognosis*. In this, the exploration or production geologists (depending on the type of well) set out the depths at which prospective horizons are to be cored, logged, or tested, and other horizons of interest are to be sampled for lithologic, micropaleontologic, or geochemical studies.

The routine, day-by-day duties of the wellsite geologist are dominated by the evaluation of mud logs and drill samples (cuttings), decisions concerning coring, testing, and logging of the well, and preliminary evaluation of the cores, formation fluids, and logs acquired. The mechanisms and techniques involved are described in Chapters 12 and 23.

19.4.1 Mud logging

The purpose of the mud logger is to evaluate oil or gas shows in the mud which has circulated through the bit and back to the surface. Oil in the mud is detected by fluorescence under ultraviolet light; gas by ignition over a hot wire. The evaluation must take account of the mud weight and additives within it (including hydrocarbons in the case of oil-base muds), and the substances monitored should include H_2S and chlorides.

19.4.2 Drill cuttings

The problems afflicting the interpretation of drill cuttings are outlined in Chapter 23. Despite the problems, cuttings are the only continuously available, visual record of the rocks passed through by the drill; their lithologies and textures must be determined as thoroughly as their condition permits. Samples should be systematically washed, dried, bagged, labelled, and shipped to the appropriate center if more detailed examination is needed. An important property which such laboratory examination may reveal is shale density, which in turn may tell the geologist and the engineer something about well pressures (Secs 11.1.4 & 14.2). The first and last appearances of any distinctive lithologies should be correlated as far as possible with the log of drilling time (rate at which the hole is deepened).

Sample descriptions, drilling breaks, and mud log data are recorded continuously on *strip logs*, using standard symbols and abbreviations so that other geologists can decipher them.

19.4.3 Cores and larger samples

The wellsite geologist should call for a core if a formation is entered which requires quantitative evaluation of any important property. The "drilling break" provides him with valuable warning. Notification from the driller, or evidence from the mud logger, tells him of any sudden increase in penetration rate; the common cause of this is entry into a horizon of abnormal fluid pressure (see Sec. 14.2). Drilling should be stopped and the bit pulled back before drilling ahead slowly. The well may "kick" on entering the permeable horizon containing the fluids. An increase of mud weight may be required, or casing may have to be set at this depth, before coring or testing begins (see Ch. 23).

The geologist must temper his natural interest in seeing a better sample of his section with recognition of the cost of coring, especially in a well already deep. The authority to order a core must be clear before the well is spudded. The geologist, the engineer, and the driller must be aware of the procedure to be followed; the weight and properties of the mud must be determined; the recovery, cleaning, marking, packing, and shipping or storage of the core must be supervised by the geologist himself. The reader is invited to recall the subject of combination sandstone reservoirs (Ch. 13), and then to imagine the possible consequences of the geologist being absent from the derrick floor when a 6 m core is brought up and recovered in four or five pieces which are then laid out, unmarked, by the roughnecks. The top or bottom of every piece of core (and the depths to both top and bottom) must not be left to anyone's memory or guess.

Detailed examination of a core is beyond the capacity of the equipment normally available at a wellsite, but before cores are shipped the wellsite geologist himself must immediately give the best scrutiny he can to the lithology, porosity, possible permeability, and fluid content (by fluorescence tests). Delay in making these observations may cause the loss of vital information. As the core has already been bathed in drilling mud in the well and then hosed with water on the derrick floor, its true fluid content has probably already been affected and is not in any event likely to remain detectable for long. Once the core has been dried, sealed in foil or plastic, packed in tins or boxes and shipped to the laboratory, only the properties of the rock and not those of its fluid content can still be determined with any assurance, despite the fact that cores are eventually studied by more individuals than any other item derived from the well.

If cuttings are of poor quality and full coring difficult to justify, the geologist calls for *sidewall samples* (Sec. 23.2). These may be samples recovered by the percussion tool, small rotary cores, or slices derived by the incision of small triangular cuts in the sidewall by a diamond saw. The exact depth from which a sidewall specimen is recovered must be known and recorded.

19.4.4 Formation tests

The geologist recommends the testing of any interval he thinks may yield hydrocarbons. As *drillstem testing* of a formation normally involves outside personnel, equipment, and a vehicle not continuously at the wellsite, the weighing of the benefits against the cost and the delay is even more important than in the case of taking a core. But many fields have been missed through failure to run tests before letting the rig go, and some great fields owed their discoveries to alert geologists who tested horizons that were not even foreseen as targets when the discovery wells were started. Hassi Messaoud in Algeria and Pembina in western Canada are among them, and both provide case histories for this volume (see Ch. 26).

The geologist chooses the horizon to be tested, and picks the packer seat which determines where the testing tool will be set. He also decides the duration of the test, which is run in open hole with drilling mud in it. At the conclusion of the test he records the fluid recovery – hydrocarbons, formation water, or drilling mud – and its amount. The latter is recorded as the height of rise of the fluid in the pipe, or the time-lapse for the fluid to reach the surface if it does this. The nature of the oil or the water is important; so also is the formation pressure under which its rise occurred (measured by pressure bomb during the test). If a brief drillstem test is unexpectedly successful, a longer test is warranted, and should be allowed to run for long enough to permit a calculation of potential initial production (IP) rate to be estimated.

Less definitive tests, called *wireline formation tests*, can be run without serious interruption of drilling. The process is similar to that for collecting sidewall samples. Only a very small sample of formation fluid is obtained, but it may be sufficient to tell the geologist what fluid the formation contains at that level and to confirm that the formation is permeable.

19.4.5 Wireline logging

The wellsite geologist's principal interest in the well logs lies in the information they give him about fluid contents of the porous formations. Their value as correlation tools will benefit others, though the wellsite geologist will be the first to report whether his well is running higher or lower than was forecast in the well prognosis. The geologist recommends the particular suite of logs to be run, though for all deep exploratory wells nowadays a company is foolish if it does not run the entire suite, and for all development and step-out wells the choice of logs to be run is determined in the exploration and production departments in advance.

The principal logs available, and their various capacities and deficiencies, are described in Chapter 12. Many operating companies have nonstandard logs of their own in addition to the standard electric, acoustic, and radioactive logs. Logs that are ineffective through casing must be run before casing is set; this becomes especially important in young clastic successions in regions like Trinidad, Indonesia, or the Gulf Coast, where casing has to be set to prevent loss of the upper part of the hole by slumping before deeper prospective horizons can be drilled.

The shallower a porous horizon lies in the well bore, the longer it is subject to invasion by mud filtrate as drilling goes deeper. If no logs are run until a deep well reaches total depth (TD), the logs may have become useless for all shallow, porous horizons hopelessly invaded by filtrate. The geologist must decide when to run logs to avoid this loss of information. In some areas in which a thick clastic section is underlain by an equally thick carbonate–evaporite section (as in much of the interior of North America), the properties of the mud must be changed for the lower section and it will commonly be advisable to run logs before making the change in case the new mud will mar the log response of the shallower section.

The physical condition of the drillhole is of immediate concern to the wellsite geologist. The driller may suspect that the hole has deviated from the vertical; the sample and mud returns may indicate extensive sloughing of weak beds into the drillhole; the geologist's own knowledge of the area may lead him to expect dipping or faulted strata. The geologist must monitor the geometry of the hole and the beds it has penetrated by running a dipmeter log, which involves caliper and deviation measurements as well as the three- or four-electrode device for determining the dip (see Sec. 12.7).

More specialized downhole devices, such as the *borehole gravimeter* and the *downhole camera*, are employed in special circumstances that normally reflect larger-scale research with objectives beyond those of the wellsite geologist.

On many exploratory wells, the geophysicists want to run *velocity surveys*, for the purpose of calibrating their seismic records with a known rock section. The wellsite geologist should notify his geophysical colleagues when he intends to run logs, so that geophones can be run also if the velocity survey is required.

19.4.6 Casing

The need to coordinate logging and testing of the well with the setting of protective casing has already been alluded to. If a drilling break is encountered, or a porous horizon worthy of testing is anticipated, logs should be run and from them the geologist chooses the depth at which the casing shoe is to be set.

19.5 The development geologist in the laboratory

In the office and laboratory, the development geologist is concerned with the rock samples and cores, fluid samples, log suites, instrumental measurements, and performance records of all the wells in an oil- or gasfield, or even all the wells in an entire basin.

Cores and trustworthy samples must be handled and

processed according to strict procedures which should be uniform throughout any one organization (and preferably throughout all comparable organizations in any one region). Lost, damaged, or mislabelled cores represent both lost information and preserved misinformation, as well as wasted time and money. Cores must be described in detail and all ascertainable properties measured. The geologist may not perform the actual analyses or measurements, but he must determine what is to be analyzed or measured, monitor the acquisition of the data and coordinate the results. Cores may be slabbed parallel to their axes and examined both before and after polishing. The greatest attention is devoted to cores of the productive or potentially productive intervals. Photographs (especially color photographs) of complete or slabbed cores are very useful, and when displayed on a laboratory wall they give a unified view of an entire reservoir rock.

Routine core analysis of potential reservoir rocks includes measurements of porosity, permeabilities to oil, gas, and water, and grain size (see Ch. 11). More specialized core analyses involve nonreservoir as well as reservoir rocks. The petrology and stratigraphy of all cores must be described from thin sections and SEM studies so that the sand and clay mineral contents, sedimentary textures and structures, and fossil content can be combined to reveal the source of the sediments, their environment of deposition (especially, in the case of marine sediments, its water depth), and their diagenetic history. Features significant to reservoir potential include far more than instrumental measurements of porosity and permeability. The nature of any clay minerals, the compositions of cements (and their paragenesis if more than one cement is detectable), the presence of fractures (and their nature and orientation), shale partings in sandstones, stylolites in carbonate rocks, or evaporites which might plug pores, may also be critical to the actual performance of the reservoir. All measurable dips must be correlated with the dipmeter data. Rocks which may be potential source sediments are analyzed by the geochemist for total organic carbon and the type and maturation level of the organic matter, and by the paleontologist and palynologist for microfossils, pollen, vitrinite fragments, and anything else that will add information about provenance, environment, diagenetic history, and organic maturity.

Log analysis may become a full-time occupation for development geologists in established areas still retaining very active drilling programs. Hundreds of wells per year may provide six or eight logs per well to a busy district office. Full-scale logs are cumbersome and their analysis is time-consuming (see Ch. 12). In densely drilled areas, logs should be digitized; log-derived data must be calibrated with core data and test data on a continuing basis. From these threefold data banks, up-to-date maps must be maintained of test results, GORs, formation water salinities (from log resistivities and analyses), well pressures, and isotherms. As production increases and then stabilizes, the geologist and the engineer regularly monitor pressure decline curves against cumulative production. Such monitoring provides the best estimates of ultimately recoverable reserves from any field.

19.5.1 *Modelling the reservoir*

The reservoirs in their oil- and gasfields are the sources of employment and income for the geologists and engineers in the oil and gas industries. It is their responsibility to treat the reservoirs in the most efficient manner possible. The development program they devise for a field must be made to fit the reservoir; if it is an inappropriate one, production will be inefficient and probably wasteful; oil and gas will remain unrecovered which could have been profitably recovered with a better program.

All data from the field – geological, geophysical, engineering, economic – are integrated so as to develop as accurate a reservoir model as possible. The model must be adjusted and updated throughout the drilling and delineation phase of the field. Geologic parameters incorporated into the model include all dip, fault, and fracture data; lithologies, porosities, stratigraphic and faunal or floral correlations; pressure and temperature measurements. Fluid dynamic data required include effective permeabilities and saturations as shown by both test measurements and well performances; production rates; production mechanisms and their maintenance or decline; changes in compositions of the fluids produced (oil, gas, and water) as determined by regular chemical and chromatographic analyses; water encroachment into the reservoir; hydrodynamic evidence for fluid migration paths and rates.

The reservoir model provides the foundation on which the geologists participate with the production and reservoir engineers in the proper development of the field. They determine the production plan, including the maximum efficient rate of production (MER); when peak production is to be achieved and at what rate; the manner of conservation or maintenance of reservoir pressures; and completion technologies. They identify obstacles in the reservoir which interfere with production: the presence of unnoticed features of the reservoir that impair its performance; formation damage; sources of water encroachment. They determine when to stimulate the reservoir, when to move from the primary recovery phase to that of secondary (assisted) recovery, and where to drill injection wells or convert producing wells to injection wells. If water injection is decided upon, the geologist may have to identify and recommend an underground source for the water. When the rate of assisted recovery declines, the reservoir model should permit a reliable estimate of the volume of extra oil that might be recovered by tertiary (enhanced) recovery (see Ch. 23). If this is decided upon, the method of such recovery and its points and rates of application must conform with the model. If enhanced recovery follows secondary recovery by waterflood, the geologist may have to identify a subsurface aquifer into which the re-produced input water can be safely injected.

Finally, all changes in ultimately recoverable reserves are constantly monitored. It is a serious mistake to imagine that all changes in the reserves of an established producing field are downwards. The reservoir may perform better than anticipated; this may lead to enhancement of the recovery by primary or secondary methods. Pools within the field may be extended; infill drilling may be economically justified by tapping unexpectedly permeable streaks in the reservoir; new fault blocks or reservoir sands may be proven commercially producible. These "extensions and revisions" to the field's reserves must be constantly brought to the attention of management so that projections of future output and income can be kept in conformity with the reservoir's capacities.

20 Surface indications and the direct detection of hydrocarbons

In our summary of the history of petroleum geology (Ch. 4), acknowledgement was made of the obvious truth that essentially all oil production before the beginning of the 20th century resulted from drilling near *seepages*. Natural seepages are the most manifest surface evidences of the existence of oil accumulations. There are, however, many other forms that surface indications of oil or gas can take, and extensive research and field testing have been devoted to these forms. The research and testing may be crudely divided among geological, geophysical, and geochemical foundations, and surface indications will be considered in this order.

20.1 Geological indications

Seepages of oil or gas merely show that petroliferous rocks are present. They do not necessarily indicate the existence of a commercial accumulation, because the mere fact that the hydrocarbons are escaping is demonstration that the trapping mechanism is at least imperfect and may be hopelessly so. Nonetheless, major seeps are the best and most direct evidence of favorability. Most geologically young oil provinces are betrayed by them. What is more, many early human settlements, and indeed entire civilizations, originated in the vicinity of oil or gas seepages, especially where the gas was burning. The natural philosophers who were the forerunners of modern scientists were not surprised that the Tigris–Euphrates valley, the Gulf of Suez, the Baku region of Russia, Sicily, northwestern Sumatra, Burma, the coastal regions of California, Ecuador, Peru, and eastern Mexico, and the mountainous northern margin of South America from Colombia to Trinidad became important producing regions for conventional oil. It is safe to say that all early-found producing regions remote from industrial populations were betrayed by seepages: the interior of Argentina, the upper Assam valley of India, and the island of Sakhalin are examples. It is also safe to say that the natural philosophers and their scientist successors have been surprised and in some ways baffled by the lack of significant production from regions of such obvious seepage as the Rhine and Indus valleys, the Dead Sea, western Japan, and the island of Barbados.

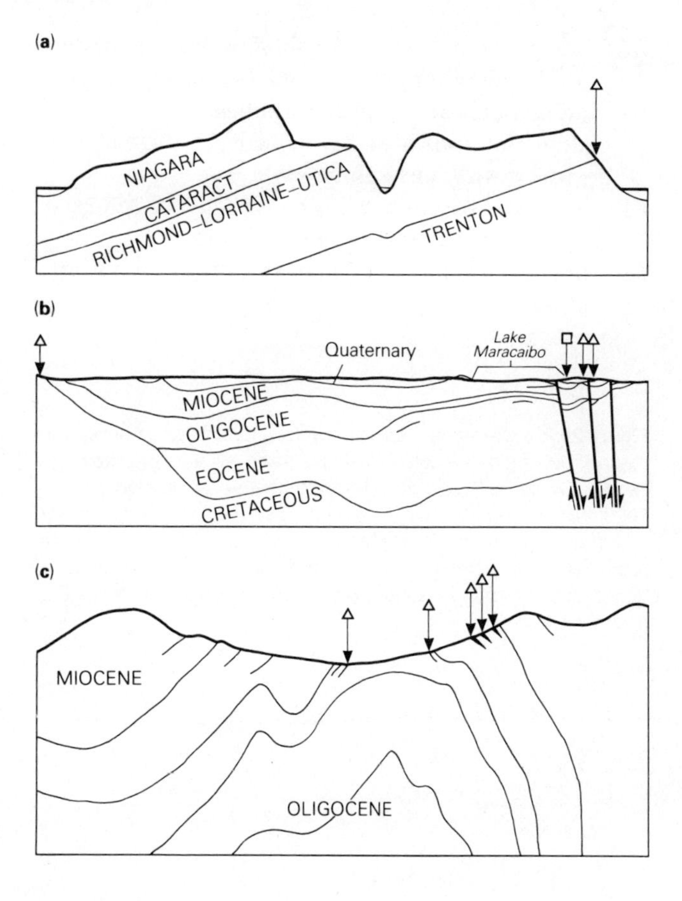

Figure 20.1 Common types of oil seepage: (a) along homocline where oil-bearing beds come to the surface (Manitoulin Island, Ontario); (b) along unconformity and eroded oil-bearing wedge (Maracaibo area, Venezuela); (c) along outcrop of oil sands in eroded anticlinal trap (Hombre Pintado, Venezuela). (From J. R. Gunter, *Asoc. Venez. Geol. Min. Petrol. Bol. Inform.*, 1961.)

Oil can move upward towards the surface via two mechanisms:

(a) *Bulk transport* or *permeation*, involving either continuous- or discontinuous-phase flow and causing *macroseepages*.
(b) *Solution-diffusion*, involving thermal or gravity flow and causing *microseepages*.

As primary migration of hydrocarbons takes place from source rock to reservoir rock, and secondary migration within the reservoir rock to a trap, movement from the trap to the surface may be thought of as *tertiary migration*. Visible macroseepages, resulting from bulk transport or upward permeation of the oil, are commonly associated with geologic structure; many are directly attributable to synchronous deformation. They represent indirect exposures of a reservoir rock, by some form of fracturing, arching, or erosion. A simple categorization of such seepage controls is due to Walter Link:

(a) along homoclines, through slow but unimpeded updip migration;
(b) along outcrops of unconformities;
(c) along outcrops of reservoir beds, through erosion with or without faulting;
(d) at intrusive contacts, against igneous plugs or dykes, salt domes, or mud volcanoes;
(e) through fracturing or shattering of exposed source rocks.

The inexperienced geologist may be surprised that no common category of seepages is attributed wholly to *faulting*. Seepages are not common along faults, unless the reservoir rocks are already at or very close to the surface. Faults are more likely to be seals than transport routes (see Sec. 16.7). Some exceptions are illustrated in Figure 20.2.

Asphalt-base oils inspissate readily to form a black substance resembling pitch. Paraffinic compounds are oxidized more readily than are the corresponding aromatic or naphthenic compounds, and asphaltenes (Sec. 5.2.5) are generated during the oxidation and biodegradation. Paraffin-base oils, however, make much less striking seepages than asphalt-base oils, lighter in color and more fluid. Heavily impregnated oil sands may be soft and plastic. Oil-saturated carbonate rock tends to acquire a dark brown color. The more compact and fine-grained the rock, the less easily are signs of oil weathered away; they may remain in *concretions* after having disappeared from the body of the rock. The presence of inspissated oil in a rock can be quickly determined by crushing the rock and letting it stand in ether in a well corked bottle for 30 min. All the hydrocarbons go into solution; pour the solvent into a white dish, and evaporate it. If oil is present, a dark ring is left around the dish, greenish or brown for a paraffinic oil, darker brown to black for an asphaltic oil.

The rocks in the immediate vicinity of active seepages undergo characteristic types of alteration. Red sandstones, in particular, become yellow or pink, and finally grey or even white; they acquire carbonate cements,

Figure 20.2 Seepages associated with faults: (a) along normal fault cutting flat-lying beds (Bothwell field, Ontario, close to the site of the original Ontario oil well, drilled beside another, similar seepage in 1858); (b) along a low-angle thrust fault some distance away from the underlying convex trap (Naft Khaneh field, Iraq); (c) along homoclinal beds overlying folded and faulted reservoir beds (Quiriquire, Venezuela). (From J. R. Gunter, *Asoc. Venez. Geol. Min. Petrol. Bol. Inform.*, 1961.)

(b)

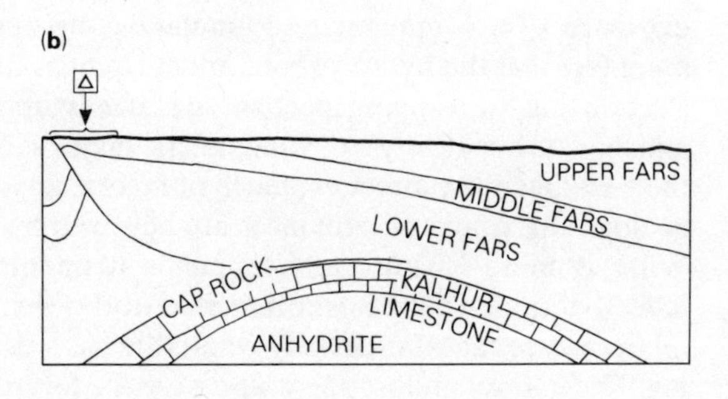

(a)

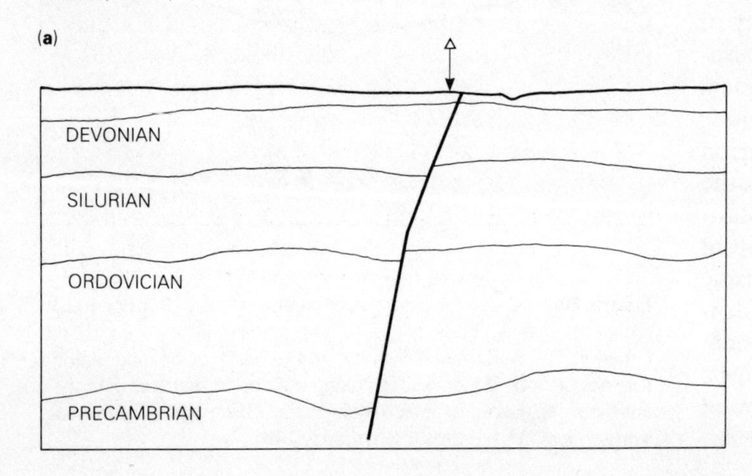

(c)

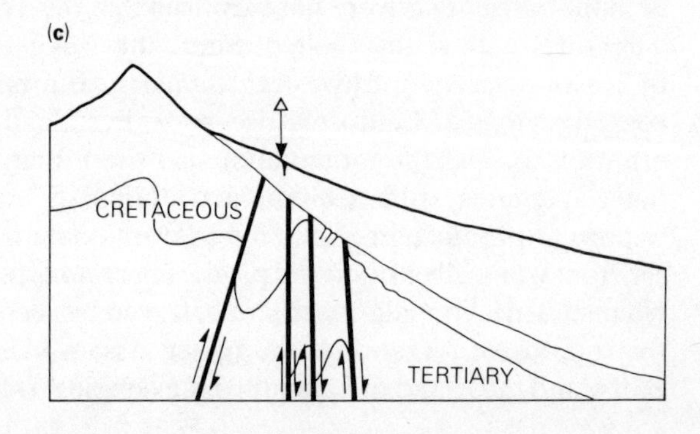

with excess contents of the light isotope of carbon and the heavy isotope of oxygen where the seepages are most active. In evaporites or black shales, gypsum (especially selenite) may be pseudomorphically replaced by calcite.

The low solubility of oil in water makes the presence of surface water conducive to the preservation of visible seepages. They are best sought in low ground and in stream beds. The film of oil on water differs from the film caused by iron hydroxides in that it does not break up when it is stirred.

Offshore oil seepages are equally unmistakable, but, with the present domination of the sea lanes by tankers and the number of drilling rigs and platforms in coastal waters, it is essential to distinguish between genuine and artificial seepages. It has been roughly estimated that 0.7×10^6 m^3 yr of oil escapes naturally from submarine seepages into the seas; the actual amount may be much greater. The highest incidence is close to tectonically active coastlines (those of subduction or transform types, with active seismicity), and the lowest close to trailing-edge coasts like those around the Atlantic Ocean. Seepages spreading to the surface of the sea take two dominant forms, vast oil slicks and floating balls of "tar." Both have been observed over many years in the Santa Barbara Channel off southern California (Sec. 25.6.4); the slick was recorded by Captain George Vancouver in 1792.

Gas may escape at the surface without associated oil. It does not, of course, form a visible deposit as escaping oil does, but it manifests itself in several ways. The most obvious arises when the gas catches fire, through lightning or brush fires. Burning gas seepages at Kirkuk, in Iraq, and Baku, in Russia, attained major religious significance as well as pointing the way to supergiant oil discoveries. Others, on the south shore of Asia Minor, at Apollonia in Thrace, and on the island of Timor, have not been shown to be associated with important commercial deposits. The second common mode of betrayal of gas seepages is the preservation of evidence of former burning, in the form of clays burnt to a brick-like color and texture. The third is the construction of mud volcanoes. These have nothing to do with true volcanic phenomena, such as solfataras or cinder cones. Gas bubbles up through a cover of weak sediments, especially where these have been pierced by clay-cored diapirs. In some prolific oil-producing regions, notably Baku and Trinidad, mud volcanoes have craters more than 100 m in diameter; smaller ones are more continuously active, and their outbursts may be violent. The fourth and simplest circumstance by which gas seepages become readily detectable is provided by water-covered ground. Rising gas bubbles are also detectable at sea, though currents may cause them to be offset from their source on the sea floor. In the northern Gulf of Mexico, for example, the bubbles create near-vertical bands of energy return within the water column, recorded on high-resolution sub-bottom profilers. They can be detected by devices called "sniffers," which are hoses towed just below the water surface behind exploration vessels; the fluids entering the hose are pumped through a shipboard analyzer system consisting of a separator and a gas chromatograph.

In addition to fluid seepages of oil or natural gas, solid and semi-solid surface hydrocarbons are very widespread. Their compositions and characteristics are described in Section 10.6. Here it is necessary only to refer to their appearances and distribution. They may be truly solid, brittle, and almost glass-like, with columnar jointing, or plasticlike highly viscous liquids. They are of two principal types, those essentially conformable with the bedding, and probably petrified oilfields in the literal sense, and those clearly cross-cutting the bedding, as dykes or vein fillings. The former type is associated with normally petroliferous, marine, sedimentary successions, like that in southern Oklahoma. The second type is peculiarly associated with oil shales and bituminous lacustrine shale sequences, like those in the Uinta Basin of the western USA and the Mesozoic pull-apart basins along the central and south Atlantic margins. They also occur in mudflow vents, as the type manjak veins of Barbados do. The veins and dykes are residues of inspissated petroleum formed below the surface.

An indirect geological route to surface indications of oil is afforded by its common association with *rock salt*. This association was recognized very early, and some of the pioneer oil discoveries were made in regions in which oil had contaminated wells drilled to recover salt. These regions include the sites of the first deliberately drilled oil well, in Pennsylvania (Fig. 16.90); the first successful Gulf Coast well, at Spindletop; the first Chinese oil wells, in the Sichuan Basin; and the early Romanian wells in the Carpathian thrust belt (see Ch. 25). The Russians and the Chinese have laid the greatest emphasis on this association, noting the control of deposition of both salt beds and oil-source sediments by subsiding, restricted basins; the similar mode of origin of salt formations and black, euxinic shales; the creation of oil-generative "sinks" through early flowage of salt; the even readier creation of later growth structures in salt, like piercement domes; and the excellence of evaporite seals to both oil and gas pools. It is also notable that, whilst some environments have clearly been conducive to the deposition of both salt and source sediments (the Permian of West Texas and of the Dnieper–Donetz trough, the Mesozoic of the Persian Gulf and Gulf Coast basins, and the Cenozoic of Romania are obvious examples among many), others seem successfully to exclude both (neither salt nor oil is a circum-Pacific phenomenon, for example; see Ch. 17).

These apparent associations are nonetheless illusory. There is a clear inverse relation in time between the deposition of evaporites and that of oil-source sediments (see Ch. 8). There is no general stratigraphic relation between the two; both source and reservoir rock may be either older or younger than the salt. There is a great deal of salt not obviously associated with any significant oil (in the Middle Devonian of Saskatchewan, for example), and where thick salt and abundant oil are spatially associated they may be clearly unassociated geologically. The oil in Iran is not associated with the salt domes, and the oil and salt in the North Sea and Algerian basins, among others, are far separated in stratigraphic age. At the same time, many prolific oil basins are wholly without salt: those of California, Venezuela, and western Siberia, are obvious examples.

20.2 Geophysical and geochemical indications

A variety of physical and chemical characteristics and concentrations may occur at the Earth's surface as consequences of hydrocarbon accumulations at depth. If these can be detected, and measured instrumentally, they may provide routes to the direct detection of petroleum deposits. Oil and gas reservoirs are dynamic systems that are not in equilibrium. At some rate which may be exceedingly slow or relatively rapid they supply surface *microseepages*, or residual hydrocarbon traces in surface sediments, soils, or waters. These microseepages may be dependent upon the transport capacity of hydrochemical plumes of ascending waters.

The detection of several such surface manifestations relies on the *halo theory* or *chimney effect*. An accumulation itself forms a barrier to the ascent of hydrocarbon-laden waters. The detectable substances therefore acquire low concentrations directly above the accumulations, with peaks, called *edge values*, above the hydrocarbon/water contacts and low or "background" concentrations outside them. The surface anomaly therefore has the form of a ring or halo over a convex trap accumulation and a half-ring over a wedge-shaped accumulation.

The most commonly claimed haloes are "intrusion" or "invasion" haloes in clayey rocks above accumulations. Positive haloes, in which concentration increases towards the pool, are shown by heavy hydrocarbons; they are likely to become displaced in the direction of water movement. Negative haloes, decreasing towards the pool, are more likely with gas microseepages. These can form by three distinct processes:

(a) Gas diffusion, the spontaneous, irreversible process in which material is transferred in the direction of lower concentration.
(b) Gas effusion, in accordance with Darcy's law, in the direction of lower pressure (commonly upwards).
(c) In aqueous solution, as the result of hydrodynamic pressure gradients.

The relative efficacies of these three processes may be briefly evaluated. Gas cannot by itself enrich the soil above deposits via *effusion* according to Darcy's law unless vertical permeability is very high (due to steep dips of the strata, or extensive vertical fracturing). *Diffusion* towards lower concentrations is governed by Fick's law, which simply states that the change in concentration of a diffused substance with time is inversely proportional to the square of the distance from the diffusing source (in our case, from the top of the gas deposit). The diffusion of methane through water-saturated rock is exceedingly slow, measured in meters per million years.

Over several geologic periods, diffusion of gas even through shale roof rocks can destroy an accumulation. Large gas accumulations are recognized as being short-lived; no present ones are likely to be older than Cenozoic, perhaps no older than Miocene. Some equilibrium must be reached between loss by diffusion through the roof rock and replenishment from the source rock. This loss does not, however, mean that significant diffusion-gas anomalies can be built up at the surface. Methane concentrations in surface waters through this mechanism cannot become much more than 1 percent of that in the pore spaces, and far lower than that in oil. When the saturation pressure of water-dissolved gases becomes equal to the rock pressure, no further gas can be dissolved. The detection of deposits through methane concentrations at the surface therefore presupposes that their principal transport has been along *hydrodynamic* gradients.

Gas microseepages formed that way are detected principally by *soil-gas analysis*. This methodology, developed independently by German and Russian scientists in the 1930s and still associated with the name of V. A. Sokolov, is used in two different ways. Methane reaching surface waters and soils is oxidized to CO_2. This enhances the normal evaporation rate of groundwater, and stable, cumulative alteration of the soils is brought about by the deposition of carbonate salts in their clay fractions. The amount of this alteration, and hence the amount of methane originally introduced into the soil, is measured by the volume of CO_2 driven off by heating to 600°C in an inert atmosphere. The second procedure utilizes the bacteria that live on the hydrocarbons in the soil. After the consumption of gas by the micro-organisms, the end-products are water and CO_2. The amount of water is proportional to the number of micro-organisms doing the consuming; it is measured by its rise in a manometer tube. Russian geochemists still using this method favor propane-oxidizing bacteria over those that consume methane; the latter are not considered diagnostic.

Micro-organisms are likely to destroy any hydrocarbon gases that find their way into *aerated* near-surface waters. An alternative measure is then the concentration of the *ammonium ion*, NH_4^+, which is a conversion-product of oil. In nonpetroliferous regions, the concentration of this ion in groundwater is normally less than 0.1 g l^{-1}. In the vicinity of oil- and gasfields it may be over 1.0 (or even 10.0) g l^{-1}. Ammonium ion measurements must be treated conservatively and with shrewd appraisal of other influences on them. For example, high concentrations are associated with acid waters of pH 4–5, such as those in volcanic regions; they also typify aquifers having very slow groundwater movement rather than the reverse. Finally, the measurements are much more reliable where the rock succession is clastic, because clastic sediments have the lowest background NH_4^+ concentrations and carbonate–evaporite sediments have the highest.

Two final cautions are necessary for the geologist or geochemist seeking macro- or microseepages of gas. It is desirable to discover whether the gas is of biogenic or thermogenic origin; the former is almost ubiquitous and tells the geologist little. The easiest determinant of this point is the carbon isotope ratio (see Sec. 6.3.1). Biogenic gases are isotopically light ($\delta\,^{13}C$ between −55 and −85); thermogenic gases are isotopically heavier ($\delta\,^{13}C$ from −35 to −50). Finally, gas seepages give no useful indication of the *depth* of any possible commercial deposit below.

20.2.1 Radiation anomalies

Anomalous amounts of radiation emanating from potassium, uranium, thorium, or radium minerals in rocks or soils, or from radon gas, are measured by the intensity of the natural gamma-ray field. Potassium minerals in surface rocks and soils are relatively abundant, in the range of 0.2–2.7 percent. Uranium and thorium concentrations are measured in small numbers $\times\ 10^{-4}$ percent. Radium (Ra) makes up about 1×10^{-10} percent, largely in the clay fraction of soils because of their higher sorption capacity. The *composition* of the soils and soil-forming rocks is the controlling factor, rather than the *type* of the soil (which reflects climate as much as anything else).

The radioactive contents of crude oils are normally much lower than those of rocks or soils; uranium in particular is very low, and concentrated in the heavy petroleum components. In subsurface waters, leaching of uranium is promoted by waters of bicarbonate–sodium type (Sulin's type *b*; Ch. 9); radium leaching is promoted by waters of chloride–sodium–potassium type (Sulin's type *d*). Radium should therefore be enriched in the reducing waters of oilfields, but any area of relatively stagnant subsurface water is likely to show this enrichment, whether oil-bearing or not. Radon (^{222}R) is derived from ^{238}U and is present in oil and gas (see Secs 5.2.4 & 10.9.5). It emits α-radiation, which is measurable; daughter isotopes emit γ-radiation.

Radiation anomalies are measured by ground or airborne scintillometer. They display the prototype halo effect: low intensity of the field over a hydrocarbon deposit, surrounded by the halo of radiation high outside which is the radiation norm. Such anomalies are considered by Lundberg and others to be caused genetically by hydrocarbons, which migrate upwards from accumulations and reduce radium sulfate to the more soluble sulfide (RaS). This postulate has been fiercely disputed. The anomaly may be related genetically to a positive (high) structure, especially if that structure has undergone uplift during the course of sedimentation, whether the structure is productive or not. Sediments tend to be coarser over the crests of growing structures than in the adjacent depressions, and anticlinal crests tend to become fractured; both circumstances promote the access of deep-seated waters to high structures, bringing higher levels of radioactivity with them.

Radiation anomalies are undoubtedly present over many hydrocarbon accumulations; moreover, they do not disappear as the accumulations are depleted by production, as hydrocarbon surface anomalies do. Gamma-ray fields have been traced through structural growth in some Russian oilfields. In the Devonian and Carboniferous fields of the Russian platform, gamma-logs read highest in oil-filled reservoir rocks and lowest in clayey roof rocks, even if no anomaly is detectable on surface radiometers. There are, of course, countless gamma-radiation anomalies in areas of sedimentary rocks not known to contain any hydrocarbon accumulations at all.

20.2.2 Temperature anomalies

Anomalous temperatures in sedimentary rocks may be expressed in terms of the *geothermal gradient* or of the *heat flow* at the surface (see Sec. 17.8). *Thermal conductivities* among sedimentary rocks are highest in evaporities (salt diapirs are geothermal highs) and lowest in shales. This variability is vastly overshadowed, however, by the variability in conductivities between reservoir fluids. They are very low in hydrocarbons; water conductivities are about five times those of oils and 14 times those of gases. Hydrocarbons accordingly insulate against continued heat flow, so a given quantity of heat raises the temperature of a hydrocarbon accumulation relative to an equivalent volume of water-filled porosity.

Anomalies mapped by contouring geothermal gradients measured in wells suggest that heat flow over hydrocarbon-bearing structures may be 10–15 percent higher than the "background" values. Again, the suspicion arises that it is the structures that cause the anomalies, not their hydrocarbon contents. Steep dips, faults, or any complexities in the structures refract the heat flow and bring about a mass transfer of heat energy into the structures whether they are petroliferous or not. In the case of the upward migration of gas, we noted that fluid transporting agents, principally waters, appeared to exert the fundamental control. If they do so in the case of heat flow also, the control by water movement is likely to obscure any effects of the hydrocarbons.

There are other pitfalls in the interpretation for exploration purposes of temperature anomalies derived from wells. Reliable data on equilibrium temperatures cannot normally be acquired from single exploratory wells. They need closely spaced wells, which are likely to be drilled only after an accumulation has been discovered. The anomalies should really be reduced to residual anomalies, attributable to the upward migration of fluids; this reduction requires the removal of any regional temperature gradient caused by areal variations in the thermal conductivities of the rocks. Correction should also be introduced for the "background" heat flow caused by the tectonic character of the region (Sec. 17.8). In a region of excessive natural flow – an immediate backarc basin such as that of Sumatra, for instance – the low thermal conductivities of oils may render the anomalies over productive structures lower, and not higher, than the background mean. A final correction should be made for the surface temperature. The temperature gradients from surface to total depth in two deep wells in central Canada will be startlingly different if one well is tested in January and the other in July.

20.2.3 Magnetic anomalies

Anomalous values of the magnetic field, commonly associated with thermal anomalies, are not thought of as influenced by hydrocarbons. Recent work has suggested, however, that common, nonmagnetic iron minerals in sedimentary rocks (hematite, goethite, siderite) may be converted to magnetite by interaction with escaping hydrocarbons in the near-surface rocks directly above oilfields. Whether such alteration in fact takes place on a scale sufficient to lead to magnetic anomalies detectable from aerial surveys is a matter for future research.

20.3 Electrical methods of direct detection

Electrical *resistivities* differ enormously among the various reservoir fluids. These differences are fundamental to electric well logging, and are discussed in Ch. 12.

Oil and gas create reduced environments, which means that the environments possess excesses of electrons available for release. Reducing and oxidizing environments are compared in terms of the *redox potential* (E_h), which is measured in millivolts and expressed thus:

$$E_h = K \log \frac{R}{O}$$

where K is a constant of value about 70, R is the concentration of the reduced substance with excess electrons, O is the concentration of the oxidized substance or electron acceptor. Oil and gas therefore create high redox potentials, the distribution or occurrence of which in any stratigraphic horizon can be mapped in several ways:

(a) From corrected SP curves on E-logs; SP is related to E_h via a function of water and mud resistivities.
(b) By surface or subsurface self-potential mapping; microseeps cause a halo of reduced rocks over accumulations.
(c) By magneto-electric surveys. Electrotelluric (ET) currents flow from reduced areas, of high E_h, to oxidized areas of low E_h outside the accumulations. The currents return via SP "sinks" in the weathered zones, directly above hydrocarbon accumulations. The geophysicist therefore tries to map areas towards which ET currents converge.

In Sylvain Pirson's view, hydrocarbon migration may be impeded by redox potential barriers, creating effective traps where no structure or pinchout exists.

Electraflex is an electromagnetic technique using electrical transients. An energizing signal is sent into the ground from electrodes at the ends of an earthed cable 0.75 km long; it gives rise to a current with very fast rise-time (in the megahertz region). The return signal is measured, in microvolts per amp, by a second pair of electrodes. The pick-up dipole is negative on the way down, and returns after reflection as positive voltage; it is in some ways analogous to seismic reflection (see Ch. 21).

Electraflex deals with the dielectric constant (specific inductive capacity), not with resistivity. Its employment requires a highly nonconductive environment, otherwise too much of the signal energy is absorbed as heat;

rocks with high conductivity are transparent to it. Proponents claim that the technique gives returns only from hydrocarbons, not merely from positive structures; detection is claimed as deep as 5000 m. The method is still in the developing stage as this is written. It may become a valuable adjunct to seismic work, especially to bright spot (see Fig. 21.20), but it is known to give a huge response from frozen ground, and a large one from some highly porous horizons above the water table.

20.4 Seismic direct detection

It is one of the dreams of the explorers for oil and gas to devise some seismic means of direct detection of their quarry. The "bright spot" technique may be considered a form of this, at least for gas. It is described in Section 21.4.

20.5 Surface indications by remote sensing

Remote sensing of the Earth's surface from satellites uses electromagnetic radiation reflected or emitted from it in both the visible and the infrared wavelength bands (from 0.3 μm to 3.0 m). The radiation is picked up by a *multispectral scanner* (MSS). The quantity and distribution of the radiation are affected by its source, the transmitting medium, the landscape cover, and the detection system. The images and tonal contrasts may be enhanced by *contrast stretching*, a computer technique that concentrates diffuse reflectance values.

LANDSAT surveys have yielded spectacular blue-green, infrared, and false-color images of the whole surface of the Earth. They provide repetitive coverage of any one area, revealing variations in climatic effects, solar illumination, soil moisture, snow cover, and other features. LANDSAT imagery was designed for agricultural and hydrologic purposes, not for geological purposes. Geological features are of course readily discernible on the images, and important information is derived on features which constantly change: ice distribution, plankton blooms, concentrations of suspended sediment, sand-dune arrays. The images may draw attention to large-scale structures, such as lineaments, not easily detected on the ground, especially where outcrop is poor. The basin now productive in the southern Sudan, for example, was initially identified on satellite images.

Some smaller-scale features of potential interest to petroleum geologists may be revealed also. Natural marine seepages are the most obvious, but those on land include subtle tonal anomalies called "hazies," due to soil alteration, and small but persistent areas in polar or mountainous regions that remain anomalously free of snow. One possible explanation for both these features is seepage of gas. Despite these possibilities, there is really no convincing evidence supporting the claims that earlier availability of LANDSAT images would have led geologists to the structures containing individual oil- or gasfields. Undoubtedly features are visible on the images that may or may not be responsible for, or caused by, petroleum deposits, but comparable features are discernible everywhere. There is no large area of true bedrock anywhere which is totally without some pattern of fractures, for example, if the examiner is determined to find them. LANDSAT images have a value and a fascination for many reasons but the remote detection of oil- and gasfields is unlikely to be one of them.

An example may be quoted, derived not from LANDSAT images but from topographic maps corrected from aerial photographs, purporting to reveal an association between petroleum deposits and topographic features detectable from space. V. A. Krotova and others conducted a statistical study of the large, mature Ural–Volga Basin in the Soviet Union, which contains about 600 producing oil- and gasfields. The fields appear to be concentrated in areas of maximum dissection of the surface, and are missing from broad interfluves. There is an array of possible explanations for this. Valleys may follow lines of weakness, especially faults, which may control hydrocarbon migration patterns, or may be responsible for lines of structural traps. Lateral migration of hydrocarbons may be facilitated by zones of groundwater discharge, which in turn may move away from neotectonic uplifts which form the interfluves. Primary exploration may be concentrated along river routes for access reasons; this was the case in the early stages of exploration in western Siberia, for example. Or the association may be fortuitous.

Modern exploration, no longer dependent upon seepages, is fundamentally governed by geophysical surveying of regions identified by geologists as sedimentary basins or the uplifts adjoining them. Anomalies will be drilled whether they are in valleys or not, though the first wells may be drilled where valleys cross anomalies.

21 Exploration seismology

21.1 Basic principles

The basic objective of all geophysical exploration is to employ the principles of physics – specifically the physical properties of rocks – to determine the disposition of rocks below the surface of the Earth. Because exploration for petroleum is concentrated upon layered sedimentary rocks, which have no great range of densities or electrical properties and little magnetic signature, petroleum geophysical exploration is practically synonymous with *seismology*.

Waves passing through the Earth during earthquakes travel with velocities that are dependent upon the elastic properties of the rocks through which they pass. The waves are reflected from, and refracted by, discontinuities in these rocks. It is from the study of the wave motions in thousands of earthquakes that the greater part of our understanding of the Earth's interior has been derived. The exploration seismologist simply creates a tiny earthquake of his own, and studies the reflection and refraction patterns of the waves he creates.

21.1.1 Seismic velocities

A compressional wave, like the wave that reaches one's ear from a sound source, consists of a series of compressions and rarefactions of the transmitting medium. The medium therefore undergoes rapid small changes both in volume and in shape. The velocity of such a wave is therefore proportional to both the *bulk modulus* of the medium (its capacity to resist change of volume) and its *rigidity* (the instantaneous resistance it offers to deformation by elastic shear). A shear or transverse wave, on the other hand, represents displacement of the particles of the medium in the direction perpendicular to the propagation direction; change of shape is imposed on the medium, but no change in volume. Thus the velocity of a shear wave is directly proportional only to the rigidity of the medium, which is the reason shear waves cannot be transmitted by fluids; rigidity is the fundamental characteristic of elastic solids.

Both compressional and shear wave velocities are also *inversely proportional to the densities* of their media. In comparing travel times for the waves through two different media, therefore, the times will be inversely affected by the densities of the media if other properties are comparable. For any one generalized medium, however – for the layered silicate and carbonate rocks of the crust, for example – the velocity of compressional waves is in linear relation to the density. The relation is expressed by *Birch's law*:

$$V_p = a + b\rho_0$$

where ρ_0 is the zero-pressure density of the medium concerned.

A contact between packets of strata having different densities and rigidities is therefore a *velocity interface*. It represents a contrast in *acoustic impedances*, acoustic impedance being simply defined as the multiple of the seismic velocity and the bulk density ($V_p\rho$). Seismic waves are therefore *reflected* from such interfaces, and returned to the surface where they can be detected and measured. They are also *refracted* at the interfaces, in accordance with *Snell's law* which relates the angles of incidence and refraction (i_i and i_r) to the velocities in the two media:

$$\frac{\sin i_i}{\sin i_r} = \frac{V_i}{V_r}$$

21.1.2 Transmission of seismic waves

The artificial energy source to create the waves is usually impulsive, though energy can also be introduced for more extended times by using coding vibratory sources. The standard impulsive source on land (and in earlier days at sea too) is the detonation of dynamite, buried in a hole between 15 and about 60 m deep. A less effective but cheaper and faster method is by the *thumper*, or very heavy weight dropped to the ground from the back of a truck. Because of the cost of dynamite, the danger it may represent, the need for cheap, quick drilling which in turn requires water and causes noise, conventional seismic sources came to be replaced, beginning in the mid-1960s, by vibratory sources.

In the *Vibroseis* technique, the vibrator is a heavy mass carried beneath a truck. The mass is hydraulically pressed to the earth, raising the back wheels of the truck and resulting in a total weight on the ground of about 8000 kg. This weight is mechanically vibrated, scanning through the range from 0.1 to 120 Hz in a period of about 7 s. The vibration is then halted and repeated, the return waves from this stationary but repetitive source being summed by a stacking procedure described later.

The seismic wave is transmitted by temporary displacements of the particles in accordance with the elastic forces between them. The energy travels in spherical wavefronts (Fig. 21.1) at the velocity appropriate to the medium. At any acoustic impedance interface, part of the energy is reflected, at the velocity of the upper medium, part is transmitted at the velocity of the lower medium. The ratio between the amplitudes of the reflected and transmitted waves is the *reflection coefficient* (R). The value of R at point A in Figure 21.1 is given by

$$R_A = \frac{V_2 \rho_2 - V_1 \rho_1}{V_2 \rho_2 + V_1 \rho_1}$$

The higher the value of R the better the resulting data.

The reflected energy is detected at the surface by arrays of geophones, popularly called "jugs," spaced along the ground and connected to amplifiers. Reflection represents systematic motion; to be usable, reflection data must possess *coherence*, or the capacity for correlation between arrivals of energy at different detectors. Coherence is hampered by *noise*, which is unwanted energy picked up by the detectors. Some noise is itself coherent, and may be due to the source of the wanted energy; for example, waves travelling directly to the geophones along the surface without having been reflected from a subsurface interface. Much noise is random and extraneous (radiation from power lines; movements of the ground due to traffic or industrial activity). The most important cause of interference is the near-surface zone of weathered rock, which causes irregular distortion of the seismic waves. The problem is most acute in areas covered by glacial drift, especially where this thickens in moraines or preglacial valleys. Because it has low velocity, a layer of such material causes time lags which must be removed from the record. All noise must be attenuated, or eliminated as far as possible.

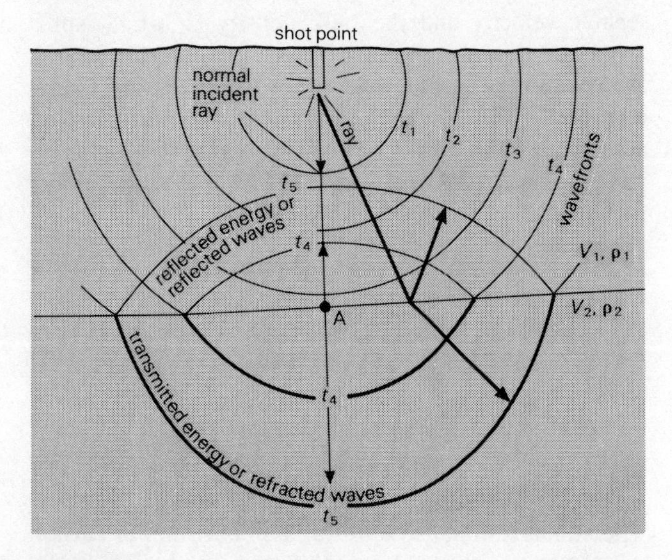

Figure 21.1 Seismic energy travelling in spherical wavefronts from the shot point.

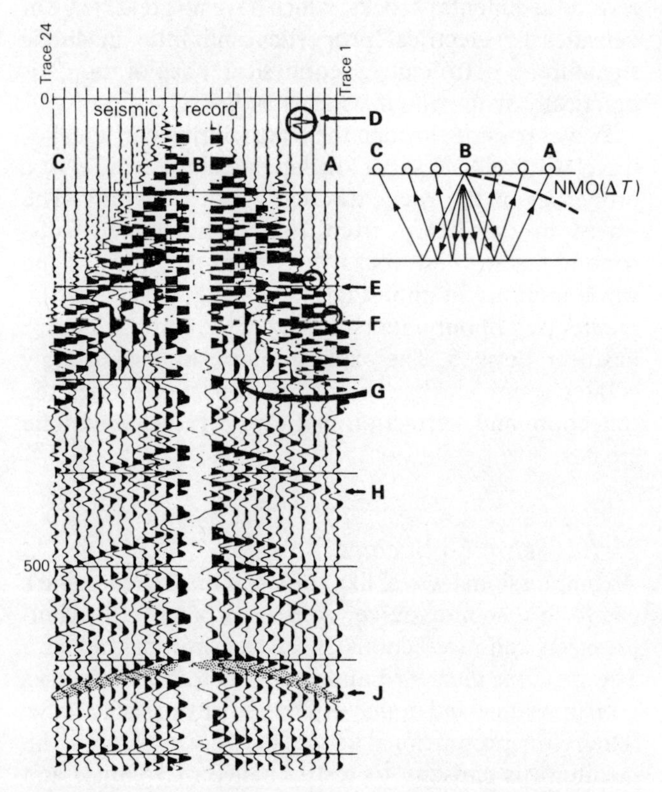

Figure 21.2 Seismic record, showing 24 traces from top (on right) to bottom (on left). Peaks emphasized in black, troughs left blank (to facilitate alignment). Note uphole time break (at D), indicating moment of shot. Reflections first recorded on traces 12 and 13, closest to shot point. First arrivals on traces 3 and 1 are indicated (at E and F). Arrival on trace 1 before that on trace 24 (in this case) indicates dip of reflector. Compare curvature of coherent line-up across record, at H and J (due to normal moveout), with incoherent "noise" generated by shot (at G).

21.2 Standard processing techniques

Uncorrected reflections do not align across the record section; instead they correlate along curved paths (Fig. 21.2), because the geophones at the ends of the array are farther from the shot-point than those at the starts. This curvature is called *normal move-out* (NMO). The amount of curvature decreases with time; the longer the travel time of the wave through the earth, the less it is deviated by thc geophone spacing correction. The curvature is therefore removed by dynamic *time shifting*, and the reflections are aligned (Fig. 21.3).

Alignments showing coherence across the record are called *events*. Some events are likely still to show residual curvature after time shifting (Fig. 21.3); these are *multiples*, caused by repeated reflections from the same interface. The interference of the two sets of reflections can be reduced by summing the traces along carefully selected straight lines, as primary reflections, already moderately aligned, add constructively better than multiple reflections that are not already aligned.

Time-shifted traces are then *stacked*. In *common depth point* (CDP) stacking, all traces having a common reflecting area in the subsurface are collected. Curvature is removed so that the reflectors are aligned in the time sense, and the traces are summed algebraically to form a single, composite trace on the CDP time section (Fig. 21.4). Individual input traces in multiples of six (by sixfold recording and processing) are summed to form each stacked trace depicted; as many as 180 traces may be shown on a single section. Normal procedure results in output traces representing sampling intervals in the subsurface of 15–60 m. The CDP processing is really one of averaging, now performed via computer programs. If the time shifts were correctly applied, the true reflection energy on the composite traces will have a common arrival time. The common signal components are therefore enhanced and the noise tends to cancel itself out. The reflection signal then has a uniform, continuous appearance from one end of the time section

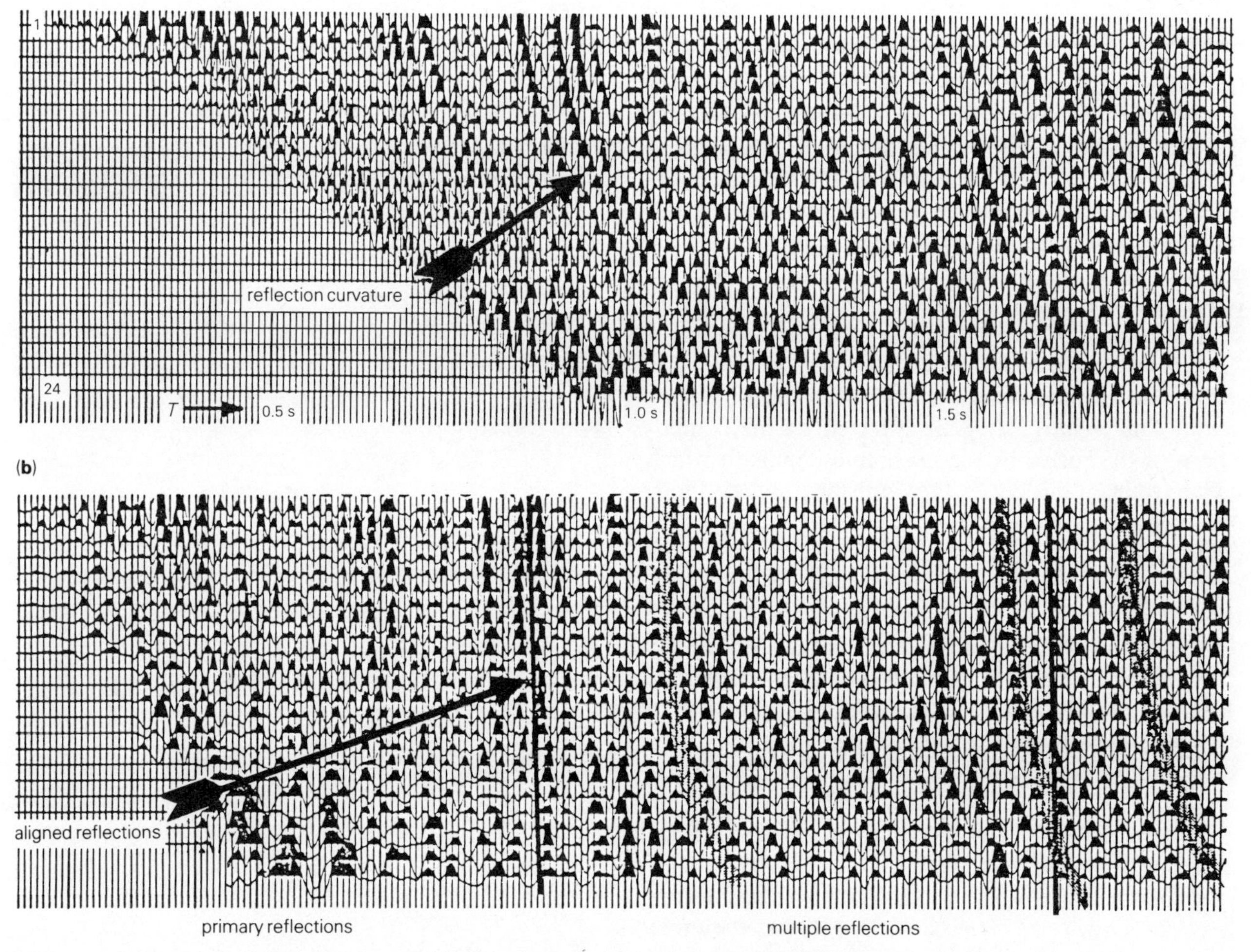

Figure 21.3 (a) Seismic recording showing reflection curvature from trace 1 (innermost) to trace 24 (outermost). Note decrease in curvature with time. (b) Same recording with curvature removed; primary reflections are now aligned. Events containing residual curvature are multiples. Interference of two sets of reflections reduced by summing traces along straight lines shown.

to the other, appearing on the record section as a dark alignment of reflections (Fig. 21.5).

The third standard correction is that made necessary by a strongly dipping reflecting horizon. The corrected seismic time-section does not display the dipping reflection in its proper position (Fig. 21.6). The data must be *migrated*, shifting the reflection up the dip with respect to its position on the time section. This process requires some knowledge of velocities in the section above the reflecting horizon.

The aim of all the editing to which the raw field data are subjected is the enhancement of the reflection signals. From the enhanced signals, the velocities of the rocks traversed by the waves are computed. From the velocity data, the display sections showing *time* are converted to display sections showing *depths*. From these, the geometric disposition of the strata is deduced.

21.3 Geologic interpretation of seismic data

From the arrival times of the reflections, the depths to the reflecting horizons are measured. Variations in arrival time from a single reflector indicate variations in the depth to that reflector and therefore some form of structure upon it. The amplitudes of the reflection signals indicate differences in acoustic impedances and therefore differences in rock types; they also, under certain conditions explained below, indicate something of the fluid contents of the rocks. Arrival times varying with location indicate a dipping reflector; converging reflectors indicate angular relations.

Because seismic wave velocities are functions of *density* and *rigidity*, they are affected by the compositions of the rocks; by the maximum depths to which they have been buried; by the porosities and the fluids occupying the pore spaces; and by the fluid pressures exerted by them. Figures 21.7, 21.8, and 21.9 show velocity–depth relations for common sedimentary rock types, for varying porosities in those types, and for differing fluid contents in the pore spaces. Among the common sedimentary rocks, the crystalline evaporites (rock salt and anhydrite) have negligible porosity and their V_p values range up to more than 6 km s^{-1}. Dolomites, limestones, sandstones, and shales have lower velocities, decreasing in that order (Fig. 21.10).

As densities normally increase, and porosities decrease, with depth, velocities increase downwards (Fig. 21.9). Velocities are also strongly temperature-dependent, and downward increase in temperature therefore enhances the increase in velocities. The final critical factors affecting seismic velocities are the fluid contents of the pore spaces of the rocks, and the fluid pressures

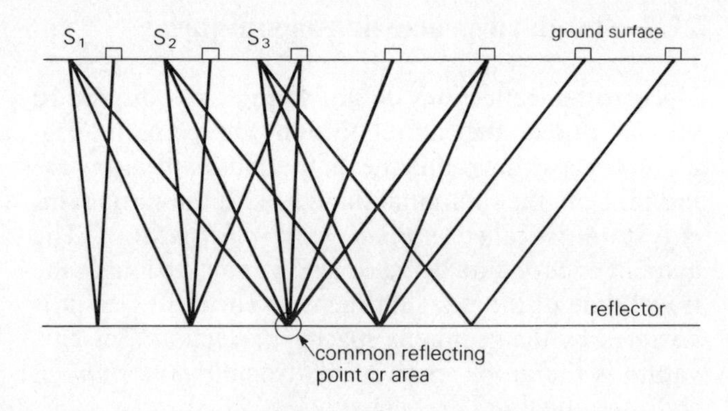

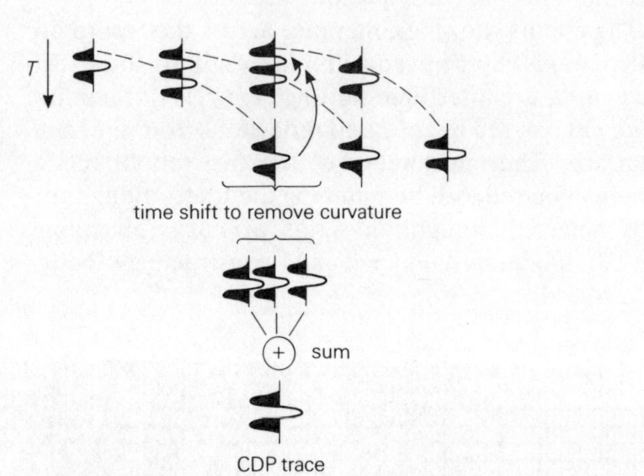

Figure 21.4 Steps in creation of common depth point (CDP) seismic trace. Reflections on all traces coming from within a small area of the reflecting horizon are summed to yield a single, enhanced trace.

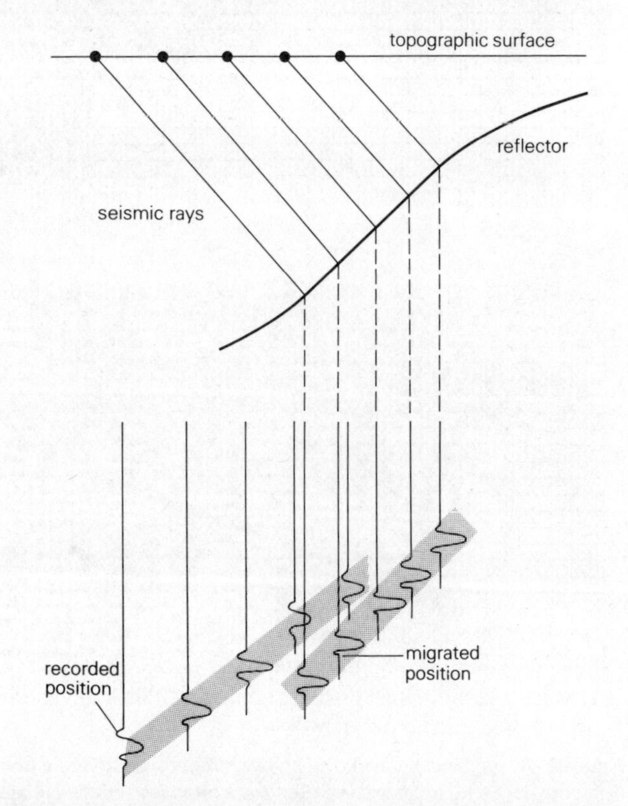

Figure 21.6 Migration of reflection data from dipping reflector.

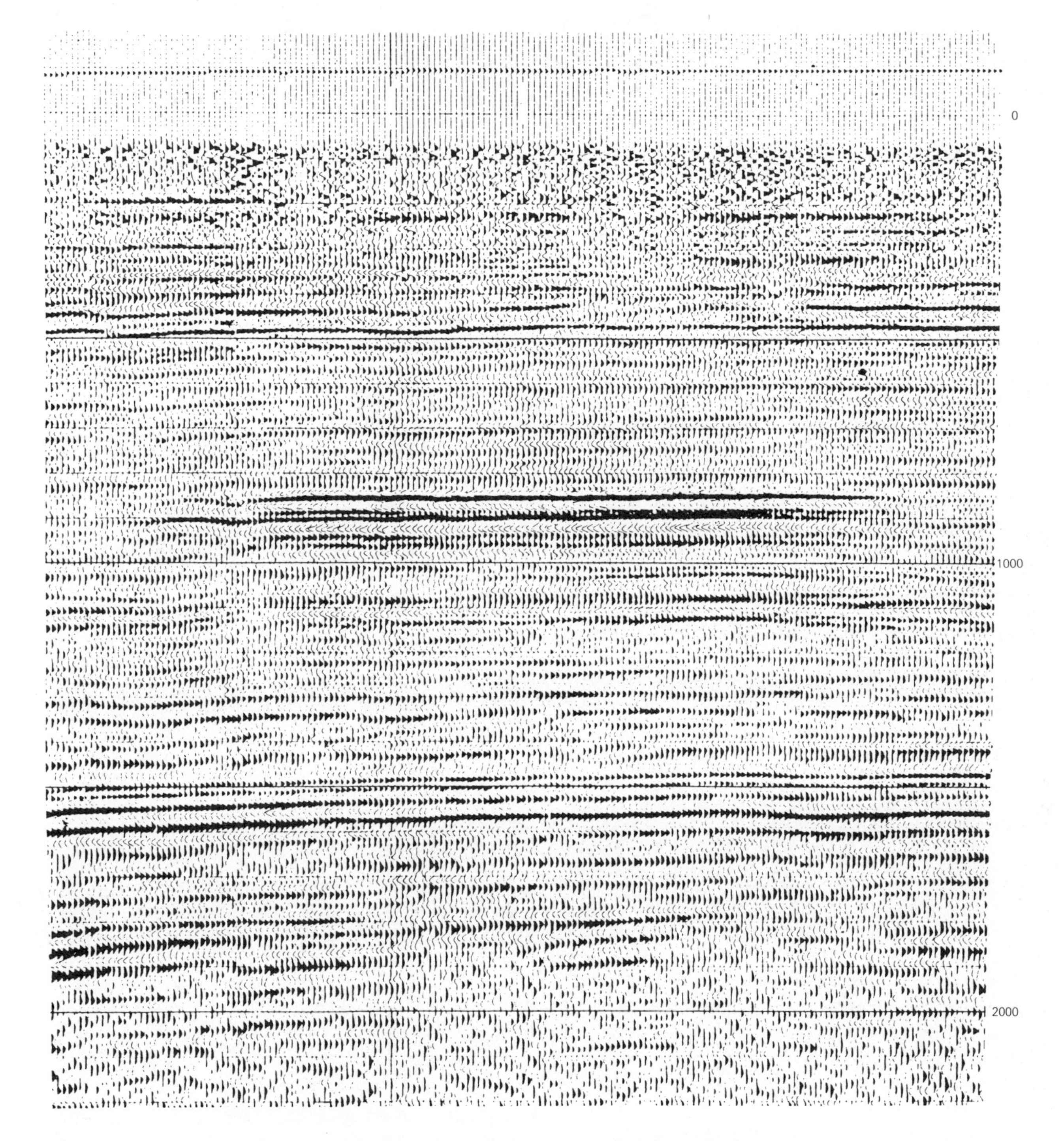

Figure 21.5 Stacked seismic record section. Six input traces (by sixfold recording and processing) have been summed to form each trace shown; there are about 180 traces on the section. Reflection data are shown by dark alignments.

exerted by them. V_p increases with fluid content, and so is higher in a fully saturated rock than in a dry or partly saturated rock. V_s on the other hand is lower in fluid-saturated rock than in dry rock, because fluids have no rigidity. The ratio of $V_p : V_s$ therefore increases with fluid content; it also increases with temperature in dry rocks.

21.3.1 Interpretation of seismic time sections

Note that the vertical axis of the stacked, corrected seismic section (Fig. 21.5) is scaled in *time*. Structure maps must show *depths*, not times, and conversion of times to depths requires knowledge of the velocities of the rocks providing the reflecting events selected for mapping. This in turn requires identification of these

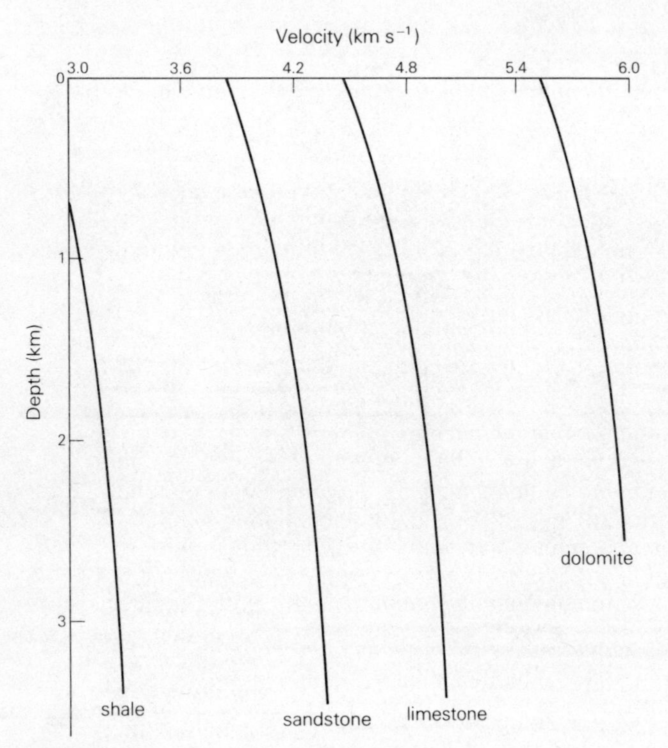

Figure 21.7 Velocity–depth relations for common sedimentary rock types, all having 10 percent porosity. (Courtesy Chevron Oil Field Research Company.)

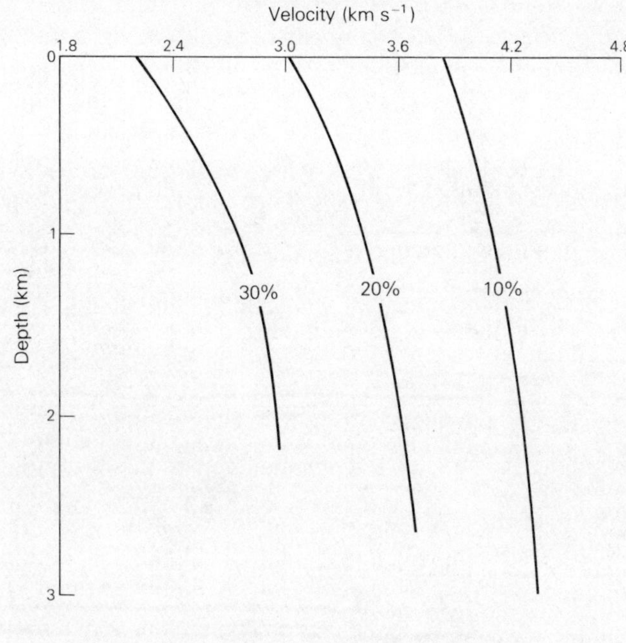

Figure 21.8 Velocity–depth relations for sandstones of varying porosity. (Courtesy Chevron Oil Field Research Company.)

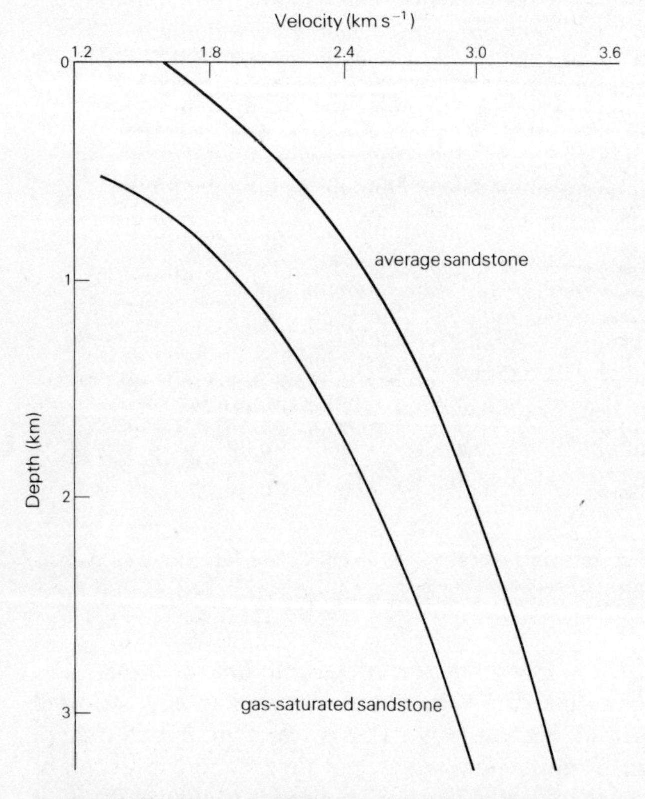

Figure 21.9 Velocity–depth curves for average water-wet sandstone and average gas-bearing sandstone, to show lowered velocity for latter. (Courtesy Chevron Oil Field Research Company.)

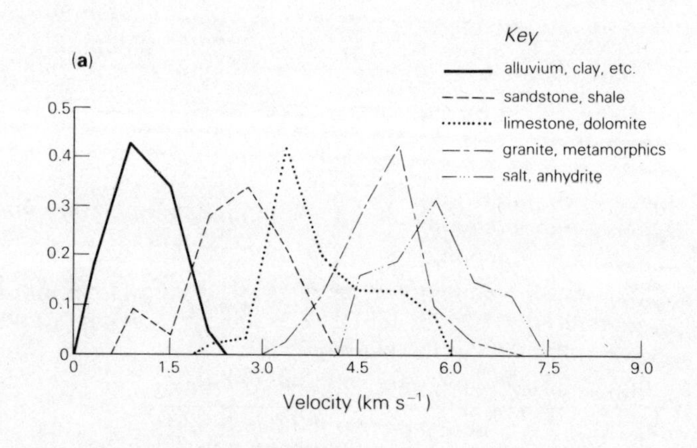

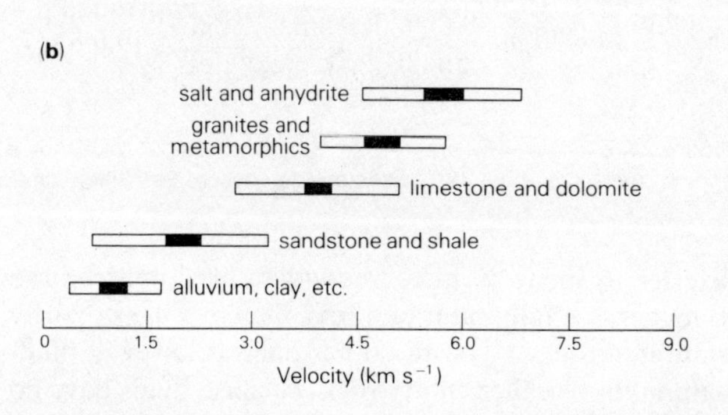

Figure 21.10 (a) Histograms of seismic wave velocities for various classes of rocks; (b) 80 percent fiducial limits for these values. (Data from F. Birch, Geological Society of America Special Paper 36, 1942.)

rocks. Some of the pitfalls inherent in this identification, in the absence of hard geologic data, may be illustrated by a few examples.

The first difficulty, and a very common one, is presented by poor or ambiguous reflection patterns (Fig. 21.11). The exploration seismologist's job would be relieved of most of its hazard if record quality was always as good as the best. Even when reflection quality is superb, uncertainty often persists. Figure 21.12 illustrates typical reflection patterns in fluviodeltaic complexes like those in the US Gulf Coast Basin. The geologist is invited to consider how many equally plausible interpretations could be made of these patterns if they were mapped in an undrilled area in which nothing was known of the subsurface geology.

21.3.2 *Examples of seismic records in complex structures*

In the seismic *time section* of Fig. 21.13, reflector A is identifiable on both sides of the section by comparably high *amplitude*. A large change in velocity above reflector A must occur between the left and right sides, perhaps across a fault downthrown to the left. The lower velocity on the left may be caused partly by the deep water, but more by a thicker and possibly younger succession of sandstones and shales there. The section must therefore be reconstructed as a *depth section*, by measuring times to A and converting them to depths by use of the velocity measurements. Reflector A now appears (lower Fig. 21.13) as probably continuous, not sharply faulted as it appears on the time section.

Figure 21.14a is a time section. Note the overlapping events on the right side. The indication of dip requires the conversion of the time section to a *migrated depth section* (Figure 21.14b). The overlapping events are thereby separated, leaving a "blank" area in the center. This may represent low-angle faulting, or simply a complication of dips too steep to record and process.

In Figure 21.15, reflections A and B appear to be the bottom and top of a salt horizon that has been gathered into domes. Unless the area is known to contain a thick salt horizon, the reflections could be attributed to some other cause – igneous bodies, perhaps, or pinnacle reefs. The reflections are lost at C. The structure there appears to extend to the surface, suggesting a piercement salt dome, but this is cast in doubt by the apparent downward continuation of the structure beyond the base of the presumed salt at A; this suggests a large fault. If the target of the drill is below reflecting horizon A, it is desirable to interpret the structure in the deeper horizons. Beds below A appear to be higher at wellsite X than at wellsite Y, but this is likely to be illusory because all rocks forming structures like these – salt, reefs, or igneous intrusives – have high velocities. They therefore produce *false structures* in underlying beds by a "velocity pull-up" (reducing the transit time through the high-velocity interval). The beds at and below A may therefore be perfectly flat at X, or even depressed.

Reefs are particularly susceptible to misinterpretation of depth through this phenomenon. If there is no recognizable reflection near the top of a reef, the reef may be mapped by false structure in pre-reef beds and assigned too great a depth. Typical reef signatures are illustrated in Figure 21.16. They may be contrasted with the much less smooth and regular convexity of a flowage structure (Fig. 21.17). The rim syncline adjacent to the salt wall is clearly delineated in this section.

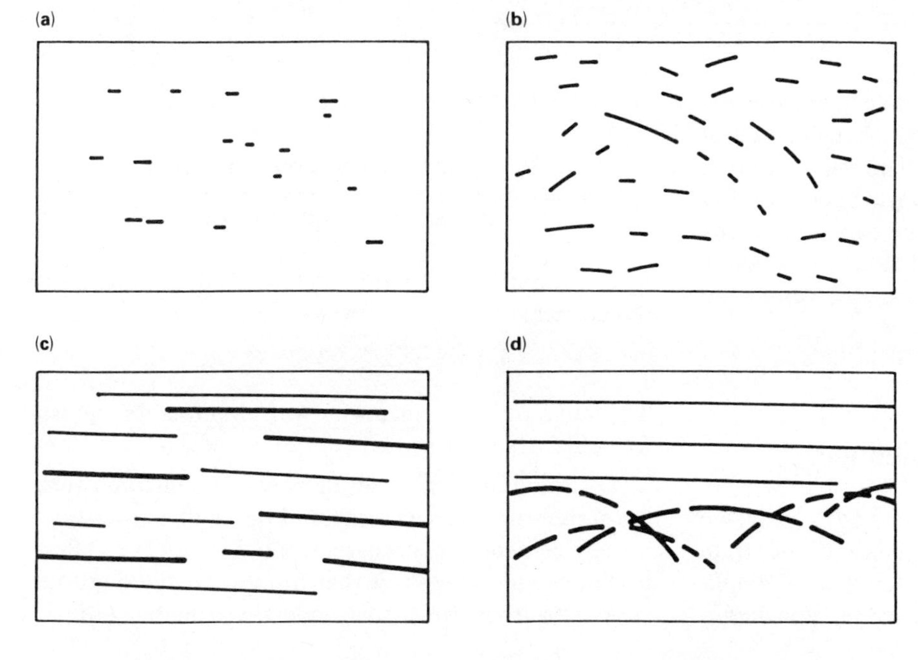

Figure 21.11 Common reflective patterns on time sections. (a) Poor reflections, due to lack of distinctive layering or velocity contrasts. May represent flowage of salt or shale, but note that these two have very different velocities. (b) Chaotic reflections due to some form of disturbance; possibly tectonic, more likely slumping, or complex buried channels. Important in offshore seismic work, because pattern may represent sea-floor conditions unsuitable for jack-up rig or production platform. (c) Discontinuous reflections, otherwise good. If all of low amplitude, probably interfingering sandstone–shale sections. If of sharply differing amplitudes, may be lake deposits. (d) Arcuate reflections usually represent rough surfaces on hard rocks (basement; lava flows; erosion surfaces).

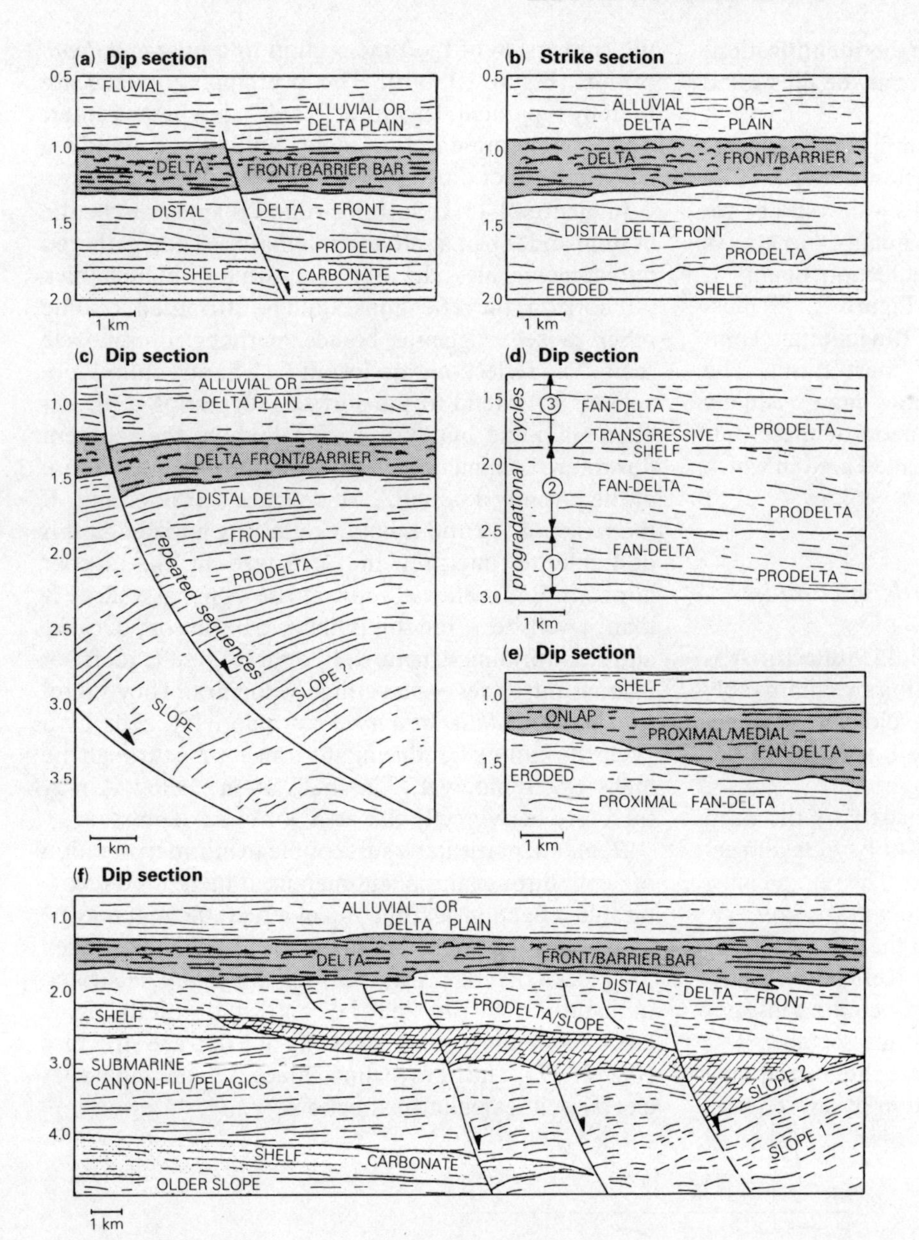

Figure 21.12 Dip and strike sections through deltaic and associated facies patterns, generalized from reflection seismic sections showing characteristic reflection attitudes and continuities. Vertical scale in seconds of two-way travel time. (From L. F. Brown, Jr and W. L. Fisher, *AAPG Memoir* 26, 1977.)

We may end this survey of seismic responses by complex structures with a dip section through a belt of overthrust faulting (Fig. 21.18). The characteristic features are the persistence of a governing direction of dip and the repetition of continuous bands of strong reflection converging down one flank, above essentially undeformed reflections representing the unmoved section below the sole fault.

21.4 "Bright spot" seismic technique

The amplitudes of reflections are governed, as we have seen, by differences in acoustic impedances of the strata at reflecting interfaces. High impedances lead to *amplitude anomalies* on integrated seismic traces. The *amplitude ratio* (R_s) is then defined by:

$$R_s = \frac{\text{peak or anomalous amplitude}}{\text{normal or background amplitude}}$$

It is proportional to the logarithm of the acoustic impedance:

$$R_s \propto \log \rho v$$

The value of $\log \rho v$ increases with depth and is higher for shales than for sands.

Acoustic impedances differ between cemented rock (nonreservoir), loosely cemented water-filled rock, and loosely cemented hydrocarbon-filled rock. The reflection of seismic waves is higher from water-filled porous rock than from dense rock, and *much* higher (10–50

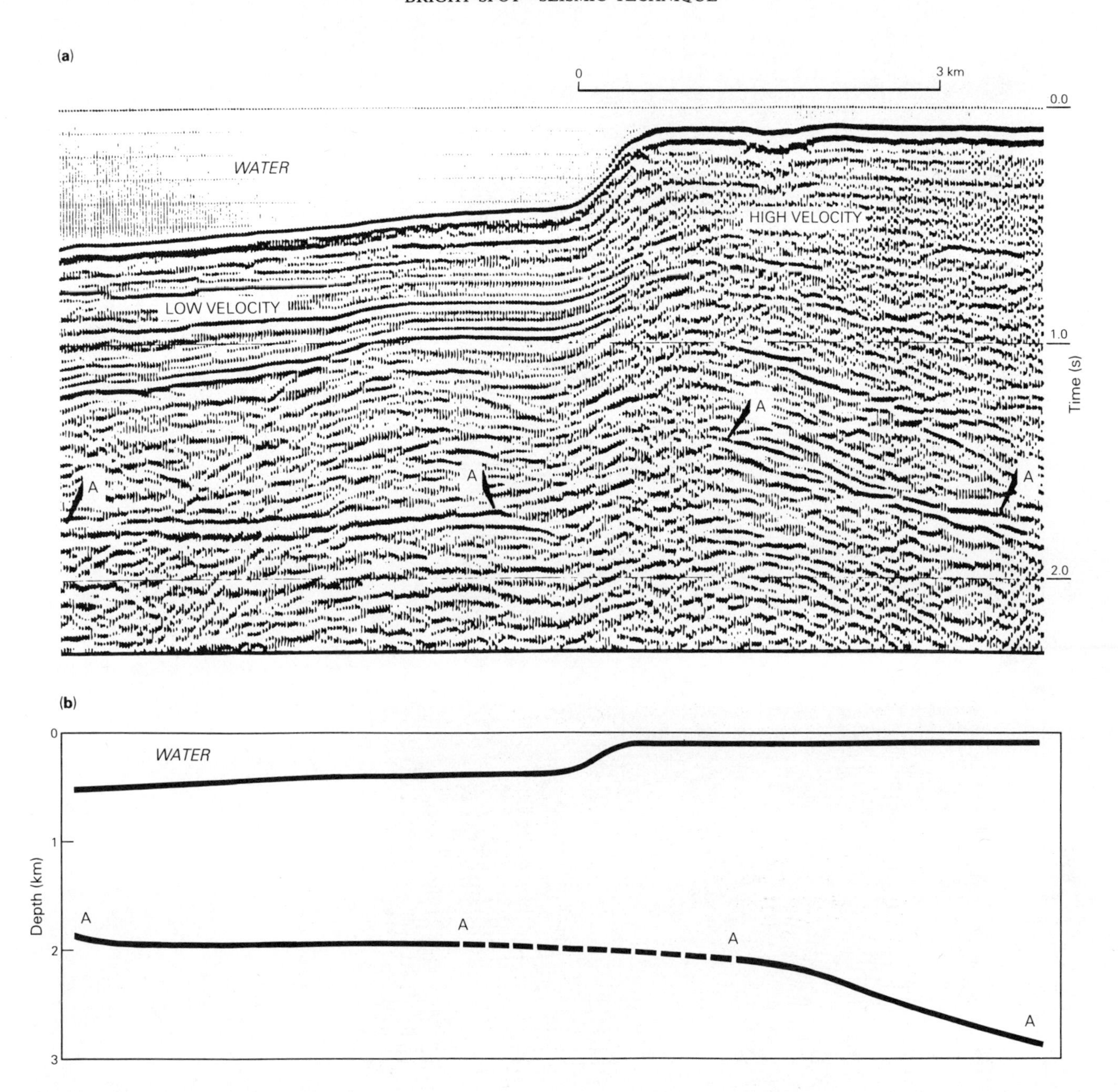

Figure 21.13 (a) Seismic time section; (b) corrected depth section. See text for explanation.

percent higher) from gas-filled porous rock. *Relative amplitude preservation* therefore permits gas-filled porous rocks to appear on integrated traces as very strong reflections or *bright spots* (Fig. 21.19).

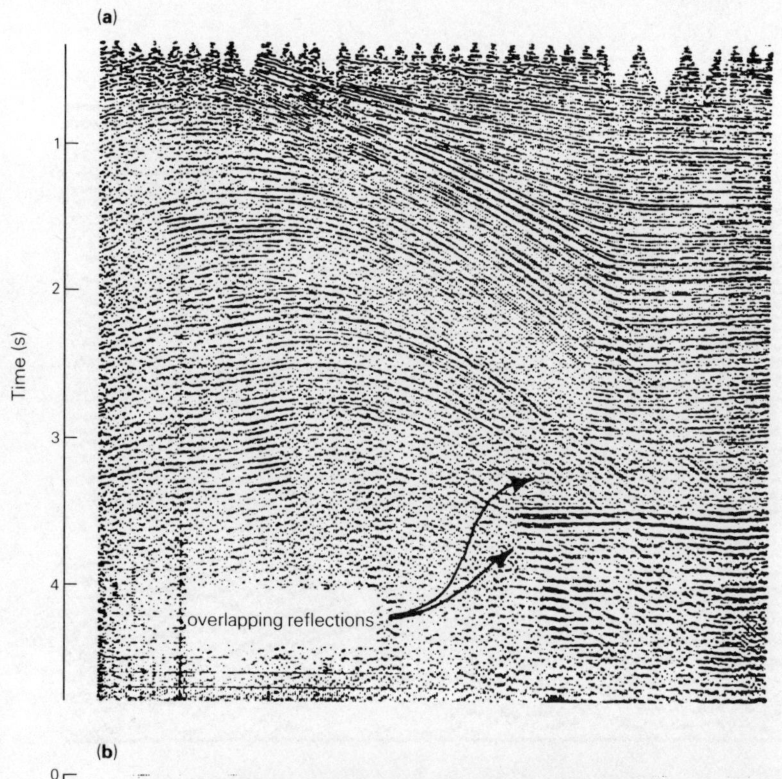

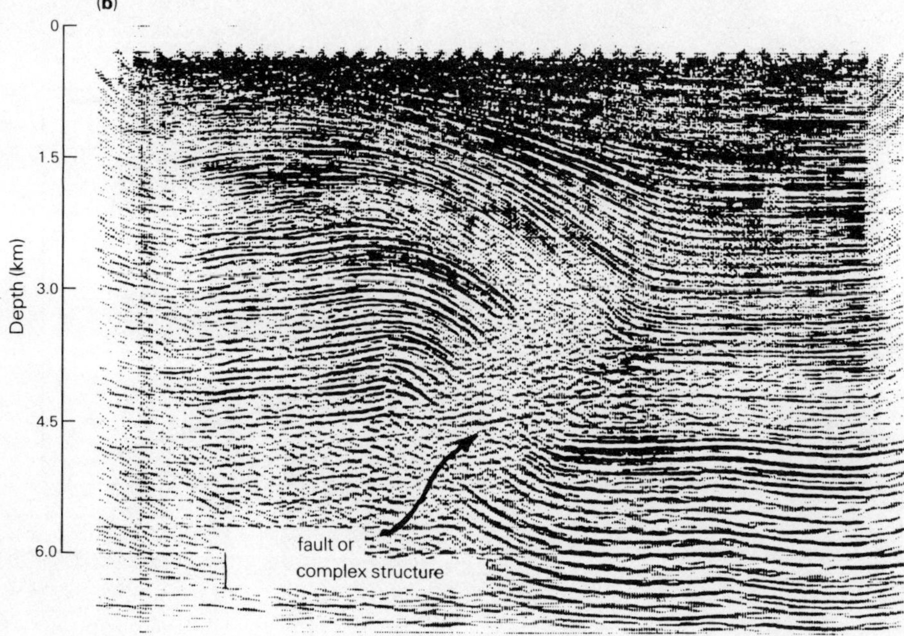

Figure 21.14 (a) Time section; (b) migrated depth section removing overlap of reflections. See text for explanation.

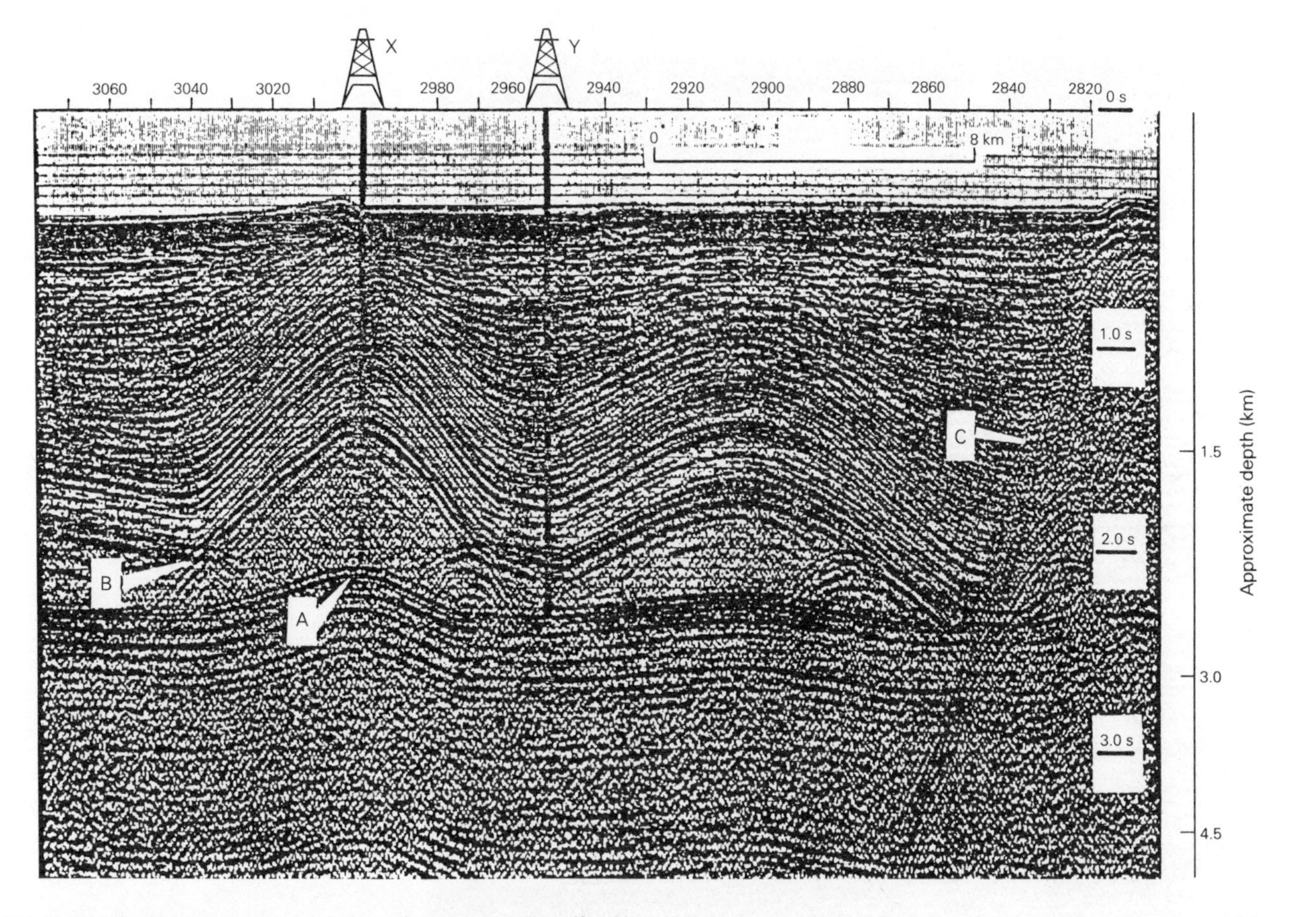

Figure 21.15 Seismic time section through apparent salt uplifts. Note that the vertical dimension is in units of *time*, requiring conversion to units of *depth* by knowledge of velocities. (Courtesy of Chevron Oil Field Research Company.)

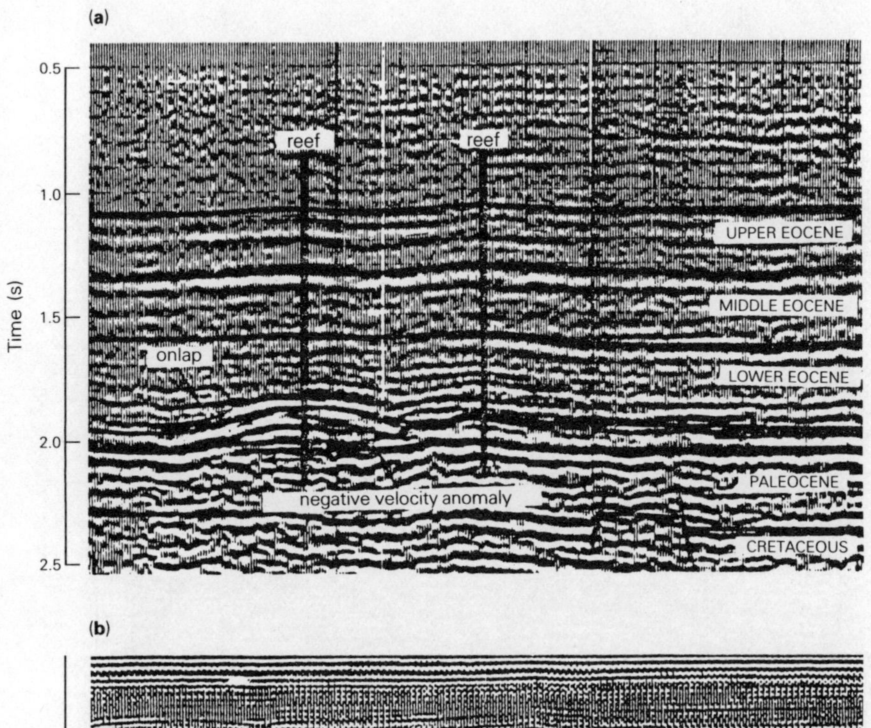

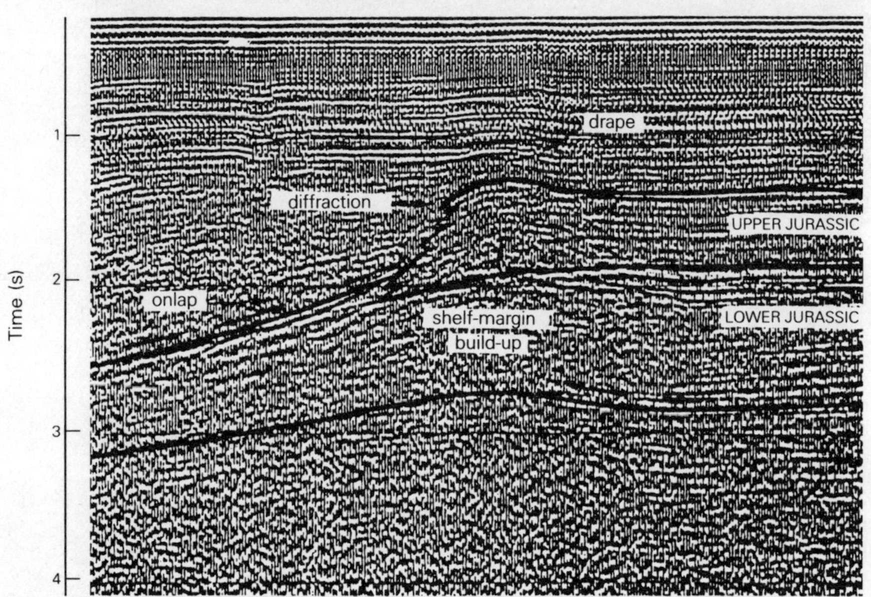

Figure 21.16 (a) Pinnacle reef in Paleocene of North Africa; length of section about 25 km. (b) Shelf-margin carbonate buildup in Jurassic strata, offshore West Africa; length of section about 15 km. In both sections, note onlapping strata seaward of the buildups, drape structures above them, and negative velocity anomalies below them. (From J. N. Bubb and W. G. Hatlelid, *AAPG Memoir* 26, 1977.)

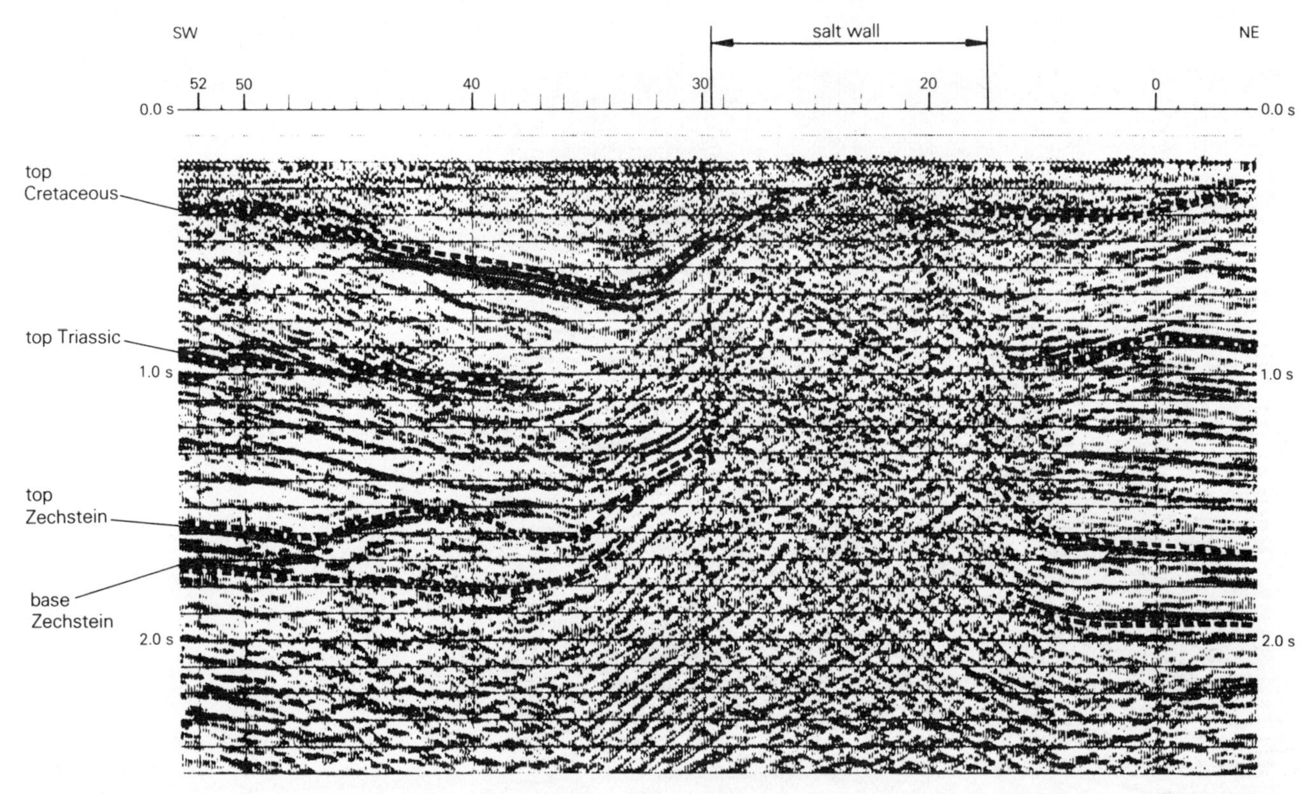

Figure 21.17 Seismic time section in the southern North Sea, showing a salt wall rising from Permian through Mesozoic strata. (From M. Gage, *AAPG Memoir* 30, 1981.)

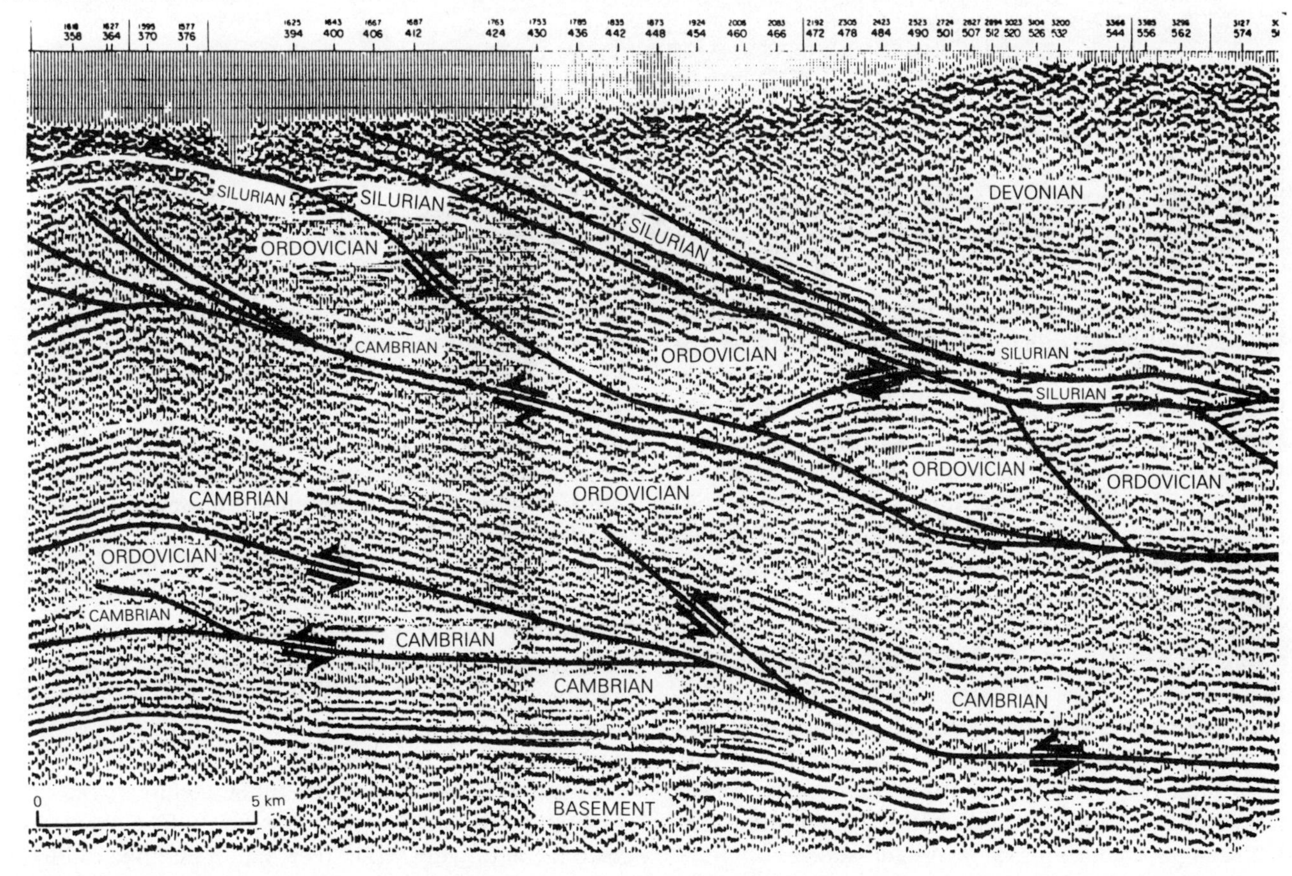

Figure 21.18 Seismic dip section through the Appalachian thrust belt, showing low-angle, east-dipping sole thrusts cutting up-section towards the west. Length of section 28 km. (From W. R. Scott and R. A. Geyer, *Oil Gas J.*, 26 July 1982.)

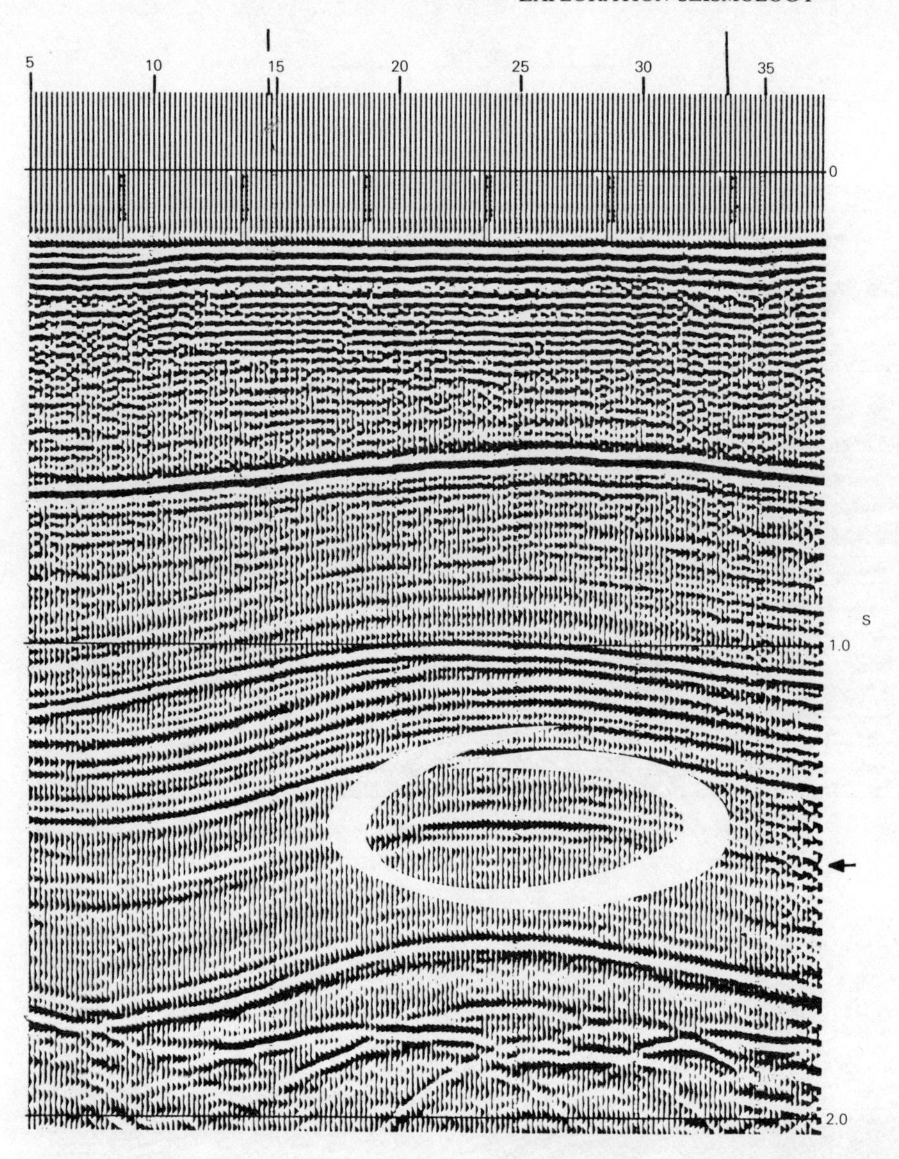

Figure 21.19 "Bright spot" on processed seismic time section. Unlike other strong reflectors, "bright spot" is not part of a continuous horizon, and represents a gas-bearing horizon at the crest of an anticline. (From Geophysical Service International, Ltd, in *Petroleum Gazette*, September 1974.)

Most of the seismic energy is reflected by the gas-bearing sands, and much of the rest is dissipated in the sands themselves if they are thick. Thus events from strata below the gas sands appear to sag downwards, because the low-energy return gives an effect opposite to that of high velocity. A porous, saturated rock (especially if it is gas-filled) is softer than a dense, barren rock of comparable lithology; it is therefore more likely to reflect compressional (P) waves as rarefactions than as compressions. This reflection shows up as a reversal of polarity at the receiving geophones from the edges of "bright spots" (Fig. 21.20); it is admittedly difficult to recognize this except on a large scale and after considerable experience. A final, invaluable characteristic of a "bright spot" anomaly is that it reveals reflections from horizontal surfaces within the fluid-filled rocks. These horizontal reflections are likely to be fluid interfaces, between gas and oil or gas or oil and water.

A combination of all these features is especially favorable (Fig. 21.21). After a number of wells have been drilled in an area, a petrophysical analysis can be made of the rock types represented, from gamma-ray and SP log data (see Sec. 13.5). From such analysis, acoustic velocities can be calculated, and hence densities can be plotted against depths of measurements. Once acoustic impedance data for the prevalent rock types are known, interpretation of seismic amplitude anomalies becomes much more reliable.

The "bright spot" technique is applicable most reliably to natural gas accumulations in young (Tertiary) sandstones, to a maximum depth of about 2500 m. Unfortunately, low-saturation gas accumulations, incapable of commercial production, also yield "bright spot" anomalies; so do lignites, some conglomerates, and a variety of volcanic rocks.

Russian seismologists call a similar technique the "diffraction conversion" of seismic records. They have also used the acronym CMDP (correlation method of direct prospecting) for the selection of anomalous absorptions of seismic energy over gas pools.

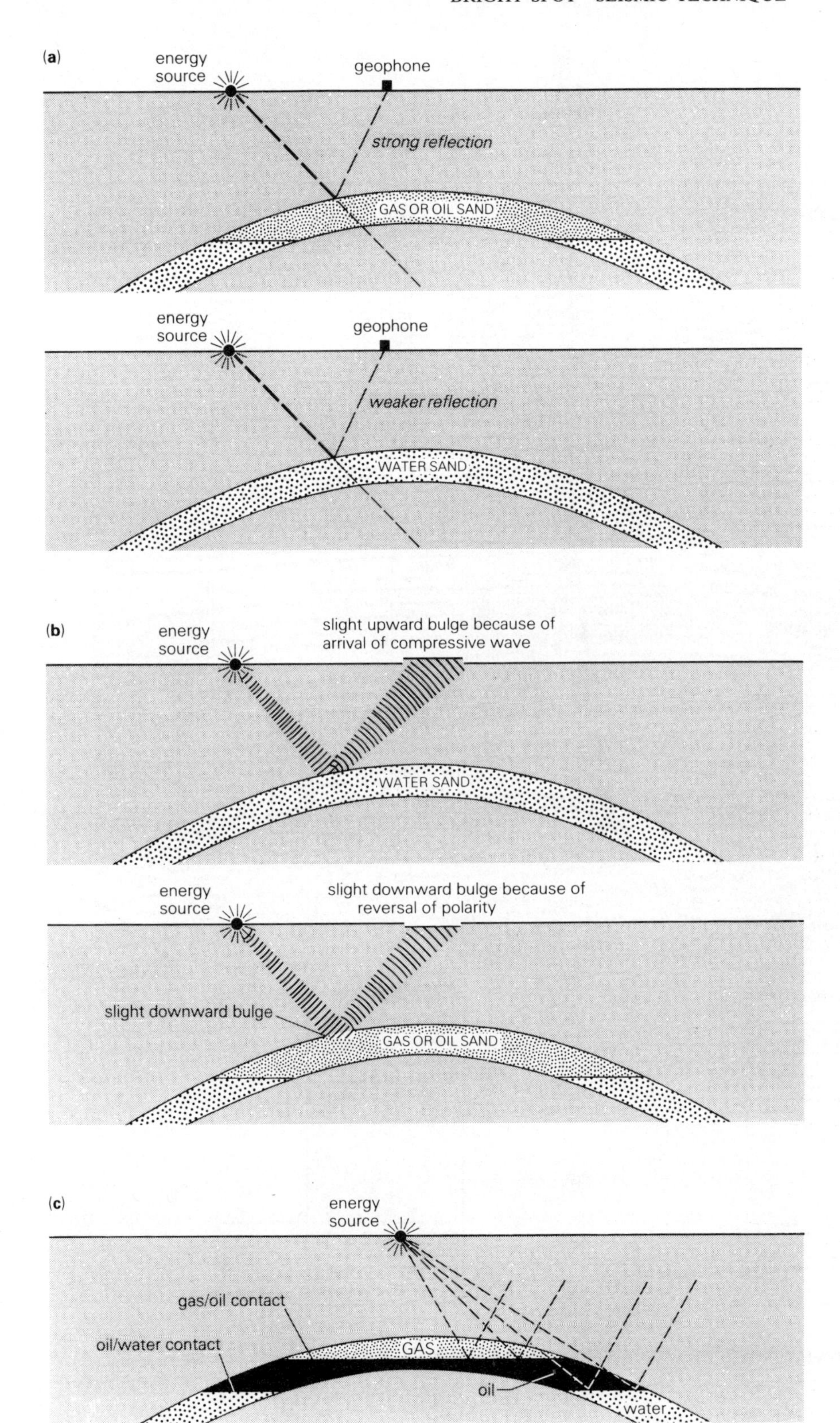

Figure 21.20 Three factors that might contribute to a "bright spot": (a) porous rock filled with oil or gas is a better reflector of seismic waves than a porous rock filled with water; (b) a gas-bearing sandstone may be sufficiently soft to cause reversal of the polarity of compressional waves; (c) in the absence of a strong hydrodynamic gradient fluid interfaces may be perfectly horizontal – no other reflecting surfaces are likely to be so. (From *Petroleum Gazette,* September 1974.)

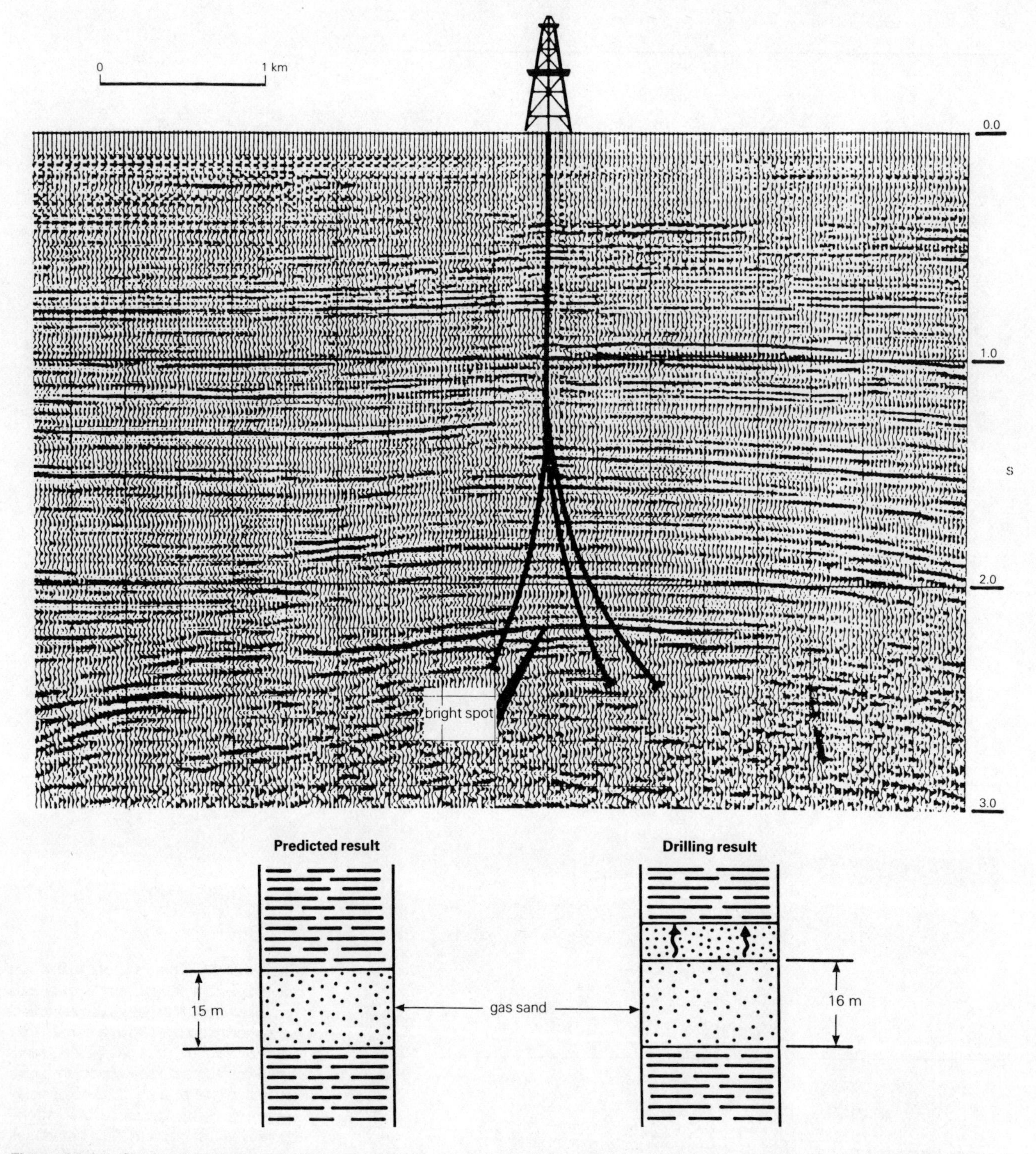

Figure 21.21 Gas-bearing structure in Neogene sandstone, Louisiana, discovered following interpretation of a "bright spot." (Courtesy of Chevron USA Inc., New Orleans.)

21.5 Seismic-geologic modelling

The geologist and the geophysicist join forces to construct *seismic models* of common geologic phenomena significant to petroleum occurrence. Once the general stratigraphic conditions in a region have been established, the average velocities and rock densities for different depth ranges can be measured. The relative sizes of the reflections to be expected from each formational contact can then be calculated. In Figure 21.22, as a simple example, a succession of shales and sandstones is modelled under different assumptions about fluid contents. If the principal sandstone contains gas, both its upper and lower contacts will provide much better reflecting events than they will if the sandstone contains water. The estimated seismic reflections are then compared with those actually obtained (Fig. 21.23).

Synthetic seismic sections, derived by converting reflections to acoustic impedance traces showing uncalibrated variations in rock properties directly, are really seismic sonic logs (see Sec. 12.2.2). They are easily calibrated with the aid of downhole sonic logs from nearby wells. As they show acoustic impedance contrasts without dependence upon convex structure, they offer especially useful aid in the search for stratigraphic traps.

21.6 Marine seismic exploration

Seismic exploration at sea is both much faster and much cheaper than equivalent coverage on land, because shipborne instruments can be operated continuously. The only drawback is the need for accurate navigation at all times, now provided by a variety of reflecting techniques.

The energy sources used on land may also be used at sea, except for the thumper. Dynamite is unpopular, however, especially in fishing grounds or near inhabited coasts; the vibrator is expensive. But three or four methods of generating seismic waves are better suited for the sea than for the land (Fig. 21.24). In the *sparker*, the waves are generated by a spark between two electrodes below the surface. High voltages are required, and salt water as the electrolyte. The *gas exploder* detonates an explosive gas mixture at the top of a vertical tube. The *air gun* introduces a pocket of compressed air about 5 m below the surface of the water.

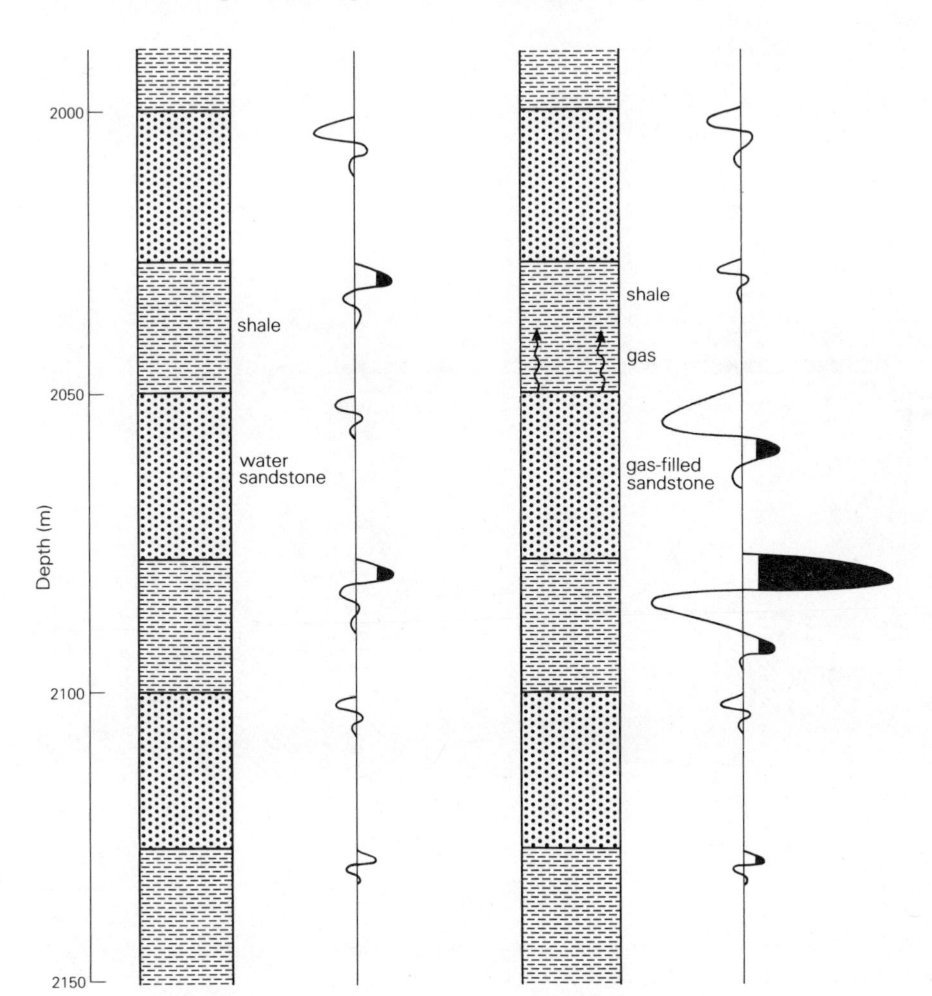

Figure 21.22 A sand–shale succession modelled under two assumptions about fluid contents, and simplified reflections expected under the two assumptions. (Courtesy of Chevron USA Inc.)

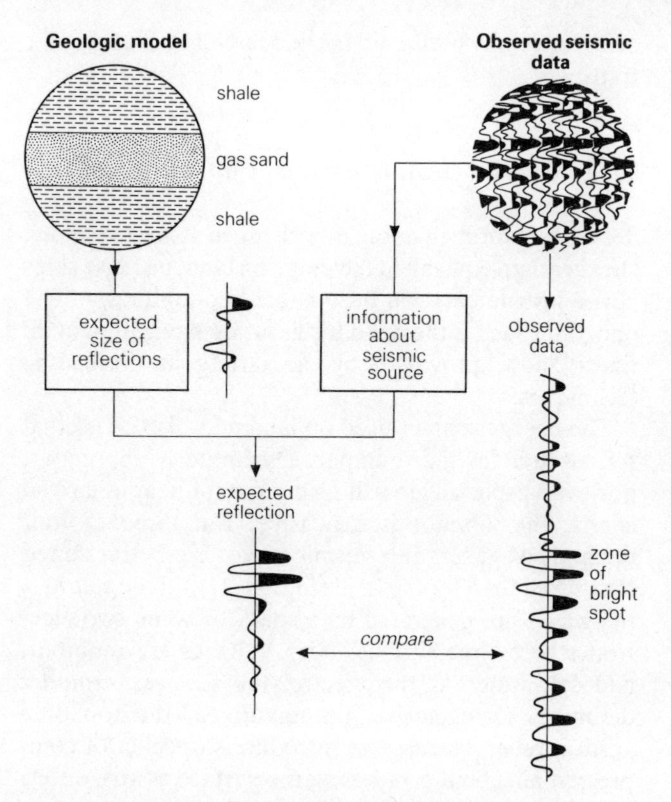

Figure 21.23 Comparison between observed seismic data and reflections expected from the geological model. (Courtesy of Chevron USA Inc.)

Sound waves penetrate the sea floor to a depth of 4 or 5 km, and their returns are picked up by hydrophones. In both the gas exploder and air gun methods, an onboard computer produces an acoustic reflection profile. Both methods are more reliable than the sparker, and fresh water is an adequate medium as no conduction of electricity is involved.

The continental shelves of the world have been extensively explored seismically, and the work is crucial to both petroleum exploration and development. No surface data are usable, and no well will be drilled without structural justification (Fig. 21.25). Only large, convex structures are suitable for development by the small numbers of development wells drillable in deep water; the structures must be found early in the exploration phase if that phase is to be extended. And development involves the positioning of expensive fixed platforms, which must be carried out accurately the first time (see Chs 23 & 26). Detailed seismic exploration must therefore be *repeated* after the discovery is confirmed. It must consist of multisensor, high-resolution profiles across the structure along several lines, preferably in three dimensions (see Sec. 21.8).

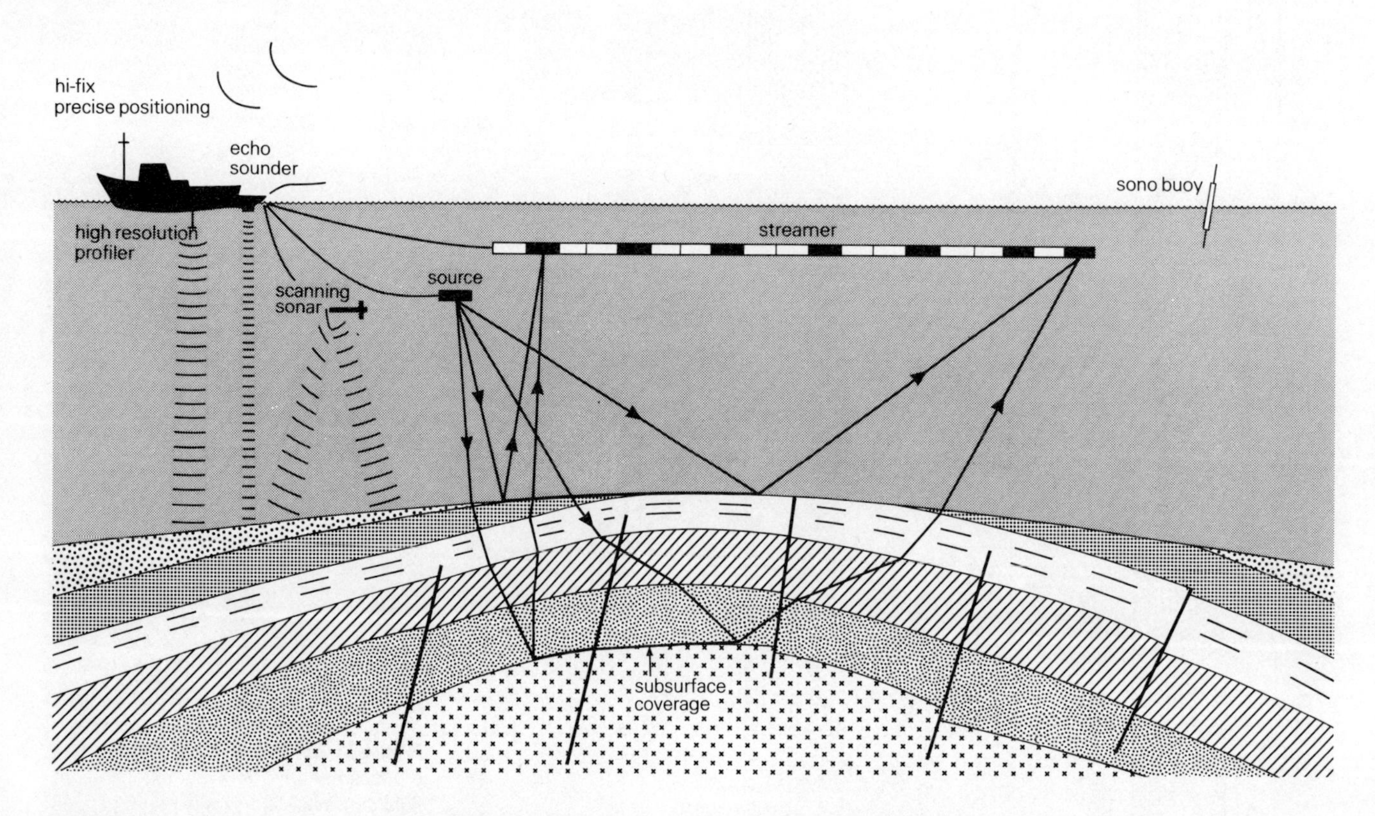

Figure 21.24 Marine seismic recording. Ship-borne recording instruments gather seismic, gravity, magnetic, and bathymetric data simultaneously. The process is much faster than its land-bound equivalent, but accurate navigation is vital.

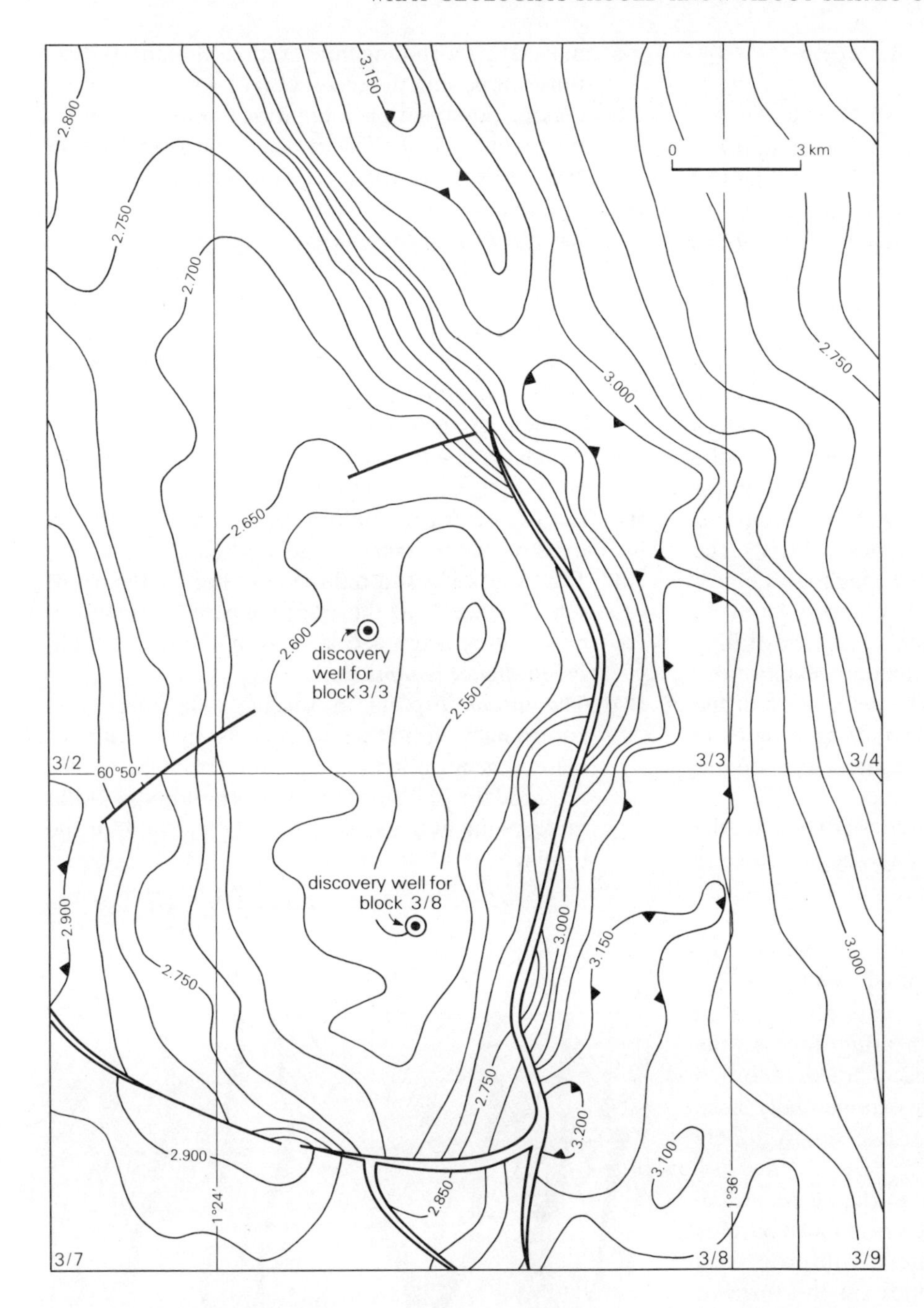

Figure 21.25 Seismic structure contours on the basal Cretaceous unconformity at Ninian in the British North Sea. The lines were shot in 1973; the two discovery wells were completed in 1974. The map was therefore prepared without detailed well control, and the contours are not depth contours but travel-time contours (interval 0.050 s two-way time). (From W. A. Albright *et al.*, *AAPG Memoir* 30, 1981.)

21.7 What geologists should know about seismic costs

Seismic exploration is expensive, though the decline in the use of dynamite was an economy. As smaller structures are sought in mature areas, more seismic effort is needed and shorter distances are therefore covered per crew-day or crew-month. The unavoidable consequence is higher cost per kilometer of coverage, because much more information is now yielded by it.

In 1970, operators paid about $600–700 per line-kilometer for conventional land crews in areas of "standard" geology and logistics. By 1984 they paid $1500–2500 per line-kilometer in the same areas. In difficult terrain, like overthrust belts, costs reach $5000–15 000 per line-kilometer. Offshore, in shallow waters, they were in 1984 about $8000 per line-kilometer. In remote, frontier areas they reached $15 000–35 000 per line-kilometer.

The geophysicists run the profiles and derive the data, but the geologists have to interpret them and try to gain the maximum benefit from them. The geologists must therefore choose the lines to be shot and the degree of detail to be sought. The costs of their choices are relatively small when compared with the costs of the leases and of deep wells, but the geologists should have an idea of what they are.

21.8 Modern advances in exploration geophysics

Great as have been the advances in our understanding of geological and geochemical processes since the end of the 1960s, advances in geophysical exploration capacities have perhaps been even greater. They have stemmed from two parallel developments, one purely technological and the other procedural and operational.

Students of geophysics can (and should) receive a much better grounding in the basic Earth sciences than their predecessors did. At the same time, it is now an inadequate university department of geology that does not provide its students with an understanding of the principles and processes of geophysics. Exploration for oil and gas (and for other Earth materials) now requires thorough integration of the work of the geologists, the geophysicists, and the geochemists throughout the program. Pre-drilling geophysical studies have to be made as complete as technologically possible when the drilling is to be carried to the depths of modern exploratory wells, especially in the offshore. The description of the discoveries and developments of two offshore fields in Chapter 26 offers a glimpse into the integrated studies now undertaken.

Technological advances in geophysics are outside the scope of a text on petroleum geology beyond a bare outline of the more important ones:

(a) The critical advance was that of *multichannel seismic recording*. In areas of geology at all complex, old-style seismic data were commonly rendered almost useless by their "smearing" across the array. The number of channels or traces recorded per shot have been increasing exponentially since the 1950s; 1024-channel recording was routinely available by 1982.

(b) The conversion from analog to digital recording was rapid during the 1960s. It was in fact a parallel development to that in computers, from processing analogs of numerical quantities (in the form of amplitudes or velocities of electrically recorded seismic waves) to processing the values directly.

(c) The introduction of seismic sources other than dynamite revolutionized the exploration process. The creation of p-waves by the vibrator (s-wave vibrators are also in use) enabled the *Vibroseis* technique to become dominant in land seismic surveys; by 1980, almost 95 percent of all such surveys involved either dynamite or Vibroseis. By that time also, *air guns* and *gas exploders* had taken over 85 percent or more of marine seismic surveys.

(d) The discrimination of individual reflected wavelets has been greatly enhanced. Minor variations in the character of the reflected signals were largely removed by shortening the durations of signal pulses, thus sharpening the resolution of the wavelets and making possible the elimination of multiple reflections (Fig. 21.3). The process, basically one of "unfiltering," is called *deconvolution*.

(e) The acquisition of high-density, high-resolution seismic data in *three-dimensional imaging* is now practical. It enables more accurate interpretation to be made of complex structures, especially if they are faulted (Fig. 21.26). It is especially valuable for offshore structures after discoveries have been made, allowing faster delineation of the field boundaries by reducing the number of step-out wells necessary, and facilitating the proper positioning of the platforms.

(f) The use of *bright spot technique* is now routine in areas of young strata (Figs 21.19 & 21).

(g) The processing and reduction of the huge mass of seismic data, and the improvement of signal-to-noise ratios, were accelerated by the use of high-speed *digital computers*.

(h) The *visual displays* of seismic data have been enormously improved, especially by the use of color. The introduction of *seismic stratigraphy* has required the presentation of lines hundreds of kilometers long in the space formerly required for the display of a local survey.

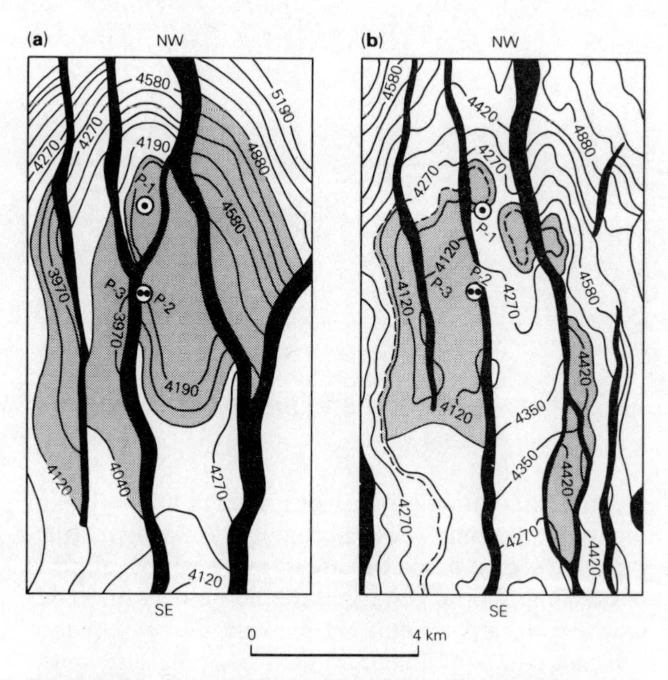

Figure 21.26 Two-dimensional (a) and three-dimensional (b) seismic structure maps made following the discovery of gas offshore from southeastern Trinidad. Note the greater detail in three dimensions, especially in the locations of faults with respect to the wells. (From J. F. Fuller and J. T. Major, *Oil Gas J.*, 16 August 1982.)

(i) There have been concurrent advances in *well logging* (see Ch. 12) and in *borehole geophysics*. Data derived from these technologies are routinely computerized and taped. The borehole gravimeter provides data from 100 m or further from the borehole. *Electromagnetic sounding* may detect hydrocarbons through their effect on the voltage decay of the introduced current pulse.

21.9 Seismic stratigraphy

Stratigraphers have recognized for many years that some unconformities are of subcontinental to continental extent. The packets of strata between these interregional breaks represent both rock units and time-rock units of higher rank than the highest employed in their respective categories, the supergroup and the system. They are, moreover, strikingly correlatable from craton to craton, though there are important exceptions. Ordovician and Silurian strata commonly occur together over wide areas of several continents, as do the Devonian and the Lower Carboniferous (Mississippian); but Silurian–Devonian continuity is much less common simply because of the worldwide effects of the redistribution of plates which we call the Caledonian orogeny.

Unconformities within systems may represent much more significant discontinuities than systemic boundaries; they are also of far greater significance to the petroleum geologist (see Sec. 16.5). Nearly all large petroleum basins provide illustrations of this important circumstance. In the Mid-Continent province of the USA, the Lower Pennsylvanian strata of the Ouachita Mountains are more closely related to the Ordovician or the Devonian than they are to the Upper Pennsylvanian molasse, from which they are separated by a major orogenic unconformity. In the North Sea Basin, there is a regional unconformity between the Jurassic and the Cretaceous, and its influence on oil distribution is critical because the principal source sediment lies immediately below it (Figs 16.79 & 80); but unconformities in the middle of the Jurassic and the middle of the Cretaceous are even more important in the trapping mechanisms (see Sec. 16.5.6). In the Middle American province, the petroliferous successions north and west of the Gulf of Mexico and south of the Caribbean are essentially Cretaceous and Tertiary; but the Paleocene and Lower Eocene, containing important reservoir rocks, are depositionally the closing components of the Cretaceous succession, and the Upper Eocene, following a regional unconformity, is the real beginning of the Tertiary succession.

L. L. Sloss in 1964 called such unconformity-bounded packets of strata *sequences*. Some prefer to call them *synthems*. Whatever they are called, they are the rock records of one of the largest rhythms in the Earth's history. The unconformities bounding a sequence above and below are not synchronous surfaces, but as they are consequences of plate movements and their attendant worldwide rises and falls of sea level the boundaries tend to be systematically heterochronous and are far from unpredictable. Most regional cycles of sedimentation, diastrophism, and erosion must be global, so that therc is good, if imperfect, correlation across whole continents and even between continents.

The great strides in seismic resolution and the development of multichannel recording made possible the identification of these interregional boundary surfaces on seismic profiles. Continent-wide seismic reflection programs like COCORP and COPSTOC now survey the entire thickness of layered rocks of the continents, including the deep Proterozoic component. This development in seismic exploration interested the petroleum geologists chiefly in showing the extensions of layered rocks far below the thick sections in deformed belts like the Appalachian and Ouachita Mountains. Much more valuable, however, was the demonstration that correlations of seismic reflection cycles are traceable over much of the ocean basins also. Correlation from prospect to prospect became possible *in advance of drilling*. Such correlation is not of course as accurate as that using drillhole data, but it has the advantage of being continuous and not subject to guesswork across the wide gaps which must inevitably remain between closely controlled patches in deep-water drilling.

Careful correlations demonstrated that the stratigraphic intervals bounded by regional seismic reflectors were of a much lower order than Sloss's sequences. Many are clearly traceable even where both upper and lower boundaries are essentially conformable. Where this is seen to be the case it is essential that the geologists and the geophysicists identify precisely what the reflectors represent. The primary exploration objective in the use of *seismic stratigraphy* is its assistance in the tracing of *facies* across wide areas. This is reliable only if the signatures of acoustic impedance (the reflectors) are tracking time-stratigraphic surfaces of deposition and not something else; this may be impossible of determination by seismic means without extensive drilling.

Use of seismic stratigraphy in exploration therefore concentrates on three features of the profiles, selected for their significance by Peter Vail and his colleagues in the Exxon group of companies (Fig. 21.27). First are the actual forms of the reflectors and the manner of their terminations between sequences; geologists need to judge whether wedge-edges of stratal packets are depositional or erosional. The tracing of successive marine margins by the onlaps of their coastal deposits, for example, itself offers a useful preliminary discrimination

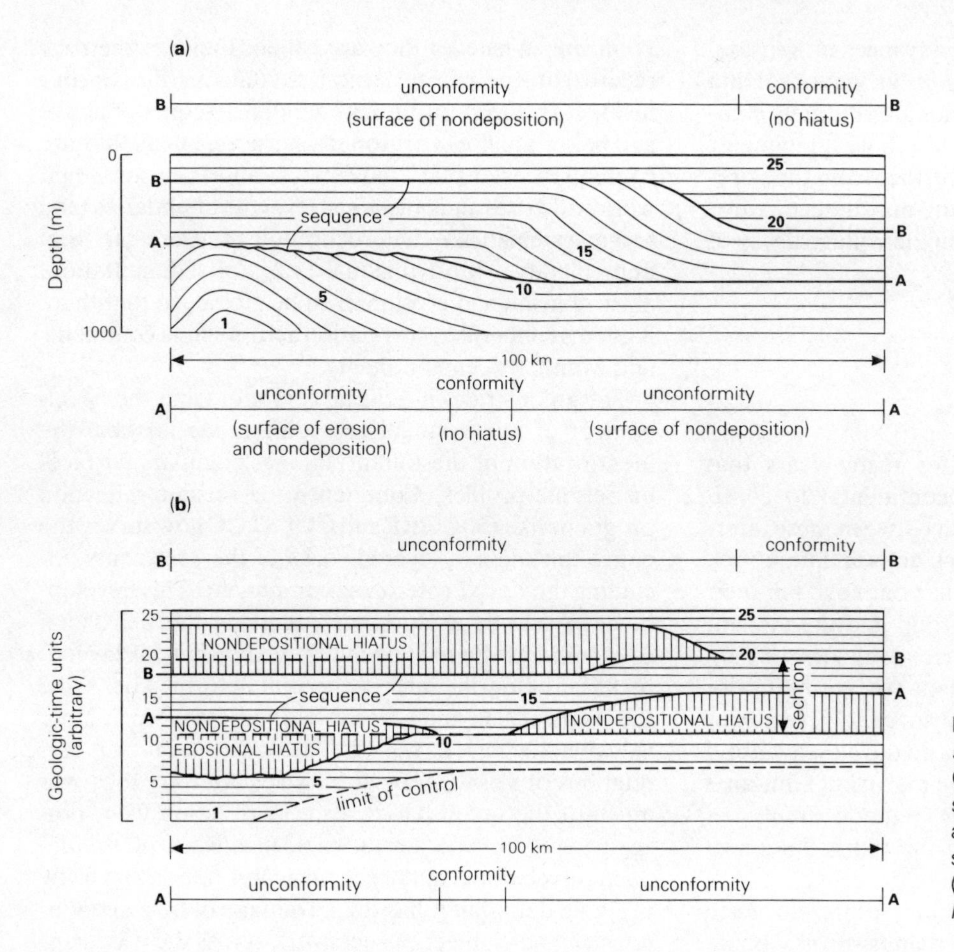

Figure 21.27 Basic characteristics of a *sequence* as defined by Vail *et al.* (a) Generalized stratigraphic section of a sequence, between two regionally traceable surfaces; (b) generalized chronostratigraphic section of a sequence. (From R. M. Mitchum Jr *et al., AAPG Memoir* 26, 1977.)

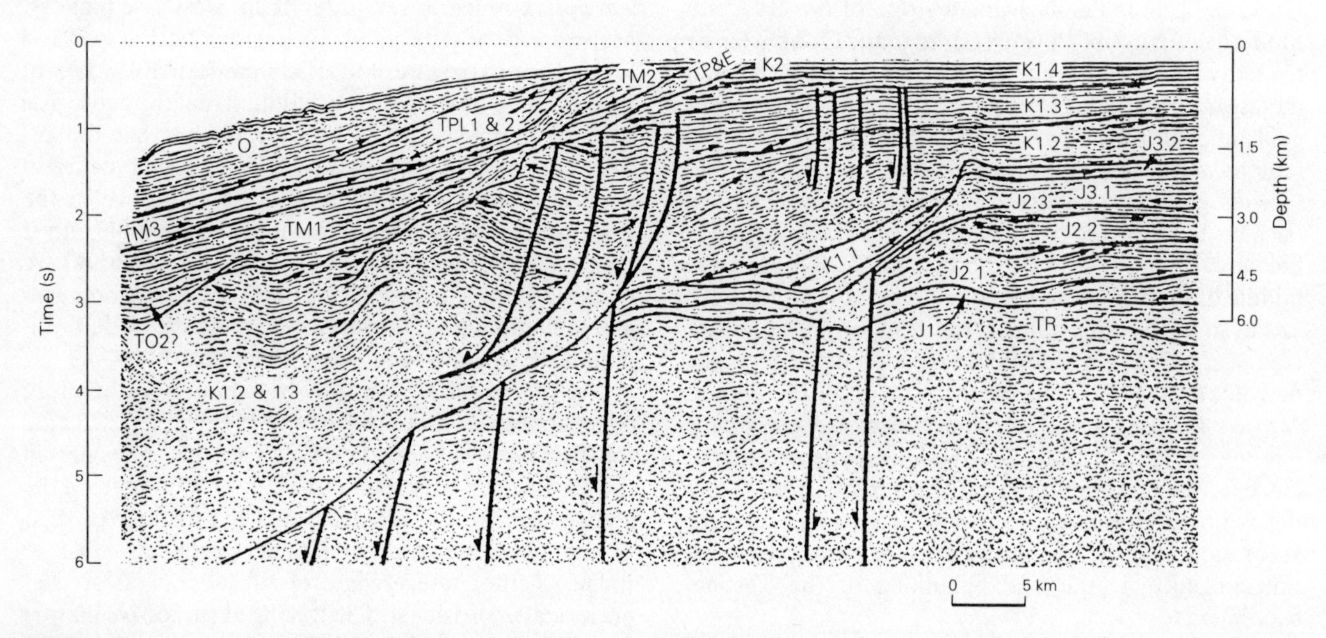

Figure 21.28 West–east seismic profile off the West African coast, showing major sequence boundaries. Stratigraphic units are of Tertiary, Cretaceous, and Jurassic ages, resting on undivided Triassic. (From R. G. Todd and R. M. Mitchum Jr, *AAPG Memoir* 26, 1977.)

between areas of maximum and minimum favorability for petroleum occurrence. Second are the three-dimensional shapes of the stratal units bounded by the seismic surfaces; they may be sheets, wedges, lenses, mounds, or variations of these, and these shapes are readily assigned to their most probable lithostratigraphic and structural foundations. Third are the configurations of secondary reflectors within the sequences. These are less conspicuous or continuous than the reflectors bounding the sequences, but any detectable pattern provides valuable information. Persistently oblique layering, or a hummocky pattern in the minor reflectors, for examples (Figs 21.18 & 28) are not likely to be attributable to the same stratigraphic environments as persistently parallel or markedly contorted patterns (Figs 21.17 & 29).

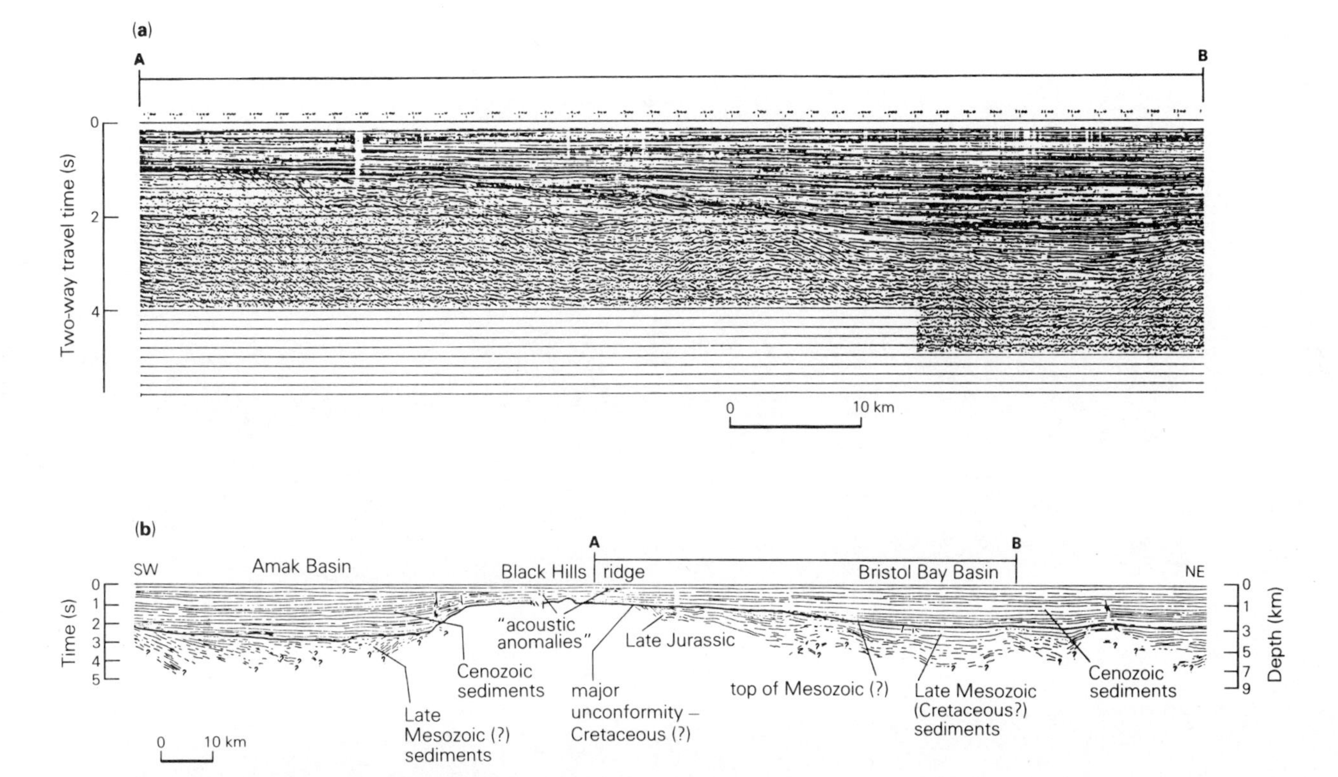

Figure 21.29 Seismic stratigraphy. (a) Example of 24-channel seismic reflection record from the southern Bering Sea shelf. (b) Interpretative drawing of seismic reflection profile, including part AB represented on seismic record above. (From M. S. Marlow and A. K. Cooper, *AAPG Bull.*, 1980.)

22 The petroleum geologist's maps and cross sections

22.1 Important generalities

Petroleum geology is first and foremost a science of the subsurface. Most of the maps made and used by petroleum geologists are therefore maps of the subsurface, but all geological and geophysical maps are part of his stockin-trade, including maps of surface geology and tectonic maps. So also may be many other kinds of surface maps – terrain maps, maps showing well locations, land holdings, land usage, or right of way. The geologist may need to combine surface and subsurface data on maps; if so, the two kinds of data should be distinguished. An elementary example of the danger of failure to make this distinction is the inclusion of both surface and subsurface thicknesses on a combined facies and thickness map of a vital stratigraphic unit. The outcrop areas have undergone an unknown amount of erosion in the present erosion cycle; the subsurface areas have not.

In this book we can deal only with those maps largely specific to the petroleum industry, and even those will necessarily omit some specialized maps devised by imaginative geologists, geochemists, geophysicists, or engineers for use in their own activities. We may preface our survey of subsurface maps with some cautions that apply to the makers and users of all maps.

First, though all easily expressed scales and contour intervals are in use, both should be appropriate to, and consistent with, the data available to the mapper. Spurious accuracy on maps or cross sections may be more misleading than spurious accuracy in tables of numbers, because it is *visual*. The desirability of converting all maps and sections to metric units is as great as that for other measures, including volumes of oil and gas and the properties of them and their host rocks. The greatest obstacle to this conversion, for the petroleum geological mapper, is the devotion of the English-speaking countries to the measurement of legal land holdings in acres.

Second, maps should not be so overloaded with data that they become confusing. Maps are working documents; they are useless if they are not informative, and their usefulness is greatly diminished if they are informative and comprehensible only to their makers. Standardization of scales, patterns, and colors is very helpful.

Third, all the geologist's maps and sections of any one area must conform with one another (and with the geophysicist's maps). Conflicts and anomalies between maps must be resolved; something valuable may be learned from the very act of resolution.

Fourth, and most important, geologists and geophysicists must learn the art of contouring. This is so critical that it deserves a short section of its own.

22.2 The matter of contouring

A 14th-century English divine bequeathed to us a Latin caution: "*Entia non sunt multiplicanda praeter necessitatem.*" Logicians call this Occam's Razor and consider it a fundamental law of their subject. Scientists use it in their habit of adopting the simplest explanation that fits the data; it is considered unscientific to draw conclusions not warranted by one's data.

This rule is a poor guide to contouring, a basic exercise of both the geologist and the geophysicist. Scores of oil- and gasfields have been discovered by people who put structures on their maps not clearly required by their control points and turned out to be right. Experienced petroleum geologists complain that neither young geologists nor young geophysicists know how to contour, and that it takes them a long time to learn the art. The reasons are simple and twofold. Young geophysicists are a species of applied mathematician; young geologists remember (or should remember) their college mathematics and probably their high-school geography. Both draw contours according to simple mathematical principles, as a computer would draw them. Lines are straight and parallel wherever they can be made so, and uniformly spaced across the intervals between control points. These are unexceptionable rules as far as they go. Unfortunately, many oil and gas pools, like many ore bodies, are preferentially located where the rules break down. Petroleum traps are formed by something interfering with stratigraphic and structural continuity; they are more likely to occur where subsurface contour lines are neither straight, nor parallel, nor continuous, nor equally spaced. Too often, these deviations from the rules are not identified until after a discovery has been made.

The second reason for the difficulty is that young people learn to contour on maps representing parts of the present topographic surface. The maps are intended to depict erosional relief, and examples are chosen to illustrate either striking relief features (hills and valleys, cliffs, volcanoes), or familiar relief features (those near home, school, tourist region). Erosional topography is dependent upon laws governing slope development, and the humid temperate regions in which most educational establishments are situated may not provide such "standard" relief as their inhabitants suppose.

Subsurface horizons should not be contoured as if they bore this kind of relief unless the geologist has legitimate reason to believe they do so. The Precambrian crystalline basement of the Russian platform, for example, may truly have the kind of erosional relief it is shown to have in Russian geology texts (Fig. 16.42); but it is nearly inconceivable that the structure on the top of the Devonian System across that platform has the same origin and the same character. Sedimentary strata, even when eroded, should not be contoured as if they were crystalline or granitic rocks. The basis of good contouring is *fidelity to the structural style* of the region. Until this is known, the geologist should try a series of interpretations, each faithful to a consistent opinion on the structural style as well as to the known stratigraphy.

Figures 22.1–5 illustrate the variety of interpretations possible if all the geologist observes are the spot values. If he is faithful to the structural style of his region, he will not make structures and isopachs in a region of simple folding (see Fig. 16.37) look on contour maps like those in a region of block faulting (Fig. 22.7), or of salt pillows and saltless sags (Fig. 22.12), or of reefs (Fig. 22.14), or of extreme erosional topography on a great unconformity (Fig. 17.1). The strongly faulted basement of the US Rocky Mountain province (Fig. 22.6) may be contrasted with the nearly unfaulted sags and swells in the cratonic interior on the opposite side of the Transcontinental Arch (Figs 17.8 & 18).

Occam's Razor is not really defied by imaginative contouring. If two control points have very different values, whether of elevation or of thickness, a uniform slope between them is no simpler or more likely geologi-

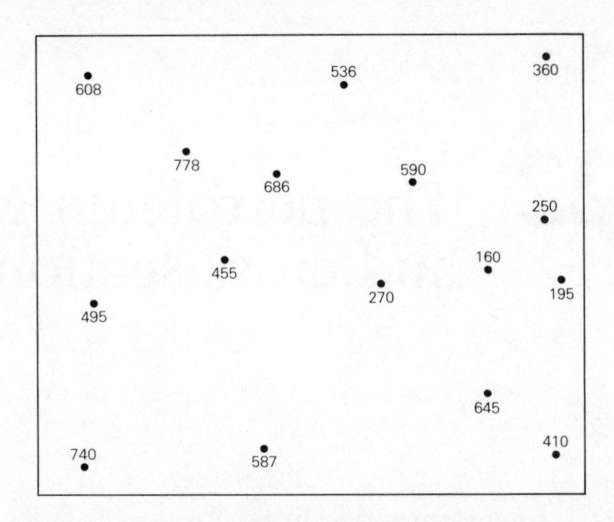

Figure 22.1 Numerical values for data points used to construct maps in Figs 22.2–22.5. Horizontal scale arbitrary.

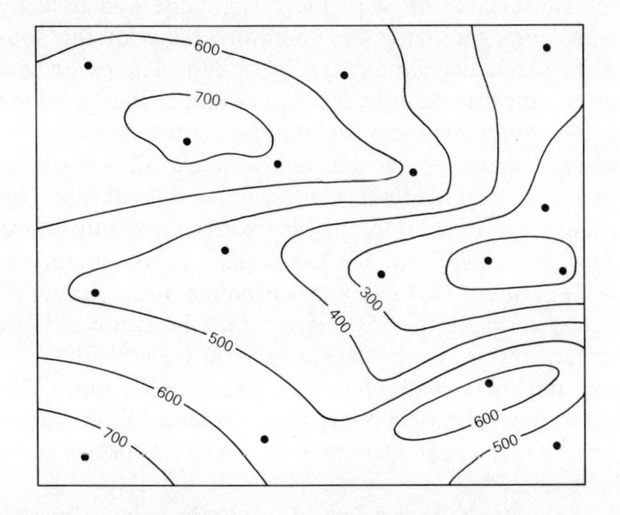

Figure 22.2 Data mapped by mechanical spacing of contours, attempting to honour all points equally.

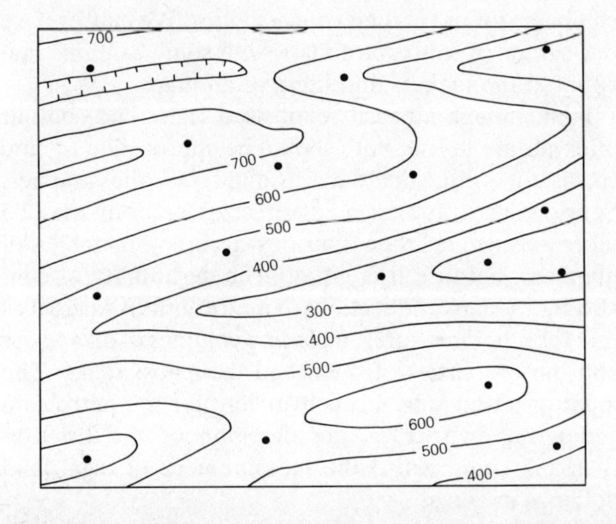

Figure 22.3 Data mapped as series of parallel ridges and valleys.

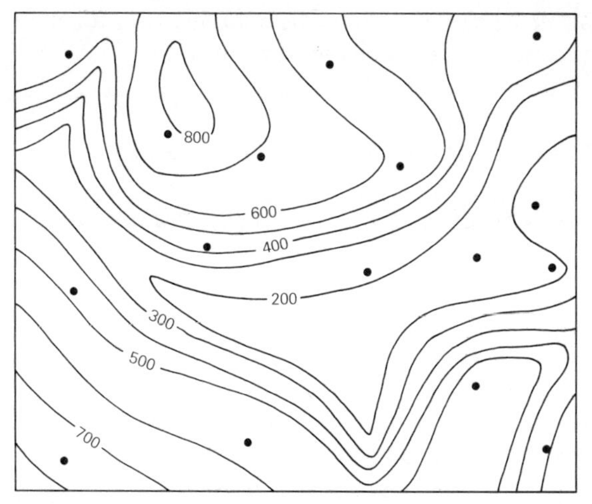

Figure 22.4 Data mapped to represent a youthful, stream-eroded landscape.

Figure 22.5 Subsurface data mapped interpretatively in region known to involve sinusoidally folded beds and dip-slip faults.

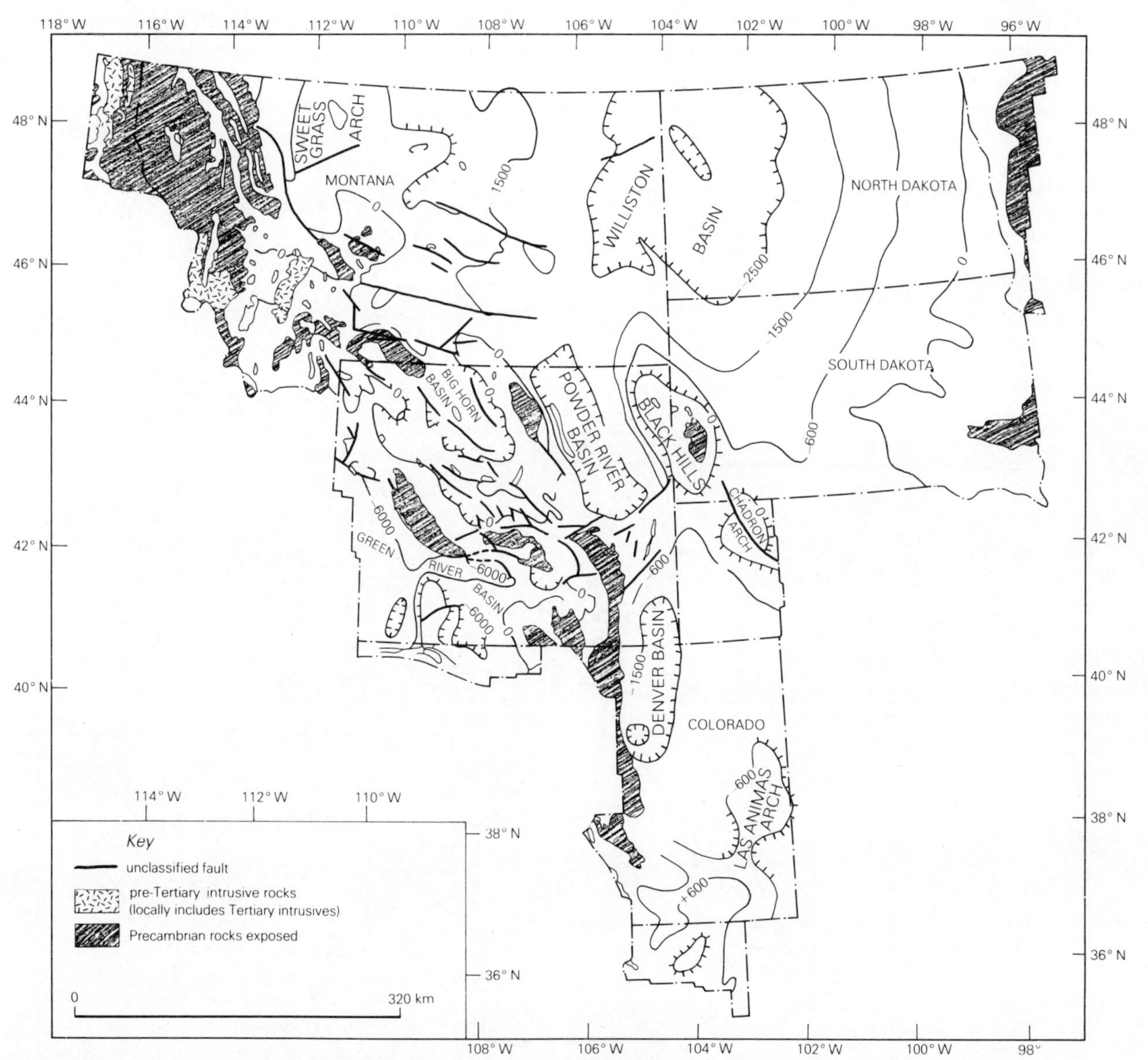

Figure 22.6 Generalized structure contours (in meters) on an easily determined horizon over a large area; Precambrian basement in northern Rocky Mountain province of the USA. (From W. H. Curry, *AAPG Memoir* 15, 1971.)

cally than a zero slope offset by a fault. If seismic or dipmeter data support the slope, or if similar slopes are already established in other parts of the region, well and good; but faults are very common phenomena and if a single fault is established within a region the likelihood of many more is manifest. The fact that the presence of a fault in the case considered above is suppositional should not deter the geologist from considering its possibility and the implications thereof. There are other ways to contour the structure shown in Figure 22.7 or the isopachs of Figure 22.8, but the ways adopted in the preparation of both maps were faithful to the geology known at the time.

Isopachs of any interval (a rock unit, biostratigraphic unit, pay zone, or whatever) do not of course represent a surface and isopach maps should not be contoured as if they did. The nature of the interval being mapped must always be clear in the geologist's mind; is it the thickness of a sedimentary rock unit, the interval between two marker horizons, a time-stratigraphic unit involving a succession of facies changes, or what?

22.3 Subsurface maps and sections

Most maps of the subsurface made and used by petroleum geologists depict either the external geometry of sedimentary strata (especially their thicknesses and the shapes of the surfaces between them), or some internal characteristic possessed by them (lithologic facies or fluid content, for examples). Many maps, and most cross sections, combine data on both the external and internal properties of the rocks.

22.3.1 Structure contour maps

Structure contour maps, often called simply *structural maps*, show contours on any subsurface geologic horizon with respect to some stated datum plane, commonly sea

Figure 22.7 Straight-line structure contour map of productive Eocene sandstone in part of La Brea–Parinas oilfield, Peru, and cross section showing abundant rectilinear normal faults. Subsea contours at 120 m intervals. (From R. B. Travis, *AAPG Bull.*, 1953.)

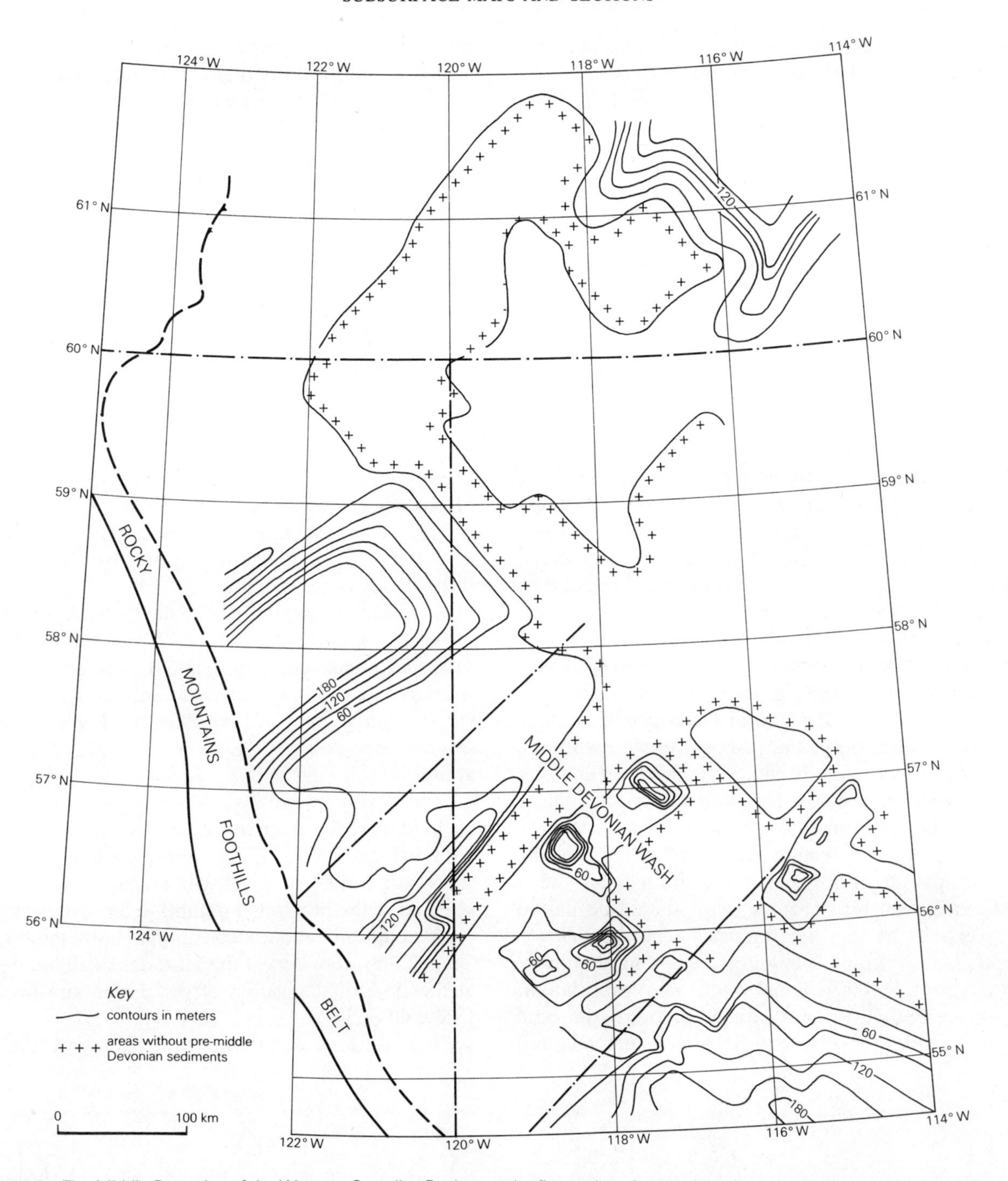

Figure 22.8 The Middle Devonian of the Western Canadian Basin was the first series of strata deposited on a platform which had undergone rectilinear faulting and deep erosion following Lower Paleozoic deposition. The series now rests on rocks ranging in age from Precambrian to Ordovician. This isopach map shows the pre-Middle Devonian sedimentary rocks (mostly thickly bedded carbonates) in the northern sector of the basin, contoured in rectilinear fashion in conformity with the known structural style. (From L. A. Sikabonyi and W. J. Rodgers, *J. Alberta Soc. Petrolm Geol.*, 1959.)

level. The geologist making the map must be clear what his chosen horizon truly represents. It may be a lithostratigraphic unit, like a formation; this may be the easiest type of contour horizon to identify when nearly all one's data are derived from well logs, but it must not be viewed as if it were a time-marker horizon. The contours may on the other hand be on what the geologist believes to be a time-marker horizon (the top of the Cretaceous, for example). If the choice of the time-horizon is well controlled by fossils (especially by time-significant microfossils), the geologist is on safer ground than if he is really contouring log picks on something given a time-stratigraphic name, like that of a stage. Structure contours on the Turonian strata may over a small area (that of an oilfield, for example) actually be on a time horizon, but they are unlikely to be so over a large area (that of a large sedimentary basin, or a country). These inaccuracies cannot be avoided if maps are to be made in the course of exploration, but the geologist must be aware that they exist.

Most structure contour maps employ sea level as the datum plane, with elevations measured in relation to those of the kelly bushings (KBs) of the wells penetrating the horizon shown on the map. In most areas of extensive drilling, only the shallow horizons are above sea level; all deep horizons (and possibly all productive horizons except in plateau and thrust-belt areas) are likely to be below sea level. Their contour values are therefore negative and large values represent greater depths, not greater elevations, than smaller values. For ease of understanding by nongeologists who help to determine programs, areas within closed or partially closed contours should be marked as positive (high) or negative (low). The contour interval should be indicated alongside the scale of the map. Faults should be shown clearly, with direction of dip if known, and with upthrown and downthrown sides distinguished. The geologist must be clear in his own mind on the reliability of his fault interpretations (Figs 16.36, 45, 94, 113 & 119).

A most useful modification of the structure contour map shows the form of the mapped horizon with any *regional dip removed*. Ideally, regional dip should be determined from the basement surface, but any widespread, fairly uniform horizon provides an adequate surface for its measurement provided that it lies below the principal target horizons. Once the regional dip is established, structure contour maps should be paired with versions having the "regional" removed. This is simply done by constructing a grid showing the divergence caused by the regional dip from a level datum, overlaying it on a true structure contour map to the same scale, and adding or subtracting the divergence values to or from the true elevations of the control points. Structural maps of the region thus returned to untilted position commonly reveal important new structural patterns which may have been responsible for trapping mechanisms.

The removal of the regional dip is especially useful for structural contours on *unconformity surfaces*. Maps of these show erosional topography, which may be considerable (especially if thick carbonate units are involved). Both erosional eminences and erosional depressions may form *paleogeomorphic traps* for hydrocarbons. The mapping procedure for identifying such traps is described in Section 16.6.

22.3.2 Cross sections

Cross sections of three types are indispensable to the petroleum geologist. Some relations are more clearly displayed, and numerous inconsistencies more starkly revealed, through the drawing of cross sections than through any other means.

Correlation cross sections are the first geologic figures to be drawn in the first phase of exploratory drilling. Their purpose is simply to enable the geologist to decide stratigraphic equivalences between the wells that provide his initial control. Until that is achieved, pre-drill seismic maps and sections show no factual stratigraphy at all.

Structural cross sections show the present structural attitudes of rocks in relation to sea level as the horizontal datum (Figs 17.11 & 22.9). *Stratigraphic cross sections* show correlations of strata with respect to one of them selected as the horizontal datum; within limits imposed by assumptions about depositional horizontality (see Sec. 22.3.3), they reveal the attitudes of all older strata at the time of deposition of the datum stratum (Figs 13.29, 40 & 50).

Cross sections should be drawn along all lines pro-

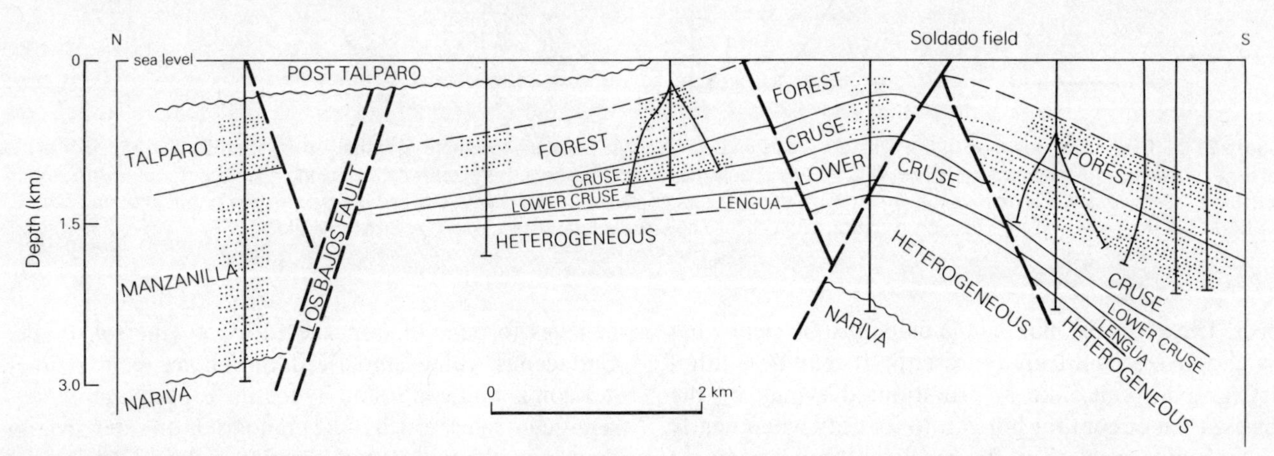

Figure 22.9 Structural cross section through Soldado oilfield, Trinidad, showing three types of fault cutting the Tertiary strata. South-dipping normal faults and north-dipping reverse fault in field area, well controlled by drillhole data despite small offsets. Large Los Bajos strike-slip fault with quite different stratigraphic sections on its two sides. (From K. Ablewhite and G. E. Higgins, 4th Caribbean Geol. Conference, (1965), 1968).

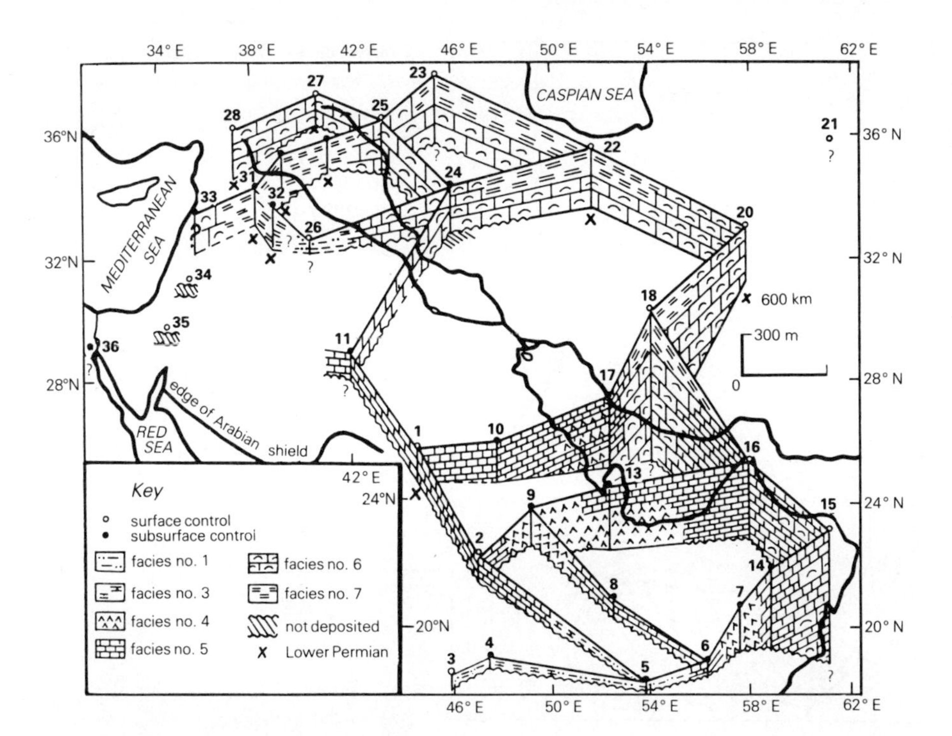

Figure 22.10 Fence diagram showing the distribution of Upper Permian lithofacies in the Persian Gulf Basin. The interval includes the Khuff Formation, the reservoir rock for supergiant gasfields on both sides of the gulf. (From F. A. Sharief, *J. Petrolm Geol.*, 1982.)

viding sufficient control points; they should not be restricted to dip- or strike-sections. Groups of cross sections in different directions are combined into *fence diagrams* (Fig. 22.10). Where control lies somewhat off a line of section and has to be projected on to it, this projection must take account of what is known of the dip (and of the plunge in the case of folded strata), and the distance of projection should be indicated on the section. Structures at all subtle may be totally obscured by uncertain or inaccurate projections which the viewer of the cross section does not know have been made. Seismic sections must be calibrated with geologic cross sections as early as possible.

All oil and gas production depends upon the presence of fluids which have migrated into their present positions and would still be capable of motion if they were not held back by some trapping mechanism. Present fluid contents of the strata, as revealed from production rates, test results, logs, GORs, must make sense on cross sections through the wells concerned. The commonest anomaly that may be revealed is fluid characteristics implying discontinuity (of reservoir, facies, structure) where no discontinuity is shown on the section. Any such anomaly needs resolution.

22.3.3 *Isopach maps*

Isopachous maps, called isopach maps throughout the industry, show lines of equal thickness of strata contained between two reference planes. The strata may constitute a lithostratigraphic unit, such as a *formation* (Fig. 22.11); a time-stratigraphic unit, such as a *system* (Figs 22.12 & 13); a packet contained between two unconformities (a *sequence* or *synthem*); or an economic unit, like the pay-zone thickness of an oilfield reservoir rock (Fig. 22.14). The geologist must choose what is to be "isopached," and he must recognize what the top and base of his chosen interval actually are. This caution is critical, because the geologist must be aware of the assumptions implied in his isopach map and of the degree to which they are likely to be valid.

Isopach maps are commonly interpreted via the assumption (often unacknowledged) that the *upper surface* of the mapped interval was originally an essentially planar surface. Variations in the thicknesses shown on the map therefore reveal structure on the lower surface and imply that it was *not* planar during the time of deposition of the rock forming the upper surface. Clearly some depositional surfaces are more likely to be planar than others. Some rock units are excellent time markers (bentonite beds, for example, or thin, persistent limestones, black shales, or coal seams); but units crossing the shelf-slope break, or those deposited along strongly embayed coastlines of submergence, near submarine canyons, around active reefs, or in a variety of other environments, do not really approximate flatness in their original attitudes (Figs 22.14 & 15).

The ascription of horizontality to the upper surface of the isopach interval of course eliminates any structure subsequently imposed upon it, whilst depicting any structure earlier imposed upon its lower surface. Isopach thicks and thins are therefore useful in determining the age and growth of structures, both of positive structures (uplifts) and negative structures (sub-

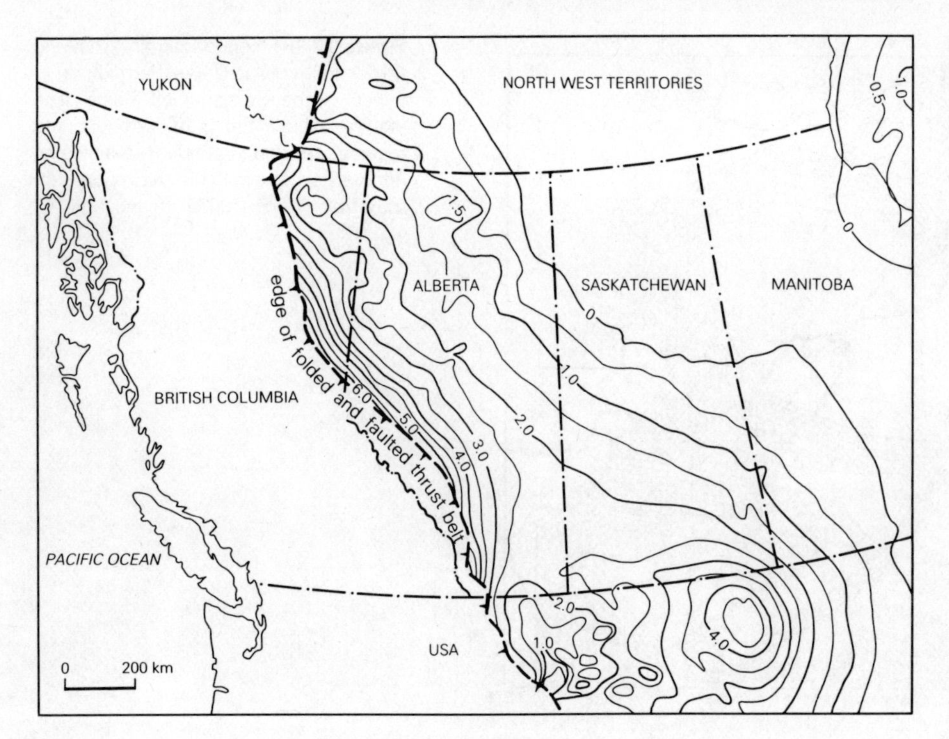

Figure 22.11 Isopach contours of a large volume of sedimentary rock in a large basin. Total preserved thickness of Phanerozoic strata in the Western Canadian Basin; contours in kilometers. (From J. W. Porter *et al., Phil. Trans. R. Soc.* 1982.)

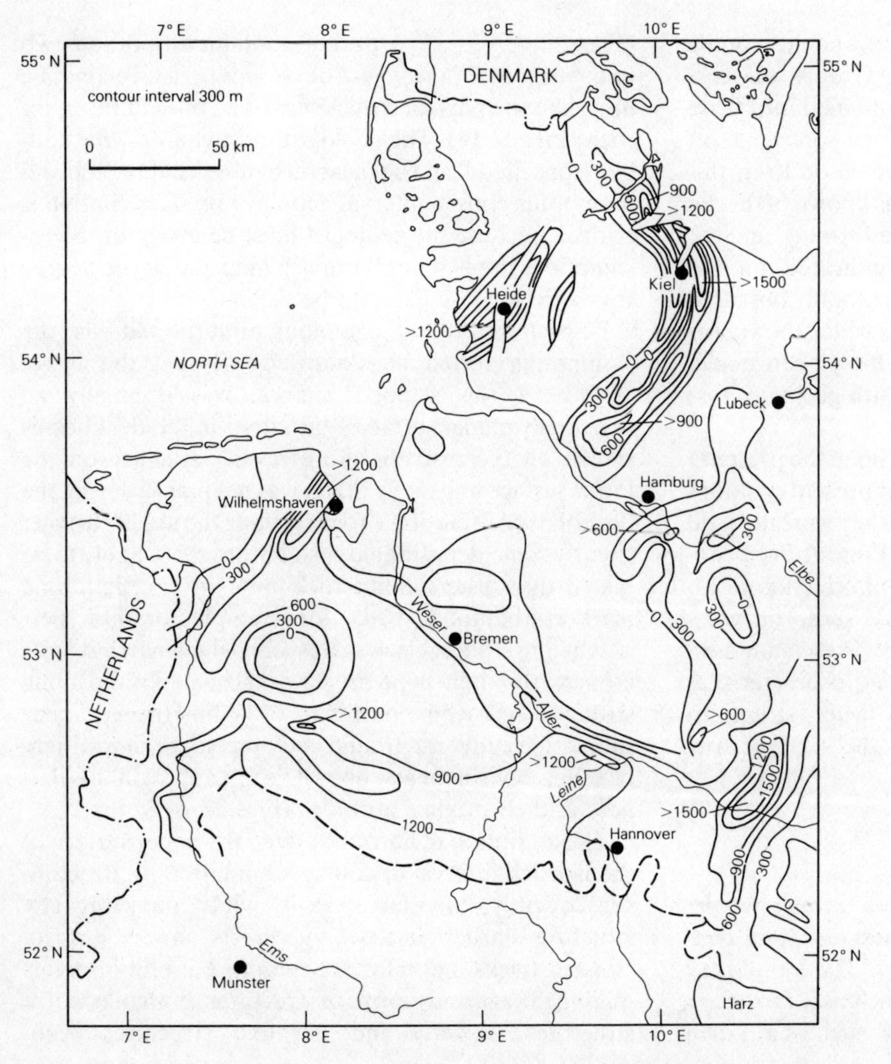

Figure 22.12 Isopach map of entire Jurassic System in Northwest German Basin. Deeply subsiding troughs of diverse strike, separated by swells over which Jurassic strata are absent by non-deposition. Troughs much influenced by movement of Permian salt below; greatest concentration of salt domes and walls lies outside deep Jurassic troughs, and southern boundary of halokinesis lies north of the largest such trough in the south. Contour interval 300 m. (From A. Bentz, in *Habitat of oil,* Tulsa, Okla: AAPG, 1958.)

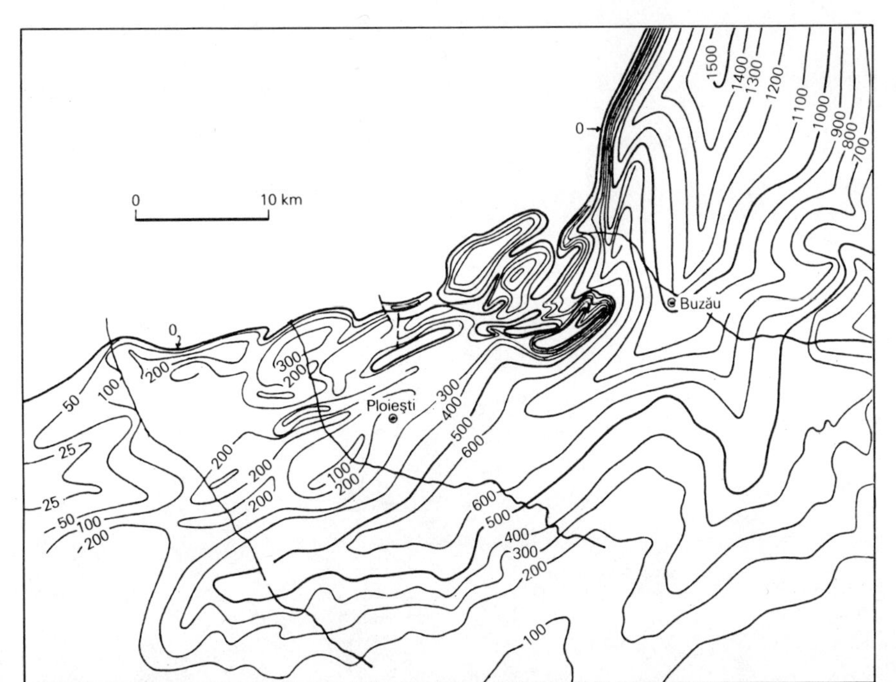

Figure 22.13 Isopach map of a foreland fold belt. Meotian Stage (Lower Pliocene), the main productive interval in the Carpathian Basin of Romania; arcuate festoons of salt-cored anticlines and intervening synclines. Contours in meters. (From D. Paraschiv and G. Olteanu, *AAPG Memoir* 14, 1970.)

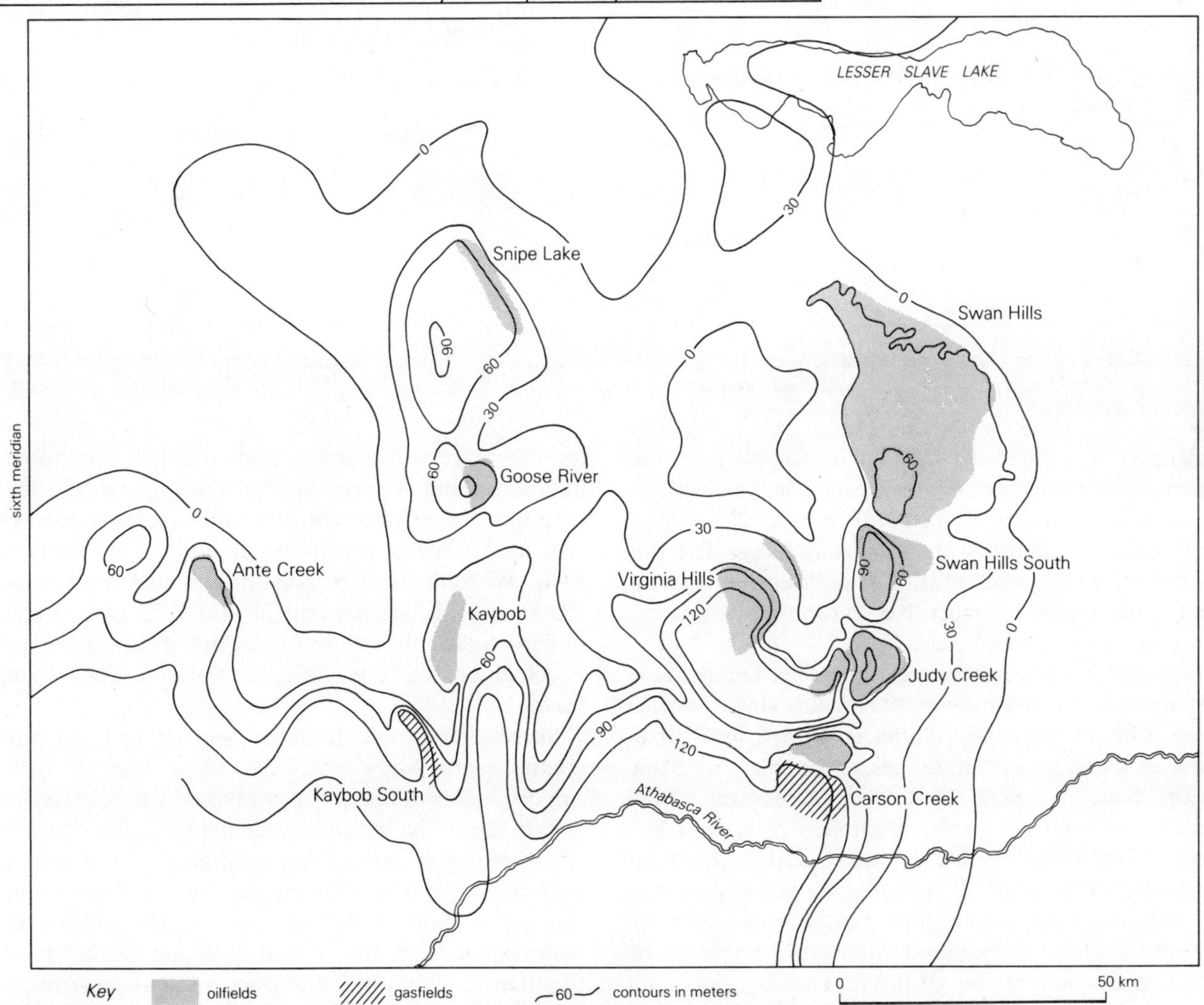

Figure 22.14 Isopach map of a productive reef-bearing interval: Swan Hills Formation, Devonian of northern Alberta, Canada. (From C. R. Hemphill *et al.*, *AAPG Memoir* 14, 1970.)

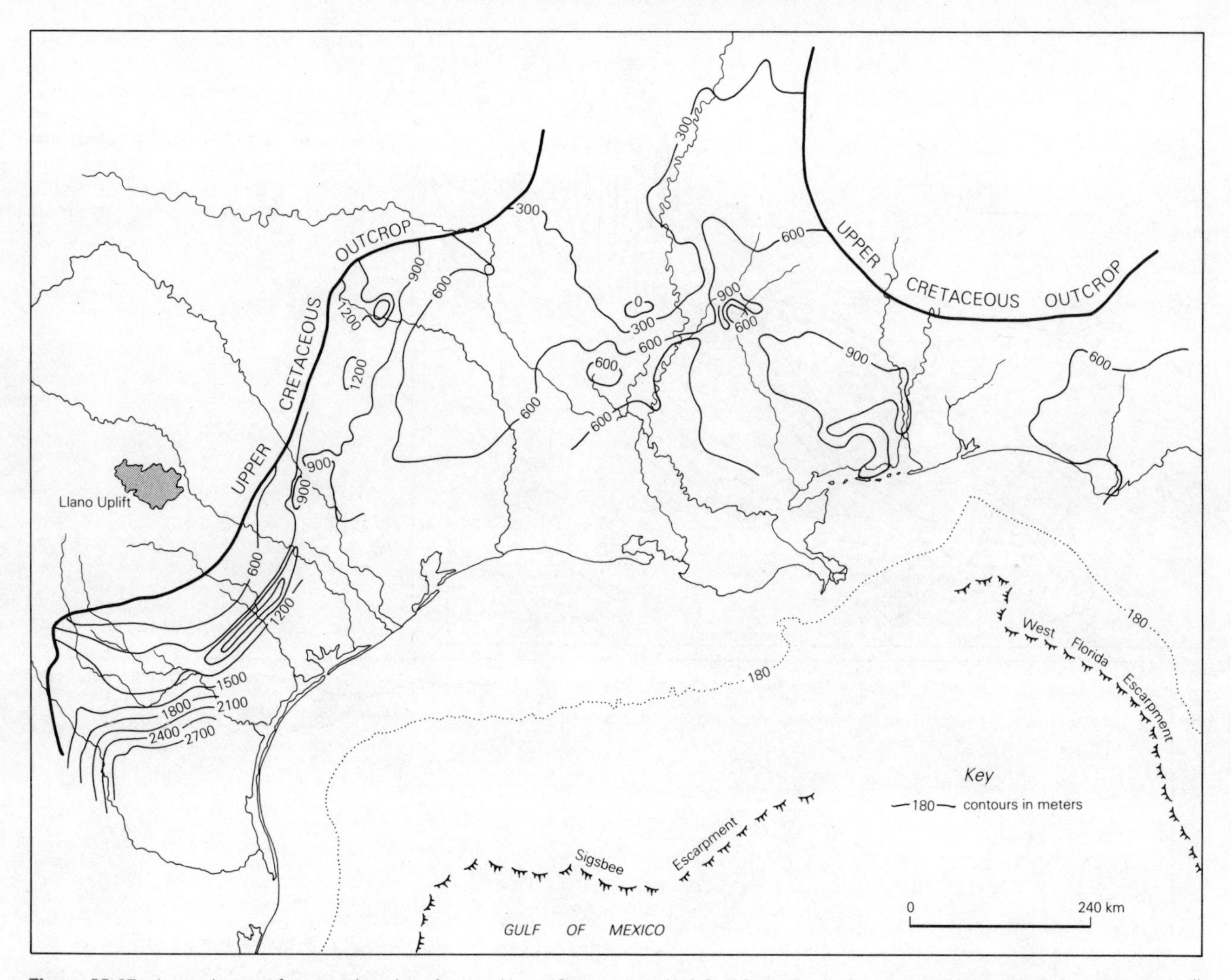

Figure 22.15 Isopach map of a coastal wedge of strata; Upper Cretaceous of US Gulf Coast Basin. Strata generally thicken seaward but actually occupy a series of depocenters – in the southwest, northwest, and to the north and northeast of the Mississippi delta. (From C. W. Holcomb, *AAPG Memoir* 15, 1971.)

sidences). Considerable caution is necessary in this usage, and only experience teaches it. The likely effects of *compaction* must be taken into account. *Net uplift* is less certainly identified than *net subsidence*. The distinction between nondeposition, deficient deposition, and erosion needs support from careful cross sections; isopachs alone are not enough.

Growth of structures through time is better illustrated by *cumulative isopach maps* than by any single isopach map. The stratigraphic column is divided into correlatable packets of strata each capable of being "isopached." Packets are added successively, from oldest to youngest, to reveal patterns of greater and lesser subsidence (which is what most "growth of structure" consists of). A still better method of revealing growth utilizes a series of *paired isopach and structure contour maps*. The geologist chooses a horizon to be illustrated – say the top of the time-stratigraphic unit containing the principal reservoir rock. He draws a structure contour map on that unit, and then constructs successive isopach maps of each younger unit, superimposing them in succession on the original structure map and thus reconstructing the horizon of that map as it was at the close of deposition of each of the overlying units. With the usual reservations about horizontality, these sequential maps reveal the time of first appearance or first closure of structures, the time of their maximum development, and any shifting of their positions during time.

Successive combinations of isopach and structure contour maps can be united in a single diagram by the *isopach triangle technique*, introduced by Russian geologists in the 1970s. It is illustrated in Fig. 22.16. The construction begins in the lower right-hand corner with an isopach map of the unit immediately overlying the zone of interest. Similar isopach maps of the successively overlying units are then added along the hypotenuse of the triangle. The left side of the triangle is constructed, from bottom to top, from structure contour maps of the bases of the units (that is, contours on the tops of the

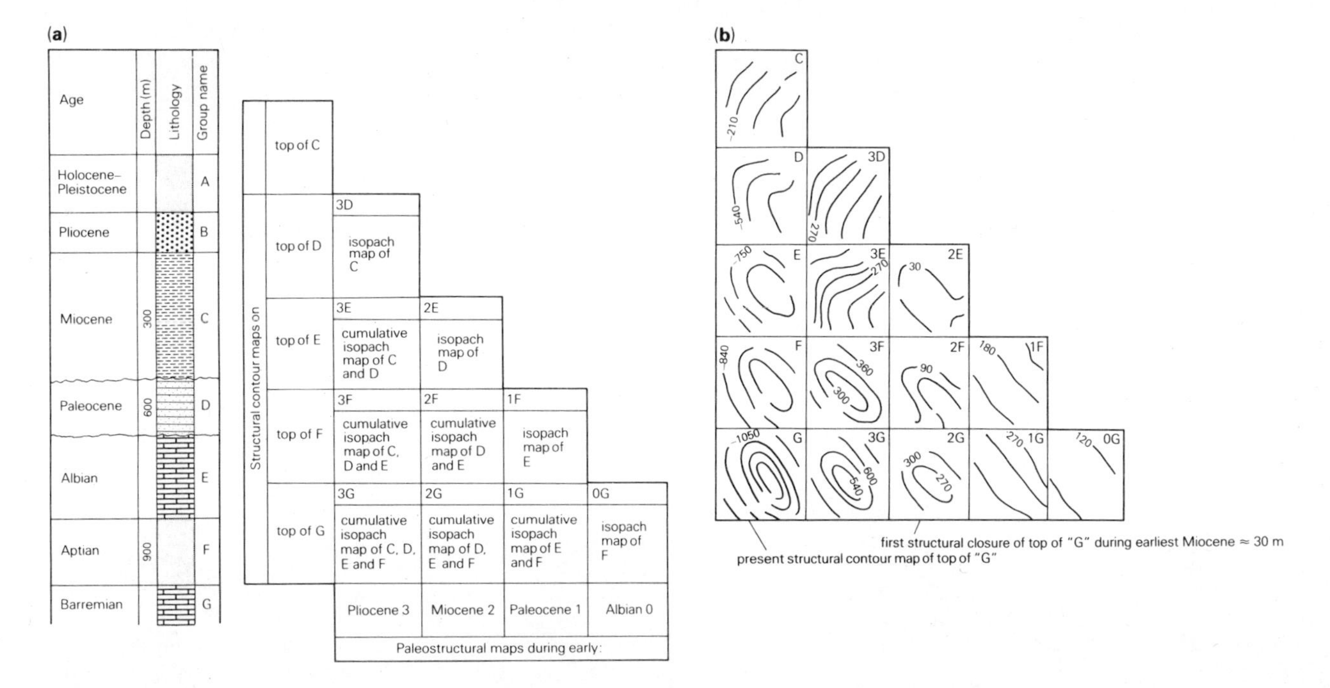

Figure 22.16 (a) Construction of the isopach triangle system for the section shown. (b) Diagrammatic isopach triangle showing first appearance of structural closure on top of zone G (Lower Cretaceous) during the earliest Miocene. Maps 0G, 1G, 2G, and 3G are paleostructural maps on top of zone G; 1F, 2F, and 3F are paleostructural maps on top of zone F; 2E and 3E are paleostructural maps on top of zone E; 3D is a paleostructural map on top of zone D. Map 0G is a paleostructural map on top of zone G during earliest Albian; 1F and 1G are paleostructural maps on top of zones G and F during earliest Miocene; 2E, 2F, and 2G are paleostructural maps on top of zones E, F, and G during earliest Miocene; and 3D, 3E, 3F, and 3G are paleostructural maps on top of zones D, E, F, and G during earliest Pliocene. (From M. W. Ibrahim *et al., AAPG Bull.*, 1981.)

underlying units). The isopach values of pairs of units forming the hypotenuse of the triangle are then added together and used to construct maps along the squares "southwest" of the hypotenuse. Further squares are then occupied by isopach maps derived by adding the values from the squares at the "northern" and "eastern" ends of their rows, until all the squares are occupied. The vertical columns thus become paleostructural maps for their particular times, revealing the appearances of structural features and changes within them through time.

Figure 22.16 depicts the growth of a single, potentially petroliferous structure through a prescribed time interval. On a much larger scale, successive isopach maps of an entire basin depict the changes in its structure and stratigraphy consequent upon the growth of much larger structures, of the dimensions of an orogen. Figures 22.17 and 22.18 show such a succession of structural changes for two petroliferous basins in the Rocky Mountain province of the western USA.

A final concern of the geologist with his isopach maps is their coordination with seismic time-interval maps. The latter are essentially without stratigraphic calibration until this coordination has been achieved. It is especially vital in the use of *seismic stratigraphy* in offshore regions (see Sec. 21.9).

22.4 Facies maps

Facies is commonly defined as the *aspect* of a three-dimensional body of rock formed during a prescribed time interval. It is a *time-stratigraphic* concept reflecting the total lithologic characteristics within a contemporaneous deposit – bedding, lithology, distinctive primary features, fossils in general.

One aspect of a sedimentary rock unit of vital concern to the petroleum geologist is its *thickness*. It is consequently natural to combine maps depicting facies with isopach maps. For both types of map, the geologist must choose sensible boundaries (upper, lower, and areal). He must also bear in mind a fundamental dichotomy between the thickness of a sedimentary rock unit and its facies. The facies depends on several factors, including water depth at the depositional site. The thickness depends on the amount of subsidence taking place during the deposition of the unit. There are undoubtedly many examples of thick sedimentary units deposited in deep water, and their thicknesses may then be correlatable with initial water depth; such units literally fill in depressions, commonly in active orogenic regions. There are, however, many more examples of thick packets of strata deposited entirely in shallow water and

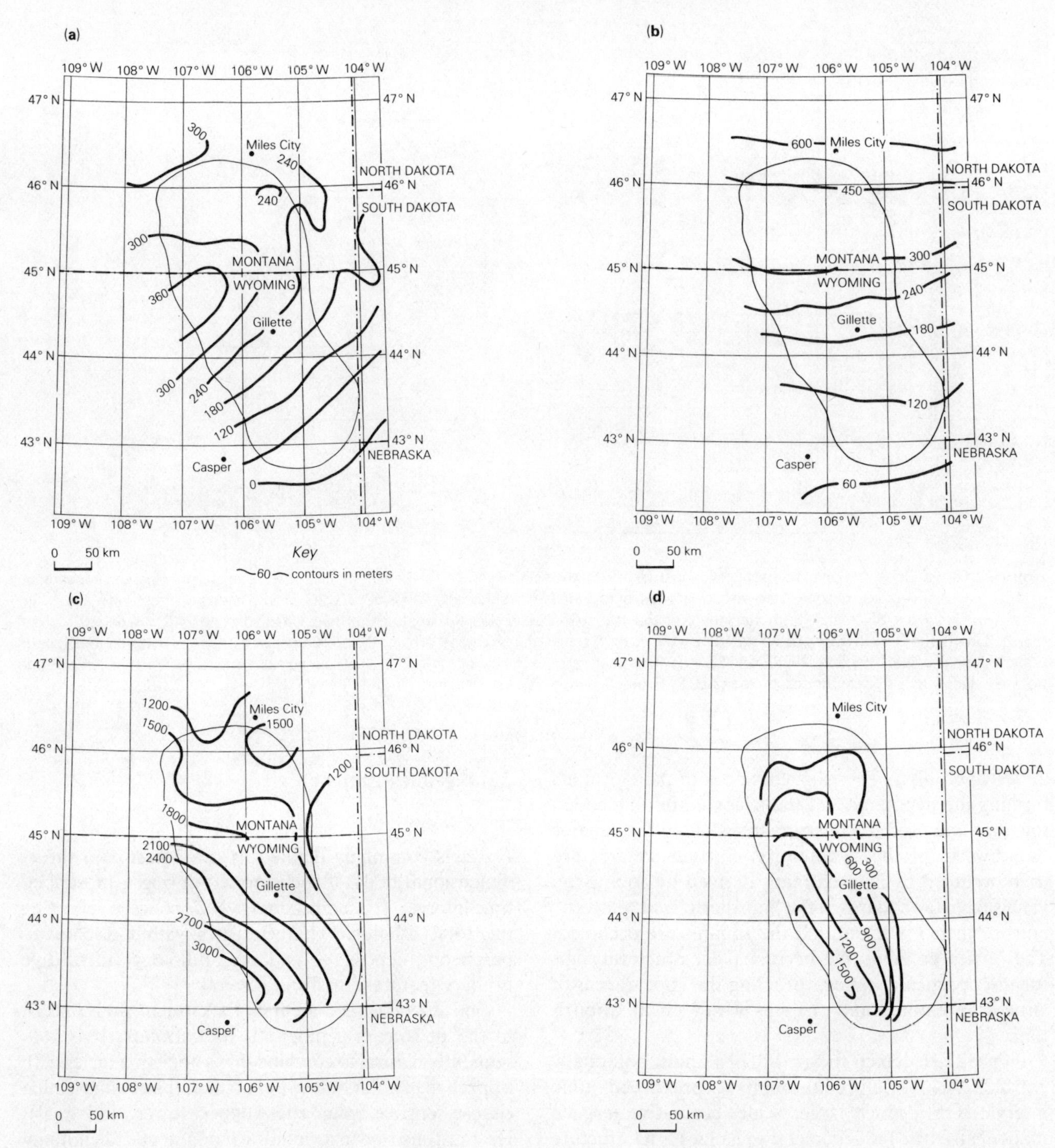

Figure 22.17 Isopach maps of four systems in the Powder River Basin of Wyoming. Cambrian and Mississippian maps (a and b) show general northward thickening unrelated to present basin. Cretaceous map (c) shows initiation of southwestward thickening as Rocky Mountain trough became dominant to the west. Tertiary map (d) shows conformity of isopachs with present basin, deepening towards Laramide Big Horn uplift to the west. (From P. T. Kinnison, *AAPG Memoir* 15, 1971.)

bearing tell-tale evidence of this throughout their thicknesses.

A second aspect of a sedimentary rock unit is its fossil assemblage; total absence of fossils is an aspect in itself and therefore contributes to the facies. A specific faunal assemblage within its appropriate sedimentary rock constitutes a *biofacies*. The biofacies is directly controlled by water depth (as well as by other characteristics such as latitude, salinity, muddiness, and degree of oxygenation). Bathymetric data derived from fossils – especially from benthonic and planktonic microfossils – can be exceedingly useful, especially within very thick successions. They not only enhance understanding of the depositional basin and help to complete the

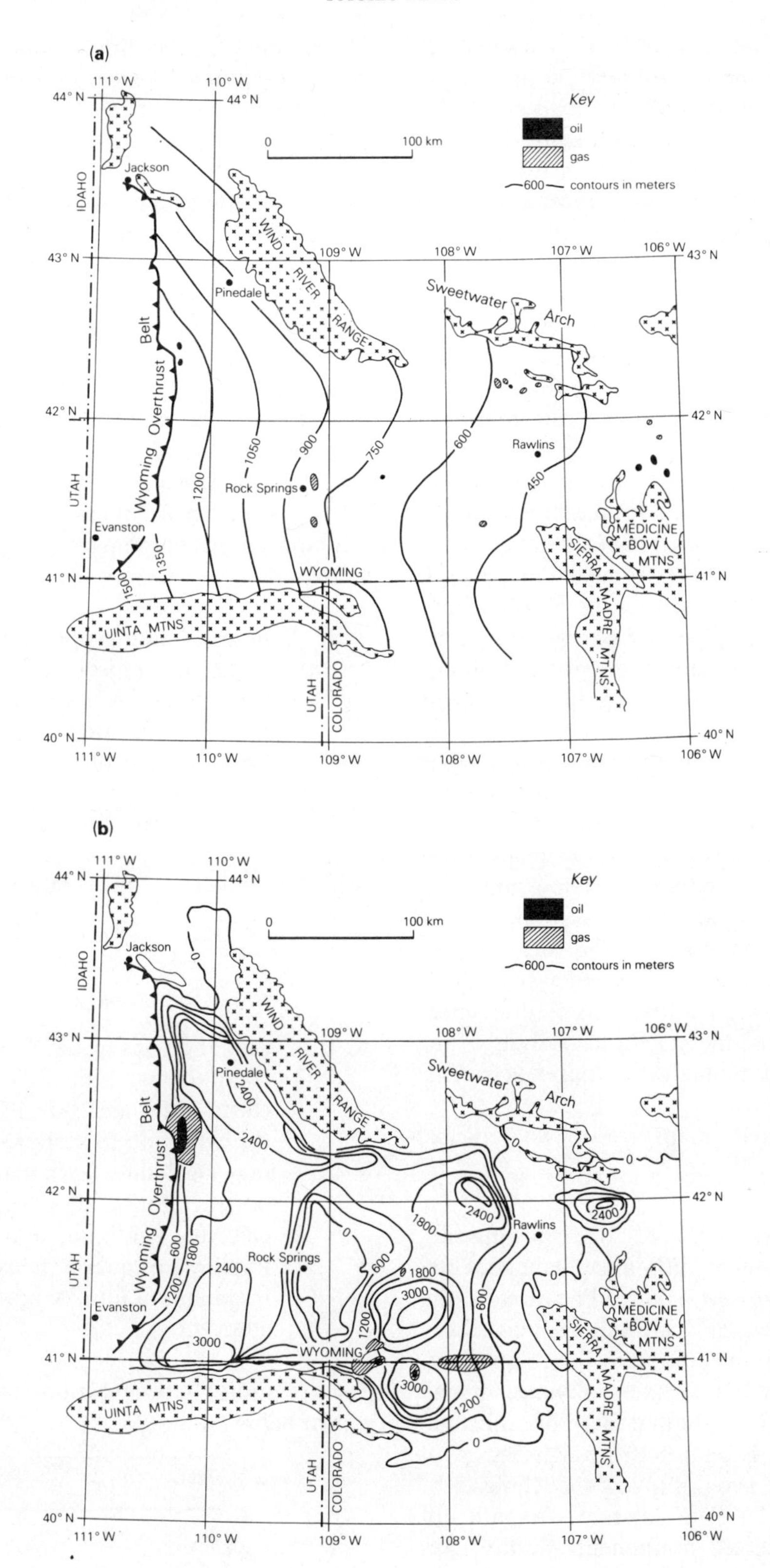

Figure 22.18 Isopach maps of two time-stratigraphic intervals in the Green River Basin of southwestern Wyoming, to illustrate the effects of intervening structural deformation. (a) Triassic–Jurassic isopach map and oil- and gas-producing areas of those ages. Strata thicken fairly uniformly westward, antedating all the structures shown. Contour interval 150 meters. (b) Tertiary isopach map and principal oil- and gas-producing areas. Rise of faulted uplifts (cross pattern) and of western overthrust belt, during the Laramide deformation, created multiple basinal subsidences (isopach thicks) with intervening arches (thins). Contour interval 600 meters. (From E. R. Keller and N. D. Thomaidis, *AAPG Memoir* 15, 1971.)

spectrum of facies represented in it; they also reveal important departures from horizontality in time-stratigraphic surfaces, thus enabling the geologists to avoid misinterpretation of their own isopach maps.

Some features of the water realm are instrumental in accumulation patterns of oil and gas. Depositional hinge belts are especially favorable, and the geologist should be on the lookout for any relations revealed between his isopach/facies maps and such reservoir characteristics as porosity, grain size, or clay mineral content.

In some thick (and thickly bedded) successions of strata, especially of shallow-water carbonates and evaporites as in much of the Paleozoic of the interior of North America, time markers may be highly uncertain or discontinuous. To avoid being restricted to the use of excessively thick intervals for isopach constructions, the geologist may use *slice maps* showing facies at arbitrary but fixed intervals above or below some datum plane. If the datum plane is well chosen, slices above it may approximate time-stratigraphic surfaces, but the geologist should be aware of the uncertainty of this and be suitably skeptical of maps depicting highly unlikely associations of facies (Fig. 22.19).

Much more abundantly available to the geologist than biofacies data are *lithofacies* data, reflecting specific lithologic assemblages within contemporaneous deposits. The most commonly used facies maps therefore show conventional lithofacies; their commonest format is a series of transparent overlays representing various lithologic characteristics of the rocks, superimposed on isopach, fluid content, or other maps. Such maps and overlays make great demands on draftsmen. In densely drilled areas, the statistical analysis of log data for lithofacies discrimination requires automatic processing techniques.

Conventional lithofacies maps are of two principal types:

(a) *Isolith maps* (Figs 22.20 & 21), showing net thicknesses of single lithologic components, especially of sandstone. A modification is the *lithologic percentage map*, showing the percentage of the total section constituted by the component, rather than its absolute thickness; obviously it may be important to show whether 50 m of sandstone represents ten beds scattered through 500 m of section or a single bed occupying the whole of it.

(b) *Ratio maps* (Figs 22.21 & 22) show the ratios of thicknesses of selected components. Ratio maps are related to percentage maps as a geometric relation to an arithmetic one.

Three lithologic components of a chosen section are readily represented by *facies triangles* (Fig. 22.23). The three components are made to add up to 100 percent of the section if there are in fact more than three present in it. Three components are adequate for the representation of most sedimentary rock sections, especially if components are combined (limestone + dolomite, for example, or conglomerate + sandstone). Simple ratios between two individual components, such as the *sand : shale ratio*, are represented by lines diverging from the corner of the triangle represented by the third member (Fig. 22.23). Ratios between one component and the sum of the other two are represented by lines parallel to the side of the triangle containing those other two. The familiar *clastic ratio*, called by some the *detrital ratio*, is that between noncarbonate clastics and nonclastics (simply carbonates plus evaporites in nearly all cases).

In constructing the standard facies triangle, the most commonly used components of sedimentary rocks are as follows.

A, B, C may be nondetrital components, sandstone, and shale respectively. Then side AB or AC represents the clastic (detrital) ratio and side BC represents the sand : shale ratio. These ratios are defined thus:

$$\text{clastic (detrital) ratio} = \frac{\text{noncarbonate clastics}}{\text{nonclastics}}$$

$$\text{sandstone : shale ratio} = \frac{\text{sandstone + conglomerate}}{\text{shale + siltstone}}$$

where 'nonclastics' = carbonates + evaporites

A, B, C may be clastic, evaporite, and carbonate components. Side AB or AC is then the clastic ratio and side BC the evaporite ratio (of evaporite to carbonate).

A, B, C may be halite, carbonate, and sulfate components, for details of evaporite sequences; or chlorite, kaolinite, and micaceous minerals for clays; or quartz, feldspar, and rock (lithic) fragments for sandstone compositional studies.

The relation between ratios and percentages is shown below:

Ratio	Percentage
0	0
¼	20
½	33
1	50
2	67
4	80
8	89
∞	100

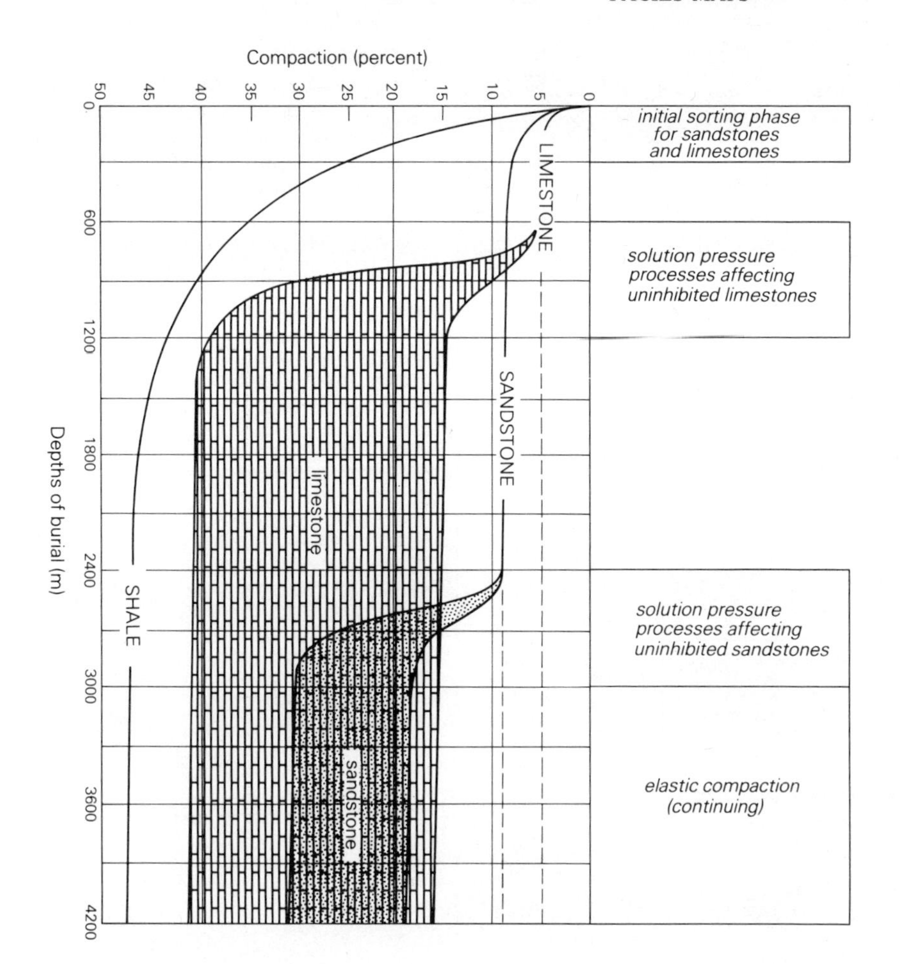

Figure 22.19 Hazards of slice maps at constant intervals of thickness, illustrated by variations in the compaction of different sedimentary rocks under different thicknesses of overburden. (From H. V. Dunnington, Iraq Petroleum Co. Ltd technical publication, 1959.)

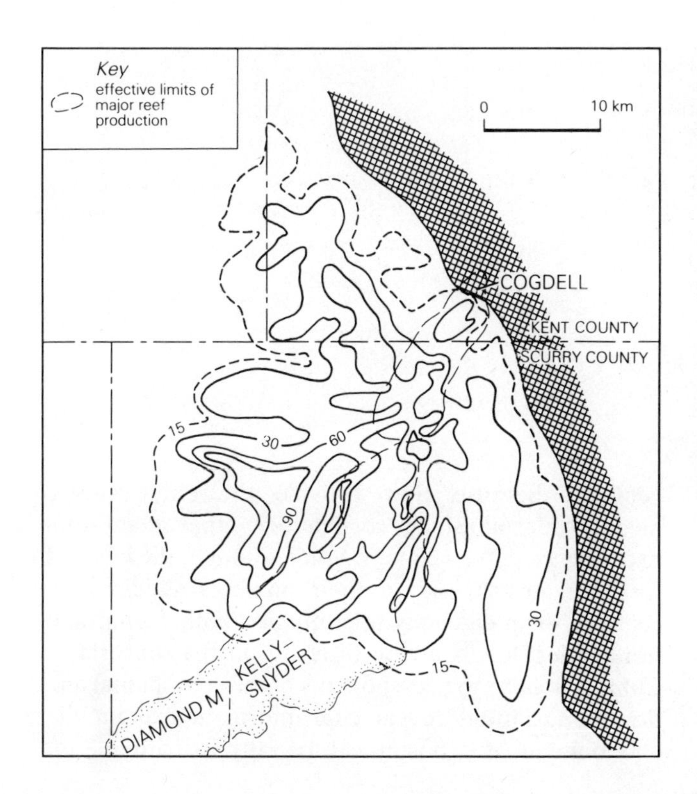

The triangle may usefully be divided into *facies areas* (commonly nine) which are colored or textured according to a standard convention for ease of transfer of data from the triangle to a map. *Triangle facies maps* normally show two sets of contours; in dominantly clastic environments, these would be the clastic ratio and the sandstone : shale ratio. This necessarily results in contours intersecting one another. For display purposes it may be desirable to avoid this, and *entropy maps* may then be constructed to display the interrelations between any three end members (Fig. 22.23). An entropy value of zero means only one end-member is present; a value of 100 means equal proportions of all three are present. The zero entropy contour therefore fits around the three corners of the facies triangle, and the 100 "contour" is the triangle's center point. Many other refinements are possible, including three-dimensional representations, but they are beyond the scope of this book.

Figure 22.20 Sandstone isolith map for part of Lower Permian clastic wedge overlying the Horseshoe Atoll complex, Midland Basin of West Texas. Map shows sandstone members thicker than 3 m. Cross-hatched area on east side is beyond limestone shelf edge. (From E. L. Vest Jr, *AAPG Memoir* 14, 1970.)

(a)

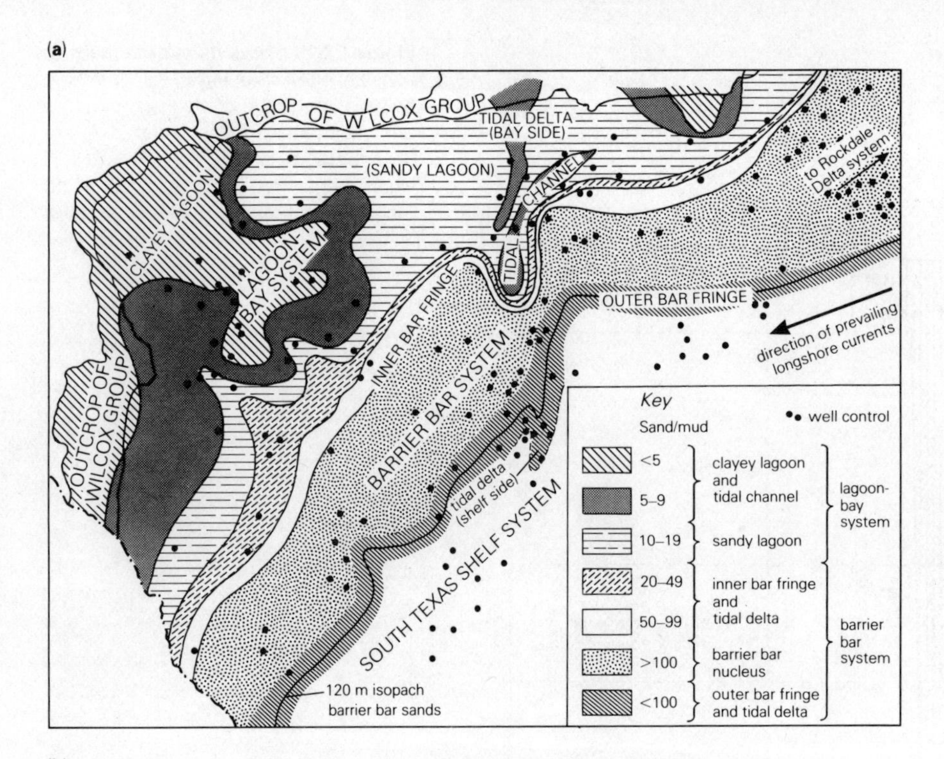

(b)

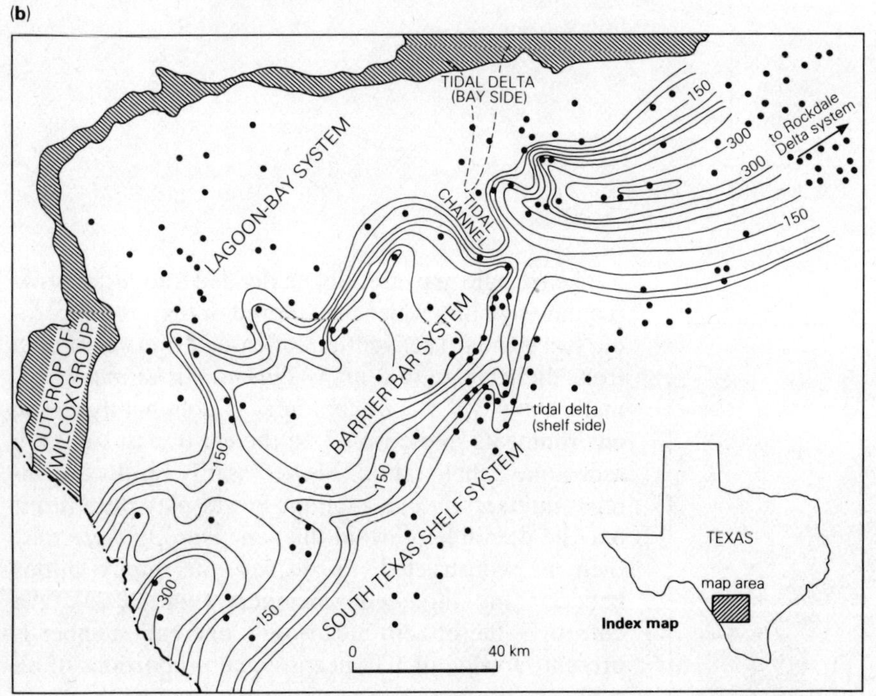

Figure 22.21 Two facies maps of a barrier-bar and bay-lagoon system in the Wilcox Group (Eocene) of the Texas Gulf Coast: (a) sandstone : shale ratio map; (b) sand isolith map – isolith interval 30 m. (From C. L. Lofton and W. M. Adams, *AAPG Memoir* 15, 1971; after W. L. Fisher and J. H. McGowen, 1967.)

Lithofacies maps should be accompanied by *lithofacies cross sections*. These are better constructed on a stratigraphic than on a structural base, using a stratigraphic horizon rather than sea level as the datum (Fig. 22.24).

22.4.1 *Biofacies maps and cross sections*

Biostratigraphic studies are critical to the petroleum geologist because they provide the only reliable means of establishing correlations other than total dependence upon well logs. Studies both of the kinds of fossil organisms and of their numbers should concentrate on indigenous microfossils, both faunal and floral. Megafossils are of little use in the subsurface. Comparisons of the components of different faunal and floral associations reveal vital information about the environment of deposition, especially its bathymetry.

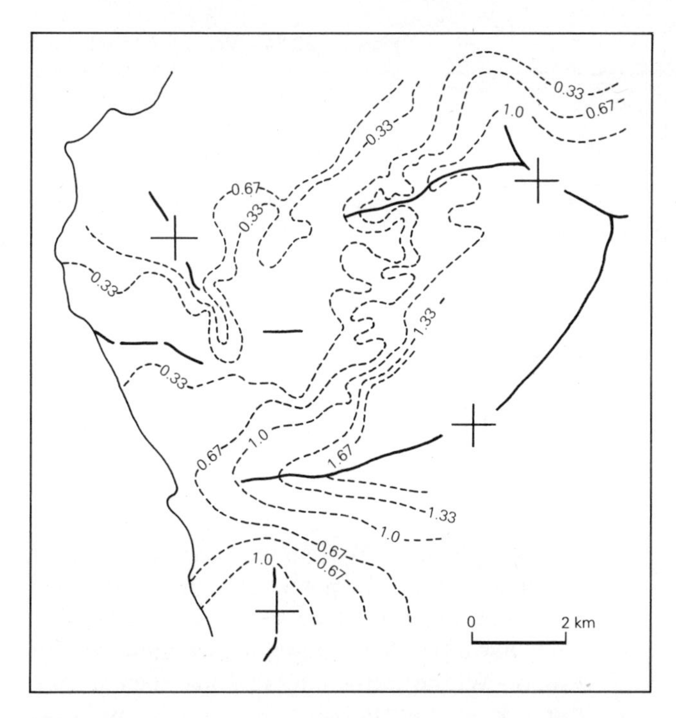

Figure 22.22 Sand : shale ratios of productive Eocene interval in La Brea–Parinas oilfield area, Peru. Note close relation between ratios and present structure (structural highs marked +, lows marked –). (From W. Youngquist, in *Habitat of oil*, Tulsa, Okla: AAPG, 1958.)

Identifications of diagnostic species (and especially of diagnostic associations of species) permit refined stratigraphic zonation, and show whether correlations are time- or facies -controlled (as, for example, by the clarity or muddiness of the water). Care must be taken to identify elements of the fauna that are not indigenous but have been displaced from their true environments, and to rely only upon fauna and flora that inhabited the environment during their lives, not upon those that ended up in it after their deaths. (Biologists distinguish the *biocoenose*, or environment of life, from the *thanatocoenose*, or environment of death; a worldwide example of the latter familiar to geologists is the graptolitic shales of the early Paleozoic.)

Biofacies maps and *correlation sections* are of little use to the petroleum geologist by themselves. The data on them should be incorporated into isopach and lithofacies maps and lithologic cross sections. An incorporation commonly employed, especially in areas of Cenozoic production, is between *paleontologic logs,* based on the bathymetric ranges and abundances of individual diagnostic species, and *electric logs*, a combination designed to reveal the environments of deposition of the sands causing the kicks on the logs.

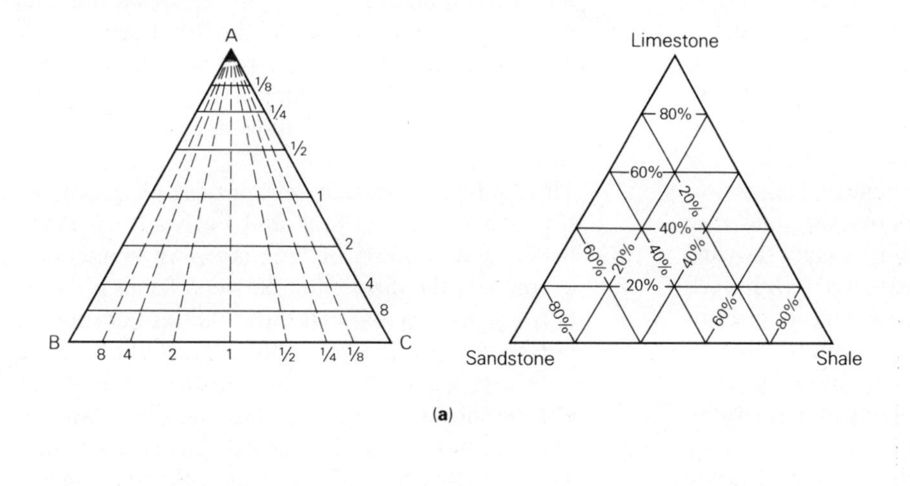

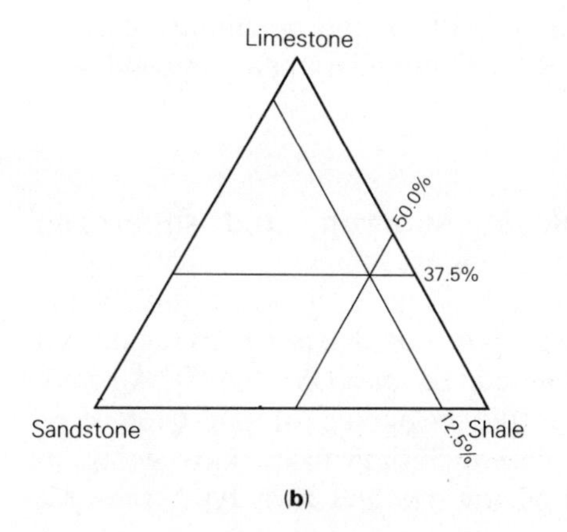

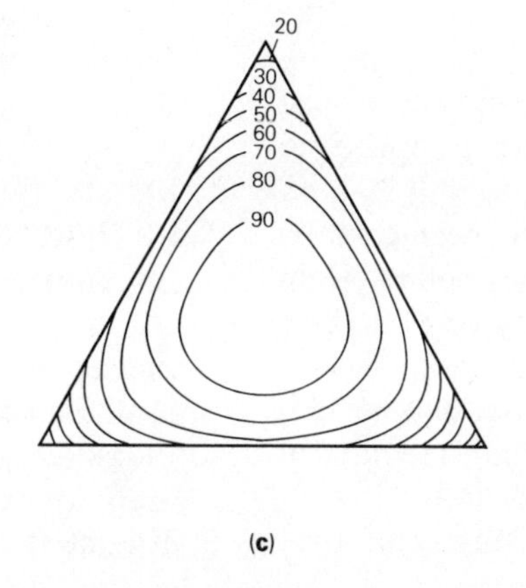

Figure 22.23 Facies triangles. (a) The basic triangle showing ratio lines (left), percentage lines (right). (b) The method of plotting points. (c) Entropy overlay (see text for explanation). (From W. C. Krumbein and L. L. Sloss, *Stratigraphy and sedimentation,* San Francisco: Freeman, 1963.)

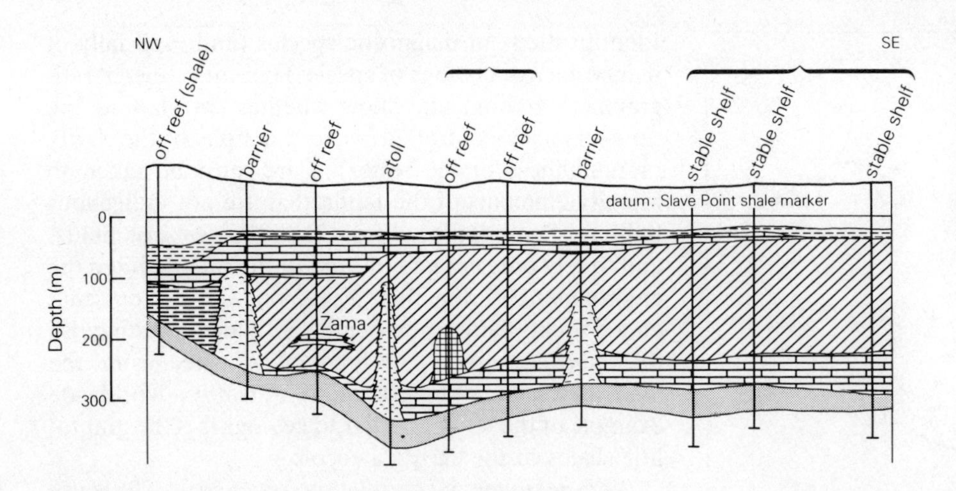

Figure 22.24 NW–SE stratigraphic section across the northern Alberta Basin, Canada, showing time and space relations in Middle Devonian strata. Facies regimes are identified at the top of the section. Diagonally ruled unit is largely evaporite, including salt. No horizontal scale is implied. (After D. L. Barss *et al.*, *AAPG Memoir* 14, 1970.)

22.4.2 Vertical variability maps

Conventional facies maps, whether based upon ratios or upon percentages, lump all representatives of each component together. It may be necessary to illustrate the internal stratigraphy of an interval in more precise detail. Sandstone may make up 50 percent of a particular section; but is there one thick sandstone unit or several thinner ones? How many sands are there that are thicker than *x* meters; whereabouts are they in the overall section?

To achieve this extra detail the geologist simply devises a convention to show what he wishes to show, and superimposes it on the conventional facies map. For example, on an isopach map showing total sand thicknesses in the chosen stratigraphic interval, he superimposes contours, patterns, or colors representing the number of sands exceeding some minimum thickness or some arbitrary log kick. For areas providing immense amounts of data, vertical variability maps showing multiple components may be constructed. One such map in common use depicts all potentially porous horizons exceeding some minimum economic thickness, regardless of their lithology. Preparation of such complex maps is difficult and time-consuming without the aid of computer programs.

Before we leave consideration of facies maps, some important cautions are necessary.

(a) The geologist must choose the components most appropriate to his data and to what he wishes to illustrate.

(b) The geologist must be clear what he means by "sand," "shale," and so on. Are they defined from logs, by cutoff values of resistivity or SP? Have they been assigned percentage limits according to grain size as determined by area counts in thin sections, or by sieve analysis? Is "sand" intended to imply "potential reservoir rock" or not?

(c) The term "carbonate" covers a variety of rock types (see Sec. 13.3). Lumping together basinal limestones which are potential source sediments and coarse reservoir-type carbonates like reefs, grainstones, or oolites may obscure more than it reveals. Within carbonate terrains such as the Middle East or the Bahamas region, it is absolutely essential to adhere strictly to a single scheme of classification (see Sec. 13.3.2) and choose the triangle components accordingly.

(d) In many areas, the presence or absence of a critical facies component may be more important than its percentage contribution to the total section. A black shale source sediment is an obvious example, as in the Appalachian or Paradox basins (Sec. 25.2.4). The geologist's judgement is on trial in these cases.

(e) The geologist should be aware of inadequacies in his map representations and be ready to make these clear to users of the maps. An elementary example is the distinction between facies changes that are gradational and those that are abrupt. A porous reef within a generally calcareous but dense rock interval must be distinguished even if the criteria chosen for the map place the whole interval in one part of the facies triangle. An oil-filled sandstone wholly enclosed in shale should be somehow shown to be so without the possibility of some other kind of facies boundary being implied.

22.5 Paleogeologic, subcrop, and supercrop maps

A *paleogeologic map* is intended to portray a region as it was at a selected time in the past (Fig. 22.25). A paleogeologic map for the beginning of the Cretaceous Period, for example, would show the surface geology as it was at that time, by age and rock type, the area

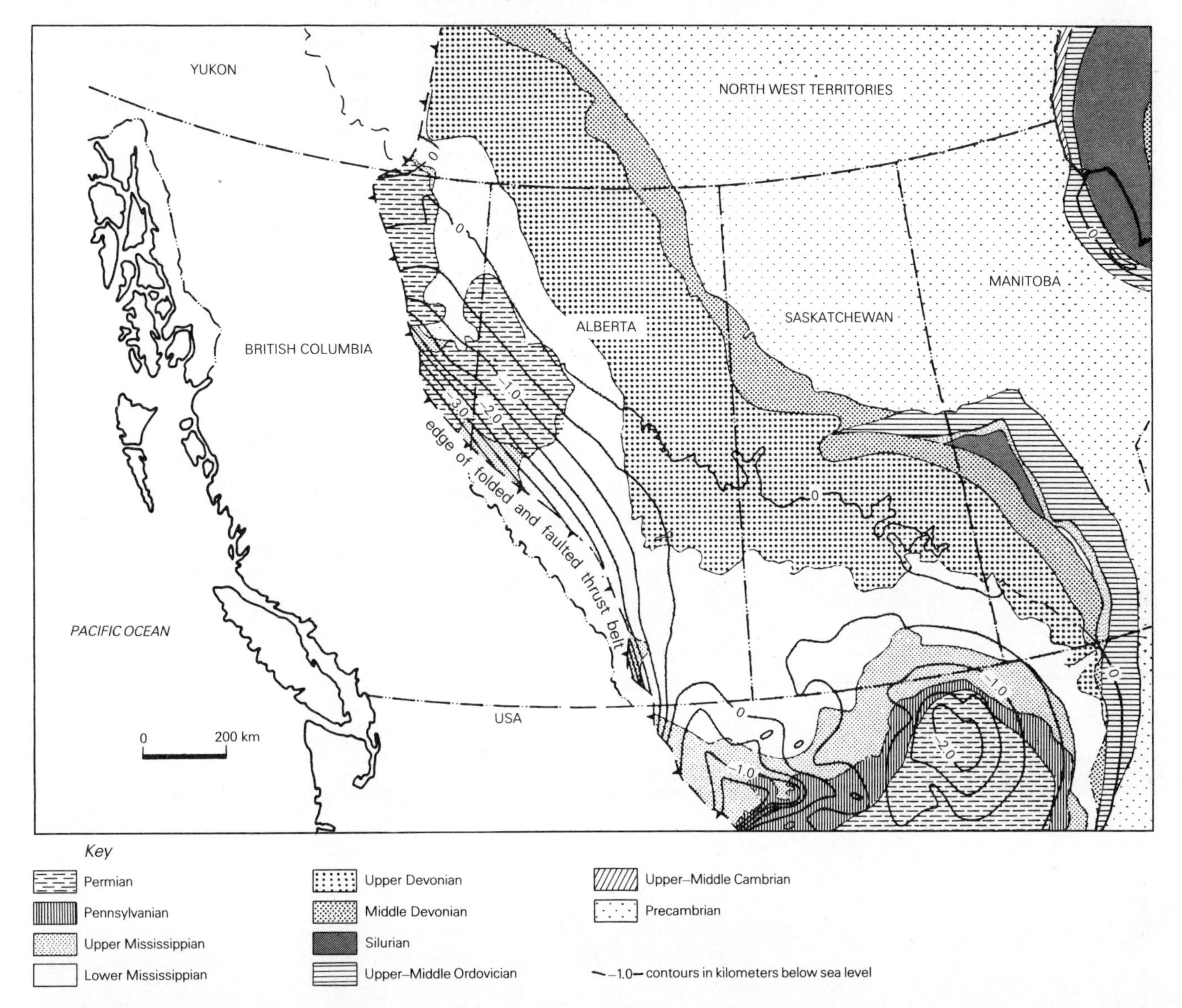

Figure 22.25 Combined structure and paleogeology of a large region: the pre-Mesozoic erosional surface of the western Canadian sedimentary basin. Strata below the unconformity become progressively younger away from the Precambrian shield, towards the Alberta Basin in the west, the Williston Basin in the south, and the Hudson Bay basin in the northeast. (From J. W. Porter *et al.*, *Phil. Trans. R. Soc.*, 1982.)

covered by early Cretaceous seas, major sources of sediment supply, and whatever other information the geologist was able to glean. The map may combine paleogeology in the simple sense of surface rock distribution with *paleotectonics*, showing structural style and age in addition. It would, in effect, be as complete a surface map as a competent geologist would have made 130 Ma ago.

Such maps may involve imaginative reconstruction. They may then be made for any time-plane or rock surface, but they are normally most easily made for times preceding unconformities. They then become *subcrop maps* for the units that later cover the unconformity surfaces. Subcrop maps are invaluable because a great variety of petroleum traps owe their existence to their subcrop positions (see Sec. 16.5). Also useful for similar reasons are *supercrop maps*, showing the rocks by age (and by type if desired) which cover the unconformity surface; subcrop traps need seals in their supercrops.

It is important to indicate the nature of zero edges on paleogeologic maps. Depositional and erosional zero edges result in different types of trapping mechanism; erosional zero edges may provide important paleogeomorphic traps (see Sec. 16.6). Both erosional and solutional zero edges may result in drape of the overlying strata over the edges; the drape structures themselves commonly form traps (Figs 16.59 & 81).

22.6 Internal property maps

Internal property maps depict selected characteristics of a single stratigraphic unit other than its shape: its porosity, average grain size, percentage of matrix, or any determinable property. Some properties may be useful for indirect purposes, as indicating something other than what is actually mapped. For instances, the percentages of grains of a particular unstable mineral (potassium feldspar, for example) may be used to indicate the direction of maturity of a clastic sediment, and the percentage of asphaltenes or other components in oils may be similarly used to indicate the direction of migration.

The most commonly used internal property maps include the following:

(a) *Isoporosity maps* show lines of equal porosity in potential reservoir rocks (Fig. 22.26). The nature of the data should be indicated; porosities may be measured in laboratory tests on cores, calculated from logs, or estimated from production data, and these sources of porosity values are not of uniform trustworthiness.

(b) *Isovolume* or *isovol maps* combine porosity values and isopachs, showing contours of equal porosity-meters or porosity-feet (products of net thicknesses and porosities).

(c) *Isoconcentration* and *isosalinity maps* show mineral contents of formation waters.

(d) *Isopotential maps* show output capacities of wells per unit of time.

(e) *Isobar maps* show any one of the pressure measurements routinely made on wells (see Sec. 23.7).

(f) *Water encroachment maps* show the "water cut," or percentage of water produced with the oil in a field undergoing depletion.

Maps such as these become out of date very quickly in producing areas. It is one of the development geologist's continuing jobs to make sure that they do not (see Secs 19.4 & 5).

Many internal property maps are now produced by organic geochemists during their investigations of

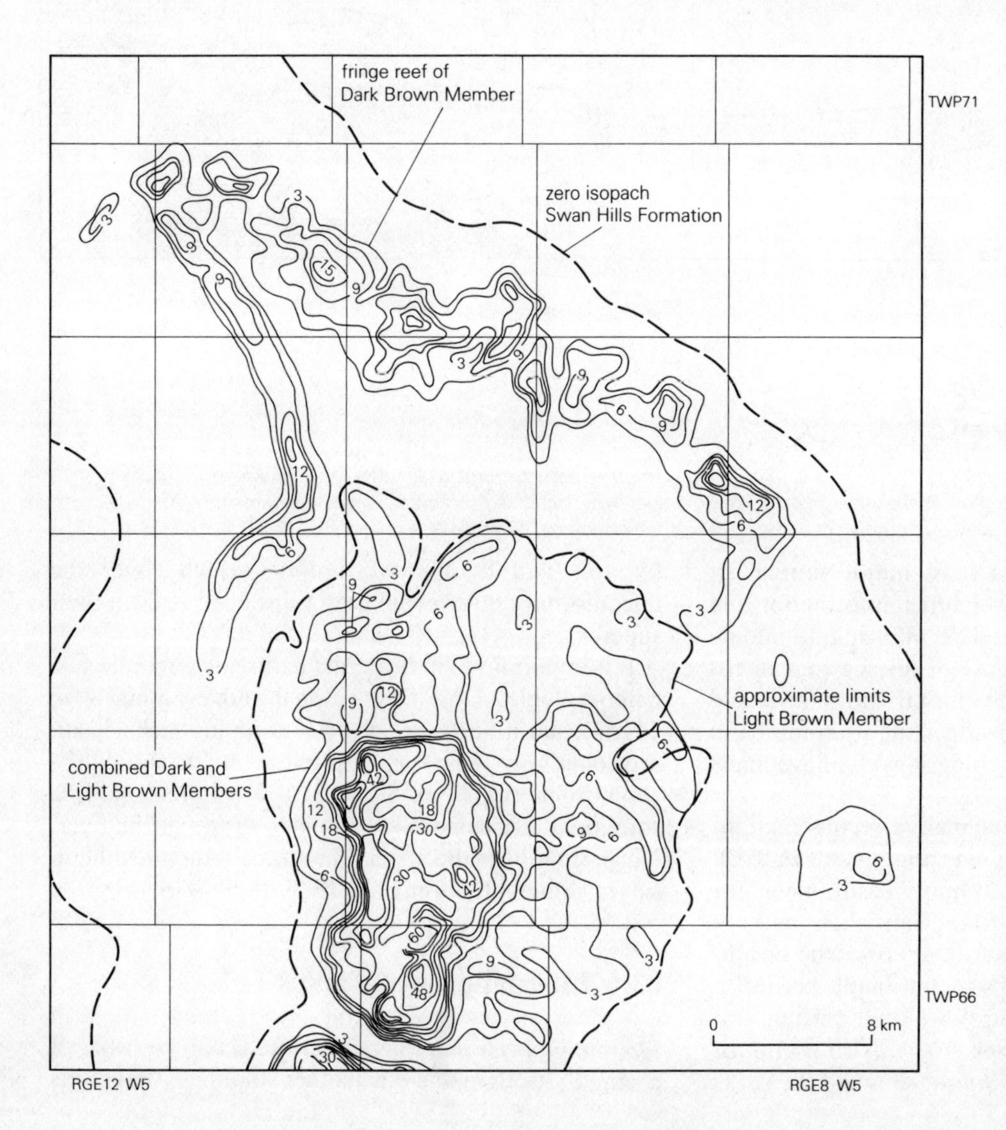

Figure 22.26 Isoporosity map of Swan Hills Formation (Devonian) in Swan Hills oil-productive region of northern Alberta, Canada. Values in meters based on logs. Areas with thick porous sections are productive reefs; in interior area, loci of porosity are patch reefs and reef outwash. (From C. R. Hemphill *et al., AAPG Memoir* 14, 1970.)

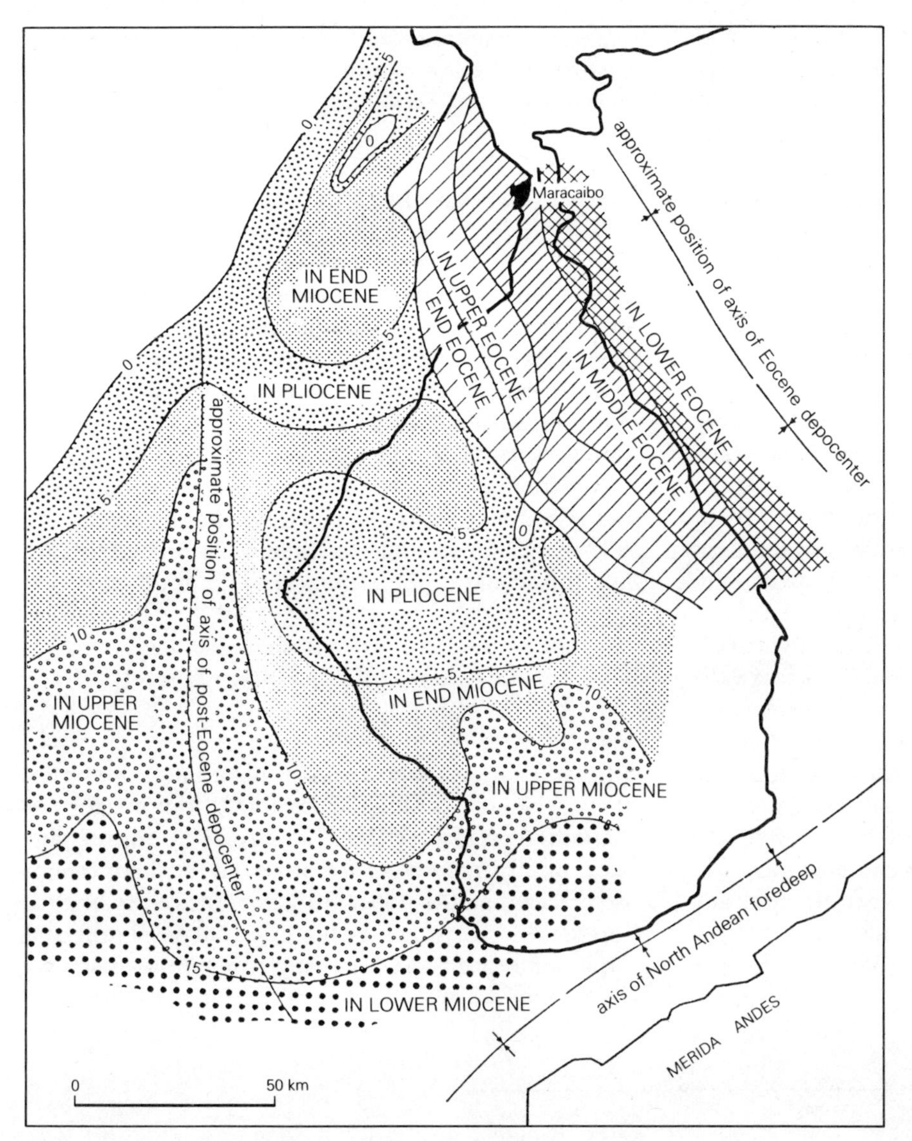

Figure 22.27 Isochron map showing time at which major source sediment of the Maracaibo Basin of Venezuela (Upper Cretaceous La Luna Formation) reached generative maturity. In a small mid-lake area, marked 0, maturity for that formation still has not been reached. (From R. Blaser and C. White, *AAPG Memoir* 35, 1984.)

source rock maturity. Figure 22.27, for example, shows maturity isochrons for the La Luna Formation in the Cretaceous of the Maracaibo Basin, Venezuela. Vitrinite reflectance and other values (see Ch. 7) were used to calculate the time before the present when the source rocks of that formation reached oil-generative maturity, and these times are contoured like any other quantifiable parameters.

22.7 Trend maps

If for purpose of illustration we consider an entire sedimentary basin to be a first-order structure, there will be a variety of lower-order structures within it. In the Persian Gulf Basin, as an example, the Zagros fold belt, the Oman Mountains, and the Arabian platform could be considered second-order structures; the Hasa Arch and Arabian sub-basin in Saudi Arabia, the Gotnian sub-basin and the Dezful embayment further north, and other swells and sags would be third-order, individual anticlinal trends fourth-order, and so on down to small individual faults and flexures.

For some purposes, it is helpful to remove the effects of lower-order structures, which may be regarded as local fluctuations, and even the effects of larger but still ancillary features in order to concentrate on the broadest, most systematic regional pattern. The more detailed a map (the larger its scale, in fact) the more the larger-scale geology is obscured by the smaller-scale, local structure. The isolation of the large-scale from the small-scale components is achieved by *trend surface analysis*. A computer analysis finds the polynomial surface best fitting the data. The value of a chosen property (the thickness of a stratigraphic unit, for example) is fitted to a linear equation involving the geographic coordinates of the control point, with a coefficient defining the slope (rate of change) of the value. Polynomial surfaces of successively higher degrees are then fitted, yielding computed values which deviate from the observed values. The differences between the observed and computed values are *residuals*, and they represent small-scale components of the structural or stratigraphic pattern which are superimposed on the larger-scale *trends*. Positive residuals represent excesses, and negative residuals deficiencies, above or below the "regional" trend (Fig. 22.28). Provided that the original contour map is first digitized, filtering can be applied either for *size* (area or thickness, for examples) or for *direction*.

Trend surface analysis is of less value to petroleum geologists than its advocates believe. Its greatest value would be in the exploration phase in new areas, when hard data are too sparse to provide a useful foundation for it. Even after considerable exploration in the Western Canadian Basin, as an example, trend surface maps for the Cretaceous System showed the regional southwesterly dip of the strata, the interruptions of this regional dip by large features like the Peace River Arch (Fig. 16.41), and *general* contours and lithofacies of the Cretaceous sandstone–shale succession. A *detailed* map of Cretaceous structure and lithofacies was significantly different, and it is the *detail* that is critical in all except the earliest phases of exploration. This is especially so in a case like the western Canadian Cretaceous; large and small traps extend the length and breadth of the basin without ever being of the structurally or stratigraphically obvious kind. Nonetheless, trend surface analysis may become a valuable adjunct to the petroleum geologist's map suite if it is extended to nonspatial variables such as pressures and temperatures in wells.

22.8 Hydrodynamic maps

The influence of subsurface water movement on hydrocarbon migration and accumulation is described in Chapters 15 and 16. A simple relation is there established between equipotential surfaces for the oil and the water in a reservoir rock; they are the surfaces normal to which the movements of the two fluids take place. The family of surfaces for the water (where v = constant) can be constructed from the shut-in pressures of wells (see Secs 15.8.2 & 16.3.9). The corresponding values for the oil (where u = constant) can then be calculated for any well from v and z (the elevation of the drillhole):

$$u = v - z \qquad \text{(if } z \text{ is above sea level)}$$
$$u = v + z \qquad \text{(if } z \text{ is below sea level)}$$

(Figs 22.29 & 30).

In an open drillhole, z represents a simple elevation (that of the kelly bushing of the hole). Contours for z = constant therefore reflect a family of horizontal surfaces. In a hydrocarbon trap, however, the fluids cannot rise higher than the upper surface of the reservoir bed. Curves for z = constant are therefore not now horizontal surfaces but are represented by structure contours on

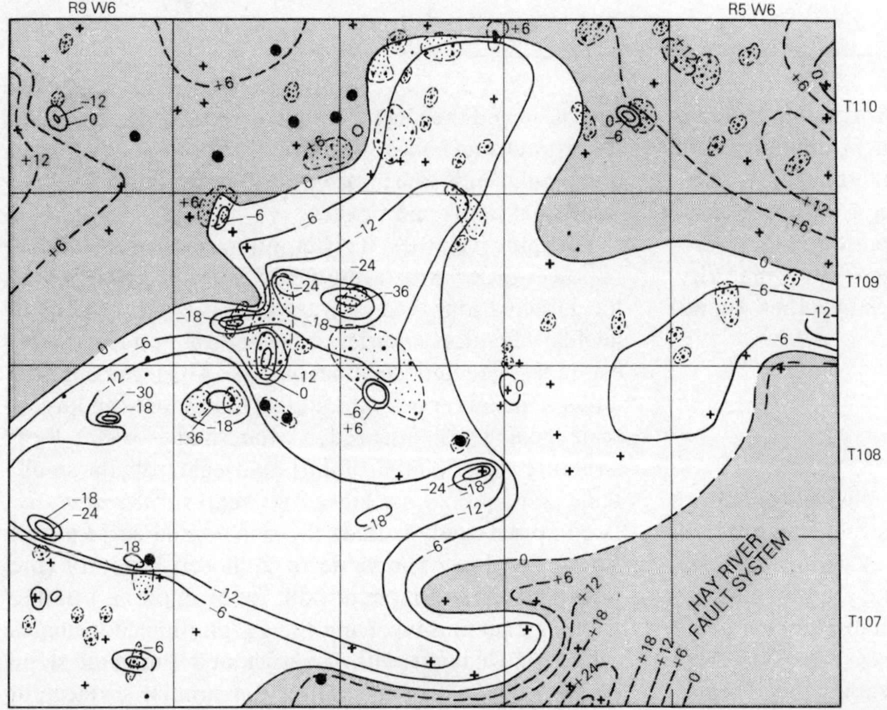

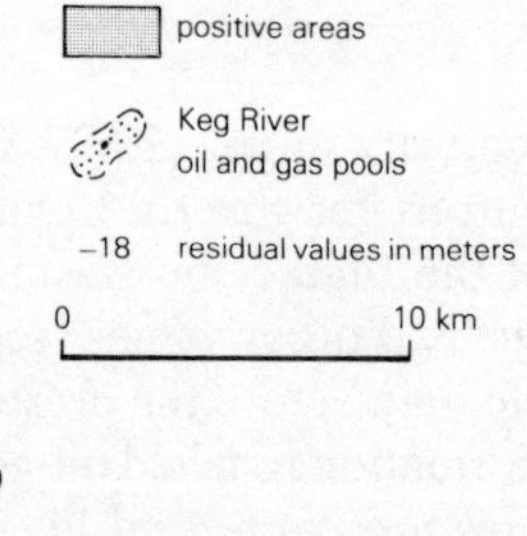

Figure 22.28 Modified first-order residual map on a Middle Devonian formation underlying the productive reefs, Rainbow area, northern Alberta, Canada. Effects of post-Devonian faulting removed, leaving only effects of Middle Devonian faults. Note general lack of correspondence between the petroliferous "thick" areas in the Keg River Formation and residual lows on the underlying surface. (From D. L. Barss *et al.*, *AAPG Memoir* 14, 1970.)

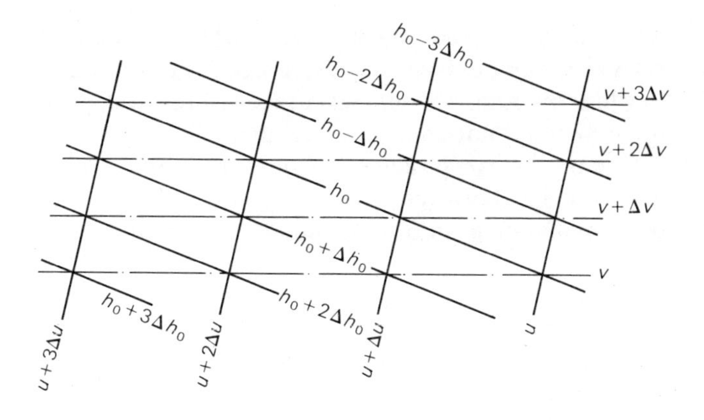

Figure 22.29 Graphical relation of surfaces along which h_o, u, and v are constants, for equal increments of each. (From M. K. Hubbert, *AAPG Bull.*, 1953.)

the top of the reservoir bed. We therefore have three sets of contours: those for constant values of v and z are known; those for constant values of u, representing the equipotential surfaces for oil, are drawn through the points of intersection of the other two contours (Fig. 22.30). Areas of closure of the u curves containing minimum values of u represent potential traps.

In an area with relatively few wells, a hydrodynamic map gives only general information about the directions of fluid movement. In areas of denser drilling and some production, values of h_w, z, ρ_w, ρ_o, and shut-in pressure become sufficiently numerous for much more meaningful maps to be constructed. These outline areas

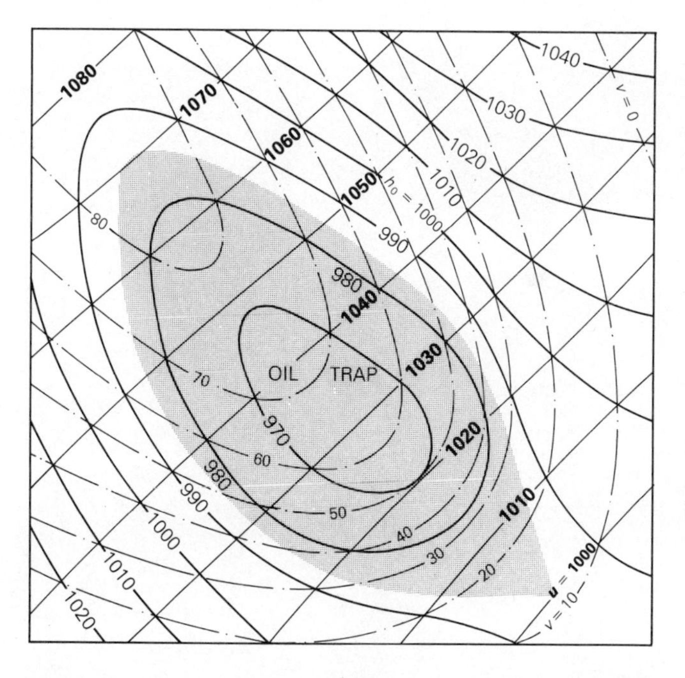

Figure 22.30 Map of traces of intersections of constant-value surfaces, h_o, u, and v, with upper surface of reservoir rock showing potential minimum and consequent trap for oil. (From M. K. Hubbert, *AAPG Bull.*, 1953.)

of favorable, downdip water flow and of closed (low-value) oil potential. Once an individual prospect has been identified, the more detailed contouring enables some estimate to be made of the height of the hydrocarbon column and the probable shape and size of the pool, and hence of its possible productivity. Examples of such hydrodynamic prospect maps are illustrated in Figures 16.65 and 66.

22.9 Computer-made maps

In all areas of dense or active drilling, well and field data must be recorded in a form suitable for storage, retrieval, and processing on magnetic tape or computer input cards. The geologists must decide what information is to be recorded in this way. It is ready for computer printout and may ultimately be plotted by machine.

The rapidly changing subject of computer mapping is beyond the scope of this book and of this author, but its fundamental aspects should be within the ken of all geologists. The essence of it is the mechanical reconstruction of a geologic surface (topographic, unconformity, stratigraphic) from data which are not of uniform spatial distribution. If the geologist understands the way in which the spatial distribution of his data exerts control over his contouring, he will be able to use his experience and judgement effectively. Otherwise his contoured maps will merely record data; they are unlikely to yield any trustworthy new insights.

For a computer to perform the contouring, it must be given a grid of points through which it then "threads" contours according to the way in which its program "searches" the data. The form of the grid may be either rectangular ("gridding") or triangular ("triangulation"). In both forms, the scale of the grid must be appropriate to the distribution of the data.

In *gridding*, a system of rectangular coordinates is superimposed on the data. For each intersection point on the grid, the closest control points are searched, a value for the intersection point is generated by least-squares fit, and contours are drawn for the intersection point values, not for the control point values. This system cannot yield a unique solution; the choice of contour is partly controlled by the orientation of the grid, which should accord with the structural grain of the area and this may be misunderstood.

In *triangulation*, the data points are linked into triangles and each triangle is solved by the computer. The surface is then fitted to the data points, not to artificial points, giving a unique solution. Contours are drawn as straight lines between points, and the pattern is consequently angular instead of smooth; smoothing requires subdivision of the triangles into smaller and smaller triangles, occupying more computer time.

Like all computer-drawn graphics, computer-contoured maps depend upon the reliability of the data. The computer can be programmed to eliminate "anomalous" data automatically, but it may be the "anomalous" data that are significant. The computer can be programmed to put faults on the map, but such programming presupposes that the geologist already knows that faults are present (and approximately where they are). Pattern or color can be added to computer-drawn maps automatically, so the best of them are not greatly different in appearance from manually drawn maps. But contouring must still be interpretative and the impression that computer contouring will find anomalies (and hence accumulations) which would be missed by even a competent geologist should be resisted. Different computer mapping packages have different merits and demerits, the principal merit they have in common being speed of regularly repeated output. Unless the geologist, knowing exactly what he wishes to discover or to illustrate, also knows which package will perform most reliably for his purpose, he must remember that the uses to which his maps are put must in the last resort depend upon him and not upon the computer.

Computers can produce wonderfully useful *derivatives* of maps, however. They can quickly convert two-dimensional maps into three-dimensional figures, with contours. These figures the computers can then tilt, rotate, or superimpose, at the command of a geologist-operator who could previously acquire such an array of viewpoints of his terrain only through exhaustive work by draftsmen skilled in descriptive geometry. The resulting *computer graphics* can be as easily produced by a microcomputer (a personal, table-top instrument) as by a mainframe computer, and with much less time, preparation, and expense. Projected as cathode-ray-tube (CRT) images, they are not merely a boon to the geo-scientists but an even greater boon to the network of communication between the scientists producing them and the committees that have to be impressed by them.

23 The drill and drillholes

23.1 Kinds of drillholes

The drilling of a hole into the outer skin of the Earth's crust is the final test of the petroleum geoscientist's role. The drill is the arbiter, and the geologist who discovers oil (or gas, nowadays) is a successful individual even if the reasoning behind his recommendation of location was faulty and if he never publishes a paper or advances our understanding of petroleum science.

There are two distinct classes of well drilled for oil or gas. They are drilled by the same kind of rig and with most drilling procedures identical for both, but their purposes and expectations differ and they are treated differently by governments, lawyers, tax collectors, and the industry in all "Western-style" jurisdictions. One class of well is drilled in the hope of discovering oil or gas where none has been discovered before; this is an *exploratory* well. The other class is drilled to permit the production of oil or gas from a field that has already been discovered by another well; this is a *development* or *production well*.

An amateur can readily understand what constitutes a development well; it is a well drilled in an existing producing oil- or gasfield. It is not so obvious what constitutes an exploratory well. How many of the wells drilled in the Pembina field in western Canada (see Sec. 26.2.1) were exploratory wells? How long after the initial discovery of the field could "exploratory" wells continue to be drilled before their locations had to be recognized as lying within the limits (enormous limits, admittedly) of the field? Clearly an "exploratory" well in the Sudan, or the upper Amazon Basin of Brazil, is not in the 1980s quite the same thing as an "exploratory" well in Oklahoma.

Inevitably the need to formalize the different views of what "exploratory" connotes arose first in the USA, where almost half of the individual states had achieved oil production before the beginning of World War II and each state applied to this production its own land regulations and tax laws. The American Association of Petroleum Geologists accordingly established, in 1948, a Committee on Statistics of Exploratory Drilling (CSED), to facilitate an appraisal of the results of oil and gas exploration, and to analyze the reasons behind the selection of exploratory well locations by the industry. Most "Western" countries adhere in whole or in part to the Committee's categorization of wells, though nowhere outside North America is the adherence so punctilious.

A well in a quite new region – in many cases in a remote and uninhabited region – has traditionally been called a *wildcat well*. How remote must it be to be called a wildcat? The Committee acknowledged a class of exploratory wells already referred to by all oilmen as *rank wildcats*; they were renamed *new field wildcats* (NFW) because they were drilled in the hope of finding new fields on geologic structures which had never produced oil. One or more *dry holes* may have been previously drilled on the same structure. Hundreds of rank wildcats, of course, have been drilled without any structure having been defined, or even having been sought. For these cases, a new field wildcat may be a hole drilled at least 2 miles (3.2 km) from the nearest previous production.

If the terms *wildcat well* and *exploratory well* had been declared synonymous, all wells drilled on structures already producing, or within 2 miles (3.2 km) of production in the absence of defined structure, would have become "development" wells. Clearly this would make the "development well" category a very large and heterogeneous one. On a very large structure, like those in the Middle East or the Permian Basin, the mere existence of production in no way establishes the areal limits it will ultimately reach. In large swaths of production without definable structure, like those of the Pembina field mentioned above or of the East Texas field, a number of wells are legitimately drilled to "explore" the limits of production long after the initial discovery well is contributing to it. Some of the very largest fields both with and without definable structure

– the Ghawar field in Saudi Arabia and the Bolivar Coastal field in western Venezuela, as obvious cases – were thought for a number of years to be four or five separate fields, and wells drilled between those putatively "separate" fields were legitimate exploratory wells. The Committee therefore recognized four categories of exploratory hole in addition to that of the new field wildcats:

(a) *New pool wildcats* (NPW) are drilled outside the area of proven production, commonly two or three locations beyond its edge. Obviously "two or three locations" means different distances in different jurisdictions, according to well-spacing regulations. It may also mean different things in different geologic circumstances; on highly complex structures like irregular salt domes or thrust-faulted anticlines (Figs 16.14, 18 & 51 illustrate such structures), continuity of reservoirs and traps may be so uncertain that a hole only one location outside the developed area may justify designation as a new pool wildcat.
(b) *Deeper pool tests* are drilled *within* the proved area of a pool wholly or partly developed, in search of another pool below the one contributing existing production. Deeper pool tests are common, but more of them are completed as field development wells than actually discover deeper pools.
(c) *Shallower pool tests* are similarly located but are searching for pools less deep than the known pools. Obviously shallower pool exploratory tests are relatively rare.
(d) *Outposts* or *extension tests* are drilled to extend known pools. They are the so-called *step-out wells*, drilled (like new pool wildcats) several locations beyond the known productive edge of the pool but designed to show that the pool, and not a separate pool, in fact extends so far. A successful extension well may of course turn out to be a new pool discovery well, and a productive new pool wildcat well may equally easily turn out to be an extension well. Successful new field wildcat discovery wells on large or ill-defined structures are invariably followed up by several extension wells in an attempt to delineate the size of the field for leasing and reserve-estimation purposes before development drilling is embarked upon.

The exact definitions of the several categories must of course be modified according to the land-measuring units used in a particular country (acres, square kilometers, hectares) and to the well-spacing regulations in force. Most producing countries outside North America, as well as nonproducing countries undergoing vigorous drilling programs, use somewhat simpler well classifications, especially that into exploratory, extension, and development wells. Some prefer to call all wells designed to extend, deepen, or develop producing fields simply *appraisal* or *confirmation wells*.

Many exploratory wells are posthumously consigned to the category of geologic successes but economic failures. Some of those so designated deserve the name, self-serving though it may be; the explorationists learn a great deal from them, and have the satisfaction of knowing that their geological or geophysical grounds for drilling the wells were correct in spite of failure to achieve commercial production. Among the fields accorded a case history elsewhere in this book are those of the West Pembina area of Alberta, Canada (see Sec. 26.2.1). If the Devonian reefs discovered by the West Pembina wells had been water-bearing, or too tight to be productive, they would have deserved to be called both geological and geophysical successes notwithstanding. Long strings of geological successes that are economic failures tend to lead managements to regard their geologists as being too easily satisfied with success. Many exploratory wells, by contrast, deserve to be called economic successes but geological (or geophysical) failures; but one seldom hears of them in those terms.

As the attraction of classical prefixes to scientific names became epidemic, during the 1940s and 1950s, an author in the journal *California Oil World* proposed that a well which was neither a real wildcat nor yet a merle development well, and was directionally drilled from the shoreline to tap oil beneath the seafloor off California, should be given the properly scientific name of "hypothalassic clinoparamesocat." He even made recommendations about pronunciations, but neither his terminology nor his pronunciations caught on, even in California.

23.2 The drill and its operation

Drilling rigs have been familiar sights to many North Americans since the days of the wooden cable-tool rigs before World War I. As this chapter is being written, there are more than 4000 rigs in use in the USA alone, drilling more than 50 000 holes each year.

In most countries outside North America, relatively few people had ever seen a full-scale rig before about 1950. Even today, there are only about 1250 rigs in all of Europe, noncommunist Asia, Africa, Australasia, and Latin America. A cursory description of a modern rotary rig is necessary for the young geologist who has yet to see one at close quarters.

Whereas the old *cable-tool rigs* were supporting frames for drilling tools that hammered their way into the ground, the *rotary rig* drills downwards with a rotating *bit.* A strong cable extends from the *draw works,* which provide the drive, to the *crown block* at the top of the high mast. The crown block must support the weight of the steel drill pipe which is suspended from it, pipe which may be 8 km long or longer in a deep well. Below the crown block is the *travelling block,* simply a pulley which moves up and down as the drill pipe is raised or lowered. From the travelling block a *swivel* is suspended, its upper half stationary, its lower half doing the rotating that turns the bit.

Below the swivel is a steel rod of square cross section called the *kelly*; it is longer than a single length of drill pipe and passes through a square hole or *kelly bushing* (KB) in the center of a steel turntable about 1.5 m in diameter. The elevation of the kelly bushing must be accurately measured when the rig is erected, because all drilling and logging depths are expressed in relation to it. The rotary turntable is driven by a motor, turning the kelly in its square hole. The drill pipe is attached below the kelly.

For many years, oilwell drill pipe has been manufactured in 30-foot (9.14 m) lengths. Because drilling downwards required the successive additions of further 30-foot lengths of pipe (called *joints*) to the *drill string,* the derrick had to stand more than 30 feet high above the *derrick floor* to enable a member of the drill crew to attach each new length of pipe between the kelly and the pipe already in the hole. With vastly improved steel and rotary equipment, and skilful operators, it became customary to add new pipe in 90-foot (27.4 m) lengths (called *stands*), three joints being left screwed together to be added as a single length. The crew member responsible for attaching this length to the kelly therefore had to stand 90 feet above the derrick floor, with the travelling and crown blocks above him, so the rig became extended to its present height – commonly 128 or 136 feet (39.0 or 41.5 m). The largest land rig used by the industry to 1980 (Delta 76) had a 147 foot mast (44.8 m) and a crown block rated at 900 tons (just over 9×10^5 kg).

Pipe is added by drawing up the string until the kelly is clear of the derrick floor, clamping the existing drill pipe there to stop it from falling back down the hole, uncoupling the kelly and pulling it upwards 90 feet, coupling in the new length of pipe with the aid of giant tongs, and returning to bottom with 90 feet of new drilling now ahead. The *bit* is a rotating arrangement of cones and wheels made of hardened steel. Bit sizes in common use range from 6 to 12¼ inches (0.15–0.31 m), with still larger sizes in less common use. Bits wear out after relatively short lives, depending upon the hardnesses of the rocks they are called upon to penetrate. To put on a new bit requires pulling all the drill pipe out of the hole, 90 feet at a time; this is known as "making a trip," merely tedious when the hole is 1500 m deep but a daunting undertaking at 7500 m. The drill string is of course under great tension, which becomes greater as the hole becomes deeper. The work of the bit, on the other hand, requires pressure, so that weight must be exerted on it whilst leaving the drill string under tension. This is made possible by the addition of a series of *drill collars* between the bit and the lowest stand of drill pipe.

The work of the bit would be impossible in dry conditions; it would be melted by frictional heat in a very short time. It is cooled by *fluid mud,* which is mixed in a large pit or *sump,* pumped through a hose to the top of the drill pipe, and forced down the inside of the pipe and through nozzles in the bit, to return to the surface through the *annulus* between the pipe and the walls of the hole. In addition to cooling and lubricating the bit, the mud serves to create a pressure head to prevent gas escape from porous formations, and it brings the rock cuttings made by the bit up to the surface. The cuttings are passed over a *shale shaker* to recover them from the mud, and the mud returns to the sump.

The *drill cuttings* are precious material for the geologist, but they are exceedingly unreliable because of three obvious failings. They may not be representative of the rock actually being drilled, because only hard rock fragments can survive the beating the cuttings receive during their journey up the hole. Even hard rock fragments may be altered by the drilling process itself. Effects such as darkening, vitrification, magnetization, and increasing insolubility have been attributed to *bit metamorphism,* best demonstrated in red beds of the North Sea and North German basins.

Furthermore, cuttings are easily contaminated by material derived by caving from any part of the hole between the bit and the bottom of the casing (described below). Finally, the journey from the bit to the shale shaker takes time. The deeper the hole, the longer the time, and consequently the less accurate the assignment of the cuttings to their proper depth. Despite these drawbacks, cuttings retain one role in which they have no substitute except at much increased expense. Cuttings derived from a fossiliferous formation may contain fossils. Macrofossils become pulverized in drill cuttings but microfossils may survive in identifiable condition. Modern micropaleontology was in fact an outgrowth of petroleum geology, and of oilwell drilling in particular (see Ch. 4). In the years between 1930 and 1960, assuredly, there were more men and women seeking and studying "bugs" around the Gulf of Mexico and the Caribbean than in all the rest of the world combined.

The unglamorous job of controlling the properties of the drilling mud is critical to the drilling process. Typical mud weights are 800–1300 kg m^{-3}, but the actual weight, consistency, and composition of the mud are varied according to the nature of the strata being drilled through, their fluid contents, and the depth and condition of the hole. The higher the formation pressure (see Sec. 14.2), the heavier the mud must be made to counteract it. Barite ($BaSO_4$) is added to increase the mud weight; it is a common mineral, easily obtained by surface mining, and adequately heavy (its relative density is about 4.5). Some 90 percent of the world's output of barite finds its use in heavy drilling muds. To increase the viscosity of the mud, bentonite (aquagel) is added; to reduce it, quebracho is added, containing tannic acid. If mud circulation is lost, because the mud invades a formation (or "thief zone") of cavernous porosity instead of returning up the hole, drilling must be suspended until it can be re-established. The mud is then made as light and dispersed as possible, to lessen its propensity for forcing its way into the walls of the hole. To serve as mud lighteners in such emergency, sacks filled with an extraordinary assortment of light refuse may be kept at the drillsite – chicken feathers, moldy cereals, ground sea shells and walnut shells, shredded banknotes, wool fiber, manure mixed with bentonite, even ground-up beer cans, are among the trash items utilized. Even the logging process (see Ch. 12) requires changes in the mud, from oil-base to water-base or vice versa, or to gyp-base by the addition of powdered gypsum when an evaporite section is to be logged.

It is not the geologist's job to control the mud, but it *is* his job to keep the mud engineer informed of the lithologies to be expected, their possible fluid contents, and their effects on the condition of the hole. He must also continuously monitor the mud for indications of hydrocarbons brought up by it (Sec. 23.2.1).

The condition of the hole must be maintained during drilling. Very weak rocks flake or crumble off into the hole, as *cavings*, interfering with the work of the bit and causing stuck drill pipe. The drilling mud itself is responsible for much of the damage done to the hole, through "mudding off" the weak formations. Oil-base muds cause less damage than water-base muds, but they are also more expensive. Shales become softened, and their expandable clay constituents become swollen or dispersed (see Sec. 15.4.1).

Caving sections must be supported behind *casing*. The weakest section to be drilled through is commonly the weathered zone or "overburden" at and shortly below the surface. The well is therefore *spudded in* with a very large bit, making a wide-diameter hole in which the *conductor pipe* is set. This may be 20 or even 30 inches in outside diameter (0.51 or 0.76 m), and serves as the funnel through which the bit is lowered into the ground. *Surface casing* is normally required, and is set through the conductor pipe; a typical outside diameter of surface casing is 13⅜ inches (0.34 m). If a succession of casings has to be set, because of weakness of the hole, the process is more than merely expensive and time-consuming; it also reduces the diameter of the hole, because each new length of casing has to be set, and cemented, through the earlier lengths which have already been cemented. Standard casing diameters for successive downhole settings include 10¾, 7⅝, and 6⅝ inches (0.27, 0.19, and 0.17 m), but other diameters are also used.

In spite of the careful setting of casing and skilful drilling practices, a variety of debris may collect at the bottom of the hole – caved rock, sand washed in from water-bearing porous zones, and cones broken from the bit. This debris is collectively called *junk* and has to be removed by specially designed recovery tools. Much more time-consuming and expensive is the recovery of twisted-off drill pipe or broken casing. The process of recovery of unwanted drilling material from the hole is called *fishing*.

Also in spite of watchful operation, the danger of a *blowout* is always present. It is caused by drilling into a well-sealed formation containing gas or oil under high pressure. The *blowout preventer* (BOP) is installed at the casinghead and enables the drillers to close the hole quickly without having to pull the drillstring.

In conventional drilling using rotary bits, an inordinate amount of energy is expended in simply turning the drillpipe and the bit, against the friction of the hole itself or the casing through which much of the drillstring may be operating. A theoretically more efficient drilling method – and potentially a much faster one – is that of *turbo-drilling*, using downhole motors powered by the mudflow through the bit. The method exerts more force against the rock and wastes less in the rotating of the drill pipe. Turbo-drilling is widely used in the Soviet Union but has not supplanted conventional drilling in regions where Western technology remains dominant.

All the material which the engineers and drillers anticipate needing to carry the well to its proposed *total depth* (TD) must be transported to the wellsite and stacked there before drilling begins – casing and drill pipe (the principal items of the oilfield equipment collectively referred to as *tubular goods*), cable, bits, cement, mud additives, sand, lubricants, and an arsenal of other items – including water for the mixing of the mud if the wellsite is in a barren region.

23.2.1 Indications during drilling

The evidence of the cuttings, the continuous mud logging, and the drilling speed may encourage the geologist to sample the rock more thoroughly. A simple route is to take *sidewall samples*. Hollow cylindrical bullets are fired electrically into the formation being penetrated, and small cores are retrieved by ingenious withdrawal wires. The samples provide some evidence of lithology and grain size of the formation, but more valuable is the evidence they may yield of the fluid content. This may be determined by retorting or by leaching with pentane, or, more conveniently, by *fluorescence tests* under ultraviolet light. Light blue fluorescence indicates gas, bright yellow or green is produced by the oil column, and spotty yellow or brown fluorescence by water. Fluorescence tests can of course also be made directly on the mud, but this should not be done if oil has already been added to the mud by the engineer. Gas in the drilling mud is easily detected by passing it over a wire just hot enough to ignite methane. The methane is burned off, raising the temperature of the wire and therefore its electric resistance, which is measured.

Any indication of porosity or of hydrocarbons, or a drilling break indicating an anticipated target horizon, may cause the geologist to call for the drilling to be supplanted by *coring*. *Wireline cores*, of small diameter, are recovered by a narrow barrel lowered and retrieved inside the drill pipe without making a trip. For a standard core, in contrast, the pipe and bit are withdrawn from the hole, and the lowest joint of pipe, with the bit, is replaced by a *core barrel*. This is a steel sleeve 20–60 feet in length (6–18 m), its bottom rim implanted with industrial diamonds or other grinding agent. The barrel cuts a cylindrical core of rock which is trapped within by a spring mechanism at the bottom. Any laboratory examination or testing that can be carried out on rock materials can be carried out on a standard core, including measurements of porosity, water saturation, and other parameters vital to reservoir engineering, possible production plans, and reserve estimations (see Chs 11, 12, 19 & 24).

An important climax of the drilling process is a successful *drillstem test*. Any favorable indication in the drilling creates the need to know what fluid or fluids the horizon contains. Instead of a core barrel, a *testing tool* is attached to the bottom of the drill string. The tool is a double steel sleeve perforated by openings called *ports* which are kept closed until the tool has been lowered opposite the formation to be tested. The tool is then isolated from the mud column by flexible *packers* expanded against the walls of the hole, and the ports are opened mechanically by rotating the inner sleeve, allowing any fluids in the formation to enter the sleeve and rise up the drill pipe. How far they rise – to the surface, or only to fill the lower part of the pipe – depends on their nature, the pressure behind them, the permeability of the formation, and the duration of the test.

The drilling process must be periodically interrupted so that the hole can be *logged*. All well logs (Ch. 12) are best run in open (uncased) hole; only some logs measuring radioactive properties, such as the pulsed neutron and gamma-ray logs, are capable of operation through casing. The battery of logs must therefore be run before casing is set (though only in rank wildcat wells are operators likely to bother running them before setting surface casing), and before each downward extension of the casing. Because well logging (like drillstem testing) is a contract operation, the logging crew with their truck and equipment remain at the drillsite only in remote or offshore locations; in all "normal" locations they are called for only when logging is necessary. Casing, logging, and testing must therefore be carefully planned and coordinated if rig time and contract expenses are not to be wasted.

23.2.2 Deviated holes

It was learned by early experience that a drillstring only 0.15–0.20 m in diameter and hundreds to thousands of meters long, pushed downwards by a drive from the surface, is easily deflected by hard rocks, especially if they have any significant dip. Many areas of dense early drilling are geologically complex, with sharp folds and numerous faults: California, southern Oklahoma, Trinidad, the Caucasus, Romania. Drillers found it impossible to maintain verticality of the drillstring, and the bottoms of holes of even modest depth were not properly located by the points on the ground on which the rigs were standing. "Wandering" or "crooked" holes had to be ingeniously surveyed to locate their positions at total depths.

Unintended deviation from the vertical is a potentially costly nuisance, called a *dogleg*. The engineers, however, quickly learned to take advantage of the possibilities of deliberate deviation. Bevelled steel tubes called *whipstocks* were designed for attachment to the drillstring to permit *directional drilling*, which under modern conditions of accurate control has numerous applications (Fig. 23.1). The commonest of these are listed here:

(a) The simplest case is that required when a fish defies recovery from the well bore. Rather than waste expensive rig time with more and more elaborate fishing methods, operators prefer to *plug back* the blocked hole with cement, and *sidetrack* it past the obstacle. After the necessary deviation has been achieved, the hole may be returned to the vertical.
(b) One of the earliest applications of directional drilling enabled an operator to obtain oil from beyond his lease boundary. Not allowed to erect his rig on the oil-bearing property, he drilled into it sideways. This is now illegal in sensible jurisdictions, but the practice is still necessary if access to a legitimate productive area is restricted by natural or man-made obstacles – buildings, airfields, parks, lakes.
(c) A pool having a thick pay zone may be of such small

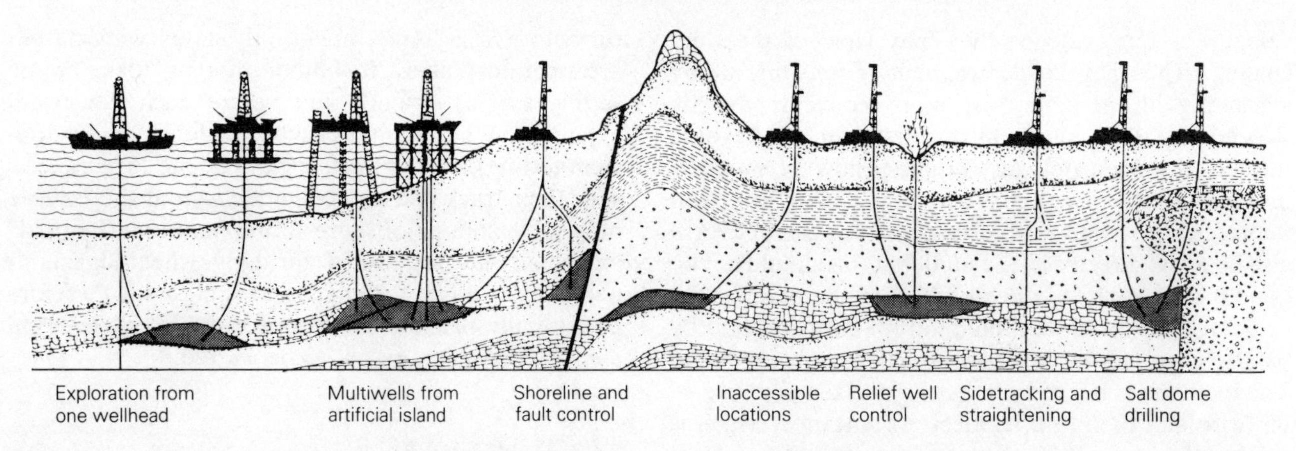

Figure 23.1 Some valuable functions made possible by directional drilling. (From the Eastman Whipstock General Catalog, 1980–1.)

area that authorized well spacing may not permit the drilling of enough vertical wells to drain it adequately (especially if output per well is also restricted). Wells drilled outside the productive limits of the pool may be drilled so as to converge into the pool at depth. Similarly, offset wells which prove to be outside the limits of a pool may be deviated towards it to save the time and expense of moving the rig (possibly into a location forbidden by well-spacing controls), and to add to the number of wells contributing an allowable volume of production. This is called *edge drilling*. It is especially convenient and common in pools buttressed against the flanks of salt domes, or confined to the crests of narrow pinnacle reefs (see Ch. 16). Edge drilling is permissible, of course, only if the production rights in the target horizon and the drilling rights at the surface location are vested in the same operator.

(d) A pool known or reasonably believed to exist in an identified structure may be missed because the well location is stepped out too far. The well may be put down into the wrong side of a fault, or too far outside the flank of a salt dome, for common examples. The well is therefore *plugged back*, by being cemented below a selected depth, and the upper part of the unsuccessful hole can then be deviated without moving the rig.

(e) It is frequently essential to tap a sizable area of an established pool without moving the rig. The commonest case is presented by offshore fields, especially those in waters deeper than a few tens of meters. The field must be exploited from a small number of fixed platforms, each tapping a sector of the field through a cluster of wells – commonly about 24 – drilled radially by a single fixed rig (Fig. 23.2). In some large western Siberian fields, drilling operations convert the permafrost terrain into hazardous swamp. Wells are directionally drilled in "superclusters" of up to 80 wells from single drillsites on artificial islands.

A less common case is that of urban oil development. Where oilfields underlie parts of cities, they must be exploited from very small and expensive bits of land, by rigs encased in structures to resemble buildings and muffled to reduce noise. Figure 23.3 shows the complex deviations of holes from only three drilling locations to tap three separate structures underlying downtown Los Angeles.

(f) The most spectacular use of directional drilling is devoted to the killing of wild wells. If a well blows out, spewing gas, oil, or both above the derrick floor – and possibly catching fire, destroying the rig, and creating a crater at the top of the hole – attempts to control it may involve the drilling of a *relief well*. From a safe distance away from the crater, a second hole is drilled and deviated so as to intersect the wild well bore above the top of the hydrocarbon column, a marvellous demonstration of pinpoint engineering. Oil or gas is drawn off from the wild well, or other fluids are pumped into it to stop the flow. Many kills of well blow-outs have been achieved, but not without loss of the lives of many drillers and safety experts.

Among hundreds of publicized blow-outs, one of the most spectacular was that of the Alborz-5 well in central Iran, near the holy city of Qum, in 1956. The well blew some 10^4 m^3 daily for 12 weeks before it "bridged" and snuffed itself out. Geologists in many countries saw, in a film made during attempts to control the well, the entire drillstring blown high into the air and contorted like spaghetti. A still more widely publicized blow-out, this one offshore, was that of the Ixtoc discovery well in Mexico's Gulf of Campeche, which spilled about 5×10^5 m^3 of oil during a 10-month rampage in 1979–80 before it was brought under control.

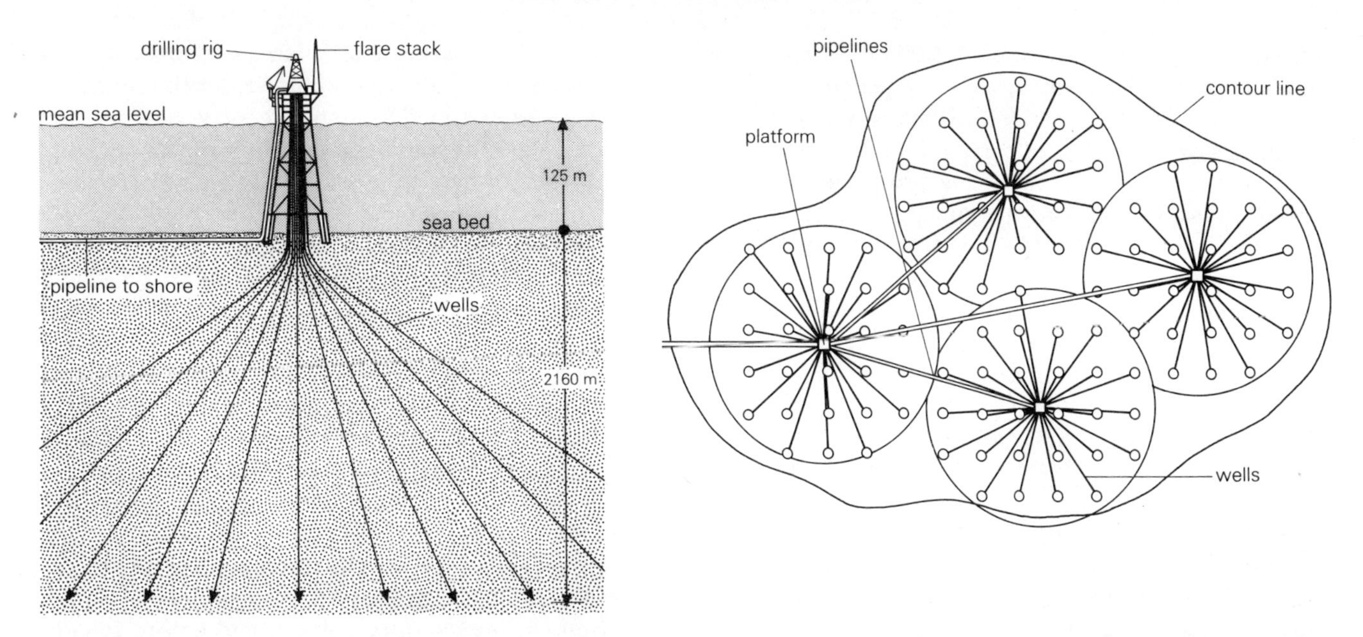

Figure 23.2 Plan of the four platforms from which the Forties field (UK North Sea) is exploited. Twenty-seven wells were directionally drilled by a single rig on each platform, as shown on the left.

The *surveying* of deviated wells, to locate their true positions at any drilled depth, is most easily done by photographing a compass needle positioned above a pendulum (actually a steel ball in a graduated hemisphere). As the drill pipe and the bit themselves would affect the compass needle, the instruments are encased within a single drill collar made of nonmagnetic metal. For highly accurate surveys, a downhole probe contains two gyroscopes, accelerometers, and a battery. The data are transmitted by cable to a surface computer, which produces a log charting the hole's path in three dimensions.

23.2.3 Small-scale drilling

A full-sized modern rig is far too costly an item to be used for shallow exploratory, experimental, or remedial operations. Shallow holes of narrower gauge than conventional drillholes (and called *slim holes*) are drilled by smaller rigs. They are seldom seen outside established producing areas, but there they serve many useful purposes for the geologist.

In many areas, structural traps at considerable depths may be subtly reflected in much shallower horizons because of draping, crestal thinning or pinchout of deposition. If accurate geophysical resolution of the deep structure is expensive, or if seismic crews are in short supply, slim-hole drilling provides a quick and relatively inexpensive supplement to it. *Structure test holes* may be put down to a few hundred meters, rather than the few thousand meters required for full-scale exploratory holes; they are merely logged, to determine

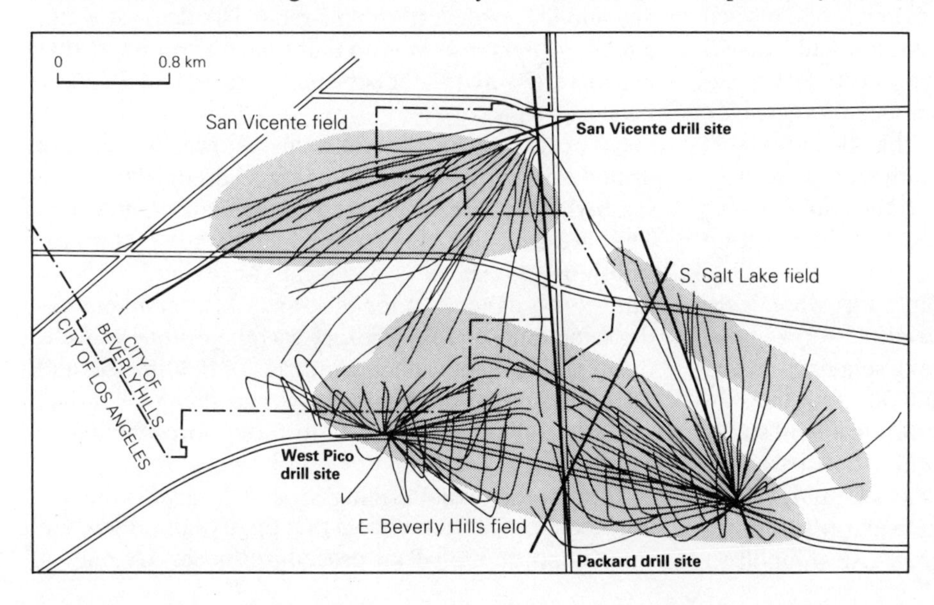

Figure 23.3 Well courses from three small drill sites in downtown Los Angeles. (From J. B. Jacobson and R. G. Lindblom, *Oil Gas J.*, 14 February 1977.)

elevations of easily recognizable marker horizons such as coal seams or bentonite beds which may reflect more pronounced structures at depth. Slim holes, not capable of full-scale testing, nor of completion, may be used to evaluate marginal or submarginal prospects, or leases about to expire; to provide quick, inexpensive exploratory step-outs in developed or semi-developed areas where target depths are shallow; or for rapid pattern or random drilling for shallow stratigraphic traps.

23.2.4 Offshore drilling

When oilwell drilling first moved offshore, it was into shallow waters that lapped against productive acreage onshore: along the eastern edge of Lake Maracaibo in Venezuela, the northern edge of the Gulf of Mexico off Louisiana, and the western edge of the Caspian Sea off Baku (Sec. 17.9). Conventional land rigs simply stood on platforms a little above the water surface.

In waters shallower than about 25–30 m, offshore drilling equipment and procedures require no more than minor modifications from land equipment and procedures. In deeper waters, however, major modifications are necessary to both. Exploratory rigs have to be mobile at sea and cannot be fixed to the sea bed. The only item set on the sea bed is the wellhead, which must be connected to the rig by a *riser* so that the hole can be re-entered after trips. At the time of writing (mid-1983), approximately 700 mobile, offshore drilling rigs are in use in the waters of the world. The rigs are divided among three quite distinct types with quite different capabilities, though the derricks themselves are not of course notably different in the three types:

(a) About 60 percent of the rigs in 1983 were *jack-ups*. Though variable in detail, all jack-ups have three tall legs which are raised well off the seafloor when the rig is under tow, and racked down on to it when the rig is in position; the process is much like that of raising a car with a ratchet-type jack. The maximum depth of water in which a jack-up rig can be used is about 100 m; legs longer than that are stable neither in planted position nor under tow. A large majority of exploratory holes off the Gulf Coast of Louisiana and Texas have been drilled by jack-up rigs.

(b) About 25 percent of the mobile rigs were *semi-submersibles*. A platform floats 30–40 m above twin horizontal hulls, with strong support bracing. A number of heavy anchors fan out from the submerged pontoons. The configuration is very stable because the structure is essentially transparent to wave-loads that would menace a jack-up rig. Yet it was a "semi" that suffered the worst offshore rig disaster (to the time of writing), off Newfoundland in 1982. "Semis" are used in water depths to about 450 m; nearly all exploratory wells in the North Sea have been drilled from them. A new generation of very large "semis" can drill in waters as deep as 2500 m or more; they cost more than $100 million apiece. True *submersibles* will no doubt provide mobile drilling rigs in the future.

(c) About 10 percent of the mobile rigs in 1983 were *drillships*, or "shipshapes" as their operators call them. In waters 30–300 m deep, the hull of the vessel is held in position by mooring lines with heavy anchors. In waters 80–1500 m deep (and deeper in the future), *dynamic positioning* is employed. No anchors are involved, and the ship is linked to the seafloor only by its riser system; it is kept in position over the hole by computer-controlled thrusters. The rig is carried in the center of the ship and drills through the *moonpool* in the hull. All exploratory holes drilled before 1983 in waters more than 450 m deep have been drilled from drillships. The ships are three to four times as expensive as jack-ups or semis, both in capital costs and daily operating costs, but they are self-propelled and can leave locations quickly if necessary. In addition to being more maneuverable than semis, drillships carry larger deck loads, but they are more sensitive to heavy weather. They also annoy environmentalists by releasing large volumes of nitrogen oxides.

Drillships have been designed to drill below 7500 m in waters deeper than 2000 m. The deepest waters in which they have actually been used are from 1300 to about 2100 m, off the Atlantic coast of North America. Other wells in waters deeper than 1000 m had been drilled by 1983 in other parts of the North and South Atlantic margins, in the western Mediterranean, in the Andaman Sea and off northwestern Australia. The deepest-water commercial oil discovery to that time was probably that in a mere 470 m of water between Spain and the Balearic Islands.

Other offshore wells, in much shallower waters, are drilled from artificial islands, of gravel dredged from the sea bed; they were pioneered in the Canadian Arctic. They are used, at huge expense, in water too shallow for drillships (15–25 m is a typical range) and with ice threats too dangerous for jack-ups. Also pioneered in the same inhospitable territory was the drilling of wells from artificially thickened platforms of floating ice, and the modification of drilling barges so that they swivel automatically about the hole to face approaching sea-ice.

Though the actual drilling process in deep waters is not essentially different from that on dry land, there are some important ancillary procedures necessary only at sea. Evacuation of a land rig by the crew and other personnel (including the geologist) is normally necessary only in the rare event of fire (though hurricanes in the Gulf of Mexico drive men off land and sea rigs alike). At sea, however, all personnel must be evacuated during storms involving very high winds and waves, at the approach of icebergs, or even in thick fog. Handling of the drilling mud may be stringently legislated; dumping of it at sea is unlikely to be permitted. Drillstem testing, cementing, plugging, and abandonment procedures should be more restrictive at sea than on land, in the interest of avoiding pollution of the sea, its inhabitants, or its coastline.

23.3 Abandonment and completion practices

The great majority of truly *exploratory wells* – those drilled outside areas known to be productive – fail to find either oil or gas in commercial quantities. They are *dry holes*, called "dusters" by many drillers. They are therefore *abandoned*, and appear in the records as "D & A." Before its abandonment, which involves attempts to retrieve the casing (which can be used again) and infilling of the hole with cement, every possible item of geological information must be extracted from a well. This requires the recording of all fluid recoveries from tests of the well, the running of a final battery of logs (most of them run in uncased hole containing drilling mud), and possibly borehole geophysical surveys to aid future exploration.

Many wells yield sufficient encouragement to warrant their *suspension* rather than their abandonment. Such wells are *capped* so that they can be re-entered later to discover whether further testing, or deepening, yields results that are economic under new conditions. If economic success already appears likely before the well reaches its total depth (TD), a quite different set of procedures becomes necessary. All horizons yielding encouraging results on test, or indicated to be hydrocarbon-bearing on the logs, are subjected to *production testing*. As this may lead to a published estimate of the well's *initial production* (IP) capacity, it should be systematic and thorough. If it is not, the well's IP may be drastically different from the volume of production it can yield on a sustained basis.

If the oil in the reservoir is capable of flowing up the well-bore to the surface under its own reservoir pressure (see Secs 23.7 & 8), its flow must be controlled by constricting valves called *chokes*, of varying sizes but all small. No well is nowadays allowed to flow freely at the surface; the legendary "gushers" are phenomena of the past, unless the drill penetrates a pool of hydrocarbons of such unexpectedly high pressure that it overcomes the blowout preventers. If the reservoir pressure is inadequate to cause a spontaneous flow of oil or gas to the surface, prospective depth intervals formerly tested are carefully isolated by packers, and re-tested; if necessary, they are *perforated*, and possibly *fractured* or *acidized* (see Sec. 23.13.1). Once commercial development is decided upon, the producing zone is cemented (to prevent the loss of either the hydrocarbons, by leakage, or the hole itself, by collapse) and the drill pipe removed. The fluid above the cement plug is removed from the hole by reverse circulation (in the days of cable-tool rigs it was removed by *bailing*, like bailing the water out of a boat), to take its weight from the face of the reservoir. The cement is then drilled out, allowing the oil to come into the borehole. If it brings loose sand or other unwanted material with it (as it is likely to do if its reservoir rock is a young, poorly consolidated sandstone), *sand control* is necessary before a wellhead or a pump is installed. The hole is finally *swabbed*, or cleaned of settled sand and debris, until the engineer is satisfied with its condition.

The *production string* or *tubing*, 2–4 inches (0.05–0.10 m) in diameter, is finally run through the *production casing* (normally 7 inches or 0.18 m in outside diameter), bottoming opposite the producing zone. An elaborate set of valves and gauges is installed at the *wellhead* to control the flow and direct the oil to the flow lines through which it will travel to the tank battery. The set of valves is called a *christmas tree* because of its appearance when illuminated at night. If the well is incapable of flowing, a *pump* is installed instead of a tree. There is a variety of means of powering modern oilwell pumps. They have no direct concern for the geologists, but the most familiar and widely used pumping device, used on scores of thousands of wells, including virtually all "old" wells, deserves brief mention. The *pumping jack* is the pivot or fulcrum device resembling a nodding horse which continuously raises and lowers *sucker rods* attached to the bottom of the production tubing. By moving up and down in the oil at the bottom of the hole, the sucker rods pump it upwards to the wellhead.

23.4 Offshore completion and production practices

The completion of successful wells offshore, in deep water, is a vastly different process. A discovery well in, say, 100 m of water is not itself completed as a producer. The directionally drilled wells required to exploit the discovery require a superstructure unlike a mobile exploratory rig, which is therefore moved off the location and replaced by a *production platform*. Large production platforms are the largest fixed objects ever moved by man. In rough seas, like the North Sea, a fixed platform that is expected to operate for 20 or 30 years must be capable of withstanding the worst expectable storm conditions – possibly 200 km h^{-1} winds and 25–30 m waves. In 125 m of water, therefore (a typical water depth for the Viking graben fields of the North Sea), the platform itself must be at least 160–170 m high, with a 50 m mast above that on the top deck. Platforms standing on the sea bed stabilized by their own weight are called *gravity structures*. The maneuvering of such a structure into position and its emplacement exactly over the intended hole position on secure bottom (with the aid of the world's largest cranes) is a marvel of modern engineering.

As the derrick drills each well of the platform's cluster, the wells are completed with seafloor wellheads, and satellite production-gathering equipment or *submerged production systems* (SPSs) also on the sea bed. If seafloor pipelines are precluded because of depth, bottom conditions, or ice scour, the production from each wellhead must be brought to the platform; this requires that the riser system contains the production lines (as well as injection lines if gas- or water-injection is to be utilized). The production is provided with temporary storage – possibly amounting to 10^5 m^3 or more – in the hollow concrete legs.

In less malign climatic conditions, less robust production platforms are installed in still deeper water. Before the end of the 1970s, for examples, a 380 m structure was installed in 312 m of water in the Cognac field southeast of New Orleans, and slightly smaller platforms were operational in 260 m of water in the Santa Barbara Channel off California. Similar platforms should be usable in waters between 350 and 450 m deep if fields requiring them are discovered. Below 450 m water depth, however, gravity structures will be supplanted by *tension leg platforms* (TLPs), floating structures tethered to the sea bed under tension by vertical mooring systems. TLPs should be operable to water depths of 2500 m and perhaps deeper.

The North Sea presented technological challenges for the engineers and drillers greater than any faced by the oil industry before, but in the Arctic, or off eastern Canada, the additional hazard of ice has to be contended with. In sea ice, as in the Beaufort Sea, *monopod platforms* may be used, the curved pedestals being designed to deflect floating ice. In "iceberg alleys," however, like that off Newfoundland in which the *Titanic* was sunk, production platforms must be *mobile*. They may be shipshapes, about the sizes of oil tankers, or semi-submersibles; the latter had already been christened FPFs, or *floating production facilities*, before any of them had been built.

23.5 Drilling and producing depths

Before the year 1920, the average depth of wells drilled for oil was about 600 m, or even less; a well deeper than 1200 m was unusual. All wells were then drilled with cable tools. Late in the 1920s, rotary rigs began to displace cable tools, permitting much deeper drilling as well as much faster drilling. By 1930, well depths of 2000–2500 m were commonplace, and in 1931 came the first wells drilled below 3000 m, in California and Mexico.

By that time, experimental temperature and pressure studies of hydrocarbons had provided evidence of their thermal destruction under conditions which would be reached (given normal geothermal gradients) at depths of about 6000 m (Ch. 7). Such depths might therefore have been visualized at that time as the practical limit of drilling-depth capability. Though oil production was established at 4500 m (in California) by 1940, the magic 20 000 feet (6096 m) was not reached until 1949 (in Wyoming) and the depth record was not beaten again until 1953.

By about that date, however, most explored basins had been drilled to the 3500–4500 m range, and the *average* depth of wells in thick Tertiary sections like the central San Joaquin Valley of California and the Mississippi delta region was more than 3000 m. In 1956, a well in southern Louisiana was still drilling in Miocene sediments at total depth of 6880 m. The depth record remained in the 7700–7800 m range from 1958 to 1972, but then came a quick succession of news-making new marks.

First, a dry hole in the Delaware Basin of West Texas bottomed at 8687 m in the Cambro-Ordovician Ellenburger dolomite. Three weeks later, the Baden well in the Anadarko Basin of Oklahoma bottomed at 9160 m in the Ordovician Viola Formation. It was the first well ever drilled to 30 000 feet; the drillstring weighed 2.7×10^5 kg; the well required the longest and heaviest string of casing ever used; and it required in addition the world's deepest fishing job at 7544 m. Drilling of the well took 543 days, by a rig 43 m high and having a crown block rated at 9.5×10^5 kg. The deepest drilling was no longer a prerogative of deep Tertiary basins. The Baden well retained the depth record for less than 18 months before losing it to another well in the thrust-faulted margin of the same basin, drilled by the same rig. The 1 Bertha Rogers bottomed at 9583 m in the Arbuckle Formation. At the bottom-hole temperature of 240 °C, a test recovered liquid sulphur. Ironically the wonder-well was eventually plugged back to the Pennsylvanian granite wash, in the topmost thrust plate, and completed as a "shallow" (!) gas producer at about 4000 m.

It is rather extraordinary that, except for a brief

interval in 1931–2 when it was held by a Mexican well, the depth record has not moved out of the USA since World War I. As a general average, 1450 m have been added to the record depth every ten years between the mid-1920s and the mid-1970s. It is nearly inconceivable that this process will continue any further, because the constraint recognized in the early 1930s is now widely accepted as real. By 1984, some 700 wells were being drilled annually, in the USA alone, to depths below 4000 m. These represent less than 1 percent of all wells drilled there. Approximately 100 deposits (less than 0.5 percent of all deposits) lie below 5000 m; nearly all of them are of gas, not of oil. The *average* depth of US wells in 1984 was 1400m; of exploratory wells, about 1800 m. Nearly 50 percent of known oil reserves lay at depths shallower than 1800 m, and more than 90 percent are shallower than 2750 m. In the Soviet Union, similarly, 90 percent of all known oil reserves lie between 1000 and 3000 m depths. There is very little oil known in the world below 4000 m. The *deepest sustained oil production* by 1980 came from 5240 m in the Bulla–More field offshore from Baku, in the Caspian Sea. Gas-condensate is produced from *Pliocene* sandstones, in the same field, at 6138–6172 m; this represents the deepest offshore production known to the author in 1982.

Natural gas is produced from significantly greater depths, but not in great quantities. The deepest gas production by 1982 came from thrust-faulted Arbuckle carbonates in the Anadarko Basin (at 7663–8082 m in the Mills Ranch field), but it was of trivial volume and exceedingly expensive. The *deepest sustained gas production* came from between 7200 and 7500 m in the Siluro-Devonian Hunton Formation of the same basin. The deepest gas production outside the USA is probably that from the Lower Cretaceous of the Lannemezan field in France's Aquitaine Basin, from 6900 m.

The deepest holes are therefore 1500 m below the deepest gas production and more than 4300 m below the deepest known producible oil. The ultimate drilling depth record may therefore be limited not by technology but by the economic constraints posed by the properties of rocks, hydrocarbons, and other reservoir fluids. If liquid hydrocarbons are gasified at the temperatures prevailing below, say, 6 or 7 km, and gases at greater depths contain increasing quantities of acid components (especially of H_2S and CO_2), the corrodability of even high-strength steels may impose practical and economic limits on drilling depths for other than experimental purposes.

The deepest drilling has in fact had precisely that objective. Beginning in 1970, Soviet geologists and drill crews began a series of "super-deep" holes drilled by turbine methods. The wells were planned for 10 km depths or more, and were expected to take 5–6 years to drill. The SG-3 "stratigraphic" test was spudded in May 1970 on the Kola Peninsula, near the Barents Sea, using a 64 m mast. The original schedule called for a total depth of 10.5 km (34 450 feet), entirely in Precambrian igneous and metamorphic rocks (in which the well was spudded). The 10 km mark was reached exactly ten years after spudding, and the 12 km mark was surpassed without apparent incident early in 1984, still in granite. Astonishingly (until one recalls the low geothermal gradients in ancient shields) the bottom-hole temperature was reported to be only about 180 °C. In 1977, a second test (SG-1) was started near Saatly, 250 km southwest of Baku and near the northern end of the Elburz Mountains. The test was planned to reach the depth of 15 km (49 212 feet), with a 70 m mast rated above 4×10^5 kg. Unlike the SG-3 test, SG-1 did not begin in Precambrian rock, but it was expected to reach it at 6–7 km (between 20 000 and 24 000 feet). The well in fact reached 8.5 km (nearly 28 000 feet) in December 1984, probably in the Jurassic.

If Thomas Gold is correct in believing that vast resources of methane lie at depths below 6 km because of diastrophic degassing of the upper mantle (see Sec. 6.2), many more wells will no doubt be drilled in the 6–10 km depth range in the future. In the much more likely event that "deep-earth gas" is not a commercially exploitable resource, drilling *by the oil and gas industries* below 10 km is never likely to become a popular pastime, and wells deeper than 8 km are unlikely to constitute any higher proportion of future wells than of past wells.

23.6 What geologists should understand about drilling costs

Costs of all exploration and exploitation, including the costs directly attributable to drilling, have increased dramatically since the 1960s along with the costs of nearly everything else. Costs change so rapidly – though inexorably in the same direction – that absolute figures become out of date annually and would be worthless in a textbook. Some generalizations about comparative and relative costs should however be known to everyone in the industry.

The *commitment* to drill a well may involve the outlay of large sums before a rig is even contracted. The amount naturally varies with the size and location of the tract, how much is already known about it, who owns the rights to it, and a host of other considerations. For a tract of about 1000 km² (250 000 acres), for example, exploration costs before a drillsite is selected may easily exceed $1 million or very much more. Contract seismic work costs in the thousands of dollars per crew-day or per square kilometer of coverage; other forms of geophysical exploration cost tens of dollars per square kilometer. Surveyors, leasers, equipment, title work, and rentals are merely the most obvious and unavoidable of the other charges that precede drilling.

The cost of the actual well includes those for site preparation, moving in equipment and supplies, rigging up, the provision of mud sumps and tanks, water supply, and communications. These costs in turn depend principally upon access (the need for roads, bridges, camps, and so on). The cost of the actual drilling then depends upon the type of rig used (it is cheapest with a standard portable rig), the target depth, lithologic or structural difficulties encountered, the amount of casing needed, the number of tests run, the need for fishing jobs, the length of time spent in logging, cementing, or abandonment procedures, and so on. One week of lost circulation, or one spent trying to recover twisted-off drill pipe, adds tens of thousands of dollars to the cost of the well. Labour; supervision; machinery rental; drill pipe; bits; mud, fuel, and water; servicing, logging, testing, and fishing; miscellaneous supplies: each of these categories of expense costs hundreds or even thousands of dollars *daily*. If the well is successful, completion costs may add a further 25–100 percent to the costs involved in reaching total depth.

Before 1960, the average well depth was less than 1500 m and a typical cost in regions of easy access was $35 per meter. Normal well costs were therefore in the order of $50 000–60 000. Each additional meter beyond about 1500 m cost about $50. If the well became very deep – say 4500 m – each extra meter cost some $250–400. In general, the cost doubles for every 1000 m of depth; at 9000 m it is 130 times what it is at 1500 m. Very deep wells therefore cost several million dollars apiece even in easily accessible locations onshore.

Between 1960 and 1980, the average depths of wells increased, but not nearly so rapidly as average costs. Costs per meter of drilling increased more than sixfold for even shallow onshore wells. By 1982, the average onshore well in North America cost nearly $500 000, or $360 per meter. Wells in the Gulf of Mexico averaged nearly $2 million; in northern waters like Cook Inlet or the Mackenzie delta, costs had come to exceed $600 per meter. All deep drilling costs at least $300 per meter, even in the interior of North America. Wells deeper than 6000 m cost $750–1000 per meter.

Offshore wells suffer a quadruple cost disadvantage – the mere fact of being offshore requires both more expensive equipment and more extensive safety precautions; the inevitability that all offshore *exploratory* wells are "deep" (it is not worth an operator's while to contract an offshore rig for a 1500 m hole); the increased amount of downtime because of adverse weather conditions; and the need to undertake the costs of enormously expensive leased platforms in the event of success. As a rough approximation, offshore wells even in benign climatic conditions cost from three to six times the cost of onshore wells of comparable depth. Mobile exploratory rigs, involving tens of millions of dollars in initial construction costs, rent for tens of thousands daily. The cost of a typical offshore exploratory well has exceeded $1 million before 2000 m depth is reached, and $2 million before 3000 m. Deep offshore exploratory wells nearly all cost in the $4–12 million range. In the event of a discovery, production platforms cost from $12–120 million, depending on the daily production facilities demanded. In the North Sea or other high latitude waters greater than 100 m deep, production platforms cost anything from $100 million to more than a billion (10^9) dollars. Development wells drilled from them cost $1–5 million apiece. The costs of operating wells, once they have been completed, run in the hundreds of dollars per month in the lower 48 United States but in thousands monthly in shallow waters and tens of thousands monthly in deeper or more hostile waters. Deep submarine producing and gathering facilities, and pipelines, are many times more expensive than their onshore equivalents.

The reported costs of very expensive wells are difficult to compare realistically; much depends on what the operator chooses to charge to the well's costs. For example, an Exxon well drilled in the state of Mississippi in 1980 and bottomed at 6950 m in Jurassic evaporites was reported to have cost $42 million because of very high bottom-hole pressures and corrosive acid gases; clearly the financial towel could have been thrown in much earlier had pure exploration, and not experimentation or experience, been the sole criterion. In contrast, the deepest wells in the Anadarko Basin, like the Baden well (see Sec. 23.5) cost about $5–6 million, so the well in Mississippi was an outrageous exception to any generalization about costs for onshore wells. But its cost would not have been so outrageous for deep offshore wells. The Issungnak 0-61 well, drilled from an artificial island in 18 m of water, off the Mackenzie delta in the Beaufort Sea, cost about $65 million (and discovered some oil and gas). The Mukluk well in the US Beaufort Sea, drilled from a gravel island during the winter of 1983–4, was probably the first well to cost more than $100 million.

The Hibernia discovery well off eastern Newfoundland cost about $40 million. As it was the first discovery off eastern North America after years of expensive failure, it was imperative that all promising, porous horizons (there were four or five of them) be exhaustively tested before the rig was released. The anticipated results justified the time and expense. The COST wells (appearing to be appropriately named, but actually being Continental Offshore Stratigraphic Tests), drilled during the 1970s by consortia of companies in preparation for outer continental shelf lease sales in the USA, are not located on expensively defined structures and they cannot be tested, treated, or completed; yet their average cost has been about $10–15 million. One test in the Gulf of Alaska cost more than $40 million. The dry holes drilled off the Atlantic seaboard of the US cost between $20 million and $35 million apiece. Completed producing wells costing over $20 million apiece had become commonplace in the North Sea by 1983.

23.7 Pressure measurements in wells

The *fluid pressure* in the reservoir is measured by a pressure gauge, popularly called a "bomb," which is lowered into the well bore inside the perforated pipe. The drilling mud is kept out of the way by *packers* which are flexibly pressed against the wall of the hole. The fluid pressure is therefore often called the *reservoir pressure, formation pressure, or bottom-hole pressure* (BHP), occasionally the *well pressure*. The original reservoir pressure of any well declines throughout the well's production. As we shall see, the behavior of the well's measured pressures is critical to the well's successful operation, and all pressures that can be measured must be continuously and carefully monitored.

A standard pressure-monitoring procedure is to halt or *shut in* production, well by well, allowing the reservoir pressure to build up to a maximum – the *shut-in pressure* (SIP) or *static* BHP – which is of course not as high as the original BHP. The SIP may be measured not at the bottom of the hole but at the top of the casing; it is then called the *casing pressure*, the value of which is therefore the reservoir pressure (or BHP) minus the weight of the fluid column in the hole between the two points of measurement. Yet a third SIP commonly recorded is the *tubing pressure*, measured at the top of the production string. Whilst a well is shut in, the casing pressure and tubing pressure are the same; there is no fluid to move through the upper part of the hole. When production is restarted, the *flowing pressure* is measured at the wellhead. It will be less than the SIP, the difference between the two being the *differential pressure*.

Pressures have traditionally been measured and recorded in *pounds per square inch* (p.s.i.) because of the dominance of American usage. They should now be measured in *kilopascals* (kPa), 1 kPa being equal to 0.01 bar and 1 p.s.i. being equal to 6.895 kPa. For some purposes requiring not merely considerable accuracy but conformity to a universally accepted convention, pressures are expressed in *absolute* units (p.s.i.a. or pounds per square inch absolute). These include the atmospheric pressure at the ground surface above the drillhole, because that is where the pressure bomb is sealed and begins its journey down the hole. The most important use of absolute pressures in an oil geologist's affairs is in the measurement of gas volumes, for expressions of reserves, production figures, MPRs, and sales volumes. Natural gas is highly compressible, and it is essential that all volumes of it are expressed in terms of accepted standards of atmospheric temperature and pressure. These standards are 15 °C (60 °F) and 101.325 kPa (14.65 p.s.i.). Pressures as measured in the well bore are converted to absolute units by adding the atmospheric pressure at the surface (corrected for elevation of the wellsite) to the gauge readings.

Actual values of the pressures encountered in wells, and the reasons for their variations, are discussed in Chapter 14.

23.7.1 *T,P conditions of reservoir fluids*

In the general case, a hydrocarbon-bearing reservoir rock contains both gas and oil, normally with significant, irreducible water saturation and with water below. These fluids store the potential energy of the reservoir, and exert the fluid or pore pressure which helps to support the weight of the superincumbent column of rock. The relations between the fluids in the reservoir, and the temperature and pressure (T,P) conditions under which they are confined and interact, govern the manner and proportion in which wells tapping that reservoir yield their fluids for recovery.

Prior to penetration by wells, the fluids in the reservoir are at the *original reservoir pressure*. At this pressure, oil under the commonest conditions is *saturated* with gas; that is, it contains all the gas in solution that it can hold under those particular T,P conditions. Any surplus of gas, beyond the volume the oil is capable of holding in solution, forms a *free gas cap* above the oil column (see Sec. 23.8.2). If the overburden pressure on the reservoir fluids is reduced, by erosion or by their updip migration, the oil's capacity to retain gas in solution is also reduced and the gas exsolves to form a *secondary gas cap*. If the reverse happens, and an oil pool saturated with solution gas becomes more deeply buried, the gas cap disappears as the gas goes into solution, until eventually the oil becomes *undersaturated* with gas.

We will consider these processes with graphical assistance. Figure 23.4 is a phase diagram of a petroleum mixture under various combinations of T and P. Consider an oil pool under the conditions represented by point A, with pressure high because of deep burial but temperature moderate because of a low thermal gradient. All the gas is in solution in the oil, which is *undersaturated*. As the pressure drops at essentially constant temperature (say through production), point A moves towards point B. At B, the pressure is no longer high enough to keep all the reservoir gas in solution in the oil, and free gas therefore bubbles out and forms a gas cap above the oil. This is the *bubble point*, representing the pressure (called the *saturation pressure*) above which only liquid is present in the reservoir. As the pressure drops still further, the amount of released gas increases and the percentage of the reservoir fluid represented by liquid oil falls progressively. The oil is now *saturated*.

If we now vary the treatment of the pool and raise the temperature whilst the pressure remains essentially constant, we will not be surprised to find that at higher temperatures the escape of solution gas is facilitated. The bubble point pressure rises to a maximum, called the *cricondenbar*, represented by point C_b, and then declines again with further increase in temperature because the liquid and gas phases become progressively less distinguishable.

We now jump to a situation opposite to that first considered. Consider a reservoir in which the temperature is so high that all the fluids have been vaporized. As this high temperature is reduced at essentially constant pressure, a point is reached at which liquid begins to condense out of the vapor. This is the *dew point*, the temperature above which only gas is present in the system. At low pressures, still further lowering them also lowers the temperature of the dew point, but at higher pressures the dew point is lowered by *increases* in the pressure because the density of the compressible gas is being increased whilst that of the oil is being decreased by the forcing of more gas into solution. Hence the dew point curve also has a maximum temperature value, the *cricondentherm* (point C_t).

The pressure, temperature, and volume conditions in the reservoir therefore dictate the state in which its particular store of fluids is held. The pressures exerted by that particular combination of fluids in turn dictate how wells penetrating that reservoir will perform. The reservoir energy provides the *drive mechanism* for wells tapping it. The drive mechanism is vital to the successful operation of any oil or gas field. It may be exerted primarily by the gas in solution in the oil, by the gas cap above the oil, or by the water below it.

23.8 Reservoir energy and drive mechanisms

23.8.1 Solution gas drive

This mechanism, also referred to as *gas expansion, dissolved gas drive,* or *depletion drive,* makes use of the energy of the gas in solution in the oil. This solution gas reduces the oil's viscosity and surface tension, facilitating its movement through the reservoir.

As reservoir pressure drops, the gas expands, driving

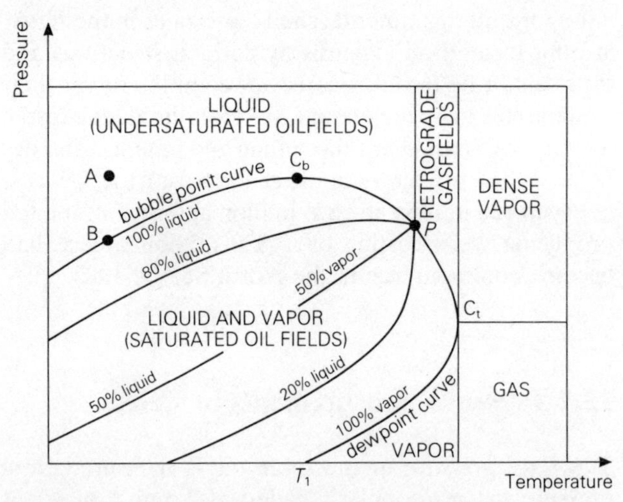

Figure 23.4 Phase diagram of petroleum mixture.

the oil in the direction of the pressure gradient (towards the well bore). When the reservoir pressure is reduced to the saturation pressure throughout the reservoir (Fig. 23.4), free gas bubbles out of solution, forming a *secondary free gas cap*. This cap is little help to production, because it simply occupies space vacated by oil. The relative permeability of the reservoir to oil declines rapidly early in the withdrawal history (Fig. 11.10). Bottom-hole pressures also decline rapidly; in general, about 50 percent of the reservoir pressure has been dissipated by the time the pool has yielded 30–40 percent of its ultimate yield. The GOR increases sharply and the well becomes primarily a gas well (Fig. 23.5).

Gas expansion alone seldom recovers more than 20–25 percent of the oil in place; recovery from carbonate reservoirs is unlikely to be higher than 20 percent and may be as low as 5 percent. Exsolved gas bubbles easily through the oil without moving it; loss of the gas quickly renders the oil more viscous and without an effective agent to keep it moving. The mechanism is

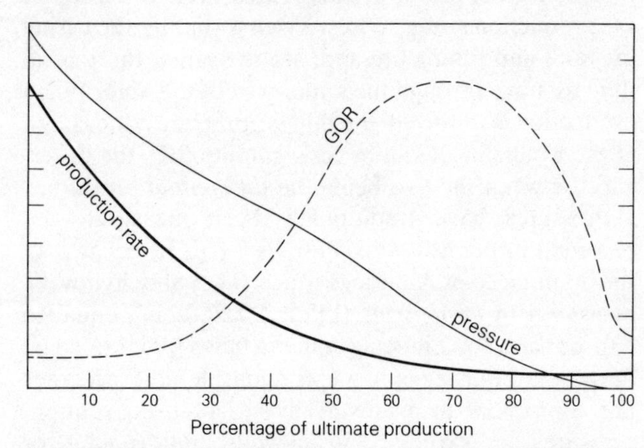

Figure 23.5 Generalized performance of solution gas-drive reservoirs, showing exponential decline in production rate and sharp increase in GOR as reservoir pressure is depleted. (From J. A. Murphy, *World Oil*, July 1952.)

especially inefficient in fracture or vuggy porosity in carbonate reservoirs. The dissolved gas flows out readily along the larger and more continuous pathways, removing oil from them but bypassing that in the intergranular pores. This phenomenon is a serious obstruction to secondary recovery in such reservoirs.

The gas saturation of the oil in the reservoir decreases downwards, but it increases with increasing thickness of the roof rock (see Sec. 11.3) for roof rocks of comparable capacity. Because methane migrates most easily through minute fractures in the roof rock, lower gas factors usually mean lower methane concentrations also. For liquid oilfields under solution gas drive, initially low GORs may rise to 350 (about 2000 cubic feet per barrel) or even to more than 850 (nearly 5000 cubic feet per barrel). For condensate fields they range from 550–600 to more than 2000 (3000–12 000 cubic feet per barrel in rough approximation).

In highly undersaturated oilfields, like those west of the Persian Gulf, great volumes of gas are necessarily produced with the oil. Unless there is a market for the gas, it must be flared or re-injected into the reservoir. As the volume of oil in the reservoir decreases, the oil/water contact rises, leaving a zone of high residual oil saturation below the new contact. This zone may constitute an impermeable tar barrier to any possible water drive from below, forcing early waterflood of the reservoir if its pressure is to be maintained (see Sec. 23.8.3).

23.8.2 Gas cap drive

A *free gas cap* may exist in the reservoir from the outset, containing the excess of gas beyond that required to saturate the oil at the reservoir *T* and *P*. The cap expands as the oil is withdrawn, exerting downward pressure on the oil and providing an effective recovery mechanism. Care must be exercised to complete wells in the oil column and not in the gas cap.

Production declines continuously from the beginning but not nearly so rapidly as under depletion drive (Fig. 23.6). Allied with good permeability, high rates of oil production may be maintained for years, eventually recovering between 30 percent and as much as 75 percent of the oil in place. GORs remain low but increase as the rate of oil extraction declines, and eventually most of the production becomes gas. The oil/water interface may remain almost static, because there is no cause for the gas/oil interface to rise.

Gas cap drive may be created artificially. The casinghead gas, produced with the oil under solution gas drive, is stripped of its NGLs (see Sec. 5.1.2), passed through a repressurizing plant, and injected under pressure at the crest of the structure or on its updip side. The artificial gas cap may increase recovery from 25–30 to 40–50 percent. The process, called *pressure maintenance*, is employed in many hundreds of oilfields, including those of Kuwait.

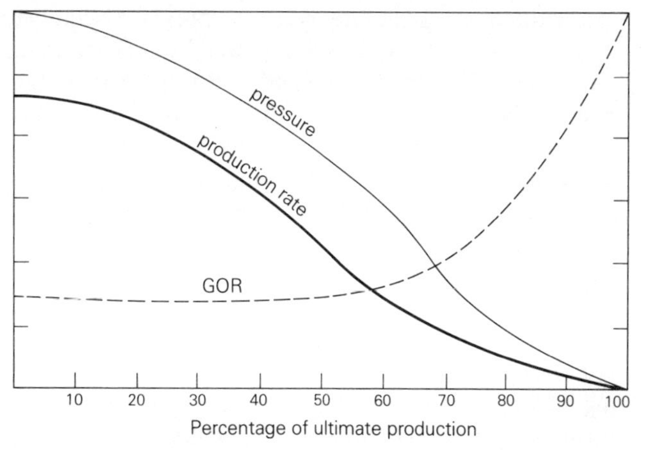

Figure 23.6 Generalized performance of gas cap drive reservoirs, showing much slower decline in oil production and reservoir pressure and slower build-up of GOR than under solution gas drive (Fig. 23.5). (From J. A. Murphy, *Petrol. Engr,* August 1952.)

Most large free gas caps are relatively shallow. With increasing depth, the density of the gas increases but that of the oil decreases because more gas is forced into solution in it. At the *critical pressure*, between 5000 and 6000 p.s.i.a. (about 38 000 kPa), gas and oil form a single phase. There is also a *critical temperature* for the gas, above which it cannot be liquefied by pressure alone. In this *T,P* range (Fig. 23.4), pressure reduction at constant temperature causes a liquid phase to appear. This is the opposite of the outcome at lower temperatures and pressures, under which vapor appears from liquid as the pressure drops through the bubble point. The anomalous reverse process is called *retrograde condensation*; at the critical temperature, a gas pool becomes a retrograde condensate pool. There is consequently a unique point on the phase diagram (Fig. 23.4) at which the bubble point curve and the dew point curve coincide, the *critical point* (P). It represents the values of *T* and *P* for that particular petroleum mixture at which all domains in which a liquid component can occur are separated from all domains in which it cannot occur.

23.8.3 Water drive

Except for a few in isolated, lenticular sandstone reservoirs, virtually all oil and gas pools contain either *bottom water* (below the hydrocarbons) or *edge water* (beyond their boundaries) or both. If new water can enter this water column in the reservoir, from a recharge area beyond the bounds of the pool, its movement along the hydrodynamic pressure gradient will maintain the reservoir pressure – provided that the entry of new water is at a rate comparable with the rate of withdrawal of the hydrocarbons. If the rate of water entry is interrupted or inadequate, reservoir pressure is likely to decline rapidly. It can be rejuvenated by shutting in the wells and allowing the head to build up to its earlier level. This process is theoretically limitless, but it may be too time-consuming to be acceptable.

The moving water sweeps through the reservoir below or behind the hydrocarbons, flushing them out of the pore space and towards the wells. Under favorable circumstances the recovery may be the highest among all drive processes, from 36 percent to as much as 80 percent of oil in place. It is lowest in oil-wet reservoirs, and is ineffective in many large carbonate reservoirs (including several of those in Saudi Arabia) because of the formation, described above, of a tar barrier at the oil/water interface.

The oil production rate may not decline significantly until some 50 percent of the recoverable volume has been extracted (Fig. 23.7). Water must by then have encroached upon much of the reservoir pore space formerly occupied by oil. If wasteful production practices have extracted the oil much faster than the water drive can be replenished, wells turn rapidly to water and much unrecovered oil is left behind. Even under careful control, the wells must ultimately become water wells, but production from the wells farthest up the dip may be maintained at reduced rates for many years. By 1970, the East Texas field, the largest in the lower 48 United States, had been in continuous production for 40 years. It was by then divided into three separate pools by two channels of water encroachment (Fig. 23.8), but it was safely forecast that the remaining wells would continue to produce oil for another 40 years.

East Texas illustrates the important circumstance that topographic variety, providing hydraulic head, is not a requisite for water drive. Nor is plentiful drainage a requisite, as demonstrated by the fields in the basal Cretaceous sandstone in Libya.

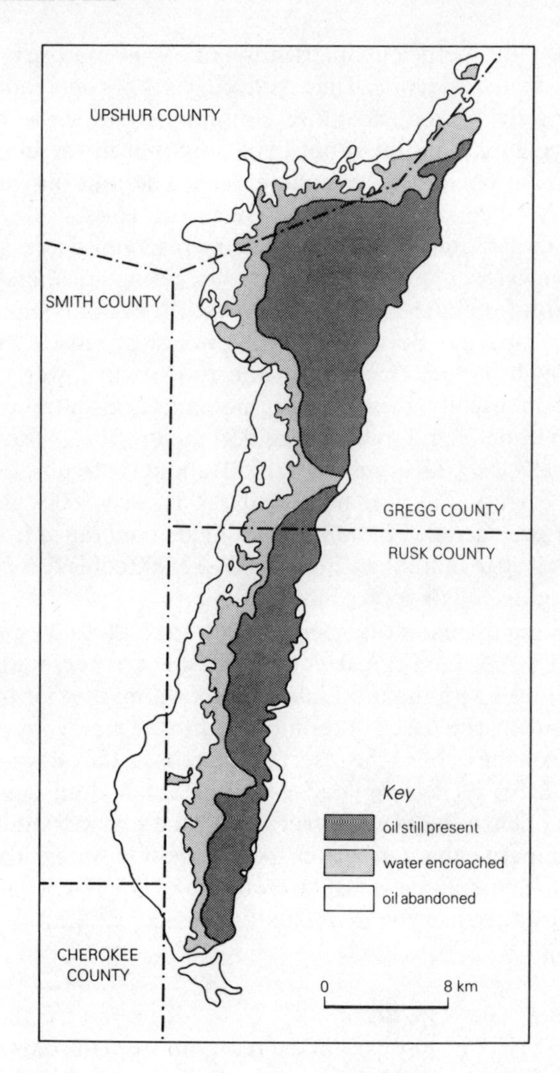

Figure 23.8 East Texas field, showing area encroached upon by edge water in 1965. Westernmost line marks original oil/water interface (1930). By 1970, the field had become separated into three fields by advance of water consequent upon withdrawal of the oil. (From K. K. Landes, *Petroleum geology of the United States*, New York: Wiley–Interscience, 1970; after Amoco Inc.)

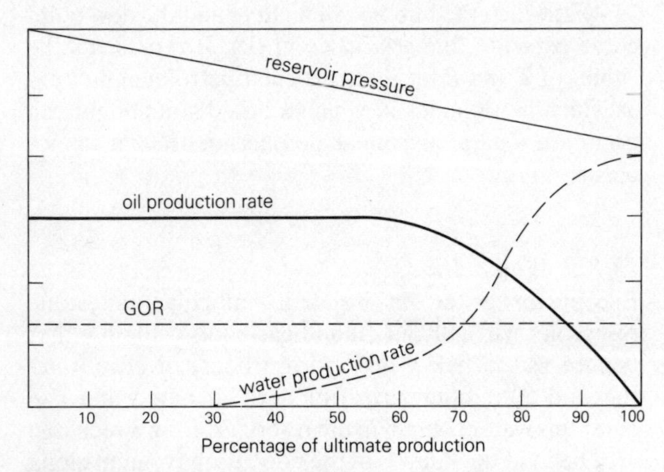

Figure 23.7 Generalized performance of water-drive reservoirs, illustrating moderate decline in reservoir pressure but supplanting of oil production by water cut after 50 percent of the recoverable oil reserves has been produced. (From J. A. Murphy, *Petrol. Engr*, August 1952.)

If water drive is the only operative mechanism, there is no essential reason for the GOR to vary much during production. Once water encroachment has increased the water cut to about 50 percent of production, however, the GOR is likely to show modest increase.

23.8.4 Gravity drainage

Gravity acts on reservoir fluids whatever their total volumes or their relative proportions. In some pools, especially those having originally been produced under solution-gas drive, a high proportion of the original oil in place – say 75 percent – may be left in the reservoir when its pressure has fallen below the effective minimum for further recovery. At the top of the reservoir the pressure may be reduced effectively to atmospheric. The secondary gas cap has disappeared from above the oil, and if there is no water drive from below the remaining oil is not acted upon by any effective force except that of gravity. It therefore seeps downwards through the reservoir, eventually finding the lowest level the permeability allows.

If wells are then completed at depths below the earlier oil/water interface, or beyond the earlier downdip limits of the pool, the oil that has seeped downwards can be recovered by pumping. The mechanism is not a "drive," because no moving agent is affecting the oil. Gravity creates a *drainage* of the reservoir; the whole "drive" is artificially exerted from the surface.

In the absence of bottom water, gravity drainage may eventually recover practically all the oil in a pool, greatly prolonging the pool's productive life. However, the recovery process is too slow to be economic unless several somewhat unusual conditions are met. In the ideal circumstance, oil must saturate at least 50 percent of the reservoir; the sand must be *oil-wet*. If the crude contains polar compounds which are adsorbed on to the grain surfaces, the grains become hydrophobic and water cannot displace the oil from the pore spaces. The reservoir thus becomes highly permeable to oil, not to water. The oil itself must be of low viscosity; if loss of the solution gas has left the oil highly viscous, it cannot seep downwards through the reservoir fast enough to justify production. Operators will then prefer some enhanced recovery mechanism, such as steam drive (see Sec. 23.13.3), as they did in San Joaquin Valley fields in California. Fields with fractured reservoirs may provide good gravity drainage. Finally, effective gravity drainage requires that the reservoir rock have substantial dip (at least 15–20°). All drive mechanisms except that by solution gas are assisted by such dips.

Clearly a wide variation in recovery rates and percentages is possible under gravity drainage. Nevertheless, many fields have been produced by this mechanism for years, especially in the USA. The most widely known big example is the Oklahoma City field. The most notable of more recent examples is provided by the central part of the Prudhoe Bay field, down-dip from the *original* gas cap (see Sec. 26.2.2).

23.8.5 Combination drives

Many fields, perhaps most, are actuated by more than one type of drive mechanism. The most commonly met combination of drives is solution-gas drive (with or without a small gas cap) augmented by a weak water drive. This combination will yield a greater recovery than solution-gas drive alone, but not as great as that from true water drive. A large gas cap or a strong water drive can effectively be regarded as the single drive mechanism even if some solution gas also contributes.

The most efficient combination drive is that associating a free gas cap and active water drive. Many fields in the interior basins of North America, in the North Sea, in North Africa, and in Indonesia are blessed with this combination.

23.9 Production rates and decline curves

The manner of decline of reservoir and other measurable pressures during the lifetime of a well depends upon the basic production mechanism and on the physical conditions of the reservoir and its fluids. Once the mechanism has been identified, and the well's early production history has been monitored, the *decline curve* can be extrapolated into the future, permitting approximate estimates of the active life of the well and of its ultimate recoverable reserves. These estimates presuppose that the historical curve is that natural to the well and has not been artificially stimulated or depressed.

The decline curve is extrapolated as an exponential curve (Fig. 23.9). Its exponential nature is actually an assumption, but one supported by abundant experience. Some engineers nonetheless prefer to use a hyperbolic curve. Two curves are extrapolated:

(a) *Production rate versus time* This is the only curve derivable during the well's early production history. The *nominal decline* (D_n) is then given by

$$D_n = \frac{q_0 - q_1}{q_0}$$

where q_0 and q_1 are the initial production rate and the rate after one year of production.

(b) *Production rate versus cumulative production* This curve is especially useful after some years of production have resulted in the rate declining below the allowable rate. This situation permits portrayal of the *true decline* (D) given by

$$D = \frac{q_0 - q_t}{Q_t}$$

where q_t is the production rate, and Q_t the cumulative production of oil, after t years. Then

$$q_t = q_0 (1 - D_n)^t = q_0 e^{-Dt}$$

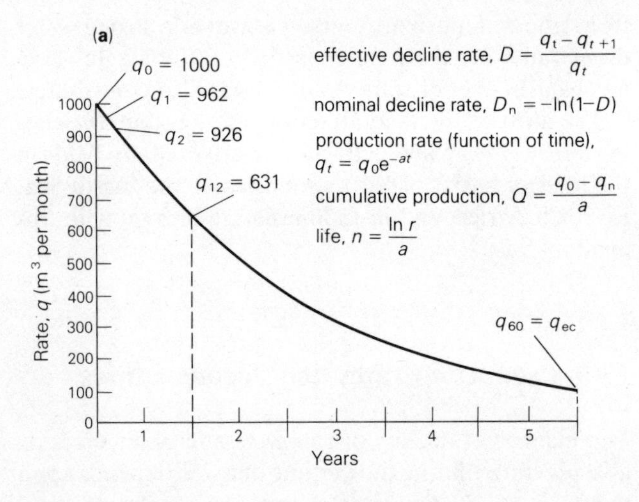

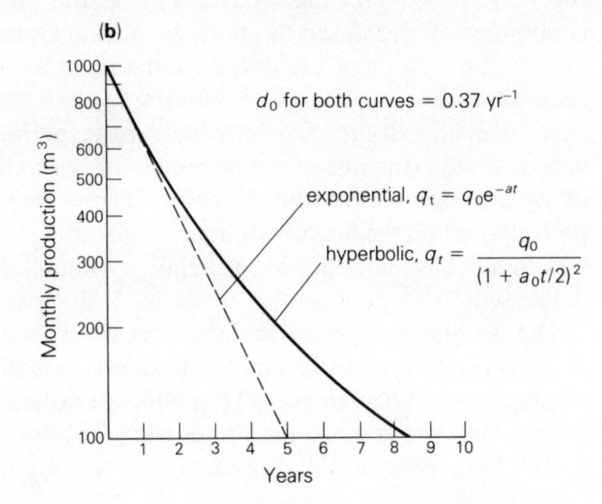

Figure 23.9 (a) Exponential decline curve; (b) hyperbolic and exponential decline curves compared.

Other decline curves that may be used are those of either reservoir pressure or cumulative gas production against cumulative oil production (these two especially if the field is produced by solution-gas or gas-cap drive) and water cut (in percent) against cumulative oil production (especially in water drive fields).

The estimates of recoverable reserves per well are summed for all the producing wells in a *pool*, and thence in a *field*, which produce under comparable recovery mechanisms. By then summing all the individual well reserves, including all pools in the field and all the recovery controls (there may of course be only one pool in the field, and a single control in common across the entire field), the field's ultimate recoverable reserves *under primary recovery* are reliably estimated. This estimate then supplants the original volumetric estimate of the field's reserves (see Ch. 24), which was made before development drilling began and indeed in most fields was the determining factor in the decision to develop. It cannot be over-emphasized that all such measurements, calculations, and estimates must be continuously maintained as current.

23.10 Field lives and MERs

For operating purposes, it is not enough to know how long a field may be able to produce nor what its ultimate output is anticipated to be. The operator needs to recoup his investment without waiting too long to do so. With the knowledge of the wells' early performances and possession of their decline curves, he must decide upon the number of development wells to be drilled and the rates at which they are to be produced. A small field will not produce for very long. A large field, however, takes time to develop, and cannot produce at its peak until an adequate number of development wells has been drilled. It will then be able to maintain that peak rate for only a relatively short time before its final decline sets in. What is to be the peak rate of production for the field, and how soon is it to be achieved? To determine these two critical operational quantities, a positively skewed bell curve must be constructed, with the value and position of the peak dependent upon the dimension of the x-axis – the life of the field.

The planned life of the field is taken to be 20 years (15 years by some operators because of political, economic, and operating conditions), and the area below the curve is the anticipated production *during that time*. The curve defines the *maximum efficient rate* of production, or MER, for the field, and therefore for the individual wells within it. The MER is the rate calculated to achieve the maximum possible recovery of the reservoir's oil and gas within the drainage areas of the wells, without unacceptable deferment of the operator's return. The MER should not be exceeded except under emergency circumstances. The actual rate of production is frequently below the MER because of restrictions dictated by political or economic circumstances (lack of pipeline, tanker, or market outlet for the production, for example). The enforced rates of output are then called MPRs, or *maximum production rates*, or, more popularly, simply *allowables*. MERs may need to be revised upwards or downwards, for a variety of reasons which will become apparent. Authority may then have to be sought to amend the MPRs. It is one of the responsibilities of the development geologist and the engineer, working as a pair, to ensure that these corrections are made.

There are absolute physical limitations on the MERs of wells, and therefore of fields. The maximum capacity of a well to produce oil is influenced by a number of factors, easily understood if one remembers that a well is simply a device to drain oil from an area of radius r_e (e for external) around it. The ability of oil to move towards the well bore declines exponentially as r_e increases. One consequence of this is the need for control of well spacing. Excessively wide spacing means too much oil is left in the ground between wells. If initially wide spacing is reduced by *infill drilling*, more of the in-place oil is recovered but of course drilling, completing, and connecting the extra wells costs money. The natural production rate of a well is therefore inversely proportional to the natural logarithm of the drainage radius (ln r_e). It is also, more obviously, inversely proportional to the viscosity of the oil (η), and directly proportional to the effective permeability of the reservoir rock to oil (k_o), to the thickness of the pay zone contributing oil to the well (h), and to the degree to which the reservoir pressure has been maintained. The maximum possible rate at which the oil can flow into the borehole is in effect directly proportional to all factors which decline during production and inversely proportional to those factors that increase, largely due to depletion of the solution gas.

In absolute quantitative terms, the maximum rate of flow of oil into the borehole (q) is given by the *Darcy equation for radial flow*. In traditional units, as formulated by Darcy, the equation is expressed thus:

$$q = \frac{7.082\ k_o\ h\ \Delta P}{\eta\ \ln\ (r_e/r_w)}$$

with q in BOPD; k_o in darcies; h in feet; and η in centipoises. P is the difference ($P_e - P_w$) between the shut-in (static) reservoir pressure and the flowing bottom-hole pressure, both in p.s.i.a. (see Sec. 23.7). The ratio r_e/r_w is that between the drainage or external boundary radius of the well and the radius of the well bore itself.

In SI units, q is in cubic meters per day; k_o is in square meters; h is in meters; P is in pascals; η is in pascal-seconds. The constant becomes 541 396, principally because of the vast change in the dimensions of the units of permeability (k_0).

The equation governs the flow of oil from the reservoir rock into the borehole. To express q in stock-tank barrels or cubic meters at the surface, the resultant should be divided by the formation volume factor (see Sec. 24.2.1), the ratio between the volume of a unit of oil in the reservoir and its volume at the surface.

We now realize that the 20 year time-frame for an oilfield's life is not arbitrary. Under *typical* reservoir mechanics, with drive mechanisms of typical efficiency, typical permeabilities and GORs and so on, and normal economic conditions, long experience of the consequences of Darcy's equation has shown that it is neither practical nor desirable to produce the first 50 percent of the total recoverable oil from the reservoir in less than about 10 years. Yearly production rates should not exceed 10 percent of the remaining reserves at any time. After the extraction of the first 50 percent of the recoverable oil, the production rate will decline exponentially at about 10 percent per year. These facts dictate the MER for each well, and for the field. Exceeding the MER – defying these rules, in effect – will normally result in decreased ultimate recovery, because of the loss of reservoir pressure through excessive gas depletion, the coning of bottom water into the wells, and the premature encroachment of edge water, leading to bypassing of the oil.

The cautionary word "normally" needs emphasis. There are fields of exceptional permeability, or exceptionally efficient drive mechanism, that perform better than "normal" fields. There are regions so productive – the Middle East fields are the prime example – that they can satisfy all output requirements while producing for years far below their MERs; their peaks of production come much later than they do in "normal" fields, and their declines are long deferred. In addition, of course, it is not imagined that a giant field will produce for only 20 years; its initial build-up of production, to a peak after a few years, is *planned* for a 20 year life, because after about 20 years its production will be far into its decline and under secondary recovery. Of the 25 most productive oilfields in the lower 48 United States in 1980, for example, 10 had been in production for 50 years or longer.

Halfway through the standard life of a "normal" oil- or gasfield – after 10 years of production – is therefore a potentially critical time for its operator. For oil and gas, as for any depleting resource, the ratio between remaining reserves and current annual production – the R : P ratio – should not be allowed to fall below 10, called the *peril point*. The ratio cannot, in practice, fall far below 10; it must level off at some minimum value (probably between 5 and 7) because reserves remaining after such advanced depletion cannot be produced fast enough to reduce the ratio further.

This circumstance should have prevented the disastrous complacency that developed within the oil industry and among the governments of some consuming countries during the "oil glut" years of the 1950s and 1960s. Unless the output of old fields is regularly augmented by that from new fields of comparable size, the overall production rate must decline after about 10 years of production at MER. In a few large countries, cyclical discovery successes materialized with marvellous regularity. In North America, the 19th century successes in the eastern USA were followed by spectacular discoveries in Oklahoma, California, and eastern Mexico between 1908 and 1925, at East Texas and Oklahoma City in 1928–30, in the Permian Basin in the 1930s, the Gulf Coast in the 1940s, western Canada in the 1950s and early 1960s, and in northern Alaska in the early 1970s. Eventually, however, expectation of further success became concentrated upon more and more remote or "frontier" regions, especially the offshore domain, and on deeper and deeper prospects.

Exploitation of these prospects, even if they prove exploitable, requires much longer lead times than are necessary in simpler, shallower prospects closer to home. The longer lead times are traditionally assessed by the oil industry at about 10 years.

The combination of the two 10-year time scales provides a warning sign that has been ignored by too many people who should have known better. Once the bulk of a nation's oil or gas production is derived from fields that have produced for more than 10 years, and the prime expectation of fields to take their places has moved to regions with long lead times, declining output cannot be avoided.

23.11 Time of accumulation of undersaturated oil pools

Undersaturated oil pools, outside the bubble point curve on the P,T phase diagram (Fig. 23.4), have no free gas cap. If an undersaturated pool was originally saturated, as is likely, its transformation to undersaturation was due to its increasing burial by sedimentary overburden. As there is no further gas to go into solution in the oil once the free gas cap has disappeared, the *saturation pressure* of the undersaturated pool should not change significantly with still deeper burial. It is therefore approximated by the initial or *original reservoir pressure* as measured at the beginning of the productive history of the pool, and it serves as an indicator of the time of disappearance of the gas cap.

Arville Levorsen used the initial reservoir pressure and the gas content of the oil in the Oklahoma City trap to calculate the minimum depth of burial necessary to create those pressure conditions. As the oil in that field is trapped beneath a major unconformity (Fig. 16.4), the approximate time of its accumulation is given by the age of the strata at that burial depth above the unconformity. William Gussow applied a similar but simpler calculation to the Devonian fields of Alberta, Canada. The originally measured reservoir pressure (presumed to represent the saturation pressure) is converted to hydrostatic head assuming the normal salt water pressure gradient of 10.5 kPa m^{-1} (0.465 p.s.i./ft), as in the following generalized example:

saturation pressure	8000 kPa (1160 p.s.i.)
reservoir datum	−500 m
hydrostatic head	8000 ÷ 10.5 = 762 m
elevation	762 − 500 = +262 m

The position in the stratigraphic column 262 m above sea-level datum dates the accumulation of sufficient overburden to destroy a secondary gas cap and convert the saturated pool to undersaturation.

The calculation should be used with extreme care. It presupposes accurate reconstruction of the pool's burial history, which may be simple if the stratigraphy is continuous and conformable but very chancy if there are several episodes of uplift, erosion, or deformation. It also presupposes that the oil and gas content of the reservoir as discovered were generated together and entered the trap together. Later recharge by new gas would invalidate the calculation. The surest approach is to combine calculations from saturation pressures with geological deductions concerning the age of formation of the trap. When was the closure completed, and the seal deposited? Better still is the check provided by applying the calculations to a series of pools in reservoirs of comparable age in a single productive basin or district. If migration always tends to move hydrocarbons up the dip, the time of flush migration is represented by the largest updip accumulation, and the end of accumulation must have occurred after the filling of the undersaturated reservoir furthest down the dip.

23.12 Temperatures in wells

Geothermal gradients under the spectrum of geologic conditions were discussed in Section 17.8. Gradients for common rock types are compared in Fig. 23.10. For many years, oil geologists and engineers have taken the typical oilfield gradient to be 1 °C per 100 feet or 1 °F per 60 feet; this is equivalent to about 32 °C km^{-1}, and somewhat higher than the gradient taken by geophysicists to be "average." A typical bottom-hole temperature (BHT) in a 3000 m (10 000 foot) well was therefore about 100–125 °C.

In most wells deeper than 6000 m the BHT is 200–240 °C (390–465 °F). The highest recorded BHT known to me is 290 °C (555 °F) at a depth of 7265 m in a well in southwestern Texas. The actual formation temperature was estimated to be 310 °C (590 °F), giving a gradient of nearly 36 °C km^{-1}. In the deep Aral Sor borehole in the Soviet Union, the temperature gradient was said to be only 24 °C km^{-1} to a depth of 6800 m.

All BHTs measured in very deep wells are suspect. Above about 200 °C (390 °F), drilling fluids in the water phase cannot be used. Conventional logging tools have been rated only to this temperature level or lower. The higher BHTs measured in very deep wells may therefore be circulating temperatures, not static (shut-in) temperatures representing true thermal equilibrium between the well bore and the surrounding rocks. Drilling fluids in the oil phase are probably usable to temperatures of about 320 °C (600 °F), but drilling into rocks significantly hotter than that will pose new problems for all existing drilling equipment.

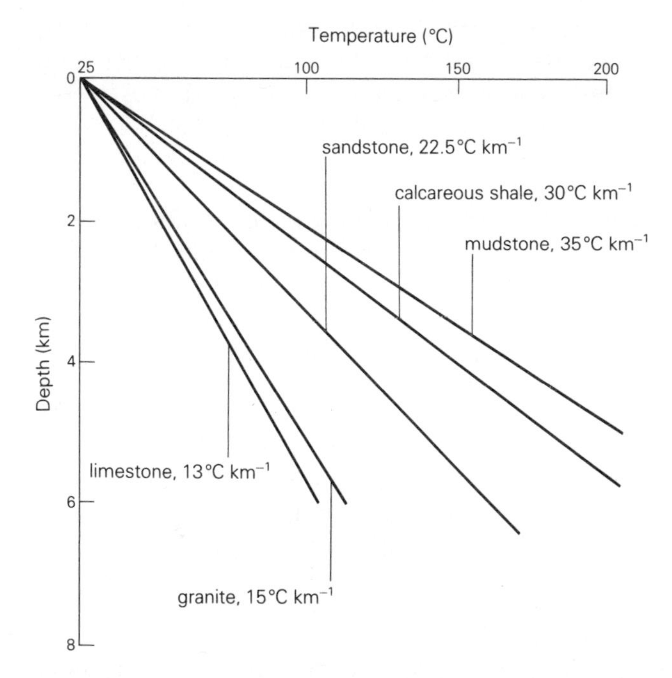

Figure 23.10 Temperature gradients from wells for the common rock types.
The gradient in degrees Celsius per kilometer is given by

$$\text{gradient} = \frac{T_{\text{depth}} - T_{\text{surface}}}{\text{depth}} \times 1000$$

where temperatures (T) are in degrees Celsius and the depth is given in meters. (From J. S. Bradley, *AAPG Bull.*, 1975.)

As temperatures and pressures both increase downwards, high temperatures and high pressures are closely related (cf. Figs 7.6, 7.10 & 14.2). The relation is in fact closer than can at first thought be obvious. Waters in overpressured zones have abnormally high temperatures (up to 270 °C in the US Gulf Coast, for example), and the temperature gradient takes a jump at the top of the overpressured zone as marked as the jump in pressure. Between 2500 and 3000 m in the geopressured "fairways" in the Gulf Coast, the temperature gradient becomes as high as 90 °C km^{-1} (5 °F per 100 feet). The quest for commercial, exploitable geothermal energy, from young, soft rocks rather than hard, igneous rocks, has been concentrated on zones having such sharp temperature increases.

An important reason for the abrupt downward increase in temperature, in such immediate association with the abrupt increase in fluid pressure, is the obstacle to the upward transmission of heat created by the undercompacted geopressured zone and its high water content; water has relatively high specific heat and low thermal conductivity. Thus geopressured hot water rises as high as it can in any available structure, and *isothermal contour maps* closely mimic structural contour maps. M. B. Kumar and others have described the construction of such maps for selected depth levels, and of isothermal surfaces for selected temperature datums or structural horizons (especially for the tops of geopressured zones). Isothermal relief, like structural relief, increases with depth, and is greatest in the geopressured zone. Geothermal "highs" correlate closely with anticlines, salt domes, and the upthrown sides of faults.

23.13 Improving recovery from the reservoir

The most valuable asset an oil operator possesses is his productive reservoir. It must be treated so as to yield as much of its hydrocarbon content as it can. The days when the speed at which the oil could be got out was more important than the percentage of the total content actually recovered were ended by all responsible operators years ago. The geologist, the engineer, the drillers, and the well service companies now cooperate to coax the maximum possible ultimate output from every reservoir.

From some wells the oil and gas flow to the surface through their own reservoir energy (see Secs 5.1.1 & 23.3). Most have to be pumped; eventually, all have to be pumped. Some reservoirs need more coaxing than others to yield their oil or gas from the beginning. Recovery from them is artificially stimulated during the *primary recovery* stage. To conserve the reservoir energy, it is eventually necessary not merely to treat the reservoir rock itself but to augment its contents by pumping some fluid – usually water – into it to maintain its internal pressure. This is the *secondary recovery* stage, to which the vast majority of the world's fields have already been consigned.

Even with the help of secondary recovery, petroleum extraction remains a very inefficient process. No more than 30–35 percent of the oil in place in the reservoir is normally recovered. This means that in the USA alone, some 50×10^9 m^3 of oil will be left in the ground in depleted fields, an amount representing the world's 1982 consumption about 15-fold. Still more drastic methods have to be adopted if the recovery factor is to exceed about 35 percent. These methods, representing the *tertiary* or *enhanced oil recovery* (EOR) stage, have been applied on a large scale in only relatively few fields, but the number of these can only increase exponentially in the future.

23.13.1 Stimulation of initial recovery

Much of the initial testing of an apparent discovery well is carried out through *perforations* in the liner. The perforations are made by charges fired from a "gun" opposite carefully selected horizons. The tests may reveal an unwillingness on the part of the reservoir to part with its oil or gas and to release it into the borehole so that it can be recovered. The porosity and permeability of the reservoir then need to be increased by some artificial means, either primarily mechanical means, by *fracturing* the rock, or primarily chemical means, by *acidizing* it. All such artificial aids to reservoir performance constitute *stimulation* of the reservoir.

A reservoir rock reluctant to release its oil or gas to the borehole is a *tight* rock; to many drillers, it is a *tight sand* even if it is in fact a carbonate reservoir (see Sec. 13.1). Tight formations nonetheless containing oil or gas had, prior to the 1980s, remained largely beyond economic exploitation except in the USA. In that country, however, an imposing array of productive formations, most of them sandstones, owe their yields almost entirely to hydraulic fracturing. Age has little to do with this reservoir phenomenon; as many tight productive formations in the USA are Cretaceous as are Carboniferous.

Since the advent of SEM studies, these should be used to determine the cause of a reservoir's tightness. Either a sandstone or a carbonate reservoir may have a calcareous cement, which is removable in solution. It may contain an excess of clay in the pore spaces, and clay is less easily removed in solution. The loss of porosity may be due to other kinds of cement or matrix, requiring other kinds of treatment. Ferruginous substances, for example, are very common in sedimentary rocks and they can form gels if treated with acid, making matters worse than before the treatment.

Following the artificial fracturing of the reservoir opposite the perforations, a carefully prepared fracturing fluid is pumped into it at desired points under pressure against the column of oil or mud. In *acid frac*, the simplest form of chemical stimulation, a 90 percent acid-in-oil emulsion or other artificially treated acid solvent, containing sand, is forced into the fractures if the reservoir rock or its cement is acid-soluble. The commonest acid in use is dilute HCl. At high temperatures in very deep, carbonate reservoirs, like those in the Smackover trend of the southeastern USA (Sec. 26.2.4), corrosion by acid gases is a menace to the downhole drilling apparatus and HCl would be too strong an acid to be introduced in addition. Formic acid, the acid of the stings of ants, bees, and nettles, is the commonest of several widely used substitutes.

If the reservoir rock and its cement are essentially unaffected by dilute acids, *sand frac* is used. Rounded sand grains are injected into the fractured reservoir in a base of thickened crude oil or diesel fuel. The sand grains remain in the fractures and prop them open to facilitate release of the oil. High quality fracturing sands, or *proppants*, should consist of uncemented, even-sized, spherical grains, free of impurities. These properties are most readily found in sandstone formations at least in part aeolian in origin. Most of the best ones are therefore of pre-Devonian age (antedating the effects of land plants on surface accumulations of sand); they are rare, and command a high price. As they become exhausted, still more expensive substitutes will include high strength glass beads, aluminum pellets, steel shot, and even crushed and rounded walnut shells.

The simplest fracturing fluid is water itself, used to *hydrafrac* the reservoirs. Fresh or salt water is injected, with or without sand. Hydrafrac is widely used in water-injection or water-supply wells in fields undergoing other methods of secondary recovery. Numerous variations on the fracturing fluids and techniques are routinely available from well-service companies, and are the subject of much expensive research.

23.13.2 Waterflooding

The universal method of *secondary recovery* from oilwells is by *waterflood* of the reservoir. Formerly adopted as a rescue process when production had gone into serious decline, it is now commonly initiated much earlier in the exploitation of the fields.

The input water must be made as compatible as is economically feasible with the reservoir's formation water (see Ch. 9). Its advance treatment may be extensive. It may for example be rendered either acidic or basic (caustic), or it may be augmented by organic polymers to improve its sweep efficiency through the reservoir. The nature of the oil in the reservoir must also be considered. In particular, water much below the oil's reservoir temperature may reduce and not enhance the producibility of oils with high paraffin contents.

The flood itself must be designed to take maximum advantage of the geometry of the reservoir in its trap. Each producing well should be essentially surrounded by injection wells, creating an inward pressure gradient from all sides and reducing the opportunity for preferential movement of the water along a single, convenient but ineffective route. The commonest design is the *pattern flood*, used where dips are moderate. The water is forced through the reservoir, from the base of the pay zone upwards, via input wells at the centers of five-spot patterns. With steeper dips, *peripheral flood* may be more effective, from the edges of the pool inwards. A rare design is the *radial flood*, from the center of the pool towards its edges. This design may provide the only effective way to force the water through a sharply convex trap containing impermeable strata of lower convexity than the trap, as in some isolated reef traps. In neither the pattern flood nor the peripheral flood would the water be able to reach the crest of the pool in a trap like that.

As impermeable barrier beds (notably shales) will confine the water flow, the flood design should use them to advantage rather than as obstacles. In a fractured reservoir, injection should be at right angles to the dominant fractures. Strong jointing may prevent any successful secondary recovery. A well designed pattern flood may lead to a 50 percent recovery from a sandstone reservoir, but a much lower recovery (perhaps 33 percent) from a carbonate. Peripheral waterflood, used almost exclusively in sandstone reservoirs, may also enable a 50 percent recovery to be achieved.

North America's first giant field, Bradford in Pennsylvania, was also the site of the first waterflood project. More oil was recovered by it than by primary methods. From that experimental beginning, waterflood has become routine. The Prudhoe Bay field in Alaska, the largest in North America, but not discovered until a century after Bradford, was prepared for seawater injection into its principal sandstone reservoir as soon as it reached its MPR in 1980. The supergiant carbonate reservoirs in Saudi Arabia, under unified operation, were put on waterflood during their flush production years in the late 1970s. By mid-1978, the six largest Saudi field reservoirs were receiving processed sea water at a rate essentially equivalent to their oil production rate, about 1.4×10^6 m^3 per day; nearly half this volume was being injected into the Ghawar field alone. The sprawling Chicontepec field of southern Mexico, in Eocene turbidite sandstone of such low permeability that every one of some 16 000 wells will need fracturing, may become the world's largest waterflood. By the end of the 1970s more than 230 Soviet oilfields had been waterflooded, including those yielding more than 90 percent of the production from the West Siberian and Ural–Volga basins.

23.13.3 Tertiary or enhanced oil recovery

Enhanced oil recovery (EOR) techniques are those employed when the efficacy of primary and secondary techniques has declined beyond economic acceptability, or when secondary recovery techniques are of no avail. Where the latter is the case, EOR is most commonly attempted by *thermal* methods. Where secondary recovery by waterflooding has already been applied, *miscible flooding* is the only proven technology for the recovery of high-viscosity crude.

Thermal recovery methods The most common cause of an oil's reluctance to leave its reservoir rock and enter the borehole is its *viscosity* (see Sec. 5.3.2). The oil may be inherently viscous, having lost its light ends through natural degradation, or it may have become viscous during the course of the field's development, because of the exhaustion of its solution gas and NGL content. The obvious route to the reduction of the viscosity of an oil is to subject it to heat. This has been carried on in a number of fields for some years, in a variety of styles. Three styles have been widely attempted, mostly on trial scales but achieving larger scales in several regions of very heavy oil production. Before the end of the 20th century, the techniques may be applied to several thousand reservoirs in North America alone.

Cyclic steam injection Cyclic steam injection is the alternating injection of high quality steam into the wells and the production of oil (plus the condensed steam) *from the same wells*. The alternation led to the name of "huff-and-puff" for the process. It may be executed on a single well basis. Wells are injected with slugs of steam at a very high rate (in the millions of kilograms) for a short time (say 10 days). The wells are allowed to "soak" in steam for a few days (providing the alternative name of "steam soak" for the process), to achieve the necessary reduction in the oil's viscosity, and are then put on production for a somewhat longer time (say 3–6 months) until the effectiveness has worn off. The process is then repeated. A refinement involves injecting a mixture of superheated steam (at 350 °C), nitrogen, and CO_2 (which dissolves in oil and further reduces its viscosity).

The technique requires considerable reservoir energy, so it would not work at the Athabasca deposit (see Sec. 10.8.4). It is most effective in weakly consolidated sand reservoirs buried at least to several hundred meter depths. These features permit compaction to bring about pressure maintenance and displace the oil towards the wells. Once reservoir compaction is about complete, oil recovery declines drastically – as it did, for example, in the fields providing 13° gravity oil from unconsolidated sands in the Magdalena Valley of Colombia. Recovery may also be aborted through the steam's reaction with clay minerals in the reservoir, forming smectite and lowering the permeability beyond economic exploitability.

It has been claimed that "steam soak" alone may recover 6–10 percent of the oil in place in a reservoir. Its longest-running demonstration is probably that in the Bolivar Coastal fields in Venezuela (Figs 9.9 & 17.30), where it was started in 1964 on great quantities of asphaltic oil as heavy as 12–15° gravity. The largest cyclic steam injection program in the near future may be in Alberta's Cold Lake heavy oil deposit (see Sec. 10.8.4). Oil of 10° gravity, with viscosity averaging about 50 000 mPa s, will be injected with nearly 5×10^6 kg of steam per hour at 325 °C and under input pressure of nearly 14×10^3 kPa.

Steam drive or steam flood This process may be used as a follow-up to steam soak or it may be employed by itself. The input wells do not become producing wells. Instead, steam is continuously injected on a pattern basis into one set of wells, called *input* or *injection wells*; it displaces the oil towards another set of wells, the *output* or *production wells*.

Because this process uses a vastly greater quantity of steam than the cyclic technique, the heat balance, or net energy gain or loss, is critical. Something must be burned to raise the steam. If crude oil is burned, the thermal efficiency is theoretically about 12 : 1; about 12 m^3 of steam are generated per cubic meter of oil burned. In practice, experience with the heavy oils in California's San Joaquin Valley fields, beginning in 1968, indicates a thermal efficiency more like 3 : 1. Steam costs may amount to half the value of the oil produced.

The largest steam flood known to this author at the time of writing is employed at the Duri field in Sumatra, where only 7 percent of the heavy, waxy, in-place crude has proved recoverable by conventional techniques. The longest-running steam drive pilot plant must surely be that in the Peace River oil sands in northern Alberta, in which the steam takes 9 years to sweep completely through a 27 m Cretaceous sandstone reservoir at a depth of about 550 m.

Fire flooding The only other thermal recovery technique with more than local and experimental application is *in situ dry combustion* or "fire flooding." The combustion of oil in the reservoir is achieved by igniting a continuous injection of air. The combustion front of heavy crude fractions moves forward through cold sand, displacing unburned oil towards the recovery wells. The oil produced is somewhat lighter than the in-place oil.

The technique requires above-average permeability in the reservoir rock. Even if that advantage is present, fire flood is a last-resort process for fields having the disadvantages of viscous oils, low GORs, low reservoir pressures, and sands too loose for waterflood (because of sand production with the oil) yet too thin for effective steam drive. One productive region suffering all these disadvantages is Lloydminster in western Canada, south of the Athabasca and Cold Lake deposits.

An imaginative but not yet very successful variation on the fireflood technique is that of *reverse combustion*. The formation is ignited, not in the injection well but *in the production well*. The combustion zone thus moves towards the oxygen supply (the injection well), in the direction opposite to the flow of air. The displaced fluids therefore move through hot sand, and the produced oil is vastly different from the in-place oil. The oil is in effect upgraded *in situ*, from (say) 10° gravity and high viscosity to (say) 25° gravity and low viscosity. If the technique worked in the field, it would revolutionize the recovery of heavy, viscous oils. Unfortunately, experience in the Orinoco belt of Venezuela showed that the continued contact of air with the in-place oil leads to spontaneous ignition near the injection well. The supply of air is thereby cut off from the combustion front and the process effectively becomes forward combustion after all.

In a further and somewhat more successful variation, called *wet combustion*, water is injected with the air and is heated by the burned sand, resulting in more efficient forward transport of heat. In effect, a type of steam drive is created. This greatly improves the thermal efficiency. In the application known as COFCAW (Amoco's combined forward combustion and water injection process), the method may become important when *in situ* recovery is developed at the Athabasca deposit.

Miscible flood techniques *Miscible flooding* is the process attempted as augmentation of recovery from reservoirs that have already been waterflooded. An injected fluid, called the *displacing fluid*, and the *displaced fluids* (oil and/or gas), are in phase continuity with no interface between them. This condition results in a great reduction in the capillary forces restraining the movement of the displaced fluid. All the water previously injected in the secondary recovery phase will be reproduced, and must be disposed of a second time.

The most widely used *miscible fluid displacement* technique in EOR uses CO_2 as the injected fluid. The CO_2 is not truly miscible with oil under reservoir conditions, but its vapor pressure is almost the same as those of the wet gases, and it is also *soluble* in oil, under the high pressures at which it is injected, to the extent of seven volume times or more. Its solution causes the oil to expand and lowers its viscosity, driving it through the reservoir. The most efficient use of CO_2 in enhanced recovery is in large carbonate reservoirs, but it has been successfully attempted on pilot scales in other types of reservoir, with oils from 15° to 45° gravities at depths from 600 to 3800 m. CO_2 and water may be injected in alternation. If CO_2 is injected alone, it has a tendency to bypass much of the oil-in-place instead of entering into solution in it.

As miscible flood is intended to restore the reservoir's pressure to a level adequate to lift the fluids from it, the reservoir must be deep enough to maintain miscibility pressures. Depending upon both the inherent and the imposed characteristics of the reservoir, from 900 to 2200 m^3 of CO_2 must be injected to recover 1 m^3 of additional oil. The CO_2 is expensive, costing $75–100 per cubic meter in 1981. It may be purchased from power, cement, or chemical plants, but it is difficult to transport over long distances and is most cheaply and conveniently used if it can be produced from gas-bearing reservoirs not far from those to be injected. Fortunately, many natural gases produced from carbonate reservoirs contain CO_2 (as well as H_2S; see Sec. 5.2). Some thick carbonate formations that have been subjected to light thermal metamorphism by igneous intrusion contain large volumes of CO_2 without hydrocarbon gas accompaniment. From such reservoirs in the Permian Basin and on the Colorado Plateau of the southwestern USA, for example, sufficient reserves of CO_2 had been established before the end of the 1970s to justify CO_2 injection into a number of large Permian Basin oilfields and the expectation of more than 300×10^6 m^3 of extra oil recovery from that basin.

Sandstone reservoirs may also be suitable for CO_2 injection once the efficacy of water and polymer flooding has worn off; several fields in Russia's Ural–Volga Basin now produce with this assistance. Furthermore, reservoirs other than carbonate reservoirs may yield commercial quantities of CO_2 for the process, especially if they lie over deep igneous intrusions or close to igneous basement. The largest commercial CO_2 production in the USA in the early 1980s has been from a Permian sandstone on the Bravo Dome in New Mexico.

Other fluids besides CO_2 are of course readily miscible with oil. Natural gas, flue gas, alcohol, or nitrogen may be used, but natural gas is too valuable in its own right and pure nitrogen is not so easily obtained in quantity as is CO_2. If LPG happens to be in surplus production in the immediate area, it may be economic to inject it as a slug driven by natural gas.

Injection of CO_2 into the reservoir may assist recovery without the CO_2 dissolving in the oil. Injected at the gas/oil contact following repressuring by water injection, CO_2 expands the gas cap and drives the oil towards the well bores. The largest *immiscible CO_2 flood* known to the author is in the sandstone reservoir of the Wilmington field in California (see Sec. 26.2.3). A zone of heavy (14° gravity) crude is to receive about 10^5 m^3 of CO_2 daily into the watered-out tar zone in a discrete fault block.

Surfactant/polymer flooding Natural surfactants cause emulsions or oil-in-water dispersions, with viscosities about the same as that of a continuous aqueous phase. Natural solubilizers of hydrocarbons are radially structured, surfactant molecules called *micelles* (Sec 15.4.2). A micellar slug can be injected into a reservoir which has already been waterflooded; the surfactant materials act as underground detergents or *scrubbers*, mobilizing the residual oil still in the reservoir. The slug is followed by *chase polymers* for mobility control. The chief polymer used is polyacrylamide, based upon $CH_2CHCONH_2$. The surfactants and the polymers "thicken" the water (increase its viscosity) and so reduce the rock's permeability to it, forcing it to flow through more channelways and so increasing its sweep efficiency.

The technique is best used in sandstone reservoirs. It is unsuitable for reservoirs of very low permeability, because the "thickened" water cannot penetrate very tight rock. With good sandstone permeability, however, the high water : oil ratios that bedevil the late stages of secondary production in many fields can be dramatically reduced. In California's Wilmington field, for example, the ratio before surfactant/polymer treatment had been nearly 100 : 1, creating a nearly insurmountable water-disposal problem. The largest oilfield accorded a pilot polymer flood by 1984 is probably Daqing, in Chinese Manchuria.

The most efficient micellar flood is by *surfactant sulphonates*, which are themselves derived from hydrocarbons. Thioalcohols have sulfur instead of oxygen in their alcohol compositions (C_2H_5SH, for example). They are insoluble in water but have a characteristic reaction with mercury, giving rise to their familiar name *mercaptans*. Mercaptans are very easily oxidized (e.g. by nitric acid), forming monobasic *sulfonic acids* which, unlike their predecessor mercaptans, are very soluble in water. A sulfonic acid, such as $C_2H_5SO_3H$, reacts with alkalis to form sulfonates, which in turn form compounds with hydrocarbons and constitute ideal surfactants for tertiary recovery.

Unfortunately, the injection and sweep processes are difficult and very costly, so their application is likely to be severely limited. To recover 1 m^3 of additional oil, 0.5 m^3 of surfactant must be injected, followed by 4.0 kg of polymer. The total cost of this unit slug amounted at the beginning of the 1980s to about 40 percent of the value of the additional oil recovered.

23.13.4 *Nuclear stimulation of reservoir rocks*

During the 1970s, several pilot attempts were made to enhance deliverability of natural gas by nuclear detonations in a group of tight reservoir sandstones in the western USA. In brief, simultaneous detonations were set off in long, narrow canisters, each provided with its own cooling system. The explosions created *rubble chimneys* 25–30 m in radius surrounding the original drillholes; permeabilities were measurably increased within a radius of about 65 m of the bores.

The choice of nuclear power as the energy behind the enhancement attempts was dictated by the geologic characteristics of the reservoir rocks. The uppermost Cretaceous and Paleocene fill of the US Rocky Mountain basins was deposited prior to the Laramide thrusting and final uplift of the mountains. The sands are of wholly freshwater origin, principally lacustrine and paludal, and carry high clay contents. Their permeabilities range from 0.1 mD to as little as 0.000 05 mD, which is about as close to total impermeability as a sedimentary rock can get. The reservoir properties of the sands are therefore beyond enhancement by any solvent treatment or any normal fracturing procedure. Like the gas adsorbed in the Paleozoic black shales in front of the Appalachians (Sec. 10.8.2), the gas content of the sands can be released only by massive rubblization of the reservoir rock. The technique requires considerable overburden (at least 300 m) to permit creation of a rubble chimney without an enormous surface crater.

The potential reward for a practicable process would be great. The average content of immature, biogenic gas in the sands is at least 3×10^9 m^3 km^{-2}, and the area of their occurrence extends over some 15° of latitude between New Mexico and northeastern British Columbia. In the event, the acceptable level of gas yield following the detonations lasted only a short time, as would be expected, and the prospect of hundreds of such experiments is not likely to commend itself to any elected authorities.

A proposal has also been made (unsuccessfully, to the time of this writing) to use underground nuclear detonation to release the oil from the Athabasca tar sands (Sec. 10.8.4). Fears of contamination aside, an obvious drawback to such a proposition is to be found in the nature of tar sands themselves. Unlike tight gas sands, tar sands are plastic rather than brittle, and do not fracture well.

The economic benefits and environmental drawbacks of any EOR (or EGR) program must be calculated very carefully in advance. In spite of this, a little simple arithmetic demonstrates the enormous potential benefit if economic criteria can be satisfied. Without EOR, worldwide oil recovery factors average about 35 percent. For each percentage point by which this factor can be increased by EOR, the noncommunist world's ultimate proven oil reserves will be increased by about 3×10^9 m^3, more than a year's consumption.

24 Estimations of reserves and resources

24.1 Reserves and resources

During the 1970s, everyone in government, industry, and scientific academia talked and wrote of the "energy crisis." Insofar as any true "crisis" existed, it was in large part the outcome of a profound confusion, in both professional and public minds, between *resources* and *reserves*. Oil was accorded unique treatment among all vital *resources*; estimates of oil volumes that might be expected to occur in the Earth's crust were published in quantitative detail and regarded as *reserves*.

Resources are accumulations of anything which is both useful and accessible to mankind. The Earth sciences are not directly concerned with such renewable resources as timber or fish. They *are* directly concerned with such apparently renewable resources as water, soil, and geothermal power.

Petroleum geologists are concerned with *nonrenewable resources,* formed by natural processes over time spans enormously long by comparison with those of human activities and now occurring as natural concentrations within the Earth's crust in such forms that their exploitation is conceivable. The particular nonrenewable resources of immediate concern to us have been called the *capital sources of energy*, "capital" because they are based on carbon which cannot be recycled once it has been oxidized (by being used as a fuel). Renewable resources, on the other hand, include the *income sources of energy* – solar, tidal, and fusion powers, for examples.

Reserves are that portion of an identified resource that is available now by being economically recoverable under existing technological conditions.

It follows that resources include reserves but are not otherwise directly comparable with them. Many resources known and investigated for years may never become reserves – deeply buried oil shales, for example, or trace metals dissolved in sea water. It is therefore desirable to insist on some restriction being placed on the use of the word "resources," in addition to very stringent restrictions on the use of the word "reserves."

The United States Geological Survey has for some years used a fourfold categorization of *resources*, illustrated by a subdivided rectangle (Fig.24.1) whose axes are the degree of geologic assurance of the existence of the resource and the degree of economic–technological feasibility of its exploitation. The category representing positive values for both parameters is known, recoverable, economic *reserves*. There are also discovered but subeconomic or marginal *resources*; undiscovered resources which will be economically extractable if and when they are discovered; and, finally, resources which are not only undiscovered but will not be economically or technologically recoverable under reasonably foreseeable circumstances even if they are discovered. It is easy to imagine that this last category may very well be enormously greater in quantity than the other three combined – but it is useless to include it in any tabulation or quantification of "future" or "potential" supplies of oil or gas or anything else.

The second and third categories, however, must be

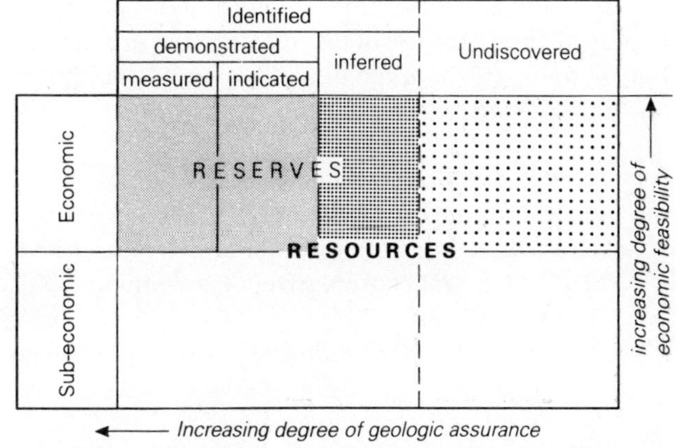

Figure 24.1 Diagrammatic representation of the petroleum resource classification used by the Geological Survey and the Bureau of Mines of the United States. (From Betty M. Miller *et al.*, US Geological Survey Circular 725, 1975; modified after V. C. McKelvey, 1973.)

recognized in any rational forecast of future supplies. We know that *reserves* will be augmented by resources currently undiscovered, but sure to be economically viable when they are discovered; such discoveries are being made all the time, *somewhere*. We also know that many known but uneconomic resources will eventually become *reserves*, because the approach of exhaustion of the exploited reserves will trigger price increases and some previously subeconomic accumulations will become economic. During the 1970s we saw this happen as the wellhead price of natural gas in North America was multiplied tenfold and the world price of uranium oxide increased sixfold.

Until about the middle of the 1970s, however, all categories of hydrocarbon resources continued to be referred to as *reserves* in the North American oil and gas industries. Terms such as probable, possible, potential, discoverable, expectable, findable, speculative, and future "reserves" were scattered through the American literature in a chaos of meaninglessness. The Russians for their part wrote of explored, indicated, predicted, and prognosticated reserves. Nearly all of these "reserves" were totally hypothetical, but energy "policies" proceeded as if they were established.

The desirability of a uniform system of reserve (and resource) terminology for any major resource material – especially for fugitive resources like oil and gas – is so obvious that it is cause for dismay that no such system has yet been adopted. A hopeful sign is the elimination, at least in North America, of the word *reserves* for any undiscovered volumes of any resource. The two principal but very different classifications of resources and reserves are discussed here.

The USA, for official statistical purposes at least, has settled upon a terminology which ostensibly eliminates the evil aspects of the earlier terminology but which is still wildly misused. The *ultimately discoverable volume* of any resource is the total quantity expected to be produced during the life of the industry that produces it. For oil and gas, the ultimately discoverable volume is made up of three components:

(a) Cumulative *production* so far.
(b) *Proved* (or *measured*) *reserves*, defined in existing fields, drilled or at least outlined (circumscribed) by drilling, and conservatively delineated.
(c) Potential *supply*:
   (i) *Probable* (or *indicated*) supply, through extensions of existing fields beyond or below the limits so far evaluated for them.
   (ii) *Possible* (or *inferred*) supply, through future discoveries in areas or formations already known to be productive.
   (iii) *Speculative* supply, through future discoveries in areas or formations not previously productive.

In effect, the formulators of this classification attempted to assess the likelihood of future exploitable discoveries being made. All "potential supply" (category (c), formerly miscalled "reserves") presupposes that exploration will proceed under acceptable economic conditions (such as adequate but reasonable prices, bearable tax levels, and so on) and improvements in technology comparable with those permitting the present state of development. As Deverle Harris expressed the intention in 1976: "Resources are the resultant of the interaction of economics and technology with a basic physical endowment." Unfortunately, characteristic optimism about technological advances and resignation to continual increases in prices (and costs) have combined to persuade many that "potential supply" is in fact assured of attainment; the deposits to provide it can be confidently assumed to exist, they merely have to be located. The three categories of "supply" are therefore habitually quantified, with no rounding-out of numbers, and their totals added together and quoted *as if they could be relied upon*. The Resource Appraisal Group of the US Geological Survey itself, for example, restricted the term *measured reserves* to those recoverable by the technologies existing in each field, and added to them the category of *indicated reserves* for volumes of oil in known reservoirs which would become available through fluid injection. The total of measured plus indicated reserves was called *demonstrated reserves*. What the parent survey called *indicated supply* was called *inferred reserves* by the Resource Appraisal Group and quantified alongside the demonstrated reserves.

Canada, because of its unique economic and trade relation to the USA, was more seriously damaged by misrepresentation of oil and gas "reserves" than any other nation. By 1978, in consequence, Canada had for official purposes abandoned all categories of unproven "reserves" and referred only to *established reserves* (the same as the American *proved reserves*) and *ultimate potential*. The latter consists of the obvious factors of cumulative production, remaining established reserves, and future additions through extensions and revisions to existing pools, plus the inevitable volumes to be discovered in new pools "at the 50 percent probability level." The potential for error in forecasting this last factor is so great that its formal quantification is assuredly a mistake. Yet in 1981 the Canadian government published a figure for the country's "conventional oil resources" (ultimate potential minus cumulative production) that was more than three times the established reserves, and included (as one instance) an amount of 3799×10^6 barrels (not what one would call a round figure) for the "eastern Arctic" in which not a single commercial oil discovery had then been made.

The Soviet Union employs a sixfold classification of "reserves" and has not at the time of this writing abandoned this ambiguous usage. Their reserve categories are as below:

Category A: reserves defined in existing deposits, through actual exploitation.
Category B: reserves in known deposits with economic recovery established in at least two wells to different depths, and drilling and geophysical indications favorable.

(Categories A and B together constitute *explored reserves*, about equivalent to the American *proved + probable reserves*.)

Category C−1: reserves proven by recovery from single wells with outline of the traps, or by being adjacent to wells in category A or B.
Category C−2: reserves inferred from indications in unexplored blocks (in the tectonic sense) or horizons (in the stratigraphic sense) of investigated deposits; also reserves in newly discovered but so far untested structures within the limits of districts known to be petroliferous.
Category D−1: *predicted* or *prognosticated* reserves, where structures have been established to exist and the stratigraphy appears favorable.
Category D−2: predicted or prognosticated reserves in apparently favorable basins in which structures have not yet been defined.

(Categories D−1 and D−2, "predicted" on general geological premises, are about equivalent to "speculative" reserves in the American classification.)

However, assessments of "predicted" reserves may be made by different procedures, or by the incorporation of different values or assumptions, from assessments of "speculative" reserves. It is not always clear what categories of "reserves" are included in any figures published by the Soviets or by Soviet *bloc* producing countries. Comparisons between "reserve" volumes reflecting different criteria and possibly calculated by different methods should be regarded with suitable caution.

Other oil- and gas-producing countries have used similar categories but with endless variations which appear to be minor but create serious confusion. A study group at the 1983 World Petroleum Congress has proposed that *all* countries adopt a simple, uniform terminology with simple, agreed definitions. The proposed categories (at the time of this writing) are:

(a) *Proved reserves*, both developed and undeveloped.
(b) *Unproved reserves*, available only from deposits already discovered; this category embraces the *probable* and *possible* reserves of the American system; it should always be expressed as a *range* of values (see Sec. 24.3.3).
(c) All volumes expected from deposits not yet discovered should be called *speculative*.
(d) The *ultimate potential recovery* is the sum of the volumes in these three categories *plus* the cumulative production.

24.2 Estimating the different categories of reserves

24.2.1. Estimation of proven reserves of oil or gas

Reliable estimation of the quantities of materials recoverable from the deposits that are (or are about to be) exploited is one of the most critical tasks faced by industrial geologists and engineers. The fundamental procedure for any resource is one of simple arithmetic, once all the factors that may influence the outcome are understood; but the procedure itself has to be modified to take account of the relation of the resource to its host rock. This relation may be of three kinds:

(a) The resource is an integral part of the rock that contains it. Obvious examples are the diamonds of primary diamond pipes, heavy metal deposits in basic or ultrabasic intrusive rocks, and the porphyry coppers. For these, the grade or tenor of the deposit within a blocked-out, minable property has to be measured by analysis or assay, and the total quantity extractable by the intended process is then calculated.
(b) The resource is an integral part of the rock assemblage or succession that contains it: iron formation, salt, coal. For this type of deposit, the volumetric calculation merely has to take into account some initial constraints on recoverable tonnage: practicable mining depth, minimum minable thickness of the deposit, percentage to be left deliberately as support within the mine, and percentage of the mined resource actually recovered in saleable form.
(c) The resource is discretely distinct from the rocks containing it. The resource may be contained in fluid, mobile rock capable of finding itself in any host rock (vein deposits, gem pegmatites, etc.), or the resource itself may be fluid and mobile, like the oil and gas of concern to us. Some of the pitfalls in reserve calculations for this category are immediately obvious; how continuous is a deposit which is or has been mobile, for example? Other pitfalls may be less obvious.

The most important derives from the contrast between fluid oil and gas, on the one hand, and virtually all the resources of categories (a) and (b) (and many of (c)) on the other. Are quantities of the resource to be expressed in terms of volume or of weight? As the two expressions are related by the density factor, any material having its reserves expressed in units of weight (especially in *tonnes*) requires some average relative density to be determined and stated. Several industrial countries (the UK and the USSR notably) traditionally express *oil* reserves, production, and consumption in terms of weight, whereas the US and Canada lead the countries using volumetric measures. We may look more closely at this nonuniformity.

The *quality* of oil is in general inversely related to its density (or directly related to its API gravity; see Sec. 5.3.1). One unit volume of 36° gravity, sweet Nigerian crude is worth much more than one unit volume of 16° gravity, sour Venezuelan crude. This is an argument in opposition to measurement by volume. It is outweighed, however, by numerous arguments in favor of such measurement. The first is economic; sale by weight provides (per unit of cost) a greater volume of an intrinsically more valuable oil to the purchaser of a light crude. The second argument is technical; the simplest expression of the reserves of any fluid within a solid host-medium is volumetric; in our present case, the porosity of the reservoir rock, its degree of saturation by water, and the gas content of its oil are all expressed in volumetric terms. The third argument is practical and domestic; gasoline for cars, heating oil for houses, and (especially) natural gas for all purposes are more easily handled by volume than by weight. Calculations of oil and gas reserves are therefore conducted in volumetric terms; the reserves themselves may or may not be later converted to tonnages.

Useful estimations of oil and gas reserves in a pool or field are impossible until several strategically spaced wells have been drilled. These wells must establish three aspects of the overall shape and size of the deposit: the elevation of the top of the deposit in its trap; the elevation of the oil/water or gas/oil or gas/water interface; and the approximate outline of the trap in plan. All other factors necessary for the calculation are measured or estimated from materials recovered from these initial wells. The calculation of *oil reserves* involves three steps:

(1) The volume of the reservoir rock between the highest point in the trap and the level of the bottom water is simply the accessible area times the average thickness of saturated rock. There are several cautions. The area must be conservatively planimetered within the structural closure (for example, from the best available seismic maps) as it is believed to be established at the time of the estimate. In the simplest of convex traps, wells on the crest of the structure will have the maximum thickness of pay, and if pre-drill exploration was very skilful these crestal wells may include the discovery well and its close step-out wells. Their pay thickness should not be taken to be the average for the pool. If the legal measuring system is nonmetric, or metric but not expressed in SI units, the appropriate units must be used. In the USA, for example, areas are expressed in acres and drilling depths in feet. Hence the estimated area of the trap *in acres* must be multiplied by 43 560 (the number of square feet in an acre, which is defined as 10 square chains or $66 \times 66 \times 10$ ft^2 and equals 4046.86 m^2). In some producing regions areas are expressed in *hectares*, 1 hectare (1 ha) being 0.01 km^2 or approximately 2.5 acres. Finally, only that portion of the trap area accessible to and drainable by wells on the allowed well spacing should be included. For example, the area of closure may include a tribal reserve or game park, an airfield, a deep lake, or inaccessible relief. In offshore regions, how many platforms are likely to be available in, say, the next 6 years? Legal well spacing may be determined by the local survey units – 80 acres, say, or 1 km^2; will the wells the operator is permitted to drill on this spacing be sufficient to drain the whole reservoir?

We now have the drainable volume of the presumably saturated reservoir in cubic meters or cubic feet. What proportion of this volume is filled with oil or gas?

(2) The *porosity* (ϕ) is determined from cores, or calculated from neutron, density, or sonic logs (Ch. 12). Some of the pore space will be occupied by water (the *irreducible water saturation*, S_w). S_w is calculated from Archie's formula (see Sec. 12.4), and needs values for the water resistivity (R_w), the true resistivity (R_t), and the formation factor(F). R_w, the most difficult of the factors to obtain with assurance, can be measured from carefully collected water samples during drillstem tests, or calculated from the SP log. R_t is calculated from the resistivity curves of focused-current logs. The formation factor F is very variable and is most conveniently determined from nomographs. In general terms, the oil content of the apparently saturated pore space which is actually capable of producing oil varies between about 35 percent and about 90 percent; below about 60 percent of *oil* saturation ($1 - S_w$) the reservoir will produce mostly water.

(3) How much of the oil or gas content of the reservoir will be recovered by the mechanism to be employed? This depends on the viscosity of the oil, the

permeability of the reservoir rock to oil, and the GOR, as well as on the mechanics of the production system (pumping, pressure maintenance, or whatever). The *recovery factor* for a heavy oil, for example, may be only 10–15 percent even with secondary recovery techniques. The figure used in the initial calculation should not be higher than about 35 percent unless deliverability of the reservoir has been seen to be exceptional. If the reservoir GOR is significant, the oil will shrink when it is brought to the surface because its gas content will bubble off. The *shrinkage factor*, commonly 15–30 percent, converts the volume of oil in the reservoir to the volume of stock-tank oil (STO); it is expressed by a factor smaller than unity (e.g. 0.82). Alternatively, the factor used may be the reciprocal of the *formation volume factor* (not to be confused with the formation factor used in log interpretation), which converts STO to reservoir oil and is therefore a number slightly greater than unity (say 1.22). Shrinkage is calculated from the temperature, pressure, and GOR of the oil (some of the shrinkage is due to cooling of the oil on leaving the reservoir for the surface); the volume is standardized as that at 15 °C (60 °F).

These six factors – area (A) and thickness (h) in SI or other appropriate units; porosity (ϕ), oil saturation (S_o or $1 - S_w$), and recovery factor (R) as decimal equivalents, and shrinkage factor, simply a number – are multiplied together to give the recoverable reserves in cubic meters or cubic feet:

$$\text{recoverable STO} = \frac{A \times h \times \phi \times S_o \times R}{\text{FVF}}$$

As oil reserves in many countries are still expressed in *barrels*, of 35 Imperial or 42 US gallons, the reserves in cubic meters should be divided by 0.159 and in cubic feet by 5.615 to provide this expression. Nomographs are published in all oilfield manuals for all factors in the reserve calculation.

The reserves should be calculated and expressed *conservatively*. The possibility of serious error is manifest otherwise. Sensibly calculated values made prior to the decision to develop, or during the early stages of development, are likely to be too low, and must be constantly revised as more data become available. Upwardly revised reserves are said to be *appreciated*. Appreciation may continue for years if the initial estimates were drastically conservative, but it is a serious mistake to project appreciation factors far into the future at any time.

Reserves of *natural gas* are calculated in the same way, at least if SI units are used, except for the addition of one new factor and the inversion of another. Gas expands when it is brought from the reservoir to the surface, instead of shrinking. The *compressibility factor* (Z) is comparable with the formation volume factor for oil, but introduces a very much larger correction which is simply read from nomographs. Furthermore, the volume of gas must be standardized to STP by introducing the *STP factor*:

$$\text{STP factor} = \frac{P_r}{P_s} \times \frac{T_s}{T_r}$$

where P_s and T_s are standard pressure and temperature (in kelvins on the absolute temperature scale), P_r and T_r are reservoir pressure and temperature.

As the STP factor involves ratios, units must merely be internally consistent. Using temperature in kelvins (i.e. degrees Celsius + 273) and pressure in pounds per square inch converted to atmospheres, the STP factor becomes

$$\text{STP factor} = \frac{P_r + 14.4}{14.4} \times \frac{15 + 273}{T_r\,(°\text{C}) + 273}$$

In the USA and some other countries, absolute temperature is commonly expressed in degrees Rankine (degrees Fahrenheit + 460), and the temperature factor becomes

$$\frac{520}{T_r\,(°\text{F}) + 460}$$

From the initial calculations of reserves in the pool or field, production rates are established as described in Chapter 23. The maximum efficient rate of production (MER) should be planned to peak at about 0.015–0.020 percent per day of the estimated recoverable reserves.

24.2.2 *Estimation of probable or possible "reserves"*

Probable and *possible* additions to reserves may justly be considered separately from the "speculative" category. *Probable* additional supply, from enlargements of known fields within, beyond, or below their currently productive limits, may be assessed by considering the historical *appreciation* of the discovered reserves in such fields. Initial estimates of recoverable reserves from newly discovered fields should be conservative, and usually are. Many producing fields have long ago yielded the volumes of oil or gas originally credited to them, and will ultimately yield several times those volumes. The appreciation "multipliers" may be projected cautiously into the future along the historically established curves, but the projections should advance in short time intervals; obviously they cannot be continued indefinitely.

The term "appreciation" in this context, like many other convenient but generalized expressions, covers a variety of actual processes. The reserves of a field may appreciate within its original areal and depth limits, because of excessively conservative early assessments, improved recovery technologies, enhanced recovery, or denser development drilling. This form of appreciation constitutes *revisions* of the reserves. Revisions may move the reserves downwards instead of upwards, of course; not all early estimates are excessively conservative. Alternatively, a field may acquire *extensions*, by having its boundaries extended by step-out wells beyond what were originally thought to be its productive limits, or by the discovery and development of *new pools* above or below the pool contributing the reserves first calculated. Lateral extensions are more likely to be made to fields having highly irregular, combination traps than to fields having simple convex or stratigraphic traps with well defined bottom- or edge-water levels. Deeper pools may reasonably be hoped for if there is known to be unexplored section below the drilled limits of the field; but they should not be expected if drilling has established *economic basement* for the field and no reservoir rocks have been found between it and the deepest established production.

Possible additions to supply, through further discoveries in areas or formations already established to be productive within the basin, are harder to assess and tabulations of them should be viewed with great caution. Some idea of their potential may be gained by extrapolating historical discovery rates in the single basin, or in clearly comparable basins elsewhere, but the exercise is largely nugatory for a simple operational reason. "Possible" reserves can only exist within regions already productive; areas within those regions having few exploratory wells, especially those having few wells penetrating the entire potentially productive section, will inevitably be drilled as soon as rigs, crews, and money are available. Operators need not wait for assessments of possible reserves. No *regional risk* is involved – the risk that the entire basin may be unproductive, as is always the dreaded possibility when a wholly new basin is being considered for exploration.

24.2.3 *Estimation of unconventional hydrocarbon resources*

Volumetric calculations of *reserves* are readily extended to unconventional *resources* of oil and gas provided that those resources consist of fluid hydrocarbons held in the pore spaces of their host rocks: heavy, viscous oils like those in the Orinoco region of Venezuela, natural gas in tight sandstones, in thermal brines in geopressured zones, or in methane hydrate deposits. One vital caution must always be emphasized when such resources are expressed in quantitative terms. The calculations include a *recovery factor*, and even the assumption of a very low factor (5–10 percent, for example) must be made under the realization that no recovery method for such resources has yet been developed within present economic and technological limits. This is why these volumes represent resources *and not reserves*.

An extreme example is the quantification of biogenic methane "reserves" in ocean floor sediments, trapped below the zone of methane hydrate formation (ZHF). In the belief that the ZHF will occupy the topmost sediments on the continental shelves in polar regions, and over the greater part of the deep ocean floor even in low latitudes, Russian geoscientists (notably Y. F. Makogon and A. A. Trofimuk) have proposed that gigantic quantities of biogenic, nonassociated methane gas will lie in the weakly cemented sediments below it. Taking the average quantities of organic matter undergoing biochemical transformation at the different oceanic depths, a rate of biogenic methane generation can be derived. This rate is then extrapolated over the time spans represented by the ocean floor sediments; a proportion is assumed for the gas lost into the ocean or irrecoverably dispersed, and a remainder value is assumed to represent the amount of gas successfully trapped in the buried sediments under a seal of hydrate. Volumes improperly referred to as "possible reserves" include about 3.5×10^{14} m^3 on the shelves alone, 10×10^{16} m^3 on the continental slopes, and 100×10^{16} m^3 on the ocean floors. If we remember that the proven plus probable *plus potential* "reserves" of conventional natural gas in the world were estimated in 1979 (by A. A. Meyerhoff) to be about 2×10^{14} m^3, the staggering size of the Russian estimates of offshore gas "reserves" may be appreciated.

A number of unconventional hydrocarbon resources, of course, are not contained in the ordinary pore spaces of their host rocks, but are physical parts of them. Natural gas contained in coal seams or in black shales is in this category (see Sec. 10.8). Some acceptable lower limit of organic carbon content must be decided upon in the case of the shales (it could scarcely be set lower than 10 percent), and some lower limit of bed or seam thickness for both sources, but otherwise the volumetric calculation is straightforward because the areas of development of the two host rocks are well understood in many countries. Oil-shale resources pose no additional problems if their recovery is to be by an *in situ* method, but if the shale is to be mined a maximum minable depth, a minimum minable bed thickness, and a proportion of the mined area to become room and not pillar, must all be established.

The most deceptive of all unconventional hydrocarbon resources for the "reserves" estimator are the Athabasca tar sands in western Canada. With hundreds of drillholes through them and reliable measurements of both the thicknesses of saturated sand and their average bitumen content, the total volume of *oil in place* is easily established. A figure of 100×10^9 m^3 is a widely accepted minimum, but it is essentially meaningless for practical purposes. The only portion of it that can be ascribed to *reserves* is that accessible to surface mining techniques (see Sec. 10.8.4); for these, a maximum acceptable thickness of overburden must be determined and a maximum depth to which surface mining can be carried (not more than, say, 75 m). The volume of oil recoverable from this accessible strip of the sands must then be further amended to take account of the upgrading processes that are necessary before the product is usable. Below and downdip from the minable strip is the greater part of the saturated sand formation, too deeply buried to be reached by mining. Much of this buried portion, however, is *not deep enough* to provide the lithostatic pressure to sustain an *in situ* recovery process. Its oil content has therefore to be disregarded in any volumetric computation even of the *resource* of the tar sands. The portions still more deeply buried would presumably be accessible to an *in situ* recovery process if a suitable one is ever developed, and they are therefore technically includable in our oil *resources*; but any such process effective at the Athabasca deposit would be so much more effective at the Orinoco, Cold Lake, and Peace River deposits that its inclusion as a resource is unlikely to have any significance for people now living.

A final warning must be heeded in connection with quantitative assessments of unconventional oil and gas resources. Measured or established *reserves* are the most important factor in the determination of production rates (Sec. 23.9). They therefore control consumption rates and hence the economics of the resource industries. Quantified, statistical statements of "reserves" of unconventional oil or gas, or of conventional but "frontier" reserves that are very difficult of access, cannot be used for this essential purpose. Deposits of unconventional oil or gas, or of conventional oil or gas in, say, the Canadian Arctic islands, cannot be produced at rates comparable with those for accessible, conventional deposits. The *recovery* system cannot usefully operate faster than the *delivery* system. The fact that there is more oil in the Athabasca deposit than in the Ghawar or Bolivar Coastal fields will not enable the tar sands to yield production at the same rate as those fields achieved and maintained for years. One tar-sand or oil-shale plant can produce only so much synthetic crude oil daily no matter how large the "reserves" its operator owns, and new plants cannot be added as fast as new oilfields can be developed if they are in accessible areas. If the Canadian Arctic islands or the Ross Sea shelf in Antarctica contained proven natural gas reserves of, say, 10^{12} m^3 (about the original size of the largest gasfield in North America), a 20 or 25 year productive life could conceivably be planned for the region because a 1200 or 1400 mm submarine pipeline or a fleet of ice-breaking LNG tankers might be able to deliver 5×10^{10} m^3 of gas annually to the markets for it. But if the same area contained proven reserves of 10×10^{12} m^3 (about the 1980 gas reserves of the whole of North America), it could not yield a production rate ten times that from the reserve of 1×10^{12} m^3. Ten "big-inch" pipelines along iceberg-scoured sea floors, or a thousand ice-breaking tankers transiting the North-West Passage, are not going to be made available during any 20 or 25 year period.

24.3 Assessments of undiscovered resource volumes

The author uses the word *assessments*, rather than *estimations*, advisedly. The estimation of recoverable reserves from an oil- or gasfield in the process of development encounters, as we have seen, certain restraints and possible pitfalls. Resource geologists, however, are confronted by the need to venture much farther, to attempt reliable forecasts – acceptable for economic planning purposes on national and international scales – of the *ultimate* recoveries of oil or gas (or other resources, of course) from entire basins, regions, or countries. These forecasts involve consideration not only of discovered deposits but of undiscovered deposits that may yet be found within productive regions, and, still more hazardous, consideration of potential contributions from basins and regions not yet known to be commercially petroliferous and possibly not yet explored at all. "Estimation" seems to me to put too much stress on the mathematical aspects of the process, and not enough on the influence of unquantified judgement.

Obviously there is a basic methodology useful for these pursuits in regions already productive but having much less obvious application to totally unproductive regions. The methodology involves the extrapolation into the future of historical experience in the region. The so-called *rate methods* of assessment require a substantial sequence of historical data on exploration and production patterns before scientifically sound forward extrapolation can be made. Where such data are not available, the resource industries have developed a variety of *volumetric methods* of assessment, assessments of the average expectable richness of estimated volumes of unexplored rock by analogy with explored regions elsewhere. Misunderstanding and misuse of both rate and volumetric methods have led to industry and even international difficulties and embarrassments in the past. It is part of the geologists' role to avoid exacerbating such difficulties and embarrassments in the future.

24.3.1 Rate methods of assessment

Rate methods fundamentally assume that production levels of any material are in large part a function of human effort or performance. The more persistent and imaginative the exploration, the greater the likelihood of achieving prolific production. Once production has been achieved, a variety of statistical data will become increasingly available: production rates; changes in proven reserves; volume of production, or number of discoveries, per unit area of exploration or per unit depth of drilling; and so on. These data can then be extrapolated into the future as evidence for predicting the ultimate number of discoveries, the number of wells necessary to make them, and the volume of their ultimate output. Of the numerous variants on the rate method, four widely quoted representatives are discussed here.

A. D. Zapp and T. A. Hendricks developed the early US Geological Survey variant, in which oil production was considered to be a direct function of the density of drilling. As exploration, other things being equal, normally proceeds from the most favorable areas to the least favorable, the volume of oil discovered per unit of drilled depth (though subject of course to considerable short-term fluctuation) necessarily declines with time (and so with cumulative drilled depth). Over a long period, measured in tens of years, the decline appears to be exponential. Zapp and his co-workers denied the validity of this conclusion so long as the areal density of drilling was highly unequal. It stands to reason that a region in which a great deal of oil has been discovered – Texas, say, or Alberta or Sumatra – will receive much more attention from the drillers than a region in which little or no oil has been discovered – like Oregon, or Quebec or Bangladesh. Zapp estimated that, for the density of drilling already achieved (by 1960) in the most productive areas of the USA to be extended over all "prospective" sedimentary areas (meaning all sedimentary areas), several times as many further wells would have to be drilled as had been drilled already, and these further wells could be assumed to result in a discovery success pattern comparable with that achieved by past drilling. The ultimate output forecast on this assumption, not surprisingly, was several times greater than that forecast by any other method (Fig.24.2).

The negative exponential as a projection into the future has been attacked on theoretical grounds as well as on grounds of mere faith. As each increment of exploration proceeds, some addition to the ultimate cumulative reserve (Q) must be made, and estimates of Q therefore keep on getting larger (as all resource experience confirms); the curve of Q against time (t) maintains a positive slope, albeit a declining one. If h represents the unit of exploratory effort (meter of drilling or cubic kilometer of explored area), the relation of dQ/dh to h overstates depletion so long as drilling depths and densities increase in successful areas whilst less successful areas remain in much less advanced stages of exploration.

There is no doubt that many areas of the world have been prematurely downgraded in the minds of exploration scientists and operators; even in areas long regarded as mature, some discoveries continue to be made. But no experience supports the contention that density of drilling is a prime determinant of ultimate production. A belief that the Gulf of Mexico is more productive of oil and gas than Hudson Bay because there

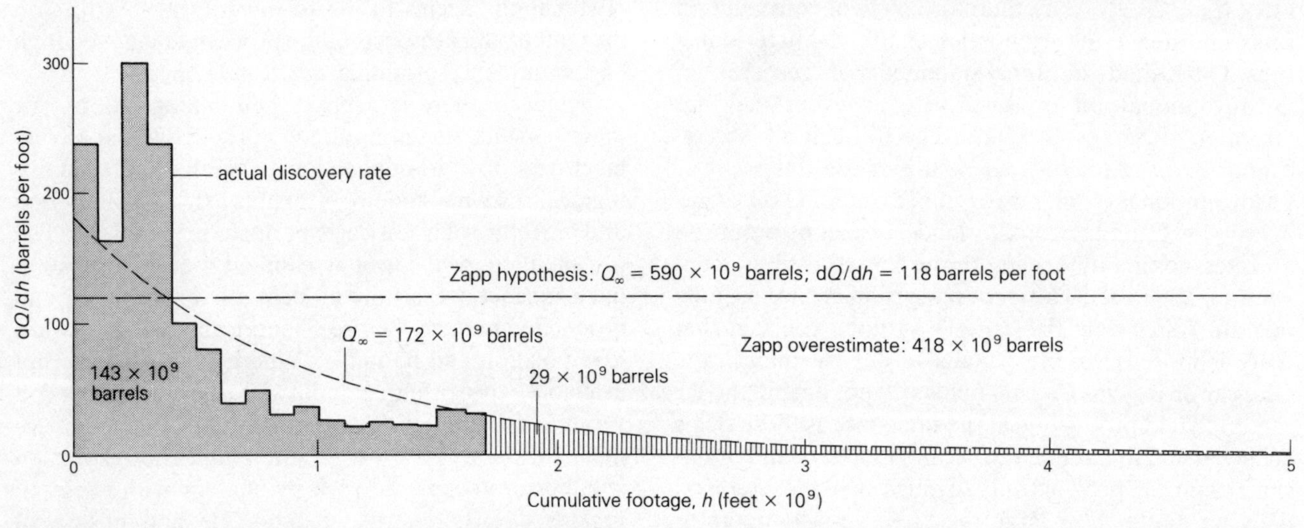

Figure 24.2 Comparison between A. D. Zapp's forecast of ultimate United States oil production, based upon discoveries per foot of drilling up to 1960 and extrapolation of that rate to a uniform future drilling density, and M. K. Hubbert's forecast derived from the area below his bell-shaped curve as the negative exponential approaches zero discoveries. (From M. K. Hubbert, *US energy resources, a review as of 1972.* Part 1. Washington, DC: US Government Printing Office, 1974.)

are more wells in the gulf than in the bay may be suspected of putting the cart before the horse. Despite this, Bernardo Grossling as late as 1976 plotted drilling densities against world areas presumed to possess oil potential and concluded that "potential reserves" were eight to ten times the quantities already produced (Fig. 24.3). The crushing weakness of such procedures is their dependence upon North American (especially United States) experience. By all rational calculations the onshore productive areas of North America are wildly overdrilled. Beyond some obviously desirable and necessary number of exploratory and development wells, their absolute numbers – the drilling densities, *in sensu stricto* – are not what matter; 100 or more Texas or Kansas or Ohio wells are drilled where two or three would yield essentially the same ultimate result. The nonproductive onshore areas of the USA and Canada are not going to become significant producers even if they are drilled as densely as Oklahoma.

Charles Moore has been a leading exponent of the use of the *Gompertz curve* to extrapolate historical production data to the ultimate life of North American oilfields. Over the lives of a large number of fields, a variety of cumulative quantities may be studied, such as numbers of discoveries per 10^3 m of drilling or cumulative additions to reserves. We will call the cumulative quantity chosen Y, and $\log Y = x$. Values of both Y and x are established at times $t_1, t_2, \ldots, t_n$ (present). Then

$$x = \frac{dx}{dt} = \log\left(\frac{dY/dt}{Y}\right)$$

Values of both Y and dY/dt obtained at any given time tend to be revised later on. This makes the concept of "measurement" of cumulative quantities complex, and leads to the possibility that one might learn from it what one wants to learn.

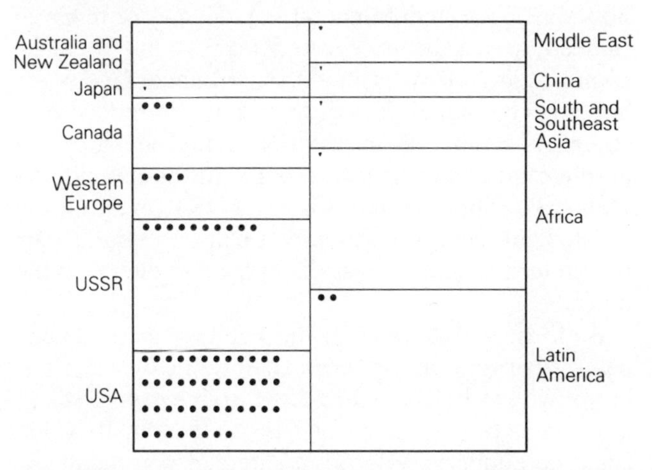

Figure 24.3 B. F. Grossling's chart showing concentrations of wells drilled in various regions of the world compared to the areas of the regions considered favorable for petroleum occurrence. Each full circle represents 50 000 wells (data from the US Geological Survey, 1976). (From H. A. Meyerhoff, *J. Petrolm Geol.*, 1979.)

However, if one regards the values of x and $\dot{x}$ as independently obtained at a number of times; they can be plotted as a regression by the method of least squares (Fig. 24.4). The equation of the line is

$$\dot{x} = a - bx$$

Extrapolated to the intercept at $\dot{x} = 0$ (i.e. as t becomes larger and larger), then $x \rightarrow a/b$ and the values of a and b are determined. A function x of t that satisfies this equation has the form

$$x = \alpha\, e^{-bt} + a/b$$

The value of α is determined for each value of t_n by making it agree with the estimate of x (= $\log Y$) at time t_{n-1} (e.g. last year).

A family of *Gompertz curves* is then constructed, plotting cumulative values of Y against t. The time t_i at which each curve passes through its inflection point is determined, and the curves are then converted to their first derivatives in which the inflection points become maxima of Y (e.g. of cumulative additions to reserves).

The critical steps in the use of the Gompertz curves are those determining the values of a, b, and α. Because the curves plotted represent *cumulative* quantities, they have no maxima and may be extended with positive slope into the indefinite future. The manner of extrapolation (itself subjective) controls the time of passage through the inflection points. Furthermore, experience with the appreciation of "old" reserves encourages the belief that all recent discoveries (those contributing the historically latest additions to the values of Y) are grossly undervalued, in spite of their appraisal by modern methods which make allowance for the experience of "appreciation." Thus the ultimate values of both Y and t tend automatically to become overestimated.

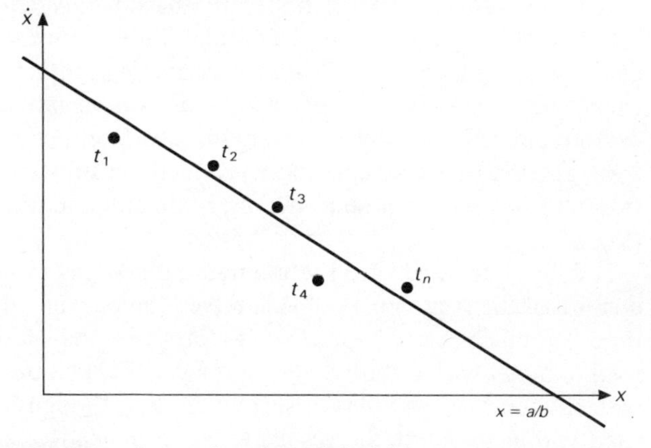

Figure 24.4 Regression of x and dx/dt ($\dot{x}$) at a succession of past times, to the present (t_n). If x is a logarithmic function of some aspect of reserve augmentation, this regression becomes the basis for use of the Gompertz curve in the extrapolation of reserve additions into the future.

King Hubbert rejected the notion that projections of past data into the future could usefully be based upon cumulative quantities only. The complete cycle of production of any nonrenewable resource, in any one region or in the entire world, must begin with production rising from zero towards a maximum; there may be several maxima separated by temporary declines; but eventually the cycle is completed by a long-continued decline back to zero via the negative exponential. The bell-shaped curve, smoothed to eliminate minor fluctuations along it, came to be referred to as "Hubbert's pimple." The area between it and the time axis is a measure of the cumulative production (Q) up to the time chosen:

$$Q_{\infty} = \int_{0}^{\infty} P.\mathrm{d}t$$

where P is the production rate.

The curve of *cumulative production* (Q_P) essentially parallels the curve of *cumulative discoveries* (or cumulative additions to reserves, Q_D) and follows it after the time lag, Δt, which is required to bring the new discoveries into the full production stage (Fig.24.5a). The curves for the *rates* of addition to reserves ($\mathrm{d}Q_D/\mathrm{d}t$) and of production ($\mathrm{d}Q_P/\mathrm{d}t$) are the first derivatives of the cumulative curves, the inflection points of which become maxima in the derivative curves (Fig.24.5b). Once the rate of addition to reserves (the rate of new discoveries) goes into decline the rate of production must follow it into decline Δt years later *unless a new surge of discovery is achieved*. At a critical time t, which can be envisaged as the half-life of the production cycle, the still-rising production rate becomes equal to the now-declining rate of addition to reserves; the first-derivative curves cross, the rate of increase of reserves becomes negative, and the remaining reserves available for future production begin to decline.

It follows that, once the first curve becomes established on the negative slope, the onset of decline for the other curves is forecastable with considerable accuracy. Each curve can then be extrapolated as a negative exponential to its future convergence with effective zero, and the expectable ultimate production estimated from the area enclosed by the complete curve (Fig.24.5b).

Hubbert's forecasts for the ultimate oil and gas production of the continental USA survived the passage of time so much better than other forecasts that his methodology was accorded almost magical capabilities by some. In fact, its utility is strictly limited, though if used within its limits the results to which it leads are crucially important. The method obviously makes no pretence of application to unexplored or little-explored areas which provide no historical data to extrapolate. It cannot be used with any reliability until the *rate of additions to reserves* ($\mathrm{d}Q_D/\mathrm{d}t$) has clearly entered the negative slope. This entry is itself critically influenced by human factors – political, economic, technological, nowadays even environmental – and it may be reversed (indeed, it has been reversed many times in many countries and many basins) if these human factors are altered. The human factors themselves introduce constraints that are not necessarily related at all to the absolute quantities of the resource in the ground. As Hubbert's critics, notably Vincent McKelvey, tirelessly pointed out, historical activities are not necessarily part of an inexorable process having only one possible outcome.

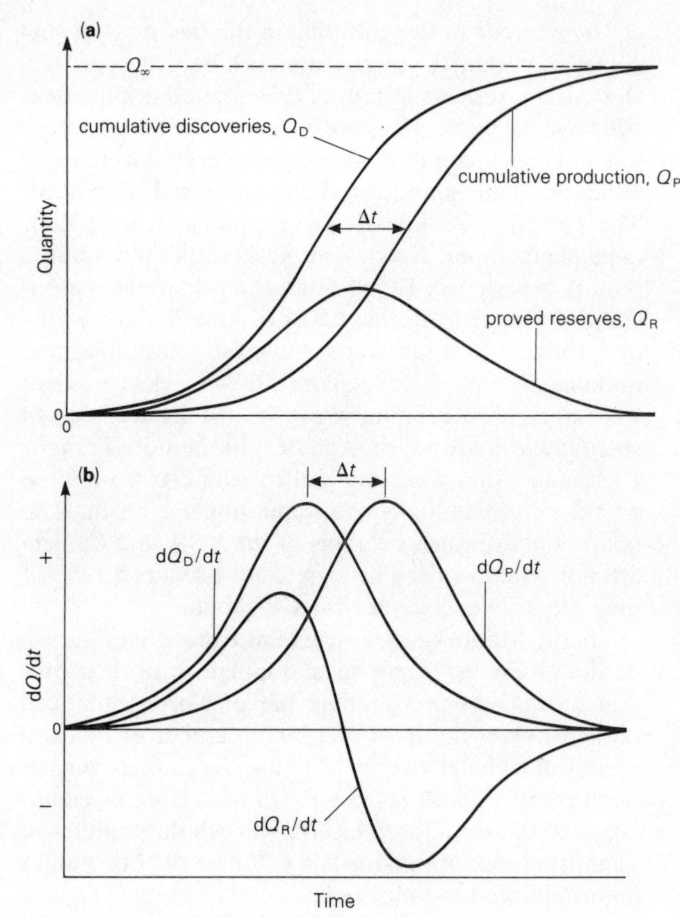

Figure 24.5 Variation with time of cumulative discoveries and production, and of proved reserves (a), and variation in rates of discovery, production, and increase in proved reserves (b) during a complete cycle of petroleum production. (From M. K. Hubbert, Natn Acad. Sci. –Natn Research Council Pub. 1000-D, 1962.)

Both users and critics of the Hubbert methodology habitually overlooked one vital point. The time lag, Δt, has proved to be about 10 years through experience in most areas of mature oil or gas production. Its value could no doubt be reduced in emergency economic or industrial conditions, but it must inevitably remain a number of years and not a small number of months – especially in "frontier" regions (see Sec. 23.10). Once the curve representing rate of production has been on the negative slope for longer than the time lag, it has become essentially linear as it trails off gradually but inexorably towards a far-future zero. It is not sufficient for further discoveries to interrupt this near-linear decline; the discoveries must be sufficiently large and sufficiently numerous *to convert the curve back to lognormality* and with dimensions comparable with those making possible the earlier production maximum.

A simple illustration may be helpful. The Prudhoe Bay field in Alaska, discovered in 1968, is the largest oilfield (and the second-largest gasfield) ever found in North America (Sec. 26.2.2). Its maximum production rate – almost 1×10^8 m^3 yr^{-1} in 1980 – was far higher than that achieved by any earlier North American field. As the achievement of this peak production involved a time lag (Δt) of about 10 years, however, the near-linear decline of the thousands of other producing fields during that lag had reduced the output of the rest of North America by more than Prudhoe Bay's contribution, which therefore created merely a short-lived bump on the negative exponential. To convert this negative slope to the positive slope of a re-created lognormal curve would require the discovery of four or five Prudhoe Bays within the span of a few months, all within the oil-productive regions of North America so that development rigs and crews, pipelines, and other facilities were instantly available and the Δt values for the new supergiant fields were reduced to very short delays. Such a surge of new discoveries, larger than any found in the long earlier history of production, actually occurred in Mexico with the opening of the Reforma and Campeche regions in the southeastern corner of the country. It is conceivable that it could happen in and offshore from, for example, Nigeria, where relatively few wells had by 1980 penetrated deeper than the productive Tertiary succession. But it is utterly inconceivable that it can happen in the long-mature continental USA or the western Canadian, eastern Venezuelan, or Ural–Volga basins. In those regions, if in few others, use of the Hubbert technique is likely to yield trustworthy assessments.

24.3.2 *Volumetric methods of assessment*

Volumetric assessments of ultimate reserves have a long history of use by mining geologists, especially for ore deposits that are integral parts of their host rocks (such as porphyry coppers). They were first adapted for undiscovered oil "reserves" by Lewis Weeks during World War II, and they have retained their appeal for the oil industry despite repeated demonstrations of the dangers of their misuse. It is not difficult to understand the reasons for this appeal. The general method is an extension of that used to calculate the proven and probable reserves in existing fields, and thus seems to provide a safe route to travel from explored to unexplored ground. The method is easy to use (and to misuse, unfortunately). It is easily modified by the introduction or attachment of variants representing imagined probability factors. It readily yields large numbers which appeal to the oilman's optimistic temperament, and, by establishing targets for both exploration and technology, these large numbers attract exploration money. Finally, for regions with no previous hydrocarbon production, no convincing, alternative, quantitative technique has been devised.

At the time the technique was adapted to assessments of oil "reserves," there were only two groups of sedimentary basins that had been drilled sufficiently extensively to provide useful data on their ultimate yields – basins in the USA and a few basins in other countries (Mexico, Venezuela, Trinidad, Iran–Iraq, Romania) which were all geologically young and all abundantly supplied with surface manifestations of oil.

Weeks first tried to assess unexplored basins by simple multiplication of their areas, the percentages of these areas anticipated (by analogy with other, drilled basins) to be oil-bearing, and the estimated yield per unit area (again by analogy with similar, drilled basins). Weeks quickly switched the basis of his assessments to the *volumes* of sedimentary rock in the basins, rather than their *areas*. The anticipated ultimate oil recoveries from mature producing basins in the USA were compared with the basins' sedimentary volumes to a maximum drillable depth of 20 000 feet (6096 m), or to the average estimated thicknesses if basinal depths were less than this. The ultimate oil yields were then expressed in barrels per cubic mile (cubem) of sedimentary rock. Weeks experimented with more refined parameters – yields per unit volume of effective reservoir facies, or calculations based on facies ratios. These, however, cannot be known until considerable subsurface exploration has been carried out; what was sought was a yardstick for use in an undrilled or little-drilled basin for which reconnaissance geophysical data (possibly through airborne surveys) provided estimates of basement depths across it.

Weeks of course recognized that the productivity of any basin depends upon a complex of geological factors – the class of basin, its structural history, its degree of mobility, the nature of its bottom environment during deposition, the development of porosity, and the portion or elements of the original basin still represented today. The reader will notice the omission that was to prove the critical weakness of the Weeks methodology. He formulated it before organic geochemistry had demonstrated that the most vital of all requirements is that of a rich, mature source sediment. Weeks did not believe in long migration for oil; where it occurs in quantity, therefore, it must have been derived from a shale formation very close, geographically and stratigraphically, to the reservoir rock. All sedimentary basins contain such shales, and therefore most sedimentary basins will be petroliferous; their richnesses will differ, but not their fundamental prospectiveness.

Weeks therefore concentrated his attention upon *basinal type* (see Ch. 17). To represent his connected interior basins, most of which were then thought to be of Paleozoic age, he chose the Michigan and Illinois basins; the open marginal type of basin was exemplified by the Texas Gulf Coast; foreland basins by those of Oklahoma; young, intermontane, mobile basins by those of California; and larger, compound regions by Texas, the USA, and European Russia *as wholes*. Weeks' estimates of "potential undiscovered reserves" of oil (not published by him, but by others) were as below, in barrels per cubem:

Michigan Basin	6 000
Illinois Basin	35 000
Texas Gulf Coast	60 000
Oklahoma	70 000
California	180 000
Texas as a whole	38 000
US basins as a whole	30 000
European Russia as a whole	50 000

The larger concentrations occur in the mobile basin types; the Californian basins, for example, reach 200 000 barrels per cubem according to Weeks (in fact, they reach several times this value). In metric terms, therefore, the range of recoverable oil contents was thought to be about 250–7500 $m^3\ km^{-3}$.

Weeks may never have intended that his method and his specimen figures should be applied by others to totally unexplored basins outside the USA. Nonetheless, they were so applied in Australia, Brazil, Canada, India, the Soviet Union, and many smaller countries, with some disastrous results. The reason for the disastrous results was simple; the method was applied without allowance for the probability factor, or, as oilmen prefer to call it, the factor of *risk*. Not all sedimentary basins are petroliferous. Weeks knew this, but of course he calculated figures only for basins that *are* petroliferous. Sedimentary basins that are not petroliferous are barren because of the absence of one or more of the critical geologic requirements, almost invariably because of the absence of a rich, mature, *source sediment*. Once the geologists and the drillers had attacked dozens of basins that bore no surface evidence of containing oil, it became apparent that only a minority of basins yield evidence that they have ever been significantly petroliferous – about one basin in three does so. Only about one explored basin out of six or seven contains even one very large oil- or gasfield. If, therefore, values such as 50 000 barrels per cubem (or even per cubic kilometer) are applied to all the different sedimentary regions in a large territory – as they were in Australia, Brazil, Canada, and India, for examples – and the resulting volumes of "potential reserves" for all the individual basins are summed arithmetically to yield the "potential reserves" of the whole territory, the outcome is *a priori* likely to overstate reality by a factor of three. If most of the basins in the territory are geographically remote or environmentally hostile, so that only very large accumulations could be logistically or economically viable, then the method is *a priori* likely to overstate reality by a factor of six. This is precisely what happened in three of the countries named here. The large basins, contributing the greater part of the sedimentary volume, were for the most part unproductive; most of the rich production came from a handful of small, coastal grabens contributing relatively minor sedimentary volumes.

Statistical averages have little meaning for any one basin. The great majority of the sedimentary regions of the world contain nothing like 30 000 or 50 000 barrels per cubem (order of 1500–2000 $m^3\ km^{-3}$); most of the world's oil comes from a relatively small number of basins (a few dozens, out of several hundreds) each from five to fifty or more times richer than this (see Secs 3.8 & 17.6). An average of, say, 30 000 barrels per cubem has a meaning comparable with an average of 10 for a group containing nine thousands and 990 ones.

A more reliable volumetric application of geologic analogy is the *average structure* method. This considers identified but undrilled structures as prospects which might be expected to enjoy a success ratio, and field sizes, comparable with those found in similar but drilled structures elsewhere (preferably elsewhere in the same general region). The method has proved very useful in some offshore areas (in the Gulfs of Mexico and Campeche, or the North Sea), and in the remote parts of large concessions (like the Empty Quarter of Saudi Arabia), where seismic delineation of structures, or their airborne reconnaissance, runs far ahead of the opportunity for drilling.

An amalgam between the "average structure" method and the bald volumetric estimation – a double analogy, in a sense – has been employed by a number of geologists, notably by Lytton Ivanhoe. Petroliferous basins normally contain a number of fields the size distribution of which, plotted on log–log paper, is essentially serial-linear for any one basin (Fig.24.6). This simply reflects the lognormal size distribution of all natural resource deposits (see Sec. 3.8). For any unexplored basin, an assumption is made about the size of the *largest field expected* – really an application of the "average structure" method. The number and size of the exploitable fields anticipated in the basin are then guesstimated by drawing a log–log line parallel to that for a key analogous basin, beginning at this largest expected size and

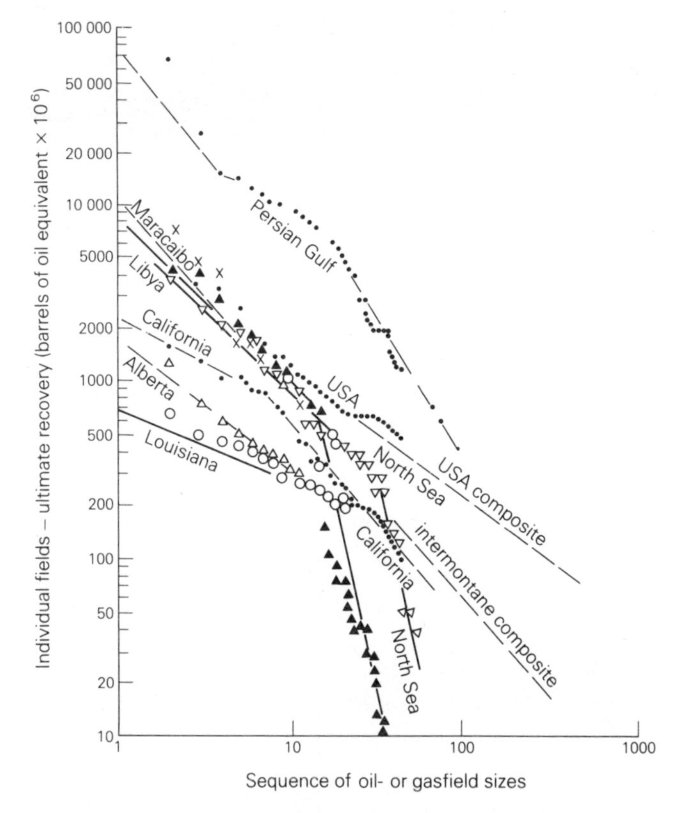

Figure 24.6 Log–log depiction of sizes of oil and/or gas fields in the world's principal producing regions, used as a means of forecasting ultimate recovery from such regions and hence from less-explored regions regarded as geologically comparable. (From L. F. Ivanhoe, *Oil Gas J.*, 6 December 1976.)

cutting off at the lower limit at the *smallest field size acceptable* under the local political, geographic, and economic restraints. In Arctic Canada's Sverdrup Basin, for example, the lower cut-off might be at 40×10^6 or 50×10^6 m^3. The method cannot escape the drawback imposed by the need to assume the size of the largest field. The larger this is assumed to be, the greater become the odds against its existence; as we have observed, only one basin in six or seven contains even one field of 100×10^6 m^3 recoverable reserves. On the opposite hand, the more conservative the appraisal of the largest anticipated field, the less likely exploration of the new basin will be.

An important variation on the volumetric assessment method utilizes the *geochemical material balance* – an item of information not available to the early practitioners of the method in its simpler form. The least elaborate version of the refined method, applicable only to oil, was described by A. N. McDowell (Fig.24.7). It requires some minimum amount of drilling, which should include holes through the entire sedimentary section regarded as potentially petroliferous. Cores from widely spaced wells are first combined with the basinal area to calculate the volume of *mature source sediment* in the basin. Geochemical study of this sediment, as described in Chapter 7, reveals its organic carbon content and the level of its maturity. These factors lead readily to an estimate of the amount of oil (Q) theoretically generated in the basin. Suitable conversion factors may be calculated from these round figure equivalences:

$$1000 \text{ p.p.m. hydrocarbons } \textit{by weight} \triangleq 3 \times 10^6 \text{ m}^3 \text{ km}^{-3}$$
$$\triangleq 82 \times 10^6 \text{ barrels per cubem}$$
$$10^6 \text{ barrels per cubem} = 38.14 \text{ p.p.m. } \textit{by volume}.$$

Of the amount Q, a smaller quantity Q_E is presumed to have been expelled from the source sediment and made available for migration and pooling. Of Q_E, some portion Q_L is inevitably lost, and some fails to find a reservoir or a trap and remains dispersed in unrecoverable form (Q_D). Only some small part (Q_T) of Q finds both reservoirs and traps. Q_T is probably less than 50 percent of Q_E and almost certainly less than 25 percent of Q. The recoverable oil (Q_R) is then no more than 30–40 percent of Q_T.

There are some obvious cautions in the use of this method, as in the use of any method, but it at least begins with quantitative, experimental data on the basin's capacity to be petroliferous. The principal cautions concern the timing of any oil migration in relation to trap formation and the length of the migration path (the larger the basin, the larger Q_L and Q_D are likely to become in relation to Q_T). Other cautions include the potential for seriously exaggerated estimates of the percentages of Q represented by Q_E and Q_T, and the inapplicability of the method in the event that the generated and pooled hydrocarbons are principally in the form of gas or gas-condensate (which have a much wider range of origin and are unpredictably mobile within the basin).

24.3.3 *The factor of risk*

It has been emphasized that the critical drawback to the volumetric assessment method in its basic form is its disregard of the factor of *risk*. Later refinement of the method tried to obviate this drawback by finding ways to introduce the risk factor, and undiscovered volumes of oil or gas were expressed as ranges of values, not as simple quantities, according to the risk assigned. The widely quoted technique used by the Resource Appraisal Group of the US Geological Survey is chosen for discussion here.

The basis of the method is the statistical treatment of a compilation of all available geologic information (including production and existing reserve data, if any), the unit of treatment being a *geologic province* (see Sec. 17.1). Underlying the procedure is the assumption, in advance, that commercial oil or gas will be found in the province being considered. An initial range of estimates

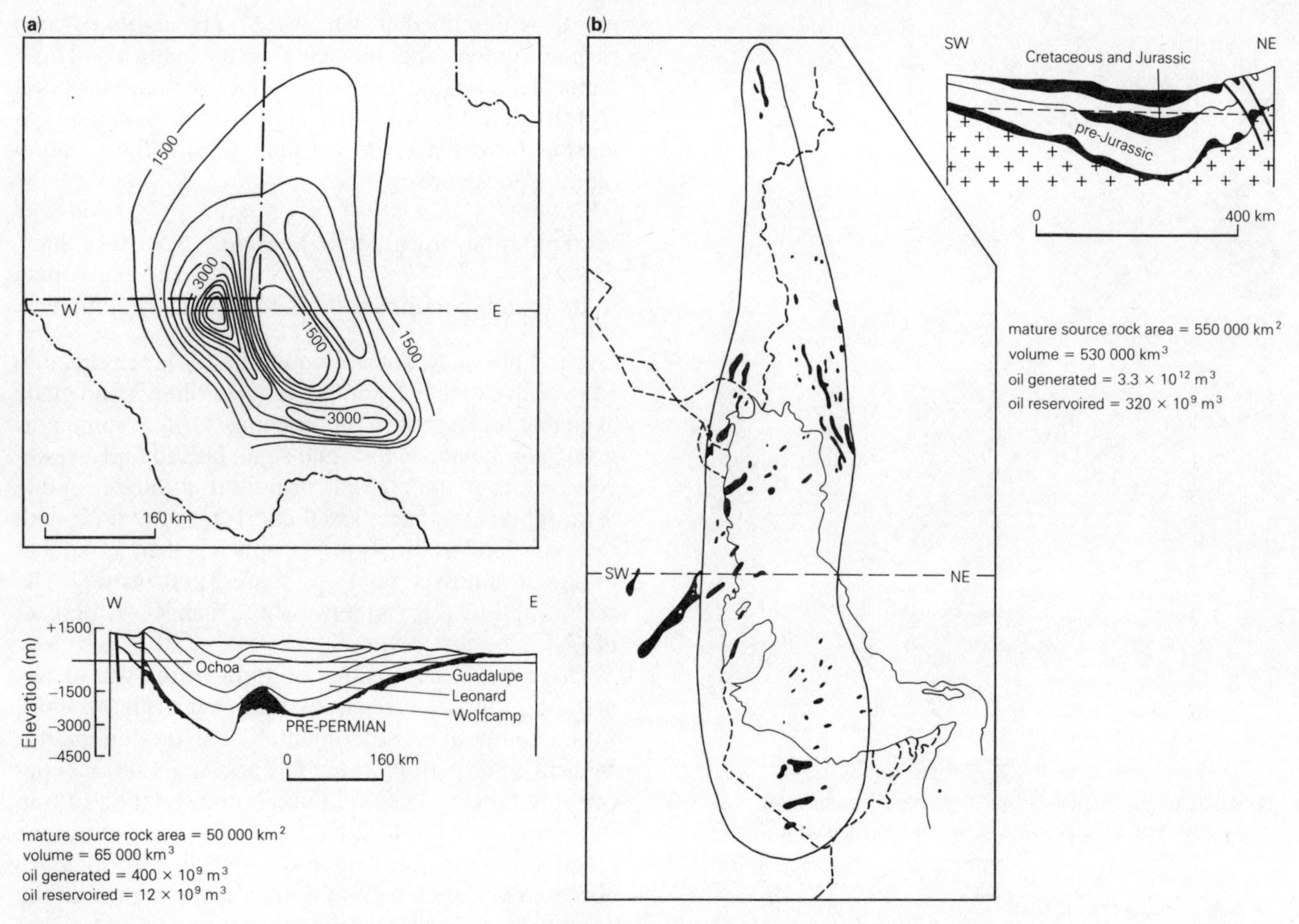

Figure 24.7 Oil reservoired (original oil in place) compared with oil generated as estimated by the geochemical "material balance" method, for two of the world's most prolific basins: (a) Permian Basin; (b) Persian Gulf Basin. (From A. N. McDowell, *Oil Gas J.*, 9 June 1975.)

is prepared for each province by a team of geologists of long experience in it. Comparison between these estimates leads to an initial appraisal of the ultimate resource potential by the assignment of essentially subjective probabilities to values within the range. This collegial or team approach, called the Delphi technique, seeks the average of the individual opinion curves, each extending from unity chance at zero yield to zero chance at the estimated maximum yield. In practice, the highest anticipated yield is assigned a 5 percent probability, the lowest anticipated yield a 95 percent probability, that *at least* that amount of oil or gas will be present. The *modal estimate* represents the highest probability of an agreed-upon amount being present. The high, low, modal, and statistical mean estimates (the last being the sum of the other three divided by three) are analyzed by Monte Carlo techniques, in which probability distributions are aggregated and displayed graphically as lognormal probability distribution curves (Fig. 24.8). These curves have single modes and positive skewness (that is, the mode is closer to the 95th than to the 5th percentile).

Steep curves with narrow ranges indicate a high degree of assurance in the assessments; wide-ranging curves indicate great disagreement or uncertainty. "Unrisked" curves must finally be "risked," or discounted by some agreed proportion reflecting geologic assessments of source rocks being present, their levels of maturity, the timing of available traps, and so on. In the event that the sum of the evidence is unfavorable to *any* commercial discovery, for example, the risked curve must not begin at the 100 percent probability level, but at some lower level representing the best assessment of the *marginal probability* of a discovery nevertheless being made.

In fact, no satisfactory method has been devised for incorporating realistic risk factors for basins so far unproductive. The problem of defining a *nonprospective region* is obviously intractable; no such definition will ever be acceptable to the explorers for oil or gas.

Despite all these cautions – but perhaps *because* of the last one – one inescapable fact remains as sustenance to the optimists. Authoritative estimates of the world's *ultimate* reserves of conventional oil and gas, published *after* World War II, include some that are lower than the present *proven* reserves. The oil industry has found more new oil and gas than its own forecasters

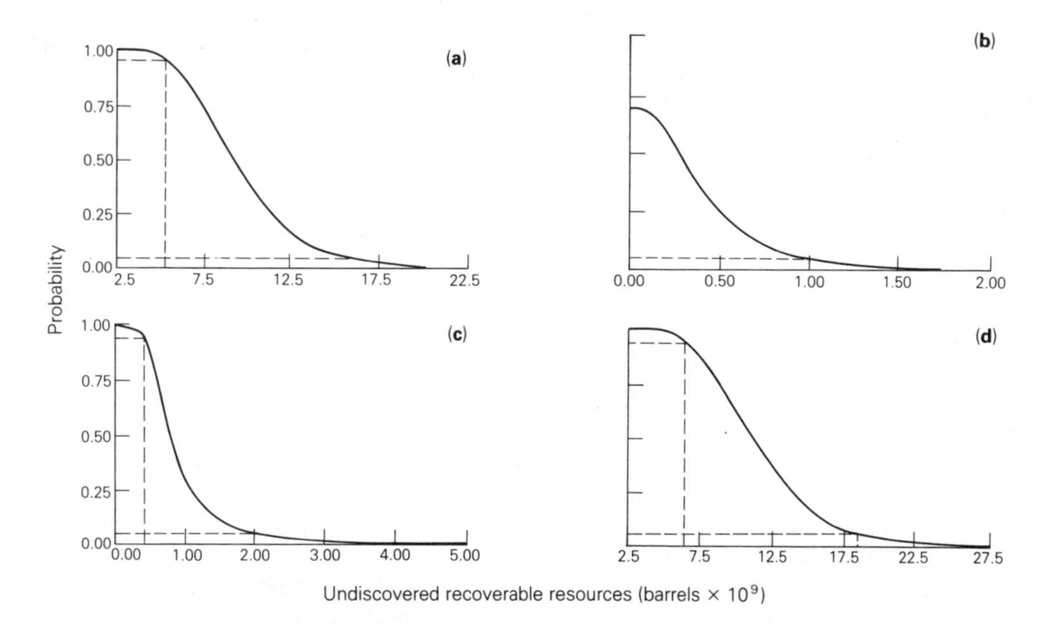

Figure 24.8 Probability distributions by Monte Carlo analysis on undiscovered recoverable resources of oil for three regions of Alaska and for all onshore Alaska: (a) northern Alaska; (b) central Alaska; (c) southern Alaska; (d) total Alaskan onshore oil. Dashed lines represent the resources anticipated at the 95th and 5th percentiles. The curve in (b) involves the assessment of a *marginal probability* of only 70 percent; at the 95th percentile no commercial discovery is anticipated. (From Betty M. Miller *et al.*, US Geological Survey Circular 725, 1975.)

had thought existed one generation earlier. There is no mystery about this; it applies to virtually all mineral and fuel resources. If we recall Deverle Harris's dictum on the nature of resources, the post-World War II period has witnessed the interaction of growing economies and burgeoning technologies with a global physical endowment the size of which could not be guessed at until this interaction took place. Much more oil and gas have been found because enormously improved exploration techniques and drilling capabilities were brought to bear on more and more places, *instead of only on the USA*.

Recognizing this manifest truth, a group at Cities Service Corporation in the late 1970s combined aspects of both the rate and the volumetric approaches with a Monte Carlo analysis to predict that the amount of oil and gas still to be discovered is at least as great as that discovered in the past, and probably considerably greater. Drillers have put down about five times as many holes in North America as in all the rest of the world combined; they were still doing so at the beginning of the 1980s. They have found five times as many oil- and gasfields in North America as in the rest of the world. One striking result has been the revelation of the significance of *nongiant fields*.

On a worldwide basis, about three-quarters of all the proven oil and gas reserves discovered have been in giant fields – those with recoverable reserves not less than 80×10^6 m^3 of oil or 10×10^{10} m^3 of gas (Sec. 3.8). In North America, however, only a little over one-third of the discovered reserves have been in such fields. The Cities Service group considered this anomaly to be a consequence of the vastly greater drilling effort expended in North America than elsewhere, an echo of Zapp and Grossling. Only dense drilling finds the small fields. No region or country can be overdrilled; everywhere except North America was in 1980 grossly underdrilled.

If the rest of the world were to be drilled to the mature level of North America, the North American statistics should (in general terms) become the worldwide statistics. Even if no further giant fields remain to be found (a postulate no petroleum scientist accepts), the reduction of the contribution of the known giants from about 75 percent to about 35 percent would mean raising the contribution of nongiants from about 25 percent (in 1979) to about 65 percent. That in turn would mean that *as yet undiscovered nongiant fields* outside North America would come to represent some 50 percent of the world's known reserves (consumed plus remaining) at "maturity;" the amount representing the other 50 percent is already known. The total of the existing world reserves and these undiscovered nongiant reserves reached nearly 34×10^{10} m^3 in 1979 (more than 2 trillion barrels) – a figure strikingly similar to those estimated years earlier by Weeks, Hendricks, and others. The Cities Service group accordingly took this volume of oil to be the mean on the Monte Carlo curve, the 95th percentile of which is of course at the known reserve (including cumulative production) of about 16×10^{10} m^3. The 5th percentile then falls at 54×10^{10} m^3 (about 3.4 trillion barrels).

Where will these undiscovered, nongiant fields be found? Some 68 percent of all known oil and gas discovered before 1980 is in basins in the interiors of present lithospheric plates. These plate interiors have seen nearly all of the drilling because they are for the most part accessible. In contrast, only about 23 percent of known oil and gas is in basins associated with currently divergent plate margins, nearly all of this percentage being around the Gulf of Mexico and in the North Sea, Nigerian, and Sirte basins. Only 9 percent of discovered oil and gas is in convergent-margin basins (nearly all of it in California and Indonesia). Unlike the plate-interior basins, the two classes of less-productive basins lie largely offshore. It is those basins that suffer the greatest deficiency in exploration, and those basins that therefore offer the greatest promise for discoveries of the nongiant fields that are found only by dense drilling. Once they are drilled to "mature" (i.e. US) density, the thousands of nongiant fields discovered by this drilling will yield the trillion barrels of new reserves.

This is a persuasive argument; in literal, mathematical terms, it may by the mid-21st century have been shown to be correct. Unfortunately, it suffers from the same drawback as all the other arguments over undiscovered "reserve" assessment. It is obsessed with American experience. It is no more necessary that the continental USA be "typical" of oil and gas occurrence patterns than that the Persian Gulf be typical; or that the USA itself be typical of gold occurrence, or southern Africa of either gold or oil occurrence. Despite drilling five to ten times as many wells as the rest of the world, and finding five to ten times as many oil- and gasfields, North America has found only one-twentieth as many supergiant oilfields and few more than one-twentieth as many supergiant gasfields as the rest of the world. For its derisorily lean density of wells, the world outside North America has found more than 40 supergiant oilfields and more than 30 supergiant gasfields; North America has found two of each.

Might it not be as logical to say that North America should have as high a proportion of giant and supergiant fields as other continents as to say that the rest of the world should have as high a proportion of modest to trivial fields as North America? The crescentic strip of the Earth's crust extending from Algeria to the Persian Gulf and around to the Caspian Sea contained 30 supergiant oilfields by 1980 in addition to several scores of merely giant fields. These are the fields primarily responsible for the discrepancy noted by the Cities Service geologists. In the Persian Gulf and Sirte basins, essentially *all* of the exploited reserves are in giant and supergiant fields. It is utterly inconceivable that these two basins can contain twice as much oil in undiscovered nongiants as they are already known to contain in giants; but this is what is implied if the contribution of the giants is to be reduced to about 35 percent. It may be protested that the US proportions do not have to apply specifically to the handful of very prolific basins (about eight of them) in which enough oil was generated and preserved to fill all the large traps, creating numerous supergiant fields. But if these supergiant-prone basins are excluded, the ratios established in North America become typical of those parts of the world that have been at all carefully explored.

It is manifestly true that North America – and the USA in particular – has made a far greater exploratory effort for oil and gas (and everything else, probably) than the rest of the world has made. But an effort like the American effort can be made only under American socioeconomic and sociopolitical conditions (including conditions of land ownership). Thousands of American oil- and gasfields and hundreds of drilling rigs are owned by independent operators who could not possibly find, develop, or operate such small finds in the jungle of the upper Amazon or the deep waters of the Gulf of Alaska. The fact that the smaller-than-desirable fields may exist in such places does not mean that they can ever be included in the *reserves*. Even if they were included, they would have no significance for the *effective* world supply.

It took more than 100 years for North America to reach its uniquely dense drillhole pattern. The "frontier" and deep-offshore regions of the world cannot be drilled to the density of wells considered "mature" in Oklahoma or Alberta, and they cannot yield oil or gas from Montana- or Manitoba-sized fields even if such fields exist (as they no doubt do). The world of concern to the consumers of oil and gas must find ways to discover more large deposits of those materials in the portions of the Earth's crust reasonably accessible to us; to recover more of them from the deposits we have found there and will find there in the future; and to utilize the equivalent materials now locked in the tar sands, oil shales, coal seams, crustal brines, and other accessible but refractory repositories.

24.3.4 *Russian forecasting of potential hydrocarbon volumes*

Russian geoscientists have carried techniques of quantitative forecasting of future oil and gas occurrences to extraordinary lengths. The principal approach has been via various versions of the volumetric estimation technique.

The use of the geochemical material balance (see Sec. 24.3.2) is called by the Russians the *volume-genetic method*. Armed with information on the quantity of disseminated organic matter in the sediments of the basin, and estimations of the generative windows by maturity studies, forecasts of the volumes of hydrocarbons *probably generated* are made with considerable confidence. Doubt arises, as we have seen, when assessments have to be made of the losses during migration, failure to find adequate traps, destruction of accumulations by unlucky geologic histories, and the consequences of other hazards. These are gauged empirically from the *coefficients of accumulation* (really coefficients of *preservation*) of well explored productive basins of different tectonic types. If the volume of oil calculated to have been generated (given the OM content and maturity index derived by the geochemical study) is taken as unity, simple empirical multipliers can be applied to estimate *pooled* contents as proportions of *calculated* contents. These proportions are said to average 0.1–0.15 for dominantly clastic basins but lower (0.05–0.07) for basins dominated by carbonates; coefficients of *gas* accumulation are significantly lower still (0.00*x*–0.015). The age factor is not as a rule taken directly into account, but the coefficients employed not unexpectedly indicate important influences of both basin age and basin size. The coefficient for the large Ural–Volga Basin, which is the Russian prototype of a Paleozoic basin with a substantial carbonate component, is said to be 0.2, but that for the even larger West Siberian Basin, wholly Mesozoic and exclusively clastic, is only 0.02. The magic of California is illustrated by the Los Angeles Basin, wholly Tertiary and wholly clastic, which is accorded a coefficient greater than unity; it has pooled more oil than it is calculated to have generated. Regions which have undergone uplifts in excess of the average depth of accumulation (about 2 km) are heavily discounted, to the extent of 0.2 times for oil and 0.1 times for gas.

The elaborately quantified volumetric assessments are called *volume-statistical methods* by the Russians. Ten geological criteria are evaluated by comparative analogies derived from all known petroliferous basins; they are then used to calculate "specific reserves" of oil, or actual volumes anticipated to be present per unit volume of sedimentary rock in the basin. The volumes are commonly expressed in tonnes of *in-place oil* per cubic kilometer. These may be converted to SI units by the factor of 1.16 $m^3\ t^{-1}$. For American comparison, 1000 t km^{-3} approximately equates with 30 000 barrels per cubic mile. The ten geological factors contributing values to the calculations are as follow:

(a) The tectonic type of the basin, considered the most important single factor. Mobile belt basins, "transitional" basins, and platformal basins contain average volumes of 21 500, 17 600, and 9600 t km^{-3} respectively.

(b) The dimensions of the basin; the larger the basin, the smaller the *specific* reserves (because more of a large basin is unproductive), but the higher the GOR (i.e. the higher the potential for gas in relation to the potential for oil). Average specific reserves range from 30 000 t km^{-3} for the smallest basins to about 9700 t km^{-3} for the largest.

(c) The maximum thickness of unfolded, Phanerozoic, sedimentary rocks in the basin; specific reserves increase from 21 000 t km^{-3} to a maximum of 32 000 t km^{-3} when the thickness reaches 8 km; they then fall off again to 11 000 t km^{-3} as the thickness becomes progressively greater than this. The GOR increases with increasing thickness.

(d) The volume of Phanerozoic sedimentary deposits, computed from (b) and (c) above. For intermontane basins of rapid subsidence (like those of California), a simple logarithmic relation has been derived:

$$\log (y + 1) = 2.07 + 0.87 \log (x + 1)$$

where y is the reserves in tonnes x 10^6 and x is the volume of sedimentary fill in cubic kilometers x 10^3.

(e) The extent of post-Paleozoic downwarping, expressed as the percentage of the total sedimentary volume of the basin that consists of Mesozoic plus Cenozoic deposits. Great variations in detail are recognized.

(f) The percentage ratio of the total sedimentary rock volume that lies below 2 km depth; this factor is clearly related to factor (c). It attains a maximum value of 28 000 t km^{-3} when about 30 percent of the sedimentary content lies deeper than 2 km, and decreases again as a greater proportion is deeply buried. Below the critical 2 km depth, the GOR increases continuously.

(g) The percentage of the total volume of sedimentary rock that consists of carbonates. Specific reserves decrease from 33 000 t km^{-3} in an ideally clastic basin to about 3500 t km^{-3} at a carbonate content of 60 percent. The GOR increases as the carbonate content increases to 15 percent but decreases again as the content increases beyond 50 percent.

(h) The percentage of the total sedimentary volume represented by marine deposits. Specific reserves decrease from 28 000 t km^{-3} at 90 percent marine content to 4300 t km^{-3} at 40 percent. The GOR on the other hand increases from 0.3 at 90 percent marine content to 0.64 at 40 percent.

(i) The age of the rocks in the principal diastrophic megacycle represented: from 26 000 t km^{-3} for the youngest (Alpine) cycle to 4500 t km^{-3} for the oldest (Caledonian) cycle. The effect of cycle age on the GOR is insignificant.

(j) The number of "megacycles" represented (complete sedimentary cycles between the major diastrophisms of (i) above). This is especially significant if the basin is shallow, so that individual "megacycles" are thin and have low specific reserves.

Russian experience has led petroleum geologists to emphasize the *interbasinal regions* (large arches or uplifts alongside or within basins) at the expense of the basins themselves (see Ch. 17). In richly petroliferous regions, it is the interbasinal regions, with relatively low values of (e) and (f), that become the principal sites of accumulation of *hydrocarbons* as prolific oil undergoes flush migration out of the basin which was the principal site of accumulation of *sediments*. The higher the ratio of (e) (for the basin) to (d) (for the interbasin "high"), the higher the specific reserves in the latter area – 1.2–6.0 times greater than in the adjoining basin. In lean regions, on the other hand, no flush migration stage is attained; the oil and gas tend to remain in smaller traps within the basin, and the interbasin regions remain essentially barren.

Non-Russian geologists may suspect that this reasoning leads to emphasis on the wrong factor, which may be the *age* of the basin. It is in the great Paleozoic basins (like the Russian Ural–Volga Basin) that the deepest parts of the original basin are now elevated into folded mountains and the largest fields consist of oil that has migrated into extensive uplifts on the forelands. In Cretaceous and younger basins, even those highly prolific, oil for the most part is still in structures within the basins.

The *specific reserves* are calculated for each of the ten indicators for the appropriate basinal type. They are in fact read directly from simple graphs expressing the values calculated for all well explored, long productive basins. The smallest value of the reserves so calculated is selected, in the so-called *weakest link assessment*. The GOR – in this context really an assessment of the likelihood that the basin will be gas-prone rather than oil-prone – may be determined as the arithmetic average of the figures derived from each indicator separately, but this is obviously an artificial procedure and the GOR is preferably calculated by regression analysis. After selection of the most informative factors available, the regression equation has the following form (where the GOR is called y and the lower case italic letters refer to the indicators enumerated above):

$$\log y = A \log d + Be^{-h} - Ce^{-1} - Df^{-1}$$
$$- Eg^2 + Fg^{-1} - Gh^2 - H$$

The multiple correlation coefficients have been calculated to be between 0.65 and 0.95.

Inevitably the Russian approach is governed even more powerfully by Russian experience than the "Western" approach is governed by American experience; North American petroleum geoscientists have collectively had a vastly greater *variety* of experience on which to base their evaluation methods. North American oilmen have explored far more barren regions than their Russian counterparts, and have drilled an enormously greater number of dry holes. The Russians have at least domestic justification for regarding all sedimentary basins as intrinsically petroliferous, even if not prolifically so; Western oilmen do not. Western geologists have vastly greater experience of areally small basins. Although Russian oil before World War II was essentially confined to narrow, compressional basins in the Caucasian province, the great subsequent developments have been in the areally huge Ural–Volga, Turkmen–Khazakh, Vilyuy–Lena, and West Siberian basins. Besides their own South Caspian (Baku) Basin, the only "areally small" basin invariably invoked by name in the Russian evaluation literature is the Los Angeles Basin, which is *sui generis* and should not be taken as typifying anything.

The Russian categorization of basins by tectonic style as "mobile," "transitional," and "platformal" is not sufficiently discriminating to assess the fundamentally different potentials of, for example, the Williston and West Siberian basins, the Ural–Volga and Great Artesian basins, or the Californian and Viennese basins. The Russian approach virtually ignores the basinal types characteristic of continental margins, and greatly undervalues carbonate–evaporite provinces. More than 90 percent of Russian and east European oil and gas is from sandstone reservoirs, and factor (g) in the sequence slotted into the reserve evaluation must read oddly to any appraiser familiar with the Persian Gulf, Sirte, Alberta, Mexican, or Permian basins.

Unlike the cases of Canada and India, very large countries for which detailed volumetric assessments were published during the early exploration phase, direct comparison between forecasts and results for Russian basins is not easily made. However, such elaborate quantification must be exceedingly risky unless the variables are reliably understood; and if they are reliably understood the basin is no longer unexplored. In the early stages of exploration, none of the parameters making up the regression equation can be known with any real assurance.

24.4 Envoi

In the first paragraph of this chapter, the point was made that the "energy crisis" of the 1970s was a direct consequence of the failure of oilmen to make sure that resources and reserves were understood by the lawmakers and the public in terms of fact and not of fantasy. Let me leave the topic by offering a few recommendations to the geologists who will have the task of performing this function properly in the future.

First, always remember that, in making the favorable assumptions that *must* be made before exploring a genuinely new area, you are acting in defiance of established odds. If your new area is a "frontier" area – as it is very likely to be – favorable assumptions are in defiance of even longer odds. The volumetric estimation techniques, even those elaborated by "risk" factors, have a very poor track record indeed, especially when the scales on which they are employed are large.

Second, do not become wedded to an analogy that proves to be fallacious. All reserve forecasting of the type we are dealing with is a processs of analogy, simple or multiple. No analogy is perfect, because no two geological basins are alike in all respects, but if the analogy chosen is a good one its appropriateness will be quickly revealed. Whoever had the Paleogene deltas of Lake Maracaibo or the Neogene deltas of Baku in mind when they chose to explore the Niger delta or the Mahakam delta in Borneo deserved their success. Even if the analogy is later shown to be in some important ways inaccurate, it is sufficient that it fairly represents the generalized, four-dimensional circumstance. The analogies that the northern North Sea Basin should be like the productive coastal basins of the South Atlantic, because all are shelf-edge half-grabens; that the North Slope of Alaska should resemble the Alberta or Ural–Volga basins because all are long-lived foreland basins in front of thrust belts; that the Gippsland Basin of Australia or the Songliao Basin of China might be akin to the productive Vienna Basin because they are all fundamentally grabens crossing older mountain structures, were all essentially correct analogies even if modern plate tectonic theory puts all the "analogous" basins into different categories.

On the other hand, exploration of the Cuba–Bahamas region by analogy with Iran–Arabia, or of the Sverdrup Basin in the Canadian Arctic with the Gulf Coast of the USA (both of which actually took place) reflected inherently poor analogies from the first and caused disastrously inflated expectations.

Third, assessing the volumes of undiscovered resources is an attempt to quantify the unknowable. It is not as lacking in foundation as guessing the distance to a star by looking at it through binoculars, but both these prognostications, performed by experienced scientists, may be wide of the mark by a comparable number of orders of magnitude. *Do not* express the results of volumetric calculations in figures that give a false impression of exactness; do not sum them for a number of separate targets as if you were summing the figures in a bank statement (or even figures of *known reserves*); your figures cannot be simply summed arithmetically because each one of them carries with it a significant but unknown risk factor. The more accurate the figures look, the more readily they are believed by the uninitiated.

Fourth, make sure that the people to whom your assessments are directed understand how they were derived and what their limitations are. After making sure, do not allow those people to use or broadcast your assessments in any misleading or unelucidated way without forceful rebuttal. If you see your prognostications (or the comparable prognostications of other geologists) published, or hear them broadcast, in such a way as to encourage belief that they represent known quantities, try to correct that impression immediately and loudly.

Fifth, and above all, do your best to ensure that the general correctness *or otherwise* of your forecasts is tested by practice as soon as is practicable; but also do your best to ensure that *no policy or commitment*, at any level, presupposes your correctness until the practical tests have been conclusively completed.

Part V
DISTRIBUTION OF OIL AND GAS

In this final part of the book, we attempt a geographic, stratigraphic, and geologic synthesis of the world's petroleum endowment as we know it at the beginning of 1984. Exploitable occurrences of oil and gas are considered by stratigraphic age, by geographic location, and by their controlling geologic mechanisms. Finally, a few notable examples are selected from the great abundance of individual fields, to illustrate the lessons and principles set out in all that has gone before.

25 Geographic and stratigraphic distribution of oil and gas

25.1 Basis of presentation

The geographic distribution of oil and gas, by countries or by regions, is described in many publications. Relative significances of the regions, as expressed in volumes of cumulative and current production and of remaining reserves, are revealed annually in single issues of the *Bulletin of the American Association of Petroleum Geologists*, the *Oil and Gas Journal*, and other professional and trade publications, and updated maps appear in the annual *International Petroleum Encyclopedia* published in Tulsa, Oklahoma.

Of more immediate interest to Earth scientists than the distribution of oil and gas by countries is the distribution by either *basins* or *geologic provinces* (see Ch. 17). A number of countries yield production from several, or even many, quite separate and geologically different basins – the USA, the USSR, Mexico, and China are obvious examples. Geologically, the Caucasus province of the Soviet Union is more sensibly considered in relation to the Carpathian province of Romania than to Russia's quite different Ural–Volga province. On the opposite hand, a few very productive basins extend across national boundaries; if the Persian Gulf Basin were the only one to do this, it alone would be ample justification for giving greater emphasis to it as a unit than to its dozen separate nations. Douglas Klemme, Lytton Ivanhoe, Richard Nehring, Bill St John, and others have published tabulations of ultimate reserves by basins, particularly of reserves contributed by giant individual oilfields.

With this geographically oriented material readily available, this chapter concentrates on the *stratigraphic* distribution of oil and gas. The control of basinal creation and extinction by tectonic processes is discussed in Chapter 17. The ideal breakdown of the stratigraphic column for our purposes would therefore take account of the great tectonic events such as the Hercynian "orogeny" and the Jura-Cretaceous onset of the latest episode of continental fragmentation. Undoubtedly the Permian Period, postdating the Hercynian events, is set off from the rest of the Paleozoic Era in critical respects, including those governing hydrocarbon incidence. The name of the system (and of the period) is enshrined in one of the greatest petroleum basins on Earth, but the Permian Basin is unique. Outside it, there is relatively little oil in Permian rocks. The six largest known oilfields in wholly Paleozoic reservoirs are all Mississippian or older. But the Permian oil is in Paleozoic basins, not in Mesozoic basins. Permian gas, on the other hand, is very much in Mesozoic basins, and its unusually high incidence carried over into Triassic and even into Jurassic rocks. In a similar way, the natural termination of most great Mesozoic basins came in the Eocene, not in coincidence with the mass extinction of life that by definition ends the Mesozoic Era at the close of the Cretaceous. Nearly all the oil in earliest Tertiary reservoirs is associated with Cretaceous and Jurassic oil, not with Oligo-Miocene oil; Libya, China, the North Sea, the circum-Atlantic basins and the Arctic Coast–Beaufort Sea province illustrate this.

Our time divisions therefore incorporate slight modifications of the standard time scale. All Paleozoic oil and all pre-Permian gas are considered together; Permian gas is considered with Triassic gas in a separate section. The entire Paleozoic, including the Permian, contributes only about 12 percent of conventional oil so far discovered; the Triassic contributes little more than 1 percent. The Permian and Triassic combined, however, contribute at least one-third of known gas accumulations. Paleocene oil and gas are considered with the Mesozoic occurrences wherever this is geologically appropriate. These occurrences then contribute some two-thirds of our total known oil endowment, and much more if tar sands are included.

Stratigraphic rather than geographic discrimination introduces its own awkwardness, of course. Individual basins which are productive from well separated strata of more than one era – like the Gulf of Mexico and Western Canadian basins, and even individually small basins in the US Rocky Mountain province – have to be treated more than once, unless there is general agreement that all the *source sediments* fall within a single era (as is the case with the Persian Gulf Basin).

Scientific textbooks are at risk of becoming out of date very quickly. The author recognizes that this one chapter is likely to become out of date even sooner than the rest of the book. He experienced a strange feeling of returning to another century when he perused again (in the process of preparing this book) the famous work by Wallace Pratt and Dorothy Good, one of the most valued references of his early days in the industry (*World geography of petroleum,* 1950). When the American Geographical Society commissioned this work and asked one of the most experienced of petroleum geologists to edit it, it was envisioned as the standard work which it became. Yet it antedated the heyday of Middle East oil, and the discoveries of the West Siberian, North Sea, and Alaskan–Arctic basins, and it had no contribution to report from Africa except for Egypt. One should not anticipate developments rivalling these in the next quarter-century, but there will certainly be enough to require extensive revision of this section of this book.

25.2 Oil and gas in Paleozoic strata

25.2.1 Paleozoic petroleum in Asia

In the heart of central Asia, between the type Baykalian ranges beside Lake Baykal and the Angara shield of Siberia, is an ancient foredeep basin containing an exceedingly thick late Proterozoic and early Paleozoic section. Russian geologists refer to it as the Irkutsk "amphitheater"; as it extends northeastward along the Lena River beyond the Tunguska River, it should properly be called the *Lena–Tunguska province*. In its geologic and geographic setting it somewhat resembles the Amadeus Basin in the center of Australia (see Sec. 25.2.6). Gas and condensate have been produced from Late Proterozoic clastic sediments, and both oil and gas from Lower Cambrian carbonates, highly fractured above salt-cored anticlines. The occurrences are more important as support for the resource optimists – who maintain that sedimentary rock of any age may be commercially petroliferous – than as economic deposits in such a location.

On the opposite side of the central Asian ranges, in the *West Siberian Basin*, small oilfields occur in Silurian and Devonian carbonate reservoirs. They are utterly insignificant by comparison with the giant oil- and gasfields in Cretaceous sandstones in other parts of the basin (Sec. 25.5.7).

In the *Sichuan Basin*, of central China, which is primarily a gas producer from Mesozoic strata (Sec. 25.5.6), oil and gas also occur in Sinian (late Proterozoic) and Carboniferous rocks and in a Permian limestone. In the *Huabei Basin*, which extends southward from Beijing, some paraffinic oil is produced from both Sinian and Paleozoic carbonates in buried hill traps below the sub-Tertiary unconformity, and deep gas occurs in Ordovician strata. The source of the oil, but less certainly of the gas, is in the overlying Tertiary.

25.2.2 Paleozoic petroleum in European Russia

Before World War II all Russian oil production was derived from Tertiary reservoir rocks in the Caucasian–south Caspian region, principally in the Baku district (Fig. 16.54). During the war, production began from an area between the Volga River and the Ural Mountains; during the late 1940s and early 1950s a succession of great discoveries enabled this region to surpass that of Baku in oil output by 1957, and thereafter it dominated Soviet production until it was itself overtaken by western Siberia in the late 1970s.

The *Ural–Volga Basin* was called the "Second Baku" by the Russians, a name which may have been economically justified but is geologically incongruous. Whereas Baku's production has come from strata less than 10 Ma old, in tightly folded traps within a very small area, the "Second Baku" produced exclusively from Paleozoic strata in huge arches scattered over a basin about 500 000 km^2 in area (Fig. 16.43). A major deformation took place in the Uralian geosyncline during Devonian time, causing wide transgression of the Russian platform by a Middle Devonian seaway. From that time to the mid-Carboniferous there were wide oscillations across the platform, with the culminating creation of a frontal foredeep and late Paleozoic folding and emergence. A consequence of this structural history was of crucial importance in the distribution of oil and gas in the basin that has extended westward from the Ural Mountain front since Permian time. Devonian strata thicken westward into the main basin, but their thickness is very variable in detail (Fig. 17.28). Carboniferous strata, in contrast, thicken continuously eastward toward their source in the mountains; much of the Carboniferous is molasse. Thick Permian is confined to the frontal foredeep, which contains the type locality of the system. Because the thickness changes between the Devonian and Carboniferous parts of the section are not in harmony, there is pronounced structural discordance between the two levels (Fig. 17.28).

The basin is an oil basin rather than an oil–gas basin – the greatest Paleozoic oil basin in the world – though some gas has been produced from both Carboniferous and Devonian reservoirs. The oil occurs principally in deltaic sandstones of Givetian and early Frasnian ages and in nonmarine Lower and Middle Carboniferous. The carbonate members, though numerous throughout the two systems and including reefs, are much less productive. The source sediment for the oil is agreed to be the famous Domanik Formation of Frasnian age (see Sec. 8.2); the traps are anticlines formed over basement uplifts superimposed on vast, regional upwarps or arches (see Sec. 16.3.4). In the Permian, the relatively minor oil (no more than about 3 percent of the basin's total recoverable reserves) occurs largely in reef reservoirs and traps. The greatest Permian carbonate field, however, is the supergiant gas accumulation at Orenburg, far to the south of the main oilfields (Fig. 16.43).

The Romashkino field in the Devonian of Tataria province is the largest wholly Paleozoic oilfield known. The province of Tataria alone had produced about 2.5×10^9 m^3 of conventional oil in just 38 years to 1982. The basin as a whole has produced more than 5×10^5 m^3 of oil *daily* since the 1960s.

North of the productive limits of the Ural–Volga Basin, the meridional Ural Mountains appear to bifurcate; the subsidiary Timan Ridge extends towards the northwest along a Proterozoic (Baykalian) trend. The ridge acts as the western foreland of the asymmetrical *Timan–Pechora Basin*, between it and the westwardly-overthrust northern Urals. Permian salt provides a *décollement* horizon for the thrusting. The Timan–Pechora Basin contains the type locality of the Domanik source-formation, and most of the oil is in multiple, lenticular, sandstone reservoirs in the Middle and Upper Devonian. A lesser amount is in Lower Paleozoic, Upper Paleozoic, and Permo-Triassic carbonates. Unlike the Ural–Volga Basin, however, the Timan–Pechora Basin is primarily a gas basin in Carbo-Permian carbonates (see Sec. 25.3). At the beginning of the 1980s the basin's oil production was about comparable with that of Australia – a little over 50 000 m^3 per day – but its gas development was overshadowed by that of the great fields of northern Siberia (and Orenburg). The basin extends offshore into the Barents Sea, where both oil and gas occur in Permo-Triassic clastic rocks and doubtless in Paleozoics also; exploitation is a matter for the future.

On the opposite side of European Russia, the area of the present Ukraine was transected by a major Paleozoic graben system, the *Dnieper–Donetz depression* (the type aulacogen). The axis of the depression contains thick evaporites in both the Devonian and the Permian; the post-Permian folds are primarily a consequence of salt diapirism. Most of the oil occurs in lower Carboniferous sandstones in folds of this sort, but gas again dominates, especially in the Permian (see Sec. 25.3).

25.2.3 Paleozoic petroleum in northern Europe

A band of northern Europe, extending from Byelorussia to eastern Britain, lies between the main Hercynian front, or Sudetic Zone, and the Baltic shield and Caledonian ranges to the north. Nearly all the oil and gas known in this belt occurs in Mesozoic or early Cenozoic reservoirs. However, the post-Hercynian extensional breakdown of the belt created a large, rift-like trough in early Permian time, extending from the Carpathian front in Poland, under Denmark, to the North Sea between Norway and Scotland. Some Paleozoic oil is preserved both within the trough and northeast of it on the Paleozoic platform.

In *Kaliningrad province* of the Soviet Union, formerly Lithuania and Latvia, light paraffinic oil occurs in Middle Cambrian and Ordovician strata. Offshore, oil occurrences extend to the island of Gothland, in small Ordovician reefs. Along the Polish coast and extending into *East Germany*, oil is produced from the Permian Zechstein dolomite, and deep gas high in nitrogen occurs in both older Permian and Carboniferous rocks. Though these occurrences are numerous, none is of more than local significance.

The *Northwest German–Netherlands Basin*, like the North Sea Basin, is a gas province in Permian strata (see Sec. 25.3) and virtually all the oil is in post-Paleozoic formations. A little oil and gas occurs in the Zechstein dolomite.

The Central and Viking grabens of the *North Sea* (Fig. 25.3) were developed during the Mesozoic over a Permian rift or aulacogen. There is a little light oil in the Permian near the southern end of the graben. However, the Permian rift itself cut across a very large Devonian (Old Red Sandstone) basin lying within the Caledonian orogenic belt. This older basin underlies the whole of the northern North Sea and extends on to its flanking coasts. At the time of writing, few wells have been carried into strata as old as Devonian, but wells in at least two fields have discovered light oil in thick, highly fractured Devonian sandstones.

The Devonian basin followed the Caledonian deformation of northern Europe much as the Permian (Zechstein) basin later followed the Hercynian deformation of central Europe. The Hercynian trough itself had become rapidly infilled during late Carboniferous time by a group of large deltas; the most familiar deposit from these deltas is the Millstone Grit of *northern England*, the Pennsylvanian foundation below the Coal Measures. A number of oilfields, very small in output but important in their industrialized setting, produced oil from both the grit units and the underlying Carboniferous limestone.

On Rona Ridge, a horst at the edge of the continental shelf west of the *Shetland Islands*, more than 0.5×10^9 m^3 of rather heavy oil has been found in nonmarine, basal detrital rocks resembling the Old Red Sandstone below the North Sea Basin and apparently of Devonian age. As the shallow reservoir is sealed by Cretaceous mudstones, a Mesozoic source for the oil is highly probable. Though the thickness of oil-saturated section is large (it extends down into fractured Precambrian basement), the reservoir permeability is so low and complex that recovery would be unlikely to reach even 10 percent of the oil in place, and enhanced recovery is unlikely to be practicable in such a hostile environment.

25.2.4 Paleozoic petroleum in North America

From some time early in the Proterozoic Eon to the Pennsylvanian Period, the structural backbone of the North American plate extended (in its present terms) from northeast to southwest. The *Transcontinental Arch* in its Paleozoic form has become completely obscured by a cover of younger sediments in the Great Plains and by much younger (early Tertiary) uplifts in the US Rocky Mountains. Throughout much of pre-Pennsylvanian time, however, it divided the continent into two gross geologic provinces, tectonically, depositionally, and faunally. One province lay to the southeast, and the other to the northwest, of an irregular landmass extending from Labrador, under the present site of Lake Superior, all the way to Arizona.

Far to the south or southeast lay the Iapetus Ocean; from the deposits of its American margin the Appalachian and Ouachita Mountains were to be raised. Far to the northwest lay the Proto-Pacific Ocean; from the Paleozoic continental shelf between it and the Canadian shield the Rocky Mountains were eventually to arise. As plate boundary adjustments occurred along the two margins, seaways were extended over the flanks of the arch and then withdrew from them again. We thus find in mainland North America today four groups of Paleozoic sedimentary basins that have yielded prolific oil and gas production. There is the Appalachian–Mid-Continent group of foredeep basins along the mountain front from New York to northern Texas; they contain largely clastic sedimentary rocks, including great thicknesses of Pennsylvanian molasse. There are the great constructional basins of West Texas–New Mexico and western Canada, developed on the craton in front of early deformed belts and containing thick successions of carbonate and evaporite rocks. And there are two sets of shallow cratonic basins formed during transgressions into re-entrants along the margins of the arch; they extend from Kansas to southwestern Ontario along the southeastern margin and from Arizona to Manitoba along the northwestern margin.

The Appalachian Basin The historic birthplace of the American oil industry yielded its first commercial production from wells, in 1858; its first giant field in a sandstone reservoir, at Bradford in Pennsylvania in 1871; and its first giant in a carbonate reservoir, in 1884 in what was to become the Lima–Indiana field southwest of Lake Erie (Fig. 13.52).

The basin is a foredeep in front (west) of the Appalachian Mountains. It extends northwestward to the broad Cincinnati Arch, of Paleozoic origin, aligned between Alabama and southwestern Ontario and separating the foredeep proper from the cratonic basins along the flank of the Transcontinental Arch. In a very general way, the basin provides an example of stratigraphic zonation of production. The stratigraphically oldest fields are in Cambro–Ordovician dolomites along the west side of the Cincinnati Arch, southeast of the Illinois Basin (Fig. 17.19); the fields also produce from younger carbonate horizons. The largest productive area is in Middle Ordovician carbonate rocks along the arch; it includes the Lima–Indiana field. Sandstones low in the Silurian section have yielded mostly gas, especially around Lake Erie; the sandstones form the base of the Niagara escarpment which rims the Michigan Basin and passes below the falls. Most of the oil and gas from the Appalachian Basin has come from Devonian strata, especially from shallow sandstones overlain by black, organic-rich shale formations. Both oil and gas are in fact produced from the shales themselves, and these may become an important future source of unconventional gas (see Sec. 10.8.2). Mississippian strata are extensively (though not prolifically) productive from shallow sandstone and limestone formations close to the mountain front and closely overlying the source sediment of the Chattanooga shale. Again, gas is greatly dominant over oil, as it is throughout the basin.

There is trivial production from thrust-related folds behind the mountain front, mostly of gas from Lower Silurian sandstones. Modern seismic exploration has revealed that the thrust belt is much wider than has traditionally been recognized, underlying not only the Valley-and-Ridge or foreland belt but extending eastward below the axis of the range. Exploration of Paleozoic strata (largely Cambro-Ordovician) below this great *décollement* surface may rejuvenate production – gas production, at least – in this oldest of conventional petroleum provinces.

A unique hydrocarbon occurrence within the Appalachian belt is that at Stony Creek, New Brunswick, in the *Canadian Maritimes*. The area was part of the Avalonian microcontinent during early Paleozoic time, lying off the African plate, and did not become part of the North American plate until the Devonian episode of plate convergence. Basal Carboniferous strata consist of bituminous shales with a rich fish fauna and sandstone lenses to provide the reservoirs, all about 1200 m thick and overlain by thick salt. The sandstones yield very light, paraffinic oil and high-quality gas, in globally trivial but locally valuable quantities.

Also unusual are the gas occurrences in Paleozoic carbonate outliers on the *Labrador shelf* off eastern Canada. The rift in which thick Cretaceous and Tertiary clastic sediments accumulated (Sec. 25.5.4) was formed across Precambrian rocks of the Canadian shield. Outlier remnants of both Ordovician limestone and Carboniferous dolomite, however, contain gas in paleogeomorphic traps and in drape structures over horst blocks. The occurrences cannot be said to belong to any definable Paleozoic petroleum province, but the presence of Carboniferous strata suggests some connection with the Appalachian foreland.

The Mid-Continent and Permian basins (Figs 17.1 & 17.3) A complex, north-facing mountain belt turns or branches westward from the Appalachians, crosses the Mississippi embayment below the Cretaceous overlap, and emerges at the surface in western Arkansas. Thence westward, the belt becomes dual, apparently through the creation of two distinct troughs during early Paleozoic time.

The *Arbuckle–Wichita–Amarillo trough* was a linear rift or aulacogen extending obliquely into the interior of the North American craton. A thick fill of Paleozoic carbonates and shales was sharply disrupted by compressional movements in mid-Carboniferous time, forming a series of closely spaced, linear structures separated by steep faults and creating or accentuating a deep basin in front of them (to the north). The high-standing structures were eroded, overlapped by thick Pennsylvanian clastic sediments, and strongly refolded and reverse-faulted in early Permian time (Fig. 16.14).

The *Ouachita–Marathon trough* was a distinctly arcuate feature lying off the edge of the Lower Paleozoic craton. It received clastic rocks, not carbonates. In mid-Carboniferous time, an episode of plate convergence caused the trough to be infilled by a vast thickness of flysch sediments, and the entire content of the trough was thrust northward and westward over the eastern end of the Arbuckle–Wichita fold-fault belt. Again, a deep frontal downbuckle was filled with enormously thick Pennsylvanian molasse.

The Ouachita front now crosses the Arbuckle uplift and its attendant frontal basin virtually at right angles (Fig. 17.1). It continues in the subsurface along the south-southwesterly course later to control the inner margin of the western Gulf Coast Basin, to re-emerge in the Marathon uplift near the Texas/Mexico border. There it abutted against yet another trend almost perpendicular to it, extending from what is now northern Mexico into the present Colorado Rocky Mountains.

Within this complex region, we therefore again have four groups of sedimentary basins, all of them with richly petroliferous representatives:

(a) Foredeep basins lying in front (north or west) of the Ouachita–Marathon thrust belt.
(b) Foreland basins lying between the Ouachita–Marathon thrust belt and the Arbuckle–Wichita–Amarillo fold-fault belt; these constitute the composite Permian Basin and its attendant shelves.
(c) Basins along and in front of the Arbuckle–Wichita–Amarillo fold-fault belt.
(d) Foreland basins between the two deformed belts and the Transcontinental Arch.

The *Val Verde* and *Fort Worth basins* lie in front of the thrust belt in west-central Texas. Deformed Mississippian and older strata, productive chiefly of deep, nonassociated gas, are unconformably overlain by thick Pennsylvanian molasse. Younger Pennsylvanian and Permian strata are lacking. In front of the Ouachita Mountains is the *Arkoma Basin*. Its production is almost entirely of nonassociated dry gas, some of it in much-faulted Lower Paleozoic carbonates but much the larger part trapped by stratigraphically controlled porosity variations in fluvial and deltaic sandstones of the enormously thick Pennsylvanian molasse. Other traps are provided by south-facing growth faults between the shelf and the deep basin.

The *Ouachita–Marathon thrust belt* itself is of daunting structural complexity and contains rocks of essentially nonporous, siliceous facies in repeated sections. Gas and some oil are produced from fracture porosity in Upper Ordovician cherts, Devonian novaculites and shales, and Mississippian–Pennsylvanian clastics of flysch aspect.

Farther to the east, in the re-entrant between the Ouachita and Appalachian mountain fronts, the *Black Warrior Basin* straddles the border between Mississippi and Alabama. Unlike the Arkoma Basin, it is not a true foredeep, but a downfaulted marginal trough protected from deformation by a basement ridge. Cambro-Ordovician platformal carbonates and thin Middle Paleozoics are largely buried by more Pennsylvanian molasse. The strata are apparently over-mature; the basin's unimportant gas production has nearly all come from upper Mississippian clastic rocks outwardly like those richly productive of oil in the Illinois Basin to the north.

The *Permian Basin* and its attendant uplifts and shelves occupy the large area between the frontal basins (which lack Permian strata) and the Arbuckle–Amarillo uplift. The structural development of this great petroleum basin is described in Section 17.4.6. Occupying much of western Texas and the southeastern corner of New Mexico, it is one of the half-dozen most prolific oil and gas basins of the world, wholly in Paleozoic rocks.

The horst-like Central Basin Platform separates a westerly (Delaware) and an easterly (Midland) basin and its eastern shelf. This last is itself separated from the foredeep of the thrust belt by the flexural Bend Arch. An east–west uplift to the north separates the entire basinal complex from a further broad but shallow basin, the Palo Duro Basin, which lies south of the much greater Arbuckle–Amarillo uplift (Fig. 17.3).

Pre-Mississippian strata in the Permian Basin are nearly all shelf carbonates, and most pre-Mississippian oil or gas accumulations are in structural traps created and accentuated by repeated episodes of deformation. A short-lived clastic regime began with the Devono-Mississippian Woodford Formation, an important source sediment over much of the southern USA. As in the other basins associated with the Ouachita–Marathon and Arbuckle–Amarillo deformed belts, Pennsylvanian and Permian rocks are here exceedingly thick, but in dominantly carbonate and evaporite facies though containing important sandstone members. Reefs are abundant in both systems, especially on and around steep-edged shelves and platforms like the one in the middle of the basin. The basin was finally filled in by a vast body of late Permian salt, the last deposit of the marine withdrawal before the whole region became stabilized.

Every structural subunit of the basin is or has been productive. Oil occurs prolifically in a Cambro-Ordovician dolomite formation (here called the Ellenburger, but equivalent to the equally productive Arbuckle Formation of Kansas); these oldest reservoirs are principally on the Central Basin Platform and the Bend Arch. Less prolific is the Siluro-Devonian section, on and around the central platform. Upper Pennsylvanian rocks, both limestones and sandstones, provide the reservoirs for hundreds of fields east of the central platform. Reservoir horizons throughout the Permian are concentrated in the Midland and Delaware basins and on the platform between them. Associated gas is very abundant in the Delaware Basin, much less so in the Midland Basin. Nonassociated gas, also abundant, occurs principally in the Cambro-Ordovician and the Pennsylvanian of the Delaware Basin. The Palo Duro Basin to the north, lying between the true Permian and true Mid-Continent basins, is the least productive subunit; created by the Pennsylvanian uplifts north and south of it, it lacks pre-Mississippian rocks and the Permian consists largely of evaporite.

There is a striking relation between the ages of the productive formations, their detailed lithologies, and the trapping mechanisms. For reasons outlined elsewhere (Sec. 13.3.6) but by no means properly understood, Pennsylvanian production is dominantly from limestones even where the traps are reefs, as in the Horseshoe Atoll or Scurry Reef in the sediment-starved Midland Basin (Fig. 17.3). Most Permian production, in contrast, has been from dolomites, and much of it has come from porosity traps consequent upon subtle facies changes rather than from conventional convex traps. Though virtually all the very large oilfields, and most of the large gasfields, are in carbonate reservoirs of one sort or the other, there are many smaller fields in stratigraphic and lithologic traps in both Pennsylvanian and Permian sandstones.

The basin as a whole had yielded more than 4×10^9 m^3 of oil and 2×10^{12} m^3 of gas by the end of 1981; it will eventually yield at least 5×10^9 m^3 of oil and nearly the equivalent in gas. Its heyday of successful exploration was undoubtedly the decade 1926–36, though the largest single accumulation in it – in the Scurry Reef – was not discovered until 1948. Despite this venerability of output, the basin still contributed almost one out of four of the 100 largest *remaining* accumulations in the continental USA in 1980. The cluster of fields around the edges and flanks of the Central Basin Platform constitutes one of the most remarkable concentrations of oil on Earth. In recent years, the principal new discoveries have been of nonassociated gas, from deep horizons in the Delaware Basin to the west of the platform.

The Arbuckle–Wichita–Amarillo belt of high-angle fault deformation is now misleadingly called the "southern Oklahoma fold belt." At its western end is the *Amarillo uplift*, the spasmodic elevation of which created a complex of horsts and grabens along its northern edge. A carbonate succession from the Cambro–Ordovician to the Mississippian gave way to thick Pennsylvanian and lower Permian clastics with granite wash sloughed off the uplift (Fig. 13.21). The clastics graded northward into shelf limestones, and the whole area was finally buried by later Permian, post-orogenic, clastic sediments. The largest gasfield in North America (Hugoton) is reservoired in Permian limestone around the western edge of the frontal basin. In the extension of the carbonate reservoir on to the Amarillo uplift, and in the granite wash associated with it, is the large oil- and gasfield of the Texas Panhandle.

Farther east, the Arbuckle–Wichita sector of the belt is a complex of sharp folds separated by deep, steep faults. The structures include both bald-headed anticlines in which eroded Ordovician carbonates are capped by Pennsylvanian sandstones, and deeply downtucked, narrow basins in which thick Ordovician–Mississippian strata are preserved in tight folds (Fig. 16.14). The most sharply downtucked strip, the *Ardmore Basin*, coincides with a zone of Cambrian silicic vulcanism in the early rift. In this subprovince generally, all the major sandstone units are prolifically productive – in the Ordovician, the late Mississippian, and the Pennsylvanian (the latter lying above the main unconformity but being itself strongly folded and faulted). In addition to sharp anticlines there is an assortment of pinchout traps along the flanks of the high structures.

The frontal basin north of the Wichita–Amarillo uplift is the *Anadarko Basin*, in which an entirely Paleozoic succession more than 12 km in thickness has been the site of many of the world's deepest drillholes (see Sec. 23.5). The ultradeep basin is a gas province *par excellence*, especially in Siluro-Devonian dolomite and Mississippian–Pennsylvanian sandstones. Much of the gas comes from depths below 5 km. Oil, on the other hand, is reservoired far shelfward towards the Transcontinental Arch, most of it in stratigraphic traps in either Ordovician or Pennsylvanian clastic strata.

The Anadarko Basin, in front of the Wichita sector of the uplift, is separated from the Arkoma Basin, in front of the Ouachita thrust belt, by the subsurface extension of the Arbuckle Mountains, called the *Seminole Arch*. From the arch, a narrow, faulted uplift of Precambrian crystalline rocks extends northward, wholly in the subsurface. Spatially, this *Nemaha Ridge* appears to link the Arbuckle Mountains to the mid-continent belt of gravity maximum which follows the band of Proterozoic mafic lavas from Lake Superior into Kansas. The ridge was elevated in early Pennsylvanian time along large faults; after extraordinarily rapid, deep erosion, mid-Pennsylvanian sediments now rest on the Precambrian on the highest parts of the uplift. The ridge constitutes an important petroleum district in its own right, having apparently impoverished its flanking basins by drawing from them oil generated by the rich Pennsylvanian (Cherokee shale) source rock.

Again, the pre-Pennsylvanian section is mostly carbonate (indeed, mostly dolomite), showing extensive diagenetic alteration. Rich oilfields (and important gas also) occurred in the ubiquitous Cambro–Ordovician dolomite, Ordovician sandstone and limestone, Middle Devonian dolomite, Upper Devonian sandstone, and Mississippian chert and limestone, especially where anticlinal structures ancillary to the main uplift enabled the Pennsylvanian source sediment to overlap the porous reservoir. The most widespread traps, however, are stratigraphic and in mid-Pennsylvanian sandstones; they include the type shoestring sands (Fig. 13.32). These highly discontinuous sand bodies, interbedded with or adjacent to the Cherokee shale source rock, illustrate the important principle of the *interorogenic time setting* of rich source sediments (see Sec. 18.2). The deposition of the Cherokee Group was immediately preceded by the Wichita disturbance and was terminated by the Arbuckle disturbance, the two together constituting this region's equivalent of the Hercynian orogeny.

Fields close to the Nemaha Ridge, in Pennsylvanian sandstones in Oklahoma, dominated world oil production at the beginning of World War I (see Ch. 4). Their presence led to the emergence of the modern city of Tulsa and its selection, in 1917, as headquarters for the American Association of Petroleum Geologists. The largest single field on the ridge (Oklahoma City, Fig. 16.4) was not discovered until 1928. Fields along its eastern flank extend northwards across Kansas almost to Kansas City.

Paleozoic petroleum along the southeast side of the Transcontinental Arch The Appalachian Mountains and the Transcontinental Arch are to all intents and purposes parallel to one another. Cratonic basins southeast of the arch are far removed from the Appalachian foredeep basin and were separated from it by wide uplifts throughout Paleozoic time. The Ouachita Mountains and the Arbuckle–Wichita–Amarillo uplift, however, head towards the arch, and the foredeep basins of the Mid-Continent merge northward, across wide shelves, towards the flanks of the arch. We will consider the Paleozoic petroleum districts along the southeast side of the Transcontinental Arch in succession from southwest to northeast.

From the southern flank of the arch, a prong of Precambrian crystalline rock extended southeastward towards the Pennsylvanian mountains. This *Central Kansas Uplift* was strongly re-elevated as a result of the foldings, and Middle Pennsylvanian strata thereafter overlapped all older rocks down to the Precambrian. All suffered weathering, fracturing, and the gamut of diagenetic changes, and all capable of acquiring porosity are now petroliferous on the uplift and along its flanks. The Precambrian basement itself provides paleogeomorphic traps (Secs 13.4.9 & 16.6); the overlying Cambro-Ordovician (Arbuckle) dolomite yielded the major share of the area's production, both of oil and of sour gas, from porosity largely due to solution. A variety of traps contain both oil and gas in overlying Ordovician sandstones, Mississippian carbonate rocks and chert, and Pennsylvanian sandstones and limestones. Not surprisingly, multipay fields – often thought of as phenomena peculiar to Tertiary orogenic basins – are the rule rather than the exception in this entirely cratonic, Paleozoic, and largely carbonate region.

The Central Kansas uplift separates two embayments into the flank of the Transcontinental Arch. The more westerly embayment, extending northward from the Anadarko Basin, contains the supergiant Hugoton gasfield in a lithologic trap in a Lower Permian dolomite (Sec. 25.3). West of the field is a residual arch separating the *Hugoton embayment* from the large Denver Basin, part of the Rocky Mountain province superimposed on the main Transcontinental Arch (Sec. 25.5.10). On the smaller arch and in its embayment, trapping mechanisms for minor oil and gas are associated with the sub-Pennsylvanian unconformity. Below it are karst topography and porosity in Mississippian carbonates; above it are stratigraphic traps in fluvial Pennsylvanian sandstones and more carbonates.

The embayment east of the Central Kansas uplift is called the *Salina Basin*, an unfortunate name which has nothing to do with the great Salina (Silurian salt) basin around the eastern Great Lakes. The relatively thin section of Paleozoic carbonates is much interrupted by unconformities and has yielded very minor oil. The next embayment to the east, the *Forest City Basin*, is also of minor significance but has yielded some oil from Ordovician and Devonian carbonates and oil and gas from mid-Pennsylvanian sandstones. The basin was really a mid-Pennsylvanian creation, and that part of the section must earlier have generated a great volume of oil, but most of what remains is so heavy that it has defied economic recovery (see Sec. 10.8.4).

Though the Appalachian province claims the birthplace of the American oil industry, the Mid-Continent province – and specifically the Oklahoma segment of it – has long laid claim to be its headquarters (see Ch. 4). In the words of Max Ball, most people considered Oklahoma and Oil to be synonymous during the years 1910–30. The Mid-Continent basins combined to form the first richly productive Paleozoic oil province. They provided a great number of fields of both oil and gas, in reservoir rocks of all Paleozoic systems and in an extraordinary assortment of traps. This variety was made possible by three episodes of Paleozoic deformation, involving three distinct kinds of basement rock, resulting in three subregional unconformities (pre-late Ordovician, pre-Mississippian, and pre-Pennsylvanian) and three rich source sediments shortly younger than the unconformities.

The Illinois and Michigan basins (Fig. 17.8) Delineation of the Illinois and Michigan basins as distinct structural entities began with the Middle Ordovician transgression of North America that followed the first major upheaval in the Appalachians. Both basins are relatively simple sags surrounded by hinge belts. The principal sagging of the Michigan Basin took place during the late Silurian, when its center accumulated a vast thickness of salt. The greatest subsidence of the Illinois Basin came considerably later, in Carboniferous time. Both basins are separated from the Appalachian foredeep by the Cincinnati–Findlay Arch; they are separated from one another by another low arch which bifurcates from the Cincinnati–Findlay Arch. All production is from Paleozoic strata, and an important oil-source rock in both basins is undoubtedly the Upper Devonian black shale that is ubiquitous in eastern North America (see Sec. 10.8.2).

The *Michigan Basin* yields some oil and gas production from the basal Ordovician sandstone and significant oil production from Middle Ordovician carbonate rocks. Both basins contain Middle Silurian reefs, which form part of the Niagara escarpment; those at the surface around the Illinois Basin include the type bioherms. Pinnacle reefs are productive of both oil and gas in the northern part of the Michigan Basin, and drape folds over other reefs yield minor production in Siluro-Devonian strata in the Illinois Basin. Most oil from the Michigan Basin, however, has come from Devonian carbonates, much of it from secondary porosity below unconformities. Most of the gas, on the other hand, is from Mississippian sandstones.

Upper Mississippian and Pennsylvanian sandstones have also provided the greater part of the oil from the *Illinois Basin*. The sands were the deposits of the channels, floodplains, and deltas of rivers draining from the new Appalachian Mountains to the east. Also contributing substantial oil from this basin are Mississippian carbonate reservoirs, especially oolitic limestones which provide popular building stones from their outcrops in the Mississippi valley along the western flank of the basin – the type locality for the Mississippian System. The commonest trapping mechanisms in the Illinois Basin fields are anticlines.

The Illinois Basin is an oil basin, not an oil–gas basin. It should ultimately produce about 600×10^6 m^3 of oil; its most prolific development came immediately before and during World War II. The Michigan Basin, in contrast, is unlikely to produce more than about a quarter of that quantity of oil, but its gas production is more important than that of its neighbor basin.

An interesting sidelight illustrates the tenuous separation between the Michigan Basin and the Appalachian foredeep. The two come together in the peninsula of southwestern Ontario (Canada), which separates the lower and upper Great Lakes (Fig. 17.8). The axis of the peninsula is formed by the shallow, plunging Cincinnati–Findlay Arch itself. The eastern edge of the Michigan Basin just gets into the western extremity of the peninsula, and has produced small quantities of oil for more than 100 years from Ordovician, Silurian, and Devonian carbonates, commonly where these are dolomitized (see Sec. 13.3.6). Structural reversals over differentially dissolved Silurian salt form common traps. A field in one such trap lays legitimate claim to being the earliest field in North America discovered by drilling; a well dug in 1857 by the spring-pole method found a prolific flow at about 60 m depth, sealed by tight glacial clay. The Appalachian Basin underlies much of Lake Erie and production has come both from below the lake bed and along its northern shore – minor oil in sandstones of both late Devonian and late Cambrian ages, and locally important gas from Lower Silurian sandstones along the Niagara peninsula.

Paleozoic petroleum along the northwest side of the Transcontinental Arch The thrust belt carrying western geosynclinal deposits over their American foreland was created in its present form during the Tertiary Laramide orogeny. We have seen that this *Western Overthrust Belt* bypassed the Permian Basin; its eastern front lies close to the Texas/Mexico border and is much buried beneath younger cover rocks. The front then swings far to the west around the southwestern prong of the Transcontinental Arch, and it was thereby enabled also to bypass the Colorado plateau. It is again obscured by younger rocks in the Basin-and-Range province, which occupies most of Nevada, but the front reappears in central and northern Utah and western Wyoming, trending almost due north. It again disappears under the volcanic blanket of Yellowstone Park, and on its further reappearance to the north, in Montana, it becomes continuous with the thrust front of the Canadian Rocky Mountains.

In Wyoming and Utah, only relatively minor discoveries have so far been made in Paleozoic strata, most of the attention being focused on the Mesozoic part of the section. Pennsylvanian sandstones and Mississippian, Devonian, and Ordovician carbonate rocks have yielded oil, gas, and condensate in several large anticlinal structures within the thrust belt. The hydrocarbons in the carbonate reservoirs are very sour, and in most of the structures the reservoir rocks lie at depths of 4 km or greater.

Between the Overthrust Belt and the Transcontinental Arch lie the northern *US Rocky Mountains*. They constitute a geologic province unique in the Americas, perhaps in the world. It consists of a complex of high but sharply discrete mountain ranges separated by deep basins, having utterly disparate trends and vergences. Within this complex, most of the basins are or have been oil- or gas-productive, and some of the production has been provided by the Paleozoic part of the section.

The southwestern end of the Transcontinental Arch became the Colorado plateau, which escaped all Phanerozoic folding episodes. During Paleozoic time it lay within the carbonate shelf, which was fragmented in the Pennsylvanian by the uplift of giant wedges of crystalline Precambrian basement rock to form the Ancestral Rocky Mountains. The wedges isolated three basins within the *Four Corners region*, so called because it surrounds the only point in the USA at which four states meet. In the *Paradox Basin*, Pennsylvanian red beds and evaporites (including thick salt) were deposited unconformably across all older rocks. A succession of thin black shales, rich in organic matter, interrupted this restricted sedimentary cycle, crossing all other Pennsylvanian lithologies (including the salt) as effective time markers. One of the black shales was intruded by an Oligocene syenite sill, which contains oil in one small field (see Sec. 13.4.9). The important oil in the basin occurs in algal bank carbonates along the hinge line between the shelf and the salt basin. Minor oil and gas also occur in Mississippian and Devonian carbonates, and heavy tar oil in the thick Permian arkose sloughed off the basement uplift and flooded across the basin. The Pennsylvanian limestone reservoir rocks also yield gas high in CO_2. Salt uplifts had begun before the end of Permian time, and constitute the most obvious structures in the basin.

The Pennsylvanian carbonates overlap to the southeast into the neighboring *San Juan Basin*, where they are trivially productive of light oil and gas. The Permian limestone similarly overlaps to the west into the *Black Mesa–Kaiparowits Basin* in Arizona and Utah; it there forms the surface rocks on either side of the Grand Canyon. It also contains a trivial amount of oil.

North of the Pennsylvanian uplift lie the *Uinta* and *Green River basins*, separated by an east–west uplift of Precambrian metasedimentary rocks. The two basins are essentially Cretaceous and Tertiary features, and contain the early Tertiary oil shales (see Sec. 10.8.1). Each basin is bounded on the east, close to the Transcontinental Arch, by a meridional uplift exposing old rock; anticlines associated with these uplifts contain oil and some gas in a Pennsylvanian sandstone of very low permeability (petrographically virtually a quartzite), and a little oil in the ubiquitous Mississippian limestone. Such occurrences are routine in most basins of the US Rocky Mountain province, but a genuine peculiarity occurs on the uplift east of the Green River Basin. The largest individual accumulations of oil there are in twin domes in the basal Cambrian sandstone, resting directly on the Precambrian basement. The Cambrian is unconformably overlapped by Carboniferous strata and the oil has probably migrated from these into the stratigraphically older but structurally higher reservoir rocks. The occurrences give the Green River Basin the unusual distinction of providing oil production from Lower Paleozoic, Upper Paleozoic, Mesozoic and Cenozoic reservoir rocks, and from every post-Devonian system; the production from both Cambrian and Tertiary strata must represent a unique combination.

The *Big Horn* and *Wind River basins* to the north are almost surrounded, as well as separated, by high blocks bounded by thrusts or reverse faults. They are oil-prone in Paleozoic rocks and somewhat gas-prone in the Mesozoic. The *Powder River Basin* is the foredeep of the northern US Rocky Mountains, bounded on its east side partly by the Transcontinental Arch and partly by a faulted anticline that separates it from the Williston Basin (Fig. 22.6). It is primarily a Cretaceous basin, its Paleozoic production having been modest.

None of these basins existed as structural entities until Cretaceous time, culminating in the early Tertiary Laramide orogeny that elevated the high-standing blocks and depressed the basins between them. During Paleozoic time, however, they lay between the Transcontinental Arch and the proto-Cordilleran geosyncline, in a belt in which mid-Permian strata were in the organic-rich facies of the Phosphoria Formation. Nearly all their Paleozoic production consequently comes from Pennsylvanian and Lower Permian sandstones.

The *Big Horn Basin* is the largest oil-producing basin in the Rocky Mountain province, and should eventually produce some 4×10^8 m^3 of oil from Paleozoic reservoirs. The *Powder River Basin* is oil-prone and seems underprovided with gas; perhaps its Phosphoria-generated oil lost most of its gas during the long time lag before the formation of the Laramide structures. Neither the Powder River nor the Wind River basin contains any giant individual fields in Paleozoic reservoirs; they will probably yield less than 1×10^8 m^3 of oil from them.

The great majority of the traps in these basins are obvious anticlines, aligned in conformity with one or both flanks of the basins. The *Wind River Basin*, for example, is a classic half-basin with its axis close to its northeastern margin. Fields in it are nearly all along its southwestern side, in anticlines having high-angle reverse faults on their southwest flanks. This basin presents a feature unusual even within the Rocky Mountain province. Its present vergence is towards the southwest, so that its Cretaceous and Cenozoic sedimentary fill thickens northeastward, opposed to the westward thickening of the underlying Paleozoic–Jurassic systems.

A noteworthy feature is the long time lag between the deposition of the Paleozoic source sediment (Permian) and the creation of the present traps (late Paleocene). The oil had probably already been generated and partially pooled before the Laramide traps were completed; much of it was no doubt lost, and what remained remigrated into the new traps where it became affected by meteoric waters invading the reservoirs from the new mountain outcrops (see Sec. 17.7.8). In the Powder River and Wind River basins, most of the Paleozoic oil is of medium to low gravity and rather high in sulfur. In the Big Horn Basin, gas and distillate occur in what had been the center of a Phosphoria sub-basin, with lower-gravity crude around its margins. The basin provided the type examples for King Hubbert's famous study of tilted oil/water interfaces.

Lying east of all these Rocky Mountain basins is the *Denver Basin*, the principal Cretaceous foredeep of the mountain belt. It lay wholly within the Transcontinental Arch and did not achieve an identity of its own until the Cretaceous. Its insignificant Paleozoic production has come from Permian and Pennsylvanian sandstones derived from the Pennsylvanian uplift of the Front Range.

The largest of all the basins associated with the northern flank of the Transcontinental Arch is the *Williston Basin*, occupying most of the state of North Dakota but extending into adjoining states and into Canada (Fig. 17.18). Though conventionally included among the basins of the Rocky Mountain province, it is a long way in front of the mountains and wholly underlain by Canadian Shield. The saucer-like basin therefore has no flanking mountain belts; it provided one of our examples of the cratonic sag type of basin in Ch. 17. Clastic formations are therefore rare in the basin, at least among those antedating the first deformation of the Rocky Mountains in Pennsylvanian time.

The basin first becomes identifiable in strata of mid-Ordovician age, and thenceforward to the end of the Mississippian its content was almost wholly in the carbonate–evaporite facies. It includes a great volume of salt and anhydrite of several ages, the most important, in the Middle Devonian, providing the potash production in central Canada. All the major carbonate units, as well as some sandstones like the one of late Cambrian age at the bottom of the column, are productive of oil, with gas or gas-condensate minor throughout. The basin was still subsiding during Jurassic time, and received yet another evaporite formation of that age. The blanket sandstone at the base of the Jurassic, unconformable on the Mississippian carbonates, has itself yielded minor oil production.

As in other cratonic-sag basins (Illinois, Michigan), the most important oil-bearing structures are related to early-formed, linear anticlines, some of them fault-controlled, transverse to the axis of the Transcontinental Arch. Though productive fields are very numerous in the Williston Basin, and exploration (especially for deep targets in the Ordovician dolomites) continued to be statistically very successful 30 years after the initial discoveries, no individually large fields are known. Among the largest are those in Mississippian carbonate reservoirs in the Saskatchewan sector of the basin. Both interbedded with and sealed above by evaporites, the traps are in effect combined lithologic, stratigraphic, and paleogeomorphic in character (Fig. 16.86).

The Western Canadian Basin (Fig. 25.1) During late Proterozoic and Paleozoic times, the western margin of the Canadian Shield (then probably its northern margin)

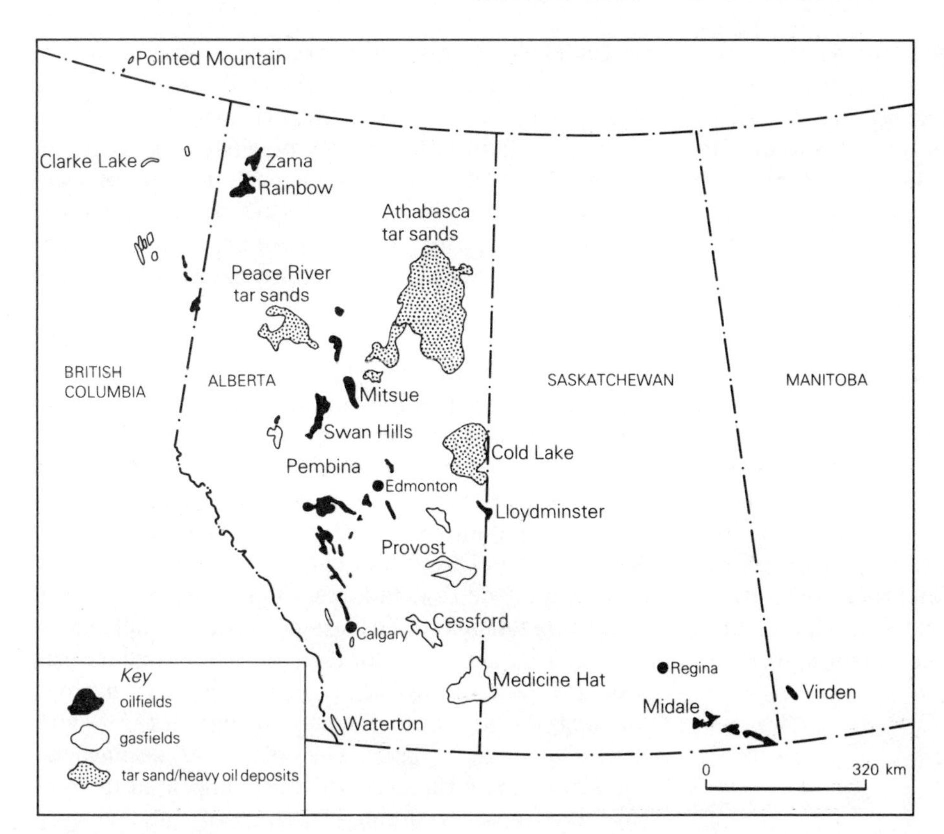

Figure 25.1 Principal oilfields, gasfields, and tar sand/heavy oil deposits in the Western Canadian Basin. (After P. W. Hay and D. C. Robertson, *AAPG Bull.*, 1983.)

had developed a thick wedge of continental shelf sediments from which the Canadian Rocky Mountains were ultimately to be raised. Because the Paleozoic equator passed through North America in a direction now nearly north–south, the seas were warm and their dominant deposits were calcareous. An intra-Devonian disturbance had interrupted the shelf-building cycle and created the first new continental crust west of the shield. The Devonian seaway was therefore closed on what is now the west, where the Cordillera now rise. The seaway extended, however, from north of what is now the Arctic Circle into the USA; once its transgression reached its maximum, it occupied both the Western Canadian Basin and the Williston Basin.

The seaway was periodically fringed by barrier reefs, dotted with pinnacle reefs, and blanketed by evaporites. At least three reef-bearing horizons of Middle Devonian age are productive of oil, and of lesser amounts of gas, in the northwestern part of the basin. Gas dominates in the far northwest. Some oil has also been produced from granite wash resting directly on the Precambrian around the edges of residual basement highs. At least four carbonate formations of Late Devonian age are much more richly productive of both light oil and gas; they include tabular and pinnacle reefs and widespread banks between and behind them. Both evaporites and shales are efficient seals.

The productive strata beneath the prairies have remained quite unfolded, but a few anticlinal traps were created by normal faulting. In the foothills of the Rocky Mountains to the west, however, the strata were thrown into parallel, linear folds by eastwardly-directed overthrusting (see Sec. 16.3). The disturbed belt is a gas province in Paleozoic rocks. Many of the anticlinal structures in its southern sector yield sour gas from Upper Devonian and Mississippian dolomites; farther north, about the British Columbia/Yukon border, the principal reservoir rocks for the gas are Middle Devonian carbonates. East of the foothills, in the western prairies, gas is fairly abundant in Mississippian dolomites, but oil is restricted to narrow traps in the updip, eroded edges of individual carbonate units; it is subordinate in significance to the oil production from equivalent Mississippian formations in the Williston Basin farther east. A little light oil is found in Pennsylvanian sandstone in an intermontane basin in the northern Yukon.

The Western Canadian Basin will probably yield about 2×10^9 m^3 of conventional oil from Upper Paleozoic reservoirs, plus a considerable quantity of natural gas. As nearly all the gas is very sour, sulfur is an exceedingly abundant byproduct. A much greater volume of very heavy oil (perhaps as much as 50×10^9 m^3) is potentially available from Upper Paleozoic carbonate reservoirs in a band across northern Alberta, underlying the Lower Cretaceous sandstones that contain the heavy oil and "tar sands" described in Section 10.8.4.

25.2.5 Paleozoic oil in the American Arctic

Nearly all the oil and gas in the North American Arctic comes from Mesozoic sandstone reservoirs (Sec. 25.5.12). At three locations, however, oil is known in Paleozoic reservoirs, and one of them is commercially productive. The oldest of the several reservoirs in the Prudhoe Bay field, on Alaska's North Slope, is a dolomitic calcarenite of Carboniferous age; it yields both oil and gas. Northeastward and offshore from Prudhoe Bay, the Lower Cretaceous unconformity (Fig. 26.11) cuts more and more deeply into the underlying section until the Cretaceous rests directly on the Lower Carboniferous. Oil then occurs not only in the carbonate reservoir but also in the basal clastic member of the Mississippian section, a conglomeratic sandstone resting on the meta-sedimentary basement complex. East of the Mackenzie delta, on the northwestern edge of the Canadian Shield, gas occurs in a Cambrian sandstone. A very small amount of oil has also been found in a Devonian reef at a single locality in Canada's Sverdrup Basin.

25.2.6 Paleozoic oil and gas in the Gondwana continents

Those pieces of the Paleozoic Gondwana continent still in the Southern Hemisphere supply only minor oil or gas production from Paleozoic strata. The present Andes Mountains of *South America* are a Tertiary phenomenon; they are superimposed, however, on both Caledonian (largely Devonian) and Hercynian (largely Permian) predecessor segments of a variety of trends. Both earlier ages of deformation are represented in Peru, Bolivia, and northern Argentina. A Devonian basin extends also into the plains region east of the mountains; in southern Bolivia, this basin is overprinted by a series of Permo-Triassic basins.

The regime was almost entirely clastic, that in post-Devonian time essentially nonmarine and with important glacial components. The small fields, of very light oil and condensate, occupy a series of narrow, meridional anticlines extending from southern Bolivia (mostly in Devonian sandstones) to the *Salta* region of northern Argentina (in Carboniferous, Permian, and Triassic sandstones, with gas and gas-condensate dominant in the Lower Devonian). Opposite the elbow-bend of the mountain front in Bolivia, in the *Santa Cruz Basin*, oil-bearing intervals occur in both the Devonian and the Carboniferous. There may be source sediments in both systems, even though no post-Devonian rocks are completely marine.

The vast Irati oil-shale deposit in the Permian of southern Brazil (see Sec. 10.8.1) is not known to have yielded any conventional oil accumulation. Gas occurs in the *Upper Amazon Basin*, in a sandstone at the base of a Pennsylvanian section resting unconformably on Devonian that includes another oil-shale formation. On the other side of the Andes, the early Tertiary production of the century-old fields of north-coastal Peru (Sec. 25.6.4) has been trivially augmented from the semi-metamorphic Pennsylvanian strata below the regional unconformity.

Australia has found small quantities of light oil and wet gas in fractured Cambrian and Ordovician sandstones in the *Amadeus Basin*, in the dead-center of the continent. Potentially larger gas deposits occur in older clastic reservoirs in the Lower Cambrian and the Upper Proterozoic. The traps are in exposed anticlines within almost mountainous topography. Some gas is also produced from Devonian sandstones in Queensland, in front of the Paleozoic mountain range which forms the eastern highlands of Australia. One of petroleum geology's greatest exploratory disappointments has been the lack of significant success in the *Fitzroy–Canning Basin* in the northern part of Western Australia. With a classic Devonian reef display and abundant large structures, the basin appeared to combine the most favorable features of the Western Canadian and Permian basins. Some Upper Devonian reef oil has been found in both the Canning and *Bonaparte Gulf* basins, but neither basin has joined the ranks of productive regions at the time of writing.

North Africa possesses the only known, large reserves of oil in Paleozoic strata outside North America and the Ural–Volga Basin. The whole of North Africa, south of the Atlas Mountains and the Mediterranean, is a vast platform of Precambrian rock formerly covered by a considerable thickness of Paleozoic clastic sediments. Differential uplift through Hercynian deformation to the north led to the total removal of these sediments from high-standing massifs, leaving varying thicknesses in the intervening depressions, and spreading a thick but discontinuous detrital blanket – the Nubian sandstone – over much of the region. The basal sandstone unit now resting on the Precambrian floor may therefore be of four distinct ages and provenances: it may be the Cambro-Ordovician sandstone, at least in part of wind-blown derivation, which represented the earliest Paleozoic cover over the basement; it may be the Nubian sandstone, of different ages in different places but commonly Carboniferous or perhaps younger; it may be a continental residuum or "granite wash" of any post-Cambrian but pre-Cretaceous age; or it may be the basal sandstone of the Mesozoic (commonly Cretaceous) transgression. All of these basal sandstones are productive in one place or another.

There are four principal areas of Paleozoic production. In northern *Algeria* is the *Ouargla* or *Eastern Erg*

Basin, containing Triassic and younger sedimentary rocks resting unconformably on the Lower Paleozoic. Nearly all the oil is found in the Cambro-Ordovician sandstone reservoir at the bottom of the section, at depths of 3–4 km. Though the area of known oil occurrence in this formation is spread over some 10° of longitude, from that of Algiers to that of Tripoli in Libya, production is dominated by the great accumulation at Hassi Messaoud, described in Chapter 26. It is the largest Lower Paleozoic oilfield known in the world, and should ultimately yield more than 10^9 m^3 of production. The oil is very light, aromatic, and undersaturated. The reservoir rock in its anticlinal trap is unconformably capped by Triassic salt. The source rock, which is either Silurian or Devonian in age, is therefore now restricted to the flanks of the structure and it seems likely that the present accumulation is the second that this source has supplied to this trap.

Farther to the southeast, close to the Algeria/Libya border, is the *Polignac* or *Illizi Basin*, another wholly cratonic basin that retained a much more complete Paleozoic section. Here most of the oil occurs in Devonian sandstones, with lesser amounts in the Carboniferous, in large, faulted anticlines. An oddity is a single field in which both oil and gas occur in an Ordovician clastic reservoir considered to be a tillite (see Sec. 13.2.7). The older oils, again probably derived from the Silurian graptolitic shales, are aromatic; the younger ones are paraffinic and were more probably derived from a Devonian source. These two Algerian basins had yielded 10^9 m^3 of oil by the end of 1982. They had also yielded some gas from Paleozoic strata, but this is totally overshadowed by the gas output from Mesozoic rocks (see Sec. 25.3).

The *Sirte Basin* of Libya is a creation of the Cretaceous and it is described with the other great petroleum basins of that age (Sec. 25.5.1). A number of its fields, however, produce oil also from fractured Precambrian basement, from the basal Cambro-Ordovician sandstone, or from other basal sandstones that may be Paleozoic in age. In several of the largest individual fields, either the Cambro-Ordovician sandstone or the fractured basement contributes the major part of the production (Fig. 16.84).

The fourth region of North African Paleozoic production is that occupied by the *Suez graben* in Egypt. The graben itself is of Oligo-Miocene origin, and by far the greater part of its oil output comes from Miocene reservoir rocks. In several fields, however, principally in those along the west side of the graben, there are traps controlled by tilted fault blocks below the sub-Tertiary unconformity, and considerable oil production has been derived from Carboniferous sandstones in these traps. The sandstones, which lie unconformably on the Precambrian basement of the Afro-Arabian platform, are stratigraphically associated with black shales, some of them manganiferous; both the sands and the shales are probably nonmarine. Oil has also been produced from the "Nubian sandstone," which has itself been called Carboniferous here but may easily be younger.

A fifth region of Paleozoic oil production in Africa may be developed in the improbable area offshore from *Ghana*. There, in a geologic province in which oil and gas are otherwise confined to Mesozoic and Cenozoic levels, oil has been found in older sandstones assigned to the Carboniferous or the Devonian.

The great oilfields of the Persian Gulf Basin all produce from late Mesozoic and Tertiary reservoir rocks. The great gasfields of the same basin are principally in Permian carbonates and are described with other gas deposits of that age in the next section. It has become apparent, however, that there are hydrocarbon accumulations in still older rocks, gathered in large domal structures probably caused by uprises of Cambrian salt. Paleozoic fields known to the time of writing are in *Oman* and *Dhofar*, at the eastern end of the Arabian peninsula and therefore in Paleozoic time close to or on the Gondwana continent. Several fields yield both oil and gas from a Carbo-Permian sequence of colored clastics and limestones; the base of this sequence is a probable tillite, lying above a major unconformity. Below this unconformity, still other, rather small oil accumulations have been found, in clastic rocks of early Paleozoic age, probably Ordovician or Silurian.

25.3 Natural gas in the Permo-Triassic

Rocks of the Permian and Triassic systems represent a low incidence of crude oil in the world. The low incidence, like an earlier low incidence in the Silurian and Early Devonian, coincides with an evaporite maximum. Outside the southern USA there are no giant oilfields producing wholly from Permian reservoir rocks (or from Silurian reservoir rocks); the only giant oilfield producing principally from a Triassic reservoir is Prudhoe Bay in northern Alaska (Sec. 26.2.2).

In contrast, there is an extraordinary assortment of giant and supergiant gasfields in Asia, Europe, North America, Africa, and Australia producing from Permian or Triassic reservoir rocks. Even where Permian or Triassic oil is prolific, as in western Texas or northern Alaska, great quantities of gas also occur. So much of the world's Permian and Triassic gas is nonassociated gas, however, that it deserves treatment independent of that given to either Paleozoic or Mesozoic oil–gas regions.

Much as the Pennsylvanian concentration of coal appears to have been consequent upon the assembly of large continental plates in subequatorial latitudes during that period, and the concentration of oil in the Jura-Cretaceous was related to initial separations in the latest continental drift episode, so the remarkable development of Permo-Triassic gas appears to have been connected with the initial breakup of the Gondwana continent (Pangaea).

The oil in the *Persian Gulf Basin* is in Jurassic and younger rocks, but an enormous volume of nonassociated gas, certainly exceeding 10×10^{12} m^3, is found in the Upper Permian Khuff limestone. An oddity is that, although large Permian gas reserves underlie giant Mesozoic oilfields like Awali in Bahrein and Umm Shaif in Abu Dhabi, even larger reserves occur in structures of their own (Northwest Dome off the coast of Qatar), some of them rather far removed from any known oil (Kangan and Pars fields in south-coastal Iran).

In North Africa, *Algeria's* Hassi R'Mel and allied fields contain more than 10^{12} m^3 of gas and condensate in Lower Triassic sandstones lying between a regional unconformity, below, and a thick salt seal, above. In northwestern Europe, gas reserves of more than 3×10^{12} m^3 occur in fields in the Lower Permian sandstone extending from the Elbe River in *Germany*, westward into *Holland* and thence northward into the British *North Sea*. About two-thirds of the reserves were contained in a single accumulation at Groningen in northern Holland (Fig. 16.10). This gas is agreed to have been generated in the underlying Pennsylvanian Coal Measures. It is high in nitrogen (up to 90 percent in the small accumulations in the Danish North Sea) and is perfectly sealed by the thick Permian evaporites. On the other side of Britain, in an arm off the *Irish Sea*, gas occurs in large quantities in a Triassic sandstone.

Though the greatest natural gas resources known are in Cretaceous sandstone reservoirs in the West Siberian Basin (Sec. 25.5.7), the Soviet Union's most productive gasfields during the 1970s had Permian carbonate reservoirs. The supergiant *Orenburg field* lies near the southern edge of the Ural–Volga Basin where the Russian platform drops off southward into the North Caspian salt basin (Fig. 16.43). Orenburg yields sour gas of very odd composition – more than 10 percent of it is either H_2S or nitrogen, and there are 28 p.p.m. of argon. The content of natural gas liquids makes the field the equivalent of a giant oilfield in addition. The field became the starting point for the world's first 1420 mm (56 inch) pipeline, the Soyuz line extending 2750 km to Czechoslovakia. Near the mouth of the Volga River in the *Emba Basin*, equally sour gas occurs in a deep limestone reservoir which is older than the Permian salt and probably of Permo-Pennsylvanian age. It is geographically close to oil in Pennsylvanian carbonates. In the *Dnieper–Donetz depression* in the Ukraine, prolific gas occurs in both sandstone and carbonate reservoirs in the Permian, associated with thick evaporites. The Shebelinka field dominated Russian gas production during the 1960s. In the Soviet Arctic, the *Timan–Pechora Basin* contains giant gasfields in Lower Permian limestones in thrust-faulted anticlines.

In Asiatic Russia, the *Vilyuy Basin* lies between the Verkhoyansk Mountains and the Angara shield. It contains giant gas reserves in sandstones of late Permian and Triassic ages; the source of the gas may again be in associated coal measures. The *Sichuan Basin* in central China contains both oil and gas in a number of horizons from late Precambrian age to Cretaceous, but it is primarily a gas province in Permian and Triassic limestones, associated with thick evaporites.

The known gas reserves of eastern *Thailand* are in a Permian limestone. *Australia* produces gas from Permian or Triassic reservoirs in four separate basins. In southeastern Queensland, gas in a Lower Permian nonmarine sandstone in the *Bowen–Surat Basin* is much more important than nearby oil. In the *Cooper Basin*, part of the Great Artesian Basin extending into the northeastern corner of South Australia, the reservoirs are in Permian and Triassic sandstones. In the *Perth Basin* along the west coast, they are in Permian sandstones and fractured limestones. On the *Dampier–Rankin shelf* off the northwest coast, gas and condensate reserves of at least 0.5×10^{12} m^3 are contained principally in horstblock traps in an Upper Triassic sandstone. The source is more likely to be Mesozoic (Jurassic?) than Paleozoic. The productive trend probably extends northeastward along the block-faulted shelf to the *Bonaparte Basin*, where gas occurs in Upper Permian rocks offshore.

In North America, half of the dozen largest gasfields are in Permian or Triassic reservoirs. In the *Sverdrup Basin* of the Canadian Arctic, the principal reservoir rocks are Upper Triassic and Lower Jurassic sandstones. At Prudhoe Bay, on the *North Slope of Alaska*, the largest oilfield in North America contains not only about 25×10^{10} m^3 of solution gas but an original gas cap of about 0.5×10^{12} m^3. The main reservoir is in a Permo-Triassic sandstone (Figs 13.43 & 16.1), and a remarkable circumstance is likely if the principal source of the *oil* is in Cretaceous shales unconformably covering the reservoir. Those shales are unlikely to be the source of the free gas cap, which would therefore consist of nonassociated gas with respect to the oil in the same trap. In the southern USA, the even larger Hugoton and Panhandle fields contain 2×10^{12} m^3 of gas in Permian dolomite and Pennsylvanian sandstone wedging out on the Amarillo uplift from the *Anadarko Basin* (see Sec. 25.2.4). In the *Permian Basin* itself, there is a great deal of gas as well as an extraordinary concentration of oil, and the largest single gasfield is in a Permian sandstone and carbonate reservoir.

25.4 The Persian Gulf Basin

The Persian Gulf Basin for the petroleum geologist occupies the two sides of the geographic gulf, and the eastern edge of the Iraq sunk land (the valley of the Tigris and Euphrates rivers), from Oman in the southeast to the northern border of Iraq in the northwest, a distance of about 2000 km. Though it takes in parts of a dozen countries, the basin is most readily described and understood as a single geologic entity. It contains more than half of all the conventionally recoverable oil ever discovered in the world, and great quantities of both associated and nonassociated gas. This concentration of riches in a single geologic province has no parallel for any other resource both valuable and common.

Reasons for this extraordinary richness have been set out by many geologists – Lees, Law, Kamen-Kaye, Murris, Tiratsoo; they are briefly outlined here. The basin contains a northeastwardly-thickening wedge of sedimentary rocks deposited on the northeastern edge of the Arabian subplate of the Gondwana continent. The subplate lay in equatorial latitudes on the southwest side of the Tethyan seaway; it therefore received sediment dominantly in the carbonate facies. Starting in Jurassic time, or possibly somewhat earlier, the subplate became a downgoing slab approaching a subduction zone beneath the Iranian subplate on the opposite side of Tethys. Though there were regional unconformities developed at the top of the Middle Jurassic, the top of the Aptian, the top of the Turonian, and the top of the Cretaceous, there was no protracted interruption of deposition, still dominantly of carbonate and evaporite sediments. On the smaller scale, however, the leading edge of the subplate became increasingly unstable as it approached the subduction zone, culminating in a major deformation of the trough ahead of it in about Turonian time. The final uplift of the new mountains (the Zagros Range of Iran) took place in Pliocene time, *before the downgoing slab with its cover of petroliferous sediments had been seriously consumed.* This is the critical factor. The basin lying west and north of the Brazilian and Guyana shields during the Cretaceous, or that between the Indian and Asian plates about the same time, may have been as petroliferous as the Persian Gulf Basin had their consumption below their bordering orogenic belts ceased earlier.

This steady progression, halted at a fortunate stage, had highly favorable consequences for both the stratigraphy and the structure of the basin. Though the carbonate–evaporite depositional regime was remarkably uniform and continuous, the episodic instability introduced just sufficient variety to permit the accumulation of three rich source sediments – one during the Oxfordian–Kimmeridgian in a long-lived depression on the Arabian foreland, one in the Albian–

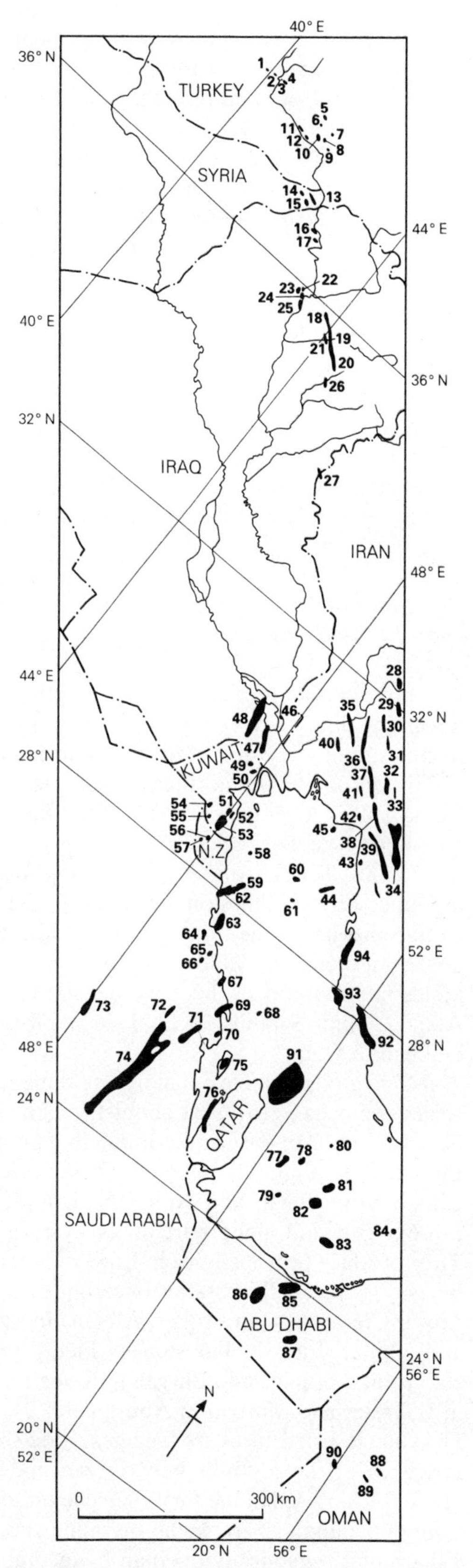

Figure 25.2 Distribution of oil- and gasfields in the Persian Gulf Basin. Ghawar is no. 74, Burgan no. 53, Kirkuk nos 18–20, Gach Saran no. 34, and North-West Dome gasfield no. 91. (From H. V. Dunnington, *J. Inst. Petrolm*, 1967.)

Cenomanian along the general axis of the present basin, and one in the Paleocene and early Eocene along what is now the mountain front and foothills belt of Iran. The episodic instability also caused the introduction of thick sandstone members into these generative basins, during the Neocomian and Albian. The thick evaporites capping the principal carbonate reservoir rocks make adequate (though imperfect) seals. Two sedimentational features critical to the prolific oil yield of the region are difficult to explain; there is scarcely any dolomitization, despite the permeable carbonates (including reefs) and the abundant evaporites; and both carbonate and sandstone reservoirs are almost free of cementation. Some of the detrital and bioclastic grainstones of the late Jurassic Arab Group in Saudi Arabia, separated by bedded anhydrites, effectively share the reservoir characteristics of the barely consolidated Albian sandstones of Kuwait, separated by shales.

The structural development was comparably favorable. Structures of three general ages and three general styles occupy three general areal belts (Fig. 25.2):

(a) Far out on the foreland, nearly north-striking linear anticlines with very gentle flank dips began to grow about at the time of the first deformation of the trough far to the northeast (Turonian–early Coniacian), probably as a result of faulting in the basement (see Sec. 16.3.2). Their growth continued until Tertiary time, trapping oil principally in Upper Jurassic limestones but also in the basal Cretaceous limestone, in Albian sandstones, and lesser amounts in Middle Jurassic limestones. Fields in structures of this style include Ghawar, Abqaiq, and Safaniya–Khafji in Arabia and Dukhan in Qatar.

(b) A second group of large but more equidimensional structures began growth at about the same time, but over uplifts of the Cambrian salt that underlies the whole stratigraphic section. These structures extend from Oman and Abu Dhabi, under the present gulf and along part of its western side. They produce principally from Lower Cretaceous limestones and Albian sandstones but also from Lower Cretaceous sandstones, Albian limestones, and Upper Jurassic limestones. Fields in this structural group include Burgan in Kuwait, Awali in Bahrein, and Murban in Abu Dhabi.

(c) The youngest structures are the linear, box-shaped anticlines of the foothills belt of Iran and Iraq. These folds, raised in the final Pliocene episode of mountain building, are highly asymmetrical and disharmonic, chaotically overlain by incompetent marls and gypsum. The principal production is from the Oligo-Miocene Asmari limestone, in Iran, and the somewhat older (Eocene–Oligocene) limestone at Kirkuk in Iraq. Oil in these reservoir rocks almost certainly reached them by late, vertical migration from older accumulations below; many of the larger structures are richly productive (commonly of heavier oils) from Cenomanian–Turonian limestones as well as from the Asmari Formation, with almost unimpeded pressure communication between the two well separated levels. Fields in this group include all the early finds of Iran and Iraq – Masjid-i-Suleiman, Agha Jari, Gach Saran, and Kirkuk (Figs 13.72, 16.13 & 16.58).

Structures of all three types may also contain enormous, nonassociated natural gas accumulations in a Permo-Triassic dolomitic limestone, commonly fractured. The Kangan and Pars gasfields, in south-coastal Iran in type (c) folds, and the Northwest Dome field off Qatar, in a type (b) fold, may be the largest gasfields in the world outside Siberia, and several others below type (a) oilfields are also very large.

The sizes of the structures of all three types – at least half a dozen of them are more than 50 km long, and closures in the thousands of meters are commonplace – permit very large individual accumulations and matchless productivity (see Sec. 13.3.10). The Ghawar field in Saudi Arabia and Burgan in Kuwait have ultimately recoverable reserves in excess of 10×10^9 m^3 of oil. Safaniya-Khafji, almost entirely offshore, Rumaila and Kirkuk in Iraq, and Berri in Saudi Arabia are certainly larger than 2×10^9 m^3 and at least a dozen other oilfields are not much below this yield level. Of the 50 largest conventional oilfields discovered in the world prior to 1984, all having recoverable reserves in excess of 0.5×10^9 m^3, no fewer than 30 are in this one basin. Fields so large inevitably contain large quantities of associated gas, but the foreland fields, especially those in type (a) structures, are nonetheless highly undersaturated.

The oils from this large basin are of remarkably constant quality, and probably all or most of them came from no more than two sources, both in Mesozoic rocks. Very little of the oil is lighter than 38° gravity, and few oils are heavier than 20° (all in Tertiary reservoirs, in northern Iraq and at the head of the gulf). Uniformly high sulfur content is a characteristic of the oils of all ages, especially of the heaviest.

At the extreme eastern edge of the basin, the Strait of Hormuz (which provides the only tanker route for the Middle East oilfields) cuts across the ophiolite thrust belt of the Oman and Makran Mountains. Within the overthrust belt, in the Trucial Coast emirates of Sharjah and Oman, a minor gas-condensate province occurs in carbonate reservoirs of the Lower Cretaceous and Upper Jurassic.

A limestone equivalent in age to the Asmari limestone

extends across central Iran into the interior *Qum Basin*, south of Tehran. This basin had begun to form in late Cretaceous time over metamorphosed basement rocks; it is now invaded by salt plugs. The Oligo-Miocene limestone is oil-productive from anticlinal traps in a few localities.

25.5 Oil and gas in Mesozoic strata

25.5.1 Mesozoic petroleum in the Mediterranean region

From the end of Paleozoic time until early in the Tertiary, southern and southeastern Europe, north Africa, and the Middle East formed the flanks of the western Tethys Ocean. During the later part of Mesozoic time, the flanks were consuming margins; the regime changed from one of extension to one of compression early in the Cretaceous. Eastward from Oman, the Mesozoic basins did not become completely closed until Tertiary time and all significant oil there is in Cenozoic strata. From the Persian Gulf to Gibraltar, however, the greater part of it is in the reservoirs of a vast, Mesozoic carbonate ramp (see Sec. 13.3.3).

Southeastern Turkey and northeastern Syria are properly parts of the Middle East oil province (Fig. 25.2) but their oil accumulations are sufficiently distinct to be better treated with the Mediterranean province. All the principal fields lie on anticlines trending along the front of the east–west Taurus Mountains, and concentrated in the re-entrant between those mountains and the NW–SE fold belt of the Zagros ranges in Iraq and Iran. In *Syria*, light oil and gas-condensate occur in Upper Triassic dolomites capped by anhydrite; some oil and gas also occur in Jurassic carbonates. Most of the production, however, has been of heavy oil from Cretaceous carbonates, including some reefs but being mostly cavernous and fissured, massive limestone. In *Turkey*, light oil is essentially restricted to dolomitized shelf carbonates below the mid-Cretaceous unconformity. The bulk of the output is of heavy, high-sulfur oil in Upper Cretaceous limestones, again including reefs. The source is probably in overlying shales of Maestrichtian–Paleocene age. There is one large field (Souedie) in Syria, but no large ones in Turkey. Turkey resembles Cuba in that most of its numerous seepages are associated with serpentinites within strongly deformed strata, but no significant accumulations have been found in such an environment.

Israel's trivial oil and gas endowment is in Jurassic limestone reservoirs, with a little in a Triassic sandstone also. In the Alamein region of *Egypt's* western desert, light oil and gas occur in a Lower Cretaceous dolomite that appears to be primary; some is found in a younger Cretaceous sandstone. Minor oil production has been derived from Aptian–Albian reefs in northern *Tunisia*, and from younger Cretaceous reefs in the *Pelagian Basin* (Gulf of Gabes), which extends into western Libya. Tunisia's largest known field is in a Triassic sandstone. Small production in *Morocco* has been derived from fractured Jurassic dolomites.

Sicily produces exceptionally heavy oil (some as heavy as 7.5° API, and containing up to 10 percent of sulfur) from thick, fractured Triassic dolomites. The source of the oil is probably in black shales immediately overlying the reservoir formation. There are several unusual and interesting aspects to the field area, which is in the southeastern corner of the island and off its south coast. It is an important center of Greek archeology; asphaltic impregnations in outcropping Tertiary limestones were known to the ancients. The Mesozoic section is unusually complete for a depositional region immediately in front of a major thrust belt and itself now bounded by large faults. Both the geometries of the fields and the nature of the oils have been markedly affected by sulfate-rich waters entering the reservoir rock downwards along faults (Fig. 16.97). Despite the extremely low API values, the fields have gas caps, with high CO_2 contents; in some offshore wells, oil heavier than water is able to flow. Several structures higher than those productive of oil are unproductive. Heavy oil has been found offshore in the Mediterranean, again in highly fractured carbonate rocks but here apparently of early Jurassic age. Until 1982, almost 80 percent of Italy's cumulative oil production of about 44×10^6 m^3 had come from the two largest Sicilian fields, Gela and Ragusa.

Partly circumscribed by the occurrences in Egypt, Tunisia, and Sicily is the *Sirte Basin* of Libya, more productive by an order of magnitude than all the others combined. Over most of the Sahara Desert the Mesozoic is largely continental in character. Its deposition followed general uplift during the Hercynian. In northern Libya, however, severe block faulting created a complex graben system with high relief; into this system a marine transgression took place in Albian–Cenomanian time. The basin became completely submerged by the end of the Cretaceous and marine sedimentation was essentially uninterrupted until the mid-Miocene.

The principal source sediments are Late Cretaceous but the reservoir rocks are of remarkably varied ages. In prominent basement highs, both fractured Precambrian basement and basal Paleozoic sandstones are richly productive (see Sec. 25.2.6) where overlapped by the Cretaceous source sediments. The basal sands of the Cretaceous transgression tend to be concentrated in troughs and absent from "bald" faulted uplifts. These Cretaceous sandstones are also important reservoir rocks and provide that at Sarir, the largest field in the basin. The commonest reservoirs, however, are in Cretaceous and Paleocene carbonates, especially Paleocene reefs. Nearly all the traps are anticlinal uplifts controlled by faulted basement.

The oil is of very high quality, lighter, sweeter, and more paraffinic than that in most basins containing so much carbonate and evaporite. There is also a great deal of associated gas. More than any other petroliferous basin this one deserves to be considered a small counterpart of the Persian Gulf Basin in the important sense that its oil is gathered into a relatively small number of individually large fields and not dispersed in hundreds of small ones. The basin contains more than a dozen giant fields (only one of which is primarily a gasfield) in an area of little more than 150 000 km^2. The fields should eventually produce at least 5×10^9 m^3 of oil and 1×10^{12} m^3 of gas.

The Tarragona basin of *Spain* lies between the mouth of the Ebro River and the Balearic Islands. Light, sweet oil occurs in a Jurassic limestone, but its source is probably in Miocene pelagic marlstones which overlie the reservoir rock with great unconformity. The most southwesterly of the fields known in 1982 (Amposta Marina) occurs in cavernous dolomitic limestone of Aptian age in the weathered zone below the same unconformity; the oil is asphaltic, sulfurous, and deeply degraded.

Oil and gas occur in Mesozoic carbonate reservoirs both north and south of the *Pyrenees*. The reservoirs are concentrated above and below the Jurassic–Cretaceous unconformity, in limestone and dolomite associated with anhydrite. The large uplifts providing the traps are cored by Triassic salt, and the resulting fracture permeability is the ingredient rendering the accumulations exploitable. In Cantabria, northern *Spain*, the fields are small; some are reservoired in Jura-Cretaceous sandstones. On the French side, the *Aquitaine Basin* south of Bordeaux is much more productive than the Spanish side, though chiefly of gas and at depths approaching 4 km. Both oil and gas are highly sulfurous. The Lacq gasfield (Fig. 16.21), one of the largest in Europe, should ultimately yield at least 0.2×10^{12} m^3 of gas, but it also yielded (in 1980) more recoverable sulfur than any other single gasfield in the world. A younger limestone (Senonian in age) yields oil from the same structure. The source of the oil in all reservoirs is probably in the Jurassic. Future production will no doubt extend offshore, off northern Spain, again from carbonate rocks below the Cretaceous unconformity. Within the Pyrenean thrust belt itself, gas occurs at great depth in a Cretaceous carbonate (and some in overthrust Eocene flysch).

Halfway up *Italy's* east coast, heavy oil of 15–16° gravity occurs in Cretaceous limestones. In the *Po Valley Basin* of northern Italy, the main oil-bearing system, as in Sicily, is the Triassic (in a Norian dolomite), with lesser amounts in a Lower Cretaceous limestone; but the contrast with Sicily could not be greater. Instead of oil heavier than 10°API, the oil from fields near Milan is lighter than 50° gravity and associated with considerable gas-condensate. In the very deep reservoir (part of it below 6 km), pressures are exceedingly high and the hydrocarbons are in the gas phase (see Chs 14 & 23). The source rock of this deep oil is believed to be a black shale of Rhaetic age, unconformably overlain by Cretaceous with the Jurassic uncharacteristically missing.

Across the northern end of the Adriatic Sea, on *Yugoslavia's* Istrian Peninsula, folded and fractured Mesozoic carbonates provide the type karst topography; offshore, they contain gas. On the other (eastern) side of the Dinaric Alps, further gas, exceedingly sour, occurs at considerable depths in the Triassic of the *Drava Basin*. The Pannonian Basin of *Hungary*, an enlargement of the Drava Basin, is a Neogene feature and only weakly petroliferous. However, Triassic and Cretaceous dolomites yield further oil from below the Mesozoic floor of the basin. On the Moesian platform of southern *Romania*, Triassic, Jurassic, and Cretaceous carbonates are again productive on a small scale. Far to the east, in *Azerbaijan* (north of the Armenian Ranges), oil occurs in an Upper Cretaceous clastic reservoir (and some in Middle Eocene sandstone and in weathered basalt also).

All the occurrences of oil and gas in Mesozoic carbonates of southern Europe, from Spain to the Caspian, are essentially alike. Their locales were parts of microplates caught between the African plate and Hercynian Europe during the closure of Tethys. Most of them lie within or close to thrust belts which contain thick Jurassic, Cretaceous, or Paleogene flysch and above which were superimposed Neogene extensional basins, many of them themselves independently petroliferous.

25.5.2 Mesozoic petroleum from the European plate

North of the Alpine ranges lies the buried Hercynian platform of western, central, and eastern Europe. There the Mesozoic succession is much less calcareous, except from the late Jurassic onwards, and relatively little of the Mesozoic oil or gas is found in carbonate reservoirs. The easternmost occurrences are in fractured Jurassic and Cretaceous carbonates in the northern foredeep of the *Caucasus Mountains*, west of the northern Caspian Sea. Some oil also occurs in Triassic, Jurassic, and Lower Cretaceous sandstones. It is utterly subordinate in historical output to that from Tertiary strata, but some of the former may have contributed to some of the latter.

The Hungarian Basin is connected to the northwest to the *Vienna Basin* of Austria. This is a half-graben that did not originate in its present form until the Miocene, and all its early production was derived from Miocene and younger strata (Sec. 25.6.1). The graben was created, however, within the zone of nappes of the Calcareous Alps, which in turn were overthrust on to the Alpine flysch, which had transgressed across autochthonous Mesozoic rocks deposited on the crystalline Bohemian massif. Deeper drilling discovered both oil and gas in Upper Triassic and Jurassic dolomites in the allochthonous Alpine nappes below the basin floor. Production from these much deeper zones may eventually surpass that from the Neogene reservoirs of the shallow graben. Mesozoic oil may indeed have provided, by leakage, much of the asphaltic oil in the Miocene Matzen field, the output of which dominated that of the whole basin before the deeper zones were discovered.

The *Molasse Zone* of Alpine detritus overlies the southern margin of the Hercynian platform from northeastern Austria, southern Germany, and northern Switzerland, to southeastern France. Minor oil and gas occur in a series of thrust sheets below the molasse: in Triassic sandstones (chiefly gas); in Middle Jurassic sandstone; and in karst zones at the tops of Upper Jurassic and Lower Cretaceous limestone formations (as well as in the more important Oligocene sandstone; see Sec. 25.6.1).

The Zechstein Basin of northwestern Europe was a post-Hercynian creation. It was succeeded and extended by a classic Jurassic basin stretching across the whole width of Europe north of the Hercynian and south of the Caledonian ranges. This basin contains the oil (and a little of the gas) of West Germany, The Netherlands, northern France, and southern England. The *Northwest German* or *Hanover Basin* extends from Schleswig-Holstein, in the "neck" joining Germany and Denmark, into northern Holland. It had a complex deformational history, beginning in the Middle Jurassic with early salt movements and creating sharply differentiated linear uplifts separated by deep sedimentary troughs (Fig. 22.12). Rich oil-source sediments accumulated in these troughs through much of the Mesozoic section, with the most important being of Liassic age. Structural movements continued into Tertiary time, raising salt domes and walls (in the eastern part of the basin), strongly folded anticlines (farther west, where the salt is not intrusive), and a variety of stratigraphic and unconformity traps (Figs 16.51 & 87). Oil occurs in Triassic, Jurassic, Lower Cretaceous, and some Upper Cretaceous sandstones; reservoirs and traps extend eastward into the corner of the Baltic and westward into the North Sea. Offshore developments will assuredly give the basin an extended productive life. Gas is produced principally from Middle and Upper Jurassic sandstones. Uppermost Jurassic, Cretaceous, and Paleocene limestones are intensely fractured over salt domes onshore in Germany, and some oil is found in such fracture porosity.

In the *Rhine graben* (Sec. 25.6.1), very minor paraffinic oil has been produced from both Triassic and Jurassic limestones; there is no Cretaceous in the graben. Westward from it, the Jurassic basin widened greatly to form the *Anglo-Paris Basin*, which contains the type localities for most Jurassic and Cretaceous stratigraphic stages. The French part of the basin has been intensively studied by petroleum geologists, and much drilled. Though it contains a multitude of petroliferous Mesozoic horizons, and a source rock of earliest Jurassic age, production has been trivial (most of it from Triassic and Lower Cretaceous sandstones; a little from Jurassic carbonates). Across the English Channel in *south-coastal England*, somewhat more prolific oil occurs at depths of only 1000–1200 m in Upper Triassic and Lower Jurassic sandstones. Elsewhere in southern England, along the western extension of the Weald which separates the Tertiary London and Hampshire Basins, light oil has been found in the oolitic limestones so characteristic of western Europe's Middle Jurassic.

25.5.3 The North Sea Basin

The present North Sea Basin is a fourth-generation depression. Following the Caledonian orogeny, a large lake-basin was created over a bifurcation in the ranges between what are now Scotland and Scandinavia. It became filled with the molasse sediments of the Old Red Sandstone. During the Permian a rift-like seaway across this older basin joined the Zechstein salt basin of Europe to the world ocean. Throughout Jurassic time, complex block faulting broke the basin down into a major graben, now forming the central axis of the North Sea, with several subsidiary fault basins (Fig. 25.3).

The complex of horsts and grabens was extensively truncated by late Jurassic unconformities, overlapped by transgressive Kimmeridgian shales, and deformed by halokinetic movements of the Permian salt. A still greater transgression in the Aptian–Albian overspread all earlier rocks, and later Cretaceous and Tertiary sediments spread far beyond the confines of the graben and created the present North Sea Basin by simple, unfaulted subsidence. The basin has not undergone any compressional deformation since early Devonian time; all later structures are either taphrogenic, halokinetic, or both. Halokinesis continued into the Tertiary.

Fault block movements during the latest Triassic and most of the Jurassic permitted the deposition of a remarkable complex of fluvial, deltaic, and shallow marine sandstones, of very high primary porosities and permeabilities. Sandstones of Late Triassic and all

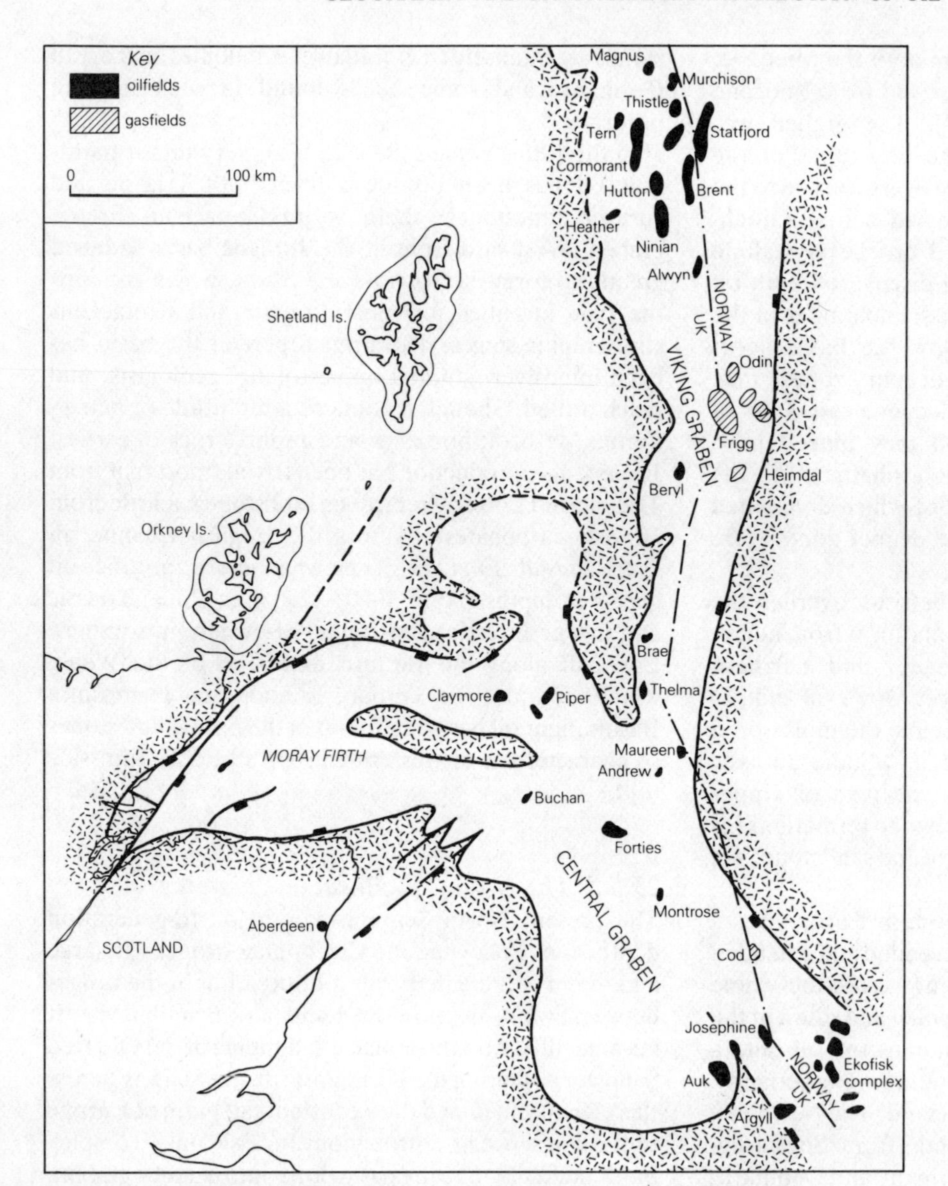

Figure 25.3 Generalized location map of Mesozoic basins and major oilfields in the northern North Sea. (From P. N. Linsley *et al., AAPG Memoir* 30, 1981.)

Jurassic ages become filled with oil (and much associated gas) from rich source sediments in the overlapping Kimmeridgian. Farther to the north, in the Norwegian North Sea, Liassic shales like those familiar in Britain and continental Europe are also source rocks; north of the 62nd parallel, still older source-type sediments are present in the Middle Triassic. The oils are light and of very high quality; some of them contain sufficient was to have high pour points. The principal traps were created by combinations of unconformities and westerly-tilted fault blocks, some of them forming true drape anticlines.

A remarkable aggregation of large oilfields, rich in associated gas, occupies a structural platform on the western side of the northern (Viking) graben, northeast of the Shetland Islands. East of this aggregation of oilfields, on the Norwegian shelf, one of the world's larger gasfields is reservoired in shallow Jurassic sandstones. Traps here are due to down-to-the-ocean growth faults formed in early Cretaceous time.

South of the oil "knot," in contrast, a group of large gasfields marks the deepest, axial part of the graben, the reservoirs being in thick Paleocene sandstones deposited in lobate submarine fans (Fig. 13.39). Still farther south, the central North Sea east of Scotland overlies an irregular basin formed by the junction of four linear grabens (Fig. 25.3). The principal reservoirs in this sector are in more Paleocene sandstones of submarine fan origin, exhibiting abrupt facies variations (see Sec. 26.2.5). Still farther south, near the geographic center of the present North Sea, comes a small area in which the large British and Norwegian sectors of the sea are joined by the narrowing Danish, German, and Dutch sectors. The area of junction happens to lie on the principal structural "high" crossing the sea from west (northeastern England) to east (Denmark). On the Norwegian side of the junction lies the first of the giant oilfields to be discovered in the North Sea (1969), now the center of the Ekofisk complex. As in the more northerly oil- and gasfields, the main reservoir at Ekofisk, and in the much smaller fields in the Danish sector, is of Paleocene age, but here it represents the top of the European Chalk and is accordingly referred to the Danian Stage. The source of the prolific oil and its associated gas, however, is still believed to be in the Upper Jurassic, which contains lesser amounts of oil in its own reservoirs. It is remarkable that a chalk should retain high primary porosity, probably to be credited to early introduction of hydrocarbons and retention of abnormally high reservoir pressures (see Secs 13.3.6 & 14.2).

The Paleocene reservoir rocks are younger than the fault phase of the basin. Their traps are due to gentle arching over deep uplifts of Permian salt. We have re-entered the main Permian (Zechstein) salt basin, and the principal hydrocarbon production in the southern North Sea is of natural gas from pre-salt (Lower Permian) desert sandstones; some comes from Triassic sandstones in addition (see Sec. 25.3).

The North Sea Basin contains more than a dozen giant oilfields in addition to many smaller ones; its ultimate oil production will be at least 4×10^9 m^3. Several of the giant oilfields are also giant gasfields. In addition there are at least seven giant, nonassociated gasfields, not counting the onshore supergiant at Groningen, near the Dutch coast. The gasfields will produce well in excess of 10^{12} m^3 of gas, and there may be as much again in the northern reaches of the basin (off Norway) which are only beginning to be explored as this is written. Large gas accumulations in Triassic and Lower and Middle Jurassic sandstones (with some oil shows also) are already established north of 66°N latitude, facing the Arctic Ocean rather than the North Sea or the Atlantic Ocean.

25.5.4 Mesozoic petroleum around the Atlantic and Indian Ocean margins

Northwards from about 10°N latitude, the North Atlantic Ocean occupies the site of an older ocean which had become closed during Paleozoic time. Numerous fold belts of a variety of ages strike towards the present continental margins. From the Bahamas to the Arctic, on the west side, and from the Bay of Biscay to the Arctic on the east side, the present continental margins lie far offshore, for the most part in hostile operating environments.

The South Atlantic, in contrast, was a brand new ocean created in the second half of Mesozoic time by the breakup of a Precambrian craton. The oldest unmetamorphosed sedimentary rocks around it are of Late Jurassic age. From the Ivory Coast and the mouth of the Amazon, in the north, to Namibia and the River Plate, in the south, narrow coastal basins extend offshore. From the River Plate southwards, the continental margin of South America now lies far offshore and the coastal basins of southern Argentina and southernmost Chile, unlike those farther north, are not continental margin basins and they have no counterparts in Africa.

The Indian and Southern Oceans had begun to open at about the same time as the Atlantic Ocean. The east coast of Africa, the west coast of Australia, and both coasts of India resemble those on the two sides of the South Atlantic, with similar coastal half-grabens bounded by growth faults and containing broadly comparable stratigraphic sections beginning with the Jurassic.

Coastal and offshore production is well established in the coastal South Atlantic basins. It seems reasonable to expect the basins around the Indian Ocean to offer comparable prospects, but this has not yet proved to be the case. Because the North Atlantic margins exhibit a far greater variety of rock ages and structural styles than do the margins of either of the other oceans, they might eventually yield the best prospects of the three; but hostile operational milieux threaten these prospects.

North Atlantic basins The *Porcupine Basin* off western Ireland contains light oil in much-faulted Jurassic sandstones. In subsidiary basins set back from the present continental margin, an important natural gasfield lies in the northeastern part of the *Irish Sea Basin*, in Lower Triassic sandstones; and a less important field occurs in Aptian–Albian sandstones closely offshore *southern Ireland*, close to the spot at which the *Lusitania* was sunk in 1915. This location is far enough to the south for the trap to be a fold created during the early Tertiary Alpine orogeny.

In the *Basque Basin*, in the angle of the Bay of Biscay, gas and condensate are produced from Cretaceous reservoirs. In the *Asturian Basin* farther west, oil occurs in Upper Cretaceous turbidite sandstones. The small *Essaouira Basin* of northwest coastal Morocco, west of Marrakesh, contains small quantities of gas and condensate in Triassic and Jurassic sandstones and oil in an Upper Jurassic limestone.

In equivalent pre-rift position on the western Atlantic margin, wet gas occurs insignificantly in the Jurassic of the *Baltimore Canyon Basin* off the US seaboard and potentially more significantly in both Cretaceous sandstone and overpressured Jurassic sandstones on the continental shelf off *Nova Scotia*. The largest single North Atlantic oil accumulation known to the time of writing occupies the edge of a narrow basin within the *Grand Banks of Newfoundland*. Light oil occurs in a rollover anticlinal trap in multiple deltaic sandstones of Early Cretaceous age. The source is probably in early rift sediments of the Upper Jurassic. The *Labrador shelf* off northeastern Canada contains another narrow, rifted basin of Early Cretaceous age. Gas and condensate occur in continental sandstones of that age, and gas alone also occurs in younger (earliest Tertiary) sandstones.

South Atlantic basins A succession of coastal and near-coastal basins, created during the early stages of continental separation during the late Mesozoic, extends along equatorial West Africa and eastern Brazil. Several of them are significantly petroliferous, the most widespread production being associated with the Upper Cretaceous part of the section and much more than half of it being derived from offshore fields. The basins are all of either rift or pull-apart type (see Ch. 17).

The most northerly oil and gas accumulations known in Brazil are off *Ceara* State, in Aptian sandstones and early Tertiary turbidites. Minor oil and gas also occur in Cretaceous sandstones in the *Potiguar Basin* at the northeast "shoulder" of Brazil. The stratigraphic section in the largely offshore basin of *Sergipe–Alagoas* typifies those of the petroliferous basins around the South Atlantic. It began with pre-rift, nonmarine Jurassic sediments deposited during the first separation of the South American and African continents; these sediments include fluvial reservoir sandstones. The succeeding Neocomian represents the rifting phase, with a remarkable assemblage of lacustrine and lagoonal clastic rocks and dolomitic marls replete with fish scales. These strata accumulated to thicknesses of several kilometers in the centers of the basins. The freshwater sediments are the principal source rocks, and freshwater to deltaic sandstones important reservoir rocks, for the highly paraffinic oils. A salt formation of Aptian age follows in nearly all the basins, associated with more euxinic shales. The section is completed by late Cretaceous and early Tertiary turbidite sandstones and conglomerates, into which oil from older Cretaceous source rocks has migrated. Some gas is also produced from these sandstones.

The longest-lived and most prolific Brazilian production has come from the *Reconcavo Basin*. The half-graben extends into the Atlantic at the city of Bahia, which under its old name of Saõ Salvador was the so-called "cultural capital" of Brazil. The Aptian salt is missing from the Reconcavo Basin section, and nearly all the oil is in Jurassic and Lower Cretaceous nonmarine sandstones. Structures are formed by a variety of drapes and offsets over horsts and tilted blocks within the half-graben (Figs 16.30 & 31). Southwest of Bahia, further minor oil occurs in the younger Cretaceous and Tertiary turbidites both on- and offshore in the state of *Espirito Santo*.

Brazil's most southerly production comes from the *Campos Basin*, offshore from Rio de Janeiro. No giant fields are known in it (nor anywhere else in Brazil) at the time of writing, but several fields have been discovered having recoverable reserves between 20×10^6 and 80×10^6 m^3 of oil. They are roughly aligned along the shelf edge close to the 200 m bathymetric contour. The oil is about equally divided between an Albian limestone reservoir and turbidite sandstones of mid-Cretaceous to early Tertiary ages, which spread over the carbonate shelf and captured oil migrating from older shales. The easily detected traps were formed by late Cretaceous halokinesis. Gas occurs in association with the oil.

On the African side of the ocean, the most northerly known oil is on the continental shelf off the Casamance River in *Senegal*; tarry oil occurs in the caprock of salt domes. At the time of writing, the age of the salt is not firmly established, but it is assuredly Mesozoic, and probably Early Cretaceous. Early exploitation of this deposit is doubtful. Both light, sweet oil and gas-condensate are derived from Cretaceous sandstones offshore from the *Ivory Coast*, western *Ghana*, and *Benin*. The localities constitute the pre-rift counterpart of Brazil's Ceara Basin. Similar Cretaceous oil in *Nigeria* and *Cameroon* is totally overshadowed by the prolific production from the Tertiary delta, which did not develop until the early Eocene (Sec. 25.6.2). However, Upper Cretaceous sandstones do contain oil to the northeast, towards the Benue graben which underwent mild later folding during the Santonian. The graben extends into the deep interior of Africa, and oil occurs in it both northeast of Lake Chad and in the subsidiary Mangara Basin of western *Chad*, south of the graben; again the reservoir rocks are Cretaceous. In the African re-entrant east of the Niger delta, gas occurs close offshore in mid-Cretaceous strata.

The coastal pull-apart basins of *Gabon* and *Angola* were the first to be opened up to Mesozoic production in West Africa. They are respectively the counterparts of Brazil's Reconcavo and Campos basins, and contain almost identical productive sections; yet familiar classifications of petroleum basins have assigned the pairs to different pigeonholes. The Aptian salt developed into pillows and piercement structures creating a variety of traps for light oil in younger Cretaceous and early Tertiary sediments, especially in Senonian sandstones. Other traps are formed by growth faults. There is also production of light oil from Lower Cretaceous dolomite, offshore; and from Lower Cretaceous sandstone, onshore; and (in Gabon) of heavy oil (some as heavy as 13° gravity) from early Eocene cherts and shales, fractured above the tops of salt intrusions.

The coastal basins of Gabon and southern Angola are widely exposed and readily identified onshore. The offshore basins between them were added to the production centers in rapid succession: in the Republic of *Congo* (Brazzaville) and in *Zaire*, each with oil, gas, and condensate in Lower and Upper Cretaceous sandstones and dolomites; and, most important, off the small Angolan enclave of *Cabinda*, which has dominated Angolan production. As in Angola proper and unlike Gabon, Cabinda has found more oil in the pre-salt section (which includes the black, lacustrine source beds) than in the post-salt section. The reservoirs are thick fluvial sandstones and algal lake biostromes and coquinas. As in Nigeria, two contrasting types of crude are produced: one from shallow, younger Cretaceous horizons averages about 25° API and is naphthenic; the other is a sweet crude of 35° gravity with a higher pour point, from a deeper reservoir formation.

On the Indian Ocean side of Africa, tar sands have long been known in the Triassic part of the nonmarine Karroo System in western *Madagascar*. Some gas and condensate occur in deltaic sandstones in the Upper Cretaceous and Paleocene in coastal *Mozambique* and both coastal and offshore *Tanzania*; in the latter country some gas occurs in an Eocene limestone reservoir also. Commerciality has not been demonstrated, to the time of writing, for any of these occurrences.

Commerciality has been demonstrated, however, for oil deposits in the swamps of the Nile valley in *southern Sudan*. They are not close to the Indian Ocean, but their structural setting is related to the rift valley system and so indirectly to the extensional phase that opened the ocean. The source and reservoir rocks are within a Cretaceous lacustrine complex; the traps are formed by normal growth faults.

The Cambay Basin on *India's* west coast was a Cenozoic creation (see Sec. 25.6.3). Minor oil and gas occur in Cretaceous and early Tertiary sandstones in the *Cauvery Basin* and Palk Strait, between India and Sri Lanka, and in the *Godavari Basin* half way up India's east coast.

25.5.5 Mesozoic oil and gas in Australasia

Off the northern and western coasts of *Western Australia* is an aborted rift zone, forming a further segment of the extensional margin of the Indian Ocean. In the *Carnarvon Basin* on the northwest corner of the continent, and on and off *Barrow Island*, oil, gas, and condensate are produced from a Lower Cretaceous sandstone of very low permeability; some occur also in the Jurassic. Much larger accumulations of gas and condensate occupy Upper Triassic sandstone reservoirs on the *Dampier–Rankin platform* off the northwest coast (see Sec. 25.3). Lesser reserves of gas, and some light oil, are available from older Triassic, Jurassic, and Lower Cretaceous sandstones; the source is probably in thick Jurassic shales off the platform, and the fault-controlled traps also originated in the Middle Jurassic. Gas and condensate occurrences in Lower Jurassic sands continue eastward into the offshore *Browse Basin*, with oil in equivalent strata still farther east, under the Timor Sea.

Immediately in front of the eastern mountain belt is the *Surat–Bowen Basin*. It yields small quantities of very light oil and gas from Triassic and Lower Jurassic sandstones; the oil may be of nonmarine source. The main foreland basin west of the mountains is the Great Artesian Basin; it is not in general petroliferous, but its southwestern extension from Queensland into South Australia yields both oil and nonassociated gas. The gas in this *Cooper–Eromanga Basin* is in Permo-Triassic rocks (see Sec. 25.3), but the exceedingly light oil (lighter than 50° API) is in Jurassic sandstones with a little in immediately older and younger systems.

The area of Australia by far the most productive of both oil and gas involves a trio of extensional basins occupying Bass Strait between the mainland and Tasmania. The easternmost of the three, the *Gippsland Basin*, has yielded all the commercial production to 1984 (Sec. 26.2.5). It originated in fault breakdown in late Jurassic time across the Paleozoic orogenic belt, and accumulated some 10 km of sediments, all clastic and coal-bearing except for a brief section in the Miocene. As in other faulted depressions originating during Mesozoic continental drift, the latest Cretaceous and Tertiary part of the sequence is post-taphrogenic and spreads far beyond the confines of the graben (Fig. 26.24).

All the oil is in Paleocene–Eocene strata which are entirely nonmarine or incompletely marine within the field area. The oil is waxy and accompanied by abundant gas and condensate. The traps are anticlines created by combinations of block faulting and erosional truncation (see Ch. 16). The basin should eventually yield at least 5×10^8 m^3 of oil and 2.5×10^{11} m^3 of gas and condensate. The source of so much petroleum in a nonmarine, coal-bearing succession has been a matter of great dispute. There are no suitable source beds within or above the productive section, and older units, where they are known in detail, appear no more favorable. However, it is difficult to believe that 10 km of sediment were accumulated in a rapidly subsiding, faulted coastal basin without any early marine phase. If such a phase exists in Upper Cretaceous strata seaward of the productive belt, it will be in favorable position both geographically and stratigraphically to provide the source sediment.

Farther west in Bass Strait, some gas is known in Cretaceous sandstones in the *Otway Basin*. At the opposite end of the continent, more occurs in similar sandstones in *Papua–New Guinea*. A very small amount of light oil is produced from Cretaceous meta-volcanic rocks in the *Barito Basin* of southernmost

Borneo, in a petroleum province otherwise wholly Tertiary (see Sec. 25.6.3).

25.5.6 Mesozoic petroleum in basins associated with the ranges of central Asia

Between the Caspian and China Seas, numerous interdeep and foredeep basins produce some oil, but more gas, from a variety of marine and nonmarine Mesozoic and Cenozoic sandstones, and some from Eocene carbonates. The region as a whole is fundamentally a vast complex of Hercynian fold-ranges, culminating in widespread Permian granitic rocks and molasse sediments with volcanics. Most of the basins began their basinal histories during the Permian, but they underwent extensive Mesozoic and Alpine restructuring and overprinting. Numerous intermont depressions contain wholly continental and lacustrine successions, responsible for most of China's oil production. We will consider the basins from west to east.

The *Mangyshlak trough* lies on the so-called Turanian plate which forms the southern margin of the Hercynian platform east of the Caspian Sea. The trough developed in Mesozoic time from the eastward extension of the Dnieper–Donetz trough of the Paleozoic (Sec. 25.2.2). Light oil very rich in paraffins is produced from nonmarine Middle Jurassic sandstones, and heavy oil in Lower Cretaceous sandstones, folded into sharp anticlines. A little oil also comes from Triassic carbonates. As both oil and gas fields continue on the opposite (western) side of the Caspian Sea (Sec. 25.5.2), offshore structures between the two areas will very likely prove productive, but to the time of writing the only large fields found have been east of the Caspian.

Wrapping around the northeast lobe of the Caspian Sea is part of the *North Caspian* or *Emba Basin*. Its numerous petroleum occurrences, none of them large, include gas in the Lower Cretaceous, heavy oil in both Lower Cretaceous and Jurassic sandstones, and some oil in a fractured syenite sill intruded into Triassic shales.

The Qizil Qum and Qara Qum deserts overlie the platform in front of the Kopet Dagh, Afghan, and Uzbek ranges south and southeast of the Aral Sea. Jurassic strata rest on the Hercynian basement. The eastward equivalent of the Mangyshlak trough is here called the *Bukhara–Khiva trough*; it contains a great volume of late Jurassic salt and anhydrite. The region is a first-order, nonassociated gas province; some of the gas is reservoired in Jurassic reef carbonates below the evaporites, but by far the greater part is in Lower Cretaceous sandstones overlying them. The Shatlyk field, in Neocomian red beds, is one of the largest gasfields in the world, and the Gazli field, in slightly younger Cretaceous sandstones, was also originally very productive. All the fields are in foreland anticlinal traps associated with deep faults propagated upwards from the Hercynian basement. Still farther east, the *Tadzhik Basin* extends into northern Afghanistan. Again, the production has been essentially all gas, in Cretaceous sandstones and Jurassic reef carbonates.

A remarkable small, intermont basin lies *en echelon* with the Tadzhik Basin to its northeast, east of Samarkand and south of Tashkent. The *Ferghana Basin* has produced oil fairly continuously since 1904, though its cumulative output is not large (perhaps 60×10^6 m^3). Its early discovery is attributable to the multitude of oil shows and solid bitumens in the metamorphosed Paleozoic rocks and their overlying Jura-Cretaceous sediments along the mountain front, phenomena familiar to the men of the early camel caravans. Oil and gas in Jurassic and Cretaceous sandstones may have migrated into them from the underlying, truncated Paleozoics, across which the Mesozoic rocks transgressed unconformably. In late Cretaceous time, the basin became a saline gulf, receiving a considerable thickness of Paleogene carbonates and gypsum capped by colored continental clastics of the Neogene. Though the most widely petroliferous part of the section is the Jurassic, most of the production has come from Middle and Upper Eocene carbonates in paleogeomorphic traps on sharp, eroded anticlines; some seeped upwards into Neogene sandstones and conglomerates.

The oil basins of northern and western *China* all lie to the north of the main Asiatic mountain ranges. They presented a paradox for petroleum geologists of an earlier generation because in a remarkable number of them the oil occurs within stratigraphic successions wholly or almost wholly of nonmarine character. Added to the further oddity that most of the basins were the products of post-orogenic block-faulting and not synorogenic downfolding, this peculiarity caused Albert Bally to classify Chinese basins in a category of their own (see Ch. 17).

The *Zhungeer* or *Wusu Basin* lies north of the Tien Shan Range in northwestern Sinkiang province. It contains basinal sediments of Permian age, which were folded during the Triassic and which may contain the principal oil source-rock of the basin. Nearly all the oil has been found in Triassic, nonmarine, clastic reservoirs and in stratigraphic traps. A little occurs in Jurassic sandstones also. The *Tarim Basin* sits astride the 40th parallel and is separated by the Tien Shan from the Ferghana Basin to the west. Small production of waxy oil has come from Jurassic lacustrine and fluvial sediments and from Oligo-Miocene nonmarine sandstones.

The *Qaidam Basin* lies on the opposite side of the Altyn Tagh Range (which contains one of the world's largest strike-slip faults) and north of the Kun Lun

Range. It was originally a compressional basin, containing marine Carboniferous–Triassic strata over 6 km thick. Nearly all the oil production, however, has come from a single field in Tertiary continental sandstones at very shallow depths; the productive strata betray their presence by numerous seepages along their outcrops. Most of the basin's other fields yield gas rather than oil.

The *Gansu Basin* is also called the Pre-Nan Shan Basin or Kan Su "corridor", it occupies a large part of the Gobi Desert. Its oil has been exploited since the 3rd century AD but the only large field known in it remained undiscovered until 1938. The reservoir is in early Tertiary alluvial red beds underlying a vast volume of Quaternary sediments. The source sediment may again have to be sought in marine Carbo-Permian rocks below the Hercynian unconformity or Jura-Cretaceous lacustrine strata above it.

The *Ordos Basin* is southwest of Beijing. Its Yenan area was exploited 2000 years ago, and modern production began in 1907. The principal oil reservoirs are in Triassic and Jurassic sandstones, part fluvial, part lacustrine; the lacustrine part may provide the source sediment, but there are oil shows in the marine Carbo-Permian sandstones below the unconformity also.

The *Sichuan Basin*, called the "Red Basin" by the Chinese, lies astride the Yangtze River, northeast of Chungking, and is almost surrounded by fold ranges. The basin is primarily a gas basin, producing from Middle Triassic limestones associated with salt and anhydrite (Sec. 25.3). The basin was first exploited for salt, recovered from brine wells by evaporation, using the associated gas as fuel. Some gas is also produced from Carboniferous strata. The relatively minor oil production comes from tight Jurassic sandstones and has a source in the ubiquitous older Jurassic lacustrine sediments. The Sichuan Basin has been the site of most of China's few very deep drillholes (below 5 km), but the production is from very shallow levels (less than 2 km).

The last of the Chinese basins within the Hercynian fold belt is the *Songliao Basin*, far to the northeast in Manchuria. Originating as a graben in the mid-Jurassic, it acquired 6 km of sediments, most of them nonmarine. The basin should be a textbook example of the influence of unconformities on petroleum source rocks, reservoir rocks, and traps (Sec. 16.5). Basinwide unconformities occur at the base of the initial Jurassic section, below the Upper Jurassic, between the Jurassic and the Cretaceous, and in the middle of the Cretaceous. The source sediment is mid-Cretaceous and nonmarine; the reservoirs are multiple lacustrine and fluvial sands in the Lower Cretaceous; the traps are of the horst-graben variety. The basin contains China's only supergiant oilfield known to date. The Daqing field had become the world's fifth most productive (in terms of current output) by 1982, and was responsible for more than 50 per cent of China's entire production.

25.5.7 The West Siberian Basin

The gigantic basin of western Siberia illustrates the pitfalls of trying to assign all petroliferous basins to clearly defined genetic categories. It is clearly an intracontinental or cratonic basin, never brought under important compression or extension, and having no associated eugeosyncline or continental-margin facies. Yet it lies not on a true shield, but on a complex of Paleozoic and Proterozoic deformed belts along the backslope of the Paleozoic Ural Mountains. The basin in fact bears a relation to the Urals very similar to that borne by the US Gulf Coast to the Paleozoic Ouachita–Marathon Mountains (see Ch. 17). Each of these Paleozoic fold belts has a major Paleozoic petroleum province on its foreland (continental) side and a major Mesozoic province on its hinterland side. Both the Gulf Coast and West Siberian basins began during Triassic time as complex, graben-like subsidences behind their respective fold belts. The Gulf Coast Basin went on to become the northern flank of a huge extensional basin, beginning with the characteristic salt fill and becoming wholly marine in character; the outer edges of the basin lie on oceanic crust. The West Siberian Basin (Fig. 25.4) never attained this condition; it remained dominantly continental in character, with marine incursions in late Jurassic, Cretaceous, and Paleogene times, but at no time was it wholly marine and only in its northern parts is a greater part of the section marine (principally in the Upper Cretaceous). Furthermore, the strata are wholly clastic and coal-bearing, there being no important carbonate or evaporite unit anywhere within the basin.

In the total absence of compressional folding since the basin's inception, all the large traps are formed by basement-controlled uplifts, many of them dome-like anticlines along wide, fault-bounded arches of early Jurassic elevation. Because so many of the large Paleozoic fields in the Ural–Volga Basin, on the other side of the mountains, are also in traps consequent upon the movement of basement arches (see Sec. 25.2.2), this circumstance in common between the two great basins helped to give rise to the characteristic Russian view of the genesis of "folds" in essentially vertical rather than tangential movements.

There is a striking areal zonation of oil and gas in the basin, nearly all of both occurring in Cretaceous sandstones. In the central part of the basin, the Middle Ob district, are the giant oilfields, including Samotlor, which have dominated Russian production since the late 1970s. The principal reservoirs are a number of Lower Cretaceous arkosic sandstones below subtle stratigraphic breaks. The source is probably the bituminous

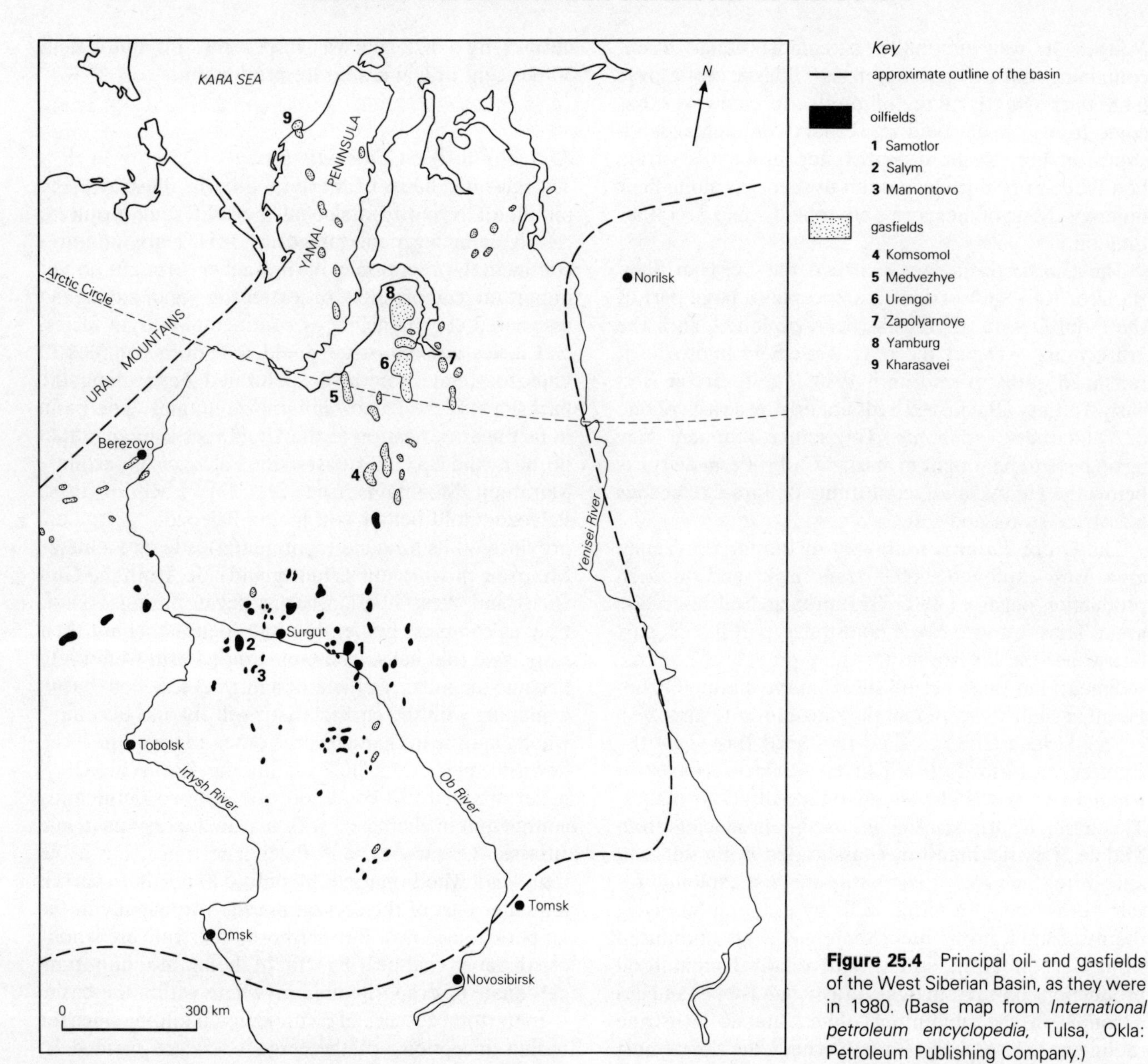

Figure 25.4 Principal oil- and gasfields of the West Siberian Basin, as they were in 1980. (Base map from *International petroleum encyclopedia,* Tulsa, Okla: Petroleum Publishing Company.)

shale of latest Jurassic age, which itself contains a great quantity of oil where it is fractured (in the Salym field). In the orthodox sandstone pools, the fluid interfaces are strongly tilted, a somewhat unexpected circumstance in the center of a symmetrical, noncompressional basin; it may be a consequence of the overall artesian nature of the basin.

Both north and south of this central oil-rich district are vast areas dominated by gasfields. That to the north is the greatest gas province known. It extends from about 64°N latitude to the shores of the Kara Sea, and undoubtedly beyond under the sea itself. An assembly of supergiant fields, the gas counterparts of the supergiant oilfields in the Persian Gulf Basin, yields gas and condensate from multipay sandstones in the Neocomian and dry gas from equally numerous reservoirs in the Cenomanian; a few fields also yield gas from sandstone reservoirs between these two levels. In the northernmost fields the gas-condensate reservoirs continue downward into the Middle and Lower Jurassic.

The dry gas, at least, is undoubtedly immature, nonassociated with oil but associated with coal. Its preservation appears to be due to a freakish circumstance; the gas pools began to form very late in Cretaceous time and passed through a clathrate (solid hydrate) stage during the Pleistocene glaciation. With the post-Pleistocene amelioration of climate, the gas began to exsolve; the pools are now trapped below permafrost, with the shallowest liberated reservoirs being the most prolific.

On the opposite (southeastern) side of the basin, in the upper reaches of the Ob and Irtysh Rivers, is another gas district, but with considerable oil in addition. The reservoir rocks are again nonmarine sandstones, of both

Jurassic and Cretaceous ages. On the western side of the basin, in contrast, the yield is almost exclusively of gas, and the reservoirs and traps reflect the position on the backslope of the Urals more than within the basin proper. Control by basement topography is very apparent; much of the gas is in Jurassic calcarenites of reef-talus derivation.

The largest Russian oilfield, Samotlor, is one of the ten largest fields known. Under politico-economic pressures to overproduce, it had become the world's second most productive oilfield (after Saudi Arabia's Ghawar) only 10 years after its discovery. By the end of 1982 it had risen to sixth place in cumulative production (with over 1.4×10^9 m^3). No other oilfield of such recent discovery and remote location had reached the top 30 in cumulative output by that time. Once provided with large-diameter pipelines, a mere half-dozen of the largest northwestern Siberian gasfields will be capable, before the end of the century, of surpassing the natural gas output of all the thousands of North American gasfields combined. At the time of writing, the Urengoi field alone has reached an annual output of 10^{11} m^3, and is probably capable of supplying 15 percent or more of the entire world's output once its delivery systems are in place.

25.5.8 Mesozoic petroleum in the Andean belt

Two Mesozoic oil basins that intersect the Atlantic coast of South America were not dealt with in Section 25.5.4. They lie so far south that their relation to continental drift phenomena is more closely tied to the early separation of Patagonia from western Antarctica than to that of South America from Africa. Furthermore, each of them lies to the east of the Austral Andean belt. Though both basins antedated the principal (Cenozoic) rise of that belt, they may appropriately be considered with the other sub-Andean basins which are petroliferous in Mesozoic strata.

The *Magallanes Basin* straddles the eastern Strait of Magellan in both Argentina and Chile. It bears some structural resemblance to the more easterly of the US Rocky Mountain basins (Sec. 25.5.10), and is similarly productive of both oil and gas from a Jura-Cretaceous sandstone. It is the most southerly commercially petroliferous basin known. Due east of the Strait is the Falkland (Malvinas) platform; between it and the mainland, oil is known to occur in the same sandstone in the southern *Malvinas Basin*.

Some 700 km further north is the semicircular Gulf of San Jorge. To its north, west, and south are the oil- and gasfields of Santa Cruz and Chubut provinces in southern Argentina; they include the longest-producing oil region south of 10°S latitude in the world, that of *Comodoro Rivadavia*. The eastern (coastal) part of the basin is fundamentally a rift, with very complex faulting. Towards the mountainous interior, simpler anticlinal traps become dominant. The multiple, lenticular sand reservoirs, derived from the Patagonian massif, are principally in the Upper Cretaceous; some are in the Paleocene immediately above an unconformity, others are in the Upper Jurassic below another unconformity. Gas occurs in both the Jurassic and the Upper Cretaceous sections also. Many of the traps are due to erratic porosity variations, especially where the sands are tuffaceous.

The *Neuquen Basin* lies between the Argentine Pampas and the Andes, and occupies a large re-entrant between the Patagonian and Brazilian platforms. The basin originated in Jurassic time, and most of its oil and gas production has come from Upper Jurassic sandstone reservoirs of very erratic distribution. A Cretaceous carbonate is also oil-bearing. Much gas is associated, and possibly nonassociated gas occurs also in early Cretaceous strata. The incidence of productive structures is powerfully controlled by an east–west structural high called the "dorsale," which arose in Jura-Cretaceous time and separated two sub-basins. That on the south contains oil principally of Lower or Middle Jurassic source, that on the north principally of Tithonian–Neocomian origin. The growth of the "dorsale" created an array of structures with varying stratigraphic gaps; on some of them the entire Jurassic section is missing below mid-Cretaceous limestones and shales.

Immediately in front of the Andes at the latitude of Buenos Aires and Santiago is the *Mendoza* or *Cuyana Basin*. It is a typical foredeep; whereas the Neuquen Basin extends eastward from the Andean front between two ancient massifs, the Mendoza Basin extends westward into the mountain front. Its western sector therefore lies within the thrust belt, and large, buried anticlinal structures are much faulted. The principal reservoir rocks are Upper Triassic sandstones and tuffs. The Triassic part of the section is essentially nonmarine in character and the oils are therefore very waxy. Much smaller anticlinal structures east of the mountain front yield small oil and gas production from Upper Jurassic and Lower Cretaceous strata (also largely continental), as well as from the Upper Triassic.

Northward from Mendoza, Mesozoic oil and gas occurrence is largely eliminated over some 20° of latitude by two factors: the eastward encroachment of the Andes towards the shield, in Bolivia, and the virtual elimination of marine sedimentary rocks from the post-Devonian succession. Some light oil and condensate are reservoired in Upper Cretaceous limestones and early Tertiary siltstones and marls in the *Salta Basin* of Argentina and the *Santa Cruz Basin* of eastern Bolivia, but their source rocks are probably Paleozoic. The same is assuredly true of the tiny but unique Pirin field, northwest of Lake Titicaca in southern *Peru*. The reservoir of Cretaceous limestone and sandstone lies between one unconformity below the Tertiary and another one at the top of the Devonian, which is the probable source. During the field's producing period, before World War I, its location in the heart of the Peruvian Andes made it the most elevated field in the world.

Oil indigenous to the Mesozoic begins again in *northern Peru*, in Cretaceous strata folded into large foreland anticlines easily visible at the surface. The anticlinal belt, involving halokinesis of Jurassic salt, continues northward, east of the Andes, through *eastern Ecuador* and into *southern Colombia* (Sub-Andean and Putumayo basins). Nearly all the oil (some of it very heavy) is in Cretaceous sandstone reservoirs, deposited on the foreland side of a classic marine basin from which the eastern Andes were eventually to arise. Some of the oil on the Colombian side of the belt is reservoired in fractured, pre-Cretaceous basement rocks; a little is trapped in latest Cretaceous and Paleocene sandstones deeply buried in the eastern thrust belt. This entire sub-Andean region between northern Peru and southern Colombia, where the Andes make their westward swing, lies in tropical jungle presenting such explorational and operational difficulties that their delayed discovery and exploitation led to strikingly bimodal output patterns for the three countries. From World War I (or much earlier in the case of Peru), coastal and interior-valley fields provided the entire output of oil for all three countries. Within a few years of the discoveries of the sub-Andean fields they had taken over the dominant production positions for all three of them.

The Cretaceous marine basin that fringed the Brazilian and Guyanan shields, between northern Peru and Trinidad, must have been comparable, as a generating site for petroleum, with that of roughly equivalent age around the Persian Gulf. Unhappily, the post-Cretaceous tectonic history of northern South America was less favorable for the preservation of hydrocarbons than was that in the Middle East. The Cretaceous basin was extensively deformed into the eastern Andes of Colombia, the Merida Andes of western Venezuela, and the coastal cordillera of eastern Venezuela and Trinidad. Only where intermontane basins survived *within* the Andean belt were great quantities of Cretaceous oil preserved.

The *Upper and Middle Magdalena Valleys* of Colombia constitute a half-graben between the eastern and central Andean ranges (Fig. 17.25). A nearly complete Cretaceous sequence of marine limestones and shales is preserved below the basin, but the thick Tertiary is almost wholly continental. A very small fraction of the valley's oil output has come from fractured Cretaceous sandstones and limestones in anticlinal and horst traps at the two ends of the basin. The oil is paraffinic.

The eastern Andes of Colombia, forming the eastern boundary of the Magdalena Valley Basin, become the western Andes in Venezuela and form the *western* boundary of the intermontane *Maracaibo Basin*. If we except the Persian Gulf Basin, which is an order of magnitude more prolific than any other basin, the Maracaibo Basin is certainly one of the two or three richest oil basins in the world. Though its production has always been dominated by Tertiary sandstone fields, fractured Cretaceous carbonates (and their underlying igneous basement) have been highly productive in sharp, faulted uplifts along the west side of Lake Maracaibo (Figs 13.73 & 79). Other Cretaceous limestones and sandstones produce lesser amounts along a north–south trend across the "lake" and from its east side, and Cretaceous sandstones are the reservoirs in several fields in tight, faulted anticlines near the southern edge of the basin (which is shared by Colombia and Venezuela). These latter structures also yield Cretaceous gas. The Cretaceous oils are unquestionably of late Cretaceous source; the Turonian La Luna Formation ranks among the oil world's classic source sediments (Sec. 8.2). There is indeed a good case for ascribing much or most of the oil from the larger Tertiary reservoirs to the same Cretaceous source.

An equally impressive source sediment of the same age is present in the coastal cordillera and adjacent deep basin of *eastern Venezuela* and *Trinidad*. Here, however, there are also rich source formations available in the mid-Tertiary section, and of the conventional oil produced in the basin scarcely any reveals any close association with Cretaceous strata. The oil which does reveal such a close association lies on the two extreme flanks of the basin. Within the deformed belt along the north flank are the Trinidad and Guanoco asphalt "lakes." Both occur in nonmarine Neogene sandstones, largely of alluvial origin, which overlap fractured and eroded Cretaceous strata of favorable source facies. Immediately in front of the mountains is the giant Quiriquire oilfield (Fig. 16.83), which may also have derived most or all of its oil from Cretaceous strata below the late Miocene unconformity. On the opposite side of the basin, the heavy oils of the Orinoco tar belt (see Sec. 10.8.4) are in sandstone reservoirs of several ages, including some assigned to the Cretaceous (here in a thin, erosional wedge-edge resting on much older rocks, including the Precambrian).

25.5.9 Mesozoic petroleum in the Gulf of Mexico province

During late Mesozoic time, the now richly productive basins of Venezuela and Trinidad flanked the forerunner of the Caribbean Sea as clearly as they flanked the Precambrian Guyana shield. They are now separated from the Caribbean basin by the early Tertiary Coastal Cordillera. Aside from very small production of exceedingly light oil and naphtha in *Cuba*, from fractured serpentinites, tuffs, and clastic limestones of Cretaceous age, no other oil or gas is known around or within the Caribbean basin proper. From Guatemala to Florida, however, the *Gulf of Mexico basin* is the second or third greatest oil and gas province on Earth.

It is seldom considered in the literature as a single basin, and indeed the productive part of it is not. From southwest to northeast (Fig. 17.4), one may distinguish a northern Guatemalan or Peten Basin; the Sound of Campeche, Chiapas–Tabasco (Reforma), Isthmus of Tehuantepec (Saline), Veracruz, Tampico–Nautla and Sabinas basins of Mexico; the northeast Mexico–south Texas Gulf Coast Basin; south Louisiana or Mississippi delta Basin; the East Texas (Tyler) Basin; the north Louisiana–Mississippi salt basin; the Apalachicola embayment of the Florida panhandle; and the south Florida Basin. The Isthmus, Mexico–Texas Gulf Coast, and Mississippi delta basins are wholly Tertiary and Quaternary from the production standpoint. All the rest are Jurassic and Cretaceous. The Gulf of Mexico basin considered as a whole, however, will ultimately produce (from fields known in 1982) a volume of oil approaching 20×10^9 m^3 and more than 12×10^{12} m^3 of natural gas and condensate. These volumes represent about 10 percent of all the conventional oil known, and more than 10 percent of all the gas. Of them, perhaps two-thirds of the oil, but less than a quarter of the gas, will come from Mesozoic reservoirs.

The inner boundaries of the productive basins are marked by the arc from the Yucatan limestone massif, the Sierra Madre of Mexico, and the great arc of down-to-the-gulf normal fault zones stretching from Texas to western Florida. Much of this boundary had become established by the end of Middle Jurassic time. It then received a thick floor of salt in discontinuous depressions around it, and a covering of marine limestone of late Jurassic and early Cretaceous age (Fig. 26.19). A gulf-encircling chain of arcuate reefs followed in Albian time, and a variety of carbonates, evaporites, shales, and some sandstones thenceforward into the early Tertiary.

The most recently discovered segment of this great basin will probably prove to be the richest in ultimate oil output. The giant fields of the *Sound of Campeche*, west of the Yucatan peninsula, produce rather heavy, sulfurous oil from deeply buried, fractured, Jurassic and Cretaceous dolomites and from early Paleocene limestone breccias derived from the bank edge to the east (Fig. 17.22). The complicated traps overlie a pattern of badly fractured horst blocks. Onshore but not on the same trend as the Campeche fields, are the *Reforma fields*, east of the Isthmus of Tehuantepec. Somewhat lighter and less sulfurous oil, plus a great deal of associated gas, is produced from depths below 3500 m in Cretaceous calcarenites, dolomitized and highly fractured over uplifts of Jurassic salt (see Chs 13, 16 & 26). The calcarenites again appear to be bank-edge talus. In the largest fields, the pay-zone extends from Upper Jurassic carbonates continuously through almost the whole of the Cretaceous, giving pay thicknesses in excess of 1000 m. The principal source for both the Campeche and Reforma oils is in the Upper Jurassic. The rapid development of the two groups of fields led to spectacular increases in Mexican oil output from 1978 onwards. An insignificant amount of oil has also been produced from the interior of the same Cretaceous carbonate bank, in northern Guatemala.

Beyond the Saline Basin to the west, small production of oil, and some gas, comes from the *Veracruz Basin*. The reservoir rocks are folded and overthrust Upper

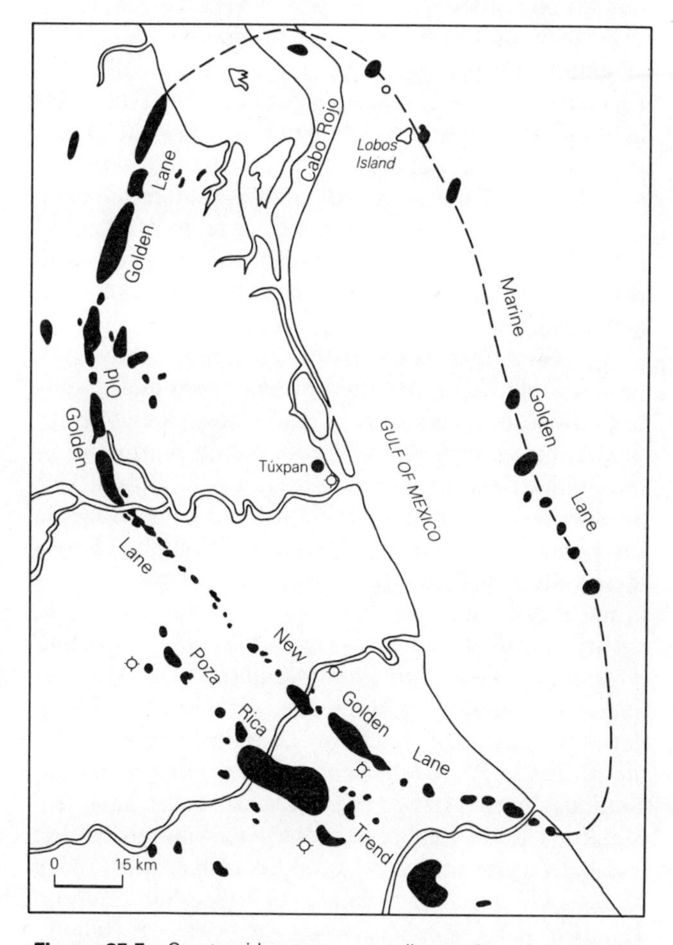

Figure 25.5 Great mid-cretaceous atoll complex of the Golden Lane, Mexico. Oil-fields along the Golden Lane in carbonate reservoirs of the El Abra (bank and reef) facies; those in the Poza Rica trend in the Tamabra (carbonate clastic) facies. (From A. H. Coogan *et al.*, *AAPG Bull.*, 1972.)

Cretaceous limestones. Beyond a coastal crystalline massif comes the *Tampico–Nautla Basin*, one of the oil industry's most spectacular and historic districts. Surface indications of oil are very abundant in it. For a brief period before and during World War I, the fields of the Golden Lane were the most productive individual oilfields in the world. By the end of 1982 the entire Golden Lane and its neighboring Poza Rica trend, discovered in 1930, had yielded more than 0.5×10^9 m^3 of oil. In its western arc, the gulf-encircling Albian reef chain was interrupted by channels in which deep-water sediments were deposited. Between two of these channels grew a giant, elliptical atoll (Figs 13.50 & 26.20). The *Old Golden Lane fields*, along the northwest edge of the atoll, yielded oil at exceptionally high rates from an Albian limestone with very high solution (moldic) porosity. The source of the oil is probably in Upper Jurassic phosphatic sediments. Later discoveries largely closed the ellipse of the giant atoll; with the rise of the Sierra Madre to the west, the eastern arc of the atoll which formerly lay in the interior of a west-facing shelf is now wholly offshore in the Gulf of Mexico (Fig. 25.5).

Farther southwest, and close to the mountain front, a remarkable paleogeomorphic depression called the *Chicontepec Basin* contains a great quantity of relatively light oil – perhaps 15×10^9 m^3 in place, plus more than 10^{12} m^3 of gas, at depths of less than 1800 m – in the turbidites and limestones of a late Paleocene–early Eocene canyon-fill (see Secs 13.2.7 & 16.6). The oil is probably of the same Jurassic source. Even with extensive artificial fracturing, however, recovery of it may never exceed about 10 percent.

At the northern end of the Tampico embayment, heavy, asphaltic oil has been produced for many years from fractured, cherty, Cretaceous limestones which are the basinal equivalents of the formation productive in the Golden Lane and at Poza Rica (see Sec. 13.4.9). This production now extends downward into Upper Jurassic carbonates. Gas occurs in fractured dolomites of both these systems in the *Sabinas Basin*, south of the big bend of the Rio Grande.

Northwards from there into the USA, a quite different geological province is entered. The Mexican Cretaceous fields lay along the *eastern* margin of a depositional trough which was later deformed into the Sierra Madre. This belt of deformation of course continues northwestward into western Texas and New Mexico. The US Gulf Coast, however, swings to the east and lies on the *southern* backslope of the much older Marathon–Ouachita mountain belt (see Sec. 25.2.4). The *US Gulf Coast Basin*, therefore, is wholly extensional, developing during Mesozoic rifting as an aspect of progressive southward tilting and normal faulting. As its sedimentary infilling has been almost continuously regressive, there is ubiquitous gulfward dip and successively younger beds offlap from north to south in that direction (Fig. 17.2).

There is a clear geologic and geographic basis for discriminating between the Mesozoic and Cenozoic basins of the US Gulf Coast. The Mesozoic basin occupied the opening Gulf of Mexico depression, beginning in the Middle Jurassic. The productive Mesozoic section lies wholly onshore and within the "inner" Gulf Coast Basin. It was much influenced by large "highs" along the backslopes of the Ouachita and Marathon Mountains. The section is dominated by carbonates, though with important sandstone interruptions formed during rapid regressions. The section below the Albian limestone contains in addition a good deal of evaporite, especially the thick salt at the base; the post-Albian section contains increasing amounts of marl and true chalk. Though many structures have been raised above deep-seated salt uplifts, there are no productive piercement structures.

The Cenozoic basin, in marked contrast, is an arcuate trough along the northern and northwestern margins of the large depression, filled with the detritus from the interior of the North American continent. The section is almost wholly clastic, thickening greatly offshore. Productive salt-piercement structures are common; scores of other traps are formed by rollover anticlines on the south sides of down-to-the-gulf growth faults.

All porous Mesozoic formations above the mid-Jurassic salt are productive, especially of oil. Jurassic production has come from the interior salt basins around the inner edge of the basin (Fig. 17.4), from northeastern Texas to western Florida. Most of the oil has come from the lower, carbonate part of the section overlying the salt (see Sec. 26.2.4); traps are principally subtle anticlines and fault-controlled folds associated with the edges of salt features. The bulk of the gas, on the other hand, is in younger, sandstone reservoirs, many of them continental and associated with evaporites, and in convex traps. Lower Cretaceous strata are especially dominated by limestones, and these contain most of the large accumulations, especially of gas. Accumulations are particularly concentrated over the Sabine uplift, between the interior salt basins, where the succession happens to contain its highest proportion of clastic sediments. The Lower Cretaceous limestone is the only reservoir horizon productive as far east as the south Florida sub-basin.

The Upper Cretaceous contains the principal oil-source sediment of the Gulf Coast Mesozoic, the Turonian Eagle Ford shale. The sandstone below it, the Woodbine Formation, is the chief oil-producing horizon, containing the oil of the East Texas field in a huge erosional wedge-out on the flank of the Sabine uplift, as well as several other large fields in faulted anticlines over deep salt uplifts (Figs 16.45 & 50). The eastern extension of the Woodbine sandstone, called the Tuscaloosa Formation, contains important gas deposits deeply buried on the gulfward sides of growth faults (see Sec. 13.2.7). The Upper Cretaceous chalk is oil-productive in southwestern Texas, where it is well fractured, and the topmost Cretaceous limestone contains prolific gas where it wedges out against another interior uplift.

25.5.10 Mesozoic and early Cenozoic petroleum in the North American cordilleran province

All the sedimentary basins within and immediately east of the US and Canadian Rocky Mountains are productive of either oil or gas, or both, in both Paleozoic and Mesozoic strata. The two broad ages of incidence are sharply distinct geologically, however. Until Pennsylvanian time, the influence of the Transcontinental Arch was dominant (Sec. 25.2.4). By mid-Jurassic time, the rising western cordillera had become the controlling influence, and through most of the Cretaceous there was established a pattern of sedimentation exceedingly complex in detail but simple in principle. Coarse molasse sediments from the rising cordillera in the west inter-fingered eastward with predominantly shaly deposits of a seaway which lay between the mountains and the eastern shield. By far the greatest proportion of both oil and gas in Mesozoic strata is in Cretaceous sandstone reservoirs and was derived from Cretaceous shale sources. The total quantity of hydrocarbons in place is very large – many thousand millions of cubic meters of oil and tens of trillions of cubic meters of gas – but *recoverable* volumes are vastly reduced by two factors: the proportion of the oil which is too heavy and viscous for any conventional recovery (this condition is at its worst in the oldest Cretaceous reservoirs), and the clay infillings of many of the sandstone reservoirs (this condition gets worse as the reservoirs become younger, especially for those containing gas).

Triassic strata are largely continental, and still much controlled by the Transcontinental Arch. Unimportant oil accumulations occur in association with the facies change from red, nonmarine sandstones to marine limestones close to complex unconformities which flank the arch. Over the Colorado plateau (which was part of the arch during the Paleozoic), and in the overthrust belt to the northwest of it, the colored Triassic strata are overlain by a thick aeolian sandstone of earliest Jurassic age, familiar in its outcrops as towering whitish cliffs in several national parks. This sandstone is the principal oil- and gas-producing horizon in the *Overthrust Belt* of southwestern Wyoming and northern Utah. The belt, along the hanging walls of huge, flat thrust sheets, is modestly productive from Triassic, younger Jurassic, and Cretaceous strata in addition. The source of the hydrocarbons is probably in Cretaceous shales below the thrusts; production is therefore dependent not only upon the presence of a structural culmination but also upon the juxtaposition across a thrust of adequate reservoir rock and the Cretaceous source rock. The same Lower Jurassic sandstone also provides minor production east of the Overthrust Belt, in the Green River and Wind River basins.

From the late Jurassic (already in molasse facies, and more productive of dinosaur remains than of hydrocarbons) through the Cretaceous and into the Paleogene, porous horizons (nearly all sandstones) are productive of either oil or gas virtually throughout the US Rocky Mountain province. The *San Juan Basin* in New Mexico, on the Transcontinental Arch until the Carboniferous and now within the Colorado plateau, yields oil from several Cretaceous sandstones but it is primarily a gas province. Production comes from both marine and nonmarine sandstones in traps caused by porosity wedge-outs. The *Uinta* and *Green River basins*, separated by the E–W-trending Uinta Mountains, are essentially Cretaceous and Paleogene constructions. The Cretaceous is largely marine, the Tertiary all continental (and including the oil shales). Both basins are markedly gas-prone, especially in Cretaceous reservoirs. The traps are structural around the edges of the basins, but stratigraphic-lithologic within the basins, commonly in deltaic or fluvial sandstones introduced into the lacustrine basins and affected by the Laramide unconformity. In the Uinta Basin, some oil has been produced from the fractured oil shale itself (see Sec. 10.8.1). On its eastern side, this basin is narrowly separated from the smaller *Piceance Basin*. Though this smaller basin contains the richest oil shales (see Ch. 10), it is not very productive of conventional oil, its small fields being in Jurassic and Cretaceous sandstones in faulted anticlines near the basin margins. Both the basin and its flanking arch, however, constitute what amounts to one huge gas accumulation, extending from the Jura-Cretaceous sandstones below and above the regional sub-Cretaceous unconformity, over the arch, to younger Cretaceous and Tertiary sandstones in the basin.

The *Wind River Basin*, next toward the north, was a Cretaceous downtuck in front of a faulted uplift on its north side. Major basement arches also flank the basin on both east and west, and oil occurs in Jurassic and Cretaceous sandstones along both arches. Most of the traps are anticlinal, but many small accumulations occur in nonconvex traps also, and some are formed by fracturing of the shales that intervene between the sandstones. The deep part of the basin, along its north flank, produces gas from multiple horizons, especially in the Upper Cretaceous. The next basin to the north, the

Big Horn Basin, is primarily a Paleozoic basin, with the rich Permian source sediment (see Sec. 25.2.4). In the Mesozoic it is gas-prone, but yields some oil from Upper Jurassic and Cretaceous sandstones in both convex and nonconvex traps. The Mesozoic oils are quite unlike those of the Paleozoic, being light and paraffinic.

The Rocky Mountain foredeep basins, east of the mountain front, extend from Colorado northwards into Canada. They are latest Jurassic and Cretaceous creations. The *Denver–Julesburg Basin*, in the south, lay originally within the Transcontinental Arch. Its present north–south axis, imposed by the uplift of the mountains, bears little relation to earlier axes, but these have powerfully influenced the distribution of hydrocarbon accumulations. Nearly all the production has come from deltaic and shallow marine Cretaceous sandstones regularly interbedded with marine shales (some of which are themselves productive where fractured). The depositional patterns of these sandstones were controlled by earlier arches and sags which show up clearly on detailed isopach maps but not on maps of present structure; the latter is a reflection of changes imposed during the Laramide upheaval. Saturated sands therefore seemed to be in a great variety of unpredictable stratigraphic traps, none of them individually large. Ultimately, however, it transpired that numerous gasfields in tight sandstones of delta front and upper shoreface facies, early Cretaceous in age, were in effect a single, extensive accumulation called the Wattenberg field.

The *Powder River Basin,* in contrast, is oil-prone, causing one to wonder whether its quota of gas has been lost because of inadequate reservoir seals. Most of its oil is notably light and sweet; it occurs chiefly in Cretaceous sandstones (nonmarine at the base, fluvial, deltaic, and shallow marine higher up), and even that in older Mesozoic strata is probably of Cretaceous origin. In conformity with the pronounced asymmetry of this typical foredeep basin, the traps along the mobile (western) side are mostly anticlinal and those along the gentle eastern flank are stratigraphic (Figs 16.70 & 94).

The basins of *northwestern Montana* lie between the disturbed belt and the large Williston Basin (Fig. 17.18). They are the southern extremity of the western Canadian Mesozoic basin, and structurally intermediate between that basin and the very different basins of the US Rocky Mountain province. Early domes of generally equidimensional plan controlled Mesozoic depositional patterns, some of them remaining bald-headed at least during the Jurassic. Some oil and gas occur in reservoir rocks of that age, mostly in nonmarine sandstones but some in limestones. Oil and gas continue into similar, nonmarine, Lower Cretaceous rocks, especially around the flanks of the early domes. Younger Cretaceous strata are largely marine and almost wholly gas-bearing, with vast volumes of shallow gas irrecoverably trapped in very tight sands of the late Cretaceous.

On the Pacific side of the US Cordillera a single basin has been richly productive of gas. The *Sacramento Valley Basin* is the northern half of the Great Valley of California, between the Coast Ranges and the Sierra Nevada. It differs totally, in its geologic make-up, from the prolifically oil-productive southern half, the San Joaquin Valley Basin, which became a deep, silled sink in mid-Tertiary time (see Sec. 25.6.4). The Sacramento Basin was part of a vast fore-arc basin during Jurassic and Cretaceous time, accumulating a thickness of sedimentary rock in excess of 10 km (the Great Valley sequence). Shales rich in radiolaria were interrupted by numerous sand bodies, some of them of submarine fan origin (see Sec. 13.2.7). The basin constitutes an isolated natural gas province, in sandstone reservoirs from late Cretaceous to Middle Eocene age; younger strata are wholly nonmarine. Most of the traps are faulted anticlines with steep flanks, along the foothills of the California Coast Range. Some of the productive structures were created by igneous intrusions, yielding gases high in CO_2.

Despite its great area, with numerous individual basins, great thicknesses of Mesozoic strata, and large overall volumes of hydrocarbons, the US Rocky Mountain province contains astonishingly few giant accumulations capable of conventional exploitation. Only two of its oilfields are giant fields. Several of the individual basins, in contrast, contain large areas of essentially continuous gas production from coalesced Cretaceous sandstone reservoirs, and these constitute truly giant accumulations: the Blanco field in the San Juan Basin and the Wattenberg field in the Denver Basin are the largest unless some recovery mechanism is developed for the "tight" gas sands.

25.5.11 The Western Canadian Basin

As we have seen, the foothills belt along the eastern front of the Canadian Rocky Mountains was a gas province in Paleozoic strata. The Triassic is poorly developed and extends eastward under the prairies only north of 54°N latitude. Again, it is principally gas-bearing, with minor oil. Reservoirs are in both dolomites and sandstones; traps are controlled both by facies and by structure. The Jurassic, by contrast, has significant extent below the prairies only in the south, where it is part of the much greater eastward extent of Jurassic sediments in the USA. In southern and western Alberta, as in Montana across the border, there is more gas than oil in the Jurassic, and traps are strongly controlled by pre-Jurassic erosional topography. Most of central Canada's Jurassic oil is pooled along the western side of the Williston Basin, where the boundary is formed by a persistent arch.

Cretaceous sedimentary rocks in the basin are wholly clastic, and similar to those in the US province in consisting essentially of coal-bearing molasse from the west and shield-derived detritus from the east, interfingering with a belt of marine shale along the basin axis. The Cretaceous basin contains great reserves of natural gas, much of it in tight sands close to the disturbed belt. Light oil occurs in the basal sandstones, resting on erosional topography and extending a short distance into the foothills. It also occurs in numerous pools in fluvial and nearshore sandstone bodies in the younger Lower Cretaceous in the central part of the basin. In the eastern part and along its eastern margin are the vast quantities of heavy oil and tar sands, both at and close to the surface and buried at depths of more than 1000 m (see Sec. 10.8.4).

The Upper Cretaceous contains the largest single oilfield (Pembina) in Canada, in a lithologic trap formed by updip wedge-out of sandstone permeability (see Sec. 16.4.1). Other sandstones contain great volumes of nonassociated gas, most of it in the immature (biogenic) facies. Between 58°N latitude and the Arctic coast, however, Mesozoic strata lack significant deposits of either oil or gas.

25.5.12 Petroliferous basins of the North American Arctic

Three richly petroliferous basins of very different types converge towards the Arctic Ocean: the North Sea–Norwegian Sea Basin, the West Siberian Basin, and the compound foredeep basin east of the American Cordillera. Wallace Pratt considered the Arctic Ocean region as a whole to be one of four great interior seas which hold the bulk of the world's petroleum; the others are the Mediterranean–Persian Gulf, Caribbean–Gulf of Mexico, and southeast Asian regions. Into the 1980s, the North American side of the Arctic, at least, has not fulfilled Pratt's expectation. A huge accumulation of oil is under exploitation in front of Alaska's Brooks Range, but otherwise the region has yielded principally gas so far.

The *Alaskan North Slope* comprises a large sedimentary basin, the Colville trough of Early Cretaceous and younger age, overlying the deformed belt in front of the folded and overthrust Brooks Range (Fig. 26.8). The slope contains the largest areally-continuous accumulation of oil in North America, nearly all of it in Triassic and Cretaceous clastic reservoirs. The older reservoir, at Prudhoe Bay, has excellent porosity and permeability (see Chs 13 & 26), but the oil is not of very high quality. The Lower Cretaceous sandstone at Kuparuk River is fine-grained and its oil is of medium gravity, but it is overlain by further, shallow sands containing a huge volume of heavy oil.

The traps represent an unusually favorable combination of stratigraphic wedge-out, broad folding, and a critical unconformity within the Lower Cretaceous succession (Figs 16.1 & 26.11). The oil and gas may be of more than one source (Sec. 26.2.2). The Prudhoe Bay–Kuparuk River complex originally contained recoverable reserves of nearly 2×10^9 m^3 of oil and 0.75×10^{12} m^3 of gas.

Lower Cretaceous sandstones are productive to a lesser degree both west and northeast of the giant producing area. In the latter direction, which passes into the *Beaufort Sea,* Paleocene sands also contain medium-gravity oil and gas-condensate and Upper Triassic sandstones rather heavy oil. Though total oil in place may be large, the geology is rather complex for a region of such operational difficulty. Where the Cretaceous source sediments come eventually to rest directly on porous Paleozoic rocks, those also contain medium-gravity oil (see Sec. 25.2.5). Further extension towards the northeast takes the play into the Canadian sector of the Beaufort Sea. Here, the Cretaceous section lacks sands, and hydrocarbons occur principally in Paleogene sandstone reservoirs (including turbidites of deep-sea fans). Medium-gravity oil occurs in the Paleocene, and both oil and gas in Eocene and Oligocene sands. Traps are controlled by growth faults and by shale-cored diapirs. Exploitation of these resources faces formidable logistic, economic, and climatic barriers.

The Canadian Beaufort Sea is encroached upon by the *delta of the Mackenzie River.* Below it, the whole Mesozoic and Cenozoic stratigraphic succession is clastic in nature and several parts of it contain wet gas. The largest accumulations lie closely on- or offshore in either the Lower Cretaceous (which rests unconformably on the Proterozoic basement) or the Paleogene (especially the Lower Eocene). The stratigraphic isolation of these reservoir horizons from the late Cretaceous source sediment has been remarked as an obstacle to oil accumulation. Another hazard to both exploration and exploitation is the complex of large growth faults crossing the delta with northeasterly strike. They control the traps that have survived.

The Canadian Arctic archipelago contains a large trapezoidal basin, the *Sverdrup Basin,* essentially confined between 75° and 80°N latitudes. The basin came into being in Pennsylvanian time and had become wholly nonmarine before the end of the Cretaceous, so it is primarily a Mesozoic feature. It contains large gas accumulations in several Mesozoic sandstones, particularly in the Upper Triassic (a nonmarine sandstone) and the Lower Jurassic. It also contains a body of tar sands of Triassic age. Light oil has been recorded from an Upper Jurassic sandstone in offshore mid-basin.

25.6 Basins essentially of Tertiary age

25.6.1 Tertiary basins of Europe

South Caspian and Caucasian province The Caspian Sea occupies a sub-meridional, fault-controlled depression of Plio-Pleistocene age. It crosses the eastern extension of the Hercynian platform which underlies most of Europe between the Alpine and Caledonian fold belts. The southern margin of this platform lies along the northern front of the Alpine belt, which is here marked by the fold line of the Apsheron "sill," the trans-Caspian junction of the Caucasus Mountains with the Kopet Dagh range in southwestern Turkmenia (Fig. 13.25).

The onshore extensions of the Apsheron "sill" form the Apsheron and Cheleken peninsulas on the western and eastern shores of the Caspian. Both are the sites of rich oilfields, which continue under the sea from the two sides and may eventually form a line of fields entirely across it. The southern Caspian Sea occupies an almost circular and very deep depression south of the sill; it contains some 22 km of sediments with no known sialic layer below them.

Until Middle Miocene time, the region was part of the Alpine geosyncline. The Oligo-Miocene part of the succession is the famous Maykop Group of bituminous claystones rich in fish remains, a classic source sediment. Middle and Upper Miocene and Lower Pliocene strata became progressively more restricted, culminating in orogenic uplift and emergence. The deep basin of the present southern Caspian was the only part of the region to remain under water. It in fact underwent sharp, continued subsidence through Middle and Late Pliocene times, as it was progressively filled from north, east, and west by the deltaic deposits of large rivers (Fig. 13.25). The delta fed from the north was the original delta of the Volga River, now driven back to its present position at Astrakhan by the enlargement of the sea.

The complex of deltaic sands and clays, partly continental, partly marine, constitutes the so-called *Productive Series,* of Middle Pliocene age. It is some 3 km thick, and the overlying Upper Pliocene and Pleistocene clastic succession, with interbeds of volcanic ash, adds sufficient additional thickness to depress the base of the Productive Series to a depth of about 6 km. Folding of the succession in late Pliocene time introduced sharp anticlines cored by abnormally-pressured pre-Pliocene clays (Fig. 16.54). Burning gas seepages created the "eternal fires" that gave rise to practices of the Zoroastrian religion. The migration (or remigration) of oil into the traps is still continuing.

The fields have produced oil at least since 1871. The *greater Baku region,* on the Apsheron Peninsula, is now largely depleted, but its cumulative production is of the order of 10^9 m^3. It totally dominated Russian oil production until World War II. In 1949, the first large offshore field was found, and in 1956 the first one on the eastern shore was added. In 1981 the production from the entire trend still reached about 75 000 m^3 daily.

The fields are unusual in many respects: the multitude of possible source sediments; the numerous productive sands (Fig. 15.3); the one-sided traps, with oil commonly pooled on only one flank of asymmetrical anticlines (Sec. 15.2.6); the hydrochemical inversion, in which alkaline formation waters tend to occur below normal salty waters; the low geothermal gradient (as low as 10°C km^{-1} in several fields).

The *pre-Caucasian foredeep* lies along the northern front of the Caucasus Mountains. The succession of strata differs from that in the Baku–Cheleken belt in two important respects; the Miocene and Pliocene strata, though well developed, are very much thinner and contain much more carbonate; the thick Mesozoic carbonates are therefore within reach of the drill and are highly productive, especially where fractured. The principal source sediment for the Mesozoic oil is probably in the Middle Jurassic, but that for the Tertiary oil (nearlyall in Middle Miocene sandstone reservoirs) is the Oligo-Miocene Maykop Group, of which this foredeep is the type locality. The principal traps are sharply asymmetrical anticlines, commonly overthrust and overturned towards the north. The Maykop field itself, however, is an oddity among European fields in being largely in shoestring sand reservoirs of Oligo-Miocene age overlapping Mesozoic rocks as a homoclinal stratigraphic trap.

Farther out on the pre-Caucasian foreland – on the so-called Scythian plate, east of the Sea of Azov – gas and some oil occur in Paleogene sandstones underlying the Maykop Group. The Caucasian foredeep continues westward across the neck of the Crimean Peninsula and the northwestern corner of the Black Sea (another deep depression within the Alpine fold belt). Here gas occurs in Oligocene and Paleocene strata, still associated with the Maykop Group.

Carpathian and Alpine basins Still farther westward from the northern Black Sea, we come to the foredeep of the *Romanian Carpathians.* Tertiary sediments some 9 km thick were severely deformed towards the south and east, along the outside of the mountain arc, during the late phases of the Alpine folding. Following a classical development of late Cretaceous–Eocene flysch, the Oligocene is dominated by richly bituminous and calcareous shales called *menilites,* resembling the Caucasian Maykop Group and associated with diatomites. The overlying thick salt is strongly diapiric (Fig. 16.53) through the Miocene and Pliocene strata, creating an extraordinary assortment of complex traps, many of them containing one-sided accumulations like those at Baku.

Nearly all the oil was in brackish to fresh water Pliocene sandstones, and it provides a classic illustration of contrasting oil types in close juxtaposition. Following a major late Miocene unconformity and considerable erosion, the Lower Pliocene sands are relatively continuous and regularly productive provided they are not structurally low or eroded; they yield paraffinic oil. The Upper Pliocene sands are exceedingly erratic; for the most part they yield asphaltic oil, in much more restricted areas containing many enigmatic barren stretches.

Southwards away from the mountain front lies the featureless plain of the Danube valley. At no great distance from the disturbed belt the salt is no longer diapiric, but has accumulated as deep-seated, pillow-like masses over which unpierced anticlines are oil-bearing. The principal reservoir sands here are somewhat older (Miocene). Proceeding around the arc of the eastern Carpathians, thrust- and fold-belt structures are even more complex than those along the southern zone. Again, reservoir sands are still older (mainly Oligocene).

Still farther around the Carpathian arc we come to the north-facing *Polish* or *external flysch Carpathians*. Small oilfields were discovered early in a zone along which Cretaceous–early Oligocene strata were closely and deeply folded and overthrust northeastwards over younger Tertiary rocks of the foreland. There is no thick Pliocene here, and the fields are much smaller and fewer than those in the Romanian sector of the arc. Nearly all the oil was in Oligocene sandstones. It is largely depleted, and the field area was incorporated into the Ukraine following World War II.

On the concave side of the Carpathian arc is the *Pannonian Basin,* which occupies most of Hungary and the northern fringe of Yugoslavia. It is a Neogene construction almost wholly surrounded by Cenozoic mountain belts. Oil occurs in small accumulations in Miocene and Lower Pliocene sandstones. In the eastern part of the basin, gas (possibly bacterial and immature) dominates over oil, especially in the topmost Miocene part of the section. An eastern extension of the large back-basin encompasses part of Transylvania in central Romania. It is another gas-producing area, from Miocene and Pliocene sandstones involved in salt-cored uplifts.

Around the flanks of the Pannonian Basin are a series of small, graben-like sub-basins. In Yugoslavia, the most important of these is the *Drava Basin,* which yields small amounts of both oil and gas from the Pliocene. By far the most important of all the sub-basins, however, is the *Vienna Basin,* which cuts across the northwestern corner of the larger basin. The present Vienna Basin, like the larger Pannonian Basin, is a construction of the early Miocene, but instead of being a back-deep behind a mountain arc it is a faulted depression superimposed on frontal mountain nappes, those of the Limestone Alps. These nappes are in turn thrust over the Alpine flysch (Fig. 16.85). As described earlier (Sec. 25.5.2), much oil is now produced from the carbonate and flysch floors of the basin, but all the earlier production came from the Neogene fill of the graben, principally from Middle and Upper Miocene sandstones. The productive part of the section is dominantly marine, with abundant marls and clays in addition to the reservoir sands. In the ancillary *Styrian Basin,* which connects the main graben to those in Yugoslavia, gas actually occurs in a Miocene reef. The higher levels of the Vienna section are wholly of brackish or fresh water origin. The Neogene basin has yielded both oil and gas, from fault-associated traps around the margins and drape folds over high basement in the basin center. Like the Romanian Carpathian Basin, the Neogene Vienna Basin yields both strongly asphaltic and strongly paraffinic oils. By the end of 1982, cumulative production of 100×10^6 m^3 of oil had been attained from an area no larger than about 7000 km^2.

North of the flysch zone, which underlies the Vienna Basin, is the *Molasse Zone,* stretching northwards from the Alpine mountain front towards the Bohemian massif (Sec. 25.5.2). Some oil occurs in Eocene sandstones, but much more is apparently pooled in underlying Mesozoic rocks, and it may be that all or most of the zone's oil content was originally of Mesozoic derivation. Cenozoic strata are dominantly gas-bearing, especially in Oligocene sandstones associated with fish-bearing shales well known to vertebrate paleontologists.

Still farther north, the *Rhine graben* cuts transversely across the Hercynian ranges and the Permo-Triassic basin which overlies them. It has been an oil-producing region since before the modern industry began, but its total output has been minuscule in modern terms. As a petroliferous basin, the graben presents two odd features. Though it contains a section ranging from the Permian to the Neogene, it totally lacks the Cretaceous strata which otherwise constitute the most widespread post-Paleozoic sedimentary rocks in the world. And though the copious oil seepages at Pechelbronn gave rise to an oil-mining industry several hundred years before the drilling era, the whole basin may ultimately revert to mining as the principal oil-recovery process.

The present Rhine graben was initiated in the early Eocene. Most of the oil has been produced from lenticular sandstone reservoirs of Oligocene age, within a thick Paleogene section otherwise consisting largely of carbonates and evaporites (including potash salts). The famous fish-bearing shales which occur throughout most of the Alpine molasse zone are present here also, and may be an oil source rock, but there is a lot of evidence that much of the oil in the Cenozoic sediments originated in the underlying Mesozoic. Some gas is also produced, much of it from younger strata (Mio-Pliocene) on the German side of the valley.

Tertiary basins in the Mediterranean region The best known and most widespread Tertiary successions in Europe are those along the Mediterranean coasts. Yet they have provided little in the way of hydrocarbons, and almost no oil. The longest productive Tertiary basin is that of the *Po Valley* in northern Italy. It is a classic, asymmetric, intermontane basin, the foredeep of the Apennines and the backslope of the Alps. The Miocene (and even the Pliocene, for which the basin provides the type locality) is again unconformable on all older strata and, again, the basin is strongly gas-prone. Miocene, Pliocene, and Pleistocene sands are all productive, from long, linear anticlines paralleling the Apenninic side of the basin. The gas occurrences extend along the east coast of Italy and into the Adriatic Sea. Minor oil occurs there also, in Lower and Middle Miocene limestones. Further Tertiary gas is found in the *Lucanian Basin* in southeastern Italy.

At the eastern end of the Mediterranean, high-sulfur oil and gas are produced from the northern extremity of the *Aegean Sea,* west of Gallipoli. The reservoirs are sandstones within a Miocene succession otherwise dominated by carbonates and evaporites; this circumstance recalls that of the Paleogene production from the Rhine graben. The Miocene rests on metamorphic basement.

Across the Mediterranean and outside Europe is a much more productive basin, that of the *Suez graben* between mainland Egypt and the Sinai Peninsula. Though the greatest marine transgression began in the Cretaceous, rifting did not become dominant until the late Oligocene. The Miocene then became even more varied than that of the Aegean, eventually becoming unconformable across all older rocks. Phosphatic marls with a rich fauna of planktonic foraminifera and fish teeth were patchily developed and followed by evaporites and then by *Lithothamnium* limestone. Most of the oil occurs in Miocene reef limestones and lagoonal dolomites; offshore in the gulf there are large accumulations in Miocene sandstones also. A lesser but still substantial amount of oil has been produced from Eocene and Upper Cretaceous limestones below the rift unconformity, and some even in the Paleozoic Nubian sandstone (Sec. 25.2.6). Most of the traps, including the reefs, are directly or indirectly controlled by the normal faults of the graben (Fig. 16.29).

A little Miocene oil also occurs in the *Gulf of Hammamet,* in a region otherwise characterized by early Mesozoic oil (in both Tunisia and Sicily). At the western extremity of the Mediterranean region, the Miocene apparently yields only gas. It has been found in small quantities in the Pre-Rif section of *Morocco* and across the Strait of Gibraltar in the *Gulf of Cadiz.*

The main basins of the deep Mediterranean contain a remarkable evaporite formation of late Miocene age, the Messinian evaporite. In the deep sinks of the eastern and western Mediterranean, thick sedimentary sections of silled basin origin, still unknown and untested, may lie below this widespread seal. Such sections provide an explorational challenge for the next generation of European searchers for petroleum.

25.6.2 The Niger Delta Basin (Fig. 17.21)

With the separation of Africa from South America, a basin originating in early Cretaceous time extended along what is now the south-facing coast of West Africa and into the Benue valley to the east. Maximum transgression occurred about Turonian time. Cretaceous oil in modest quantities occurs in many segments of this basin (see Sec. 25.5.4). Heavy oil seepages are common and bituminous sands crop out in the eastern outcrop belt. From the beginning of Eocene time, however, the main depocenter was established in the angle of the West African coastline with the first growth of the modern Niger delta. The total sedimentary section in the depocenter probably exceeds 12 km; the Neogene section alone exceeds 4 km in the wedge of greatest subsidence.

The seaward advance of the delta has resulted in a very characteristic configuration of the deltaic prism. Beneath continental sandstones of Plio-Pleistocene age, a deltaic or paralic facies of alternating sandstones and shales overlies prodelta marine shales, penetrated by diapirs and cut by a great number of down-to-the-coast growth faults (Figs 16.102 & 103). Nearly all the principal oilfields are in rollover anticlinal traps in Miocene deltaic sandstones. In the eastern part of the prism, only the prodelta shales have reached sufficient maturity to serve as source sediments, but in the central and western parts the lower part of the true deltaic section has also yielded its oil for pooling.

Like the American Gulf Coast Basin on the two sides of the Mississippi delta, the Niger Delta Basin exhibits the characteristic age-zonation, the reservoir rocks becoming younger towards the basin (Sec. 17.4.3). The oil produced is of a great range of gravities, considering the geologic unity of the productive area and its stratigraphic restriction (Sec. 17.7.3). As in most deltaic accumulations, there is a great deal of associated gas (the incidence of which extends eastward into *Cameroon*). The delta alone put Nigeria, by 1981, into about twelfth position in recoverable gas reserves of all the countries of the world. It undoubtedly occupies an equally elevated rank in ultimate oil reserves. Yet –

again characteristic of deltas, with their festoons of small traps – the basin contains very few giant individual fields, and no supergiants.

25.6.3 Tertiary petroleum in southern and south-eastern Asia

The modestly producing basins of northern China are essentially Mesozoic in age, and were considered with other Mesozoic basins (Sec. 25.5.6). Nevertheless, at least three of them yield oil also from Tertiary reservoirs, doubtless from pre-Tertiary sources. The *Tarim Basin* north of Tibet yields some waxy oil from the Oligo-Miocene. The *Qaidam Basin* northeast of the Kun Lun provides numerous seepages from Tertiary sediments and some shallow production from Oligocene to Pleistocene continental sandstones. In both these cases the probable source of the oil is in the Jurassic. The *Gansu Basin* of the Gobi Desert has provided small production from nonmarine, early Tertiary red beds; the source is undoubtedly older (Sec. 25.5.6).

The fault-controlled west coast of *India* was developed during the Cretaceous breakup of the Gondwana continent. It is interrupted by a major fault-bounded embayment north of Bombay. The onshore portion of this graben, the *Cambay Basin,* was established during the Eocene within the late Cretaceous Deccan Trap plateau. Eocene black shales of estuarine or lagoonal character are the source rock for oil in several Eocene and Oligocene sandstones which represent fluvial and deltaic infillings from the north. A basinwide unconformity at the base of the Miocene forms an upper limit of oil occurrence. The fields are mostly on complex domal anticlines set *en echelon* along the graben; most of the oils are very light and waxy.

The offshore continuation of the basin extends past Bombay, becoming increasingly gas-prone. An easily-identified "high," developed during the Paleogene, contains prolific oil in a limestone reservoir of Middle Miocene age. Both west and east of the "high" are down-faulted clastic basins.The basin between the high and the coast contains oil and gas in a Paleogene limestone, their source probably in prodelta shales and silts.

Pakistan's *Indus Basin* lies between the north-trending Baluchistan ranges and the desert of north-western India. Paleocene and Eocene limestones and sandstones are well developed but apparently exclusively gas-bearing. The fields, in large simple anticlines on top of a larger arch, have large in-place reserves. Much of the gas is sour, however, and in some of the fields it contains high percentages of nitrogen and CO_2 (see Sec. 5.2). If any commercial oil occurs in this outwardly attractive foredeep basin, it is likely to be in pre-Tertiary strata; light oil has been reported from Cretaceous sandstones.

In northern Pakistan the intermontane *Potwar Basin* lies between the Hindu Kush Mountains and the Salt Range. Paleogene sediments rest unconformably on Jurassic sandstones and other units of the Gondwana System; they are largely in calcareous and saline facies like their equivalents in Russia's Ferghana Basin on the opposite side of the Hindu Kush (Sec. 25.5.6). Some of the beds contain enormous numbers of fish remains or of foraminifera. Most of the oil production has come from erratic permeability patches in Lower Eocene limestone; some also occurred in Miocene continental sandstones which overlie the Eocene unconformably and which initiated the great thickness of final basinal fill.

Oil production from young sandstones in mountainous *southeast Asia* is of considerable antiquity. Numerous seepages, especially around clay diapirs, prompted exploitation of the oil long before the drilling era and lured the early drillers also. Several Burmese fields, Digboi in Assam, both northern and southern Sumatra, Ledok in Java, Sanga Sanga in eastern Borneo, and northern Ceram in the Moluccas were all productive before the end of the 19th century. Tarakan Island off eastern Borneo and Miri in Sarawak were added before World War I. Sumatra, easily the most productive part of the region, provides an interesting sidelight on the economic consequences of geography. Oil seepages in northern Sumatra gave birth to the Royal Dutch–Shell group of companies during the 1880s; the Stanvac fields in southern Sumatra were discovered and exploited before 1900. The jungles of central Sumatra, bisected by the Equator, were the last to be explored and developed, during and after World War II, and they turned out to contain the two largest oilfields in southeast Asia (Minas and Duri fields).

An Eocene flysch trough fringed the western margin of the Sunda platform within which the islands and mountains of southeast Asia were to be developed. This geosyncline was first impinged upon by India, forming the syntaxis of *Upper Assam* between the Himalayan and Burmese ranges. The depression contained fresh water from the mid-Miocene onwards; late Miocene uplift and erosion followed, and still later thrusting from both sides apparently took place along the sites of earlier down-to-the-basin normal faults. Most of the oil occurs at or near the unconformity between the Oligocene and the Lower Miocene, a level which is also close to the main thrust plane. There are numerous productive sands, over a stratigraphic range of more than 1000 m. The source of the oil is below the unconformity and largely in the footwalls of the thrusts; the traps are obvious linear anticlines, principally in the hanging walls. The crude is light, of mixed base and high wax content.

The *Bengal Basin* in Bangladesh, like its Indus

counterpart on the opposite side of the subcontinent, is gas-prone in the Tertiary section. The principal reservoirs are in Miocene sandstones. To the east, the Irrawaddy River valley in *Burma* occupies an orogenic intermontane basin. In it, a series of narrow, elongate anticlines are oil-productive. As in Assam, the Oligo-Miocene unconformity is critical to the trapping. It lies in the middle of the oil-bearing section and oil occurs in both subcrop and supercrop; again, the source rocks are probably in the subcrop. The reservoir sandstones, largely of deltaic environment, are numerous but markedly inconsistent. Some deeper limestones also now provide reservoirs. The crudes, as in Assam, are light and waxy but more paraffinic, becoming heavier and more asphaltic towards the eastern foreland.

The axis of a series of back-arc basins continues southward, east of the *Andaman Islands* (where gas has been found), into northeastern Sumatra and the Malacca Strait. Subsidence progressed southward, so that *Sumatra* has much less Paleogene section than has Burma or Assam but much more Neogene. The subsidence was brought about by block-faulted break-up of the pre-Tertiary basement; the first marine transgression over it occurred in the earliest Miocene, only shortly before the seas withdrew entirely from the Upper Assam basin. The Miocene orogeny greatly accentuated a series of linear basinal cells between the mobile zone on the southwest (now the Barisan Mountain chain) and a hinged stable shelf to the northeast. The anticlinal oilfields of Sumatra are concentrated along or closely basinward of this hinge-line in three of these cells (Fig. 16.40).

Both source rocks and reservoir rocks are in the transgressive Miocene succession of sandstones and shales. Some carbonate reef reservoirs of early and middle Miocene age were later discovered along the northern shelf. Sumatra should eventually attain a cumulative output approaching 2×10^9 m^3 of oil; the Minas field alone had initial recoverable reserves of about half this amount.

The basin continues through the Java Sea. Important fields of both oil and gas lie largely offshore from southeastern Sumatra and western Java, with smaller fields scattered along the north coast of *Java* in Miocene and Pliocene sandstones folded into north-facing, thrust-faulted anticlines. The only Tertiary oil so far in production between Java and the open Pacific Ocean comes from the Salawati Basin, which occupies the western end of the Vogelkop in *Irian Jaya* (western New Guinea). Oil occurs in pinnacle reefs of Middle Miocene age, in another basin not initiated until early Tertiary time.

On the inner, concave side of this great sweep of back-arc basins comes a second, more closely curvilinear belt from Thailand, through Borneo to the Philippine Islands. It is separated from the larger arc of basins by the mountain ranges of Burma, the Malay Peninsula, the Celebes, and the Philippines. In *Thailand,* sweet oil and deep gas occur in Miocene sandstones. Offshore, the *Gulf of Thailand* covers a complex graben in which deltaic sandstones of early and middle Miocene ages contain oil, gas, and condensate. Gas and condensate are also known in the Red River delta (Hanoi trough) of *Viet Nam.* In the *Malay Basin* offshore from the peninsula, oil and a little gas occur in Oligocene and Miocene sandstones. More oil occurs in Miocene deltaic sandstones on the *Natuna Islands,* between Malaya and Borneo.

Sarawak, Brunei, and *Sabah* make up the northwestern flank of the island of Borneo. Again, pre-Eocene rocks are basement, here represented by the ophiolite–chert association. The Eocene is a flysch, as in Assam and Burma, and a late Eocene orogeny created the present structural grain with northwesterly vergence. The post-orogenic sequence comprises 10 km of strata, mostly but not entirely clastic. The principal oil-source sediment is Lower Miocene, the reservoir rocks are Miocene and Pliocene sandstones in molasse facies, and the traps are sharp, much faulted anticlines raised during a late Pliocene deformation. Yet again, the oils produced show rather sharp distinction between heavy, asphaltic crudes at very shallow depths and paraffinic crudes below.

In all the Sumatra, Java, and northwestern Borneo basins, offshore exploitation has been as much of gas and condensate as of oil. Off Sarawak, in particular, oil tends to dominate in sandstone reservoirs and gas in Middle and Upper Miocene carbonate buildups. Like many offshore areas, these are cut by great numbers of normal faults.

The tiny enclave of Brunei, remnant of an ancient, much larger sultanate, scarcely rivals Kuwait in its ratio of oil reserves to area but it probably surpasses the more famous Trinidad. At least two of its fields – Ampa (offshore) and Seria (onshore) – should eventually yield more than 150×10^6 m^3 of oil apiece; only the Minas and Duri fields of central Sumatra exceed them in recoverable reserves in all of southeast Asia. In 1980, moreover, Brunei had the world's largest LNG terminal, designed largely for the Japanese market.

Between Brunei and the Philippine Islands is the island of *Palawan.* It differs from the established producing areas just described in having received important carbonate deposition from late Eocene to late Miocene time. Very light, sweet oil occurs in Oligo-Miocene limestone, including reefs, lying offshore to the west of the island. The reservoirs lie between two major unconformities. Some oil has also been found in Miocene turbidites.

In *eastern Kalimantan* (Borneo), the stratigraphic succession is again wholly Tertiary and nearly 10 km thick, representing a single cycle of sedimentation but structurally separated into three foreland sub-basins. Following paralic sedimentation during the Eocene, and thick, bathyal Oligocene, a series of large deltas began to form in the Miocene and their formation continues today. Late Pliocene folding intervened and continued into the Pleistocene. Thick, multiple pay sands in the Miocene and Pliocene contain very aromatic oil with a rather high pour point, and prolific associated gas. Minor oil also occurs in Eocene sandstones.

On the opposite side of the *South China Sea* from Palawan Island is the island of Hainan. West of it, the *Beibu Gulf Basin* (Gulf of Tonkin) yields light oil and gas from Lower Tertiary sandstones. East of it, the *Pearl River Basin* west of Hong Kong is a future producer of very waxy oil.

25.6.4 Tertiary petroleum in the circum-Pacific belt

The oil- and gasfields of Irian Jaya and Palawan Island lie close to the western margin of the Pacific plate; the margin follows the trenches and transform faults of the island arc system. Most other oil and gas deposits known along this western margin (all of them small) are more accurately associated with the coast of the Asian mainland than with the true margin of the Pacific plate.

On the west side of *Taiwan,* a little paraffinic oil, and some gas, is produced from Miocene sandstones. In the *East China Sea* off Shanghai, thick Tertiary strata appear to be richly organic and will probably yield both oil and gas. At the northwestern extremity of the Yellow Sea, between China and Korea, several large deltas (including that of the Yellow River) are being built into the *Gulf of Bo Hai.* Their rivers cross the low-lying *Huabei (Hopeh) Basin,* superimposed during the Tertiary on a deformed Paleozoic succession. The basin contains exceedingly thick Paleogene strata, including rich hydrocarbon source sediments. Some oil is produced from late Cretaceous and early Tertiary marine beds, but much more comes from continental and paralic sandstones (and some reef limestones) of Miocene age. Inland from the coastal basins are extensive oil-shale deposits, some of which are Tertiary in age (see Sec. 10.8.1).

Japan has been an oil-producing country for many years, but never to more than a trivial extent. Oil and gas are apparently restricted to the back-arc Akita Basin along the west coast of northern Honshu; nearly all the reservoir rocks are of Late Miocene or Early Pliocene age and of partly volcanic derivation (tuffaceous sandstones and sandy tuffs). *Sakhalin* possesses a number of asphalt lakes aligned along major fault zones. Heavy, asphaltic oil, plus considerable gas, is produced from Miocene and Lower Pliocene multiple sandstones in an enormously thick Neogene succession along the northeast coast. Production extends offshore into the Sea of Okhotsk. In southern and western *Kamchatka,* gas and very minor oil occur in Miocene sandstones, with a little in earlier Tertiary and Upper Cretaceous sandstones in addition. More gas, with a little oil, is found in Miocene sandstones in the *Anadyr Basin,* at the northeastern corner of Siberia.

On the American Pacific margin, the only commercial oil known north of California is in the *Cook Inlet Basin* of southern Alaska. It provides one of the case histories for this book (Sec. 26.2.2). All the oil and gas is found in sandstone and conglomerate reservoirs of the thick Tertiary succession – the oil in Oligo-Miocene fluvial sandstones and gas throughout the Oligocene–Pliocene section wherever there are suitable reservoir rocks. The oil, at least, is believed to be of Mesozoic source. Small quantities of gas are produced in *northwestern Oregon,* from Middle Eocene sandstones. The *Great Basin,* so called because no big river flows out of it despite its elevation, is largely in Nevada and so not particularly close to the Pacific margin. Nor is it of any significance as a petroleum province, but the circumstances of its tiny oil production are sufficiently unusual to warrant brief comment. The oil occurs in an Eocene lacustrine limestone and an overlying fractured ignimbrite of Oligocene age. These two odd reservoir units are overlain and sealed by very thick fill in a graben within the famous Basin-and-Range province; they are underlain with great unconformity by Carboniferous limestone, which also yields oil shows and might be imagined to be the most probable source of the oil. The oil, however, is sweet and paraffinic, making it more probably indigenous to the Eocene lake beds.

Oil production in the circum-Pacific region has been totally dominated by that from three of the six productive basins in *California* (Figs 16.27 & 37). All are individually small, intermontane, fault-bounded basins; all are essentially Tertiary in age; all produce from sandstone reservoirs almost exclusively. The basins in their present form owe their origin to the conversion of a triple plate junction (involving the trench off North America and a ridge–ridge transform fault in the Pacific Ocean) to a sliding margin now governed by great strike-slip faults. This conversion caused transtensional (oblique) opening of rhomb-shaped basins, their abrupt depression, and subsequent transpressional (oblique) deformation. The basins now lie on the two sides of the San Andreas fault where it crosses the boundary between continental (granitic) and oceanic (Franciscan) basement and the east–west Transverse Ranges. The conversion to sliding boundary took place near the end of Oligocene time. Two of the basins (the Los Angeles and Cuyama–Salinas basins) were created during the Miocene and contain no pre-Miocene Tertiary sediments; all the others except the Sacramento Valley

Basin were greatly modified (indeed, re-created) at this time. The Late Miocene, diatomaceous Monterey Formation is the principal source sediment of all the oil in three of the basins and most of it in the other two that yield oil. A strong Pleistocene deformation is clearly manifested in the mountain belts that border all the basins.

The development of California's basins within the regime of an active plate boundary resulted in multiple sandstone reservoirs in very narrow basins. California accordingly became synonymous, in the American oil industry, with very high rates of oil production per unit area; the Los Angeles Basin is still commonly quoted as having produced more oil per acre (or per square kilometer) than any other basin in the world. At least ten individual Californian oilfields (four of them in the Los Angeles Basin) have already produced more than 100×10^6 m^3 of oil, and many more have produced the volume (about 16×10^6 m^3) required to qualify as giant fields within the US pantheon. This great output has been achieved despite the heavy, asphaltic character of much of the oil and its consequent recalcitrance both of recovery and of refining.

The Californian oil and gas province is split in two by

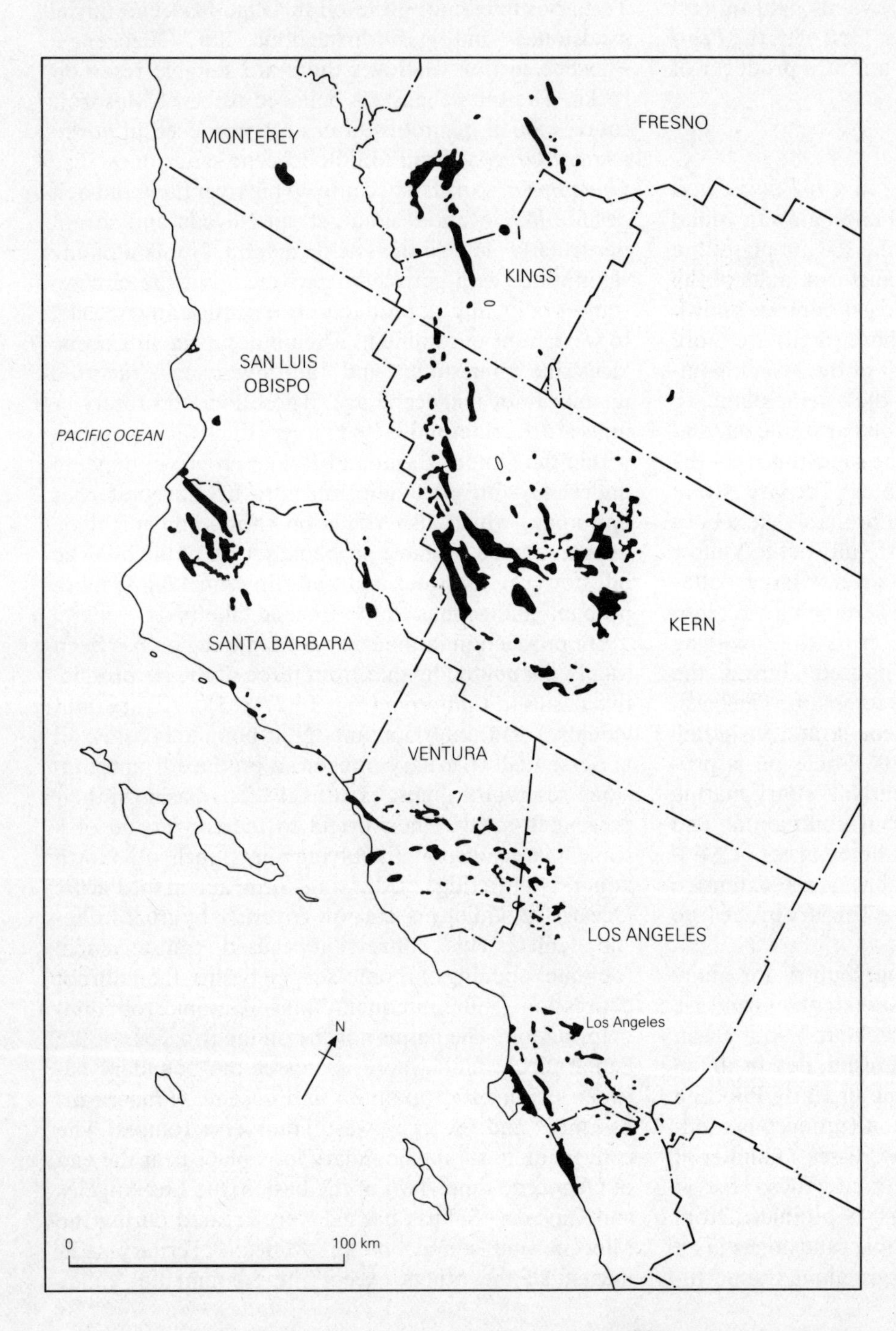

Figure 25.6 Principal oil- and gasfields of California, exclusive of the gasfields of the Sacramento Basin, north of the area shown. (After R. C. Blaisdell and T. W. Dignes, *AAPG Bull.*, 1980.)

the present San Andreas fault (Fig. 25.6). Northeast of the fault, the *Sacramento Valley Basin* is wholly a gas province and its output from Eocene and Upper Cretaceous sandstones is included among the Mesozoic basins (Sec. 25.5.10). Its southern extension through the Pleistocene and present Great Valley of California provides a very different circumstance. The *San Joaquin Valley Basin* became a rapidly subsiding depression during early Tertiary time and had become a quite distinct basin by the Miocene. Considerable oil has been produced from Paleogene sandstones, but by far the most prolific production has come from the Miocene and Pliocene. The Miocene had the advantage of the introduction of several bodies of deep-water sandstone into the basin in which the diatomaceous source sediment was being deposited. Where the displaced sandstone is not present, the diatomaceous unit (commonly a shale or chert) is itself productive from fractures (of very heavy oil in some instances). Pliocene sandstones are productive principally on the two flanks of the basin, in traps at least partly stratigraphic (Figs 16.8 & 38). Valuable quantities of gas have been associated with the oilfields; much less obviously associated gas occurs in anticlinal traps in the central basin.

Southwest of the San Andreas fault a succession of narrow, fault-bounded basins has yielded oil (and some gas) from mid-Tertiary and younger sandstones. The *Cuyama–Salinas basins* are fault troughs containing only Neogene sediments; oil (some of it heavier than 15°API) is produced from Miocene sandstones.The *Santa Maria Basin,* north of the Transverse Ranges, is productive chiefly from fractured siliceous shale and chert and a little dolomite in the Middle and Upper Miocene diatomaceous succession, and from sandstones immediately above and below this (Fig. 13.75). The *Ventura Basin* (Fig. 17.12) lies wholly within the Transverse Ranges, and was compressional throughout its Neogene history. Prolific oil occurs in multiple sandstones in high-amplitude anticlines associated with (and in many cases cut by) reverse faults. The sandstones occur throughout the Oligocene–Pleistocene succession, and are unusual in becoming more marine upwards; the productive Pliocene sands were among the earliest identified turbidites (see Ch. 4).

The Ventura Basin extends offshore along the central axis of the Transverse Ranges; here it is familiarly referred to as the *Santa Barbara Channel.* It was the site of a notorious spill from a drilling platform in 1969 and has been at the center of political and environmental controversy ever since. Oil, gas, and condensate are produced from Eocene and Oligo-Miocene sandstones, but the greatest production is of heavy, high-sulfur oil from fractured Miocene Monterey rocks (cherts, shales, diatomites) and from Pliocene sandstones which again include turbidites. The *Los Angeles Basin* ranks among the world's most famous and oft-quoted oil-producing regions (Fig. 16.27). Nearly all its output has come from Upper Miocene and Lower Pliocene sands; a little has come from fractured schist and schist-conglomerate forming the floor of the present basin.

Offshore developments are likely along much of the coastal waters of both California and Baja California, though obstacles to these are more determined and more publicized than those in any other jurisdiction. No commercial oil or gas is known from the Pacific margin of North America south of 33°N except for small quantities of gas and condensate in Miocene and younger deltaic sandstones in the Gulf of California (which belongs to Mexico). None is known along the Pacific coasts of Central America or Colombia. *Southwestern Ecuador* exposes another fore-arc basin of the type that is otherwise productive only in southern Alaska and California; it has long been productive of minor oil from Middle and Upper Eocene sandstones, associated with abundant chert and a complex of gravitational slump deposits. Some gas is known in Miocene sandstones in the Gulf of Guayaquil. On the opposite side of the gulf, in *north-coastal Peru,* oil has been produced in small quantities for many years from Oligocene and Miocene sandstones and conglomerates along a down-to-the-coast rift zone. For many more years, a much greater amount of oil has come from the *Talara region* of Peru. Extensional breakdown over a late Paleozoic orogenic belt (a precursor of the present northern Andes) began in Cretaceous time and reached a climax in the Eocene with the development of a complex mosaic of normal faulting. Oil occurs in Eocene sandstones (especially Lower Eocene) in a large number of separate pools within this mosaic; it extends offshore (Fig. 16.114).

No commercial oil is known along the Pacific coast of South America south of northern Peru. A little gas occurs on- and offshore from *Chile* about at the latitude of Argentina's Neuquen Basin (see Sec. 25.5.8). The principal reservoir rock is an Oligocene sandstone, but the source may well be Cretaceous. Other gas (and perhaps some oil) is reasonably to be expected somewhere along the great length of Chile's Pacific strip.

Across the South Pacific, the Taranaki Basin extends into Cook Strait between the north and south islands of *New Zealand.* As in Australia's Bass Strait basins (Sec. 25.5.5), Paleocene and Eocene nonmarine sandstones are coal-bearing and very thick. Unlike their Australian equivalents, however, the Taranaki Basin Paleogene strata rest directly on basement without any Cretaceous, and they provide a gas-condensate province, not an oil province (though a little oil occurs in Eocene and Oligocene sandstones in a thrust belt). The source of the gas is therefore more likely to be in the Eocene coal measures than in any pre-Tertiary strata.

The offshore Maui field and smaller fields onshore should yield nearly 0.2×10^{12} m^3 of gas and condensate.

25.6.5 Tertiary petroleum in northern South America

The second greatest assembly of Tertiary oilfields on Earth extends along the northern Andean and coastal Cordilleran belt from northern Colombia to Trinidad. Much of the oil is heavy and asphaltic; nearly all of it is contained in sandstone reservoirs which extend through the whole Tertiary succession.

The outer margin of the Cretaceous basin (Sec. 25.5.8) was severely deformed between Turonian and Eocene time, isolating two intermont basins (the Magdalena Valley of Colombia and the Maracaibo Basin of Venezuela) behind major mountain belts. The climatic uplift of the northern Andes, in the Mio–Pliocene, completed this isolation.

The folded Eastern Andes of Colombia separate the Middle Magdalena Valley from the Llanos Basin on the internal side. The *Middle Magdalena Valley* (Fig. 17.25) is a half-graben in which a thick wedge of nonmarine Tertiary clastics is superimposed on basinal Cretaceous strata containing rich source sediments.

Most of the oil (some of it heavier than 15°API) occurs in Paleogene arkosic sandstones, many of which are scarcely consolidated. The commonest traps are heavily faulted anticlines. Smaller oilfields extend southward into the upper and northward into the lower Magdalena Valley also; in the latter direction they contain more gas, and gas alone occurs in the delta region offshore. The *Llanos Basin* is a foredeep, with stratigraphic section and unconformities virtually a mirror image of those in the valley. Some oil occurs in uppermost Cretaceous sands, but the greater part is in Eocene and Oligocene sands of very high permeabilities. The range of oil gravities is extreme – from 7° to more than 40°. The Llanos Basin extends into Venezuela as the Barinas Basin, with minor production from Eocene sandstone reservoirs.

The *Maracaibo Basin* (Fig. 17.30) is shared by Colombia and Venezuela, with the latter country having by far the larger share of it. Fields in the southwestern corner of the basin lie between two branches of the Andes and are pooled in prominent fold traps, principally in Eocene sandstones. Much thicker Eocene and Paleocene sandstones are productive from under Lake Maracaibo, the oil being associated with major unconformities below (with Cretaceous source sediments underneath) and above (with further reservoir sands in the Miocene). Again, much of the oil is very heavy, but there is a good deal of associated gas. Until late in the 1970s, the great complex of pools called the Bolivar Coastal Field, below the "lake" and along its eastern side, maintained the highest cumulative production of any oilfield in the world – a quantity surpassed by only six *countries* (including Venezuela itself). The greater part of this huge output has come from Miocene sandstones in a gigantic stratigraphic trap above the regional unconformity (Fig. 16.2).

Oligo-Miocene sandstones are productive again in the *Falcon Basin* northeast of the "lake," but only in a very minor way; structural complexity there is much greater than it is in the Maracaibo Basin. South of the coastal Cordillera, however, a huge foredeep basin extends across almost 10° of longitude. The *Orinoco* or *Eastern Venezuelan–Trinidadian Basin* has been progressively tilted towards the east, and production extends in that direction into younger and younger marine strata. Virtually all the oil is in Oligocene and younger sandstones, which became nonmarine during the late Miocene except in the extreme east. The productive section rests unconformably upon the Cretaceous, which has undoubtedly contributed some of the source sediment. The principal source is in Oligo-Miocene deepwater shales, especially well developed in southern Trinidad.

The oils display a remarkable range of gravities, from 10° API or heavier in the tar belt along the north side of the Orinoco River to lighter than 50° (Figs 10.16 & 17.29); wax contents also range from high to negligible. An essentially complete range of trapping mechanisms is also present: stratigraphic traps along the margins, prominent anticlines (many of them with thrust faults) along the deeper, northern flank of the basin, growth-fault traps in its center, and diapiric folds in the delta area and in Trinidad.

Oil and gas in *Trinidad* are restricted to the southern basin, which is simply the eastern end of the Orinoco Basin and extends offshore both west and east of the island. The basin is characterized by notoriously complex stratigraphy and by youthful, active structure. Both features led to the early and continuing prominence of this small sample of the Earth as a standard for the microfaunal zonation of much of the Tertiary System. Oil and considerable associated gas are produced largely from Miocene sandstones of highly lenticular nature. There is no significant production from older horizons, and oil in latest Miocene and younger strata appears to be partly controlled by faults and to be migrating from rocks below. Though there are festoons of large anticlines, many with piercing clay cores, crestal folds are mostly barren, oil accumulations being more commonly found in flank positions. Southeast of the island, production is from progressively younger strata, oil occurring in multiple sands in the Pliocene and gas even in the Pleistocene.

In the Caribbean region proper, commercial oil is

negligible. Gas occurs in Eocene sandstone north of Trinidad and of eastern Venezuela. West of the island of Margarita, light oil occurs with the gas but its commerciality is uncertain. Trivial quantities of oil and gas have been produced from the complicated Tertiary rocks of *Barbados*.

The basins of Colombia, Venezuela, and Trinidad have yielded a combined cumulative production of almost 7×10^9 m^3 of oil to 1982. Some 95 percent of this output has come from Tertiary sandstone reservoirs, and 90 percent of it has been provided by Venezuela. The three countries should ultimately yield nearly 12×10^9 m^3 of oil and more than 2×10^{12} m^3 of gas.

25.6.6 *Tertiary petroleum in the Gulf of Mexico basins*

A great Tertiary basin encloses the Gulf of Mexico on its northern and western sides. It differs from the earlier, Jura-Cretaceous basin (Sec. 25.5.9) in being almost wholly clastic and in consisting of a series of wedges thickening gulfward over major flexures. Each wedge effectively consists of a regressive unit dominated by thick sandstones, incompletely marine, enclosed in marine shales. As the overall sedimentational history during the Cenozoic was regressive, each successively younger unit offlaps the one below it. The consequence is a series of arcuate productive trends becoming younger towards the Gulf (Fig. 17.10).

Virtually all sizable sandstone units are productive of either oil or gas at some points around their trends; the major sand formations are prolifically productive along most of their extents. By far the greater part of the output has come from such sands in the Lower Eocene (Wilcox Group), Middle Oligocene (Frio Formation), and Lower Miocene. Younger sands, up to and including those in the Pleistocene, lie progressively farther offshore; the Plio-Pleistocene basin lies almost wholly offshore from the present Mississippi delta.

The larger oilfields are trapped in anticlines, domes, or fault traps over deep-seated salt domes; some of them are extraordinarily complex (Fig. 16.52). Others occur in rollover anticlines on the downdropped (gulfward) sides of growth faults, which extend in festoons around the northern and northwestern margins of the basin. The largest gas and condensate fields are in Miocene and older sandstones arched into anticlines over salt masses. Scores of smaller gas or condensate fields occur in later Miocene–Pleistocene sands in a variety of fault, diapiric, and stratigraphic traps.

Many petroleum geologists think of the Gulf Coast Tertiary basin as belonging essentially to *Texas* and *Louisiana*. Its Jura-Cretaceous predecessor, of course, is magnificently developed into *eastern Mexico* (Sec. 25.5.9), and the Tertiary basin also extends onshore in parts of that country. The Oligocene Frio trend is gas-bearing both in the *Burgos Basin* of northeastern Mexico (where it is continuous from southern Texas) and in *Veracruz* (where most of the oil is in the Cretaceous). The Lower Miocene trend of Texas heads offshore about at the Mexican border, but reappears onshore in the *Saline Basin* of the Isthmus of Tehuantepec. Here the Oligocene is the oldest Tertiary series present, and Lower Miocene sandstones yield both oil and gas from structures brought on by salt intrusions. Still farther to the east, beyond the great Reforma oilfields in Mesozoic carbonates (Sec. 26.2.4) is the areally small *Macuspana Basin* (Fig.26.21). Large quantities of gas and condensate are produced there from anticlinal traps in more Miocene sandstones.

Associated with the obvious geological distinctions between the Mesozoic and Cenozoic Gulf Coast basins is a striking difference in hydrocarbon characteristics. The Cenozoic basin should ultimately yield $6 \times 10^9 - 7 \times 10^9$ m^3 of oil and at least 10×10^{12} m^3 of gas and condensate. It is therefore notably less oil-prone than the Mesozoic carbonate basin and very much more gas-prone (see Sec. 25.5.9). Of the ten largest oilfields in the entire Gulf Coast Basin, from Florida to Yucatan, only one is a Cenozoic field (the Bay Marchand complex; Fig. 16.52). Of the nine fields producing from Mesozoic rocks, only the East Texas field has a sandstone reservoir. The three largest gasfields (one, the Bermudez complex in the Reforma area of Mexico, also a supergiant oilfield) also produce from Mesozoic carbonate reservoirs, but by far the greater part of the Cenozoic gas is in sandstone reservoirs and in relatively small individual accumulations.

25.6.7 *Tertiary petroleum in the interior of North America*

We have seen that basins on the continental side of the northern Andes and Coastal Cordillera of South America are among the world's most prolific producers of oil from Tertiary strata. The equivalent basins east of the North American Cordillera do not approach their output from reservoirs of any age (see Secs 25.2.4 & 25.5.10). The last marine sedimentation there was in the Cretaceous, not in the Miocene; the last compressional deformation in the Paleocene, not the Neogene. Petroleum output from Tertiary strata between New Mexico at 32°N latitude and the delta of the Mackenzie River at 68°N has been trivial.

The Uinta–Piceance Basin in Utah and Colorado and the Green River Basin in Wyoming have yielded a very little oil from fractured Eocene oil shales, and somewhat more gas from nonmarine sandstones associated with them. The Powder River Basin has yielded insignificant gas from the Paleocene. Tertiary production from the Western Canadian Basin has been nil.

In the outer delta of the Mackenzie River, and in the Beaufort Sea beyond it, both oil and gas are known in Paleogene sandstones. As they lie in an otherwise Mesozoic province, the occurrences are described with the larger Mesozoic fields (Sec. 25.5.12).

26 Case histories of selected fields

26.1 Explication

We have completed our survey of all the aspects of a petroleum geologist's work. All the aspects are crucial to the conduct of the oil and gas industries, and all are best handled by teams of specialists, as we recognized at the outset of our survey in Chapter 1.

It remains the case that the aspects most dependent upon individual geologists, and occupying the attentions of the great majority of petroleum geologists the world over, are the discovery and development of oil- and gasfields. Some fields were discovered very easily and cheaply; others only after exhaustive and expensive detective work. Some fields are very easily developed and operated; others present formidable hazards. Because the most easily found fields were naturally found first, and the most easily developed fields were developed when only simple technologies were available, we can be fairly confident that future fields will present the geologists (and their colleagues) with more and greater challenges.

To conclude this book with an attempt to illustrate some of the challenges that have already been faced and overcome, I have chosen the artifice of pairing important fields which provide either comparisons or contrasts to this end.

26.2 The fields paired

26.2.1 Pembina and West Pembina fields, western Canada

After years of disappointment, yielding a handful of trivially small fields for the drilling of more than 100 wildcat wells, the Western Canadian Basin gave up its first giant discovery at Leduc, near Edmonton, early in 1947. In the ensuing 6 years, all the large fields in Devonian reefs of Leduc age were discovered in rapid succession; their combined recoverable reserves were some 635×10^6 m^3 of oil.

The western limit of these discoveries trended approximately north–south up the center of Alberta, following the edge of a Devonian carbonate shelf. Westward of the shelf lay an unpromising shale basin, apparently too deep for either reefs or platform despite the appearance of more Devonian reefs in the Rocky Mountains to the west. In the basin, the Devonian is buried below 2 km of Cretaceous sediments, most of them in nonmarine molasse facies and offering little promise of their own. A wry aphorism in the industry referred to the geologists who undertook to drink all the oil that would be found west of the fifth meridian (the land survey boundary that approximately coincides with the productive reef trend).

By 1953, companies holding land in this deep part of the basin began to farm it out to other operators willing to drill in it. One company, accepting a farmout, drilled a rank wildcat well on what was said to be a seismic anomaly, interpreted as another Devonian reef. The well had shows of oil in Lower Cretaceous sandstones, but there was no reef or carbonate platform to total depth of about 3000 m (at that time a very deep well for Canada). Complete logs were run, to extract the maximum geologic information from the dry hole before abandonment, and an alert geologist spotted an unforeseen sandstone in the upper part of the hole (at about 1600 m) that appeared to be hydrocarbon-bearing. The recovery of oil on a drillstem test heralded the discovery of a new oilfield, the Pembina field.

The sandstone was in the stratigraphic position of the so-called Cardium sandstone in outcrops of the Rocky Mountain foothills to the west. This thin, quartzose sandstone and conglomerate, isolated within a thick succession of Upper Cretaceous shales, was a familiar outcrop marker bed in easily traced ridges. The Cardium sandstone of the outcrops would not have been an encouraging target for oil exploration in the subsurface because it is essentially a quartzite. Certainly no well had ever been drilled with it as its target, because the sandstone formation was not known to extend far eastward from its outcrop and it had never been certainly identified in the subsurface.

Following one of the most hectic land plays in Canadian oil history, rapid development of the new field showed it to be in an updip wedge-out (stratigraphic) trap surrounded by shales (Fig. 26.1). Its original attitude must have been downdip, as the sand was derived from rising cordillera in the west; it was reversed during Laramide deformation and uplift of those cordillera, with concomitant depression of the Alberta Basin below frontal thrust faults (Sec. 16.4.1). Oil occurs in the formation only along the edges of eastwardly-projecting lobes, like that at Pembina. Obviously the fluids that deposited the materials impairing the porosity down-dip were unable to enter the updip parts that contained the oil. Either the oil was pooled there early, by buoyancy, and the downdip porosity was then plugged; or the oil was driven into its trap by the porosity-plugging fluids themselves.

Pembina became the largest oilfield in Canada, with about 3×10^8 m^3 of recoverable light oil. Though the pay section is thin (maximum about 12 m, average little over 5 m), the field is areally one of the largest in the world (about 1600 km^2; Fig. 26.2). Between discovery of the

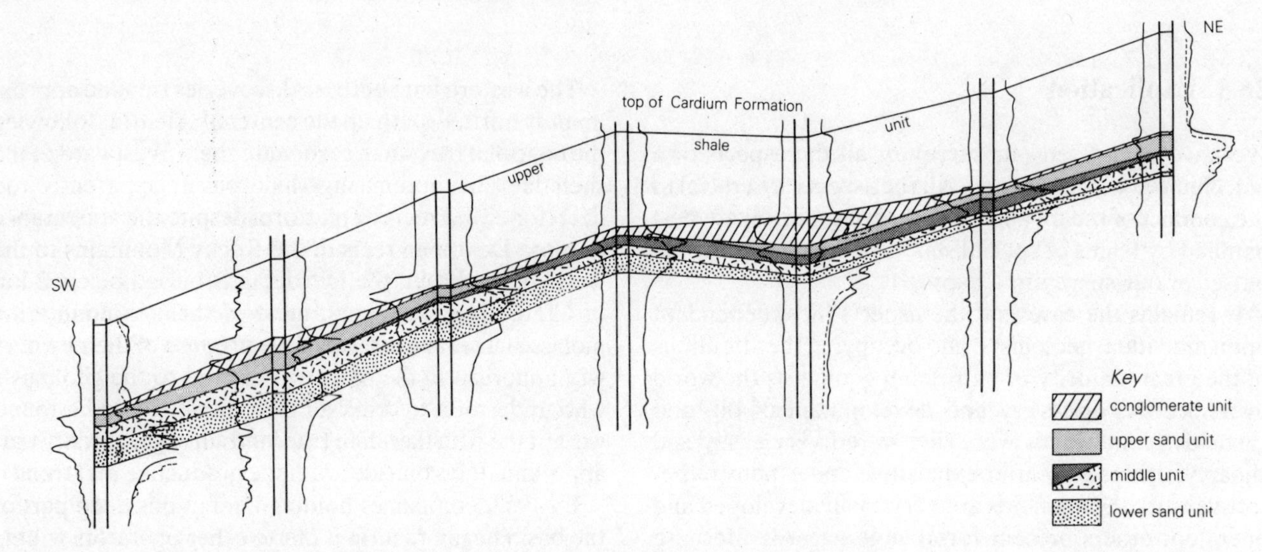

Figure 26.1 Pembina oilfield, Alberta, Canada: structural cross section from southwest to northeast, showing the nonconvex trap in the Cretaceous Cardium Formation. (From A. R. Nielsen, *J. Alberta Soc. Petrolm Geol.*, 1957.)

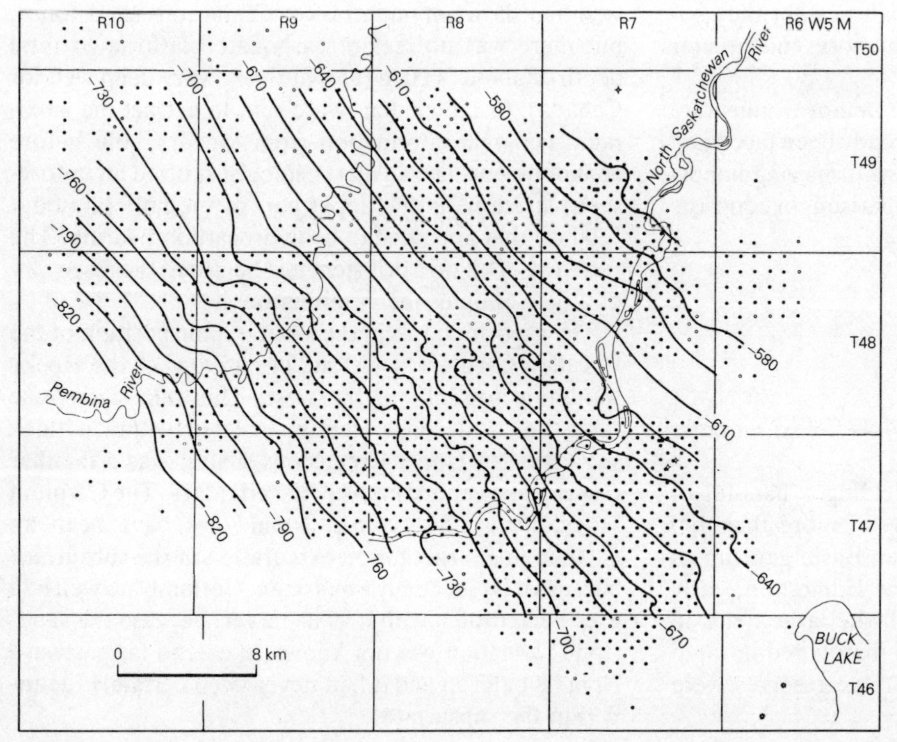

Figure 26.2 Structure contours on the Cardium sandstone reservoir in the Pembina oilfield, western Canada. Note the uninterrupted southwesterly dip. The land survey squares are 6 x 6 miles (9.6 x 9.6 km); contour interval 15 m. (From A. R. Nielsen, *J. Alberta Soc. Petrolm Geol.*, 1957.)

field, in 1953, and 1968, exploration was widely extended across western and northern Alberta. Extensive reef development was proven in at least five distinct Devonian horizons, three of them richly petroliferous (Sec. 25.2.4). By the early 1970s, perceptive geologists had realized that the sprawling Pembina field, with nearly 5000 wells drilled to its Upper Cretaceous target, covered up the largest area in the Alberta Basin in which no Devonian reefs were known. Yet productive reefs existed both southeast and northwest of it, and thrust-shortened outcrops in the mountains to the west showed that these two trends had undoubtedly been continuous in the southwest also. Reefs of *some* Devonian age were therefore legitimate expectations somewhere below the Pembina field.

Very detailed seismic shooting over more than 2 years, by a company having little representation in the Pembina field and virtually none in its extensive western half, was carefully processed by common depth point techniques (Sec. 21.2). These detected the western edge of a carbonate shelf long known from the Devonian producing area to the east (an edge related to, but younger than, the shelf edge referred to in the second paragraph of this history; Fig. 26.3). West of this shelf edge, small, isolated, pimple-like anomalies were interpreted as pinnacle reefs of uncertain age, but resembling those already familiar in other parts of western Canada in both Middle and Upper Devonian strata (Fig. 26.4).

What followed illustrates the oil industry's concern for security of geological and geophysical data. Because the company in possession of the encouraging new data had no significant land holdings in the area, it had to acquire acreage over the deep anomalies from other operators whose concern was the shallow Cretaceous field. Having succeeded in putting together all the acreage available, the company drilled a new discovery well at the beginning of 1977 and followed it with a succession of new pools in isolated pinnacle reefs. Once a few wells had been drilled, velocity logs from them permitted the construction of synthetic seismograms (Sec. 21.5), which in turn facilitated more accurate interpretation of the seismic data and led to the identification and discovery of still more of the tiny targets.

West Pembina's recoverable reserves are probably less than one-fifth of those of the big Pembina field above; some of the reefs are so small that they represent one-well pools. The oil has high H_2S content and pore pressures are both high and erratic; drilling and completion of wells has consequently been difficult and expensive. But the contrast between the discovery histories of the two Pembina fields provides an enlightening exercise. The first illustrated the luck of an unpredicted discovery in a previously unexplored region prematurely downgraded. The second reflected the skill and care devoted to the successful prediction of new fields in a mature area thought by many to have nothing more to offer the explorers.

26.2.2 *The Cook Inlet and North Slope fields of Alaska*

It has been remarked by commentators on the general American productivity phenomenon that the USA is the only nation that established unified government over a country of semi-continental proportions lying entirely within the Earth's temperate climatic zone. When Alaska became one of the united states, it brought with it a contrasting distinction; it is the only part of the

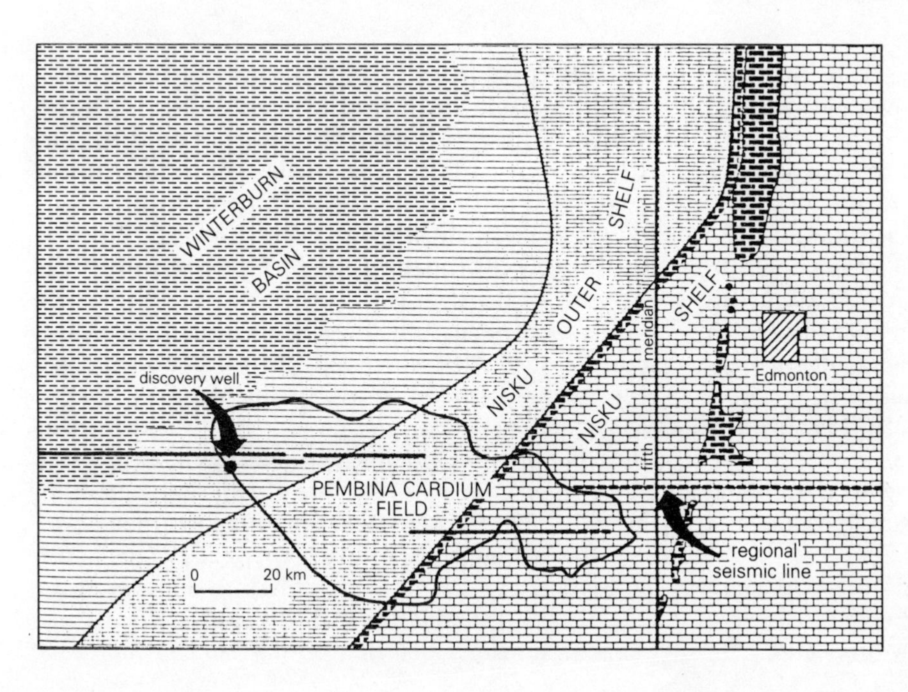

Figure 26.3 West Pembina field, Alberta, Canada, in relation to the Pembina field and to two carbonate shelf edges. (From Chevron Standard Limited, *Bull. Can. Petrolm Geol.*, 1979.)

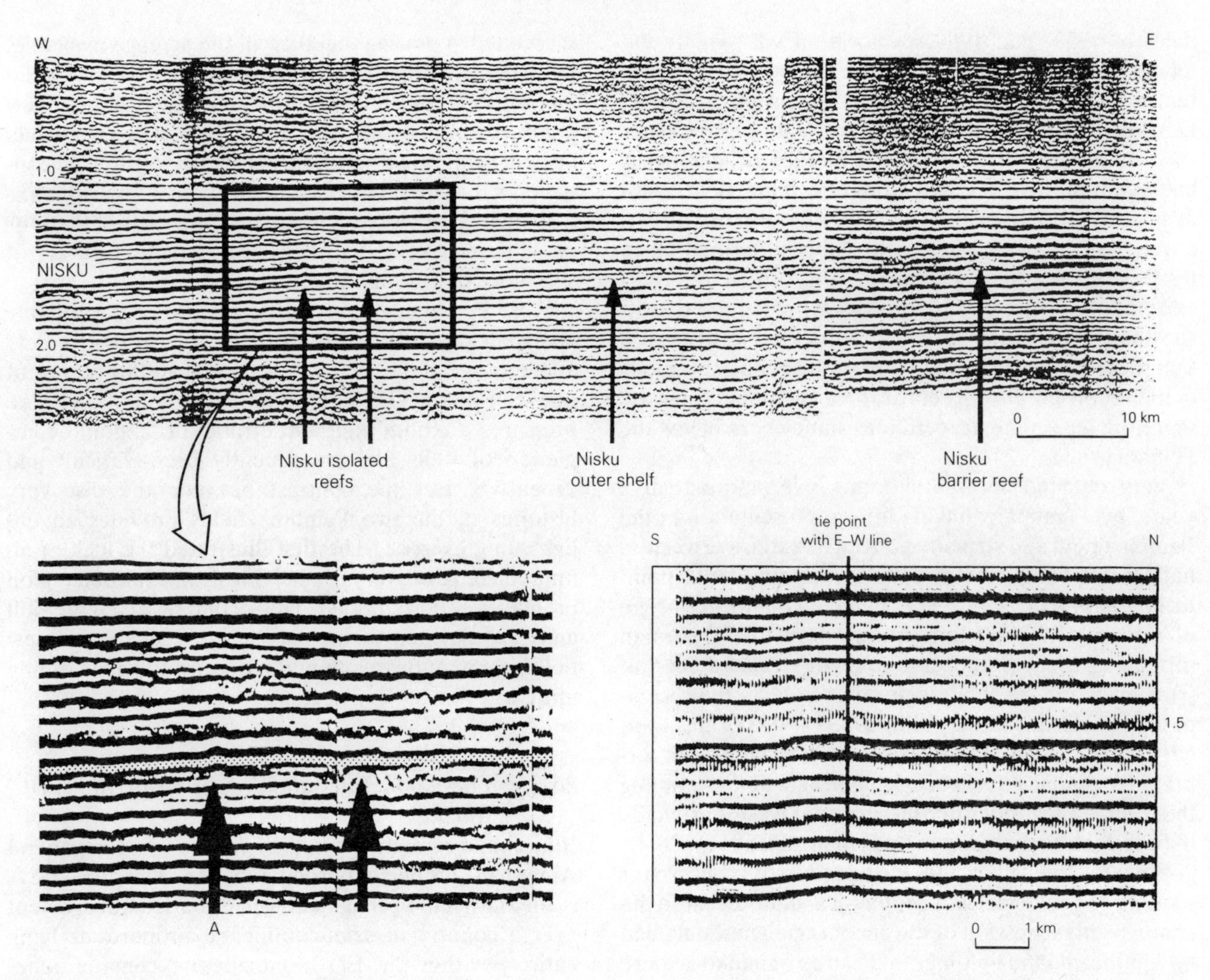

Figure 26.4 West Pembina field, Alberta, Canada: regional seismic section showing carbonate bank and isolated reefs. Lower diagrams show enlargement of the reef anomalies, and a north–south tie-line on the anomaly marked A. (From Chevron Standard Limited, *Bull. Can. Petrolm Geol.*, 1979.)

Western Hemisphere to possess three long coastlines formed principally of sedimentary rocks and lying entirely within a hostile climatic zone. Exploration for oil and gas in Alaska has proceeded from the margins inward.

Most of the Pacific coast of Alaska is a classic arc/trench structural system now called a subduction orogenic zone. Prior to 1955, only three areas productive of oil were known anywhere around the true Pacific Ocean margin. The only very important one, southern California, was already recognized as unique in its direct association with major strike-slip faults; it is now described as a converted triple junction (Sec. 25.6.4). Neither of the other two areas, Japan and coastal Ecuador–Peru, is significant enough to be viable in a polar environment.

Both the Arctic and Pacific coasts of Alaska are dotted with oil and gas seepages, and both coasts were investigated early for possible shallow production. On the north coast, indeed, early encouragement had led to the setting aside, in 1923, of a large area as a US Naval Petroleum Reserve (NPR). Neither it nor its equivalent stretch on the south coast yielded more than trivial discoveries, and it was to be a relatively small private company, the Richfield Oil Corporation, that enjoyed the distinction of making the first giant discovery on each side of the state.

Cook Inlet is a narrow, fjord-like bay interrupting the south coast of Alaska (see Fig. 26.7). Most oil and gas seepages are found in Jurassic strata along its western side, where reflection seismic surveys were begun in 1955. The basin within which the inlet lies is bounded by major reverse faults, so it is not a typical graben. Furthermore, the structures detected within it turned out to be not horsts, as in productive grabens elsewhere in the world (Sec. 17.4.2), but complex narrow anticlines with reverse or thrust faults along their margins. The largest of these anticlines detected on dry land in the basin was on its eastern side (the Kenai Peninsula), not on the western side where most of the seepages were found.

Most of the section in the basin consists of alluvial Tertiary clastics some 8 km thick, called the Kenai Group. The rocks include numerous massive sands of fluvial, estuarine, and deltaic origin and some of them conglomeratic, with floodplain shales and thick coal seams (Fig. 13.12). Correlations have to be made by palynologic techniques. This thick pile of sediment lies on a major unconformity reflecting long erosion (see Fig. 26.6). The Cretaceous is very incomplete, but Jurassic strata are thick and provide the most probable source for the oil. Lower Jurassic volcanic rocks constitute "economic basement" for exploratory wells, as they do in much of the circum-Pacific rim.

The Swanson River field was discovered on the easternmost large anticline in 1957, in Oligocene conglomeratic sands low in the Kenai Group at a depth of more than 3000 m. Development revealed several unexpected features. Structure contours on successive marker horizons were so close to being parallel that the structural growth must have been exceedingly young. The uplift was in fact Plio-Pleistocene and Recent; the city of Anchorage, at the head of the inlet, remains aware that the movements still continue. Carbon isotope studies could not unequivocally relate the oil either to the nonmarine shales within the Kenai reservoir group as their source or to the underlying Jurassic sediments. Later opinion has accepted the Jurassic source (see Fig. 26.6). As these older strata are buried below 3600 m, and the oil they generated cannot have become pooled in its present traps before the Pleistocene, both generation and migration were remarkably long delayed by comparison with the equivalent processes in California, or across the Pacific in Indonesia, where very prolific pooled oil originated in source sediments that cannot be older than Miocene. The Cook Inlet basin proved to have a very low geothermal gradient, probably explaining the delayed maturation of the source material. The very light, paraffinic oil, markedly undersaturated with gas, is accompanied by pools of both biogenic and thermogenic dry methane gas. The dry biogenic gas, which constitutes the major gas reserve of the basin, is nonassociated and occurs in shallow sandstones younger than the oil reservoirs. The thermogenic gas accompanies the oil in the older, conglomeratic sandstone reservoirs; it is of minor economic significance.

The basin surface is occupied almost wholly by Quaternary alluvium, glacial deposits, and the waters of the inlet. Exploration quickly moved offshore, where a group of closely spaced anticlines yielded their first oil discovery in 1963 (Middle Ground Shoal field; Fig. 26.5). By 1981, there were five large oilfields and a dozen gasfields in production. None of the large structures is filled to its spill point, demonstrating that the present traps were the final ones.

The first offshore field antedated by 5 years the first big oil discoveries in the North Sea, and development proceeded in a highly experimental fashion in the absence of any earlier industry experience in such rough water and weather. Floating rigs proved useless in an

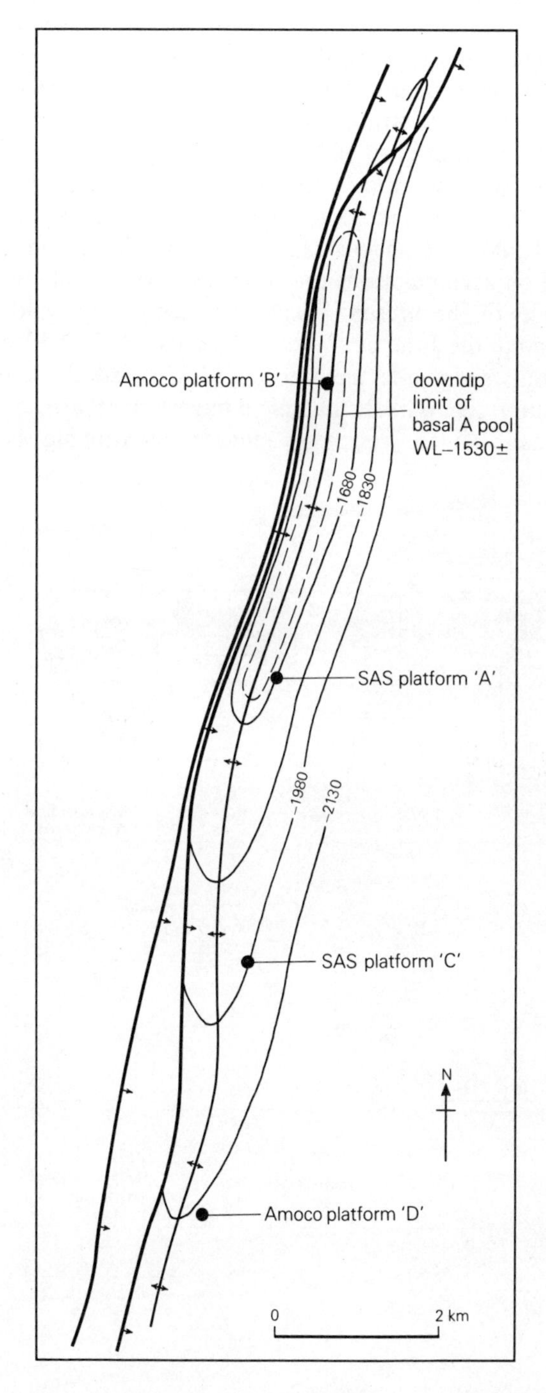

Figure 26.5 Middle Ground Shoal field, Alaska. A typical field of the Cook Inlet Basin (actually an offshore field), showing the elongate, faulted anticlinal structure. (From R. F. Boss *et al.*, *AAPG Memoir* 24, 1976.)

inlet with a tidal range of nearly 10 m, very strong currents, and severe winter ice. The multiplicity of hydrocarbonbearing sands created a succession of individual pools in each field, each pool having its own oil/water contact necessitating expensive coring to enable each pool to be treated as a productive unit in itself.

Industry expectations from *lower* Cook Inlet were not fulfilled. The explanation for this disappointment may lie in the age difference between the source sediment and the structural traps. Southward, down the plunge of the basin structure, the Kenai Group rests on progressively younger strata which contain neither source nor reservoir rocks.

It is apparent that, although the productive anticlines are very young, there was a critical earlier deformation in the late Cretaceous or very early Tertiary. Commercial oil accumulations are restricted to strips along the flanks of the present basin on which erosion had cut down to the Jurassic source sediments before the Kenai Group sediments unconformably covered the entire basin. It seems likely that initial migration of oil from the Jurassic source rocks was into supercropping sands above the unconformity (Fig. 26.6), and that the only part of this early accumulation remaining is that captured by the present, very young traps.

The *North Slope* region on the opposite side of Alaska provides a classic setting for a great oil basin, if geologic conditions alone are considered and geographic conditions ignored. A continental margin of Atlantic type received sediment from a northern source until early in the Cretaceous. Crustal extension then occurred, leaving an elongate remnant of the northern landmass adjoining the depositional basin. A Cretaceous collision episode then raised a folded mountain belt, overthrust northwards towards the continental remnant. Between the range and the remnant a linear foredeep, the Colville trough, contains 10 km of sedimentary rocks. The progradational molasse succession overstepped the remnant in late Cretaceous time (Fig. 26.8).

The remnant now constitutes the Barrow Arch (Fig. 26.9), which is responsible for the present north coast of Alaska eastward from Point Barrow. Point Barrow is a large cape caused by a northwardly projecting structural high, and numerous oil seepages on it drew

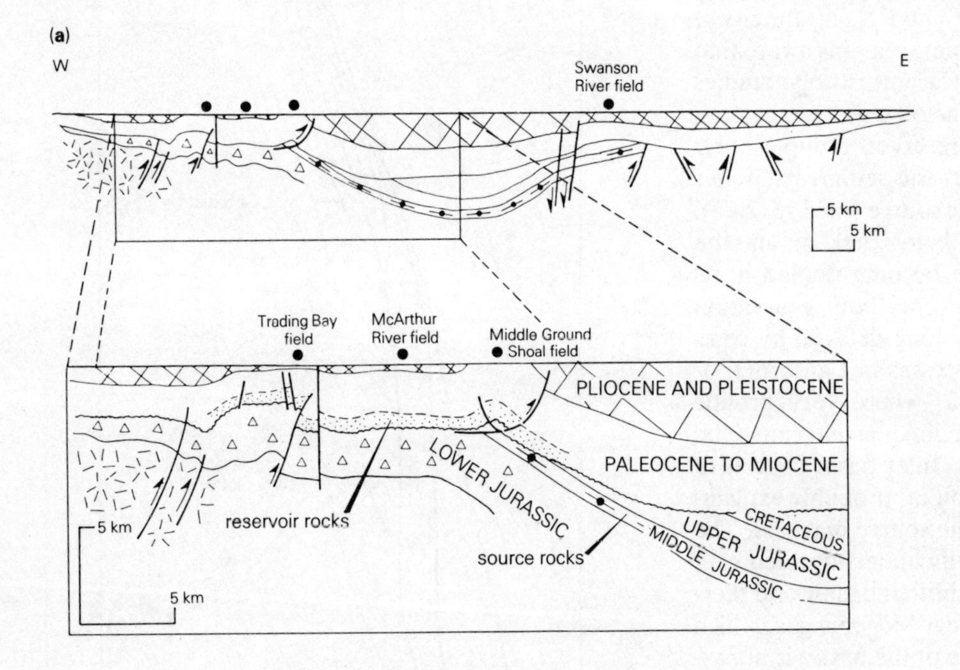

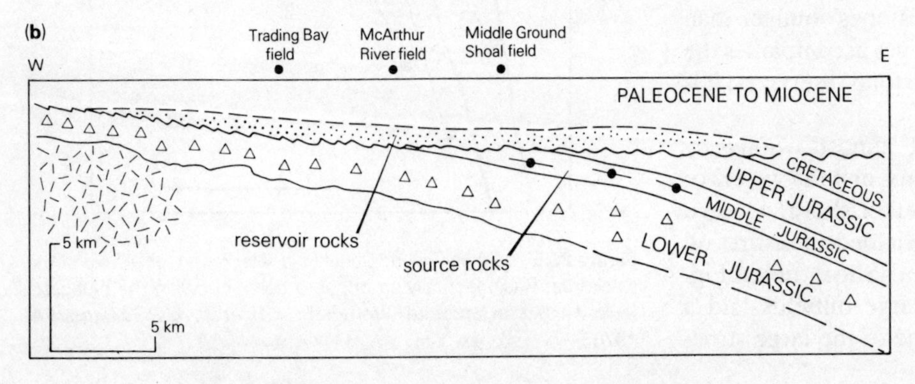

Figure 26.6 (a) Cross section through upper Cook Inlet, and detail of the west flank of the basin. Note relation between Middle Jurassic source formation and Tertiary reservoir rocks. (b) Geologic reconstruction of the west flank of the basin at the end of Miocene time, showing restored relation between source and reservoir rocks in a stratigraphic trap. (From L. B. Magoon and G. E. Claypool, *AAPG Bull.*, 1981.)

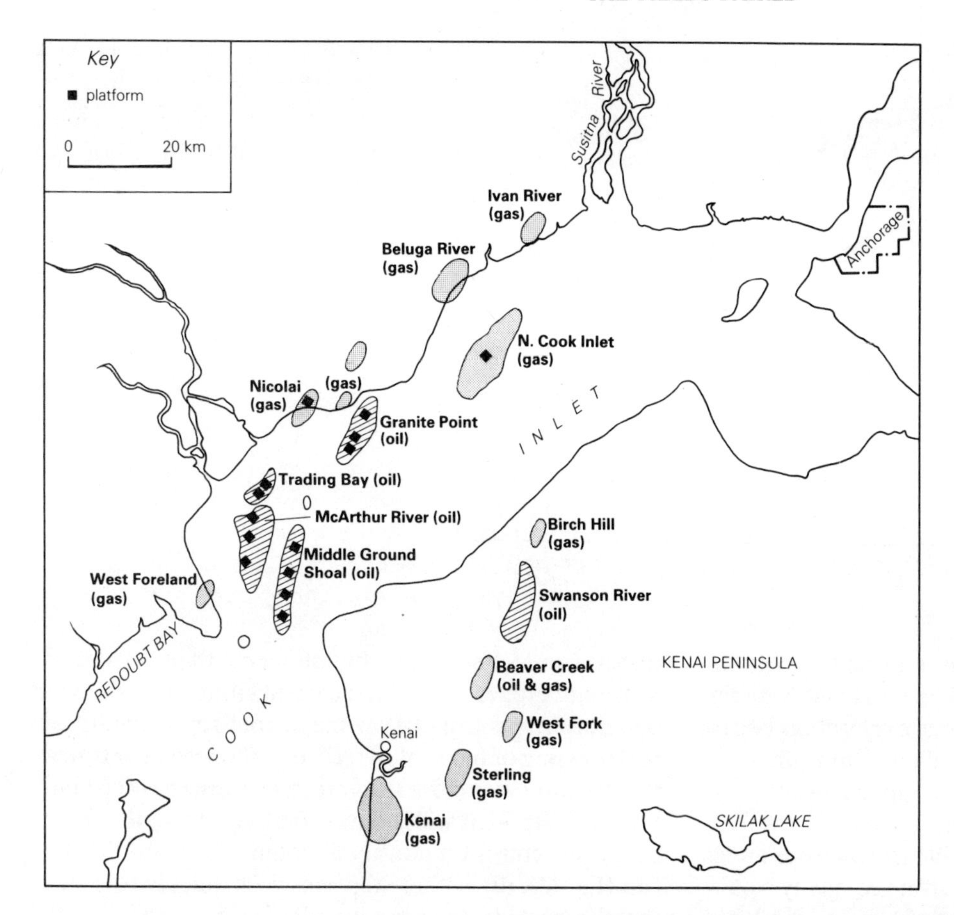

Figure 26.7 Oil- and gasfields in Cook Inlet Basin, Alaska. (From R. F. Boss *et al.*, *AAPG Memoir* 24, 1976.)

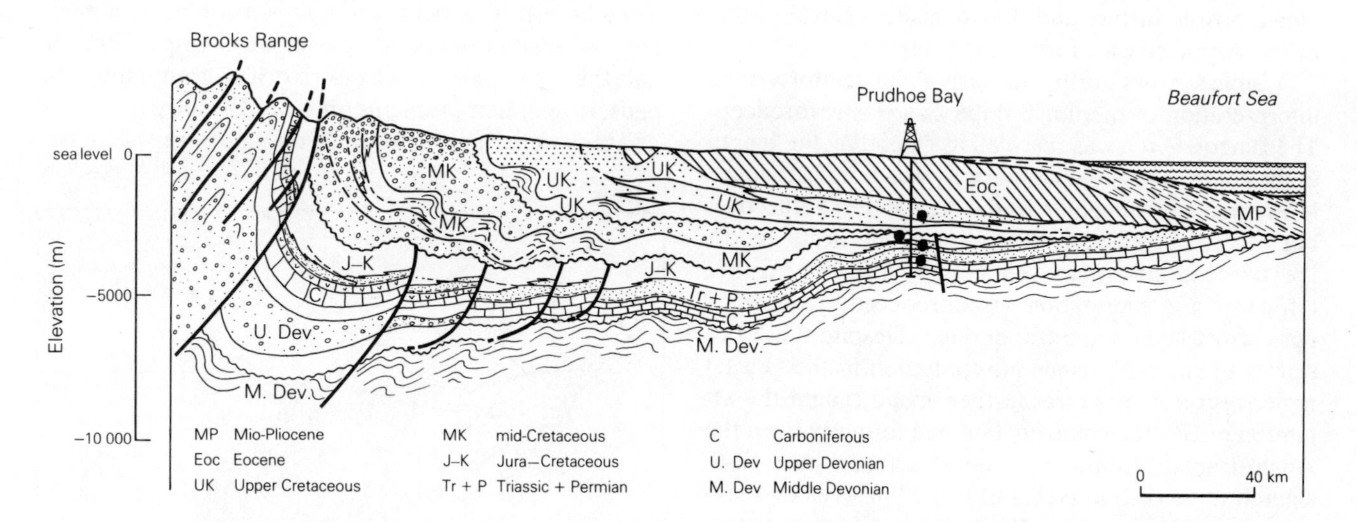

Figure 26.8 Cross section through the Arctic Slope Basin of northern Alaska, showing the Brooks Range, the Colville trough, and the Barrow Arch. Note the two major unconformities within the Cretaceous section. (From A. Bouillot, 8th World Petroleum Congress, 1971.)

early attention. Several small oil and gas finds were made, mostly in Cretaceous strata, and these led during the national protection program that followed World War I to the creation there of one of four US Naval Petroleum Reserves (NPR 4). A 10-year drilling program by the Navy in its reserve, begun in 1944, found only trivial accumulations.

Farther to the east, the nation established a formal Wildlife Range, which extends to the Alaska/Canada border. Except for a small state-held area along the

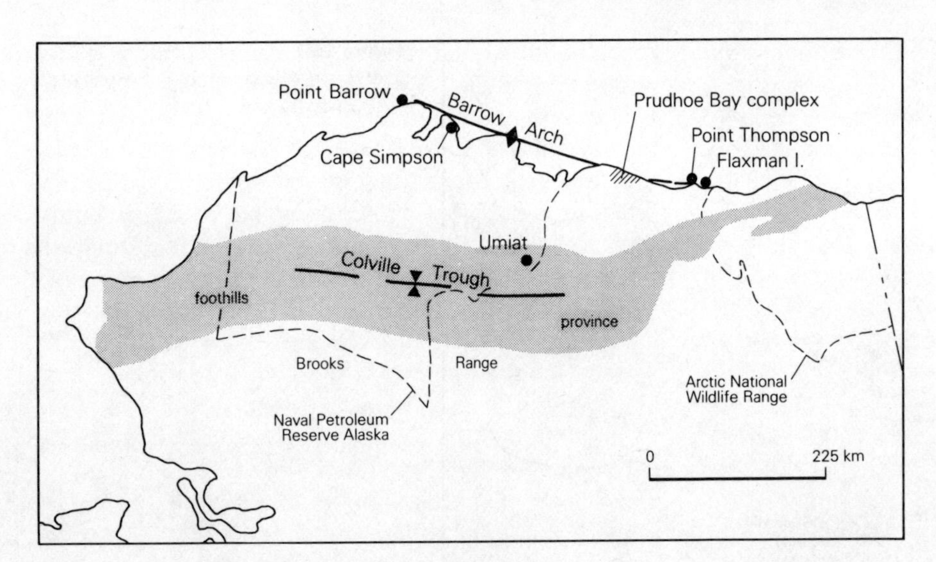

Figure 26.9 Major structural elements of the Alaska North Slope. (From H. C. Jamison *et al.*, *AAPG Memoir* 30, 1981.)

coast, between the Petroleum Reserve and the Wildlife Range, the whole of the North Slope was and is under federal jurisdiction. Though seismic exploration began in 1962, only federal land was drilled. Unfortunately, the Cretaceous and Tertiary strata dipped steadily towards the northeast without clearly definable closures; pre-Cretaceous strata in the Colville trough were much too deep to be explored; and no private company would have been interested in the small blocks of land available under federal regulations. The remote locale and its bleak, hostile surface and climate made the exploration of the Arctic coast a forbidding prospect.

Seismic surveys during the early 1960s reinforced the interpretation of the north slope as a classic foredeep. The Barrow Arch was revealed as extending the length of the coast. A major unconformity was manifested, at a level originally thought to be the base of the Cretaceous, but multiple reflections ("noise") from strata below the unconformity made resolution difficult. Further difficulty in interpretation was introduced by the thick permafrost layer some 650 m deep. Despite these obstacles to routine seismic interpretation in the coastal region, several structures farther inland caught the attention of BP, the company that had formerly been the Anglo-Iranian Company. The structures were considered to be of Iran type and scale, but wells on them were dry. However, by 1965 a very large structure, apparently anticlinal, had been certainly delineated seismically.

In that year, the state opened its share of the coastal area to competitive leasing, and ARCO (the successor to the Richfield Oil Corporation which had already been responsible for Alaska's other oilfields) put together a large block of drillable land. The first well was drilled near the crest of the large reflection-seismic structure having nearly 300 m of closure. The seismic resolution turned out to be remarkably efficient, thanks to good reflecting properties and the careful elimination of multiple reflections from below the unconformity by digital reprocessing of old analog records. The well penetrated the structure very close to its crest, in a large gas accumulation. A step-out well in mid-1968, far down the flank of the structure, found the oil column beneath the gas cap (Fig. 26.10). Other step-out wells quickly revealed that the field was the largest oilfield ever found in North America (it turned out to be the second largest gasfield in addition). The field was unitized for development purposes, with two operating companies responsible for half the area apiece. Wells were drilled from multiwell pads, to minimize problems presented by the permafrost and then were capped whilst the operators waited for the

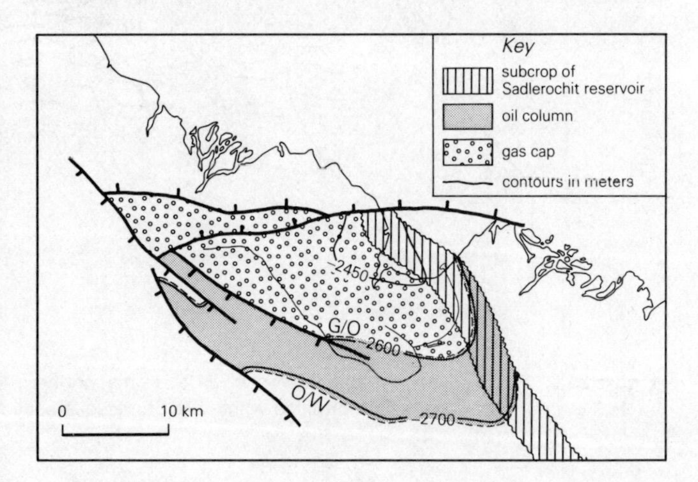

Figure 26.10 Prudhoe Bay oil- and gasfield, Arctic Alaska, in a giant combination trap. Structure contours (subsea) and fluid contacts at the top of the main reservoir. Oil column extends beneath gas cap, except at the very top of the structure (where the words "Prudhoe Bay" occur). Faults hachured on downthrown sides, (From D. L. Morgridge and W. B. Smith Jr, *AAPG Memoir* 16, 1972.)

decision on a delivery system and its subsequent construction.

The productive Prudhoe Bay structure is the result of a most favorable combination of circumstances. Pre-Cretaceous sediments were derived from the north, where there is now an Arctic Ocean. The remnant of this source area, now the Barrow Arch, had undergone sporadic uplift for at least 200 Ma before the end of Jurassic time; both "Caledonian" and "Hercynian" unconformities are regionally developed. In early Cretaceous time, the arch was tilted westward and deeply eroded, so that formations are systematically truncated from east to west. The Brooks Range was elevated in the south, and thrust northwards, switching the principal source of sediment to the south for the first time. During subsequent molasse deposition, starting in the mid-Cretaceous, the plunge of the arch has been progressively reversed, so that the latest Mesozoic and all Cenozoic formations thicken greatly eastward while resting on older and older truncated Paleozoic formations (Fig. 26.11).

The early Cretaceous uplift and eastward truncation brought all older reservoir rocks into subcrop trap position. The Brooks Range thrusting raised the linear anticlinal structure that was detected seismically and drilled successfully. The trap is therefore of the type described as "combination" (Sec. 16.2), basically stratigraphic in nature but with critical structural closure superimposed upon it (Fig. 16.1). The hydrocarbons would be in the same subcrop reservoirs and effectively in the same location if the present closure had never been created.

There are several possible source sediments for the oil at Prudhoe Bay. The marine shales of both the Jurassic (below the unconformity) and the Cretaceous (above it) are particularly rich in organic matter. The Cretaceous shales constitute the only horizon which is in direct contact with all four reservoir horizons in the field; all of them – ranging in age from Carboniferous to early Cretaceous – are truncated at the unconformity and overlapped by the Cretaceous. It is logical to conclude that the Cretaceous shales, in addition to providing the seal for the traps, were also the source of the oil; but it does not follow that they were the *only* source. Detailed geochemical studies, concentrated upon "fingerprint" components of the oils (Sec. 5.1), indicate that the greater part of the oil had more than one source, possibly as many as three, in the Triassic, Jurassic, and Cretaceous.

Important principles of petroleum geology illuminated by the Prudhoe Bay and Cook Inlet fields are many.

(a) The geologic settings of the two basins are utterly different. Cook Inlet is in a fault-bounded depression within a compressional fore-arc basin. The North Slope is an orogenic foredeep superimposed on a trailing margin.

(b) The versatile role played by unconformities in hydrocarbon accumulation is pointedly illustrated (Sec. 16.5). In the Cook Inlet fields, the reservoir rocks are all above the unconformity and the source rock below it; the oil's primary migration was stratigraphically and structurally upward from marine strata into nonmarine. On the North Slope, the reservoir rocks are below the unconformity and at least the principal source rock is above it; the oil therefore migrated stratigraphically downward (though possibly structurally upward; see (e) below). As in so many other areas of the world, both these unconformities were the results of Cretaceous events.

(c) The reservoir rocks in both areas illustrate the generalization that most large reservoir sandstones

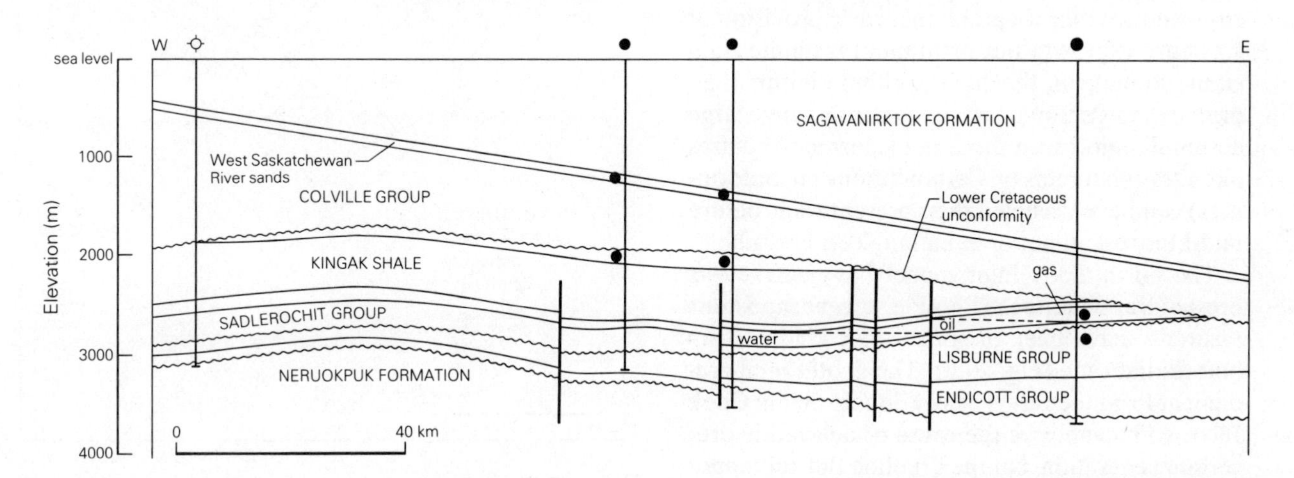

Figure 26.11 West–east cross section, Prudhoe Bay complex, showing eastward truncation of strata below the Lower Cretaceous unconformity, and great eastward thickening of younger strata. Compare the north–south section in Fig. 16.1. (After H. C. Jamison *et al.*, *AAPG Memoir* 30, 1981.)

are complexes of fluvial, deltaic, and beach deposits (Sec. 13.2.7). The Cook Inlet reservoir succession is an assemblage of coarse sands and conglomerates of the fluvial and estuarine regime, with a middle sequence of lower-energy sediments of floodplain origin. The lower part of the main Prudhoe Bay reservoir (the Permo-Triassic sandstone) was deltaic in origin, the upper part a river channel complex formed as regression permitted the river to overrun its own delta. The sands prograded southwards over offshore shales, leading to upward (and northward) coarsening of grain size (Fig. 13.43).

Both productive successions contain sands of excellent reservoir quality despite the abundance of nonquartz constituents (Sec. 13.2.1). This is more surprising (and significant) than it may seem. Alaska, which possesses no shield terrain, lacks sources of supply of clean, quartzose sand, and the consequent dearth of good reservoir rocks militates seriously against the state's ultimate petroleum potential. The Cook Inlet sands were derived from Jurassic granites west of the inlet, but the principal source for the Sadlerochit sands at Prudhoe Bay must have been metasedimentary rocks north of the present Barrow Arch.

(d) The hydrocarbon accumulations in Cook Inlet are in relatively small traps wholly structural in character. The giant accumulation at Prudhoe Bay affords a prime illustration of the axiom that few of the very greatest oilfields in the well explored Western Hemisphere are in traps that are wholly structural. This point is forcefully emphasized in Ch. 16.

(e) Both basins illustrate the crucial influence of the *time of accumulation* of oil and gas pools. Two apparently conflicting principles have been stressed in earlier chapters: that early provision of structure ("growth through time") facilitates efficient pooling of the hydrocarbons during their primary migration, but that hardly any large accumulations (even those in Paleozoic structures like Devonian reefs or Carboniferous unconformities) can have occupied their present traps before mid-Mesozoic time, or even mid-Tertiary time.

The oil in Cook Inlet cannot have entered its present traps before the Plio-Pleistocene, and must assuredly have been in some other kind of trap before that time (Fig. 26.6). The geothermal gradient at Prudhoe Bay is nearly double that in Cook Inlet, so it cannot be the cause of delayed hydrocarbon generation; but the Prudhoe Bay oil cannot have been emplaced before mid-Cretaceous time and cannot have acquired its present distribution in its trap until some time during the Tertiary. The overlapping Cretaceous shales are generationally immature in the immediate field area, and the portion of them which is at once mature and oil-prone does not extend far southward into the Colville trough. In the deep trough, the shales are both overmature and of gas-prone organic content.

Furthermore, *secondary migration* of the oil at Prudhoe Bay must have taken place more than once, first towards the east and north into the truncated sands, later back towards the west as final eastward tilting progressed (Figs 26.12 & 26.13).

(f) Both basins contain both biogenic (immature) and thermogenic (post-mature) gas in addition to their large oil accumulations. Prudhoe Bay may be unique among all the fields of the world in being both a supergiant oilfield and a supergiant gasfield *in the same reservoir*. The huge gas cap is unlikely to have migrated downwards into the trap from overlying source sediments, as much of the oil is agreed to have done. The gas cap may in fact be of Permo-Triassic source, like so many other great gas accumulations (Sec. 25.3), and represent the startling circumstance of the gas cap of a giant oil accumulation consisting of nonassociated gas.

(g) The Prudhoe Bay field illustrates a constraint of enormous significance to all future oil exploration and production. A field, no matter how large, that is discovered in a remote or operationally difficult region, especially one in polar latitudes and more especially still if it is offshore, can only be brought to market after a long *lead time*. An oil industry

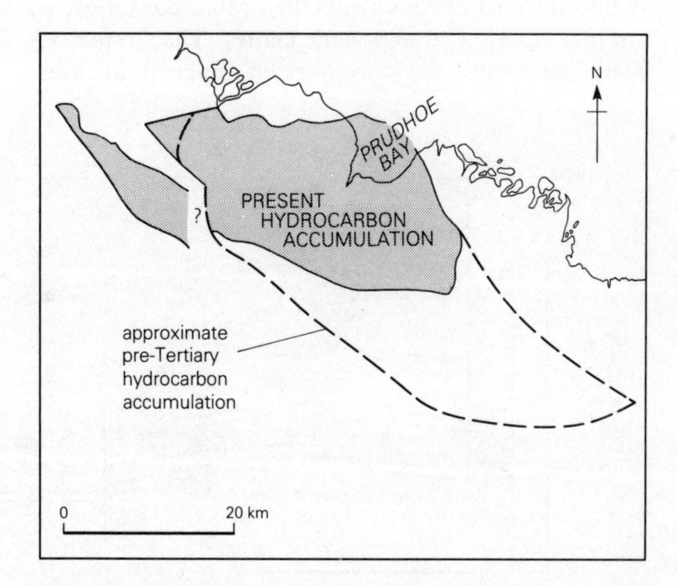

Figure 26.12 Contrast between present and original (pre-Tertiary) hydrocarbon accumulations in Prudhoe Bay structure; the redistribution was the result of post-Cretaceous tilting. (From H. P. Jones and R. G. Speers, *AAPG Memoir* 24, 1976.)

rule of thumb since the 1960s has considered the lead time to be at least 10 years (Sec. 23.10). Ten years is also a fair approximation of the "half-life" of most oilfields, and during the lead time required for a remote field to make its full contribution, production from existing fields elsewhere may easily decline by more than the amount contributed by the new field (Sec. 24.3.1).

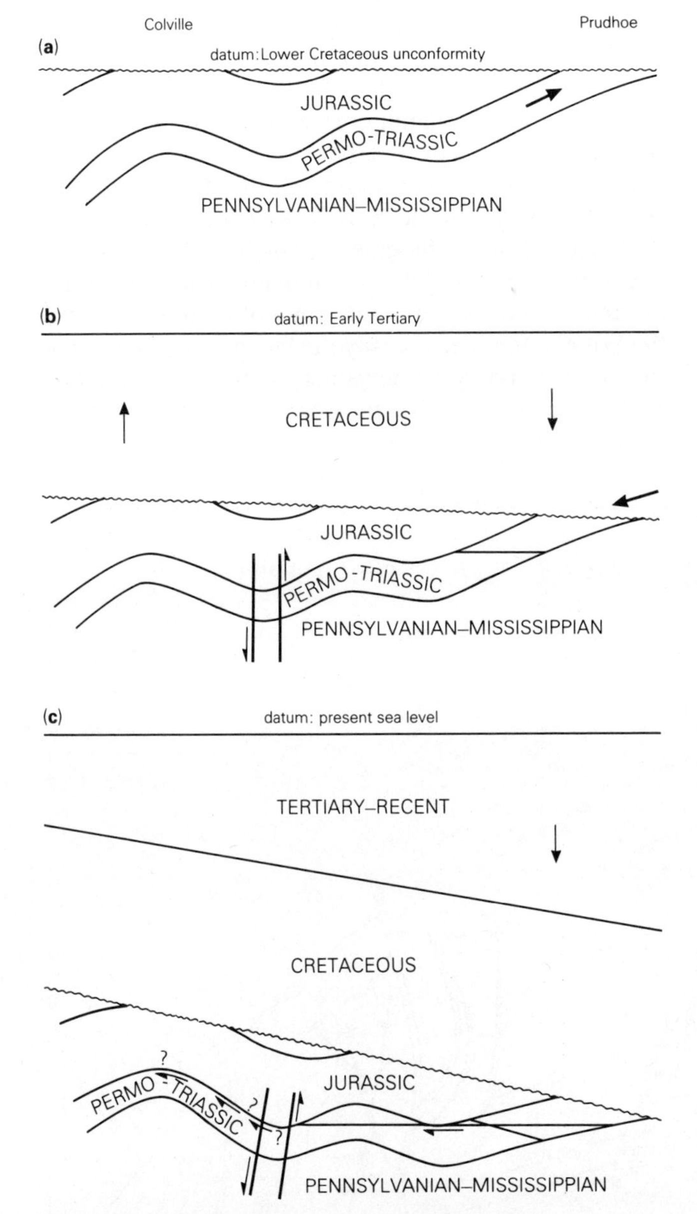

Figure 26.13 Postulated multistage fluid migration in Permo-Triassic reservoir sandstone in Prudhoe Bay field. (a) Updip water migration preceding Cretaceous eastward tilting. (b) Oil accumulation in subcrop below pre-Cretaceous unconformity; oil from overlapping Cretaceous source trapped by downdip water movement following initial eastward tilting. (c) Remigration of hydrocarbons within reservoir following Tertiary tilting to present position. (From H. P. Jones and R. G. Speers, *AAPG Memoir* 24, 1976.)

The longer the lead time, the greater the expense caused by it. Between the discovery of Prudhoe Bay and the completion of its pipeline across Alaska, a group of only three operating companies undertook construction and development drilling costing millions of dollars *daily* without a penny of return. Few organizations, public or private, are able or willing to sustain that level of outlay, and few oilfields, gasfields, or mineral deposits are large enough to justify it.

(h) Just as early drilling in the waters of Cook Inlet led to enormous and rapid strides in drilling and production technology, so the coastal position of Prudhoe Bay led to great advances in offshore seismic surveys and test drilling in the ice-bound waters of the Beaufort Sea, underlain by soft sediments partially bound by permafrost.

26.2.3 Wilmington, California, and Hassi Messaoud, Algeria

The Wilmington oilfield in California and the Hassi Messaoud oilfield in Algeria seem an incongruous pair for the purpose of any illustrative lesson at all. The first has one of the youngest reservoirs of any giant oilfield in the world (principally Pliocene); the second has the oldest (Cambrian). One is a very shallow field; the other very deep. One produces heavy crude difficult to refine; the other a crude so light and sweet that it approaches natural gasoline in constitution. The regulatory and fiscal environments in which the two fields were discovered, and have been operated, could hardly present a greater contrast.

Wilmington and Hassi Messaoud, however, provide perfect illustrations of two fundamental points. Each is easily the largest field within its jurisdiction, yet each escaped discovery whilst much smaller fields were routinely found. Despite the encouragement from modern exploratory techniques for the belief that the largest fields in any basin should be discovered early, these two enormous fields remind the geologists that a single faulty premise can forestall a giant discovery for years. The second point is basically political. Oilmen do not like politicians, but sometimes politicians contribute to the discovery of oil that would not be discovered (or not discovered so early) without them; and politicians very frequently determine how the geologists and engineers develop it once it has been discovered.

For much of the first half of the 20th century, California was the most important oil-producing area in the world. Until the end of World War I, the state had been responsible for nearly 15 percent of cumulative world production. The San Joaquin Basin, in the interior of the state, had made the major contribution to this share, every one of its early fields having been discovered either through surface evidence (including copious seepages) or by random drilling.

The areally small Los Angeles Basin had also produced oil prior to 1920, from a line of fields along the faulted foothills of its northeast side (Fig. 16.27). Between 1920 and 1925, a series of spectacular discoveries was made along a second fault zone close to the southwest margin of the basin (and extending into the Pacific Ocean at both ends). These discoveries enabled California to increase its share of world oil output to 25 percent by 1925. Given a single one of the discoveries, the rapid achievement of the others is not surprising. The basin contains no pre-Miocene sediments, and its young sedimentary fill was strongly folded during the Pleistocene Pasadenan orogeny. Relief is so young that anticlines rise as prominent ridges or knobs above the flat floor of the basin.

Still farther southwest, a third fault zone separates the basin from the Palos Verdes Hills, an uplifted promontory interrupting the smooth arc of the southern California coast. A modest oilfield (Torrance) had been discovered on the eastern (downthrown) side of this fault in 1922, during the basin's discovery heyday, but this segment of the basin was so flat and featureless that it was understandably deemed to be also structureless. In 1935, however, 10 years after the latest major discovery in the basin, a geophysical survey revealed a prominent closed anomaly, also on the downthrown side of the Palos Verdes fault, southeast of the Torrance field and essentially on strike with it. The anomaly turned out to reflect a major anticline, raised by mid-Pliocene uplift over rejuvenated basement (Fig. 26.14). Because the post-uplift Plio-Pleistocene sediments were still as flat as when they were deposited, the structure was invisible at the surface. It is the only large productive structure in the Los Angeles Basin of which this can be said.

The new field, Wilmington, enjoyed successive discoveries of deeper sand reservoirs (a hallmark of the Los Angeles Basin fields) for more than 10 years, the reservoirs extending from the mid-Pliocene unconformity to the fractured schist basement (which is of Jurassic or Cretaceous age; Fig. 26.15). Actually continuous with the smaller Torrance field in some of these reservoirs, the Wilmington field is easily the largest oilfield in California (twice as large as any other). Until 1980, it was the fourth largest field in North America and had remained

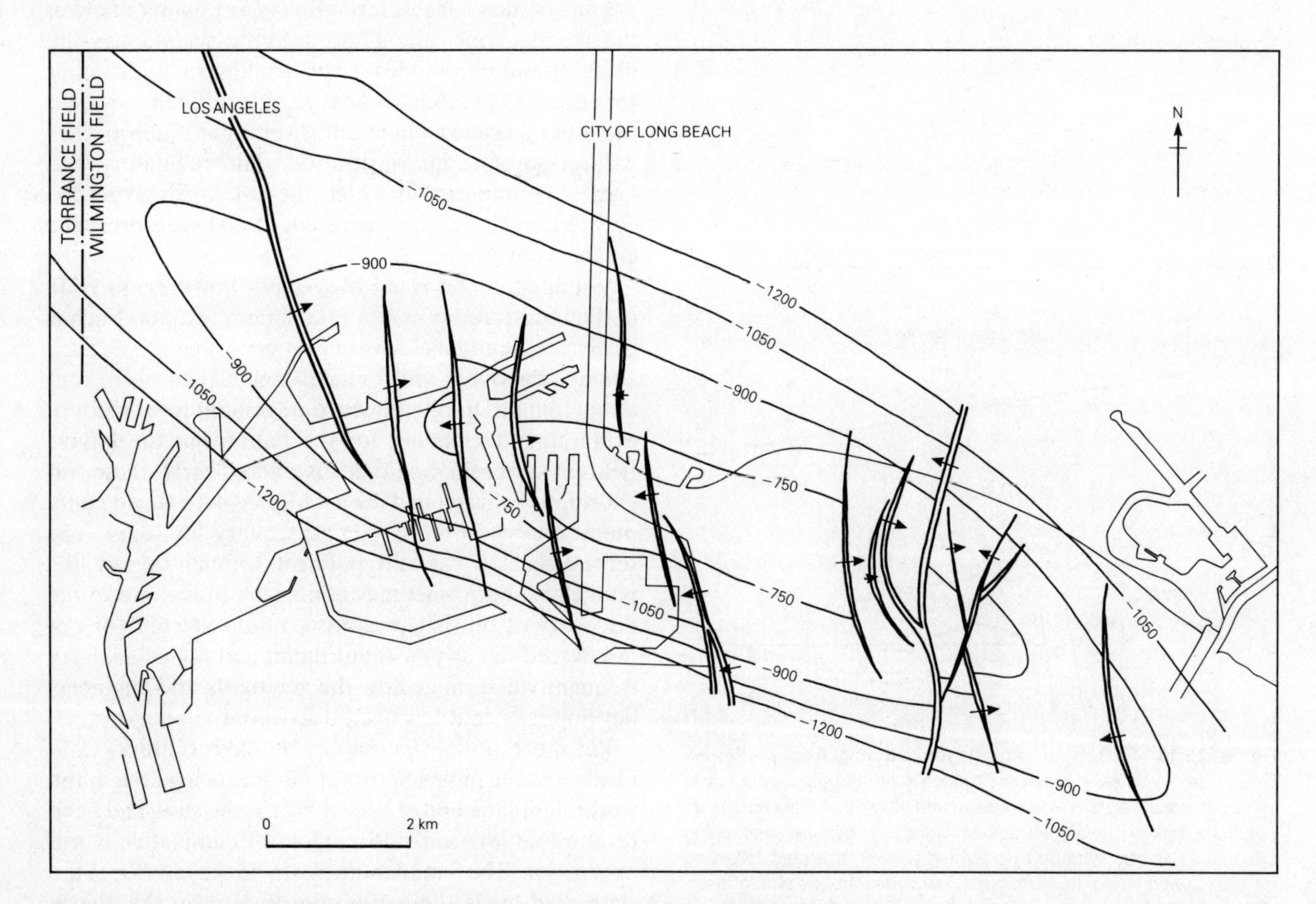

Figure 26.14 Wilmington oilfield, California: structure contours on the principal Pliocene producing zone. Note the arbitrary division from the Torrance field at the northwest end of the structure. (From M. N. Mayuga, *AAPG Memoir* 14, 1970.)

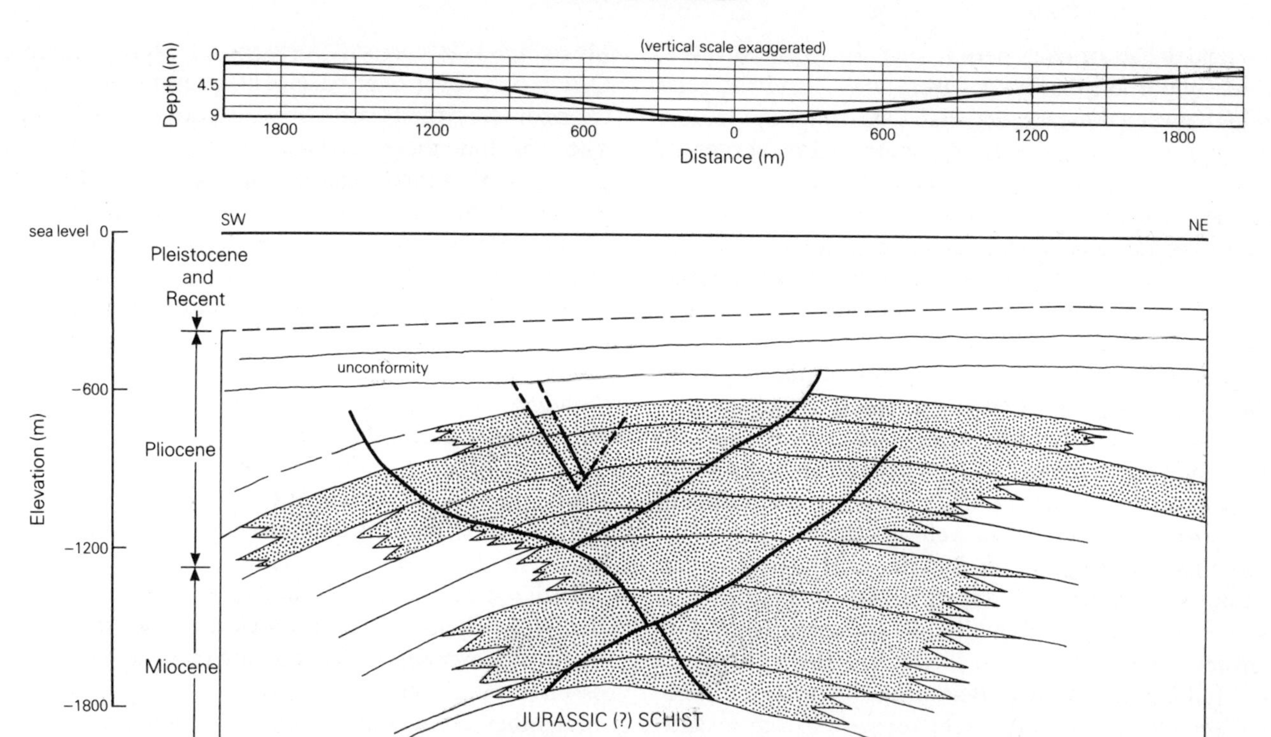

Figure 26.15 Wilmington oilfield, California. Cross section through the central part of the field, showing multiple sand reservoirs in relation to underlying basement and overlying unconformity. Top section shows subsidence bowl above the anticline. (From M. N. Mayuga, *AAPG Memoir* 14, 1970.)

among the top three or four producing fields in the USA for 40 years. Yet it was the last of California's many giant fields to be discovered.

The field entered a new phase of its history following World War II. Surface subsidence had been detected over its crest, and became so great in the early 1950s that it seriously damaged the facilities of the US Naval Yard at Long Beach (under which the field extends at its southeastern end). Following a pilot program, water was injected into the reservoir on an enormous scale, successfully halting the subsidence as oil continued to be withdrawn, but of course incapable of reversing the subsidence already suffered. One might wonder why salt water was injected into the depleted reservoir, when the field area was dotted with large refineries importing oil from overseas; why not pump this imported oil into the reservoir as a strategic reserve? The revelation that California's law of capture would entitle householders above the replenished reservoirs to sue for ownership of the injected oil frustrated any such sensible course.

The field had yet more lessons for its developers. A new seismic survey in 1954 revealed that its southeastern extension, already responsible for the subsidence in the naval yard, went much farther into the sea than had been realized. The devising of drilling islands, in the shallow Pacific waters off a heavily populated shoreline that could accommodate neither unsightliness nor environmental damage, involved remarkable ingenuity. And the need for all these defensive enterprises to be conducted as efficiently and cheaply as possible forced the pooling of a great number of small unit holdings and demonstrated that the oil industry could be cooperative instead of competitive when it had to be.

As in many regions of the Eastern Hemisphere, systematic oil exploration in *Algeria* began following the end of World War II. Between then and 1958, this exploration was conducted by two companies controlled by the Algerian and French governments. They undertook a great deal of geophysical exploration, but drilling was confined to the mountainous northern strip of the country until 1952. It met with trivial success.

Following the 1951 awards of large concession areas in the northern Sahara desert, drilling emphasis shifted to them despite difficulties of access and operation. Several dozen exploratory wells were drilled, mostly to Devonian or Silurian strata. Several had shows, mostly of gas, but nothing commercial was discovered.

As these first concession permits were valid for only 5 years, their expiration became due at the end of 1956. The two original concessionaires would then have had to surrender half of each permit area. In 1956, however, three important discoveries were made, and the relations between them were crucial. One company, holding a concession in the extreme southeastern corner of Algeria, discovered oil early in the year by drilling on a surface structure (the Edjeleh anticline) in an Upper Paleozoic basin having fundamental geological affinities more with western Libya than with the rest of Algeria.

The two original companies, which had drilled nearly all Algeria's exploratory wells prior to that date, had lost confidence in the Paleozoic strata of the northern Sahara, and intended to relinquish most of their acreage there when their permits expired. Their Algerian and French government overseers granted them approval to do this with the proviso that, before relinquishment areas were determined, a well was drilled *to the Precambrian basement* on the highest detectable structure within their concession. Identification of this structure was made by a combination of surface geology and *refraction* seismology (Fig. 26.16). The Hassi Messaoud test fulfilled the proviso. Below the anticipated Triassic salt section, it recovered light, sweet oil from a sandstone formation at about 3300 m depth. The well was completed on the last day of the permit year.

The productive sandstone, resting directly on granite basement, was regarded as the basal sandstone of the Triassic transgression. This interpretation was reinforced by the third discovery well of 1956, at Hassi R'Mel, in which gas was found in a basal Triassic sandstone beneath a seal of late Triassic salt and shale. Within 1 year, however, the reservoir sandstone at Hassi Messaoud was correctly recognized as the Cambro-Ordovician cover over basement, preserved beneath a regional Hercynian (late Carboniferous) unconformity. The sandstone therefore became a completely new target, not earlier recognized as such because it was older than any other prolific oil reservoir then known in the world, and it was moreover essentially a nonprospective quartzite wherever else it was known. Recognition of this new target paved the way for newly optimistic forecasts. A year later, promulgation of the first Sahara Petroleum Code led to numerous new concessions and the entry of foreign companies to the region. But no other comparable field has yet been found.

The Hassi Messaoud field is one of the three or four largest oilfields in Africa. If fields around the Persian Gulf are excluded, it is undoubtedly one of the 15 or 20 largest fields in the world. Throughout the 1960s and 1970s it maintained its position among the 20 most productive fields also. Even on a world scale, the field is unique in a number of respects, and presents a variety of lessons to exploration geologists.

No other known oilfield remotely comparable in size has its whole oil accumulation in so old a reservoir rock. In no other wholly Paleozoic oilfield of any significant size is the highest point in the oil column at so great a depth. In no other large oilfield is the age difference between reservoir rock and roof-rock so large (more than 250 Ma). In every other oilfield yielding considerable production from basement or from a basal clastic reservoir rock, the oil is undoubtedly derived from a source somewhere in the roof-rock; moreover, the roof-

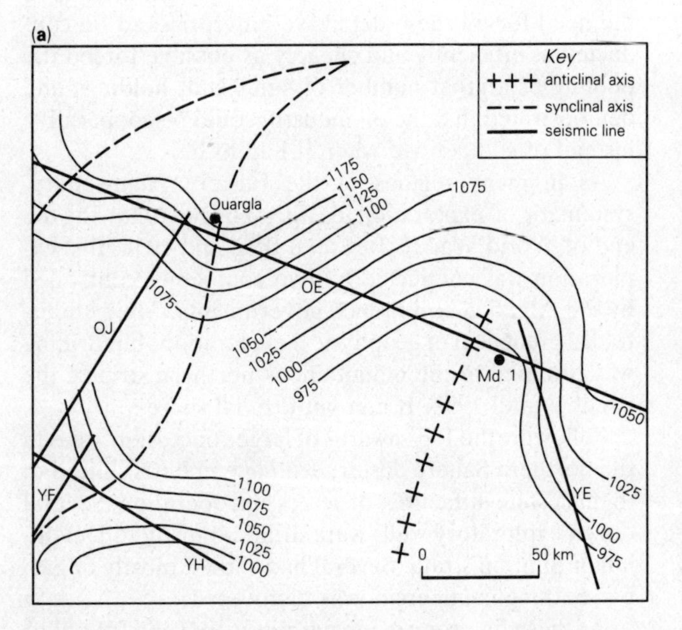

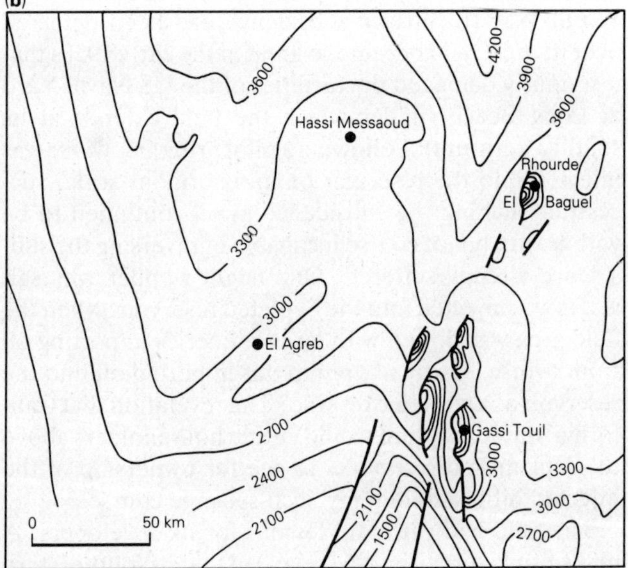

Figure 26.16 Hassi Messaoud field, Algeria. (a) Original seismic refraction map on which drilling location was based; contour interval 25 ms. Lettered lines are seismic lines. (b) Structure contour map on top of the Paleozoic. Contour interval 300 m; elevations below sea level. (From A. Balducchi and G. Pommier, *AAPG Memoir* 14,1970.)

rock in all cases contains an important producing reservoir in its own right. At Hassi Messaoud, erosion following Hercynian uplift must have removed the whole of the Devonian, Silurian, and Upper Ordovician strata (and possibly some Lower Carboniferous in addition) from the crest of the structure, and this removal had to be achieved between late Pennsylvanian time and the mid-Triassic (Figs 15.1 & 26.17).

A general consensus among petroleum geologists and geochemists in North Africa is that the source of the Hassi Messaoud oil lay in Silurian shales, which form a broad apron around the lower flanks of the structure. The geologists must choose between two possibilities. The first is that the source rocks, rich as they are, remained immature during 125 Ma of late Paleozoic sedimentation, and a further 100 Ma, at least, of post-Hercynian erosion and the Triassic retransgression of the region. Burial depths and P,T conditions must have been inadequate to bring on generative maturity (see Sec. 7.4). The second possibility is that a Paleozoic oil accumulation was destroyed during Permian erosion, probably with the removal of Lower Carboniferous or Upper Devonian reservoir rocks. At some time in the Mesozoic, and probably not before Cretaceous time, the remaining Silurian source sediments around the flanks of the uplift were brought to generative maturity a second time, and the oil reached its stratigraphically older but structurally higher reservoir rock by updip migration below the seal of Triassic shale and salt.

The phenomenon of long deferment of source rock maturation is established or inferred in several basins, most of them Paleozoic (for the Devonian source sediments in the Western Canadian Basin and their Ordovician counterparts in the Williston Basin, for example; see Sec. 23.11). The alternative phenomenon, of successive long-separated spasms of generation and migration is also known or inferred in several producing regions; but in few is the time interval between spasms so large as at Hassi Messaoud.

Had the precise stratigraphic and structural configuration of this trap been revealed through divine vision in advance of the selection of the site, few experienced petroleum geologists would have risked their own money to drill it. Given the history of its discovery, it is intriguing to speculate how long the field might have remained undiscovered had a government not required the drilling of one final hole in 1956, or had the government's geologists chosen a different site for it when the requirement was passed on to them.

26.2.4 Jay field, Florida, and the Reforma fields, Mexico

The Gulf of Mexico came into being as an aspect of the opening of the central and northern Atlantic Ocean, in the late Triassic and early Jurassic (215–165 Ma). As in many parts of the initial oceanic rift systems, two facies were very conspicuous: a rapidly deposited saline facies,

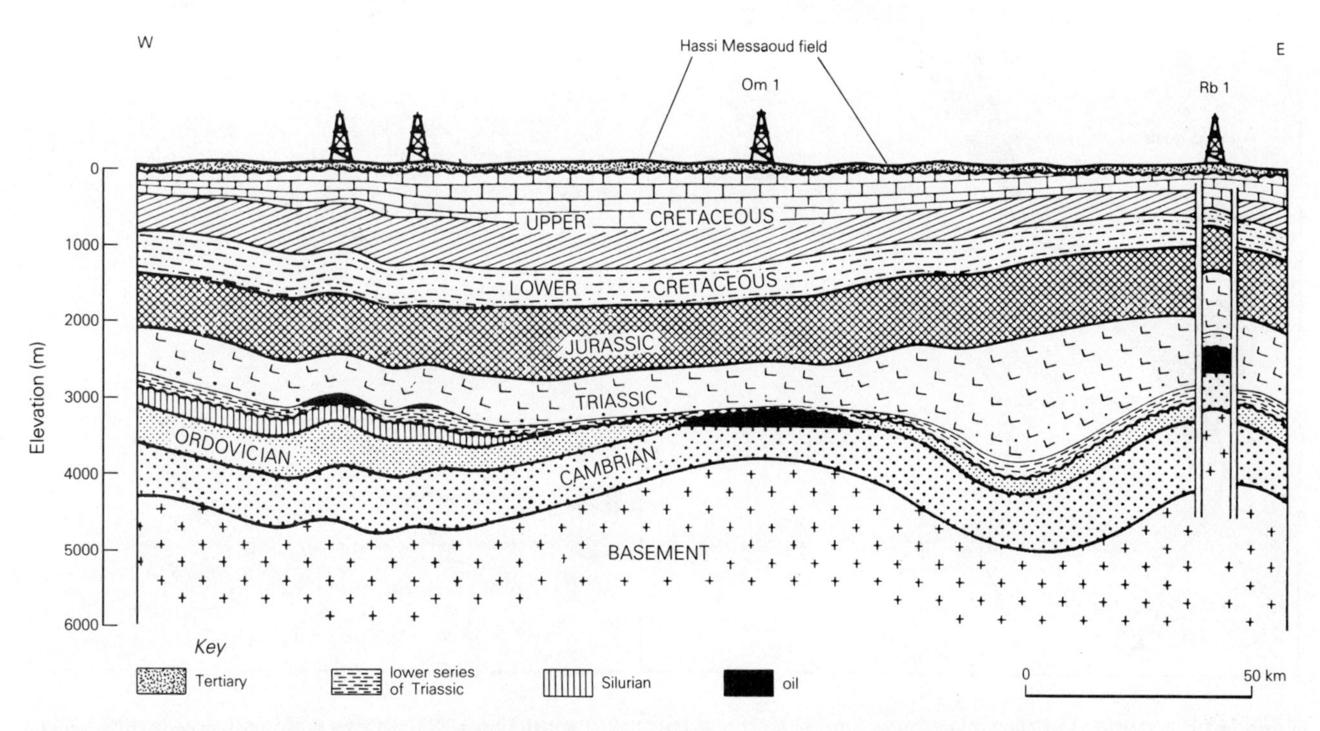

Figure 26.17 West–east cross section of the central part of the Ouargla Basin, showing the Hassi Messaoud (Om) and Rhourde el Baguel (Rb) fields. (From A. Balducchi and G. Pommier, *AAPG Memoir* 14, 1970.)

and a facies of bank carbonates with fringing reefs. The salt in the Gulf of Mexico is Jurassic in age. The principal carbonates extended over a longer interval, from the Oxfordian to the Cenomanian, inclusively, representing a great marine transgression of the continental margins (Sec. 25.5.9).

During the late Jurassic, the best developed segment of the carbonate bank stretched across the northern margin of the gulf, from eastern Texas through Mississippi and Alabama into western Florida and on into the Bahamas Banks and north-coastal Cuba (Fig. 26.18). Another segment of it wrapped around the Yucatan promontory on the opposite side of the gulf. During the early Cretaceous, the bank was extended gulfward as the shelf grew in width, and by Albian time one of the largest bank-barrier reef complexes in the geologic record almost completely encircled the gulf (Figs 17.4 & 26.19). In the southern USA, the Cretaceous representative is commonly called the Edwards limestone; in Mexico it is called the El Abra limestone (Fig. 13.50).

The El Abra limestone yielded spectacular early discoveries in the so-called Faja de Oro (Golden Lane), making Mexico the world's second leading oil-producing country in the years preceding and during World War I. The towns of Tampico and Tuxpan became the Latin American equivalents of Tulsa and Baku. In 1922, an important discovery was made on the American side, at Smackover in southern Arkansas, in the Upper Jurassic limestone promptly given the field's name. The Golden Lane and Smackover discoveries were the largest carbonate-reservoired oilfields in the world at that time except for the single field of Masjid-i-Suleiman in Persia, the field that opened up the Middle Eastern petroleum province; these three were also the first large oilfields discovered in the world to occur in carbonate reservoirs not Paleozoic in age.

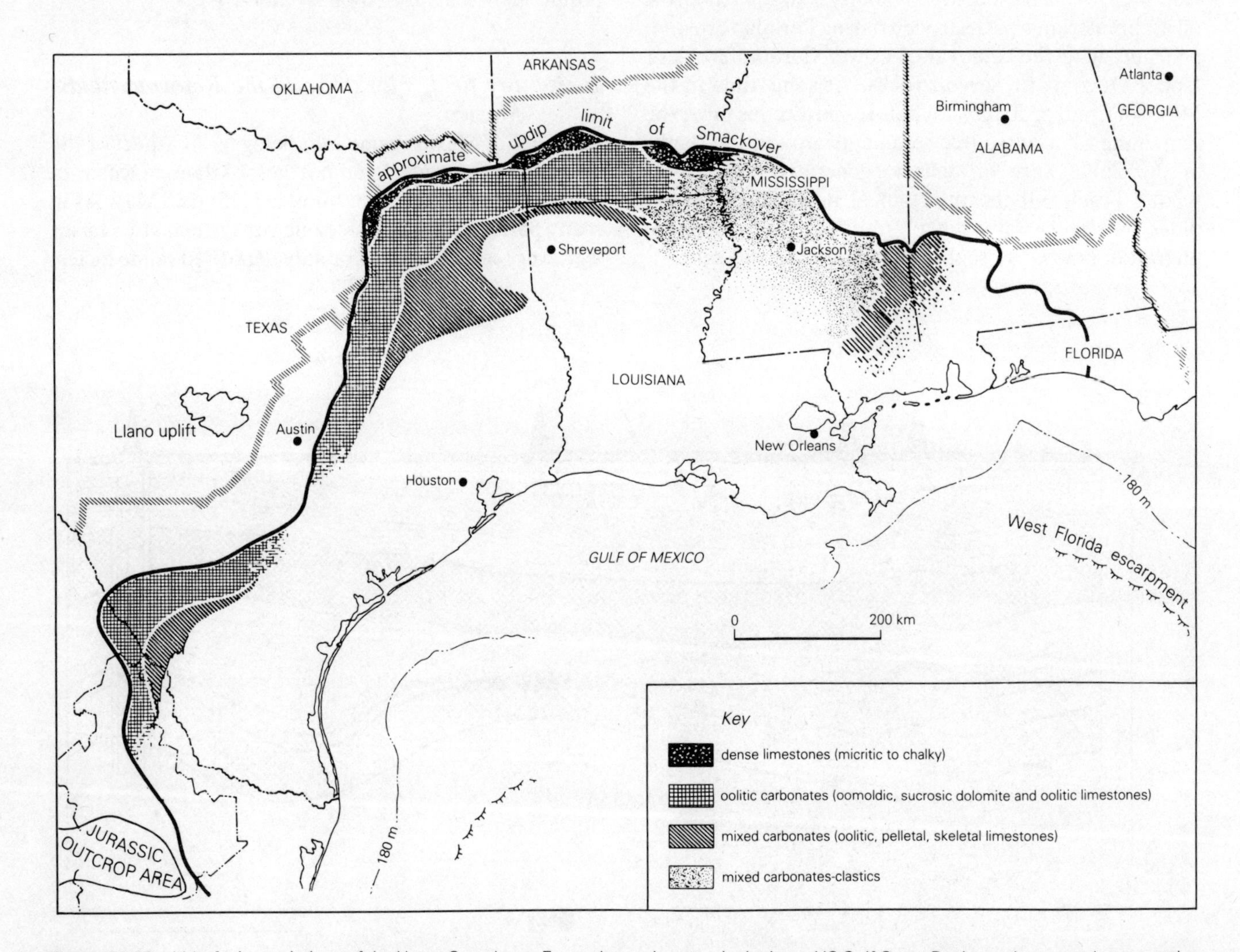

Figure 26.18 Lithofacies variations of the Upper Smackover Formation carbonates in the inner US Gulf Coast Basin, as they were known at the beginning of drilling of the Jay discovery well in the "blind" area at the eastern end of Smackover development. The trap is formed by the juxtaposition of the "mixed carbonate" and "dense limestone" lithofacies. (From T. F. Newkirk, *AAPG Memoir* 15, 1971.)

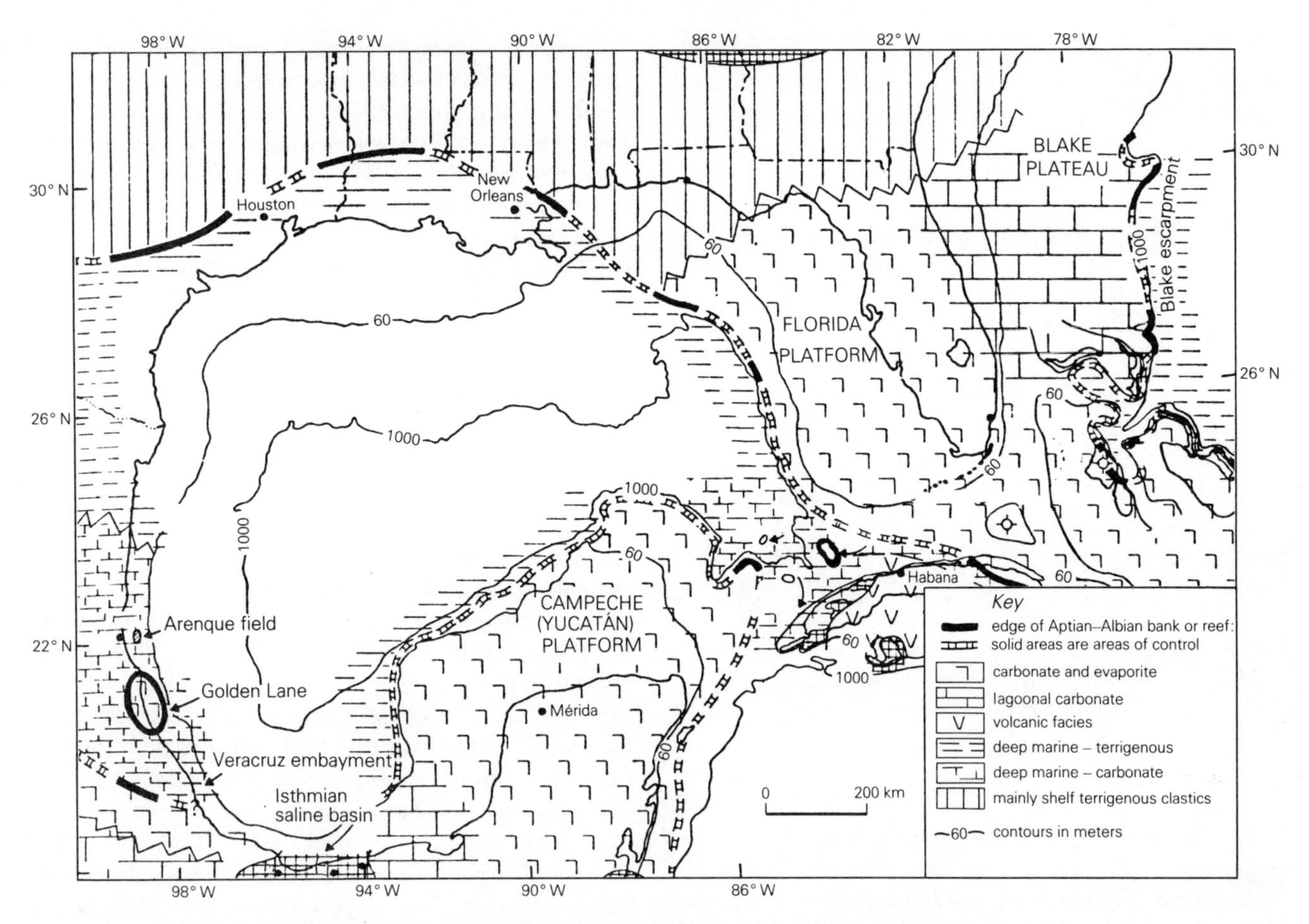

Figure 26.19 The bank and reef complex circumscribing the early Gulf of Mexico during mid-Cretaceous (Aptian–Albian) time. Note positions of the platform edges in Florida and southeastern Mexico. (From W. R. Bryant *et al.*, *AAPG Bull.*, 1969.)

It was not until 30 years after their discovery that the Golden Lane fields were recognized as occupying the reefs of one side of an elliptical atoll (Fig. 25.5). It was 15 years after the Smackover discovery before a trend of similar, but much smaller, fields identified the extent of the carbonate bank marking the northern rim of the late Jurassic Gulf Coast Basin, basinward of a major, down-to-the-basin, normal fault zone. The Smackover limestone represents the deposit on shallow-water structures separating sink-like salt basins that were to become concentrations of salt domes. The thickly bedded carbonates and evaporites were frustrating to the efforts of early geophysicists, and it was not until the development of common depth point seismic studies, in the 1960s (Ch. 21), that the eastward extension of the Smackover trend was shown to reach the deep basin of western Florida.

There, a large subsurface structure, having no outcrop representation and no surface relief, proved to be gas-bearing in a rather dense, micritic limestone or lime mud. However, in a southward-plunging nose, extending from the "high" towards the deep basin, the shallow-water lime muds contained ooidal grainstones which had been dolomitized. They were thereby provided with uniform rhombohedral porosity, which had later been enhanced by solution leaching. These dolomitized grainstones contained light oil which yielded high recoveries per unit volume. The Jay field, discovered in 1970, was the first important oilfield discovered in the continental USA since the 1950s. At a depth of over 4500 m, it was also one of the deepest productive oilfields in the world. As in all deep carbonate reservoirs, the oil and gas have high contents of H_2S, CO_2, and nitrogen.

The trap in which the Jay field occurs combines structure and stratigraphy in such a way that both are vital to it (Fig. 26.20). With one of the Gulf Coast's bounding faults immediately to the northeast, the trap resembles a buried combination of dip-slope and scarp-face, so that it is in a genuine sense a stratigraphic trap with a steep northeast flank and southwesterly dip. But the porosity pinches out to the northwest, essentially at right angles to the strike. What appears to have happened is that meteoric waters invaded along the updip side of the nose, from the outcrop, creating early solution porosity. To the south there was no outcrop, but solution by

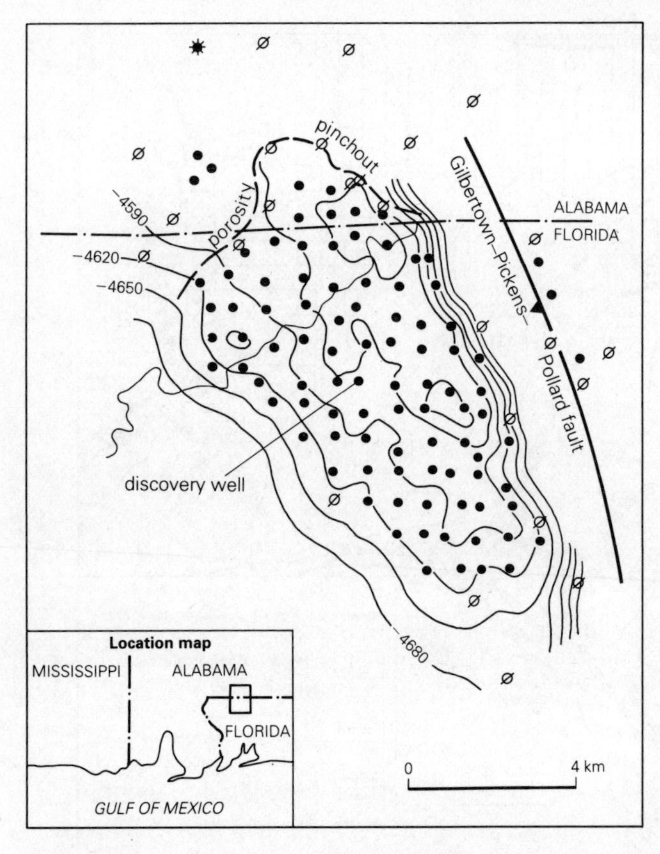

Figure 26.20 Jay field, western Florida; structure contours on top of the Smackover Formation, the carbonate reservoir rock. Contours at 30 m intervals. Note the depositional margin to the northeast and the porosity pinchout to the northwest, forming a combined structural-lithologic trap. (From R. D. Ottmann *et al., AAPG Memoir* 24, 1976.)

marine waters there enhanced the primary intergranular porosity. All the porosity is thus due largely to dissolution; dolomitization created the permeability within originally calcitic micrites. The dolomitization may have been brought about by hypersaline waters percolating downward from a second salt and evaporite section overlying the reservoir.

The trap is therefore dependent upon both *structure* and *facies*; structure alone would not be sufficient, and the facies, changing as the shoreline is approached, forms with it a combination very difficult to predict. The area apparently lay within stable, shallow flats along the protected, inner side of a line of Jurassic bank islands. A great range of carbonate types, comparable with the range now found in equally restricted areas of the Bahaman Banks, included pelletal or fecal, grain-supported calcarenites virtually free of mud.

The Jay field has endured production difficulties – narrow production strings at the great depths, and corrosion problems caused by the action of acid gases at high temperatures on bits and tubular goods; but it remained one of the dozen or so leading producing oilfields in North America throughout the 1970s.

Diagonally across the Gulf of Mexico, the carbonate-reservoir fields of *southeastern Mexico* had languished for almost as long after World War I as the fields of the American trend. The largest single field, Poza Rica, had been discovered in 1930, but it remained the southeastern limit of important carbonate production for 40 years. True, there was production farther to the southeast, in the Isthmus of Tehuantepec, but it came from Miocene sandstones within a wholly clastic Tertiary succession overlying another Jurassic salt basin. The thick Mesozoic carbonates of the Yucatan Peninsula and Guatemala, like those of peninsular Florida and the Bahamas, remained no more than trivially productive.

Yet the Jurassic salt was known to be present, and to have become intrusive into domes and ridges. These had already been identified beneath the waters of the southwestern gulf, and they extended on to shore in both the Isthmian (Saline) Basin and the Macuspana Basin, west of the Mexico/Guatemala border (Fig. 26.21). The gap in the explorers' information lay in the flat, tropical coastal plain of the states of Tabasco and Chiapas, facing the Gulf of Campeche. Careful seismic work carried out by Pemex across this strip of country revealed a number of vaguely defined anomalies. Because these were broad and somewhat diffuse, and clearly lay at considerable depths within a succession of strata having the velocity characteristics of carbonate rocks, it was tentatively concluded that they represented deep-seated salt intrusions within the Mesozoic carbonate succession. The carbonates were known to exist directly to the east and southeast; the salt intrusions were known to exist to the east, west, and northwest; and salt intrusions into thick carbonates were known to result in deep uplifts and not in piercement structures that approach or reach the surface (Sec. 16.3.5).

Drilling began on the crests of the more southerly (inland) structures, presumed to be the shallowest, in 1971–2. They proved to be very broad domes with exceedingly gentle flanks; two of them yielded both oil and gas from fractured calcarenites of late Cretaceous age, with net pay intervals measured in hundreds of meters. In the shallower field, called *Cactus*, the productive interval was below 3740 m; in the deeper field, *Sitio Grande*, it was below 4120 m, approaching the depth of the pay zone in the Jay field in Florida.

Seismic exploration quickly identified some 150 drillable structures in the new field area, the North Chiapas or *Reforma* province (Fig. 26.22). Within 5 years of the initial discoveries, about 16 fields had been found, all in Mesozoic carbonate reservoirs; every structure drilled had shows of oil or gas if nothing more; none was condemned by its first testing. The average productivity per well was close to 1000 m^3 per day; some wells came in at over 3000 m^3 per day. Total output for the new province reached 150 000 m^3 per day before the end of 1978, the

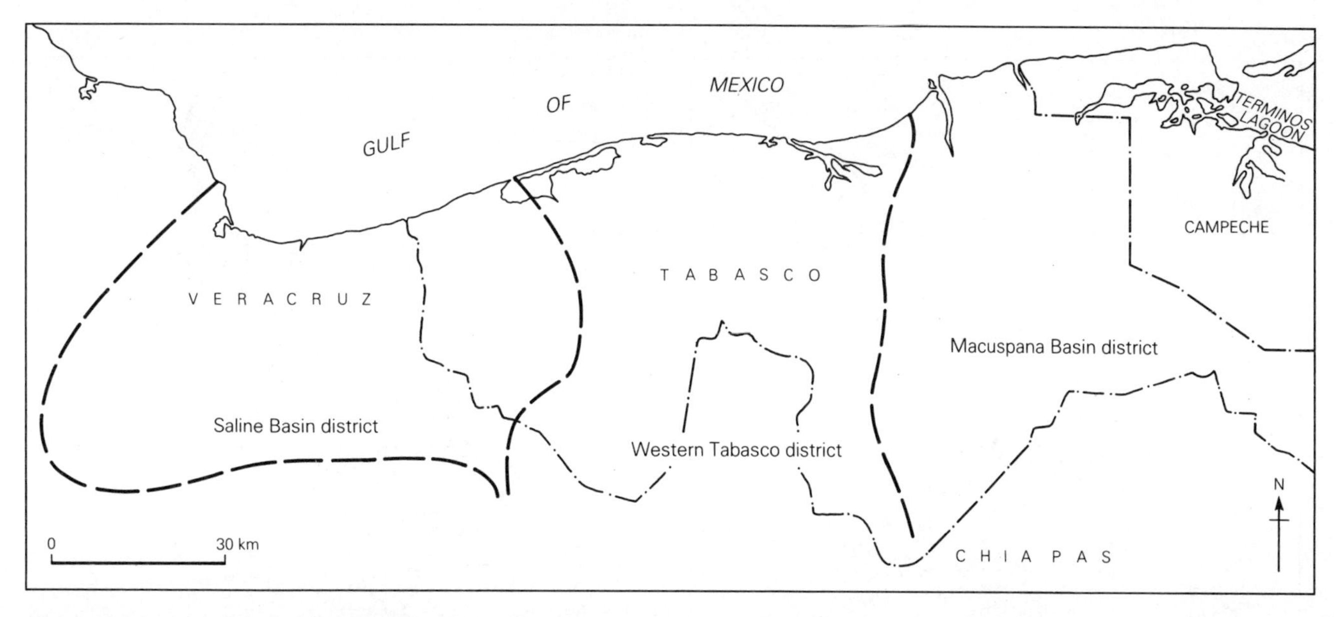

Figure 26.21 Map of the Isthmus–Tabasco region, southeastern Mexico. In the Saline Basin (west), oil is produced from Miocene sandstones in salt-formed traps. In the Macuspana Basin (east), oil in Miocene sandstones is trapped in anticlines over the edge of a west-facing carbonate bank–reef complex. The Reforma fields, in deeply buried Mesozoic carbonates, lie over the Villahermosa "horst" between the two basins. (From E. J. Guzman, *AAPG Bull.*, 1959.)

earliest discoveries proving to be very much larger than indicated by their first tests and outputs. Unfortunately, a crucial phenomenon quickly made its influence felt. As each field reached a production level of about 20 000 m^3 per day, well pressures declined severely and the wells quickly turned to water. Though the larger fields had relatively high initial GORs (nearly 200), they were in fact undersaturated for their depths. The early rapid loss of reservoir pressure soon stabilized, demonstrating that the short-lived gas expansion recovery mechanism (see Sec. 23.8) was readily replaced by an active water drive in the underlying aquifers. The combination of primary porosity, solution porosity, and intense microfracturing produced exceptionally high and continuous intercommunication both along and across the beds. The rapid loss of pressure would therefore have resulted in very poor primary recovery of the oil content – possibly as little as 10–12 percent – unless the natural water drive could quickly be augmented. The first field, Sitio Grande, was artificially waterflooded 5 years after its initial discovery, at a rate of about 25 000 m^3 of water per day. One year later, an even larger waterflood – about 160 000 m^3 of water per day – was initiated in the largest of the fields, A. J. Bermudez, north of Sitio Grande and Cactus (Fig. 16.46).

The stratigraphic and structural settings of the Reforma fields provide an ideal combination for oil generation and accumulation; the only drawbacks are consequences of the great drilling depths. As in the Golden Lane, there are three facies of Cretaceous limestone in addition to the thick, underlying Upper Jurassic limestone. The reef, "mixed," and basinal facies are described in Section 16.3.7 and illustrated in Figure 13.50. The "mixed" or Tamabra facies was no doubt in ideal stratigraphic trap position from the beginning, to receive oil from the basinal facies to the west and north. The belt of facies change, however, was superimposed upon (and doubtless in part caused by) a complex uplift extending northwards from the anticlinorium which forms the axial backbone of southeastern Mexico and Guatemala. The uplift had an initial northward plunge, away from its axial zone; but repeated rejuvenations resulted in a reversal of the substructure so that the Jurassic now has a general southerly and westerly dip. The Cretaceous units wedge out in succession towards the north in consequence of successive unconformities.

The uplift thus came to constitute a complex horst, the Villahermosa horst, with numerous subsidiary horsts and grabens, separating not only the Cretaceous shelf and basin but also the two younger basins penetrated by piercement salt structures (Fig. 26.21). During the Middle Eocene (in the Mexican equivalent of North America's Laramide deformation), salt pillowing was instigated within the uplift. Because the post-salt section is composed essentially of carbonates several thousand meters in thickness, no salt piercement was achieved (Sec. 16.3.5). Instead, the beds over the pillows were intensely fractured, and the structures were transected by a set of left-slip faults trending NW–SE across the horst.

All the principal domes along the axis of the Villahermosa horst contain oil. The smaller structures along

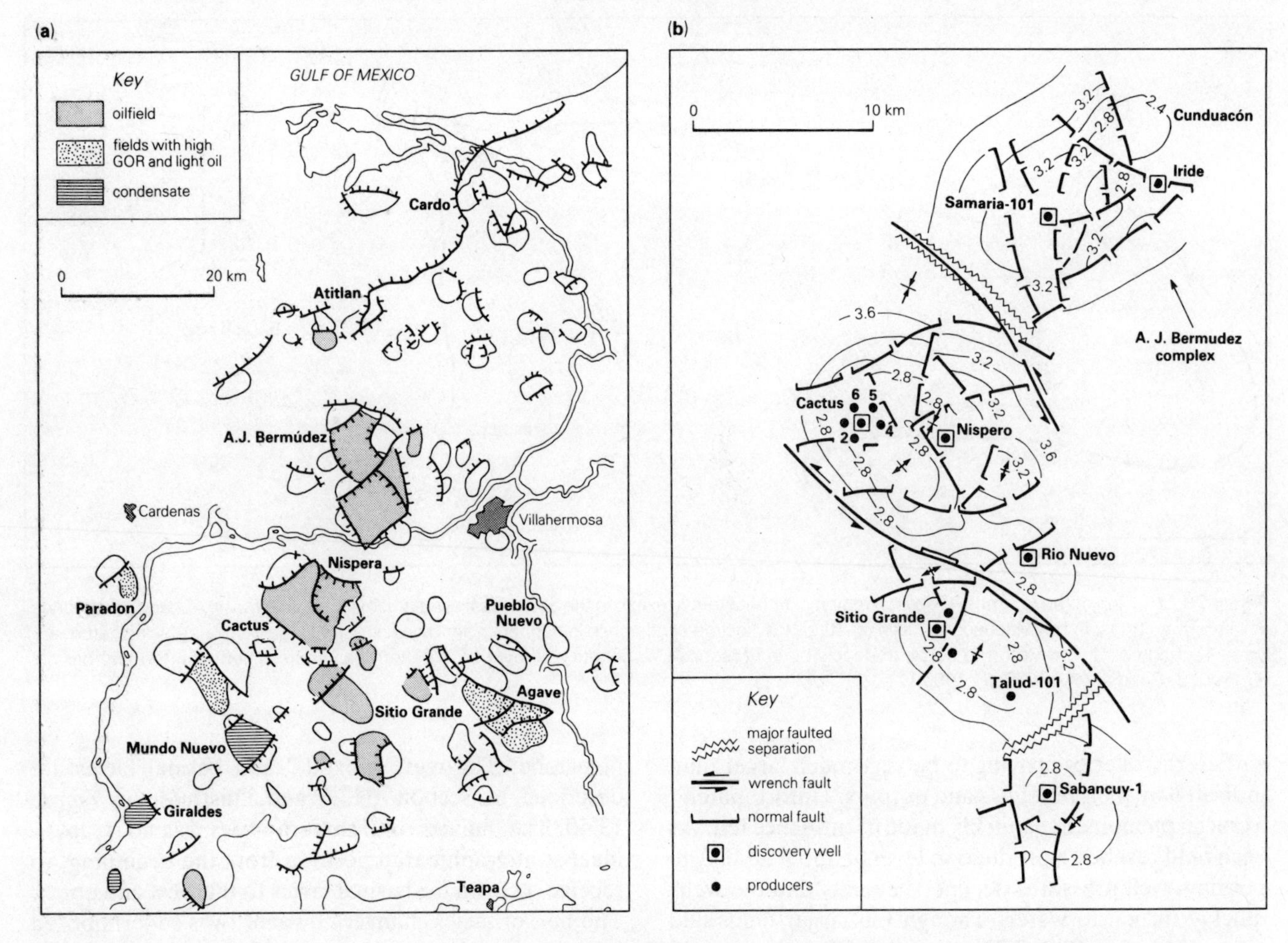

Figure 26.22 Principal oil and condensate fields of Reforma area, southeastern Mexico. (a) Map showing concentration of largest oilfields along axis of north–south arch. (b) Structure contours (in kilometers below sea level) on the main Cretaceous carbonate producing horizon in the central producing area, showing equidimensional structures above deep salt uplifts, cut by numerous normal faults and by northwest-striking left-slip faults. (From A. A. Meyerhoff, *Oil Gas J.*, 21 April 1980; after F. Viniegra, 1975.)

the eastern and western flanks mostly contain gas and condensate, with GORs of 500–1200. The largest field, almost in the center of the onshore portion of the horst, is actually a complex of fault blocks (Figs 16.46 & 26.22) that can be called a single field (A. J. Bermudez) because the continuous Jura-Cretaceous pay section demonstrated excellent interconnection of reservoir permeabilities over a vertical extent of 600–2000 m and horizontal extent exceeding 150 km^2. The combined, original, recoverable reserves of the four or five largest fields alone were nearly 2×10^9 m^3 of oil and about 3×10^{11} m^3 of wet gas. Oil gravities are between 20° and 40°, the commonest range being 28–33° API (see Sec. 17.7.3). As in all deep carbonate reservoirs associated with evaporites, sulfur content is relatively high, increasing downwards from 1 to about 3 percent by weight. By the end of 1980, production from the trend had reached 300 000 m^3 daily, vaulting Mexico into the leading group of oil-producing countries. Within 2 years, equally large fields in the offshore Gulf of Campeche had augmented production still further, but the geology of the Campeche fields is unlike that of the Reforma fields in important respects (see Sec. 25.5.9).

26.2.5 *Halibut field, Australia, and Forties field, UK North Sea*

Until the very late 1960s, all offshore oil production was achieved under two critical controls:

(a) The offshore regions were direct extensions of areas of prolific and long-established onshore production (see Sec. 17.9.1).

(b) The productive areas lay in shallow waters, less than 50 m deep, and exploration and development were conducted via elementary modifications of technologies used onshore.

Other areas that had by 1970 achieved offshore production, or were shortly thereafterwards about to do so, included a number of grabens along continental margins created by Mesozoic seafloor spreading. These strips of seafloor were either downthrown along faults controlling the present coastlines (as in the Gabon and Angola basins offshore from West Africa), or enclosed within downfaulted blocks extending seaward at angles to the coastal scarps (Reconcavo Basin in Brazil; Cambay Basin in India).

During the 1960s, a surge of new offshore exploration was extended into areas to which neither of the earlier conditions applied; they were remote from any commercial onshore oil production, and they involved drilling in significantly deeper waters. Two such areas were destined to become major producers. The comparisons and contrasts between them provide useful lessons which form the subject of this final section.

The late Lewis Weeks had recognized that another of the Mesozoic transverse coastal grabens underlies the strait between Australia and Tasmania. The grabens are transverse in the sense that they cut obliquely across ancient structures that control the present continental margins; they may be parallel and not transverse to the present *coastlines* (Fig. 26.23). Because the grabens were formed by extension during continental separation, their bounding faults are normal faults. Australia resembles the other countries containing petroliferous grabens of this sort – Brazil, west-central Africa, and India (and Scotland and Scandinavia, as will become apparent shortly) – in that it contains relatively little young, marine, sedimentary rock. The northern margins of the Bass Strait grabens form much of coastal Victoria, and both seepages and numerous shows in onshore wells had given encouragement over several decades. But no commercial oil had been found, probably because of flushing of potential traps by fresh waters.

At the beginning of the 1960s, an aeromagnetic grid had outlined *three* fault-controlled basins, and had provided some estimates of depths to basement within them. These estimates turned out to be seriously understated. Follow-up seismic surveys identified a number of drillable structures, and drilling in the offshore part of the most easterly basin (the Gippsland Basin) began in 1964. Within less than 3 years, the basin's two largest oilfields, and two commercial gasfields, had been discovered. The discovery and development of one of the oilfields is described here.

The Halibut field was discovered in August 1967, by a hole drilled in 72 m of water. The oil-bearing Paleocene–Eocene reservoir succession, nearly 6 km thick, is wholly deltaic or fluvial in origin, principally the massive sands of braided streams and numerous coal seams (Fig. 13.11). About the end of the Eocene, the area was uplifted, tilted, and considerably eroded, and the truncated, tilted strata form a succession of erosional wedge-outs beneath a very uneven erosional surface. This surface was transgressed by Oligocene marine

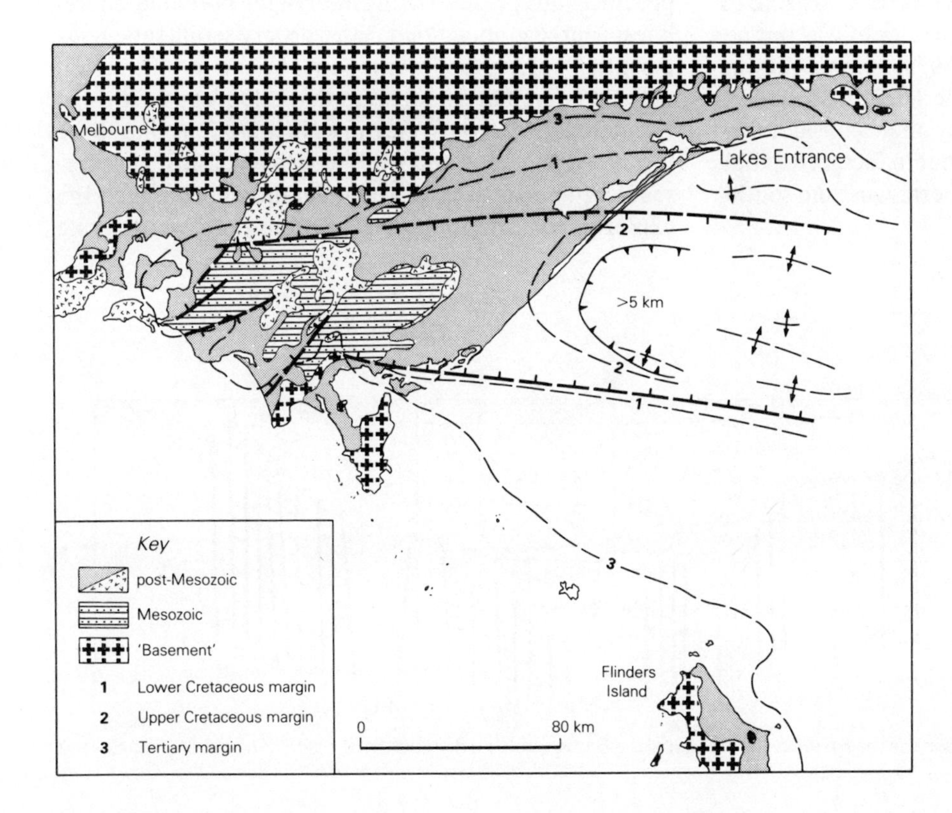

Figure 26.23 Structural map of the Gippsland Basin, Australia, according to general data available at the time of the Halibut discovery. The contour labelled 3 is the zero edge of the Tertiary fill of the graben. (From L. G. Weeks and B. M. Hopkins, *AAPG Bull.*, 1967.)

mudstones and buried by still younger rocks (Fig. 26.24).

Although the structure is basically a very simple one, the erosion surface at the top of the reservoir succession introduced critical complications (Fig. 26.25). The surface is an excellent seismic reflector, but the sediments overlying it suffer severe variations in seismic velocity. In particular, deep erosional channels were filled by thick, high-velocity calcareous rocks of Miocene age, creating velocity pull-ups (see Sec. 21.3.2). A well apparently in a crestal structural position was maddeningly likely to be drilled into a deep channel and might enter the reservoir beds below the oil/water interface.

This cause of unexpected dry holes is an embarrassing nuisance to the developers of onshore fields, but it is far more than that in deep water offshore. Development wells in deep water are drilled from expensive and immovably fixed platforms; they cannot afford wastage of this kind. So it is not a matter of simple infill drilling; careful interpretation of the details of every well location, in advance of its drilling, is essential.

Successive seismic surveys of the Halibut structure were made following each drilling venture, utilizing improved offshore techniques (including common depth point shooting by air gun), in an attempt to define the productive structure before a platform was put into place. Despite this succession of more and more detailed seismic refinements, including the mandatory migration of all sections, the crest of the structure was still incorrectly located. It was not until March 1969, 18 months after the field's discovery, that development was able to begin, and a year after that before the field was put on stream. A further 8 years later, in 1978, a new field was discovered on the west flank of the Halibut structure, separated from the main field (by then considerably depleted) only by a base seal. Halibut in fact turned out to have important extensions both northward and southward as well.

Diametrically on the opposite side of the Earth, exploration of the North Sea began on a different expectation. Westward progression of drilling from Germany into The Netherlands, in search of natural gas, had led to the 1959 discovery of the largest gasfield in Europe, at Groningen (Fig. 16.10). The reservoir rock for the field was a basal Permian sandstone, the Rotliegendes Formation (Fig. 13.7). It forms part of one of the most famous stratigraphic successions in the world: the Pennsylvanian Coal Measures of western Europe below the sandstone, the Küpferschiefer (copper shales) above it, and the Zechstein evaporites, including the Stassfurt potash deposits, above those. This general succession was known to underlie at least part of the North Sea, offshore from the Groningen field, so exploration for more Permian gasfields was extended into the southern reaches of the sea, between Holland and East Anglia. The exploration was strikingly successful, and a number of large counterparts of the Groningen field were discovered; but nearly all of them were in the British sector of the sea. The Dutch, German, and Danish sectors proved to be less well endowed with large gas accumulations.

Among the political consequences of the exploration was careful delineation of the international boundaries in North Sea waters. By far the longest and most critical of these boundaries was that between Britain and Norway. During its settlement, reconnaissance seismic surveys were progressively extended northwards along the axis of the sea, and they revealed a major geologic province unrepresented on either of the bounding shorelines: a large graben filled with Tertiary sediments, lying essentially along the international boundary. Though it was not immediately apparent that this graben provided an analogy with transverse marginal grabens like those forming the Gippsland or Reconcavo basins, the presence of an axis of thick sedimentation encouraged the explorers to move their investigations northwards along

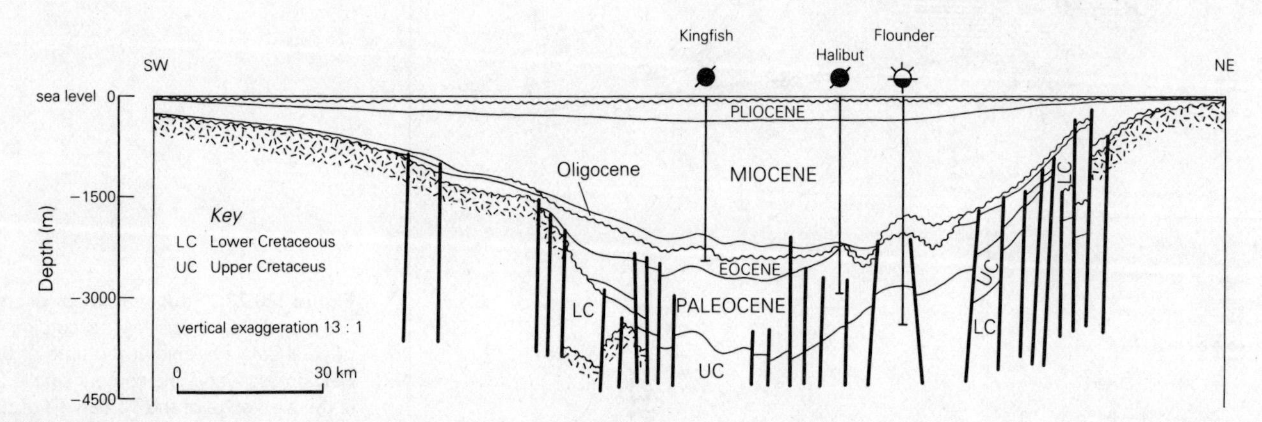

Figure 26.24 SW–NE cross section through the main oilfields of the Gippsland Basin, showing distribution of strata. (From E. H. Franklin and B. B. Clifton, *AAPG Bull.*, 1971.)

it. Uncertain of the nature or age of their targets, the geologists and geophysicists had perforce to drill on anything that looked like an anomaly. The first finds were small. Such finds are insufficient to promote development in any remote or difficult environment, but at the end of 1969 a more promising discovery was made on the Norwegian side of the international median line. Intensive testing and step-out drilling in 1970 confirmed that the find was indeed a major one; though not the first oil find in the North Sea, it was the one that precipitated the onslaught of leasing and exploration that followed.

Lacking any geographic reference point in the middle of the North Sea, from which to name their field, the operators half-facetiously christened it Ekofisk. The reservoir, at a depth of over 3000 m, was a Danian limestone, effectively the top of the Cretaceous chalk succession, fractured above what appeared to be a deep intrusion of Permian salt. Armed with this intelligence,

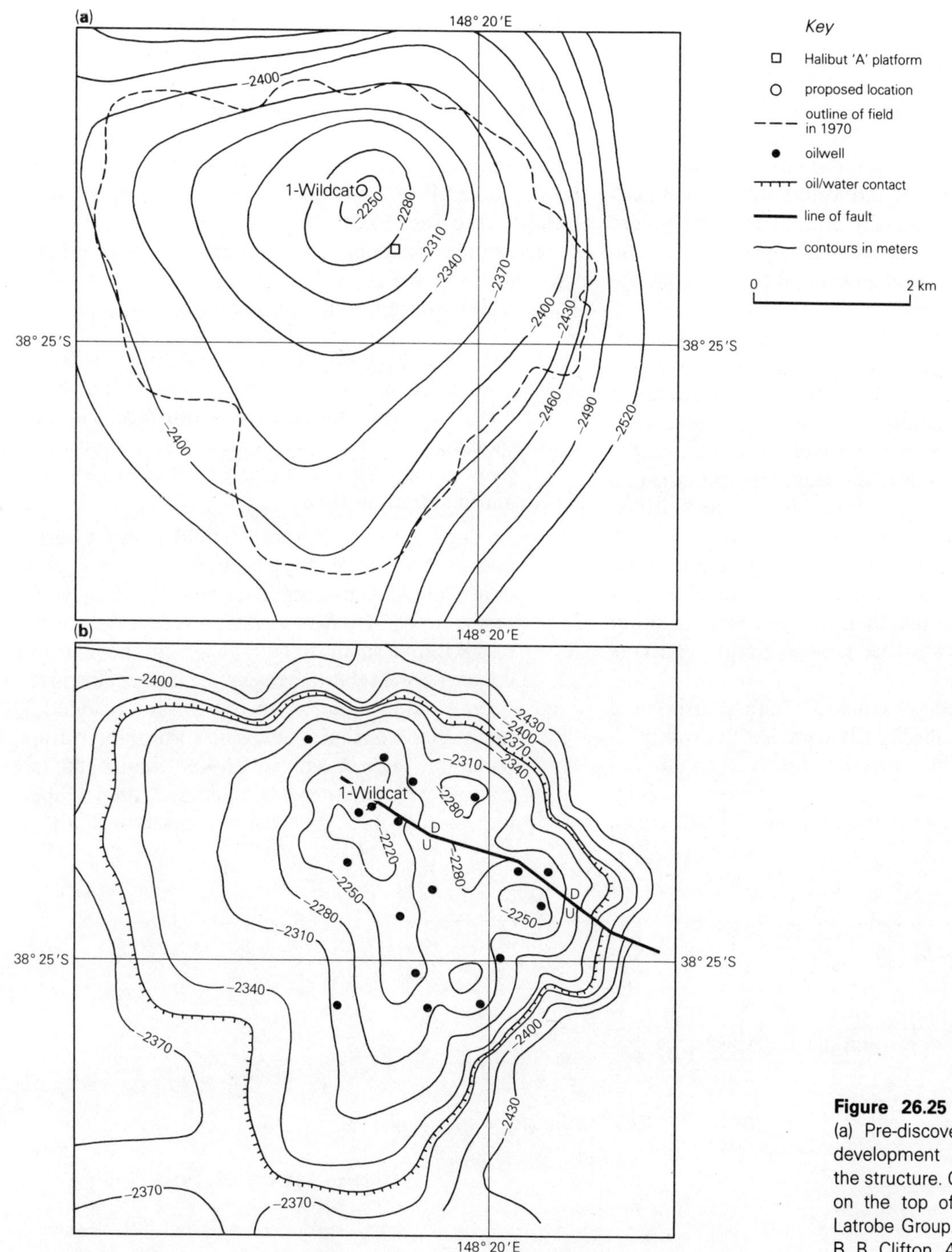

Figure 26.25 Halibut field, Australia. (a) Pre-discovery (1966) and (b) post-development (1970) interpretations of the structure. Contours at 30 m intervals on the top of the reservoir-containing Latrobe Group. (After E. H. Franklin and B. B. Clifton, *AAPG Bull.*, 1971.)

operators quickly found further fields, but with an unexpected twist. In October 1970, the first comparably large field in British waters was discovered by drilling a low-amplitude closure on a general structural "nose" – essentially an outward deflection of the structural contours towards a low region, itself without closure. The reservoir here was slightly younger than that at Ekofisk, a Paleocene sandstone overlying the Danian and uppermost Cretaceous chalk succession. The field was given the name Forties, in at least oblique obedience to the rule of geographic identification.

Step-out wells revealed that the productive area was so large (Fig. 16.11) that its efficient development would require no fewer than four platforms, positioned at the corners of a parallelogram (Fig. 23.2). With water depths varying from 90 to 130 m (deeper than those in Bass Strait), each platform would stand 170 m high from the seabed to the top deck, and would carry a 54 m mast on top of that. With a soft clay bottom and very severe weather conditions, the design and positioning of these platforms presented a formidable engineering challenge in themselves.

Fortunately, the difficulties encountered at Halibut were not encountered at Forties, because the top of the reservoir succession had not been eroded (Fig. 26.26). It was still necessary to undertake detailed seismic surveys *after* the discovery, however. To avoid positioning what were then the world's largest and most expensive drilling platforms in such a way that those of their wells drilled directionally outwards might enter the reservoir horizon below the oil/water interface, that interface had to be defined as accurately as possible in advance. The four platforms at Forties, and their counterparts in other large fields in the North Sea, now constitute veritable marine cities.

The Paleocene reservoir sandstone in the Forties field turned out to be extraordinarily complex. Its two most salient features were identified early: first, the sands are dominated by turbidites of submarine fan origin (Fig. 13.34); second, there are several distinct sand bodies, not all of the same nature or provenance. The stratigraphy is not of the layer-cake variety; instead, the geometry and facies of each sand body vary abruptly across the field. The pre-reservoir succession includes major sandstone formations of Mesozoic age (several of them richly productive elsewhere in the basin; see Sec. 25.5.3), as well as the Old Red Sandstone below them all. Yet these older sandstones appear not to have provided the sources for the Forties reservoir sandstones. Instead, the principal source lay in Carboniferous sandstones formerly blanketing the East Shetland platform (Figs 13.39, 16.5 & 16.26), and these in turn had been derived from the metamorphic basement of the Scottish mainland. By the time of Paleocene deposition, the graben had become greatly widened (see Sec. 17.7.4), and source areas for clastic detritus were controlled by the vagaries of intrabasinal tectonism. The Forties reservoir exemplifies the value of detailed petrophysical studies (see Sec. 13.5).

The similarities between the Halibut and Forties fields are more striking than the differences, but both similarities and differences illustrate some important points:

(a) In both fields (indeed in both offshore basins) the productive sections are unknown on either shoreline and known only in the subsurface. Halibut, though not Forties, illustrates the circumstance that not all fields now underlying the seafloor have marine strata in them.

(b) In deep waters offshore, a field needs a large, convex trap, to make possible a thick pay zone exploitable from a small number of fixed platforms. Wells from the platforms must drain the trap within a small, fixed radius from the platform center (say 2000 m). Because development rigs in deep waters cannot be moved about ("skidded"), many giant onshore fields in stratigraphic traps would be uneconomic offshore; they need too many development wells to drain a pool shaped like a blanket rather than a cushion. Only in

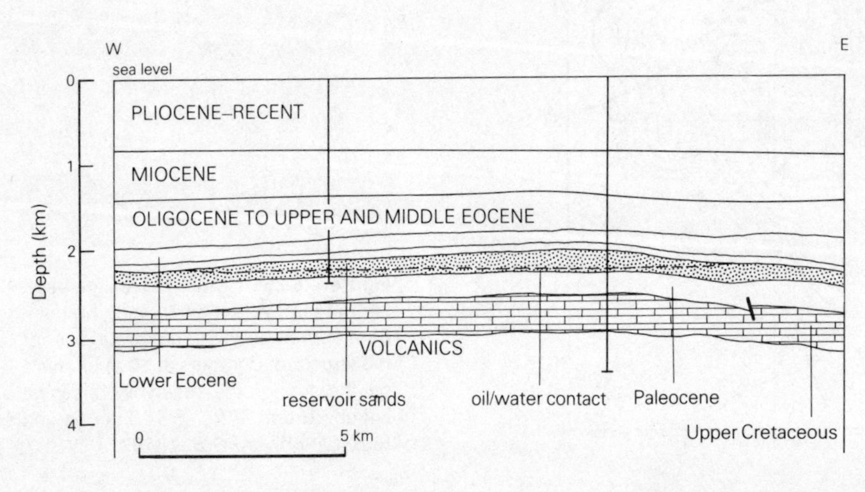

Figure 26.26 West–east structural cross section of the Forties field, showing the uneroded reservoir formation. (From A. N. Thomas *et al.*, *AAPG Bull.*, 1974.)

shallow waters of the US continental shelf is it yet practicable and economic to erect three or four platforms to develop a field of $15 \times 10^6 - 30 \times 10^6$ m^3 ($100 \times 10^6 - 200 \times 10^6$ barrels). The existence of this familiar exception suggests that the time will come, before the end of the 20th century, when this obstacle has to be removed from all offshore basins except perhaps those in polar waters.

(c) For any field in deep waters, it is therefore imperative that both the limits of the structure and the periphery of the oil/water interface be defined in detail, by exhaustive seismic surveys, before proceeding with development. It would be far too expensive and time-consuming to do this by drilling. The discovery wells themselves, drilled from floating or jack-up structures unsuited for development procedures, are not normally completed as producers as they would be on dry land.

(d) Given points (b) and (c), above, it is not surprising that the largest exploitable fields are found early in the exploration phase. If the structures needed to contain such fields exist in the basin, modern seismic techniques normally identify them at least in general outline with little delay. In the absence of outcrop, the earliest drilling is inevitably carried out on the largest structures identified. If these prove dry, exploration is unlikely to continue for long except very perfunctorily. Deep offshore basins are not given the opportunity to duplicate the history of some of the world's most productive onshore basins – the Permian, Los Angeles, Alberta, and Ouargla basins are examples – in which exploration has been conducted for years, with varying degrees of success, before the very large fields were found.

(e) The level of commerciality of deep-water fields is quite different from the levels acceptable onshore. The time gap between initial discovery and first production is far longer, and capital expenditures during that preparatory interval may be huge (for platforms and pipelines in particular). Forties, for example, did not begin production until 5 years after its discovery, and peak production was not attained until a further 3 years had elapsed. The recovery factor must be high. Heavy, viscous oils are likely to be unexploitable even if they are found in large volumes; enhanced recovery processes from offshore platforms would be exceedingly expensive and dangerous. Waterflood, on the other hand, may be necessary almost from the beginning, especially if the natural reservoir energy is low (as it is at Forties). Rapid declines and inefficient recoveries are unacceptable with operating costs so high; at least there is no difficulty finding a source of water.

(f) Despite these parallels between the two fields, they differ in some fundamental respects. Though both occur in basins controlled by faulting, neither is in a fault trap. The Halibut field's reservoir is involved in the faulting, but that at Forties postdates it and reflects the post-faulting subsidence that affected Mesozoic basins on every continental plate (Figs 16.26, 26.24 & Ch. 17). The Gippsland Basin is like other transverse grabens at continental margins, and its discovery stemmed from a correct analogy with those grabens. The North Sea was not known to contain a major graben until exploration in it had begun. The Ekofisk and Forties fields, the first and second "giants" discovered within it, are quite untypical of the other giant fields, in which the reservoirs are in much older strata involved in the taphrogenic or rifting stage, and in traps much more directly consequent upon the faults themselves.

(g) The Halibut field, and its neighboring fields in the Gippsland Basin, provide a striking parallel with the fields of Cook Inlet Basin, Alaska, the discovery and development of which are also discussed here (Sec. 26.2.2). The two basins are much alike in both structural setting and structural form. Both productive sections are very thick, paralic to nonmarine Paleogene clastic successions with abundant coal. In neither case are the overlapping strata regarded as adequate source sediments, and the underlying strata at Halibut are as nonmarine and coal-bearing as the reservoir itself. For both basins, therefore, the nonmarine, carbonaceous, reservoir succession was in the first instance regarded as providing its own source material. This view is still maintained by many geologists and geochemists familiar with the basins; but the Cook Inlet reservoir rocks are underlain by suitably rich Jurassic marine shales more widely accepted as the source rocks, and it may be speculated that the deeper parts of the Gippsland Basin will likewise prove to contain pre-reservoir (Cretaceous or Upper Jurassic) strata in marine facies capable of providing the source of the oil.

Author index

Subject index